Algebraic Curves over a Finite Field

PRINCETON SERIES IN APPLIED MATHEMATICS

TITLES IN THE SERIES

Chaotic Transitions in Deterministic and Stochastic Dynamical Systems: Applications of Melnikov Processes in Engineering, Physics, and Neuroscience by Emil Simiu

Selfsimilar Processes by Paul Embrechts and Makoto Maejima

Self-Regularity: A New Paradigm for Primal-Dual Interior-Point Algorithms by Jiming Peng, Cornelis Roos, and Tams Terlaky

Analytic Theory of Global Bifurcation: An Introduction by Boris Buffoni and John Toland

Entropy by Andreas Greven, Gerhard Keller, and Gerald Warnecke

Auxiliary Signal Design for Failure Detection by Stephen L. Campbell and Ramine Nikoukhah

Thermodynamics: A Dynamical Systems Approach by Wassim M. Haddad, VijaySekhar Chellaboina, and Sergey G. Nersesov

Optimization: Insights and Applications by Jan Brinkhuis and Vladimir Tikhomirov

Max Plus at Work by Bernd Heidergott, Geert Jan Olsder, and Jacob van der Woude

Impulsive and Hybrid Dynamical Systems: Stability, Dissipativity, and Control by Wassim M. Haddad, VijaySekhar Chellaboina, and Sergey G. Nersesov

Wave Scattering by Time-Dependent Perturbations: An Introduction by Gary C. Roach

The Traveling Salesman Problem: A Computational Study by David L. Applegate, Robert E. Bixby, Vasek Chvatal, and William J. Cook

Positive Definite Matrices by Rajendra Bhatia

Genomic Signal Processing by Ilya Shmulevich and Edward Dougherty

The Princeton Series in Applied Mathematics publishes high quality advanced texts and monographs in all areas of applied mathematics. Books include those of a theoretical and general nature as well as those dealing with the mathematics of specific applications areas and real-world situations.

Algebraic Curves over a Finite Field

J.W.P. Hirschfeld
G. Korchmáros
F. Torres

PRINCETON UNIVERSITY PRESS
PRINCETON AND OXFORD

Published by Princeton University Press, 41 William Street, Princeton, New Jersey 08540

In the United Kingdom: Princeton University Press, 3 Market Place, Woodstock, Oxfordshire OX20 1SY

Library of Congress Control Number: 2007940767

ISBN: 978-0-691-09679-7 (acid-free paper)

British Library Cataloging-in-Publication Data is available

This book has been composed in LaTeX

The publisher would like to acknowledge the authors of this volume for providing the camera-ready copy from which this book was printed.

Printed on acid-free paper. ∞

press.princeton.edu

Printed in the United States of America

10 9 8 7 6 5 4 3 2 1

For Lucca

born 9th June 2006

JWPH

For Adriana, Annachiara and Gabriella

GK

To the memory of Eduardo W. Torres Orihuela

6th March 1970 to 27th December 1989

FT

Contents

Preface

BACKGROUND

Algebraic curves defined over a finite field have been much studied in recent years, both for their intrinsic interest and for their applications in other areas, such as finite geometry, number theory, error-correcting codes, and cryptology. Conversely, applications in these areas have focused attention on particular problems for curves.

This book is a self-contained introduction to this subject, including topics of current research interest. It is divided into three parts.

The first is a general theory of algebraic curves over an algebraically closed field K of arbitrary characteristic p. To keep the treatment as simple as possible, Part 1 starts with the classical idea of a plane algebraic curve $\mathcal{F}$ as the set of zeros of a polynomial $f(X, Y)$ or a homogeneous polynomial $F(X_0, X_1, X_2)$, and develops its subject in the spirit of the geometric approach to the theory of algebraic curves, such as the treatises of Enriques, Lefschetz, Seidenberg, Semple and Kneebone, Semple and Roth, Severi, van der Waerden, and Walker. A rigorous treatment requires concepts and results from algebra; this is supplied by means of the function field $K(\mathcal{F})$ obtained by adjoining to $K(x)$ an element y satisfying $f(x, y) = 0$.

Part 2 is the heart of the book. It deals to a great extent with the case where K is the algebraic closure $\overline{\mathbf{F}}_q$ of a finite field $\mathbf{F}_q$ giving many properties of curves peculiar to finite fields.

Part 3 collects together a number of more advanced results on curves over finite fields, on automorphism groups of curves, and on applications of curves in coding theory and in the combinatorics of finite projective spaces. It also contains a collection of the most important families of curves.

The last section in each chapter consists of notes and references.

Appendix B is a list of notation used in the book.

Throughout the book numerous examples are worked out to illustrate the theory. The reader may also benefit from the exercises, the second last section in each chapter, and Appendix A reviewing basic facts required from algebra. The exercises are not supplied with hints for their solutions, but for many of them there are references in the notes.

The algebraic side of the theory of curves can also be developed further with a modern approach demanding considerable background in algebra, in the style of the theory of compact Riemann surfaces or of the theory of algebraic functions of one variable.

The book developed from a two-year course taught by the second author at the University of Basilicata and a series of lectures by the first author on the Stöhr–Voloch Theorem at various summer schools. Advanced topics from Part 3 were also the subject of talks at international conferences, delivered by the authors.

OUTLINE OF THE BOOK

The field K of coordinates is assumed to be algebraically closed except in the preliminary Section 1.1. This is not a restriction as every field is a subfield of an algebraically closed field. No assumption on the characteristic p of K is made. Hence either $p = 0$ or $p \geq 2$. Since all curves considered in the book are algebraic, the adjective 'algebraic' is mostly omitted.

Part 1

Classical algebraic geometry deals with plane projective curves in terms of their projective invariants. These are the properties of the curves which do not change under any projective transformation or, equivalently, under any change of the homogeneous coordinate system. For their study, the classical definition in its projective formulation is adequate: a plane projective curve $\mathcal{F}$ is the set of all points in the projective plane $\mathrm{PG}(2, K)$ satisfying a homogeneous polynomial $F(X_0, X_1, X_2)$ with coefficients in K; it is denoted by $\mathcal{F} = \mathbf{v}(F(X_0, X_1, X_2))$. If $\mathcal{F}$ is not the line $\mathbf{v}(X_0)$, the notation $\mathcal{F} = \mathbf{v}(f(X, Y))$ is used, where $f(X, Y) = F(1, X, Y)$ is the associated non-homogeneous polynomial. Calculations are often simplified by using the canonical form of a curve that goes back at least to Descartes in 1637 for conics, and to Newton in 1710 for cubics.

The relevant projective invariants, notably that of degree, inflexion, k-fold point, ordinary singularity, intersection number, bitangents, are considered in Chapter 1. Many problems on curves ultimately reduce to questions about the intersections of two plane curves. So the most important projective invariant is the intersection number $I(P, \mathcal{F} \cap \ell)$ of a plane curve $\mathcal{F}$ and a line ℓ at a point P. It is defined in such a way that the number of common points, each counted according to the intersection number at that point, is exactly the degree of $\mathcal{F}$, unless ℓ is a component of $\mathcal{F}$. Bézout's Theorem extends this formula from ℓ to any plane curve $\mathcal{G}$ having no common component with $\mathcal{F}$. Essentially, $I(P, \mathcal{F} \cap \ell)$ is generalised to $I(P, \mathcal{F} \cap \mathcal{G})$, the intersection number of $\mathcal{F}$ with another plane curve $\mathcal{G}$ at P, such that the number of common points, each counted according to the intersection number at that point, is exactly the product of the degrees of $\mathcal{F}$ and $\mathcal{G}$.

Such a generalisation is not immediate. The method for calculating $I(P, \mathcal{F} \cap \mathcal{G})$ requires knowing what condition the coefficients of two polynomials must satisfy in order that the polynomials have a common root of a given multiplicity. This is done by elimination theory relying on the resultant in Chapter 2. The geometric properties including the projective invariance of $I(P, \mathcal{F} \cap \mathcal{G})$, together with the proof of Bézout's Theorem, are in Chapter 3.

In investigating curves, an analysis of the singularities is essential. Singularities can be complicated and hence intractable for computations. Since projective transformations do not change the nature of the singularities, more general transformations are required. In Chapter 3, it is shown that locally quadratic transformations suffice to transform any plane curve, with possibly bad singularities, to one with only ordinary singularities. Then, certain properties of the curve can be retrieved from its transform, which is a plane curve with singularities in as simple a form as possible.

In such an analysis, as in many other contexts in algebraic geometry, the idea of a branch of a plane curve $\mathcal{F}$ plays an important role. The centre of a branch of $\mathcal{F}$ is a point of $\mathcal{F}$. There is exactly one branch of $\mathcal{F}$ centred at every non-singular point of $\mathcal{F}$, but an ordinary r-fold point is the centre of exactly r branches of $\mathcal{F}$. In general, an r-fold point of $\mathcal{F}$ is the centre of at least one and at most r branches of $\mathcal{F}$. Also, two distinct irreducible plane curves $\mathcal{F}$ and $\mathcal{G}$ have no common branch. For any branch γ of $\mathcal{G}$ centred at a point P, the intersection number $I(P, \mathcal{F} \cap \gamma)$ of $\mathcal{F}$ and γ is defined in such a way that $I(P, \mathcal{F} \cap \mathcal{G})$ is the sum of the $I(P, \mathcal{F} \cap \gamma)$ with γ ranging over all branches of $\mathcal{G}$ centred at P. This intersection number is suitable for use in Bézout's Theorem, and provides an efficient way for computing. The technical tools are formal power series rather than polynomials. Chapter 4 is dedicated to the theory of branches, including classical results on power series, such as Hensel's Lemma and Weierstrass' Preparatory and Division Theorems.

Locally quadratic transformations are particular birational transformations. This suggests investigating the effect of birational transformations on a plane curve, and studying its unaffected properties, that is, its birational invariants. However, the intersection number, like other projective invariants considered in Chapter 1, is not a birational invariant. To understand why this happens, it is enough to observe that the nodal cubic and the cuspidal cubic are birationally equivalent and that any birational transformation mapping one to the other is not defined at singular points. So the classical idea of a plane curve is not sufficient to deal with birational transformations. A plane curve needs to be considered as the set of its branches rather than of its points. Then the idea of a generic point of a curve can be viewed as a straightforward generalisation of the idea of a branch. Another actor is the function field $K(\mathcal{F})$ of an irreducible plane curve $\mathcal{F}$, which is a field of transcendency degree 1 over K. Importantly, birationally equivalent curves have the same function field, up to a K-isomorphism, and the geometric idea of a branch of $\mathcal{F}$ can be translated into $K(\mathcal{F})$ by the purely algebraic concept of a place. In particular, the valuation of an element ξ of $K(\mathcal{F})$ at a place $\mathcal{P}$ leads to the consideration of the zeros and the poles of ξ.

In Chapter 5, after a detailed presentation of these concepts, the most important birational invariant, namely the *genus* g of an irreducible plane curve $\mathcal{F}$, is investigated. The genus is a non-negative integer which can be calculated from the equation of $\mathcal{F}$ when $\mathcal{F}$ has only ordinary singularities. If $\mathcal{F}$ is non-singular, then $g = \frac{1}{2}(n-1)(n-2)$ where $n = \deg \mathcal{F}$. In particular, irreducible plane cubics are elliptic curves, that is, they have genus 1. Elliptic curves have many peculiar properties; their study requires a particular theory that is not included in the book. However, some of their structure is given in Chapter 6 and a summary of the isomorphism and projective equivalence classes is given in Chapter 8.

An important formula to compute the genus g of any irreducible plane curve $\mathcal{F}$ is established in terms of differentials of $K(\mathcal{F})$. Higher differentials are also considered, with some changes in the familiar definition of higher derivatives to avoid undesirable behaviour in positive characteristic. Such Hasse derivatives are the technical tool for the study of dual curves as well as in several contexts in later chapters.

Other important birational invariants are the parameters of linear series, namely the order and dimension. A divisor is any formal sum of places; the principal divisor $\operatorname{div}\xi$ of $\xi \in K(\mathcal{F})$, $\xi \neq 0$, is the sum of all zeros and poles of ξ, each counted with multiplicity. Two divisors A, B are equivalent if $B = A + \operatorname{div}\xi$ for some $\xi \in K(\mathcal{F})$. A divisor A is effective or positive if no place appears in A with negative weight. A linear series is a set of certain effective equivalent divisors. A complete linear series $|A|$ consists of all effective divisors equivalent to A. Chapter 6 develops the theory of linear series based on the idea of adjoint curves. For a complete linear series of order n and projective dimension r, the important Riemann–Roch Theorem states that $r = n - g + i$, where i is another birational invariant, the index of speciality. In particular, $i = 0$ when $n > 2g - 2$. The Riemann–Roch Theorem not only provides an alternative definition of the genus g, but is also the essential ingredient in many fundamental results on curves, such as the Reciprocity Theorem of Brill and Noether, Clifford's Theorem, and Weierstrass' Gap Theorem. The last concerns gaps at a place $\mathcal{P}$, which are positive integers k such that $k\mathcal{P}$ is not the divisor of all poles of any function in $K(\mathcal{F})$; it states that there exist exactly g gaps at any place of $K(\mathcal{F})$. Chapter 6 also includes some basic facts on the zero-divisor class group, the Picard group, and on its factor group, $\operatorname{Pic}_0(K(\mathcal{F}))$, together with an important result on the structure of certain subgroups of $\operatorname{Pic}_0(K(\mathcal{F}))$. The study of the divisor class groups is expanded in Part 3 for the case $K = \overline{\mathbf{F}}_q$.

Even using birational transformations, an irreducible plane curve can rarely be transformed into a non-singular plane curve. To obtain a non-singular model for any irreducible plane curve, space curves need to be considered. The idea of a space curve in a higher-dimensional space as an algebraic variety of dimension one, that is, the set of common zeros of a finite set of polynomials, is not suitable for this. An alternative approach, developed in Part 2, provides a different but equivalent definition: an irreducible space curve Γ is the image of an irreducible plane curve $\mathcal{F}$ under a birational transformation ω. In particular, the function field $K(\Gamma)$ of Γ is $K(\mathcal{F})$. The approach is based on the idea of branches in a higher-dimensional space; those which are the images of the branches of $\mathcal{F}$ under ω are viewed as the branches of Γ. The points of Γ are the centres of its branches. Here, care is required, as some point of Γ may not be the image of a point of $\mathcal{F}$ under ω. This theory of space curves begins in Chapter 7.

The geometry of a space curve Γ of $\mathrm{PG}(r, K)$ can be extracted from its intersections with hyperplanes, and studied in terms of the linear series g_n^r on Γ cut out by hyperplanes. Here n is the degree of Γ, that is, the number of common points P of Γ with a hyperplane H, each counted according to the intersection number $I(P, \Gamma \cap H)$. In particular, the singular points of Γ are those points $P \in \Gamma$ for which $I(P, \Gamma \cap H) \geq 2$ for every hyperplane H through P. For sufficiently large n

compared to r, no point of Γ is singular; this shows the existence of a non-singular model of Γ. For $g > 1$, the canonical curve $\mathcal{K}$ of Γ arising from the canonical linear series g_{2g-2}^{g-1} is a non-singular curve of $\mathrm{PG}(g-1, K)$, provided that Γ is not hyperelliptic. Hyperelliptic curves occur as exceptions since their canonical linear series are composed of an involution of order two; in other words, $\mathcal{K}$ is doubly covered by Γ; equivalently, $[K(\Gamma) : K(\mathcal{K})] = 2$.

Another important geometric idea in Chapter 7 comes from classical differential geometry. For a non-singular point P of Γ, there are exactly $r + 1$ Hermitian P-invariants; they are the non-negative integers j for which some hyperplane H intersects Γ at P with multiplicity j, that is, $I(P, \Gamma \cap H) = j$. The Hermitian P-invariants written in increasing order $j_0 = 0 < j_1 < \cdots < j_r$ define the order sequence $(j_0, j_1 \ldots, j_r)$ of Γ at P. Let L_i be the intersection of all hyperplanes H in $\mathrm{PG}(r, K)$ such that $I(P, \Gamma \cap H) \geq j_{i+1}$. Then, L_0 is the point P, and L_1 is the tangent line to Γ at P. The projective subspace L_i is the i-th osculating space, and the flag $L_0 \subset L_1 \subset \cdots \subset L_{r-1}$ is the algebraic analogue of the Frenet frame. For a singular point P, all this remains valid for each branch of Γ centred at P. Apart from finitely many points of Γ, the order sequence of Γ at P is the same and it defines the order sequence $(\epsilon_0, \epsilon_1, \ldots, \epsilon_r)$ of Γ. In zero characteristic, every curve is classical, the order sequence being $(0, 1, \ldots, r)$. This does not hold true for $p > 0$, although non-classical curves are rare and often have unusual behaviour. For instance, certain non-classical curves are projectively equivalent to their Gauss duals, and, up to a birational transformation, are the curves with many automorphisms. When $K = \overline{\mathbf{F}}_q$, these are the curves lying in $\mathrm{PG}(r, q)$ with many points. So non-classical curves are important and germane to several research areas of current interest, such as coding theory and cryptology. Chapter 7 contains the theory of non-classical curves.

The existence of non-classical curves is a real obstacle to extending classical results from zero to positive characteristic. Nevertheless, Castelnuovo's Bound, Halphen's Theorem, and several other relevant results on curves do hold true for $p > 0$ under some reasonable extra conditions, notably the Linear General Position Principle and the Uniform Position Principle. This is the final topic in Part 2, together with the explanation of how the concept of a place, as used in this book, is related to the concept of a place arising in valuation theory. Also, the proof of the equivalence between the definition of an irreducible space curve and the concept of an algebraic variety of dimension one is described.

Part 2

With $K = \overline{\mathbf{F}}_q$, the field K has a set of finite subfields containing $\mathbf{F}_q$,

$$\mathbf{F}_q, \mathbf{F}_{q^2}, \ldots, \mathbf{F}_{q^i}, \ldots,$$

such that $K = \cup_{i \geq 1} \mathbf{F}_{q^i}$. So, for any r, the projective space $\mathrm{PG}(r, K)$ contains the finite projective spaces $\mathrm{PG}(r, q^i)$ with $i \geq 1$.

A plane curve $\mathcal{F} = \mathbf{v}(f(X, Y))$ is defined over $\mathbf{F}_q$ when $f(X, Y) \in \mathbf{F}_q[X, Y]$. Irreducible plane curves defined over $\mathbf{F}_q$ are also called $\mathbf{F}_q$-rational. This natural concept extends to higher dimensions: an irreducible space curve Γ is $\mathbf{F}_q$-rational if it arises from an $\mathbf{F}_q$-rational plane curve by an $\mathbf{F}_q$-birational transformation ω. This means that ω is defined over the subfield $\mathbf{F}_q(\mathcal{F})$ of $K(\mathcal{F})$, the $\mathbf{F}_q$-rational function field of $\mathcal{F}$.

Given an $\mathbf{F}_q$-rational irreducible curve Γ of $\mathrm{PG}(r, K)$, the aim is to extract useful information from the sequence

$$M_1, M_2, \ldots, M_i \ldots,$$

with M_i the number of points of Γ in $\mathrm{PG}(r, q^i)$. However, such a sequence may not be the same for another curve $\mathbf{F}_q$-birationally equivalent to Γ.

Again, the idea of an irreducible curve as a set of its branches is useful for placing the theory within birational equivalence: it is enough to count the $\mathbf{F}_{q^i}$-rational branches of Γ, and replace M_i by the number N_i of such branches, and consider the sequence

$$N_1, N_2, \ldots, N_i \ldots.$$

From the function field point of view, the corresponding concept of an $\mathbf{F}_q$-rational branch is an $\mathbf{F}_q$-rational place of $K(\mathcal{F})$. It should be noted that a singular point of Γ lying in $\mathrm{PG}(r, q)$ may not be the centre of any $\mathbf{F}_q$-rational branch of Γ. This shows that a bijection between the points of Γ lying in $\mathrm{PG}(r, q)$ and the $\mathbf{F}_q$-rational places of $K(\Gamma)$ can only be ensured for non-singular curves. If this is the case, points of Γ lying in $\mathrm{PG}(r, q)$ are the $\mathbf{F}_q$-rational points.

The foundation of the theory of irreducible curves over finite fields is given in Chapter 8. Gauss in 1801 considered cubic and quartic congruences in two variables modulo a prime, and so calculated the number of rational points for particular cubic and quartic curves. Here, a central result is the Stöhr–Voloch Theorem, which highlights several peculiar geometric properties of curves with many $\mathbf{F}_q$-rational branches. Such curves are mostly non-classical. The Stöhr–Voloch Bound derived from this theorem provides an accurate estimate of the number S_q of $\mathbf{F}_q$-rational branches for large families of curves. Chapter 8 explains all the background and details of the Stöhr–Voloch Theorem, and gives an exposition of the theory of Frobenius non-classical curves. An elementary proof of the Stöhr–Voloch Bound for non-classical plane curves is also presented.

Another bound on S_q, depending only on g and q, is the famous Hasse–Weil Bound, proved by Hasse for $g = 1$ in 1934 and by Weil in 1948 for $g > 1$:

$$q + 1 - 2g\sqrt{q} \leq S_q \leq q + 1 + 2g\sqrt{q}.$$

This bound is deduced, via zeta functions, from the Riemann hypothesis for function fields over finite fields.

The Hasse–Weil Bound and its refinements, especially the Serre Bound, together with the Stöhr–Voloch Bound, are widely used in investigating curves over finite fields, in various areas such as finite geometry, number theory, coding theory, and cryptology. In coding theory, the asymptotic behaviour of $N_q(g)/g$ is important, where $N_q(g)$ is the maximum number of $\mathbf{F}_q$-rational points that a non-singular $\mathbf{F}_q$-rational curve of genus g can have.

The Hasse–Weil Theorem leads to other results on curves over finite fields, on their Picard groups, p-ranks and Frobenius linear series.

These themes are treated in Chapter 9. A few results, such as the fundamental equation of a Frobenius linear series, are discussed without proofs but with an eye to further reading on algebraic number theory and abelian varieties. Modular curves are also important, but they are not included.

Part 3

For $q = p^r$, the Hermitian curve $\mathcal{H}_q = \mathbf{v}(Y^q - Y - X^{q+1})$ is a non-singular plane curve of genus $g = \frac{1}{2}(q^2 - q)$ with exactly $q^3 + 1$ points in $\mathrm{PG}(2, q^2)$. So the number of $\mathbf{F}_{q^2}$-rational points of $\mathcal{H}_q$ is equal to $q^3 + 1 = q^2 + 1 + 2gq$, the maximum allowed in the Hasse–Weil upper bound. Other important $\mathbf{F}_{q^2}$-maximal curves are the DLS and DLR curves. Curves of genus zero are lines and so the lines of $\mathrm{PG}(2, q^2)$ are maximal. For each square q, there is also a maximal elliptic curve. Maximal curves are naturally used in algebraic-geometry codes. The theory of $\mathbf{F}_{q^2}$-maximal curves is covered in Chapter 10.

Various characterisations of $\mathbf{F}_{q^2}$-maximality in both algebraic and geometric terms are given. An irreducible curve $\mathcal{F}$ defined over $\mathbf{F}_{q^2}$ is $\mathbf{F}_{q^2}$-maximal if and only if one of the following holds.

(i) The numerator in the zeta function of $\mathcal{F}$, that is, the $L_{q^2}(t)$-polynomial of $\mathcal{F}$, is $(t + q)^{2g}$.

(ii) Let $\mathcal{X}$ be a non-singular model of $\mathcal{F}$, defined over $\mathbf{F}_{q^2}$ and embedded in $\mathrm{PG}(r, K)$. For any $\mathbf{F}_{q^2}$-rational point $P_0 \in \mathcal{X}$, the Frobenius linear series $|(q + 1)P_0|$ has fundamental equation

$$qP + \Phi(P) \equiv (q + 1)P_0$$

satisfied by all points $P \in \mathcal{X}$, where Φ is the Frobenius collineation of $\mathrm{PG}(r, K)$ fixing $\mathrm{PG}(r, q^2)$ point-wise.

(iii) The curve $\mathcal{F}$ is $\mathbf{F}_{q^2}$-birationally equivalent to an irreducible curve Γ of degree $q + 1$ embedded in $\mathrm{PG}(r, K)$ and lying on a non-degenerate Hermitian variety of $\mathrm{PG}(r, q^2)$.

The curve Γ in (iii) is a non-singular model defined over $\mathbf{F}_{q^2}$ such that the Frobenius linear series is cut out by hyperplanes of $\mathrm{PG}(r, q^2)$. Therefore Γ plays a central role in the study of the geometry of $\mathbf{F}_{q^2}$-maximal curves. The orders at any $\mathbf{F}_{q^2}$-rational point $P \in \Gamma$ are determined by the Weierstrass non-gaps at P less than or equal to $q + 1$. The same kind of computation also works for other points giving all but one of these orders. From these results and basic facts on Weierstrass semigroups, Γ is shown to be Frobenius non-classical.

Another relevant property of maximal curves, originally pointed out by Serre, is related to covers: given an $\mathbf{F}_{q^2}$-maximal curve $\mathcal{F}$, every curve $\mathcal{F}'$ which is covered by $\mathcal{F}$ over $\mathbf{F}_{q^2}$ is also $\mathbf{F}_{q^2}$-maximal. Notably, the quotient curve $\mathcal{F}' = \mathcal{F}/G$ arising from any non-trivial $\mathbf{F}_{q^2}$-automorphism group G of $\mathcal{F}$ is covered by $\mathcal{F}$. Since the Hermitian, DLS, and DLR curves have very large automorphism groups, Serre's result provides an efficient construction technique. The Garcia–Stichtenoth curve over $\mathbf{F}_{3^6}$ is an example of a curve which is not a quotient curve of the Hermitian curve. Very recently, Giulietti and Korchmáros constructed a family of $\mathbf{F}_{q^2}$-maximal curves which are not covered over $\mathbf{F}_{q^2}$ by any of the Hermitian, DLS, and DLR curves.

For a given q, the highest value of g for which an $\mathbf{F}_{q^2}$-maximal curve $\mathcal{F}$ of genus g exists is $\frac{1}{2}(q^2 - q)$ and equality holds if and only if $\mathcal{F}$ is the Hermitian curve, up to $\mathbf{F}_{q^2}$-birational equivalence. The upper part in the spectrum of genera of $\mathbf{F}_{q^2}$-maximal curves contains only few values. The classification of some of these curves and the known results are given. Although the whole spectrum depends on the nature of q, the highest three genera in the spectrum have the same positions independently of q.

Particular $\mathbf{F}_{q^2}$-maximal curves, such as those of Hurwitz type, are also investigated, and some non-isomorphic $\mathbf{F}_{q^2}$-maximal curves with the same genus are considered.

Chapter 10 also describes some examples of $\mathbf{F}_q$-optimal curves, that is, curves defined over $\mathbf{F}_q$ whose number of $\mathbf{F}_{q^2}$-rational points is $N_q(g)$. By definition, $\mathbf{F}_{q^2}$-maximal curves are $\mathbf{F}_{q^2}$-optimal. The DLS and DLR curves are both $\mathbf{F}_q$-optimal, but the DLS curve is $\mathbf{F}_{q^4}$-maximal and the DLR curve is $\mathbf{F}_{q^6}$-maximal. It is shown that the DLS curve is characterised as the unique $\mathbf{F}_q$-optimal curve with genus $g = q_0(q-1)$ for $p = 2$, $q_0 = 2^s$ and $q = 2q_0^2$. A similar characterisation for the DLR curve may be possible.

Chapter 11 gives the theory of automorphisms of an irreducible curve. To establish the concept of an automorphism as a birational invariant, the theory begins with the concept of a field automorphism applied to a function field Σ of transcendency degree one over K or, equivalently, to the function field $K(\mathcal{F})$ of an irreducible plane curve $\mathcal{F}$. Here, only automorphisms fixing every element in K are admitted. Such a K-automorphism α of Σ defines a K-birational transformation of Σ, from which a geometric interpretation of α in terms of $\mathcal{F}$ can be deduced.

The theory develops on the algebraic side with the study of the K-automorphism group $\mathrm{Aut}_K(\Sigma)$ of $\Sigma = K(\mathcal{F})$ in terms of its abstract structure and its actions on the sets of places and divisors. Nevertheless, K-automorphism is a truly geometric concept. In fact, $\mathcal{F}$ is birationally equivalent to an irreducible non-singular curve $\mathcal{X}$ of some $\mathrm{PG}(r, K)$, and $\mathcal{X}$ can be chosen in such a way that every K-automorphism of Σ is represented as a linear collineation of $\mathrm{PG}(r, K)$ preserving $\mathcal{X}$. Then $\mathrm{Aut}_K(\Sigma)$ is the K-automorphism group of $\mathcal{X}$, and the term 'K-automorphism group of Γ' is adopted for any curve Γ birationally equivalent to $\mathcal{X}$.

The major result is the finiteness of $\mathrm{Aut}_K(\Sigma)$ for curves of genus $g > 1$, which is a consequence of the finiteness of the stabiliser of any place $\mathcal{P}$ of Σ. This result on stabilisers also holds for elliptic curves.

Let G be a finite non-trivial subgroup of $\mathrm{Aut}_K(\Sigma)$. The elements in Σ fixed by every K-automorphism in G is the subfield Σ^G of Σ. Importantly, Σ^G is larger than K and Σ/Σ^G is a Galois extension over K with Galois group G. In particular, $\Sigma^G = K(\mathcal{G})$ is the function field of a quotient curve $\mathcal{G}$ of $\mathcal{F}$, defined up to birational equivalence. As a matter of fact, $\mathcal{G}$ inherits important properties from $\mathcal{F}$: for instance, $\mathbf{F}_{q^2}$-maximality as considered in Chapter 10.

Derived from Hurwitz's Theorem, linking g to the genus g' of Σ^G, is the important equation

$$2g-2 = |G|(2g'-2)+d,$$

where d is the degree of the different divisor $D(\Sigma/\Sigma^G)$. To compute d, two cases are distinguished according as G is tame or non-tame (wild). In the tame case, $p \nmid |G_{\mathcal{P}}|$ for every place $\mathcal{P}$ of Σ; this occurs for instance for $p=0$ and for $p \nmid |G|$.

If G is tame, the classical argument due to Hurwitz works and gives a formula for d depending only on the lengths of the short G-orbits of places of Σ. From this, the bound, $|G| \leq 84(g-1)$, follows for $g \geq 2$.

If G is non-tame, then $|G|$ and hence $\mathrm{Aut}_K(\Sigma)$ may be much larger, even exceeding g^4. A major result is the classification of all curves with $|\mathrm{Aut}_K(\Sigma)| \geq 8g^3$. The long and delicate proof not only uses Hurwitz's work with Hilbert's different formula and Serre's results on ramification groups but also exploits many results on finite permutation groups, especially the classification of all 2-transitive groups whose 2-point stabilisers are cyclic. There are four such curves, namely two hyperelliptic curves, the Hermitian curve and the DLS curve.

On the other hand, $|G|$ is bounded above by $4g+4$ for abelian groups and p-subgroups of curves with positive p-rank, and by $24g^2$ for certain soluble groups of curves with zero p-rank. Also, Hurwitz's bound holds for ordinary curves with p-rank equal to g. There remains the problem of understanding better to what extent the Hurwitz bound holds in positive characteristic.

Chapter 12 is dedicated to curves defined over finite fields with unusual properties that a complex curve cannot have. Artin–Schreier curves and, in particular, Hermitian curves are of this type. A family of such plane curves arises from separated polynomials. It consists of curves $\mathcal{F} = \mathbf{v}(A(Y) - B(X))$ where $p \nmid m$ with $m = \deg B(X) \geq 2$ and $A(Y)$ is any additive separable polynomial.

The main properties of $\mathcal{F}$ are extracted from the local analysis of its unique singular point P_∞. The exposition describes the genus, the Weierstrass gap sequence at P_∞ and the ramification groups of its translation automorphism group fixing P_∞. The full K-automorphism group of $\mathcal{F}$ fixes P_∞ except in two cases, namely, when $\mathcal{F}$ is the Hermitian curve $\mathbf{v}(Y^{p^n} - Y - X^{p^n+1})$ or the curve,

$$\mathcal{F} = \mathbf{v}(Y^{p^n} + Y - X^m) \quad \text{with } m < p^n, \text{ and } p^n \equiv -1 \pmod{m}.$$

For $p > 2$ and $m = 2$, the latter curve is hyperelliptic. Notably for $p > 2$, these hyperelliptic curves and the Hermitian curves are the only curves whose K-automorphism groups have order larger than $8g^3$.

Deligne–Lusztig varieties of dimension one provide other examples of significant curves over finite field, namely the DLS curves of Suzuki type and the DLR curves of Ree type. They are characterised by their genera and K-automorphism groups. For $p = 2$, the Hermitian curves, the DLS curves, and the hyperelliptic curves $\mathcal{F} = \mathbf{v}(Y^2+Y+X^{2^h+1})$ are the only curves with K-automorphism groups of order larger than $8g^3$.

A further interesting example worked out in Chapter 12 is defined over $\mathbf{F}_2$ and attains the Serre Bound for $q = 2^{11}$. The proof illustrates how the theory presented in Chapter 9 may be used to investigate sporadic curves.

Chapter 13 looks at applications of curves in coding theory and in the combinatorics of finite geometry. A brief exposition of the main construction due to Goppa and a few illustrative examples are presented. Coding theory is also connected with curves via finite geometry, since complete arcs in $\mathrm{PG}(r, q)$ are the geometric counterpart of maximum distance separable (MDS) codes, which are linear codes correcting the greatest number of errors with respect to the field size and their dimension. An account of this subject is given, especially of the Main Conjecture for MDS codes.

ACKNOWLEDGEMENTS

We would like to thank the following colleagues: Massimo Giulietti for sharing the writing of Chapter 11 and also for proof-reading; Marien Abreu, Vincenzo Giordano, Lucia Indaco for proof-reading Part 1; Angelo Sonnino and Fabio Pasticci for technical help.

Discussions on a miriad of aspects of algebraic curves with colleagues in many countries were of great assistance and we are most grateful to all of them; they are Miriam Abdón, Angela Aguglia, Antonio Campillo, John Cannon, Antonio Cossidente, Rainer Fuhrmann, Arnaldo Garcia, Paolo Maroscia, Nicola Melone, Ruud Pellikaan, Alessandro Siciliano, Henning Stichtenoth, Karl Stöhr, Felipe Voloch, and Mike Zieve.

The Mathematics Departments of the Universities of Basilicata, Campinas, Essen, Perugia, Rome 'La Sapienza', Sussex, Sydney and Valladolid, as well as IMPA in Rio de Janeiro and the Abdus Salam International Centre for Theoretical Physics in Trieste have supported us.

Finally, this book could not have been produced without the love and support of our families.

PART 1
General theory of curves

Chapter One

Fundamental ideas

In this chapter, basic facts about curves are presented. The exposition also highlights some of the peculiarities that occur for positive characteristic, such as the existence of strange curves, that is, curves whose tangent lines at non-singular points have a point in common.

1.1 BASIC DEFINITIONS

Over the real numbers, $\mathbf{R}$, consider the parabola $\mathcal{F}$ given by $F = Y - X^2$; its points form, as in Figure 1.1, the set

$$\{(t, t^2) \mid t \in \mathbf{R}\}.$$

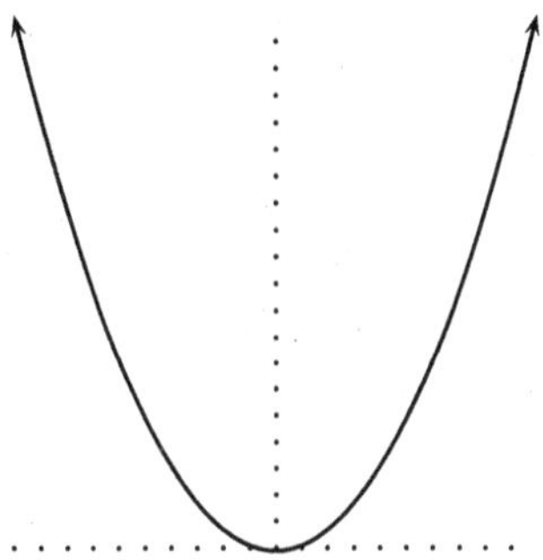

Figure 1.1 The parabola $y = x^2$ in the real affine plane

However, there are two other types of points associated with $\mathcal{F}$, namely, (a) those at infinity and (b) those with coordinates in $\mathbf{C}$, the algebraic closure of $\mathbf{R}$. For example, regarding (b), the line with equation $y + 1 = 0$ meets $\mathcal{F}$ in the two points $(i, -1), (-i, -1)$, where $i^2 = -1$. Regarding (a), if F is homogenised to $F^* = X_0X_2 - X_1^2$, with $X = X_1/X_0$, $Y = X_2/X_0$, then the line with equation $X_0 = 0$ meets the corresponding projective curve $\mathcal{F}^*$ at the point $(0, 0, 1)$.

All these ideas need to be considered for a general curve and a general field. First, some notation and fundamental definitions for the spaces that appear are explained.

DEFINITION 1.1 (i) For a field K, let $K^n = \{(x_1, x_2, \ldots, x_n) \mid x_i \in K\}$, the *$n$-fold Cartesian product of K.*

(ii) Let $V(n,K)$ be *n-dimensional vector space over K*, which may be regarded as $(K^n,+,\cdot)$, where, for $x_i, y_i, \lambda \in K$,

$$(x_1,x_2,\ldots,x_n)+(y_1,y_2,\ldots,y_n)=(x_1+y_1,x_2+y_2,\ldots,x_n+y_n),$$
$$\lambda(x_1,x_2,\ldots,x_n)=(\lambda x_1,\lambda x_2,\ldots,\lambda x_n).$$

(iii) The *affine plane* $\mathrm{AG}(2,K)=\mathbf{A}^2(K)$ is a pair $(\mathcal{P},\mathcal{L})$, where

$$\mathcal{P}=\{P=(x,y)\mid x,y\in K\},$$
$$\mathcal{L}=\{\ell=aX+bY+c\mid a,b,c\in K,(a,b)\neq(0,0)\},$$

and a *point* $P=(x,y)$ lies on a *line* $\ell=aX+bY+c$ if $ax+by+c=0$.

(iv) More generally, *affine space of n-dimensions* is $\mathrm{AG}(n,K)=\mathbf{A}^n(K)$ with *points* $x=(x_1,x_2,\ldots,x_n)$ and *r-dimensional subspaces* $x+S$, for r-dimensional subspaces S of $V(n,K)$.

(v) The *projective plane* $\mathrm{PG}(2,K)=\mathbf{P}^2(K)$ is a pair $(\mathcal{P},\mathcal{L})$, where

$$\mathcal{P}=\{P=(x,y,z)=(\lambda x,\lambda y,\lambda z)\mid (x,y,z)\in K^3\backslash\{(0,0,0)\},\ \lambda\in K\backslash\{0\}\},$$
$$\mathcal{L}=\{\ell=aX+bY+cZ=\lambda aX+\lambda bY+\lambda cZ\mid a,b,c,\lambda\in K,\ (a,b,c)\neq(0,0,0),\ \lambda\neq 0\},$$

and a *point* $P=(x,y,z)$ lies on a *line* $\ell=aX+bY+cZ$ if $ax+by+cz=0$.

(vi) More generally, *projective space of n-dimensions* is $\mathrm{PG}(n,K)=\mathbf{P}^n(K)$ with *points*,

$$\mathbf{x}=(x_0,x_1,x_2,\ldots,x_n)=(\lambda x_0,\lambda x_1,\lambda x_2,\ldots,\lambda x_n),$$

$(x_0,x_1,x_2,\ldots,x_n)\neq(0,0,0,\ldots,0)$, $\lambda\neq 0$, and *r-dimensional subspaces* S, for $(r+1)$-dimensional subspaces S of $V(n+1,K)$.

In each type of space, it is important to consider the structure-preserving transformations.

DEFINITION 1.2 (i) A *linear transformation* $T: V(n,K)\to V(n,K)$, is given as follows:

$$T(x)=x' \quad \text{where } {}^tx'=A\,{}^tx \text{ for a suitable non-singular matrix } A,$$

with $x=(x_1,x_2,\ldots,x_n)$, $x'=(x_1',x_2',\ldots,x_n')$, and ${}^tx$ the transpose of x. The linear transformations of $V(n,K)$ constitute the *the general linear group* $\mathrm{GL}(n,K)$.

A *semilinear transformation* $T: V(n,K)\to V(n,K)$, is given as follows:

$$T(x)=x', \quad \text{where } {}^tx'=A\,{}^t\sigma(x) \text{ for a suitable non-singular matrix } A,$$

with $\sigma(x)=(\sigma(x_1),\sigma(x_2),\ldots,\sigma(x_n))$ for some automorphism σ of K.

The semilinear transformations of $V(n,K)$ constitute its *general semilinear group* $\Gamma L(n,K)$.

(ii) An *affine transformation* $S : \mathrm{AG}(n,K) \to \mathrm{AG}(n,K)$ is given as follows:

$$S(x) = x' = T(x) + b,$$

where T is a linear transformation and $b = (b_1, b_2, \ldots, b_n)$ The affine transformations of $\mathrm{AG}(n,K)$ constitute its *affine group* $\mathrm{AGL}(n,K)$.

An *affine collineation* $S : \mathrm{AG}(n,K) \to \mathrm{AG}(n,K)$ is given as follows:

$$S(x) = x' = T(\sigma(x)) + b,$$

with T as above and $\sigma(x)$ as in (i).

(iii) A *projectivity* $T : \mathrm{PG}(n,K) \to \mathrm{PG}(n,K)$ is given as follows:

$$T(\mathbf{x}) = \mathbf{x}', \qquad \text{where} \quad {}^t\mathbf{x}' = A\,{}^t\mathbf{x},$$

with

$$\mathbf{x} = (x_0, x_1, \ldots, x_n), \quad \mathbf{x}' = (x'_0, x'_1, \ldots, x'_n),$$

and A a suitable non-singular matrix. It is also called a *projective transformation* or *linear collineation*. The projectivities of $\mathrm{PG}(n,K)$ constitute its *projective general linear group* $\mathrm{PGL}(n+1,K)$.

A *collineation* $T : \mathrm{PG}(n,K) \to \mathrm{PG}(n,K)$ is given as follows:

$$T(\mathbf{x}) = \mathbf{x}', \qquad \text{where} \quad {}^t\mathbf{x}' = A\,{}^t\sigma(\mathbf{x}),$$

with A as above and $\sigma(\mathbf{x}) = (\sigma(x_0), \ldots, \sigma(x_n))$.

The collineations of $\mathrm{PG}(n,K)$ constitute its *projective semilinear group* $\mathrm{P\Gamma L}(n+1,K)$.

(iv) When K contains the finite field $\mathbf{F}_q$, the mapping

$$\Phi : \mathbf{F}_q \to \mathbf{F}_q,$$
$$x \mapsto x^q,$$

is the *Frobenius automorphism*. The *n-th Frobenius automorphism of K* is the map Φ^n that takes $x \in K$ to $x^{q^n} \in K$. Then $\mathbf{F}_{q^n}$ consists of all elements in K which are fixed by Φ^n.

The *Frobenius collineation* associated to the Frobenius automorphism is the collineation of $\mathrm{PG}(n,K)$ with $\sigma = \Phi$; that is,

$$\mathbf{x} \mapsto \mathbf{x}', \quad {}^t\mathbf{x}' = A\,{}^t\Phi(\mathbf{x}),$$

with $\Phi(\mathbf{x}) = (x_0^q, x_1^q, \ldots, x_n^q)$, for some non-singular matrix A.

REMARK 1.3 (i) When $K = \mathbf{F}_q$, it is customary to replace K by q in the notation for all the spaces and groups; so $V(n,q)$ means $V(n,K)$, and similarly for $\mathrm{AG}(r,q)$, $\mathrm{PG}(r,q)$, $\mathrm{GL}(r,q)$, $\Gamma\mathrm{L}(r,q)$, $\mathrm{AGL}(r,q)$, $\mathrm{PGL}(r,q)$, and $\mathrm{P\Gamma L}(r,q)$.

(ii) When K is algebraically closed and has characteristic $p > 0$, then K contains the finite field $\mathbf{F}_q$ for every power of q of p.

1.2 POLYNOMIALS

DEFINITION 1.4 (i) A polynomial f in the ring $K[X_1, X_2, \ldots, X_n]$ of polynomials in the indeterminates $X_1, X_2, \ldots, X_n$ is *reducible* if there exist non-constant f_1, f_2 in $K[X_1, X_2, \ldots, X_n]$ with $f = f_1 f_2$; otherwise, f is *irreducible*.

(ii) The *degree* of a monomial $X_1^{r_1} X_2^{r_2} \ldots X_n^{r_n}$ is $r_1 + r_2 + \cdots + r_n$.

(iii) A polynomial is *homogeneous* if all its terms have the same degree.

(iv) The *degree* of a polynomial f is the largest degree of all its terms; write $\deg f$.

1.3 AFFINE PLANE CURVES

In the first instance, a curve is associated both to a set of points and to a polynomial. Let K be an algebraically closed field, and let $F \in K[X, Y]$. Then an affine curve is viewed as a set of points.

DEFINITION 1.5 (i) The *plane affine curve*

$$\mathcal{F} = \mathbf{v}_a(F) = \{P = (x, y) \in \mathrm{AG}(2, K) \mid F(x, y) = 0\}.$$

(ii) The *degree of* $\mathcal{F}$, written $\deg \mathcal{F}$, is $\deg F$.

Any affine transformation sends an affine curve to another having the same degree. Therefore $\deg \mathcal{F}$ of an affine curve $\mathcal{F}$ is an affine invariant.

DEFINITION 1.6 (i) A *component* of the affine curve $\mathcal{F} = \mathbf{v}_a(F)$ is an affine curve $\mathcal{G} = \mathbf{v}_a(G)$ such that G divides F.

(ii) The affine curve $\mathcal{F} = \mathbf{v}_a(F)$ is *irreducible* when it has no proper component, that is, when F is irreducible.

Components are covariant, that is, the diagram below is commutative for any affine transformation T of $\mathrm{AG}(2, K)$.

$$\begin{array}{ccc}
\mathcal{F} = \mathbf{v}(F) & \xrightarrow{\;T\;} & \mathcal{F}' = \mathbf{v}_a(F') \\
\text{component of } \mathcal{F} \Big\downarrow & & \Big\downarrow \text{component of } \mathcal{F}' \\
\mathcal{G} = \mathbf{v}_a(G) & \xrightarrow{\;T\;} & \mathcal{G}' = \mathbf{v}_\alpha(G')
\end{array}$$

Any line containing at least $n + 1$ points from an affine curve $\mathcal{F}$ of degree n is a component of $\mathcal{F}$. To show this, it may be assumed by covariance that $\ell = \mathbf{v}_a(Y)$. Let $\mathcal{F} = \mathbf{v}_a(F(X, Y))$. Then $|\ell \cap \mathbf{v}_a(F)| \geq n + 1$ implies that $F(X, 0)$ has more than n roots. Therefore $F(X, 0) = 0$, and hence X divides $F(X, Y)$.

Let $\mathcal{F} = \mathbf{v}_a(F)$ be an affine curve with $\deg F = d$, and let $\ell = bX - aY + c$ be a line containing the point $P_0 = (x_0, y_0)$ on $\mathcal{F}$. Then, for any point $P = (x, y) \in \ell$,

$$\begin{aligned} bx - ay &= bx_0 - ay_0, \\ b(x - x_0) &= a(y - y_0) = abt, \\ x = x_0 + at, &\quad y = y_0 + bt \end{aligned}$$

for some $t \in K$. Then

$$\begin{aligned} F(x, y) = F(x_0 + at, y_0 + bt) = G(t) &= G_0 + G_1 t + G_2 t^2 + \cdots + G_d t^d \\ &= G_m t^m + \cdots + G_d t^d, \end{aligned} \tag{1.1}$$

with $G_m \neq 0,\ G_d \neq 0$.

LEMMA 1.7 *The two irreducible curves $\mathcal{F}_1 = \mathbf{v}_a(F_1)$ and $\mathcal{F}_2 = \mathbf{v}_a(F_2)$ are the same if and only if $F_2 = \lambda F_1$ for some $\lambda \in K \backslash \{0\}$.*

Proof. This is a consequence of Theorem 2.10. □

DEFINITION 1.8 If $F \in K[X, Y]$ satisfies

$$F = F_1^{n_1} F_1^{n_2} \cdots F_s^{n_s}$$

with each F_i irreducible, then $\mathcal{F} = \mathbf{v}_a(F)$ has *components* $\mathcal{F}_i = \mathbf{v}_a(F_i)$ with *multiplicity* n_i for $i = 1, \ldots, s$.

The multiplicity of a component is an affine invariant.

DEFINITION 1.9 Let ℓ be a line which is not a component of $\mathcal{F}$.

(i) The integer m of (1.1) is the *intersection number of ℓ and $\mathcal{F}$ at P_0*: write

$$m = I(P_0, \ell \cap \mathcal{F});$$

(ii) if $m = 1$ for some line ℓ through P_0, then P_0 is a *simple* or *non-singular* point of $\mathcal{F}$;

(iii) if $m \geq 2$ for all lines ℓ through P_0, then P_0 is a *singular* or *multiple* point of $\mathcal{F}$;

(iv) if $m_0 = \min\{m \mid \ell \text{ a line through } P_0\}$, then m_0 is the *multiplicity of P_0 on $\mathcal{F}$*, or P_0 is an *m_0-fold point of $\mathcal{F}$*, and write

$$m_0 = m_{P_0}(\mathcal{F}) = m_{P_0}(F);$$

(v) if $m > m_0$ for a line ℓ, then ℓ is a *tangent to $\mathcal{F}$ at P_0*.

The intersection number and the multiplicity of a point are affine invariants.

DEFINITION 1.10 If $m_P(\mathcal{F}) = 2$, then P is a *double* point of $\mathcal{F}$. A double point P with two distinct tangents to $\mathcal{F}$ at P is a *node*, and with only one tangent to $\mathcal{F}$ at P is a *cusp*. If $m_P(\mathcal{F}) = 3$, then P is a *triple* point of $\mathcal{F}$.

REMARK 1.11 Let M be a subfield of K and suppose that $\mathcal{F}$ is defined over M, that is, $\mathcal{F} = \mathbf{v}(f(X,Y))$ with $f(X,Y) \in M[X,Y]$. If P is a double point with two distinct tangents, neither of them defined over M, then P is an *isolated double point over* M.

LEMMA 1.12 *If P_0 is a simple point of $\mathcal{F}$, then, in (1.1),*

$$G_1 = \left.\frac{\partial F}{\partial X}\right|_{P_0} a + \left.\frac{\partial F}{\partial Y}\right|_{P_0} b.$$

COROLLARY 1.13 *The tangent to $\mathcal{F}$ at a simple point $P = (x,y)$ is*

$$\ell_P = \left.\frac{\partial F}{\partial X}\right|_{P} (X - x) + \left.\frac{\partial F}{\partial Y}\right|_{P} (Y - y).$$

Note the meaning of this corollary: the line ℓ_P has intersection multiplicity at least 2 with $\mathcal{F}$ at P.

DEFINITION 1.14 A non-singular point P of $\mathcal{F}$ is a *point of inflexion* of $\mathcal{F}$ if

$$I(P, \ell_P \cap \mathcal{F}) \geq 3.$$

Here, P is also called an *inflexion* or, in some sources, a *flex*; the tangent ℓ_P at P is the *inflexional tangent*. Tangents and inflexional tangents are covariant.

REMARK 1.15 The behaviour of $P = (0,0)$ for an affine curve $\mathcal{F} = \mathbf{v}_a(F)$ follows simply from the form of F. Write

$$F(X,Y) = F_m + F_{m+1} + \cdots + F_d,$$

where F_i is homogeneous of degree i in X and Y, and $F_m \neq 0$. Then

(i) if $m > 0$, the point P lies on $\mathcal{F}$;

(ii) if $m = 1$, the point P is simple and F_1 is the tangent at P;

(iii) if $m \geq 2$, the term $F_m = \prod \ell_i$, where $\ell_1, \ldots, \ell_m$ are the tangents at P;

(iv) if $\ell_1, \ldots, \ell_m$ are distinct, then P is an *ordinary* multiple point.

EXAMPLE 1.16 (i) If $F = Y - X^3$, then $\mathcal{F} = \mathbf{v}_a(F)$ has no singular points but $(0,0)$ is an inflexion.

(ii) If $F = Y^2 - X^3$, then $\mathcal{F} = \mathbf{v}_a(F)$ has a singular point $(0,0)$ of multiplicity 2 with the repeated tangent Y; so it is a cusp.

See Figure 1.2.

DEFINITION 1.17 An affine curve $\mathcal{F} = \mathbf{v}_a(f(X,Y))$ is *rational* if there exist rational functions $a(T)/c(T)$ and $b(T)/c(T)$ with $a(T), b(T), c(T) \in K[T]$ such that $f(a(T)/c(T), b(T)/c(T)) = 0$.

Lines and conics, that is, curves of degree 1 and irreducible curves of degree 2, are rational curves.

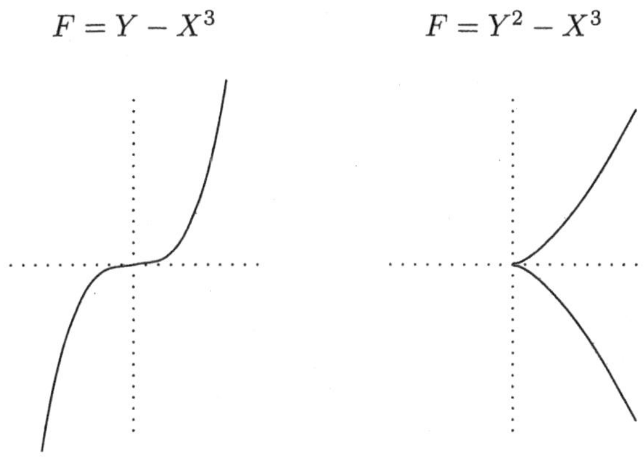

Figure 1.2 Plane cubics

1.4 PROJECTIVE PLANE CURVES

Let K be an algebraically closed field. For any polynomial $F \in K[X, Y]$ of degree d, associate the homogeneous polynomial $F^* \in K[X_0, X_1, X_2]$, given by

$$X = X_1/X_0, \quad Y = X_2/X_0, \quad F^*(X_0, X_1, X_2) = X_0^d F(X_1/X_0, X_2/X_0).$$

Similarly, for any homogeneous polynomial $F \in K[X_0, X_1, X_2]$, associate the polynomial $F_* \in K[X, Y]$, given by

$$F_*(X, Y) = F(1, X, Y).$$

DEFINITION 1.18 Given $F \in K[X, Y]$, the *projective plane curve* of affine equation $F(X, Y) = 0$, or homogeneous equation $F^*(X_0, X_1, X_2) = 0$, is

$$\mathcal{F} = \mathbf{v}(F) = \mathbf{v}(F^*) = \{(x_0, x_1, x_2) \in \mathrm{PG}(2, K) \mid F^*(x_0, x_1, x_2) = 0\}.$$

This definition implies that the projective curve consists of the affine points plus the points at infinity; that is,

$$\mathbf{v}(F) = \{(1, x, y) \mid (x, y) \in \mathbf{v}_a(F)\} \cup \{(0, x, y) = (0, \lambda x, \lambda y) \mid F^*(0, x, y) = 0\}.$$

REMARK 1.19 (i) For a linear form $L = a_0X_0 + a_1X_1 + a_2X_2$, the corresponding line is indicated both by L and $\mathbf{v}(L)$.

(ii) The notions of irreducibility, component, and degree extend in a natural way from affine curves to projective curves. It also follows, for any projective curve $\mathcal{F} = \mathbf{v}(F(X, Y))$ not containing the line $\mathbf{v}(X_0)$ as a component, that $\mathcal{F}$ is irreducible if and only if the affine curve $\mathcal{F}' = \mathbf{v}_a(F(X, Y))$ is irreducible.

DEFINITION 1.20 (i) With F homogeneous, a point $P = (x_0, x_1, x_2)$ of $\mathcal{F}$ is *singular* if

$$\frac{\partial F}{\partial X_0} = \frac{\partial F}{\partial X_1} = \frac{\partial F}{\partial X_2} = 0$$

at P.

(ii) Otherwise, P is simple and the *tangent at* P is

$$\left.\frac{\partial F}{\partial X_0}\right|_P X_0 + \left.\frac{\partial F}{\partial X_1}\right|_P X_1 + \left.\frac{\partial F}{\partial X_2}\right|_P X_2.$$

Throughout, write

$$X_\infty = (0,1,0), \quad Y_\infty = (0,0,1), \quad O = (1,0,0), \quad E = (1,1,1).$$

Here, X_∞ is the *point at infinity* on the X-axis, Y_∞ is the *point at infinity* on the Y-axis, and O is the *origin*.

In Example 1.16 (i), Y_∞ is a cusp on $\mathcal{F}$ and $F^* = YZ^2 - X^3$; in (2), Y_∞ is an inflexion and $F^* = Y^2Z - X^3$. So the curves are projectively equivalent.

In general, to determine the behaviour of a point, translate it to the origin, and use Remark 1.15.

Sometimes, it is convenient to use the notation

$$\mathbf{U}_0 = (1,0,0), \quad \mathbf{U}_1 = (0,1,0), \quad \mathbf{U}_2 = (0,0,1), \quad \mathbf{U} = (1,1,1).$$

REMARK 1.21 The properties of projective curves such as the degree, multiplicity of singular points, multiplicity of contact of a tangent are *covariant*, that is, invariant under projective transformations.

Let $\mathcal{F} = \mathbf{v}(F(X_0, X_1, X_2))$ be a projective plane curve of degree d.

DEFINITION 1.22 For any point $P = (x_0, x_1, x_2)$, if

$$G(X_0, X_1, X_2) = \frac{\partial F}{\partial X_0} x_0 + \frac{\partial F}{\partial X_1} x_1 + \frac{\partial F}{\partial X_2} x_2 \tag{1.2}$$

is not the zero polynomial, the plane curve $\mathcal{G} = \mathbf{v}(G(X_0, X_1, X_2))$ of degree $d-1$ is the *polar curve of* P with respect to $\mathcal{F}$. Otherwise, the polar curve of P is *vanishing*.

THEOREM 1.23 *The polar curve is covariant.*

Proof. If (X_0', X_1', X_2') is a new homogeneous coordinate system, the link between the old frame and the new one is given by a linear substitution $X_i = \sum_{j=0}^{2} a_{ij} X_j'$, $i = 0, 1, 2$, such that the matrix (a_{ij}) is non-singular. Under this change, the image of $\mathcal{F}$ is the curve $\mathcal{F}' = \mathbf{v}(F')$ with

$$F'(X_0', X_1', X_2') = F(\textstyle\sum_{j=0}^{2} a_{0j} X_j',\ \sum_{j=0}^{2} a_{1j} X_j',\ \sum_{j=0}^{2} a_{2j} X_j'),$$

while the image of P is the point $P' = (\sum_{j=0}^{2} a_{0j} x_j, \sum_{j=0}^{2} a_{1j} x_j, \sum_{j=0}^{2} a_{2j} x_j)$. Covariance means that the diagram below is commutative, which is now shown.

$$\begin{array}{ccc}
\mathcal{F} = \mathbf{v}(F) & \xrightarrow{(a_{ij})^{-1}} & \mathcal{F}' = \mathbf{v}(F') \\
\text{polar of } P \Big\downarrow & & \Big\downarrow \text{polar of } P' \\
\mathcal{G} = \mathbf{v}(G) & \xrightarrow{(a_{ij})^{-1}} & \mathcal{G}' = \mathbf{v}(G')
\end{array}$$

By the usual rules of the derivatives of composite functions,

$$\frac{\partial F'}{\partial X'_k} = \sum_{i=0}^{2} a_{ik} \frac{\partial F}{\partial X_i}, \qquad k = 0, 1, 2.$$

Therefore

$$\sum_{k=0}^{2} \frac{\partial F'}{\partial X'_k} x'_k = \sum_{i=0}^{2} \frac{\partial F}{\partial X_i} x_i.$$

□

THEOREM 1.24 *Suppose that the polar curve $\mathcal{F}'$ of P with respect to $\mathcal{F}$ is not vanishing.*

(i) *If A is a non-singular point of $\mathcal{F}$, then $A \in \mathcal{F}'$ if and only if the tangent to $\mathcal{F}$ at A passes through P.*

(ii) *Let A be an r-fold point of $\mathcal{F}$, with $r > 1$. Then*

 (a) *A is at least an $(r-1)$-fold point of $\mathcal{F}'$;*

 (b) *if there are infinitely many points P for which A is at least an r-fold point of $\mathcal{F}'$, every tangent to $\mathcal{F}$ at A has multiplicity divisible by p.*

Proof. Let $A = (a_0, a_1, a_2)$. The tangent line to $\mathcal{F}$ at A is

$$\frac{\partial F}{\partial X_0} X_0 + \frac{\partial F}{\partial X_1} X_1 + \frac{\partial F}{\partial X_2} X_2,$$

where the partial derivatives are evaluated at (a_0, a_1, a_2). Thus (i) is a consequence of the definition of a polar curve.

To show (ii), note that the covariance of the polar curve allows A to be mapped to the origin. Write

$$F = \Phi_0 X_0^d + \Phi_1 X_0^{d-1} + \cdots + \Phi_i X_0^{d-i} + \cdots + \Phi_d,$$

where Φ_i is a homogeneous polynomial in X_1 and X_2 of degree i. Since A is an r-fold point, $\Phi_0 = \cdots = \Phi_{r-1} = 0$, but $\Phi_r \neq 0$. Then, with G as in (1.2),

$$G = X_0^{d-r} G_1 + G_2,$$

where

$$G_1 = \frac{\partial \Phi_r}{\partial X_1} x_1 + \frac{\partial \Phi_r}{\partial X_2} x_2,$$

and where the terms in G_2 have degree at least r in X_1 and X_2. Since G_1 has degree at least $r-1$, so (a) follows.

If there are infinitely many points P for which A is an s-fold point of $\mathcal{F}'$ with $s \geq r$, then both $\partial\Phi_r/\partial X_1$ and $\partial\Phi_r/\partial X_2$ vanish. This can occur only when Φ_r is a p-th power of some polynomial Ψ_r. But then every tangent to $\mathcal{F}$ at A has multiplicity divisible by p. □

THEOREM 1.25 *If $\mathcal{F}$ is irreducible of degree d, then there is at most one point P with vanishing polar curve $\mathcal{F}'$.*

Proof. Assume, on the contrary, that the polar curves of two distinct points are vanishing. By the covariance of polar curves, let these points be X_∞ and Y_∞. Then $\partial F/\partial X_1$ and $\partial F/\partial X_2$ are both zero. Hence the general term in $F(X_0, X_1, X_2)$ is

$$a_{ij} X_0^{d-(i+j)} X_1^i X_2^j$$

with both i and j divisible by p. Then p divides d, as otherwise X_0 divides F, contradicting the irreducibility of $\mathcal{F}$. So p also divides $d-(i+j)$. Now define the coefficients b_{ij} by $b_{ij}^p = a_{ij}$, and the polynomial

$$L = \sum b_{ij} X_0^{(d-(i+j))/p} X_1^{i/p} X_2^{j/p}.$$

Then $F = L^p$ and $\mathcal{F}$ is reducible, a contradiction. □

Points with vanishing polar curves are characterised by a purely geometric property.

THEOREM 1.26 *If $\mathcal{F}$ is irreducible, then the polar curve of P is vanishing if and only if the tangents to $\mathcal{F}$ at non-singular points pass through P.*

Proof. By the covariance, suppose that P is the point Y_∞. With $\mathcal{F}$ in its inhomogeneous form $F(X,Y) = \sum a_{ij} X^i Y^j$, the polar curve of P is vanishing if and only if $a_{ij} = 0$ whenever $j \not\equiv 0 \pmod p$. This occurs if and only if there is no X_2-term in the form of the tangent to $\mathcal{F}$ at any non-singular point. □

This condition can be weakened.

THEOREM 1.27 *If $\mathcal{F}$ is irreducible, then the polar curve of P is vanishing if and only if infinitely many tangents to $\mathcal{F}$ pass through P.*

Proof. If the polar curve $\mathcal{F}'$ of P is vanishing, the assertion follows from Theorem 1.26. If $\mathcal{F}$ is not vanishing, then the weak form of Bézout's Theorem 3.13 applied to $\mathcal{F}$ and $\mathcal{F}'$ implies that there are finitely many common points. Since $\mathcal{F}$ has a finite number of singular points, P lies only on finitely many tangents to $\mathcal{F}$. □

As becomes apparent later, vanishing polar curves can cause serious difficulties in certain situations, especially when the point lies on the curve. Here, some examples are given after formalising these concepts.

DEFINITION 1.28 (i) A point is the *nucleus* of a projective irreducible curve $\mathcal{F}$ if it is the common point of all tangents to $\mathcal{F}$ at non-singular points.

(ii) A curve $\mathcal{F}$ is *strange* if it admits a nucleus N.

Following this definition, Theorem 1.26 states that a point P is a nucleus of $\mathcal{F}$ if and only if the polar curve of P with respect to $\mathcal{F}$ is vanishing. Theorem 1.25 has the following corollary.

THEOREM 1.29 *An irreducible curve has at most one nucleus.*

The simplest example of a strange curve is an irreducible conic in characteristic 2. Note that the nucleus does not lie on the conic.

EXAMPLE 1.30 An example of a strange curve $\mathcal{F}$ whose nucleus lies on the curve is the following. Let $p = 2$ and $q = 2^r$. For an integer $k < q$ with $\gcd(k, q-1) = 1$, let $k = 2^s u$ with $1 < s < r$, $u > 1$ and u odd. Let $\mathcal{F}$ be the irreducible curve $\mathbf{v}(f(X,Y))$, where

$$\begin{aligned}(X^k + 1)(Y + 1) + (Y^k + 1)(X + 1)\\ = f(X,Y)(X+1)(Y+1)(X+Y).\end{aligned} \tag{1.3}$$

To show that $N = (1,1,1)$ is a nucleus of $\mathcal{F}$, write (1.3) as follows:

$$\begin{aligned}(X^u)^{2^s} Y + (Y^u)^{2^s} X + (X^u + Y^u)^{2^s} + X + Y\\ = f(X,Y)(X+1)(Y+1)(X+Y).\end{aligned} \tag{1.4}$$

Using this form, the tangent to $\mathcal{F}$ at a non-singular point $A = (a_0, a_1, a_2)$ of $\mathcal{F}$ is

$$(a_1^k + a_2^k)X_0 + (a_0^k + a_2^k)X_1 + (a_0^k + a_1^k)X_2.$$

This shows that the tangent passes through the point N.

EXAMPLE 1.31 An example in odd characteristic is the following.

Let $p > 2$, and $\mathcal{F} = \mathbf{v}(Y^m - g(X))$ with $m \not\equiv 0 \pmod p$. Then $\mathcal{F}$ is strange if and only if one of the following hold:

(i) $g(X) = f(X^p) + cX$, where $f \in K[X]$ and $c \in K$ with $c = 0$ for $m \neq 1$;

(ii) $g(X) = (X + a)^r f(X^p)$ where $a \in K$, $f \in K[X]$ and $1 \leq r < p$ with $r \equiv m \pmod p$.

Strange curves have exceptional behaviour with respect to duality; this is treated in Section 5.11. They also cause difficulties in the resolution of singularities; see the last part of the proof of Theorem 3.27.

1.5 THE HESSIAN CURVE

Let $\mathcal{F} = \mathbf{v}(F(X_0, X_1, X_2))$ be a projective curve of degree d. Write

$$F_i = \frac{\partial F}{\partial X_i}, \quad F_{ij} = \frac{\partial^2 F}{\partial X_i \partial X_j}. \tag{1.5}$$

DEFINITION 1.32 If the determinant

$$H(X_0, X_1, X_2) = \begin{vmatrix} F_{00} & F_{01} & F_{02} \\ F_{01} & F_{11} & F_{12} \\ F_{02} & F_{12} & F_{22} \end{vmatrix}$$

is not vanishing, then the projective curve $\mathcal{H} = \mathbf{v}(H(X_0, X_1, X_2))$ is the *Hessian* curve of $\mathcal{F}$; it has degree $3(d-2)$. Otherwise, the Hessian curve is *vanishing*.

Strange curves have vanishing Hessian; see Exercise 3. Other relevant examples of curves with vanishing Hessian are the Hermitian curves.

THEOREM 1.33 *The Hessian curve is covariant.*

Proof. Arguing as in the proof of Theorem 1.23, for $0 \leq i, j \leq 2$,

$$\frac{\partial^2 F'}{\partial X_i' \partial X_j'} = \sum_{k=0}^{2} \sum_{m=0}^{2} a_{ki} a_{mj} \frac{\partial^2 F}{\partial X_k \partial X_m}.$$

This shows that

$$H'(X_0', X_1', X_2') = H(X_0, X_1, X_2) \cdot \det(a_{ij})^2.$$

Since $\det(a_{ij}) \neq 0$, the assertion follows. □

THEOREM 1.34 *If* $d \equiv 1 \pmod p$, *then the Hessian curve is vanishing.*

Proof. Let $G \in K[X_0, X_1, X_2]$ be a homogeneous polynomial of degree m. By Euler's formula,

$$\frac{\partial G}{\partial X_0} X_0 + \frac{\partial G}{\partial X_1} X_1 + \frac{\partial G}{\partial X_2} X_2 = mG.$$

When $G = F$, this becomes

$$\frac{\partial F}{\partial X_0} X_0 + \frac{\partial F}{\partial X_1} X_1 + \frac{\partial F}{\partial X_2} X_2 = d\,F, \tag{1.6}$$

while for $G = \partial F / \partial X_i$ with $i = 0, 1, 2$,

$$\frac{\partial^2 F}{\partial X_i \partial X_0} X_0 + \frac{\partial^2 F}{\partial X_i \partial X_1} X_1 + \frac{\partial^2 F}{\partial X_i \partial X_2} X_2 = (d-1) \frac{\partial F}{\partial X_i}. \tag{1.7}$$

Then, with the notation of (1.5),

$$X_0 H(X_0, X_1, X_2) = \begin{vmatrix} (d-1)F_0 & F_{01} & F_{02} \\ (d-1)F_1 & F_{11} & F_{12} \\ (d-1)F_2 & F_{12} & F_{22} \end{vmatrix}.$$

Therefore $H(X_0, X_1, X_2) = 0$ when $d \equiv 1 \pmod p$. □

The fundamental property of the Hessian curve is stated in the following theorem.

THEOREM 1.35 *Assume that* $\mathcal{H}$ *is not vanishing.*

(i) *Let* $p \neq 2$. *Then a non-singular point of* $\mathcal{F}$ *is an inflexion if and only if it is a common point of* $\mathcal{F}$ *and* $\mathcal{H}$.

(ii) *Every singular point of* $\mathcal{F}$ *lies on* $\mathcal{H}$.

Proof. Let P be a non-singular point of $\mathcal{F}$. Again by the covariance, suppose that P is the origin and that the tangent line ℓ to $\mathcal{F}$ at P is the X-axis. Then

$$F(X_0, X_1, X_2) = X_0^{d-1} X_2 + X_0^{d-2}(a_{20} X_1^2 + a_{11} X_1 X_2 + a_{02} X_2^2) + \cdots,$$

where the other terms are powers of X_0 with exponent at most $d-3$. A straightforward calculation shows that

$$H(X_0, X_1, X_2) = -2(d-1)^2 a_{20} X_0^{3(d-2)} + \cdots,$$

where the other terms are powers of X_0 with exponent less than $3(d-2)$. Since $p \neq 2$ and $d \not\equiv 1 \pmod p$ by Theorem 1.34, $\mathcal{H}$ passes through P if and only if $a_{20} = 0$; that is, P an inflexion. If P is singular, then the above computation shows that no term containing only X_0 appears in $H(X_0, X_1, X_2)$. Hence $P \in \mathcal{H}$. □

In affine form, when $\mathcal{F} = \mathbf{v}(f(X,Y))$, write

$$f_X = \frac{\partial f}{\partial X},\ f_Y = \frac{\partial f}{\partial Y},\ f_{XX} = \frac{\partial^2 f}{\partial X^2},\ f_{XY} = \frac{\partial^2 f}{\partial X \partial Y},\ f_{YY} = \frac{\partial^2 f}{\partial Y^2}.$$

Lemma 1.36 *Let $d \not\equiv 1 \pmod p$. If $\mathcal{F} = \mathbf{v}(f(X,Y))$, then $\mathcal{H} = \mathbf{v}(h(X,Y))$, where*

$$\begin{aligned} h(X,Y) = {} & (f_X)^2 f_{YY} + (f_Y)^2 f_{XX} - 2 f_X f_Y f_{XY} \\ & - d(d-1)^{-1} \left(f_{XX} f_{YY} - (f_{XY})^2 \right) f. \end{aligned}$$

Proof. As in the proof of Theorem 1.34, equations (1.6) and (1.7) are used. Substituting in $H(X_0, X_1, X_2)$, the expression reduces to the following:

$$X_0^2 H(X_0, X_1, X_2) = \begin{vmatrix} (d-1)dF & F_1 & F_2 \\ (d-1)^2 F_1 & F_{11} & F_{12} \\ (d-1)^2 F_2 & F_{12} & F_{22} \end{vmatrix}.$$

In inhomogeneous coordinates, this becomes the determinant

$$h(X,Y) = \begin{vmatrix} (d-1)d f & f_X & f_Y \\ (d-1)^2 f_X & f_{XX} & f_{XY} \\ (d-1)^2 f_Y & f_{XY} & f_{YY} \end{vmatrix}.$$

Expansion of the determinant and division by $-(d-1)^2$ give the result. □

When studying the inflexion points of $\mathcal{F}$ or, more generally, the intersection of $\mathcal{H}$ and $\mathcal{F}$, the last term of the polynomial in Theorem 1.36 can be omitted. So it is also possible to define the Hessian curve as $\mathcal{H}' = \mathbf{v}(h'(X,Y))$, with

$$h'(X,Y) = (f_X)^2 f_{YY} + (f_Y)^2 f_{XX} - 2 f_X f_Y f_{XY}. \tag{1.8}$$

An advantage is that the Hessian curve in this form is no longer necessarily vanishing when $\mathcal{F}$ has degree $d \equiv 1 \pmod p$; see Exercise 2. Therefore curves with such degrees can be considered in the study of inflexion points. With this approach, just two remarks are needed.

(1) Theorem 1.33 holds true if $\mathcal{H}$ is given by (1.8).

(2) Rewording the proof of Theorem 1.35 in terms of inhomogeneous coordinates, $h(X,Y) = -2a_{20} + \cdots$, which shows that condition $d \not\equiv 1 \pmod p$ depending on Theorem 1.34 disappears.

Therefore the following result is established.

Theorem 1.37 *Let $p \neq 2$. If either (1.8) is identically zero or the Hessian $\mathcal{H}$ of $\mathcal{F}$ as in (1.8) contains all points of $\mathcal{F}$, then every non-singular point of $\mathcal{F}$ is an inflexion. The converse also holds.*

REMARK 1.38 The form (1.8) of the Hessian curve is inadequate to deal with the even characteristic case. What is actually needed is to modify the definition of the second partial derivatives, as suggested by Hasse. If $f = \sum a_{ij}X^iY^j$ then the Hasse second partial derivatives are as follows:

$$\frac{\partial^{(2)} f(X,Y)}{\partial X^{(2)}} = \sum a_{ij} \binom{i}{2} X^{i-2}Y^j, \qquad \frac{\partial^{(2)} f(X,Y)}{\partial Y^{(2)}} = \sum a_{ij} \binom{j}{2} X^i Y^{j-2}.$$

Put $f = f(X,Y)$ for brevity. If $p \neq 2$, then

$$\frac{\partial^2 f}{\partial X^2} = 2\frac{\partial^{(2)} f}{\partial X^{(2)}}, \qquad \frac{\partial^2 f}{\partial Y^2} = 2\frac{\partial^{(2)} f}{\partial Y^{(2)}}.$$

Now the Hessian curve $\mathcal{H} = \mathbf{v}(\widetilde{h})$ with inhomogeneous form $\widetilde{h}$ is defined to be the curve with

$$\widetilde{h}(X,Y) = \left(\frac{\partial f}{\partial X}\right)^2 \frac{\partial^{(2)} f}{\partial Y^{(2)}} + \left(\frac{\partial f}{\partial Y}\right)^2 \frac{\partial^{(2)} f}{\partial X^{(2)}} - \frac{\partial f}{\partial X}\frac{\partial f}{\partial Y}\frac{\partial^2 f}{\partial X \partial Y}. \tag{1.9}$$

With this definition, $\mathcal{H}$ is still covariant, and Theorem 1.35 holds true for $p = 2$.

REMARK 1.39 The vanishing of the Hessian is related to exceptional behaviour of the dual curve in positive characteristic. This is treated in Section 5.11. The Hasse derivatives are useful in several other contexts; see Sections 5.10 and 7.6.

EXAMPLE 1.40 Let n be a positive integer which is not divisible by p. The curve

$$\mathcal{F} = \mathbf{v}(X_0^n + X_1^n + X_2^n)$$

is the *Fermat curve of degree* n.

When $K = \overline{\mathbf{F}}_q$ with q a power of p and $n = q+1$, then the Fermat curve is the *Hermitian curve* and denoted by $\mathcal{H}_q$; that is,

$$\mathcal{H}_q = \mathbf{v}(X_0^{q+1} + X_1^{q+1} + X_2^{q+1}).$$

The properties of the Hermitian curve are further developed in Section 12.3.

The Fermat curve is the set of its inflexion points if and only if (a) $n \equiv 1 \pmod{p}$ for $p \neq 2$ or (b) $n \equiv 1 \pmod{2^2}$ when $p = 2$. In particular, this holds for the Hermitian curve $\mathcal{H}_q$ when $p \neq 2$.

EXAMPLE 1.41 Let $p \neq 0$. Choose a power q of p such that $q \equiv -1 \pmod 3$. Let $\mathcal{F} = \mathbf{v}(F)$ be the plane curve with homogeneous form

$$F(X_0, X_1, X_2) = X_0^q X_2 + X_1^q X_0 + X_2^q X_1 - 3(X_0X_1X_2)^{(q+1)/3}.$$

It is first shown that $\mathcal{F}$ is an irreducible curve with only ordinary singularities.

Each of the fundamental lines meets $\mathcal{F}$ in only two points. If $\mathcal{G}$ is a component of $\mathcal{F}$ then $\mathcal{G}$ meets each fundamental line in at least one point, and so at least two of the vertices of the fundamental triangle are on $\mathcal{G}$. If $\mathcal{G}_1$ is another component of $\mathcal{F}$, then it also passes through two of the vertices of the fundamental triangle. So one of these points is common to $\mathcal{G}$ and $\mathcal{G}_1$, and hence is a singular point of $\mathcal{F}$. However, this is not the case, since a first partial derivative of F is not zero at each of these points. More precisely, $F_{X_0}(0,1,0) = F_{X_1}(0,0,1) = F_{X_2}(1,0,0) = 1 \neq 0$.

Next, it is shown that $\mathcal{F}$ has exactly $(q^2-q+1)/3$ singular points, each of which is an ordinary double point. This requires some typical computations involving polynomials and their partial derivatives.

Since no vertex of the fundamental triangle is a singular point of $\mathcal{F}$, it suffices to consider affine points. So put $X = X_1/X_0$, $Y = X_2/X_0$; then $\mathcal{F} = \mathbf{v}(f(X,Y))$ with

$$f(X,Y) = Y + X^q + XY^q - 3(XY)^{(q+1)/3}.$$

Let $A = (a,b)$ be a singular point of $\mathcal{F}$. Then $ab \neq 0$ and $f_X(a,b) = f_Y(a,b) = 0$. A direct calculation shows that this implies that

$$1 = a^{(q-2)/3}b^{(1-2q)/3}, \tag{1.10}$$

$$1 = a^{(q+1)/3}b^{(q-2)/3}. \tag{1.11}$$

Thus $a^{(q-2)/3}b^{(1-2q)/3} = a^{(q+1)/3}b^{(q-2)/3}$, whence $a = b^{1-q}$. Now (1.11) implies the following:

$$b^{(q^2-q+1)/3} = 1, \quad a = b^{-q^2}.$$

Conversely, if $a, b \in K$ satisfy these conditions, then $f_X(a,b) = f_Y(a,b) = 0$, showing that every point $A = (a,b)$ such that $b^{(q^2-q+1)/3} = 1$, $a = b^{1-q} = b^{-q^2}$ is a singular point of $\mathcal{F}$. Also,

$$\begin{aligned}
f_{XX}(a,b) &= \tfrac{2}{3}\, a^{(q-5)/3}b^{(q+1)/3},\\
f_{YY}(a,b) &= \tfrac{2}{3}\, a^{(q+1)/3}b^{(q-5)/3},\\
f_{XY}(a,b) &= \tfrac{1}{3}\, (ab)^{(q-2)/3},\\
f_{XX}(a,b)f_{YY}(a,b) - f_{XY}(a,b)^2 &= \tfrac{1}{3}\, (ab)^{(2q-4)/3}.
\end{aligned}$$

It follows that, if $p \neq 2$, then A is a double point, more precisely, a node. The same holds true for $p = 2$, but the proof requires Hasse partial derivatives, as explained in Remark 1.38.

Finally, the Hessian $\mathcal{H}$ of $\mathcal{F}$ is $\mathbf{v}((X_0X_1X_2)^{(2q-4)/3})$, which shows that $\mathcal{H}$ splits into three linear components, each counted $(2q-4)/3$ times.

EXAMPLE 1.42 For a divisor $k \geq 2$ of $q-1$, and $u, v \in K$ with $uv \neq 1$, the plane curve $\mathcal{F} = \mathbf{v}(vX^kY^k - X^k - Y^k + u)$ is irreducible and has the following properties.

(i) $\mathcal{F}$ has two singular points, namely X_∞ and Y_∞, both ordinary k-fold points.

(ii) The Hessian $\mathcal{H} = \mathbf{v}(H)$ of $\mathcal{F}$ has

$$H = k^3X^{k-2}Y^{k-2}(X^ku-1)(Y^ku-1)(2X^kY^ku + (k-1)(X^k+Y^k)).$$

In particular, $\mathcal{F}$ is not the locus of its inflexions.

(iii) For each $a \in K$ with $a^k = u$, the points $A_1 = (1,0,a)$ and $A_2 = (1,a,0)$ of $\mathcal{F}$ are inflexions. For $i = 1,2$, the tangent to $\mathcal{F}$ at A_i is $\ell_i = \mathbf{v}(X_i - aX_0)$. Also, $I(A_i, \mathcal{F} \cap \ell_i) = k$.

1.6 PROJECTIVE VARIETIES IN HIGHER-DIMENSIONAL SPACES

DEFINITION 1.43 (i) Given a homogeneous $F \in K[X_0, X_1, \ldots, X_n]$, the *projective hypersurface*

$$\mathcal{F} = \mathbf{v}(F) = \{(x_0, x_1, \ldots, x_n) \in \mathrm{PG}(n, K) \mid F(x_0, x_1, \ldots, x_n) = 0\}.$$

When $\deg F = 1$, the hypersurface is a *hyperplane*.

(ii) Given homogeneous $F_1, \ldots, F_r \in K[X_0, X_1, \ldots, X_n]$, the *projective variety*

$$\begin{aligned}\mathcal{F} = \mathbf{v}(F_1, \ldots, F_r) = \{\mathbf{x} = (x_0, x_1, \ldots, x_n) \in \mathrm{PG}(n, K) \mid \\ F_1(\mathbf{x}) = \cdots = F_r(\mathbf{x}) = 0\}.\end{aligned}$$

When $n = 2$, a projective hypersurface is just a projective plane curve. Varieties can also be curves in a higher-dimensional space; this is explained in Section 7.17.

1.7 EXERCISES

1. Let $p > 2$ and let $\mathcal{C} = \mathbf{v}(F)$, where

$$F(X_0, X_1, X_2) = X_0^{q+1} + X_1^{q+1} + X_2^{q+1} + X_0^{q-1} X_1 X_2.$$

Show that the Hessian curve $\mathcal{H}$ of $\mathcal{C}$ in its inhomogeneous form (1.8) is non-vanishing.

2. Let K have characteristic 2. Show that the Hessian curve $\mathcal{H}$ of the Hermitian curve $\mathcal{H}_2$ in its inhomogeneous form (1.9) is non-vanishing.

3. If $p > 3$, show that the Hessian of a strange curve vanishes.

4. Prove that a plane curve $\mathcal{F}$ of degree $k \geq 2$ is irreducible if it has both the following properties.

   (a) $\mathcal{F}$ contains a point A such that $I(A, \mathcal{F} \cap \ell) = k$ for the tangent ℓ of $\mathcal{F}$ at A when A is a simple point, and for at least one tangent ℓ to $\mathcal{F}$ at A when A is a singular point. In the latter case, ℓ is required to have multiplicity 1.

   (b) $\mathcal{F}$ contains no linear component through the point A.

5. Prove that an irreducible plane cubic curve is rational if and only if it is singular.

6. Prove that an irreducible plane curve of degree n with an $(n-1)$-fold point is rational.

7. Prove that any plane curve of degree at least 3 whose tangent lines at collinear points are concurrent is either strange or projectively equivalent to the Hermitian curve.

8. With $\mathcal{F}$ as in Example 1.30, show that the necessary and sufficient condition for N to lie on $\mathcal{F}$ is that $s > 1$. Also, show that the singular points of $\mathcal{F}$ are as follows:

   (a) N is an ordinary $(2^s - 2)$- fold point with tangents
   $$(m+1)X_0 + mX_1 + X_2$$
   for $m^{2^s-1} = 1$ but $m \neq 1$;

   (b) for each b with $b^u = 1$ but $b \neq 1$, the point $P = (1, b, 1)$ is a $(2^s - 1)$-fold point with a single tangent;

   (c) for each c with $c^u = 1$ but $c \neq 1$, the point $P = (1, 1, c)$ is a $(2^s - 1)$-fold point with a single tangent;

   (d) for all b, c with $b^u = c^u = 1$ but $b \neq 1$, $c \neq 1$, $b \neq c$, the point $P = (1, b, c)$ is a 2^s-fold point with a single tangent;

   (e) for b with $b^u = 1$ but $b \neq 1$, the point $P = (1, b, b)$ is a $(2^s - 1)$-fold point with a single tangent.

1.8 NOTES

There are many works on the elements of algebraic curves: Abhyankar [8], Baker [26], Bertini [49], Coolidge [85], Enriques [122], Fischer [128], Fulton [135], Lefschetz [301], Hartshorne [193], Hilton [212], Kirwan [269], Reid [368], Salmon [380], Segre [399], Seidenberg [400], Semple and Kneebone [402], Semple and Roth [403], Severi [408], Walker [497].

For the historical development of algebraic geometry, see Coolidge [86] and Dieudonné [104, 105].

Example 1.31 comes from [230]. For Exercise 7, see [236].

For strange curves, as in Section 1.4, see Hartshorne [192, Chapter IV].

In [37], strange curves invariant under a cyclic projective group fixing a triangle are investigated.

A survey paper on Fermat curves in positive characteristic is [429].

In the classical literature, polar curves and their generalisations, the m-ic polar curves, together with Hessian curves, are used to establish the Plücker formulas and their generalisations. These are equations linking numerical parameters of curves, such as the number of nodes, cusps, inflexion points, singularities of higher multiplicities, and other constants; see [85, Chapter 9]. Polar curves in positive characteristic are studied in [202]; see also [203].

Hessian curves are also important in investigating the number of points of curves defined over a finite field. For instance, let $\mathcal{F} = \mathbf{v}(F)$ be a projective irreducible curve of degree d, where $F \in \mathbf{F}_q[X_0, X_1, X_2]$. If its Hessian curve is non-vanishing, then the number of points of $\mathcal{F}$ lying in $\mathrm{PG}(2, q)$ is bounded above by $\frac{1}{2}d(q + d - 1)$; see Theorem 8.41. For q square and $d = \sqrt{q} + 1$, this upper bound is $\frac{1}{2}\sqrt{q}(\sqrt{q} + 1)^2$.

On the other hand, the Hermitian curve $\mathcal{H}_{\sqrt{q}}$ has vanishing Hessian and has $q\sqrt{q} + 1$ points lying in $\mathrm{PG}(2, q)$; see Example 8.67. Also, the polar curve of any point with respect to $\mathcal{H}_{\sqrt{q}}$ is reducible, being a line counted $\sqrt{q}$ times. See Chapter 12 for more examples of plane curves in positive characteristic with properties that a complex algebraic curve cannot have.

Chapter Two

Elimination theory

The intersection number $I(P, \ell \cap \mathcal{F})$ of an algebraic plane curve $\mathcal{F}$ and a line ℓ at a point P was used in Chapter 1 for the study of local projective invariants, such as singularities and inflexions. In this and subsequent chapters, a natural extension of $I(P, \ell \cap \mathcal{F})$ to the intersection number $I(P, \mathcal{G} \cap \mathcal{F})$ of $\mathcal{F}$ and another algebraic plane curve $\mathcal{G}$ at P is investigated. The main outcome, Bézout's Theorem 3.14, is an equation linking the degrees of $\mathcal{F}$ and $\mathcal{G}$ to the intersection numbers at their common points.

Consider $\mathcal{F} = \mathbf{v}(f(X, Y))$, $\ell = \mathbf{v}(uY - X)$ and $P = (0, 0)$, such that ℓ is not a component of $\mathcal{F}$. In this special situation, the intersection number $I(P, \ell \cap \mathcal{F})$ is obtained by eliminating X from the system

$$\begin{aligned} f(X,Y) &= 0, \\ uY - X &= 0; \end{aligned}$$

that is, $I(P, \ell \cap \mathcal{F}) = m$ if $D(Y) = F(uY, Y) = D_m Y^m + \cdots + D_d Y^d$ with $D_m \neq 0$.

When ℓ is replaced by $\mathcal{G} = \mathbf{v}(g(X, Y))$, the elimination of X or Y is no longer straightforward; the technical tool to deal with it is the resultant, treated in Sections 2.1 and 2.2.

From Section 2.3, the polynomial $D(Y)$ obtained by elimination of X in the system, $f = 0$, $g = 0$, is the resultant $R_X(f, g)$ of f and g. Therefore the natural definition of the intersection number of $\mathcal{F}$ and $\mathcal{G}$ at $P = (0, 0)$ is the degree of the smallest non-zero term in $D(Y)$. That this definition is satisfactory from a geometric point of view is shown in Section 3.1.

2.1 ELIMINATION OF ONE UNKNOWN

The *resultant* decides whether or not two given polynomials have a root in common. Two polynomials $f(X)$ and $g(X)$ with coefficients in a field L, not necessarily algebraically closed, have a root in common in L or in its closure $\overline{L}$ if and only if they have a factor in common.

Consider the two polynomials

$$\begin{aligned} f(X) &= a_0 X^n + a_1 X^{n-1} + \cdots + a_{n-1} X + a_n, \\ g(X) &= b_0 X^s + b_1 X^{s-1} + \cdots + b_{s-1} X + b_s. \end{aligned} \tag{2.1}$$

Let $\alpha_1, \ldots, \alpha_n$ be the roots of $f(X)$ and let $\beta_1, \ldots, \beta_s$ be the roots of $g(X)$; they may lie in $\overline{L}$ if L is not algebraically closed.

DEFINITION 2.1 The *resultant* of the polynomials $f(X)$ and $g(X)$ is

$$R(f,g) = ({a_0}^s{b_0}^n)\prod_{i=1}^n \prod_{j=1}^s (\alpha_i - \beta_j).$$

It is a symmetric polynomial in the two sets of indeterminates α_i, β_j.

LEMMA 2.2 *The resultant $R(f,g)$ is zero if and only if there exists a common root of the polynomials $f(X)$ and $g(X)$.*

LEMMA 2.3 *The following properties hold:*

(i) $R(f,g) = {a_0}^s\prod_{i=1}^n g(\alpha_i)$;

(ii) $R(f,g) = {b_0}^n\prod_{j=1}^s f(\beta_j)$;

(iii) $R(f,g) = (-1)^{ns}R(g,f)$;

(iv) $R(f_1f_2,g) = R(f_1,g)\cdot R(f_2,g)$.

Proof. (i) By definition,

$$R(f,g) = {a_0}^s{b_0}^n\prod_{i=1}^n \prod_{j=1}^s (\alpha_i - \beta_j).$$

Therefore

$$g(X) = b_0(X-\beta_1)\cdots(X-\beta_s) = b_0\prod_{j=1}^s (X-\beta_j).$$

Now substitute in place of X the value α_i:

$$g(\alpha_i) = b_0(\alpha_i-\beta_1)\cdots(\alpha_i-\beta_s) = b_0\prod_{j=1}^s (\alpha_i-\beta_j).$$

Then

$$\begin{aligned}\prod_{i=1}^n g(\alpha_i) &= b_0(\alpha_1-\beta_1)\cdots(\alpha_1-\beta_s)b_0(\alpha_2-\beta_1)\cdots(\alpha_2-\beta_s)\cdots\\ &\qquad b_0(\alpha_n-\beta_1)\cdots(\alpha_n-\beta_s)\\ &= b_0^n\prod_{i=1}^n \prod_{j=1}^s (\alpha_i-\beta_j).\end{aligned}$$

So

$$\prod_{i=1}^n g(\alpha_i) = b_0^n\prod_{i=1}^n \prod_{j=1}^s (\alpha_i-\beta_j).$$

Multiplying both sides of this equation by a_0^s and using Definition 2.1 shows that

$$R(f,g) = a_0^s\prod_{i=1}^n g(\alpha_i).$$

(ii) The proof of this is analogous.

(iii) This follows immediately.

(iv) Let $\deg g = s$, $\deg(f_1f_2) = m+n$, with $m = \deg f_1$ and $n = \deg f_2$. The leading coefficient of f_1f_2 is $a_{10}a_{20}$, where a_{10} and a_{20} are the leading coefficients of f_1 and f_2. The roots of f_1f_2 are given by $x_1,\ldots,x_n$ and $y_1,\ldots,y_m$, which are just the roots of f_1 and f_2; write these roots as α_i for $i = 1,\ldots,m+n$. Then

$$\begin{aligned}R(f_1f_2,g) &= (a_{10}a_{20})^s\prod_{i=1}^{m+n} g(\alpha_i)\\ &= a_{10}^s\prod_{i=1}^n g(\alpha_i)\cdot a_{20}^s\prod_{i=n+1}^{n+m} g(\alpha_i)\\ &= R(f_1,g)\cdot R(f_2,g).\end{aligned}$$

□

COROLLARY 2.4 *The product rule for resultants is as follows:*

$$R\left(\prod_{i=1}^{n} f_i, g\right) = \prod_{i=1}^{n} R(f_i, g).$$

The resultant $R(f, g)$ can be expressed as a polynomial in the indeterminates $a_0, \ldots, a_n, b_0 \ldots, b_s$. To show this, the theory of symmetric polynomials is used. From Definition 2.1,

$$R(\alpha_1, \ldots, \alpha_n, \beta_1, \ldots, \beta_s) = ({a_0}^s {b_0}^n)^{-1} R(f, g) = \prod_{i=1}^{n} \prod_{j=1}^{s} (\alpha_i - \beta_j)$$

is a symmetric polynomial with respect to both the set $\{\alpha_1, \ldots, \alpha_n\}$ and the set $\{\beta_1, \ldots, \beta_s\}$.

Let $\sigma_1, \ldots, \sigma_n$ be the elementary symmetric polynomials in the indeterminates $\alpha_1, \ldots, \alpha_n$ and let $\tau_1, \ldots, \tau_s$ be the polynomials in $\beta_1, \ldots, \beta_s$. By Newton's theorem on symmetric functions, $R(\alpha_1, \ldots \beta_s)$ can be expressed in terms of the σ_i and τ_j. Up to sign, $\sigma_1, \ldots, \sigma_n$ and $\tau_1, \ldots, \tau_s$ are the coefficients of the polynomials $a_0^{-1} f(X)$ and $b_0^{-1} g(X)$. So $R(\alpha_1, \ldots, \beta_s)$ can be expressed as a polynomial in the indeterminates

$$\frac{a_1}{a_0}, \ldots, \frac{a_n}{a_0}, \frac{b_1}{b_0}, \ldots, \frac{b_s}{b_0}.$$

The factor ${a_0}^s {b_0}^n$ makes the denominator in $R(\alpha_1, \ldots, \beta_s)$ disappear.

Now it is shown that $R(f, g)$, as given by Definition 2.1, is equal to a determinant D of order $n + s$.

THEOREM 2.5 *Given*

$$f(X) = a_0 X^n + a_1 X^{n-1} + \cdots + a_{n-1} X + a_n,$$
$$g(X) = b_0 X^s + b_1 X^{s-1} + \cdots + b_{s-1} X + b_s,$$

then $R(f, g) = D$, *where*

$$D = \begin{vmatrix} a_0 & a_1 & \ldots & a_n & 0 & 0 & & \ldots & 0 \\ 0 & a_0 & a_1 & \ldots & a_n & 0 & & \ldots & 0 \\ \vdots & & \vdots & & & & & \vdots & \\ 0 & 0 & \ldots & 0 & a_0 & a_1 & & \ldots & a_n \\ b_0 & b_1 & & \ldots & & b_s & 0 & \ldots & 0 \\ 0 & b_0 & b_1 & & \ldots & & b_s & 0 \ldots & 0 \\ \vdots & & & & & \vdots & & \vdots & \\ 0 & \ldots & 0 & b_0 & b_1 & & & \ldots & b_s \end{vmatrix} \begin{matrix} \left.\vphantom{\begin{matrix}a\\a\\a\\a\end{matrix}}\right\} s \text{ rows} \\ \left.\vphantom{\begin{matrix}a\\a\\a\\a\end{matrix}}\right\} n \text{ rows} \end{matrix} . \qquad (2.2)$$

Proof. Let $\alpha_1, \ldots, \alpha_n$ be the roots of $f(X)$ and $\beta_1, \ldots, \beta_s$ the roots of $g(X)$. Consider another determinant of order $n + s$:

$$M = \begin{vmatrix} \beta_1^{n+s-1} & \beta_2^{n+s-1} & \cdots & \beta_s^{n+s-1} & \alpha_1^{n+s-1} & \cdots & \alpha_n^{n+s-1} \\ \beta_1^{n+s-2} & \beta_2^{n+s-2} & \cdots & \beta_s^{n+s-2} & \alpha_1^{n+s-2} & \cdots & \alpha_n^{n+s-2} \\ \vdots & \vdots & & \vdots & \vdots & & \vdots \\ \beta_1^2 & \beta_2^2 & \cdots & \beta_s^2 & \alpha_1^2 & \cdots & \alpha_n^2 \\ \beta_1 & \beta_2 & \cdots & \beta_s & \alpha_1 & \cdots & \alpha_n \\ 1 & 1 & \cdots & 1 & 1 & \cdots & 1 \end{vmatrix} . \qquad (2.3)$$

The product $(a_0{}^s b_0{}^n)DM$ is now calculated in two different ways. First, M is a Vandermonde determinant , whose value is given by the following formula:

$$M = \prod_{1\leq i<j\leq s}(\beta_i - \beta_j) \prod_{j=1}^{s}\prod_{i=1}^{n} (\beta_j - \alpha_i) \prod_{1\leq i<j\leq n} (\alpha_i - \alpha_j).$$

Now

$$\begin{aligned} & a_0^s b_0^n \, DM \\ & = D\, a_0^s b_0^n \prod_{j=1}^{s}\prod_{i=1}^{n} (\beta_j - \alpha_i) \cdot \prod_{1\leq i<j\leq s} (\beta_i - \beta_j) \prod_{1\leq i<j\leq n}(\alpha_i - \alpha_j) \\ & = D \cdot R(g,f) \prod_{1\leq i<j\leq s} (\beta_i - \beta_j) \prod_{1\leq i<j\leq n} (\alpha_i - \alpha_j). \qquad (2.4) \end{aligned}$$

The product of the matrices associated to the determinants D and M is the matrix

$$N = \begin{bmatrix} B & \mathbf{0} \\ \mathbf{0} & A \end{bmatrix},$$

where

$$B = \begin{bmatrix} \beta_1^{s-1} f(\beta_1) & \beta_2^{s-1} f(\beta_2) & \cdots & \beta_s^{s-1} f(\beta_s) \\ \beta_1^{s-2} f(\beta_1) & \beta_2^{s-2} f(\beta_2) & \cdots & \beta_s^{s-2} f(\beta_s) \\ \vdots & \vdots & \cdots & \vdots \\ f(\beta_1) & f(\beta_2) & \cdots & f(\beta_s) \end{bmatrix},$$

$$A = \begin{bmatrix} \alpha_1^{n-1} g(\alpha_1) & \cdots & \alpha_n^{n-1} g(\alpha_n) \\ \alpha_1^{n-2} g(\alpha_1) & \cdots & \alpha_n^{n-2} g(\alpha_n) \\ \vdots & \cdots & \vdots \\ g(\alpha_1) & \cdots & g(\alpha_n). \end{bmatrix}.$$

Expanding the determinant $\det N$ of N by Laplace's method and taking out column and row factors, and calculating the corresponding Vandermonde determinants gives the following:

$$\begin{aligned} & (a_0{}^s b_0{}^n) \det N \\ & = (a_0^s\, b_0^n) \prod_{j=1}^{s} f(\beta_j) \prod_{i=1}^{n} g(\alpha_i) \prod_{1\leq k<l\leq s} (\beta_k - \beta_l) \prod_{1\leq r<t\leq n} (\alpha_r - \alpha_t). \end{aligned}$$

Regarding this as a polynomial relation between the indeterminates $\alpha_1, \ldots, \alpha_n$, $\beta_1, \ldots, \beta_s$, and using the fact that

$$R(f,g) = a_0^s \prod_{i=1}^{n} g(\alpha_i), \; R(g,f) = b_0^n \prod_{j=1}^{s} f(\beta_j),$$

it follows that

$$(a_0^s\, b_0^n) \det N = R(f,g) \cdot R(g,f) \prod_{1\leq i<j\leq s}(\beta_i - \beta_j) \prod_{1\leq i<j\leq n}(\alpha_i - \alpha_j). \quad (2.5)$$

Equating the expressions (2.4) and (2.5), and simplifying, gives the required result that $R(f,g) = D$. □

PROPOSITION 2.6 *The determinant D is isobaric of weight $n \cdot s$ in the indeterminates $a_0, \ldots, a_n, b_0, \ldots, b_s$.*

Proof. Given a general term of D, namely

$$c a_0^{i_0} a_1^{i_1} \cdots a_n^{i_n} b_0^{j_0} b_1^{j_1} \cdots b_s^{j_s},$$

define the weight of such a term as the following quantity:

$$\sum_{h=1}^{n} hi_h + \sum_{k=1}^{s} kj_k.$$

Let λ be a non-zero element of L. Consider, other than the polynomials $f(X)$ and $g(X)$, two new polynomials:

$$\begin{aligned} f_\lambda(X) &= a_0X^n + a_1\lambda X^{n-1} + \cdots + a_{n-1}\lambda^{n-1}X + a_n\lambda^n, \\ g_\lambda(X) &= b_0X^s + b_1\lambda X^{s-1} + \cdots + b_{s-1}\lambda^{s-1}X + b_s\lambda^s. \end{aligned}$$

If α_i is a root of $f(X)$, then $\lambda\alpha_i$ is a root of $f_\lambda(X)$ for each $i = 1, \ldots, n$. In fact,

$$\begin{aligned} f_\lambda(\lambda\alpha_i) &= a_0\lambda^n\alpha_i^n + a_1\lambda\lambda^{n-1}\alpha_i^{n-1} + \cdots + a_n\lambda^n \\ &= \lambda^n(a_0\alpha_i^n + \cdots + a_n) \\ &= \lambda^n f(\alpha_i) \\ &= 0. \end{aligned}$$

In a similar way, if β_j is a root of $g(X)$, then $\lambda\beta_j$ is a root of $g_\lambda(X)$ for each $j = 1, \ldots, s$.

The resultant $R(f_\lambda, g_\lambda)$ equals the determinant

$$D_\lambda = \left|\begin{array}{ccccccccc} a_0 & \lambda a_1 & \cdots & & & \lambda^n a_n & & & \\ & a_0 & \lambda a_1 & \cdots & & & \lambda^n a_n & & \\ & & & \vdots & & \vdots & & & \\ & & & a_0 & \lambda a_1 & \cdots & & & \lambda^n a_n \\ b_0 & \lambda b_1 & \cdots & & \lambda^s b_s & & & & \\ & b_0 & \lambda b_1 & \cdots & & \lambda^s b_s & & & \\ & & & \vdots & & \vdots & & & \\ & & & & b_0 & \lambda b_1 & \cdots & & \lambda^s b_s \end{array}\right|, \quad \begin{array}{l} \left.\right\} s \text{ rows} \\ \left.\right\} n \text{ rows} \end{array}$$

whose general term is

$$ca_0^{i_0}(\lambda a_1)^{i_1}(\lambda^2 a_2)^{i_2}\cdots(\lambda^n a_n)^{i_n} b_0^{j_0}(\lambda b_1)^{j_1}\cdots(\lambda^s b_s)^{j_s}.$$

In terms of λ, this becomes

$$c\lambda^{i_1+2i_2+\cdots+ni_n+j_1+2j_2+\cdots+sj_s} \cdot (a_0^{i_0}a_1^{i_1}\cdots a_n^{i_n} b_0^{j_0} b_1^{j_1}\cdots b_s^{j_s}).$$

So D and D_λ differ solely by the factor

$$\lambda^{i_1+2i_2+\cdots+ni_n+j_1+2j_2+\cdots+sj_s}.$$

Again, by Definition 2.1,

$$R(f, g) = {a_0}^s {b_0}^n \prod_{i=1}^{n}\prod_{j=1}^{s}(\alpha_i - \beta_j).$$

So

$$\begin{aligned} R(f_\lambda, g_\lambda) &= {a_0}^s {b_0}^n \prod_{i=1}^{n}\prod_{j=1}^{s}(\lambda\alpha_i - \lambda\beta_j) \\ &= {a_0}^s {b_0}^n \lambda^{ns} \prod_{i=1}^{n}\prod_{j=1}^{s}(\alpha_i - \beta_j) = \lambda^{ns} R(f, g). \end{aligned}$$

Hence

$$D_\lambda = R(f_\lambda, g_\lambda) = \lambda^{ns} R(f, g) = \lambda^{ns} D,$$

from which it follows that

$$\lambda^{i_1+2i_2+\cdots+j_1+\cdots+sj_s} = \lambda^{ns},$$

as required. □

The expansion of the determinant (2.2) gives a homogeneous polynomial in the $n+s+2$ indeterminates $a_0, \ldots, a_n, b_0, \ldots, b_s$, whose general term has degree $n+s$. Now some properties of the polynomial D and then of $R(f, g)$ are described.

THEOREM 2.7 *The determinant $D(a_0, \ldots, a_n, b_0, \ldots, b_s)$ is homogeneous both in $a_0, \ldots, a_n$ and in $b_0, \ldots, b_s$.*

Proof. The general term of the determinant D is

$$a_0^{i_0} a_1^{i_1} \cdots a_n^{i_n} b_0^{j_0} b_1^{j_1} \cdots b_s^{j_s},$$

where the factors a_h, each counted with multiplicity i_h, are in number $\sum i_h = s$, since they are precisely the factors coming from the first s rows of (2.2). Since $D(a_0, \ldots, b_s)$, viewed as a polynomial only in the indeterminates $a_0, \ldots, a_n$, is a sum of monomials all of the same degree s. So such a polynomial is homogeneous in $a_0, \ldots, a_n$. The result is established analogously for the indeterminates $b_0, \ldots, b_s$. □

THEOREM 2.8 *The polynomial $D(a_0, \ldots, a_n, b_0, \ldots, b_s)$ is irreducible over K.*

Proof. To prove the result, the relation between D and $R(f, g)$ is used.

Suppose that D is reducible; that is, there exist non-constant homogeneous polynomials P, $Q \in K[a_0, \ldots, a_n, b_0, \ldots, b_s]$ such that $D = P \cdot Q$. Now introduce new indeterminates $a_1', a_2', \ldots, a_n'$ and $b_1', b_2', \ldots, b_s'$, with $a_i' = a_i/a_0$ and $b_j' = b_j/b_0$. Then

$$D(a_0, a_1, \ldots, a_n, b_0, b_1, \ldots, b_s) = D(a_0, a_0a_1', \ldots, a_0a_n', b_0, b_0b_1', \ldots, b_0b_s').$$

By the homogeneity of D,

$$\begin{aligned} &D(a_0, a_0a_1', \ldots, a_0a_n', b_0, b_0b_1', \ldots, b_0b_s') \\ &\quad = {a_0}^s {b_0}^n D(1, a_1', \ldots, a_n', 1, b_1', \ldots, b_s'). \end{aligned}$$

Put

$$D(1, a_1', \ldots, a_n', 1, b_1', \ldots, b_s') = D'(a_1', \ldots, a_n', b_1', \ldots, b_s');$$

the polynomial D' is not constant, since ${a_0}^s {b_0}^n$ does not divide D. Hence

$$\begin{aligned} &D'(a_1', \ldots, a_n', b_1', \ldots, b_s') \\ &\quad = P'(a_1', \ldots, a_n', b_1', \ldots, b_s') \cdot Q'(a_1', \ldots, a_n', b_1', \ldots, b_s'). \end{aligned}$$

Consider

$$R'(f, g) = R(f, g)/({a_0}^s {b_0}^n) = D/({a_0}^s {b_0}^n) = D' = P'Q'.$$

Expressing the a_i' and b_j' as elementary symmetric functions in the α_i and β_j gives that

$$P' \cdot Q' = \prod_{i=1}^{n} \prod_{j=1}^{s} (\alpha_i - \beta_j).$$

Since the factor $(\alpha_i - \beta_j)$ divides $P' \cdot Q'$, it must divide one of the two polynomials. Suppose P' contains it as a factor; then, as P' is a symmetric polynomial in the α_i and in the β_j, it contains them all. Therefore Q' is a constant, whence the irreducibility of D' and so of D follows. □

THEOREM 2.9 *Given two polynomials $f(X)$, $g(X)$ with coefficients in L, there exist polynomials $A(X)$, $B(X)$ with $\deg A(X) \leq s-1$ and $\deg B(X) \leq n-1$ such that*

$$R(f,g) = A(X)f(X) + B(X)g(X).$$

Proof. Write

$$\begin{aligned} f(X) &= a_0X^n + \cdots + a_{n-1}X + a_n, \\ g(X) &= b_0X^s + \cdots + b_{s-1}X + b_s, \end{aligned}$$

and consider the following identities:

$$\begin{aligned} X^{s-1}f(X) &= a_0X^{n+s-1} + \cdots + a_{n-1}X^s + a_nX^{s-1}, \\ X^{s-2}f(X) &= \quad a_0X^{n+s-2} + \cdots + a_{n-1}X^{s-1} + a_nX^{s-2}, \\ &\;\vdots \\ f(X) &= \qquad\qquad a_0X^n + \cdots + a_{n-1}X + a_n, \\ X^{n-1}g(X) &= b_0X^{s+n-1} + \cdots + b_{s-1}X^n + b_sX^{n-1}, \\ X^{n-2}g(X) &= \quad b_0X^{s+n-2} + \cdots + b_{s-1}X^{n-1} + b_sX^{n-2}, \\ &\;\vdots \\ g(X) &= \qquad\qquad b_0X^s + \cdots + b_{s-1}X + b_s. \end{aligned}$$

The determinant of the coefficients of the right-hand side is the same as the determinant (2.2).

Let A_i for $i = 0, \ldots, s-1$ and B_j for $j = 0, \ldots, n-1$ be the cofactors of the determinant (2.2) with respect to the last column. Multiplying the i-th of the first s rows by A_i and, similarly, the i-th of the remaining n rows by B_i, and summing gives

$$A(X)f(X) + B(X)g(X) = R(f,g),$$

with $A(X) = A_0X^{s-1} + \cdots + A_{s-1}$ and $B(X) = B_0X^{n-1} + \cdots + B_{n-1}$. □

THEOREM 2.10 (Study) *Let f and g be elements of $K[X_1, \ldots, X_r]$, and let f be irreducible over K. If each root $\xi_1, \ldots, \xi_r$ of f is a root of g, then f divides g.*

Proof. Put

$$\begin{aligned} f(X_1, \ldots, X_r) = a_0(X_1, \ldots, X_{r-1})X_r^n &+ a_1(X_1, \ldots, X_{r-1})X_r^{n-1} \\ &+ \cdots + a_n(X_1, \ldots, X_{r-1}), \end{aligned}$$

and consider it as a polynomial in the single variable X_r. Suppose that the indeterminate X_r does not occur in g; then there exist $\xi_1, \ldots, \xi_{r-1} \in K$ such that

$$g(\xi_1, \ldots, \xi_{r-1}) \cdot a_0(\xi_1, \ldots, \xi_{r-1}) \neq 0,$$

since, otherwise,

$$g(X_1,\dots,X_{r-1})\cdot a_0(X_1,\dots,X_{r-1})=0,$$

for all $X_1,\dots,X_{r-1}$. So one of the two terms of the product must be zero; but this is impossible by hypothesis.

Since K is algebraically closed, the polynomial $f(\xi_1,\dots,\xi_{r-1},X_r)$ has a root $X_r=\xi_r$. From this, it follows that the r-ple $(\xi_1,\dots,\xi_r)$ are roots of f but not of g, contrary to hypothesis.

Therefore write g as a polynomial in X_r :

$$\begin{aligned}&g(X_1,\dots,X_r)\\&=b_0(X_1,\dots,X_{r-1})X_r^s+b_1(X_1,\dots,X_{r-1})X_r^{s-1}+\cdots+b_s(X_1,\dots,X_{r-1}).\end{aligned}$$

Consider the indeterminates $X_1,\dots,X_{r-1}$ as parameters so that f and g become polynomials in the single variable X_r; that is, $f=f(X_r)$ and $g=g(X_r)$. By Theorem 2.9, there exist polynomials $A(X_r)$ and $B(X_r)$ such that

$$A(X_r)f(X_r)+B(X_r)g(X_r)=R(f,g),$$

where $R(f,g)$ is a polynomial in $X_1,\dots,X_{r-1}$, and $A=A(X_r)$ and $B=B(X_r)$ depend on the parameters $X_1,\dots,X_{r-1}$.

If ξ_r is a root of $f(X_r)$, then ξ_r is also a root of $g(X_r)$ by hypothesis; therefore

$$A(X_1,\dots,X_r)f(X_1,\dots,X_r)+B(X_1,\dots,X_r)g(X_1,\dots,X_r)=0.$$

Hence

$$A(X_1,\dots,X_r)f(X_1,\dots,X_r)=-B(X_1,\dots,X_r)g(X_1,\dots,X_r).$$

So $f(X_1,\dots,X_r)$ divides $-B(X_1,\dots,X_r)g(X_1,\dots,X_r)$, and since f is irreducible, it must divide one of the two polynomials. However, as $\deg B<\deg f$, so f divides g. □

The polynomials f and g, as in (2.1), with $a_0,b_0\neq 0$, have a root in common if and only if $R(f,g)=0$.

If a_0 or b_0 or both are zero, the expression in Definition 2.1 for the resultant is no longer usable. Instead, D can be used, keeping in mind that $D=R(f,g)$ when $a_0b_0\neq 0$. It can no longer be affirmed that the existence of roots common to f and g is equivalent to the resultant being zero. In fact, if $a_0=b_0=0$, then, in each case, $R(f,g)=0$, independently of the existence of common roots; but this is the only case in which the resultant being zero does not guarantee that there are common roots.

THEOREM 2.11 *If the leading coefficients of the polynomials f and g given by* (2.1) *are not both zero, then $R(f,g)=0$ if and only if the two polynomials have a root in common.*

Proof. Suppose that $a_0\neq 0$, $b_0=b_1=\cdots=b_{k-1}=0$, but $b_k\neq 0$. Put

$$\overline{g}(X)=b_kX^{s-k}+b_{k+1}X^{s-(k+1)}+\cdots+b_{s-1}X+b_s.$$

Since the leading coefficients of f and $\overline{g}$ are both non-zero, $R(f,\overline{g})=0$ if and only if f and $\overline{g}$ have a root in common.

Substituting zero for the elements $b_0, b_1, \ldots, b_{k-1}$ in (2.2) and expanding by Laplace's Theorem, it follows that $R(f,g) = a_0^k R(f,\overline{g})$. The polynomials $g(X)$ and $\overline{g}(X)$ have the same roots, whence $R(f,g) = 0$ implies that $a_0^k R(f,\overline{g}) = 0$. □

These results extend to homogeneous polynomials. As in Section 1.4, with f a polynomial of degree n,

$$f(X) = a_0 X^n + a_1 X^{n-1} + \cdots + a_{n-1} X + a_n,$$

let

$$\begin{aligned} F(X,Y) = f^*(X,Y) &= Y^n f(X/Y) \\ &= a_0 X^n + a_1 X^{n-1} Y + \cdots + a_{n-1} X Y^{n-1} + a_n Y^n. \end{aligned}$$

Theorem 2.11 immediately gives the comparable result for homogeneous polynomials.

THEOREM 2.12 *The homogeneous polynomials*

$$\begin{aligned} F(X,Y) &= a_0 X^n + a_1 X^{n-1} Y + \cdots + a_{n-1} X Y^{n-1} + a_n Y^n, \\ G(X,Y) &= b_0 X^s + b_1 X^{s-1} Y + \cdots + b_{s-1} X Y^{s-1} + b_s Y^s \end{aligned}$$

have a non-constant common factor if and only if the determinant D in (2.2) is zero.

Proof. If $a_0 = b_0 = 0$, then $D = 0$ and X is a common factor of F and G. Otherwise, with $f(X) = F(X,1)$ and $g(X) = G(X,1)$, Theorem 2.11 gives the result. □

THEOREM 2.13 *Given the homogeneous polynomials*

$$\begin{aligned} F(X_0,\ldots,X_r) &= A_0 X_r^n + A_1 X_r^{n-1} + \cdots + A_{n-1} X_r + A_n, \\ G(X_0,\ldots,X_r) &= B_0 X_r^s + B_1 X_r^{s-1} + \cdots + B_{s-1} X_r + B_s, \end{aligned}$$

where A_i and B_i are homogeneous in $K[X_0,\ldots,X_{r-1}]$, with $A_0 B_0 \neq 0$, the resultant $R(F,G)$ is a homogeneous polynomial of degree ns if F and G have no common factor.

Proof. Let $A_i = A_i(X_0,\ldots,X_{r-1})$. Then $R_{X_r}(F,G) = D(X)$, with the notation $X = (X_0,\ldots,X_{r-1})$, is given by the determinant

$$\left| \begin{array}{cccccccccc} A_0 & A_1 & & \cdots & \cdots & A_n & 0 & \cdots & 0 \\ 0 & A_0 & A_1 & \cdots & \cdots & & A_n & \cdots & 0 \\ & \vdots & & \vdots & & \vdots & & & \\ 0 & \cdots & 0 & A_0 & A_1 & \cdots & & \cdots & A_n \\ B_0 & B_1 & \cdots & & B_s & 0 & \cdots & & 0 \\ 0 & B_0 & B_1 & \cdots & & B_s & 0 & \cdots & 0 \\ & \vdots & & \vdots & & \vdots & & & \\ 0 & & \cdots & 0 & B_0 & B_1 & \cdots & & B_s \end{array} \right| \begin{array}{l} \left.\vphantom{\begin{array}{c}1\\1\\1\\1\end{array}}\right\} s \text{ rows} \\ \left.\vphantom{\begin{array}{c}1\\1\\1\\1\end{array}}\right\} n \text{ rows} \end{array}$$

Replace X by tX. Since A_i and B_j are homogeneous polynomials, $D(tX)$ is equal to the determinant

$$\left.\begin{vmatrix}
A_0 & tA_1 & & \cdots & \cdots & t^n A_n & 0 & \cdots & 0 \\
0 & A_0 & tA_1 & \cdots & \cdots & & t^n A_n & \cdots & 0 \\
& \vdots & & \vdots & & \vdots & & & \\
0 & \cdots & 0 & A_0 & tA_1 & \cdots & & \cdots & t^n A_n \\
B_0 & tB_1 & \cdots & & t^s B_s & 0 & \cdots & & 0 \\
0 & B_0 & tB_1 & \cdots & & t^s B_s & 0 & \cdots & 0 \\
& \vdots & & \vdots & & \vdots & & & \\
0 & & \cdots & 0 & B_0 & tB_1 & \cdots & & t^s B_s
\end{vmatrix}\begin{array}{l} \left.\begin{array}{l} \\ \\ \\ \\ \end{array}\right\} s \text{ rows} \\ \left.\begin{array}{l} \\ \\ \\ \\ \end{array}\right\} n \text{ rows} \end{array}\right. .$$

Multiply the i-th row in A by t^i for $1 \leq i \leq s$, and the j-th row in B by t^j for $1 \leq j \leq n$. This shows that $t^u D(tX)$ is equal to the determinant

$$\begin{vmatrix}
tA_0 & t^2A_1 & \cdots & \cdots & t^{n+1}A_n & 0 & \cdots & 0 \\
0 & t^2A_0 & t^3A_1 & \cdots & t^{n+1}A_{n-1} & t^{n+2}A_n & \cdots & 0 \\
& \vdots & & \vdots & & \vdots & & \\
0 & \cdots & 0 & t^sA_0 & t^{s+1}A_1 & \cdots & & t^{n+s}A_n \\
tB_0 & t^2B_1 & \cdots & t^{s+1}B_s & 0 & \cdots & & 0 \\
0 & t^2B_0 & t^3B_1 & \cdots & t^{s+1}B_{s-1} & t^{s+2}B_s & \cdots & 0 \\
& \vdots & & \vdots & & \vdots & & \\
0 & \cdots & 0 & t^nB_0 & t^{n+1}B_1 & \cdots & & t^{s+n}B_s
\end{vmatrix},$$

where $u = \frac{1}{2}s(s+1) + \frac{1}{2}n(n+1)$.

The last determinant is equal to $t^v D(X)$ with $v = \frac{1}{2}(s+n)(s+n+1)$. Thus $t^u D(tX) = t^v D(X)$ and hence $D(tX) = t^{v-u} D(X)$. Therefore

$$D(tX_0, \ldots, tX_{r-1}) = t^{ns} D(X_0, \ldots, X_{r-1}),$$

which shows that $D(X_0, \ldots, X_{r-1})$ is a homogeneous polynomial of degree ns. □

2.2 THE DISCRIMINANT

The condition that a polynomial should have a multiple root is considered.

As in (2.1), let

$$f(X) = a_0X^n + a_1X^{n-1} + \cdots + a_{n-1}X + a_n, \quad a_0 \neq 0,$$

have roots $\alpha_1, \ldots, \alpha_n$. If $f'(X)$ is identically zero, then $p \neq 0$ and there exists $g(X) \in K[X]$ such that $f(X) = g(X)^p$. In this case, every zero of $f(X)$ is multiple, at least p-fold.

Let $f'(X)$ be not identically zero.

THEOREM 2.14 *The resultant of f and its derivative f' is*

$$R(f, f') = a_0^{2n-1} \prod_{i=1}^{n} \prod_{j \neq i} (\alpha_i - \alpha_j). \tag{2.6}$$

Proof. From Lemma 2.3 (i),

$$R(f, f') = a_0^{n-1} \prod_{i=1}^{n} f'(\alpha_i). \tag{2.7}$$

Taking the derivative of

$$f(X) = a_0 \prod_{k=1}^{n} (X - \alpha_k)$$

gives

$$f'(X) = a_0 \sum_{k=1}^{n} \prod_{j \neq k} (X - \alpha_j),$$

whence

$$f'(\alpha_i) = a_0 \prod_{j \neq i} (\alpha_i - \alpha_j).$$

Substituting in (2.7) gives the result. □

DEFINITION 2.15 The *discriminant* of $f(X)$ is

$$D(f) = a_0^{2n-2} \prod_{1 \leq j < i \leq n} (\alpha_i - \alpha_j)^2.$$

COROLLARY 2.16 *The discriminant can be calculated from the formula*

$$R(f, f') = (-1)^{n(n-1)/2} a_0 D(f). \tag{2.8}$$

Proof. In (2.7), for every pair i, j with $i \neq j$, there are two factors, $\alpha_i - \alpha_j$ and $\alpha_j - \alpha_i$. As there are $\frac{1}{2}n(n-1)$ pairs, so

$$R(f, f') = (-1)^{n(n-1)/2} a_0^{2n-1} \prod_{1 \leq j < i \leq n} (\alpha_i - \alpha_j)^2,$$

and the result follows. □

PROPOSITION 2.17 *The polynomial $f(X)$ has a multiple zero if and only if its discriminant $D(f)$ is zero.*

EXAMPLE 2.18 Let $f(X) = aX^2 + bX + c$. Then $f'(X) = 2aX + b$ and

$$R(f, f') = \begin{vmatrix} a & b & c \\ 2a & b & 0 \\ 0 & 2a & b \end{vmatrix} = a(-b^2 + 4ac).$$

From (2.8), $R(f, f') = -aD(f)$; so

$$D(f) = b^2 - 4ac.$$

2.3 ELIMINATION IN A SYSTEM IN TWO UNKNOWNS

Let L be any field, and let $f, g \in L[X, Y]$ be written as follows:

$$\begin{aligned} f(X,Y) &= a_0(Y)X^n + a_1(Y)X^{n-1} + \cdots + a_{n-1}(Y)X + a_n(Y), \\ g(X,Y) &= b_0(Y)X^s + b_1(Y)X^{s-1} + \cdots + b_{s-1}(Y)X + b_s(Y). \end{aligned} \tag{2.9}$$

Then $f,\ g \in L(Y)[X]$ and previous results can be applied for $L(Y)[X]$.

Now let $R_X(f, g)$ be the resultant of the polynomials f and g considered as polynomials in X. By the property of the resultant, $R_X(f, g)$ is a polynomial $D(Y)$ in Y with coefficients in K as follows:

$$D(Y) = \qquad (2.10)$$

$$\begin{vmatrix}
a_0(Y) & a_1(Y) & \dots & a_n(Y) & 0 & 0 & & \dots & 0 \\
0 & a_0(Y) & a_1(Y) & \dots & a_n(Y) & 0 & & \dots & 0 \\
\vdots & & \vdots & & & & & \vdots & \\
0 & 0 & \dots & 0 & a_0(Y) & a_1(Y) & & \dots & a_n(Y) \\
b_0(Y) & b_1(Y) & & \dots & & b_s(Y) & 0 & \dots & 0 \\
0 & b_0(Y) & b_1(Y) & & \dots & & b_s(Y)\ 0 & \dots & 0 \\
\vdots & & & & & \vdots & & \vdots & \\
0 & \dots & 0 & b_0(Y) & b_1(Y) & & & \dots & b_s(Y)
\end{vmatrix};$$

here, the determinant has entries $a_i(Y)$ in its first s rows and entries $b_j(Y)$ in its next n rows.

If the two polynomials (2.9) have a common root, $(X, Y) = (\alpha, \beta)$, substituting $Y = \beta$ in (2.9) gives polynomials in the indeterminate X, namely, $f(X, \beta)$, $g(X, \beta)$. These polynomials have a common root α, whence their resultant is zero. So β is a root of $R_X(f, g)$.

Conversely, if the resultant $R_X(f, g)$ of the polynomials (2.9) admits a root β, then the resultant of the polynomials $f(X, \beta)$, $g(X, \beta)$ is zero. Thus the polynomials $f(X, \beta)$ and $g(X, \beta)$ have a common root or their leading coefficients $a_0(\beta)$ and $b_0(\beta)$ are both zero.

The determination of the common roots of the polynomials (2.9) is reduced to determining the roots of the polynomial (2.10) in Y; this is expressed by saying that the indeterminate X has been eliminated from the system of the two equations or that the equation $D(Y) = 0$ is the result of eliminating X from (2.9).

EXAMPLE 2.19 Let $L = \mathbf{Q}$. To determine the common roots of

$$f(X, Y) = X^2Y + 3XY + 2Y + 3, \quad g(X, Y) = 2XY - 2X + 2Y + 3,$$

the resultant is calculated as $R_X(f, g) = 2Y^2 + 11Y + 12$; its roots are $y_1 = -4$, $y_2 = -\frac{3}{2}$. The leading coefficients of $f(X, y_i)$ and $g(X, y_i)$ are non-zero; hence, for each y_i, there is a common root in X. In fact,

$$f(X, -4) = -4X^2 - 12X - 5, \quad g(X, -4) = -10X - 5,$$

and they have the common root $x_1 = -\frac{1}{2}$. Similarly, $x_2 = 0$ is the common root of $f(X, \frac{3}{2}) = -\frac{3}{2}X^2 - \frac{9}{2}X$ and $g(X, \frac{3}{2}) = -5X$. Therefore the solution of the problem is given by the two pairs $(-\frac{1}{2}, -4)$ and $(0, -\frac{3}{2})$.

THEOREM 2.20 *The polynomials f and g given by (2.9) have a non-constant factor in common if and only if $R_X(f, g) = D(Y)$ is identically zero.*

Proof. From Theorem 2.9, there exist, for each fixed y, two polynomials $A_y(X)$, $B_y(X)$ such that

$$R_X(f, g) = A_y(X)f(X, y) + B_y(X)g(X, y)$$

with $\deg A_y(X) \le s-1$ and $\deg B_y(X) \le n-1$. In the proof, the coefficients of $A_y(X)$ and $B_y(X)$ are cofactors of the determinant D; it follows that there exist polynomials $A(X,Y)$, $B(X,Y)$ such that $A_y(X) = A(X,y)$, $B_y(X) = B(X,y)$ for each y. Hence

$$R_X(f,g) = A(X,Y)f(X,Y) + B(X,Y)g(X,Y)$$

with $A(X,Y)$ and $B(X,Y)$ of degree at most $s-1$ and $n-1$ in Y.

If $R_X(f,g) = 0$ for every $Y = y$, then $A(X,Y)f(X,Y) = -B(X,Y)g(X,Y)$. Since $f(X,Y)$ cannot divide $B(X,Y)$, some non-constant factor of $f(X,Y)$ must divide $g(X,Y)$, as required. Conversely, if $h(X,Y)$ is common factor of $f(X,Y)$ and $g(X,Y)$, for each y there exists at least one x such that $h(x,y) = 0$. It follows that $R_X(f,g) = 0$ for each y, whence $R_X(f,g)$ is the zero polynomial. □

THEOREM 2.21 *If $R_X(f,g)$ is not identically zero, then* $\deg R_X(f,g) \le NS$, *where* $N = \deg f(X,Y)$ *and* $S = \deg g(X,Y)$.

This theorem can be reformulated as follows: if $f(X,Y)$ and $g(X,Y)$ have no common factors, then the number of solutions of the system

$$f(x,y) = 0, \quad g(x,y) = 0$$

is at most NS.

Proof. Rewrite $f(X,Y)$ and $g(X,Y)$ in decreasing powers of X:

$$f(X,Y) = a_0(Y)X^n + \cdots + a_n(Y), \qquad g(X,Y) = b_0(Y)X^s + \cdots + b_s(Y).$$

This gives the following inequalities:

$$\begin{array}{ll} \deg a_0(Y) \le N-n, & \deg b_0(Y) \le S-s, \\ \deg a_1(Y) \le N-(n-1), & \deg b_1(Y) \le S-(s-1), \\ \vdots & \vdots \\ \deg a_h(Y) \le N-(n-h), & \deg b_k(Y) \le S-(s-k), \\ \vdots & \vdots \\ \deg a_n(Y) \le N, & \deg b_s(Y) \le S. \end{array} \tag{2.11}$$

The general term in the expansion of the determinant in (2.10) is

$$c\, a_0(Y)^{i_0} a_1(Y)^{i_1} \cdots a_n(Y)^{i_n} b_0(Y)^{j_0} b_1(Y)^{j_1} \cdots b_s(Y)^{j_s},$$

whose degree is, by (2.11), at most

$$\begin{aligned} &(N-n)i_0 + \cdots + (N-(n-h))i_h + \cdots + Ni_n \\ &\quad + (S-s)j_0 + \cdots + (S-(s-k))j_k + \cdots + Si_s. \end{aligned}$$

This can be written as

$$(N-n)\sum_{i=0}^n i_h + (S-s)\sum_{k=0}^s j_k + \sum_{h=1}^n h i_h + \sum_{k=1}^s k j_k. \tag{2.12}$$

Since

$$\sum_{i=0}^n i_h = s, \quad \sum_{k=0}^s j_k = n, \quad \sum_{h=1}^n h i_h + \sum_{k=1}^s k j_k = ns,$$

the sum (2.12) becomes $(N-n)s + (S-s)n + ns$. Since $n \le N$ and $s \le S$, the degree of the general term of the determinant $D(Y)$ is at most NS. □

THEOREM 2.22 *If $f(X,Y)$ and $g(X,Y)$ are two polynomials containing no terms of degree less than u and v respectively, then $y=0$ is a root of $R_X(f,g)=D(Y)$ of multiplicity at least uv.*

Proof. As an example, consider first the case $u=2, v=2$, with $\deg f(X,Y)=3$, $\deg g(X,Y)=2$:

$$f(X,Y)=a_{20}X^2+a_{11}XY+a_{02}Y^2+a_{30}X^3+a_{21}X^2Y+a_{12}XY^2+a_{03}Y^3,$$
$$g(X,Y)=b_{20}X^2+b_{11}XY+b_{02}Y^2.$$

It must be shown that $R_X(f,g)=D(Y)$ is divisible by Y^4. First,

$$D(Y)=$$
$$\begin{vmatrix} a_{30} & a_{21}Y+a_{20} & a_{11}Y+a_{12}Y^2 & a_{02}Y^2+a_{03}Y^3 & 0 \\ 0 & a_{30} & a_{21}Y+a_{20} & a_{11}Y+a_{12}Y^2 & a_{02}Y^2+a_{03}Y^3 \\ b_{20} & b_{11}Y & b_{02}Y^2 & 0 & 0 \\ 0 & b_{20} & b_{11}Y & b_{02}Y^2 & 0 \\ 0 & 0 & b_{20} & b_{11}Y & b_{02}Y^2 \end{vmatrix}.$$

In each entry of the determinant, consider only the terms of lowest degree and delete the others. This gives a new determinant $E(Y)$ as follows:

$$E(Y)=\begin{vmatrix} a_{30} & a_{20} & a_{11}Y & a_{02}Y^2 & 0 \\ 0 & a_{30} & a_{20} & a_{11}Y & a_{02}Y^2 \\ b_{20} & b_{11}Y & b_{02}Y^2 & 0 & 0 \\ 0 & b_{20} & b_{11}Y & b_{02}Y^2 & 0 \\ 0 & 0 & b_{20} & b_{11}Y & b_{02}Y^2 \end{vmatrix}.$$

Then Y^4 divides $D(Y)$ if and only if Y^4 divides $E(Y)$. Now multiply rows $1,2,4,5$ by Y,Y^2,Y,Y^2 respectively, giving a new determinant $F(Y)$, which is

$$F(Y)=\begin{vmatrix} a_{30}Y & a_{20}Y & a_{11}Y^2 & a_{02}Y^3 & 0 \\ 0 & a_{30}Y^2 & a_{20}Y^2 & a_{11}Y^3 & a_{02}Y^4 \\ b_{20} & b_{11}Y & b_{02}Y^2 & 0 & 0 \\ 0 & b_{20}Y & b_{11}Y^2 & b_{02}Y^3 & 0 \\ 0 & 0 & b_{20}Y^2 & b_{11}Y^3 & b_{02}Y^4 \end{vmatrix}.$$

Here, $F(Y)=Y^6E(Y)$.

So it must be shown that $F(Y)$ is divisible by Y^{10}. In $F(Y)$, Columns $2,3,4,5$ are divisible by Y,Y^2,Y^3,Y^4 respectively, whence Y^{10} divides $F(Y)$, as required.

Now consider the general case. The two polynomials have the following form:

$$f(X,Y)=a_{u,0}X^u+a_{u-1,1}X^{u-1}Y+\cdots+a_{1,u-1}XY^{u-1}+a_{0,u}Y^u+\cdots,$$
$$g(X,Y)=b_{v,0}X^v+b_{v-1,1}X^{v-1}Y+\cdots+b_{1,v-1}XY^{v-1}+b_{0,v}Y^v+\cdots.$$

Let $a(i)$ be the lowest exponent in the term $a_i(Y)$, and $b(i)$ the lowest in $b_i(Y)$. Then

$$f(X,Y)=a_0(Y)X^n+a_1(Y)X^{n-1}+\cdots+a_{n-1}(Y)X+a_n(Y),$$
$$a(n-i)\geq u-i;$$
$$g(X,Y)=b_0(Y)X^s+b_1(Y)X^{s-1}+\cdots+b_{s-1}(Y)X+b_s(Y),$$
$$b(s-i)\geq v-i.$$

Since $R_X(f,g) = D(Y)$, consider, as in the example, a new determinant $E(Y)$ obtained from $D(Y)$ by retaining only the terms of lowest degree in each row.

For the first s rows, multiply the $(s-i+1)$-th row by Y^{v-i+1} for $1 \le i \le v$. For the remaining n rows multiply the $(n-j+1)$-th row by Y^{u-j+1} for $1 \le j \le u$. After these operations, the $(n+s-k)$-th column is divisible by Y^{u+v-k}, for $0 \le k \le u+v$. Therefore $R_X(f,g)$ is divisible by Y^t, where

$$\begin{aligned} t &= \textstyle\sum_{k=1}^{u+v} k - \sum_{i=1}^{u} i - \sum_{j=1}^{v} j \\ &= \tfrac{1}{2}(u+v)(u+v+1) - \tfrac{1}{2}u(u+1) - \tfrac{1}{2}v(v+1). \end{aligned}$$

Hence $t = uv$, completing the proof. □

2.4 EXERCISES

1. Let

$$\begin{aligned} f(X) &= a_0X^n + a_1X^{n-1} + \cdots + a_n, \\ g(X) &= b_0X^s + b_1X^{s-1} + \cdots + b_s. \end{aligned}$$

Show that $R(f,g)$ is equal to the determinant whose entries are the coefficients of the polynomials

$$\begin{aligned} h_k(X) = &(a_0X^{k-1} + a_1X^{k-1} + \cdots + a_{k-1})g(X) \\ &-(b_0X^{k-1} + b_1X^{k-1} + \cdots + b_{k-1})f(X), \end{aligned}$$

where $k = 1, \ldots, n$ and $\deg h_k \le n-1$.

2. Assume that either $p = 0$ or $p \nmid n, m$, and let $F_n(X)$ be the n-th cyclotomic polynomial. Show that the resultant of the polynomials $F_n(X)$ and $X^m - 1$ is

$$R(F_n(X), X^m - 1) = \begin{cases} 0 & \text{when either } m \mid n \text{ or } n_1 = 1, \\ e^{\phi(n)/\phi(n_1)} & \text{when } n_1 = e^k \text{ with e prime}, \\ 1 & \text{otherwise}, \end{cases}$$

where ϕ is the Euler function, $d = \gcd(n,m)$, and $n_1 = n/d$.

3. Assume that either $p = 0$ or $p \nmid n, m$. Show that the resultant of the cyclotomic polynomials $F_m(X)$ and $F_n(X)$, with $m \ge n > 1$, is

$$R(F_m(X), F_n(X)) = \begin{cases} 0 & \text{when } m = n, \\ e^{\phi(n)} & \text{when } m = ne^k \text{ with e prime}, \\ 1 & \text{otherwise}. \end{cases}$$

4. Show that the discriminant of

$$f(X) = X^3 + aX^2 + bX + c$$

is

$$D(f) = a^2b^2 - 4b^3 - 4a^3c + 18abc - 27c^2.$$

5. Solve the following system over $\mathbf{Q}$:

$$Y^2 - 7XY + 4X^2 + 13X - 2Y - 3 = 0,$$
$$Y^2 - 14XY + 9X^2 + 28X - 4Y - 5 = 0.$$

6. If $p \nmid n$, show that the discriminant of $f(X) = X^n + uX + v$ is

$$D(f) = (-1)^{n(n-1)/2} n^n v^{n-1} + (-1)^{(n-1)(n-2)/2} (n-1)^{n-1} u^n.$$

7. Using Study's Theorem 2.10, prove Hilbert's Nullstellensatz:

   Let I be an ideal of $K[X_1, \dots, X_n]$. If each root of $F \in K[X_1, \dots, X_n]$ is a root of all polynomials in I, then there is positive integer m such that $F^m \in I$.

2.5 NOTES

For the basic theory of resultants, see Hodge and Pedoe [225, Chapter IV], Rédei [366], Kurosh [285]. The theory has been expanded to any finite number of polynomials; see Gelfand, Kapranov, and Zelevinsky [165], Lang [293, Chapter 9], Segre [399], and van der Waerden [484].

Related to Exercises 2 and 3, the resultant of two cyclotomic polynomials was investigated by Apostol, Cheng, Diederichsen, Lehmer, Louboutin, McKay and Wang; see [75].

For a proof of Hilbert's Nullstellensatz in Exercise 7, see [512, Chapter VII].

Chapter Three

Singular points and intersections

The aim of Section 3.1 is to define the intersection number $I(P, \mathcal{F} \cap \mathcal{G})$ of two plane curves at a point P by means of purely geometric postulates. The main results, Theorems 3.8 and 3.9, ensure that such a number not only exists but is also uniquely determined. Also, for $P = (0, 0)$, this definition coincides with that arising from elimination theory as described at the beginning of Chapter 2.

In Section 3.2, the concept of intersection number is extended to include projective curves $\mathcal{F}$ and $\mathcal{G}$ and points P at infinity. With this, Bézout's Theorem 3.14 can be formulated and proved.

The major obstacle in the study of a plane curve $\mathcal{F}$ is the presence of badly-behaved singular points, especially with several multiple tangents. So it is of interest to obtain a plane curve $\mathcal{G}$ with only ordinary singular points that is equivalent to $\mathcal{F}$. Such an equivalence cannot be projective; that is, $\mathcal{G}$ cannot be the image of $\mathcal{F}$ by a linear transformation. The idea is to allow more general transformations expressed by polynomials; how to do this by using birational transformations is explained in Section 3.3.

One type of birational transformations is a quadratic transformation. The effect of the standard quadratic transformation on an irreducible plane curve $\mathcal{F}$ is described in Section 3.4.

Section 3.5 is devoted to the classical result that $\mathcal{F}$ can be transformed to an irreducible plane curve $\mathcal{G}$ with only ordinary singularities by a finite sequence of quadratic transformations. The main ingredients in this process are the virtual genus and the index of an irreducible plane curve. In positive characteristic, the concept of a strange curve also occurs.

3.1 THE INTERSECTION NUMBER OF TWO CURVES

In this section, a *curve* in $\mathrm{AG}(2, K)$ is simply an equivalence class of polynomials in $K[X, Y]$ under multiplication by a non-zero scalar. Given two curves F and G, the *intersection number* $I(P, F \cap G)$ of F and G at the point $P = (a, b)$ is defined by the following axioms:

(I1) $I(P, F \cap G)$ is a non-negative integer when F and G have no common component through P;

(I2) $I(P, F \cap G) = \infty$ if F and G have a common component through P;

(I3) $I(P, F \cap G) = 0$ if and only if $P \notin F \cap G$;

(I4) $I(P, F \cap G) = 1$ if F and G are two distinct lines through P;

(I5) $I(P, F \cap G) = I(P, G \cap F)$;

(I6) $I(P, F \cap (G + AF)) = I(P, F \cap G)$ for any $A \in K[X, Y]$;

(I7) $I(P, F \cap GH) = I(P, F \cap G) + I(P, F \cap H)$ for any $H \in K[X, Y]$.

The first result shows the relation between the geometry and the algebra.

THEOREM 3.1 *The intersection number $I(P, F \cap G)$ depends only on the ideal (F, G) generated by F and G, in the sense that, if F' and G' are curves such that $(F, G) = (F', G')$, then*

$$I(P, F \cap G) = I(P, F' \cap G'). \tag{3.1}$$

Proof. If $(F, G) = (F', G')$, then

$$F' = AF + BG, \tag{3.2}$$
$$G' = CF + DG, \tag{3.3}$$

with $A, B, C, D \in K[X, Y]$. There exist $A', B', C', D' \in K[X, Y]$ such that

$$F = A'F' + B'G', \tag{3.4}$$
$$G = C'F' + D'G'. \tag{3.5}$$

If H divides both F and G, then it also divides both F' and G', and vice versa. By (I2), this says that, if $I(P, F \cap G) = \infty$, then $I(P, F' \cap G') = \infty$.

So it may be supposed that F' and G' have no common factor. It is first shown that, if P is a common point of the curves F and G, then $\Delta(P) \neq 0$, where $\Delta = AD - BC$. A direct calculation, using (3.2), (3.3), (3.4), (3.5), gives

$$F'(1 - (AA' + BC')) = (AB' + BD')G',$$
$$G'(1 - (CB' + DD')) = (CA' + DC')F'.$$

Suppose first that $AB' + BD'$, and hence also $(1 - (AA' + BC'')$, is not the zero polynomial. Then F' and G' divide both these. Since $F(P) = G(P) = 0$ implies that $F'(P) = G'(P) = 0$, it follows that, with

$$R_1 = AB' + BD', \quad R_2 = AA' + BC',$$
$$R_1(P) = 0, \quad R_2(P) = 1. \tag{3.6}$$

Similarly, with

$$R_3 = CA' + DC', \quad R_4 = CB' + DD',$$
$$R_3(P) = 0, \quad R_4(P) = 1, \tag{3.7}$$

provided that R_3, and so $1 - R_4$, is not the zero polynomial.

It follows immediately that (3.6) and (3.7) remain true when R_1 and R_3 are the zero polynomials. Therefore

$$R_2(P)R_4(P) - R_1(P)R_3(P) = 1.$$

Now

$$\begin{aligned}&R_2R_4 - R_1R_3\\&= (AA' + BC')(CB' + DD') - (AB' + BD')(CA' + DC')\\&= (AD - BC)(A'D' - B'C').\end{aligned}$$

Hence $\Delta(P) = A(P)D(P) - B(P)C(P) \neq 0$, as required.

From (I 3), it follows that

$$I(P, \Delta \cap F) = 0. \tag{3.8}$$

Now the result can be deduced, using (I 5), (I 6), (I 7), and (3.7):

$$\begin{aligned}&I(P, F' \cap G')\\&= I(P, [AF + BG] \cap [CF + DG])\\&= I(P, [AFD + BGD] \cap [CF + DG]) - I(P, D \cap [CF - DG])\\&= I(P, [AFD - BCF + BCF + BGD] \cap [CF + DG]) - I(P, D \cap CF)\\&= I(P, [(AD - BC)F + B(CF + DG)] \cap [CF + DG]) - I(P, D \cap CF)\\&= I(P, \Delta F \cap [CF + DG]) - I(P, D \cap CF)\\&= I(P, \Delta \cap [CF + DG]) + I(P, F \cap [CF + DG]) - I(P, D \cap CF)\\&= I(P, F \cap DG) - I(P, D \cap CF)\\&= I(P, F \cap D) + I(P, F \cap G) - I(P, D \cap C) - I(P, D \cap F)\\&= I(P, F \cap G) - I(P, C \cap D).\end{aligned}$$

It remains to see that $I(P, C \cap D) = 0$. If not, then $C(P) = D(P) = 0$, and so $A(P)D(P) - B(P)C(P) = 0$, contradicting that $\Delta \neq 0$. □

The following useful formula for reducible curves is a consequence of (I 7) and (I 5).

THEOREM 3.2 *If* $F = \prod_i F_i^{r_i}$ *and* $G = \prod_j G_j^{s_j}$, *then*

$$I(P, F \cap G) = \sum_{i,j} r_i s_j \, I(P, F_i \cap G_j).$$

Now some propositions are established that are required later.

PROPOSITION 3.3 *For every* $F \in K[X, Y]$ *not having* Y *as a component, if*

$$f(X) = F(X, 0) = X^r (X - a_1)^{r_1} \cdots (X - a_m)^{r_m},$$

then $I(O, F \cap Y) = r$, *where* $O = (0, 0)$ *is the origin.*

Proof. Since $F(X, Y) = f(X) + YH(X, Y)$, from (I 7) it follows that

$$I(O, F \cap Y) = I(O, f(X) \cap Y).$$

So, by Theorem 3.2,

$$\begin{aligned}&I(O, F \cap Y)\\&= rI(O, X \cap Y) + r_1 I(O, (X - a_1) \cap Y) + \cdots + r_m I(O, (X - a_m) \cap Y).\end{aligned}$$

Then the result follows from (I 3) and (I 4). □

Now define the homomorphism

$$\begin{aligned}\Phi : K[X,Y] &\to K[X], \\ H(X,Y) &\mapsto h(X) = H(X,0).\end{aligned} \tag{3.9}$$

Given two non-zero polynomials $F, G \in K[X,Y]$, let $f = \Phi(F), g = \Phi(G)$, and let $d \in K[X]$ be the greatest common divisor of f and g. Then there exist polynomials $a, b, c, e \in K[X]$ such that $d = af + bg, f = cd, g = ed$. Therefore

$$ac + be = 1. \tag{3.10}$$

Now put

$$D = aF + bG, \quad H = -eF + cG. \tag{3.11}$$

With O the origin, calculate $I(O, D \cap H)$. Since $\Phi(D) = d$ and $\Phi(H) = 0$, it follows that $H = YH'$ with $H' \in K[X,Y]$. From the above, the ideals (F, G) and (D, H) are the same. Then, from (I 7) and Theorem 3.1,

$$I(O, F \cap G) = I(O, D \cap H) = I(O, D \cap Y) + I(O, D \cap H'). \tag{3.12}$$

PROPOSITION 3.4 *If $\bar{I}(P, F \cap G)$ is a function satisfying* (I 1),…,(I 7), *then, for every pair (F, G) with $F, G \in K[X,Y]$,*

$$\bar{I}(O, F \cap G) = I(O, F \cap G). \tag{3.13}$$

Proof. Put $i = I(O, F \cap G)$; if $i = 0$ the proposition follows from (I 3). Now proceed by induction on i.

Suppose that $I(O, U \cap V) = \bar{I}(O, U \cap V)$ for all $U, V \in K[X,Y]$ under the hypothesis that $I(O, U \cap V) < i$. First, let G be linear; that is, $G = \lambda X + \mu Y$ with $\lambda, \mu \in K$ and $\lambda \neq 0$. With the above notation, $g(X) = \lambda X$. Also, $f(X)$ is divisible by X, since $f(0) = F(0,0) = 0$. Therefore

$$f(X) = cX,\ a = 0,\ b = \lambda^{-1},\ d = X,\ e = \lambda,\ D = \lambda^{-1}G = X + \mu\lambda^{-1}Y.$$

From (I 4) and (I 6),

$$\begin{aligned}I(O, D \cap Y) &= I(O, X + \mu\lambda^{-1}Y \cap Y) = I(O, X \cap Y) = 1 \\ &= \bar{I}(O, X \cap Y) = \bar{I}(O, X + \mu\lambda^{-1}Y \cap Y) = \bar{I}(O, D \cap Y).\end{aligned}$$

Using (3.12), it follows that

$$\begin{aligned}I(O, D \cap H') &= I(O, D \cap H) - I(O, D \cap Y) \\ &= I(O, F \cap G) - 1 = i - 1.\end{aligned}$$

By the inductive hypothesis, $I(O, D \cap H') = \bar{I}(O, D \cap H')$. Therefore

$$\begin{aligned}I(O, F \cap G) &= I(O, D \cap Y) + I(O, D \cap H') \\ &= \bar{I}(O, D \cap Y) + \bar{I}(O, D \cap H') \\ &= \bar{I}(O, D \cap H') = \bar{I}(O, F \cap G),\end{aligned}$$

as required.

It remains to consider the case that G is not linear. Next, by (3.12), the inequality, $I(O, D \cap Y) \leq i$, holds. By (I 2), Y does not divide D. Also, Proposition 3.3 asserts that

$$I(O, D \cap Y) = \bar{I}(O, D \cap Y). \tag{3.14}$$

Since $F(0,0) = G(0,0) = 0$, also $D(0,0) = 0$. Hence $I(O, D \cap Y) \geq 1$ by (I3).
Next, $I(O, D \cap H') < i$ by (3.12); by the inductive hypothesis,

$$I(O, D \cap H') = \bar{I}(O, D \cap H').$$

Using (3.14),

$$\begin{aligned} I(O, F \cap G) &= I(O, D \cap Y) + I(O, D \cap H') \\ &= \bar{I}(O, D \cap Y) + \bar{I}(O, D \cap H') \\ &= \bar{I}(O, D \cap H) = \bar{I}(O, F \cap G). \end{aligned}$$

This completes the proof of Proposition 3.4. □

PROPOSITION 3.5 *If $D \in K[X, Y]$, then*

$$I(O, D \cap Y) \geq m_O(D),$$

with equality if and only if Y is not tangent to D at O.

Proof. By (I2), it may be supposed that Y is not a component of D. Write

$$D(X, Y) = d(X) + Y D'(X, Y), \tag{3.15}$$

where

$$d(X) = X^m(\lambda_m + \lambda_{m+1} X + \cdots), \quad \lambda_m \neq 0. \tag{3.16}$$

With $m_O(D) = n$,

$$D(X, Y) = D_n(X, Y) + D_{n+1}(X, Y) + \cdots, \tag{3.17}$$

where $D_i(X, Y)$ is a homogeneous polynomial of degree i. From (3.15), (3.16), (3.17), it follows that $m \geq n$ and

$$D_n(X, Y) = \lambda_m X P_{n-1}(X, Y) + Y P'_{n-1}(X, Y),$$

where $\lambda_m = 0$ for $m > n$ and P'_{n-1} is a homogeneous polynomial of degree $n-1$. Therefore $n = m$ if and only if Y does not divide $D_n(X, Y)$; that is, if Y is not tangent to D at O. From (I5), (3.15), (I6), (3.16), (I7), (I3), (I4),

$$\begin{aligned} I(O, D \cap Y) &= I(O, Y \cap (d + Y D')) \\ &= I(O, Y \cap d) \\ &= I(O, Y \cap X^m(\lambda_m + \lambda_{m+1} X + \cdots) \\ &= I(O, Y \cap X^m) + I(O, Y \cap (\lambda_m + \lambda_{m+1} X + \cdots)) \\ &= m I(O, Y \cap X) + 0 \\ &= m. \end{aligned}$$

□

The next result is important for calculating the intersection number at a point.

PROPOSITION 3.6 *The intersection number satisfies the inequality*

$$I(O, F \cap G) \geq m_O(F) \cdot m_O(G)$$

with equality if and only if the tangents at O to F are all distinct from the tangents at O to G.

Proof. By (I3), the proposition is true when $I(O, F\cap G) = 0$. Suppose, by induction, that the proposition has been shown for $U, V \in K[X,Y]$ such that $I(O, U\cap V) < i$. To show that $I(O, F\cap G) = i$, with the notation of (3.4), from Proposition 3.5 it follows that $I(O, D\cap Y) > 0$. Now, by (3.12), $I(O, D\cap H') < i$. By the inductive hypothesis, $I(O, D\cap H') \geq m_O(D)\cdot m_O(H')$, with equality if and only if D and H' do not have the same tangent at O. In any case, Proposition 3.5 together with (3.12) implies that

$$\begin{aligned} I(O, F\cap G) &= I(O, D\cap Y) + I(O, D\cap H') \\ &\geq m_O(D) + m_O(D)m_O(H') \\ &= m_O(D)m_O(H), \end{aligned} \tag{3.18}$$

with equality if and only if D has no tangent in common either with Y or H', and so also not with H.

Now a relationship is established between $m_O(D)m_O(H)$ and $m_O(F)m_O(G)$. Suppose first that Y divides neither $F_m(X,Y)$ nor $G_n(X,Y)$, where

$$\begin{aligned} F(X,Y) &= F_m(X,Y) + F_{m+1}(X,Y) + \cdots, \\ G(X,Y) &= G_n(X,Y) + G_{n+1}(X,Y) + \cdots, \end{aligned}$$

with $F_i(X,Y)$ and $G_i(X,Y)$ homogeneous polynomials of degree i. Multiplying F and G by suitable scalars ensures that F_m and G_m have the following form:

$$\begin{aligned} F_m(X,Y) &= X^m + Y\cdot F'_{m-1}(X,Y), \\ G_n(X,Y) &= X^n + Y\cdot G'_{n-1}(X,Y), \end{aligned}$$

where F'_{m-1} and G'_{n-1} are homogeneous polynomials of degrees $m-1$ and $n-1$. With $f(X) = F_m(X,0)$, $g(X) = G_n(X,0)$,

$$f(X) = X^m(1 + f_1(X)), \quad g(X) = X^n(1 + g_1(X)).$$

Without loss of generality, let $m \leq n$. Then $d = \gcd(f, g)$ is divisible by X^m but not by X^{m+1}. Therefore

$$\begin{aligned} d &= X^m + d_{m+1}X^{m+1} + \cdots, \\ c &= 1 + c_1X + \cdots, \\ e &= X^{n-m} + e_{n-m+1}X^{n-m+1} + \cdots. \end{aligned} \tag{3.19}$$

Put

$$\begin{aligned} a &= \alpha_r X^r + \alpha_{r+1}X^{r+1} + \cdots, \\ b &= \beta_s X^s + \beta_{s+1}X^{s+1} + \cdots. \end{aligned}$$

Then, from (3.10),

$$\begin{aligned} 1 &= (\alpha_r X^r + \alpha_{r+1}X^{r+1} + \cdots)(1 + c_1X + \cdots) \\ &\quad + (\beta_s X^s + \beta_{s+1}X^{s+1} + \cdots)(X^{n-m} + e_{n-m+1}X^{n-m+1} + \cdots) \\ &= (\alpha_r X^r + \cdots) + (\beta_s X^{n-m+s} + \cdots). \end{aligned}$$

Hence

$$\begin{cases} r = 0,\ \alpha_0 = 1, & \text{if } n + s > m; \\ \beta_0 = 1, & \text{if } n = m, s = 0, r > 0; \\ \alpha_0 + \beta_0 = 1,\ \alpha_0\beta_0 \neq 0, & \text{if } n = m, r = s = 0. \end{cases} \tag{3.20}$$

Since $D = aF + bG$ and $H = -eF + cG$, as in (3.11),

$$\begin{aligned} D &= (\alpha_r X^r + \alpha_{r+1} X^{r+1} + \cdots)(F_m + F_{m+1} + \cdots) \\ &\qquad + (\beta_s X^s + \beta_{s+1} X^{s+1} \cdots)(G_n + G_{n+1} + \cdots) \\ &= (\alpha_r X^r F_m + \cdots) + (\beta_s X^s G_n + \cdots); \end{aligned} \tag{3.21}$$

$$\begin{aligned} H &= -(X^{n-m} + e_{n-m+1} X^{n-m+1} + \cdots)(F_m + F_{m+1} + \cdots) \\ &\qquad + (1 + c_1 X + \cdots)(G_n + G_{n+1} + \cdots) \\ &= (-X^{n-m} F_m + G_n) + \cdots . \end{aligned} \tag{3.22}$$

It follows that $m_O(H) \geq n$ and $m_O(D) \geq m$.

On the other hand, (3.15) and (3.19) imply that

$$D = d + YD' = (X^m + d_{m+1} X^{m+1} + \cdots) + YD'(X,Y).$$

Therefore $m_O(D) \leq m$; hence $m_O(D) = m$. Now write $D = D_m + \cdots$, with D_m a homogeneous polynomial of degree m. From (3.21) and (3.20),

$$D_m = \begin{cases} F_m, & \text{if } n + s > m; \\ G_m, & \text{if } n = m, s = 0, r > 0; \\ \alpha_0 F_m + \beta_0 G_m, & \text{if } n = m, r = s = 0. \end{cases} \tag{3.23}$$

Now the proof can be completed. If $m_O(H) > n$, then (3.22) implies that $G_n = X^{n-m} F_m$, so that each tangent to F at O is also tangent to G at O. By (3.18),

$$I(O, F \cap G) \geq m_O(D) \cdot m_O(H) > mn.$$

If $m_O(H) = n$, then $m_O(D) \cdot m_O(H) = mn$, and

$$H_n = -X^{n-m} F_m + G_n \neq 0. \tag{3.24}$$

It remains to show that F and G have tangents in common if and only if D and H have; in other words, H_n and D_m have a non-constant divisor if and only if F_m and G_n have. But this follows immediately from the fact that H_n and D_m are linear combinations of F_m and G_n with coefficients in $K[X]$, and vice versa; see (3.24), (3.23), (3.20). So the result is true for $m_O(H) = n$.

It remains to show that the assumption that Y divides neither F_m nor G_n is unnecessary. Given $a, b, c, d \in K$ with $ad - bc \neq 0$, define a new function

$$\bar{I}(P, \bar{F} \cap \bar{G}) = I(P, F \cap G),$$

where

$$\begin{aligned} \bar{F}(X,Y) &= F(aX + bY, cX + dY), \\ \bar{G}(X,Y) &= G(aX + bY, cX + dY). \end{aligned}$$

It follows immediately that $\bar{I}(P, \bar{F} \cap \bar{G})$ also satisfies the postulates (I 1),...,(I 7). Now choose $a, b, c, d \in K$ so that, as a sum of homogeneous polynomials,

$$\begin{aligned} \bar{F}(X,Y) &= \bar{F}_m(X,Y) + \bar{F}_{m+1} + \cdots , \\ \bar{G}(X,Y) &= \bar{G}_n(X,Y) + \bar{G}_{n+1} + \cdots , \end{aligned}$$

the terms $\bar{F}_m$ and $\bar{G}_n$ are not divisible by Y. All the same, $\bar{F}_m$ and $\bar{G}_n$ have a common divisor $cX + dY$. By the preceding result applied to $\bar{I}$, it follows that

$\bar{I}(O, \bar{F} \cap \bar{G}) > m_O(\bar{F}) m_O(\bar{G})$. Since $m_O(\bar{F}) = m_O(F), m_O(\bar{G}) = m_O(G)$, it also follows that $I(O, F \cap G) > m_O(F) m_O(G)$, as required. □

The results so far for $I(P, F \cap G)$ have $P = O$, the origin. Now they are extended to the case that $P \neq O$. With $P = (x_0, y_0)$, the translation

$$\tau : (X, Y) \mapsto (X - x_0, Y - y_0)$$

sends P to the origin. The same τ sends the curves F and G to the curves F' and G', where

$$F'(X, Y) = F(X - x_0, Y - y_0),$$
$$G'(X, Y) = G(X - x_0, Y - y_0).$$

Define the function $\bar{I}(P, F \cap G)$ by

$$\bar{I}(P, F \cap G) = I(O, F' \cap G').$$

Then it follows immediately that $\bar{I}(P, F \cap G)$ satisfies all of (I 1),..., (I 7). Conversely, if a function $I(P, F \cap G)$ satisfies these properties for a fixed point P, then also the function $\bar{I}(O, F' \cap G') = I(P, F \cap G)$ enjoys the same properties with respect to the origin. By Propositions 3.4 and 3.6, the following theorems are finally obtained.

THEOREM 3.7 *$I(P, F \cap G) \geq m_P(F) \cdot m_P(G)$, with equality if and only if the curves F and G have no common tangent at P.*

THEOREM 3.8 (Uniqueness Theorem) *For two curves F and G, there is at most one function $I(P, F \cap G)$ that satisfies the seven postulates* (I 1),..., (I 7).

The following result completes the theory of the intersection number.

THEOREM 3.9 (Existence Theorem) *There exists a function $I(P, F \cap G)$ that satisfies the seven postulates* (I 1),..., (I 7).

Proof. Suppose that F and G have no common component. let $R_Y(F, G)$ be their resultant when F and G are considered as polynomials in Y. Then $R_Y(F, G)$ is a polynomial $D(X) \in K[X]$. Now choose coordinates so that

$$\left.\begin{array}{rl} \text{(i)} & Y_\infty = (0,0,1) \text{ is not on } F \text{ or } G; \\ \text{(ii)} & Y_\infty \text{ is not on any line through two distinct points of } F \cap G; \\ \text{(iii)} & Y_\infty \text{ is not on the tangent to } F \text{ or } G \text{ at any point of } F \cap G. \end{array}\right\} \quad (3.25)$$

It is now shown that, for $P = (x_0, y_0) \in F \cap G$,

$$I(P, F \cap G) = \max\{k \in \mathbf{N} \mid (X - x_0)^k \text{ divides } R_Y(F, G)\}.$$

Let

$$D(X) = c \textstyle\prod_i (X - \alpha_i)^{r_i}, \quad c \in K;$$

then, for each point $P = (x_0, y_0)$ such that $F(P) = G(P) = 0$, the abscissa x_0 is one of the roots α_i of $D(X)$. In accord with this, put

$$I(P, F \cap G) = \begin{cases} \infty, & \text{if } F, G \text{ have a common factor } H \text{ and } H(P) = 0; \\ 0, & \text{if } F(P) \neq 0 \text{ or } G(P) \neq 0; \\ r_i, & \text{if } F(P) = G(P) = 0 \text{ and } x_0 = \alpha_i. \end{cases} \quad (3.26)$$

The truth of postulates (I 1)...(I 7) is a consequence of results on the resultant and elimination theory, as can now be verified.

The conditions (I 1),(I 2),(I 3),(I 5) follow from the definition. For (I 4), let

$$F = A_1(X - x_0) + B_1(Y - y_0),$$
$$G = A_2(X - x_0) + B_2(Y - y_0).$$

Then

$$R_Y(F, G) = (B_2A_1 - B_1A_2)(X - x_0),$$

with $B_2A_1 - B_1A_2 \neq 0$. So $I(P, F \cap G) = 1$.

For (I 6), note that the resultant $R_Y(F, G + AF)$ in the form of (2.2) equals $R_Y(F, G)$, since the former is obtained from the latter by adding linear combinations of the first m rows to the last n rows, where $n = \deg F,\ m = \deg G$.

It follows from Lemma 2.3 that $R_Y(F, GH) = R_Y(F, G)R_Y(F, H)$, from which (I 7) follows.

Thus (I 1),..., (I 7) are satisfied. □

REMARK 3.10 For another proof of this theorem in terms of the local ring at P, see Exercise 6.

These results lead to the following geometric and covariant concept.

DEFINITION 3.11 For the affine curves, $\mathcal{F} = \mathbf{v}_\alpha(F)$, $\mathcal{G} = \mathbf{v}_\alpha(G)$, and the point $P = (a, b)$, the *intersection number* is

$$I(P, \mathcal{F} \cap \mathcal{G}) = I((a, b), F \cap G).$$

3.2 BÉZOUT'S THEOREM

For a projective curve, local properties are the same as for the corresponding affine curve.

DEFINITION 3.12 For the projective curves, $\mathcal{F} = \mathbf{v}(F)$, $\mathcal{G} = \mathbf{v}(G)$, and the point $O = (1, 0, 0)$ the *intersection number* is

$$I(O, \mathcal{F} \cap \mathcal{G}) = I((0, 0), F_*, G_*),$$

where F_* and G_* are the polynomials associated to F and G as in Section 1.4.

Intersection numbers of $\mathcal{F} = \mathbf{v}(F)$ and $\mathcal{G} = \mathbf{v}(G)$ at another point P are calculated, as in the affine case, by using covariant properties; that is, a projectivity is applied to change P to Y_∞.

The covariance of the intersection number of projective curves at a point can be shown by using Theorem 3.9 and (3.26). This important property also follows from the proof of Bézout's Theorem 3.14 below. Now some global results are obtained.

THEOREM 3.13 *If the projective plane curves $\mathcal{F}$ and $\mathcal{G}$ have degrees m and n, and no common component, then they have at most mn points in common.*

Proof. Suppose, on the contrary, that the curves have more than mn points in common. Choose any $mn+1$ of the common points and join every pair by a line. There is a point P neither on any of these lines nor on $\mathcal{F}$ or $\mathcal{G}$, and there is a line ℓ through P distinct from these lines. Take a frame for which $P = \mathbf{U}_1$ and $\ell = \mathbf{U}_1\mathbf{U}_2$. Then every one of the $mn+1$ points chosen has a different value for y, since the join of any two does not contain $\mathbf{U}_1$. Let $\mathcal{F} = \mathbf{v}(f)$, $\mathcal{G} = \mathbf{v}(g)$. Then the system of equations

$$\begin{aligned} f(X,Y) &= a_0(Y)X^m + a_1(Y)X^{m-1} + \cdots + a_{m-1}(Y)X + a_m(Y), \\ g(X,Y) &= b_0(Y)X^n + b_1(Y)X^{n-1} + \cdots + b_{n-1}X + b_n(Y) \end{aligned} \tag{3.27}$$

has at least $mn+1$ solutions (x,y) with different values for y. As explained in Section 2.3, this implies that the polynomial $D(Y) = R_X(f,g)$ has at least $mn+1$ distinct roots. From Theorem 2.21, $R_X(f,g)$ is identically zero. By Theorem 2.20, $\mathcal{F}$ and $\mathcal{G}$ have a common factor, a contradiction. □

THEOREM 3.14 (Bézout's Theorem) *If the projective plane curves $\mathcal{F} = \mathbf{v}(F)$ and $\mathcal{G} = \mathbf{v}(G)$ have degrees m and n, and no common component, then*

$$\sum I(P, F \cap G) = mn.$$

Proof. By Theorem 3.13, $\mathcal{F}$ and $\mathcal{G}$ have at most nm common points. Let $h \geq 0$ denote the exact number of such points.

Let $R = (\xi_0, \xi_1, \xi_2)$ and $Q = (\eta_0, \eta_1, \eta_2)$ be two distinct points, and let ℓ be the line through R and Q. A point $P = (x_0, x_1, x_2)$ lies on ℓ if and only if there exist $\lambda, \mu \in K$ such that

$$P = (\lambda\xi_0 + \mu\eta_0, \lambda\xi_1 + \mu\eta_1, \lambda\xi_2 + \mu\eta_2).$$

A necessary and sufficient condition for P to be a point of $\mathcal{F}$ is

$$F(\lambda\xi_0 + \mu\eta_0, \lambda\xi_1 + \mu\eta_1, \lambda\xi_2 + \mu\eta_2) = 0.$$

Since $F(\lambda\xi_0 + \mu\eta_0, \lambda\xi_1 + \mu\eta_1, \lambda\xi_2 + \mu\eta_2)$ can be viewed as a homogeneous polynomial $F'(\lambda, \mu)$ in the indeterminates λ and μ, write

$$F'(\lambda, \mu) = a_0\lambda^m + a_1\lambda^{m-1}\mu + \cdots + a_{m-1}\lambda\mu^{m-1} + a_m\mu^m.$$

To calculate a_s, let

$$F(X_0, X_1, X_2) = \textstyle\sum_{i+j+k=m} a_{ijk} X_0^i X_1^j X_2^k.$$

Then

$$a_s = \sum_{u+v+w=m-s} \; \sum_{i+j+k=m} \binom{i}{u}\binom{j}{v}\binom{k}{w} a_{ijk} (\xi_0^u \xi_1^v \xi_2^w)(\eta_0^{i-u}\eta_1^{j-v}\eta_2^{k-w}).$$

It may be noted that a_s is a homogeneous polynomial of degree m in the indeterminates $\xi_0, \xi_1, \xi_2, \eta_0, \eta_1, \eta_2$. If η_0, η_1, η_2 are considered to be constants, then a_s is a homogeneous polynomial of degree $m-s$ in the indeterminates ξ_0, ξ_1, ξ_2. Similarly, if ξ_0, ξ_1, ξ_2 are constants, then a_s is a homogeneous polynomial of degree s in the indeterminates η_0, η_1, η_2. The non-trivial roots of $F'(\lambda, \mu)$ correspond to the common points of the curve $\mathcal{F}$ with the line ℓ.

A similar result holds for the curve $\mathcal{G}$. The line ℓ through R and Q meets $\mathcal{G}$ in the points P such that (λ, μ) is a non-trivial root of the polynomial

$$G'(\lambda, \mu) = b_0\lambda^n + b_1\lambda^{n-1}\mu + \cdots + b_{n-1}\lambda\mu^{n-1} + b_n\mu^n.$$

Therefore $\mathcal{F}$ and $\mathcal{G}$ have a non-trivial common root; equivalently,

$$R(F'(\lambda, \mu), G'(\lambda, \mu)) = 0$$

if and only if $\mathcal{F}$ and $\mathcal{G}$ have a common point

$$P_i = (x_0^{(i)}, x_1^{(i)}, x_2^{(i)})$$

on the line through R and Q. It follows that $R(F'(\lambda, \mu), G'(\lambda, \mu)) = 0$ if and only if at least one of the determinants

$$D(\xi, \eta, x^{(i)}) = \begin{vmatrix} \xi_0 & \xi_1 & \xi_2 \\ \eta_0 & \eta_1 & \eta_2 \\ x_0^{(i)} & x_1^{(i)} & x_2^{(i)} \end{vmatrix} \tag{3.28}$$

is zero.

Now fix the point R and consider Q as a variable point; that is, leave the homogeneous triple (ξ_0, ξ_1, ξ_2) fixed and consider η_0, η_1, η_2 as indeterminates. Then the determinant $D(\xi, \eta, x^{(i)})$ in (3.28) is a linear homogeneous polynomial while the resultant $R(F'(\lambda, \mu), G'(\lambda, \mu))$ is a homogeneous polynomial of degree mn.

Every root of $D(\xi, \eta, x^{(i)}))$ is also a root of $R(F'(\lambda, \mu), G'(\lambda, \mu))$. By Study's Theorem 2.10, $D(\xi, \eta, x^{(i)})$ divides $R(F'(\lambda, \mu), G'(\lambda, \mu))$. Conversely, every root of $R(F'(\lambda, \mu), G'(\lambda, \mu))$ is a root of some $D(\xi, \eta, x^{(i)})$. Therefore, for certain positive integers r_i,

$$R(F'(\lambda, \mu), G'(\lambda, \mu)) = c \prod_{i=1}^{k} D(\xi, \eta, x^{(i)})^{r_i}, \tag{3.29}$$

where c is a non-zero constant and $k \leq h$. This implies that

$$nm = \sum_{i=1}^{k} r_i. \tag{3.30}$$

As a consequence, $k \geq 1$ and hence $h \geq 1$.

Next, it is shown that r_i is covariant. To do this, the notation in the proof of Theorem 1.23 is used. It is a straightforward to show that

$$D(\xi, \eta, x^{(i)}) = D(\xi', \eta', (x'^{(i)}))\ \det(a_{ij}),$$

which shows the covariance of $D(\xi, \eta, x^{(i)})$. Also, from the definition of $F'(\lambda, \mu)$ and $G'(\lambda, \mu)$, it follows that their resultant is a covariant.

Therefore $\mathcal{F}$ and $\mathcal{G}$ may be assumed to have no common point on the line $\mathbf{v}(X_0)$. Then their intersection is in the affine plane with coordinates (x, y) where, as usual, $x = x_1/x_0$ and $y = x_2/x_0$. So take the curves in their the affine forms, with $\mathcal{F} = \mathbf{v}(F_*(X, Y))$ and $\mathcal{G} = \mathbf{v}(G_*(X, Y))$. It may also be supposed that no two such common points lie on the same line $\mathbf{v}(X - c)$.

If $R = (u, 0)$ and $Q = Y_\infty$, then

$$D(\xi, \eta, x^{(i)}) = X - u_i,$$

where u_i is the abscissa of the point R_i. Also, $R(F'(\lambda,\mu), G'(\lambda,\mu))$ is the resultant $R_Y(X)$ of $F_*(X,Y)$ and $G_*(X,Y)$, both considered as polynomials in the indeterminate Y. Now (3.29) reads as follows:

$$R_Y(X) = c\prod_{i=1}^{k}(X-u_i)^{\nu_i}.$$

This shows that, if $P_i = (u_i, v_i)$ is a common point of the curves $\mathcal{F}$ and $\mathcal{G}$, then r_i is equal to the multiplicity of the factor $X - u_i$ in the resultant $R_Y(X)$. Finally,

$$r_i = I(P_i, \mathcal{F}\cap\mathcal{G}) \tag{3.31}$$

from Theorem 3.9 and (3.26). □

EXAMPLE 3.15 With $K = \mathbf{C}$ and (X,Y,Z) homogeneous coordinates, let

$$F = Y^5 - X(Y^2 - XZ)^2,$$
$$G = Y^4 - X^2Z^2 + Y^3Z.$$

Then

$$\begin{aligned} ZF - Y^2G &= -XZ(Y^2-XZ)^2 - Y^2(Y^2-XZ)(Y^2+XZ)\\ &= (Y^2-XZ)(-Y^2XZ + X^2Z^2 - Y^4 - Y^2XZ)\\ &= (Y^2-XZ)(X^2Z^2 - 2Y^2XZ - Y^4).\end{aligned}$$

If $\mathcal{F} = \mathbf{v}(F), \mathcal{G} = \mathbf{v}(G)$ and $P = (x,y,z) \in \mathcal{F}\cap\mathcal{G}$, then

$y^2 - xz = 0 \Rightarrow y = 0 \Rightarrow P = P_1$ or P_2

with $P_1 = (1,0,0),\ P_2 = (0,0,1)$;

$y^4 - x^2z^2 = -2y^2xz \Rightarrow -2y^2xz + y^3z = 0$

$\Rightarrow y = 2x \Rightarrow 16x^4 - x^2z^2 + 8x^3z = 0$

$\Rightarrow 16x^2 + 8xz - z^2 = 0$

$\Rightarrow P = P_3$ or P_4

with $P_3 = (-1+\sqrt{2}, -2+2\sqrt{2}, 4),\ P_4 = (-1-\sqrt{2}, -2-2\sqrt{2}, 4)$.

The points P_3 and P_4 are simple on both the curves $\mathcal{F}$ and $\mathcal{G}$, which intersect transversally at both points, whence

$$I(P_3, F\cap G) = I(P_3, F\cap G) = 1.$$

For $P_2 = (0,0,1)$, put

$$f(X,Y) = F(X,Y,1), \quad g(X,Y) = G(X,Y,1).$$

Then

$$\begin{aligned} f &= Y^5 - X(Y^2-X)^2 = -X^3 + 2X^2Y^2 - XY^4 + Y^5,\\ g &= Y^4 - X^2 + Y^3 = -X^2 + Y^3 + Y^4,\\ h &= f - Xg = 2X^2Y^2 - XY^3 - 2XY^4 + Y^5\\ &\qquad = Y^2(2X-Y)(X-Y^2);\end{aligned}$$

$$\begin{aligned} I(P_2, F\cap G) &= I(P_2, G\cap F) = I(P_2, g\cap f) = I(P_2, g\cap h)\\ &= I(P_2, g\cap Y^2) + I(P_2, g\cap(2X-Y)) + I(P_2, g\cap(X-Y^2))\\ &= 4 + 2 + I(P_2, (X-Y^2)\cap Y^3)\\ &= 4+2+3 = 9.\end{aligned}$$

For $P_1 = (1, 0, 0)$, put

$$f(Y, Z) = F(1, Y, Z), \quad g(Y, Z) = G(1, Y, Z).$$

Then

$$\begin{aligned} f &= -Z^2 + 2Y^2Z - Y^4 + Y^5, \\ g &= -Z^2 + Y^3Z + Y^4, \\ h &= g - f = -2Y^2Z + Y^3Z + 2Y^4 - Y^5 \\ &= -Y^2(2 - Y)(Z - Y^2); \end{aligned}$$

$$\begin{aligned} I(P_1, F \cap G) &= I(P_1, f \cap g) = I(P_1, f \cap h) \\ &= I(P_1, f \cap Y^2) + I(P_1, f \cap (2 - Y)) + I(P_1, f \cap (Z - Y^2)) \\ &= 4 + 0 + I(P_1, (Z - Y^2) \cap f) \\ &= 4 + I(P_1, (Z - Y^2) \cap (f + (Z - Y^2)^2)) \\ &= 4 + I(P_1, (Z - Y^2) \cap Y^5) \\ &= 4 + 5 = 9. \end{aligned}$$

Hence

$$\textstyle\sum_{i=1}^{4} I(P_i, F \cap G) = 1 + 1 + 9 + 9 = 20 = \deg \mathcal{F} \cdot \deg \mathcal{G}.$$

3.3 RATIONAL AND BIRATIONAL TRANSFORMATIONS

DEFINITION 3.16 Given two projective planes π and π' over K, then the mapping $\omega : \pi \to \pi'$ is a *rational transformation* if there exist independent homogeneous polynomials $\varphi_0, \varphi_1, \varphi_2 \in K[X_0, X_1, X_2]$ of the same degree n and, for any point $P = (a_0, a_1, a_2) \in \pi$,

$$\begin{aligned} \omega(P) &= P' = (b_0, b_1, b_2), \\ \rho b_0 &= \varphi_0(a_0, a_1, a_2), \\ \rho b_1 &= \varphi_1(a_0, a_1, a_2), \\ \rho b_2 &= \varphi_2(a_0, a_1, a_2), \end{aligned} \tag{3.32}$$

where $\rho \in K$. Also, n is the *order* of the transformation.

Denote by ζ the net of curves $\mathbf{v}(\mu_0\varphi_0 + \mu_1\varphi_1 + \mu_2\varphi_2)$. As P' traverses the line $\mathbf{v}(\lambda_0 Y_0 + \lambda_1 Y_1 + \lambda_2 Y_2)$ of π', its pre-image P describes in π the curve

$$\mathbf{v}(\lambda_0\varphi_0(X_0, X_1, X_2) + \lambda_1\varphi_1(X_0, X_1, X_2) + \lambda_2\varphi_2(X_0, X_1, X_2)).$$

Here, to a point $P' = (\eta_0, \eta_1, \eta_2)$, centre of a pencil of lines, corresponds the set of base points of a pencil of curves φ of the net ζ. Therefore to the point P' correspond the points common to three curves:

(1) $\mathbf{v}(\varphi_1,\ \varphi_2)$, for $(\eta_0, \eta_1, \eta_2) = (c, 0, 0),\ c \neq 0$;

(2) $\mathbf{v}(\eta_0\varphi_1 - \eta_1\varphi_0,\ \varphi_2)$, for $\eta_1 \neq 0, \eta_2 = 0$;

(3) $\mathbf{v}(\eta_2\varphi_0 - \eta_0\varphi_2, \eta_2\varphi_1 - \eta_1\varphi_2)$, for $\eta_2 \neq 0$.

To ensure that ω is bijective, $n^2 - 1$ of these points must be fixed: denote them $P_1, \ldots, P_{n^2-1}$. Then the n^2-th point $P = (\xi_0, \xi_1, \xi_2)$ is the pre-image of P', and X_0, X_1, X_2 can be obtained as rational functions of Y_0, Y_1, Y_2:

$$\begin{aligned} &\omega^{-1} : \pi' \to \pi, \\ &\rho X_0 = f_0(Y_0, Y_1, Y_2), \\ &\rho X_1 = f_1(Y_0, Y_1, Y_2), \\ &\rho X_2 = f_2(Y_0, Y_1, Y_2). \end{aligned} \tag{3.33}$$

Here f_0, f_1, f_2 are independent homogeneous polynomials of the same degree. In this case, the transformation ω is *birational*; it is also called a *Cremona* transformation. Then through the points $P_0, P_1, \ldots, P_{n^2-1}$ pass all curves of the net ζ. This is therefore a *homaloidal* net; that is, every two of its curves have a single variable intersection.

PROPOSITION 3.17 (i) *The transformation ω changes the curves φ of degree n of the homaloidal net ζ passing through the points $P_0, P_1, \ldots, P_{n^2-1}$ in π into the lines of π'.*

(ii) *In π, the points $P_0, P_1, \ldots, P_{n^2-1}$ constitute the set of* fundamental *points of ω.*

(iii) *The image P' of a point P other than a fundamental point is the image under ω of the unique pencil in ζ passing through P.*

(iv) *When there are two curves C_1, C_2 in ζ with a common component C, every P on C different from a fundamental point has the same image P', which is a fundamental point for ω^{-1}.*

Of particular interest among birational transformations are those of order two, that is, the *quadratic* transformations. If ω is a quadratic transformation from π to π', to the set $\mathcal{L}$ of lines of π correspond in π' the set of conics of a homaloidal net ζ', that is, the set of conics passing through three fixed points, Q_0, Q_1, Q_2, the *fundamental points* of ω in π'.

Also the inverse transformation ω^{-1} is quadratic, since to the lines of π' correspond in π the conics passing through three fixed points P_0, P_1, P_2, the fundamental points in π, and constitute the homaloidal net ζ.

In the general case, the points P_0, P_1, P_2 are distinct and not collinear, as are the points Q_0, Q_1, Q_2. If the reference system is chosen so that

$$P_0 = Q_0 = \mathbf{U}_0,\ P_1 = Q_1 = \mathbf{U}_1,\ P_2 = Q_2 = \mathbf{U}_2,\ \omega(\mathbf{U}) = \mathbf{U},$$

then

$$\begin{aligned} &\omega : \pi \to \pi', \\ &\rho Y_0 = X_1 X_2, \\ &\rho Y_1 = X_0 X_2, \\ &\rho Y_2 = X_0 X_1. \end{aligned} \tag{3.34}$$

The transformation ω is not defined at the points P_0, P_1, P_2. The lines joining pairs of these points are the *exceptional lines*. Every non-fundamental point on one of these lines $\mathbf{v}(X_j)$ is transformed to the point $(\delta_{0j}, \delta_{1j}, \delta_{2j})$, where δ_{ij} is the Kronecker symbol. So, for example, under ω the line P_1P_2 goes to the point Q_0, while, under ω^{-1} the point Q_0 dilates to P_1P_2. Similarly, ω takes the lines P_0P_2 and P_0P_1 to the points Q_1 and Q_2, and ω^{-1} reverses these.

If P is any point off the exceptional lines, its image P' is also off the exceptional lines of ω^{-1}, and $\omega^{-1}(P') = P$. Therefore, with

$$\Delta = \{P = (x_0, x_1, x_2) \in \pi \mid x_0x_1x_2 \neq 0\},$$
$$\Delta' = \{Q = (y_0, y_1, y_2) \in \pi' \mid y_0y_1y_2 \neq 0\},$$

the restriction $\omega_{|\Delta} : \Delta \longrightarrow \Delta'$ is bijective. The inverse ω^{-1} of ω has the following equation:

$$\begin{aligned} &\omega^{-1} : \pi' \to \pi, \\ &\tau X_0 = Y_1Y_2, \\ &\tau X_1 = Y_0Y_2, \\ &\tau X_2 = Y_0Y_1. \end{aligned} \tag{3.35}$$

DEFINITION 3.18 The transformation ω is the *standard quadratic transformation.*

3.4 QUADRATIC TRANSFORMATIONS

Given a projective plane curve $\mathcal{F} = \mathbf{v}(F)$ with $F = F(X_0, X_1, X_2)$, the points $Q = (y_0, y_1, y_2)$ on the image of $\mathcal{F}$ under ω lie on the curve $\mathcal{G} = \mathbf{v}(G)$, with

$$G(Y_0, Y_1, Y_2) = F(Y_1Y_2, Y_0Y_2, Y_0Y_1). \tag{3.36}$$

The curve $\mathcal{G}$ given by (3.36) is the *algebraic transform* of $\mathcal{F}$.

To better understand the geometric relation between $\mathcal{F}$ and $\mathcal{G}$, consider first the case that $\mathcal{F}$ is a line with $F = a_0X_0 + a_1X_1 + a_2X_2$. There are three possibilities.

(1) *$\mathcal{F}$ does not pass through any fundamental point.* In this case, $a_0a_1a_2 \neq 0$ and the algebraic transform of $\mathcal{F}$ is the conic

$$\mathcal{C}_2 = \mathbf{v}(a_0Y_1Y_2 + a_1Y_0Y_2 + a_2Y_0Y_1).$$

This conic $\mathcal{C}_2$ passes through each of the fundamental points in π' but has no further intersection with any exceptional line. The fundamental points Q_0, Q_1, Q_2 correspond to the points of intersection S_0, S_1, S_2 of $\mathcal{F}$ with the exceptional lines in π. So ω establishes a bijection between $\mathcal{F}\backslash\{S_0, S_1, S_2\}$ and $\mathcal{C}_2\backslash\{Q_0, Q_1, Q_2\}$.

(2) *$\mathcal{F}$ is a line through P_0 other than an exceptional line.* Hence $F = X_1 + \lambda X_2$, and its algebraic transform under ω splits into two distinct lines. They are the exceptional line $\mathbf{v}(Y_0)$ and the line $\mathcal{F}' = \mathbf{v}(F')$ with $F' = Y_2 + \lambda Y_1$. As in (1), ω induces a bijection between $\mathcal{F}$ and $\mathcal{F}'$.

(3) *$\mathcal{F}$ coincides with one of the exceptional lines.* Now the algebraic transform of $\mathcal{F}$ splits into two exceptional lines; they are $\mathbf{v}(Y_1)$ and $\mathbf{v}(Y_2)$. Every point of $\mathcal{F}$

for which ω is defined has as its image the point Q_0 that can be considered as the transform of $\mathcal{F}$.

Given a curve $\mathcal{F} = \mathbf{v}(F(X_0, X_1, X_2))$ containing no exceptional line as a component, let $\mathcal{G} = \mathbf{v}(G(Y_0, Y_1, Y_2)) = \mathbf{v}(F(X_1X_2, X_0X_2, X_0X_1))$ be its algebraic transform. Then

$$G(Y_0, Y_1, Y_2) = m(Y)F'(Y_0, Y_1, Y_2),$$

where $m(Y) = {Y_0}^{r_0}{Y_1}^{r_1}{Y_2}^{r_2}$ but $F'(Y_0, Y_1, Y_2)$ is not divisible by any Y_i.

DEFINITION 3.19 The curve $\mathcal{F}' = \mathbf{v}(F')$ is the *geometric transform* of $\mathcal{F}$ under ω.

THEOREM 3.20 (i) *If $\mathcal{F}'$ is the geometric transform of $\mathcal{F}$ for ω, then $\mathcal{F}$ is the geometric transform of $\mathcal{F}'$ for ω^{-1}.*

(ii) *Apart from a finite number of points of $\mathcal{F}$ and $\mathcal{F}'$, the remaining points of $\mathcal{F}$ and $\mathcal{F}'$ are in bijective correspondence.*

Proof. By definition,

$$\begin{aligned} F(Y_1Y_2, Y_0Y_2, Y_0Y_1) &= M_1(Y)F'(Y_0, Y_1, Y_2), \\ F'(X_1X_2, X_0X_2, X_0X_1) &= M_2(X)F''(X_0, X_1, X_2), \end{aligned} \tag{3.37}$$

where $\mathcal{F}'' = \mathbf{v}(F'')$ is the geometric transform of $\mathcal{F}'$ for ω^{-1}. Put $Y_i = X_jX_k$ in (3.37) for $\{i, j, k\} = \{1, 2, 3\}$:

$$\begin{aligned} F(X_0^2X_1X_2, X_0X_1^2X_2, X_0X_1X_2^2) &= M_3(X)F'(X_1X_2, X_0X_2, X_0X_1) \\ &= M_4(X)F''(X_0, X_1, X_2), \end{aligned}$$

where M_3, M_4 are monomials in X_0, X_1, X_2. Since F has degree n, it follows that

$$F(X_0^2X_1X_2, X_0X_1^2X_2, X_0X_1X_2^2) = (X_0X_1X_2)^nF(X_0, X_1, X_2).$$

Since no X_i divides F or F'', so $F = F''$. Further, both $\mathcal{F}$ and $\mathcal{F}''$ have a finite number of points in common with the exceptional lines, and the bijection between the components of $\mathcal{F}$ and $\mathcal{F}''$ is a consequence of the formulas that define ω and ω^{-1}. □

Now consider the behaviour of a singular point of a plane curve under a quadratic transformation. Here, use the following correspondence between the non-exceptional lines through P_0 and the non-fundamental points of the line $\mathbf{v}(Y_0)$:

$$\mathbf{v}(X_1 + \lambda X_2) \longleftrightarrow (0, 1, -\lambda).$$

As $P = (1, -\lambda u, u)$ describes the line ℓ on π, the image $P' = (-\lambda u, 1, -\lambda)$ describes the line $\mathbf{v}(Y_0 - uY_2)$ on π'. As $P \mapsto P_0$, that is, the parameter $u \mapsto 0$, then $P' \mapsto Q = (0, 1, -\lambda)$.

THEOREM 3.21 *Let $\mathcal{F} = \mathbf{v}(F(X_0, X_1, X_2))$ be a curve of degree d such that*

(a) *the fundamental point P_i has multiplicity m_i on $\mathcal{F}$, $0 \leq i \leq 2$;*

(b) *no exceptional line through a fundamental point is tangent to $\mathcal{F}$ at that point.*

Then

(i) *the algebraic transform $\mathcal{G}$ of $\mathcal{F}$ admits the fundamental line* $\mathbf{v}(Y_i)$ *as a component of multiplicity* m_i;

(ii) *the geometric transform* $\mathcal{F}'$ *has degree* $2d - m_0 - m_1 - m_2$;

(iii) *there is a multiplicity-preserving bijection between the tangents to* $\mathcal{F}$ *at* P_i *and the intersections of* $\mathcal{F}'$ *with the exceptional line* $\mathbf{v}(Y_i)$, *apart from the two fundamental points of* $\mathbf{v}(Y_i)$;

(iv) $I(R, \mathcal{F}' \cap \mathbf{v}(Y_i)) = j$, *where R is the point corresponding to a line counted j times among the tangents to $\mathcal{F}$ at P_i*;

(v) *the curve $\mathcal{F}'$ has a point of multiplicity $d - m_j - m_k$ at the fundamental point Q_i, where $\{i, j, k\} = \{1, 2, 3\}$*;

(vi) *the tangents to $\mathcal{F}'$ at Q_i are not exceptional lines and correspond to the intersections of $\mathcal{F}$ with the line $\mathbf{v}(X_i)$, apart from the fundamental points.*

Proof. (i),(ii) Here the discussion is limited to considering what happens to the point P_0. The polynomial F of degree d is written in the form

$$F(X_0, X_1, X_2) = \textstyle\sum_{j=m_0}^{d} X_0^{d-j} F_j(X_1, X_2),$$

where F_j is a homogeneous polynomial of degree j in X_1, X_2 and F_{m_0}, F_d are both non-zero. The algebraic transform of $\mathcal{F}$ is $\mathcal{G} = \mathbf{v}(G)$ with

$$\begin{aligned} G(Y_0, Y_1, Y_2) &= F(Y_1Y_2, Y_0Y_2, Y_1Y_2) \\ &= \textstyle\sum_{j=m_0}^{d} (Y_1Y_2)^{d-j} F_j(Y_0Y_2, Y_0Y_1) \\ &= Y_1^{d-m_0} Y_2^{d-m_0} Y_0^{m_0} F_{m_0}(Y_2, Y_1) + \cdots + Y_0^d F(Y_2, Y_1). \end{aligned}$$

So $Y_0^{m_0}$ divides G but $Y_0^{m_0+1}$ does not, and similarly for $Y_1^{m_1}$ and $Y_2^{m_2}$. Therefore $G = Y_0^{m_0} Y_1^{m_1} Y_2^{m_2} F'$, where F' is the geometric transform. It follows that F' has degree $2d - (m_0 + m_1 + m_2)$.

(iii) On the one hand, the tangents to $\mathcal{F}$ at P_0 correspond to the linear factors of

$$F_{m_0}(X_1, X_2) = \textstyle\prod_{k=1}^{k=n} (\mu_k X_1 + \nu_k X_2)^{\alpha_k},$$

with $\alpha_1 + \cdots + \alpha_n = m_0$; on the other hand, $I(Q, \mathcal{F}' \cap \mathbf{v}(Y_0))$ at points Q is given by the multiplicity of the roots of the polynomial

$$\begin{aligned} &Y_1^{d-m_0-m_1} Y_2^{d-m_0-m_2} F_{m_0}(Y_2, Y_1) \\ &\qquad = Y_1^{d-m_0-m_1} Y_2^{d-m_0-m_2} \textstyle\prod_{k=1}^{k=n} (\mu_k Y_2 + \nu_k Y_1)^{\alpha_k}, \end{aligned}$$

with $\alpha_1 + \cdots + \alpha_n = m_0$.

(iv) Note that

$$F' = Y_1^{d-m_0-m_1} Y_2^{d-m_0-m_2} F_{m_0}(Y_1, Y_2) + \cdots + Y_0^{d-m_0} Y_1^{-m_1} Y_2^{-m_2} F_d(Y_2, Y_1);$$

then the multiplicity of $\mathcal{F}'$ at $Q_0 = (1, 0, 0)$ is given by the degree of

$$B(Y_1, Y_2) = Y_1^{-m_1} Y_2^{-m_2} F_d(Y_1, Y_2),$$

and this is $d - m_1 - m_2$.

(v) The tangents to $\mathcal{F}'$ at Q_0 are given by the factors of $B(Y_1, Y_2)$. If one of these were Y_1 or Y_2, then $F_d(Y_2, Y_1)$ would be divisible by a power $m > m_1$ of Y_1 or $m > m_2$ of Y_2. But, if that happens, then $I(P_1, \mathcal{F} \cap \mathbf{v}(X_0)) = m$ or $I(P_2, \mathcal{F} \cap \mathbf{v}(X_0)) = m$. Therefore one of the tangents to $\mathcal{F}$ at P_1 or at P_2 would be the exceptional line $\mathbf{v}(X_0)$, a contradiction.

(vi) Applying (v) to ω^{-1} gives the result. □

THEOREM 3.22 (i) *To each point P external to the fundamental triangle of π corresponds a point Q external to the fundamental triangle of π'.*

(ii) *If such a point P is an m-ple point of the curve $\mathcal{F}$, then its image Q is an m-ple point of the geometric transform $\mathcal{F}'$ of $\mathcal{F}$.*

Proof. Take $P = (1, 1, 1)$, and so $Q = (1, 1, 1)$. Since the algebraic transform $\mathcal{G}$ and the geometric transform $\mathcal{F}'$ differ by a component not passing through P, it suffices to consider $\mathcal{G}$ instead of $\mathcal{F}'$. In π, take a new system of coordinates:

$$\begin{aligned} \rho Z_0 &= X_0, \\ \rho Z_1 &= X_1 - X_0, \\ \rho Z_2 &= X_2 - X_0. \end{aligned} \tag{3.38}$$

The point P is now $(1, 0, 0)$, and

$$\begin{aligned} F(X_0, X_1, X_2) &= F(Z_0, Z_0 + Z_1, Z_0 + Z_2) \\ &= \bar{F}(Z_0, Z_1, Z_2) \\ &= Z_0^{d-m} \bar{F}_m(Z_1, Z_2) + \cdots + \bar{F}_d(Z_1, Z_2) \\ &= X_0^{d-m} \bar{F}_m(X_1 - X_0, X_2 - X_0) + \cdots + \bar{F}_d(X_1 - X_0, X_2 - X_0). \end{aligned}$$

Therefore the algebraic transform $\mathcal{G} = \mathbf{v}(G(Y_0, Y_1, Y_2))$ is given by

$$\begin{aligned} G = {} & (Y_1 Y_2)^{d-m} \bar{F}_m(Y_0 Y_2 - Y_1 Y_2, Y_0 Y_1 - Y_1 Y_2) \\ & + \cdots + \bar{F}_d(Y_0 Y_2 - Y_1 Y_2, Y_0 Y_1 - Y_1 Y_2). \end{aligned}$$

In π', use a similar change of coordinates:

$$\begin{aligned} \rho W_0 &= Y_0, \\ \rho W_1 &= Y_0 - Y_1, \\ \rho W_2 &= Y_0 - Y_2. \end{aligned} \tag{3.39}$$

This takes $Q = (1, 1, 1)$, the image of P under ω, to $Q' = (1, 0, 0)$ in π'. Also,

$$G = G(Y_0, Y_1, Y_2) = G(W_0, W_0 - W_1, W_0 - W_2) = \bar{G}(W_0, W_1, W_2).$$

Now

$$\begin{aligned} & \bar{G}(W_0, W_1, W_2) \\ & = [(W_0 - W_1)(W_0 - W_2)]^{d-m} \bar{F}_m(W_0 W_1 - W_1 W_2, W_2 W_0 - W_1 W_2) \\ & \quad + [(W_0 - W_1)(W_0 - W_2)]^{d-m-1} \bar{F}_{m-1}(W_0 W_1 - W_1 W_2, W_2 W_0 - W_1 W_2) \\ & \quad + \cdots + \bar{F}_d(W_0 W_1 - W_1 W_2, W_2 W_0 - W_1 W_2) \\ & = W_0^{2d-m} \bar{F}_m(W_1, W_2) + \cdots + W_0^{2d-m-1} \bar{F}_{m+1}(W_1, W_2) \\ & \quad + \cdots + W_0^d \bar{F}_d(W_1, W_2) + \cdots, \end{aligned}$$

in descending powers of W_0. The highest power of W_0 appears in the first term. Therefore $\bar{F}_d(W_1, W_2)$ determines the configuration of the tangents to $\mathcal{G}$ at Q', and the theorem is proved. □

3.5 RESOLUTION OF SINGULARITIES

This section is devoted to the classical result that every projective plane curve is birationally equivalent to a curve with only ordinary singularities. Here, a point P of a curve is *ordinary* if the tangents to the curve at P are distinct.

DEFINITION 3.23 If $\mathcal{C}^n$ is an irreducible projective plane curve of degree n, its *virtual genus* is

$$g^*(\mathcal{C}^n) = \tfrac{1}{2}(n-1)(n-2) - \sum \tfrac{1}{2}r(r-1), \tag{3.40}$$

where the sum is over all singular points P of $\mathcal{C}^n$, with r the multiplicity of P.

LEMMA 3.24 *The virtual genus of a curve is non-negative.*

Proof. Let $P_1, P_2, \ldots, P_k$ be the singular points of the curve $\mathcal{C}^n$, with multiplicities $r_1, r_2, \ldots, r_k$. First, it is shown that the integer

$$s = \tfrac{1}{2}(n-1)(n+2) - \textstyle\sum_{i=1}^{k} \tfrac{1}{2}r_i(r_i-1) \tag{3.41}$$

is non-negative.

If $\mathcal{C}^n = \mathbf{v}(F(X_0, X_1, X_2))$, then the polar curve of $\mathcal{C}^n$ with respect to the point $\mathbf{U}_0 = (1, 0, 0)$ is $\mathcal{C}_0^n = \mathbf{v}(\partial F/\partial X_0)$. When $p > 0$, it is possible that $\partial F/\partial X_0$ is the zero polynomial. In that case, Theorem 1.25 ensures that $\partial F/\partial X_1$ is not the zero polynomial.

Thus it may be assumed that $\mathcal{C}_0^n$ exists.

A point P such that $m_P(\mathcal{C}^n) = r$ satisfies $m_P(\mathcal{C}_0^n) \geq r - 1$. By Bézout's Theorem 3.14,

$$\textstyle\sum_{i=1}^{k} r_i(r_i - 1) \leq n(n-1).$$

Since $n(n-1) < (n-1)(n+2)$, so $s = 0$ for $n = 1$ and $s > 0$ otherwise.

Now choose s points $Q_1, \ldots, Q_s$ on $\mathcal{C}^n$. First, it is shown that there exists a curve $\mathcal{C}^{n-1}$ of degree $n-1$ passing through each point P_i with multiplicity at least $r_i - 1$ and through each Q_j. If $\mathcal{G} = \mathbf{v}(G)$ is a general plane curve of degree $n-1$, then G has $n(n+1)/2$ coefficients. Its passage through an $(r-1)$-ple point P requires at most $\frac{1}{2}r(r-1)$ linear conditions on its coefficients. In fact, if $P = \mathbf{U}_0$ and $G = \sum g_i(X_1, X_2)X_0^{n-1-i}$, the condition that P is an $(r-1)$-ple point is that

$$g_0 = g_1 = \cdots = g_{r-2} = 0;$$

these terms contain $\frac{1}{2}r(r-1)$ coefficients. It follows that the passage through the k singular points requires at most $\sum_{i=1}^{k} \frac{1}{2}r(r-1)$ linear conditions on its coefficients.

Therefore the number of conditions on $\mathcal{C}^{n-1}$ to pass through the P_i and Q_j is at most

$$s + \textstyle\sum_{i=1}^{k} \tfrac{1}{2}r_i(r_i-1) \leq \tfrac{1}{2}(n-1)(n+2),$$

by the definition of s in (3.41). Hence, such a $\mathcal{C}^{n-1}$ exists.

Now apply Bézout's Theorem 3.14 to the curves $\mathcal{C}^n$ and $\mathcal{C}^{n-1}$:

$$\begin{aligned} n(n-1) &= \sum I(P, \mathcal{C}^n \cap \mathcal{C}^{n-1}) \\ &\geq \sum I(Q_j, \mathcal{C}^n \cap \mathcal{C}^{n-1}) + \sum I(P_i, \mathcal{C}^n \cap \mathcal{C}^{n-1}) \\ &= s + \textstyle\sum_{i=1}^k r_i(r_i - 1) \\ &= \tfrac{1}{2}(n-1)(n+2) - \textstyle\sum_{i=1}^k \tfrac{1}{2} r_i(r_i-1) + \textstyle\sum_{i=1}^k r_i(r_i-1) \\ &= n(n-1) - \tfrac{1}{2}(n-1)(n-2) + \textstyle\sum_{i=1}^k \tfrac{1}{2} r_i(r_i-1) \\ &= n(n-1) - g^*(\mathcal{C}^n). \end{aligned}$$

Hence $g^*(\mathcal{C}^n) \geq 0$. □

DEFINITION 3.25 The *index* of the irreducible plane projective curve $\mathcal{F} = \mathbf{v}(F)$ is the non-negative integer

$$I(\mathcal{F}) = I(F) = \textstyle\sum_P \sum_T (m_T - 1), \tag{3.42}$$

where the outer sum is over all singular points P of $\mathcal{F}$ and the inner sum is over all distinct tangents T to $\mathcal{F}$ at P, with m_T the multiplicity of the tangent.

It is worth noting explicitly the following property of the index.

LEMMA 3.26 *The index $I(\mathcal{F})$ is zero if and only if each singular point of $\mathcal{F}$ is ordinary.*

Now a start is made on the proof of the famous theorem on the resolution of singularities.

THEOREM 3.27 *Through a finite sequence of quadratic transformations, an irreducible projective plane curve can be transformed into a curve with only ordinary singularities.*

Proof. Given a finite sequence of quadratic transformations, $\omega_1, \ldots, \omega_r$, if $\mathcal{F}$ is an irreducible plane curve, the symbol $\mathcal{F}^{(i)}$ indicates the geometric transform of $\mathcal{F}^{(i-1)}$ under ω_i, for $i = 1, \ldots, r-1$ with $\mathcal{F}^{(0)} = \mathcal{F}$.

Now let $\mathcal{F}$ be an irreducible plane projective curve of degree d with a non-ordinary singularity; it suffices to exhibit a finite sequence $\omega_1, \ldots, \omega_r$ of quadratic transformations such that $I(\mathcal{F}^{(k)}) \leq I(\mathcal{F}^{(k-1)})$ and, if equality holds, then

$$g^*(\mathcal{F}^{(k)}) < g^*(\mathcal{F}^{(k-1)}).$$

If $\mathcal{F}^{(k)}$ still has non-ordinary singularities, then, with a new sequence of quadratic transformations, another curve is obtained with strictly lower index or virtual genus. This procedure is repeated till a curve is obtained with index zero, which necessarily has only ordinary singularities by Lemma 3.26.

To construct such a sequence, a particular system of coordinates is used that satisfies the following conditions for a given m-ple point of $\mathcal{F}$:

(i) the m-ple point is $P_0 = (1, 0, 0)$;

(ii) the curve $\mathcal{F}$ does not pass through the other two fundamental points, and $\mathcal{F}$ intersects each of the two exceptional lines through P_0 in $d-m$ points other than P_0;

(iii) the exceptional line $\mathbf{v}(X_0)$ meets $\mathcal{F}$ in d distinct points.

The existence of such a system of coordinates does not always happen, for in positive characteristic there exist strange curves containing their own nucleus, for which condition (ii) does not hold. This possibility is treated below. □

The utility of such a system of coordinates lies in the following result.

LEMMA 3.28 *If ω is the quadratic transformation relative to a triangle satisfying conditions $(1), (2), (3)$ and, if $\mathcal{F}'$ is the geometric transform of $\mathcal{F}$ under ω, then the following properties hold:*

(i) $g^*(\mathcal{F}') \le g^*(\mathcal{F}), \quad I(\mathcal{F}') \le I(\mathcal{F})$;

(ii) *if P_0 is not an ordinary singularity, then*

$$I(\mathcal{F}') = I(\mathcal{F}) \Longrightarrow g^*(\mathcal{F}') < g^*(\mathcal{F}).$$

Proof. Denote by $\ell_1, \ldots, \ell_s$ the tangents to $\mathcal{F}$ at P_0, and by $\alpha_1, \ldots, \alpha_s$ their respective multiplicities. From Theorem 3.22, every singular point of $\mathcal{F}$ other than P_0 is transformed into a point of $\mathcal{F}'$ with the same multiplicity; an ordinary singularity of $\mathcal{F}$ external to the fundamental triangle is transformed into an ordinary singularity of $\mathcal{F}'$, also external to the fundamental triangle.

By part (v) of Theorem 3.21, $\mathcal{F}'$ has three 'new' singular points: Q_0 of multiplicity d, and Q_1, Q_2 both of multiplicity $d-m$. From part (iv) of the same theorem, it follows that $\mathcal{F}'$ has s points $R'_1, \ldots, R'_s$ on the line $\mathbf{v}(Y_0)$ distinct from the fundamental points, with r'_i the multiplicity of R'_i; then

$$r'_i \le I(R'_i, \mathcal{F}' \cap \mathbf{v}(Y_0)) = \alpha_i$$

for $1 \le i \le s$. Also suppose that $\mathcal{F}'$ has degree $2d-m$ as in (ii) of Theorem 3.21.

To prove (i), it suffices to see that

$$g^*(\mathcal{F}') = g^*(\mathcal{F}) - \textstyle\sum_{j=1}^{s} \frac{1}{2} r'_j(r'_j - 1). \tag{3.43}$$

If $\mathcal{F}$ has k singular points other than P_0 with multiplicities $r_1, \ldots, r_k$, then

$$\begin{aligned} g^*(\mathcal{F}') = {} & \tfrac{1}{2}(2d-m-1)(2d-m-2) - \tfrac{1}{2}d(d-1) - (d-m)(d-m-1) \\ & - \textstyle\sum_{i=1}^{k} \frac{1}{2} r_i(r_i-1) - \sum_{j=1}^{s} \frac{1}{2} r'_j(r'_j-1). \end{aligned}$$

This expression can also be written in the form

$$\begin{aligned} & \tfrac{1}{2}(d^2-3d+2) - \tfrac{1}{2}m(m-1) - \textstyle\sum_{i=1}^{k} \frac{1}{2} r_i(r_i-1) - \sum_{j=1}^{s} \frac{1}{2} r'_j(r'_j-1) \\ & \quad = g^*(\mathcal{F}) - \textstyle\sum_{j=1}^{s} \frac{1}{2} r'_j(r'_j-1), \end{aligned}$$

as required.

For part (ii), note that, if $I(\mathcal{F}) = I(\mathcal{F}')$, then

$$\textstyle\sum_{i=1}^{s} (\alpha_i - 1) - \sum_{i=1}^{s} \sum_{T_i} (m'_{T_i} - 1) = 0,$$

where, in the second summation, T_i varies in the set of tangents to $\mathcal{F}'$ at R_i'. It follows that

$$\sum_{i=1}^{s} [\alpha_i - \sum_{T_i} m'_{T_i}] + S - s = 0,$$

where S is the total number of tangents to $\mathcal{F}'$ at the points $R_1', \ldots, R_s'$. Since

$$\alpha_i = I(R_i, \mathcal{F}' \cap \mathbf{v}(Y_0)) \geq r_i' = \sum_{T_i} m'_{T_i}$$

and $S \geq s$, this implies that $\alpha_i = r_i'$ and $S = s$.

It follows that if, as in the proof of (ii) of Lemma 3.28, some $\alpha_i > 1$, then also $r_i' > 1$. Note also that $T = s$ if and only if $\mathcal{F}'$ has a unique tangent at each point R_i'. Hence the sum $\sum_{i=1}^{s} \frac{1}{2} r_i'(r_i' - 1)$ is positive, which proves (ii). □

By virtue of Lemma 3.28, the procedure described immediately after the enunciation of Theorem 3.27 carries over to an irreducible plane curve with only ordinary singularities once it is possible to find, for each curve $\mathcal{F}^{(i)}$ of the sequence, a system of coordinates for which conditions (i),(ii),(iii) hold. To finish the proof of Theorem 3.27, it is necessary to construct the sequence $\mathcal{F}^{(1)} = \mathcal{F}', \mathcal{F}^{(2)}, \ldots$ in such a way that points not satisfying (ii) are avoided.

To do this, use non-homogeneous coordinates $X = X_1/X_0,\ Y = X_2/X_0$, and let $\mathcal{F} = \mathbf{v}(f(X,Y))$. To see that condition (ii) can be satisfied, the intersection of $\mathcal{F}$ with a line $\ell = \mathbf{v}(Y - tX)$, $t \in K^*$, must be found:

$$f(X,Y) = 0, \quad Y - tX = 0. \tag{3.44}$$

Write

$$f(X,Y) = F_m(X,Y) + F_{m+1}(X,Y) + \cdots + F_d(X,Y),$$

where $F_i(X,Y)$ is a homogeneous polynomial of degree i with $F_m \neq 0,\ F_d \neq 0$.

Now (3.44) gives

$$F_m(X,tX) + F_{m+1}(X,tX) + \cdots + F_d(X,tX) = 0,$$

which becomes

$$X^m(F_m(1,t) + XF_{m+1}(1,t) + \cdots + X^{d-m}F_d(1,t)) = 0. \tag{3.45}$$

If ℓ is not a tangent to $\mathcal{F}$ at P_0, then $F_m(t) \neq 0$.

Suppose this is the case, and put $F_i(t) = F_i(1,t)$. Then (ii) is satisfied by the line ℓ since

$$P_t(X) = F_m(t) + XF_{m+1}(t) + \cdots + X^{d-m}F_d(t)$$

has no multiple roots. This is equivalent to saying that $P_t(X)$ and its derivative $Q_t(X)$ have no common root. Now

$$Q_t(X) = F_{m+1}(t) + 2XF_{m+2}(t) + \cdots + (d-m)X^{d-m-1}F_d(t).$$

To see that this happens, distinguish two cases according as $Q_t(X)$ is identically zero or not.

In the second case, since $F_j(t)$ is a polynomial in t, so $P_t(X)$ can be written as a polynomial in the indeterminates X, t as $P(X,t)$. Similarly, define $Q(X,t)$. Using the resultant, the system

$$P(X,t) = 0, \quad Q(X,t) = 0 \tag{3.46}$$

has a finite number of solutions (x, λ) unless the two polynomials have a common factor. This possibility can be ruled out as follows.

First, it is shown that $P(X, t)$ is irreducible. Otherwise,

$$P(X,t) = U(X,t)V(X,t),$$

with U, V non-constant; therefore

$$P(X,Y/X) = U(X,Y/X)V(X,Y/X).$$

Now, if both sides are multiplied by X^m, the left becomes $f(X,Y)$ and the right the product of $X^iU(X,Y/X)$ and $X^jV(X,Y/X)$ with $i + j = m$. But this is a contradiction since both are non-constant in that they contain Y; otherwise, X divides $P(X,t)$, which is impossible since $F_m(t) \neq 0$ for all t.

To show that the two polynomials in (3.46) have no common factors, it suffices to show that $\deg P(X,t) > \deg Q(X,t)$. To see this, note that, as $F_j(t)$ has degree at most j, so $2i + m - 1$ is an upper limit for the degree of $F_{m+i}(t)X^{i-1}$. This number reaches its maximum for the largest value of i allowed that equals $d - m$. Therefore the degree of $Q(X,t)$ cannot be greater than

$$2(d - m) + m - 1 = 2d - m - 1.$$

Since $\deg P(X,t) = 2d - m$, the result is proved. So it has been shown that if $Q_t(X)$ is not identically zero, then (3.46) has a finite number of solutions (x, λ). Therefore there is an infinite number of values of λ for which (x, λ) is not a solution; so the line $\ell = \mathbf{v}(Y - \lambda X)$ satisfies condition (ii).

Now consider the case that $Q_t(X)$ is the zero polynomial. This can only happen when p is positive, and $p \mid (d - m)$. Also $p \mid i$ for each i, with $1 \leq i \leq d - m$, for which $F_{m+i}(t)$ is not identically zero. This gives a necessary and sufficient condition for $Q_t(X)$ to be zero. The following example shows that this can occur.

EXAMPLE 3.29 Let $f(X,Y) = Y - X^{p+1}$. The plane curve $\mathcal{F} = \mathbf{v}(f(X,Y))$ is non-singular. The intersection of $\mathcal{F}$ with $\ell = \mathbf{v}(Y - tX)$, apart from P_0, is the single point $P = (x, x^{p+1})$, where x is the unique root of $X^p - t$. Therefore there is no line through P_0 satisfying condition (ii).

DEFINITION 3.30 The point P is a *terrible* point of $\mathcal{F}$ if, when it is transformed to P_0, it does not satisfy condition (ii).

It follows that a point of $\mathcal{F}$ is terrible if and only if $\mathcal{F}$ is strange and P is its unique nucleus. Summarising so far, if none of the singular points of $\mathcal{F}$ is terrible, Theorem 3.27 is proved.

Now it is shown how to get rid of the obstacle of terrible points. Suppose P is a terrible m-ple point of $\mathcal{F}$; then, as above, $d \equiv m \pmod{p}$.

Take a triangle with sides not containing P and with vertices not lying on the curve if $d \not\equiv 0 \pmod{p}$, but with a single vertex on the curve at a non-singular point if $d \equiv 0 \pmod{p}$. Apply the quadratic transformation ω relative to this triangle as fundamental triangle; it is shown that ω takes P to a non-terrible point Q of the geometric transform $\mathcal{F}'$. Note that $\mathcal{F}'$ has degree $2d$ or $2d - 1$ according as the vertices of the triangle are all external to $\mathcal{F}$ or as one is a non-singular point of $\mathcal{F}$.

If Q were terrible, in the first case $d \not\equiv 0 \pmod p$ and $2d - m \equiv 0 \pmod p$; in the second case, $d \equiv 0 \pmod p$ and $2d - 1 - m \equiv 0 \pmod p$. But both cases are impossible in that they contradict $d \equiv m \pmod p$. Therefore Q is not terrible.

So it has been shown that, if there is a singular terrible point P on $\mathcal{F}$, the strategy is to first apply the above quadratic transformation ω to give a new curve $\mathcal{F}'$, the geometric transform of $\mathcal{F}$, for which the corresponding point Q is no longer terrible. Using this, Theorem 3.27 is finally proved.

EXAMPLE 3.31 Let $q = 2^s$ and let $K = \overline{\mathbf{F}}_q$. For $u < 2^s$, let $\mathcal{D}$ be the irreducible plane curve $\mathbf{v}(g(X,Y))$ such that

$$\begin{aligned}(X^u+1)(Y^{2^s}+1)+(Y^u+1)(X^{2^s}+1)\\ =g(X,Y)(X+1)(Y+1)(X+Y).\end{aligned} \tag{3.47}$$

Let $g_1(X,Y) = g(X+1, Y+1)$. Then

$$\begin{aligned}((X+1)^u+1)Y^{2^s}+((Y+1)^u+1)X^{2^s}\\ =g_1(X,Y)XY(X+Y).\end{aligned} \tag{3.48}$$

The analysis of the singular points of $\mathcal{D}$ is as follows:

(i) $E = (1,1)$ is an ordinary (2^s-2)-ple point, with the appropriately distinct tangents $\mathbf{v}(Y+mX+m+1)$, where $m^{2^s-1} = 1$ but $m \neq 1$;

(ii) for each b with $b^u = 1$ but $b \neq 1$, the point $P = (b,1)$ is a (2^s-1)-ple point with a single tangent;

(iii) for each c with $c^u = 1$ but $c \neq 1$, the point $P = (1,c)$ is a (2^s-1)-ple point with a single tangent;

(iv) for all b, c with $b^u = c^u = 1$ but $b \neq 1,\ c \neq 1$, the point $P = (b,c)$ is a 2^s-ple point with a single tangent.

Let ω be the standard quadratic transformation $(X,Y) \mapsto (X^{-1},Y^{-1})$. Then the geometric transform $\mathcal{D}'$ of $\mathcal{D}$ is the curve $\mathbf{v}(g_2(X,Y))$ of degree 2^s-2, where

$$\begin{aligned}(X^{u-1}+\cdots+1)X^{2^s-u}+(Y^{u-1}+\cdots+1)Y^{2^s-u}\\ =g_2(X,Y)(X+Y).\end{aligned} \tag{3.49}$$

Then the curve $\mathcal{D}'$ has only ordinary singularities. Also, the genus of $\mathcal{D}'$, and so of $\mathcal{C}$, is $(u-2)(2^s-u-2)$.

The analysis of the singular points of $\mathcal{D}'$ is as follows:

(1) $O = (0,0)$ is an ordinary (2^s-u-1)-ple point;

(2) $E = (1,1)$ is an ordinary $(u-1)$-ple point.

No point at infinity is a singular point. So, if $P = (b,c)$ is a singular point of $\mathcal{D}'$, then $\partial g_2/\partial X = 0$ at P, and so

$$(1+b)^{u-1}b^{2^s-u-1} = 0.$$

Then $b = 0$ or 1; similarly, $c \in \{0, 1\}$. So neither $P = (0, 1)$ nor $Q = (1, 0)$ lies on $\mathcal{D}'$, whence $\mathcal{D}'$ has only two singular points, $O = (0, 0)$ and $E = (1, 1)$. More precisely, $O = (0, 0)$ is an ordinary $(2^s - u - 1)$-ple point with tangents $\mathbf{v}(Y + \lambda X)$, where $\lambda^{2^s - u} = 1$ but $\lambda \neq 1$.

To find the tangents to $\mathcal{D}'$ at $E = (1, 1)$, consider the translation

$$(X, Y) \mapsto (X + 1, Y + 1)$$

that takes E to O. Then $\mathcal{D}'$ becomes $\mathcal{D}'' = \mathbf{v}(g_3(X, Y))$, where

$$X^u(X+1)^{2^s-u} + X^{2^s} + Y^u(Y+1)^{2^s-u} + Y^{2^s} = g_3(X, Y)(X + Y).$$

This shows that E is a $(u - 1)$-ple point with tangents $\mathbf{v}(Y + \lambda X + \lambda + 1)$, where $\lambda^u = 1$ but $\lambda \neq 1$.

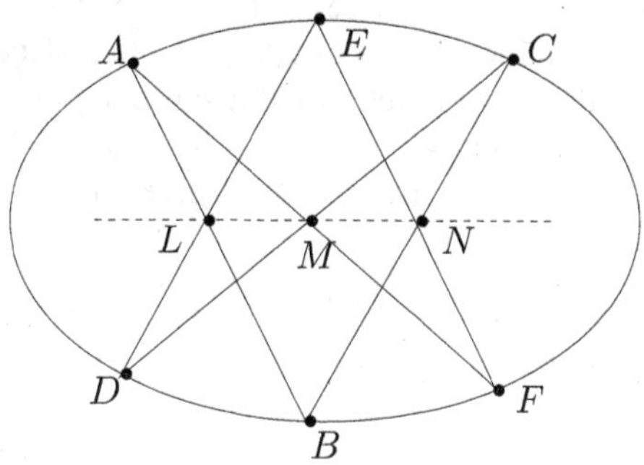

Figure 3.1 The Pascal hexagon

3.6 EXERCISES

1. Let $\mathcal{F} = \mathbf{v}(X_1X_0^2 - X_2^3)$ and $\mathcal{G} = \mathbf{v}(X_1X_0^4 - X_2^5)$. Show that the common points of $\mathcal{F}$ and $\mathcal{G}$ are

$$P_1 = (1, 0, 0),\ P_2 = (1, 1, 1),\ P_3 = (1, -1, -1), P_4 = (0, 0, 1),$$

where

$$I(P_i, \mathcal{F} \cap \mathcal{G}) = \begin{cases} 3 & \text{when } i = 1, \\ 1 & \text{when } i = 2, 3, \\ 10 & \text{when } i = 4. \end{cases}$$

2. (Pascal's Theorem) Assume that the points of intersection L, M, N of the opposite sides of a hexagon $ABCDEF$ in PG$(2, K)$ are distinct. Show that these points, the diagonal points, are collinear if and only if the hexagon is inscribed in a conic.

3. (Maclaurin's Theorem) Let $\mathcal{F}$ be an irreducible projective plane cubic curve, and let P be any non-singular point of $\mathcal{F}$. If P is not an inflexion point of $\mathcal{F}$, the *tangential point* P' of P is the unique point of $\mathcal{F}$ on the tangent line

to $\mathcal{F}$ at P other than P. If P is an inflexion point, let $P' = P$. If P, Q, R are three non-singular collinear points of $\mathcal{F}$, show that their tangential points P', Q', R' are also collinear.

4. (Lamé's Theorem) Let ℓ_1, ℓ_2, ℓ_3 and r_1, r_2, r_3 be two triples of distinct lines in $\mathrm{PG}(2, K)$. Assume that no line from one triple passes through the common point of two lines from the other triple. For $1 \leq j, k \leq 3$, let R_{jk} denote the common point of the lines ℓ_j and r_k. There are nine such common points. Show that if eight of these points lie on a cubic curve then the ninth also does.

   This is a particular case of the Theorem of the Nine Associated Points that two projective plane cubic curves with eight common points have a ninth common point; it is used in Section 9.9.

5. Generalise the preceding exercise. Let $N = \frac{1}{2}n(n+3)$, and consider a set $\mathcal{S}$ of N distinct common points of two projective plane curves of degree n. Then either $\mathcal{S}$ contains a point P and a subset $\mathcal{S}'$ of $N-2$ points distinct from P such that any curve of degree n through $\mathcal{S}'$ also passes through P, or, for every point $P \in \mathcal{S}$, any curve of degree n through $\mathcal{S} \backslash \{P\}$ also passes through P.

6. For two polynomials $F, G \in K[X, Y]$ and a point P, the ideal generated by F and G in the local ring $\mathcal{O}_P$ is $(F, G) \cdot \mathcal{O}_P$. Show that the residue class ring $\mathcal{O}_P/((F, G) \cdot \mathcal{O}_P)$ is a finite-dimensional vector space over K whose dimension $\dim_K \mathcal{O}_P/((F, G) \cdot \mathcal{O}_P)$ satisfies the postulates $(\mathrm{I}\,1), \ldots, (\mathrm{I}\,7)$ studied in Section 3.1. Note that, by Theorem 3.8,

$$\dim_K \mathcal{O}_P/((F, G) \cdot \mathcal{O}_P) = I(P, F \cap G).$$

3.7 NOTES

Section 3.1 is based on Scherk [382], which in turn is a variant of Fulton [135, Chapter 3]. See also the latter for more on the local ring approach as in Remark 3.10.

The proof of Bézout's Theorem 3.14 is based on van der Waerden [484, Chapter 3]; see also [166, Chapter 2]. For other proofs, see [135], [269], [400], [497].

For Sections 3.3, 3.4, 3.5, see Fulton [135, Chapter 7] and Moreno [336].

Chapter Four

Branches and parametrisation

From now on, only projective curves are considered, and the adjective 'projective' is omitted.

Concepts such as the degree of a curve, the multiplicity of a point on a curve, and the number of singular points of a curve are all projective invariants.

The resolution of singularities in Chapter 3 required a broader class of transformations, namely birational transformations: projective transformations were not sufficient. Any projective invariant is also a birational invariant. So the question arises of the existence of birational invariants; that is, non-negative integers that are constant under a birational transformation that does not necessarily fix a projective invariant. Such an invariant must have the property that some points in the plane have no image under a transformation.

The aim of this chapter is to introduce the concept of a branch of a curve $\mathcal{F}$ and show that every branch of $\mathcal{F}$ is transformed to a branch of $\mathcal{F}'$, the geometric transform of $\mathcal{F}$ under a birational transformation. So this concept can be viewed as a birational invariant; but, more importantly, it leads to the idea of the genus of a curve, which is the most useful birational invariant; see Chapters 5 and 6.

Every point of the curve defines exactly one branch, but finitely many branches can arise from a singular point. This, and other properties of branches, are investigated by means of formal power series rather than polynomials.

4.1 FORMAL POWER SERIES

In this section, the relevant facts on formal power series are presented.

Let $K[[X,Y]]$ be the set of elements of the form

$$a_{00} + a_{10}X + a_{01}Y + a_{20}X^2 + \cdots, \ a_{ij} \in K,$$

in which two operations, the sum and the product, are defined as follows:

$$\begin{aligned}
&(a_{00} + a_{10}X + a_{01}Y + \cdots) + (b_{00} + b_{10}X + b_{01}Y + \cdots) \\
&\quad = (a_{00} + b_{00}) + (a_{10} + b_{10})X + (a_{01} + b_{01})Y + \cdots; \\
&(a_{00} + a_{10}X + a_{01}Y + \cdots) \times (b_{00} + b_{10}X + b_{01}Y + \cdots) \\
&\quad = a_{00}b_{00} + (a_{00}b_{10} + a_{10}b_{00})X + \cdots.
\end{aligned}$$

Then $(K[[X,Y]], +, \times)$ is a commutative ring whose elements are called *formal power series*. The identity is represented by the element $1 + 0X + 0Y + \cdots$ and the zero by the element $0 + 0X + 0Y + \cdots$. A formal power series can be written in the form

$$F = F_0 + F_1 + F_2 + \cdots ,$$

where F_i is a homogeneous polynomial in X and Y of degree i. With this convention, if

$$G = G_0 + G_1 + G_2 + \cdots ,$$

then

$$F + G = (F_0 + G_0) + (F_1 + G_1) + \cdots , \tag{4.1}$$

$$F\,G = (F_0G_0) + (F_0G_1 + F_1G_0) + \cdots . \tag{4.2}$$

If $F = F_r + F_{r+1} + \cdots$, with $F_r \neq 0$, then r is called the *order* or *subdegree* of F. The element $0 = 0 + 0X + 0Y + \cdots$ does not have an order since there is no first non-zero term; the convention is to write ord $0 = \infty$. Then, for $F, G \in K[[X,Y]]$,

$$\text{ord}\,(F+G) \geq \min\{\text{ord}\,F, \text{ord}\,G\},$$

$$\text{ord}\,(FG) = \text{ord}\,F + \text{ord}\,G.$$

Since $K[[X,Y]]$ is an integral domain, it has a quotient field

$$K((X,Y)) = \{F/G \mid F, G \in K[[X,Y]],\ G \neq 0\},$$

called the *quotient field of formal power series*.

Given $F_1, F_2, \ldots \in K[[X,Y]]$, the definition of $\sum_{i=0}^{n} F_i$ follows from the definition of the sum of two power series, (4.1). However, to define $\sum_{i=0}^{\infty} F_i$, the following procedure suffices. Let $F_i = F_{i0} + F_{i1} + \cdots + F_{ij} + \cdots$, and suppose that ord $F_i \to \infty$ for $i \to \infty$; then define

$$\textstyle\sum_{i=0}^{\infty} F_i = \sum_{i=0}^{\infty} F_{i0} + \sum_{i=0}^{\infty} F_{i1} + \cdots .$$

This make sense since each sum $\sum_{i=0}^{\infty} F_{ij}$ is well-defined. The following properties hold:

$$\textstyle\sum_{i=0}^{\infty} F_i + \sum_{i=0}^{\infty} G_i = \sum_{i=0}^{\infty} (F_i + G_i); \tag{4.3}$$

$$\textstyle F\sum_{i=0}^{\infty} G_i = \sum_{i=0}^{\infty} FG_i. \tag{4.4}$$

THEOREM 4.1 *The units of $K[[X,Y]]$ are the formal power series of order zero.*

Proof. Let F be a unit in $K[[X,Y]]$. Then there exists $G \in K[[X,Y]]$ such that $FG = 1$; hence $0 = \text{ord}\,(1) = \text{ord}\,(FG) = \text{ord}\,F + \text{ord}\,G$. Since both ord F and ord G are non-negative, then ord F = ord $G = 0$.

To show the converse, let ord $F = 0$; then, with $a_{00} \neq 0$,

$$\begin{aligned} F &= a_{00} + a_{10}X + a_{01}Y + \cdots \\ &= a_{00}\left(1 + \frac{a_{10}}{a_{00}}X + \frac{a_{01}}{a_{00}}Y + \cdots\right) \\ &= a_{00}(1 - G(X,Y)), \end{aligned}$$

with ord $G(X,Y) > 0$; that is,

$$F = a_{00}(1 - G). \tag{4.5}$$

Now note that, if $\operatorname{ord} G = r$, then $\operatorname{ord} G^i = ir$ and so $\operatorname{ord} G^i \to \infty$ for $i \to \infty$. Hence $\sum_0^\infty G^i$ makes sense. So multiplying (4.5) by this gives the following:

$$\begin{aligned} F(1 + G + G^2 + \cdots + G^i + \cdots) &= a_{00}(1 - G)(1 + G + G^2 + \cdots + G^i + \cdots) \\ &= a_{00}(1 - G)\textstyle\sum_{i=0}^\infty G^i \\ &= a_{00}\textstyle\sum_{i=0}^\infty (1 - G)G^i \\ &= a_{00}\textstyle\sum_{i=0}^\infty (G^i - G^{i+1}) \\ &= a_{00}(1 - G + G - G^2 + \cdots) \\ &= a_{00}. \end{aligned}$$

Dividing by a_{00} gives

$$F({a_{00}}^{-1}(1 + G + G^2 + \cdots + G^i + \cdots)) = 1;$$

that is, F is invertible with inverse, ${a_{00}}^{-1}(1 + G + G^2 + \cdots + G^i + \cdots)$. □

The preceding ideas and results for two indeterminates can be extended to any finite number of indeterminates, including the case of a single indeterminate. Some further properties of

$$K[[t]] = \{F(t) = a_r t^r + a_{r+1}t^{r+1} + \cdots,\ a_i \in K\}$$

are now given. First, if $\operatorname{ord} F = r$, write

$$F = t^r(a_r + a_{r+1}t + \cdots) = t^r E(t),$$

with $E(t)$ invertible.

Theorem 4.2 *The ring $K[[t]]$ is a unique factorisation domain.*

Proof. Every element has the form $f = t^r E(t)$, with $E(t)$ invertible and $r \geq 1$. If t were reducible, then $t = t^{r_1}E_1(t)t^{r_2}E_2(t)$ with $r_1, r_2 \geq 1$. But this implies that $1 = \operatorname{ord} t = r_1 + r_2$, a contradiction. If f is irreducible, then $r = 1$; further, $tE(t)$ and t are associates. Therefore every complete factorisation of $f \in K[[t]]$ has exactly $r = \operatorname{ord} f$ factors of the form t or $tE(t)$. Hence the only complete factorisation of f is $t \cdots t \cdot tF(t)$, for a unit $F(t)$. □

Definition 4.3 The field

$$\begin{aligned} K((t)) &= \{F/G \mid F, G \in K[[t]],\ G \neq 0\} \\ &= \left\{ \frac{t^s E_1(t)}{t^r E_2(t)} = t^{s-r}E(t) = t^m E(t) \right\}, \end{aligned}$$

with $E_1(t), E_2(t), E(t)$ invertible in $K[[t]]$ and m an integer, is the *field of rational functions of the formal power series* in the indeterminate t. The definition of $\operatorname{ord} F$ extends to $K((t))$: when $F = t^m E(t) \in K((t))$ with $E(t)$ invertible in $K[[t]]$, then $\operatorname{ord} F = m$.

Now all the K-monomorphisms and K-automorphisms of $K[[t]]$ are determined. Here, a *K-homomorphism* of $K[[t]]$ is a mapping of the ring to itself preserving addition and multiplication, and fixing every element of K: it is a *K-monomorphism* if it is injective, that is, it has zero kernel. A *K-automorphism* is a surjective K-monomorphism, that is, a bijective *K-homomorphism.* If $\tau \in K[[t]]$, $\tau \neq 0$ with $\mathrm{ord}_t\, \tau \geq 1$, then the *substitution* $t \mapsto \tau$ is the mapping

$$\begin{gathered} K[[t]] \to K[[t]], \\ \textstyle\sum_{i=0}^{\infty} c_i t^i \mapsto \sum_{i=0}^{\infty} c_i \tau^i. \end{gathered} \tag{4.6}$$

The *order* of the substitution is the order of τ. It is immediate that such a mapping is a K-monomorphism of $K[[t]]$. The following theorem shows that every K-monomorphism of $K[[t]]$ is of this type. All this extends to $K((t))$.

THEOREM 4.4 (i) *The K-monomorphisms of $K[[t]]$ are precisely the substitutions $t \mapsto \tau$.*

(ii) *A K-monomorphism of $K[[t]]$ is a K-automorphism if and only if the associated substitution has order* 1.

(iii) *Any K-monomorphism of $K((t))$ induces a K-monomorphism of $K[[t]]$, and conversely.*

Proof. It is first shown that, if the substitution $t \mapsto \tau$ has order 1, then it is surjective. Given $g = g_0 + g_1 t + g_2 t^2 + \cdots$ in $K[[t]]$, it is opportune to ensure the existence of an element $f = f_0 + f_1 t + f_2 t^2 + \cdots$ in $K[[t]]$ such that

$$f_0 + f_1 \tau + f_2 \tau^2 + \cdots = g_0 + g_1 t + g_2 t^2 + \cdots .$$

Equivalently, with $\tau = d_1 t + d_2 t^2 + d_3 t^3 + \cdots$, the coefficients $f_0, f_1, \cdots$ can be chosen with $f_0 = g_0$ and successively

$$\begin{aligned} f_1 d_1 &= g_1, \\ f_2 d_1^2 + f_1 d_2 &= g_2, \\ f_3 d_1^3 + f_2 2 d_1 d_2 + f_1 d_3 &= g_3, \\ &\vdots . \end{aligned}$$

Since $f_i d_1^i$ appears in the i-th row, but there is no other term in this row containing f_i, these equations determine the coefficients $f_0, f_1, f_2, \ldots$ uniquely, and so f is obtained.

It remains to consider a K-monomorphism σ of $K[[t]]$, and show that it is obtained by a substitution $t \mapsto \tau$, for a suitable element $\tau \in K[[t]]$ of order at least 1. First, note that σ transforms every invertible element of $K[[t]]$ to another invertible element of $K[[t]]$. In fact, if $f \in K[[t]]$ is invertible, there exists g such that $fg = 1$; then $\sigma(f)\sigma(g) = \sigma(1) = 1$, and so $\sigma(f)$ is invertible. The invertible elements have order zero; so σ sends elements of order zero to elements of order zero.

Now it is shown that $\sigma(t)$ has positive order. If it were otherwise, put

$$\sigma(t) = c_0 + c_1 t + \cdots ;$$

then

$$\sigma(-c_0 + t) = c_1 t + \cdots$$

with $c_0 \neq 0$, while $\sigma(-c_0 + t)$, being of order 0, must be invertible; this is a contradiction. It follows that σ sends non-invertible elements to non-invertible elements in $K[[t]]$; in other words, σ sends elements of positive order to elements of positive order. In fact, if $f = t^n E(t)$ with $n \geq 1$ and $E(t)$ is invertible, then $\sigma(f) = \tau^n F(t)$ with $\mathrm{ord}_t\, \tau \geq 1$, and $F(t) = \sigma(E(t))$ is invertible. More precisely, σ sends an element of order v to one of order $v \cdot \mathrm{ord}_t\, \tau$, and fixes the orders if and only if $\mathrm{ord}_t\, \tau = 1$.

This last observation also shows that, if σ is an automorphism, then $\sigma(t)$ must have order 1. Now, since σ is a K-monomorphism, then, for every positive integer r,

$$\sigma(\textstyle\sum_{i=0}^r c_i t^i) = \sum_{i=0}^r c_i \sigma(t^i) = \sum_{i=0}^r c_i (\sigma(t))^i = \sum_{i=0}^r c_i (\tau)^i.$$

More must be shown, namely that

$$\sigma(\textstyle\sum_{i=0}^\infty c_i t^i) = \sum_{i=0}^\infty c_i \sigma(t)^i = \sum_{i=0}^\infty c_i (\tau)^i.$$

For this, write

$$\textstyle\sum_{i=0}^\infty c_i t^i = (\sum_{i=0}^r c_i t^i) + R_r,$$

with $R_r = \sum_{i=r+1}^\infty c_i t^i \in K[[t]]$ and $\mathrm{ord}\, R_r > r$. Therefore

$$\sigma(\textstyle\sum_{i=0}^\infty c_i t^i) = \sigma(\sum_{i=0}^r c_i t^i) + \sigma(R_r) = \sum_{i=0}^r c_i (\tau)^i + \sigma(R_r). \qquad (4.7)$$

Since $\mathrm{ord}\, \sigma(R_r) \geq \mathrm{ord}\, R_r > r$, it follows from (4.7) that

$$\mathrm{ord}_t \left(\sigma(\textstyle\sum_{i=0}^\infty c_i t^i) - \sum_{i=0}^r c_i \tau^i\right) > r.$$

Then, since r can be chosen arbitrarily, this implies that

$$\sigma(\textstyle\sum_{i=0}^\infty c_i t^i) - \sum_{i=0}^\infty c_i \tau^i = 0,$$

as required.

To prove that every K-monomorphism σ of $K((t))$ induces a K-monomorphism of $K[[t]]$, it suffices to add just a remark to the above proof, namely that $\sigma(t)$ has non-negative order. Assume on the contrary that $\nu = \mathrm{ord}_t\, \sigma(t) < 0$. Then $\mathrm{ord}_t\, \sigma(t + c) = \mathrm{ord}_t(\sigma(t) + c) = \nu$. On the other hand, $\mathrm{ord}_t\, \sigma(t + c) = 0$ for $c \in K\backslash\{0\}$, as $t + c$ is invertible in $K[[t]]$ for $c \in K\backslash\{0\}$, and hence $\sigma(t + c)$ is also invertible in $K[[t]]$. But this contradicts the hypothesis that $\nu < 0$.

The converse follows from (i) since any substitution defines a K-monomorphism of $K((t))$. □

It is worth noting explicitly that the preceding proof also shows the following result.

THEOREM 4.5 *Let $f(t) \in K[[t]]$ and $g(\tau) \in K[[\tau]]$. If $f(t) = g(\tau)$ for some $\tau \in K[[t]]$, then*

$$\mathrm{ord}_t\, f = \mathrm{ord}_\tau\, g \cdot \mathrm{ord}_t\, \tau. \qquad (4.8)$$

The following result relates polynomials and power series.

THEOREM 4.6 *Let $F(X, Y) \in K[X, Y]$ with $F(0, 0) = 0$. Let $\partial F/\partial Y \neq 0$ at $(0, 0)$. Then there is a unique $f(X) = c_1 X + c_2 X^2 + \cdots$ in $K[[X]]$ such that*

$$F(X, c_1 X + c_2 X^2 + \cdots) = 0.$$

Proof. Without loss of generality, assume that $\partial F/\partial Y = 1$ at $(0,0)$, Then

$$F(X,Y) = Y + a_{10}X + \cdots \in K[X,Y],$$
$$F(X,Y+Z) = F(X,Y) + \frac{\partial F}{\partial Y}Z + \cdots,$$

where the partial derivatives are calculated with respect to X and Y. Since

$$\frac{\partial F}{\partial Y} = 1 + G(X,Y)$$

with ord $G(X,Y) > 0$, it follows that

$$F(X,Y+Z) = F(X,Y) + [1 + G(X,Y)]Z + G_2Z^2 + \cdots. \qquad (4.9)$$

First, the uniqueness of f is shown. Suppose that both the formal power series

$$c_1X + \cdots + c_{i-1}X^{i-1} + c_iX^i + \cdots,$$
$$c_1X + \cdots + c_{i-1}X^{i-1} + d_iX^i + \cdots$$

satisfy the conditions of the theorem. Substitute

$$Y = c_1X + \cdots + c_{i-1}X^{i-1},\ Z = c_iX^i + \cdots$$

in (4.9):

$$\begin{aligned} 0 &= F(X, c_1X + \cdots + c_{i-1}X^{i-1} + c_iX^i + \cdots) \\ &= F(X, c_1X + \cdots + c_{i-1}X^{i-1}) + [1 + G(X, c_1X + \cdots + c_{i-1}X^{i-1})] \\ &\quad \times (c_iX^i + \cdots) + X^{i+1}H(X). \end{aligned}$$

Hence it follows that c_i is uniquely determined by the coefficients $c_1, \ldots, c_{i-1}$. But this implies that $c_i = d_i$ for all i.

For the existence of F, construct inductively for each i a polynomial

$$c_1X + \cdots + c_iX^i$$

such that ord $F(X, c_1X + \cdots + c_iX^i) > i$. For $i = 1$, put $c_1 = -a_{10}$. Suppose then that a polynomial $c_1X + \cdots + c_{i-1}X^{i-1}$ has been determined with

$$\operatorname{ord} F(X, c_1X + \cdots + c_{i-1}X^{i-1}) > i - 1.$$

Write

$$\begin{aligned} &F(X, c_1X + \cdots + c_{i-1}X^{i-1} + c_iX^i) \\ &= F(X, c_1X + \cdots + c_{i-1}X^{i-1}) + [1 + G(X, c_1X + \cdots + c_{i-1}X^{i-1})]c_iX^i \\ &\qquad + X^{i+1}H(X) \\ &= (dX^i + \cdots) + c_iX^i + X^{i+1}H_1(X). \end{aligned}$$

Now choose $c_i = -d$ to ensure that $F(X, c_1X + \cdots + c_iX^i + \cdots) = 0$. In fact, by (4.9),

$$\begin{aligned} &F(X, c_1X + \cdots + c_{i-1}X^{i-1} + c_iX^i + \cdots) \\ &\qquad = F(X, c_1X + \cdots + c_{i-1}X^{i-1}) + X^iH_2(X). \end{aligned}$$

By the inductive hypothesis, ord $F(X, c_1X + \cdots + c_{i-1}X^{i-1}) > i - 1$, whence ord $F(X, c_1X + \cdots + c_iX^i) > i$. Since this is true for all i, the only possibility is that $F(X, c_1X + c_2X^2 + \cdots) = 0$. □

It may be noted that the condition in this theorem can be expressed geometrically: the algebraic curve $\mathcal{F} = \mathbf{v}(F(X,Y))$ has a non-singular point at the origin and the tangent there is not the Y-axis. Since $Y = c_1X + c_2X^2 + \cdots$ is a solution in $K[[X]]$ of the equation $F(X,Y) = 0$, it follows from Theorem 4.6 that

$$F(X,Y) = [Y - (c_1X + c_2X^2 + \cdots)]F_1(X,Y),$$

with $F_1(X,Y) \in K[[X]][Y]$. Also, $F_1(0,0) \neq 0$, since the origin is a simple point of the curve; that is, ord $F_1 = 0$. It follows that

$$\text{ord}\, F(X,0) = \text{ord}\,(c_1X + c_2X^2 + \cdots).$$

So the following theorem has been proved.

Theorem 4.7 *Let $\mathcal{F}$ be a plane curve passing through the origin O. If O is a non-singular point of $\mathcal{F}$ and the tangent to $\mathcal{F}$ at O is not the Y-axis, then the intersection multiplicity of $\mathcal{F}$ and $\mathbf{v}(Y)$ at O is*

$$I(O, \mathcal{F} \cap \mathbf{v}(Y)) = \text{ord}\,(c_1X + c_2X^2 + \cdots).$$

This theorem is generalised as follows.

Theorem 4.8 *Let $\mathcal{F} = \mathbf{v}(F(X,Y))$ and $\mathcal{G} = \mathbf{v}(G(X,Y))$ be two plane curves with no common components, and suppose that $\mathcal{F}$ has a non-singular point at the origin with tangent different from the Y-axis. If $c_1X + c_2X^2 + \cdots$ is the unique formal power series such that $F(X, c_1X + c_2X^2 + \cdots) = 0$, then*

$$I(O, \mathcal{F} \cap \mathcal{G}) = \text{ord}\, G(X, c_1X + c_2X^2 + \cdots).$$

When O is a singular point of $\mathcal{F}$, the study of $I(O, \mathcal{F} \cap \mathcal{G})$ is more involved. It is postponed to Section 4.4 because it requires the idea of a branch, or branch point, as developed in Section 4.2. Meanwhile, some more results on power series must be established; these are essential for the study of $I(O, \mathcal{F} \cap \mathcal{G})$, as well. Interestingly, these technical results are also the main ingredients in the generalisation of Theorem 4.2 to any number of indeterminates; see Theorem 4.16.

Lemma 4.9 (Hensel) *Let*

$$F(X,Y) = Y^n + a_1(X)Y^{n-1} + \cdots + a_n(X) \in K[[X]][Y]$$

be a monic polynomial of degree $n > 0$ in Y with coefficients $a_1(X), \ldots, a_n(X)$ in $K[[X]]$. Suppose that $F(0,Y) = \overline{G}(Y)\overline{H}(Y)$, where

$$\overline{G}(Y) = Y^r + \overline{b}_1Y^{r-1} + \cdots + \overline{b}_r,$$
$$\overline{H}(Y) = Y^s + \overline{c}_1Y^{s-1} + \cdots + \overline{c}_s,$$

are monic polynomials of degree $r > 0$ and $s > 0$ in Y with coefficients in K such that $\gcd(\overline{G}(Y), \overline{H}(Y)) = 1$. Then there exist two unique monic polynomials

$$G(X,Y) = Y^r + b_1(X)Y^{r-1} + \cdots + b_r(X),$$
$$H(X,Y) = Y^s + c_1(X)Y^{s-1} + \cdots + c_s(X)$$

in $K[[X]][Y]$ of degree r, s in Y with coefficients $b_1(X), \ldots, b_r(X)$ and $c_1(X), \ldots, c_s(X)$ in $K[[X]]$ such that

$$G(0,Y) = \overline{G}(Y),\ H(0,Y) = \overline{H}(Y),$$

$$F(X,Y) = G(X,Y)H(X,Y). \tag{4.10}$$

Proof. Write $F = F(X, Y)$ as a power series in X whose coefficients are polynomials in Y:

$$F = F_0(Y) + F_1(Y)X + \cdots + F_m(Y)X^m + \cdots,$$

where $F_0(Y) = F(0, Y) \in K[Y]$ is monic with $\deg F_0 = n$ and $F_m(Y) \in K[Y]$ with $\deg F_m < n$ for all $m > 0$.

It is necessary to find

$$\begin{aligned} G &= G_0(Y) + G_1(Y)X + \cdots + G_i(Y)X^i + \cdots, \\ H &= H_0(Y) + H_1(Y)X + \cdots + H_j(Y)X^j + \cdots, \end{aligned}$$

with

$$\begin{aligned} &G_0 = G_0(Y) = \overline{G}(Y) \in K[Y] \quad \text{monic of degree } r, \\ &H_0 = H_0(Y) = \overline{H}(Y) \in K[Y] \quad \text{monic of degree } s, \\ &G_i = G_i(Y) \in K[Y] \quad \text{and } \deg G_i(Y) < r \text{ for each } i > 0, \\ &H_j = H_j(Y) \in K[Y] \quad \text{and } \deg H_j(Y) < s \text{ for each } j > 0, \end{aligned}$$

such that $F = GH$. It should be noted that $F = GH$ means that

$$F_m = \textstyle\sum_{i+j=m} G_i H_j \quad \text{for each } m \geq 0.$$

The proof proceeds by induction on m. For $m = 0$, the result is that

$$\begin{aligned} &F_0 = \textstyle\sum_{i+j=0} G_i H_j, \\ &F_0 = F_0(Y) = F(0, Y) = \overline{G}(Y)\overline{H}(Y) = G_0(Y)H_0(Y). \end{aligned}$$

Let $m > 0$. Assume that there exist G_i, $H_j \in K[Y]$, with $\deg G_i < r$ and $\deg H_j < s$ for $1 \leq i, j < m$ satisfying the equation (4.10) for values smaller than m. Then it is necessary to find G_m, H_m in $K[Y]$ with $\deg G_m < r$, $\deg H_m < s$ satisfying

$$F_m = \textstyle\sum_{i+j=m} G_i H_j.$$

This equation may be written as follows:

$$G_0 H_m + H_0 G_m = U_m;$$

here,

$$U_m = F_m - \sum_{\substack{i+j=m \\ i,j<m}} G_i H_j$$

is known, since it contains only terms already calculated by the inductive hypothesis.

It may be observed that $\deg U_m < n$; in fact,

$$\deg F_m < n, \quad \deg G_i < r < n, \quad \deg H_j < s < n,$$

from which it follows that $\deg(G_i H_j) = \deg(G_i) + \deg(H_j) < r + s \leq n$ and hence $\deg U_m < n$.

Since $\gcd(G_0, H_0) = 1$, so

$$G_0 H^* + H_0 G^* = 1$$

with $G^*,\ H^* \in K[Y]$. Multiplying both sides by U_m gives that

$$G_0 H_m^* + H_0 G_m^* = U_m$$

with $G_m^* = U_m G^*$ and $H_m^* = U_m H^*$. Using the division algorithm to divide H_m^* by H_0, the polynomial H_m is obtained:

$$H_m^* = E_m H_0 + H_m,$$

where $E_m,\ H_m \in K[Y]$ with $\deg H_m < s$. Substituting this expression for H_m^* in

$$G_0 H_m^* + H_0 G_m^* = U_m = G_0 H_m + H_0 G_m$$

gives that

$$G_0(E_m H_0 + H_m) + H_0 G_m^* = G_0 H_m + H_0 G_m,$$

from which, simplifying and dividing by H_0,

$$G_m = G_0 E_m + G_m^*.$$

Also, since $\deg U_m < n$, it follows that $\deg G_m < r$. By induction, G_i and H_j have been calculated so that $F_m = \sum_{i+j=m} G_i H_j$. Thus the existence of G and H with the desired properties has been established.

To prove the uniqueness, suppose there exist two other polynomials g_m and h_m that satisfy the hypothesis. Then

$$\begin{aligned}
&G_0 H_m + G_m H_0 = G_0 h_m + g_m H_0 \\
&\Rightarrow G_0(H_m - h_m) = (G_m - g_m) H_0 \\
&\Rightarrow G_0 \mid (G_m - g_m), \quad \text{since } G_0 \text{ and } H_0 \text{ are coprime,} \\
&\Rightarrow G_m - g_m = 0, \quad \text{as } \deg(G_m - g_m) < \deg G_0, \\
&\Rightarrow G_m = g_m,\ H_m = h_m.
\end{aligned}$$

The lemma has thus been proved. □

An equivalent way of stating Hensel's Lemma 4.9 is the following.

Let $F(X,Y) \in K[[X]][Y]$ be monic of degree n in Y and such that $F(0,Y)$ has at least two roots. Then $F(X,Y)$ is reducible in $K[[X]][Y]$. Also, if

$$F(0,Y) = G_0(Y) H_0(Y)$$

with $\deg G_0 = r,\ \deg H_0 = (n-r)$ *and* $\gcd(G_0, H_0) = 1$, *then*

$$F(X,Y) = G(X,Y) H(X,Y)$$

with $G(X,Y), H(X,Y) \in K[[X]][Y]$ of degrees $r, n-r$ in Y, where

$$G(0,Y) = G_0(Y),\ H(0,Y) = H_0(Y).$$

THEOREM 4.10 (Weierstrass Preparation Theorem) *Let*

$$F(X,Y) = \sum a_{ij} X^i Y^j \in K[[X,Y]]$$

with $a_{ij} \in K$ and $F(0,Y) \neq 0$. If $d =$ ord $F(0,Y)$, then there exist unique $G(X,Y),\ H(X,Y) \in K[[X,Y]]$ such that

$$F(X,Y) = G(X,Y) H(X,Y),$$

where $G(0,0) \neq 0$, and

$$H(X,Y) = Y^d + c_1(X) Y^{d-1} + \cdots + c_d(X),$$

with $c_i(X) \in K[[X]]$ and $c_i(0) = 0$ for $1 \le i \le d$.

Note that

(i) $G(X, Y)$ is a unit;

(ii) $F(X, Y)$ has no independent term;

Proof. The proof is analogous to that of Hensel's Lemma 4.9. Write

$$F(X, Y) = F_0(Y) + F_1(Y)X + F_2(Y)X^2 + \cdots + F_m(Y)X^m + \cdots,$$

with $F_m(Y) \in K[[Y]]$. Then it is necessary to find power series

$$G(X, Y) = G_0(Y) + G_1(Y)X + \cdots + G_i(Y)X^i + \cdots,$$
$$H(X, Y) = H_0(Y) + H_1(Y)X + \cdots + H_j(Y)X^j + \cdots,$$

where

$$G_0(Y) \in K[[Y]] \text{ with } G_0(0) \neq 0 \text{ and } G_i(Y) \in K[[Y]] \text{ for } i > 0,$$
$$H_0(Y) = Y^d \text{ and } H_j(Y) \in K[Y] \text{ with } \deg H_j(Y) < d \text{ for } j > 0,$$

such that $F(X, Y) = G(X, Y)H(X, Y)$.

Now $F(X, Y) = G(X, Y)H(X, Y)$ means that

$$F_m(Y) = \textstyle\sum_{i+j=m} G_i(Y)H_j(Y) \quad \text{for each } m \geq 0. \tag{4.11}$$

Proceed by induction on m. Since $\operatorname{ord} F(0, Y) = d$, with $H_0(Y) = Y^d$, there exists a unique $G_0(Y) \in K[[Y]]$ with $G_0(0) \neq 0$ such that

$$F_0(Y) = F(0, Y) = G_0(Y)H_0(Y).$$

So the case that $m = 0$ has been established.

Let $m > 0$, and suppose that there exist $G_i(Y) \in K[[Y]]$ for $1 \leq i < m$ and $H_j(Y) \in K[Y]$ with $\deg H_j(Y) < d$ for $1 \leq j < m$ that satisfy (4.11). So the task is to find $G_m(Y) \in K[[Y]]$ and $H_m(Y) \in K[Y]$ with $\deg H_m(Y) < d$ that satisfy the equation

$$F_m(Y) = \textstyle\sum_{i+j=m} G_i(Y)H_j(Y).$$

This equation may be written as follows:

$$\begin{aligned} &G_0(Y)H_m(Y) + H_0(Y)G_m(Y) \\ &\quad = U_m(Y) = F_m(Y) - \sum_{\substack{i+j=m \\ i,j<m}} G_i(Y)H_j(Y). \end{aligned}$$

Since $G_0(Y)$ is a unit, define $H_m(Y)$ to be the sum of the terms of degree less than d in $U_m(Y)/G_0(Y)$ with $\deg H_m(Y) < d$, from which it follows that

$$G_m(Y) = \{U_m(Y)/G_0(Y) - H_m(Y)\}Y^{-d}G_0(Y) \quad \in K[[Y]].$$

Thus $G_i(Y)$, for each $i > 0$, and $H_j(Y)$, with $\deg H_j(Y) < d$, that satisfy (4.11) have been calculated.

To prove the uniqueness, again proceed by induction. Since $G_0(Y)$ is unique, suppose that $G_i(Y)$ and $H_j(Y)$ are unique for $i < m$ and $j < m$. Then, by the construction of $G_m(Y)$ and $H_m(Y)$, their uniqueness follows. □

A generalisation of Theorem 4.10 to more variables may be stated as follows.

THEOREM 4.11 *If* $F(X_1,\ldots,X_n) \in K[[X_1,\ldots,X_n]]$ *and*

$$F(0,\ldots,0,X_n) = c_\nu X_n^\nu + c_{\nu+1}X_n^{\nu+1} + \cdots, \; c_\nu \neq 0,$$

then there exists a unique unit G *in* $K[[X_1,\ldots,X_n]]$ *such that*

$$GF = X_n^\nu + A_1(X_1,\ldots,X_{n-1})X_n^{\nu-1} + \cdots + A_\nu(X_1,\ldots,X_{n-1}) \qquad (4.12)$$

with $A_i(X_1,\ldots,X_{n-1}) \in K[[X_1,\ldots,X_{n-1}]]$ *and* $A_i(0,\ldots,0) = 0$ *for each* $i = 1,\ldots,\nu$.

An element of $K[[X_1,\ldots,X_n]]$ that contains a term in which X_n appears is *regular* in X_n; hence F is *regular* in X_n. A polynomial in X_n with form (4.12) is *special.*

THEOREM 4.12 (Weierstrass Division Theorem) *Given* $F(X_1,\ldots,X_n,Y)$ *and* $H(X_1,\ldots,X_n,Y)$ *in* $K[[X_1,\ldots,X_n,Y]]$, *assume that* $F(0,\ldots,0,Y) \neq 0$ *and* $d = \operatorname{ord}_Y F(0,\ldots,0,Y)$. *Then there exist unique* A, B *such that*

$$H = AF + B,$$

with $A \in K[[X_1,\ldots,X_n,Y]]$, $B \in K[[X_1,\ldots,X_n]][Y]$, *and* $\deg_Y B < d$.

Proof. Proceed by induction on n. The case $n = 1$ coincides with Theorem 4.10. For $n > 1$, write

$$\begin{aligned} H &= H_0(X_2,\ldots,X_n,Y) + H_1(X_2,\ldots,X_n,Y)X_1 + \cdots, \\ A &= A_0(X_2,\ldots,X_n,Y) + A_1(X_2,\ldots,X_n,Y)X_1 + \cdots, \\ F &= F_0(X_2,\ldots,X_n,Y) + F_1(X_2,\ldots,X_n,Y)X_1 + \cdots, \\ B &= B_0(X_2,\ldots,X_n,Y) + B_1(X_2,\ldots,X_n,Y)X_1 + \cdots. \end{aligned}$$

Then the required equation $H = AF + B$ is equivalent to the following:

$$\begin{aligned} H_0 &= A_0F_0 + B_0, \\ H_1 &= A_0F_1 + A_1F_0 + B_1, \\ &\;\;\vdots \\ H_i &= A_0F_i + \cdots + A_{i-1}F_1 + A_iF_0 + B_i, \\ &\;\;\vdots \end{aligned}$$

Since $F(0,\cdots,0,Y) \neq 0$, so $F_0(0,\cdots,0,Y) \neq 0$. Also,

$$\operatorname{ord}_Y F(0,\ldots,0,Y) = d$$

implies that $\operatorname{ord}_Y F_0(0,\ldots,0,Y) = d$. Since

$$B(X_1,\ldots,X_n,Y) \in K[[X_1,\ldots,X_n]][Y]$$

and $\deg_Y B < d$, then also

$$B_i(X_1,\ldots,X_n,Y) \in K[[X_1,\ldots,X_n]][Y]$$

and $\deg_Y B_i < d$ for all $i \geq 0$.

By induction, A_i and B_i are found successively; so they exist and are unique by construction. Therefore the same is true for A and B. □

The proof of Theorem 4.11 follows from Theorem 4.12 by taking $H = Y^d$.

The following corollaries are a consequence of the general Weierstrass Preparation Theorem 4.10.

COROLLARY 4.13 *Given $F \in K[[X_1, \ldots, X_n]]$ with*

$$F(0, \ldots, 0, X_n) = c_\nu X_n^\nu + c_{\nu+1} X_n^{\nu+1} + \cdots, \; c_\nu \neq 0,$$

and $A'F \in K[[X_1, \ldots, X_{n-1}]][X_n]$ of degree $< \nu$, with $A' \in K[[X_1, \ldots, X_n]]$, then $A' = 0$.

Proof. The proof follows from Theorem 4.12. Putting $H = 0$, it follows that $A'F = B'$ and $F \neq 0$ by the uniqueness of A and B. Hence $A' = 0$. □

COROLLARY 4.14 *Let F be a special polynomial in $K[[X_1, \ldots, X_{n-1}]][X_n]$ of degree ν and let $A' \in K[[X_1, \ldots, X_n]]$. If $A'F = B' \in K[[X_1, \ldots, X_{n-1}]][X_n]$ has degree ν, then $A' \in K[[X_1, \ldots, X_{n-1}]]$.*

Proof. Let $B' = CX_n^\nu + \cdots$, with $C \in K[[X_1, \ldots, X_{n-1}]]$. Then

$$CX_n^\nu - A'F = CX_n^\nu - B'$$

has degree less than ν. Put $H = CX_n^\nu$. Then the relation $H - AF = B$ is satisfied not only by $A = A'$ and $B = CX_n^\nu - B'$ but also by $A = C$ and $B = H - CF$. From Theorem 4.12, since A and B are unique, it follows that $A' = C$. □

THEOREM 4.15 *If F is special in X_n and irreducible in $K[[X_1, \ldots, X_n]]$, then F is irreducible in $K[[X_1, \ldots, X_{n-1}]][X_n]$.*

Proof. Suppose that F is reducible in $K[[X_1, \ldots, X_{n-1}]][X_n]$ with $F = G_1G_2$, where $\deg_{X_n} G_i < \deg_{X_n} F$. Since F is irreducible in $K[[X_1, \ldots, X_n]]$, then one of the G_i is a unit in this ring, If it is G_1, then $G_1^{-1}F = G_2$ and, by Corollary 4.13, $G_2 = 0$, which is impossible. □

THEOREM 4.16 *The ring $K[[X_1, \ldots, X_n]]$ is a unique factorisation domain.*

Proof. For $n = 1$, the assertion is Theorem 4.2. The proof is by induction on n; so $K[[X_1, \ldots, X_{n-1}]][X_n]$ is assumed to be a unique factorisation domain. If $F = G_1 \cdots G_m$, then $\operatorname{ord} F = \sum \operatorname{ord} G_i$. This shows that any non-unit F has a factorisation into irreducible non-units.

Let

$$E_1F_1 \cdots F_s = E_2G_1 \cdots G_t$$

be two complete factorisations of F in $K[[X_1, \ldots, X_n]]$, where E_1, E_2 are units but F_i and G_i are non-units. Then the following occur.

(i) Applying a linear homogeneous substitution $\mathbf{X} = A(\mathbf{X}')$ with

$$\mathbf{X} = (X_1, \ldots, X_n) \mapsto \mathbf{X}' = (X_1', \ldots, X_n'), \quad A \in \mathrm{PGL}(n+1, K),$$

it may be assumed that F_i and G_i are regular in X_n.

(ii) By Weierstrass's Preparation Theorem 4.10 , and absorbing the extra units into E_1 and E_2, it may be assumed that F_i and G_i are special in X_n.

Suppose that $F = G_1G_2$ with $\deg_{X_n} G_i < \deg_{X_n} F$. Since F is irreducible in $K[[X_1, \ldots, X_n]]$, one of the G_i is a unit, say G_1. Then $G_1^{-1}F = G_2$ and, by

Theorem 4.13, $G_2 = 0$. This is impossible, and hence the following has been shown.

(iii) If F is special in X_n and irreducible in $K[[X_1, \ldots, X_n]]$, then F is also irreducible in $K[[X_1, \ldots, X_{n-1}]][X_n]$.

This has the following consequence.

(iv) Multiplying both sides of $E_1F_1 \cdots F_s = E_2G_1 \cdots G_t$ by E_1^{-1}, it may be assumed that $E_1 = 1$.

Theorem 4.13 shows that $\deg F_1 \cdots F_s < \deg G_1 \cdots G_t$ is not possible. On the other hand,

$$E_2^{-1}F_1 \cdots F_s = G_1 \cdots G_t$$

shows that $\deg F_1 \cdots F_s > \deg G_1 \cdots G_t$ is impossible, as well. Hence the degrees are equal, and $E_2 = 1$ from Theorem 4.14. Now the unique factorisation in $K[[X_1, \ldots, X_{n-1}]][X_n]$ implies that $s = t$ and that, up to relabelling of subscripts, F_i and G_i are associates in $K[[X_1, \ldots, X_{n-1}]][X_n]$. As the units in this ring are also units in $K[[X_1, \ldots, X_n]]$, it follows that F_i and G_i are also associates in $K[[X_1, \ldots, X_n]]$. □

4.2 BRANCH REPRESENTATIONS

DEFINITION 4.17 A *branch representation* is a point of the plane $\mathrm{PG}(2, K((t)))$ not belonging to $\mathrm{PG}(2, K)$.

Given a point $(x_0(t), x_1(t), x_2(t))$ of $\mathrm{PG}(2, K((t)))$, it is therefore a branch representation if $\rho(t)x_i(t) \notin K$ for at least one i and for all $\rho(t) \in K((t))\backslash\{0\}$. If $x_0(t) \neq 0$, put $x(t) = x_1(t)x_0(t)^{-1}$, $y(t) = x_2(t)x_0(t)^{-1}$; then the branch representation becomes the point $(x(t), y(t))$ of the affine plane

$$\mathrm{AG}(2, K((t))) = \mathrm{PG}(2, K((t)))\backslash\ell_\infty,$$

where ℓ_∞ is the line at infinity of $\mathrm{AG}(2, K((t))$.

Let $(x_0(t), x_1(t), x_2(t))$ be a branch representation and let

$$m = \min\{\mathrm{ord}_t\, x_i(t)\}.$$

Then, with $\xi_i(t) = t^{-m}x_i(t)$, the branch representation $(\xi_0(t), \xi_1(t), \xi_2(t))$ has the order of one of its components zero and the other two non-negative; it is a *special* branch representation and its coordinates are also *special*. In this case, let

$$x_k(t) = c_k + c_{k1}t + \cdots + c_{ki}t^i + \cdots, \quad k = 0, 1, 2;$$

the point $P = (c_0, c_1, c_2)$ of $\mathrm{PG}(2, K)$ is the *centre* of the branch representation.

Now the further notion of the *order of a branch representation* is introduced. With $(x_0(t), x_1(t), x_2(t))$ already in special coordinates, the *order* is the positive integer

$$\min\{\mathrm{ord}_t(a_0x_0(t) + a_1x_1(t) + a_2x_2(t))\},$$

for all points $(a_0, a_1, a_2) \in \mathrm{PG}(2, K)$ with $a_0c_0+a_1c_1+a_2c_2 = 0$. If $x_0(t), x_1(t)$, $x_2(t)$ are linearly independent over K, then $\mathrm{ord}_t(a_0x_0(t) + a_1x_1(t) + a_2x_2(t))$ is zero or positive according as $a_0c_0 + a_1c_1 + a_2c_2$ is non-zero or zero. Also, it takes one of two positive values j_1 and j_2, where $j_1 < j_2$; the value is j_1 or j_2 according as $a_0c_{01} + a_1c_{11} + a_2c_{21}$ is non-zero or zero. In fact,

$$\mathrm{ord}_t(a_0x_0(t) + a_1x_1(t) + a_2x_2(t)) = j_2$$

for a unique point (a_0, a_1, a_2), in which case the line $\mathbf{v}(a_0X_0 + a_1X_1 + a_2X_2)$ of $\mathrm{PG}(2, K)$ is the *tangent* of the branch representation. The triple $(0, j_1, j_2)$ is the *order sequence* of the branch representation and j_1 is its *order*. When $x_0(t), x_1(t), x_2(t)$ are linearly dependent over K, the same holds true provided that j_2 is defined to be ∞.

The notion of branch representation is projective in the sense that every projective transformation of $\mathrm{PG}(2, K)$ transforms a branch representation into a branch representation; also, special coordinates become special coordinates. Similarly, the notions of order and tangent are projective invariants.

If one of the special coordinates is 1, then $\xi_0(t)$ is invertible. Hence every branch representation can be expressed in *special affine coordinates* $(x(t), y(t))$ with

$$x(t) = u + u_1t + \cdots, \qquad y(t) = v + v_1t + \cdots,$$

with centre the point (u, v). This notion is preserved under a linear transformation of the affine plane $\mathrm{AG}(2, K)$. In what follows, the branch representations mostly considered are those in special affine coordinates, as these simplify the exposition.

DEFINITION 4.18 (i) Two branch representations in special affine coordinates

$$(x(t), y(t)) \quad \text{and} \quad (\xi(t), \eta(t))$$

are *equivalent* if there exists a K-automorphism σ of $K[[t]]$ such that

$$x(t) = \sigma(\xi(t)) \quad \text{and} \quad y(t) = \sigma(\eta(t)).$$

(ii) Two branch representations are *equivalent* if they are so when put in the form of special affine coordinates.

In other words, $(x(t), y(t))$ and $(\xi(t), \eta(t))$ are equivalent branch representations if there exists a substitution $t \mapsto \tau$ of order 1 such that $x(t) = \xi(\tau)$ and $y(t) = \eta(\tau)$.

DEFINITION 4.19 (i) A branch representation in special affine coordinates

$$(x(t), y(t))$$

is *imprimitive* if there exists a branch representation in special affine coordinates

$$(\xi(t), \eta(t))$$

such that

$$x(t) = \sigma(\xi(t)) \quad \text{and} \quad y(t) = \sigma(\eta(t)),$$

with σ a K-monomorphism of $K[[t]]$ that is not a K-automorphism.

(ii) A branch representation is *imprimitive* if it is so when put in the form of special affine coordinates.

This definition can also be stated in terms of substitutions: $(x(t), y(t))$ is imprimitive if there exists a branch representation in special affine coordinates $(\xi(t), \eta(t))$ such that $x(t) = \xi(\tau)$ and $y(t) = \eta(\tau)$ for a suitable $\tau \in K[[t]]$ with $\text{ord}_t\, \tau > 1$. If this is the case, then $(x(t), y(t))$ has ramification index at least $\nu = \text{ord}_t\, \tau > 1$.

The reason for this definition is that $\xi(t), \eta(t)$ have, in general, simpler expressions than $x(t), y(t)$. Now, if $(\xi(t), \eta(t))$ is also imprimitive, then a further branch representation $(\xi_1(t), \eta_1(t))$ with ramification index $\nu_1 > 1$ is obtained, and so on until a primitive branch representation is obtained. As this happens in a finite number of steps, the following result is reached.

THEOREM 4.20 *Let $(x(t), y(t))$ and $(\xi(t), \eta(t))$ be two branch representations in special affine coordinates. If*

(a) $x(t) = \sigma(\xi(t)),\ y(t) = \sigma(\eta(t))$,

(b) *σ is a K-monomorphism of $K[[t]]$,*

(c) $\nu = \text{ord}_t\, \sigma(t)$,

then

$$\text{ord}_t(x(t), y(t)) = \nu \cdot \text{ord}_t(\xi(t), \eta(t)).$$

Proof. If (u, v) is the centre of $(x(t), y(t))$, then, since σ is a K-monomorphism of $K[[t]]$, the same point (u, v) is also the centre of $(\xi(t), \eta(t))$. Therefore, given a non-trivial triple (a_0, a_1, a_2), with $a_i \in K$ satisfying $a_0 + a_1 u + a_2 v = 0$, then $(x(t), y(t))$ has order $\text{ord}_t(a_0 + a_1 x(t) + a_2 y(t))$ and $(\xi(t), \eta(t))$ has order $\text{ord}_t(a_0 + a_1\xi_(t) + a_2\eta(t)))$. Now

$$\begin{aligned}\text{ord}_t(a_0 + a_1 x(t) + a_2 y(t)) &= \text{ord}_t(a_0 + a_1\sigma(\xi(t)) + a_2\sigma(\eta(t))) \\ &= \text{ord}_t\, \sigma(a_0 + a_1\xi(t) + a_2\eta(t)).\end{aligned}$$

By Theorem 4.5, this is the required result. □

Since every product of K-monomorphisms of $K[[t]]$ is also such a K-monomorphism, for each imprimitive branch representation in special affine coordinates $(x(t), y(t))$ there exists a *primitive* branch representation in special affine coordinates $(\xi(t), \eta(t))$ such that

$$x(t) = \sigma(\xi(t)),\ y(t) = \sigma(\eta(t)),$$

where σ is a K-monomorphism of $K[[t]]$ that is not a K-automorphism. To show the uniqueness, three further results are required that are interesting in themselves.

THEOREM 4.21 *Let $(x(t), y(t))$ be a branch representation in special affine coordinates.*

(i) *If τ is of positive minimum order in the subfield $K(x(t), y(t))$ of $K((t))$, then the order of each element $z \in K(x(t), y(t))$ is divisible by $g = \text{ord}_t\, \tau$.*

(ii) *The element z can be expanded in a formal power series in τ with coefficients in K.*

(iii) *In the particular case that $(x(t), y(t))$ is primitive, there is an element of order 1 in $K(x(t), y(t))$.*

Proof. (i) If μ is the order of z, the Euclidean algorithm gives integers q, r such that $\mu = qg + r$ and $0 \leq r < g$. Hence z/τ^q is an element of order r belonging to $K(x(t), y(t))$. Since τ has minimum order, then $r = 0$, as required.

(ii) Now write $z/\tau^q = c + z_1$ with $c \in K,\ c \neq 0$ and $z_1 \in K[[t]]$; then z_1 is also in $K(x(t), y(t))$. As in the previous argument, there is a positive integer q_1 such that $z_1/\tau^{q_1} = c_1 + z_2$. Hence

$$z = c\tau^q + c_1\tau^{q+q_1} + \tau^{q+q_1} z_2.$$

Continuing this procedure gives the required formal power series.

(iii) In particular, the same coordinates $x(t)$ and $y(t)$ become formal power series in τ, so that, if $(x(t), y(t))$ is primitive, τ is necessarily of order 1. □

THEOREM 4.22 *Given a branch representation $(x(t), y(t))$ in special affine coordinates, let $(x_1(t), y_1(t))$ and $(x_2(t), y_2(t))$ also be branch representations in special affine coordinates, of which the first is primitive. If the relations*

$$x(t) = x_1(\tau_1) = x_2(\tau_2), \quad y(t) = y_1(\tau_1) = y_2(\tau_2), \tag{4.13}$$

with $\tau_1, \tau_2 \in K[[t]]$, are satisfied, then τ_1 can be written as a formal power series in τ_2. Further, if $(x_2(t), y_2(t))$ is also primitive, then $\mathrm{ord}_t\, \tau_1 = \mathrm{ord}_t\, \tau_2$.

Proof. Let τ be an element of minimum positive order in $K(x_1(\tau_1), y_1(\tau_1))$. By Theorem 4.21 and the primitivity of $(x_1(t), y_1(t))$, this order must be 1. Therefore $\tau = c\tau_1 + \cdots$ with $c \neq 0$, whence $\tau = \sigma(\tau_1)$ for some σ that is a K-automorphism of $K[[t]]$. Since σ^{-1} is also such an automorphism, so also $\tau_1 = d\tau + \cdots$ with $d \neq 0$.

Since τ is a rational function of the elements $x_1(\tau_1)$ and $y_1(\tau_1)$, there exist polynomials $a(X,Y), b(X,Y) \in K[X,Y]$ such that

$$\tau = \frac{a(x_1(\tau_1), y_1(\tau_1))}{b(x_1(\tau_1), y_1(\tau_1))}.$$

From (4.13), it follows that

$$\tau = \frac{a(x_2(\tau_2), y_2(\tau_2))}{b(x_2(\tau_2), y_2(\tau_2))}.$$

Therefore $\tau = e\,\tau_2{}^i + \cdots$ with $i \geq 0$. Hence $\tau_1 = de\,\tau_2{}^i + \cdots$; that is, τ_1 is a formal power series in τ_2 of positive order. When $(x_2(t), y_2(t))$ is primitive, then their roles may be interchanged, giving that τ_2 is a formal power series in τ_1. But this is possible if and only if τ_1 is an element of order 1 in $K[[\tau_2]]$. □

From this theorem and Theorem 4.20, the following result is obtained.

THEOREM 4.23 *Let $(x(t), y(t))$ be an imprimitive branch representation in special affine coordinates. If $(x_1(t), y_1(t))$ and $(x_2(t), y_2(t))$ are branch representations in special affine coordinates satisfying the relations*

$$x(t) = x_1(\tau_1) = x_2(\tau_2), \quad y(t) = y_1(\tau_1) = y_2(\tau_2),$$

with $\tau_1, \tau_2 \in K[[t]]$, then $(x_1(t), y_1(t))$ and $(x_2(t), y_2(t))$ have the same order.

Now the idea of ramification index can be introduced.

DEFINITION 4.24 (i) If $(x(t), y(t))$ is an imprimitive branch representation in special affine coordinates and if $(\xi(t), \eta(t))$ is a primitive branch representation in special affine coordinates with $x(t) = \sigma(\xi(t)), y(t) = \sigma(\eta(t))$, then $\nu = \operatorname{ord}_t \sigma(t)$ is the *ramification index* of $(x(t), y(t))$.

(ii) The *ramification index of a branch representation* is its ramification index in its special affine coordinate form.

Theorem 4.23 thus has the following corollary.

COROLLARY 4.25 *Equivalent imprimitive branch representations have the same ramification index. Further, the primitive representations to which they give rise are also equivalent.*

A sufficient condition for imprimitivity is reducibility, where a branch representation in special affine coordinates $(x(t), y(t))$, and any equivalent one, is *reducible* if, in Definition 4.19, $\sigma(t) = t^m$; that is, $x(t), y(t) \in K[[t^m]]$ for an integer $m > 1$. In fact, the adjective 'reducible' signifies 'imprimitivity for a substitution of type $t \mapsto \tau$ with $\tau^m = t$'. A useful criterion for reducibility is given by the following theorem.

THEOREM 4.26 *Let $p = 0$ or $p > 0$ with $p \nmid n$. A branch representation in special affine coordinates, where $a_1 \neq 0$ and $0 < n_1 < n_2 < \cdots$,*

$$\begin{aligned} x(t) &= u + t^n, \\ y(t) &= v + a_1 t^{n_1} + a_2 t^{n_2} + \cdots, \end{aligned} \tag{4.14}$$

is reducible if and only if $\gcd(n, n_1, n_2, \ldots) > 1$.

Proof. That the condition is sufficient is immediate.

To show the necessity, suppose that the representation is reducible. Then there exists a substitution $t \mapsto \tau$ of order 1 such that $x(\tau), y(\tau) \in K[[t^m]]$ with $m > 1$.

Next it is shown that $\tau/t \in K[[t^m]]$. Suppose, by way of contradiction, that

$$\tau = t(b_0 + b_1 t^m + \cdots + b_h t^{hm} + ct^s + \cdots)$$

with $c \neq 0$ but with m not dividing s. It follows that

$$\begin{aligned} x(\tau) &= u + t^n(b_0 + b_1 t^m + \cdots + b_h t^{hm} + ct^s + \cdots)^n \\ &= u + t^n(b_0 + \cdots + b_h t^{hm})^{n-1} + \cdots. \end{aligned}$$

Since $x(\tau) \in K[[t^m]]$, it is immediate that $m \mid n$. Further, $m \mid (n + s)$ since $n \not\equiv 0 \pmod{p}$. Hence $m \mid s$, a contradiction.

Therefore write $\tau = zt$ with $z = u_0 + u_1 t^m + \cdots \in K[[t^m]]$ and $u_0 \neq 0$. Now suppose that at least one of the exponents $n_1, n_2, \ldots$ is not divisible by m, and let n_{k+1} be the smallest such exponent. Then

$$\begin{aligned} &y(t) - (v + a_1 t^{n_1} z^{n_1} + \cdots + a_k t^{n_k} z^{n_k}) \\ &= a_{n_k+1} t^{n_k+1} (u_0 + u_1 t^m + \cdots)^{n_k+1} + \cdots \\ &= a_{n_k+1} u_0^{n_k+1} t^{n_k+1} + \cdots . \end{aligned}$$

This shows that the first expression, but not the second, belongs to $K[[t^m]]$. This is a contradiction, which shows that m must divide not only n but also each exponent $n_1, n_2, \ldots$; this finishes the proof. □

THEOREM 4.27 *Let $p = 0$ or $p \nmid n$. Every branch representation of order n has a special affine coordinate form of the type*

$$x(t) = u + t^n, \quad y = v + \eta(t) \tag{4.15}$$

with $\operatorname{ord}_t \eta(t) \geq n$.

Proof. Let

$$\begin{aligned} x(t) &= u + u_n t^n + u_k t^k + \cdots , \\ y(t) &= v + y_1(t), \end{aligned}$$

with $u_n \neq 0$, $\operatorname{ord}_t y_1(t) \geq n$, be a branch representation. First, the equation

$$\tau^n = u_n t^n + u_k t^k + \cdots \tag{4.16}$$

is solved for $\tau \in K[[t]]$. The case $n = 1$ already provides a solution; so suppose that $n > 1$. Dividing both sides of (4.16) by t^n gives the equation

$$z^n = u_n + u_k t^{k-n} + \cdots \tag{4.17}$$

in the unknown $z \in K[[t]]$. Since $p \nmid n$, the polynomial $Z^n - u_n$ certainly does not have multiple roots, and therefore has at least two distinct roots. By Hensel's Lemma 4.9, $Z^n - (u_n + u_k t^{k-n} + \cdots)$ is reducible in $K[[t]][Z]$; that is,

$$Z^n - (u_n + u_k t^{k-n} + \cdots) = (Z^r + \rho(Z))(Z^s + \sigma(Z)),$$

where $\rho(Z), \sigma(Z) \in K[[t]][Z]$. Write

$$Z^r + \rho(Z) = Z^r + \rho_1(t) Z^{r-1} + \cdots + \rho_r(t), \ \rho_i \in K[[t]].$$

The polynomial $g(Z) = Z^r + \rho_1(0) Z^{r-1} + \cdots + \rho_r(0)$, with coefficients in K, divides $Z^n - u_n$. Hence $g(Z)$ has no multiple roots and so at least two distinct roots. Applying Hensel's Lemma 4.9 iteratively shows that the polynomial

$$Z^n - (u_n + u_k t^{k-n} + \cdots)$$

splits into linear factors in $K[[t]][Z]$. Let $Z - \zeta(t)$ be one of these factors; note that $\operatorname{ord}_t \zeta(t) = 0$. Then

$$\zeta(t)^n - (u_n + u_k t^{k-n} + \cdots) = 0,$$

whence $(\zeta(t)t)^n = u_n t^n + u_k t^k \cdots$. Putting $\tau = \zeta(t)t$ gives $x = u + \tau^n$. Since $\operatorname{ord}_t \zeta(t)t = 1$, the relation $\tau = \zeta(t)t$ gives t; that is, there exists an element $f(\tau) \in K[[\tau]]$ such that $t = f(\tau)$. Finally, putting $\eta(\tau) = y_1(f(\tau))$ gives

$$x = u + \tau^n, \quad y = v + \eta(\tau),$$

with $\operatorname{ord}_\tau \eta(\tau) \geq n$, as required. □

THEOREM 4.28 *Let $p = 0$ or $p \nmid n$. A branch representation of order n is imprimitive if and only if it is reducible.*

Proof. It suffices to show that the imprimitivity implies the reducibility. By Theorem 4.27, the branch representation in special affine coordinates $(x(t), y(t))$ may be taken in the form (4.15). If $(x(t), y(t))$ is imprimitive, then there exists a representation $(\xi(\tau), \eta(\tau))$ with $\xi(\tau), \eta(\tau) \in K[[\tau]]$ such that, for a suitable element $\tau = c_i t^i + \cdots$, $i > 1$,

$$x(t) = \xi(\tau), \quad y(t) = \eta(\tau).$$

By Theorem 4.20, the order of the primitive representation $(\xi(\tau), \eta(\tau))$ is n/i, which is coprime to p; write $m = n/i$. Therefore, still by Theorem 4.27, it may be supposed that (ξ, η) is of type (4.15), that is,

$$\xi(\tau) = u + \tau^m, \quad \eta(\tau) = v + b_1\tau^{m_1} + b_2\tau^{m_2} + \cdots .$$

Hence

$$t^n = \tau^m, \tag{4.18}$$

$$a_1 t^{n_1} + a_2 t^{n_2} + \cdots = b_1\tau^{m_1} + b_2\tau^{m_2} + \cdots . \tag{4.19}$$

From (4.18) it follows that $(t^i/\tau)^m = 1$, whence $\tau = ct^i$ since c is an m-th root of unity; in particular $c \in K\backslash\{0\}$. Now (4.19) implies that $n_j = im_j$ for every j. This shows that $\gcd(n, n_1, n_2, \ldots) > 1$; hence $(x(t), y(t))$ is reducible. □

Finally, in this section, the definition of a *branch* is given; it is one of the central concepts in the following sections.

DEFINITION 4.29 A *branch* is an equivalence class of primitive branch representations. The *centre*, the *order*, and the *order sequence* of a branch are the centre, order, and the order sequence of any branch representation.

4.3 BRANCHES OF PLANE ALGEBRAIC CURVES

The next definition is the foundation for developing the study of algebraic plane curves based on the idea of branch; in fact, branches constitute an important tool for the local study of algebraic curves.

DEFINITION 4.30 A *branch of a plane curve* $\mathcal{F} = \mathbf{v}(F(X_0, X_1, X_2))$ is a branch whose representations $(x_0(t), x_1(t), x_2(t))$ are zeros of $F(X_0, X_1, X_2)$.

THEOREM 4.31 *The centre of a branch of a plane curve is a point of the curve.*

Proof. Let $\mathcal{F} = \mathbf{v}(F)$ be a plane curve and let $(\xi_0(t), \xi_1(t), \xi_2(t))$ be any special branch representation; then

$$F(\xi_0(t), \xi_1(t), \xi_2(t)) = 0.$$

Putting $t = 0$ gives $F(\xi_0(0), \xi_1(0), \xi_2(0)) = 0$. Therefore the point

$$P = (\xi_0(0), \xi_1(0), \xi_2(0)),$$

which is the centre of the branch representation, is a point of the curve $\mathcal{F}$. □

THEOREM 4.32 *There exists a unique branch of a plane curve $\mathcal{F}$ centred at a simple point of $\mathcal{F}$.*

Proof. Without loss of generality, let the point be the origin and let the tangent there be any line other than $\mathbf{v}(X)$. If $\mathcal{F} = \mathbf{v}(F(X, Y))$, then, by Theorem 4.26, the representation

$$x = t, \quad y = c_1 t + c_2 t^2 + \cdots$$

is primitive and represents a branch of $\mathcal{F}$ centred at the origin.

To show the uniqueness, let $(\xi(t), \eta(t))$ be a primitive branch representation of $\mathcal{F}$ centred at the origin; then $\operatorname{ord} \xi > 0$, $\operatorname{ord} \eta > 0$ and $F(\xi, \eta) = 0$. If the origin O is a simple point of $\mathcal{F}$ and the tangent O to $\mathcal{F}$ is not $\mathbf{v}(X)$, by Theorem 4.6, there exists a unique formal power series

$$c_1 X + c_2 X^2 + \cdots$$

such that

$$F(X, c_1 X + c_2 X^2 + \cdots) = 0.$$

Then $F(X, Y)$ factorises in $K[[X]][Y]$:

$$F(X, Y) = (Y - (c_1 X + c_2 X^2 + \cdots))H(X, Y),$$

with $\operatorname{ord} H(X, Y) = 0$. In $K[[t]]$,

$$F(\xi, \eta) = (\eta - (c_1 \xi + c_2 \xi^2 + \cdots))H(\xi, \eta) = 0.$$

Hence $\eta = c_1 \xi + c_2 \xi^2 + \cdots$.

Since (ξ, η) is a primitive representation, ξ must be of order 1. The substitution $\sigma : t \to \xi$ is of order 1, and so, by Theorem 4.4, is an automorphism of $K[[t]]$; its inverse σ^{-1} is also an automorphism. Also, σ^{-1} sends ξ to t and $c_1 \xi + \cdots$ to $(t, c_1 t + c_2 t^2 + \cdots)$. It follows that $(\xi, c_1 \xi + \cdots)$ is equivalent to the branch representation $(x(t), y(t))$ introduced at the beginning of the proof; so the uniqueness is established. □

EXAMPLE 4.33 Let $\mathcal{H}_q = \mathbf{v}(Y^q + Y - X^{q+1})$, with q a power of p, be the Hermitian curve. Then the origin is a simple point and the tangent to $\mathcal{H}$ at the origin is $\mathbf{v}(Y)$. A primitive representation of the unique branch of $\mathcal{H}_q$ at the origin is $(x = t,\ y = y(t))$ with

$$y(t) = \textstyle\sum_{i=1}^{\infty} (-1)^{i+1} t^{q^{i+1}+q^i}.$$

DEFINITION 4.34 Let $\mathcal{G} = \mathbf{v}(G)$ be a plane curve and γ a branch centred at the point P. If $(\xi_0(t), \xi_1(t), \xi_2(t))$ is a representation of γ in special coordinates, then the *intersection multiplicity* is

$$I(P, \mathcal{G} \cap \gamma) = \begin{cases} \operatorname{ord}_t G(\xi_0(t), \xi_1(t), \xi_2(t)) & \text{if } \gamma \notin \mathcal{G}, \\ \infty & \text{if } \gamma \in \mathcal{G}. \end{cases}$$

Note that $I(P, \mathcal{G} \cap \gamma)$ depends only on γ, and not on the representation chosen.

THEOREM 4.35 *The intersection number $I(P, \mathcal{G} \cap \gamma)$ is covariant.*

Proof. For $i = 0, 1, 2$, let $X_i = \sum_{j=0}^{2} a_{ij} X'_j$ be the equation of the change of the coordinate system (X_0, X_1, X_2) into (X'_0, X'_1, X'_2). If

$$G'(X_0, X_1, X_2) = G(\sum_{j=0}^{2} a_{0j} X'_j, \sum_{j=0}^{2} a_{1j} X'_j, \sum_{j=0}^{2} a_{2j} X'_j)$$

and $x_i(t) = \sum_{j=0}^{2} a_{ij} x'_j(t)$ for $i = 0, 1, 2$, then

$$\text{ord}_t\, G(x_0(t), x_1(t), x_2(t)) = \text{ord}_t\, G'(x'_0(t), x'_1(t), x'_2(t)).$$

If $G(x_0(t), x_1(t), x_2(t)) = 0$, then $I(P, \mathcal{G} \cap \gamma) = \infty$. □

The following result shows how Definition 4.34 appears naturally in the study of the intersection multiplicity at a common point of two curves.

THEOREM 4.36 (i) *If the branch γ of the curve $\mathcal{F}$ is centred at a simple point P of $\mathcal{F}$, and $\mathcal{G}$ is any other curve, then*

$$I(P, \mathcal{G} \cap \gamma) = I(P, \mathcal{G} \cap \mathcal{F}).$$

(ii) *More generally, if P is a singular point of the irreducible curve $\mathcal{F}$ and $\mathcal{G}$ is a plane curve not containing $\mathcal{F}$ as a component, then*

$$I(P, \mathcal{G} \cap \mathcal{F}) = \textstyle\sum_{\gamma} I(P, \mathcal{G} \cap \gamma),$$

where γ runs over all branches of $\mathcal{F}$ centred at P.

(iii) *If P is an m_P-fold point of $\mathcal{F}$ then m_P is a bound on the number of branches of $\mathcal{F}$ with centre P.*

Before dealing with Theorem 4.36 it needs to be shown that every point of $\mathcal{F}$ is the centre of at most a finite number of branches. This ensures that the definition $\sum_{\gamma} I(P, \mathcal{G} \cap \gamma)$ is meaningful. The proof requires some more results on the behaviour of branches under local quadratic transformations. These results are stated in the next section after proving the following result.

THEOREM 4.37 *Two distinct irreducible plane curves have no branches in common.*

Proof. Suppose, by way of contradiction, that the two curves $\mathcal{C}^n$ and $\mathcal{C}^m$ have a branch γ in common. Let $(\xi(t), \eta(t))$ be a branch representation; it is a point of the plane over $K((t))$ and so is a common point of the two curves regarded as curves of the plane over $K((t))$. Applying Bézout's Theorem 3.14 over the field K, the two curves being irreducible, gives that $\sum I(P, \mathcal{C}^n \cap \mathcal{C}^m) = nm$. Similarly, Bézout's Theorem 3.14 over $K((t))$ gives the same result. Hence the two curves have all their intersections in the smaller field, contradicting that, over $K((t))$, they have the point $(\xi(t), \eta(t))$ in common. □

COROLLARY 4.38 *Every branch (ξ, η) with $\xi = 0$ is a branch of the line $\mathbf{v}(X)$, but of no other irreducible curve.*

4.4 LOCAL QUADRATIC TRANSFORMATIONS

The study of branches centred at a singular point of a plane curve is helped by local quadratic transformations, that is, transformations of the form

$$\begin{array}{rcl} \sigma(X,Y) & = & (X',Y'), \\ X' & = & X, \\ Y' & = & Y/X. \end{array} \tag{4.20}$$

It is not defined for points on the line $\mathbf{v}(X)$, but the point $Y_\infty = (0,1,0)$ is taken to be the image of these points, other than the origin. Let $(\xi(t),\eta(t))$ be a branch representation with centre the origin and $\xi(t) \neq 0$. Then the branch representation

$$(\xi(t),\eta(t)/\xi(t))$$

is the image of the branch $(\xi(t),\eta(t))$ under σ, and so a branch representation corresponds to another branch representation. This gives the following result.

THEOREM 4.39 *The local quadratic transformation σ given by (4.20) defines a bijective correspondence between branches centred at the origin with tangent not $\mathbf{v}(X)$ and those centred at any affine point on $\mathbf{v}(X)$.*

For a point $P = (a,b)$, with $a \neq 0$, on a plane curve $\mathcal{F} = \mathbf{v}(F(X,Y))$, the image of P under σ is on the curve $\mathbf{v}(\bar{F}(X,Y))$, where $\bar{F}(X,Y) = F(X,XY)$. This gives the following idea.

DEFINITION 4.40 The *algebraic transform* of a curve $\mathcal{F} = \mathbf{v}(F(X,Y))$ is the curve $\bar{\mathcal{F}} = \mathbf{v}(\bar{F}(X,Y))$, where $\bar{F}(X,Y) = F(X,XY)$.

A useful property is the following relation: given a branch γ centred at P and an algebraic plane curve $\mathcal{G}$, let γ' and $\bar{\mathcal{G}}$ be their images under a local quadratic transformation. If $\bar{P}$ is the centre of γ', then

$$I(P,\mathcal{G}\cap\gamma) = I(\bar{P},\bar{\mathcal{G}}\cap\gamma'). \tag{4.21}$$

To show (4.21), put $\mathcal{G} = \mathbf{v}(G(X,Y))$, $\bar{\mathcal{G}} = \mathbf{v}(\bar{G}(X,Y))$, and let $(x'(t),y'(t))$ be a primitive representation of γ'. Then

$$\operatorname{ord}_t G(x(t),y(t)) = \operatorname{ord}_t \bar{G}(x(t),y(t)/x(t)) = \operatorname{ord}_t \bar{G}(x'(t),y'(t)),$$

showing (4.21).

Every polynomial $F(X,Y) \in K[X,Y]$ can be written in the form

$$F(X,Y) = F_r(X,Y) + F_{r+1}(X,Y) + \cdots, \ F_r \neq 0,$$

with $F_i(X,Y)$ homogeneous of degree i. The curve $\mathcal{F} = \mathbf{v}(F(X,Y))$ passes through the origin O if and only if $r \geq 1$; in this case, O is a point of $\mathcal{F}$ of multiplicity r. Since

$$\begin{aligned} \bar{F}(X,Y) &= F(X,XY) \\ &= F_r(X,XY) + F_{r+1}(X,XY) + \cdots \\ &= X^r(F_r(1,Y) + XF_{r+1}(1,Y) + \cdots), \end{aligned}$$

so the algebraic transform of $\mathcal{F}$ has X^r as a component. For this reason, it is appropriate to eliminate X^r and introduce the notion of the geometric transform.

DEFINITION 4.41 The *geometric transform* of a curve $\mathcal{F} = \mathbf{v}(F(X,Y))$ with an r-ple point, with $r \geq 0$, at the origin is the curve $\mathcal{F}' = \mathbf{v}(F'(X,Y))$ given by

$$F'(X,Y) = \bar{F}(X,Y)/X^r = F(X,XY)/X^r.$$

Now the first, important property of the geometric transform is established.

THEOREM 4.42 *If $\mathcal{F}$ is irreducible and is not the line $\mathbf{v}(X)$, then the geometric transform $\mathcal{F}'$ is also irreducible.*

Proof. Let $F(X,Y)$ be irreducible and different from cX, $c \in K$. It must be shown that $F'(X,Y) = F(X,XY)/X^r$ is irreducible.

Suppose that $F'(X,Y) = G(X,Y)H(X,Y)$ with $G(X,Y), H(X,Y)$ non-constant polynomials. First, consider the possibility that $H(X,Y) = h(X)$. Then $F(X,XY) = X^r h(X)G(X,Y)$. Putting $Y = Z/X$ gives

$$F(X,Z) = X^r h(X)G(X,Z/X).$$

Let d be the minimum value for which $X^d G(X,Z/X) \in K[X,Z]$; then

$$X^{r-d}h(X) \in K[X],$$

whence

$$h(X) = h_1 X^{d-r} + h_2 X^{d-r+1} + \cdots.$$

Therefore

$$F(X,Z) = (h_1 + h_2 X + \cdots)(X^d G(X,Z/X)).$$

Since $F(X,Y)$ contains Y, then $F(X,Z)$ must contain Z and consequently is not a constant. The irreducibility of $F(X,Z)$ therefore implies that $h(X) = h_1 X^{d-r}$. This means that $F(X,XY) = X^d G(X,Y)$. Then, from the definition of r, necessarily $d = r$; that is, $h(X)$ is a constant, as required.

Finally, the possibility that $H(X,Y)$ or $G(X,Y)$ contains Y must be excluded. If it were so, then

$$\begin{aligned} F(X,Z) &= X^r G(X,Z/X)H(X,Z/X) \\ &= [X^m G(X,Z/X)][(X^{r-m} H(X,Z/X)], \end{aligned}$$

where both factors in square brackets are polynomials. But they are not constant polynomials since they contain Z and that contradicts the irreducibility of both $F(X,Z)$ and $F(X,Y)$. □

As a corollary, there is the following result.

THEOREM 4.43 *Let $\mathcal{F}_1 = \mathbf{v}(F_1), \ldots, \mathcal{F}_s = \mathbf{v}(F_s)$ be irreducible curves, with each F_i in $K[X,Y]$ and $\mathcal{F}_i \neq \mathbf{v}(X)$ for all i. Then the geometric transform of $\mathcal{F} = \mathbf{v}(F_1^{m_1} \cdots F_s^{m_s})$ is $\mathcal{F}' = \mathbf{v}(F_1'^{m_1} \cdots F_s'^{m_s})$.*

Now Theorem 4.39 may be made more precise.

THEOREM 4.44 *If $\mathcal{F}$ is a plane curve such that the line $\mathbf{v}(X)$ is not tangent to $\mathcal{F}$ at O and $\mathcal{F}'$ is the geometric transform of $\mathcal{F}$ by the local quadratic transformation σ as in (4.20), then there exists a bijection between the branches of $\mathcal{F}$ centred at O and the branches of $\mathcal{F}'$ centred at an affine point of $\mathbf{v}(X)$.*

Proof. Considering Corollary 4.38, it follows from Theorem 4.37 that the branches of $\mathcal{F}$ and those of $\mathcal{F}'$ admit a representation $(\xi(t), \eta(t))$ with $\xi_(t) \neq 0$. Further, if $(\xi(t), \eta(t))$ satisfies $F(X,Y) = 0$, then $(\xi(t), \eta(t)/\xi(t))$ satisfies $F(X, XY) = 0$ and also $F(X, XY)/X^r = 0$; and conversely. This gives a bijective correspondence between the branches of $\mathcal{F}$ and $\mathcal{F}'$.

To show that the branches of $\mathcal{F}'$ obtained in this way are centred at affine points of $\mathbf{v}(X)$, it suffices to see that $\text{ord}\, \xi(t) \leq \text{ord}\, \eta(t)$; that is,

$$\xi(t) = a_i t^i + \cdots, \quad \eta(t) = b_j t^j + \cdots$$

with $i \leq j$. Suppose instead that $j < i$. The hypothesis that $\mathbf{v}(X)$ is not tangent to $\mathcal{F}$ at O implies that Y^r is present in $F(X,Y)$, whence, with $c_r \neq 0$,

$$F_r(X,Y) = c_0 X^r + c_1 X^{r-1} Y + \cdots + c_r Y^r.$$

So $\text{ord}\, F_r(\xi(t), \eta(t)) = jr$, but $\text{ord}\, F_{r+k}(\xi(t), \eta(t)) \geq (r+1)j$ for $k > 0$. Therefore $\text{ord}\, F(\xi(t), \eta(t)) = rj$, which contradicts that $F(\xi(t), \eta(t)) = 0$. This finishes the proof. □

THEOREM 4.45 *If O is an ordinary singular point on the curve $\mathcal{F} = \mathbf{v}(F(X,Y))$ of multiplicity r, then*

(i) *it is the centre of exactly r branches of $\mathcal{F}$;*

(ii) *the branches are all linear, and their tangents are precisely the tangents to $\mathcal{F}$ at O.*

Proof. With O at the origin and no vertical tangent to $\mathcal{F}$ at O,

$$F(X,Y) = \textstyle\prod_{i=1}^r (Y - m_i X) + \cdots.$$

Then the geometric transform $\mathcal{F}' = \mathbf{v}(F'(X,Y))$ of $\mathcal{F}$ under the local quadratic transformation (4.20) is given by

$$F'(X,Y) = \textstyle\prod_{i=1}^r (Y - m_i) + XH(X).$$

The line $\mathbf{v}(X)$ meets $\mathcal{F}'$ in the points $Q_i = (0, m_i)$ for $i = 1, \ldots, r$.

Now it is shown that these points are all simple on $\mathcal{F}'$. Choose Q_1, say, and take it to $(0,0)$ via the translation $(X,Y) \mapsto (X, Y - m_1)$. The transformed curve is $\mathcal{G} = \mathbf{v}(G(X,Y))$ with

$$\begin{aligned} G(X,Y) &= F'(X, Y + m_1) \\ &= \textstyle\prod_{j=1}^r ((Y + m_1) - m_j) + XH_1(X) \\ &= Y(Y - (m_2 - m_1)) \cdots (Y - (m_r - m_1)) + XH_2(X) \\ &= (-1)^{r-1} Y \textstyle\prod_{i=2}^r (m_i - m_1) + \cdots. \end{aligned}$$

Since all the $m_i - m_1$ are non-zero, it follows that $G(X,Y)$ contains a single term of degree 1 given by $(-1)^{r-1} Y \prod_{i=2}^r (m_i - m_1)$; this shows that Q_1 is simple for $\mathbf{v}(\mathcal{G})$, and so also for $\mathcal{F}'$. From Theorem 4.32 it now follows that there are precisely r branches of $\mathcal{F}'$ with centre at an affine point on $\mathbf{v}(X)$. By the previous theorem, $\mathcal{F}$ has exactly r branches centred at the origin. By Theorem 4.32 the branch of $\mathcal{F}'$ centred at Q_i has a representation $(t, m_i + \cdots)$. The corresponding branch of $\mathcal{F}$ therefore has a representation $(t, m_i t + \cdots)$. Since the tangent of such a branch is the line $\mathbf{v}(Y - m_i X)$, the theorem is proved. □

THEOREM 4.46 *Each point P of $\mathcal{F}$ is the centre of at least one and at most a finite number of branches of $\mathcal{F}$.*

Proof. Since each branch of $\mathcal{F} = \mathbf{v}(F)$ must annul at least one irreducible factor $\mathcal{F}_i$, the polynomial F may be supposed irreducible. In fact, let

$$\mathcal{F} = \mathcal{F}_1 \cup \mathcal{F}_2 \cup \cdots \cup \mathcal{F}_l$$

with each $\mathcal{F}_i = \mathbf{v}(F_i)$ an irreducible component of $\mathcal{F}$; then

$$F(X,Y) = F_1(X,Y) \cdots F_l(X,Y).$$

Let $\gamma = (\xi(t), \eta(t))$ be a branch of $\mathcal{F}$; so $F(\xi(t), \eta(t)) = 0$, or

$$F_1(\xi(t), \eta(t)) \cdots F_l(\xi(t), \eta(t)) = 0.$$

Hence $F_i(\xi(t), \eta(t)) = 0$ for at least one i and so γ is a branch of $\mathcal{F}_i$ for that i.

It may also be supposed that P, of multiplicity r, is at the origin and that the line $\mathbf{v}(X)$ is not tangent to $\mathcal{F}$ at P. Then $F = \prod_{i=1}^{r}(Y - m_i X) + \cdots$ with the m_i not necessarily distinct. Applying σ to $\mathcal{F}$ gives $\mathcal{F}' = \mathbf{v}((F'(X,Y')/X^r)$ with

$$F'(X,Y')/X^r = \prod_{i=1}^{r} (Y' - m_i) + XH(X) = 0.$$

Therefore the line $\mathbf{v}(X)$ meets $\mathcal{F}'$ in the points $Q_i'(0, m_i)$, and the sum of the intersection multiplicities at these points is r; that is,

$$\sum_{i=1}^{s} I(Q_i', \mathbf{v}(X) \cap \mathcal{F}') = r. \tag{4.22}$$

If at least two of the m_i are distinct, each point $(0, m_j)$ of $\mathcal{F}'$ has multiplicity less than r. By an induction argument, with the assumption of the existence of at least one and at most a finite number of branches centred at the point $(0, m_j)$, then $\mathcal{F}'$ has at least one and at most a finite number of branches centred at an affine point on $\mathbf{v}(X)$. By Theorem 4.44, $\mathcal{F}$ has at least one and at most a finite number of branches centred at O.

This leaves the case that all the m_i are equal:

$$F(X,Y) = (Y - m_1 X)^r + \cdots, \ r > 1.$$

In this case, $\mathcal{F}' = \mathbf{v}(F'(X,Y')/X^r)$ with

$$F'(X,Y')/X^r = (Y' - m_1)^r + XH(X).$$

Now the translation $X = X$, $Y = Y' - m_1$, which takes the point $(0, m_1)$ to $(0,0)$, transforms $\mathcal{F}'$ to $\mathbf{v}(Y^r + XG(X,Y))$. Hence $\mathcal{F}'$ may be assumed to have the singular point of multiplicity r at the origin.

With a similar argument to the above, by induction on r, it may be supposed that the curve $\mathcal{F}' = \mathbf{v}(F'(X,Y))$ with

$$F'(X,Y) = (Y - m_2 X)^r + H(X,Y).$$

Then applying σ gives a plane curve $\mathcal{F}^{(2)} = \mathbf{v}(F^{(2)}(X,Y))$, which may be taken to have a point of multiplicity r at $(0, m_2)$, whence

$$F^{(2)}(X,Y) = (Y - m_2)^r + XH(X,Y).$$

Is it possible that an infinite sequence of points of multiplicity r is obtained? The answer is, in fact, in the negative, as is shown below. Therefore this procedure gives rise to a point that is not of multiplicity r. Induction now finishes the proof.

To show that the answer is negative, introduce $m_3, m_4, \ldots$ through the curves $\mathcal{F}^{(2)}, \mathcal{F}^{(3)}, \ldots$ exactly as m_1 and m_2 were obtained in relation to $\mathcal{F}$ and $\mathcal{F}'$. For each $i \geq 1$, write $F(X,Y)$ in the form

$$F(X,Y) = \textstyle\sum_{b>0} c_{ab} X^a (Y - m_1 X - m_2 X^2 - \cdots - m_i X^i)^b + X^{h_i} P_i(X), \tag{4.23}$$

where $c_{ab} \in K$, $P_i(X) \in K[X]$, $P_i(0) \neq 0$; this can be done uniquely.

In the expansion of the sum, the term $c_{ab} X^a \eta^b$ is not eliminated by other terms. Since $\mathcal{F}$ has a point of multiplicity r at the origin, it follows that $a + b \geq r$. Also, $h_i \geq r$. Using the form (4.23) for $\mathcal{F}$, the algebraic transform $\bar{\mathcal{F}} = \mathbf{v}(\bar{F}(X, Y'))$ is as follows:

$$\begin{aligned}
&\bar{F}(X,Y') \\
&= F(X, XY') \\
&= \textstyle\sum c_{ab} X^a (XY' - m_1 X - m_2 X^2 - \cdots - m_i X^i)^b + X^{h_i} P_i(X) \\
&= \textstyle\sum c_{ab} X^{a+b} (Y' - m_1 - m_2 X - \cdots - m_i X^{i-1})^b + X^{h_i} P_i(X) = 0.
\end{aligned}$$

Since, by assumption, the geometric transform $\mathcal{F}'$ of $\mathcal{F}$ has a point of multiplicity r at the origin, so $\mathcal{F}' = \mathbf{v}(F')$ has

$$\begin{aligned}
F'(X,Y') &= \bar{F}(X,Y')/X^r \\
&= \textstyle\sum c_{ab} X^{a+b-r} (Y' - m_1 - m_2 X - \cdots - m_i X^{i-1})^b \\
&\quad + X^{h_i - r} P_i(X) = 0.
\end{aligned}$$

Since $h_i(\mathcal{F}) = h_i(\mathcal{F}') + r = (h_i(\mathcal{F}'') + r) + r$, repeating the process i times gives $h_i(\mathcal{F}) = h_i(\mathcal{F}^i) + ir$, from which $h_i(\mathcal{F}) \geq ir$. It follows that $h_i \to \infty$ as $i \to \infty$.

Substituting $Y = m_1 X + m_2 X^2 + \cdots$ in (4.23) gives

$$F(X, m_1 X + m_2 X^2 + \cdots) = \textstyle\sum c_{ab} X^a (m_{i+1} X^{i+1} + \cdots)^b + X^{h_i} P_i(X).$$

Hence

$$\operatorname{ord}_x F(X, m_1 X + m_2 X^2 + \cdots) \geq \min\{i + 1, h_i\}$$

for each i. Therefore $\operatorname{ord}_x F(X, m_1 X + m_2 X^2 + \cdots) = \infty$; in other words, there is a branch $\gamma = (\xi = t, \eta = m_1 t + m_2 t^2 + \cdots)$ centred at O, which is the first part of the theorem. The next proposition completes the proof of Theorem 4.46. □

Proposition 4.47 *The sequence of quadratic transformations appearing in the above proof does not give rise, for $r > 1$, to an infinite sequence of points of multiplicity r.*

Proof. Assume at first that the polynomial $\partial F/\partial Y$ does not vanish. Then the polar curve $\mathcal{G}$ of $Y_\infty = (0,0,1)$ with respect to $\mathcal{F}$ exists. Since $\mathcal{F}$ is irreducible, the branch $\gamma = (\xi = t, \eta = m_1 t + m_2 t^2 + \cdots)$ cannot be a branch of $\mathcal{G}$ by Theorem 4.37. Let $\mathcal{G}'$ be the polar curve of Y_∞ with respect to the geometric transform $\mathcal{F}'$ of $\mathcal{F}$ under the local quadratic transformation. Then $\mathcal{G}' = \mathbf{v}(\partial F'/\partial Y')$, and

$$\begin{aligned}
\left(\frac{\partial F(X,Y)}{\partial Y} \Big/ X^{r-1} \right)_{Y = XY'} &= \left(\frac{\partial F(X, XY')}{\partial Y'} \frac{\partial Y'}{\partial Y} \right) \Big/ X^{r-1} \\
&= \frac{\partial}{\partial Y'} \left(\frac{1}{X^r} F(X, XY') \right),
\end{aligned}$$

whence

$$\frac{1}{X^{r-1}}\frac{\partial F}{\partial Y} = \frac{\partial F'}{\partial Y'}.$$

With $O = (0,0)$ the centre of γ and $O' = (0, m_1)$ the centre of

$$\gamma' : (\xi' = t, \eta' = \eta/\xi = m_1 + m_2 t + \cdots),$$

then

$$\begin{aligned} I(O, \mathcal{G}\cap\gamma) &= \mathrm{ord}_t \left(\frac{\partial F}{\partial Y}\right)_{X=t, Y=m_1 t+\cdots} \\ &= \mathrm{ord}_t \left(\frac{\partial F'}{\partial Y'}\right)_{X'=t, Y'=m_1+\cdots} + \mathrm{ord}_t(X^{r-1})_{X=t} \\ &= I(O', \mathcal{G}'\cap\gamma') + \mathrm{ord}_t(t^{r-1}) \\ &= I(O', \mathcal{G}'\cap\gamma') + r - 1. \end{aligned}$$

Therefore, if $r > 1$,

$$I(O, \mathcal{G}\cap\gamma) > I(O', \mathcal{G}'\cap\gamma') > 0.$$

Continuing this argument gives an infinite sequence of decreasing positive integers, a contradiction which proves the assertion under the initial hypothesis.

If this hypothesis does not hold, then choose another non-tangent line to $\mathcal{F}$ to be the Y-axis. Then the polynomial $\partial F/\partial Y$ does not vanish any more. Now the polar curve of $\mathcal{F}$ for Y_∞ exists. By (4.24), the polar curve of $\mathcal{F}'$ for Y_∞ also exists, and hence this proof still works. □

Now Theorem 4.36 can be proved. To do this, define

$$J(P, \mathcal{G}\cap\mathcal{F}) = \textstyle\sum_\gamma I(P, \mathcal{G}\cap\gamma),$$

where γ ranges over all branches of $\mathcal{F}$ centred at P. Now check that $J(P, \mathcal{G}\cap\mathcal{F})$ satisfies the seven postulates in Chapter 3.1. By Theorem 4.35, $J(P, \mathcal{G}\cap\mathcal{F})$ is covariant; so it may be assume that P coincides with the origin and that the Y-axis is not tangent to $\mathcal{F}$ at P.

If $\mathcal{F}$ and $\mathcal{G}$ have no common component through P, then the branches of $\mathcal{F}$ centred at P are distinct from those of $\mathcal{G}$ centred at P. Hence $J(P, \mathcal{G}\cap\mathcal{F})$ is a non-negative integer for every branch γ of $\mathcal{F}$. This shows the validity of (I 1).

If $\mathcal{F}$ and $\mathcal{G}$ have a common component, say $\mathcal{C}$, through P, then any branch γ of $\mathcal{C}$ centred at P is a common branch of $\mathcal{F}$ and $\mathcal{G}$. This implies that $J(P, \mathcal{G}\cap\mathcal{F}) = \infty$ showing (I 2).

If P is not a point of $\mathcal{F}$, no branch of $\mathcal{F}$ is centred at P; so $J(P, \mathcal{G}\cap\mathcal{F}) = 0$. If P is a point of $\mathcal{F}$ and γ is a branch of $\mathcal{F}$ centred at P given by the primitive representation $(x(t), y(t))$, then $\mathrm{ord}_t\, x(t) > 0$, $\mathrm{ord}_t\, y(t) > 0$. Thus

$$G(x(t), y(t)) = G(0,0) + tH(t).$$

If P is not a point of $\mathcal{G}$, then $G(0,0) \neq 0$. Hence

$$J(P, \mathcal{G}\cap\mathcal{F}) = \mathrm{ord}_t\, G(x(t), y(t)) = 0,$$

and this shows (I 3). Also, (I 4) follows from Theorem 4.36 (i).

To show (I 5), the following technical result is required.

THEOREM 4.48 *Let the origin P be an r-fold point of $\mathcal{F} = \mathbf{v}(F)$ and an s-fold point of $\mathcal{G} = \mathbf{v}(G)$ such that $\mathbf{v}(X)$ is not a tangent at P to $\mathcal{F}$ or $\mathcal{G}$. Let $\mathcal{F}'$ and $\mathcal{G}'$ be the geometric transforms of $\mathcal{F}$ and $\mathcal{G}$ under a local quadratic transformation. If $P'_1, \ldots, P'_n$ are the common affine points of $\mathcal{F}'$ and $\mathcal{G}'$ on $\mathbf{v}(X)$, then*

$$J(P, \mathcal{F} \cap \mathcal{G}) = rs + \textstyle\sum_{i=1}^{n} J(P'_j, \mathcal{F}' \cap \mathcal{G}').$$

Proof. Let $\gamma_1, \ldots, \gamma_m$ be the branches of $\mathcal{G}$ centred at P, and $(x_i(t), y_i(t))$ a primitive representation of γ_i for $i = 1, \ldots, m$. Let the branches of $\mathcal{G}'$ centred at affine points of $\mathbf{v}(X)$ are $\gamma'_1, \ldots, \gamma'_m$, and $(x'_i(t) = x_i(t), y'_i(t) = y_i(t)/x_i(t))$ is a primitive representation of γ'_i, for $i = 1, \ldots, m$. Hence

$$\begin{aligned} J(P, \mathcal{F} \cap \mathcal{G}) &= \textstyle\sum_{i=1}^{m} \operatorname{ord}_t F(x_i(t), y_i(t)) \\ &= \textstyle\sum_{i=1}^{m} \operatorname{ord}_t x_i(t)^r F'(x'_i(t), y'_i(t)) \\ &= r\textstyle\sum_{i=1}^{m} \operatorname{ord}_t x_i(t) + \sum_{i=1}^{m} \operatorname{ord}_t F'(x'_i(t), y'_i(t)) \\ &= rs + \textstyle\sum_{i=1}^{n} J(P'_i, \mathcal{F}' \cap \mathcal{G}'). \end{aligned}$$

□

Now (I5) is shown by induction on $k = J(P, \mathcal{F} \cap \mathcal{G})$. Since $J(P, \mathcal{F} \cap \mathcal{G}) = 0$ if and only if P is not a common point of $\mathcal{F}$ and $\mathcal{G}$, the assertion is true for $k = 0$. So assume it is true for every integer less than k, and let $J(P, \mathcal{F} \cap \mathcal{G}) = k$. By Theorem 4.48,

$$J(P, \mathcal{F} \cap \mathcal{G}) = rs + \textstyle\sum_{i=1}^{n} J(P'_i, \mathcal{F}' \cap \mathcal{G}').$$

As $rs \geq 1$, so $J(P'_i, \mathcal{F}' \cap \mathcal{G}') < k$. Hence $J(P'_i, \mathcal{F}' \cap \mathcal{G}') = J(P'_i, \mathcal{G}' \cap \mathcal{F}')$, and so $J(P, \mathcal{F} \cap \mathcal{G}) = J(P, \mathcal{G} \cap \mathcal{F})$.

The validity of (I6) and (I7) follows from the next lemma.

LEMMA 4.49 *Let γ be a branch of $\mathcal{F} = \mathbf{v}(F)$ centred at the point P, and let $\mathcal{G} = \mathbf{v}(G)$. If $\Phi = \mathbf{v}(G + FH)$ with $H \in K[X, Y]$, then*

$$J(P, \mathcal{G} \cap \gamma) = J(P, \Phi \cap \gamma).$$

If Λ is a reducible curve that splits into the components $\mathcal{G}$ and Ψ, then

$$I(P, \Lambda \cap \gamma) = I(P, \mathcal{G} \cap \gamma) + I(P, \Psi \cap \gamma).$$

Proof. Let $(x(t), y(t))$ be a primitive representation of γ. Then

$$\begin{aligned} J(P, \Phi \cap \gamma) &= \operatorname{ord}_t(G(x(t), y(t)) + F(x(t), y(t))H(x(t), y(t))) \\ &= \operatorname{ord}_t G(x(t), y(t)), \end{aligned}$$

whence the first assertion follows.

Let $\Psi = \mathbf{v}(M)$ and $\Lambda = \mathbf{v}(GM)$. Then

$$\begin{aligned} I(P, \Lambda \cap \gamma) &= \operatorname{ord}_t GM(x(t), y(t)) \\ &= \operatorname{ord}_t G(x(t), y(t))M(x(t), y(t)) \\ &= \operatorname{ord}_t G(x(t), y(t)) + \operatorname{ord}_t M(x(t), y(t)) \\ &= I(P, \mathcal{G} \cap \gamma) + I(P, \Psi \cap \gamma), \end{aligned}$$

whence the second assertion follows. □

The uniqueness of the intersection number, as in Theorem 3.8, gives the following important result.

THEOREM 4.50 *Let $\mathcal{F}$ and $\mathcal{G}$ be curves. Then, for any point P,*

$$I(P, \mathcal{F} \cap \mathcal{G}) = J(P, \mathcal{F} \cap \mathcal{G}).$$

It is possible to show directly that $J(P, \mathcal{F} \cap \mathcal{G}) = J(P, \mathcal{G} \cap \mathcal{F})$ when $\mathcal{G}$ is a line. To do this, fix the coordinate system in such a way that $P = (0,0)$, $\mathcal{G} = \mathbf{v}(Y)$ and that no tangents to $\mathcal{F}$ at P is $\mathbf{v}(X)$. By Theorem 4.36, if $\mathcal{F} = \mathbf{v}(F)$, then $J(P, \mathcal{F} \cap \mathbf{v}(Y)) = \mathrm{ord}_x F(x, 0)$. So the following result is required:

$$J(P, \mathbf{v}(Y) \cap \mathcal{F}) = \mathrm{ord}_x F(x, 0). \tag{4.24}$$

To prove this, suppose P to be an r-fold point of $\mathcal{F}$ with $r > 1$, as the assertion is true for $r = 1$ by Theorem 4.36. Then

$$F(X, Y) = \prod_{i=1}^{r} (Y - m_i X) + G(X, Y),$$

with ord $G > r$.

Let γ_i, for $i = 1, \ldots, k$, be the branches of $\mathcal{F}$ centred at P and let $(x_i(t), y_i(t))$ be a primitive representation of γ_i. Under a local quadratic transformation, γ_i is transformed into a branch γ_i' of the geometric quadratic transform $\mathcal{F}'$ of $\mathcal{F}$.

With γ_i' centred at $P_i' = (0, m_i)$, let $(x_i'(t), y_i'(t))$ be a primitive representation of γ_i'. Since the algebraic quadratic transform of $\mathcal{G} = \mathbf{v}(Y)$ is the conic $\mathbf{v}(XY)$, from (4.21),

$$J(P, \mathbf{v}(Y) \cap \mathcal{F}) = \sum_{i=1}^{k} \mathrm{ord}_t\, y(t) = \sum_{i=1}^{k} \mathrm{ord}_t\, x_i'(t) + \sum_{i=1}^{k} \mathrm{ord}_t\, y_i'(t).$$

By (4.22),

$$J(P, \mathbf{v}(Y) \cap \mathcal{F}) = r + \sum_{i=1}^{k} \mathrm{ord}_t\, y_i'(t).$$

In fact, as $y_i'(t) = y_i(t)/x_i(t)$, the branches giving a positive contribution to the sum are those centred at P. If their number is s, relabel the indices i so that the branches of $\mathcal{F}'$ centred at P have indices $1, \ldots, s$. Then

$$J(P, \mathbf{v}(Y) \cap \mathcal{F}) = r + \sum_{i=1}^{s} \mathrm{ord}_t\, y_i'(t) = r + J(P, \mathbf{v}(Y) \cap \mathcal{F}').$$

Now, if a least two of $P_i' = (0, m_i)$ are distinct, then each point $P_i' = (0, m_i)$ is an r_i'-fold point of $\mathcal{F}'$ with $r_i' < r$, and induction can be applied on r. Hence

$$\sum_{i=1}^{k} \mathrm{ord}_t\, y_i'(t) r + \mathrm{ord}_x F'(x, 0).$$

On the other hand, $r = \mathrm{ord}_x\, x^r$. Thus

$$r + \mathrm{ord}_x F(x, 0) = \mathrm{ord}_x (x^r F'(x, 0)) = \mathrm{ord}_x F(x, 0).$$

Even if there is only one point P_i', this can be done when $r_i' < r$. Otherwise, there is only one P_i', say P', and $r_i' = r$. In this case, take P' to the origin by a translation, and apply the local quadratic transformation. Perhaps r is reduced. If not, repeat the procedure. As in the proof of Theorem 4.46, this procedure ends in a finite number of steps. This completes the proof.

4.5 NOETHER'S THEOREM

Let $\mathcal{F} = \mathbf{v}(F(X,Y))$ and $\mathcal{G} = \mathbf{v}(G(X,Y))$ be two curves with no common components. By Bézout's Theorem 3.14, $\mathcal{F} \cap \mathcal{G} = \{P_1, \ldots, P_s\}$. Given a polynomial $H(X,Y)$, it is proposed to find conditions for it to belong to the ideal generated by F and G; that is, $H = AF + BG$ for some polynomials A and B. A necessary condition is that H vanishes at the points $P_1, \ldots, P_s$. This condition is also sufficient if the multiplicity of $\mathcal{F}$ and $\mathcal{G}$ at each point P_i is equal to 1.

The result is not proved directly, but follows from other results.

DEFINITION 4.51 For a point $P = (x_0, y_0)$, the *local ring $\mathcal{O}_P$ at the point P* is the ring of all rational functions defined at P; that is,

$$\mathcal{O}_P = \{A/d \mid A, d \in K[X,Y],\ d(P) \neq 0\}.$$

DEFINITION 4.52 Given any point P of the plane, let

$$\mathfrak{q}_P = \{H \in K[X,Y] \mid H = (AF + BG)/d;\ d(P) \neq 0;\ d, A, B \in K[X,Y]\}.$$

PROPOSITION 4.53 *The set $\mathfrak{q}_P$ is an ideal.*

Proof. It must be shown that

(i) if H_1 and H_2 are in $\mathfrak{q}_P$, then $(H_1 + H_2) \in \mathfrak{q}_P$;

(ii) if $H \in \mathfrak{q}_P$ and $D \in K[X,Y]$, then $HD \in \mathfrak{q}_P$.

Let $H_1 = (A_1F + B_1G)/d_1,\ H_2 = (A_2F + B_2G)/d_2$ with $d_1(P), d_2(P) \neq 0$. Then

$$H_1 + H_2 = \frac{(A_1F + B_1G)}{d_1} + \frac{(A_2F + B_2G)}{d_2} = \frac{CF + BG}{h}$$

with $C = d_2A_1 + d_1A_2,\ D = d_2B_1 + d_1B_2$ and $0 \neq h(P) = d_1(P)d_2(P)$. Hence $H_1 + H_2 \in \mathfrak{q}_P$. Also,

$$DH = D\left(\frac{AF + BG}{d}\right) = \frac{(DA)F + (DB)G}{d},$$

and so $DH \in \mathfrak{q}_P$. □

The following lemma is a consequence of Hilbert's Nullstellensatz; see Exercise 7 in Chapter 2.

LEMMA 4.54 *Let $\{d_i(X,Y)\}$ be a set of polynomials with no common zeros. Then the ideal generated by the $d_i(X,Y)$ is (1); that is, $1 = \sum A_i d_i$ for some polynomials A_i.*

DEFINITION 4.55 The polynomial H satisfies *Noether's conditions* for F, G and P if $H \in \mathfrak{q}_P$.

THEOREM 4.56 *If H satisfies Noether's conditions for F, G and for every point P, then $H \in (F, G)$.*

Proof. For every point P, there is a polynomial d_P such that $d_P(P) \neq 0$ and $d_P \cdot H \in (F, G)$. From the previous lemma, it follows that the ideal generated by the d_P is equal to (1), whence $1 = \sum A_P d_P$.

Since (F, G) is an ideal and $d_P \cdot H \in (F, G)$, so also $A_P \cdot (d_P H) \in (F, G)$ and hence $H = 1H = (\sum A_P d_P)H \in (F, G)$. □

THEOREM 4.57 *The ideal $\mathfrak{q}_P$ is (1) except for $P = P_1, \ldots, P_s$.*

Proof. Let $P \neq P_i$ for $i = 1, \ldots, s$. Then one of the polynomials F, G, say F, is not zero at P; that is, $F(P) \neq 0$. Since $1 = F/F$, so $1 \in \mathfrak{q}_P$, where $A = 1$ and $B = 0$ in $(AF + BG)/F$. But, if $1 \in \mathfrak{q}_P$, then $\mathfrak{q}_P = (1)$.

Now let $P = P_i$ and suppose that $1 \in \mathfrak{q}_P = \mathfrak{q}_{P_i}$. By the definition of $\mathfrak{q}_{P_i}$,

$$1 = (AF + BG)/d$$

with $d(P_i) \neq 0$. From this, $d = AF + BG$. However, as F and G pass through P_i, so $d(P_i) = 0$, a contradiction. So $\mathfrak{q}_{P_i} \neq (1)$. □

COROLLARY 4.58 (Lasker–Noether) *The ideal generated by F and G can be decomposed as follows*:

$$(F, G) = \mathfrak{q}_{P_1} \cap \cdots \cap \mathfrak{q}_{P_s}.$$

Proof. If $H \in (F, G)$ then $H = AF + BG = (AF + BG)/1$. Take $d = 1$; then $d(P_i) = 1 \neq 0$ for all P_i. So $H \in \mathfrak{q}_{P_i}$ for each i; that is, $H \in \bigcap_{i=1}^{s} \mathfrak{q}_{P_i}$. Hence

$$(F, G) \subset \mathfrak{q}_{P_1} \cap \cdots \cap \mathfrak{q}_{P_s}.$$

Conversely, if $H \in \mathfrak{q}_{P_1} \cap \cdots \cap \mathfrak{q}_{P_s}$, then there exists a polynomial d such that $d(P_i) \neq 0$ for each $i = 1, \ldots, s$ and $H = (AF + BG)/d$. So H satisfies Noether's conditions by Theorem 4.56, giving that $H \in (F, G)$; that is,

$$\mathfrak{q}_{P_1} \cap \cdots \cap \mathfrak{q}_{P_s} \subset (F, G).$$

From the two inclusions, it follows that $(F, G) = \bigcap_{i=1}^{s} \mathfrak{q}_{P_i}$ □

COROLLARY 4.59 *If $P_1 = (0, 0)$, then all powers of sufficiently high degree of X, Y and the ideal (X, Y) are in $\mathfrak{q}_{P_1}$.*

Proof. Write F and G in descending powers of Y, and calculate the resultant with respect to Y, that is, taking X as parameter. This gives a polynomial in X, namely

$$R_Y(F, G) = AF + BG = X^\rho(c_0 + c_1 X + \cdots + c_m X^m),$$

with $c_i \in K$, $c_0 \neq 0$ and $A, B \in K[X, Y]$. Since $P_1 = (0, 0)$ belongs to both F and G, so $\rho > 0$. Then

$$X^\rho = (AF + BG)/(c_0 + c_1 X + \cdots + c_m X^m) \in \mathfrak{q}_{P_1};$$

so also all powers greater than ρ belong to $\mathfrak{q}_{P_1}$.

If, instead of considering X as unknown and Y as parameter, it is the other way round, then there exists $\sigma > 0$ such that $Y^\sigma \in \mathfrak{q}_{P_1}$. Let $\tau = \rho + \sigma - 1$. It is shown that $(X,Y)^\tau \subset \mathfrak{q}_{P_1}$.

Now $(X,Y)^\tau = (X,Y)\cdots(X,Y)$; if $u \in (X,Y)^\tau$ then $u = u_1 \cdots u_\tau$ with $u_i = A_iX + B_iY \in (X,Y)$ for $i = 1,\ldots,\tau$. So

$$\begin{aligned} u &= (A_1X + B_1Y)\cdots(A_\tau X + B_\tau Y) \\ &= C_\tau X^\tau + X^{\tau-1}YC_{\tau-1} + \cdots + C_1XY^{\tau-1} + C_0Y^\tau. \end{aligned}$$

It remains to see that $X^iY^j \in \mathfrak{q}_{P_1}$. Then, since $\mathfrak{q}_{P_1}$ is an ideal, it follows that $u \in \mathfrak{q}_{P_1}$ and so $(X,Y)^\tau \subset \mathfrak{q}_{P_1}$.

Since, in the powers of the product X^iY^j with $i+j=\tau$, either $i \geq \rho$ or $j \geq \sigma$, since, if $i \leq (\rho - 1)$ and $j \leq (\sigma - 1)$, then $\tau = i + j \leq \rho + \sigma - 2$; but this is impossible by the definition of τ. Hence $X^iY^j \in \mathfrak{q}_{P_1}$. In fact, if $i \geq \rho$, then it has been shown above that $X^i \in \mathfrak{q}_{P_1}$. Since $\mathfrak{q}_{P_1}$ is an ideal, so $X^iY^j \in \mathfrak{q}_{P_1}$. □

THEOREM 4.60 *Let $P_1 = (0,0)$. Then H is in the ideal $\mathfrak{q}_{P_1}$ if and only if there exist $A, B \in K[[X,Y]]$ such that $H = AF + BG$.*

Proof. Let $H \in \mathfrak{q}_{P_1}$. Then $H = (A_1F + B_1G)/d$ with $A_1,\ B_1,\ d \in K[X,Y]$ and

$$d = d_{0,0} + d_{1,0}X + d_{0,1}Y + \cdots,$$

with $d_{0,0} \neq 0$ because $d(P_1) \neq 0$. Hence d is a unit in $K[[X,Y]]$ and also $A = A_1d^{-1}$, $B = B_1d^{-1}$ are the required elements in $K[[X,Y]]$.

Conversely, suppose that $H = AF + BG$ with $A,\ B \in K[[X,Y]]$. Now write $A = A_r + A'_r$ and $B = B_r + B'_r$, where A_r and B_r are the sums of the terms of degree less than or equal to r in A and B. So

$$H = A_rF + B_rG + A'_rF + B'_rG = A_rF + B_rG + T_{r+1},$$

where T_{r+1} contains only terms of degree greater than r. Since H and $A_rF + B_rG$ are polynomials, their difference T_{r+1} is also a polynomial. By Corollary 4.59, $T_{r+1} \in \mathfrak{q}_{P_1}$ for r sufficiently large. Further, $A_rF + B_rG \in \mathfrak{q}_{P_1}$, by taking $d = 1$, and so $H \in \mathfrak{q}_{P_1}$. □

Consider the ring $K[[X]][Y]$, and let $G,\ H,\ f$ be three of its elements, where

$$f = \prod_{i=1}^{r}(Y - m_iX - \cdots) \quad \text{with } m_i \neq m_j \text{ for } i \neq j.$$

THEOREM 4.61 *If, for $i = 1,\ldots,r$,*

$$\operatorname{ord}_X H(X, m_iX + \cdots) \geq \operatorname{ord}_X G(X, m_iX + \cdots) + r - 1,$$

then

$$H = Af + BG \qquad \text{with } A,\ B \in K[[X]][Y].$$

Also, B can be chosen so that $\operatorname{ord}_X B \geq r - 1$.

Proof. The theorem is proved by induction on r. For $r = 1$,

$$\operatorname{ord}_X H(X, m_iX + \cdots) \geq \operatorname{ord}_X G(X, m_iX + \cdots).$$

Then

$$H(X, m_1X + \cdots) = B(X)G(X, m_1X + \cdots);$$

so $Y = m_1X + \cdots$ is a root of the polynomial $H(X,Y) - B(X)G(X,Y)$, whence

$$H(X,Y) - B(X)G(X,Y) = A(X,Y)(Y - m_1X - \cdots).$$

So the case $r = 1$ has been proved.

Suppose the theorem is true in the following case:

$$f_1 = \prod_{i=2}^{r} (Y - m_iX - \cdots), \quad G_1 = XG.$$

By the inductive hypothesis, $H = A_1f_1 + B_1XG$ with $\text{ord}_X B_1 \geq r - 2$. Since $m_i \neq m_j$ for $i \neq j$,

$$\begin{aligned} &\text{ord}_X A_1(X, m_1X + \cdots) + r - 1 \\ &= \text{ord}_X (H(X, m_1X + \cdots) - B_1(X, m_1X + \cdots)XG(X, m_1X + \cdots)) \\ &\geq \text{ord}_X G(X, m_1X + \cdots) + r - 1. \end{aligned}$$

Here, use has been made of the following two inequalities:

$$\begin{aligned} &\text{ord}_X H(X, m_1X + \cdots) \geq \text{ord}_X G(X, m_1X + \cdots) + r - 1; \\ &\text{ord}_X (B_1(X, m_1X + \cdots)XG(X, m_1X + \cdots)) \\ &\qquad\qquad \geq \text{ord}_X G(X, m_1X + \cdots) + r - 1. \end{aligned}$$

It has therefore been shown that

$$\text{ord}_X A_1(X, m_1X + \cdots) \geq \text{ord}_X G(X, m_1X + \cdots);$$

so

$$A_1(X,Y) = A_2(X,Y)(Y - m_1X - \cdots) + B_2(X,Y)G(X,Y).$$

Substituting the expression for A_1 in $H = A_1f_1 + B_1XG$ gives that

$$\begin{aligned} H(X,Y) &= [A_3(Y - m_1X - \cdots) + B_3G(X,Y)]f_1 + B_1XG(X,Y) \\ &= A_4f + [A_5f_1 + B_1X]G(X,Y) \\ &= Af + BG, \end{aligned}$$

where $A_3, A_4, A_5, B_1, B_3 \in K[X,Y]$ and

$$\text{ord}_X B = \text{ord}_X B_1 + \text{ord}_X X \geq r - 2 + 1 = r - 1. \qquad \square$$

Let $\mathcal{F} = \mathbf{v}(F(X,Y))$ be a plane curve for which $O = (0,0)$ is an ordinary singular point with multiplicity r. Suppose that $\mathbf{v}(X)$ is not a tangent to $\mathcal{F}$ at O. Then

$$F(X,Y) = \prod_{i=1}^{r} (Y - m_iX) + \cdots,$$

with $m_i \neq m_j$ for $i \neq j$.

From the previous chapter, the curve $\mathcal{F}$ has r branches at O that are of the following type:

$$x = t, \quad y = m_it + \cdots.$$

Then, by the definition of a branch, the point $(X, m_iX + \cdots)$ is a root of $F(X,Y)$; that is, $F(X, m_iX + \cdots) = 0$. Hence $F(X,Y)$ is divisible by $(Y - m_iX - \cdots)$ in $K[[X]][Y]$ and, if $m_i \neq m_j$ for $i \neq j$, then $(Y - m_iX - \cdots)$ and $(Y - m_jX - \cdots)$ are not associated. This proves the following result.

COROLLARY 4.62 *The polynomial*

$$F(X,Y) = \prod_{i=1}^{r} (Y - m_i X - \cdots) \cdot g(X,Y),$$

with $g \in K[[X]][Y]$ *and* $g(0,0) \neq 0$, *where* $\operatorname{ord}_X F = r$.

COROLLARY 4.63 *Let* $\mathcal{F} = \mathbf{v}(F(X,Y))$ *be a plane curve for which* P_1 *is an ordinary singular point with multiplicity* r, *and let* $\gamma_1, \ldots, \gamma_r$ *be the branches of* $\mathcal{F}$ *centred at* P_1. *If* $\mathcal{G} = \mathbf{v}(G(X,Y))$ *and* $\mathcal{H} = \mathbf{v}(H(X,Y))$ *are two curves such that*

$$I(P_1, \mathcal{H} \cap \gamma_j) \geq I(P_1, \mathcal{G} \cap \gamma_j) + r - 1 \quad \text{for } j = 1, \ldots, r,$$

then $H \in \mathfrak{q}_{P_1}$.

Proof. By the previous corollary,

$$F(X,Y) = \prod_{i=1}^{r} (Y - m_i X - \cdots) g(X,Y)$$

with $g(0,0) \neq 0$. With $\prod_{i=1}^{r}(Y - m_i X - \cdots) = f$, then $F = fg$. Since g is a unit, $f = g^{-1}F$. Let $\gamma_j = (x, y)$, with

$$\begin{aligned} x &= t, \\ y &= m_i t + \cdots, \end{aligned}$$

be a branch of $\mathcal{F}$. Then

$$I(P_1, \mathcal{G} \cap \gamma_j) = \operatorname{ord}_t G(t, m_j t + \cdots) = \operatorname{ord}_X G(X, m_j X + \cdots);$$

analogously,

$$I(P_1, \mathcal{H} \cap \gamma_j) = \operatorname{ord}_t H(t, m_j t + \cdots) = \operatorname{ord}_X H(X, m_j X + \cdots).$$

Since $\operatorname{ord}_X H \geq \operatorname{ord}_X G + r - 1$ by Theorem 4.61,

$$H = Af + BG \quad \text{with } A, B \in K[[X,Y]].$$

Since $f = g^{-1}F$, it follows that $H = (Ag^{-1})F + BG$ and, by Theorem 4.60, $H \in \mathfrak{q}_{P_1}$. □

Let $\{\gamma\}$ be the set of all branches of an irreducible curve $\mathcal{F}$.

DEFINITION 4.64 A formal sum $\sum n_\gamma \gamma$, where each n_γ is an integer and $n_\gamma \neq 0$ only for a finite number of γ, is a *divisor* of $\mathcal{F}$.

The set of divisors is an abelian group under addition:

$$\sum n_\gamma \gamma + \sum m_\gamma \gamma = \sum (n_\gamma + m_\gamma)\gamma.$$

Its identity is the *zero divisor*, that is, the divisor for which every n_γ is zero.

DEFINITION 4.65 (i) A divisor $\sum n_\gamma \gamma$ is *effective* if $n_\gamma \geq 0$ for each branch γ; write $\sum n_\gamma \gamma \succ 0$.

(ii) A non-effective divisor is *virtual*.

(iii) Given two divisors D and D', define $D \succ D'$ if $D - D' \succ 0$.

Let $\mathcal{F}$ be a plane curve with only ordinary singularities $P_1, \ldots, P_s$ having respective multiplicities $r_1, \ldots, r_s$. Also, let $\gamma_{i,j}$, for $j = 1, \ldots, r_i$, be the branches of $\mathcal{F}$ centred at P_i. Then the divisor

$$D = \sum_{i,j} (r_i - 1)\gamma_{i,j}$$

is the *double-point divisor* of $\mathcal{F}$.

For two curves $\mathcal{F}$ and $\mathcal{G}$, denote by $\mathcal{G} \cdot \mathcal{F}$ the divisor

$$\mathcal{G} \cdot \mathcal{F} = \sum_{\gamma} I(P, \mathcal{G} \cap \gamma)\gamma,$$

where γ varies over all branches of $\mathcal{F}$. Then $\mathcal{G} \cdot \mathcal{F}$ represents the divisor cut out on $\mathcal{F}$ by $\mathcal{G}$.

To formulate Noether's Theorem, let $\mathcal{F} = \mathbf{v}(F(X,Y))$ be a plane curve with only ordinary singularities and suppose that $\mathcal{G} = \mathbf{v}(G(X,Y))$, $\mathcal{H} = \mathbf{v}(H(X,Y))$ are two other curves, where $\mathcal{F}$ and $\mathcal{G}$ have no common components.

THEOREM 4.66 (Noether) *If $\mathcal{H} \cdot \mathcal{F} \succ D + \mathcal{G} \cdot \mathcal{F}$, then*

$$H = AF + BG,$$

with $A,\ B \in K[X,Y]$.

Proof. Let γ be a branch of $\mathcal{F}$ with centre a point P of multiplicity $r \geq 1$ for $\mathcal{F}$. By hypothesis,

$$I(P, \mathcal{H} \cap \gamma) \geq I(P, \mathcal{G} \cap \gamma) + r - 1.$$

By Corollary 4.63, $H \in \mathfrak{q}_P$ and so H satisfies Noether's conditions. Hence, by Theorem 4.56, H belongs to the ideal (F, G). □

Now the projective formulation of this theorem is given. Let

$$\mathcal{F} = \mathbf{v}(F^*(X_0, X_1, X_2), \ \mathcal{G} = \mathbf{v}(G^*(X_0, X_1, X_2)), \ \mathcal{H} = \mathbf{v}(H^*(X_0, X_1, X_2))$$

be the corresponding projective curves.

THEOREM 4.67 (Noether) *If $\mathcal{H} \cdot \mathcal{F} \succ D + \mathcal{G} \cdot \mathcal{F}$, then*

$$H^* = A^*F^* + B^*G^*,$$

with $A^*,\ B^*$ homogeneous polynomials in $K[X_0, X_1, X_2]$.

Proof. Applying a projectivity if necessary, assume that $\mathcal{G}$ and $\mathcal{F}$ do not meet on $\mathbf{v}(X_0)$ and that $\mathbf{v}(X_0)$ is a component of neither $\mathcal{G}$ nor $\mathcal{F}$; therefore

$$G^*(0, X_1, X_2) \neq 0, \quad F^*(0, X_1, X_2) \neq 0.$$

Consider the affine polynomials of the curves:

$$\begin{aligned} F(X,Y) &= f_{0,0} + f_{1,0}X + f_{0,1}Y + \cdots + f_{n,0}X^n + f_{0,n}Y^n, \quad \deg F = n; \\ G(X,Y) &= g_{0,0} + g_{1,0}X + f_{0,1}Y + \cdots + g_{m,0}X^m + g_{0,m}Y^m, \quad \deg G = m; \\ H(X,Y) &= h_{0,0} + h_{1,0}X + h_{0,1}Y + \cdots + h_{s,0}X^s + h_{0,s}Y^s, \quad \deg H = s. \end{aligned}$$

By the preceding theorem,

$$H^*(1, X, Y) = A(X,Y)F^*(1, X, Y) + B(X,Y)G^*(1, X, Y), \qquad (4.25)$$

with $\deg A = n_0$, $\deg B = m_0$. Putting $X = X_1/X_0$, $Y = X_2/X_0$ in (4.25) gives

$$\frac{1}{X_0^s} H^*(X_0, X_1, X_2) = \frac{1}{X_0^{n_0+n}} A^*(X_0, X_1, X_2) F^*(X_0, X_1, X_2) + \frac{1}{X_0^{m_0+m}} B^*(X_0, X_1, X_2) G^*(X_0, X_1, X_2)$$

with A^* and B^* homogeneous. Put $t = \max\{s, n_0 + n, m_0 + m\}$; then

$$X_0^{t-s} H^* = X_0^{t-(n_0+n)} A^* F^* + X_0^{t-(m_0+m)} B^* G^*,$$

which can be re-written as follows:

$$X_0^\rho H^*(X_0, X_1, X_2) = A_1^*(X_0, X_1, X_2) F^*(X_0, X_1, X_2) + B_1^*(X_0, X_1, X_2) G^*(X_0, X_1, X_2) \qquad (4.26)$$

with A_1^* and B_1^* homogeneous. Suppose that $\rho > 0$, and put $X_0 = 0$:

$$A_1^*(0, X_1, X_2) F^*(0, X_1, X_2) + B_1^*(0, X_1, X_2) G^*(0, X_1, X_2) = 0.$$

Noting that $F^*(0, X_1, X_2)$ is a homogeneous polynomial in the indeterminates X_1, X_2, it may be factorised as

$$F^*(0, X_1, X_2) = \prod_{i=1}^n (\alpha_i X_1 - \beta_i X_2)$$

with $(0, \alpha_i, \beta_i) \in \mathcal{F}$. Analogously,

$$G^*(0, X_1, X_2) = \prod_{j=1}^m (\gamma_j X_1 - \delta_j X_2)$$

with $(0, \gamma_j, \mathcal{G}_j) \in \mathcal{G}$.

Since the curves $\mathcal{F}$ and $\mathcal{G}$ have no common component, so points on $\mathbf{v}(X_0)$, namely of type $(0, \beta_i, \alpha_i)$, are in $\mathbf{v}(B_1^*)$. Hence the terms of the product

$$\prod_{i=1}^n (\alpha_i X_1 - \beta_i X_2)$$

divide B_1^*, whence

$$A_1^*(0, X_1, X_2) = -\frac{B_1^*(0, X_1, X_2)}{F^*(0, X_1, X_2)} G^*(0, X_1, X_2) = A_2^*(X_1, X_2) G^*(0, X_1, X_2).$$

Therefore

$$A_1^*(X_0, X_1, X_2) = A_2^*(X_1, X_2) G^*(X_0, X_1, X_2) + X_0 A_3^*(X_0, X_1, X_2)$$

with A_2^*, A_3^* homogeneous. Substituting this expression for A_1^* in (4.26) gives

$$X_0^\rho H^*(X_0, X_1, X_2) = X_0 A_3^*(X_0, X_1, X_2) F^*(X_0, X_1, X_2) + B_2^*(X_0, X_1, X_2) G^*(X_0, X_1, X_2).$$

Now put $X_0 = 0$; hence $B_2^*(0, X_1, X_2) = 0$ and so

$$B_2^*(X_0, X_1, X_2) = X_0 B_3^*(X_0, X_1, X_2).$$

Eliminating X_0 gives

$$X_0^{\rho-1} H^*(X_0, X_1, X_2) = A_3^*(X_0, X_1, X_2) F^*(X_0, X_1, X_2) + B_3^*(X_0, X_1, X_2) G^*(X_0, X_1, X_2).$$

Repeating this procedure reduces ρ to 0, which proves the theorem. □

4.6 ANALYTIC BRANCHES

In this section, another definition of the intersection multiplicity of two algebraic curves is given, and it is shown that the new definition coincides with the previous one.

For a curve $\mathcal{F} = \mathbf{v}(F)$ with $F(X,Y) \in K[X,Y]$, the idea of a branch was introduced. Now this idea is extended to the case that $F(X,Y)$ is a formal power series; that is, $F(X,Y) \in K[[X,Y]]$.

DEFINITION 4.68 An *analytic branch centred at* O is a branch centred at O in the sense of Definition 4.29.

The branch *belongs* to $F(X,Y) \in K[[X,Y]]$ if, for any representation $(x(t), y(t))$, then $F(x(t), y(t)) = 0$.

It is shown below that there is a unique branch that belongs to a given irreducible power series other than 1 and that, given $F_1, F_2 \in K[[X,Y]]$, both irreducible and different from 1, if a branch belongs to both of them, then they are associates. So it is possible to introduce the idea of analytic branch as an irreducible formal power series different from 1 or, more precisely, as an equivalence class of irreducible power series other than 1 with respect to the equivalence relation: $F_1 \sim F_2$ if $F_1 = EF_2$ with E a unit.

Given $F(X,Y) \in K[[X,Y]]$ with $F = F_r + F_{r+1} + \cdots$, perform a transformation of the type

$$X = aX' + bY', \; Y = cX' + dY',$$

with $ad - bc \neq 0$, in such a way that F_r is not divisible by X.

Now, since F_r is not divisible by X, then

$$F(0,Y) = a_r Y^r + a_{r+1} Y^{r+1} + \cdots, \qquad \text{with } a_r \neq 0.$$

The Weierstrass Preparation Theorem 4.10 applied to $F(X,Y)$ determines a polynomial in Y with coefficients in $K[[X]]$ which is an associate of $F(X,Y)$; that is, there exists a unit $E(X,Y) \in K[[X,Y]]$ such that

$$F(X,Y) = E(X,Y)(Y^r + A_1(X)Y^{r-1} + \cdots + A_r(X)),$$

with $A_i(X) \in K[[X]]$ and $A_i(0) = 0$ for $i = 1, \ldots, r$. This allows the substitution, for each formal power series $F(X,Y)$, of its associated polynomial in $K[[X]][Y]$.

DEFINITION 4.69 (i) An *analytic curve* is an element $F(X,Y) \in K[[X,Y]]$ that is not a unit and that has no repeated factors.

(ii) If $F(X,Y)$ is a formal power series with $F \neq 0$, then F is an *analytic cycle* and $F(X,Y) = 0$ is its equation.

DEFINITION 4.70 An analytic cycle F is *centred at* $O = (0,0)$ if $F(0,0) = 0$. More generally, $F(X,Y)$ *is centred at* $(0,m)$ if $F(0,m) = 0$.

DEFINITION 4.71 The centre $(0,0)$ is *r-ple* for $F(X,Y)$ if

$$F = F_r + F_{r+1} + \cdots, \; F_r \neq 0,$$

and is *simple* if $r = 1$.

The factors that appear in F_r represent the tangents of $F(X, Y)$ at its centre.

REMARK 4.72 Both F and EF, where E is a unit, refer to the same analytic curve.

THEOREM 4.73 *Let* $F(X, Y) \in K[[X, Y]]$ *with* $\operatorname{ord} F = 1$. *Then there is one and only one branch* $(x(t), y(t))$ *such that* $F(x(t), y(t)) = 0$.

Proof. The proof is similar to that of Theorem 4.32, considering a formal power series instead of a polynomial. The existence of the branch is obtained as in Theorem 4.6 and, if $F = Y - c_1X - \cdots$, then the branch is $(t, c_1t + \cdots)$. □

DEFINITION 4.74 If $(x(t), y(t))$ is a primitive representation of the branch γ and $\mathcal{G}$ is an analytic cycle with equation $G(X, Y) = 0$, then put

$$I(P, \mathcal{G} \cap \gamma) = \operatorname{ord}_t G(x(t), y(t)).$$

The idea of multiplicity of a branch at its centre is defined as in the case of a branch of a plane curve.

THEOREM 4.75 *Two distinct irreducible analytic curves* $\mathcal{F} = \mathbf{v}(F(X, Y))$ *and* $\mathcal{G} = \mathbf{v}(G(X, Y))$ *cannot have a branch in common.*

Proof. The transformation

$$X = aX' + bY', \quad Y = cX' + dY',$$

with $ad - bc \neq 0$, can make F and G regular in Y and substitute the corresponding special polynomials F_1 and G_1 obtained from the Weierstrass Preparation Theorem 4.10.

The special polynomials are irreducible in $K[[X]][Y]$ and also in $K((X))[Y]$. By hypothesis, F and G are not associates in $K[[X, Y]]$; so F_1 and G_1 are also associates neither in $K[[X]][Y]$ nor in $K((X))[Y]$. Therefore their greatest common divisor in $K((X))[Y]$ is 1; so $1 = aF_1 + bG_1$ with $a,\ b \in K((X))[Y]$.

Note that

$$a(Y) = a_n(X)Y^n + \cdots + a_i(X)Y^i + \cdots + a_0(X)$$

with $a_i(X) \in K((X))$; that is,

$$a_i(X) = \frac{s_i(X)}{t_i(X)} = X^{k_i}E_i(X),$$

where $E_i(X)$ is a unit for $i = 0, \ldots, n$. Also,

$$b(Y) = b_m(X)Y^m + \cdots + b_j(X)Y^j + \cdots + b_0(X),$$

with $b_j(X) \in K((X))$; that is,

$$b_j(X) = \frac{r_j(X)}{u_j(X)} = X^{h_j}H_j(X),$$

with $H_j(X)$ a unit for $j = 0, \ldots, m$.

It may happen that some k_i and h_j are negative; write $\rho = \max\{|k_i|,\ |h_j|\}$ for $i = 0, \ldots, n$ and $j = 0, \ldots, m$. Hence multiplying by X^ρ gives

$$X^\rho = AF_1 + BG_1, \tag{4.27}$$

with A, $B \in K[[X]][Y]$.

Suppose that there is a branch $(x(t), y(t))$ common to the two analytic curves. Substituting in (4.27) gives that $x^\rho = 0$, whence $x = 0$. Hence F_1 is special; that is, $F_1 = Y^r + c_1(X)Y^{r-1} + \cdots + c_r(X)$ and $(x(t), y(t))$ is a branch of F_1, as it is an associate of F. Therefore $0 = F_1(0, y(t)) = y(t)$, whence $y = 0$. This contradicts the definition of a branch; so they cannot have branches in common. □

COROLLARY 4.76 *Let $(\xi(t), \eta(t))$ be a branch with $\xi(t) = 0$; then it is a branch of the line $\mathbf{v}(X)$ and of no other irreducible analytic curve.*

Proof. The branch considered satisfies the equation $X = 0$ and no other irreducible analytic curve can have such a branch. □

THEOREM 4.77 *Let $\mathcal{F}$ be an analytic cycle centred at $O = (0,0)$ not containing $\mathbf{v}(X)$ as a component. Then $\mathcal{F}$ has a unique special equation $F = 0$, where F is reducible in $K[[X]][Y]$ if and only if it is reducible in $K[[X, Y]]$.*

Proof. Suppose that F is regular in Y; otherwise, it can be made so by an affine transformation. Then, by the Weierstrass Preparation Theorem 4.10, there exist unique G and H such that $F = GH$ with H a special polynomial which is an associate of F. By Theorem 4.15 the second part follows. □

DEFINITION 4.78 An *analytic cycle along* $\mathbf{v}(X)$ is a finite number of irreducible analytic cycles centred on $\mathbf{v}(X)$ together with their associated multiplicity. The notation is $m_1\mathcal{F}_1 + \cdots + m_s\mathcal{F}_s$.

If $F \in K[[X]][Y]$ is monic and irreducible, Hensel's Lemma 4.9 shows that the polynomial $F(0, Y)$ has only one root λ and F defines an irreducible analytic cycle centred at $(0, \lambda)$ and different from $\mathbf{v}(X)$. If $\lambda = 0$, then F is special; in general, F is special relative to $Y_1 = Y - \lambda$.

Conversely, every irreducible analytic cycle centred on $\mathbf{v}(X)$, but not this line, is represented by a unique monic irreducible element $F \in K[[X]][Y]$.

If $\mathcal{F}_1, \ldots, \mathcal{F}_s$ are such cycles with monic irreducible series $F_1, \ldots, F_s$, then by definition $F_1^{m_1} \cdots F_s^{m_s}$ is the monic series $m_1\mathcal{F}_1 + \cdots + m_s\mathcal{F}_s$. Conversely every monic element $F \in K[[X]][Y]$ has a unique factorisation into monic irreducible factors: $F = F_1^{m_1} \cdots F_s^{m_s}$. It defines the cycle $m_1\mathcal{F}_1 + \cdots + m_s\mathcal{F}_s$.

Also, in the theory of analytic branches, it is useful to consider the local quadratic transformation

$$X' = X, \quad Y' = Y/X.$$

Let $\mathcal{F} = \mathbf{v}(F(X, Y))$ be an analytic cycle centred at $O = (0, 0)$, and define the transform of $\mathcal{F}$ only in the case that $\mathbf{v}(X)$ is not tangent to $\mathcal{F}$. Defining $\mathcal{F}'$ as the special series of $\mathcal{F}$, it may be assumed directly that $\mathcal{F}$ is special.

DEFINITION 4.79 The curve given by $\mathcal{F}' = \mathbf{v}(F(X, XY')/X^r)$, where $(0, 0)$ is r-ple for $\mathcal{F}$, is the *transform* $\mathcal{F}'$ of $\mathcal{F}$.

Theorems 4.42 to 4.45 that connect $\mathcal{F}$ and $\mathcal{F}'$ in the context of branches continue to hold for analytic curves $\mathcal{F}$ centred at $(0, 0)$ and that do not have the line $\mathbf{v}(X)$

as a tangent. The proofs are the same, apart from taking the special series $F(X,Y)$ of $\mathcal{F}$.

Corresponding to Theorem 4.46 is the following result.

THEOREM 4.80 *Let $\mathcal{F} = \mathbf{v}(F(X,Y))$ be an analytic curve centred at $O = (0,0)$. Then there is at least one and at most a finite number of branches $(x(t), y(t))$ that satisfy F.*

Proof. The proof is similar to that of Theorem 4.46 with some small changes. Let $F = F_r(X,Y) + F_{r+1}(X,Y) + \cdots$; then, by a transformation

$$X' = aX + bY, \quad Y' = cX + dY,$$

it can be ensured that F_r is not divisible by X. Applying Weierstrass Preparation Theorem 4.10 again gives the series

$$Y^r + a_1(X)Y^{r-1} + \cdots + a_r(X),$$

which is the special series of $\mathcal{F}$ associated to $F(X,Y)$. Suppose further that $\mathcal{F}$ is irreducible; then the transform $\mathcal{F}'$ of $\mathcal{F}$ having polynomial

$$F'(X,Y') = F(X,XY')/X^r$$

is still irreducible. By Hensel's Lemma 4.9, $F'(0,Y')$ has only one root λ. By a translation $Y'' = Y' - \lambda$, it may be assumed that $\lambda = 0$; then

$$F(X,Y') = Y'^r + b_1(X)Y'^{r-1} + \cdots.$$

If r does not decrease, then F' is still special and the axis $\mathbf{v}(X)$ is not tangent to $\mathcal{F}'$. The difference from Theorem 4.46 is the need to make transformations apart from translations on the analytic curves before applying the local quadratic transformations. □

It remains to prove, as in Theorem 4.46, that r decreases. This then gives the following result.

COROLLARY 4.81 *The sequence of quadratic transformations and translations that occur in the preceding proof cannot produce a sequence of r-ple points for $r > 1$.*

Proof. Let $\mathcal{F}$ be irreducible with $F(X,Y) = 0$ as its special equation, and assume that $\mathcal{F}$ is not tangent to the line $\mathbf{v}(X)$. As in the proof of Theorem 4.47, if the sequence of quadratic transformations does not produce a reduction in r, then

$$F(X,Y) = (Y - m_1X - m_2X^2 - \cdots)G(X,Y).$$

Since F is irreducible and monic, it follows that

$$F(X,Y) = Y - m_1X - m_2X^2 - \cdots,$$

whence $r = 1$. □

THEOREM 4.82 *An irreducible analytic cycle $\mathcal{F}$ has only one branch.*

Proof. The proof is by induction on r. Let $O = (0,0)$ be r-ple for $\mathcal{F}$. If $r = 1$, then, by Theorem 4.73, there is one and only one branch of $\mathcal{F}$.

Suppose therefore that $r > 1$ and ensure by a suitable transformation that $\mathbf{v}(X)$ is not tangent to $\mathcal{F}$. Consider the special equation $F(X,Y) = 0$ and apply the local quadratic transformation. If there is a reduction in r, the result is obtained by the inductive hypothesis. If r does not reduce, first make a translation and then another quadratic transformation; by Corollary 4.81, r reduces and the result follows. □

Let $F(X,Y) = Y^e + a_1(X)Y^{e-1} + \cdots + a_e(X) \in K[[X]][Y]$ be monic and irreducible. By Hensel's Lemma 4.9, $F(0,Y)$ has only one root λ, which may be taken as $\lambda = 0$. Let

$$x = ct^d + \cdots, \; y = y(t), \; c \neq 0,$$

be a branch of $\mathbf{v}(F(X,Y))$. It is shown that $d = e$.

LEMMA 4.83 *If $(x(t), y(t))$ is a primitive branch representation, then the field $K(x(t), y(t))$ contains an element of order* 1.

Proof. The proof is given in Theorem 4.21. □

DEFINITION 4.84 An element of order 1 is a *uniformising parameter*.

LEMMA 4.85 *Let $x = ct^d + \cdots$, $c \neq 0$. Then*

$$K[[x]][t] = K[[t]] \quad \textit{and} \quad [K((x))(t) : K((x))] = d.$$

Proof. Let $f \in K[[x]][t]$. Then f is a polynomial in t with coefficients in $K[[x]]$; that is,

$$f = a_h(x)t^h + a_{h-1}(x)t^{h-1} + \cdots + a_0(x),$$

with $a_i(x) = a_{0,i} + a_{1,i}x + a_{2,i}x^2 + \cdots$ for $i = 0, \ldots, h$. Substituting in $a_i(x)$ the value of x, in f only powers of t are obtained. So $f \in K[[t]]$; this shows that $K[[x]][t] \subset K[[t]]$.

Conversely, consider

$$1, t, \ldots, t^{d-1}, x, tx, \ldots, t^{d-1}x, x^2, tx^2, \ldots;$$

they have degrees $0, 1, 2, \ldots$ in t and constitute a basis for $K[[t]]$.

Take an element $h \in K[[t]]$; it can be expressed as a linear combination of the basis elements

$$h = h_0 + h_1 t + h_2 t^2 + \cdots + h_{d-1}t^{d-1} + h_d x + \cdots.$$

So h can be written as a polynomial in t of degree $d-1$ with coefficients in $K[[x]]$. In fact,

$$h = h'_0 + h'_1 t + \cdots + h'_{d-1}t^{d-1},$$

with

$$\begin{aligned} h'_0 &= h_0 + h_d x + h_{2d}x^2 + \cdots + h_{id}x^i + \cdots, \\ h'_1 &= h_1 + h_{d+1}x + h_{2d+1}x^2 + \cdots + h_{id+1}x^i + \cdots, \\ &\vdots \quad \vdots \\ h'_j &= h_j + h_{d+j}x + h_{2d+j}x^2 + \cdots + h_{id+j}x^i + \cdots. \end{aligned}$$

Hence $h \in K[[x]][t]$ and so $K[[x]][t] = K[[t]]$.

For the second part of the theorem, consider an element $F \in K[[X,T]]$ with $F = X - cT^d - \cdots$ and $c \neq 0$. Note that F is irreducible with subdegree 1. Since $F(0,T) = -cT^d + \cdots$ with $c \neq 0$, the Weierstrass Preparation Theorem 4.10 shows the existence of a unit $E(X,T)$ such that

$$E(X,T)F(X,T) = T^d + a_1(X)T^{d-1} + \cdots .$$

The polynomial associated to F is irreducible in $K[[X]][T]$. As in the proof of Theorem 4.15, it is also irreducible in $K((X))[T]$. By the choice of F, the point $(x(t),t)$ with $F(x(t),t) = 0$ also satisfies the equation

$$T^d + a_1(X)T^{d-1} + \cdots = 0.$$

Hence t is a root of the polynomial

$$T^d + a_1(x(t))T^{d-1} + \cdots ,$$

which is a polynomial of degree d with coefficients in $K((x(t)))$. So it has been shown that $[K((x(t)))(t) : K((x(t)))] = d$. □

COROLLARY 4.86 *If $(x = ct^d + \cdots , y = et^f + \cdots)$ is a primitive branch representation centred at $(0,0)$ and $x \neq 0$, then y is algebraic over $K((x))$.*

Proof. As in Lemma 4.83, choose an element $\tau = at + \cdots , a \neq 0$ of order 1 in $K(x,y)$. Then $x = \bar{c}\tau^d + \cdots , y = \bar{e}\tau^f + \cdots$ and, further, $\tau = a(x,y)/b(x,y)$ with $a(X,Y), b(X,Y) \in K[X,Y]$. If $\tau \in K(x)$, then, *a fortiori*, $\tau \in K((x))$, whence $y \in K(x)$. If, instead, $\tau \notin K(x)$, write $\tau b(x,y) = a(x,y)$ and then re-order it in powers of y, giving

$$a_1(x,\tau)y^m + a_2(x,\tau)y^{m-1} + \cdots = 0,$$

with $m \geq 1$ and $a_i(x,\tau) \in K(x,\tau)$. Therefore y is algebraic over $K((x))(\tau)$. By Lemma 4.85, the result follows. □

THEOREM 4.87 *Given a monic, irreducible polynomial*

$$F = Y^e + a_1(X)Y^{e-1} + \cdots \in K[[X]][Y],$$

if $(x = ct^d + \cdots , y = y(t))$, with $c \neq 0$, is a branch representation of F, then $d = e$.

Proof. By Lemma 4.85,

$$y(t) \in K[[t]] = K[[x(t)]][t] \quad \text{and} \quad [K((x(t)))(t) : K((x(t)))] = d.$$

It follows that y is a root of a polynomial F of degree $\ell \leq d$ with coefficients in $K((x))$. Then F is a multiple of $Y^e + a_1(X)Y^{e-1} + \cdots$. This shows that $e \leq \ell \leq d$. To see that also $d \leq e$, choose an element τ of order 1 in $K(x,y)$. Then $\tau \in K((x))(y)$ and $[K((x))(y) : K((x))] = e$, from which it follows that τ is a root of a polynomial of degree at most e. But, then $d \leq e$, as required. □

THEOREM 4.88 *With $p = 0$, consider a monic, irreducible polynomial*

$$F = Y^d + a_1(X)Y^{d-1} + \cdots \in K[[X]][Y]$$

and one of its branches $\gamma = (x(t), y(t))$. Then $F(x(t), Y)$ splits completely into linear factors in $K((t))$.

Proof. Let $(x = ct^d + \cdots, y = y(t))$ with $c \neq 0$ be the branch representation γ. By applying Theorem 4.27 it is possible to transform such a pair into an equivalent branch representation $(x = t^d, y = \eta(t))$. Hence it may be assumed that $x = t^d$.

Let

$$y = \sum c_{i_j} t^{i_j} \quad \text{with } c_{i_j} \neq 0. \tag{4.28}$$

Write only the terms that have non-zero coefficients. Then, with

$$e = \gcd(d, \ldots, i_j, \ldots),$$

$e = 1$. For, if $e > 1$, then $(x(t), y(t))$ can be re-written in terms of t^e, but this contradicts the primitive character of $(x(t), y(t))$.

Before continuing with the proof, some algebraic arguments need to be made. Suppose that $K = \mathbf{C}$; then the d-th roots of unity are the d complex numbers

$$\zeta_{d,0}, \ldots, \zeta_{d,d-1},$$

where

$$\zeta_{d,k} = \cos(2k\pi/d) + i\sin(2k\pi/d),$$

$0 \leq k < d$.

A d-th root of unity ζ is *primitive* if $\zeta^k \neq 1$ for $1 \leq k \leq d-1$ and $\zeta^d = 1$; then $\cos(2\pi/d) + i\sin(2\pi/d)$ is primitive in $\mathbf{C}$. Similarly, every algebraically closed field of characteristic 0 has a primitive d-th root of 1.

Let ζ be a primitive d-th root of 1 and consider the d substitutions

$$t \mapsto \zeta^i t, \quad i = 0, \ldots, d-1. \tag{4.29}$$

Such a substitution sends t^d to t^d and $y(t)$ to $y_i(t)$ for $i = 0, \ldots, d-1$. Since such substitutions are K-automorphisms of $K[[t]]$, the $y_i(t)$ are roots of $F(t^d, Y)$. Hence $(x(t), y(t))$ is a branch of $\mathcal{F}$; that is, $F(x(t), y(t)) = 0$, whence $y(t)$ is a root of $F(x(t), Y) = 0$. By (4.29), $y(t) \mapsto y_i(t)$ by an automorphism; if $y(t)$ is a root, so also is $y_i(t)$.

If it is shown that the y_i are distinct, then they are the d roots of $F(t^d, Y)$ and the proof is complete. The expression for y_λ is

$$y_\lambda = \sum c_{i_j} \zeta^{\lambda i_j} t^{i_j}, \tag{4.30}$$

where the c_{i_j} are the same as in (4.28). If $y_\lambda = y_\mu$, then $\zeta^{\lambda i_j} = \zeta^{\mu i_j}$, whence $\zeta^{(\lambda-\mu)i_j} = 1$ and $(\lambda - \mu)i_j \equiv 0 \pmod{d}$. Put $e = \gcd(\ldots, i_j, \ldots)$; then

$$(\lambda - \mu)e \equiv 0 \pmod{d}.$$

Since $\gcd(d, e) = 1$, it follows that $\lambda - \mu \equiv 0 \pmod{d}$, so $\lambda = \mu$, as $0 \leq \lambda \leq d-1$ and $0 \leq \mu \leq d-1$. Since the y_i are distinct, the result is proved. □

There is a more general result valid in any characteristic.

THEOREM 4.89 *Let*

$$F = Y^d + a_1(X)Y^{d-1} + \cdots \in K[[X]][Y]$$

be a monic polynomial. Then there exists $x(v) \in K[[v]]$ such that $F(x(v), Y)$ splits completely into linear factors in $K[[v]][Y]$.

Proof. If $(x(t), y(t))$ is a branch representation of $\mathcal{F}$, then $y(t)$ is a root of the polynomial $F(x(t), Y) \in K[[t]][Y]$. Therefore $F(x(t), Y) = (Y - y(t))F_1(t, Y)$, with

$$F_1(t, Y) = Y^{d-1} + b_1(t)Y^{d-2} + \cdots \in K[[t]][Y].$$

If $d > 1$, there is a branch representation $(t(t_1), y_1(t_1))$ of $\mathcal{F}_1 = \mathbf{v}(F_1)$ and the argument can be repeated to obtain that

$$F_1(t_1, Y) = (Y - t(t_1))(Y^{d-2} + c_1(t_1)Y^{d-3} + \cdots) \in K[[t_1]][Y].$$

After this procedure has been repeated $d - 1$ times,

$$F(x(t), Y) = (Y - y(t))(Y - y_1(t_1)) \cdot (Y - y_{d-1}(t_{d-1})).$$

Put $v = t_{d-1}$, and express t_{d-2} in terms of v; successively, do this for each t_i with $i = d - 3, \ldots, 2, 1$, whence the result follows. □

THEOREM 4.90 *Given a monic, irreducible polynomial*

$$F = Y^d + a_1(X)Y^{d-1} + \cdots \in K[[X]][Y],$$

let $\mathcal{F}$ be an analytic cycle with special equation $F = 0$ centred at $O = (0, 0)$ and let $\gamma = (x(t), y(t))$ be one of its branches. If $\mathcal{G} = \mathbf{v}(G)$ is another analytic cycle also centred at O, then

$$I(P, \mathcal{G} \cap \gamma) = \operatorname{ord}_X R_Y(F(X, Y), G(X, Y));$$

that is,

$$\operatorname{ord}_t G(x(t), y(t)) = \operatorname{ord}_X R_Y(F(X, Y), G(X, Y)).$$

Proof. First, a proof is given in the case of characteristic zero; the proof in the general case is similar.

Recall that $R_Y(F, G)$ is a polynomial in X. By a preceding theorem,

$$F(x(t), Y) = \prod_{i=1}^{d} (Y - y_i(t))$$

and there are K-automorphisms of $K[[t]]$ that send x to x and y to y_i. Then

$$\operatorname{ord}_t G(x(t), y_i(t)) = \operatorname{ord}_t G(x(t), y(t)).$$

From Section 4.1,

$$R_Y(\prod_{i=1}^{d} (Y - y_i), G) = \prod_{i=1}^{d} G(x(t), y_i(t)),$$

whence

$$\operatorname{ord}_t R_Y(F(x, Y), G(x, Y)) = d \operatorname{ord}_t G(x, y).$$

Recall that $\operatorname{ord}_t x = d$; so it follows that

$$\operatorname{ord}_t R_Y(F(x, Y), G(x, Y)) = d \operatorname{ord}_x R_Y(F(x, Y), G(x, Y)),$$

and the theorem is proved. □

In the general case, write

$$F(x(u), Y) = \prod_{i=1}^{d} (Y - y_1(u)),$$

where $(x(u), y(u))$ and $(x(u), y_i(u))$ are not-necessarily-primitive representations. By a substitution of the type $t_i = c_i u^{\rho_i} + \cdots$, the pair $(x(u), y_i(u))$ is transformed to a primitive representation $(x'_i(t_i), y'_i(t_i))$. By Theorem 4.87, $\mathrm{ord}_{t_i} x'_i(t_i) = d$; so $\mathrm{ord}_u x(u) = \rho_i d$. Thus all the redundancies ρ_i are equal: write $\rho_i = \rho$.

Now $(x'_i(t_i), y'_i(t_i))$ and $(x'_j(t_j), y'_j(t_j))$ are primitive branch representations of $\mathcal{F}$. The substitution $t_i \mapsto t_j$ transforms $(x'_i(t_i), y'_i(t_i))$ to $(x'_j(t_j), y'_j(t_j))$, which still represents a branch of $\mathcal{F}$. Since such a substitution is an automorphism, then

$$\mathrm{ord}_{t_i} G(x'_i(t_i), y'_i(t_i)) = \mathrm{ord}_{t_j} G(x'_j(t_j), y'_j(t_j)),$$

whence

$$\mathrm{ord}_u G(x(u), y_i(u)) = \mathrm{ord}_u G(x(u), y_j(u)).$$

The remainder of the proof proceeds as before.

Now an alternative proof can be given for the following:

$$J(P, \mathcal{G} \cap \mathcal{F}) = I(P, \mathcal{G} \cap \mathcal{F}).$$

COROLLARY 4.91 *Let $\mathcal{F}$ and $\mathcal{G}$ be plane curves with no common components, and let P be a point of the plane. Then the intersection multiplicities defined by the resultant coincide with the intersection multiplicities defined using the branches of $\mathcal{F}$ centred at P; that is, $I(P, \mathcal{G} \cap \mathcal{F}) = J(P, \mathcal{G} \cap \mathcal{F})$.*

Proof. If P does not belong to both curves, then, from the definitions, both intersection multiplicities are zero. Therefore suppose that the point belongs to both curves and that it is the origin; suppose also that $(0,1,0)$ does not belong to both curves and that the curves meet $\mathbf{v}(X)$ only in $P = (0,0) = (0,0,1)$.

Let $\mathcal{F} = \mathbf{v}(F(X,Y))$ and $\mathcal{G} = (G(X,Y))$; if $\mathcal{F}$ does not pass through $(0,1,0)$, then $F(X,Y)$ is monic in Y. Let $F = F_1 F_2 \cdots F_s$ be the complete factorisation of F in $K[[X]][Y]$ with F_i monic and irreducible for each i. Let $F_1, \ldots, F_k$ the polynomials centred at $(0,0)$,. The polynomial F_i defines the branch γ_i of $\mathcal{F}$ centred at the origin. Note that $R_Y(F_i, G) = P_i(X)$ is a unit in $K[[X]]$ for $i > k$, since $G(0,Y)$ and $F_i(0,Y)$ could have a root in common only if the two curves met $\mathbf{v}(X)$ at a point other than the origin. Then

$$\begin{aligned} I(P, \mathcal{G} \cap \mathcal{F}) &= I(P, \mathcal{F} \cap \mathcal{G}) \\ &= \mathrm{ord}_X R_Y(F, G) \\ &= \mathrm{ord}_X R_Y(\textstyle\prod_{i=1}^s F_i, G) = \mathrm{ord}_X \prod_{i=1}^s R_Y(F_i, G) \\ &= \textstyle\sum_{i=1}^s \mathrm{ord}_X R_Y(F_i, G) = \sum_{i=1}^k \mathrm{ord}_X R_Y(F_i, G) \\ &= \textstyle\sum_{i=1}^k I(P, \mathcal{G} \cap \mathcal{F}) = J(P, \mathcal{G} \cap \mathcal{F}). \end{aligned}$$

□

4.7 EXERCISES

1. For $p > 0$, show that the branch representation

$$x(t) = u + t^p + t^{p+1}, \ y(t) = v + t^p + t^{p+2}$$

has order p, and that it is not equivalent to a branch representation of type

$$x(t) = u + t^p, \ y(t) = v + \eta(t), \quad \text{with } \mathrm{ord}_t \eta(t) \geq p.$$

2. If $\mathcal{R}$ denotes the set of all branch representations in special affine coordinates, for each K-monomorphism σ of $K[[t]]$, define the mapping $\sigma^\star : \mathcal{R} \to \mathcal{R}$ as follows:

$$\sigma^\star : (x(t), y(t)) \mapsto (\sigma(x(t), \sigma(y(t)).$$

Let $\mathcal{T} \subset \mathcal{R}$ be the union of the images $\sigma^\star(\mathcal{T})$ as σ varies in the set of K-monomorphisms of $K[[t]]$ that are not automorphisms. Show that a branch representation $(x(t), y(t))$ in special affine coordinates is primitive if and only if it is in $\mathcal{R}\backslash\mathcal{T}$.

3. Over $K = \mathbf{C}$, the irreducible plane curve

$$\mathcal{F} = \mathbf{v}((X^2 + Y^2)^2 - X(5X^2 + 4Y^2) + 4X^2)$$

has a cusp at the origin. After the linear substitution

$$X' = X + Y,\ Y' = Y,$$

apply the quadratic transformation (3.34) and show that the geometric transform of $\mathcal{F}$ is

$$\mathcal{F}' = \mathbf{v}((X^2 + 1)^2 + Y(X - 1)(5X^2 + 4Y^2) + 4Y^2(Y - X)^2).$$

Also, show that $\mathcal{F}'$ has a node at the infinite point $(0, 2, -1)$. Deduce from this that $\mathcal{F}$ has two linear branches centred at the origin.

4. For $u \in K((t))$ with $\mathrm{ord}_t\, u = m \geq 1$, the field $K((u))$ can be viewed as a subfield of $K((t))$. Show that $[K((t)) : K((u))] = m$.

5. Let $p = 3$. Show that the irreducible plane curve

$$\begin{aligned}\mathcal{F} = \mathbf{v}(&X^3Y^3 + 2X^2Y^4 + X^4Y + X^3Y^2 + 2XY^4 + 2X^4\\ &+2X^2Y^2 + 2XY^3 + 2X^3 + X^2Y + XY + Y^2)\end{aligned}$$

has five singular points, namely,

$$\begin{aligned}&O = (1, 0, 0), \quad X_\infty = (0, 1, 0), \quad Y_\infty = (0, 0, 1),\\ &E = (1, 1, 1), \quad P = (1, 2, 1),\end{aligned}$$

each the centre of exactly two branches.

6. Give a direct proof of Noether's Theorem 4.66 in its simplest case. If the curves $\mathcal{F} = \mathbf{v}(F)$ and $\mathcal{G} = \mathbf{v}(G)$ intersect in mn distinct points, where $m = \deg F$, $n = \deg G$, show that any curve $\mathcal{H} = \mathbf{v}(H)$ of degree r which passes through all common points of $\mathcal{F}$ and $\mathcal{G}$ has $H = AF + BG$, where A and B are polynomials of degrees $r - m$ and $r - n$.

7. Show that the result in Exercise 6 holds true if each common point P of $\mathcal{F}$ and $\mathcal{G}$ is a simple point of $\mathcal{F}$ and $I(P, \mathcal{H} \cap \mathcal{F}) \geq I(P, \mathcal{G} \cap \mathcal{F})$.

8. Suppose that

   (a) $\mathcal{F} = \mathbf{v}(F)$ and $\mathcal{G} = \mathbf{v}(G)$ are plane curves of degree m and n;

(b) at every common point P the tangents to $\mathcal{F}$ are distinct from the tangents to $\mathcal{G}$;

(c) $\mathcal{H} = \mathbf{v}(H)$ is a plane curve of degree r which has an $(h+k-1)$-fold point at P whenever P is h-fold on $\mathcal{F}$ and k-fold on $\mathcal{G}$.

Then $H = AF + BG$, where A and B are polynomials of degrees $r-m$ and $r-n$ such that the curves $\mathcal{A} = \mathbf{v}(A)$ and $\mathcal{B} = \mathbf{v}(B)$ have points of multiplicity at least $k-1$ and $h-1$ at P.

9. Let the curves $\mathcal{F} = \mathbf{v}(F)$ and $\mathcal{G} = \mathbf{v}(G)$ intersect in mn distinct points, where $m = \deg F$, $n = \deg G$. Show that, if nr of these points lie on a plane curve of degree r, then the remaining $(m-r)n$ lie on a plane curve of degree $m-r$. Deduce Pascal's Theorem, Chapter 3, Exercise 2, from this result.

10. (Cayley–Bacharach Theorem) Let S be the set of mn points of intersection of two plane curves of degrees m and n. Show that that every curve of degree $m+n-3$ containing all but one point of S must contain S.

4.8 NOTES

Chapter 4 is based on Seidenberg [400, Chapters 11,12,15].

The theory of branches in positive characteristic is treated in Campillo [64], and includes Puiseux and Hamburger–Noether expansions, characteristic exponents, and Newton coefficients; see also [201].

For generalisations of the Cayley–Bacharach Theorem, see [403, Chapter 5] and [115].

Chapter Five

The function field of a curve

A generalisation of the idea of a branch is a generic point of an irreducible plane curve $\mathcal{F} = \mathbf{v}(F)$. The coordinates of a generic point are in some proper extension of K, but still are assumed to satisfy the polynomial F. As in Section 4.2, branch representations are points in this sense.

The coordinates of a generic point of $\mathcal{F}$ together with K generate a field Σ of transcendence degree 1 over K. Distinct generic points of $\mathcal{F}$ give rise to the same field Σ, the function field $K(\mathcal{F})$ associated to $\mathcal{F}$, determined up to a K-isomorphism. It may be noted that no generic point of $\mathcal{F}$ has a special property that the other generic points do not have. This justifies the adjective 'generic' as the term for a generic point in the classical literature used to indicate a point without any particular property.

The idea of branches of $\mathcal{F}$ can be transferred to $K(\mathcal{F})$ via the concept of a place in such a way that birationally equivalent branches correspond to the same place.

The aim of this chapter is to show how function fields equipped with their places are used in the study of the birational properties of an irreducible plane curve. Almost all results are similar to those for curves over the complex numbers. However, the presence of inseparable variables in $K(\mathcal{F})$ when K has positive characteristic requires the concept of a Hasse derivative, which is an efficient technical tool but must be carefully used. This becomes apparent in Section 5.11, where the behaviour of dual curves in positive characteristic is considered.

5.1 GENERIC POINTS

An (affine) *point* now means any point $P = (x, y)$ with coordinates x, y belonging to some extension of the field K. As in Section 4.2, branch representations are points in this sense. A point is *constant* if both its coordinates x, y belong to K; in the other case, the point is *variable*.

DEFINITION 5.1 A point $P = (x, y)$ is *generic* on the curve $\mathcal{F} = \mathbf{v}(F(X, Y))$ if, for every $G(X, Y)$ in $K[X, Y]$ with $G(x, y) = 0$, it follows that $G(a, b) = 0$ for all points $Q = (a, b)$ of $\mathcal{F}$. In other words, the point $P = (x, y)$ is generic if $G(x, y) = 0$ implies that $G \equiv 0 \pmod{F}$.

REMARK 5.2 It follows from this definition that two distinct irreducible curves cannot have a common generic point.

THEOREM 5.3 *A reducible curve has no generic points.*

Proof. Let $\mathcal{C}$ be a curve with two components $\mathcal{F} = \mathbf{v}(F)$ and $\mathcal{G} = \mathbf{v}(G)$, with $\mathcal{F} \not\subset \mathcal{G}$, $\mathcal{G} \not\subset \mathcal{F}$. Then $\mathcal{C} = \mathbf{v}(FG)$. So, if $P = (a, b)$ is a point, constant or variable, such that $F(a, b)G(a, b) = 0$; then $F(a, b) = 0$ or $G(a, b) = 0$.

Suppose that $\mathcal{C}$ has a generic point $P = (x, y)$; then $F(x, y) = 0$ or $G(x, y) = 0$. Suppose the former holds; that is, $F(x, y) = 0$. Then F is zero at all points of $\mathcal{C}$ and so also at all points of $\mathcal{G}$. So $\mathcal{G}$ must be a component of $\mathcal{F}$, a contradiction. □

THEOREM 5.4 *A variable point $P = (x, y)$ of an irreducible curve $\mathcal{F}$ is a generic point.*

Proof. Let $\mathcal{F} = \mathbf{v}(F(X, Y))$ and suppose that $G(x, y) = 0$ but $G(X, Y) \not\equiv 0 \pmod{F(X, Y)}$, with $P = (x, y)$ a variable point of $\mathcal{F}$. Taking $\mathcal{F}$ to be irreducible, it follows from Bézout's Theorem that the common points of the curves $\mathcal{F}$ and $\mathcal{G} = \mathbf{v}(G(X, Y))$ are finite in number and so belong to the plane $\mathrm{AG}(2, K)$. In other words, each of the common points of $\mathcal{F}$ and $\mathcal{G}$ is constant. In particular, since $P = (x, y)$ is a variable point of $\mathcal{F}$, it cannot belong to the curve $\mathcal{G}$, and so $G(x, y) \neq 0$, as required. □

From now on, the term 'point' of an irreducible curve $\mathcal{F}$ only indicates any of its constant points, while 'generic point' of $\mathcal{F}$ means any of its variable points.

The above discussion extends to the projective plane. Let $\mathcal{F} = \mathbf{v}(F)$ with homogeneous $F = F(X_0, X_1, X_2) \in K[X_0, X_1, X_2]$. A *generic point* of $\mathcal{F}$ is defined to be a point $P = (x_0, x_1, x_2)$ with $F(x_0, x_1, x_2) = 0$, whose coordinates x_0, x_1, x_2 come from an extension field of K such that, if G is a homogeneous polynomial in $K[X_0, X_1, X_2]$ that vanishes at P, then G vanishes at every point of $\mathcal{F}$. This amounts to saying that G is divisible by F. As in the affine case, a curve has a generic point if and only if it is irreducible. Using homogeneous polynomials, the function field of $\mathcal{F}$ can be described as the subfield in $K(x_0, x_1, x_2)$ consisting of all elements $\xi = u(x_0, x_1, x_2)/v(x_0, x_1, x_2)$, where $u(X_0, X_1, X_2)$, $v(X_0, X_1, X_2) \in K[X_0, X_1, X_2]$ are homogeneous of the same degree and $v(x_0, x_1, x_2) \neq 0$. In fact, if $x_0 \neq 0$, this subfield is K-isomorphic to $K(1, x_1/x_0, x_2/x_0)$ which is just $\Sigma = K(x, y)$.

If all coordinates x_0, x_1, x_2 are constant, then P is a *constant* point; otherwise, it is a *variable* point. As before, a point of an irreducible curve is generic if and only if it is variable.

REMARK 5.5 With the notion of a universal domain Ω over K, all points are in the plane over Ω. Since Ω is an algebraically closed field, a curve $\mathcal{F}$ over K can be considered as a curve over Ω, meaning that the coefficients in the equation of $\mathcal{F}$ are in Ω. All theorems previously proved remain true as stated, although they may have a different content because of the change of meaning for point.

DEFINITION 5.6 The generic points $P = (x, y)$ and $P_1 = (x_1, y_1)$ of an irreducible curve $\mathcal{F}$ are *K-isomorphic* if there exists an isomorphism

$$\sigma : K[x, y] \to K[x_1, y_1]$$

such that $\sigma(x) = x_1$ and $\sigma(y) = y_1$.

THEOREM 5.7 *Any two generic points of an irreducible curve are K-isomorphic.*

Proof. Let $P = (x, y)$ and $P_1 = (x_1, y_1)$ be generic points of the irreducible curve $\mathcal{F}$. Now let $\sigma : K[x, y] \to K[x_1, y_1]$ be a map such that $\sigma(G(x, y)) = G(x_1, y_1)$ for every $G(X, Y) \in K[X, Y]$. It must be shown that $G(x, y) = H(x, y)$ implies $G(x_1, y_1) = H(x_1, y_1)$ for every $H(X, Y) \in K[X, Y]$. In other words, if $G(x, y) - H(x, y) = 0$, then also $G(x_1, y_1) - H(x_1, y_1) = 0$. But this follows immediately from the notion of generic point. Therefore the map σ is well-defined. Both the following relations hold:

$$\sigma(G_1(x, y) + G_2(x, y)) = \sigma(G_1(x, y)) + \sigma(G_2(x, y));$$
$$\sigma(G_1(x, y)G_2(x, y)) = \sigma(G_1(x, y))\sigma(G_2(x, y)).$$

This shows that σ is a homomorphism from $K[x, y]$ to $K[x_1, y_1]$. Further, the kernel of σ is zero since $G(x_1, y_1) = 0$ implies that $G(x, y) = 0$, since, by hypothesis, $P = (x_1, y_1)$ is a generic point of $\mathcal{F}$. □

COROLLARY 5.8 *An irreducible curve $\mathcal{F}$ has infinitely many generic points.*

Proof. To prove the assertion, it is enough to show that $\mathcal{F}$ has infinitely many points $P = (x, y)$ such that $x, y \in K((t))$. Given a constant point Q of $\mathcal{F}$, let $(x(t), y(t))$ be any representation of a branch centred at Q. Then $P = (x(t), y(t))$ is a variable point of $\mathcal{F}$. By Theorem 5.4, $P = (x(t), y(t))$ is a generic point of $\mathcal{F}$. For any K-automorphism σ of $K[[t]]$, the point $P' = (x'(t), y'(t))$ with $x'(t) = \sigma(x(t))$ and $y'(t) = \sigma(y(t))$ is also a variable point of $\mathcal{F}$. By Theorem 4.4, each $\tau \in K[[t]]$ defines a K-automorphism of $K[[t]]$. Since $K[[t]]$ has infinitely many such K-automorphisms, each of them defines a generic point of $\mathcal{F}$. □

DEFINITION 5.9 In the case that $P = (x, y)$ is a generic point of an irreducible curve $\mathcal{F} = \mathbf{v}(F(X, Y))$, then $\Sigma = K(x, y)$ is the *field of rational functions* or, more briefly, the *function field* of $\mathcal{F}$.

Note that, by Theorem 5.7, this field is uniquely determined up to K-isomorphism. Also, its transcendence degree over K is 1. In other words, the following theorem holds.

THEOREM 5.10 *For any two elements $\xi, \eta \in \Sigma$, there exists an irreducible polynomial $g(X, Y) \in K[X, Y]$ such that $g(\xi, \eta) = 0$; that is, there is a polynomial relation between any two elements in Σ.*

An immediate corollary is the following result.

COROLLARY 5.11 *Every point $P = (\xi, \eta)$ of $\mathrm{AG}(2, \Sigma)$ not in $\mathrm{AG}(2, K)$ is a generic point of an irreducible curve $\mathcal{F}$.*

5.2 RATIONAL TRANSFORMATIONS

In classical literature on projective geometry, a rational transformation is a mapping

$$X' = \frac{f(X, Y)}{h(X, Y)}, \; Y' = \frac{g(X, Y)}{h(X, Y)},$$

with $f, g, h \in K[X, Y]$.

Since $h(X, Y)$ could vanish at some points, these formulas may not allow the images of such points to be calculated. Further, if the domain of the transformation is to be non-empty and be contained in a given algebraic curve $\mathcal{F} = \mathbf{v}(F(X, Y))$, then it is necessary that $h(X, Y) \not\equiv 0 \pmod{F(X, Y)}$. In this case, $h(a, b)$ is zero only at a finite number of points of $\mathcal{F}$. Then, for any $P = (a, b) \in \mathcal{F}$ and $h(a, b) \neq 0$, the following transformation can be defined:

$$\mathfrak{T}(a, b) = \left(\frac{f(a, b)}{h(a, b)}, \frac{g(a, b)}{h(a, b)} \right).$$

In particular, $\mathfrak{T}$ is defined at every generic point of $\mathcal{F}$ and transforms such a generic point $P = (x, y)$ into a point $P' = (x', y')$ of $\mathrm{AG}(2, \Sigma)$. Now the proper definition can be given.

DEFINITION 5.12 A *rational transformation* $\mathfrak{T}$ of an algebraic curve $\mathcal{F}$ is given by a generic point $P = (x, y)$ of $\mathcal{F}$ and a point $P' = (x', y')$, $x', y' \in K(x, y)$. Then

$$x' = \frac{f(x, y)}{h(x, y)}, \; y' = \frac{g(x, y)}{h(x, y)},$$

with $f, g, h \in K[X, Y]$ and $h(x, y)$ not vanishing identically on $\mathcal{F}$. From this, for each $P = (a, b) \in \mathcal{F}$ such that $h(a, b) \neq 0$,

$$\mathfrak{T}(a, b) = \left(\frac{f(a, b)}{h(a, b)}, \frac{g(a, b)}{h(a, b)} \right).$$

REMARK 5.13 The expressions for x', y' as rational functions of x, y are not uniquely determined; but $\mathfrak{T}(a, b)$ does not depend on the particular functions chosen.

Proof. Suppose that

$$x' = \frac{f_1(x, y)}{h_1(x, y)}, \; y' = \frac{g_1(x, y)}{h_1(x, y)}$$

and

$$\mathfrak{T}'(a, b) = \left(\frac{f_1(a, b)}{h_1(a, b)}, \frac{g_1(a, b)}{h_1(a, b)} \right),$$

with $h_1(a, b) \neq 0$. Then it must be shown that $\mathfrak{T}'(a, b) = \mathfrak{T}(a, b)$; that is, if

$$\frac{f(x, y)}{h(x, y)} = \frac{f_1(x, y)}{h_1(x, y)},$$

then

$$\frac{f(a, b)}{h(a, b)} = \frac{f_1(a, b)}{h_1(a, b)}.$$

Thus the mapping is well-defined and does not depend on the representative functions. From the hypothesis that

$$\frac{f(x, y)}{h(x, y)} = \frac{f_1(x, y)}{h_1(x, y)},$$

it follows that

$$f(x,y)h_1(x,y) - f_1(x,y)h(x,y) = 0.$$

So the polynomial

$$G(X,Y) = f(X,Y)h_1(X,Y) - f_1(X,Y)h(X,Y)$$

vanishes at the generic point $P = (x,y)$ of $\mathcal{F}$. Hence $G(a,b) = 0$ for every point $Q = (a,b) \in \mathcal{F}$, whence

$$f(a,b)h_1(a,b) - f_1(a,b)h(a,b) = 0.$$

Since $h(a,b)h_1(a,b) \neq 0$, it may be concluded that

$$f(a,b)/h(a,b) = f_1(a,b)/h_1(a,b).$$

□

The mapping $\mathfrak{T}$ is *defined at* $P = (a,b)$ if, for some expression in $h(x',y')$, the value $h(a,b) \neq 0$. If there exists no such expression, then $\mathfrak{T}$ is *not defined* at $P = (a,b)$.

EXAMPLE 5.14 Let $\mathcal{C} = \mathbf{v}(Y - X^2)$ be a conic and $P = (x,y)$ one of its generic points. Let $\mathfrak{T}$ be defined by the generic point $P' = (x',y')$ where

$$x' = x,\ y' = y/x.$$

Then $\mathfrak{T}((0,0)) = (0,0)$ is defined as, although $0/0$ is not defined, $y/x = x$. In fact, $x',\ y'$ can be written in the form $x' = y,\ y' = x$, whence $\mathfrak{T}((0,0)) = O$. On the other hand, consider the transformation $\mathfrak{T}$ determined by $x' = y,\ y' = x/y$. The expression x/y cannot be rewritten in the form $g(x,y)/h(x,y)$ with $h(0,0) \neq 0$. In fact, since $h(x,y)x - g(x,y)y = 0$, so

$$h(X,Y)X - g(X,Y)Y = d(X,Y)(Y - X^2)$$

with $d(X,Y) \in K[X,Y]$, which is impossible, as the terms of least degree in $h(X,Y)X - g(X,Y)Y$ contain a term cX with $c \in K,\ c \neq 0$, which cannot be the case in the right-hand side.

In terms of homogeneous coordinates, a *rational transformation* $\mathfrak{T}$ of $\mathcal{F}$ is defined by a generic point $P = (x_0,x_1,x_2)$ of $\mathcal{F}$ and a point $P' = (x'_0,x'_1,x'_2)$ such that $K(P) \supset K(P')$. Then

$$x'_0 : x'_1 : x'_2 = f_0(x_0,x_1,x_2) : f_1(x_0,x_1,x_2) : f_2(x_0,x_1,x_2),$$

with f_0, f_1, f_2 homogeneous polynomials of the same degree. If $Q = (a_0,a_1,a_2)$ is a point of $\mathcal{F}$, and f_0, f_1, f_2 do not vanish at Q, then $\mathfrak{T}(Q)$ is defined to be $(f_0(Q), f_1(Q), f_2(Q))$. The triple (f_0, f_1, f_2) is not unique. But, if

$$\begin{aligned} f_0(x_0,x_1,x_2) &: f_1(x_0,x_1,x_2) : f_2(x_0,x_1,x_2) \\ &= g_0(x_0,x_1,x_2) : g_1(x_0,x_1,x_2) : g_2(x_0,x_1,x_2), \end{aligned}$$

where the g_i are also homogeneous of the same degree, and if g_0, g_1, g_2 do not all vanish at Q, then $(f_0(Q), f_1(Q), f_2(Q))$ and $(g_0(Q), g_1(Q), g_2(Q))$ define the same point $Q' = \mathfrak{T}(Q)$. In this way, $\mathfrak{T}$ is defined at all but a finite number of points of $\mathcal{F}$.

EXAMPLE 5.15 In Example 5.14 in which $\mathcal{C} = \mathbf{v}(Y - X^2)$ and $\mathfrak{T}$ is given by the equations $x' = y,\ y' = x/y$, the mapping is not defined at the point $P = (0,0)$. In $\mathrm{PG}(2,K)$, however,

$$\begin{aligned}(1, x', y') &= (y, y^2, x) = (x^2, yx^2, x) = (x, yx, 1)\\ &= (f_0(1,x,y), f_1(1,x,y), f_2(1,x,y)),\end{aligned}$$

where

$$f_0 = X_0X_1,\ f_1 = X_1X_2,\ f_2 = X_0^2.$$

Then, with $P = (1,0,0)$,

$$\mathfrak{T}(P) = P' = (0,0,1).$$

In other words, $\mathfrak{T}$ is not defined at P viewed as a point of $\mathrm{AG}(2,K)$, but it is defined at P when that point is regarded as a point in $\mathrm{PG}(2,K)$.

For an example of a rational transformation in the projective plane not everywhere defined on a given curve, consider the curve $\mathcal{C}' = \mathbf{v}(Y^2 - X^2 - X^3)$ and the transformation $\mathfrak{T}$ given by $x' = x,\ y' = y/x$. It can be shown that $\mathfrak{T}$ is not defined at $P = (1,0,0)$.

Returning to the affine case, two cases occur for $x', y' \in K(x,y)$.

(i) If $x',\ y' \in K$, then all points of $\mathcal{F}$ are transformed into the same point, namely $P' = (x', y')$. This is the *trivial* case.

(ii) If at least one of x', y' is not in K, then the field extension $K(x', y')/K$ has still transcendence degree 1. Hence $P' = (x', y')$ is a generic point of an irreducible curve $\mathcal{F}'$, the *image curve* of $\mathcal{F}$ under $\mathfrak{T}$. This is the *non-trivial* case, and it is usually tacitly assumed that this is the case.

These definitions give the following two results.

THEOREM 5.16 *All transformed points of $\mathcal{F}$ under $\mathfrak{T}$ are on the image curve.*

THEOREM 5.17 *A rational transformation $\mathfrak{S}$ of $\mathcal{F}$ is always defined on generic points of $\mathcal{F}$ and the image of a generic point of $\mathcal{F}$ is a generic point of the image curve $\mathcal{F}'$.*

As a special case, take a branch representation of $\mathcal{F}$ as a generic point.

THEOREM 5.18 *The image of a branch representation of a curve $\mathcal{F}$ under a rational transformation is a branch representation of the image curve $\mathcal{F}'$.*

The image of a primitive branch representation need not be primitive. Still, equivalent branch representations yield equivalent branch representations, so that a given branch γ of $\mathcal{F}$ gives rise, under $\mathfrak{T}$, to a definite branch γ' of $\mathcal{F}'$. The branch γ' is the *image* of γ under $\mathfrak{T}$.

Let $\mathfrak{T}$ be given by a generic point $P' = (x', y')$ of $\mathcal{F}$, and define a map $\mathfrak{T}^*$ as follows: if $P = (a, b)$ is a point of $\mathcal{F}$ and γ is a branch of $\mathcal{F}$ centred at $P(a, b)$, set $\mathfrak{T}^*(P) = P'$ where P' is the centre of the image γ' of γ under $\mathfrak{T}$. Here $\mathfrak{T}^*$ need not be a *univalent* map, as $\mathcal{F}$ may have several branches centred at P. Thus $\mathfrak{T}^*(P)$

should be defined to be the set of all centres of the branches of $\mathcal{F}'$ arising from the branches of $\mathcal{F}$ centred at P. It will also not be univalent, in general. The centre of the transformed branch γ' of γ need not be at finite distance. However, from the discussion on the projective plane, $\mathfrak{T}^*$ will be defined at every point of $\mathcal{F}$, since there is at least one branch of $\mathcal{F}$ centred at any point of $\mathcal{F}$.

Let $P = (a_0, a_1, a_2)$ be a point of $\mathcal{F}$ and assume that $\mathfrak{T}$ is defined at P. Then

$$1 : x' : y' = f_0(1, x, y) : f_1(1, x, y) : f_2(1, x, y),$$

where f_0, f_1, f_2 are homogeneous of the same degree and $f_i(a_0, a_1, a_2) \neq 0$ for some i. Let

$$x_0(t) = a_0 + ty_0(t),\ x_1(t) = a_1 + ty_1(t),\ x_2(t) = a_2 + ty_2(t)$$

be a special representation of a branch of $\mathcal{F}$. Put $t = 0$ in $f_i(x_0(t), x_1(t), x_2(t))$; this shows that γ' is centred at the point

$$P' = (f_0(a_0, a_1, a_2), f_1(a_0, a_1, a_2), f_2(a_0, a_1, a_2)).$$

In other words, $\mathfrak{T}^*$ extends $\mathfrak{T}$, and it has the advantage of being defined everywhere on $\mathcal{F}$. Nevertheless, $\mathfrak{T}$ cannot be defined at a point where $\mathfrak{T}^*$ is multivalued.

DEFINITION 5.19 A rational transformation $\mathfrak{S} : \mathcal{F} \to \mathcal{F}'$, given by the generic points $P = (x, y) \in \mathcal{F}$ and $P' = (x', y') \in \mathcal{F}'$, is *birational* if

$$K(x', y') = K(x, y).$$

In this case, the generic point $P' = (x', y')$ of $\mathcal{F}'$ together with the pair of elements (x, y) yield a rational transformation $\bar{\mathfrak{S}} : \mathcal{F}' \to \mathcal{F}$.

THEOREM 5.20 *If a birational transformation $\mathfrak{S}$ is defined at the point P and $\bar{\mathfrak{S}}$ is defined at the point $\mathfrak{S}(P)$, then $\bar{\mathfrak{S}}(\mathfrak{S}(P)) = P$.*

Proof. Let $P = (a, b)$ and let $r, s, t \in K[X, Y]$ such that

$$\mathfrak{S} : x' = \frac{r(x, y)}{t(x, y)},\ y' = \frac{s(x, y)}{t(x, y)},$$

with $t(a, b) \neq 0$. For $P' = (a', b') = \mathfrak{S}(P)$, let $r', s', t' \in K[X, Y]$ be such that

$$\bar{\mathfrak{S}} : x = \frac{r'(x', y')}{t'(x', y')},\ y = \frac{s'(x', y')}{t'(x', y')},$$

with $t'(a', b') \neq 0$. Now

$$x = \frac{r'(r(x, y)/t(x, y), s(x, y)/t(x, y))}{t'(r(x, y)/t(x, y), s(x, y)/t(x, y))},$$
$$y = \frac{s'(r(x, y)/t(x, y), s(x, y)/t(x, y))}{t'(r(x, y)/t(x, y), s(x, y)/t(x, y))}.$$

There exists an integer $h \geq 1$ such that

$$w(X, Y) = (t(X, Y))^h \left[Xt'\left(\frac{r(X, Y)}{t(X, Y)}, \frac{s(X, Y)}{t(X, Y)}\right) - r'\left(\frac{r(X, Y)}{t(X, Y)}, \frac{s(X, Y)}{t(X, Y)}\right)\right]$$

is a polynomial with coefficients in K. It follows simply that $w(x,y) = 0$, and so the point $P = (x,y)$ is generic point of $\mathcal{F}$. Hence $w(a,b) = 0$ for each point $Q = (a,b) \in \mathcal{F}$. Since $t(a,b) \neq 0$, $t'(a',b') \neq 0$,

$$a = \frac{r'(r(a,b)/t(a,b), s(a,b)/t(a,b))}{t'(r(a,b)/t(a,b), s(a,b)/t(a,b))}$$

and, similarly,

$$b = \frac{s'(r(a,b)/t(a,b), s(a,b)/t(a,b))}{t'(r(a,b)/t(a,b), s(a,b)/t(a,b))}.$$

Therefore

$$a = \frac{r'(a',b')}{t'(a',b')}, \; b = \frac{s'(a',b')}{t'(a',b')},$$

and $\bar{\mathfrak{S}}(a',b') = (a,b)$. However, $(a',b') = \mathfrak{S}(P)$ and $(a,b) = P$, from which it follows that $\bar{\mathfrak{S}}(\mathfrak{S}(P)) = P$, as required. □

The rational transformation $\bar{\mathfrak{S}}$ is the *inverse* of $\mathfrak{S}$ and so is written $\bar{\mathfrak{S}} = \mathfrak{S}^{-1}$.

Since $\mathfrak{S}$ is defined at every generic point P of $\mathcal{F}$ and since $\mathfrak{S}(P)$ is generic point of $\mathcal{F}'$, the following corollary is immediate.

COROLLARY 5.21 *A birational transformation $\mathfrak{S}$ of a curve $\mathcal{F}$ defines a bijective correspondence between the generic points of $\mathcal{F}$ and those of the image curve $\mathcal{F}'$.*

DEFINITION 5.22 Let Σ be the function field of a curve; equivalently, let K have an extension $\Sigma = K(x,y)$ of transcendence degree 1 over K. A *model of* Σ is given by a point $P = (\xi,\eta)$ such that $\Sigma = K(\xi,\eta)$ and by the curve $\mathcal{F}$ having P as generic point; in symbols, write $(\mathcal{F}; (\xi,\eta))$.

It should be noted that two models $(\mathcal{F}; (x,y))$ and $(\mathcal{F}'; (x',y'))$ are distinct if $(x,y) \neq (x',y')$. It may, however, happen that $\mathcal{F} = \mathcal{F}'$ but $(x,y) \neq (x',y')$. So a given curve $\mathcal{F}$ may give rise to more than one model, consistent with the fact that a curve can have non-trivial automorphisms, that is, can be birational images of themselves under a non-identical birational transformation. Automorphisms are investigated in Chapter 11. Here two examples illustrate this situation.

EXAMPLE 5.23 Let $\mathcal{C}$ and $P = (x,y)$ be as in Example 5.14. If $\mathfrak{S}$ is the map

$$\begin{aligned} x' &= x + c, \\ y' &= 2cx + y + c^2 \end{aligned}$$

with $c \in K$, then $\mathfrak{S}^{-1}$ is as follows:

$$\begin{aligned} x &= x' - c, \\ y &= y' - 2cx' + c^2. \end{aligned}$$

Hence

$$y - x^2 = (y' - 2cx' + c^2) - (x' - c)^2 = y' - x'^2.$$

This shows that $\mathfrak{S}$ transforms $\mathcal{C}$ into itself, as $y - x^2 = 0$ implies that $y' - x'^2 = 0$. Therefore $(\mathcal{C}; (x,y))$ is transformed into the model $(\mathcal{C}; (x',y'))$ by $\mathfrak{S}$.

EXAMPLE 5.24 Let $q_0 = 2^e$ and $q = 2q_0^2$. The DLS curve is the irreducible plane curve

$$\mathcal{F} = \mathbf{v}(X^{2q_0}(X^q + X) + Y^q + Y).$$

It has only one singular point, namely Y_∞, which is a $2q_0$-fold point. For a generic point $P = (x, y)$, let $K(\mathcal{F}) = K(x, y)$ be the function field of $\mathcal{F}$. Now put $h = xy + x^{2q_0+2} + y^{2q_0}$. A straightforward computation shows that the rational transformation

$$\mathfrak{S} : x' = \frac{y}{h}, \quad y' = \frac{x}{h}$$

leaves $\mathcal{D}$ invariant. In particular, the associated map $\mathfrak{T}$ on $\mathcal{F}$ is defined on all points of $\mathcal{F}$ apart from Y_∞ and O. Further, the branch γ centred at O is transformed to a branch centred at Y_∞. Also, $\mathfrak{S}$ has order two as it coincides with its inverse. Therefore Y_∞ is the centre of only one branch, say γ', which is the image of γ under $\mathfrak{T}$. Since $(x(t), y(t))$, with

$$x(t) = t, \quad y(t) = t^{2q_0} + t^{q+2q_0} + t^{2qq_0} + t^{q(q+2q_0)} + \cdots,$$

is a primitive representation of γ, so $(\xi(t), \eta(t))$, with

$$\xi(t) = \frac{1 + t^q + t^{2q_0(q-1)} + \cdots}{t(1 + t + t^{2q-2q_0-1} + \cdots)}, \quad \eta(t) = \frac{1}{t^{2q_0}(1 + t + t^{2q-2q_0-1} + \cdots)},$$

is a primitive representation of γ'.

If $(\mathcal{F}; (\xi, \eta))$ and $(\mathcal{F}'; (\xi', \eta'))$ are two models of Σ, then $\Sigma = K(\xi, \eta)$ and $\Sigma = K(\xi', \eta')$. Hence $K(\xi, \eta) = K(\xi', \eta')$ and so, by Definition 5.19, the points $P = (\xi, \eta)$ and $P' = (\xi', \eta')$ determine a birational transformation from $\mathcal{F}$ to $\mathcal{F}'$.

THEOREM 5.25 *Any two models of a given field Σ are birationally equivalent.*

A *property* of the field Σ is a birationally invariant property of any one of its models.

Now consider the irreducible curves $\mathcal{F} = \mathbf{v}(F(X, Y))$ and $\mathcal{G} = \mathbf{v}(G(X, Y))$, and take generic points $P = (x, y)$ of $\mathcal{F}$ and $Q = (\xi, \eta)$ of $\mathcal{G}$. Assume that their function fields $\Sigma = K(x, y)$ and $\Sigma' = K(\xi, \eta)$ are isomorphic over K. Let

$$\varphi : \Sigma \to \Sigma'$$

be a K-isomorphism, and put $x' = \varphi(\xi)$, $y' = \varphi(\eta)$. Then $\Sigma' = K(x', y')$ and $\varphi F(x, y) = 0$ implies $F(x', y') = 0$. Hence $(\mathcal{F}; (x', y'))$ is a model of Σ', in particular a birationally equivalent model to the model $(\mathcal{G}; (\xi, \eta))$.

This discussion motivates the following definition.

DEFINITION 5.26 Two irreducible plane curves are *birationally isomorphic* or *birationally equivalent* if their function fields are isomorphic over K.

5.3 PLACES

Let $(\mathcal{F}, (x, y))$ be a model of a field Σ of transcendence degree 1. If $(\xi(t), \eta(t))$ is a not-necessarily-primitive branch representation of $\mathcal{F}$, then, by Theorem 5.7, there is a K-monomorphism σ of Σ into $K((t))$ such that $\sigma(x) = \xi(t), \sigma(y) = \eta(t)$. For an automorphism ρ of $K[[t]]$, extended to $K((t))$ in a natural way, put $\sigma' = \rho\circ\sigma$. Then σ' is also a K-monomorphism of Σ into $K((t))$ such that the branch representations $(\sigma'(x) = \xi'(t),\ \sigma'(y) = \eta'(t))$ and (ξ, η) are equivalent, as in this diagram:

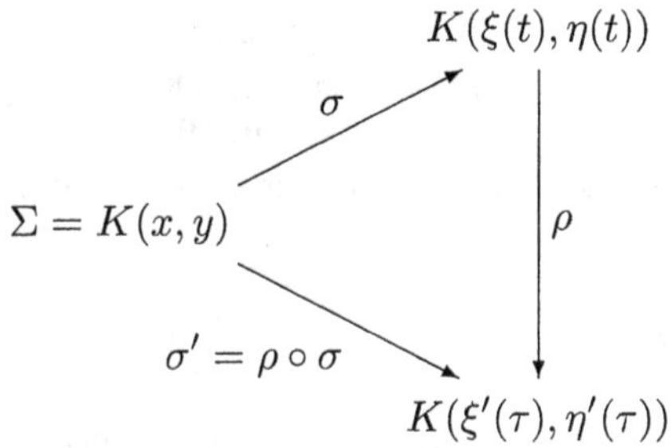

This leads to the following fundamental concepts.

DEFINITION 5.27 (i) A K-monomorphism $\sigma : \Sigma \to K((t))$ is a *place representation.*

(ii) A place representation σ is *primitive* if $(\sigma(x), \sigma(y))$ for a pair, and hence for any pair, of generators x, y of Σ is a primitive branch representation of $\mathcal{F}$.

(iii) Two place representations σ and σ' are *equivalent* if there is a K-automorphism ρ of $K((t))$ such that $\sigma' = \rho \circ \sigma$.

(iv) A *place* is an equivalence class of primitive place representations.

(v) The set of all places of Σ is denoted by $\mathfrak{P}(\Sigma)$.

A place is merely the field-theoretic or birational counterpart of the geometric concept of a branch. The following two theorems explain this point.

THEOREM 5.28 *The places of Σ and the branches of any model of Σ are in one-to-one correspondence, in a natural way.*

Proof. Let $(\mathcal{F}; (x, y))$ be a model of Σ. It has been shown how a primitive branch representation $(\xi(t), \eta(t))$ leads to a primitive place representation σ in a natural way, and also how to go back to a primitive branch representation $(\xi(t), \eta(t))$ from a primitive place representation σ by $\sigma(\xi) = \xi(t)$, $\sigma(\eta) = \eta(t)$. In this way, a branch representation and a place representation give rise to one other. Also, equivalent branch representations correspond to equivalent place representations; so branches and places are in one-to-one correspondence. □

THEOREM 5.29 *Let $(\mathcal{F}; (x, y))$ and $(\mathcal{F}; (x', y'))$ be two models of Σ, and let $\mathfrak{S}$ be the birational transformation sending $P = (x, y)$ into $P' = (x', y')$. Then $\mathfrak{S}$ induces a one-to-one correspondence between the branches of $\mathcal{F}$ and those of $\mathcal{F}'$, two branches corresponding if and only if they correspond to the same place of Σ.*

Proof. Let the place $\mathcal{P}$, that is, one of its primitive representations, send both (x, y) to $(x(t), y(t))$ and (x', y') to $(x'(t), y'(t))$, as in the following diagram:

$$\begin{array}{ccc} (x, y) & \xrightarrow{\mathcal{P}} & (x(t), y(t)) \\ \tau \downarrow & & \vdots \\ (x', y') & \xrightarrow{\mathcal{P}} & (x'(t), y'(t)) \end{array}$$

There exist $r(X, Y), s(X, Y) \in K[X, Y]$ such that $x' = r(x, y)/s(x, y)$. Applying $\mathcal{P}$ gives $x'(t) = r(x(t), y(t))/s(x(t), y(t))$, and similarly for y'. This shows that $\mathfrak{S}(P(x(t), y(t))) = (x'(t), y'(t))$, justifying the fourth arrow in the diagram and proving one part of the theorem. A similar argument gives the converse. □

5.4 ZEROS AND POLES

Let $\mathcal{F} = \mathbf{v}(F(X, Y))$ be an irreducible curve with a generic point $P = (x, y)$. For a place $\mathcal{P}$ of the field $\Sigma = K(x, y)$, let $\tau : \Sigma \to K((t))$ be a primitive representation of $\mathcal{P}$. If ξ is an element of Σ, then $\text{ord}_t\, \tau(\xi)$ does not depend on the choice of τ. To show this, take another primitive representation $\tau' : \Sigma \to K((t))$ of $\mathcal{P}$. Then $\tau' = \rho \circ \tau$ with an automorphism ρ of $K[[t]]$ naturally extended to $K((t))$. Hence $\text{ord}_t\, \tau'(\xi) = \text{ord}_t\, \rho(\tau(\xi)) = \text{ord}_t\, \tau(x_i)$ by Theorem 4.5. This leads to the following concept.

DEFINITION 5.30 Let τ be any primitive representation of a place $\mathcal{P}$ of Σ.

(i) The *order of* ξ *at* $\mathcal{P}$ is $\text{ord}_{\mathcal{P}}\, \xi = \text{ord}_t\, \tau(\xi)$.

(ii) If $\text{ord}_{\mathcal{P}}\, \xi = 1$, then ξ is a *uniformising element* or *local parameter* at $\mathcal{P}$.

Let γ be the branch of $\mathcal{F}$ associated to the place $\mathcal{P}$. The *centre* of $\mathcal{P}$ is the centre of γ. Since, by Theorem 5.28, there is a bijection between the branches of $\mathcal{F}$ and the places of Σ, and by Theorem 4.46 there is a only a finite number of branches centred at the same point, there is a only a finite number of places of Σ with the same centre. Put

$$\text{ord}_\gamma\, \xi = \text{ord}_{\mathcal{P}}\, \xi.$$

In particular, $\text{ord}_\gamma\, x = \text{ord}_t\, x(t)$, $\text{ord}_\gamma\, y = \text{ord}_t\, y(t)$ which can be thought of as alternative notation for $\text{ord}_t\, x(t)$ and $\text{ord}_t\, y(t)$. From part (iii) of Theorem 4.21 the following result appears.

LEMMA 5.31 *The function field* Σ *has a uniformising element at every place.*

DEFINITION 5.32 (i) The place $\mathcal{P}$ is a *zero* of ξ if $\text{ord}_{\mathcal{P}}\, \xi > 0$; if $\mathcal{P}$ is a zero of ξ, it is a zero of *multiplicity* $\text{ord}_{\mathcal{P}}\, \xi$.

(ii) The place $\mathcal{P}$ is a *pole* of ξ if $\text{ord}_{\mathcal{P}}\, \xi < 0$; if $\mathcal{P}$ is a pole of ξ, it is a pole of *multiplicity* $-\text{ord}_{\mathcal{P}}\, \xi$.

(iii) If the place $\mathcal{P}$ is not a pole of ξ, then ξ is *regular* at $\mathcal{P}$.

Henceforth, ζ denotes a function of Σ, that is, an element of $\Sigma\backslash K$.

THEOREM 5.33 *The function ζ has only finitely many zeros. The number of zeros counted with their multiplicity is $[\Sigma : K(\zeta)]$.*

Proof. Since ζ is transcendental over K, then Σ is a finite algebraic extension of $K(\zeta)$ and $[\Sigma : K(\zeta)]$ is an integer. If L is a finite algebraic extension of a field M, then L is a simple extension of M. By the Theorem of Primitive Element, there exists $\eta \in \Sigma$ such that $\Sigma = K(\zeta, \eta)$. Write $\zeta = x$, $\eta = y$; in other words, consider ζ as the first coordinate of a generic point of a model $\mathcal{F}$ of Σ.

Let

$$F(X, Y) = c_0(X)Y^m + c_1(X)Y^{m-1} + \cdots + c_m(X)$$

be an irreducible polynomial with coefficients in K of which (x, y) is a zero. Then

$$\begin{aligned} c_0(X)^{m-1}F(X,Y) &= (c_0(X)Y)^m + c_1(X)(c_0(X)Y)^{m-1} + \\ &\qquad \cdots + c_m(X)c_0(X)^{m-1} \\ &= Z^m + c_1(X)Z^{m-1} + \cdots, \end{aligned}$$

having put $c_0(X)Y = Z$. For this polynomial in $K[X, Z]$, the pair $(x, c_0(x)y)$ is a zero. Since $c_0(x)y$ is a generator of Σ over $K(x)$, then $c_0(x)y$ may be substituted for y; in other words, it may be supposed that $c_0(X) = 1$. Then $[\Sigma : K(x)] = m$.

Let $(\mathcal{F}; (x; y))$ be a model of $\Sigma = K(x, y)$ with (x, y) as a generic point. Then the branches γ of $\mathcal{F}$ such that $\text{ord}_\gamma\, x > 0$ are to be counted each with its order. These branches have centres on the Y-axis, but *a priori* possibly at Y_∞. This can happen but not when $c_0(X) = 1$. To check it, assume γ is a branch centred at Y_∞. Note that if $(x(t), y(t))$ is a primitive branch representation of γ, then $\text{ord}_\gamma\, y = \text{ord}\, y(t) < 0$. So

$$\text{ord}_\gamma\, (y^m + c_1(x)y^{m-1} + \cdots) = \text{ord}_\gamma\, y^m < 0,$$

which contradicts that

$$y^m + c_1(x)y^{m-1} + \cdots = 0, \quad \text{ord}_\gamma\, 0 = \infty.$$

Let $\gamma_1, \gamma_2, \ldots, \gamma_s$ be the branches for which $\text{ord}_\gamma\, x > 0$; they are precisely the branches centred at points $(0, b, 1)$ on the y-axis. Consider now such a point P_j on $\mathcal{F}$. Let $\gamma_{j1}, \ldots, \gamma_{jk}$ be the branches of $\mathcal{F}$ centred at P_j. Then

$$I(P_j, \mathbf{v}(X) \cap \mathcal{F}) = \text{ord}_{\gamma_{j1}}\, x + \cdots + \text{ord}_{\gamma_{jk}}\, x.$$

Summing for all such points P_j, it follows that

$$\textstyle\sum I(P_j, \mathbf{v}(X) \cap \mathcal{F}) = \sum \text{ord}_t\, x(t);$$

the right-hand side counts the zeros of x with appropriate multiplicity. To finish the proof, it suffices to note that, if $P_j = (0, y_j, 1)$ and y_j is an m_j-ple root of the polynomial $Y^m + c_0(0)Y^{m-1} + \cdots$, then $I(P_j, \mathbf{v}(X) \cap \mathcal{F}) = m_j$; hence $\sum_j I(P_j, \mathbf{v}(X) \cap \mathcal{F}) = m$. □

THEOREM 5.34 *The function ζ has finitely many poles. The number of poles each counted with its multiplicity is $[\Sigma : K(\zeta)]$.*

Proof. Poles of ζ are zeros of $1/\zeta$,. Hence there are only a finite number of them, and this number, counting with multiplicity, is $[\Sigma : K(1/\zeta)] = [\Sigma : K(\zeta)]$. □

COROLLARY 5.35 *The number of zeros of ζ equals the number of its poles. Equivalently,*

$$\sum_{\mathcal{P}\in\mathfrak{P}(\Sigma)} \operatorname{ord}_{\mathcal{P}} \zeta = 0,$$

where the summation extends over all places $\mathcal{P}$.

EXAMPLE 5.36 Suppose that $p = 5$, and let $K = \overline{\mathbf{F}}_5$. Take $\mathcal{F} = \mathbf{v}(f(X,Y)$ with $f(X,Y) = Y^5 + Y - X^3$; then $K(\mathcal{F}) = \Sigma = K(x,y)$ with $y^5 + y - x^3$. The irreducible plane curve $\mathcal{F}$ has a unique singular point, namely the double point $X_\infty = (0,1,0)$. It is the centre of a unique branch. Therefore there is a bijection between branches and points of $\mathcal{F}$; so branches and points may be identified.

To illustrate Corollary 5.35, let $g(x) = x^2(1-x^{12})$ and compute its zeros and poles. The roots ξ of the polynomial $g(X) = X^2(1-X^{12})$ are 0, counted twice, together with the 12 non-zero square elements of $\mathbf{F}_{25}$, the quadratic extension of $\mathbf{F}_5$ in K. If w is a primitive element of $\mathbf{F}_{25}$, then $w_i = w^{2i}$ with $i = 1, \ldots, 12$ are the non-zero square elements. Therefore the zeros of $g(x)$ are the places associated with the points $P = (a,b)$ such that $b^5 + b = a^3$ and either $a = 0$, counted twice, or a is a non-zero square element of $\mathbf{F}_5$. The former points are five, say $P_1, \ldots, P_5$, while the latter ones are 60, say $Q_1, \ldots, Q_{60}$. These 65 points together with X_∞ are those points of $\mathcal{F}$ which lie in PG(2, 25). If $\mathcal{P}_i$ and $\mathcal{Q}_i$ are the corresponding places, then the zero divisor of $g(x)$ is

$$\operatorname{div}(x^2(1-x^{12}))_0 = 2\sum_{i=1}^{5} \mathcal{P}_i + \sum_{i=1}^{60} \mathcal{Q}_i.$$

Also, $g(x)$ has a unique pole, namely the place $\mathcal{P}_\infty$ arising from X_∞. Since $(x(t) = t^{-5} + \cdots,\ y(t) = t^{-3} + \cdots)$ is a primitive representation of the unique branch of $\mathcal{F}$ centred at X_∞, the pole divisor of $g(x)$ is

$$\operatorname{div}(x^2(1-x^{12}))_\infty = 70\mathcal{P}_\infty.$$

5.5 SEPARABILITY AND INSEPARABILITY

Any field of transcendence degree 1 over K can be obtained by a transcendental extension $K(x)/K$ followed by an algebraic extension $\Sigma/K(x)$. In positive characteristic p, the latter extension $\Sigma/K(x)$ can be split into a separable step $\Sigma_s/K(x)$ and a purely inseparable step Σ/Σ_s; see Section A.1. Separable extensions have similar properties to algebraic extensions in characteristic zero. Purely inseparable extensions are related to certain subfields of Σ containing K, namely, the subfields $\Sigma^q = \{\xi^q \mid \xi \in \Sigma\}$, where q denotes a power of p.

THEOREM 5.37 *For every q, the extension Σ/Σ^q is purely inseparable of degree q. Conversely, if $K \subset \Sigma' \subset \Sigma$ and Σ/Σ' is purely inseparable of degree q, then $\Sigma' = \Sigma^q$.*

Proof. Since K is perfect, it follows that Σ^q is a field that contains K. The extension Σ/Σ^q is inseparable, as $\xi^q \in \Sigma^q$ for every $\xi \in \Sigma$. To calculate the degree of Σ/Σ^q, choose a local parameter $x \in \Sigma$ at some place $\mathcal{P}$ of Σ.

By the Theorem of the Primitive Element, Theorem A.1, there exists $y \in \Sigma$ such that $\Sigma = K(x,y)$. Also, $\Sigma = K(x,y^q)$. In fact, $\Sigma = K(x,y^q)(y)/K(x,y^q)$ is a purely inseparable extension since y satisfies the equation $X^q - y^q = 0$ over $K(x,y^q)$. On the other hand, $K(x) \subset K(x,y^q) \subset \Sigma$, and therefore $\Sigma/K(x,y^q)$ is also a separable extension. This is only possible when $\Sigma = K(x,y^q)$.

Now $\Sigma^q = K(x^q,y^q)$ and $\Sigma = K(x,y^q)$ imply that $\Sigma = \Sigma^q(x)$. Since x is a root of the polynomial $X^q - x^q$ with coefficients in Σ^q, it follows that $[\Sigma : \Sigma^q] \leq q$.

To prove the reverse inequality, let $a_0^q X^m + \cdots + a_m^q$ be a minimal polynomial of the extension $\Sigma = \Sigma^q(x)/\Sigma^q$. It is enough to show that $m \geq q$. Assume that $\mathrm{ord}_{\mathcal{P}}\, a_i^q x^i$ is minimal; that is, $\mathrm{ord}_{\mathcal{P}}\, a_i^q x^i \leq \mathrm{ord}_{\mathcal{P}}\, a_j^q x^j$ for every $j = 0,\ldots,m$, Then $a_0^q x^m + \cdots + a_m^q = 0$ yields that $\mathrm{ord}_{\mathcal{P}}\, a_i^q x^i = \mathrm{ord}_{\mathcal{P}}\, a_k^q x^k$ for some k, with $0 \leq k \leq m$ and $i \neq k$. Since $\mathrm{ord}_{\mathcal{P}}\, x = 1$, it follows that $i \equiv k \pmod{q}$. Therefore $m \geq q$.

For the converse, assume that the extension Σ/Σ' is purely inseparable of degree q. Then $\xi^q \in \Sigma'$ for every $\xi \in \Sigma$. Hence $\Sigma^q \subset \Sigma' \subset \Sigma$. Since the degree of Σ/Σ^q is q, so it follows that $\Sigma^q = \Sigma'$. □

Let Σ' be a subfield of Σ which contains K and has transcendence degree 1 over K. The extension Σ/Σ' is finite. Also, there is a unique intermediate field Σ'', with $\Sigma' \subset \Sigma'' \subset \Sigma$ such that Σ''/Σ' is separable and Σ/Σ'' is purely inseparable. Then, $[\Sigma : \Sigma'] = [\Sigma : \Sigma''][\Sigma'' : \Sigma']$, where $[\Sigma'' : \Sigma']$ is the *separable degree* and $[\Sigma : \Sigma'']$ is the *inseparable degree* of the extension $[\Sigma : \Sigma']$. An element $\xi \in \Sigma$ is a *separable* or *inseparable variable* of Σ according as the finite extension $\Sigma/K(\xi)$ is separable or inseparable.

A separable variable ζ of Σ is characterised in the following manner. For any element $\xi \in \Sigma\backslash K$, the irreducible polynomial $f(X,Y) \in K[X,Y]$ linking ξ and ζ, that is $f(\xi,\zeta) = 0$, has the property that the polynomial $f(X,\zeta) \in K(\zeta)[X]$ is separable; that is, its derivative $df(X,\zeta)/dX$ is not identically zero. Note that $df(X,\zeta)/dX = \partial f(X,Y)/\partial X$ with $Y = \zeta$.

A useful criterion for separability in Σ is given in the following lemma.

Lemma 5.38 *An element $\zeta \in \Sigma$ is a separable variable if and only if $\zeta \notin \Sigma^p$, that is, ζ is not a p-th power of an element in Σ.*

Proof. If ζ is a separable variable, $K(\zeta) \not\subset \Sigma^p$, because Σ/Σ^p is purely inseparable of degree greater than 1. Conversely, if $\zeta \in \Sigma\backslash K$ is not separable, there is an intermediate field Σ' with $K(\zeta) \subset \Sigma' \subset \Sigma$ such that Σ/Σ' is purely inseparable of degree p. By Theorem 5.37, $\Sigma' = \Sigma^p$, whence it follows that $\zeta \in \Sigma^p$. □

5.6 FROBENIUS RATIONAL TRANSFORMATIONS

In this section, K has positive characteristic p, and q denotes a power of p. Also, if $g(X,Y) \in K[X,Y]$ and $g(X,Y) = \sum c_{ij} X^i Y^j$, then $g^q(X,Y)$ denotes the polynomial $\sum c_{ij}^q X^i Y^j$. The previous notation is maintained: $\mathcal{F} = \mathbf{v}(f(X,Y))$ is an irreducible curve; $P = (x,y)$ is a generic point of $\mathcal{F}$; the field $\Sigma = K(x,y)$ is the associated function field; ω is a rational transformation of Σ; the subfield $\omega(\Sigma) = \{\omega(\xi) \mid \xi \in \Sigma\}$ is the image of Σ under ω.

DEFINITION 5.39 The rational transformation

$$\omega_q : x' = x^q, \quad y' = y^q$$

of Σ is the *q-th Frobenius* transformation.

Let $f(X,Y) = \sum a_{ij} X^i Y^j$. Note that $f^q(X,Y)$ is also irreducible. Also, the image curve of $\mathcal{F}$ under ω_q is $\mathcal{F}' = \mathbf{v}(f^q(X,Y))$. In fact, as $x' = x^q$, $y' = y^q$, so $f^q(x',y') = \sum a_{ij}^q x'^i y'^j = (\sum a_{ij} x^i y^j)^q = 0$. In particular, $P' = (x',y')$ is a generic point of $\mathcal{F}'$, and the subfield $\Sigma' = K(x',y')$ of Σ can be viewed as the function field associated with $\mathcal{F}'$.

THEOREM 5.40 *The mapping ω_q is not birational. However, $\Sigma \cong \Sigma'$.*

Proof. If ω_q were birational, then both x and y, and hence every element in Σ, would be a q-th power; but this is impossible. Now it is shown that the mapping

$$\rho : \Sigma \longrightarrow \Sigma', \qquad \frac{U(x,y)}{V(x,y)} \longmapsto \frac{U^q(x',y')}{V^q(x',y')},$$

with $U(X,Y), V(X,Y) \in K[X,Y]$, $V(x,y) \neq 0$, is well-defined and does not depend on the representative functions. To do this, choose U_1, $V_1 \in K[X,Y]$ so that

$$\frac{U_1(x,y)}{V_1(x,y)} = \frac{U(x,y)}{V(x,y)}.$$

Then $U(x,y)V_1(x,y) - U_1(x,y)V(x,y) = 0$. As $P = (x,y)$ is a generic point of $\mathcal{F}$, there exists $h(X,Y) \in K[X,Y]$ such that

$$f(X,Y)h(X,Y) = U(X,Y)V_1(X,Y) - U_1(X,Y)V(X,Y).$$

It follows that

$$f^q(X,Y)h^q(X,Y) = U^q(X,Y)V_1^q(X,Y) - U_1^q(X,Y)V^q(X,Y).$$

As $f^q(x,'y') = 0$, this shows that

$$U(x',y')V_1(x',y') - U_1(x',y')V(x',y') = 0,$$

whence

$$\frac{U_1^q(x',y')}{V_1^q(x',y')} = \frac{U^q(x',y')}{V^q(x',y')}.$$

Since the mapping ρ induces an automorphism of K, so ρ is a homomorphism from Σ to Σ'. Actually, ρ is injective, as $\xi^q = 0$ implies $\xi = 0$, and surjective as $\sum b_{ij} x'^i y'^j$ is the image of $\sum a_{ij} x^i y^j$ when $a_{ij}^q = b_{ij}$ for every i and j. Therefore ρ is an isomorphism. □

DEFINITION 5.41 A rational transformation ω of Σ is either *separable* or *inseparable* according as the field extension $\Sigma/\omega(\Sigma)$ is separable or inseparable.

THEOREM 5.42 (i) *Every inseparable rational transformation ω of Σ is the product of a separable rational transformation by a q-th Frobenius transformation.*

(ii) *The inseparability degree of ω is equal to q.*

Proof. Let ω be given by the generic points $P = (x, y)$ of $\mathcal{F}$ and $P' = (x', y')$ of $\mathcal{F}'$. If either x or y is not a p-th power, $\Sigma/\omega(\Sigma)$ is separable. Otherwise, let q denote the highest power of p such that $x' = \xi^q$ and $y' = \eta^q$. By Theorem 5.37, $\omega(\Sigma)$ is purely inseparable and its inseparability degree is equal to q. Further, the rational transformation ω_s given by the generic point (ξ, η) is separable. Hence ω is the product of ω_s and the q-th Frobenius transformation. □

5.7 DERIVATIONS AND DIFFERENTIALS

DEFINITION 5.43 A *derivation* D in a field L is a mapping $D : L \to L$ given by $a \mapsto a' = D(a)$ such that, for all $a, b \in L$,

(i) $(a + b)' = a' + b'$;

(ii) $(ab)' = a'b + ab'$.

LEMMA 5.44 (i) $0' = 0$;

(ii) $1' = 0$;

(iii) $(a/b)' = (a'b - ab')/b^2$ *for all* $a, b \in L$ *with* $b \neq 0$.

Proof. (i) Here, $0' = (0 + 0)' = 0' + 0'$, whence $0' = 0' - 0' = 0$.
(ii) In this case, $1' = (1 \cdot 1)' = 1' \cdot 1 + 1 \cdot 1' = 1' + 1'$, whence $1' = 1' - 1' = 0$.
(iii) Finally, $a' = (b \cdot a/b)' = b' \cdot a/b + b \cdot (a/b)'$. So

$$(a/b)' = (a' - b' \cdot a/b)/b = a'/b - b' \cdot a/b^2 = (a'b - ab')/b^2.$$ □

DEFINITION 5.45 Given a subfield M of L, a derivation D of L is said to be *over* M if $D(c) = 0$ for each $c \in M$.

Let D be a derivation of L over M and let $\theta, \phi \in L$; then

$$D(\theta^i \phi^j) = i\theta^{i-1}\phi^j D\theta + j\theta^i \phi^{j-1} D\phi.$$

So

$$D(F(\theta, \phi)) = \frac{\partial F}{\partial \theta} D\theta + \frac{\partial F}{\partial \phi} D\phi,$$

where $F \in M[X, Y]$. Here, $\partial F/\partial \theta$ means $\partial F/\partial X$ evaluated for $X = \theta,\ Y = \phi$.

Consider now the particular case in which $L = K((t))$. In $K((t))$, a derivation is given by the following rule:

$$\frac{d}{dt}\sum c_i t^i = \sum i c_i t^{i-1},$$

with $c_i \in K$. Take dt as an indeterminate and consider the field $K((t))(dt)$. If $\theta \in K((t))$, define the *differential* of θ in the following way:

$$d\theta = \left(\frac{d\theta}{dt}\right) dt.$$

Let $\phi \in K((t))\backslash K$ and, when $p > 0$, suppose also that $d\phi/dt \neq 0$; that is, $\phi \notin K((t^p))$. Define, in the field $K((t))(dt)$,

$$\frac{d\theta}{d\phi}$$

as $d\theta$ divided by $d\phi$. From the rule that

$$\frac{d}{dt}(\theta_1\theta_2) = \theta_1\frac{d\theta_2}{dt} + \theta_2\frac{d\theta_1}{dt},$$

multiplication by dt gives

$$d(\theta_1\theta_2) = \theta_1 d\theta_2 + \theta_2 d\theta_1$$

and similarly

$$d(\theta_1 + \theta_2) = d\theta_1 + d\theta_2.$$

Dividing by $d\phi$ gives

$$\frac{d(\theta_1\theta_2)}{d\phi} = \theta_1\frac{d\theta_2}{d\phi} + \theta_2\frac{d\theta_1}{d\phi},$$

$$\frac{d(\theta_1 + \theta_2)}{d\phi} = \frac{d\theta_1}{d\phi} + \frac{d\theta_2}{d\phi}.$$

Therefore

$$\theta \mapsto \frac{d\theta}{d\phi}$$

is a derivation of $K((t))$, which coincides with d/dt for $\phi = t$. Such a derivation is denoted by $d/d\phi$.

As

$$\frac{d}{dt}F(\theta,\phi) = \frac{\partial F}{\partial \theta}\frac{d\theta}{dt} + \frac{\partial F}{\partial \phi}\frac{d\phi}{dt},$$

multiplying by dt gives

$$\left[\frac{d}{dt}F(\theta,\phi)\right] dt = \frac{\partial F}{\partial \theta}\frac{d\theta}{dt}dt + \frac{\partial F}{\partial \phi}\frac{d\phi}{dt}dt,$$

and hence

$$dF(\theta,\phi) = \frac{\partial F}{\partial \theta}d\theta + \frac{\partial F}{\partial \phi}d\phi,$$

where F is a rational function f/g with $f, g \in K[X,Y]$ and $g(\theta,\phi) \neq 0$.

To deal with Σ, a characterisation for separability in terms of derivatives is needed.

THEOREM 5.46 *The element $\zeta \in \Sigma$ is a separable variable if and only if $d\bar{\zeta} \neq 0$ for each place $\mathcal{P}$ of Σ.*

Proof. By the Theorem of the Primitive Element, there exists $\xi \in \Sigma$ such that $\Sigma = K(\xi, \zeta)$. Note that for any place $\mathcal{P}$, either $d\bar{\xi}$ or $d\bar{\zeta}$ does not vanish. In fact, for a primitive place representation τ of $\mathcal{P}$, the equality $d\tau(\xi) = 0$ occurs if and only if $\tau(\xi) \in K((t))$ has no term $c_i t^i$ with $i \not\equiv 0 \pmod{p}$. If $d\tau(\zeta) = 0$ also happens then neither $\tau(\zeta) \in K((t))$ has a term $c_i t^i$ with $i \not\equiv 0 \pmod{p}$. Then $(\tau(\xi), \tau(\zeta))$ is an imprimitive representation of the branch of the model $(\mathcal{F}; (\xi, \zeta))$. But this would mean that τ is an imprimitive place representation.

Let $U(X,Y) \in K[X,Y]$ be an irreducible polynomial for which $U(\xi, \zeta) = 0$, and $\mathcal{F} = \mathbf{v}(U)$ the corresponding irreducible curve having $P = (\xi, \zeta)$ as a generic point. For any place $\mathcal{P}$ of Σ, the equation $U(\bar{\xi}, \bar{\zeta}) = 0$ holds; hence, by derivation,

$$(\partial U/\partial \bar{\xi}) d\bar{\xi} + (\partial U/\partial \bar{\zeta}) d\bar{\zeta} = 0.$$

Assume the existence of a place $\mathcal{P} \in \mathfrak{P}(\Sigma)$ such that $d\bar{\zeta} = 0$. Then $\bar{\zeta} = \tau(\zeta)$ for a primitive representation τ of $\mathcal{P}$, and so

$$(\partial U/\partial \bar{\xi}) d\bar{\xi} = 0.$$

Since $d\bar{\xi} \neq 0$, it follows that

$$\partial U/\partial \bar{\xi} = 0.$$

Therefore either $\partial U(X,Y)/\partial X = 0$ and ζ is inseparable or $\partial U(X,Y)\partial X \neq 0$ and $P = (\xi, \zeta)$ is also a generic point of the curve $\mathbf{v}(\partial U(X,Y)\partial X)$. The latter case cannot actually occur, as $\deg \partial U(X,Y)\partial X < \deg U(X,Y)$, while $U(X,Y)$ is irreducible.

Conversely, suppose ζ is inseparable; then there exist ξ in Σ and $U(X,Y)$ in $K[X,Y]$ such that

$$U(\xi, \zeta) = 0, \quad \partial U(X, \zeta)/\partial X = 0.$$

Therefore, for each place $\mathcal{P}$ of Σ,

$$U(\bar{\xi}, \bar{\zeta}) = 0, \quad \partial U(\bar{\xi}, \bar{\zeta})/\partial \bar{\xi} = 0.$$

It follows that also

$$(\partial U(\bar{\xi}, \bar{\zeta})/\partial \bar{\zeta}) d\bar{\zeta} = 0.$$

Here,

$$\partial U(\bar{\xi}, \bar{\zeta})/\partial \bar{\zeta} \neq 0$$

by the irreducibility of $U(X,Y)$. So $d\bar{\zeta} = 0$, as required. □

Given a place $\mathcal{P}$ of Σ, let $\tau : \Sigma \to K((t))$ be a primitive representation of $\mathcal{P}$. Having fixed a separable variable ζ of Σ, let ξ be any element of Σ. Then there exists a polynomial $f(X,Y) \in K[X,Y]$ such that $f(\xi, \zeta) = 0$. it follows that $f(\bar{\xi}, \bar{\zeta}) = 0$, where, with the usual conventions, $\bar{\xi}(t) = \tau(\xi)$ and $\bar{\zeta}(t) = \tau(\zeta)$. Therefore $df(\bar{\xi}, \bar{\zeta}) = 0$, and so

$$\frac{\partial f}{\partial \bar{\xi}} d\bar{\xi} + \frac{\partial f}{\partial \bar{\zeta}} d\bar{\zeta} = 0.$$

Note that $\partial f/\partial\bar{\xi} \neq 0$. If this were not true, then $\partial f/\partial X = 0$ is satisfied by $(\bar{\xi}, \bar{\zeta})$; but this implies that the polynomial $\partial f/\partial X$ is identically zero, simply from the fact that such a polynomial, if not zero, would have degree less than that of $f(X, Y)$, contradicting the irreducibility of $f(X, Y)$. Since $\partial f/\partial X = 0$, so $\partial f/\partial\xi = 0$; but this is impossible with ζ a separable variable of Σ. By the previous theorem, $d\bar{\zeta} \neq 0$.

Therefore it makes sense to write

$$\frac{d\bar{\xi}}{d\bar{\zeta}} = -\frac{\partial f/\partial\bar{\zeta}}{\partial f/\partial\bar{\xi}} \in K((t)), \tag{5.1}$$

and to make the following definition.

DEFINITION 5.47

$$\frac{d\xi}{d\zeta} = -\frac{\partial f/\partial\zeta}{\partial f/\partial\xi} \in \Sigma. \tag{5.2}$$

In a diagram,

$$\begin{array}{ccc} \xi & \xrightarrow{d/d\zeta} & d\xi/d\zeta \\ \tau \downarrow & & \downarrow \tau \\ \bar{\xi} & \xrightarrow{d/d\bar{\zeta}} & d\bar{\xi}/d\bar{\zeta} \end{array}$$

The mapping $\xi \mapsto d\xi/d\zeta$ is a derivation of Σ, since $\bar{\xi} \mapsto d\bar{\xi}/d\bar{\zeta} \in K((t))$ and the mapping $\tau : \Sigma \to K((t))$ is a K-monomorphism.

Now the idea of the differential of a separable variable of Σ is introduced in a similar way to the analogous idea for $K((t))$.

DEFINITION 5.48 (i) For a separable variable ζ of Σ, a *differential* is any element $x d\zeta$, with $x \in \Sigma$, of the field $\Sigma(d\zeta)$, a transcendental extension of Σ by the symbol $d\zeta$.

(ii) The *differential* $d\xi$ of $\xi \in \Sigma$ is defined by the expression

$$d\xi = \frac{d\xi}{d\zeta}\, d\zeta. \tag{5.3}$$

(iii) The differentials with respect to the fixed element ζ form a vector space of dimension 1 over Σ, the *vector space of exact differentials*.

It should be noted that the differential does not depend on the choice of ζ in the following sense. If ζ_1 is another separable variable of Σ, then $d\zeta_1/d\zeta \neq 0$, and if the new indeterminate $d\zeta_1$ is identified with $(d\zeta_1/d\zeta)d\zeta$, then $d\xi$ as defined from $d\zeta_1$ can be considered as the same as $d\xi$ defined from $d\zeta$. If this is done, then

$$\frac{d\xi}{d\zeta}\, d\zeta = \frac{d\xi}{d\zeta_1}\, d\zeta_1,$$

by the following theorem.

THEOREM 5.49 *If $\zeta, \zeta_1 \in \Sigma$ are both separable, then, for each $\xi \in \Sigma$,*

$$\frac{d\xi}{d\zeta_1} \cdot \frac{d\zeta_1}{d\zeta} = \frac{d\xi}{d\zeta}. \tag{5.4}$$

Proof. Let $\mathcal{P}$ be a place of Σ. If σ is a primitive representation of $\mathcal{P}$, then

$$\sigma\left(\frac{d\xi}{d\zeta_1} \cdot \frac{d\zeta_1}{d\zeta}\right) = \sigma\left(\frac{d\xi}{d\zeta_1}\right) \cdot \sigma\left(\frac{d\zeta_1}{d\zeta}\right) = \frac{d\bar{\xi}}{d\bar{\zeta_1}} \cdot \frac{d\bar{\zeta_1}}{d\bar{\zeta}} = \frac{d\bar{\xi}}{d\bar{\zeta}}.$$

Since σ is a K-monomorphism from Σ into $K((t))$, the result is proved. □

The above definition of differential is functional but awkward in that it singles out a separable variable ζ for a special role. An alternative introduction to the concept of the vector space of exact differentials which avoids such an inconvenience is possible. To do this, define for every $\zeta \in \Sigma$, including inseparable variables and elements of K, a symbol $d\zeta$ and consider the infinite-dimensional vector space over Σ whose elements are the formal finite sums

$$\alpha_1 d\zeta_1 + \alpha_2 d\zeta_2 + \cdots,$$

with $\zeta_1, \zeta_2, \ldots \in \Sigma$ and $\alpha_1, \alpha_2, \ldots \in K$; then these symbols satisfy the equations

$$\frac{\partial F}{\partial \zeta_1}\, d\zeta_1 + \frac{\partial F}{\partial \zeta_2}\, d\zeta_2 + \cdots = 0,$$

where $F(X_1, X_2, \ldots)$ is a rational function over K satisfied by $\zeta_1, \zeta_2, \ldots$. This one-dimensional vector space over Σ is the desired vector space of exact differentials.

Familiar results on derivatives, such as the first assertion in the following theorem, hold true in zero characteristic. Extensions to positive characteristic are also possible but inseparability needs to be considered carefully.

THEOREM 5.50 (i) *If $p = 0$, then $d\xi = 0$ only for $\xi \in K$.*

(ii) *If $p > 0$, then ξ is a separable variable of Σ if and only if $d\xi \neq 0$.*

Proof. Let $f(X, Y) \in K[X, Y]$ be irreducible and such that $f(\xi, \zeta) = 0$. If $f(X, Y)$ is written in the usual canonical form

$$f(X, Y) = a_{00} + a_{10}X + a_{01}Y + \cdots + a_{ij}X^iY^j + \cdots,$$

then, with $g(X, Y) = \partial f(X, Y)/\partial Y$,

$$g(X, Y) = a_{01} + a_{11}X + 2a_{02}Y + \cdots + ja_{ij}X^iY^{j-1} + \cdots.$$

Suppose that $d\xi/d\zeta = 0$. From (5.2) it follows that (ξ, ζ) satisfies the equation $g(X, Y) = 0$. Since $P = (\xi, \zeta)$ is a generic point of the curve $\mathbf{v}(f(X, Y))$, that can happen only when $g(X, Y)$ is the zero polynomial.

Then, in zero characteristic,

$$f(X, Y) = f(X, 0) = a_{00} + a_{10}X + \cdots + a_{n0}X^n,$$

whence $f(\xi, 0) = 0$, which implies that $\xi \in K$. Instead, in positive characteristic p,

$$f(X, Y) = \sum a_{ik}X^iY^{pk}.$$

Since $\partial f(\xi, Y)/\partial Y = 0$, so ξ is inseparable. Finally, as $dx = 0$ for any inseparable variable in Σ, so (ii) holds. □

By (5.2), $d\xi/d\zeta$ is an element in Σ. Hence it makes sense to consider the second derivative of ξ, that is,

$$\frac{d^2\xi}{d\zeta^2} = \frac{d(d\xi/d\zeta)}{d\zeta},$$

and iteratively the higher derivatives of ξ:

$$\frac{d^{i+1}\xi}{d\zeta^{i+1}} = \frac{d(d^i\xi/d\zeta^i)}{d\zeta}.$$

Hence

$$\frac{d^n(\xi\eta)}{d\zeta^n} = \sum_{j=0}^{n} \binom{n}{j} \frac{d^j\xi}{d\zeta^j} \cdot \frac{d^{n-j}\eta}{d\zeta^{n-j}}, \tag{5.5}$$

which is essentially a consequence of the product rule:

$$\frac{d(\xi\eta)}{d\zeta} = \frac{d\xi}{d\zeta}\eta + \frac{d\eta}{d\zeta}\xi.$$

5.8 THE GENUS OF A CURVE

Given a place $\mathcal{P}$ of Σ and a primitive representation τ, if ξ is a separable variable of Σ and $\tau(\xi) = \bar{\xi}$, the order of $d\xi$ at $\mathcal{P}$ is defined as follows.

DEFINITION 5.51

$$\mathrm{ord}_{\mathcal{P}}\, d\xi = \mathrm{ord}_t \frac{d\bar{\xi}(t)}{dt}. \tag{5.6}$$

This concept is well defined by Theorem 5.46 and does not depend on the choice of the representative τ.

DEFINITION 5.52 Let ξ be a separable variable of Σ.

(i) A place $\mathcal{P}$ with $\mathrm{ord}_{\mathcal{P}}\, d\xi > 0$ is a *zero* of $d\xi$; if $\mathcal{P}$ is a zero, then it is a *zero of order* $\mathrm{ord}_{\mathcal{P}}\, d\xi$.

(ii) A place $\mathcal{P}$ with $\mathrm{ord}_{\mathcal{P}}\, d\xi < 0$ is a *pole* of $d\xi$; if $\mathcal{P}$ is a pole, then it is a *pole of order* $-\mathrm{ord}_{\mathcal{P}}\, d\xi$.

LEMMA 5.53 *If $x \in \Sigma$ is separable, then $\mathrm{ord}_{\mathcal{P}}\, dx = 0$ except for a finite number of places; that is, dx has only a finite number of zeros and poles.*

Proof. Let y be an element of Σ such that $\Sigma = K(x, y)$. Consider the irreducible curve $\mathcal{F} = \mathbf{v}(F(X, Y))$ for which $P = (x, y)$ is a generic point. Now it is shown that the affine simple points of $\mathcal{F}$ with tangent $\mathbf{v}(X)$ are finite in number. In fact, such a point $P = (a, b)$ of $\mathcal{F}$ also belongs to the polar curve $\mathcal{F}'$ of $\mathcal{F}$ at the point $Q = (0, 1, 0)$. Note that, since x is a separable variable of Σ, so $\partial F(X, Y)/\partial Y$ is not identically zero; hence the polar curve $\mathcal{F}'$ is $\mathbf{v}(\partial F(X, Y)/\partial Y)$.

Then the number in question is at most $(\deg F)(\deg F - 1)$ by Bézout's Theorem. It remains to calculate $\mathrm{ord}_{\mathcal{P}}\, dx$ at the places $\mathcal{P}$ of Σ centred at affine simple points P with tangent distinct from $\mathbf{v}(X)$. As before, this includes all places except perhaps for a finite number. If $P = (a, b)$ is such a point, then $x = a + ct + \cdots$, where $c \neq 0$; otherwise, $\mathcal{F}$ either has $\mathbf{v}(X)$ as tangent at P or P is a singular point. It follows that $dx/dt = c$, whence $\mathrm{ord}_{\mathcal{P}}\, dx = 0$. This proves the result. □

Lemma 5.53 shows that the sum $\sum_{\mathcal{P} \in \mathfrak{P}(\Sigma)} \mathrm{ord}_{\mathcal{P}}\, d\zeta$ is meaningful, since all but a finite number of terms are zero.

A fundamental result that enables the genus of a curve to be defined is the following.

THEOREM 5.54 *For every two separable variable ξ, η of Σ,*

$$\textstyle\sum_{\mathcal{P} \in \mathfrak{P}(\Sigma)} \mathrm{ord}_{\mathcal{P}}\, d\xi = \sum_{\mathcal{P} \in \mathfrak{P}(\Sigma)} \mathrm{ord}_{\mathcal{P}}\, d\eta,$$

where the summation is over all places of Σ.

Proof. Since $d\xi = (d\xi/d\zeta)d\zeta$, so

$$\textstyle\sum_{\mathcal{P} \in \mathfrak{P}(\Sigma)} \mathrm{ord}_{\mathcal{P}}\, d\xi = \sum_{\mathcal{P} \in \mathfrak{P}(\Sigma)} \mathrm{ord}_{\mathcal{P}}(d\xi/d\zeta) + \sum_{\mathcal{P} \in \mathfrak{P}(\Sigma)} \mathrm{ord}_{\mathcal{P}}\, d\zeta.$$

Since $d\xi/d\zeta \in \Sigma$, it follows from Corollary 5.35 that $\sum_{\mathcal{P} \in \mathfrak{P}(\Sigma)} \mathrm{ord}_{\mathcal{P}}(d\xi/d\zeta) = 0$, and so $\sum_{\mathcal{P} \in \mathfrak{P}(\Sigma)} \mathrm{ord}_{\mathcal{P}}\, d\xi = \sum_{\mathcal{P} \in \mathfrak{P}(\Sigma)} \mathrm{ord}_{\mathcal{P}}\, d\zeta$, as required. □

DEFINITION 5.55 Let g be the integer for which

$$\textstyle\sum_{\mathcal{P} \in \mathfrak{P}(\Sigma)} \mathrm{ord}_{\mathcal{P}}\, d\zeta = 2g - 2.$$

Then g, necessarily non-negative, is the *genus of* Σ.

It is shown in the next section that the genus of Σ is less than or equal to the virtual genus of each curve $\mathcal{F}$ which is a model of Σ; equality holds if and only if $\mathcal{F}$ has only ordinary singularities. The following theorem provides a formula for $\mathrm{ord}_{\mathcal{P}}$ for a place centred at a simple point.

THEOREM 5.56 *Let P be a simple point of an irreducible curve $\mathcal{F} = \mathbf{v}(F(X, Y))$ such that the vertical line ℓ through P is the tangent to $\mathcal{F}$ at P, and let $\mathcal{P}$ be the unique place of $\mathcal{F}$ centred at P. Given a generic point (x, y) of $\mathcal{F}$ such that x is separable, then*

$$\mathrm{ord}_{\mathcal{P}}\, dx = \mathrm{ord}_{\mathcal{P}} \left(\frac{\partial F(x, y)}{\partial y} \right).$$

Proof. Without loss of generality, take P to be the origin. Then

$$F(X,Y) = X + \Phi_2(X,Y) + \cdots + \Phi_n(X,Y).$$

Suppose that Y^r, with $r \geq 2$, is the term of least degree not containing X. Write

$$F(X,Y) = X + cY^r + \cdots,$$

where all other terms are divisible by X^2, XY or Y^{r+1}. Then $I(P, \mathcal{F} \cap \mathbf{v}(X)) = r$ and a primitive representation of the branch of $\mathcal{F}$ centred at P is

$$\begin{aligned} x(t) &= -ct^r + \cdots, \\ y(t) &= t. \end{aligned}$$

Hence $dx(t)/dt = -crt^{r-1} + \cdots$; that is, $\text{ord}_{\mathcal{P}}\, dx = r - 1$. Also,

$$\frac{\partial F(X,Y)}{\partial Y} = cr\, Y^{r-1} + \cdots,$$

where all other terms are divisible by X or Y^r. Hence

$$\text{ord}_{\mathcal{P}}(\partial F(x,y)/\partial y) = r - 1,$$

as required.

This proof fails when $p > 0$ and $r = np$ with n a positive integer.

To prove the result in this case, take the derivative of $F(x(t), y(t)) = 0$. Then

$$\left(\frac{\partial F(x(t),y(t))}{\partial X}\right) x'(t) + \left(\frac{\partial F(x(t),y(t))}{\partial Y}\right) y'(t) = 0.$$

Hence

$$\frac{\partial F(x(t),y(t))}{\partial X} x'(t) = -\frac{\partial F(x(t),y(t))}{\partial Y}.$$

Since $\text{ord}_t\, y'(t) = 0$, so

$$\text{ord}_t(\partial F(x(t),y(t))/\partial Y) = \text{ord}_t(\partial F(x(t),y(t))/\partial X) + \text{ord}_t\, x'(t).$$

To obtain the result, it is necessary to show that $\text{ord}_t(\partial F(x(t),y(t))/\partial X) = 0$.

Suppose, therefore, that $\text{ord}_t(\partial F(x(t),y(t))/\partial X) > 0$ and consider the polar curve $\mathcal{F}' = \mathbf{v}(\partial F(X,Y)/\partial X)$ at $(1,0,0)$. Hence $\partial F(x(0),y(0))/\partial X = 0$, from which it follows that $\mathcal{F}'$ passes through P. So the tangent to $\mathcal{F}$ at P passes through $(1,0,0)$. This gives a contradiction since the tangent to $\mathcal{F}$ at P is $\mathbf{v}(X)$. Therefore

$$\text{ord}_t(\partial F(x(t),y(t))/\partial X) = 0,$$

as required. □

THEOREM 5.57 *If $P_1, \ldots, P_k$ are the singular points of an irreducible curve $\mathcal{F}$, each of them ordinary, then the virtual genus of $\mathcal{F}$ is equal to g; that is,*

$$g = \tfrac{1}{2}(d-1)(d-2) - \textstyle\sum_{i=1}^{k} \tfrac{1}{2} m_i(m_i - 1)$$

where d is the degree of $\mathcal{F}$ and m_i is the multiplicity of the point P_i.

Proof. Choose a reference system (X_0, X_1, X_2) in the plane so that $\mathbf{v}(X_0)$ is the line at infinity and so that the following conditions are satisfied:

(i) the points at infinity of $\mathcal{F}$ are d distinct, simple points, none of which is $Y_\infty = (0,0,1)$;

(ii) no tangent to $\mathcal{F}$ at a singular point is the Y-axis.

Let $P = (x, y)$ be a generic point of $\mathcal{F}$. Now calculate $\sum_{\mathcal{P}} \text{ord}_{\mathcal{P}}\, dx$ that, by definition, equals $2g - 2$.

Given these assumptions, the following properties are demonstrated:

(a) dx has poles only at places centred on the line at infinity and $\text{ord}_{\mathcal{P}}\, dx = -2$ for each of these places;

(b) dx has zeros only at places centred at simple points of $\mathcal{F}$ with vertical tangent and $\text{ord}_{\mathcal{P}}\, dx = \text{ord}_{\mathcal{P}}\, \partial F(X,Y)/\partial Y$;

(c) if P is a point of multiplicity m of $\mathcal{F}$ and $\mathcal{F}' = \mathbf{v}(\partial F(X,Y)/\partial Y)$, then $I(P, \mathcal{F} \cap \mathcal{F}') = m(m-1)$.

To show (a), a primitive representation of a branch γ of $\mathcal{F}$ centred at the simple point $P = (0,1,a)$, with $a \neq 0$, must be obtained. Interchanging X_0 and X_1, the point P becomes the affine point $(1,0,a)$ and the branch γ is centred at $(0,a)$ with tangent $\mathbf{v}(X_2 - aX_0 - bX_1)$, with $b \in K$; in non-homogeneous coordinates, this becomes $\mathbf{v}(Y - bX - a)$, which, it should be noted, is not $\mathbf{v}(X)$ for $(0,a)$. A primitive representation of the branch γ has the form

$$x(t) = t,$$
$$y(t) = a + ct^r + \cdots$$

with $r \geq 1$; in homogeneous coordinates,

$$x_0(t) = 1,$$
$$x_1(t) = t,$$
$$x_2(t) = a + ct^r + \cdots .$$

It follows that a representation of γ centred at $(0,1,a)$ has the form

$$x_0(t) = t,$$
$$x_1(t) = 1,$$
$$x_2(t) = a + ct^r + \cdots ;$$

this, in non-homogeneous coordinates, becomes

$$x(t) = t^{-1},$$
$$y(t) = (a + ct^r + \cdots)/t.$$

Hence $dx(t)/dt = -1/t^2$ and $\text{ord}_{\mathcal{P}}\, dx = -2$.

To prove (b), consider a simple affine point $P = (a,b)$ of $\mathcal{F}$. If the line $\mathbf{v}(X - a)$ through P is not the tangent to $\mathcal{F}$ at P, a branch representation centred at P is given by

$$x(t) = a + t,$$
$$y(t) = b + ct^r + \cdots ,$$

whence $\mathrm{ord}_{\mathcal{P}}\, dx = 0$.

If, instead, the tangent to $\mathcal{F}$ at P is $\mathbf{v}(X - a)$, a branch representation is given by

$$\begin{aligned} x(t) &= a + dt^r + \cdots, \\ y(t) &= b + t, \end{aligned}$$

with ≥ 2. By Theorem 5.56,

$$\mathrm{ord}_{\mathcal{P}}\, dx = \mathrm{ord}_{\mathcal{P}} \left(\frac{\partial F(x(t), y(t))}{\partial Y} \right).$$

This means that $\mathrm{ord}_{\mathcal{P}}\, dx = I(P, \mathcal{F} \cap \mathcal{F}')$, where $\mathcal{F}' = \mathbf{v}(\partial F / \partial Y)$.

To finish the proof, it remains to show (c); that is, it is necessary to calculate $I(P, \mathcal{F} \cap \mathcal{F}')$, where P is an m-fold singular point of $\mathcal{F}$. Note that $\mathcal{F}'$ has an $(m-1)$-fold singularity at P.

Now it is shown that the tangents at P for $\mathcal{F}$ are all distinct from those for $\mathcal{G}$, and hence that $I(P, \mathcal{F} \cap \mathcal{F}') = m(m-1)$.

Without loss of generality, let $P = (0,0)$. Then

$$F(X, Y) = \Phi_m(X, Y) + \cdots + \Phi_d(X, Y),$$

with

$$\Phi_m(X, Y) = a_{m,0} X^m + \cdots + a_{i,m-i} X^i Y^{m-i} + \cdots + a_{0,m} Y^m.$$

Note that $a_{0,m} \neq 0$, as otherwise $\mathbf{v}(X)$ is a tangent.

Let $t_1, \ldots, t_m$ be the roots of the polynomial

$$\Phi_m(Z) = \Phi_m(1, Z) = a_{m,0} + \cdots + a_{i,m-i} Z^{m-i} + \cdots + a_{0,m} Z^m.$$

Since P is an ordinary singular point of $\mathcal{F}$, the roots of $\Phi_m(Z)$ are distinct; so $\Phi_m(Z)$ and its derivative $\Phi'_m(Z)$ have no common roots.

To determine the tangents to $\mathcal{F}'$, let

$$\frac{\partial F(X, Y)}{\partial Y} = \Psi_i(X, Y) + \cdots + \Psi_{d-1}(X, Y),$$

where

$$\Psi_i(X, Y) = \frac{\partial \Phi_m(X, Y)}{\partial Y};$$

also $\partial \Phi_m(X, Y)/\partial Y$ cannot be identically zero, as otherwise $\Phi_m(Z)$ would be the p-th power of a polynomial and its roots would have multiplicity p. Hence the tangents to $\mathcal{F}'$ at P are given by the roots of $\Psi_i(Z) = \Psi_i(1, Z)$, except possibly for tangents equal to $\mathbf{v}(X)$.

The curves $\mathcal{F}$ and $\mathcal{F}'$ have a tangent at P in common if and only if $\Phi_m(Z)$ and $\Psi_i(Z)$ have a root in common; but that cannot happen since $\Psi_i(Z) = \partial \Phi_m(Z)/\partial Z$ and $\Phi_m(Z)$ has no multiple roots. This proves (c).

Now, if γ is a branch centred at a singular point of $\mathcal{F}$, it must be linear since P is an ordinary singularity. Hence a primitive representation of γ has the form $x(t) = a + t,\ y(t) = b + ct^r + \cdots$. It follows that $\mathrm{ord}_\gamma\, dx = 0$.

As a consequence,

$$\textstyle\sum_{\mathcal{P}\in\mathfrak{P}(\Sigma)} \mathrm{ord}_{\mathcal{P}}\, dx = \sum I(P, \mathcal{F}\cap\mathcal{F}') - 2d,$$

where the second summation is over all simple points of $\mathcal{F}$ with tangent $\mathbf{v}(X-a)$. Applying Bézout's Theorem, the intersection of $\mathcal{F}$ and $\mathcal{F}'$ can be determined. With $d = \deg\mathcal{F}$ and $d-1 = \deg\mathcal{F}'$,

$$d(d-1) = \sum_{i=1}^{k} m_i(m_i-1) + \sum_{\mathcal{P}\in\mathfrak{P}(\Sigma)} \mathrm{ord}_{\mathcal{P}}\left(\frac{\partial F(x,y)}{\partial y}\right),$$

where the second summation is over all places centred at affine non-singular points of $\mathcal{F}$ with vertical tangent. From (b),

$$d(d-1) = \textstyle\sum_{i=1}^{k} m_i(m_i-1) + \sum_{\mathcal{P}\in\mathfrak{P}(\Sigma)} \mathrm{ord}_{\mathcal{P}}\, dx,$$

and so

$$\sum_{\mathcal{P}\in\mathfrak{P}(\Sigma)} \mathrm{ord}_{\mathcal{P}}\left(\frac{\partial F(x,y)}{\partial y}\right) = d(d-1) - \sum_{i=1}^{k} m_i(m_i-1).$$

Therefore

$$\textstyle\sum_{\mathcal{P}\in\mathfrak{P}(\Sigma)} \mathrm{ord}_{\mathcal{P}}\, dx = d(d-1) - \sum_{i=1}^{k} m_i(m_i-1) - 2d.$$

Finally,

$$d(d-3) - \textstyle\sum_{i=1}^{k} m_i(m_i-1) = 2g-2,$$

whence

$$g = \tfrac{1}{2}(d-1)(d-2) - \tfrac{1}{2}\textstyle\sum_{i=1}^{k} m_i(m_i-1). \qquad \square$$

It has been shown that the genus of an irreducible algebraic curve with only ordinary singularities coincides with its virtual genus. This leads to the following definition.

Definition 5.58 The *genus* of an irreducible algebraic curve is the genus of its function field Σ, whence it is the genus of each model $(\mathcal{F}; (x,y))$ of Σ, and so is the virtual genus of a plane model with only ordinary singular points.

However, the genus of an irreducible curve with some non-ordinary singularities is in general smaller than its virtual genus; more precisely,

$$g \le \tfrac{1}{2}(n-1)(n-2) - \textstyle\sum_P \tfrac{1}{2}r(r-1),$$

where $n = \deg\mathcal{F}$, where P ranges over the singular points of $\mathcal{F}$, and where r is the multiplicity of P.

This follows from Theorem 3.27 and (i) of Lemma 3.28. The following example illustrates this case.

EXAMPLE 5.59 Let $p \neq 2$ and, for a positive integer $h \geq 4$, let

$$F(X,Y) = Y^2 - f(X), \quad \text{with} \quad f(X) = \prod_{i=1}^{h}(X - \alpha_i),$$

where $\alpha_i \neq \alpha_j$ for $1 \leq i < j \leq h$. Let $\mathcal{F}$ be the irreducible curve $\mathbf{v}(F(X,Y))$ and let $P = (x,y)$ be a generic point of $\mathcal{F}$.

First, all the affine points of $\mathcal{F}$ are simple, since all the roots of $f(X)$ are distinct, and Y_∞ is a singular point of multiplicity $h-2$. So the virtual genus g_v of $\mathcal{F}$ is equal to $h-2$.

To determine the genus g of $\mathcal{F}$, the sum $\sum_{\mathcal{P}\in\mathfrak{P}(\Sigma)} \mathrm{ord}_{\mathcal{P}}\, dx$ is calculated. For this, the simple points of $\mathcal{F}$ with vertical tangent and the unique point at infinity must be considered. Among the simple points of $\mathcal{F}$ the only ones with vertical tangent are the points $Q_i = (\alpha_i, 0)$ for $i = 1, \ldots, h$. If $P = (x_0, y_0)$ is a simple point of $\mathcal{F}$, its tangent is $\mathbf{v}(G)$ with

$$G = F_X(x_0,y_0)\,(X - x_0) + F_Y(x_0,y_0)\,(Y - y_0);$$

this is vertical when $F_Y(x_0,y_0) = 0$, that is, if and only if $y_0 = 0$.

A primitive branch representation centred at Q_i is

$$\begin{aligned} x(t) &= \alpha_i + ct^2 + \cdots, \; c \neq 0, \\ y(t) &= t. \end{aligned}$$

Then $dx(t)/dt = 2ct + \cdots$ and so $\mathrm{ord}_t(dx(t)/dt) = 1$. Hence, if $\mathcal{P}$ is the place corresponding to Q_i, then $\mathrm{ord}_{\mathcal{P}}\, dx = 1$.

Actually, $\mathcal{F}$ has only one point at infinity, which is $Y_\infty = (0,0,1)$. Now it is shown that are one or two branches centred at Y_∞, according as h is odd or even. The corresponding places are poles whose order is 3 in the former case, and 2 for both places in the latter case.

Take Y_∞ to the origin $O = (0,0)$ by interchanging the X-axis and the line at infinity. Then the transformed curve is $\mathcal{F}'$, given by

$$Y^{h-2} - X^h + f(X,Y), \quad \text{with} \quad f(X,Y) = X^h - \prod_{i=1}^{h}(X - \alpha_i Y),$$

which shows that X_∞ is an $(h-2)$-ple point of $\mathcal{F}$ with only the one tangent, given by ℓ_∞. A primitive representation $(x'(t), y'(t))$ of a branch γ' of $\mathcal{F}'$ centred at $P = (0,0)$ has the form

$$\begin{array}{rcl} x'(t) & = & t^i + \cdots, \\ y'(t) & = & a\,t^j + \cdots \end{array} \tag{5.7}$$

with $i < j$. From the equation $y(t)^{h-2} - x(t)^h + y(t)^h = 0$, it follows that $(h-2)j = hi$, whence one of the following occurs:

(i) $i = h-2$, $j = h$;

(ii) h is even and $i = \frac{1}{2}(h-2)$, $j = \frac{1}{2}h$.

The two cases are investigated separately.

In case (i), $\mathcal{F}'$ has only one branch centred at $P = (0,0)$. In fact,

$$I(P, \mathbf{v}(X) \cap \mathcal{F}') = h - 2,$$

while $\text{ord}_t\, x'(t) = h - 2$ for every branch γ'. By Theorem 4.50,

$$h - 2 = \sum \text{ord}_t\, x'(t),$$

where the summation is for all branches γ' of $\mathcal{F}'$ centred at $P(0,0)$, whence the assertion follows. Now (5.7) becomes

$$\begin{array}{rcl} x'(t) & = & t^{h-2} + \cdots, \\ y'(t) & = & a\, t^{h} + \cdots. \end{array} \tag{5.8}$$

Going back to $\mathcal{F}$, the corresponding branch γ of $\mathcal{F}$ centred at Y_∞ has a primitive representation

$$\begin{array}{rcl} x(t) & = & t^{-2}E_1(t), \\ y(t) & = & t^{-h}E_2(t)), \end{array} \tag{5.9}$$

with invertible $E_1(t),\ E_2(t) \in K[[t]]$. So $\text{ord}_t\, dx(t)/dt = -3$, since $p \neq 2$. Hence $\text{ord}_{\mathcal{P}_\infty}\, dx = -3$, where $\mathcal{P}_\infty$ is the place coming from γ. This implies that $2g - 2 = h - 3$, whence $g = \frac{1}{2}(h-1)$ and h is odd.

In case (ii), h is even. Now Theorem 4.50 together with $x'(t) = \frac{1}{2}(h-2)$ shows that $\mathcal{F}'$ has exactly two branches centred at $O = (0,0)$, each having a primitive representation

$$\begin{array}{rcl} x'(t) & = & t^{(h-2)/2} + \cdots, \\ y'(t) & = & a\, t^{h/2} + \cdots. \end{array} \tag{5.10}$$

Each of the two corresponding branches of $\mathcal{F}$ centred at Y_∞ has primitive representation

$$\begin{array}{rcl} x(t) & = & t^{-1}F_1(t)), \\ y(t) & = & t^{-h/2}F_2(t), \end{array} \tag{5.11}$$

with invertible $F_1(t),\ F_2(t) \in K[[t]]$. It follows that $\text{ord}_t\, dx(t)/dt = -2$. If $\mathcal{P}_\infty^{(i)}$ with $i = 1, 2$ are the corresponding places, $\text{ord}_{\mathcal{P}_\infty^{(i)}}\, dx = -2$. Therefore

$$2g - 2 = \textstyle\sum_{\mathcal{P}} \text{ord}_{\mathcal{P}}\, dx = h - 4,$$

and so the genus of $\mathcal{F}$ is $g = \frac{1}{2}(h-2)$ for h even.

THEOREM 5.60 *Let m be the smallest non-gap of Σ at a place $\mathcal{P} \in \mathcal{P}_{K(\mathcal{X})}$. Suppose that $\xi \in K(\mathcal{X})$ has a pole of order m at $\mathcal{P}$ but is regular elsewhere. If F is a subfield of Σ containing $K(\xi)$ such that $[F : K(\xi)] > 1$, then F has positive genus.*

Proof. Suppose F is rational, and let $F = K(\eta)$. Then $\xi = f(\eta)/g(\eta)$, where $f(X),\ g(X) \in K[X]$ and the roots of $f(X)$ are distinct from those of $g(X)$. Assume first that $g(X)$ is constant. Then $f(X)$ has positive degree k; in fact, $k > 1$ as $[F : K(\xi)] > 1$. Since any pole of η is also a pole of ξ, the hypothesis implies that η has a unique pole, that is, $\text{div}(\eta)_\infty = -n\mathcal{P}$. Now $m = nk$; but this contradicts the definition of ξ.

Assume now that $g(X)$ is divisible by $X - a$. Then $\mathcal{P}$ is a zero of $\eta - a$. Actually, $\mathcal{P}$ is the unique zero of $\eta - a$, because $\mathcal{P}$ is the unique pole of ξ. Thus $(\eta - a)^{-1}$ has a pole at $\mathcal{P}$ but is regular elsewhere; let $n = \text{div}\,((\eta - a)^{-1})_\infty$. The fact that $\mathcal{P}$

is the unique pole of ξ also implies that $g(X) = c(X-a)^r$ with $r \geq 1$. As $\mathcal{P}$ is a pole of order $n = mr$ of $(\eta - a)^{-l}$, the definition of m implies that $r = 1$.

If $f(X)$ is constant, then $\xi = c(\eta - a)^{-1}$ with $c \in K$, $c \neq 0$, which shows that $F = K(\xi)$, a contradiction.

If $f(X)$ has degree $k > 0$, then every pole of η is a pole of order k of ξ. Again, by the definition of ξ, this can happen only when $k = 1$. But then $c(\eta - b)(\eta - a)^{-1} = \xi$ with $a, b, c \in K$, $c \neq 0$, whence $K(\xi) = F$, a contradiction. □

THEOREM 5.61 *Suppose that* $\Sigma = K(x, y)$. *If* Σ *has genus* g, *then*

$$g \leq ([\Sigma : K(x)] - 1)([\Sigma : K(y)] - 1).$$

Proof. Put $n = [\Sigma : K(x)]$ and $m = [\Sigma : K(y)]$. Let $\mathcal{F} = \mathbf{v}(f(X, Y))$ be the irreducible plane curve with $P = (x, y)$ as a generic point. Let d denote the degree of $\mathcal{F}$. If $Y_\infty \in \mathcal{F}$, choose a non-tangent line $\ell = \mathbf{v}(X - a)$ to $\mathcal{F}$ at Y_∞ and transform X to $X - a$, and similarly for X_∞. Then $\mathcal{F} = \mathbf{v}(g(X, Y))$ with $g(X, Y) = f(X + a, Y + b)$, and $P = (\xi, \eta)$ with $\xi = x - a, \eta = y - b$ is a generic point of $\mathcal{F}$. Note that $K(\xi) = K(x)$, $K(\eta) = K(y)$. Also, no place arising from a branch of $\mathcal{F}$ centred at Y_∞ is a zero of ξ. By Theorem 5.33, Y_∞ is a $(d - n)$–fold point of $\mathcal{F}$. Similarly, X_∞ is a $(d - m)$–fold point of $\mathcal{F}$. Hence the virtual genus

$$g^* \leq \tfrac{1}{2}(d^2 - 3d + 2 - (d - n)(d - n - 1) - (d - m)(d - m - 1)).$$

Since $g \leq g^*$, it suffices to check that

$$\tfrac{1}{2}(d^2 - 3d + 2 - (d - n)(d - n - 1) - (d - m)(d - m - 1)) \leq (n - 1)(m - 1).$$

A straightforward calculation also shows that equality occurs in two cases, namely, for $d = n + m$ and $d = n + m - 1$. Since $g^* = g$ only happens when $\mathcal{F}$ has only ordinary singularities, it follows that $g = (n - 1)(m - 1)$ if and only if each of the following occurs:

(i) either $d = n + m$ or $d = n + m - 1$;

(ii) $\mathcal{F}$ has at most two singular points, namely X_∞ and Y_∞;

(iii) $\mathcal{F}$ has only ordinary singularities.

□

5.9 RESIDUES OF DIFFERENTIAL FORMS

If $y \in K((t))$ with $y = \sum a_k t^k$, then a_{-1}, that is, the coefficient of t^{-1} is the *residue of* y *with respect to* t, denoted by $\text{Res}_t\, y$. Note that the residue is a function, $K((t)) \to K$, which is K-linear. Also, $\text{Res}_t\, y = 0$ when $y \in K[[t]]$.

LEMMA 5.62 *If* $x, y \in K((t))$, *and if* $\text{ord}_t(u) = 1$, *that is,* u *is a local parameter of* $K((t))$, *then*

$$\text{Res}_u \left(y \frac{dx}{du} \right) = \text{Res}_t \left(y \frac{dx}{dt} \right).$$

Proof. To simplify the computations, note that it suffices to prove the lemma in the particular case that $x = t$ and $y = t^n$ with n any integer. Also, the lemma remains true when t is replaced by at with $a \neq 0$. Hence it may be assumed that

$$t = u + a_2u^2 + \cdots, \quad dt/du = 1 + 2a_2u + \cdots.$$

The following must be shown:

$$\mathrm{Res}_u\left(t^n \frac{dt}{du}\right) = \begin{cases} 1 & \text{for } n = -1, \\ 0 & \text{otherwise.} \end{cases}$$

For $n \geq 0$, the assertion follows, as $t^n dt/du$ has no term in negative powers of t.

If $n = -1$, then

$$\frac{1}{t}\frac{dt}{du} = \frac{1 + 2a_2u + \cdots}{u + a_2u^2 + \cdots} = \frac{1}{u} + \cdots,$$

and the residue is equal to 1.

When $n < -1$, the case $p = 0$ is considered first. Here,

$$\mathrm{Res}_u\left(t^n \frac{dt}{du}\right) = \mathrm{Res}_u\left(\frac{d}{du}\left(\frac{1}{n+1}t^{n+1}\right)\right),$$

which is zero for $n \neq -1$. When $p \geq 0$, put $m = -n$ for a fixed $n < -1$. Then there is an integer N and a polynomial $F(X_1, \ldots, X_N)$ with integer coefficients such that

$$\frac{1}{t^m}\frac{dt}{du} = \frac{1 + 2a_2u + \cdots}{u^m(1 + a_3u + \cdots)} = \frac{F(a_2, a_3, \ldots, a_{N+1})}{u} + \cdots.$$

Actually, $F(X_1, \ldots, X_N)$ is the same for any field K. Hence the truth of the assertion in positive characteristic is a formal consequence of the result in zero characteristic, because in that case the assertion holds, showing that $F(X_1, \ldots, X_N)$ must be the zero polynomial. This completes the proof. □

Note that, by Lemma 5.62, $\mathrm{Res}_t(ydx)$, that is, the *residue of the differential form* ydx is meaningful as it is independent of the choice of t.

Next, the case that $\mathrm{ord}_t\, u = m > 1$ is investigated. Up to a multiplication by a constant,

$$u = t^m + b_1t^{m+1} + b_2t^{m+2} + \cdots = t^m(1 + b_1t + b_2t^2 + \cdots).$$

If $K((u))$ is viewed as subfield of $K((t))$, then $[K((t)) : K((u))] = m$; see Exercise 4 in Chapter 4. So the trace function T from $K((t))$ to $K((u))$ is defined.

Lemma 5.63 *Let $u \in K((t))$ be a non-zero element of order $m \geq 1$ such that $m \not\equiv 0 \pmod p$. If $y \in K((t))$, then*

$$\mathrm{Res}_t\left(y\frac{du}{dt}dt\right) = \mathrm{Res}_u(\mathrm{T}(y)du). \tag{5.12}$$

Proof. By Exercise 4 in Chapter 4, the powers $1, t, \ldots, t^{m-1}$ are linearly independent over $K((u))$. Hence they form a basis for $K((t))$ over $K((u))$, and the trace of an element y of $K((t))$ can be computed from the matrix (a_{ij}) representing y with respect to this basis, as in (III) of Section A.3.

Multiplying u by a non-zero constant does not alter the validity of the lemma. Hence it may assumed that

$$u = t^m + b_1 t^{m+1} + \cdots = t^m(1 + b_1 t + b_2 t^2 + \cdots).$$

with $b_1, b_2, \ldots \in K$.

Hence there exist $f_0(u), f_1(u), \ldots, f_{m-1}(u) \in K((u))$ such that

$$t^m = f_0(u) + f_1(u)t + \cdots + f_{m-1}(u)t^{m-1}.$$

These elements can be found recursively and the coefficients of $f_i(u)$ are polynomials in $b_1, b_2, \ldots$ with integer coefficients; that is,

$$f_i(u) = \sum F_\alpha^{(i)}(b) u^\alpha,$$

where each $F_v^{(i)}(b)$ is a polynomial with integer coefficients involving finitely many b_j. Therefore the matrix representing an arbitrary element

$$g_0(u) + g_1(u)t + \cdots + g_{m-1}(u)t^{m-1}$$

in $K((t))$ is of type

$$\begin{bmatrix} G_{0,1}(u) & \cdots & G_{0,m-1}(u) \\ \vdots & \cdots & \vdots \\ G_{m-1,1}(u) & \cdots & G_{m-1,m-1}(u) \end{bmatrix},$$

with $G_{\nu,\mu}(u) \in K((u))$, where the coefficients of $G_{\nu,\mu}(u)$ are polynomials with integer coefficients in the b_j and in the coefficients of the $g_i(u)$. This means that (5.12), if it is true, is a formal identity. The proof of (5.12) may therefore be carried out in zero characteristic; the case of positive characteristic is then a corollary, since if two expression only involving integers are equal then they are also equal $(\text{mod } p)$. So take $p = 0$.

From Theorem 4.27, $u = v^m$ where $v = t + c_2 t^2 + \cdots$. By Lemma 5.62, to prove (5.12) it suffices to show that

$$\operatorname{Res}_v \left(y \frac{du}{dv} dv \right) = \operatorname{Res}_u(\mathrm{T}(y)du). \tag{5.13}$$

Now (5.13) holds if $\operatorname{ord}_t y$ is large enough, in which case both sides are equal to zero. Actually, it suffices to consider the case $y = v^j$ with $j \in \mathbf{Z}$. This simplification is possible by linearity since y can be written as a sum of a finite number of terms $a_j v^j$ and an element whose order is as high as required.

Write $j = ms + r$ with $0 \leq r < m$. Then $v^j = u^s v^r$ and $\mathrm{T}(v^j) = u^s \mathrm{T}(v^r)$. Also

$$\mathrm{T}(v^r) = \begin{cases} 0 & \text{for } r \neq 0, \\ m & \text{for } r = 0, \end{cases}$$

whence

$$\mathrm{T}(v^j) = \begin{cases} mu^s & \text{for } j = ms, \\ 0 & \text{otherwise.} \end{cases}$$

Thus

$$\operatorname{Res}_u(\mathrm{T}(v^j)du) = \begin{cases} m & \text{for } j = -m, \\ 0 & \text{otherwise.} \end{cases} \tag{5.14}$$

For $y = v^j$, the left-hand side in (5.13) is $\mathrm{Res}_v(v^j m v^{m-1} dv)$, which is equal to the right-hand side in (5.14). Note that, if $p \mid m$, both sides of the equation are zero. This completes the proof. □

As before, let Σ denote a field of transcendence degree 1 over K. In the same way that the derivation d/dt of $K((t))$ gives rise to a derivation in Σ, it is possible to define a residue on differential forms. To do this, choose a differential form ydx and a place $\mathcal{P}$ of Σ. If $\sigma : \Sigma \to K((t))$ is a primitive representation of $\mathcal{P}$, put $\bar{z} = \sigma(z)$ for $z \in \Sigma$. Then $\mathrm{Res}_t(\bar{y}d\bar{x})$ is the residue of the differential form $\bar{y}d\bar{x}$ at $\mathcal{P}$. As already noted after the proof of Lemma 5.62, this is independent of the choice of σ.

DEFINITION 5.64 Let $\mathcal{P}$ be a place of Σ. Then the *residue of the differential form ydx at $\mathcal{P}$* is

$$\mathrm{Res}_{\mathcal{P}}(ydx) = \mathrm{Res}_t(\bar{y}d\bar{x}),$$

where $\bar{x} = \sigma(x)$ and $\bar{y} = \sigma(y)$ for a primitive representation σ of $\mathcal{P}$.

The next lemma, which is an essential ingredient in the proof of the main result on residue of differential forms, requires some preliminary observations.

Suppose $y \in \Sigma$ is a primitive element satisfying an irreducible monic polynomial $g(Y) \in K[x][Y]$. The idea is to view $g(Y)$ as a polynomial over $K((x))$. This can be done as follows. Let $K((\tilde{x}))$ be a field of power series. If

$$g(Y) = Y^n + a_1(x)Y^{n-1} + \cdots + a_n(x),$$

with $a_1(X), \ldots, a_n(X) \in K[X]$, put

$$\tilde{g}(Y) = Y^n + a_1(\tilde{x})Y^{n-1} + \cdots + a_n(\tilde{x}).$$

Since the subfield $K(\tilde{x})$ of $K((\tilde{x}))$ is naturally K-isomorphic to $K(x)$, the rational field $K(x)$ may be assumed to be embedded in $K((\tilde{x}))$. This makes it possible to replace $\tilde{x}$ by x. With this convention, $\tilde{g}(Y) = g(Y)$ and $g(Y) \in K[[x]][Y]$.

Since $K[[x]][Y]$ is a unique factorisation domain, $g(Y)$ splits into monic and irreducible factors over $K[[x]]$:

$$g(Y) = g_1(Y) \cdots g_r(Y). \tag{5.15}$$

Put $d_j = \deg g_j(Y)$ for $j = 1, \ldots, r$. The factors are distinct, x being a separable variable of Σ.

Write

$$\begin{aligned} g_j(Y) &= Y^{d_j} + \alpha_1(x)Y^{d_j-1} + \cdots, \\ g_j(X, Y) &= Y^{d_j} + \alpha_1(X)Y^{d_j-1} + \cdots, \end{aligned}$$

with $\alpha_1(X), \ldots \in K[[X]]$. By Hensel's Lemma 4.9, $g_j(0, Y) = 0$ has only one solution, say b_j. Then

$$h_j(X, Y) = g_j(X, Y + b_j) \in K[[X]][Y],$$

with $h_j(0, 0) = 0$, and $h_j(X, Y)$ is still monic and irreducible. By Theorem 4.89 there is a finite extension $K((x))(v_j)/K((x))$ such that $h_j(x(v_j), Y)$ splits completely into monic linear factors in $K[[v_j]][Y]$. Each such factor $Y - f(v_j)$ defines a branch representation $(x = x(v_j), y = f(v_j))$ of the irreducible analytic cycle $h_j(X,Y) = 0$.

From the proof of Theorem 4.89, one of these branch representations is primitive. By Theorem 4.87, it is of the form

$$(x = cv_j^{d_j} + \cdots, \; y = f(v_j)),$$

with $c \neq 0$, and $f(X) \in K[[X]]$. In particular, $K((x))(v_j)/K((x))$ is a finite extension of degree d_j. Then

$$(x = x(v_j), \; y_j = -b_j + f(v_j))$$

defines the unique branch of the analytic cycle $g_j(X,Y) = 0$. The same primitive branch representation

$$(x = x(v_j), y_j = -b_j + f(v_j))$$

defines a branch γ of the irreducible plane curve $\mathcal{F} = \mathbf{v}(g(X,Y))$. Note that γ is a centred at the point $P_j = (0, -b_j)$. Also, by Theorem 4.82, all these d_j factors give rise to the same branch of $g_j(X,Y)$ and hence the same branch γ of $\mathcal{F}$. If $\mathcal{P}_j$ is the associated place of Σ, the subfield $K((v_j))$ of the algebraic closure $\overline{K((x))}$ of $K((x))$ is denoted by $\Sigma_{\mathcal{P}_j}$. Note that, if σ_i is the primitive representation arising from this primitive representation of γ, that is, $\sigma_i(x) = x$, $\sigma_i(y) = y_j$, then σ_i is a K-monomorphism from Σ into $\Sigma_{\mathcal{P}_j}$.

Now each of the d_j elements $-b_j + f(v_j)$ is a root of $g(Y)$ Also, two such elements arising from distinct values of j are different, by Theorem 4.75. Therefore every root of $g(Y)$ is one of these elements for some j with $1 \leq j \leq r$.

Finally, every place $\mathcal{P}$ of $\mathcal{F}$ over $\mathcal{P}'$ is obtained in this way, where $\mathcal{P}'$ is the place of $K(x)$ arising from the origin viewed as a point of the line $\Gamma = \mathbf{v}(Y)$, a non-singular model $(\Gamma; (x,0))$ of $K(x)$.

For each $j = 1, \ldots, r$, let T_j be the trace from $\Sigma_{\mathcal{P}_j}$ to $K((x))$, and T the trace from Σ to $K(x)$.

LEMMA 5.65 *With this notation,*

$$\mathrm{T}(y) = \textstyle\sum_{i=1}^{r} \mathrm{T}_j(\sigma_j(y)). \tag{5.16}$$

Proof. If $[\Sigma : K(x)] = n$, then $\mathrm{T}(y)$ is the coefficient of Y^{n-1} in the irreducible polynomial $g(Y)$ as above. The same holds for the local traces in $\Sigma_{\mathcal{P}_j}$. Hence the equation follows from (5.15). □

REMARK 5.66 Lemma 5.65 holds true for any $y \in \Sigma$. In fact, if y is not a primitive element of Σ over $K(x)$, let z be such a primitive element. By Theorem A.1, which is the Theorem of the Primitive Element in its strong form, there exists $c \in K$ such that $w = y + cz$ is a primitive element. Note that the formula (5.16) is valid for both cz and w. Since (5.16) is linear on both sides, it follows that (5.16) also holds for y.

LEMMA 5.67 *Let x be separable variable of Σ. If $\mathcal{P}'$ is a place of $K(x)$ and $\mathcal{P}_1, \ldots, \mathcal{P}_r$ are the places of Σ lying over $\mathcal{P}$, then*

$$\mathrm{Res}_{\mathcal{P}'}(\mathrm{T}(y)dx) = \textstyle\sum_{i=1}^{r} \mathrm{Res}_{\mathcal{P}_i}(ydx). \tag{5.17}$$

Proof. Without loss of generality, suppose that $\mathcal{P}'$ arises from the origin as in the preceding discussion. By definition,

$$\mathrm{Res}_{\mathcal{P}_i}(y dx) = \mathrm{Res}_{v_i}\left(\sigma_i(y)\frac{\sigma_i(x)}{dv_i}dv_i\right).$$

By Lemma (5.12), the right-hand side is equal to $\mathrm{Res}_x(\mathrm{T}_i(\sigma_i(y))dx)$. Taking (5.16) into account, this gives

$$\textstyle\sum_{i=1}^r \mathrm{Res}_{\mathcal{P}_i}(y dx) = \mathrm{Res}_x(\sum_{i=1}^r \mathrm{T}_i(\sigma_i(y))dx) = \mathrm{Res}_x(\mathrm{T}(y)dx),$$

whence (5.17) follows. □

The main result on residues of differential forms in Σ is the following.

THEOREM 5.68 *Let Σ be a field of transcendence degree* 1 *over K. If $y dx$ is a differential form, then*

$$\sum \mathrm{Res}_{\mathcal{P}}(y dx) = 0. \tag{5.18}$$

Note that the sum is taken over all places $\mathcal{P}$, but it is a finite sum since any differential form has only a finite number of poles, by Lemma 5.53.

Proof. Consider first the rational case; that is, suppose $\Sigma = K(x)$. There exist $h(X)$, $k(X) \in K[X]$ with $k(X) \neq 0$ such that $y = h(x)/k(x)$. Places of $K(x)$ are in one-to-one correspondence with branches of the line $\Gamma = \mathbf{v}(Y)$. If $\mathcal{P}$ is a place of $K(x)$ corresponding to the branch γ centred at the point $P = (c, 0)$ and having primitive representation $(x = c + t, y = 0)$, then

$$\mathrm{Res}_{\mathcal{P}}(y dx) = \mathrm{Res}_t\left(\frac{h(c+t)}{k(c+t)}\right).$$

Expressing the rational function $h(X)/k(X)$ as partial fractions and replacing x by $t + b$ gives the equation

$$\frac{h(x)}{k(x)} = \sum \frac{f_\mu(x)}{(x - b_\mu)^{i_\mu}} + f(x) \tag{5.19}$$

where $f(X)$, $f_\mu(X) \in K[X]$, and $\deg f_\mu(X) < i_\mu$. Write

$$f_\mu(X) = c_{\mu,1}X^{i_\mu - 1} + c_{\mu,2}X^{i_\mu-2} + \cdots + c_{\mu,i_\mu}.$$

To calculate $\mathrm{Res}_{\mathcal{P}}(y dx)$, only the coefficient of x^{-1} matters. With this notation, the sum of the residues taken over all $\mathcal{P}$ whose corresponding branches are centred at affine points is equal to $\sum_\mu c_{\mu,1}$.

Suppose that the branch γ arising from $\mathcal{P}$ is centred at the infinite point of Γ. Then $(x = t^{-1}, y = 0)$ is a primitive representation of γ. Since $dx/dt = -t^{-2}$, the coefficient of t^{-1} in (5.19) can be calculated. The residue of $-f(1/t)/t^2$ at t is zero. The other terms in (5.19) can be written as follows:

$$\begin{aligned}
&-t^{-2}\sum \frac{f_\mu(t^{-1})}{(t^{-1} - b_\mu)^{i_\mu}}\\
&= -t^{-2}\sum \left(c_{\mu,1}(t^{-1})^{i_\mu-1} + c_{\mu,2}(t^{-1})^{i_\mu-2} + \cdots + c_{\mu,i_\mu}\right)\frac{t^{i_\mu}}{(1 - tb_\mu)^{i_\mu}}\\
&= -\sum(c_{\mu,1}t^{-1} + c_{\mu,2} + c_{\mu,3}t + \cdots + c_{\mu,i_\mu}t^{i_\mu-2})(1 + tb_\mu + \cdots)^{i_\mu}.
\end{aligned}$$

This shows that the contribution to the residue only comes from the coefficient of the first term, which is precisely $-\sum_\mu c_{\mu i}$. Therefore $\sum \mathrm{Res}_{\mathcal{P}}(ydx) = 0$.

In the general case, the assertion follows from the rational case together with Lemma 5.67. □

REMARK 5.69 In the classical case, $K = \mathbf{C}$, any irreducible curve has the natural structure of a Riemann surface, and $\mathrm{Res}_P(\omega) = \oint_P \omega/(2\pi i)$. Hence Theorem 5.68 for $K = \mathbf{C}$ follows from Stokes' Theorem.

EXAMPLE 5.70 For an odd integer n not divisible by p, with p odd, let

$$F(X,Y) = Y^{n-2} - X^n + Y^n.$$

Then $\Sigma = K(x,y)$, with $F(x,y) = 0$, has x as a separable variable and ydx as a non-trivial differential form. The place $\mathcal{P}$ of Σ corresponds to a branch γ on the model $(\mathcal{F},(x,y))$ of Σ, where $\mathcal{F} = \mathbf{v}(F)$,

Assume first that γ is centred at an affine point. Then, for any primitive representation σ, both $\sigma(x)$ and $\sigma(y)$ are in $K[[t]]$. Hence $\mathrm{Res}_{\mathcal{P}}(ydx) = 0$. The points of $\mathcal{F}$ at infinity are $P_i = (\epsilon^i, 1, 0)$, where ϵ is a primitive n-th root of unity and $i = 0, 1, \ldots, n-1$. If the centre of γ is $P = (0, \epsilon^i, 1)$, then σ can be chosen such that

$$\sigma(x) = \bar{x} = \frac{\epsilon^i + (1/n)\epsilon^i t^2 + \cdots}{t}, \qquad \sigma(y) = \bar{y} = \frac{1}{t},$$

is a primitive representation of γ. Then

$$\bar{y}d\bar{x} = -\frac{\epsilon^i + (1/n)\epsilon^i t^2 + \cdots}{t^3}.$$

This shows that $\mathrm{Res}_t(\bar{y}d\bar{x}) = -(1/n)\epsilon^i$. Therefore, apart from the constant factor $-(1/n)$, the right-hand side in (5.18) is $\sum_{i=0}^{n-1} \epsilon^i$, that is, the sum of all n-th root of unity. Since this sum is equal to zero, (5.18) follows.

5.10 HIGHER DERIVATIVES IN POSITIVE CHARACTERISTIC

Formula (5.5) in Section 5.7 shows that $d^n\zeta^n/d\zeta^n = n!$. However, this implies that $d^n\zeta^n/d\zeta^n = 0$ for any separable variable $\zeta \in \Sigma$ when p is less than or equal to n. To avoid all the negative consequences of this unusual behaviour, a modification to higher derivatives is made by replacing factorials with binomial coefficients.

The binomial coefficient $\binom{n}{m}$ is usually defined for non-negative integers n and m with $n \geq m$. Here, it is convenient to extend this definition to all integers n and all non-negative integers m by putting

$$\binom{n}{m} = n(n-1)\cdots(n-(m-1))/[m(m-1)\cdots 1], \qquad \binom{n}{0} = 1.$$

Note that $\binom{n}{m} = 0$ when $0 \leq n < m$.

Higher derivatives in $K((t))$ are considered first.

DEFINITION 5.71 For $i = 0, 1, 2, \ldots$, the *i-th Hasse derivative* of $\sum a_k t^k$ is

$$D_t^{(i)}(\sum a_k t^k) = \sum \binom{k}{i} a_k t^{k-i}.$$

Note that $D_t^{(i)}$ for $i \geq 1$ is over K, that is $D_t^{(i)} c = 0$ for $c \in K$ and $i \geq 1$. Also,

$$D_t^{(i)} f = (1/i!) d^i f / dt^i,$$

for $f \in K((t))$ provided that $p = 0$, or $i < p$. In particular, $D_t^{(0)} f = f$ for every $f \in K((t))$, while $D_t^{(1)} f$ coincides with the usual derivative df/dt of f. Further,

$$D_t^{(i)} t^j = \begin{cases} 1 & \text{for } i = j, \\ 0 & \text{for } i > j \geq 0. \end{cases} \tag{5.20}$$

Now some fundamental equations for Hasse derivatives are established.

LEMMA 5.72 *Let $f, g \in K((t))$ and $s \in K$. Then*

$$D_t^{(i)}(f+g) = D_t^{(i)} f + D_t^{(i)} g, \quad D_t^{(i)}(sf) = s D_t^{(i)} f; \tag{5.21}$$

$$D_t^{(i)}(\sum_{k=0}^{\infty} c_k f^k) = \sum_{k=0}^{\infty} D_t^{(i)}(c_k f^k) \quad \textit{for } \mathrm{ord}_t\, f > 0; \tag{5.22}$$

$$D_t^{(i)}(fg) = \sum_{j=0}^{i} D_t^{(j)} f\, D_t^{(i-j)} g; \tag{5.23}$$

$$D_t^{(i)} D_t^{(j)} f = \binom{i+j}{i} D_t^{(i+j)} f; \tag{5.24}$$

$$D_t^{(i)}(f_1 \cdots f_r) = \sum D_t^{(k_1)} f_1 \cdots D_t^{(k_r)} f_r, \tag{5.25}$$

where the last sum is over all r-ples of non-negative integers $(k_1, \cdots, k_r)$ with $k_1 + \cdots + k_r = i$.

Proof. Both equations in (5.21) are immediate. As a consequence, for every positive integer m,

$$D_t^{(i)}(\sum_{k=0}^{m} c_k f^k) = \sum_{k=0}^{m} D_t^{(i)}(c_k f^k).$$

If $\mathrm{ord}_t\, f > 0$, then all terms of a given order m in $D_t^{(i)}(\sum_{k=0}^{\infty} c_k f^k)$ are already present in $D_t^{(i)}(\sum_{k=0}^{m+i} c_k f^k)$, and the same holds for $\sum_{k=0}^{\infty} D_t^{(i)}(c_k f^k)$ and $\sum_{k=0}^{m+i} D_t^{(i)}(c_k f^k)$. Thus (5.22) follows from a finite number of repeated applications of (5.21).

To show (5.23), let $f = \sum c_v t^v$, $g = \sum d_v t^v$. Then the coefficient of t^v in fg is $\sum c_k d_{v-k}$. Therefore $D_t^{(i)}(fg) = \sum e_v t^{v-i}$ with $e_v = \binom{v}{i} \sum c_k d_{v-k}$. On the other hand, from $D_t^{(j)} f = \sum \binom{k}{j} c_k t^{k-j}$ and $D_t^{(i-j)} g = \sum \binom{w}{i-j} d_w t^{w-(i-j)}$,

$$D_t^{(j)} f D_t^{(i-j)} g = \sum_k \sum_w \binom{k}{j} \binom{w}{i-j} c_k d_w t^{k+w-i}.$$

Hence the coefficient of t^{v-i} in $\sum_{j=0}^{i} D_t^{(j)} f D_t^{(i-j)} g$ is

$$\sum_{j=0}^{i} \sum_k \binom{k}{j} \binom{v-k}{i-j} c_k d_{v-k}.$$

Since

$$\sum_{j=0}^{i} \binom{k}{j}\binom{v-k}{i-j} = \binom{v}{i},$$

which may be established by comparing the coefficient of X^i on both sides of $(1+X)^v = (1+X)^k(1+X)^{v-k}$, so (5.23) follows.

Now

$$D_t^{(i)} D_t^{(j)} f = \sum c_v \binom{v}{j}\binom{v-j}{i} t^{v-(i+j)}.$$

This, together with the identity,

$$\binom{v}{j}\binom{v-j}{i} = \binom{i+j}{i}\binom{v}{i+j},$$

proves (5.24). Finally, (5.25) follows from (5.23) by induction on r. □

For any $b \in K$ and non-negative integers i, m, since

$$D_t^{(j)}(b+t) = \begin{cases} b+t & \text{if } j = 0, \\ 1 & \text{if } j = 1, \\ 0 & \text{if } j \geq 2, \end{cases}$$

so (5.25) implies that

$$D_t^{(i)}(b+t)^m = \binom{m}{i}(b+t)^{m-i}. \tag{5.26}$$

Another fundamental equation is the chain rule.

LEMMA 5.73 *Let* $f, g \in K[[t]]$ *and* $\operatorname{ord}_t g > 0$. *Then*

$$D_t^{(i)} f(g(t)) = \left[(D_t^{(1)} g(t))^i D_T^{(i)} f(T) + \textstyle\sum_{j=1}^{i-1} g_j D_T^{(j)} f(T)\right]\Big|_{T=g(t)}, \tag{5.27}$$

where

$$g_j = \sum D_t^{(r_1)} g(t) \cdots D_t^{(r_j)} g(t),$$

and the sum is over all j-ples of positive integers $(r_1, \ldots, r_j)$ with $r_1 + \cdots + r_j = i$.

Proof. By (5.22), it is enough to prove (5.27) for $f(t) = t^n$. From (5.25),

$$D_t^{(i)} g(t)^n = \textstyle\sum_{k_1+\cdots+k_n=i} D_t^{(k_1)} g(t) \cdots D_t^{(k_n)} g(t).$$

In each summand, among the k_i some positive integers occur; let $(s_1, \ldots, s_m)$ be the m-ple arising from $(k_1, \ldots, k_n)$ by omitting the zeros. Then $1 \leq m \leq i$, $s_1 + \cdots + s_m = i$, and

$$D_t^{(k_1)} g(t) \cdots D_t^{(k_n)} g(t) = g(t)^{n-m}(D_t^{(s_1)} g(t) \cdots D_t^{(s_m)} g(t)).$$

Note that each term $g(t)^{n-m}(D_t^{(s_1)} g(t) \cdots D_t^{(s_m)} g(t))$ comes from exactly $\binom{n}{m}$ terms $D_t^{(k_1)} g(t) \cdots D_t^{(k_n)} g(t)$. Hence

$$D_t^{(i)} g(t)^n = \sum_{m=1}^{i} \sum_{s_1+\cdots+s_m=i} \binom{n}{m} g(t)^{n-m} D_t^{(s_1)} g(t) \cdots D_t^{(s_m)} g(t). \tag{5.28}$$

Since

$$\binom{n}{m} g(t)^{n-m} = D_T^{(m)} T^n \Big|_{T=g(t)},$$

the assertion follows. □

EXAMPLE 5.74 Let $f(t) = t^3$ and $g(t) = t^2 + 1$. If $i = 3$, then $g_1 = 0$, $g_2 = 4t$ and $D_t^{(3)} = 6t^3 + 12t^2$.

Another basic relation is given in Lemma 5.75. The proof requires the number-theoretical result given in Lemma A.6.

LEMMA 5.75 *Let $q = p^h$. For $f \in K((t))$ with $\mathrm{ord}_t f > 0$, and $0 \le j, k < q$,*

$$D_t^{(q)} f^{jq+k} = f^{jq} D_t^{(q)} f^k + f^k (D_t f^j)^q. \tag{5.29}$$

Proof. From (5.23),

$$D_t^{(q)} f^{jq+k} = \textstyle\sum_{m=0}^{q} D_t^{(m)} f^{jq} D_t^{(q-m)} f^k.$$

By Lemma A.6, this sum consists of two terms only, namely the first and the last. It remains to show that $D_t^{(q)} f^{jq} = (D_t^{(1)} f^j)^q$.

Put $f^j = \sum c_n t^n$. Then

$$\begin{aligned} D_t^{(1)} f^j &= \textstyle\sum n c_n t^{n-1}, \\ (D_t^{(1)} f^j)^q &= \textstyle\sum n^q c_n^q t^{nq-q}, \\ f^{jq} = (f^j)^q &= \textstyle\sum c_n^q t^{qn}, \\ D_t^{(q)} f^{jq} &= \textstyle\sum c_n^q \binom{nq}{q} t^{nq-q}. \end{aligned}$$

As $\binom{nq}{q} \equiv n \pmod p$, by Lemma A.6, so the result follows. □

Using similar arguments, the following lemma is established.

LEMMA 5.76 *Let $i = rp^m + s$ with $1 \le r < p$ and $0 \le s < p^m$. Then, for all $f \in K((t))$,*

$$D_t^{(i)} f = \frac{1}{r!} \binom{i}{s}^{-1} D_t^{(s)} \underbrace{\left(D_t^{(p^m)} \cdots \left(D_t^{(p^m)} f\right)\right)}_{r \text{ times}}.$$

Proof. By (5.24),

$$D_t^{(p^m)} \cdots \left(D_t^{(p^m)} f\right) = \binom{i}{s} \binom{rp^m}{p^m} \cdots \binom{2p^m}{p^m} D_t^{(i)} f,$$

where, as in Lemma A.6, $\binom{np^m}{p^m} \equiv n \pmod p$ for all $1 \le n \le r$. Since $r < p$ and $s < i$, so $r! \not\equiv 0 \pmod p$ and $\binom{i}{s} \not\equiv 0 \pmod p$. Hence the result follows. □

Now Hasse derivatives in Σ are introduced using a similar idea that led from d/dt to $d/d\zeta$ via any place of Σ; see Section 5.7. This time, however, not all places are suitable, or admissible, as it may happen at some places that the idea does not work. As a matter of fact, computational and some theoretical complications arise when either $\mathrm{ord}_{\mathcal{P}}\, \zeta < 0$, or $\mathrm{ord}_{\mathcal{P}}\, \zeta \ge 0$ but $\mathrm{ord}_{\mathcal{P}}(\zeta - b) > 1$ with $\sigma(\zeta) = b + \cdots$ for a primitive representation of σ of $\mathcal{P}$. This obstacle can be overcome by using the argument below.

By Lemma 5.53, all but finitely many places of Σ are neither zeros nor poles of $d\zeta$. Choose one of these infinite places, say $\mathcal{P}$. Let σ be a primitive representation of $\mathcal{P}$, and put $\sigma(\zeta) = \overline{\zeta}$. Then $\overline{\zeta} = b + \phi(t)$ with $\phi(t) \in K[[t]]$ and $\operatorname{ord}_t \phi(t) = 1$. Since $t \mapsto \phi(t)$ is an automorphism of $K((t))$, it may be assumed that $\overline{\zeta} = b + t$. In other words, $\zeta - b$ is a local parameter at $\mathcal{P}$.

For an element $\xi \in \Sigma$, let $f(X, Y) \in K[X, Y]$ be an irreducible polynomial such that $f(\xi, \zeta) = 0$. Let $\mathcal{F} = \mathbf{v}(f(X, Y))$ be the corresponding irreducible curve. All but finitely many branches of $\mathcal{F}$ are centred at simple affine points such that the corresponding places are neither zeros nor poles of $d\zeta$. Hence $\mathcal{P}$ may be assumed to come from a branch γ of $\mathcal{F}$ with a primitive representation in the form $(\overline{\xi}, \overline{\zeta})$ with $\overline{\xi} = a + \psi(t)$, $\overline{\zeta} = b + t$, and $\psi(t) \in K[[t]]$. Such a place $\mathcal{P}$ is *admissible* with respect to ζ and ξ.

Now Hasse derivatives are generalised to more than one variable. As pointed out in Remark 1.38, in positive characteristic the usual partial derivatives of a polynomial $f(X, Y) = \sum a_{m,k} X^m Y^k$, that is,

$$\frac{\partial^{\alpha+\beta} f(X, Y)}{\partial X^{\alpha} \partial X^{\beta}} = \sum a_{m,k}\, \alpha! \binom{m}{\alpha} \beta! \binom{k}{\beta} X^{m-\alpha} Y^{k-\beta}$$

are changed to Hasse partial derivatives

$$\frac{\partial^{(\alpha+\beta)} f(X, Y)}{\partial X^{(\alpha)} \partial Y^{(\beta)}} = \sum a_{m,k} \binom{m}{\alpha} \binom{k}{\beta} X^{m-\alpha} Y^{k-\beta}.$$

It may be noted that, if u and v are indeterminates, then

$$f(X + u, Y + v) = \sum_{\mu,\nu \geq 0} \frac{\partial^{(\mu+\nu)} f(X, Y)}{\partial X^{(\mu)} \partial Y^{(\nu)}} u^{\mu} v^{\nu}.$$

Lemma 5.77 *The i-th Hasse derivative of $\overline{\xi}$ with respect to t can be computed as follows*:

$$D_t^{(i)} \overline{\xi} = -\frac{1}{\partial f(\overline{\xi}, \overline{\zeta}) / \partial \overline{\xi}} \left[\frac{\partial^{(i)} f(\overline{\xi}, \overline{\zeta})}{\partial \overline{\zeta}^{(i)}} + \sum_{j=1}^{i-1} \frac{\partial^{(i-j+1)} f(\overline{\xi}, \overline{\zeta})}{\partial \overline{\xi} \partial \overline{\zeta}^{(i-j)}} D_t^{(j)} \overline{\xi} + \right.$$
$$\left. \sum_{n=2}^{i} \sum_{j=n}^{i} \sum_{r_1 + \cdots + r_n = j} \frac{\partial^{(i-j+n)} f(\overline{\xi}, \overline{\zeta})}{\partial \overline{\xi}^{(n)} \partial \overline{\zeta}^{(i-j)}} D_t^{(r_1)} \overline{\xi} \cdots D_t^{(r_n)} \overline{\xi} \right], \tag{5.30}$$

where $r_1, \ldots, r_n > 0$ *and*

$$\frac{\partial^{(i-j+n)} f(\overline{\xi}, \overline{\zeta})}{\partial \overline{\xi}^{(n)} \partial \overline{\zeta}^{(i-j)}} = \frac{\partial^{(n)} f(\overline{\xi}, \overline{\zeta})}{\partial \overline{\xi}^{(n)}}$$

when $j = i$.

Proof. Let $f(X, Y) = \sum a_{km} X^k Y^m$. Then $\sum a_{km} \overline{\xi}^k \overline{\zeta}^m = 0$. From (5.21), $\sum a_{km} D_t^{(i)} (\overline{\xi}^k \overline{\zeta}^m) = 0$. By (5.26), $D_t^{(j)} \overline{\zeta}^m = \binom{m}{j} \overline{\zeta}^{m-j}$. From (5.23),

$$\begin{aligned} D_t^{(i)} (\overline{\xi}^k \overline{\zeta}^m) &= \textstyle\sum_{j=0}^{i} D_t^{(j)} \overline{\xi}^k D_t^{(i-j)} \overline{\zeta}^m \\ &= \binom{m}{i} \overline{\zeta}^{m-i} \overline{\xi}^k + \sum_{j=1}^{i} \binom{m}{i-j} \overline{\zeta}^{m-(i-j)} D_t^{(j)} \overline{\xi}^k. \end{aligned}$$

By (5.28), for every j with $1 \le j \le i$,

$$D_t^{(j)}\overline{\xi}^k = k\overline{\xi}^{k-1}D_t^{(j)}\overline{\xi} + \sum_{n=2}^{j}\sum_{r_1+\cdots+r_n=j}\binom{k}{n}\overline{\xi}^{k-n}D_t^{(r_1)}\overline{\xi}\cdots D_t^{(r_n)}\overline{\xi}.$$

Hence

$$\begin{aligned}\sum a_{km}D_t^{(i)}(\overline{\xi}^k\overline{\zeta}^m) = \sum a_{mk}\Bigg[&\binom{m}{i}\overline{\zeta}^{(m-i)}\overline{\xi}^k \\ &+\sum_{j=1}^{i-1}\binom{m}{i-j}k\overline{\zeta}^{m-(i-j)}\overline{\xi}^{k-1}D_t^{(j)}\overline{\xi} + k\overline{\zeta}^m\overline{\xi}^{k-1}D_t^{(i)}\overline{\xi} \\ &\cdot\sum_{n=2}^{i}\sum_{j=n}^{i}\sum_{r_1+\cdots+r_n=j}\binom{m}{i-j}\binom{k}{n}\overline{\zeta}^{m-i+j}\overline{\xi}^{k-n}D_t^{(r_1)}\overline{\xi}\cdots D_t^{(r_n)}\overline{\xi}\Bigg] = 0.\end{aligned} \tag{5.31}$$

Since $\partial f(\overline{\xi},\overline{\zeta})/\partial\overline{\xi} \neq 0$, the assertion now follows by dividing the last equation by $\partial f(\overline{\xi},\overline{\zeta})/\partial\overline{\xi}$. □

Although the above proof is constructive, for higher values of i the computation of the coefficients of $c^{(i)}(X)$ is usually long, since it depends on the number of partitions of a natural number. For $j = n$, the condition that $r_1 + \cdots + r_n = j$ implies that $r_1 = \cdots = r_n = 1$ and the corresponding term is

$$\frac{\partial^{(i)} f(\overline{\xi},\overline{\zeta})}{\partial\overline{\xi}^{(n)}\partial\overline{\zeta}^{(i-n)}}(D_t^{(1)}\overline{\xi})^n;$$

but, for $n = 2$ and $j > n$, there are already $\lfloor j/2 \rfloor$ summands with coefficient

$$\frac{\partial^{(i-j+2)} f(\overline{\xi},\overline{\zeta})}{\partial\overline{\xi}^{(2)}\partial\overline{\zeta}^{(i-j)}}.$$

For small values of i,

$$D_t^{(1)}\overline{\xi} = D_t\overline{\xi} = -\frac{\partial f(\overline{\xi},\overline{\zeta})/\partial\overline{\zeta}}{\partial f(\overline{\xi},\overline{\zeta})/\partial\overline{\xi}}; \tag{5.32}$$

$$\begin{aligned}&D_t^{(2)}\overline{\xi} \\ &= -\frac{1}{\partial f(\overline{\xi},\overline{\zeta})/\partial\overline{\xi}}\left(\frac{\partial^{(2)} f(\overline{\xi},\overline{\zeta})}{\partial\overline{\xi}^{(2)}}(D_t\overline{\xi})^2 + \frac{\partial^{(2)} f(\overline{\xi},\overline{\zeta})}{\partial\overline{\xi}\partial\overline{\zeta}}D_t\overline{\xi} + \frac{\partial^{(2)} f(\overline{\xi},\overline{\zeta})}{\partial\overline{\zeta}^{(2)}}\right).\end{aligned} \tag{5.33}$$

To check (5.33), the procedure in the proof of Lemma 5.30 is used. It suffices to calculate $D_t^{(2)}(\overline{\xi}^k\overline{\zeta}^m)$, and repeat the final argument:

$$\begin{aligned}D_t^{(2)}(\overline{\xi}^k\overline{\zeta}^m) &= \overline{\zeta}^m D_t^{(2)}\overline{\xi}^k + \binom{m}{2}\overline{\zeta}^{m-2}\overline{\xi}^k + m\overline{\zeta}^{m-1}D_t\overline{\xi}^k \\ &= k\overline{\xi}^{k-1}\overline{\zeta}^m D_t^{(2)}\overline{\xi} + \binom{k}{2}\overline{\zeta}^m\overline{\xi}^{k-2}(D_t\overline{\xi})^2 + \binom{m}{2}\overline{\zeta}^{m-2}\overline{\xi}^k \\ &\quad + mk\overline{\zeta}^{m-1}\overline{\xi}^{k-1}D_t\overline{\xi}.\end{aligned}$$

Similarly, the third and fourth derivatives are calculated as follows:

$$
\begin{aligned}
D_t^{(3)}\bar{\xi} = -\frac{1}{\partial f(\bar{\xi},\bar{\zeta})/\partial\bar{\xi}} \Bigg(& 2\frac{\partial^{(2)} f(\bar{\xi},\bar{\zeta})}{\partial\bar{\xi}^{(2)}}(D_t^{(2)}\bar{\xi})(D_t\bar{\xi}) + \frac{\partial^{(3)} f(\bar{\xi},\bar{\zeta})}{\partial\bar{\xi}^{(3)}}(D_t\bar{\xi})^3 \\
& + \frac{\partial^{(2)} f(\bar{\xi},\bar{\zeta})}{\partial\bar{\xi}\partial\bar{\zeta}}(D_t^{(2)}\bar{\xi}) + \frac{\partial^{(3)} f(\bar{\xi},\bar{\zeta})}{\partial\bar{\xi}^{(2)}\partial\bar{\zeta}}(D_t\bar{\xi})^2 \\
& + \frac{\partial^{(3)} f(\bar{\xi},\bar{\zeta})}{\partial\bar{\xi}\partial\bar{\zeta}^{(2)}}(D_t\bar{\xi}) + \frac{\partial^{(3)} f(\bar{\xi},\bar{\zeta})}{\partial\bar{\zeta}^{(3)}} \Bigg);
\end{aligned}
\tag{5.34}
$$

$$
\begin{aligned}
D_t^{(4)}\bar{\xi} = -\frac{1}{\partial f(\bar{\xi},\bar{\zeta})/\partial\bar{\xi}} \Bigg(& \frac{\partial^{(2)} f(\bar{\xi},\bar{\zeta})}{\partial\bar{\xi}^{(2)}}(D_t^{(3)}\bar{\xi})(D_t\bar{\xi}) + \frac{\partial^{(2)} f(\bar{\xi},\bar{\zeta})}{\partial\bar{\xi}}(D_t^{(2)}\bar{\xi})^2 \\
& + 3\frac{\partial^{(3)} f(\bar{\xi},\bar{\zeta})}{\partial\bar{\xi}^{(3)}}(D_t^{(2)}\bar{\xi})(D_t\bar{\xi})^2 + \frac{\partial^{(4)} f(\bar{\xi},\bar{\zeta})}{\partial\bar{\xi}^{(4)}}(D_t\bar{\xi})^4 + \frac{\partial^{(2)} f(\bar{\xi},\bar{\zeta})}{\partial\bar{\xi}\partial\bar{\zeta}}(D_t^{(3)}\bar{\xi}) \\
& + \frac{\partial^{(3)} f(\bar{\xi},\overline{gz})}{\partial\bar{\xi}^{(2)}\partial\bar{\zeta}}(D_t^{(2)}\bar{\xi})(D_t\bar{\xi}) + \frac{\partial^{(4)} f(\bar{\xi},\bar{\zeta})}{\partial\bar{\xi}^{(3)}\partial\bar{\zeta}}(D_t\bar{\xi})^3 + \frac{\partial^{(3)} f(\bar{\xi},\bar{\zeta})}{\partial\bar{\xi}\partial\bar{\zeta}^{(2)}}(D_t^{(2)}\bar{\xi}) \\
& + \frac{\partial^{(4)} f(\bar{\xi},\bar{\zeta})}{\partial\bar{\xi}^{(2)}\partial\bar{\zeta}^{(2)}}(D_t\bar{\xi})^2 + \frac{\partial^{(4)} f(\bar{\xi},\bar{\zeta})}{\partial\bar{\xi}\partial\bar{\zeta}^{(3)}}(D_t\bar{\xi}) + \frac{\partial^{(4)} f(\bar{\xi},\bar{\zeta})}{\partial\bar{\zeta}^{(4)}} \Bigg).
\end{aligned}
\tag{5.35}
$$

As in Lemma 5.77, higher derivatives can be defined using the same approach that produced D_ζ from D_t via a place of Σ.

DEFINITION 5.78 Let ζ be a separable variable of Σ. For $i = 0, 1, 2, \ldots$, the *i-th Hasse derivative* $D_\zeta^{(i)}\xi$ of $\xi \in \Sigma$ is defined iteratively: $D_\zeta^{(0)}\xi = \xi$, and

$$
\begin{aligned}
D_\zeta^{(i)}\xi = -\frac{1}{\partial f(\xi,\zeta)/\partial\xi} \Bigg[& \frac{\partial^{(i)} f(\xi,\zeta)}{\partial\zeta^i} + \sum_{j=1}^{i-1} \frac{\partial^{(i-j+1)} f(\xi,\zeta)}{\partial\xi\partial\zeta^{(i-j)}} D_t^{(j)}\xi \\
& + \sum_{n=2}^{i}\sum_{j=n}^{i}\sum_{r_1+\cdots+r_n=j} \frac{\partial^{(i-j+n)} f(\xi,\zeta)}{\partial\xi^n\partial\zeta^{(i-j)}} D_t^{(r_1)}\xi \cdots D_t^{(r_n)}\xi \Bigg].
\end{aligned}
\tag{5.36}
$$

Diagrammatically,

$$
\begin{array}{ccc}
\xi & \xrightarrow{\;D_\zeta^{(i)}\;} & D_\zeta^{(i)}\xi \\
\tau\big\downarrow & & \big\downarrow\tau \\
\bar{\xi} & \xrightarrow{\;D_t^{(i)}\;} & D_t^{(i)}\bar{\xi}
\end{array}
$$

Some previous remarks also hold true for $D_\zeta^{(i)}$. For $i \geq 1$, the Hasse derivative $D_\zeta^{(i)}$ is over K and, for $i < p$,

$$i!\, D_\zeta^{(i)}\xi = d^i\xi/d\zeta^i.$$

REMARK 5.79 By definition, $\sigma(D_\zeta^{(i)}\xi) = D_t^{(i)}\sigma(\xi)$ for every admissible place $\mathcal{P}$ of Σ, where σ is a primitive representation of $\mathcal{P}$ such that $\sigma(\zeta) = b+t$ with $b \in K$. Since σ is a K-monomorphism from Σ onto $K((t))$, the previous results on $D_t^{(i)}$ extend to $D_\zeta^{(i)}$.

LEMMA 5.80 *All equations but (5.22) in Lemma 5.72 as well as (5.20) and Lemma 5.76 remain true when $K((t))$ and t are replaced by Σ and ζ.*

EXAMPLE 5.81 (i) Suppose that $p = 2$. Let $f(X,Y) = Y + 1 + Y^5X^{23}$ and $\Sigma = K(x,y)$ with $y + 1 = y^5x^{23}$. Then x is a separable variable of Σ. By straightforward calculations,

$$\frac{\partial f(x,y)}{\partial x} = x^{22}y^5;\ \frac{\partial f(x,y)}{\partial y} = \frac{1}{y};$$
$$\frac{\partial^{(2)} f(x,y)}{\partial x^{(2)}} = x^{21}y^5;\ \frac{\partial^{(2)} f(x,y)}{\partial x\partial y} = x^{22}y^4;\ \frac{\partial^{(2)} f(x,y)}{\partial y^{(2)}} = 0;$$
$$\frac{\partial^{(3)} f(x,y)}{\partial x^{(3)}} = x^{20}y^5;\ \frac{\partial^{(3)} f(x,y)}{\partial x^{(2)}\partial y} = x^{21}y^4;\ \frac{\partial^{(3)} f(x,y)}{\partial x\partial y^{(2)}} = \frac{\partial^{(3)} f(x,y)}{\partial y^{(3)}} = 0;$$
$$\frac{\partial^{(4)} f(x,y)}{\partial x^{(4)}} = x^{11}y^5;\ \frac{\partial^{(4)} f(x,y)}{\partial x^{(3)}\partial y} = x^{20}y^4;$$
$$\frac{\partial^{(4)} f(x,y)}{\partial x^{(2)}\partial y^{(2)}} = \frac{\partial^{(4)} f(x,y)}{\partial x\partial y^{(3)}} = 0;\ \frac{\partial^{(4)} f(x,y)}{\partial y^{(4)}} = yx^{23}.$$

From (5.32),

$$D_xy = \frac{x^{22}y^5}{1+x^{23}y^4} = x^{22}y^6.$$

From (5.33), (5.34) and (5.35),

$$D_x^{(2)}y = x^{21}y^7;\ D_x^{(3)}y = x^{20}y^8;\ D_x^{(4)}y = y^6(y^4+y^3+1)x^{19}.$$

(ii) In Example 5.36, x is a separable variable of Σ, and

$$\begin{aligned}
&D_xx = 1,\ D_xy = -2x^2,\ D_xy^2 = x^2y;\\
&D_x^{(2)}x = 1,\ D_x^{(2)}y = -2x,\ D_x^{(2)}y^2 = -xy - x^4;\\
&D_x^{(3)}x = 0,\ D_x^{(3)}y = 1,\ D_x^{(3)}y^2 = -2x^3 + 2y;\\
&D_x^{(4)}x = 0,\ D_x^{(4)}y = 0,\ D_x^{(3)}y^2 = 0;\\
&D_x^{(5)}x = 0,\ D_x^{(5)}y = 2x^{10},\ D_x^{(5)}y^2 = x - x^{10}y;\\
&D_x^{(6)}x = 0,\ D_x^{(5)}y = 0,\ D_x^{(6)}y^2 = 2x^{12} + 1.
\end{aligned}$$

THEOREM 5.82 *Let ζ and η be separable variables of Σ. Then there are elements $d_1,\ldots,d_{i-1}$ in Σ that are polynomials in the indeterminates $D_\eta^{(j)}\zeta$ with integer coefficients for $1 \le j \le i$ such that, for any $\xi \in \Sigma$,*

$$D_\eta^{(i)}\xi = (d\zeta/d\eta)^iD_\zeta^{(i)}\xi + \textstyle\sum_{j=1}^{i-1} d_jD_\zeta^{(j)}\xi. \tag{5.37}$$

Proof. Let $\mathcal{P}$ be an admissible place of Σ with respect to the pair (ξ, ζ). To compare D_η and D_ζ, admissibility with respect to the pair (ξ, η) also enters into considerat-ion. Since there remain infinitely many places with the required properties, $\mathcal{P}$ may be assumed admissible with respect to both pairs (ξ, ζ) and (ξ, η).

Let σ be a primitive representation σ of $\mathcal{P}$. Then

$$\sigma(\zeta) = b + t, \qquad \sigma(\xi) = \overline{\xi} = a + f \text{ with } \mathrm{ord}_t f > 0.$$

Hence $\sigma(D_\zeta^{(j)}\xi) = D_t^{(i)}\overline{\xi} = D_t^{(j)} f$ for every $j \geq 0$. Also, $\sigma(\eta) = \overline{\eta} = c + h$ with $\mathrm{ord}_t\, h = 1$. By Theorem 4.4, the substitution $t \to h$ is a K-automorphism of $K[[t]]$. Its inverse μ also operates as a substitution, namely $t \to g$ with $g \in K[[t]]$. Then $\mu(\overline{\eta}) = c + t$, $\mu(\overline{\xi}) = a + \mu(f) = a + f(g(t)))$, and $\mu(\overline{\zeta}) = b + g(t)$. Let

$$\xi^* = a + f(g(t)),\ \eta^* = c + t,\ \zeta^* = b + g(t).$$

Since $\mu\sigma$ is another primitive representation of $\mathcal{P}$, for every $j \geq 0$,

$$(\mu\sigma)(D_\eta^{(j)}\xi) = D_t^{(j)}\xi^* = D_t^{(j)} f(g(t));$$
$$(\mu\sigma)(D_\eta^{(j)}\zeta) = D_t^{(j)}\zeta^* = D_t^{(j)} g(t).$$

Also,

$$(\mu\sigma)(D_\zeta^{(j)}\xi) = \mu(D_t^{(j)}\bar{\xi}) = D_T^{(i)} f(T))|_{T=g(t)},$$

Since $\mu\sigma$ is a monomorphism from Σ into $K((t))$, Theorem 5.82 follows from Lemma 5.73. □

Take $p > 0$. Again, let ζ be a separable variable of Σ. For any positive integer m, put

$$\Sigma_m = \{\xi \in \Sigma \mid D_\zeta^{(i)}\xi = 0,\ 1 \leq i < p^m\},$$
$$\Sigma_\infty = \{\xi \in \Sigma \mid D_\zeta^{(i)}\xi = 0,\ i = 1, 2, \ldots\}.$$

Then $\Sigma_m \supset K$ and $\Sigma_\infty \supset K$.

LEMMA 5.83 *For any positive integer m,*

(i) *Σ_m is a subfield of Σ containing K;*

(ii) *$D_\zeta^{(p^m)}$ is a derivation of Σ_m;*

(iii) *$\Sigma_{m+1} = \{\xi \in \Sigma_m \mid D_\zeta^{(p^m)}\xi = 0\}$.*

Proof. (i) This follows from Lemma 5.72.

(ii) From (5.23),

$$D_\zeta^{(p^m)}(\xi\eta) = \eta D_\zeta^{(p^m)}\xi + \xi D_\zeta^{(p^m)}\eta$$

for $\xi, \eta \in \Sigma_m$. This, together with (5.21), shows that $D_\zeta^{(p^m)}$ is a derivation of Σ_m.

(iii) Put $T = \{\xi \in \Sigma_m \mid D_\zeta^{(p^m)}\xi = 0\}$. If it is shown that $T = \Sigma_{m+1}$, then the lemma follows by induction. By definition, it suffices to show that $D_\zeta^{(i)}\xi = 0$ for

$\xi \in T$ and $p^m < i < p^{m+1}$. Now write $i = rp^m + s$ with $1 \le r \le p-1$ and $0 \le s \le p^m - 1$. From Lemmas 5.76 and 5.80,

$$D_\zeta^{(s)} \left(D_\zeta^{(p^m)}\right)^r \xi = r!\, D_\zeta^{(s)} D_\zeta^{(rp^m)} \xi = r! \binom{i}{s} D_\zeta^{(i)} \xi.$$

From Lemma A.6, $\binom{i}{s} \not\equiv 0 \pmod{p}$. Hence, for $\xi \in T$,

$$D_\zeta^{(i)} \xi = \frac{1}{r!} \binom{i}{s}^{-1} D_\zeta^{(s)} \left(D_\zeta^{(p^m)}\right)^r \xi = 0.$$

□

LEMMA 5.84 *For any positive integer m,*

$$\Sigma_m = \Sigma^{p^m} = \{u^{p^m} \mid u \in \Sigma\}.$$

Proof. Let $\eta \in \Sigma$. Choose an admissible place $\mathcal{P}$ with respect to ζ and η. For a primitive representation τ of $\mathcal{P}$, $\tau(\eta) = \overline{\eta} = \sum c_k t^k$. As $\tau(\overline{\eta}^{p^m}) = \sum c_k^{p^m} t^{kp^m}$, it follows that

$$D_t^{(i)} \overline{\eta}^{p^m} = \sum c_k^{p^m} \binom{kp^m}{i} t^{kp^m - i}, \tag{5.38}$$

which equals zero for $1 \le i < p^m$. Since τ is a monomorphism from Σ into $K((t))$, this only occurs when $D_\zeta^{(i)} \eta^{p^m} = 0$. Thus $\eta^{p^m} \in \Sigma_m$ for every $\eta \in \Sigma$.

To show the converse, note that every $\xi \in \Sigma_1$ is either a constant or an inseparable variable of Σ, by Theorem 5.50. Hence the assertion for $m = 1$ follows from Lemma 5.38. Now the proof is by induction on m.

Assume that every element in Σ_m is a p^m-th power of an element in Σ. Then ξ is in Σ_{m+1} when $\xi = \eta^{p^m}$ and $D_\zeta^{(i)} \xi = 0$ for $p^m \le i < p^{m+1}$. Again, choose an admissible place $\mathcal{P}$ with respect to ζ and η, and write $\tau(\eta) = \overline{\eta} = \sum c_k t^k$ with τ a primitive representation of $\mathcal{P}$. Arguing as before, (5.38) follows, and this is zero since $p^m < i < p^{m+1}$ is assumed. By the p-adic criterion of Lemma A.6, this implies that every k in the expansion of $\overline{\eta}$ is divisible by p. Thus $D_t \overline{\eta} = 0$. By Theorem 5.50 η is an inseparable variable of Σ. Hence Lemma 5.38 implies that $\eta = u^p$ for some $u \in \Sigma$. Therefore $\xi = \eta^{p^m} = u^{p^{m+1}}$. □

LEMMA 5.85 $\Sigma_\infty = K$.

Proof. Let ξ be any non-zero element in $\in \Sigma_\infty$. For a place $\mathcal{P}$ of Σ, let $k = \mathrm{ord}_{\mathcal{P}}\, \xi$. From the definition, $\Sigma_\infty \subset \Sigma_m$ for any positive integer m. Choose m such that $p^m > k$. By Lemma 5.84, $\xi = \eta^{p^m}$ for some $\eta \in \Sigma$. Hence $\mathrm{ord}_{\mathcal{P}}\, \xi = p^m \mathrm{ord}_{\mathcal{P}}\, \eta$. But this only occurs when $\mathrm{ord}_{\mathcal{P}}\, \xi = 0$ and $\mathrm{ord}_{\mathcal{P}}\, \eta = 0$. Therefore $\mathrm{ord}_{\mathcal{P}}\, \xi$ has no zeros. By Theorem 5.33, $\xi \in K$. □

THEOREM 5.86 *Let $x_0, \ldots, x_r \in \Sigma$. Then $x_0, \ldots, x_r$ are linearly independent over Σ_m if and only if there exist integers $\epsilon_0, \ldots, \epsilon_r$, with*

$$0 = \epsilon_0 < \epsilon_1 < \cdots < \epsilon_r < p^m,$$

such that $\det(D_\zeta^{(\epsilon_i)} x_j) \neq 0$.

Proof. Suppose that $x_0, \ldots, x_r \in \Sigma$ are linearly dependent over Σ_m; that is, there exist $z_0, \ldots, z_r \in \Sigma_m$ not all zero for which $\sum_{j=0}^{r} z_j x_j = 0$. By (5.23) and the definition of Σ_m,

$$\textstyle\sum_{j=0}^{r} z_j D^{(i)} x_j = 0.$$

This proves the 'if' part of the theorem.

The converse is proved by induction on r. The case $r = 0$ is trivial. Let $x_0, \ldots, x_r$ be linearly independent over Σ_m, and assume that the assertion of the theorem does not hold; that is, the $r+1$ vectors

$$u_j = (x_j, D_\zeta^{(1)} x_j, D_\zeta^{(2)} x_j, \ldots, D_\zeta^{(p^m-1)} x_j)$$

are linearly dependent over Σ. Then there exist $z_0, \ldots, z_r \in \Sigma$ such that

$$\textstyle\sum_{j=0}^{r} z_j D_\zeta^{(i)} x_j = 0 \quad \text{for} \quad 0 \le i \le p^m - 1. \tag{5.39}$$

Since $x_0, \ldots, x_{r-1}$ are linearly independent over Σ_m, from the inductive hypothesis it follows, for $0 \le j \le r-1$, that the above r vectors are linearly independent over Σ. Hence $z_r \ne 0$. Therefore it may be assumed that $z_r = 1$. It suffices to show that $z_j \in \Sigma_m$ for $j = 0, \ldots, r-1$. In fact, this, together with (5.39) for $i = 0$, gives a contradiction that completes the proof.

To prove that $z_j \in \Sigma_m$, it is enough to show by Lemma 5.83 that $D_\zeta^{(n)} z_j = 0$ for $1 \le n \le p^{m-1}$. This is done by induction on n.

For $n = 1$, from (5.39),

$$D_\zeta^{(1)} \textstyle\sum_{j=0}^{r} z_j D_\zeta^{(i)} x_j = (i+1) \sum_{j=0}^{r} z_j D^{(i+1)} x_j + \sum_{j=0}^{r-1} D_\zeta^{(1)} z_j D_\zeta^{(i)} x_j = 0.$$

From (5.39), $(i+1) \sum_{j=0}^{r} z_j D_\zeta^{(i)} x_j = 0$ for $i < p^m - 1$. This holds true for $i = p^m - 1$ since in this case $i + 1 \equiv 0 \pmod{p}$. Hence

$$\textstyle\sum_{j=0}^{r-1} D_\zeta^{(1)} z_j D_\zeta^{(i)} x_j = 0$$

for $0 \le i \le p^m - 1$. Since the above r vectors for $0 \le j \le r-1$ are linearly independent over Σ, the conclusion is that $D_\zeta^{(1)} z_j = 0$ for $j = 0, \ldots, r-1$.

Assume that $D_\zeta^{(s)} z_j = 0$ for $1 \le s < n \le p^{m-1}$ and for $0 \le j \le r-1$. From (5.39) by derivation,

$$\binom{i+n}{i} \sum_{j=0}^{r} z_j D_\zeta^{(i+n)} x_j + \sum_{j=0}^{r-1} D_\zeta^{(n)} z_j D_\zeta^{(i)} z_j = 0.$$

As in the case $n = 1$, to prove that $D_\zeta^{(n)} z_j = 0$, it suffices to show that

$$\binom{i+n}{i} \sum_{j=0}^{r} z_j D_\zeta^{(i+n)} x_j = 0 \quad \text{for} \quad 0 \le i \le p^m - 1.$$

For $i + n < p^m$, this follows from (5.39). For $i + n \ge p^m$,

$$p^m \le i + n < p^m + p^{m-1},$$
$$p^m > i \ge p^m - n \ge (p-1)p^{m-1}.$$

Hence the coefficient of p^{m-1} in the p-adic expansion of $i+n$ is zero and that of i is $p - 1$. From Lemma A.6, $\binom{i+n}{i} \equiv 0 \pmod{p}$, and this completes the proof. □

The following result is a consequence of Theorem 5.86.

THEOREM 5.87 *Let* $x_0, \ldots, x_r \in \Sigma$. *Then* $x_0, \ldots, x_r$ *are linearly independent over* Σ_∞ *if and only if there exist integers* $\epsilon_0, \ldots, \epsilon_r$, *with* $0 = \epsilon_0 < \epsilon_1 < \cdots < \epsilon_r$, *such that* $\det(D_\zeta^{(\epsilon_i)} x_j) \neq 0$.

5.11 THE DUAL AND BIDUAL OF A CURVE

Let $\mathcal{F}$ be an irreducible plane curve of degree $n > 1$. To each non-singular point of $\mathcal{F}$ there corresponds a unique tangent line to $\mathcal{F}$ at the point. In the dual plane, tangent lines determine points and, at least intuitively, the point set obtained is a curve. A rigorous proof is the main purpose of the present section, in which some unexpected behaviour of the dual curves in positive characteristic is also shown.

DEFINITION 5.88 The *dual curve* of $\mathcal{F}$ is an irreducible plane curve $\mathcal{F}'$ containing all but finitely many points $P' = (b_0, b_1, b_2)$ such that the line $b_0X_0+b_1X_1+b_2X_2$ in $\mathrm{PG}(2, K)$ is tangent to $\mathcal{F}$.

Assume that $\mathcal{F} = \mathbf{v}(F)$ for a homogeneous $F \in K[X_0, X_1, X_2]$ of degree n. With the notation of (1.5), Euler's formula gives that $nF = X_0F_0+X_1F_1+X_2F_2$. Since $\mathcal{F}$ is irreducible, one of the partial derivatives, say F_0, is not vanishing; see the proof of Theorem 1.25. Then, with coordinates $x = (x_0, x_1, x_2)$ for the generic point P,

$$x_0F_0(x) + x_1F_1(x) + x_2F_2(x) = 0. \tag{5.40}$$

One of the coordinates of P, say x_0, is distinct from 0. Without loss of generality, $\xi = x_1/x_0$ is a separable variable of the function field Σ of $\mathcal{F}$.

DEFINITION 5.89 The rational transformation

$$\omega_{\mathcal{F}} : \Sigma \to \Sigma, \quad (x_0', x_1', x_2') = (F_0(x), F_1(x), F_2(x)) \tag{5.41}$$

is the *rational Gauss map* associated to $\mathcal{F}$.

First, it is shown that the rational Gauss map is non-trivial. Without loss of generality, suppose that $F_0(x) \neq 0$. In the trivial case, both polynomials $F_1 - sF_0$ and $F_2 - tF_0$ are divisible by F. Since they have both degree less than n, the trivial case only occurs when $F_1 = sF_0$ and $F_2 = tF_0$. Then Euler's formula gives that

$$nF = X_0F_0 + X_1F_1 + X_2F_2 = (X_0 + sX_1 + tX_2)F_0.$$

Since F is irreducible and $F_0 \neq 0$, this only happens for $n = 1$ and F_0 is constant, contradicting the hypothesis $n > 1$. So the transform of $\mathcal{F}$ under the rational Gauss map is an irreducible curve $\mathcal{F}'$ which is actually the dual curve of $\mathcal{F}$.

Now some important properties of the dual curve are established. Choose any separable variable ζ of Σ, and derive $F(x) = 0$. Then

$$\frac{dx_0}{d\zeta}F_0(x) + \frac{dx_1}{d\zeta}F_1(x) + \frac{dx_2}{d\zeta}F_2(x) = 0. \tag{5.42}$$

So the triple $(F_0(x), F_1(x), F_2(x))$ provides a non-trivial solution of the following system of homogeneous linear equations in the unknowns u_0, u_1, u_2:

$$x_0u_0 + x_1u_1 + x_2u_2 = 0, \tag{5.43}$$

$$\frac{dx_0}{d\zeta}u_0 + \frac{dx_1}{d\zeta}u_1 + \frac{dx_2}{d\zeta}u_2 = 0. \tag{5.44}$$

There is a unique non-trivial solution, up to a non-zero factor, since the matrix of the system has rank 2. To show this, assume that $x_0 = 1$. Then $dx_0/d\zeta = 0$, and the rank is 1 if and only if $dx_1/d\zeta = 0$ and $dx_2/d\zeta = 0$. But x_1/x_0 is a separable variable, and thus $dx_1/d\zeta \neq 0$ by Theorem 5.50(ii). Hence the rank is equal to 2. Taking derivatives and subtracting, a further equation from (5.40) and (5.42) is obtained:

$$x_0\frac{dF_0(x)}{d\zeta} + x_1\frac{dF_1(x)}{d\zeta} + x_2\frac{dF_2(x)}{d\zeta} = 0. \tag{5.45}$$

Now consider the dual curve $\mathcal{F}''$ of $\mathcal{F}'$, that is, the *bidual* of $\mathcal{F}$. In the real and complex planes, $\mathcal{F}''$ coincides with $\mathcal{F}$. But this is not true for every curve in positive characteristic, and this behaviour is now investigated.

Put $y_i = F_i(x)$ for $i = 0, 1, 2$, and write $y = (y_0, y_1, y_2)$. If $\mathcal{F}' = \mathbf{v}(F')$ for an irreducible homogeneous polynomial F' with coefficients in K, then $F'(y) = 0$; that is, $Q = (y_0, y_1, y_2)$ is a generic point of $\mathcal{F}'$. Let $\Sigma' = K(\mathcal{F}')$; then Σ' is a subfield of Σ, and the rational Gauss map associated to $\mathcal{F}'$ is

$$\omega_{\mathcal{F}'} : \Sigma' \to \Sigma', \quad (y_0', y_1', y_2') = (F_0'(y), F_1'(y), F_2'(y)), \tag{5.46}$$

where F_i' is the partial derivative of F' with respect to X_i. If $\mathcal{F}'$ is not a line, that is, $\mathcal{F}$ is not a strange curve, then $\mathcal{F}'$ is transformed into the curve $\mathcal{F}''$, and $R = (F_0'(y), F_1'(y), F_2'(y))$ is a generic point of $\mathcal{F}''$. Since $\mathcal{F}''$ arises from $\mathcal{F}'$ in the same way as $\mathcal{F}'$ from $\mathcal{F}$, the triple $(F_0'(y), F_1'(y), F_2'(y))$ is a solution of the system of homogeneous linear equations in the unknowns v_0, v_1, v_2:

$$y_0v_0 + y_1v_1 + y_2v_2 = 0, \tag{5.47}$$

$$\frac{dy_0}{d\zeta}v_0 + \frac{dy_1}{d\zeta}v_1 + \frac{dy_2}{d\zeta}v_2 = 0. \tag{5.48}$$

Another solution is the triple (x_0, x_1, x_2). This follows from (5.45) and (5.40). In this system, two cases occur according to whether the matrix

$$\begin{bmatrix} y_0 & y_1 & y_2 \\ \dfrac{dy_0}{d\zeta} & \dfrac{dy_1}{d\zeta} & \dfrac{dy_2}{d\zeta} \end{bmatrix} = \begin{bmatrix} F_0(x) & F_1(x) & F_2(x) \\ \dfrac{F_0(x)}{d\zeta} & \dfrac{F_1(x)}{d\zeta} & \dfrac{F_2(x)}{d\zeta} \end{bmatrix} \tag{5.49}$$

has maximal rank or not. If the rank is 2, then there is a unique solution up to a non-zero factor, and $\mathcal{F}$ and $\mathcal{F}''$ share a generic point. Since both $\mathcal{F}$ and $\mathcal{F}''$ are irreducible, this is only possible when they coincide. Therefore $\omega_{\mathcal{F}'}\omega_{\mathcal{F}}$ is a birational transformation leaving $\mathcal{F}$ and also $\mathcal{F}'$ invariant. In terms of separability, see Theorems 5.40 and 5.42, this implies that both Gauss maps $\omega_{\mathcal{F}}$ and $\omega_{\mathcal{F}'}$, and hence also their product, are separable; that is, $\mathcal{F}$ is a *reflexive* curve.

Now the possibility that the matrix in (5.49) has rank 1 is examined. For this purpose, let $x_0 = 1$ and $\zeta = x_1$. Note that $F_2(x) \neq 0$, since x_1 is a separable variable. If the rank is 1, then

$$F_1(x)\frac{dF_2(x)}{dx_1} = F_2(x)\frac{dF_1(x)}{dx_1}. \tag{5.50}$$

Also, from (5.42),

$$F_1(x) + \frac{dx_2}{dx_1}F_2(x) = 0. \tag{5.51}$$

By taking derivatives,

$$\frac{dF_1(x)}{dx_1} = -\frac{d^2x_2}{dx_1^2}F_2(x) - \frac{dx_2}{dx_1}\frac{dF_2(x)}{dx_1}.$$

This, together with (5.50) and (5.51), gives

$$\frac{d^2x_2}{dx_1^2} = 0. \tag{5.52}$$

Note that, if $p = 2$, then (5.52) holds trivially. For $p \neq 2$, (5.52) implies that all non-singular points of $\mathcal{F}$ are inflexions. To show this, derive (5.51). In terms of F,

$$\frac{\partial^2 F(x)}{\partial x_1^2} + 2\frac{dx_2}{dx_1}\frac{\partial^2 F(x)}{\partial x_1 \partial x_2} + \left(\frac{dx_2}{dx_1}\right)^2 \frac{\partial^2 F(x)}{\partial x_2^2} + \frac{d^2x_2}{dx_1^2}\frac{\partial F(x)}{\partial x_2} = 0.$$

Writing $f(X,Y) = F(1,X,Y)$, this, together with (5.51) and (5.52), yields

$$\left(\frac{\partial f}{\partial x_1}\right)^2 \frac{\partial^2 f}{\partial x_2^2} + \left(\frac{\partial f}{\partial x_2}\right)^2 \frac{\partial^2 f}{\partial x_1^2} - 2\frac{\partial f}{\partial x_1}\frac{\partial f}{\partial x_2}\frac{\partial^2 f}{\partial x_1 \partial x_2} = 0.$$

Since $f(x_1, x_2) = 0$, by Theorem 1.37 the assertion holds. Another consequence of (5.52) for $p > 2$ is that the Gauss map $\omega_{\mathcal{F}}$ is inseparable; that is, $\mathcal{F}$ is a non-reflexive curve. In fact, both $F_0(x)/F_2(x)$ and $F_1(x)/F_2(x)$ are in Σ^p. This depends on the above equations: from (5.52),

$$\frac{d(F_1(x)/F_2(x))}{dx_1} = 0,$$

and this, together with (5.40), implies that

$$\frac{d(F_0(x)/F_2(x))}{dx_1} = 0.$$

Let p^m be the largest power of p for which

$$F_0(x)/F_2(x) = z_0^{p^m}, \quad F_1(x)/F_2(x) = z_1^{p^m},$$

for some $z_0, z_1 \in \Sigma$. Then (5.40) becomes $z_0^{p^m} + z_1^{p^m}x_1 + x_2 = 0$. Therefore

$$H(X,Y)f(X,Y) = Z_0(X,Y)^{p^m} + Z_1(X,Y)^{p^m}X + Z_2(X,Y)^{p^m}Y, \tag{5.53}$$

with $Z_0, Z_1, Z_2, H \in K[X,Y]$. In summary, the following theorem is established.

THEOREM 5.90 *Let* $p \neq 2$.

(i) *A sufficient condition for an irreducible plane curve $\mathcal{C}$ to coincide with its bidual curve is that $\mathcal{C}$ is reflexive.*

(ii) *The curve $\mathcal{F} = \mathbf{v}(f(X,Y))$ is non-reflexive if and only if $f(X,Y)$ satisfies (5.53) with p^m the inseparability degree of the Gauss map of $\mathcal{C}$.*

(iii) *$\mathcal{F}$ is non-reflexive if and only if it is the locus of its inflexion and singular points.*

For $p = 2$, the situation is different.

THEOREM 5.91 *Let $p = 2$.*

(i) *Every irreducible plane curve $\mathcal{F} = \mathbf{v}(f(X,Y)$ is non-reflexive, and $f(X,Y)$ satisfies (5.53) with p^m the inseparability degree of the Gauss map of $\mathcal{F}$.*

(ii) *If $p^m > 2$, then $\mathcal{F}$ is the locus of its inflexion and singular points.*

Proof. The hypothesis $p = 2$ implies that $d(F_1(x)F_2(x))/dx_1 = 0$. To show this, let $F(1,X,Y) = \sum a_{ij}X^iY^j$, and, for every non-negative integer n, define $\chi(n)$ to be 0 or 1 according as n is even or odd. Then

$$F_1 = \sum \chi(i)a_{ij}X^{i-1}Y^j, \quad F_2 = \sum \chi(j)a_{ij}X^iY^{j-1}.$$

Therefore

$$\frac{dF_1(x)}{dx_1} = \frac{dx_2}{dx_1}\sum \chi(i)\chi(j)x^{i-1}y^{j-1} = \frac{F_1(x)}{F_2(x)}\sum \chi(i)\chi(j)x^{i-1}y^{j-1},$$

$$\frac{dF_2(x)}{dx_1} = \sum \chi(i)\chi(j)x^{i-1}y^{j-1},$$

whence

$$\frac{F_1(x)F_2(x)}{dx_1} = \frac{dF_1(x)}{dx_1}F_2(x) + \frac{dF_2(x)}{dx_1}F_1(x) = 0.$$

Similarly, $d(F_0(x)F_1(x))/dx_1 = 0$.

From Theorem 5.50 and Lemma 5.38, $F_1(x)F_2(x)$ is in Σ^2; that is, it is a square in Σ. Since $F_2(x) \neq 0$, this gives that $F_1(x)/F_2(x) \in \Sigma^2$. Similarly, $F_1(x)F_0(x)$ is in Σ^2, and hence $F_0(x)/F_2(x) = (F_0(x)F_1(x))/(F_1(x)F_2(x)) \in \Sigma_2$. This means that $d(F_0(x)F_1(x))/dx_1 = 0$, and hence the matrix (5.49) has rank 1. In particular, the rational Gauss map of $\mathcal{F}$ is inseparable. By (5.40), this shows that (5.53) holds for $p = 2$, where 2^m is the inseparability degree of the rational Gauss map.

Assertion (ii) follows from (1.9) and (5.50). □

REMARK 5.92 For $p^m = 2$, Theorem 5.90 (iii) does not hold true. A counterexample is the Hermitian curve $\mathcal{H}_2 = \mathbf{v}(X_0^3 + X_1^3 + X_2^3)$ which is non-reflexive, but has only nine inflexion points. A necessary and sufficient condition for a non-reflexive curve $\mathcal{F}$ to have only finitely many inflexion points is that the matrix

$$\begin{bmatrix} F_0(x) & F_1(x) & F_2(x) \\ D^{(2)}_{x_1}F_0(x) & D^{(2)}_{x_1}F_1(x) & D^{(2)}_{x_1}F_0(x) \end{bmatrix}$$

has rank 2. This is shown in Section 7.7.

REMARK 5.93 It is possible for a non-reflexive curve to coincide with its dual curve, up to a Frobenius collineation. An example of this is the Hermitian curve $\mathcal{H}_q = \mathbf{v}(X_0^{q+1} + X_1^{q+1} + X_2^{q+1})$ for any power q of p. In fact, the Gauss map is

$$\omega_{\mathcal{H}_q} : \mathcal{H}_q \to \mathcal{H}_q, \quad (x_0, x_1, x_2) \mapsto (x_0^q, x_1^q, x_2^q),$$

and hence is the product of the identity rational transformation and the q-th Frobenius collineation.

REMARK 5.94 The dual curve of a strange curve is a line, and hence strange curves have no bidual curve.

For the dual of a non-singular plane curve the following results hold.

THEOREM 5.95 (i) *If $\mathcal{F}$ is a non-singular plane curve of degree n, and if its dual curve $\mathcal{F}'$ has degree n', then $qn' = n(n-1)$, where q is the inseparable degree of the Gauss map.*

(ii) *If $\mathcal{F}$ and $\mathcal{G}$ are two distinct non-singular plane curves, then their duals $\mathcal{F}'$ and $\mathcal{G}'$ coincide if and only if $p = 2$ and they are conics with the same nucleus.*

5.12 EXERCISES

1. Let $\Sigma = K(x, y)$. It follows from Lemma 5.38 that either x or y is a separable variable of Σ. Give another proof here by using Theorem 1.25.

2. With Σ a field of transcendence degree 1, give a new proof for Bézout's Theorem using Theorem 5.35.

3. Show that a uniformising element at any place is a separable variable of Σ.

4. Show that, if $p = 0$, then $d\xi/d\zeta = c$ if and only if $\xi = c\xi + d$, with $c, d \in K$. This result extends to positive characteristic provided that ξ is separable.

5. Prove that the image of a rational curve under a rational transformation is rational.

6. Prove that an irreducible plane curve has genus zero if and only if it is rational.

7. Let γ be a branch of $\mathcal{F}$ with order sequence $(0, j_1, j_2)$. For $p > 0$, assume that $p \geq 3$ but that p does not divide either j_1 or $j_2 - j_1$. Show that the corresponding branch γ' of the dual curve $\mathcal{F}'$ of $\mathcal{F}$ has order $j_2 - j_1$.

8. Prove the following generalisation of Theorem 5.61. Let Σ_1 and Σ_2 be two subfields of Σ both of transcendence degree 1. For $i = 1, 2$, put $n_i = [\Sigma : \Sigma_i]$ and denote by g_i the genus of Σ_i. If $\Sigma = \Sigma_1\Sigma_2$ is the compositum of Σ_1 and Σ_2, that is, the smallest subfield of Σ containing both Σ_1 and Σ_2, then

$$g \leq n_1 g_1 + n_2 g_2 + (n_1 - 1)(n_2 - 1).$$

9. Let ζ be a separable variable of Σ. Take $x, y \in \Sigma$ and let $G = G(x, y)$ with $G(X, Y) \in K[X, Y]$. Show that the higher derivatives of G may be computed from those of x and y using the following formula:

$$D_\zeta^{(i)}(G)$$
$$= \sum \frac{\partial^{(r_1+r_2+\cdots+s_1+s_2)} G(x,y)}{\partial x^{(r_1+r_2+\cdots)}\, \partial y^{(s_1+s_2+\cdots)}} \binom{r_1+r_2+\cdots}{r_1, r_2, \ldots} \times$$
$$\binom{s_1+s_2+\cdots}{s_1, s_2, \ldots} (D_\zeta^{(1)} x)^{r_1} (D_\zeta^{(2)} x)^{r_2} \cdots (D_\zeta^{(1)} x)^{s_1} (D_\zeta^{(1)} x)^{s_2} \cdots,$$

where the summation is for all non-negative integers $r_1, r_2, \ldots, s_1, s_2, \ldots$ such that $r_1 + 2r_1 + \cdots + s_1 + 2s_2 + \cdots = i$.

5.13 NOTES

The first four sections and Sections 5.7, 5.8 are based on [400]. Introductory works dealing with function fields of transcendence degree 1 over an algebraically closed field are [385], [387], [77], [102], [114], [84], [489].

Section 5.9 is based on [291], and has applications in coding theory; see [178]. Another proof of Theorem 5.68 is found in [196].

Hasse derivatives were introduced in [198]. Lemmas 5.72 and 5.73 come from [451]. The other lemmas and theorems in Section 5.10 are due to Garcia and Voloch; see [161]. Theorem 5.87 is also found in [386].

For Theorem 5.82, see [197] and [176]. The curve in Example 5.81 is birationally equivalent to that in Section 12.5.

The exceptional behaviour of duals and biduals of irreducible plane curves in positive characteristic was noted by Wallace [499]. For a detailed investigation of this topic, see Hefez [200], in which some generalisations of Theorems 5.90 and 5.91 are also found. See also [254], [271], [272], [357], [229], [230], [232], [231]. For a proof of Theorem 5.95, see Homma [235].

A generalisation of the concept of dual curve to higher-dimensional projective spaces is treated in Section 7.9.

Exercise 8 comes from [427]; see also [428, Theorem 10.3].

For Exercise 9, see [197].

Chapter Six

Linear series and the Riemann–Roch Theorem

In the investigation of the birational invariants of an irreducible plane curve, a central role is played by divisors, which are finite formal sums of places, and by linear series, which are certain sets of effective divisors. This chapter develops the theory of linear series; this includes the Riemann–Roch Theorem 6.61, the Weierstrass Gap Theorem 6.89 and the study of divisor class groups, which are also called Picard groups.

6.1 DIVISORS AND LINEAR SERIES

As in the previous chapter, the field $\Sigma = K(x, y)$ is associated to an irreducible algebraic curve $\mathcal{F}$ equipped with a generic point $P = (x, y)$. From an algebraic point of view, Σ is a function field of transcendence degree 1 whose places are in a one-to-one correspondence with branches of $\mathcal{F}$.

A natural generalisation of a place is a divisor.

DEFINITION 6.1 (i) A *divisor* of Σ is a formal sum of places:

$$D = \sum_{\mathcal{P} \in \mathfrak{P}(\Sigma)} n_{\mathcal{P}} \mathcal{P},$$

with $n_{\mathcal{P}}$ integers such that $n_{\mathcal{P}} = 0$ except for a finite number of places.

(ii) The coefficient $n_{\mathcal{P}}$ is the *multiplicity* or *weight* of $\mathcal{P}$ in D.

(iii) The *support* of D, denoted by Supp D, is the set of all places appearing with non-zero coefficients in D.

LEMMA 6.2 *The set of all divisors of Σ is an abelian group with respect to the addition, the* divisor group $\text{Div}(\Sigma)$ *of* Σ.

Proof. The sum $D + E$ of two of divisors $D = \sum n_{\mathcal{P}}\mathcal{P}$ and $E = \sum m_{\mathcal{P}}\mathcal{P}$ is the divisor $\sum(n_{\mathcal{P}} + m_{\mathcal{P}})\mathcal{P}$. The identity element (0) is the *zero divisor* which has no non-zero weight. □

DEFINITION 6.3 (i) If $n_{\mathcal{P}} \geq 0$ for every place $\mathcal{P}$, the divisor $D = \sum n_{\mathcal{P}}\mathcal{P}$ is *effective* or *positive*.

(ii) A non-effective divisor is *virtual*.

(iii) Given two divisors $D = \sum n_{\mathcal{P}}\mathcal{P}$ and $E = \sum m_{\mathcal{P}}\mathcal{P}$, the notation $D \succ E$ is used when $D - E$ is effective.

The most important divisors arise directly from elements of Σ, due to Theorem 5.33. For $\xi \in \Sigma\backslash\{0\}$, these are the following:

(i) $\operatorname{div}(\xi) = \sum_{\mathcal{P}\in\mathfrak{P}(\Sigma)} n_{\mathcal{P}}\mathcal{P}$ with $n_{\mathcal{P}} = \operatorname{ord}_{\mathcal{P}} \xi$ is the *principal divisor of* ξ;

(ii) $\operatorname{div}(\xi)_0 = \sum_{\mathcal{P}\in\mathfrak{P}(\Sigma)} n_{\mathcal{P}}\mathcal{P}$ with $n_{\mathcal{P}} = \operatorname{ord}_{\mathcal{P}} \xi$ and $n_{\mathcal{P}} \geq 0$ is the *divisor of zeros* of ξ;

(iii) $\operatorname{div}(\xi)_\infty = \sum_{\mathcal{P}\in\mathfrak{P}(\Sigma)} (-n_{\mathcal{P}})\mathcal{P}$ with $n_{\mathcal{P}} = \operatorname{ord}_{\mathcal{P}} \xi$ and $n_{\mathcal{P}} \leq 0$ is the *divisor of poles* of ξ.

Note that $\operatorname{div}(\xi) = \operatorname{div}(\xi)_0 - \operatorname{div}(\xi)_\infty$.

THEOREM 6.4 *Let* $\xi \in \Sigma\backslash\{0\}$. *Then* $\operatorname{div}(\xi) = (0)$ *if and only if* $\xi \in K\backslash\{0\}$.

Proof. If $\xi \in K\backslash\{0\}$, then ξ has neither zeros nor poles. Assume $x \notin K$. From the proof of Theorem 5.33, Σ has a model $(\mathcal{F}; (\xi, \eta))$ such that some branch of $\mathcal{F}$ is centred at an affine point on the X-axis. The corresponding place of Σ is a zero of $\operatorname{div}(\xi)$. Hence $\operatorname{div}(\xi)_0 \neq (0)$. Since zeros of $\zeta \in \Sigma\backslash\{0\}$ are poles of its inverse ζ^{-1}, so $\operatorname{div}(\xi)_\infty \neq (0)$ also holds. Hence $\operatorname{div}(\xi) \neq (0)$. □

LEMMA 6.5 *The elements in* $\operatorname{Div}(\Sigma)$ *which are principal divisors constitute a subgroup* $\operatorname{Prin}(\Sigma)$.

Proof. If $f, g \in \Sigma\backslash\{0\}$, then $\operatorname{div}(fg) = \operatorname{div}(f) + \operatorname{div}(g)$. □

DEFINITION 6.6 (i) The quotient group $\operatorname{Pic}(\Sigma) = \operatorname{Div}(\Sigma)/\operatorname{Prin}(\Sigma)$ is the *divisor class group* of Σ.

(ii) For a divisor D of Σ, the corresponding element in the factor group $\operatorname{Pic}(\Sigma)$ is denoted by $[D]$, the *divisor class* of D.

DEFINITION 6.7 Two divisors A and B of Σ are *equivalent*, written $A \equiv B$, if $[A] = [B]$, that is if there exists $\xi \in \Sigma\backslash\{0\}$ for which $A - B = \operatorname{div}(\xi)$.

LEMMA 6.8 (i) *The relation of Definition 6.7 is an equivalence relation.*

(ii) *For divisors* A, B, C, D *of* Σ,

(a) *if* $A \equiv B$, *then* $A + C \equiv B + C$;

(b) *if* $A \equiv B$, *then* $-A \equiv -B$;

(c) *if* $A \equiv B$, $C \equiv D$, *then* $A + C \equiv B + D$.

Proof. (ii) (a) Since $A \equiv B$, there exists $\xi \in \Sigma\backslash\{0\}$ for which $A - B = \operatorname{div}(\xi)$. Since $A - B = (A + C) - (B + C)$, the assertion follows.

(ii) (b) If $A \equiv B$, then $A - B = \operatorname{div}(\xi)$ with $\xi \in \Sigma\backslash\{0\}$, whence it follows that $-A - (-B) = \operatorname{div}(\xi^{-1})$.

(ii) (c) If $C \equiv D$ also holds, that is, $C - D = \operatorname{div}(\eta)$ with $\eta \in \Sigma\backslash\{0\}$, then $A + C - (B + D) = A - B + (C - D) = \operatorname{div}(\xi) + \operatorname{div}(\eta) = \operatorname{div}(\xi\eta)$. □

DEFINITION 6.9 The *degree* of a divisor $D = \sum_{\mathcal{P}\in\mathfrak{P}(\Sigma)} n_{\mathcal{P}}\mathcal{P}$ is $\deg D = \sum n_{\mathcal{P}}$.

Note that $\varphi : A \mapsto \deg A$ is an additive map. In other words, φ is a homomorphism from the divisor group to the additive group of integers.

THEOREM 6.10 *If* $A \equiv B$, *then* $\deg A = \deg B$.

Proof. Let $A - B = \operatorname{div}(\xi)$ with $\in \Sigma \backslash \{0\}$. If $\xi \in K$, then $A = B$. Otherwise, as $\operatorname{div}(\xi) = 0$ by Corollary 5.35, $\deg(A - B) = 0$, whence the assertion follows. □

Theorem 6.10 shows that $\deg[A]$ is well-defined. Hence the map

$$\varphi : [A] \mapsto \deg[A]$$

is a homomorphism from the divisor class group to the additive group of integers.

DEFINITION 6.11 The *subgroup* $\operatorname{Div}_0(\Sigma)$ *of divisors of degree zero* of $\operatorname{Div}(\Sigma)$ consists of all divisors of Σ of degree zero.

Under the natural homomorphism $\operatorname{Div}(\Sigma) \mapsto \operatorname{Pic}(\Sigma)$, the subgroup $\operatorname{Div}_0(\Sigma)$ corresponds to the subgroup $\operatorname{Pic}_0(\Sigma)$ of divisor classes of degree zero; that is,

$$\operatorname{Pic}_0(\Sigma) = \{[A] \in \operatorname{Pic}(\Sigma) \mid \deg A = 0\}.$$

Certain subsets of divisors, not necessarily subgroups, play an important role in the study of Σ. The most relevant are the linear series.

Let $x_0, x_1, \ldots, x_r$ be elements of Σ, not all zero, and let B be a fixed divisor of Σ. An ordered $(r+1)$-ple $\mathbf{c} = (c_0, c_1, \ldots, c_r)$ of elements in K is *permissible* when $c_0x_0 + c_1x_1 + \cdots + c_rx_r \neq 0$. For such an $(r+1)$-ple $\mathbf{c}$, consider the divisor

$$A_{\mathbf{c}} = \operatorname{div}(c_0x_0 + c_1x_1 + \cdots + c_rx_r) + B. \tag{6.1}$$

DEFINITION 6.12 The set $\mathcal{L}$ of divisors $A_{\mathbf{c}}$ for all permissible $(r+1)$-ples $\mathbf{c}$ is a *virtual linear series.*

Note that there may be a finite number of places occurring in every divisor of $\mathcal{L}$. Such places are the *fixed* places of $\mathcal{L}$. A sum of fixed places is a *fixed* divisor of $\mathcal{L}$.

Generally, it is supposed that $x_0, x_1, \ldots, x_r$ are linearly independent over K. Then the condition of permissibility on $\mathbf{c}$ is that $\mathbf{c} \neq \mathbf{0} = (0, 0, \ldots, 0)$. The next two results follow readily.

THEOREM 6.13 *All divisors of a virtual linear series* $\mathcal{L}$ *are equivalent to* B *and hence are equivalent to each other.*

DEFINITION 6.14 The common degree of the divisors of a virtual linear series is the *order* or *degree* of the virtual linear series.

THEOREM 6.15 *For every* $\zeta \neq 0$,

$$A_{\mathbf{c}} = \operatorname{div}(c_0x_0\zeta + c_1x_1\zeta + \cdots + c_rx_r\zeta) - \operatorname{div}(\zeta) + B.$$

COROLLARY 6.16 *If* $B \equiv B'$, *then divisors in* $\mathcal{L}$ *can be represented in the form*

$$\operatorname{div}(c_0x_0 + c_1x_1 + \cdots + c_rx_r) + B'$$

for some $x_0, x_1, \ldots, x_r$ *of* Σ *linearly independent over* K.

Proof. If $B \equiv B'$, then $B - B' = \mathrm{div}(\zeta)$ for an element $\zeta \in \Sigma\backslash\{0\}$. Then Theorem 6.15 applies. □

THEOREM 6.17 *Let $y_0, \ldots, y_s \in \Sigma$ be linearly independent elements over K such that the divisors $A'_{\mathbf{d}} = \mathrm{div}(d_0y_0 + d_1y_1 + \cdots + d_sy_s) + B'$, with a fixed divisor B', are exactly the divisors of $\mathcal{L}$ arising from $x_0, x_1, \ldots, x_r$, linearly independent elements over K. Then $r = s$ and, for some non-zero element $\zeta \in \Sigma$,*

$$y_i = \zeta \sum c_{ij} x_j,$$

with $\det(c_{ij}) \neq 0$.

Proof. Since $\mathrm{div}(y_i) + B' \in \mathcal{L}$, there exists $\mathbf{c}_i$ such that $\mathrm{div}(y_i) + B' = A_{\mathbf{c}_i}$. Thus $A_{\mathbf{c}_i} \equiv B'$, whence $B \equiv B'$. Therefore $B - B' = \mathrm{div}(\zeta)$ for a non-zero $\zeta \in \Sigma$. Hence

$$\begin{aligned} \mathrm{div}(y_i) &= \mathrm{div}(\textstyle\sum_{j=0}^r c_{ij}x_j) + B - B' = \mathrm{div}(\textstyle\sum_{j=0}^r c_{ij}x_j) + \mathrm{div}(\zeta) \\ &= \mathrm{div}(\textstyle\sum_{j=0}^r (c_{ij}x_j)\zeta) = \mathrm{div}(\textstyle\sum_{j=0}^r c_{ij}x_j\zeta). \end{aligned}$$

Thus $\mathrm{div}(\sum_{j=0}^r (c_{ij}x_j)/y_i) = 0$. By Theorem 6.4, there exists $k \in K\backslash\{0\}$ such that $y_i = k\sum_{j=0}^r c_{ij}x_j\zeta$. Absorbing the constant k into the coefficients c_{ij}, this becomes $y_j = \sum_{j=0}^r c_{ij}x_j\zeta$. As a consequence, $s \leq r$. Similarly, $r \leq s$, whence $r = s$. □

COROLLARY 6.18 *The integer r depends only on the series $\mathcal{L}$ and not on its representations.*

DEFINITION 6.19 The integer r is the *dimension* of the virtual linear series $\mathcal{L}$.

It should be noted that the virtual linear series $\mathcal{L}$ is a projective space over K whose points are the divisors of $\mathcal{L}$. The dimension of $\mathcal{L}$ is equal to the dimension of the projective space. Hence $\mathrm{PG}(r, K)$ can be regarded as a *representative* of $\mathcal{L}$.

If $\mathbf{d} = k\mathbf{c}$, with $k \in K\backslash\{0\}$, then $A_{\mathbf{d}} = A_{\mathbf{c}}$. Conversely, if

$$\mathrm{div}(\textstyle\sum_{i=0}^r c_ix_i) + B = \mathrm{div}(\textstyle\sum_{i=0}^r d_ix_i) + B,$$

then $\mathrm{div}(\sum_{i=0}^r c_ix_i) = \mathrm{div}(\sum_{i=0}^r d_ix_i)$, whence, by Theorem 6.4,

$$\textstyle\sum_{i=0}^r c_ix_i = k\sum_{i=0}^r d_ix_i,$$

with $k \in K\backslash\{0\}$. Assuming the elements x_i to be linearly independent over K, this only occurs when $\mathbf{d} = k\mathbf{c}$. Thus the set of divisors $A_{\mathbf{c}}$ of the linear series $\mathcal{L}$ is in one-to-one correspondence with the points of the r-dimensional projective space $\mathrm{PG}(r, K)$ over K, the divisor $A_{\mathbf{c}}$ corresponding to the point with coordinates $(c_0, c_1, \ldots, c_r)$.

A change of representation (6.1) of divisors in $\mathcal{L}$ changes $\mathbf{c}$ by a projectivity of $\mathrm{PG}(r, K)$. To check this, use the same notation as in the proof of Theorem 6.17. If $A'_d = A_{\mathbf{c}}$, that is,

$$\mathrm{div}(\textstyle\sum_{i=0}^r d_iy_i) + B = \mathrm{div}(\textstyle\sum_{i=0}^r c_ix_i) + B',$$

then $\mathrm{div}(\sum_{i=0}^r d_iy_i) = \mathrm{div}(\sum_{i=0}^r c_iy_i)$. Since $y_i = \zeta\sum_{j=0}^r c_{ij}x_j$, so

$$\mathrm{div}(\textstyle\sum_{i=0}^r \sum_{j=0}^r d_ic_{ij}x_j) = \mathrm{div}(\textstyle\sum_{i=0}^r c_ix_i).$$

By Corollary 5.35, for a non-zero $k \in K$,

$$k\, c_j = \sum_{i=0}^{r} c_{ij} d_i.$$

Thus $\mathbf{c} = (c_0, \ldots, c_r)$ and $\mathbf{d} = (d_0, \ldots, d_r)$ are related in the way stated. Put $C = (c_{ij})$ and let ${}^tC$ be the transpose of the matrix C; then the relation can also be written as $\mathbf{c} = k^{-1}\, {}^tC\mathbf{d}$ and $\mathbf{d} = k\, {}^tC^{-1}\mathbf{c}$.

DEFINITION 6.20 A *subseries of a virtual linear series* $\mathcal{L}$ is the set of all divisors corresponding to a projective subspace of $\mathrm{PG}(r, K)$, the projective space representative of $\mathcal{L}$.

The following result shows that $\mathrm{PG}(r, K)$ is a good representative of $\mathcal{L}$.

THEOREM 6.21 *Every subseries of a virtual linear series is a virtual linear series.*

Proof. Let Π_s be an s-dimensional subspace of $\mathrm{PG}(r, K)$ whose points are in one-to-one correspondence with the divisors of a linear series $\mathcal{L}$. Given $s + 1$ linearly independent points in Π_s, say,

$$\mathbf{c}_0 = (c_{00}, \ldots, c_{0r}), \ldots, \mathbf{c}_s = (c_{s0}, \ldots, c_{sr}),$$

the subseries corresponding to Π_s consists of the divisors

$$\begin{aligned} &\mathrm{div}((\textstyle\sum_{i=0}^{s} \lambda_i c_{i0})x_0 + \cdots + (\sum_{i=0}^{s} \lambda_i c_{ir})x_r) + B \qquad (6.2)\\ &= \mathrm{div}(\lambda_0(c_{00}x_0 + \cdots + c_{0r}x_r) + \cdots + \lambda_s(c_{s0}x_0 + \cdots + c_{sr}x_r)) + B. \end{aligned}$$

Put $y_i = c_{i0}x_0 + \cdots + c_{ir}x_r$ for $i = 0, 1 \ldots, s$. Then the divisor (6.2) can be rewritten as $\mathrm{div}(\sum_{i=0}^{s} y_i) + B$, that is, as the divisor of a linear series $\mathcal{L}'$. A straightforward calculation shows that $y_0, \ldots, y_s$ are linearly independent if and only if the matrix $C = (c_{ij})$ has maximum rank $s + 1$. This condition is satisfied here, since $\mathbf{c}_0, \ldots, \mathbf{c}_s$ have been chosen to represent linearly independent points in $\mathrm{PG}(r, K)$. □

Let $\mathcal{L}$ be a virtual linear series consisting of all divisors (6.1). If there exists $\mathbf{c}$ such that $\sum_{i=0}^{r} c_i x_i = 1$, then $\mathcal{L}$ is *normalised.*

THEOREM 6.22 *A virtual linear series $\mathcal{L}$ is normalised if and only if its fixed divisor B belongs to $\mathcal{L}$.*

Proof. If $\mathcal{L}$ is normalised, take the $(r + 1)$-tuple $\mathbf{c}$ so that $\sum_{i=0}^{r} c_i x_i = 1$; then $A_{\mathbf{c}} = \mathrm{div}(1) + B = B$, showing that $B \in \mathcal{L}$.

Conversely, if $B \in \mathcal{L}$, then there exists $\mathbf{c} \neq \mathbf{0}$ such that $B = A_{\mathbf{c}}$. For such an $(r + 1)$-ple $\mathbf{c}$, the divisor $\mathrm{div}(\sum_{i=0}^{r} c_i x_i)$ is equal to 0. Therefore $\sum_{i=0}^{r} c_i x_i$ has no zeros or poles. By Corollary 5.35, $\sum_{i=0}^{r} c_i x_i$ is equal to a non-zero constant $k \in K$. Putting $d_i = c_i/k$, this sum can be rewritten in the form $\sum_{i=0}^{r} d_i x_i = 1$, as required. □

Actually, it may be supposed that $\mathcal{L}$ is given in normalised form. In fact, as in Theorem 6.17, in (6.1) the elements x_i may be replaced by $x_i\zeta$ with an arbitrary $\zeta \in \Sigma\backslash\{0\}$ and B by a suitable B'. Now put $\zeta = \sum_{i=0}^{r} c_i x_i$, and $y_i = x_i/\zeta_i$, bearing in mind that $\zeta \neq 0$ as $x_0, \ldots, x_r$ are linearly independent over K. This gives that $A_{\mathbf{c}} = \mathrm{div}(\sum_{i=0}^{r} c_i y_i) + B'$, with $\sum_{i=0}^{r} c_i y_i = 1$.

THEOREM 6.23 *If $\mathcal{L}$ and $\mathcal{L}'$ are two virtual linear series and an element of $\mathcal{L}$ is equivalent to one of $\mathcal{L}'$, then $\mathcal{L}$ and $\mathcal{L}'$ are both contained in a virtual linear series.*

Proof. Let

$$\mathcal{L} = \{A_{\mathbf{c}} \mid A_{\mathbf{c}} = \operatorname{div}(\textstyle\sum_{i=0}^{r} c_i x_i) + B,\ \mathbf{c} \neq \mathbf{0}\},$$
$$\mathcal{L}' = \{A'_{\mathbf{d}} \mid A'_{\mathbf{d}} = \operatorname{div}(\textstyle\sum_{j=0}^{s} d_j y_j) + B',\ \mathbf{d} \neq \mathbf{0}\}.$$

In $\mathcal{L}'$, the divisor B' may be replaced by any divisor equivalent to it, in particular B. Hence $B = B'$ may assumed. Now the virtual linear series

$$\mathcal{L}_0 = \operatorname{div}(\textstyle\sum_{i=0}^{r} c_i x_i + \sum_{j=0}^{s} d_j y_j) + B$$

contains both $\mathcal{L}$ and $\mathcal{L}'$. □

In this proof, $x_0, \ldots, x_r, y_0, \ldots, y_s$ are not necessarily linearly independent. Hence $\dim \mathcal{L}_0 \leq 1 + \dim \mathcal{L} + \dim \mathcal{L}'$.

THEOREM 6.24 *An arbitrary divisor C can be added to or subtracted from the divisors of a virtual linear series to obtain another virtual linear series of the same dimension.*

THEOREM 6.25 *The multiplicities with which a place $\mathcal{P}$ appears in the divisors of a virtual linear series $\mathcal{L}$ are bounded above and below.*

Proof. The divisor B in $\mathcal{L}$ does not affect the question of boundedness; so assume that $B = (0)$. Let $\mathcal{P}$ be a place of Σ. For $i = 0, \ldots, r$, let $-\nu_i = \operatorname{ord}_{\mathcal{P}} x_i$. Since $\operatorname{ord}_{\mathcal{P}}(x + y) \geq \min\{\operatorname{ord}_{\mathcal{P}} x, \operatorname{ord}_{\mathcal{P}} y\}$, so $\operatorname{ord}_{\mathcal{P}}(\sum_{i=0}^{r} c_i x_i) \geq n_{\mathcal{P}}$, where

$$n_{\mathcal{P}} = \begin{cases} 0 & \text{if } \mathcal{P} \text{ is not a pole of any } x_i, \\ \min\{-\nu_i\} & \text{if } \mathcal{P} \text{ is a pole of some } x_i, \end{cases}$$

which gives a lower bound. This also shows that

$$-n_{\mathcal{Q}} \geq -\operatorname{ord}_{\mathcal{Q}}(\textstyle\sum_{i=0}^{r} c_i x_i)$$

for each place $\mathcal{Q}$ of Σ. Hence

$$\textstyle\sum -n_{\mathcal{Q}} \mathcal{Q} \succ \operatorname{div}(\sum_{i=0}^{r} c_i x_i)_\infty,$$

with the summation over all places $\mathcal{Q}$ of Σ. Since

$$\operatorname{div}(\textstyle\sum_{i=0}^{r} c_i x_i) = \operatorname{div}(\sum_{i=0}^{r} c_i x_i)_0 - \operatorname{div}(\sum_{i=0}^{r} c_i x_i)_\infty,$$

so

$$\textstyle\sum -n_{\mathcal{Q}} \mathcal{Q} + \operatorname{div}(\sum_{i=0}^{r} c_i x_i) \succ \operatorname{div}(\sum_{i=0}^{r} c_i x_i)_0.$$

Hence

$$\textstyle\sum -n_{\mathcal{Q}} + \sum \operatorname{ord}_{\mathcal{Q}}(\sum_{i=0}^{r} c_i x_i) \geq \sum m_{\mathcal{Q}},$$

where $m_{\mathcal{Q}}$ is the *zero number of* $\sum_{i=0}^{r} c_i x_i$; that is, $m_{\mathcal{Q}} = \operatorname{ord}_{\mathcal{Q}}(\sum_{i=0}^{r} c_i x_i)_0$. On the other hand, $\sum \operatorname{ord}_{\mathcal{Q}}(\operatorname{div} \sum_{i=0}^{r} c_i x_i) = 0$, by Corollary 5.35. This means that $\sum n_{\mathcal{Q}} \geq \sum m_{\mathcal{Q}}$, whence $\sum n_{\mathcal{Q}} \geq m_{\mathcal{P}}$. Since $\sum n_{\mathcal{Q}}$ depends only on $\mathcal{L}$ but is independent of the particular choice of the coefficients $c_0, \ldots, c_r$, it follows that $\operatorname{ord}_{\mathcal{P}}(\sum_{i=0}^{r} c_i x_i)$ is bounded above by a number only depending on $\mathcal{L}$. □

THEOREM 6.26 *Let $\mathcal{P}$ be a place of Σ and ν the smallest multiplicity with which $\mathcal{P}$ occurs in the divisors of $\mathcal{L}$. Then the divisors of $\mathcal{L}$ in which $\mathcal{P}$ occurs with multiplicity at least $\nu + 1$ make up a subseries of $\mathcal{L}$ of dimension* $\dim \mathcal{L} - 1$.

Proof. Let (6.1) be a divisor of $\mathcal{L}$. Choose $B \in \mathcal{L}$ in which $\mathcal{P}$ occurs in B with multiplicity ν; that is, $n_{\mathcal{P}}(B) = \nu$. Then $\mathcal{P}$ is not a pole of any x_i, for otherwise $\mathcal{P}$ would occur with multiplicity less than ν in some $\operatorname{div}(x) + B$. In fact,

$$n_{\mathcal{P}}(\operatorname{div}(x_i) + B) = \operatorname{ord}_{\mathcal{P}}(x_i) + n_{\mathcal{P}}(B) = \operatorname{ord}_{\mathcal{P}}(x_i) + \nu,$$

and, if $\operatorname{ord}_{\mathcal{P}}(x_i) < 0$ were true, then $n_{\mathcal{P}}(\operatorname{div}(x_i) + B) < \nu$ would also be true, contradicting the minimality of ν. Also, $\mathcal{P}$ is not a zero for every x_i. If this were not true, then

$$n_{\mathcal{P}}(\operatorname{div}(\textstyle\sum_{i=0}^{r} c_i x_i) + B) = \operatorname{ord}_{\mathcal{P}}(\textstyle\sum_{i=0}^{r} c_i x_i) + \nu > \nu,$$

since

$$\operatorname{ord}_{\mathcal{P}}(\textstyle\sum_{i=0}^{r} c_i x_i) \geq \min\{\operatorname{ord}_{\mathcal{P}}(x_i)\} \geq \nu$$

and $\operatorname{ord}_{\mathcal{P}}(x_i) > 0$ for $0 \leq i \leq r$.

Let $\sigma : \Sigma \to K((t))$ be a primitive representation of the place $\mathcal{P}$. Then

$$\sigma(x_i) = a_{i0} + a_{i1}t + \cdots .$$

Since $\operatorname{ord}_{\mathcal{P}}(x_i) = 0$ for some i, the coefficient a_{i0} does not vanish for every i. Therefore

$$\begin{aligned} A_c &= \operatorname{div}(c_0 x_0 + \cdots + c_r x_r) + B, \\ &\operatorname{div}(c_0(a_{00} + a_{01}t + \cdots) + \cdots + c_r(a_{r0} + a_{r1}t + \cdots)) + B \\ &\quad = \operatorname{div}((c_0 a_{00} + \cdots + c_r a_{r0}) + tM_1 + \cdots) + B, \end{aligned}$$

whence

$$\begin{aligned} n_{\mathcal{P}}(A_c) &= n_{\mathcal{P}}(\operatorname{div}(c_0 a_{00} + \cdots + c_r a_{r0}) + tL_1 + \cdots) + n_{\mathcal{P}}(B) \\ &= n_{\mathcal{P}}(\operatorname{div}(c_0 a_{00} + \cdots + c_r a_{r0}) + tL_1 + \cdots) + \nu. \end{aligned}$$

Now $n_{\mathcal{P}}(A_c) \geq \nu + 1$ if and only if

$$n_{\mathcal{P}}(\operatorname{div}(c_0 a_{00} + \cdots + c_r a_{r0}) + tL_1 + \cdots) \geq 1,$$

whence $c_0 a_{00} + \cdots + c_r a_{r0} = 0$.

This means that, in the projective space $\mathrm{PG}(r, K)$ representing $\mathcal{L}$, the hyperplane $H = \mathbf{v}(a_{00}X_0 + \cdots + a_{0r}X_r)$ passes through the point $A = (c_0, \ldots, c_r)$. Hence the divisors of $\mathcal{L}$ in which $\mathcal{P}$ appears with multiplicity at least $\nu + 1$ correspond to those points A lying in H, and hence they constitute a virtual linear series of dimension $\dim \mathcal{L} - 1$. □

COROLLARY 6.27 *The divisors of $\mathcal{L}$ in which a given place $\mathcal{P}$ appears with multiplicity at least μ constitute a subseries of $\mathcal{L}$.*

Proof. If $\mu \leq \nu$, then $\mathcal{L}$ itself is the subseries. If $\mu = \nu + 1$, then the corollary is just Theorem 6.26. If $\mu = \nu + 2$, apply Theorem 6.26 to the subseries consisting of all divisors in which $\mathcal{P}$ occurs with multiplicity at least $\nu + 1$; continue this process. □

REMARK 6.28 With the notation of Theorem 6.13, if $\nu < \mu$, the subseries is empty. Also, the divisors of $\mathcal{L}$ in which $\mathcal{P}$ appears with multiplicity at least $\nu + 2$ is a subseries $\mathcal{L}'$ of $\mathcal{L}$ whose dimension is at least $\dim \mathcal{L} - 2$. In fact, $\mathcal{L}$ may happen not to contain divisors in which $\mathcal{P}$ appears with multiplicity $\nu + 1$. If this is the case, Theorem 6.13 does not apply to $\mathcal{L}'$ and $\nu + 1$.

COROLLARY 6.29 *If $\mathcal{P}_1, \ldots, \mathcal{P}_s$ is a set of places of Σ and $\mu_1, \ldots, \mu_s$ are integers, then the divisors of $\mathcal{L}$ in which each $\mathcal{P}_i$ appears with multiplicity at least μ_i constitute a subseries of $\mathcal{L}$.*

THEOREM 6.30 *The effective divisors of $\mathcal{L}$ constitute a subseries.*

Proof. There are only finitely many places of Σ in which the divisors (6.1) can have negative degree, namely, the places occurring in B and the poles of the elements x_i. The theorem now follows from the previous corollary when $\mathcal{P}_1, \ldots, \mathcal{P}_s$ are defined to be these places and $\mu_i = 0$ for $i = 1, \ldots, s$. □

DEFINITION 6.31 A *linear series* is a virtual linear series consisting of effective divisors; it is denoted by g_n^r, where n is its order and r is its dimension.

THEOREM 6.32 *In a linear series g_n^r, the order and dimension satisfy $r \leq n$.*

Proof. If $r = 0$, then $r \leq n$. Let $r > 0$, and take a place $\mathcal{P}$ that does not occur in every divisor of g_n^r. Then the divisors of g_n^r in which $\mathcal{P}$ occurs constitute a subseries g_n^{r-1}. Subtracting $\mathcal{P}$ from each of these divisors, a subseries g_{n-1}^{r-1} is obtained. Applying induction on r shows that $r - 1 \leq n - 1$, whence the assertion follows. □

DEFINITION 6.33 Let g_m^s and g_n^r be two linear series of Σ. Then g_m^s is *contained* in g_n^r if every divisor $D \in g_m^s$ is contained in some divisor $E \in g_n^r$; that is, $E \succ D$.

Here,

$$g_m^s = \{\mathrm{div}(c_0x_0 + \cdots + c_mx_m) + B \mid \mathbf{c} = (c_0, \ldots, c_m) \in \mathrm{PG}(m, K)\}$$

is contained in g_n^r if and only if there exist $x_{s+1}, \ldots, x_r \in \Sigma$, and $F \succ B$ such that

$$g_n^r = \{\mathrm{div}(c_0x_0 + \cdots + c_rx_r) + F \mid \mathbf{c} = (c_0, \ldots, c_r) \in \mathrm{PG}(r, K)\}.$$

THEOREM 6.34 *If g_n^r and $g_{n'}^{r'}$ are linear series of Σ such that a divisor of one of them is equivalent to a divisor of the other, then $n = n'$ and both are contained in a linear series.*

Proof. Theorem 6.25 together with Theorem 6.13 yield $n = n'$. From Theorem 6.23, g_n^r and $g_n^{r'}$ are both contained in a virtual linear series $\mathcal{L}$. Let $\mathcal{L}'$ be the set of all effective divisors in $\mathcal{L}$. By Theorem 6.30, $\mathcal{L}'$ is a linear series, and both g_n^r and $g_{n'}^{r'}$ are subseries of $\mathcal{L}'$. □

THEOREM 6.35 *The effective divisors equivalent to a divisor C constitute a linear series.*

Proof. It may be that no effective divisor is equivalent to C, for instance when $\text{ord}_{\mathcal{P}}\, C < 0$. If this is the case then such effective divisors constitute the empty linear series. Otherwise there is at least one effective divisor B equivalent to C, and C may be replaced by B since the divisors equivalent to B are the same as those equivalent to C.

Let $\text{ord}_{\mathcal{P}}\, B = n$. The virtual linear series $\mathcal{L}_0 = \text{div}(x_0)+B$ consists of effective divisors, more precisely, $\mathcal{L}_0 = g_n^0$. Every divisor E equivalent to B can be written as $E = \text{div}(f) + B$ for some $f \in \Sigma$. Let Γ the set of such elements $f \in \Sigma$ with E ranging over the set of all effective divisors equivalent to B. Note that Γ is a subspace of Σ regarded as a vector space over K. To show this, take a place $\mathcal{P}$ of Σ and two elements $f_1, f_2 \in \Gamma$ together with two constants $c_1, c_2 \in K$. Since

$$\text{ord}_{\mathcal{P}}(c_1 f_1 + c_2 f_2) \geq \min\{\text{ord}_{\mathcal{P}}\, f_1, \text{ord}_{\mathcal{P}}\, f_2\},$$

it follows that

$$\text{div}(c_1 f_1 + c_2 f_2) \succ \textstyle\sum_{\mathcal{P}\in\Sigma_{\mathcal{P}}} n_{\mathcal{P}}\mathcal{P},$$

with $n_{\mathcal{P}}$ defined to be $\min\{\text{ord}_{\mathcal{P}} f_1,\ \text{ord}_{\mathcal{P}} f_2\}$. Since $f_1, f_2 \in \Gamma$ implies $n_{\mathcal{P}} \geq 0$, the assertion follows.

To show that Γ has dimension at most n, suppose on the contrary that Γ contains a K-independent set $\{x_0, \ldots, x_r\}$ with $r > n$. Then $\mathcal{L} = \{A_{\mathbf{c}} \mid \mathbf{c} \neq \mathbf{0}\}$ is a linear series with dimension $r > n$; but this contradicts Theorem 6.32. □

Definition 6.36 The linear series consisting of all effective divisors equivalent to a divisor C is a *complete* linear series and is denoted by $|C|$.

Example 6.37 (i) Given an irreducible non-singular cubic $\mathcal{C}$ and a fixed point $P_0 \in \mathcal{C}$, the lines through P_0 cut out a complete g_2^1 on $\mathcal{C}$.

(ii) Given $P_1 \notin \mathcal{C}$, the lines through P_1 cut out a g_3^1 on $\mathcal{C}$ contained in the complete g_3^2 cut out by all lines of the plane.

Lemma 6.38 *Let $|A|$, $|B|$ be two complete linear series, with A and B not necessarily effective. If $A_1 \equiv A$ and $B_1 \equiv B$, then*

$$|A_1 + B_1| = |A + B|, \quad |A_1 - B_1| = |A - B|.$$

Proof. By Lemma 6.8 (iii), $A_1 + B_1 \equiv A + B$, whence $|A_1 + B_1| = |A + B|$. The same holds when '+' is replaced by '−'. □

Definition 6.39 (i) The *sum* of two complete linear series $|A|$ and $|B|$ consists of all effective divisors equivalent to $A + B$, is denoted by $||A| + |B||$.

(ii) The *difference* of $|A|$ and $|B|$ is the linear series of all effective divisors equivalent to $A - B$.

Theorem 6.40 *Let A and B be two effective divisors. Then $||A| - |B||$ consists of all divisors $D - B$, where $D \in |A|$ and $D \succ B$.*

Proof. Suppose $C \in ||A|-|B|| = |A-B|$. Then $C \equiv A-B$. Hence $C+B \equiv A$, whence $C+B \in |A|$. Therefore $C+B = A_1$ with $A_1 \in |A|$, which yields $C = A_1 - B$. Conversely, if $C = A_1 - B$ with $A_1 \in |A|$ and $A_1 \geq B$, then

$$C \in |A_1 - B| = ||A_1| - |B|| = ||A| - |B||. \qquad \square$$

In Section 5.7 the concept of a differential of the form dx for a separating variable $x \in \Sigma$ is defined and investigated. In the context of divisors, associate with dx the *divisor of the differential* dx defined by

$$\operatorname{div}(dx) = \sum \operatorname{ord}_{\mathcal{P}}\, dx\, \mathcal{P}.$$

Every differential is of the form ξdx with $\xi \in \Sigma$. Hence $(\xi dx) = \operatorname{div}(\xi) + \operatorname{div}(dx)$, which shows that all divisors ξdx belong to the same equivalence class of divisors, the *canonical class*. A canonical divisor has degree $2g-2$, by Definition 5.55.

DEFINITION 6.41 The complete linear series $|dx|$ is the *canonical series.*

The canonical series has order $2g-2$. In the next section, an explicit description of the canonical series is given which demonstrates that its dimension is $g-1$.

6.2 LINEAR SYSTEMS OF CURVES

The aim of this section is to describe the construction of a linear series from a linear system of curves by means of intersection divisors.

Let Σ be a function field of transcendence degree 1 and $(\mathcal{F}; (x,y))$ a model of Σ, where $\mathcal{F}$ is an irreducible curve and $P = (x,y)$ is a generic point of $\mathcal{F}$. A divisor of Σ arises from every curve $\mathcal{G}$ that contains $\mathcal{F}$ as a component. Such a divisor is the *intersection divisor* cut out on $\mathcal{F}$ by Φ and defined to be

$$\mathcal{F} \cdot \mathcal{G} = \sum I(P, \mathcal{G} \cap \gamma)\mathcal{P},$$

where the summation is over all places of Σ, and γ is the branch of $\mathcal{F}$ corresponding to $\mathcal{P}$.

THEOREM 6.42 *Suppose that*

(a) $\mathcal{F} = \mathbf{v}(F(X_0, X_1, X_2))$ *is an irreducible curve with generic point* $(1, x, y)$*;*

(b) $\mathcal{G} = \mathbf{v}(G(X_0, X_1, X_2)$ *is any curve of degree* m*;*

(c) $\mathcal{F} \neq \ell_\infty = \mathbf{v}(X_0)$*, the line at infinity;*

(d) F *does not divide* G*.*

Then

$$\operatorname{div} G(1,x,y) = \mathcal{G} \cdot \mathcal{F} - m(\ell_\infty \cdot \mathcal{F}).$$

Proof. For a place $\mathcal{P}$ of Σ, let $(x_0(t), x_1(t), x_2(t))$ be a primitive representation in special coordinates of the branch of $\mathcal{F}$ corresponding to $\mathcal{P}$. Note that $\ell_\infty \neq \mathcal{F}$ implies $x_0(t) \neq 0$. As usual, let $x(t) = x_1(t)/x_0(t)$ and $y(t) = x_2(t)/x_0(t)$. Then

$$\text{ord}_{\mathcal{P}}\, G(x_0(t), x_1(t), x_2(t)) = \text{ord}_{\mathcal{P}}\, x_0(t)^m + \text{ord}_{\mathcal{P}}\, G(1, x, y).$$

Multiplying by $\mathcal{P}$ and summing over all places $\mathcal{P}$, it follows that

$$\sum \text{ord}_{\mathcal{P}}\, G(x_0(t), x_1(t), x_2(t))\, \mathcal{P} = m \sum \text{ord}_{\mathcal{P}}\, x_0(t)\, \mathcal{P} + \sum \text{ord}_{\mathcal{P}}\, G(1, x, y)\, \mathcal{P}.$$

By (4.34), the left-hand side is $\mathcal{D} \cdot \mathcal{F}$, while $\sum \text{ord}_{\mathcal{P}}\, x_0(t) \mathcal{P} = \ell_\infty \cdot \mathcal{F}$. Therefore

$$\mathcal{G} \cdot \mathcal{F} = m(\ell_\infty \cdot \mathcal{F}) + \text{div}\, G(1, x, y),$$

whence the assertion follows. □

COROLLARY 6.43 *The curve $\mathcal{G}$ cuts out on $\mathcal{F}$ the divisor $\mathcal{G} \cdot \mathcal{F}$ equivalent to $m(l_\infty \cdot \mathcal{F})$.*

COROLLARY 6.44 *If $\mathcal{G}_1 = \mathbf{v}(G_1(X_0, X_1, X_2))$ is another curve of degree m that does not contain $\mathcal{F}$ as a component, then $\mathcal{G}_1$ and $\mathcal{G}$ cut out equivalent divisors on $\mathcal{F}$.*

DEFINITION 6.45 Let $\Phi_0 = \mathbf{v}(\varphi_0(X_0, X_1, X_2)), \ldots, \Phi_r = \mathbf{v}(\varphi_r(X_0, X_1, X_2))$ be curves of the same degree m, with $\varphi_0, \ldots, \varphi_r$ linearly independent over K.

(i) The set of curves $\mathcal{F} = \mathbf{v}(\sum_{i=0}^r c_i \varphi_i)$, with $\mathbf{c} = (c_0, \ldots, c_r)$ ranging over all non-trivial $(r+1)$-ples of elements of K, is a *linear system of dimension* r, and is *the linear system generated by* $\Phi_0, \ldots, \Phi_r$;

(ii) when $r = 1$, the system is a *pencil*;

(iii) when $r = 2$, the system is a *net*.

There is a one-to-one correspondence between the curves of a linear system of dimension r and the points of $\text{PG}(r, K)$. It is worth noting that, if the polynomials are written in their inhomogeneous form, that is, $\varphi'(X, Y) = \varphi(1, X, Y)$, then some of them could have degree less than m.

THEOREM 6.46 *The curves in a linear system of degree m that do not contain $\mathcal{F}$ as a component cut out on $\mathcal{F}$ the divisors of a linear series. The converse also holds, up to a fixed divisor.*

Proof. For the converse, let $\mathcal{L}$ be a linear series consisting of all divisors (6.1). Since $\Sigma = K(x, y)$ with $P = (x, y)$ a generic point of $\mathcal{F}$, there are polynomials $\varphi_i(X, Y), d(X, Y) \in K[X, Y]$ such that $x_i = \varphi_i(x, y)/d(x, y)$ for $i = 0, \ldots, r$. Hence

$$\textstyle\sum_{i=0}^r c_i x_i = \sum_{i=0}^r c_i \varphi_i(x, y)/d(x, y).$$

Also,

$$\begin{aligned} A_c &= \text{div}\,(\textstyle\sum_{i=0}^r c_i \varphi_i(x, y)/d(x, y)) + B \\ &= \text{div}\,(\textstyle\sum_{i=0}^r c_i \varphi_i(x, y)) - \text{div}(d(x, y)) + B. \end{aligned}$$

If m is the maximum degree of the polynomials $\varphi_i(X, Y)$, define the homogeneous polynomial $\varphi_i(X_0, X_1, X_2)$ to be $X_0^m \varphi(X_1/X_0, X_2/X_0)$, for $i = 0, \ldots, r$. Let $\mathcal{F}_i = \mathbf{v}(\varphi(X_0, X_1, X_2))$ be the corresponding curve, and let $\mathcal{D} = \mathbf{v}(D)$ with $D(X_0, X_1, X_2)$ a homogeneous polynomial such that $D(1, X, Y) = d(X, Y)$.

Now the linear system Γ generated by the curves $\mathcal{F}_i$ is well defined. Let $\Phi_{\mathbf{c}}$ be the curve $\mathbf{v}(\sum_{i=0}^r c_i\varphi_i(X_0, X_1, X_2))$. From Theorem 6.42,

$$\operatorname{div}\left(\sum_{i=0}^r c_i\varphi(x, y)\right) = \Phi_{\mathbf{c}} \cdot \mathcal{F} - m(\ell_\infty \cdot \mathcal{F}).$$

Therefore

$$\operatorname{div}\left(\sum_{i=0}^r \frac{c_i\varphi_i(x, y)}{d(x, y)}\right) + B = \Phi_{\mathbf{c}} \cdot \mathcal{F} - m(\ell_\infty \cdot \mathcal{F}) - \operatorname{div} d(x, y) + B.$$

Since $C = -m(\ell_\infty \cdot \mathcal{F}) - (d(x, y)) + B$ does not depend on $\mathbf{c}$, the linear system Γ cuts out on $\mathcal{F}$ the linear series consisting of the divisors (6.1), up to the fixed divisor C.

The direct part of Theorem 6.46 can be proved by reversing these steps. □

REMARK 6.47 Theorem 6.46 establishes a close relationship between linear series and linear systems. However, the dimension $\dim \Gamma$ of a linear system Γ differs from the dimension of the linear series $\mathcal{L}$ cut out by Γ on $\mathcal{F}$. In fact, $\dim \Gamma \geq \dim \mathcal{L}$ and equality only holds when no two distinct curves in Γ cut out the same divisor in $\mathcal{L}$.

To eliminate this difference, linear dependency modulo F is needed.

DEFINITION 6.48 If $\mathcal{F} = \mathbf{v}(F(X_0, X_1, X_2))$ and

$$F(X_0, X_1, X_2) \text{ divides } \phi(X_0, X_1, X_2) - \psi(X_0, X_1, X_2),$$

then $\phi \equiv \psi \pmod{F}$.

Given a linear system $\mathbf{v}(\sum_{i=0}^r c_i\varphi_i(X_0, X_1, X_2))$, if one of the φ_i, say φ_0, is linearly dependent on the others modulo F, then φ_0 can be eliminated; that is, the linear system $\mathbf{v}(\sum_{i=1}^r c_i\varphi_i)$ cuts out the same linear series on $\mathcal{F}$ as $\mathbf{v}(\sum_{i=0}^r c_i\varphi_i)$ does. So the φ_i may be supposed linearly independent modulo F. Note that, if $m < \deg F$, then linear independence means linear independence modulo F.

To obtain deeper results on linear series and especially on the canonical series, a model $(\mathcal{F}; (x, y))$ of Σ is chosen such that the irreducible curve $\mathcal{F}$ has only ordinary singularities.

Let $P_1, \ldots, P_k$ denote the singular points of $\mathcal{F}$ and let r_i be the multiplicity of P_i for $i = 1, \ldots, k$. Let $\{\gamma_{ij} \mid j = 1, \ldots, r_i\}$ be the set of all branches of $\mathcal{F}$ centred at P_i.

DEFINITION 6.49 (i) The *double-point divisor* of $\mathcal{F}$, more precisely of Σ, with respect to the model $(\mathcal{F}; (x, y))$, is

$$D = \sum_{i=1}^k \sum_{j=1}^{r_i} (r_i - 1)\mathcal{P}_{ij},$$

where $\mathcal{P}_{ij}$ is the place of Σ corresponding to γ_{ij}.

(ii) A curve $\mathcal{G}$ is an *adjoint* of $\mathcal{F}$ if $\mathcal{G} \cdot \mathcal{F} \succ D$.

In this case, $\mathcal{G} \cdot \mathcal{F} = D + E$ with E an effective divisor, and E is the *divisor cut out by* $\mathcal{G}$ on $\mathcal{F}$, up to D. Also, $\mathcal{G}$ *passes through* E' for any effective divisor E' such that $E \succ E'$.

THEOREM 6.50 *A curve $\mathcal{G}$ is an adjoint of $\mathcal{F}$ if and only if $\mathcal{G}$ has at least an $(r-1)$-fold point at every r-fold point of $\mathcal{F}$.*

Proof. Let P be an r-fold point of $\mathcal{F}$. If P is an $(r-1)$-fold point of $\mathcal{G}$ and γ is a branch of $\mathcal{F}$ centred at P, then $I(P, \mathcal{G} \cap \gamma) \geq r - 1$. Suppose now that $\mathcal{G}$ has only an s-fold point, with $s \leq r - 2$, at a point of multiplicity r of $\mathcal{F}$. Since P is ordinary singularity, $\mathcal{G}$ has at most s tangents at P. Since $s < r$, some tangent to $\mathcal{F}$ at P, say ℓ, is not tangent to $\mathcal{G}$ at P. Let γ be the branch whose tangent is ℓ. Since γ is linear, $I(P, \mathcal{G} \cap \gamma) = s$. As $s \leq r - 2$, so $\mathcal{G}$ is not an adjoint. □

THEOREM 6.51 *The adjoints of a given degree cut out a complete linear series on the curve $\mathcal{F}$.*

Proof. Let Φ be an adjoint of degree m. Then $\Phi \cdot \mathcal{F} = D + E$, with $E \succ (0)$. If E' is another effective divisor with $E' \equiv E$, the existence of an adjoint Φ' of $\mathcal{F}$ for which $\Phi' \cdot \mathcal{F} = D + E'$ needs to be established.

First it is shown that $E \equiv E'$ implies the existence of two curves Ψ and Ψ' such that

$$\Psi \cdot \mathcal{F} = E + F, \quad \Psi' \cdot \mathcal{F} = E' + F.$$

Since $E = E' + \operatorname{div}(u)$ for an element $u = a(x,y)/b(x,y)$ with $a(X,Y), b(X,Y)$ both in $K[X,Y]$, so

$$E + \operatorname{div} b(x,y) = E' + \operatorname{div} a(x,y).$$

Applying Lemma 6.42 to the curves $\mathcal{C}_1 = \mathbf{v}(a(X,Y))$ and $\mathcal{C}_2 = \mathbf{v}(b(X,Y))$ gives

$$\begin{array}{rcl} \operatorname{div} a(x,y) & = & \mathcal{C}_1 \cdot \mathcal{F} - (\deg \mathcal{C}_1)(\ell_\infty \cdot \mathcal{F}), \\ \operatorname{div} b(x,y) & = & \mathcal{C}_2 \cdot \mathcal{F} - (\deg \mathcal{C}_2)(\ell_\infty \cdot \mathcal{F}). \end{array} \tag{6.3}$$

Hence

$$E + \mathcal{C}_2 \cdot \mathcal{F} - (\deg \mathcal{C}_2)(\ell_\infty \cdot \mathcal{F}) = E' + \mathcal{C}_1 \cdot \mathcal{F} - (\deg \mathcal{C}_1)(\ell_\infty \cdot \mathcal{F}).$$

Now suppose that the reference system is chosen with the following two properties:

(i) $\ell_\infty \cap \mathcal{F}$ is disjoint from both $\mathcal{C}_1 \cap \mathcal{F}$ and $\mathcal{C}_2 \cap \mathcal{F}$;

(ii) if $\mathcal{P}$ appears either in E or in E', then the centre of $\mathcal{P}$ is not a point of ℓ_∞.

Then (6.3) implies that $\deg \mathcal{C}_1 = \deg \mathcal{C}_2$. Therefore

$$E + \mathcal{C}_2 \cdot \mathcal{F} = E' + \mathcal{C}_1 \cdot \mathcal{F},$$

whence $\mathcal{C}_1 \cdot \mathcal{F} - E = \mathcal{C}_2 \cdot \mathcal{F} - E'$. Putting $F = \mathcal{C}_1 \cdot \mathcal{F} - E$, the assertion follows.

Note that F need not to be effective even if this happens in some cases. For instance, if E and E' do not have places in common. Nevertheless, a slight change is enough to ensure that F is effective. In fact, write $F = F_0 - F_\infty$ with effective divisors F_0 and F_∞. If $F_\infty = n_0\mathcal{P}_0 + \cdots + n_t\mathcal{P}_t$, choose a line ℓ_i through the centre of $\mathcal{P}_i$ and define the totally reducible curve $\mathcal{C}$ whose components are the lines ℓ_i for $i = 1, \ldots t$, each counted n_i times. Put $U = \mathcal{C} \cdot \mathcal{F}$. Then U is an effective divisor such that $U \succ F_\infty$. Then also $U + F \succ (0)$. Replacing Ψ by $\mathcal{C}\Psi$ and Ψ' by $\mathcal{C}\Psi'$ ensures that

$$(\mathcal{C}\Psi) \cdot \mathcal{F} = \mathcal{C} \cdot \mathcal{F} + \Psi \cdot \mathcal{F} = U + E + F = U + F,$$
$$(\mathcal{C}\Psi') \cdot \mathcal{F} = \mathcal{C} \cdot \mathcal{F} + \Psi' \cdot \mathcal{F} = U + E' + F = E' + U + F.$$

This shows that $F \succ (0)$ can be assumed. Then

$$(\Phi\Psi') \cdot \mathcal{F} = D + E + E' + F \succ D + \Psi \cdot \mathcal{F}.$$

Put

$$\Phi = \mathbf{v}(\varphi), \quad \Psi = \mathbf{v}(\psi)), \quad \Psi' = \mathbf{v}(\psi').$$

By Theorem 4.66,

$$\varphi\psi' = AF + \varphi'\psi,$$

where A and φ' are homogeneous polynomials, and $\varphi' \not\equiv 0 \pmod F$). Thus

$$\begin{aligned} D + E + E' + F &= \Phi \cdot \mathcal{F} + \Psi' \cdot \mathcal{F} \\ &= (\Phi\Psi') \cdot \mathcal{F} \\ &= (AF + \varphi'\psi) \cdot \mathcal{F} \\ &= (\varphi'\psi) \cdot \mathcal{F} \\ &= \Phi' \cdot \mathcal{F} + \Psi \cdot \mathcal{F} \\ &= \Phi' \cdot \mathcal{F} + E + F. \end{aligned}$$

Hence $\Phi' \cdot \mathcal{F} = D + E'$. □

THEOREM 6.52 *Every complete linear series is cut out, up to a fixed divisor, by the adjoints of some degree m.*

Proof. Let $|E|$ be a complete linear series. Apart from the trivial case when $|E|$ is empty, it may be assumed that E is effective. Choose an adjoint passing through E. Such adjoints certainly exist; for example, take lines passing through the centres of the branches corresponding to the places occurring in E, as well as, for every r-fold point of $\mathcal{F}$, some $r - 1$ lines passing through it. The totally reducible curve whose components are these lines is an adjoint of $\mathcal{F}$ passing through E. Any adjoint Φ of $\mathcal{F}$ through E satisfies $\Phi \cdot \mathcal{F} = D + E + E'$ with $E' \succ (0)$.

Let m denote the degree of Φ, By Theorem 6.51, the adjoints of degree m cut out the complete linear series $|E + E'|$ on $\mathcal{F}$. The divisors of $|E + E'|$ containing E' are cut out by the adjoints of degree m which pass through E'. By Theorem 6.40, the linear series $||E + E'| - |E'||$ consists of all divisors $H - E'$, where H is cut out by the adjoints of degree m passing through E'. To complete the proof it is enough to note that $|E| = ||E + E'| - |E'||$. □

THEOREM 6.53 *An irreducible plane curve $\mathcal{F}$ has genus 0 if and only if $\mathcal{F}$ is birationally equivalent to a line.*

Proof. From Theorem 5.57, lines have genus 0. For the converse, $\mathcal{F} = \mathbf{v}(f(X,Y))$ is assumed to have only ordinary singularities. Also, $\deg f(X,Y) \geq 2$ may be supposed. From the proof of Theorem 3.24, the complete linear series cut out on $\mathcal{F}$ by the adjoint curves of $\mathcal{F}$ of degree $n-1$ has dimension

$$r \geq \tfrac{1}{2}(n-1)(n+2) - \tfrac{1}{2}\textstyle\sum_{i=1}^{k} r_i(r_i - 1) - (2n-3) = 1.$$

Suppose $r > 1$. Then there is an adjoint curve $\mathcal{G}$ of degree 1 containing two non-singular points of $\mathcal{F}$. By Bézout's Theorem applied to $\mathcal{F}$ and $\mathcal{G}$, it turns out that they must have a common component, since

$$\textstyle\sum_{i=1}^{k} r_i(r_i - 1) + (2n-3) + 2 > n(n-1).$$

This is impossible since $\mathcal{F}$ is irreducible and $\mathcal{G}$ has lower degree than $\mathcal{F}$. Therefore $r = 1$. So adjoint curves of $\mathcal{F}$ degree $n-1$, with one exception, can be expressed in the form $\mathcal{G}_\lambda = \mathbf{v}(g_\lambda(X,Y))$, with $\lambda \in K$ and

$$g_\lambda(X,Y) = g_1(X,Y) + \lambda g_2(X,Y),$$

where $\mathcal{G}_1 = \mathbf{v}(g_1(X,Y)$ and $\mathcal{G}_2 = \mathbf{v}(g_2(X,Y))$ are two fixed, distinct adjoints of degree $n-1$.

Choose a coordinate system so that

$$f(X,Y) = uY^n + \cdots,\ g_1(X,Y) = v_1Y^{n-1} + \cdots,\ g_2(X,Y) = v_2Y^{n-1} + \cdots,$$

with $u, v_1, v_2 \neq 0$, and so that no common point of $\mathcal{F}$ and $\mathcal{G}_1$ is at infinity. Let $R(X,\lambda)$ be the resultant with respect to Y of $\mathcal{F}$ and $\mathcal{G}$. Then

$$R(X,\lambda) = b_0(\lambda) + b_1(\lambda)X + \cdots + b_s(\lambda)X^s,$$

where $s = n(n-1)$. Now $R(X,0)$ is the resultant of $\mathcal{F}$ and $\mathcal{G}_1$, and it has degree $n(n-1)$. Hence $b_s(0) \neq 0$, and so $b_s(\lambda_0) = 0$ for at most a finite set of values of λ_0. Ignoring this finite set, $\mathcal{F}$ and $\mathcal{G}_\lambda$ have no common point at infinity. The polynomial $R(X,\lambda) \in K[X]$ has a root of a_i of multiplicity at least $r_i(r_i-1)$ at each r_i-fold point $P_i = (a_i, b_i)$, and has a simple root c_i at each of the $2n-3$ chosen simple points $Q_i = (c_i, d_i)$ of $\mathcal{F}$. Since

$$\textstyle\sum_{i=1}^{k} r_i(r_i - 1) + (2n-3) = n(n-1) - 1,$$

all but one of the roots of $R(X,\lambda) = 0$ have been accounted for. The remaining root is, therefore,

$$\varphi(\lambda) = \frac{b_{s-1}(\lambda)}{b_s(\lambda)} - \textstyle\sum_{i=1}^{k} r_i(r_i-1)a_i - \sum_{i=1}^{k} c_i.$$

Note that $\varphi(\lambda)$ is a rational function. Similarly, from the resultant with respect to X, another rational function $\psi(\lambda)$ is obtained.

For each value λ which has not been ruled out, the point $Q = (\varphi(\lambda), \psi(\lambda))$ is on $\mathcal{F}$. If $Q = (x_0, y_0)$ is a point of $\mathcal{F}$ not on $\mathcal{G}_2$, there is a unique λ_0 such that $Q \in \mathcal{G}_{\lambda_0}$, namely $\lambda_0 = -g_1(x_0,y_0)/g_2(x_0,y_0)$. Hence x_0 is a root of the polynomial $R(\lambda_0, X) \in K[X]$, and so $x_0 = \varphi(\lambda_0)$. Similarly, $y_0 = \psi(\lambda_0)$. There is only a finite set of points of $\mathcal{F}$ which are on $\mathcal{G}_2$. Let Σ be the function field of the line ℓ of equation $Y = 0$. Then $\Sigma = K(\xi)$.

Let ω be the rational transformation

$$x' = \varphi(\xi), \quad y' = \psi(\xi)$$

of Σ. Then the image curve $\mathcal{F}'$ of ℓ by ω is an irreducible plane curve which has infinitely many common points with $\mathcal{F}$. By Bézout's Theorem, they coincide. Hence $\mathcal{F}$ is birationally equivalent to the line ℓ. □

DEFINITION 6.54 For a curve $\mathcal{F}$ of degree $n > 3$, the adjoints of degree $n-3$ are the *canonical* or *special* adjoints of $\mathcal{F}$.

THEOREM 6.55 *If $\mathcal{F}$ has degree $n > 3$ and genus g, there are g linearly independent canonical adjoints of $\mathcal{F}$.*

Proof. The curves of degree $n-3$ form a linear system of dimension $\frac{1}{2}n(n-3)$. If $\mathcal{F}$ is non-singular, all these curves are adjoints and the result follows since

$$g = \tfrac{1}{2}(n-1)(n-2) = \tfrac{1}{2}n(n-3) + 1.$$

Suppose P is an r-fold point of $\mathcal{F}$. If $\mathcal{G}$ is an adjoint, P is an $(r-1)$-fold point of $\mathcal{G}$ by Theorem 6.50. To find a canonical adjoint $\mathcal{G}$ of $\mathcal{F}$, take any polynomial $G(X,Y) \in K[X,Y]$ of degree $n-3$. Arguing as in the proof of Theorem 3.24, a curve $\mathcal{G} = \mathbf{v}(G(X,Y))$, with

$$G(X,Y) = a_{00} + a_{10}X + \cdots + a_{0,n-3}Y^{n-3},$$

is an adjoint of $\mathcal{F}$ if and only if $(a_{00}, a_{10}, \ldots, a_{0,n-3})$ is a solution of a certain system of linear equations. The number of the equations is the sum of the number of singular points of $\mathcal{F}$, each counted $1 + 2 + \cdots + (r-1) = \sum_{i=1}^{k} \frac{1}{2}r_i(r_i - 1)$ times, where $P_1, \ldots, P_k$ are the singular points of $\mathcal{F}$ and r_i is the multiplicity of P_i. These equations can be interpreted as hyperplane equations in $\mathrm{PG}(N,K)$ with $N = \frac{1}{2}n(n-3)$. In this way, $\frac{1}{2}\sum r(r-1)$ is the total number of hyperplanes arising from singular points of $\mathcal{F}$. The intersection of these hyperplanes is a subspace Π_M of dimension $M \geq N - \frac{1}{2}\sum r(r-1)$. To $M+1$ linearly independent points in Π_M there correspond the same number of linearly independent canonical adjoints. Now the assertion follows from the fact that $M \geq g-1$, a consequence of Theorem 5.57. □

THEOREM 6.56 *If $\mathcal{F}$ has degree $n > 3$, then the canonical series $|(dx)|$ is cut out by the canonical adjoints.*

Proof. With a change of coordinates, take $\mathcal{F}$ as in the proof of Theorem 5.57. Then the polar curve $\mathcal{F}' = \mathbf{v}(\partial(F(X,Y)/\partial Y)$ of $\mathcal{F}$ at Y_∞ exists and, if $P = (x,y)$ denotes a generic point of $\mathcal{F}$, then x is a separable variable of Σ. From the proof of Theorem 5.57,

$$\mathcal{F}' \cdot \mathcal{F} = D + \mathrm{div}(dx)_0, \quad \mathrm{div}(dx)_\infty = 2(\ell_\infty \cdot \mathcal{F}).$$

If E is a canonical adjoint, then $(E + 2\ell_\infty) \cdot \mathcal{F} \equiv \mathcal{F}' \cdot \mathcal{F}$, and so it follows that $E \cdot \mathcal{F} \equiv D + \mathrm{div}(dx)$. Therefore every canonical adjoint cuts out a divisor equivalent to $\mathrm{div}(dx)$ on $\mathcal{F}$. By Theorem 6.51, the result is established provided that there exists at least one canonical divisor. However, this follows from Theorem 6.55 for $g \geq 1$. For $g = 0$, the canonical series is empty, and the result holds. □

REMARK 6.57 The above approach to the canonical series does not work when the condition that $\mathcal{F}$ has only ordinary singularities is dropped. However, in some cases, the canonical linear series may have a nice explicit description in terms of linear systems of curves, as the following example shows.

EXAMPLE 6.58 Assume h to be odd in Example 5.59, and consider the linear system Δ consisting of all plane curves of degree $\frac{1}{2}(n-3)$ for which Y_∞ is an $\frac{1}{2}(n-1)$-fold point. In other words, Δ consists of all completely reducible plane curves split into $\frac{1}{2}(n-1)$ lines through Y_∞. Since Y_∞ is an $(n-2)$-fold point of $\mathcal{F}$, so Δ cuts out on $\mathcal{F}$ a linear series L for which Y_∞ is a fixed place of multiplicity $\frac{1}{2}(n-2)(n-3)$. It then follows that L has dimension $\frac{1}{2}(n-3)$. Discarding $\frac{1}{2}(n-2)(n-3)Y_\infty$ from each divisor of L, the resulting linear series is a g^d_{2d} with $d=\frac{1}{2}(n-3)=g-1$; by Theorem 6.72 (i), this is the canonical series.

6.3 SPECIAL AND NON-SPECIAL LINEAR SERIES

Now the Riemann–Roch Theorem can be proved; it is the main result concerning complete linear series. Again, a model $(\mathcal{F};(x,y))$ of Σ is used with the property that $\mathcal{F}$ is an irreducible algebraic curve with only ordinary singularities. The degree of $\mathcal{F}$ is denoted by m, and W stands for a canonical divisor.

If $m \leq 3$, take a triangle meeting $\mathcal{F}$ in $3m$ distinct points. Then the image curve of $\mathcal{F}$ under a quadratic transformation associated to the triangle is birationally equivalent to $\mathcal{F}$, and it has degree $3m$, which is larger than 3 for $m>1$. If $m=1$, repeating this process, an irreducible curve of degree greater than 3 is obtained. Thus it may be assumed that $m>3$.

DEFINITION 6.59 For a complete linear series g^r_n,

(i) the *index of speciality* i is the number of linearly independent canonical divisors through a given divisor;

(ii) the series is *special* if $i>0$ and *non-special* if $i=0$.

THEOREM 6.60 (Riemann) *If g^r_n is a complete linear series on the curve $\mathcal{F}$ of genus g, then $r \geq n-g$.*

THEOREM 6.61 (Riemann–Roch) *If g^r_n is a complete linear series $|D|$ on the curve $\mathcal{F}$ of genus g, with D an effective divisor and W a canonical divisor, then*

(i) $r=n-g+i$;

(ii) $r=g-1$ *and* $n=2g-2$ *when* $D=W$;

(iii) $r=n-g$ *when* $n>2g-2$.

EXAMPLE 6.62 Let $\mathcal{F}$ be as in Example 5.59. Put $P_i=(a_i,0)$ for $i=1,\dots,h$, and denote by $\mathcal{P}_i$ the place associated to the unique branch of $\mathcal{F}$ centred at P_i. Let $D=\sum_{i=1}^n \epsilon_i\mathcal{P}_i$ with $\epsilon_i=0$ or $\epsilon_i=1$. By Example 6.58, every canonical divisor is cut out on $\mathcal{F}$ by a completely reducible curve split into $\frac{1}{2}(h-1)$ lines through Y_∞ minus the fixed divisor $\frac{1}{2}(n-2)(n-3)\mathcal{P}_\infty$, where $\mathcal{P}_\infty$ is the place centred at Y_∞. From this, the index of speciality i of D is calculated:

$$i = \begin{cases} \frac{1}{2}(h-1) - \deg D & \text{for } \deg D \leq \frac{1}{2}(h-1), \\ 0 & \text{for } \deg D > \frac{1}{2}(h-1). \end{cases} \tag{6.4}$$

The Riemann–Roch Theorem shows that $\dim |D| = 0$ for $\deg D \leq \frac{1}{2}(h-1)$.

In the proof of Theorem 3.24 it is shown that the maximum number of linearly independent adjoints of degree $m-2$ of $\mathcal{F}$ is $\frac{1}{2}m(m-1) - \sum \frac{1}{2}r_i(r_i-1) + e$ with $e \geq 0$. These cut out divisors of degree

$$m(m-2) - \sum r_i(r_i-1) = 2g - 2 + m$$

on $\mathcal{F}$. Thus the adjoints of degree $m-2$ cut out out a complete linear series $g_{2g-2+m}^{g-2+m+e}$. *A priori*, e may be positive, as the $\sum r_i(r_i-1)$ conditions imposed on a curve of degree $m-2$ to be adjoint may be linearly independent. *A posteriori*, by the Riemann–Roch Theorem, $e = 0$. For the present it may only be asserted that $e \geq 0$. However, the assertion that $e = 0$ can be formulated as a consequence of the Riemann–Roch Theorem.

COROLLARY 6.63 *The maximum number of linearly independent adjoints of degree $m-2$ of $\mathcal{F}$ is $\frac{1}{2}m(m-1) - \sum \frac{1}{2}r_i(r_i-1)$.*

Similarly, the maximum number of linearly independent adjoints of degree $m-3$ of $\mathcal{F}$ is $\frac{1}{2}(m-1)(m-2) - \sum \frac{1}{2}r_i(r_i-1) + e$ with $e \geq 0$. Here, again, $e = 0$ follows from the Riemann–Roch Theorem.

To prove Theorem 6.61, a particular case is first examined.

LEMMA 6.64 *If a complete linear series g_n^r has order $n = g+1$, then $r \geq 1$.*

Proof. Since $m > 2$, so $2g + m - 2 \geq g + 1$. The adjoints of degree $m-2$ cut out on $\mathcal{F}$ a series $g_{2g-2+m}^{g-2+m+e}$. Given a divisor $C \in g_n^r$, then it is first shown that there exists $D_0 \in \mathcal{L}$ with $D_0 \succ C$. By Theorem 6.40, the divisors $H - C$, with $H \in \mathcal{L}$ and $H \succ C$, is a $g_{n'}^{r'}$ with

$$\begin{aligned} r' &= g - 2 + m + e - (g+1) + e' = m - 3 + e + e', \\ n' &= 2g - 2 + m - (g+1) = g - 3 + m, \end{aligned}$$

with $e' \geq 0$. Since $r' \geq 0$, so F exists in $\mathcal{L}$ with $F \succ C$.

Consider now the divisors in $\mathcal{L}$ which contain $F - C$; they form a series of dimension,

$$s = (g - 2 + m + e) - (2g - 2 + m - (g+1)) + e'' = 1 + e + e'' \geq 1.$$

This series contains C and has dimension at least one. □

COROLLARY 6.65 *If a complete linear series g_n^r has $n > g+1$, then $r \geq 1$.*

Proof. Let $C + E \in g_n^r$, with C, E effective divisors and $\deg C = g+1$. By Lemma 6.64, $\dim |C| \geq 1$; hence there exists an effective divisor C' such that $C' \equiv C$ but $C \neq C'$. So $C' + E \equiv C + E$ but $C + E \neq C' + E$. Thus the dimension $r \geq 1$. □

THEOREM 6.66 (Noether's Reduction Theorem) *Given a place $\mathcal{P}$ centred at the simple point P of $\mathcal{F}$, if there exists a special adjoint through the divisor C but not through $C+\mathcal{P}$, then $\mathcal{P}$ is a fixed place of $|C+\mathcal{P}|$.*

Proof. The original proof of Noether which is reproduced here does not work in positive characteristic when P is a terrible point. To deal with terrible points, the same argument as after Definition 3.30 does work; a quadratic transformation is applied to give a curve, the geometric transform of $\mathcal{F}$, whose singular points are ordinary, for which the corresponding point Q of P is no longer terrible. Therefore P may be assumed not to be terrible. Then there is a line ℓ through the centre P of the branch corresponding to $\mathcal{P}$ that meets $\mathcal{F}$ in m distinct points, viewed as the centres of the branches corresponding to the places $\mathcal{P}=\mathcal{P}_1,\ldots,\mathcal{P}_m$. Then

$$(\mathcal{F}_{m-3}\,\ell)\cdot\mathcal{F}=D+C+E+\mathcal{P}_1+\cdots+\mathcal{P}_m.$$

The adjoint curves $\mathcal{F}_{m-2}$ cut out on $\mathcal{F}$ the complete linear series

$$|C+E+\mathcal{P}_1+\cdots+\mathcal{P}_m|.$$

As above, the adjoint curves $\mathcal{F}_{m-2}$ through $E+\mathcal{P}_2+\cdots+\mathcal{P}_m$ form a linear subsystem Γ' which cuts out on $\mathcal{F}$ the complete series $|C+\mathcal{P}|$; that is,

$$|C+\mathcal{P}|=\big||C+E+\mathcal{P}_1+\cdots+\mathcal{P}_m|-|E+\mathcal{P}_2+\cdots+\mathcal{P}_m|\big|.$$

The curves of Γ' meet ℓ in $m-1$ points, and therefore contain ℓ as a component. It follows that each curve of Γ' passes through the centre of $\mathcal{P}$. Since $\mathcal{P}$ is not in $E+\mathcal{P}_1+\cdots+\mathcal{P}_m$, so $\mathcal{P}$ is a fixed place of $|C+\mathcal{P}|$. □

REMARK 6.67 In the theorem, the case $C\succ P$ is permissible, since places are counted with multiplicity.

Proof of Theorem 6.61. The proof works by induction on i.

Suppose first that $i=0$; in this case, the proof is by induction on r. The first case is $r=0$, and it must be shown that $n=g$. From Lemma 6.64 and Corollary 6.65 it follows that $n\le g$. As above, the canonical curves cut out a g_{2g-2}^{g-1+e} on $\mathcal{F}$; therefore through any $g-1$ places there is at least one canonical divisor. Since $i=0$, so $n\ge g$. Therefore $n=g$ and the theorem is proved for $i=r=0$.

Suppose, still with $i=0$, that $r>0$. Given a place $\mathcal{P}$ centred at a simple point that is not a fixed place for g_n^r, Theorem 6.13 shows that the divisors of g_n^r, with $\mathcal{P}$ removed, form a g_{n-1}^{r-1}. Note that the index of speciality for the latter series is zero. In fact, if it were otherwise, there would exist a special adjoint through a divisor C of g_{n-1}^{r-1}, which cannot pass through $\mathcal{P}$ since $i=0$. By Theorem 6.66, $\mathcal{P}$ would be a fixed place of g_n^r, a contradiction. By induction, $r-1=n-1-g$, whence $r=n-g$.

Finally, suppose that $i>0$, and choose a divisor C in g_n^r and a canonical curve $\mathcal{C}'$ through C. Let P be a simple point of $\mathcal{F}$ not lying on $\mathcal{C}'$; then the place $\mathcal{Q}$ with centre Q is fixed for $|C+\mathcal{P}|$, by Theorem 6.66. Therefore this series $|C+\mathcal{P}|$ is a g_{n+1}^r, and its index of speciality is one less than that of $|C|$. Hence

$$r=n+1-g+(i-1)=n-g+i.$$

□

6.4 REFORMULATION OF THE RIEMANN–ROCH THEOREM

Theorem 6.61 can be written in the following form:

$$\dim |C| = \deg C - g + 1 + \dim |W - C|; \tag{6.5}$$

here, C is an effective divisor and W a canonical divisor. This has been established in the case that $|C|$ contains at least one effective divisor; that is, $r \geq 0$. Now it is shown that (6.5) is still valid when $|C|$ contains no effective divisor; that is, $r = -1$.

If $\dim |W - C| \geq 0$, then, with $W - C$ for C, equation (6.5) takes the form

$$\begin{aligned} \dim |W - C| &= \deg(W - C) - g + 1 + \dim |W - (W - C)| \\ &= 2g - 2 - \deg C - g + 1 + \dim |C|, \end{aligned} \tag{6.6}$$

since $\deg W = 2g - 2$. But this implies (6.5); so considering C or $W - C$ gives equivalent results.

If $\dim |W - C| = -1$, then $W - C$ contains no effective divisors. Since $|C|$ also contains no effective divisors, it must be shown that $\deg C = g - 1$ or, equivalently, that $\deg(W - C) = g - 1$.

Suppose that $\deg C \geq g - 1$ and let $C = A - B$ with both A and B effective divisors. Then

$$\begin{aligned} \dim |C + B| &\geq \deg(C + B) - g \\ &= \deg C + \deg B - g \\ &\geq \deg B. \end{aligned}$$

So $\deg B$ conditions can be imposed on the divisors of $|C + B|$; therefore there exists $A' \in |C + B|$ such that $A' \succ B$. But then $A' - B \equiv C$, and $A' - B$ is effective, contradicting that $\dim |C| = -1$.

Now suppose that $\deg C < g - 1$. Then $\deg(W - C) > g - 1$, and the above argument holds, with $W - C$ in place of C. Thus (6.5) is valid for any divisor C on $\mathcal{F}$.

The Riemann–Roch Theorem can be reformulated in terms of functions on $\mathcal{F}$.

DEFINITION 6.68 For any divisor $C = \sum n_{\mathcal{P}} \mathcal{P}$, effective or not,

$$\begin{aligned} \mathcal{L}(C) &= \{f \in \Sigma \mid \operatorname{ord}_{\mathcal{P}}(f) \geq -n_{\mathcal{P}} \text{ for all } \mathcal{P} \in \mathfrak{P}(\Sigma)\} \\ &= \{f \in \Sigma \mid \operatorname{div}(f) + C \succ 0\} \cup \{0\}; \end{aligned}$$

this is the *Riemann–Roch space* of C.

LEMMA 6.69 *The set $\mathcal{L}(C)$ is a vector space over K with the following properties*:

(i) *if $C \succ D$, then $\mathcal{L}(D)$ is a subspace of the vector space $\mathcal{L}(C)$;*

(ii) *$\mathcal{L}(0) = K$, and $\mathcal{L}(C) = \{0\}$ when* $\deg C < 0$;

(iii) $\dim \mathcal{L}(C)$ *is finite, and* $\dim \mathcal{L}(C) \leq \deg C + 1$ *when* $\deg C \geq 0$;

(iv) *if $C \equiv D$, then* $\dim \mathcal{L}(C) = \dim \mathcal{L}(D)$.

Proof. (i) Let $C = D + E$ with $E \succ 0$. If $f \in \mathcal{L}(D)$, then $\mathrm{div}(f) + D \succ 0$, whence $\mathrm{div}(f) + D + E \succ 0$. Thus $\mathcal{L}(C) \supset \mathcal{L}(D)$ and (i) follows.

(ii) If $f \in \Sigma$ is a function, then $\mathrm{div}(f) = \mathrm{div}(f)_0 - \mathrm{div}(f)_\infty$ with $\mathrm{div}(f)_\infty \neq 0$. Therefore $\mathrm{div}(f) \not\succ 0$, whence $\mathcal{L}(0) = K$. When $\deg C < 0$, then there is no function f such that $\mathrm{div}(f) + C \succ 0$. Hence $\mathcal{L}(C) = \{0\}$.

(iii) A more general assertion is proved. Let A, B be divisors of Σ. If $B \succ A$, then

$$\dim \mathcal{L}(B)/\mathcal{L}(A) \leq \deg B - \deg A.$$

It is enough to prove it for $B = A + \mathcal{P}$, as the general case follows from the special case by induction. It may be assumed that $\mathcal{L}(B) \neq \mathcal{L}(A)$. Choose an element ξ in $\mathcal{L}(B)$ that does not belong to $\mathcal{L}(A)$. If $A = \sum n_\mathcal{Q} \mathcal{Q}$, then $\mathrm{ord}_\mathcal{P}\, \xi = n_\mathcal{P} + 1 = m$, and hence $\mathrm{ord}_\mathcal{P}\, \xi^{-1} = -m$. Now $\mathrm{ord}_\mathcal{P}\, \xi^{-1}\eta \geq 0$ for any element $\eta \in \mathcal{L}(B)$.

Given a primitive representation τ of $\mathcal{P}$, let

$$\begin{aligned} \tau(\xi^{-1}) &= ct^{-m} + \cdots, \ c \neq 0, \\ \tau(\eta) &= dt^m + \cdots. \end{aligned}$$

Let $V_1(K)$ be one-dimensional vector space. Then the mapping

$$\begin{aligned} \varphi : \mathcal{L}(B) &\to V_1(K), \\ \eta &\mapsto cd \end{aligned}$$

is K-linear, and its kernel is $\mathcal{L}(A)$. Thus $\dim \mathcal{L}(B)/\mathcal{L}(A) \leq 1$. It follows that $\dim \mathcal{L}(B)/\mathcal{L}(A) = 1$.

(iv) Let $D = C + \mathrm{div}(g)$. The mapping

$$\begin{aligned} \varphi : \mathcal{L}(C) &\to \mathcal{L}(D), \\ f &\mapsto fg \end{aligned}$$

is K-linear, and its kernel is the zero divisor (0). So φ is an isomorphism. □

THEOREM 6.70 *For any divisor C on the curve $\mathcal{F}$,*

$$\dim \mathcal{L}(C) = \deg C - g + 1 + \dim \mathcal{L}(W - C). \tag{6.7}$$

Proof. Let $r = \dim \mathcal{L}(C) - 1$ and suppose that $f_0, f_1, \ldots, f_r$ is a basis for $\mathcal{L}(C)$. Let $\mathcal{A}_C = \{\sum c_i f_i \mid c_i \in K\}$; then $\mathcal{A}_C$ is a linear series, and $\sum c_i f_i \in \mathcal{L}(C)$ if and only if the divisor $\mathrm{div}(f) + C$ is effective; that is, $\mathrm{div}(f) + C \in \mathcal{A}_C$. Hence $|C|$ is a linear series and has dimension r. Now Theorem 6.5 gives the required result. □

EXAMPLE 6.71 Consider $F = X_0^3 + X_1^3 + X_2^3$ defined over $K = \overline{\mathbf{F}}_4$, with

$$\mathbf{F}_4 = \{0, 1, \omega, \bar{\omega} = \omega^2 = \omega + 1\}.$$

The curve $\mathcal{F} = \mathbf{v}(F)$ is non-singular, and its points are identified with the places of its function field $\Sigma = K(x, y)$ with $x^3 + y^3 + 1 = 0$. There are nine points $(1, x, y) = (x_0, x_1, x_2)$ in PG$(2, 4)$ which belong to $\mathcal{F}$; namely,

$$\begin{array}{lll} P_0 = (0,1,1), & P_3 = (1,0,1), & P_6 = (1,1,0), \\ P_1 = (0,1,\omega), & P_4 = (1,0,\omega), & P_7 = (1,\omega,0), \\ P_2 = (0,1,\bar{\omega}), & P_5 = (1,0,\bar{\omega}), & P_8 = (1,\bar{\omega},0). \end{array}$$

Let $f_1 = 1/(x+y) = x_0/(x_1+x_2)$. Then $\operatorname{div} f_1 = P_1 + P_2 - 2P_0$. Note that

$$\frac{1}{f_1} = \frac{x_1+x_2}{x_0} = \frac{1}{x_1^2+x_1x_2+x_2^2}\, x_0^2 = u\, x_0^2,$$

where $u = 1/(x_1^2+x_1x_2+x^2)$ has no pole at P_0, and x_0 is a uniformising parameter; so f_1 has a pole of order 2 at P_0. Similarly, $f_2 = x/(x+y) = x_1/(x_1+x_2)$ has a pole of order 3 at P_0, since

$$\frac{1}{f_2} = \frac{x_1+x_2}{x_1} = \frac{1}{x_1(x_1^2+x_1x_2+x_2^2)}\, x_0^3.$$

Here, $\operatorname{div} f_2 = P_3 + P_4 + P_5 - 3P_0$. Let $C = 3P_0$; then f_1 and f_2 are both in $\mathcal{L}(C)$. Since $g = 1$, $\deg C = 3$, $\mathcal{L}(W - C) = \{0\}$, so $\dim \mathcal{L}(C) = 3$. Therefore a basis for $\mathcal{L}(C)$ is $\{1, f_1, f_2\}$.

6.5 SOME CONSEQUENCES OF THE RIEMANN–ROCH THEOREM

THEOREM 6.72 (i) *The canonical series is the only series on a curve of genus g that has order $2g-2$ and dimension $g-1$.*

(ii) *The canonical series has no fixed place.*

(iii) *A divisor of a complete special g_n^r puts $n-r$ conditions on a canonical divisor to contain it.*

Proof. (i) A g_{2g-2}^{g-1} with dimension $r = g-1$ and $r > n-g$ is special; so it is totally contained in the canonical series, and hence coincides with it, having the same dimension.

(ii) If the canonical series had a fixed point P, the residual series would be a g_{2g-3}^{g-1}. The addition of another place Q to the divisors of this series would give a g_{2g-2}^{g-1} different from the canonical series.

(iii) If i is the index of speciality of the g_n^r, then $i = g - (n-r)$, whence $i - 1 = g - 1 - (n-r)$. □

EXAMPLE 6.73 Let $\mathcal{C}^6 = \mathbf{v}(F)$ be a sextic curve with one singular point P_0 that is an ordinary quadruple point; for example, if char $K \neq 2$, let

$$F = XY(X-Y)(X+Y)Z^2 + X^6 + Y^6.$$

Then a special adjoint is a cubic curve with a triple point at P_0, and is therefore a set S of three lines through P_0. The canonical series therefore has order 6, since each line of S meets $\mathcal{C}^6$ in two points apart from P_0, and dimension 3, since the pencil of lines through P_0 has dimension 1. Since $\mathcal{C}^6$ has genus

$$g = \tfrac{1}{2}(6-1)(6-2) - \tfrac{1}{2}4(4-1) = 4,$$

this is consistent with the canonical series being a g_6^3.

Now the converse of Theorem 6.66 is established.

THEOREM 6.74 *Given a linear series g_n^r on $\mathcal{F}$, if there exists a place $\mathcal{P}$ such that the complete linear series $g_n^r + \mathcal{P} = \{D + \mathcal{P} \mid D \in g_n^r\}$ has $\mathcal{P}$ as a fixed place, then the g_n^r is complete and special, and there exists a canonical divisor containing a divisor of g_n^r but not containing $\mathcal{P}$.*

Proof. If $g_n^r + \mathcal{P}$ has the fixed place $\mathcal{P}$, since between the order and dimension of this series there is the inequality $r \geq (n+1) - g$, so $r > n - g$; hence the series g_n^r is special. Further, g_n^r is complete, since otherwise $g_n^r + \mathcal{P}$ would not be complete. Now, if each canonical divisor containing a divisor of g_n^r also contained $\mathcal{P}$, then $g_n^r + \mathcal{P}$ would also be special, with the same index of speciality as g_n^r; that is, a divisor of g_n^r and one of $g_n^r + \mathcal{P}$ impose the same number of conditions on a canonical divisor. This contradicts Theorem 6.72. □

COROLLARY 6.75 *The complete series $g_n^r + \mathcal{P}$ has $\mathcal{P}$ as a fixed place if and only if the series g_n^r is special and there exists a canonical divisor containing a divisor of g_n^r but not containing $\mathcal{P}$. In other words, for any effective divisor A and any place $\mathcal{P}$, one and only one of the following equations holds:*

(i) $\dim |A + \mathcal{P}| = \dim |A|$;

(ii) $i(|A + \mathcal{P}|) = i(|A|)$.

DEFINITION 6.76 The *sum* of two linear series g_n^r, $g_{n'}^{r'}$ is

$$g_n^r + g_{n'}^{r'} = \{D + D' \mid D \in g_n^r,\ D' \in g_{n'}^{r'}\}.$$

COROLLARY 6.77 *A complete series g_n^r is non-special if $n > 2g-2$ or $r > g-1$, whereas it is special if $n < g$.*

THEOREM 6.78 (Reciprocity Theorem of Brill and Noether). *If g_n^r and $g_{n'}^{r'}$ are residual with respect to the canonical series $|W|$, that is, $|g_n^r + g_{n'}^{r'}| = |W|$, then $n - 2r = n' - 2r'$.*

Proof. The speciality indices i and i' of g_n^r and $g_{n'}^{r'}$ satisfy $i = r'+1$ and $i' = r+1$. Thus $r = n - g + i = n - g + r' + 1$ and, similarly, $r' = n' - g + r + 1$; subtracting the two equations gives the result. □

THEOREM 6.79 (Clifford's Theorem) *If a linear series g_n^r is special, then $n \geq 2r$.*

Proof. It may be supposed that the series g_n^r is complete. If H is the residual of a divisor G of g_n^r with respect to a canonical divisor, then G imposes r conditions on the canonical divisors through H, while it imposes $n - r$ conditions on an arbitrary canonical divisor containing it, by Theorem 6.72 (iii). As it cannot happen that independent conditions in the general case become dependent for a subseries, so $n - r \geq r$, whence $n \geq 2r$. Equality can only occur when the series is complete.

An alternative view is as follows. Let Φ be the canonical linear system of $\mathcal{F}$. If the g_n^r is complete, then it is cut out by a subsystem Φ' of dimension r of Φ. More precisely, for a given divisor G of g_n^r and a canonical adjoint $\mathcal{G}$ through G, let H

be the residual of G in the canonical divisor C cut out by $\mathcal{G}$; that is, $C = G + H$. Then Φ' consists of all special adjoints through H.

Now consider the subsystem Φ'' of all the canonical adjoints through G. If $i-1$ is the dimension of Φ'', then $r = n - g + i$ by the Riemann–Roch Theorem. Also, Φ' and Φ'' have $\mathcal{G}$ in common but not a pencil of curves, as otherwise $\mathcal{F}$ would be a component of a curve in such a pencil. Hence the system containing both Φ' and Φ'' has dimension at most $r + i - 1$. Since $\dim C = g - 1$, so $r + i - 1 \leq g - 1 = n - r + i - 1$, whence $n \geq 2r$. □

EXAMPLE 6.80 A non-singular quartic curve C^4 has genus 3. The lines of the plane are the special adjoints and cut out the canonical series g_4^2. The lines through a point Q off C^4 cut out a special g_4^1.

A complement to Theorems 6.32 and 6.79 is the following result.

THEOREM 6.81 (i) *A necessary and sufficient condition for a curve to be rational is that it possesses at least one linear series* g_n^n.

(ii) *The complete linear series of a rational curve are* g_n^n *for* $n \in \mathbf{N}$.

Proof. A series g_n^n is complete as its dimension is the largest possible for its order; see Theorem 6.32.

Suppose the curve is rational; that is, it is the line at infinity $\ell_\infty = \mathbf{v}(X_0)$ up to a birational transformation. The linear system of all plane curves of degree n cuts out on ℓ_∞ a linear series of degree n. Since the $n+1$ monomials X^iY^{n-i} for $i = 0, \ldots, n$ are linearly independent over K, the linear series has dimension n, and hence it is a g_n^n. Since there is a curve of degree n cutting out any pre-assigned divisor of degree n, so g_n^n consists of all divisors of ℓ_∞ of degree n. Thus everything apart from the sufficiency condition is proved.

Suppose now that the curve possesses a series g_n^n, with $n > 0$. By Theorem 6.32, g_n^n is complete. Choose a simple point P on the curve such that the corresponding place $\mathcal{P}$ is not a fixed place of g_n^n. The divisors $C - \mathcal{P}$, with $C \in g_n^n$ and C containing $\mathcal{P}$, constitute a linear subseries g_{n-1}^{n-1}. By repeating this argument, the existence of a g_1^1 follows. Let

$$g_1^1 = \{\operatorname{div}(c_0 + c_1 x) + B \mid c_0, c_1 \in K, (c_0, c_1) \neq (0,0)\}.$$

By Theorem 7.36, g_1^1 is not composed of an involution. Thus $[\Sigma : K(x)] = 1$, whence $\Sigma = K(x)$. □

6.6 THE WEIERSTRASS GAP THEOREM

In Section 5.4, it is shown that every function $z \in \Sigma$ has a finite number of poles, giving rise to the effective divisor of poles $\operatorname{div}(z)_\infty$. The converse of this theorem is generally false, as not every effective divisor is the pole divisor of a function in Σ. This is explained in the following theorem.

THEOREM 6.82 *An effective divisor A is the divisor of poles* $\operatorname{div}(z)_\infty$ *of some function of Σ if and only if $|A|$ has no fixed place.*

Proof. If A is the divisor of poles of z, then $A \equiv \text{div}(z)_\infty$; that is, $|A|$ contains $\text{div}(z)_0$. Since A and $\text{div}(z)_0$ do not share any place, so $|A|$ has no fixed place.

Conversely, let $|A|$ have no fixed places. It is shown first that $B \equiv A$ for some divisor B such that A and B have no common place. Write $A = \sum_{i=1}^{k} n_{\mathcal{P}_i} \mathcal{P}_i$. By assumption, for every j with $1 \leq j \leq k$, there is a divisor $A_j \equiv A$ such that A_j does not contain $\mathcal{P}_j$. If $\mathcal{F}_j = \mathbf{v}(\varphi_j(X_0, X_1, X_2))$ cuts out A_j, then the curves of the linear system $\mathbf{v}(\sum_{i=1}^{k} \lambda_j \varphi_j)$ cut out divisors in $|A|$. Such a divisor contains $\mathcal{P}_j$ if and only if the k-ple $(\lambda_1, \ldots, \lambda_k)$ is a solution of a certain system of linear equations. Choosing coefficients λ_j that satisfy none of the equations in that system, a divisor B of the desired type is obtained. Now the argument in the first part can be reversed. □

As a corollary, the following theorem is obtained.

THEOREM 6.83 *Every effective divisor is the divisor of poles of some function of Σ if and only if Σ is a rational function field.*

REMARK 6.84 Theorem 6.83 remains true if the condition is only required for divisors consisting of one place.

For the rest of this section, the genus g of Σ is assumed to be positive.

THEOREM 6.85 *Let $A_s = \mathcal{P}_1 + \cdots + \mathcal{P}_s$ be an effective, non-special divisor with not-necessarily-distinct places $\mathcal{P}_i$, and let $A_0 = 0, A_n = \mathcal{P}_1 + \cdots + \mathcal{P}_n$, for $1 \leq n \leq s$. Then*

(i) *there exist at least g values of n for which A_n is not the divisor of poles of any function in Σ;*

(ii) *the indices of the $\mathcal{P}_i$ can be relabelled such that this occurs precisely for g values of n.*

Proof. Put $i_n = i(|A_n|)$; in particular, $i_0 = g$, $i_s = 0$. By Theorem 6.75,

$$\begin{array}{ll} i_n = i_{n-1} & \text{if } \dim |A_{n-1}| < \dim |A_n|, \\ i_n = i_{n-1} - 1 & \text{if } \dim |A_{n-1}| = \dim |A_n|. \end{array}$$

This implies the existence of exactly g values of n for which the relationship $\dim |A_{n-1}| = \dim |A_n|$ holds, that is, for which $\mathcal{P}_n$ is a fixed place in $|A_n|$. From Theorem 6.82, there are k values of n, with $k \geq g$, for which A_n is not the divisor of poles of a function in Σ.

To show (ii), merely relabel the indices of the $\mathcal{P}_i$ as follows. Let n_1 be the largest values of n for which $|A_n|$ has a fixed place. Relabel the places of A_{n_1} so that $\mathcal{P}_{n_1}$ is such a fixed place. Proceeding similarly for the new labelling, let n_2 be the largest value of n less than n_1 for which $|A_n|$ has a fixed place, and relabel the places of A_{n_2} so that $\mathcal{P}_{n_2}$ is fixed. By continuing this procedure, an arrangement of the places $\mathcal{P}_i$ is eventually obtained so that whenever A_n has a fixed place, $\mathcal{P}_n$ is fixed. From the preceding argument it then follows that $k = g$. □

Now the case $\mathcal{P}_1 = \mathcal{P}_2 = \cdots = \mathcal{P}_s$ is considered.

DEFINITION 6.86 For any place $\mathcal{P}$, a non-negative integer n is a *pole number* of $\mathcal{P}$ if there is a function z such that $\text{div}(z)_\infty = n\mathcal{P}$; otherwise, n is a *gap number* of $\mathcal{P}$.

LEMMA 6.87 *The set of pole numbers of $\mathcal{P}$ is a sub-semigroup $H(\mathcal{P})$ of the additive semigroup of the non-negative integers.*

Proof. If $\text{div}(z_1)_\infty = n_1\mathcal{P}$ and $\text{div}(z_2)_\infty = n_2\mathcal{P}$, then $\text{div}(z_1z_2)_\infty = (n_1+n_2)\mathcal{P}$. □

DEFINITION 6.88 The set $H(\mathcal{P})$ is the *Weierstrass semigroup at $\mathcal{P}$.*

The integer n is a pole number of $\mathcal{P}$ if and only if

$$\dim(|n\mathcal{P}|) > \dim|(n-1)\mathcal{P}|.$$

By Theorem 6.85, this occurs if and only if $i(n\mathcal{P}) = i((n-1)\mathcal{P})$. In particular, every $n > 2g-1$ is a pole number.

On the other hand, 1 is a gap number. Suppose, on the contrary, that 1 is a pole number of $\mathcal{P}$. Then $H(\mathcal{P})$ is the whole semigroup of non-negative integers showing that there are no gaps at all, contradicting Theorem 6.85. Therefore, for $\mathcal{P}_1 = \cdots = \mathcal{P}_n$, Theorem 6.85 reads as follows.

THEOREM 6.89 (Weierstrass Gap Theorem) *For any place $\mathcal{P}$, there exist exactly g gap numbers, $n_1 < n_2 < \cdots < n_g$, with $n_1 = 1$ and $n_g \leq 2g-1$.*

It is shown in Section 7.6, see Corollary 7.57, that all but finitely many places have the same sequence of gap numbers. These are the *ordinary places* of Σ. If the gap sequence of Σ at an ordinary point is $(1, 2, \ldots, g)$ then Σ has *classical gap sequence at a general point.* The non-ordinary places are the *Weierstrass points,* and the i-th gap of an ordinary point is greater than or equal to the i-th gap of any Weierstrass point. Also, if $g \geq 2$, then Σ contains at least one Weierstrass point.

Two illustrative examples are now given.

EXAMPLE 6.90 Let $\mathcal{F} = \mathbf{v}(f(X,Y))$ be the irreducible plane curve in Example 5.59. Assume that h is odd. Then $\mathcal{F}$ has genus $g = \frac{1}{2}(h-1)$. It is now shown that the gap numbers of $K(\mathcal{F})$ at ordinary places are $1, 2, \ldots, g$, and that $\mathcal{F}$ has exactly $2g+2$ Weierstrass points where the gap numbers are the odd integers less than or equal to $2g-1$.

With same notation employed in Example 5.59,

$$\text{div}(x-a_i)^{-1} = \begin{cases} -2Q_i + 2P_\infty & \text{for } h \text{ odd,} \\ -2Q_i + P_\infty^{(1)} + P_\infty^{(2)} & \text{for } h \text{ even.} \end{cases}$$

This shows that 2 is a pole number at $\mathcal{Q}_i$. Since $g \neq 0$, it follows that the semigroup $H(\mathcal{Q}_i)$ consists of all even numbers. Therefore the odd integers less than or equal to $2g-1$ are gap numbers. This holds true for $\mathcal{P}_\infty$. These $2g+2$ places are the only Weierstrass points of $\mathcal{F}$.

To show the latter assertion, it suffices to prove that $1, 2, \ldots, g$ are gap numbers at any place $\mathcal{P}$ corresponding to a branch of $\mathcal{F}$ centred at a point $P = (a,b)$ with

$b \neq 0$. This occurs if and only if $i(n\mathcal{P}) = i((n-1)\mathcal{P}) - 1$ for $n = 1, \ldots, g$. To do this, the canonical series of $K(\mathcal{F})$ is first determined. Let $Q = (x, y)$ be a generic point of $\mathcal{F}$. The condition $p \nmid h = \deg f(X, Y)$ ensures that x is a separating variable. From the calculations carried out in Example 5.59,

$$\begin{aligned}\operatorname{div}(dx) &= \mathcal{Q}_1 + \cdots + \mathcal{Q}_h - 3\mathcal{P}_\infty,\\ \operatorname{div} y &= \mathcal{Q}_1 + \cdots + \mathcal{Q}_h - h\mathcal{P}_\infty.\end{aligned}$$

Therefore $(y^{-1}dx) = (h-3)\mathcal{P}_\infty$. This shows that the canonical series of $K(\mathcal{F})$ is $|(h-3)\mathcal{P}_\infty|$. Also,

$$\operatorname{div}(x^i) = i(\mathcal{R}_1 + \mathcal{R}_2) - 2i\mathcal{P}_\infty,$$

where $\mathcal{R}_1$ and $\mathcal{R}_2$ are the places corresponding to the branches centred at the points of $\mathcal{F}$ lying at finite distance on the Y-axis. The case $\mathcal{R}_1 = \mathcal{R}_2$ is possible, and this occurs when $f(0) = 0$, that is, the Y-axis meets $\mathcal{F}$ at finite distance only in the origin. Hence, if $i = 0, 1, \ldots, (h-3)/2$, then $\operatorname{div}(x^i) + (h-3)P_\infty$ is a canonical divisor. Since $1, x, \ldots, x^{(h-3)/2}$ are linearly independent over K, it follows in terms of g that canonical series of $K(\mathcal{F})$ consists of all divisors

$$\operatorname{div}(c_0 + c_1x_1 + \cdots + c_{g-1}x^{g-1}) + (2g-2)\mathcal{P}_\infty,$$

as $\mathbf{c} = (c_0, \ldots, c_{g-1})$ ranges over the points of $\mathrm{PG}(g-1, K)$.

For $n = 1, \ldots, g-1$, this implies that a canonical divisor W contains $n\mathcal{P}$ if and only if

$$W = \operatorname{div}(c_0 + c_1x_1 + \cdots + c_{g-1}x^{g-1}) + (2g-2)\mathcal{P}_\infty$$

such that the polynomial $c_0 + c_1X + \cdots + c_{g-1}X^{g-1}$ has an n-fold root at a. Therefore $i(n\mathcal{P}) = i((n-1)\mathcal{P}) - 1$, for $n = 1, \ldots, g$.

EXAMPLE 6.91 Take $p = 3$, and let $\mathcal{H}_3 = \mathbf{v}(Y^3 + Y - X^4)$, the Hermitian curve. Since $\mathcal{H}_3$ has genus 3, it is the canonical curve of its function field $K(\mathcal{H}_3)$. Hence the canonical divisors are cut out on $\mathcal{H}_3$ by lines.

Let $\mathcal{P}$ be the place arising from the point $P = (a, b)$ of $\mathcal{H}_3$. Assume at first that both a, b are in the prime field $\mathbf{F}_3$ of K. There are 27 such points. If ℓ is the tangent to $\mathcal{H}_3$ at P, then $I(P, \ell \cap \mathcal{H}_3) = 4$. Thus

$$i(\mathcal{P}) = 2, \quad i(2\mathcal{P}) = i(3\mathcal{P}) = i(4\mathcal{P}) = 1, \quad i(5\mathcal{P}) = 0.$$

This shows that the gap numbers of $K(\mathcal{H}_3)$ at $\mathcal{P}$ are $1, 2, 5$. The same holds for the place $\mathcal{P}_\infty$ centred at the infinite point of the Y-axis. Now assume that either a or b does not belong to $\mathbf{F}_3$. Then $I(P, \ell \cap \mathcal{H}_3) = 3$, and hence

$$i(\mathcal{P}) = 2, \quad i(2\mathcal{P}) = i(3\mathcal{P}) = 1, \quad i(4\mathcal{P}) = 0.$$

Therefore the gap numbers of $K(\mathcal{H}_3)$ at P are $1, 2, 4$. To sum up, the gap sequence at ordinary places is $(1, 2, 4)$, while there are 28 Weierstrass points with gap sequence $(1, 2, 5)$.

Let $\mathcal{P}$ be any place of Σ, and let m denote the smallest pole number of $\mathcal{P}$. Choose an element $x \in \Sigma$ such that $\operatorname{div}(x)_\infty = mP$. Every integer n in the Weierstrass semigroup $H(P)$ can be written in the form $n = am + i$ where a and i are non-negative integers such that $a > 0$ and $i < m$. Since there are finitely many gaps at $\mathcal{P}$, all the values $i = 0, 1, \ldots, m-1$ arise. Let $a_i m + i$ be the least pole number congruent to i modulo m. Choose elements $z_0, z_1, \ldots, z_{m-1} \in \Sigma$ such that $\operatorname{div}(z_i)_\infty = (a_i m + i)\mathcal{P}$. Note that z_0 may be taken to be x.

It is shown that every $y \in \Sigma$ which has a pole at $\mathcal{P}$ but is regular elsewhere can be written in the form

$$y = A(x) + B(x)z_1 + \cdots + C(x)z_{m-1}, \tag{6.8}$$

with $A(X), B(X), \ldots, C(X) \in K[X]$.

Let n be the pole number arising from y; that is, $\operatorname{div}(y)_\infty = nP$. If $n \equiv i \pmod{m}$, write $n = am + i$. Since n cannot be less then $a_i m + i$, there is a constant c so that $z' = y - cx^{a-a_i} z_i$ is either zero or $\operatorname{div}(z')_\infty = n'\mathcal{P}$ with $n' < n$. Proceeding then similarly with z', an equation (6.8) is obtained.

Applying this equation to the first $m-1$ powers of y gives the following:

$$\begin{array}{rcl} B_1(x)z_1 + \cdots + C_1(x)z_{m-1} &=& y - A_1(x), \\ B_2(x)z_1 + \cdots + C_2(x)z_{m-1} &=& y^2 - A_2(x), \\ \vdots && \vdots \qquad \vdots \\ B_{m-1}(x)z_1 + \cdots + C_{m-1}(x)z_{m-1} &=& y^{m-1} - A_{m-1}(x). \end{array} \tag{6.9}$$

Note that $(z_1, \ldots, z_{m-1})$ can be viewed as the solution of a system of linear equations over $K(x)$.

If these equations are linearly independent, that is, the $(B_i(x), \ldots, C_i(x))$ are $m-1$ linear independent vectors over the field $K(x)$, then Cramer's rule gives the result

$$\begin{array}{l} z_i = Q_{i,1}(x)(y - A_1(x)) + Q_{i,2}(x)(y^2 - A_2(x)) + \cdots \\ \qquad + Q_{i,m-1}(x)(y^{m-1} - A_{m-1}(x)), \end{array} \tag{6.10}$$

with $Q_{i,1}(X), Q_{i,2}(X), \ldots, Q_{i,m-1}(X) \in K(X)$.

Otherwise,

$$P(x) + P_1(x)y + \cdots + P_{m-1}(x)y^{m-1} = 0, \tag{6.11}$$

with $P(X), P_1(X), \ldots, P_{m-1}(X) \in K[X]$. Since

$$\operatorname{ord}_{\mathcal{P}}(P_k(x)y^k) = m \deg P_k(X) + kn$$

from (6.11), so

$$m \deg P_k(X) + kn = m \deg(P_{k'}(X)) + k'n,$$

for some k, k' with $1 \leq k < k' \leq m-1$. This implies that

$$\frac{n}{m} = \frac{\deg P_k(X) - \deg P_{k'}(X)}{k - k'}.$$

Now k and k' are both less than m. Therefore m and n have a common divisor.

Now take n prime to m. Then the equations in (6.9) must be linearly independent, and (6.10) holds. This together with (6.8) applied to y^m gives an irreducible equation of the form

$$U_0(x)y^m + U_1(x)y^{m-1} + \cdots + U_{m-1}(x)y + U_m(x) = 0, \tag{6.12}$$

with $U_i(X) \in K[X]$, and $U_0(X) \neq 0$. More precisely, $U_0(X)$ is a non-zero constant polynomial as both x and y have a unique pole $\mathcal{P}$. Further, if $i < j$, then

$$\begin{aligned} m \deg U_i(X) + (m-i)n &= \mathrm{ord}_{\mathcal{P}}(U_i(x)y^{m-i}) \\ &= \mathrm{ord}_{\mathcal{P}}(U_j(x)y^{m-j}) \\ &= m \deg U_j(X) + (m-j)n \end{aligned}$$

only for $j \equiv i \pmod{m}$. As this congruence is only valid for $i = 0, j = m$, it follows that $\deg U_m(X) = n$ and that

$$\deg U_i(X) < in/m, \quad \text{for } i = 1, \ldots, m-1. \tag{6.13}$$

Since $\gcd(m,n) = 1$ implies that $\Sigma = K(x,y)$, the following result is obtained.

THEOREM 6.92 (Weierstrass Normal Form) *Let $\mathcal{F}$ an irreducible curve of genus $g \geq 1$. For a place $\mathcal{P}$ of $K(\mathcal{F})$, let m be the first non-gap at $\mathcal{P}$ and let n be the least non-gap which is prime to m. Then*

(i) *$\mathcal{F} = \mathbf{v}(f(X,Y))$ with*
$$f(X,Y) = Y^m + U_1(X)Y^{m-1} + \cdots + U_{m-1}(X)Y + U_m(X), \tag{6.14}$$
where $U_i(X) \in K[X]$, $\deg U_m(X) = n$ and (6.13) holds;

(ii) *$P = (x,y)$ is a generic point of $\mathcal{F} = \mathbf{v}(f(X,Y))$ for $\mathrm{div}(x)_\infty = m\mathcal{P}$ and $\mathrm{div}(y)_\infty = n\mathcal{P}$;*

(iii) *the branch of $\mathcal{F}$ associated to $\mathcal{P}$ is the unique branch of $\mathcal{F}$ with centre at Y_∞.*

A consequence is the following result.

PROPOSITION 6.93 *Let $\mathcal{F} = \mathbf{v}(f(X,Y))$ be given in Weierstrass normal form (6.14). Then the genus g has the following properties:*

(i) *$g \leq \frac{1}{2}(n-1)(m-1)$;*

(ii) *if no point of $\mathcal{F}$ other than Y_∞ is singular, then $g = \frac{1}{2}(n-1)(m-1)$.*

Proof. (i) The idea is to apply the Riemann–Roch Theorem 6.61 to the complete linear series $|(mn)\mathcal{P}|$. For each ordered pair (i,j) of non-negative integers i,j for which $im + jn < mn$, the divisor $\mathrm{div}(x^iy^j) + (mn)\mathcal{P}$ belongs to $|(mn)\mathcal{P}|$. This also holds for $\mathrm{div}(x^n) + (mn)\mathcal{P}$. As in the discussion on (6.9), the hypothesis $\gcd(m,n) = 1$ implies that $1, y, \ldots, y^{m-1}$ are linearly independent over $K(x)$. Therefore x^n and the functions x^iy^j form a linearly independent set in the vector space $L(\mathcal{P})$. So

$$\begin{aligned} \dim|(mn)\mathcal{P}| &\geq n + m - 1 + \textstyle\sum_{j=1}^{m-1} \lfloor n(m-j)/m \rfloor \\ &= n + m - 1 + \tfrac{1}{2}(m-1)(n-1). \end{aligned} \tag{6.15}$$

In particular, $\dim|(mn)\mathcal{P}| > \frac{1}{2}nm = \frac{1}{2}\deg(mn)\mathcal{P}_0$. This and Clifford's Theorem 6.79 imply that $|(mn)\mathcal{P}|$ is not special. Then, from (6.15) and Theorem 6.61, $g \leq \frac{1}{2}(n-1)(m-1)$. □

(ii) Both m and n cannot be divisible by p. Suppose first that $m \not\equiv 0 \pmod{p}$. Then x is a separable variable of $K(\mathcal{F})$ and

$$g(X,Y) = \partial f/\partial Y = mY^{m-1} + \cdots + (m-i)U_i(X)Y^{m-i} + \cdots + U_{m-1}(X).$$

A primitive representation of the branch of $\mathcal{F}$ associated to $\mathcal{P}$ is

$$x(t) = t^{-m}, \qquad y(t) = t^{-n} + \cdots .$$

From (6.13),

$$\begin{aligned} \mathrm{ord}_t(m-1)y(t)^{m-1} &= -(m-1)n \\ &< -\lfloor in/m \rfloor m - (m-i-1)n \\ &\leq \mathrm{ord}_t(m-i)U_i(x(t))y(t)^{m-i-1}. \end{aligned}$$

Hence $\mathrm{ord}_t g(x(t), y(t)) = -n(m-1)$. From this and Corollary 5.35,

$$\textstyle\sum_{\mathcal{Q}} \mathrm{ord}_{\mathcal{Q}}\, g(x,y) = n(m-1),$$

where the summation is over all places of $K(\mathcal{F})$ other than $\mathcal{P}$. On the other hand, since the branch of $\mathcal{F}$ associated to such a place $\mathcal{Q}$ is centred at a non-singular point of $\mathcal{F}$, from Theorem 5.56,

$$\mathrm{ord}_{\mathcal{Q}}\, g(x,y) = \mathrm{ord}_{\mathcal{Q}}\, dx.$$

Also,

$$\mathrm{ord}_{\mathcal{P}}\, dx = -(m+1).$$

Now, by Definition 5.55, $2g - 2 = n(m-1) - (m+1)$, whence (ii) follows.

If $n \not\equiv 0 \pmod{p}$, a similar argument can be used, this time with

$$g(X,Y) = \partial f/\partial X = nX^{n-1} + \cdots$$

and dy.

6.7 THE STRUCTURE OF THE DIVISOR CLASS GROUP

In Section 6.1, basic facts on divisors of a function field Σ of transcendence degree 1 are discussed. Now a fundamental result on the divisor class group $\mathrm{Pic}(\Sigma)$ is stated. As the group $\mathrm{Pic}(\Sigma)$ is written additively, that is, $[A+B] = [A] + [B]$ for $[A], [B] \in \mathrm{Pic}(\Sigma)$, the meaning of $m \cdot [A]$ is just $[mA]$, for any positive integer m and divisor A of Σ. So the set $\{[A] \mid m \cdot [A] = 0,\ [A] \in \mathrm{Pic}(\Sigma)\}$ is a subgroup $\mathrm{Pic}(\Sigma, m)$ of $\mathrm{Pic}(\Sigma)$. Therefore

$$\mathrm{Pic}_0(\Sigma, m) = \mathrm{Pic}(\Sigma, m) \cap \mathrm{Pic}_0(\Sigma)$$

is a subgroup of $\mathrm{Pic}_0(\Sigma)$, which plays an important role in the study of the structure of the divisor class group, as its order only depends on g.

THEOREM 6.94 *For any positive integer m not divisible by p,*

$$\mathrm{Pic}_0(\Sigma, m) \cong (\mathbf{Z}_m)^{2g},$$

where g is the genus of Σ and $(\mathbf{Z}_m)^{2g}$ is the direct product of $2g$ copies of the additive group of integers modulo m.

Theorem 6.94 is a special case of a more general result on abelian varieties. The idea is to prove the theorem first for $K = \mathbf{C}$ and then use reduction theory to extend it from $\mathbf{C}$ to any K. No more detail is given here. A proof of Theorem 6.94, valid in the case that $K = \overline{\mathbf{F}}_q$, is given in Section 9.7. The following example illustrates how to determine $\mathrm{Pic}_0(\Sigma, 2)$ in a particular case.

EXAMPLE 6.95 Let $\mathcal{F}$ and Σ be as in Example 5.59, and suppose that h is odd. With the notation of Example 6.62, let $\mathcal{D}_1$ be the set of all divisors D for which $\epsilon_1 = 0$. Then $\deg D < h - 1$.

For two distinct divisors D_1 and D_2 in $\mathcal{D}_1$, write $E_1 - E_2 = D_1 - D_2$ with E_1 and E_2 effective divisors. Note that both E_1 and E_2 are in $\mathcal{D}_1$. Thus

$$\deg E_1 = \deg E_2 < \tfrac{1}{2}(h-1).$$

Then $\dim |E_1| = 0$ by (6.4), and hence $D_1 \not\equiv D_2$. Now, for any $D \in \mathcal{D}_1$, let $F = m\mathcal{P}_1 - D$ with $m = \deg D$; then $[F] \neq [0]$. Since $2\mathcal{P}_1 \equiv \mathcal{P}_i$, as shown in Example 6.62, so $2m\mathcal{P}_1 \equiv 2D$; that is, $[2F] = [0]$. Since divisors D in $\mathcal{D}_1$ can be chosen in 2^{h-1} different ways, the total number of inequivalent divisors F amounts to 2^{h-1}. Since $g = \frac{1}{2}(h-1)$, Theorem 6.94 shows that such divisors F constitute a full representative system of cosets in the divisor class group $\mathrm{Pic}(\Sigma, 2)$.

The hypothesis $p \nmid m$ cannot be omitted in Theorem 6.94, even though the group $\mathrm{Pic}_0(\Sigma, m)$ is finite. In particular, $\mathrm{Pic}_0(\Sigma, p) \cong (\mathbf{Z}_p)^\gamma$ for a non-negative integer γ, the *p-rank of* Σ. This important particular case appears in Theorem 11.62. Here the main result is stated, which can be viewed as describing the essence of Theorem 6.94.

THEOREM 6.96 *The p-rank γ is less than or equal to the genus g.*

A proof of Theorem 6.96 valid in the case that $K = \overline{\mathbf{F}}_q$ is given in Section 9.7.

DEFINITION 6.97 A plane non-singular cubic curve $\mathcal{C}$ is *supersingular* if the p-rank of its function field $K(\mathcal{C})$ is equal to 0.

An alternative definition of the p-rank is the Hasse–Witt invariant and relies on the Hasse–Witt matrix of Σ. For this purpose, assume that $p \neq 0$ and choose a separable variable ζ of Σ. From Theorem 5.37, every differential ω of Σ can be written as follows:

$$\omega = (u_0^p + u_1^p\zeta + \cdots + u_{p-1}\zeta^{p-1})d\zeta, \text{ with } u_0, u_1, \ldots, u_{p-1} \in \Sigma.$$

The *Cartier operator* C is defined by the relation,

$$C(\omega) = u_{p-1}d\zeta.$$

Some fundamental properties of the Cartier operator are the following:

(i) $C(\omega)$ is independent of the choice of ζ;

(ii) $C(y_1^p\omega_1 + y_2^p\omega_2) = y_1C(\omega_1) + y_2C(\omega_2)$ for every $y_1, y_2 \in \Sigma$;

(iii) if ω is holomorphic, that is, $\mathrm{div}\,\omega$ is effective, then $C(\omega)$ is also holomorphic;

(iv) $C(\omega)$ is independent of the choice of ζ.

Choose $x_0, x_1, \ldots, x_{g-1} \in \Sigma$ such that the canonical series $|d\zeta|$ consists of all divisors

$$A_{\mathbf{c}} = \operatorname{div}(\textstyle\sum_{i=0}^{g-1} c_i x_i) + \operatorname{div}(d\zeta).$$

Let $\omega_i = x_i d\zeta$. Then the representation matrix (h_{ij}) over K of C is the *Hasse–Witt matrix*, or *Cartier–Manin matrix*, with respect to $\zeta, x_0, \ldots, x_{g-1}$ such that

$$(C(\omega_0), C(\omega_1), \ldots, C(\omega_{g-1}))^T = (h_{ij})\,(\omega_0, \omega_1, \ldots, \omega_{g-1})^T.$$

The rank of the matrix

$$M = (h_{ij})(h_{ij}^p)\cdots({h_{ij}}^{p^{g-1}})$$

is the *Hasse–Witt invariant* and is independent of $\zeta, x_0, \ldots, x_{g-1}$.

A deep result which implies Theorem 6.96 is stated in the theorem below and is followed by two illustrative examples.

THEOREM 6.98 *The p-rank is equal to the Hasse–Witt invariant.*

EXAMPLE 6.99 For $p = 2$ and $b \in K$, let $\mathcal{F} = \mathbf{v}(Y^4 + b^2Y^2 + Y + X^5)$ be the non-singular plane curve belonging to the family of curves investigated in Section 12.1. Then $\mathcal{F}$ has genus $g = 6$ and its function field $\Sigma = K(x, y)$, with $y^4 + b^2y^2 + y = x^5$, contains a separable variable x. From Lemma 12.1 (iii), $\operatorname{div}(dx) = 10P_\infty$, where P_∞ is the unique infinite point of $\mathcal{F}$. The canonical linear series of Σ consists of all divisors

$$A_{\mathbf{c}} = \operatorname{div}(c_0 + c_1x + c_2x^2 + c_3y + c_4y^2 + c_5xy) + 10P_\infty,$$

with $\mathbf{c} = (c_0, \ldots, c_5)$ ranging over the points of $\mathrm{PG}(5, K)$. Let

$$\omega_0 = dx,\ \omega_1 = xdx,\ \omega_2 = x^2dx,\ \omega_3 = ydx,\ \omega_4 = y^2dx,\ \omega_5 = xydx.$$

Then

$$(C(\omega_0), C(\omega_1), C(\omega_2), C(\omega_3), C(\omega_4), C(\omega_5)) = (0, 1, 0, x^2, 0, y^2 + by)dx,$$

whence the Hasse–Witt matrix (h_{ij}) is equal to

$$\begin{bmatrix} 0 & 0 & 0 & 0 & 0 & 0 \\ 1 & 0 & 0 & 0 & 0 & 0 \\ 0 & 0 & 0 & 0 & 0 & 0 \\ 0 & 0 & 1 & 0 & 0 & 0 \\ 0 & 0 & 0 & 0 & 0 & 0 \\ 0 & 0 & 0 & b & 1 & 0 \end{bmatrix}.$$

Further, the matrix, $(h_{ij})(h_{ij}^2)\cdots(h_{ij}^{2^{r-1}})$, has rank 3, 1, or 0 according as $r = 1, 2$, or $r > 2$. This shows that the Hasse–Witt invariant is equal to zero.

EXAMPLE 6.100 For $p \nmid n$, let $\Sigma = K(\mathcal{D}_n) = K(x, y)$, with $x^n + y^n + a = 0$, be the function field of the Fermat curve $\mathcal{D}_n = \mathbf{v}(X^n + Y^n + a)$, which has genus

$g = \frac{1}{2}(n-1)(n-2)$, where $n \neq 3$, and $a \in K\backslash\{0\}$. For any pair (k,i) with integers $0 \le k \le m-3$ and $0 \le i \le k$, let

$$x_{ki} = x^i y^{k-(i+m)+1}.$$

Then the canonical series $|dx|$ consists of all divisors

$$A_{\mathbf{c}} = \operatorname{div}(\textstyle\sum_{k,i} c_{ki} x_{ki}) + \operatorname{div}(dx).$$

Let $\omega_{ki} = x_{ki}dx$. To calculate $C(\omega_{ki})$, two number-theoretic results are required.

LEMMA 6.101 *For any pair (k,i) with integers satisfying $0 \le k \le m-3$ and $0 \le i \le k$, there exists exactly one pair (u,v) of integers such that*

$$pu + nv = (p-1)(n-1) + k - i,\ 0 \le u \le n-2 \text{ and } 0 \le v \le p-1. \quad (6.16)$$

For these pairs (k,i) and (u,v), there exists exactly one pair (j,s) of integers such that

$$nj + i = p(s+1) - 1,\ 0 \le j \le p-1 \text{ and } 0 \le s \le n-2. \quad (6.17)$$

Further, if $j \le v$ then $u+s \le n-3$ and if $j > v$ then $u+s \ge n-1$.

From this, for any such pair (k,i), let (u,v) be the solution of (6.16) and (j,s) the solution of (6.17). Then

$$C(\omega_{k,i}) = \begin{cases} (-1)^j \binom{v}{j} \left(a^{1/p}\right)^{v-j} \omega_{u+s,s} & \text{if } j \le v, \\ 0 & \text{if } j > v. \end{cases}$$

Therefore the Hasse–Witt matrix (h_{ij}) can be calculated. This gives the following results.

PROPOSITION 6.102 (i) *The Hasse–Witt matrix (h_{ij}) of Σ has*

(a) *maximum rank g if and only if $p \equiv 1 \pmod{n}$;*

(b) *minimum rank 0 if and only if $p \equiv -1 \pmod{n}$.*

(ii) *If $p = 2$ and $n > 4$, then $1 \le \operatorname{rank}(h_{ij}) \le g-1$.*

Another equivalent definition of the p-rank of Σ follows from Theorem 11.64.

In the rest of this section, the special case when $\Sigma = K(\mathcal{F})$ for a non-singular plane cubic curve $\mathcal{F}$ is worked out.

The places of $\Sigma = K(\mathcal{F})$ and points of $\mathcal{F}$ are identified, and P_0 denotes a given point of $\mathcal{F}$.

PROPOSITION 6.103 *Every zero divisor class of $K(\mathcal{F})$ contains a unique divisor $P - P_0$ with $P \in \mathcal{F}$.*

Proof. If $P - P_0 \equiv Q - P_0$ with $P, Q \in \mathcal{F}$, then $P \equiv Q$. Since $K(\mathcal{F})$ has genus 1, from Theorem 6.81 (i) $P = Q$, and the uniqueness is established.

To show the existence, first the following assertion is shown. For any effective divisor D, there exists $P \in \mathcal{F}$ and $m \in \mathbf{N}_0$ such that

$$D \equiv P + mP_0. \quad (6.18)$$

If $\deg D = 1$, then (6.18) holds with $m = 0$. If $\deg D > 1$, then there exists an effective divisor D' together with a point $R \in \mathcal{F}$ such that $D = D' + R$ with $\deg D = \deg D' + 1$. By induction, (6.18) may be assumed true for D'. Then $D' \equiv P + m'P_0$ and hence $D' \equiv R + P + m'P_0$. If there is a point $Q \in \mathcal{F}$ for which

$$R + P \equiv Q + P_0, \tag{6.19}$$

then (6.18) follows. If $R \neq P$, the line ℓ through R and P is either a tangent, or meets $\mathcal{F}$ in a third point. In both cases, there exists $S \in \mathcal{F}$ such that the intersection divisor $\ell \cdot \mathcal{F}$ is equal to $P + R + S$. If $P_0 \neq S$, let r be the line through P_0 and S. If $Q \in \mathcal{F}$ is the point such that $r \cdot \mathcal{F} = Q + P_0 + S$, then $R + P + S \equiv Q + P_0 + S$, whence (6.18) follows. This argument still works for $R = P$ and $S = P_0$, when the lines ℓ and r are the tangents to $\mathcal{F}$ at R and at S.

In the remaining case, $\deg D = 0$, that is, $D = D_1 - D_2$ with two effective divisors of the same degree m. From (6.18) applied to D_1 and to D_2,

$$D_1 \equiv P_1 + mP_0, \quad D_2 \equiv P_2 + mP_0.$$

Therefore $D \equiv P_1 - P_2$, and the assertion follows when $P_2 = P_0$. For the case $P_0 \neq P_2$, the above geometric argument may be used. Let s be the line through P_0 and P_1, or the tangent at P_0 when $P_0 = P_1$. Then there is a point $U \in \mathcal{F}$ such that $s \cdot \mathcal{F} = U + P_0 + P_1$. If u is the line through U and P_2, which is the tangent at U when $U = P_2$, there exists a point $P \in \mathcal{F}$ such that $u \cdot \mathcal{F} = U + P + P_2$. Then $U + P + P_2 \equiv U + P_0 + P_1$ whence $P_1 - P_2 \equiv P - P_0$. Thus $D \equiv P - P_0$. □

The group $\mathrm{Pic}_0(K(\mathcal{F}))$ has a nice geometric representation. To show this, the following result is needed.

THEOREM 6.104 *The points of a non-singular plane cubic curve $\mathcal{F} = \mathbf{v}(F)$ can be equipped with an operation to form an abelian group.*

Proof. For any point $T \in \mathcal{F}$, let ℓ_T denote the tangent at T to $\mathcal{F}$. For $P, Q \in \mathcal{F}$, let ℓ be the line through P and Q; take $\ell = \ell_P$ when $P = Q$. Let $A = \phi(P, Q)$ be the point of $\mathcal{F}$ such that

$$\ell \cdot \mathcal{F} = P + Q + A.$$

Now define

$$P \oplus Q = \phi(P_0, A);$$

see Figure 6.1. Then, under this operation, $\mathcal{F}$ is an abelian group $G_{\mathcal{F}}$ with identity P_0.

First, the operation is well defined and abelian. Next, P_0 is the identity, since if for a point P the point P' is defined by $\phi(P_0, P) = P'$, then $\phi(P_0, P') = P$; that is, $P \oplus P_0 = P$.

To show that P has an inverse, let P'_0 be point where ℓ_{P_0} meets $\mathcal{F}$ again; that is, $\phi(P_0, P_0) = P'_0$. Then $\phi(P'_0, P) = -P$; see Figure 6.2. This incidentally means that, when $P'_0 = P_0$, then $-P = P'$.

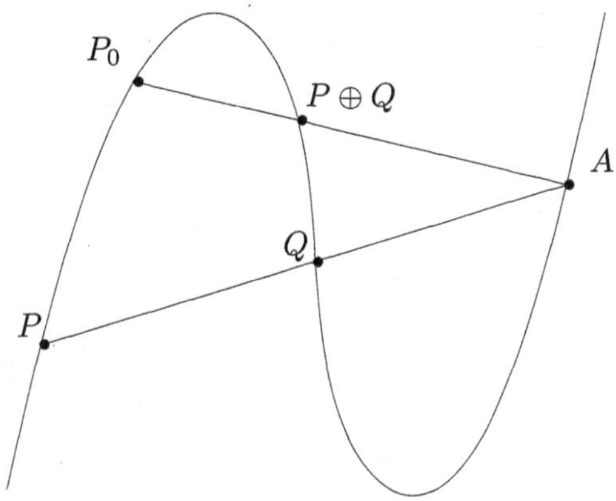

Figure 6.1 Abelian group law on an elliptic curve

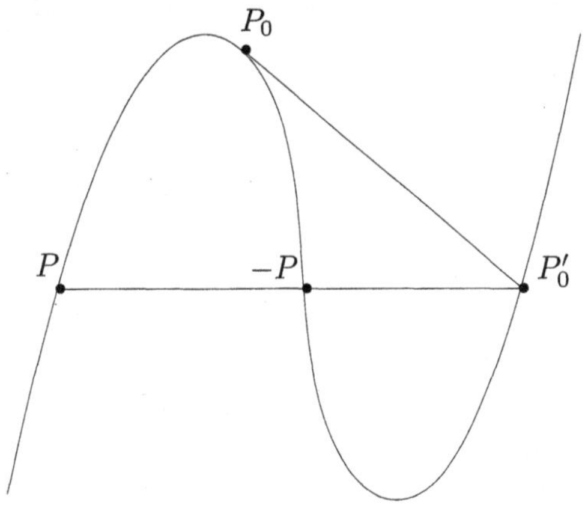

Figure 6.2 Inverse of an element

It remains to show that the associative law holds. For $P, Q, R \in \mathcal{F}$, define A, B, C, D, $P \oplus Q$, $Q \oplus R$ as follows:

$$\phi(P, Q) = A,\ \phi(A, P_0) = P \oplus Q,\ \phi(P \oplus Q, R) = B,$$
$$\phi(Q, R) = C,\ \phi(C, P_0) = Q \oplus R,\ \phi(P, Q \oplus R) = D.$$

The vertical and horizontal lines of Figure 6.3 make up two cubic curves with eight common points on $\mathcal{F}$. By the Theorem of the Nine Associated Points, see Exercise 4 in Section 9.9, they have a ninth point in common, namely $B = D$; hence

$$(P \oplus Q) \oplus R = \phi(P_0, B) = \phi(P_0, D) = P \oplus (Q \oplus R). \qquad \square$$

COROLLARY 6.105 (i) *P, Q, R are collinear if and only if $P \oplus Q \oplus R = P_0'$, where $\ell_{P_0}.\mathcal{F} = 2P_0 + P_0'$.*

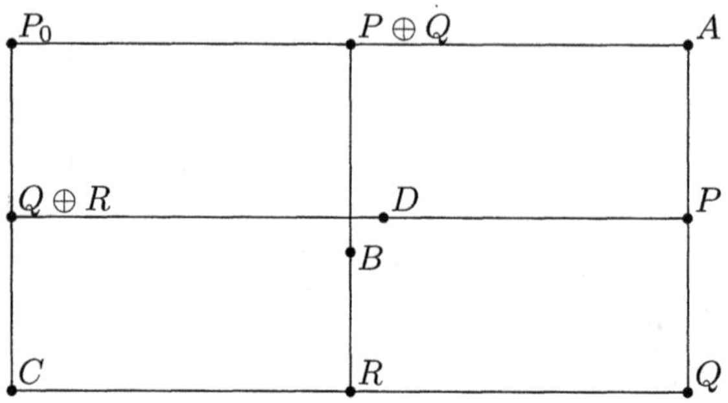

Figure 6.3 The associative law

(ii) *If P_0 is an inflexion, then P, Q, R are collinear if and only if $P \oplus Q \oplus R = P_0$.*

COROLLARY 6.106 (i) *Any inflexion in $\mathcal{F}$ other than P_0 has order 3 in the group.*

(ii) *If P_0 is an inflexion, then the points of order 2 are the points of contact of the tangents through P_0 other than the inflexional tangent.*

For $P \oplus P \oplus \cdots \oplus P$ (m times), write $[m]P$.

THEOREM 6.107 *If $G_{\mathcal{F}}$ be the group in Theorem 6.104, then $G_{\mathcal{F}} \cong \mathrm{Pic}_0(K(\mathcal{F}))$.*

Proof. By Proposition 6.103, every zero class is uniquely represented by $[P - P_0]$ with $P \in \mathcal{F}$. In particular, there is a one-to-one correspondence between points of $\mathcal{F}$ and zero divisor classes of $K(\mathcal{F})$). To show that the map $G_{\mathcal{F}} \to \mathrm{Pic}_0(K(\mathcal{F}))$ which sends $P \in \mathcal{F}$ to the zero divisor class $[P - P_0]$ is an isomorphism, it must be shown that $P \oplus Q = R$ implies that $[P - P_0] + [Q - P_0] = [R - P_0]$.

Let ℓ be the line through P and Q (the tangent at P when $P = Q$). There exists a point $S \in \mathcal{F}$ such that $\ell \cdot \mathcal{F} = P + Q + S$. Let r be the line through S and P_0 (the tangent at S when $S = P_0$). Then $R = P \oplus Q$ is the point for which $r \cdot \mathcal{F} = P_0 + R + S$. Therefore

$$P - P_0 + Q - Q_0 = P + Q + S - S - 2P_0 \equiv S + R + P_0 - S - 2P_0 = R - P_0,$$

showing that $[P - P_0] + [Q - P_0] = [R - P_0]$. □

6.8 EXERCISES

1. (Bertini's Theorem) Assume that K has characteristic zero. Prove that, if any curve in a pencil has a multiple point, this must be a base point.

   Show that Bertini's Theorem is not necessarily true in positive characteristic by considering the counter-example of the pencil consisting of all curves $\mathcal{F}_\lambda = \mathbf{v}(X^p + Y^2 - 2\lambda Y)$.

2. For any two effective divisors A, B,

$$\dim|A| + \dim|B| \leq \dim|A+B|.$$

3. Let $\mathcal{P}$ be a place of Σ. Show that the following conditions are equivalent:

   (a) $\mathcal{P}$ is a Weierstrass point;

   (b) $g\mathcal{P}$ is a special divisor;

   (c) $m\mathcal{P}$ is a special divisor for some $m \geq g$.

4. Show that the curve in Example 6.91 is the only irreducible plane curve of genus 3 with non-classical gap sequence.

5. Let $p = 3$. For function fields of genus 3 show that ordinary places have the classical gap sequence $(1, 2, 3, 4)$.

6. For q odd, let $\mathcal{F} = \mathbf{v}(X^{(q+1)/2} + Y^{(q+1)/2} + 1)$ be the Fermat curve of degree $\frac{1}{2}(q+1)$. Show that the Weierstrass semigroup of $\mathcal{F}$ at each inflexion point of $\mathcal{F}$ is $\langle \frac{1}{2}(q-1), \frac{1}{2}(q+1)\rangle$.

7. For $q \equiv 3 \pmod 4$, let $\mathcal{F} = \mathbf{v}(Y^q + Y - X^{(q+1)/4})$. Show that the Weierstrass semigroup of $\mathcal{F}$ at the place $\mathcal{P}_\infty$ arising from the unique branch centred at Y_∞ is $\langle q, \frac{1}{4}(q+1)\rangle$.

8. Let $\mathcal{F}$ be an irreducible curve of genus $g \geq 0$. For a place $\mathcal{P}$ of $K(\mathcal{F})$, let m be the first non-gap at $\mathcal{P}$ and n another non-gap at $\mathcal{P}$. Choose $x, y \in \Sigma$ such that $\text{div}(x)_\infty = m\mathcal{P}$ and $\text{div}(y)_\infty = n\mathcal{P}$. Show that there exists an irreducible polynomial $g(X, Y) \in K[X, Y]$ with the following properties:

   (a) $g(x, y) = 0$;

   (b) $g(X, Y) = Y^\ell + U_1(X)Y^{\ell-1} + \cdots + U_{\ell-1}(X)Y + U_\ell(X)$, where $U_i(X) \in K[X]$;

   (c) if $n/m = n'/m'$ with $\gcd(n', m') = 1$, then $m' \mid \ell \mid m$;

   (d) $\deg U_\ell(X) = (n\ell)/m$;

   (e) $\deg U_i(X) \leq \lfloor (ni)/m \rfloor$ for $1 \leq i \leq \ell - 1$;

   (f) the subfield $K(x, y)$ of $K(\mathcal{F})$ is a separable extension of $K(x)$;

   (g) if m does not divide n, then $g > 0$;

   (h) if $\gcd(m, n) = 1$, then $K(\mathcal{F}) = K(x, y)$.

9. Let $\mathcal{H}_3$ be the Hermitian curve, as in Example 6.91. After showing the properties below, find all the $2^{2g} = 64$ divisor classes that represent an involution in $\text{Pic}_0(K(\mathcal{H}_3))$, where $K = \overline{\mathbf{F}}_9$. This gives an explicit presentation of $\text{Pic}_0(K(\mathcal{H}_3), 2)$ as in Example 6.95.

   (a) The set $\mathcal{H}_3(\mathbf{F}_9)$ of points of $\mathcal{H}_3$ lying in $\text{PG}(2, 9)$ has size 28.

(b) For any three non-collinear points $P_0, P_1, P_2 \in \mathcal{H}_3(\mathbf{F}_9)$, there is a unique irreducible conic $\mathcal{C}_2$ such that the intersection divisor $\mathcal{C}_2 \cdot \mathcal{H}_3$ has the following property:

$$\mathcal{C}_2 \cdot \mathcal{H}_3 \succ 2P_0 + 2P_1 + 2P_2. \tag{6.20}$$

(c) If (6.20) holds, then $\mathcal{C}_2$ is a 4-tangent irreducible conic to $\mathcal{H}_3$; that is, $\mathcal{C}_2 \cdot \mathcal{H}_3 = 2P_0 + 2P_1 + 2P_2 + 2P_3$ with a fourth point $P_4 \in \mathcal{H}_3(\mathbf{F}_9)$ distinct from P_1, P_2, P_3.

(d) For a fixed point $P_0 \in \mathcal{H}_3(\mathbf{F}_9)$, the total number of such 4-tangent irreducible conics to $\mathcal{H}_3$ is 36.

(e) For each of these thirty-six 4-tangent irreducible conics $\mathcal{C}_2$ to $\mathcal{H}_3$, the divisor class $[P_1 + P_2 + P_3 - 3P_0]$ is an involution in $\text{Pic}_0(K(\mathcal{H}_3))$ such that no two of them lie in the same divisor class.

(f) For a fixed point $P_0 \in \mathcal{H}_3(\mathbf{F}_9)$, each of the twenty-seven divisor classes $[2P - 2P_0]$ with $P \in \mathcal{H}_3(\mathbf{F}_9)$ is an involution in $\text{Pic}_0(K(\mathcal{H}_3))$.

(g) Divisor classes arising from distinct points $P \in \mathcal{H}_3(\mathbf{F}_9)$ are distinct.

(h) No two of the $63 = 36+27$ divisor classes in $\text{Pic}_0(K(\mathcal{H}_3), 2)$ coincide.

6.9 NOTES

The idea of a linear series arose in classical accounts of the subject, such as [122], as a set of effective divisors cut out on a plane algebraic curve by a linear system of plane algebraic curves. Later, it assumed a more general form, so that linear equivalence of non-effective divisors was also considered, as in Section 6.1 based on Seidenberg's book [400]. It should be noted however that other authors, notably Lefschetz [301] and Fulton [135], only allowed effective divisors in a linear series in order to ensure the equivalence between linear series and Riemann–Roch spaces. This has now become prevalent in the literature. In the present book, the name *virtual linear series* is adopted for what Seiedenberg called linear series, while *linear series* is reserved for linear series of effective divisors in Seidenberg's terminology.

Weierstrass points in positive characteristic were introduced by Schmidt [387]; see also Laksov [288] and Neeman [346].

The proof of Theorem 6.92 is adapted from Baker's proof in [25, Chapter 5]; see also [208], [260], [261] and [502]. For a different proof of Proposition 6.93 (i), see [252].

For direct proofs of Theorem 6.94, see [338] and [130]. For a survey on p-ranks of a curves, see [58].

Fundamental papers on the Hasse–Witt matrix and invariant are Hasse and Witt [199], Manin [321], and Stöhr-Voloch [433]. Examples 6.99 and 6.100 come from [431] and [275]; see also [329], [276], [351], [177].

For Example 6.91 and Exercises 4, 5, see [277].

Exercise 8 can be viewed as a generalisation of Theorem 6.92; see [383].

Chapter Seven

Algebraic curves in higher-dimensional spaces

Algebraic curves in higher-dimensional spaces are usually introduced in the more general context of arbitrary algebraic varieties. However, an equivalent *ad hoc* definition which fits in with the plan of this book is also possible. Essentially, an irreducible algebraic curve of $\mathrm{PG}(n, K)$ is a set of points which can be put in a birational correspondence with the points of an irreducible algebraic plane curve. The idea of deriving the concept of an algebraic curve in a higher-dimensional space directly from that of an algebraic plane curve has at least two advantages. On the one hand, it allows the development of a theory of algebraic curves depending only on the basic concepts that have been investigated so far. On the other hand, it makes it possible to go directly to important results on linear series and rational transformations.

7.1 BASIC DEFINITIONS AND PROPERTIES

A *point* in the n-dimensional projective space $\mathrm{PG}(n, K)$ is a homogeneous $(n+1)$-tuple $(a_0, a_1, \ldots, a_n)$ with $a_i \in K$. As before, define a *branch representation* to be a point in $\mathrm{PG}(n, K(t))$ but not in $\mathrm{PG}(n, K)$. All the basic definitions from Chapter 5 extend to projective spaces. For example, if

$$\begin{aligned} x_0(t) &= a_0 + b_0 t^{i_0} + \cdots, \quad & b_0 \neq 0, \\ &\vdots \qquad \vdots & \\ x_n(t) &= a_n + b_n t^{i_n} + \cdots, \quad & b_n \neq 0, \end{aligned}$$

is a special representation of a branch, then its centre is the point $A = (a_0, \ldots, a_n)$, and its order is $\min\{i_0, \ldots, i_n\}$.

Let $Q = (x_0, \ldots, x_n)$ be a point in an n-dimensional projective space whose coordinates x_k belong to an extension field of K. Then $x_i \neq 0$ for some $0 \leq i \leq n$. The field $K(x_0/x_i, \ldots, x_n/x_i)$ is the *field of the point* $Q = (x_0, \ldots, x_n)$ and denoted by $K(x_0, \ldots, x_n)$ or $K(Q)$. Here,

$$K(x_0/x_j, \ldots, x_n/x_j) = K(x_0/x_i, \ldots, x_n/x_i)$$

provided that also $x_j \neq 0$. In fact, $x_k/x_j = (x_k/x_i)(x_i/x_j)$ holds for every $k = 0, \ldots, n$. Hence $K(Q)$ is independent of the choice of the non-zero coordinate x_i of P.

As before, $\Sigma = K(x, y)$ stands for a finite extension of K of transcendence degree 1. So x is transcendental over K and y is related to x by an irreducible polynomial equation over K:

$$F(x,y)=0$$

with $F(X,Y)\in K[X,Y]$. As usual, consider Σ as the function field $K(\mathcal{F})$ of an irreducible plane curve $\mathcal{F}=\mathbf{v}(F)$ arising from its generic point $P=(x,y)$. With the terminology of Chapter 5, $(\mathcal{F};(x,y))$ is a birational model of Σ.

From now on, assume that $K(\mathcal{F})=K(Q)$. In particular, $x_0,\ldots,x_n\in K(\mathcal{F})$ and, if $x_i\neq 0$, then some $x_j/x_i\notin K$.

An immediate consequence of the hypothesis is that some coordinate x_j/x_i must be a separable variable in $K(\mathcal{F})$, otherwise both x and y would be inseparable which is impossible; see Exercise 1 in Chapter 5.

Let γ be a branch of $\mathcal{F}$ and $\mathcal{P}$ the corresponding place of $K(\mathcal{F})$. If

$$\tau:\quad K(\mathcal{F})\to K((t))$$

is a primitive representation of $\mathcal{P}$, then the *corresponding branch in* $\mathrm{PG}(n,K)$ *with respect to the point* $Q=(x_0,\ldots,x_n)$ is defined, namely the branch with primitive representation $(\tau(x_0),\ldots,\tau(x_n))$. The centres of all such branches form a point set in $\mathrm{PG}(n,K)$ which gives the idea of a curve. The aim here is to develop the theory of space curves based on this idea.

DEFINITION 7.1 An *irreducible algebraic curve* Γ in $\mathrm{PG}(n,K)$ consists of all points which are centres of branches corresponding to the places of $K(\mathcal{F})$ with respect to a point $Q=(x_0,\ldots,x_n)$. These branches are the *branches of* Γ, and

$$(\Gamma,(x_0,\ldots,x_n))=(\Gamma,Q)$$

is a *model* of $K(\mathcal{F})$ in $\mathrm{PG}(n,K)$. The x_j are the *coordinate functions* of Γ.

An advantage of Definition 7.1 is that it is a straightforward extension of that of an irreducible plane curve. Several results, especially the following theorems, are higher-dimensional versions of the analogous results for plane curves.

THEOREM 7.2 *Each point of an irreducible curve* Γ *is the centre of finitely many branches of* Γ.

Proof. Let $A=(a_0,\ldots,a_n)$ be a point of Γ, and γ a branch centred at A. Choose a special representation $(x_0(t),\ldots,x_n(t))$ of γ. Suppose that $a_i\neq 0$; then $x_i\neq 0$. Put $y_k=x_k/x_i$ for $k=0,\ldots,n$. One of them, say y_j, is not a constant. Then $y_j-a_j\notin K$, and hence $\mathcal{P}$ is a zero of y_j-a_j. By Theorem 5.33, any function in $K(\mathcal{F})$ has only finitely many zeros, and hence there are only finitely many branches centred at A. □

As a finite number of points can be the centres for only a finite number of branches, Theorem 7.2 has the following corollary.

THEOREM 7.3 *An irreducible curve* Γ *has infinitely many points.*

DEFINITION 7.4 (i) A *hypersurface* Γ in $\mathrm{PG}(n,K)$ is the set of all points satisfying a polynomial equation

$$F(X_0,\ldots,X_n)=0,$$

where $F\in K[X_0,\ldots,X_n]$ is a homogeneous non-constant polynomial; write $\Gamma=\mathbf{v}(F)$.

(ii) The *degree* of the hypersurface Γ is $\deg F$.

(iii) A *hyperplane* H is a hypersurface of degree 1; that is,

$$H = \mathbf{v}(a_0X_0 + \cdots + a_nX_n).$$

Most of the material in Chapter 6.2 extends from curves to hypersurfaces. In particular, two hypersurfaces $\Phi_1 = \mathbf{v}(F_1)$ and $\Phi_2 = \mathbf{v}(F_2)$ have the same set of points if and only if the polynomials F_1 and F_2 have the same irreducible factors.

If γ is a branch with centre A and a primitive representation $(x_0(t), \ldots, x_n(t))$ in special coordinates, and Δ is a hypersurface $\mathbf{v}(G)$, then the *intersection multiplicity* is

$$I(A, \Delta \cap \gamma) = \operatorname{ord}_t G(x_0(t), \ldots, x_n(t)).$$

The branch γ is *linear* if $I(A, H \cap \gamma) = 1$ for some hyperplane H.

THEOREM 7.5 *Let Γ be the irreducible curve of $\mathrm{PG}(n, K)$ given by the point $Q = (x_0, \ldots, x_n)$.*

(i) *The hypersurface $\Delta = \mathbf{v}(G)$ contains Γ if and only if $G(x_0, \ldots, x_n) = 0$.*

(ii) *If Δ does not contain Γ, then they have only finitely many common points.*

Proof. (i) If $G(x_0, \ldots, x_n) = 0$, then it is immediate that Δ contains Γ.

For the converse, suppose $G(x_0, \ldots, x_n) \neq 0$ and show that Γ is not contained in Δ. By relabelling the indices, assume that $x_0 \neq 0$ and put $y_i = x_i/x_0$. Then

$$G(1, y_1, \ldots, y_n) \neq 0,$$

and hence $G(1, y_1, \ldots, y_n)$ has a finite number of zeros. If $A = (1, a_1, \ldots, a_n)$ is a point of Γ and

$$(1,\ y_1(t) = a_1 + \cdots,\ \ldots,\ y_n(t) = a_n + \cdots)$$

is a primitive representation of a branch γ of Γ centred at A, then the corresponding place $\mathcal{P}$ is a zero of $G(1, y_1, \ldots, y_n)$. Hence Δ can contain only a finite number of affine points of Γ. On the other hand, Γ has infinitely many affine points, as in the proof of Theorem 7.3. Hence Γ cannot be contained in Δ.

(ii) It is only necessary to show that the hyperplane $\mathbf{v}(X_0)$ meets Γ in a finite number of points. But this follows since x_0 has only a finite number of zeros. □

THEOREM 7.6 *Let γ be a branch of Γ centred at A, and assume that the hypersurface Δ does not contain Γ. Then $I(A, \Delta \cap \gamma)$ is finite; that is,*

$$G(x_0(t), \ldots, x_n(t)) \neq 0,$$

where $(x_0(t), \ldots, x_n(t))$ is a primitive representation of γ and $\Delta = \mathbf{v}(G)$.

Proof. If σ is a primitive representation of the place associated to γ, then σ maps $K(Q)$ isomorphically onto $K(x_0(t), \ldots, x_n(t))$ with x_i mapped to $x_i(t)$. Thus the $x_i(t)$ satisfy the same polynomial relations over K as do the x_i. It follows that $G(x_0, \ldots, x_n) \neq 0$ implies $G(x_0(t), \ldots, x_n(t)) \neq 0$. □

This theorem, together with Theorem 7.2, provides the extension of the concept of *intersection divisor*: if Δ does not contain Γ, then

$$\Delta \cdot \Gamma = \textstyle\sum_\gamma I(A, \Delta \cap \gamma)\,\mathcal{P},$$

where $\mathcal{P}$ is the place associated to γ and γ ranges over all branches of Γ.

Some slight changes in the wording and proof of Theorem 6.42 are sufficient to obtain the following result.

THEOREM 7.7 *Let the curve Γ be given by the point $P = (x_0, \dots, x_n)$, and let $\Delta = \mathbf{v}(G(X_0, \dots, X_n))$ be a hypersurface of degree m. If Δ does not contain Γ, and H_∞ is the hyperplane $\mathbf{v}(X_0)$, then*

$$\text{div}\,(G(x_0, \dots, x_n)) = \Delta \cdot \Gamma - m(H_\infty \cdot \Gamma).$$

As in Section 6.2, $\deg(\Delta \cdot \Gamma)$ stands for the degree of the intersection divisor of Δ and Γ.

COROLLARY 7.8 *For any hyperplane not containing Γ,*

$$\deg(\Delta \cdot \Gamma) = \deg \Delta \cdot \deg(H_\infty \cdot \Gamma), \quad \deg(H \cdot \Gamma) = \deg(H_\infty \cdot \Gamma).$$

The positive integer $\deg(H \cdot \Gamma)$ is the *degree* of Γ. Then the first equation in Corollary 7.8 can be viewed as the extension of Bézout's Theorem 3.14 to space curves.

COROLLARY 7.9 (Bézout) *If a hypersurface Δ does not contain Γ, then*

$$\deg(\Delta \cdot \Gamma) = \deg \Delta \cdot \deg \Gamma.$$

Also, Theorem 6.43 and Corollaries 6.44 and 6.46 extend to space curves with similar proofs.

THEOREM 7.10 *Let Δ be a hypersurface of degree m not containing Γ. Then Δ cuts out on Γ a divisor equivalent to $m(H_\infty \cdot \Gamma)$.*

THEOREM 7.11 *If Δ_1 is another surface of degree m not containing Γ, then $\Delta_1 \cdot \Gamma$ is equivalent to $\Delta \cdot \Gamma$.*

THEOREM 7.12 *Every linear system of hypersurfaces of a given degree cuts out a linear series on Γ. Conversely, every linear series of $K(\mathcal{F})$, except for a fixed divisor, is cut out on $K(\mathcal{F})$ by a linear system of hypersurfaces of the same degree.*

Let $\Delta_0 = \mathbf{v}(\varphi_0), \dots, \Delta_s = \mathbf{v}(\varphi_n)$ be hypersurfaces of degree m, where

$$\varphi_0, \dots, \varphi_s \in K[X_0, \dots, X_n].$$

It is possible that some hypersurface in the linear system $\mathbf{v}(\sum_{i=0}^{s} c_i \varphi_i)$ contains Γ even though none of the Δ_i does. In the case of plane curves, there was the concept of linear independence modulo F, where $\Gamma = \mathbf{v}(F)$; see Remark 6.47. Here a similar concept is required.

DEFINITION 7.13 The hypersurfaces $\Delta_0, \ldots, \Delta_s$ are *linearly dependent modulo* Γ if some hypersurface in the linear system $\sum_{i=0}^{s} c_i\varphi_i$ vanishes over Γ; equivalently, if

$$\textstyle\sum_{i=0}^{s} c_i\varphi_i(x_0, \ldots, x_n) = 0,$$

where Γ is given by the point $P = (x_0, \ldots, x_n)$.

If this happens, one or more of the Δ_i can be eliminated and in this way the linear system becomes linearly independent modulo Γ.

DEFINITION 7.14 A *singular* point of an curve Γ is the centre of either at least two branches or just one non-linear branch of Γ. A non-singular point is also called *simple*.

THEOREM 7.15 *Every irreducible curve has only finitely many singular points.*

Proof. The notation of the proof of Theorem 7.2 is used. There are finitely many branches centred on H_∞, and these are put aside.

Let γ be one of the branches centred at $A = (1, a_1, \ldots, a_n)$. One of the x_i, say x_1, is separable. Let

$$x_1 = a_1 + ct^v + \cdots, \; c \neq 0.$$

If $v > 1$, then the place corresponding to γ is a zero of dx_1. By Lemma 5.53, this only happens for a finite number of places. Hence all but finitely many branches of Γ are linear. Eliminating the non-linear branches, it may be supposed that the branches centred at A are linear. Since $K(\mathcal{F}) = K(Q)$,

$$x = \frac{f(x_1, \ldots, x_n)}{d(x_1, \ldots, x_n)}, \quad y = \frac{g(x_1, \ldots, x_n)}{d(x_1, \ldots, x_n)}.$$

with

$$d, \; f, \; g \in K[X_1, \ldots, X_n].$$

Since $d(x_1, \ldots, x_n) \neq 0$, the hypersurface $\mathbf{v}(d)$ does not contain Γ. By Theorem 7.5, this hypersurface meets Γ in only a finite number of points. All branches centred at these points are eliminated. Any branch γ centred at $A = (1, a_1, \ldots, a_n)$ goes into a branch of the plane curve $\mathcal{C}$ centred at the point

$$A' = \left(\frac{f(a_1, \ldots, a_n)}{d(a_1, \ldots, a_n)}, \frac{g(a_1, \ldots, a_n)}{d(a_1, \ldots, a_n)}\right).$$

If A' is a singular point of $\mathcal{C}$, eliminate the branch γ. In this way, finitely many branches are eliminated. Then there remains at most one branch of $\mathcal{C}$ centred at A', and hence with at most one branch of Γ centred at A. □

7.2 RATIONAL TRANSFORMATIONS

Rational transformations are defined similarly to Section 5.2 when considering points $Q = (x_0, x_1, \ldots, x_n)$, $Q' = (x'_0, x'_1, \ldots, x'_{n'})$ satisfying $K(Q') \subset K(Q)$.

Up to relabelling indices, neither x_0 nor x'_0 is zero; so it is possible to assume that $x_0 = x'_0 = 1$. Let Γ and Γ' denote the irreducible curves in $\mathrm{PG}(n, K)$ associated with Q and Q'.

DEFINITION 7.16 A *rational transformation* $\omega : K(Q) \to K(Q')$ is given by

$$x'_i = f_i(x_1, \dots, x_n)/d(x_1, \dots, x_n) \quad \text{for } i = 1, \dots, n',$$

where $f_i,\ d \in K[X_1, \dots, X_n]$, and the hypersurface $H = \mathbf{v}(d)$ does not contain Γ.

Basic facts on rational transformations stated in Section 5.2 for $n = 2$ hold true in any dimension $n \geq 3$. The image of a branch γ of Γ under ω is a branch of Γ'. The image A' of a point $A \in \Gamma$ exists and $A' \in \Gamma'$ provided that A is an affine point and that A does not lie in the hypersurface $H = \mathbf{v}(d)$. Actually, both conditions can be dropped when A is the centre of only one branch, in particular when A is a non-singular point of Γ. In fact, computing the coordinates of the image A' of such a point $A \in \Gamma$ is possible using the fact that the coordinate functions are determined up to a non-zero factor.

Now it is shown how to carry out the calculation after replacing $(1, \dots, x'_{n'})$ by $(z, \dots, zx'_{n'})$ for a suitably chosen $z \in K(Q)$. Putting

$$m_{\mathcal{P}} = \min\{\mathrm{ord}_{\mathcal{P}} f_1/d, \dots, \mathrm{ord}_{\mathcal{P}} f_n/d\},$$

take an index k for which $m_{\mathcal{P}} = \mathrm{ord}_{\mathcal{P}} f_k/d$. In the definition of ω, replace $(1, f_1/d, \dots, f_n/d)$ by $(1/f_k, f_1/f_k, \dots, f_n/f_k)$. Note that $x'_k = 1$. Now all $f'_j(A) = f'_j(a_0, \dots, a_n)$ exist. Therefore

$$A' = \phi(A) = (f'_0(A), \dots, f'_{n'}(A))$$

provided that A is the centre of a unique branch of Γ. In particular, if Γ is an irreducible non-singular curve, then the rational transformation ω defines the *morphism* $\phi : \Gamma \longrightarrow \Gamma'$.

If $K(Q) = K(Q')$, then Γ is *birationally equivalent* or *birationally isomorphic* to the curve Γ' given by Q', and branches correspond to each other when derived from the same place of $K(Q)$. A non-singular point of Γ and a non-singular point of Γ' *correspond* if they are the centres of corresponding branches. This definition extends to singular points provided that they are the centre of only one branch. If both Γ and Γ' are non-singular curves, the morphism ϕ is an *isomorphism*.

Since every field of transcendence degree 1 over K has a pair of generators, the following result is obtained.

THEOREM 7.17 *Any irreducible curve is birationally equivalent to a plane curve.*

EXAMPLE 7.18 The Klein quartic is the non-singular plane curve

$$\mathcal{F} = \mathbf{v}(X + Y^3 + X^3Y)$$

of genus 3. If $P = (x, y)$ is a generic point of $\mathcal{F}$, then $K(\mathcal{F}) = K(x, y)$, with

$$x + y^3 + x^3y = 0,$$

is the function field of $\mathcal{F}$. Put

$$x_0 = x,\ x_1 = x^2y,\ x_2 = y^2,\ x_3 = -3xy.$$

Let Γ be the irreducible curve in $\mathrm{PG}(3, K)$ given by $P = (x_0, x_1, x_2, x_3)$. The associated rational transformation ω defines a morphism $\phi : \mathcal{F} \to \Gamma$ such that, if $A = (a, b)$, then

$$A' = \phi(A) = (a, a^2b, b^2, -3ab),$$

provided that A is not a vertex of the triangle of reference in $\mathrm{PG}(2, K)$. So, as in Section 7.2, the image of such a vertex is also uniquely determined.

Assume that $p = 5$. First, $A = (1, 0, 0)$ is considered. To carry out the computation, choose a primitive representation σ of the unique place of $K(\mathcal{F})$ whose corresponding branch of $\mathcal{F}$ centred at A. If $\sigma(x) = x(t)$, $\sigma(y) = y(t)$, then

$$\begin{aligned} x(t) &= a_1t^i + a_2t^j + \cdots, \quad a_1a_2 \neq 0, \\ y(t) &= t. \end{aligned}$$

From $f(a_1t^i + a_2t^j + \cdots, t) = 0$,

$$x(t) = -t^3 + t^{10} + \cdots, \quad y(t) = t.$$

Let $\sigma(x_i) = x_i(t)$ for $i = 0, 1, 2, 3$. Then

$$\begin{aligned} x_0(t) &= -t^3 + t^{10} + \cdots, & x_1(t) &= (-t^3 + t^{10} + \cdots)^2 t, \\ x_2(t) &= t^2, & x_3(t) &= -3(-t^3 + t^{10} + \cdots)t. \end{aligned}$$

So $e_{\mathcal{P}} = -2$ and $\phi(A) = B_2$ with $B_2 = (0, 0, 1, 0)$. The place $\mathcal{P}$ corresponds to the branch of Γ which is centred at B_2 and has the following primitive branch representation:

$$\begin{aligned} y_0(t) &= -t + t^8 + \cdots, & y_1(t) &= t^5 - 2t^{12} + \cdots, \\ y_2(t) &= 1, & y_3(t) &= 3t^2 - 3t^9 + \cdots. \end{aligned}$$

Similar calculations show that

$$\begin{aligned} \phi(X_\infty) &= B_0 \quad \text{where } B_0 = (1, 0, 0, 0), \\ \phi(Y_\infty) &= B_1 \quad \text{where } B_1 = (0, 1, 0, 0). \end{aligned}$$

The intersection divisor of the plane $H = \mathbf{v}(X_0)$ with Γ is $H \cdot \Gamma = 5\mathcal{P}_1 + \mathcal{P}_2$, where the places $\mathcal{P}_1$ and $\mathcal{P}_2$ correspond to the branches of $\mathcal{C}$ centred at the points Y_∞ and O.

It is opportune to investigate rational transformations that are not birational. In the *non-trivial case*, the transcendence degree of $K(Q')$ is 1; in the trivial case, $K(Q') = K$, and Γ is transformed into a point. Unless specifically mentioned, only the non-trivial case is considered.

If $\mathcal{P}$ is a place of $K(Q)$, given by a monomorphism $\tau : K(Q) \to K((t))$, then τ induces a monomorphism $\tau' : K(Q') \to K((t))$, giving a place $\mathcal{P}'$ of $K(Q')$; then $\mathcal{P}$ *lies over* $\mathcal{P}'$. In terms of curves, if Q, Q' define the curves Γ, Γ', then a branch of Γ gives rise to a branch of Γ'. The point A of Γ, which is the centre of the first branch, is taken to the point A' of Γ', which is the centre of the second. As before, A' is well defined if just one branch is centred at A. This is the case when A is a non-singular point, and hence for all but finitely many points by Theorem 7.15. From a geometric point of view, Γ *covers* Γ', as the images of the non-singular points of Γ cover all but finitely many points of Γ'.

To pursue the discussion on rational transformations and morphisms, a purely field-theoretic question is first considered.

Let $\mathcal{P}'$ be a place of $K(Q')$ and $\mathcal{P}$ a place of $K(Q)$ over $\mathcal{P}'$. Let $\mathcal{P}$ be given by τ and let τ induce τ'. Then, even if τ is primitive, which is assumed, τ' need not be.

LEMMA 7.19 *The redundancy of τ' is independent of the choice of τ.*

Proof. As usual, $\Sigma = K(x, y)$, $x(t) = \tau(x)$, $y(t) = \tau(y)$, and $f \in K[X, Y]$ is an irreducible polynomial for which $f(x, y) = 0$. Choose $\xi \in \Sigma$, and write $\xi = u(x, y)/v(x, y)$ with u, $v \in K[X, Y]$, and $f \nmid v$. Then

$$\mathrm{ord}_{\mathcal{P}}\xi = \mathrm{ord}_t\, u(x(t), y(t))/v(x(t), y(t)).$$

Let $\Sigma' = K(x', y')$, and $f'(x', y') = 0$ for an irreducible $f' \in K[X, Y]$. The pair $(x'(t), y'(t))$ with $x'(t) = \tau'(x'), y'(t) = \tau'(y')$ can be viewed as a representation of the branch corresponding to $\mathcal{P}'$. If τ' has *redundancy* $\nu \geq 1$, then there exists $s(t) \in K[[t]]$ such that

$$s(t) = c_\nu t^\nu + c_{\nu+1}t^{\nu+1} + \cdots, \quad c_\nu \neq 0$$

and $x'(t) = \bar{x}(s(t))$, $y'(t) = \bar{y}(s(t))$ for a primitive representation $(\bar{x}(s), \bar{y}(s))$ of $\mathcal{P}'$ in $K[[s]]$.

Suppose that $\xi \in \Sigma'$. Let $\xi = u'(x', y')/v(x', y')$ with u', $v' \in K[X, Y]$, and $f' \nmid v'$. Then

$$\tau(\xi) = u(\tau(x'), \tau(y'))/v(\tau(x'), \tau(y')).$$

Now, since

$$u(\tau(x), \tau(y))/v(\tau(x), \tau(y)) = u'(\tau'(x'), \tau'(y'))/v'(\tau'(x'), \tau'(y')),$$

it follows that $\mathrm{ord}_{\mathcal{P}}\xi = \nu \cdot \mathrm{ord}_{\mathcal{P}'}\xi$, which proves the assertion. □

From Lemma 7.19, the notation $e_{\mathcal{P}} = \mathrm{ord}_{\mathcal{P}}\,\xi/\mathrm{ord}_{\mathcal{P}'}\xi$ is meaningful. Hence the proof of Lemma 7.19 gives the following formula:

$$\mathrm{ord}_{\mathcal{P}}\,\xi = e_{\mathcal{P}} \cdot \mathrm{ord}_{\mathcal{P}'}\,\xi. \tag{7.1}$$

Now $e_{\mathcal{P}}$ is the *ramification index of* $\mathcal{P}$. Also, if $e_{\mathcal{P}} > 1$, say that ω *ramifies at* $\mathcal{P}$.

LEMMA 7.20 *Over every place $\mathcal{P}'$ of Σ', there lies at least one and at most a finite number of places of Σ.*

Proof. Consider the linear series $|m\mathcal{P}'|$. By the Riemann–Roch Theorem 6.61,

$$\dim |m\mathcal{P}'| \geq m - g',$$

where g' denotes the genus of Σ'. In other words, $|m\mathcal{P}'| = g^r_m$. Then $r \geq 1$ for sufficiently large m. So there exists an effective divisor $B \equiv m\mathcal{P}'$ such that $B \neq m\mathcal{P}'$. If $B \succ \mathcal{P}'$, cancel $\mathcal{P}'$ as often as possible; in the end, $B \not\succ \mathcal{P}'$. Hence there exists an element $x \in \Sigma'$ such that $\mathrm{div}(x) = m\mathcal{P}' - B$.

Now check that $\mathcal{P}'$ is a zero of x. Since $B \not\succ \mathcal{P}'$, so $\mathrm{div}(x)_0 = m\mathcal{P}$ and $\mathrm{div}(x)_\infty = B$. It follows that $\mathrm{ord}_{\mathcal{P}'}x = m$. Actually, it has been shown that x

has $\mathcal{P}'$ but no other place as a zero. By (7.1), the places $\mathcal{P}$ of Σ lying over $\mathcal{P}'$ are also zeros of x. As the number of zeros of x is equal to $[\Sigma : K(x)]$, this number gives not only an upper bound for the number of places of Σ lying over $\mathcal{P}'$, but also ensures the existence of at least one place $\mathcal{P}$ of Σ at which x vanishes. Such a place $\mathcal{P}$ lies over a place $\mathcal{Q}'$ of Σ'. By (7.1) $\mathcal{Q}'$ is a zero of x; but then $\mathcal{Q}' = \mathcal{P}'$. This completes the proof. □

Let $\mathcal{P}_1, \ldots, \mathcal{P}_j$ be the places of Σ lying over $\mathcal{P}'$, and let $e_{\mathcal{P}_1}, \ldots, e_{\mathcal{P}_j}$ be their ramification indices, so that $\text{ord}_{\mathcal{P}_i} x = e_{\mathcal{P}_i} \cdot \text{ord}_{\mathcal{P}_i'} x$. Then $\mathcal{P}'$ *corresponds to*

$$D(\mathcal{P}') = e_{\mathcal{P}_1}\mathcal{P}_1 + \cdots + e_{\mathcal{P}_j}\mathcal{P}_j. \tag{7.2}$$

Also,

$$[\Sigma : K(x)] = \textstyle\sum_{i=1}^{j} \text{ord}_{\mathcal{P}_i} x = \sum_{i=1}^{j} e_{\mathcal{P}_i}\text{ord}_{\mathcal{P}_i'}\, x = \left(\sum_{i=1}^{j} e_{\mathcal{P}_i}\right) \cdot [\Sigma' : K(x)],$$

whence

$$\textstyle\sum_{i=1}^{j} e_{\mathcal{P}_i} = [\Sigma : \Sigma']. \tag{7.3}$$

As shown in Lemma 7.28, $e_{\mathcal{P}_1} = \cdots = e_{\mathcal{P}_j} = 1$ in general and, if this occurs, then each of the $[\Sigma : \Sigma']$ places over $\mathcal{P}'$ is an *unramified place*. On the other hand, $\mathcal{P}_1$ *ramifies completely* when $j = 1$ or, equivalently, when $e_{\mathcal{P}_1} = [\Sigma : \Sigma']$.

An interpretation of (7.3) is given in the following theorem.

THEOREM 7.21 *To each place $\mathcal{P}'$ of Σ' there correspond $[\Sigma : \Sigma']$ places of Σ, counted with multiplicity.*

Geometrically speaking, $\Gamma \to \Gamma'$ is a $[\Sigma : \Sigma']$-*fold covering*.

In a rational transformation, multiplicity is counted in a similar way. Hence the following criterion for a rational transformation to be birational is obtained.

COROLLARY 7.22 *A rational transformation from Γ to Γ' is birational if and only if to each branch of Γ' there corresponds just one branch of Γ; equivalently, if and only if to some branch of Γ' there corresponds just one branch of Γ.*

THEOREM 7.23 *Let Γ be given by $P = (1, x_1, \ldots, x_n)$. Then, for some pair of elements*

$$y_1 = a_{01} + a_{11}x_1 + \cdots + a_{n1}x_n,\; y_2 = a_{02} + a_{12}x_1 + \cdots + a_{n2}x_n,$$

with $a_{ij} \in K$, the curve Γ' having $Q = (1, y_1, y_2)$ as a generic point is birationally equivalent to Γ.

Proof. Let $\mathcal{P}$ be a place of $K(Q)$ centred at a non-singular point A of Γ. To avoid poles of the coordinate functions x_i, suppose that A is an affine point; that is, A does not lie in the hyperplane $\mathbf{v}(X_0)$. Then, for some $a_{j1} \in K$, the hyperplane

$$H_1 = \mathbf{v}(a_{01} + a_{11}X_1 + \cdots + a_{n1}X_n)$$

passes through A and $\text{ord}_{\mathcal{P}}(a_{01} + a_{11}x_1 + \cdots + a_{n1}x_n) = 1$ at the place $\mathcal{P}$ centred at A. Take

$$y_1 = a_{01} + a_{11}x_1 + \cdots + a_{n1}x_n;$$

this element y_1 has only a finite number of zeros. Choose a hyperplane

$$H_2 = \mathbf{v}(a_{02} + a_{12}X_1 + \cdots + a_{n2}X_n)$$

through A but not through the other zeros of y_1. Take

$$y_2 = a_{02} + a_{12}x_1 + \cdots + a_{n2}x_n,$$

and let $\Sigma' = K(y_1, y_2)$. Since $\Sigma' \subset \Sigma$, there is a place $\mathcal{P}'$ lying under $\mathcal{P}$. Then $\mathcal{P}'$ is a common zero of y_1 and $\mathcal{P}$ is the only place lying over $\mathcal{P}'$. Also, the multiplicity is 1. This follows from $\text{ord}_{\mathcal{P}}\, y_1 = 1$ together with (7.1). Corollary 7.22 completes the proof. □

REMARK 7.24 A geometric interpretation of Theorem 7.23 can be given in terms of projection. With H_1 and H_2 as above and (a_{ij}), $j = 1, \ldots n+1$, $i = 0, \ldots, n$, a matrix such that $\det(a_{ij}) \neq 0$, apply to $\text{PG}(n, K)$ the projectivity associated to the matrix (a_{ij}). This can be done so that $y_1 = x_1,\ y_2 = x_2$ in Theorem 7.23. The mapping

$$\begin{aligned} \pi : \text{AG}(n, K) &\longrightarrow \text{AG}(2, K), \\ (a_1, a_2, \ldots, a_n) &\mapsto (a_1, a_2) \end{aligned} \tag{7.4}$$

is a projection. Thus the *rational transformation given in Theorem 7.23 determined by* (P, Q) *is induced by a projection.*

Using projective coordinates, π can be viewed as induced by the transformation $(a_0, a_1, \ldots, a_n) \mapsto (a_0, a_1, a_2)$, which is defined at points outside the subspace $\Pi_{n-3} = \mathbf{v}(X_0, X_1, X_2)$, and which in fact is the projection from Π_{n-3} as vertex to the plane $\Pi_2 = \mathbf{v}(X_3, \ldots, X_n)$. The join of $(a_0, a_1, \ldots, a_n)$ and Π_{n-3} is a subspace Π_{n-2} intersecting Π_2 in the point $(a_0, a_1, a_2, 0, \ldots, 0)$ or, omitting the last coordinates as superfluous, in (a_0, a_1, a_2). It is to be noted that Γ may intersect the vertex Π_{n-3} and, if so, the transform, or transforms, of the points of intersection are defined via the branches centred at those points.

In summary, *every irreducible curve can be transformed by means of a projection into a birationally equivalent plane curve.*

7.3 HURWITZ'S THEOREM

This section establishes a basic theorem of Hurwitz concerning the genera of subfields Σ' of a function field Σ of transcendence degree 1 over K. For a finite separable extension Σ/Σ' of degree n, Hurwitz's Theorem states the numerical relation

$$2g - 2 = n(2g' - 2) + d,$$

where g and g' are the genera of Σ and Σ', and d is a non-negative integer; it also gives an interpretation for d together with a method for computing it. By Hurwitz's Theorem applied to $\Sigma = K(\Gamma)$ and $\Sigma' = K(\Gamma')$, certain properties of Γ' can be derived from those of Γ; in many cases the degree of the covering $\Gamma \to \Gamma'$ can be computed.

If Σ/Σ' is a Galois extension, then the corresponding covering $\Gamma \to \Gamma'$ is a *Galois covering*. In terms of automorphism groups of $K(\Gamma)$, the covering $\Gamma \to \Gamma'$ is Galois if and only if $K(\Gamma')$ is the set of all fixed elements of an automorphism group of $K(\Gamma)$. A detailed treatment of Hurwitz's Theorem for Galois covers is given in Chapter 11.

A key lemma is a formula similar to (7.1).

LEMMA 7.25 *If $\xi \in \Sigma'$ is a separable variable of Σ,*

$$\mathrm{ord}_{\mathcal{P}}\, d\xi = \mathrm{ord}_t\, \frac{ds}{dt} + \mathrm{ord}_{\mathcal{P}'}\, d\xi. \tag{7.5}$$

Proof. From the proof of (7.1), $\xi(t) = \xi'(s(t))$, whence $d\xi/dt = ds/dt \cdot d\xi'/ds$ by derivation. Note that $ds/dt \neq 0$, since otherwise $d\xi/dt = 0$ contradicting the separability of ξ. This gives the result. □

By (7.5),

$$\sum \mathrm{ord}_t\, \frac{ds}{dt} = \sum \mathrm{ord}_{\mathcal{P}}\, d\xi - \sum \mathrm{ord}_{\mathcal{P}'}\, d\xi,$$

where the summation is over the places of S. From (7.1),

$$\sum \mathrm{ord}_{\mathcal{P}'}\, d\xi = [\Sigma : \Sigma'] \cdot {\sum}' \mathrm{ord}_{\mathcal{P}'}\, d\xi,$$

the latter summation being over the places of Σ'. Since $d\xi$ has a finite number of zeros and poles, it follows that $\sum \mathrm{ord}_t\, (ds/dt)$ is a finite sum. Also, $\sum \mathrm{ord}_t\, (ds/dt)$ is independent of the choice of ξ in Σ'. This leads to the following definition.

DEFINITION 7.26 The *different* of a finite separable extension Σ/Σ' is the divisor $D(\Sigma/\Sigma') = \sum d_{\mathcal{P}}\mathcal{P}$ with $d_{\mathcal{P}} = \mathrm{ord}_t(ds/dt)$, where the summation is over all places $\mathcal{P}$ of Σ.

THEOREM 7.27 (Hurwitz) *Let Σ be a finite separable extension of Σ', and let g and g' be the genera of Σ and Σ'. Then*

$$2g - 2 = [\Sigma : \Sigma'] \cdot (2g' - 2) + \deg D(\Sigma/\Sigma'). \tag{7.6}$$

Proof. By Definition 5.55, $\sum \mathrm{ord}_{\mathcal{P}}\, d\xi = 2g - 2$ and ${\sum}' \mathrm{ord}_{\mathcal{P}'}\, d\xi' = 2g' - 2$. Now the result follows from Lemma 7.25. □

If $e_{\mathcal{P}}$ is prime to p, then (7.5) becomes

$$\mathrm{ord}_{\mathcal{P}}\, d\xi = (e_{\mathcal{P}} - 1) + \mathrm{ord}_{\mathcal{P}'}\, d\xi.$$

Otherwise, it can only be asserted that $\mathrm{ord}_{\mathcal{P}}\, d\xi \geq e_{\mathcal{P}} + \mathrm{ord}_{\mathcal{P}'}\, d\xi$. In any case,

$$\mathrm{ord}_{\mathcal{P}}\, d\xi \geq e_{\mathcal{P}} - 1 + \mathrm{ord}_{\mathcal{P}'}\, d\xi.$$

Hence

$$\sum (e_{\mathcal{P}} - 1) \leq \sum \mathrm{ord}_t(ds/dt) = \deg D(\Sigma/\Sigma'), \tag{7.7}$$

where the summation is for all places of Σ at which ω ramifies. In particular, $\sum(e_{\mathcal{P}} - 1)$ is a finite sum. This gives the following.

LEMMA 7.28 *If Σ be a finite separable extension of Σ', then $\omega : \Sigma \to \Sigma'$ ramifies at a finite number of places.*

Let Σ/Σ_1 and Σ_1/Σ_2 be finite separable extensions. Then Σ/Σ_2 is also a finite separable extension, and it is straightforward to prove the equation

$$d_{\mathcal{P}}(\Sigma/\Sigma_2) = d_{\mathcal{P}}(\Sigma/\Sigma_1) + [\Sigma : \Sigma_1]\, d_{\mathcal{P}'}(\Sigma_1/\Sigma_2), \tag{7.8}$$

where $\mathcal{P}$ is any place of Σ and $\mathcal{P}'$ is the place of Σ_1 lying under $\mathcal{P}$.

The following example shows how this machinery works.

EXAMPLE 7.29 Let $\mathcal{H}_q = \mathbf{v}(Y + X^q + XY^q)$ be the Hermitian curve. For a generic point $P = (x, y)$ of $\mathcal{H}_q$, let $\Sigma = K(x, y)$ the function field of $\mathcal{H}_q$. Since $dx \neq 0$, so x is a separable variable of Σ by Theorem 5.50. Now fix a divisor m of $q^2 - q + 1$, and consider the rational transformation

$$\omega : \xi = x^m,\ \eta = y^m.$$

Let Γ' be the curve with generic point $Q = (\xi, \eta)$, and let $\Sigma' = K(\xi, \eta)$. Since $p \nmid m$, so ξ is a separable variable of Σ, by Lemma 5.38. The different $D(\Sigma/\Sigma')$ is calculated explicitly.

Take the point $P_1 = (1, 0, 0)$ of Γ. There is a unique branch γ of Γ centred at P_1. Since the tangent to Γ at P_1 is the X-axis, there is a primitive branch representation of γ of type $(x = t, y = y(t))$, with $y(t) \in K[[t]]$. Note that

$$f(t, y(t)) = y(t) + t^q + ty(t)^q = 0$$

shows that

$$y(t) = -(t^q + t^{q^2+1} + \cdots + a_j t^{q+\alpha_j(q^2-q+1)} + \cdots).$$

Let $\mathcal{P}$ be the place of Σ corresponding to the branch γ, given by the primitive place representation $\tau : \Sigma \to K((t))$. Let $\tau' : \Sigma' \to K((t))$ be the monomorphism induced by τ on Σ', and let $\mathcal{P}'$ be the place of Σ' arising from τ'. From the definition of ω, the place $\mathcal{P}$ is the only place of Σ lying over $\mathcal{P}'$. Now define $\tau'(\xi) = \xi(t)$, $\tau'(\eta) = \eta(t)$. Then $(\xi(t), \eta(t))$ is a representation of the branch γ' of Γ' corresponding to $\mathcal{P}'$. From the definition of ω,

$$\begin{aligned} \xi(t) &= t^m, \\ \eta(t) &= (-1)^m (t^m)^q \left[1 + (t^m)^{(q^2-q+1)/m} + \cdots\right]^m. \end{aligned}$$

Therefore it is a reducible representation with ramification index m. Replacement of t^m by s provides a primitive representation of γ':

$$\begin{aligned} \xi(s) &= s, \\ \eta(s) &= (-1)^m s^q [1 + s^{(q^2-q+1)/m} + \cdots]. \end{aligned}$$

This shows that $\mathrm{ord}_{\mathcal{P}'}\eta = q$. Since $\mathrm{ord}_{\mathcal{P}}\eta = mq$, and no other place of Σ lies over $\mathcal{P}'$, so $e_{\mathcal{P}} = m$; that is, $[\Sigma : \Sigma'] = m$. Also, as $s(t) = t^m$ yields $ds/dt = mt^{m-1}$, the place $\mathcal{P}$ contributes $(m-1)\mathcal{P}$ to the different divisor $D(\Sigma/\Sigma')$.

Similar arguments can be used in doing computations for the other two fundamental points $P_2 = (0, 1, 0)$ and $P_3 = (0, 0, 1)$. The results for $\mathcal{P}_1$ hold true for both P_2 and P_3, which can be shown directly. The change of the coordinate system by $(X_0, X_1, X_2) \to (X_1, X_2, X_0)$ leaves Γ invariant and takes P_1 to P_2 and P_2 to P_3, Therefore the contribution of the three fundamental points to $D(\Sigma/\Sigma')$ is $(m-1)\mathcal{P}_1 + (m-1)\mathcal{P}_2 + (m-1)\mathcal{P}_3$.

Each of the remaining points P of Γ is affine: $P = (a, b)$ with $a, b \neq 0$. The tangent to Γ at P is not the horizontal line $\mathbf{v}(Y - a)$. Hence a primitive representation of the unique branch γ of Γ centred at P is of type

$$x = a + t + \cdots, \quad y = b + t.$$

Therefore

$$\xi(t) = a^m + \cdots, \quad \eta = (b+t)^m.$$

Now choose $\eta' \in K[[s]]$ and $s(t) \in K[[t]]$ such that $\eta(t) = \eta'(s(t))$. Then

$$m \cdot \eta(t)/dt = m(b+t)^{m-1} = m \cdot \eta'/dt \cdot ds(t)/dt,$$

whence $\text{ord}_t\, ds/dt = 0$. Therefore $e_{\mathcal{P}} = 1$ for the place $\mathcal{P}$ corresponding to γ. This shows that P does not contribute to $D(\Sigma/\Sigma')$. Summing up, the different divisor $D(\Sigma/\Sigma')$ of the rational transformation ω is equal to

$$(m-1)\mathcal{P}_1 + (m-1)\mathcal{P}_2 + (m-1)\mathcal{P}_3,$$

and its degree is equal to $3(m-1)$. By Hurwitz's Theorem 7.6,

$$(2g-2) = m(2g'-2) + 3(m-1),$$

where $g = \frac{1}{2}(q^2 - q)$ is the genus of Γ. Hence the genus g' of Γ' is equal to

$$g' = \tfrac{1}{2}\left(\frac{q^2 - q + 1}{m} - 1\right).$$

7.4 LINEAR SERIES COMPOSED OF AN INVOLUTION

Let Γ' be a curve given by the point $Q = (y_0, \ldots, y_r)$ with $y_i \in \Sigma$ and $K(Q) \subset \Sigma$. There are two basic ideas related to Q: the rational transformation

$$\tau : (1, x, y) \mapsto (y_0, \ldots, y_r),$$

and the virtual linear series L consisting of all divisors

$$A_{\mathbf{c}} = \text{div}\left(\textstyle\sum c_i y_i\right) + B, \quad \mathbf{c} = (c_0, \ldots, c_r) \in \text{PG}(r, K), \tag{7.9}$$

where B is a fixed divisor. The aim now is to give a geometric representation of L on Γ'. The simplest case occurs when τ is birational and, in this case, the divisors of L are the hyperplane sections on Γ'.

The curve Γ' lies in $\text{PG}(r, K)$. However, if $y_0, \ldots, y_r$ are linearly dependent over K, that is, if $c_0 y_0 + \cdots + c_r y_r = 0$ with some $c_i \in K$ not all zero, the hyperplane $H = \mathbf{v}(c_0 Y_0 + \cdots + c_r Y_r)$ contains Γ' by Theorem 7.5. Conversely, if Γ' lies in H, then the same theorem yields that $c_0 y_0 + \cdots + c_r y_r = 0$.

The significance of the usual assumption when discussing virtual linear series, that $y_0, \ldots, y_r$ are linearly independent, is now clear: this property holds if and only if Γ' is not contained in a proper subspace of $\text{PG}(r, K)$.

If a fixed divisor C is added to the divisors of L, a new virtual linear series is obtained, but both Γ' and the rational transformation τ remain the same. Hence it is assumed that *the virtual linear series L in question is a linear series and has no fixed place.* Also, it is helpful to make the following purely field-theoretic definition.

Let $\mathcal{P}'$ be a place of Σ' and let

$$D(\mathcal{P}') = e_{\mathcal{P}_1}\mathcal{P}_1 + \cdots + e_{\mathcal{P}_s}\mathcal{P}_s$$

be the divisor as in (7.2). The set of all divisors $D(\mathcal{P}')$ obtained as $\mathcal{P}'$ ranges over the places of Σ' is an *involution* of Σ of *order*

$$[\Sigma : \Sigma'] = e_{\mathcal{P}_1} + \cdots + e_{\mathcal{P}_s}.$$

Also, write $\nu = [\Sigma : \Sigma']$; then γ_ν indicates an involution of order ν. Also, by abuse of notation, write $\tau(\mathcal{P}) = \mathcal{P}'$ when $\mathcal{P}$ lies over $\mathcal{P}'$, and extend this to divisors by writing $\tau(\sum \mu_i \mathcal{P}_i) = \sum \mu_i \tau(\mathcal{P}_i)$. In other words, τ is a homomorphism from the divisor group of Σ into the divisor group of Σ'.

THEOREM 7.30 *For $x \in \Sigma'$, let $\operatorname{div}(x)_0$ and $\operatorname{div}(x)'_0$ denote the divisor of zeros of x for Σ and Σ'. Then*

$$\tau(\operatorname{div}(x)_0) = [\Sigma : \Sigma'] \cdot \operatorname{div}(x)'_0.$$

Proof. Let

$$\operatorname{div}(x)'_0 = \mu_1 \mathcal{P}'_1 + \cdots + \mu_s \mathcal{P}'_s,$$

with $\mathcal{P}'_i \neq \mathcal{P}'_j$, for $i \neq j$. If $\sum_j e_{\mathcal{P}_{ij}} \mathcal{P}_{ij}$ is the divisor corresponding to $\mathcal{P}'_i$, then

$$\tau(\textstyle\sum_j e_{\mathcal{P}_{ij}} \mathcal{P}_{ij}) = [\Sigma : \Sigma'] \cdot \mathcal{P}'_i.$$

Also, $\operatorname{ord}_{\mathcal{P}_{ij}} x = e_{\mathcal{P}_{ij}} \operatorname{ord}_{\mathcal{P}'_j} x = e_{\mathcal{P}_{ij}} \mu_j$. Hence

$$\operatorname{div}(x)_0 = \textstyle\sum_{i,j} \mu_i (\operatorname{ord}_{\mathcal{P}_{ij}} x) \mathcal{P}_{ij},$$

and

$$\begin{aligned}\tau(\operatorname{div}(x)_0) &= \tau(\textstyle\sum_{ij} \mu_i (\operatorname{ord}_{\mathcal{P}_{ij}} x)\, \mathcal{P}_{ij}) = \sum_{ij} \mu_i (\operatorname{ord}_{\mathcal{P}_{ij}} x) \tau(\mathcal{P}_{ij}) \\ &= \textstyle\sum_i \mu_i \left(\sum_j \operatorname{ord}_{\mathcal{P}_{ij}} x\right) \mathcal{P}'_i = \sum_i \left(\sum_j e_{\mathcal{P}_{ij}}\right) \mu_i \mathcal{P}'_i \\ &= [\Sigma : \Sigma'] \textstyle\sum_i \mu_i \mathcal{P}'_i = [\Sigma : \Sigma'] \operatorname{div}(x)'_0.\end{aligned}$$

□

COROLLARY 7.31 *Let $\operatorname{div}(x)'_0 = \mathcal{P}'_1 + \cdots + \mathcal{P}'_t$. Then $\operatorname{div}(x)_0$ is the sum of the divisors corresponding to the $\mathcal{P}'_i$. Hence, it is made up of the divisors of an involution.*

PROPOSITION 7.32 *The divisors of a linear series L are made up of divisors of an involution.*

Proof. If the representation (7.9) of L is not normalised, change it to a normalised one; see Theorem 6.22. This involves replacing y_i by a multiple zy_i, $z \in \Sigma$, and hence it does not change Γ'. By Theorem 6.22, B is then a divisor of L.

Let $\mathcal{P}_1, \ldots, \mathcal{P}_s$ be the places occurring in B. The condition that the divisor $A_{\mathbf{c}} \in L$ contains $\mathcal{P}_i$ is a linear condition $\sum \lambda_{ij} c_j = 0$ on the c_j; see Theorem 6.26. This condition is not trivial; that is, some $\lambda_{ij} \neq 0$, since $\mathcal{P}_i$ is not a fixed divisor of L. Thus the polynomial $\sum \lambda_{ij} X_j$ is not identically zero, and hence $H = \mathbf{v}(\sum \lambda_{ij} X_j)$ is a hyperplane of $\mathrm{PG}(r, K)$. Since finitely many hyperplanes cannot cover all points of $\mathrm{PG}(r, K)$, there are points off the hypersurface $\Delta = \mathbf{v}(\prod_{i=1}^{s}(\sum \lambda_{ij} X_j))$. Choose such a point $P = (c_0, \ldots, c_m)$. Then the corresponding divisor $A_{\mathbf{c}}$ contains none of the $\mathcal{P}_i$. Thus $A_{\mathbf{c}}$ is the set of zeros and B is the set of poles of $\sum_{i=0}^{r} c_i y_i$. Hence B is the set of zeros of an element in Σ. By Corollary 7.31, B is made up of elements of an involution. □

Thus the linear series L, or the rational transformation τ, defines an associated involution, $\gamma(L)$ or $\gamma(\tau)$. If $\gamma(L) = \gamma_\nu$ with $\nu > 1$, then $\{A_{\mathbf{c}}\}$ is *composed of the involution* γ_ν; otherwise $\{A_{\mathbf{c}}\}$ is *simple*. Then τ is birational if and only if the corresponding series is simple. In fact, $\Sigma = \Sigma'$ if and only if $\nu = 1$. Applying τ to (7.9) gives

$$\tau(A_{\mathbf{c}}) = \nu \cdot \mathrm{div}(\textstyle\sum c_i y_i) + \tau(B).$$

Each place in $\tau(B)$ occurs as many times as a multiple of ν, and the same holds for $\tau(A_{\mathbf{c}})$. Dividing by ν gives

$$\nu^{-1}\,\tau(A_{\mathbf{c}}) = \mathrm{div}(\textstyle\sum c_i y_i)' + \nu^{-1}\,\tau(B), \quad \nu = [\Sigma : \Sigma']. \tag{7.10}$$

The divisors on the right-hand side cut out a linear series L' on Γ'. If $\mathcal{P}_1' + \cdots + \mathcal{P}_t'$ is a divisor in L', then the sum of the divisors corresponding to the $\mathcal{P}_i'$ is an $A_{\mathbf{c}} \in L$. Actually, each $A_{\mathbf{c}} \in L$ is obtained in this way.

The linear series L' has no fixed place. To prove this, suppose otherwise and that $\mathcal{P}'$ is a fixed point. Let $\mathcal{P}_1, \ldots, \mathcal{P}_s$ be the places over $\mathcal{P}'$. As seen before, there is a divisor $A_{\mathbf{c}}$ containing none of the $\mathcal{P}_i$ and such that $\tau(A_{\mathbf{c}})$ does not contain $\mathcal{P}'$, a contradiction. Also, L' is effective. The hyperplanes of $\mathrm{PG}(r, K)$ cut out on Γ' the linear series

$$\mathrm{div}(\textstyle\sum c_i y_i)' + D.$$

As this has no fixed place and is effective, the multiplicity λ with which any place $\mathcal{P}$ enters D is $-\min_{\mathbf{c}}\{(\mathrm{ord}_{\mathcal{P}}(\sum c_i y_i)\}$. Since $(1/\nu)\tau(B)$ can be described in the same way, this also shows that $D = (1/\nu)\tau(B)$. Hence (7.10) *is the linear series cut out on* Γ' *by hyperplanes.* From this the following result is obtained.

THEOREM 7.33 *Let* Γ' *be a curve arising from a rational transformation* τ *of* Σ. *If* τ *is birational, then the corresponding linear series* L *is cut out on* Γ' *by hyperplanes.*

REMARK 7.34 Definition 6.33 can be interpreted geometrically. With a slight change in the wording in Remark 7.24, the projection π_s from $\mathrm{PG}(r, K)$ to its s-dimensional subspace $\Pi_s = \mathbf{v}(X_{s+1}, \ldots, X_r)$, for $2 \leq s < r$, is defined, with the vertex of π_s being $\Pi_{r-(s+1)} = \mathbf{v}(X_0, \ldots, X_s)$.

Let g_m^s and g_n^r be two simple linear series of Σ such that $m \leq n$, $s \leq r-1$. Suppose that g_m^s is contained in g_n^r. Then

$$g_m^s = \{\mathrm{div}\,(c_0 x_0 + \cdots + c_s x_s) + B \mid \mathbf{c} = (c_0, \ldots, c_s) \in \mathrm{PG}(s, K)\},$$
$$g_n^r = \{\mathrm{div}\,(c_0 x_0 + \cdots + c_r x_r) + F \mid \mathbf{c} = (c_0, \ldots, c_r) \in \mathrm{PG}(r, K)\}$$

with $F \succ B$.

Let $Q = (x_0, \ldots, x_s)$ and $Q' = (x_0, \ldots, x_r)$. If Γ and Δ are the irreducible curves determined by $K(Q)$ and $K(Q')$, then Δ is the projection of Γ with vertex $\Pi_{r-(s+1)}$. The converse also holds. An irreducible curve Γ is *normal* if it cannot be obtained in this way by a projection of another curve. Equivalently, if Γ is determined by $K(Q)$ with $Q = (x_0, \ldots, x_r)$ then the associated simple, fixed-place-free linear series

$$\{\text{div}\,(c_0x_0 + \cdots + c_rx_r) + B \mid \mathbf{c} = (c_0, \ldots, c_r) \in \text{PG}(r, K)\}$$

is complete.

EXAMPLE 7.35 Let $\Sigma = K(x)$, and $Q = (1, x, \ldots, x^r)$. The algebraic curve Γ of $\text{PG}(r, K)$ determined by Q is normal, and it is the *normal rational curve of* $\text{PG}(r, K)$.

Another fundamental result is the following.

THEOREM 7.36 *Let g_n^r be a linear series without fixed points but composed of an involution. If $\mathcal{P}$ is a place of Σ, then every divisor of g_n^r containing $\mathcal{P}$ also contains at least another place $\mathcal{Q}$, where $\mathcal{Q}$ depends only on $\mathcal{P}$, and conversely. The case $\mathcal{P} = \mathcal{Q}$ can occur when a divisor of g_n^r contains $\mathcal{P}$ with multiplicity at least* 2.

Proof. Let g_n^r be composed of the involution γ_ν with $\nu > 1$. Take any divisor $\mathcal{P} + \mathcal{P}_2 + \cdots + \mathcal{P}_s$ of γ_ν. Then every divisor containing $\mathcal{P}$ contains each of the remaining places $\mathcal{P}_2, \ldots, \mathcal{P}_\nu$. This proves the direct part.

For the converse, assume that g_n^r is simple, so that it defines a birational transformation, and hence g_n^r is cut out on Γ' by hyperplanes. Let $\mathcal{P}$ be any place centred at a simple point of Γ'. If $\mathcal{Q}$ another place of Σ, then there is a divisor in g_n^r containing $\mathcal{P}$ but $\mathcal{Q}$. In fact, if the centre Q of $\mathcal{Q}$ is distinct from the centre P of $\mathcal{P}$, then there is a hyperplane through P but not through Q, and hence the divisor cut out has the required property. For $P = Q$, such a divisor arises from any hyperplane containing $\mathcal{P}$ but not $2\mathcal{P}$, or geometrically, from any non-tangent hyperplane to Γ' at P. □

A characterisation for a point of a space curve to be non-singular is given in the following theorem.

THEOREM 7.37 *Let g_n^r be the linear series cut out on Γ by hyperplanes. Then a point P of Γ is non-singular if and only if the subseries of g_n^r containing a place $\mathcal{P}$ centred at P, minus $\mathcal{P}$, has no fixed place.*

Proof. Let $\gamma_1, \ldots, \gamma_k$ be the branches of Γ centred at P, and $\mathcal{P}_1, \ldots, \mathcal{P}_k$ the corresponding places of Σ. For $i = 1, \ldots, k$, put $n_{\mathcal{P}_i} = \text{ord}_{\mathcal{P}}\, \gamma_i$. Then the divisor $E = \sum_{i=1}^k n_{\mathcal{P}_i}\mathcal{P}_i$ is the fixed divisor of the linear subseries L' of g_n^r cut out on Γ by the hyperplanes through P. Also, $\deg E \geq 1$ and equality holds if and only if P is a non-singular point of Γ. On the other hand, $\deg E = 1$ occurs if and only if L minus $\mathcal{P}_1$ has no fixed divisor. □

DEFINITION 7.38 A singular point of Γ is an *s-fold point* if $\deg E = s$.

REMARK 7.39 A point P of Γ is an s-fold point if and only if the fixed divisor of the linear series cut out on Γ by the hyperplanes containing P has degree s.

As a consequence of Theorem 7.37, there is the following result.

THEOREM 7.40 *Let g_n^r be a complete linear series of Σ. If $n > 2g$, then the curve Γ, arising from g_n^r via the associated rational transformation, is free of singularities; that is, it is a non-singular model of Σ.*

Proof. By the Riemann–Roch Theorem 6.61, $r = n - g$ since $n > 2g > 2g - 2$ implies that g_n^r is not special. Also, g_n^r has no fixed point; otherwise, a complete non-special linear series g_{n-1}^r would be obtained by subtracting a fixed point. However, then $r = n - 1 - g$, again by the Riemann–Roch Theorem 6.61, contradicting the previous equation, $r = n - g$.

Now let P be any point of Γ, and let $\mathcal{P}$ be a place centred at P. The subseries of all divisors of g_n^r containing $\mathcal{P}$ minus $\mathcal{P}$ is a linear series g_{n-1}^{r-1} which has no fixed point. The latter property can be shown as before. The hypothesis that g_{n-1}^{r-1} has a fixed point would imply the existence of a complete non-special linear series g_{n-2}^{r-1}, contradicting the Riemann–Roch Theorem 6.61. □

For $g > 1$, the most important linear series is the *canonical series* g_{2g-2}^{g-1}. It is possible that g_{2g-2}^{g-1} may be composed of an involution. However, the following theorem shows that this happens only in a very special situation.

THEOREM 7.41 *Let $g > 1$. The canonical series is composed of an involution if and only if Σ has a complete g_2^1. If this is the case, then Σ has a subfield Σ' such that $[\Sigma : \Sigma'] = 2$.*

Proof. Assume first that Σ admits a complete g_2^1, and take one of the divisors in g_2^1, namely $C = \mathcal{P}_1 + \mathcal{P}_2$. Since the canonical series $|W|$ has no fixed point, its subseries of divisors containing $\mathcal{P}$ has dimension 1. If $|W|$ were not composed of an involution, then requiring a $W \in |W|$ to contain C would impose two linear conditions. So this would imply that $\dim |W - C| = (g-1) - 2 = g - 3$ and hence $i(C) = g - 2$, contradicting the Riemann–Roch Theorem 6.61 applied to g_2^1. Actually, it has been shown that g_{2g-2}^{g-1} is made up of divisors of g_2^1.

Conversely, let $|W|$ be composed of an involution γ_ν. By Theorem 7.36, the rational transformation associated to the canonical series gives rise to a curve Γ' in $\mathrm{PG}(g-1, K)$ of degree $(2g-2)/\nu$ such that the hyperplanes cut out on Γ' a linear series $g_{(2g-2)/\nu}^{g-1}$. Since $r \le n$ for any g_n^r, this is only possible for $\nu \le 2$. As $|W|$ is assumed to be composed, so $\nu = 2$, showing that $|W|$ is composed of an involution γ_2 of order 2.

It remains to show that γ_2 is a linear series, that is, its divisors are the divisors of a g_2^1. As $\nu = 2$, the linear series cut out on Γ' by hyperplanes is g_{g-1}^{g-1}. Hence Γ' has genus 0 by Theorem 6.81. Therefore the field Σ' of Γ' is $K(\xi)$ for some $\xi \in \Sigma$, and $[\Sigma : K(\xi)] = 2$. Let $\mathcal{P}$ be a place of Σ which is not a pole of x. For a primitive representation σ of $\mathcal{P}$,

$$\sigma(\xi) = a + bt^i + \cdots .$$

Then $\mathcal{P}$ is a zero of $\xi - a$ with multiplicity i. Let $\mathcal{P}'$ be the place of Σ' under $\mathcal{P}$. Hence (7.1) implies that $\mathcal{P}'$ is a zero of $\xi - a$ regarded as an element in Σ'. If $i = 1$, just one more place of Σ , say $\mathcal{Q}$, lies over $\mathcal{P}'$. By (7.1), $\mathcal{Q}$ is also a zero of $\xi - a$. Therefore $\text{div}(\xi - a)_0 = \mathcal{P} + \mathcal{Q}$. Now the case $i > 1$ is examined. By (7.1), $i = e_{\mathcal{P}}\, \text{ord}_{\mathcal{P}'}(\xi - a)$. Since Γ' is rational, $\text{ord}_{\mathcal{P}'}(x - a) = 1$; thus $i = e_{\mathcal{P}}$. Since $e_{\mathcal{P}} \le \nu = 2$, so $i = e_{\mathcal{P}} = 2$. This shows that $\mathcal{P}$ is the only place of Σ lying over $\mathcal{P}'$. Hence $2\mathcal{P} \in \gamma_2$, and $\text{div}(\xi - a)_0 = 2\mathcal{P}$.

Thus it has been shown that γ_2 consists of all divisors

$$\text{div}(c_0 + c_1\xi) + \text{div}(\xi)_\infty. \tag{7.11}$$

The completeness of g_2^1 depends on the hypothesis that $g > 0$. □

THEOREM 7.42 *If Σ has more than one g_2^1, then its genus is 0 or 1; that is, Σ is either rational or elliptic.*

Proof. Assume that Σ admits a g_2^1. From the proof of Theorem 7.41, the degree $[\Sigma : K(\xi)] = 2$ for some ξ, and g_2^1 consists of all divisors (7.11). If Σ has another g_2^1, then also $[\Sigma : K(\eta)] = 2$ with some $\eta \notin K(\xi)$. Therefore $\Sigma = K(\xi, \eta)$. From Theorem 5.61, the assertion $g \le 1$ follows. □

DEFINITION 7.43 A function field of genus $g > 1$ that has a complete, necessarily unique, linear series g_2^1 is *hyperelliptic*. An irreducible curve Γ is *hyperelliptic* if its function field is *hyperelliptic*.

THEOREM 7.44 *Let $K(x)$ be the unique subfield of a hyperelliptic Σ such that $[\Sigma : K(x)] = 2$. If $[\Sigma : K(y)] \le g$, with g the genus of Σ, then $K(y)$ is contained in $K(x)$.*

Proof. Suppose that $[\Sigma : K(z)] \le g$ but $z \notin K(x)$. Then $\Sigma = K(x, z)$. From Theorem 5.61,

$$g \le ([\Sigma : K(x)] - 1)([\Sigma : K(y)] - 1) \le g - 1,$$

a contradiction. □

Curves of genus 2 are examples of hyperelliptic curves. For a more detailed account of hyperelliptic curves, see Section 7.10. For non-hyperelliptic curves, the relevant results are stated in the following section.

7.5 THE CANONICAL CURVE

In this section, $\mathcal{C}$ denotes a non-hyperelliptic curve of genus g. Since, the canonical series on $\mathcal{C}$ is a simple g_{2g-2}^{g-1} with no fixed points, the projective image of such a series is a curve Γ of degree $2g-2$ in $\text{PG}(g-1, K)$, birationally equivalent to $\mathcal{C}$ and unique up to projectivities; see Theorem 7.45 (ii) below. It is the *canonical curve of genus g*. Conversely, if Γ is a curve of degree $2g - 2$ in $\text{PG}(g - 1, K)$ which is birationally equivalent to $\mathcal{C}$, then the series g_{2g-2}^{g-1} cut out by the hyperplanes is the canonical series, and so Γ is the canonical curve.

THEOREM 7.45 (i) *The canonical curve is non-singular.*

(ii) *Two canonical curves that are birationally equivalent are projectively equivalent.*

(iii) *A divisor* $D = \mathcal{P}_1 + \cdots + \mathcal{P}_n$ *defines a complete special series* $|D| = g_n^r$ *if and only if the centres of the* n *branches corresponding to the places* $\mathcal{P}_i$, $i = 1, \ldots, n$, *span an* $(n - r - 1)$*-dimensional subspace of* $\mathrm{PG}(g - 1, K)$.

(iv) *The canonical curve is normal.*

Proof. (i) Assume on the contrary that M is a singular point of Γ, that is, a $(k+2)$-fold point with k non-negative. By Theorem 7.37, the hyperplanes through M cut out on Γ a linear series g_{2g-4-k}^{g-2} without fixed places. If $\gamma_1, \ldots, \gamma_s$ are the branches of Γ centred at M and $\mathcal{P}_1, \ldots, \mathcal{P}_s$ are the corresponding places of Σ, the hyperplanes cut out in addition a fixed divisor $D = \sum_{i=1}^{s} n_{\mathcal{P}}\mathcal{P}$, where the degree $\sum_{i=1}^{s} n_{\mathcal{P}} = k + 2$. Hence a fixed divisor $D' = \sum_{i=1}^{s} n'_{\mathcal{P}}\mathcal{P}$, with $0 \leq n'_{\mathcal{P}} \leq n_{\mathcal{P}}$ and $\sum_{i=1}^{s} n'_{\mathcal{P}} = k$, can be added to g_{2g-4-k}^{g-2}, so as to produce a special series

$$g_{2g-4}^{g} = |D' + g_{2g-4-k}^{g-2}|.$$

The dimension of this series is less than $g - 1$, but also at least $g - 2$, and hence it is a special series g_{2g-1}^{g-2}. The residual divisors relative to the canonical series constitute a g_2^s. By the Reciprocity Theorem 6.78 of Brill and Noether,

$$2g - 6 = 2(g - 2 - s),$$

whence $s = 1$, and the series is a g_2^1. This is ruled out, however, since Γ is supposed not to be hyperelliptic.

(ii) This follows from Theorem 6.17 together with Theorem 6.72 (ii).

(iii) If G is a special divisor of n points on the canonical curve Γ that defines a complete g_n^r, the linear system of hyperplanes of $\mathrm{PG}(g - 1, K)$ through G has dimension $g - 1 - n + r$; so G is in a subspace Π_{n-r-1} . Conversely, if a divisor G of n points is in a subspace Π_{n-r-1} and not in any lower dimension, the index of speciality of G is $g - n + r$, and so defines a complete g_n^r.

(iv) This follows from Remark 7.34, since the canonical series is complete. □

7.6 OSCULATING HYPERPLANES AND RAMIFICATION DIVISORS

The idea is to consider for any point P of a non-singular space curve the different intersection multiplicities of hyperplanes with the curve at P. There is only a finite number of these intersection multiplicities, this number being equal to the dimension of the space. There is a unique hyperplane, called the *osculating hyperplane*, with the maximum intersection multiplicity. The concept of osculating hyperplane will also be defined for singular curves.

Let Γ be an irreducible curve of degree n in $\mathrm{PG}(r, K)$. The hyperplanes cut out on Γ a simple, fixed-point-free, not-necessarily-complete, linear series L. Note that

L has degree n and dimension r, and hence it is a g_n^r. Let Γ be given by the point $P = (x_0, \ldots, x_n)$. Without loss of generality, suppose that $x_0 \neq 0$; then

$$K(\Gamma) = K(x_1/x_0, \ldots, x_r/x_0)$$

is the function field of Γ. Further, L consists of all divisors

$$A_{\mathbf{c}} = \operatorname{div}(c_0x_0 + c_1x_1 + \cdots + c_rx_r) + B$$

with $\mathbf{c} = (c_0, \ldots, c_r) \in \mathrm{PG}(r, K)$ and $B = \sum e_{\mathcal{P}}\mathcal{P}$, where

$$e_{\mathcal{P}} = -\min\{\operatorname{ord}_{\mathcal{P}} x_0, \ldots, \operatorname{ord}_{\mathcal{P}} x_r\}.$$

Here $\operatorname{ord}_{\mathcal{P}} B = n$.

Geometrically speaking, $A_{\mathbf{c}}$ is cut out on Γ by the hyperplane

$$H = \mathbf{v}(c_0X_0 + \cdots + c_rX_r);$$

that is, $A_{\mathbf{c}}$ is the intersection divisor $H \cdot \Gamma = \sum I(P, H \cap \gamma)\mathcal{P}$, where

$$I(P, H \cap \gamma) = \operatorname{ord}_t(\textstyle\sum_{i=0}^{r} c_ix_i(t)) + n_{\mathcal{P}}.$$

Here $x_i(t) = \tau(x_i)$, $i = 0, \ldots, r$, is a primitive representation τ of $\mathcal{P}$, and the $(r+1)$-tuple $(x_0(t), \ldots, x_r(t))$ is a primitive representation of the branch γ of Γ corresponding to $\mathcal{P}$, with P the centre of γ.

To develop local properties of Γ, a finite sequence of integers is defined.

Definition 7.46 An integer j is a *Hermitian $\mathcal{P}$-invariant* or an *$(L, \mathcal{P})$-order* if there exists a hyperplane H such that $I(P, H \cap \gamma) = j$.

In Example 7.18, if $P = (0, 0, 1, 0)$ then the $(L, \mathcal{P})$-orders are $0, 1, 2, 5$.

In the case that L is the canonical series, it follows from the Riemann–Roch Theorem 6.61 that j is a $(L, \mathcal{P})$-order if and only if $j + 1$ is a Weierstrass gap; that is, there exists no rational function in Σ which has a pole of order $j + 1$ at $\mathcal{P}$ but is regular at all places different from $\mathcal{P}$; see Section 6.6.

For any non-negative integer i, consider the set of all hyperplanes H of $\mathrm{PG}(r, K)$ for which $I(P, H \cap \gamma) \geq i$. Such hyperplanes correspond to the points of a subspace $\overline{\Pi}_i$ in the dual space of $\mathrm{PG}(r, K)$. Then

$$\mathrm{PG}(r, K) = \overline{\Pi}_0 \supset \overline{\Pi}_1 \supset \overline{\Pi}_2 \supset \cdots.$$

An integer j is an $(L, \mathcal{P})$-order if and only if $\overline{\Pi}_j \neq \overline{\Pi}_{j+1}$, in which case $\overline{\Pi}_{j+1}$ has codimension 1 in $\overline{\Pi}_j$. Since $\deg L = n$, so Π_i is empty as soon as $i > n$. The number of $(L, \mathcal{P})$-orders is exactly $r+1$; they are $j_0, j_1, \ldots, j_r$ in increasing order. Since $\mathcal{P}$ is not a base point of L, so $j_0 = 0$. Note also that $j_1 = 1$ if and only if the branch γ is linear.

In the case that the linear series L is complete, from the Riemann–Roch Theorem 6.61,

$$\dim \overline{\Pi}_{j_i} = n - j_i - g,$$

for $j_i < n - 2g + 2$; in particular, $j_i = i$ when $i \leq n - 2g$.

Consider the intersection Π_i of hyperplanes H of $\mathrm{PG}(r, K)$, for which

$$I(P, H \cap \gamma) \geq j_{i+1}.$$

Then Π_0 is simply P, the centre of the branch γ, and Π_1 is the tangent line to the branch γ at P. Also, Π_i is the *i-th osculating space* or the *osculating i-space*; in particular, Π_{r-1} is the *osculating hyperplane*. The flag

$$\Pi_0 \subset \Pi_1 \subset \cdots \subset \Pi_{r-1} \subset \mathrm{PG}(r, K)$$

can be viewed as the algebraic analogue of the Frenet frame in differential geometry.

Now an equation of the osculating hyperplane is given in terms of the coordinate functions x_i. The essential tool is a generalisation of the Wronskian determinant whose entries are Hasse higher derivatives of the coordinate functions.

Choose a local parameter ζ at $\mathcal{P}$, and divide x_k by $\zeta^{e_\mathcal{P}}$ for $k = 0, 1, \ldots, r$. Then $\mathrm{ord}_\mathcal{P}\, D_\zeta^{(i)} x_k \geq 0$; that is,

$$\tau(D_\zeta^{(i)} x_k) = c_{0k}^{(i)} + \cdots,$$

with $c_{0k}^{(i)} \in K$. To write the Wronskian determinant, it is better to use simpler and more expressive notation. For this purpose, in the rest of this section, points of Γ are thought of as *branch points*; that is, the term *a point P of* Γ is used to indicate the branch of Γ associated to the place $\mathcal{P}$. In this spirit, $D_\zeta^{(i)} x_k(P)$ is an appropriate notation for $c_{0k}^{(i)}$. In particular, the centre of P is the point $(x_0(P), \ldots, x_r(P))$ of Γ. Also, $I(P, \Gamma \cap \Phi)$ is used to indicate $I(A, \gamma \cap \Phi)$ for a hypersurface Φ of $\mathrm{PG}(r, K)$.

THEOREM 7.47 *The $(L, \mathcal{P})$-orders can be computed iteratively. If the $(L, \mathcal{P})$-orders $j_0, \ldots, j_{i-1}$ are known, then the successive order j_i is the smallest integer such that, in* $\mathrm{PG}(r, K)$, *the $i + 1$ points*

$$P_m = (D_\zeta^{(j_m)} x_0)(P),\ (D_\zeta^{(j_m)} x_1)(P),\ \ldots,\ (D_\zeta^{(j_m)} x_r)(P)),$$

for $m = 0, 1, \ldots, i$, are linearly independent. The i-th osculating space at P is spanned by these points.

Proof. Up to a projectivity of $\mathrm{PG}(r, K)$, for every $i = 0, \ldots, r$ the subspace L_i is the intersection of the $r - i$ hyperplanes $\mathbf{v}(X_{i+1}), \ldots, \mathbf{v}(X_r)$. Then

$$j_{i+1} = \min\{\mathrm{ord}_\mathcal{P}(a_{i+1}x_{i+1} + \cdots + a_r x_r) \mid a_{i+1}, \ldots, a_r \in K\}.$$

Hence, considering the suffixes in descending order,

$$j_r = \mathrm{ord}_\mathcal{P}\, x_r,\ j_{r-1} = \mathrm{ord}_\mathcal{P}\, x_{r-1},\ \ldots,\ j_0 = \mathrm{ord}_\mathcal{P}\, x_0.$$

Now define the $(r + 1) \times (r + 1)$ matrix $A = (a_{mi})$, where

$$a_{mi} = (D_t^{(j_m)} x_i)(P),$$

for $0 \leq r \leq n$, $0 \leq i \leq n$. Since, by Lemma 5.80 and (5.20), $D_\zeta^{(j)} \zeta^j = 1$ but $D_\zeta^{(k)} \zeta^j = 0$ for $k > j$, so A is lower-triangular with every diagonal element non-zero. Hence L_i is spanned by the points

$$P_m = (D_\zeta^{(j_m)} x_0(P), \ldots, D_\zeta^{(j_m)} x_r(P)),$$

for $m = 0, 1, \ldots, i$. □

The minimality of the j_i holds in an even stronger sense.

COROLLARY 7.48 *If $m_0, m_1, \ldots, m_k$ are integers with $0 \le m_0 < \cdots < m_k$ such that the points $P_i = (D_\zeta^{(m_i)} x_0(P), \ldots, D_\zeta^{(m_i)} x_r(P))$, for $i = 0, 1, \ldots, k$, are linearly independent, then $j_i \le m_i$ for $i = 0, \ldots, k$.*

Proof. Since the points $P_s = (D_t^{(s)} x_0(P), \ldots, D_t^{(s)} x_r(P))$, with $s = 0, \ldots, j_i - 1$ span a subspace of dimension i, while the $i+1$ points with $s = m_0, \ldots, m_i$ are linearly independent, so $j_i - 1 < m_i$; that is, $j_i \le m_i$. □

COROLLARY 7.49 *The osculating hyperplane at P has equation*

$$\begin{vmatrix} X_0 & X_1 & \cdots & X_r \\ D_\zeta^{(j_0)} x_0(P) & D_\zeta^{(j_0)} x_1(P) & \cdots & D_\zeta^{(j_0)} x_r(P) \\ D_\zeta^{(j_1)} x_0(P) & D_\sigma^{(j_1)} x_1(P) & \cdots & D_\zeta^{(j_1)} x_r(P) \\ \vdots & \vdots & \cdots & \vdots \\ D_\zeta^{(j_{r-1})} x_0(P) & D_\zeta^{(j_{r-1})} x_1(P) & \cdots & D_\zeta^{(j_{r-1})} x_r(P) \end{vmatrix} = 0.$$

The point P is an *osculation point* or, more precisely, an *L-osculation point* if $j_r > r$, that is, if there is a hyperplane intersecting Γ at P with multiplicity greater than r.

EXAMPLE 7.50 (i) Let $\Sigma = K(x)$ be the rational function field. Choose the X-axis $\mathcal{C} = \mathbf{v}(Y)$ as a model $(\mathcal{C}; (x, 0))$ of Σ. Let $s \ge 1$ be any integer; then the elements $1, x, x^{p^s}, x^{p^s+1}$ are linearly independent over K. Let Γ be the associated curve of $\mathrm{PG}(3, K)$. Note that the divisors $A_\mathbf{c} + B$, with

$$A_\mathbf{c} = \mathrm{div}\,(c_0 + c_1 x + c_2 x^{p^s} + c_3 x^{p^s+1})$$

and $B = \sum e_\mathcal{P} \mathcal{P}$, constitute a simple, fixed-point-free linear series L. Then

$$(j_0, j_1, j_2, j_3) = (0, 1, p^s, p^{s+1})$$

for any P of Γ, and the osculating plane H_P has equation

$$\begin{vmatrix} X_0 & X_1 & X_2 & X_3 \\ 1 & t & t^{p^s} & t^{p^s+1} \\ 0 & 1 & 0 & t^{p^s} \\ 0 & 0 & 1 & t \end{vmatrix} = 0;$$

that is,

$$H_P = \mathbf{v}(-t^{p^s+1} X_0 + t^{p^s} X_1 + t X_2 - X_3).$$

(ii) Let $p = 5$. Take $\mathcal{F} = \mathbf{v}(f(X, Y))$ as in Examples 5.36 and 5.81 (ii). Let $r = 3$ and $x_0 = 1$, $x_1 = x$, $x_2 = y$, $x_3 = y^2$. If

$$R_i = (D_x^{(i)} x_0,\ D_x^{(i)} x_1,\ D_x^{(i)} x_2,\ D_x^{(i)} x_3),$$

then

$$\begin{aligned} R_0 &= (1, x, y, y^2), & R_1 &= (0, 1, -2x^2, x^2 y), \\ R_2 &= (0, 0, -2x, xy - x^4), & R_3 &= (0, 0, 1, -2x^3 + 2y), \\ R_4 &= (0, 0, 0, 0), & R_5 &= (0, 0, 2x^{10}, x - x^{10} y), \\ R_6 &= (0, 0, 0, 2x^{12} + 1). \end{aligned}$$

Let $A = (1, 0, 0) \in \mathcal{F}$. Then the corresponding point $P = (1, 0, 0, 0)$ is in Γ. The order sequence of Γ at P is $(0, 1, 3, 6)$, and the osculating plane H_P to Γ at P has equation

$$\begin{vmatrix} X_0 & X_1 & X_2 & X_3 \\ 1 & 0 & 0 & 0 \\ 0 & 1 & 0 & 0 \\ 0 & 0 & 1 & 0 \end{vmatrix} = 0;$$

that is, $H_P = \mathbf{v}(X_3)$. Let $A = (0, 1, 0) \in \mathcal{F}$. Then the corresponding point $P_\infty = (0, 0, 0, 1)$ and a primitive representation of the unique branch of Γ centred at P_∞ is as follows:

$$x_0 = t^6 + \cdots, \ x_1 = t + \cdots, \ x_2 = t^3 + \cdots, \ x_3 = 1.$$

The order sequence of Γ at P_∞ is $(0, 1, 3, 6)$.

Let

$$R_i(P_\infty) = (D_x^{(i)} x_0(P_\infty), D_x^{(i)} x_1(P_\infty), D_x^{(i)} x_2(P_\infty), D_x^{(i)} x_3(P_\infty)).$$

Then

$$\begin{array}{lll} R_0(P_\infty) = (0, 0, 0, 1), & R_1(P_\infty) = (0, 1, 0, 0), & R_2(P_\infty) = (0, 0, 0, 0), \\ R_3(P_\infty) = (0, 0, 1, 0), & R_4(P_\infty) = (0, 0, 0, 0), & R_5(P_\infty) = (0, 0, 0, 0), \\ R_6(P_\infty) = (1, 0, 0, 0). & & \end{array}$$

The osculating plane H_P to Γ at P_∞ has equation

$$\begin{vmatrix} X_0 & X_1 & X_2 & X_3 \\ 0 & 0 & 0 & 1 \\ 0 & 1 & 0 & 0 \\ 0 & 0 & 1 & 0 \end{vmatrix} = 0;$$

that is, $H_P = \mathbf{v}(X_0)$.

To investigate global properties of Γ, the condition on ζ to be a local parameter is weakened to be a separable variable. However, the idea of considering a determinant such as that in Theorem 7.47 is still useful. This depends on the existence of a strictly increasing sequence of non-negative integers $\epsilon_0, \ldots, \epsilon_n$, the *orders*, such that the generalised Wronskian

$$W = \det(D_\zeta^{(\epsilon_i)} x_j) \tag{7.12}$$

with $0 \le i, j \le r$ does not vanish. To determine such a sequence the idea is to choose $\epsilon_0, \ldots, \epsilon_r$ in strictly increasing order such that $\epsilon_0 = 0$ and define ϵ_i inductively so that, if $\epsilon_0, \ldots, \epsilon_{i-1}$ are known, then ϵ_i is the smallest integer such that the points

$$P_k = (D_\zeta^{(\epsilon_k)} x_0(P), \ldots, D_\zeta^{(\epsilon_k)} x_r(P)),$$

$k = 0, \ldots, i$, are linearly independent in $\mathrm{PG}(r, \Sigma)$. As in Corollary 7.48, the ϵ_i are minimal in a stronger sense: if $m_0, \ldots, m_i$ are integers with $0 \le m_0 < \cdots < m_i$ such that the points $Q_k = (D_\zeta^{(m_k)} x_0(P), \ldots, D_\zeta^{(m_k)} x_r(P))$, $k = 0, 1, \ldots, i$, are linearly independent in $\mathrm{PG}(r, \Sigma)$, then $\epsilon_k \le m_k$ for $k = 0, \ldots, i$. To show that this inductive procedure works properly, the following lemma is required.

LEMMA 7.51 (i) *If* $y_i = \sum a_{ij} x_j$ *with* $(a_{ij}) \in \mathrm{GL}(r+1, K)$, *then*

$$\det(D_\zeta^{(\epsilon_i)} y_j) = \det(a_{ij}) \det(D_\zeta^{(\epsilon_i)} x_j).$$

(ii) *If* $h \in \Sigma$, *then*

$$\det(D_\zeta^{(\epsilon_i)}(h x_j)) = h^{r+1} \det(D_\zeta^{(\epsilon_i)} x_j).$$

(iii) *If* η *is another separable variable, then*

$$\det(D_\eta^{(\epsilon_i)} x_j) = (d\zeta/d\eta)^{\epsilon_1 + \cdots + \epsilon_r} \det(D_\zeta^{(\epsilon_i)} x_j).$$

Proof. (i) This comes from projective geometry, and does not depend on the minimality of the ϵ_i.

(ii) By Leibniz' rule, as in Lemma 5.80 and (5.23),

$$D_\zeta^{(\epsilon_i)}(h x_j) = h \cdot D_\zeta^{(\epsilon_i)} x_j + \textstyle\sum_{s=1}^{\epsilon_i} (D_\zeta^{(s)} h) \cdot (D_\zeta^{(\epsilon_i - s)} x_j).$$

In the first row of W, every element has h as a factor. The second row is

$$h \cdot D_\zeta^{(\epsilon_1)}(x_0) + \cdots, \ \ldots, h \cdot D_\zeta^{(\epsilon_1)}(x_k) + \cdots,$$

where the ellipses in each entry indicate terms that are the components of a vector w which is a linear combination of vectors $(D_\zeta^{(k)} x_0(P), \ldots, D_\zeta^{(k)} x_r(P))$ with $0 \leq k < \epsilon_1$. This vector w is a multiple of $(D_\zeta^{(\epsilon_0)} x_0(P), \ldots, D_\zeta^{(\epsilon_0)} x_r P))$, by the minimality of ϵ_1. Hence, in the determinant, w can be omitted, and so h is again a factor. Proceeding inductively on the other rows of W gives the result.

(iii) The same argument can be employed if the chain rule (5.77) is used in place of Leibniz' rule. Here, $D_\eta^{(\epsilon_i)} x_j$ in W may be replaced by

$$D_t^{(\epsilon_i)} x_j \cdot (d\zeta/d\eta)^{\epsilon_i}.$$

□

From Lemma 7.51, it follows that $\epsilon_0, \ldots, \epsilon_n$ depend only on the linear series L, and therefore can be called the *L-orders* or, simply, the *orders* of the curve Γ.

To show the actual existence of the orders, choose a place $\mathcal{P}$ and let ζ be a local parameter of Σ at $\mathcal{P}$, and suppose that $e_{\mathcal{P}} = 0$. By Theorem 7.47, there exist $j_0, \ldots, j_r$, the $(L, \mathcal{P})$-*orders*, for which

$$\det(D_\zeta^{(j_i)} x_j(P)) \neq 0.$$

This implies that $\det(D_\zeta^{(j_i)} x_j)) \neq 0$. Hence there exists the *order sequence*, that is, the strictly increasing minimal sequence $\epsilon_0, \ldots, \epsilon_r$ for which $\det(D_\zeta^{(\epsilon_i)} x_j)) \neq 0$ and $\epsilon_i \leq j_i$ for $i = 0, \ldots, r$.

DEFINITION 7.52 The *ramification divisor* of L is

$$R = \mathrm{div}\,(\det(D_\zeta^{(\epsilon_i)} x_j)) + (\epsilon_0 + \cdots + \epsilon_r)\mathrm{div}\,(d\zeta) + (r+1)E,$$

where $E = \sum e_{\mathcal{P}} \mathcal{P}$ and $e_{\mathcal{P}} = -\min\{\mathrm{ord}_{\mathcal{P}}\, x_0, \ldots, \mathrm{ord}_{\mathcal{P}}\, x_r\}$.

EXAMPLE 7.53 Let $p = 5$. With the notation in Example 7.50, the order sequence of Γ is $(0, 1, 2, 5)$, and

$$\det(D_x^{(\epsilon_i)}) = \begin{vmatrix} 1 & x & y & y^2 \\ 0 & 1 & 3x^2 & x^2y \\ 0 & 0 & 3x & xy - x^4 \\ 0 & 0 & 2x^{10} & x - x^{10}y \end{vmatrix} = 3x^2(1 - x^{12}).$$

The unique branch of Γ centred at $P_\infty = (0, 0, 0, 1)$ gives rise to the place $\mathcal{P}_\infty$. Then $\operatorname{div}(dx) = 6\mathcal{P}_\infty$ and $E = 6\mathcal{P}_\infty$; see Lemma 12.1. Therefore the ramification divisor is

$$R = \operatorname{div}(x^2(1 - x^{12})) + 72\mathcal{P}_\infty = 2\textstyle\sum_{i=1}^{5} \mathcal{P}_i + \sum_{i=1}^{60} \mathcal{Q}_i + 2\mathcal{P}_\infty,$$

where $\mathcal{P}_i$ and $\mathcal{Q}_i$ are as in Example 5.36.

LEMMA 7.54 *The divisor R depends only on the linear series L.*

Proof. From Theorem 6.17, suppose that the divisors in L can also be written as $A_{\mathbf{d}} + B$ where $A_{\mathbf{d}} = \operatorname{div}(\sum_{k=0}^{r} d_k y_k) + B$ with $\mathbf{d} = (d_0, \ldots, d_r)$ and $P = \mathbf{d}$ in $\mathrm{PG}(r, K)$. From Lemma 7.51 (i),

$$\operatorname{ord}_{\mathcal{P}}(\operatorname{div}(\det(D_\zeta^{(\epsilon_i)} x_j)) = \operatorname{ord}_{\mathcal{P}}(\operatorname{div}(\det(D_\zeta^{(\epsilon_i)} y_j)),$$

for every place $\mathcal{P}$ of Σ. Given any non-zero element $h \in \Sigma$, let

$$e_{\mathcal{P}}^* = -\min\{\operatorname{ord}_{\mathcal{P}}(hx_i)\}, \quad E^* = \textstyle\sum e_{\mathcal{P}}^* \mathcal{P}.$$

Then $E^* + \operatorname{div}(h) = E$ and

$$\begin{aligned}
&\operatorname{div}(\det(D_\zeta^{(\epsilon_i)}(hx_j))) + (\epsilon_0 + \cdots + \epsilon_r)\operatorname{div}(d\zeta) + (r+1)E^* \\
&= (r+1)\operatorname{div}(h) + \operatorname{div}(\det(D_\zeta^{(\epsilon_i)} x_j)) \\
&\qquad + (\epsilon_0 + \cdots + \epsilon_r)\operatorname{div}(d\zeta) + (r+1)E^* \\
&= \operatorname{div}(\det(D_\zeta^{(\epsilon_i)} x_j)) + (\epsilon_0 + \cdots + \epsilon_r)\operatorname{div}(d\zeta) + (r+1)E.
\end{aligned}$$

Also, if η is another separable variable in Σ, then

$$\begin{aligned}
&\operatorname{div}(\det(D_\eta^{(\epsilon_i)} x_j)) + (\epsilon_0 + \cdots + \epsilon_r)\operatorname{div}(d\eta) + (r+1)E \\
&= (\epsilon_0 + \cdots + \epsilon_r)(\operatorname{div}(d\zeta/d\eta) + \operatorname{div}(\det(D_\zeta^{(\epsilon_i)} x_j)) \\
&\qquad + (\epsilon_0 + \cdots + \epsilon_r)(\operatorname{div}(d\eta) + (r+1)E \\
&= \operatorname{div}(\det(D_\zeta^{(\epsilon_i)} x_j)) + (\epsilon_0 + \cdots + \epsilon_r)\operatorname{div}[(d\eta) \cdot (d\zeta/d\eta)] + (r+1)E \\
&= \operatorname{div}(\det(D_\zeta^{(\epsilon_i)} x_j)) + (\epsilon_0 + \cdots + \epsilon_r)\operatorname{div}(d\zeta) + (r+1)E.
\end{aligned}$$

This completes the proof. □

From Corollary 5.35 and the definition of the genus (5.55),

$$\deg R = (\epsilon_1 + \cdots + \epsilon_r)(2g - 2) + (r+1)n. \tag{7.13}$$

THEOREM 7.55 *If $j_0, \ldots, j_r$ are the $(L, \mathcal{P})$-orders, then*

$$v_{\mathcal{P}}(R) \geq \textstyle\sum_{i=0}^{r} (j_i - \epsilon_i)$$

with equality if and only if

$$\det\left(\binom{j_i}{\epsilon_k}\right) \not\equiv 0 \pmod{p}.$$

Proof. Suppose again that $e_{\mathcal{P}} = 0$. For a local parameter ζ at $\mathcal{P}$,

$$v_{\mathcal{P}}(R) = \text{ord}_{\mathcal{P}}(\det(D_t^{\epsilon_k} x_i)).$$

Arguing as in Theorem 7.47, it is possible to take $y_i \in \Sigma$ such that

$$\tau(y_i) = c_i t^{j_i} + \cdots \text{ for } i = 0, \ldots, n,$$

where, as usual, ellipses indicate terms of higher orders. Then

$$\begin{aligned}\det(D_t^{(\epsilon_k)}\tau(y_i)) &= \det\left(\binom{j_i}{\epsilon_r} t^{j_i-\epsilon_r} + \cdots\right)\\ &= \det\left(\binom{j_i}{\epsilon_r} t^{j_i} + \cdots\right) t^{-\epsilon_0-\cdots-\epsilon_n}\\ &= \det\left(\binom{j_i}{\epsilon_r}\right) t^{j_0+\cdots+j_n-\epsilon_0-\cdots-\epsilon_n} + \cdots.\end{aligned}$$

Thus $\text{ord}_{\mathcal{P}}(\det(D_\zeta^{(\epsilon_k)} y_i) \geq \sum(j_i - \epsilon_i)$, with equality for

$$\det\left(\binom{j_i}{\epsilon_k}\right) \not\equiv 0 \pmod p.$$

□

By Theorem 7.55, the ramification divisor R is effective, and $n_{\mathcal{P}} = 0$ in the divisor $R = \sum n_{\mathcal{P}}\mathcal{P}$ if and only if $j_i = \epsilon_i$ all i. Hence $\epsilon_0, \epsilon_1, \ldots, \epsilon_r$ are the $(L, \mathcal{P})$-orders for almost all places $\mathcal{P}$ of Σ; so these places are *L-ordinary*. The other places, finite in number, satisfy the inequality

$$(j_0, \ldots, j_r) \neq (\epsilon_0, \ldots, \epsilon_r)$$

and are *L-Weierstrass points*; also, $n_{\mathcal{P}}$ is the *weight* of $\mathcal{P}$, and is denoted by $v_{\mathcal{P}}(R)$. Note that, if ζ is a local parameter at $\mathcal{P}$, then

$$v_{\mathcal{P}}(R) = \begin{vmatrix} x_0 & x_1 & \cdots & x_r \\ D_\zeta^{(\epsilon_1)} x_0 & D_\zeta^{(\epsilon_1)} x_1 & \cdots & D_\zeta^{(\epsilon_1)} x_r \\ \vdots & \vdots & \cdots & \vdots \\ D_\zeta^{(\epsilon_r)} x_0 & D_\zeta^{(\epsilon_r)} x_1 & \cdots & D_\zeta^{(\epsilon_r)} x_r \end{vmatrix}. \tag{7.14}$$

In this terminology, the number of *L-Weierstrass* points, each counted according to its weight, is

$$\sum v_{\mathcal{P}}(R) = (\epsilon_0 + \cdots + \epsilon_r)(2g-2) + (r+1)n.$$

REMARK 7.56 In the particular case that L is the canonical linear series of Σ, the notion of *L-Weierstrass* point is the same as in Section 6.6. An example for this particular case is worked out in Section 12.5.

This has the following important corollary.

COROLLARY 7.57 *Every curve which is neither rational, nor elliptic, nor hyperelliptic has finitely many Weierstrass points.*

Note that this corollary holds true for hyperelliptic curves; see Theorem 7.103.

DEFINITION 7.58 The linear series L, or equivalently the curve Γ, is *classical* if the order sequence $(\epsilon_0, \epsilon_1, \ldots, \epsilon_r)$ is $(0, 1, \ldots, r)$.

Now some criteria are given that are useful in deciding whether or not a given linear series is classical. First, a proposition follows that refines the preceding inequality, $\epsilon_i \leq j_i$.

PROPOSITION 7.59 *Let $j_0, \ldots, j_r$ be the $(L, \mathcal{P})$-orders. If $m_0, \ldots, m_r$ are integers such that $0 \leq m_0 < \cdots < m_r$ and*

$$\det\left(\binom{j_i}{m_k}\right) \not\equiv 0 \pmod p,$$

then $\epsilon_i \leq m_i$ for each $i = 0, \ldots, r$.

Proof. As in the proof of Theorem 7.55,

$$\det(D_t^{(m_k)}\tau(y_i)) = \det\left(\binom{j_i}{m_k}\right) t^{j_0 - m_0 + \cdots + j_n - m_n} + \cdots.$$

Therefore $\epsilon_i \leq m_i$ for all i by the minimality of the ϵ_i. □

COROLLARY 7.60 *Let $j_0, \ldots, j_r$ be the $(L, \mathcal{P})$-orders. If*

$$\prod_{i>s} (j_i - j_s)/(i - s)$$

is not divisible by the characteristic p of K, then L is classical and

$$v_{\mathcal{P}}(R) = \sum_{i=0}^{r} (j_i - i).$$

Proof.

$$\det\left(\binom{j_i}{k}\right)$$
$$= \det({j_i}^k/i! + \cdots) = \det({j_i}^k)/(1!\,2! \cdots n!) = \prod_{i>s} (j_i - j_s)/(i - s).$$

Hence $\epsilon_i = i$ by Proposition 7.59 and the weight of $\mathcal{P}$ in R equals $\sum_{i=0}^{r}(j_i - i)$ by Theorem 7.55. □

COROLLARY 7.61 *If the characteristic p of K is 0, or larger than n, then L is classical and $v_{\mathcal{P}}(R) = \sum_{i=0}^{r}(j_i - i)$.*

Proof. The condition in Corollary 7.60 is satisfied if $j_i \not\equiv j_s \pmod p$ for all $i \neq s$; in particular, this holds when $j_n \leq n$. □

LEMMA 7.62 *If ϵ is an L-order and μ is such that $\binom{\epsilon}{\mu} \not\equiv 0 \pmod p$, then μ is also an L-order. In particular, if $\epsilon < p$, then $0, 1, \ldots, \epsilon - 1$ are also L-orders.*

Proof. Since $\binom{\epsilon}{\mu} \neq 0$, so $0 \leq \mu \leq \epsilon$. It may be supposed that $\mu > 0$. Let k be the largest integer such that $\epsilon_k < \mu$. The matrix

$$\begin{bmatrix} \binom{\epsilon_0}{\epsilon_0} & \cdots & \binom{\epsilon_k}{\epsilon_0} & \binom{\epsilon}{\epsilon_0} \\ \vdots & & \vdots & \vdots \\ \binom{\epsilon_0}{\epsilon_k} & \cdots & \binom{\epsilon_k}{\epsilon_k} & \binom{\epsilon}{\epsilon_r} \\ \binom{\epsilon_0}{\mu} & \cdots & \binom{\epsilon_k}{\mu} & \binom{\epsilon}{\mu} \end{bmatrix}$$

is triangular with $1, \ldots, 1, \binom{\epsilon}{\mu}$ on the main diagonal. Hence the rows are linearly independent over K. Thus $\epsilon_{k+1} \leq \mu$ by Proposition 7.59, and so $\mu = \epsilon_{k+1}$ by the definition of k. □

LEMMA 7.63 *Assume that $p \geq r$, and that the L-order $\epsilon_i = i$ for $0, 1, \ldots, r-1$. If L is non-classical, then ϵ_r is a power of p.*

Proof. From Lemma 7.62, $\epsilon_r \geq p$. Assume, on the contrary, that ϵ_r is not a power of p. By Lemma A.6, there is an integer k such that $p \leq k < \epsilon_r$ and also $\binom{\epsilon_r}{k} \not\equiv 0 \pmod{p}$. Then k is an L-order. But this contradicts that $\epsilon_{r-1} < k < \epsilon_r$. □

In the classical case, the L-Weierstrass points are precisely the L-osculating points. In zero characteristic, every curve Γ is classical.

THEOREM 7.64 *Let N be the number of Weierstrass points of a function field Σ which is neither rational, nor elliptic, nor hyperelliptic. If either $p = 0$ or $p \geq 2g-2$, then*

$$2g + 3 \leq N \leq (g-1)g(g+1). \tag{7.15}$$

Proof. By Corollary 7.61, the canonical curve Γ of Σ is classical, and

$$\sum v_{\mathcal{P}}(R) = (g-1)g(g+1).$$

For a point $P \in \Gamma$, let $j_0 = 0, j_1 = 1, \ldots, j_{g-1}$ be the orders of Γ at P. By Corollary 7.55,

$$v_{\mathcal{P}}(R) = \textstyle\sum_{i=0}^{g-1}(j_i - i).$$

Since $v_{\mathcal{P}}(R) \geq 1$ for any Weierstrass point $\mathcal{P}$, the upper bound follows.

An upper bound for j_i is

$$j_i \leq 2i. \tag{7.16}$$

This comes from Clifford's Theorem 6.79 applied to the fixed-point-free special linear series cut out on Γ by the hyperplanes H through P with intersection multiplicity $I(P, H \cap \Gamma)$ at least j_i. So $j_i - i \leq i$ for $i = 2, \ldots, g-1$ and $0 = j_1 - 1 < 1$. Hence

$$v_{\mathcal{P}}(R) < 1 + 2 + \cdots + g - 1 = \tfrac{1}{2}g(g-1).$$

Since the Weierstrass points are the points $P \in \Gamma$ with $v_{\mathcal{P}}(R) > 0$, it follows that $\sum v_{\mathcal{P}}(R) < \frac{1}{2}g(g-1)N$. Thus $(g-1)g(g+1) < \frac{1}{2}g(g-1)N$. Therefore $2g+3$ is a lower bound for N. □

If L is non-classical, every point is L-osculating. However, it should be noted that non-classical linear series are rare.

THEOREM 7.65 *Let $\epsilon_0 < \epsilon_1 < \cdots < \epsilon_r$ be the L-orders.*

(i) *If $\epsilon_r \geq p^m$, then there exist $z_0, \ldots, z_r \in \Sigma$ such that*

$$z_0^{p^m} x_0 + z_1^{p^m} x_1 + \cdots + z_r^{p^m} x_r = 0. \tag{7.17}$$

(ii) *Let* $\epsilon_r = p^m$ *and assume that (7.17) holds. Then the osculating hyperplane at all but finitely many points* P *is*

$$\mathbf{v}(z_0^{p^m}(P)X_0 + z_1^{p^m}(P)X_1 + \cdots + z_r^{p^m}(P)X_r). \qquad (7.18)$$

(iii) *If* $\epsilon_{r-1} < p^m$, *and (7.17) holds, then* $\epsilon_r = p^k$ *with* $k \geq m$. *Equality holds if and only if some* z_i *is a separable variable of* Σ.

Proof. Since $\det(D_\zeta^{(\epsilon_i)} x_j) \neq 0$, (i) follows from Theorem 5.86 and Lemma 5.84. Choose a point $P \in \Gamma$ or, more precisely, a branch of Γ centred at P such that P is neither a pole of any z_k, nor a zero of all the z_k. For a primitive representation τ of the place corresponding to P,

$$\tau(z_k) = a_{k0} + a_{k1}t + \cdots,$$

for $k = 0, \ldots, r$, and at least one term a_{u0} is distinct from 0. Now (7.17) can also be written as follows:

$$a_{k0}^{p^m} x_0 + \cdots + a_{r0} z_r^{p^m} x_r + (z_0 - a_{k0})^{p^m} x_0 + \cdots + (z_r - a_{kr})^{p^m} x_r = 0.$$

Then

$$\operatorname{ord}_t(a_{k0}^{p^m} x_0(t) + \cdots + a_{r0} z_r^{p^m} x_r(t)) \geq p^m.$$

As $\epsilon_r = p^m$, this shows that

$$H_P = \mathbf{v}(a_{k0}^{p^m} X_0 + \cdots + a_{r0}^{p^m} X_r)$$

is the osculating hyperplane at P. Since, by definition, $z_k(P) = a_{k0}$ for $0 \leq k \leq r$, assertion (ii) holds.

The previous argument also shows that, if $\epsilon_{r-1} < p^{m-1}$ and (7.17) holds, then $\epsilon_r \geq p^m$. From Lemma 7.62, $\epsilon_r = p^k$ with $k \geq m$. If z_i is inseparable, then Lemma 5.38 provides u_i such that $z_i = u_i^p$. If no z_i is separable, then

$$u_0^{p^{m+1}} x_0 + \cdots + u_r^{p^{m+1}} x_r = 0,$$

and hence $\epsilon_r \geq p^{m+1}$.

Conversely, if $\epsilon_r = p^k$ with $k > m$ and $\epsilon_{r-1} < p^m$, then $x_0, \ldots, x_r$ are linearly dependent over Σ_k but independent over Σ_m. Hence, if (7.17) holds, then each $z_i^{p^m}$ is Σ_k. By Lemma 5.84, $z_i = u_i^p$ with $u_i \in \Sigma$ for $0 \leq i \leq r$. So the last part of Theorem 7.65 follows from Lemma 5.38. □

REMARK 7.66 If $\epsilon_r = p^m$ in Theorem 7.65, then the $(r+1)$-ple $(z_0, \ldots, z_r)$ is determined up to a non-zero factor $u \in \Sigma$. This follows from (ii) by the uniqueness of the osculating hyperplane.

EXAMPLE 7.67 Let $p = 5$. Take $\mathcal{F}$ and Γ as in Example 7.50(ii). Then $\epsilon_i = i$ for $i = 0, 1, 2$ but $\epsilon_3 = 5$. The hypotheses of Theorem 7.65 are satisfied for $m = 1$. Since

$$(Y^5 + Y - X^3)(Y^5 + Y + X^3) = (Y^2)^5 + 2YY^5 + Y^2 - XX^5,$$

so (7.17) holds for $z_0 = y^2$, $z_1 = -x$, $z_2 = 2y$, $z_3 = 1$.

7.7 NON-CLASSICAL CURVES AND LINEAR SYSTEMS OF LINES

In this section, classical and non-classical curves with respect to lines are considered.

Let $\mathcal{F} = \mathbf{v}(f(X,Y))$ be an irreducible plane curve. Choose any positive integer s smaller than the degree n of $f(X,Y)$. The linear system of all plane curves of degree s cuts out on $\mathcal{F}$ a g^r_d that has no fixed place. Here $d = ns$ and $r = \frac{1}{2}s(s+3)$. Theorem 7.65 together with Lemma 7.63 has the following corollary.

THEOREM 7.68 *Suppose $ns > p \geq r = \frac{1}{2}s(s+3)$. Then $\mathcal{F}$ is non-classical with respect to the linear series cut out on $\mathcal{F}$ by all curves of a given degree s if and only if there exist $z_0(X,Y), \ldots, z_r(X,Y), h(X,Y) \in K[X,Y]$ and an integer $m \geq 1$ such that*

$$h(X,Y)f(X,Y) = z_0(X,Y)^{p^m} + \cdots + z_r(X,Y)^{p^m}Y^s. \tag{7.19}$$

REMARK 7.69 Rewording Theorem 7.68 in its homogeneous form, (7.19) reads as follows:

$$h(X_0,X_1,X_2)f(X_0,X_1,X_2) = \textstyle\sum_{j=0}^r z_j(X_0,X_1,X_2)^{p^m} X_0^u X_1^v X_2^w;$$

here the $z_j(X_0,X_1,X_2)$ are homogeneous of the same degree. This shows that interchanging X_0 with X_1 or X_2 produces a similar equation. Therefore, in the local study of $\mathcal{F}$, it is enough to consider branches of $\mathcal{F}$ centred at an affine point. For this purpose, the inhomogeneous equation (7.19) is used, as it is more manageable than the homogeneous one.

The simplest case occurs when the linear system consists of all lines, that is, the linear series L_1 cut out on $\mathcal{F} = \mathbf{v}(f(X,Y))$ consists of all divisors

$$A_{\mathbf{c}} = \text{div}\,(c_0 + c_1x + c_2y) + B, \quad \mathbf{c} = (c_0, c_1, c_2) \in \text{PG}(2,K).$$

Without loss of generality, take x to be a separable variable. So $\partial f(X,Y)/\partial Y$ does not vanish. Let $R = \sum n_{\mathcal{P}}\mathcal{P}$ be the ramification divisor.

First, the classical case is considered; that is, the order sequence of L is assumed to be $(0,1,2)$. Then

$$\deg R = 3(2g-2) + 3n. \tag{7.20}$$

Assume $p \neq 2$. Let P be a a non-singular point of $\mathcal{F}$. Then P is an L-Weierstrass point if and only if it is inflexion point of $\mathcal{F}$. If this is the case, $j_2 = I(P, \ell \cap \mathcal{F})$ with ℓ the inflexional tangent to $\mathcal{F}$ at P, and $n_{\mathcal{P}} = j_2 - 2$ where $\mathcal{P}$ is the place arising from the unique branch centred at P. Now let P be a singular point of $\mathcal{F}$, and γ a branch of Γ centred at P. Then γ, namely the branch point P of $\mathcal{F}$, is an L-Weierstrass point if and only if P is non-linear and, when this occurs, $n_{\mathcal{P}} = j_1 + j_2 - 3$ for the place arising from P.

All these results are straightforward consequences of the previous general results, especially Lemmas 7.62 and 7.61. A direct, elementary argument is also possible. Again let $p \neq 2$. Since $\epsilon_2 = 2 \neq p$, the Hessian curve of $\mathcal{F}$ does not vanish; write its equation in inhomogeneous form

$$\frac{\partial f^2}{\partial^2 X}\left(\frac{\partial f}{\partial Y}\right)^2 - 2\frac{\partial f^2}{\partial X \partial Y}\frac{\partial f}{\partial X}\frac{\partial f}{\partial Y} + \frac{\partial f^2}{\partial^2 Y}\left(\frac{\partial f}{\partial X}\right)^2 = 0.$$

Note that

$$\frac{d^2y}{dx^2} = \frac{\partial f^2}{\partial^2 x}\left(\frac{\partial f}{\partial y}\right)^2 - 2\frac{\partial f^2}{\partial x \partial y}\frac{\partial f}{\partial x}\frac{\partial f}{\partial y} + \frac{\partial f^2}{\partial^2 y}\left(\frac{\partial f}{\partial x}\right)^2.$$

Let P be a branch point of $\mathcal{F}$. If $(\bar{x}, \bar{y})$ is a primitive representation of P, put $\bar{x}' = d\bar{x}/dt$ and $\bar{x}''(t) = d^2\bar{x}/dt^2$, and similarly for $\bar{y}$. Then

$$\frac{d^2\bar{y}}{d\bar{x}^2} = \frac{(\bar{x}'\bar{y}'' - \bar{x}''\bar{y}')}{\bar{x}'^3}.$$

By Theorem 7.51 (i), $\bar{x}'\bar{y}'' - \bar{x}''\bar{y}'$ is covariant. Therefore

$$\begin{aligned}\mathrm{ord}_t &\left(\frac{\partial \bar{f}^2}{\partial^2 \bar{x}}\left(\frac{\partial \bar{f}}{\partial \bar{y}}\right)^2 - 2\frac{\partial \bar{f}^2}{\partial \bar{x}\partial \bar{y}}\frac{\partial \bar{f}}{\partial \bar{x}}\frac{\partial \bar{f}}{\partial \bar{y}} + \frac{\partial \bar{f}^2}{\partial^2 \bar{y}}\left(\frac{\partial \bar{f}}{\partial \bar{x}}\right)^2\right)\\ &= \mathrm{ord}_t(\bar{x}'\bar{y}'' - \bar{x}''\bar{y}') + 3\left(\mathrm{ord}_t \frac{\partial \bar{f}}{\partial \bar{y}} - \mathrm{ord}_t \bar{x}'\right).\end{aligned}$$

Since the Hessian has degree $3(n-2)$, Bézout's Theorem 3.14 gives the equation

$$3n(n-2) = \deg R + 3\left(\mathrm{ord}_t \frac{\partial \bar{f}}{\partial \bar{y}} - \mathrm{ord}_t \bar{x}'\right).$$

Bézout's Theorem applied to the polar curve at Y_∞ shows that

$$\sum \mathrm{ord}_t \frac{\partial \bar{f}}{\partial \bar{y}} = n(n-1),$$

while $\sum \mathrm{ord}_t \bar{x}' = 2g - 2$ by definition. Therefore (7.20) holds.

In the non-classical case, the order sequence is $(0, 1, p^m)$ with $m \geq 1$ for p odd, and $m \geq 2$ for $p = 2$. In particular, $\mathcal{F}$ is a non-reflexive curve whose Gauss map has inseparability degree p^m; this is the *General Order of Contact Theorem.* Theorem 7.65 together with Lemma 7.63 shows that

$$z_0^{p^m} + z_1^{p^m} x + z_2^{p^m} y = 0,$$

where $z_0, z_1, z_2 \in K(\mathcal{F})$ and at least one of them is separable. By Remark 7.66, the triple (z_0, z_1, z_2) is determined up to a non-zero factor in $K(\mathcal{F})$. After clearing denominators,

$$h(X,Y)f(X,Y) = z_0(X,Y)^{p^m} + z_1(X,Y)^{p^m} X + z_2(X,Y)^{p^m} Y, \qquad (7.21)$$

with $z_0, z_1, z_2, h \in K[X, Y]$.

PROPOSITION 7.70 *The factor $h(X,Y)$ is not required in (7.21) for non-singular curves; that is, $h(X,Y) = 1$ may be assumed.*

THEOREM 7.71 *If $\deg \mathcal{F} = q$, then $\mathcal{F}$ is a rational strange curve and, up to a change of the coordinate system,*

$$\mathcal{F} = \mathbf{v}(X_2^q - (X_1^{q-1}X_0 + b_2 X_1^{q-2}X_0^2 + \cdots + b_{q-2}X_1^2X_0^{q-2} + X_1X_0^{q-1})),$$

with $b_2, \ldots, b_{q-2} \in K$.

THEOREM 7.72 *If* $\deg \mathcal{F} = q+1$, *then, up to a change of the coordinate system,* $\mathcal{F}$ *is the Hermitian curve* $\mathcal{H}_q$, *the rational nodal cubic or a strange curve*

$$\mathbf{v}(X_0X_2^q + X_0^{q-1} + a_1X_0^qX_1 + a_2X_0^{q-1}X_2^2 + \cdots + a_qX_0X_1^q + X_1^{q+1})$$

with suitable $a_1, \ldots, a_q \in K$.

EXAMPLE 7.73 The DLS curve $\mathcal{F}$ given in Example 5.24 is a non-classical curve with order sequence $(0, 1, 2q_0)$. This follows from Theorem 7.65 for

$$z_0(X,Y) = X^{q_0+1} + Y^{q_0}, \quad z_1(X,Y) = X, \quad z_2(X,Y) = 1.$$

7.8 NON-CLASSICAL CURVES AND LINEAR SYSTEMS OF CONICS

In this section, classical and non-classical curves with respect to conics are considered. So the case $s = 2$ in Theorem 7.68 is examined. The linear system of all conics cuts out on $\mathcal{F}$ a simple, fixed-point-free linear series $L_2 = g_{2n}^5$. This gives rise to the irreducible curve Γ of $\mathrm{PG}(5,K)$ of degree $2n$ which is the image curve of $\mathcal{F}$ under the *Veronese map*, that is, the birational transformation

$$\varphi: \quad (1,x,y) \mapsto (1,x,y,x^2,xy,y^2).$$

Some or all L_2-orders may be calculated from the L_1-orders, where $L_1 = g_n^2$ is cut out on $\mathcal{F}$ by all lines.

To do this, two possibilities are considered separately according as L_1 is classical or non-classical. Take any non-singular point $P \in \Gamma$ which is not an L_1-Weierstrass point of $\mathcal{F}$, and let $\mathcal{P}$ be the corresponding place. If $\mathcal{F}$ is non-classical for L_1, then the $(L_1, \mathcal{P})$-orders are $(j_0 = 0, j_1 = 1, j_2 = p^m)$ with $p^m > 2$. Choose three lines ℓ_0, ℓ_1, ℓ_2 such that $I(P, \ell_i \cap \Gamma) = \epsilon_i$, for $0 \le i \le 2$. There are six degenerate conics whose components are two of these three lines. They provide six different intersection multiplicities with $\mathcal{F}$ at P, namely $0, 1, 2, p^m, p^m+1, 2p^m$. Therefore these integers are the $(L_2, \mathcal{P})$-orders, and also the L_2-orders due to the general choice of P.

From now on, let $\mathcal{F}$ be classical with respect to L_1. The above argument shows that $0, 1, 2, 3, 4$ are $(L_2, \mathcal{P})$-orders, and hence L_2-orders. Since the hypotheses of Theorem 7.68 are satisfied for $p \ge 5$, so ϵ_5 is a power of p. The condition $p \ge 5$ cannot be omitted; see Exercise 3. For $p \ge 5$, Theorem 7.65 shows that $\epsilon_5 \ge p^m$ if and only if there exist $z_0, \ldots, z_5, h \in K[X,Y]$ such that

$$\begin{aligned} &h(X,Y)f(X,Y) \\ &= z_0(X,Y)^{p^m} + z_1(X,Y)^{p^m}X + z_2(X,Y)^{p^m}Y \\ &\quad + z_3(X,Y)^{p^m}X^2 + z_4(X,Y)^{p^m}XY + z_5(X,Y)^{p^m}Y^2. \end{aligned} \tag{7.22}$$

Unlike the case $r = 2$, the non-singularity condition on $\mathcal{F}$ is not enough to ensure that $h(X,Y)$ can be assumed in this equation. An example is given in Exercise 5.

Non-classicality of L_2 gives rise to several interesting properties of the curve. Here, consider the case that $p > 2$, the curve $\mathcal{F}$ is classical with respect to L_1, and the L_2-orders of $\mathcal{F}$ are $0, 1, 2, 3, 4, p^m$. Put $\mathcal{F} = \mathbf{v}(f(X, Y))$, and let $Q = (x, y)$ be a generic point of $\mathcal{F}$. Then (7.22) holds, and at least one of the elements $z_i(x, y)$ in $K(\mathcal{F})$ is a separable variable.

Let $\mathcal{P}$ be a place of $K(\mathcal{F})$ with $(L_2, \mathcal{P})$-orders $0, j_1, j_2$. Choose an index j such that $\text{ord}_{\mathcal{P}}\, z_j(x, y) \le \text{ord}_{\mathcal{P}}\, z_i(x, y)$, and put $v_i = z_i(x, y)/z_j(x, y)$. From (7.22),

$$v_0^{p^m} + v_1^{p^m} x + v_2^{p^m} y + v_3^{p^m} x^2 + v_4^{p^m} xy + v_5^{p^m} y^2 = 0. \qquad (7.23)$$

Let γ be the branch of $\mathcal{F}$ corresponding to $\mathcal{P}$. Without loss of generality, assume that γ is centred at an affine point, say at the point $P = (a, b)$. Given a primitive representation $(x(t), y(t))$ of γ, write the formal power series $v_i(t)$ explicitly in the following form:

$$v_i(t) = \mu_i^{(0)} + \mu_i^{(1)} t + \cdots + \mu_i^{(k)} t^k + \cdots. \qquad (7.24)$$

So

$$\begin{aligned}
&(\mu_0^{(0)})^{p^m} + (\mu_1^{(0)})^{p^m} x(t) + (\mu_2^{(0)})^{p^m} y(t) + (\mu_3^{(0)})^{p^m} x(t)^2 + \\
&(\mu_4^{(0)})^{p^m} x(t)y(t) + (\mu_5^{(0)})^{p^m} y(t)^2 + t^{p^m}[(\mu_0^{(1)})^{p^m} + (\mu_1^{(1)})^{p^m} x(t) + \\
&(\mu_2^{(1)})^{p^m} y(t) + (\mu_3^{(1)})^{p^m} x(t)^2 + (\mu_4^{(1)})^{p^m} x(t)y(t) + (\mu_5^{(1)})^{p^m} y(t)^2] + \cdots + \\
&t^{kp^m}[(\mu_0^{(k)})^{p^m} + (\mu_1^{(k)})^{p^m} x(t) + (\mu_2^{(k)})^{p^m} y(t) + (\mu_3^{(k)})^{p^m} x(t)^2 + \\
&(\mu_4^{(k)})^{p^m} x(t)y(t) + (\mu_5^{(k)})^{p^m} y(t)^2] + \cdots = 0.
\end{aligned}$$

For $k = 0, 1, \ldots$, put

$$\begin{aligned}
s_k(X, Y) = {} & (\mu_0^{(k)})^{p^m} + (\mu_1^{(k)})^{p^m} X + (\mu_2^{(k)})^{p^m} Y \\
& + (\mu_3^{(k)})^{p^m} X^2 + (\mu_4^{(k)})^{p^m} XY + (\mu_5^{(k)})^{p^m} Y^2.
\end{aligned}$$

Then, with $s_k(t) = s_k(x(t), y(t))$, it follows that

$$s_0(t) + t^{p^m} s_1(t) + \cdots + t^{kp^m} s_k(t) + \cdots = 0. \qquad (7.25)$$

As $s_0(x, y)$ contains a non-zero coefficient, $\mathbf{v}(s_0(X, Y))$ is a conic, possibly reducible. Putting $\mu_0^{(i)} = \mu_i$, the following result is obtained.

LEMMA 7.74 (i) *The conic $\mathcal{C}_0 = \mathbf{v}(s_0(X, Y))$, with*

$$s_0(X, Y) = \mu_0^{p^m} + \mu_1^{p^m} X + \mu_2^{p^m} Y + \mu_3^{p^m} X^2 + \mu_4^{p^m} XY + \mu_5^{p^m} Y^2,$$

intersects the branch γ of $\mathcal{F}$ with multiplicity at least p^m.

(ii) *This multiplicity is greater than p^m if and only if $s_1(a, b) = 0$, where*

$$\begin{aligned}
s_1(a, b) = {} & (\mu_0^{(1)})^{p^m} + (\mu_1^{(1)})^{p^m} a + (\mu_2^{(1)})^{p^m} b + (\mu_3^{(1)})^{p^m} a^2 \\
& + (\mu_4^{(1)})^{p^m} ab + (\mu_5^{(1)})^{p^m} b^2.
\end{aligned}$$

(iii) *If (a, b) is not a common zero of the six polynomials $z_i(X, Y)$, then*

$$\begin{aligned}
\mathcal{C}_0 = \mathbf{v}(& z_0(a, b)^{p^m} + z_1(a, b)^{p^m} X + z_2(a, b)^{p^m} Y \\
& + z_3(a, b)^{p^m} X^2 + z_4(a, b)^{p^m} XY + z_5(a, b)^{p^m} Y^2). \qquad (7.26)
\end{aligned}$$

In particular, for all but finitely many points $P = (a, b)$ of $\mathcal{F}$, the osculating conic of $\mathcal{F}$ at P is $\mathcal{C}_0$.

REMARK 7.75 It may happen that a particular branch does not have $\mathcal{C}_0$ as osculating conic, as the following example shows. The Fermat curve $\mathcal{F} = \mathbf{v}(F)$, with

$$F = X^{p-1} + Y^{p-1} + 1,$$

satisfies (i), since

$$XY(X^{p-1} + Y^{p-1} + 1) = Y^p X + X^p Y + XY.$$

Now take a point $P = (0, c)$ with $c^{p-1} = -1$ of $\mathcal{F}$ on the Y-axis. Then P is an inflexion of $\mathcal{F}$; thus the osculating conic of $\mathcal{F}$ at P is degenerate and is $\mathbf{v}((Y-c)^2)$, consisting of the horizontal line through P counted twice. On the other hand, the conic $\mathcal{C}_0$ is also degenerate but is $\mathbf{v}(X(Y-c))$ consisting of the horizontal and vertical lines through P, since $c^p X + XY = X(Y-c)$.

Looking back through this section, or alternatively looking forward to Section 8.7, the conic $\mathcal{C}_0$ plays a central role in the study of non-classical curves with respect to conics. When the symbol $\mathcal{C}_0$ is used it should be understood that $\mathcal{C}_0$ denotes the conic with equation (7.26).

Introduce a new system of reference taking the centre (a, b) of the branch γ to the origin and the tangent ℓ of γ to the $X-$axis.

This change of coordinates from (X, Y) to $(\bar{X}, \bar{Y})$ is given by

$$\begin{array}{rcl} X & = & v_{11}\bar{X} + v_{12}\bar{Y} + a, \\ Y & = & v_{21}\bar{X} + v_{22}\bar{Y} + b, \end{array} \tag{7.27}$$

with $v_{11}v_{22} - v_{12}v_{21} \neq 0$. Correspondingly, let $K(\mathcal{F}) = K(\bar{x}, \bar{y})$ such that

$$\begin{array}{rcl} x & = & v_{11}\bar{x} + v_{12}\bar{y} + a, \\ y & = & v_{21}\bar{x} + v_{22}\bar{y} + b. \end{array} \tag{7.28}$$

Then $\text{ord}_{\mathcal{P}}\, \bar{x} = j_1$, $\text{ord}_{\mathcal{P}}\, \bar{y} = j_2$, Equation (7.22) is invariant under this transformation. To see this, put

$$\begin{array}{lclcl} z_i(X, Y) & = & z_i(v_{11}\bar{X} + v_{12}\bar{Y} + a,\ v_{21}\bar{X} + v_{22}Y + b) & = & \bar{z}_i(\bar{X}, \bar{Y}), \\ f(X, Y) & = & f(v_{11}\bar{X} + v_{12}\bar{Y} + a,\ v_{21}\bar{X} + v_{22}\bar{Y} + b) & = & F(\bar{X}, \bar{Y}), \\ h(X, Y) & = & h(v_{11}\bar{X} + v_{12}\bar{Y} + a,\ v_{21}\bar{X} + v_{22}\bar{Y} + b) & = & H(\bar{X}, \bar{Y}), \\ \multicolumn{5}{l}{a = c^{p^m}, \quad b = d^{p^m}, \quad v_{ij} = n_{ij}{}^{p^m},\ i, j = 1, 2,} \end{array}$$

and write $\bar{z}_i = \bar{z}_i(\bar{X}, \bar{Y})$. Then, with

$$\begin{array}{l} Z_0(\bar{X}, \bar{Y}) = \bar{z}_0 + c\bar{z}_1 + d\bar{z}_2 + c^2\bar{z}_3 + cd\bar{z}_4 + d^2\bar{z}_5, \\ Z_1(\bar{X}, \bar{Y}) = n_{11}\bar{z}_1 + n_{21}\bar{z}_2 + 2cn_{11}\bar{z}_3 + (cn_{11} + dn_{21})\bar{z}_4 + 2dn_{21}\bar{z}_5, \\ Z_2(\bar{X}, \bar{Y}) = n_{12}\bar{z}_1 + n_{22}\bar{z}_2 + 2cn_{12}\bar{z}_3 + (cn_{22} + dn_{12})\bar{z}_4 + 2dn_{22}\bar{z}_5, \\ Z_3(\bar{X}, \bar{Y}) = n_{11}{}^2\bar{z}_3 + n_{11}n_{21}\bar{z}_4 + n_{21}{}^2\bar{z}_5, \\ Z_4(\bar{X}, \bar{Y}) = 2n_{11}n_{12}\bar{z}_3 + (n_{12}n_{21} + n_{11}n_{22})\bar{z}_4 + 2n_{21}n_{22}\bar{z}_5, \\ Z_5(\bar{X}, \bar{Y}) = n_{12}{}^2\bar{z}_3 + n_{12}n_{22}\bar{z}_4 + n_{22}{}^2\bar{z}_5, \end{array}$$

equation (7.22) becomes

$$\begin{array}{l} H(\bar{X}, \bar{Y})F(\bar{X}, \bar{Y}) \\ = Z_0(\bar{X}, \bar{Y})^{p^m} + Z_1(\bar{X}, \bar{Y})^{p^m}\bar{X} + Z_2(\bar{X}, \bar{Y})^{p^m}\bar{Y} \\ \quad + Z_3(\bar{X}, \bar{Y})^{p^m}\bar{X}^2 + Z_4(\bar{X}, \bar{Y})^{p^m}\bar{X}\bar{Y} + Z_5(\bar{X}, \bar{Y})^{p^m}\bar{Y}^2. \end{array} \tag{7.29}$$

Since the relationship between Z_i and z_i is linear and invertible,

$$\min\{\mathrm{ord}_{\mathcal{P}} z_i(x,y))\} = \min\{\mathrm{ord}_{\mathcal{P}} Z_i(\bar{x},\bar{y})\}. \tag{7.30}$$

Also, if $w_0, \ldots, w_5 \in K[X,Y]$, put

$$\Delta(w_0,\ldots,w_5) = \begin{bmatrix} w_0 & \frac{1}{2}w_1 & \frac{1}{2}w_2 \\ \frac{1}{2}w_1 & w_3 & \frac{1}{2}w_4 \\ \frac{1}{2}w_2 & \frac{1}{2}w_4 & w_5 \end{bmatrix}.$$

With this notation,

$$\mathrm{ord}_{\mathcal{P}}\, \Delta(z_0(x,y),\ldots,z_5(x,y)) = \mathrm{ord}_{\mathcal{P}}\, \Delta(Z_0(\bar{x},\bar{y}),\ldots,Z_5(\bar{x},\bar{y})). \tag{7.31}$$

Put $W_i = Z_i(\bar{X},\bar{Y})/Z_j(\bar{X},\bar{Y})$ and $V_i = W_i(\bar{x},\bar{y})$, where j is chosen such that

$$\mathrm{ord}_{\mathcal{P}}\, Z_j(\bar{x},\bar{y}) = \min\{\mathrm{ord}_{\mathcal{P}}\, Z_i(\bar{x},\bar{y})\}.$$

Then, in the new reference system $(\bar{X},\bar{Y})$, the conic

$$\begin{aligned} \mathcal{C}_0 = \mathbf{v}(&W_0(a,b)^{p^m} + W_1(a,b)^{p^m}\bar{X} + W_2(a,b)^{p^m}\bar{Y} \\ &+ W_3(a,b)^{p^m}\bar{X}^2 + W_4(a,b)^{p^m}\bar{X}\bar{Y} + W_5(a,b)^{p^m}\bar{Y}^2), \end{aligned} \tag{7.32}$$

which shows the covariance of $\mathcal{C}_0$. From (7.29),

$$V_0^{p^m} + V_1^{p^m}\bar{x} + V_2^{p^m}\bar{y} + V_3^{p^m}\bar{x}^2 + V_4^{p^m}\bar{x}\bar{y} + V_5^{p^m}\bar{y}^2 = 0.$$

Let $k_i = \mathrm{ord}_{\mathcal{P}}\, V_i$. Then the left-hand side is the sum of six elements from $K(\mathcal{F})$ whose orders are as follows:

$$\begin{array}{lllll} \mathrm{ord}_{\mathcal{P}}\, V_0^{p^m} & = & k_0p^m, & \mathrm{ord}_{\mathcal{P}}\, V_1^{p^m}\bar{x} & = & k_1p^m + j_1, \\ \mathrm{ord}_{\mathcal{P}}\, V_2^{p^m}\bar{y} & = & k_2p^m + j_2, & \mathrm{ord}_{\mathcal{P}}\, V_3^{p^m}\bar{x}^2 & = & k_3p^m + 2j_1, \\ \mathrm{ord}_{\mathcal{P}}\, V_4^{p^m}\bar{x}\bar{y} & = & k_4p^m + j_1 + j_2, & \mathrm{ord}_{\mathcal{P}}\, V_5^{p^m}\bar{y}^2 & = & k_5p^m + 2j_2. \end{array}$$

At least two of these orders must be equal, and they are less than or equal to the remaining four. Hence one of the following relations must hold:

(a) $(k_0 - k_1)p^m = j_1$, (b) $(k_1 - k_3)p^m = j_1$,
(c) $(k_2 - k_4)p^m = j_2$, (d) $(k_0 - k_3)p^m = 2j_1$,
(e) $(k_0 - k_2)p^m = 2j_2$, (f) $(k_2 - k_5)p^m = j_2$,
(g) $(k_0 - k_4)p^m = j_1 + j_2$, (h) $(k_0 - k_5)p^v = 2j_2$,
(i) $(k_1 - k_2)p^m = j_2 - j_1$, (j) $(k_3 - k_4)p^m = j_2 - j_1$,
(k) $(k_1 - k_4)p^m = j_2$, (l) $(k_4 - k_5)p^m = j_2 - j_1$,
(m) $(k_1 - k_5)p^m = 2j_2 - j_1$, (n) $(k_2 - k_3)p^v = 2j_1 - j_2$,
(o) $(k_3 - k_5)p^m = 2(j_2 - j_1)$.

If $n < p^m$, then $j_2 < p^m$. This leaves just three possibilities; namely, (n) $k_2 = k_3$; (m) $k_1 = k_5 + 1$; (g) $k_0 = k_4 + 1$. In particular, the following result is obtained.

THEOREM 7.76 *Let $\mathcal{F} = \mathbf{v}(f(X,Y))$ be an irreducible curve of degree n such that the following conditions are satisfied:*

(a) $3 \le n < p^m$;

(b) *$f(X,Y)$ satisfies (7.22);*

(c) *γ is a branch of $\mathcal{F}$ whose centre is the origin and whose tangent is the X-axis;*

(d) *the $(L_1, \mathcal{P})$-orders of a branch point P of $\mathcal{F}$ are $0, j_1, j_2$.*

Then one of the following holds:

(i) $$j_2 = 2j_1, \text{ and } \mathcal{C}_0 = \mathbf{v}(\bar{Y} + c\bar{X}^2),\ c \neq 0; \tag{7.33}$$

(ii) $$j_2 = \tfrac{1}{2}(j_1 + p^m), \text{ and } \mathcal{C}_0 = \mathbf{v}(\bar{Y}^2); \tag{7.34}$$

(iii) $$j_2 = p^m - j_1, \text{ and } \mathcal{C}_0 = \mathbf{v}(\bar{X}\bar{Y}). \tag{7.35}$$

Let $G(\bar{X}, \bar{Y}) = H(\bar{X}, \bar{Y})F(\bar{X}, \bar{Y}))$. Then

$$G_{\bar{X}} = H_{\bar{X}}F + HF_{\bar{X}} = Z_1^{p^m} + 2Z_3^{p^m}\bar{X} + z_4^{p^m}\bar{Y}, \tag{7.36}$$

$$G_{\bar{Y}} = H_{\bar{Y}}F + FF_{\bar{Y}} = Z_2^{p^m} + Z_4^{p^m}\bar{X} + 2Z_5^{p^m}\bar{Y}. \tag{7.37}$$

Note that both $F_{\bar{X}}$ and $F_{\bar{Y}}$ are not zero polynomials, otherwise $\mathcal{F}$ would be non-classical with order sequence $(0, 1, p^v)$ where $v \leq m$. The Hessian $\mathbf{v}(G^*(X, Y))$ of the plane curve $\mathcal{G} = \mathbf{v}(G(X, Y))$, which may be reducible, is given by

$$\begin{aligned}
G^* &= G_{\bar{X}\bar{X}}(G_{\bar{Y}})^2 - 2G_{\bar{X}\bar{Y}}G_{\bar{X}}G_{\bar{Y}} + G_{\bar{X}\bar{Y}}(G_{\bar{X}})^2\\
&= 2Z_3^{p^m}(Z_2^{p^m} + Z_4^{p^m}\bar{X} + 2Z_5^{p^m}\bar{Y})^2 - 2Z_4^{p^m}(Z_2^{p^m} + Z_4^{p^m}\bar{X} + 2Z_5^{p^m}\bar{Y})\\
&\qquad (Z_1^{p^m} + 2Z_3^{p^m}\bar{X} + z_4^{p^m}\bar{Y}) + 2Z_5^{p^m}(Z_1^{p^m} + 2Z_3^{p^m}\bar{X} + Z_4^{p^m}\bar{Y})^2\\
&= 2\{(Z_3Z_2^2 - Z_1Z_2Z_4 + Z_5Z_1^2)^{p^m}\\
&\qquad + (4Z_3Z_5 - Z_4^2)^{p^m}(Z_1^{p^m}\bar{X} + Z_2^{p^m}\bar{Y} + Z_3^{p^m}\bar{X}^2 + Z_4^{p^m}\bar{X}\bar{Y} + Z_5^{p^m}\bar{Y}^2)\}\\
&\equiv 2\{(Z_3Z_2^2 - Z_1Z_2Z_4 + Z_5Z_1^2)^{p^m} + (4Z_3Z_5 - Z_4^2)^{p^m}Z_0^{p^m}\}\\
&\qquad\qquad\qquad\qquad\qquad\qquad\qquad\qquad (\text{mod } F(\bar{X}, \bar{Y}))\\
&\equiv -\Delta(Z_0, \ldots, Z_5)^{p^m} \quad (\text{mod } F(\bar{X}, \bar{Y})).
\end{aligned}$$

On the other hand,

$$\begin{aligned}
&G_{\bar{X}\bar{X}}(G_{\bar{Y}})^2 - 2G_{\bar{X}\bar{Y}}G_{\bar{X}}G_{\bar{Y}} + G_{\bar{Y}\bar{Y}}(G_{\bar{X}})^2\\
&\equiv H^3[F_{\bar{X}\bar{X}}(F_{\bar{Y}})^2 - 2F_{\bar{X}\bar{Y}}F_{\bar{X}}F_{\bar{Y}} + F_{\bar{Y}\bar{Y}}(F_{\bar{X}})^2]\ (\text{mod } F(\bar{X}, \bar{Y})).
\end{aligned}$$

It follows that

$$\begin{aligned}
&H^3[F_{\bar{X}\bar{X}}(F_{\bar{Y}})^2 - 2F_{\bar{X}\bar{Y}}F_{\bar{X}}F_{\bar{Y}} + F_{\bar{Y}\bar{Y}}(F_{\bar{X}})^2]\\
&\qquad\qquad \equiv -\Delta(Z_0, \ldots, Z_5)^{p^m}\ (\text{mod } F(\bar{X}, \bar{Y})).
\end{aligned} \tag{7.38}$$

It may be noted that the left-hand side, apart from H^3, gives the Hessian of $\mathcal{F}$. Since $F_{\bar{Y}}$ is not the zero polynomial, $\bar{x}$ is a separable variable of $K(\mathcal{F})$. Hence

$$\frac{d^2\bar{y}}{d\bar{x}^2} = -\frac{F_{\bar{X}\bar{X}}F_{\bar{Y}}^2 - 2F_{\bar{X}\bar{Y}}F_{\bar{X}}F_{\bar{Y}} + F_{\bar{Y}\bar{Y}}F_{\bar{X}}^2}{F_{\bar{Y}}^3}, \tag{7.39}$$

where the polynomials on the right-hand side are evaluated at $(\bar{x}, \bar{y})$. Therefore

$$\frac{d^2\bar{y}}{d\bar{x}^2} = \Delta(Z_0(\bar{x}, \bar{y}), \ldots, Z_5(\bar{x}, \bar{y}))^{p^m} / (H(\bar{x}, \bar{y})F_{\bar{Y}}(\bar{x}, \bar{y}))^3. \tag{7.40}$$

As $\mathcal{F}$ is classical with respect to L_1, this shows that

$$\Delta(Z_0(\bar{x}, \bar{y}), \ldots, Z_5(\bar{x}, \bar{y})) \neq 0.$$

If R is the ramification divisor with respect to the linear series cut out on $\mathcal{F}$ by lines, the following result is obtained.

LEMMA 7.77

$$v_{\mathcal{P}}(R) = p^m \mathrm{ord}_{\mathcal{P}}\, \Delta(Z_0(\bar{x},\bar{y}),\ldots,Z_5(\bar{x},\bar{y})) + 3\,\mathrm{ord}_{\mathcal{P}}\, d\bar{x}$$
$$-3[\mathrm{ord}_{\mathcal{P}}\,(Z_2(\bar{x},\bar{y})^{p^m} + Z_3(\bar{x},\bar{y})^{p^m}\bar{x} + 2Z_5(\bar{x},\bar{y})^{p^m}\bar{y})]. \quad (7.41)$$

Now a geometric interpretation of (7.41) is given. For this purpose, the following conditions are assumed satisfied:

(i) $z_0(x,y),\ldots,z_5(x,y)$ are linearly independent over K;

(ii) the linear system,

$$\lambda_0 z_0(X,Y) + \cdots + \lambda_5 z_5(X,Y),$$

cuts out a simple linear series L_2'.

It may be noted that both conditions are satisfied when $n < p^m$, but the proof is postponed to the end of the section. As usual, it is also assumed that L_2' is fixed-point-free.

Let $\mathcal{Z}$ be the image curve of $\mathcal{F}$ under the rational transformation

$$\varphi' : (1,x,y) \mapsto (z_0(x,y), z_1(x,y), z_2(x,y), z_3(x,y), z_4(x.y), z_5(x,y)).$$

By condition (ii), $\mathcal{Z}$ is an irreducible curve of $\mathrm{PG}(5,K)$, birationally equivalent to $\mathcal{F}$ over K. It may be noted that $\mathcal{Z}$ can be viewed as the dual curve of Γ. In fact, $\mathcal{Z}$ is birationally equivalent to the dual curve of Γ as the Gauss map associated to Γ is the product of φ' by the Frobenius transformation of $\mathrm{PG}(5,K)$. For more details on dual curves, see Section 7.9.

As in Section 7.6, points of $\mathcal{Z}$ are considered as branch points, so that the term 'point P of $\mathcal{Z}$' stands for a branch γ of $\mathcal{Z}$ associated to the place $\mathcal{P}$. If γ is centred at A and Φ is a hypersurface of $\mathrm{PG}(5,K)$, then $I(P,\mathcal{Z}\cap\Phi)$ is used to indicate $I(A,\gamma\cap\Phi)$.

Apart from $\mathcal{Z}$, it is appropriate to consider the cubic hypersurface

$$\Phi = \mathbf{v}(\det(\Delta(X_0,\ldots,X_5))).$$

This depends on the fact that

$$I(P,\mathcal{Z}\cap\Phi) = \mathrm{ord}_{\mathcal{P}}\ \det\Delta(z_0(x,y),\ldots,z_5(x,y) - \min\mathrm{ord}_{\mathcal{P}}(z_i(x,y)).$$

In particular, $\mathcal{Z}$ is not contained in Φ, and Bézout's Theorem 3.14 gives that

$$3\deg\mathcal{Z} = \sum I(P,\mathcal{Z}\cap\Phi). \quad (7.42)$$

From (7.30) and (7.31), the computation of the orders on the right-hand side can be carried out in the new reference system. Let

$$\mathrm{ord}_{\mathcal{P}}\ Z_j(\bar{x},\bar{y}) = \min\mathrm{ord}_{\mathcal{P}}(z_i(x,y));$$

then

$$v_{\mathcal{P}}(R) = p^m\, I(P,\mathcal{Z}\cap\Phi) + 3\,\mathrm{ord}_{\mathcal{P}}\, d\bar{x} - 3\mathrm{ord}_{\mathcal{P}}\, B(\bar{x},\bar{y}) \quad (7.43)$$

with

$$B(\bar{x},\bar{y}) = \left(\frac{Z_2(\bar{x},\bar{y})}{Z_j(\bar{x},\bar{y})}\right)^{p^m} + \left(\frac{Z_3(\bar{x},\bar{y})}{Z_j(\bar{x},\bar{y})}\right)^{p^m}\bar{x} + 2\left(\frac{Z_5(\bar{x},\bar{y})}{Z_j(\bar{x},\bar{y})}\right)^{p^m}\bar{y}.$$

Orders on the right-hand side can be directly computed for each of the 15 possibilities listed above. As in Theorem 7.76, only the case $n < p^m$ is treated.

Summing (7.43) gives the result

$$p^m \deg \mathcal{Z} = c - n + \sum \deg B, \tag{7.44}$$

where c denotes the class of $\mathcal{F}$.

If (7.33) holds, then $\operatorname{ord}_{\mathcal{P}} Z_2(\bar{x}, \bar{y}) = \operatorname{ord}_{\mathcal{P}} Z_3(\bar{x}, \bar{y})$ is strictly less than the orders of the four other $Z_i(\bar{x}, \bar{y})$. Therefore

$$B(\bar{x}, \bar{y}) = 1 + [Z_4(\bar{x}, \bar{y})/Z_2(\bar{x}, \bar{y})]^{p^m} + [Z_5(\bar{x}, \bar{y})/Z_2(\bar{x}, \bar{y})]^{p^m} \bar{x}.$$

Hence $\operatorname{ord}_{\mathcal{P}} B(\bar{x}, \bar{y}) = 0$.

Assume that (7.34) holds. This time, $\operatorname{ord}_{\mathcal{P}} Z_5(\bar{x}, \bar{y})$ is strictly less than the orders of the other $\operatorname{ord}_{\mathcal{P}} Z_i(\bar{x}, \bar{y})$. Therefore

$$B(\bar{x}, \bar{y}) = [Z_2(\bar{x}, \bar{y})/Z_5(\bar{x}, \bar{y})]^{p^m} + [Z_4(\bar{x}, \bar{y})/Z_5(\bar{x}, \bar{y})]^{p^m} \bar{x} + 2\bar{y}.$$

Since $\operatorname{ord}_{\mathcal{P}} \bar{y} < p^m$, so $\operatorname{ord}_{\mathcal{P}} B(\bar{x}, \bar{y}) = j_2$.

Finally, if (7.35) holds, then $\operatorname{ord}_{\mathcal{P}} Z_4(\bar{x}, \bar{y})$ is strictly less than the orders of the other $\operatorname{ord}_{\mathcal{P}} Z_i(\bar{x}, \bar{y})$. Therefore

$$B(\bar{x}, \bar{y}) = [Z_2(\bar{x}, \bar{y})/Z_4(\bar{x}, \bar{y})]^{p^m} + \bar{X} + 2[Z_5(\bar{x}, \bar{y})/Z_4(\bar{x}, \bar{y})]^{p^m} \bar{y}.$$

Since $\operatorname{ord}_{\mathcal{P}} \bar{x} < p^m$, so $\operatorname{ord}_{\mathcal{P}} B(\bar{x}, \bar{y}) = j_1$.

As $\operatorname{ord}_{\mathcal{P}} d\bar{x} \geq j_1 - 1$, the following result is obtained.

LEMMA 7.78 *Let $n < p^m$. Then*

$$v_{\mathcal{P}}(R) \geq p^m I(P, \mathcal{Z} \cap \Phi) + \begin{cases} 3(j_1 - 1) & \text{when (7.33) holds;} \\ -3(j_2 - j_1 + 1) & \text{when (7.34) holds;} \\ -3 & \text{when (7.35) holds.} \end{cases}$$

Since $v_{\mathcal{P}}(R) \geq j_1 + j_2 - 3$, this yields the following inequality:

$$4v_{\mathcal{P}}(R) \geq p^m I(P, \mathcal{Z} \cap \Phi) - \begin{cases} 6 & \text{when } j_1 = 1 \text{ and (7.34) holds;} \\ 0 & \text{otherwise.} \end{cases} \tag{7.45}$$

If $j_1 < p$, then $\operatorname{ord}_{\mathcal{P}} d\bar{x} = j_1 - 1$, and $(j_1 - 0)(j_2 - 0)(j_2 - j_1) \not\equiv 0 \pmod{p}$ in each of the three cases in Theorem 7.76. Hence, by Corollary 7.60, $v_{\mathcal{P}} = j_1 + j_2 - 3$, and equality holds in Lemma 7.78. Therefore the following result holds.

LEMMA 7.79 *If $j_1 < p$, then*

$$I(P, \mathcal{Z} \cap \Phi) = \begin{cases} 0 & \text{when (7.33) holds;} \\ 2 & \text{when (7.34) holds;} \\ 1 & \text{when (7.35) holds.} \end{cases}$$

So far, places corresponding to branches centred at affine points have been considered. However, these results hold true when the centre is an infinite point, as in Remark 7.69.

Now summing the inequality (7.45) as P ranges over all branch points of $\mathcal{F}$ centred at an affine point,

$$12(2g - 2) + 12n \geq 3p^m \deg \mathcal{Z} - 6N,$$

where N denotes the number of linear branches P satisfying (7.34). Since

$$v_{\mathcal{P}}(R) \geq j_1 + j_2 - 3 = \tfrac{1}{2}(p^m - 3),$$

for any P counted in N, an upper bound on N is

$$N \leq 2\,\frac{3(2g-2)+3n}{p^m - 3.}$$

Therefore the following theorem is established.

THEOREM 7.80 *Let $3 \leq n < p^m$. Then*

$$\deg \mathcal{Z} \leq \frac{4}{p^m - 3}[(2g-2)+n]. \tag{7.46}$$

Equality holds if and only if $\mathcal{F}$ is non-singular. In this case, (7.46) becomes

$$\deg \mathcal{Z} = \frac{4n(n-2)}{p^m - 3}. \tag{7.47}$$

EXAMPLE 7.81 Let q be an even power of the characteristic p of K. When the degree $n = \frac{1}{2}(\sqrt{q}+1)$, the Fermat curve $\mathcal{F} = \mathbf{v}(X^n + Y^n + 1)$ is a non-classical curve with respect to L_2, as the L_2-orders are $(0,1,2,3,4,\sqrt{q})$. In fact, (7.22) holds for

$$h = (X^n + Y^n - 1)(X^n - Y^n + 1)(X^n + Y^n - 1),$$

$$z_0 = 1,\ z_1 = -2X,\ z_2 = 2Y,\ z_3 = X^2,\ z_4 = -2XY,\ z_5 = Y^2.$$

Since the six curves $\mathbf{v}(Z_i(X,Y))$ have no common points, $\deg \mathcal{Z} = 2n$. On the other hand, $\mathcal{F}$ has exactly $3n$ inflexion points, and its branch points satisfying (7.34) are precisely its inflexion points.

As mentioned before, conditions (i) and (ii) are satisfied when $n < p^m$. Now a proof is given by using some of the above results. To prove (i), assume on the contrary that a non-trivial linear combination of $z_i(x,y)$ is zero; that is,

$$\lambda_0 z_0(x,y) + \cdots + \lambda_5 z_5(x,y) = 0, \tag{7.48}$$

with $\lambda_i \in K$ not all zero. From Bézout's Theorem 3.14, Φ and $\mathcal{Z}$ have non-trivial intersection. Choose a branch point P of $\mathcal{Z}$ whose centre $(z_0(P),\ldots,z_5(P))$ is a point of Φ. Suppose, without loss of generality, that P arises from the branch γ under φ.

From (7.48) it follows that neither (7.34) nor (7.35) can actually occur in the Theorem 7.76. To show this, note first that since the relationship between $Z_i(\bar{x},\bar{y})$ and $z_i(x,y)$ are linear and invertible, (7.48) implies that

$$\rho_0 Z_0(\bar{x},\bar{y}) + \cdots + \rho_5 Z_5(\bar{x},\bar{y}) = 0, \tag{7.49}$$

with $\rho_i \in K$ not all zeros. Therefore at least two of the six numbers $\mathrm{ord}_{\mathcal{P}}\, Z_i(\bar{x},\bar{y})$ are equal and less than the remaining four. This means that just one case, namely (n), can occur from the previous long list (a), ..., (o). Hence (7.33) is the only possibility in Theorem 7.76. In particular, $\mathcal{C}_0$ is a non-degenerate conic, $j=2$ or $j=3$, and $\mathrm{ord}_{\mathcal{P}}\, \det \Delta(Z_0(\bar{x},\bar{y}),\ldots,Z_5(\bar{x},\bar{y})) - \mathrm{ord}_{\mathcal{P}}\, Z_j(\bar{x},\bar{y}) = 0$. But then $I(P, \mathcal{Z} \cap \Phi) = 0$, a contradiction.

If L_2' is composed of an involution of order r, then $n > rp^m$, which is impossible for $n < p^m$. Thus (ii) is satisfied when $b < p^m$.

Hence Theorem 7.46 holds for any non-classical curve $\mathcal{F}$ with order sequence $(0,1,2,3,4,p^m)$ with respect to the linear series L_2 cut out on $\mathcal{F}$ by all conics.

7.9 DUAL CURVES OF SPACE CURVES

In Section 5.11, the dual curve of a plane curve is investigated. Now duality for space curves is considered.

Let Γ be a non-degenerate irreducible curve of $\mathrm{PG}(r, K)$.

DEFINITION 7.82 *The dual curve of Γ is the irreducible curve Γ' of $\mathrm{PG}(r, K)$ containing all but finitely many points $P' = (b_0, \ldots, b_r)$ such that the hyperplane $\mathbf{v}(b_0X_0 + \cdots + b_rX_r)$ in $\mathrm{PG}(r, K)$ is an osculating hyperplane of Γ.*

Corollary 7.49 shows how to obtain the dual curve of a space curve as its image under a rational transformation. For every $i = 0, \ldots, r$, delete the last row and the i-th column in the Wronskian determinant (7.12), and denote by F_i the resulting $(r-1) \times (r-1)$ determinant multiplied by $(-1)^{r+i}$. Then the rational transformation ω_C given by the formula

$$\omega_C(x_0', x_1', \ldots, x_r') = (F_0, F_1, \ldots, F_r)$$

is the *Gauss map* associated to Γ, and the curve Γ' given by the point

$$P' = (F_0, F_1, \ldots, F_r)$$

is the *dual curve* of Γ. Note that, from the definition of the F_i,

$$\begin{aligned} F_0x_0 + \cdots + F_rx_r &= 0, \\ F_0D_\zeta^{(1)}x_0 + \cdots + F_rD_\zeta^{(1)}x_r &= 0. \end{aligned}$$

The geometric meaning of these equations is that the osculating hyperplane contains the tangent line. As for plane curves, the *bidual* Γ'' of Γ is the dual of Γ'. Also, ω' stands for the Gauss map associated to Γ'.

The argument used in Section 5.11 can be adapted to the study of duality of space curves. However, the present situation is much more involved in positive characteristic and no details are given here. Nevertheless, the main results are stated together with illustrative examples. Let $\epsilon_0 < \epsilon_1 < \cdots < \epsilon_r$ be the orders of a non-degenerate irreducible curve Γ of $\mathrm{PG}(r, K)$. Denote by $\deg_i \omega$ the inseparability degree of the Gauss map.

THEOREM 7.83 *The highest power of p dividing ϵ_r is equal to* $\deg_i \omega$.

THEOREM 7.84 *Suppose that the dual curve Γ' of Γ is also non-degenerate and let $\epsilon_0' < \epsilon_1' < \cdots \epsilon_r'$ be its orders. For a rational transformation $\delta : \Gamma \to \Gamma'$ whose inseparability degree is equal to $q \deg_i \omega$ with $q \geq 1$, the following conditions are equivalent:*

(i) $\Gamma = \Gamma''$ *and* $\delta = \omega' \circ \omega$;

(ii) *q divides ϵ_r' and, for a general point $P \in \Gamma$,*

$$L_{\omega(P)}^{(d)}(\Gamma) \subset L_P^{(r-1)}(\Gamma),$$

where the integer d is defined by $\epsilon_d < \epsilon_r'/q \leq \epsilon_{d+1}$.

Also, if this is the case, then $\epsilon_{d+1} = \epsilon'_r/q$, *and* $\epsilon_{d+1} \not\equiv 0 \pmod p$.

EXAMPLE 7.85 Let the irreducible plane curve $\mathcal{F} = \mathbf{v}(f(X,Y))$ be non-classical with orders $0, 1, q'$, where $q' = p^m > 2$. Then (7.21) holds, and hence the tangent to $\mathcal{F}$ at a general point Q is

$$\ell_Q = \mathbf{v}(z_0(Q)^{q'} + z_1(Q)^{q'}X + z_2(Q)^{q'}Y). \tag{7.50}$$

Suppose that $s < q'$ and let g_d^r be the linear series introduced before Theorem 7.68. The corresponding rational transformation τ is given by $x'_{ij} = x^i y^j$ with $0 \le i+j \le s$, and it defines the irreducible curve $\Gamma = \tau(\mathcal{F})$ in $\mathrm{PG}(r,K)$. The orders of Γ are all integers in the set $\{u + vq' \mid u + v \le s\}$. In particular, $\epsilon_r = sq'$, and the equation of the osculating hyperplane of Γ at the general point $\delta(Q)$ is obtained formally raising the linear polynomial in (7.50) to the s-th power and replacing X^iY^j with X_{ij}. Here the X_{ij} with $0 \le i+j \le s$ denote the non-homogeneous coordinates in $\mathrm{PG}(r,K)$.

Let δ be the qq'-th Frobenius transformation. If ℓ_Q contains $\delta(Q) = Q^{qq'}$ for a general point Q, then

$$L_{\tau(P)}^{(s-1)}(\Gamma) \subset L_P^{(r-1)}(\Gamma), \quad L_{\tau(P)}^{(s)}(\Gamma) \not\subset L_P^{(r-1)}(\Gamma),$$

for a generic point P. Additionally, Γ' is not degenerate provided that $p > s$. Also, Γ' coincides with Γ up to a linear collineation associated to a diagonal matrix. In particular, Γ and Γ' have the same degrees.

THEOREM 7.86 *Suppose the following conditions hold:*

(a) $\epsilon_0 < \epsilon_1 < \cdots < \epsilon_r$ *are the orders of a non-degenerate irreducible curve* Γ *of* $\mathrm{PG}(r,K)$;

(b) *the dual curve* Γ' *of* Γ *is non-degenerate*;

(c) $\epsilon'_0 < \epsilon'_1 < \cdots \epsilon'_r$ *are the orders of* Γ';

(d) $\epsilon_r \cdot \epsilon'_r \not\equiv 0 \pmod p$.

Then $\Gamma'' = \Gamma$ *if and only if* $\epsilon'_r \le \epsilon_r$. *If this is the case, then* $\omega' \circ \omega$ *is the identity map and* $\epsilon_r = \epsilon'_r$.

EXAMPLE 7.87 Let $q > 3$. The rational curve Γ given by the point

$$Q = (1, x, x^2, x^3, x^q + x^{2q+3}, x^{q+1} + x^{2q+2})$$

is a non-degenerate curve in $\mathrm{PG}(5,K)$ with order sequence $(0,1,2,3,q,q+1)$. The dual curve Γ' is also non-degenerate and its orders are $(0,1,q,q+1,2q,2q+1)$. This gives an example of a curve with $\epsilon_r\epsilon'_r \not\equiv 0 \pmod p$. Thus, by Theorem 7.86, $\Gamma \ne \Gamma''$.

Another generalisation of the concept of dual curve of a plane curve is the Gaussian dual, which is an irreducible curve Δ in $\mathrm{PG}(r',K)$ with $r' = \binom{r+1}{2} - 1$, obtained from Γ by the following Gauss map, via the classical Plücker embedding of the lines of $\mathrm{PG}(r,K)$ in $\mathrm{PG}(r',K)$.

Let Γ be given by the point $Q = (1, x_1, \dots, x_r)$, where $x = x_1$ is a separable variable of $K(\Gamma)$. The 2×2 minors of the matrix

$$\begin{bmatrix} 1 & x & x_2 & \cdots & x_r \\ 0 & 1 & D_x^{(1)} x_2 & \cdots & D_x^{(1)} x_r \end{bmatrix}$$

can be arranged in a sequence of $r' + 1$ elements, say, $y_0 = 1, y_1, \dots, y_{r'}$. Then the irreducible curve Δ given by the point

$$R = (1, y_1, \dots, y_{r'})$$

is the *Gaussian dual* of Γ. Also, the points $Q = (1, x, \dots, x_r)$ and R define a rational transformation ω of $K(\Gamma)$. Put $\Sigma = K(Q) = K(\Gamma)$ and $\Sigma' = K(R)$.

The first of the above minors is equal to 1, while the others are

$$\begin{aligned} \alpha_i &= D_x^{(1)} x_i, \quad 2 \le i \le r; \\ \beta_i &= x_i - x\alpha_i, \quad 2 \le i \le r; \\ \gamma_{ij} &= x_i \alpha_j - x_j \alpha_i, \quad 2 \le i < j \le r. \end{aligned}$$

Since

$$\gamma_{ij} = x_i \alpha_j - x_j \alpha_i = (\beta_i + x\alpha_i)\alpha_j - (\beta + xa_j)\alpha_i = \beta_i \alpha_j - \beta_j \alpha_i,$$

it follows that

$$\Sigma' = K(\alpha_2, \dots, \alpha_r, \beta_2, \dots, \beta_r).$$

On the other hand, since $\Sigma = K(x, x_2, \dots, x_r)$ and $x_i = x\alpha_i + \beta_i$ for every $2 \le i \le r$, so

$$\Sigma = \Sigma'(x) = K(\alpha_2, \dots, \alpha_r, \beta_2, \dots, \beta_r, x).$$

The following result is a generalisation of Theorems 5.90 and 5.91 to space curves.

THEOREM 7.88 *Let $\epsilon_0 = 0, \epsilon_1 = 1, \epsilon_2, \dots, \epsilon_r$ be the orders of Γ. Then $\epsilon_2 = q'$ is a power of p if and only Σ/Σ' is inseparable with inseparability degree q'.*

Proof. Let q' be a power of p, and let d be any power of p such that $1 \le d < q'$. By Lemmas 5.83 (iii), 5.84, and Theorem 5.37, It suffices to show that the following conditions are equivalent:

(i) for all $2 \le i \le r$ and $i < j \le r$,

$$D_x^{(d)} \alpha_i = D_x^{(d)} \beta_i = D_x^{(d)} \gamma_{ij} = 0; \tag{7.51}$$

(ii) for all $2 \le i \le r$,

$$D_x^{(1)} D_x^{(1)} x_i = D_x^{(d)} x_i = 0. \tag{7.52}$$

Now

$$D_x^{(d)} \beta_i = D_x^{(d)} x_i - \textstyle\sum_{k=0}^{d} D_x^{(d-k)} \alpha_i D_x^{(k)} x = \begin{cases} -xD_x^{(1)} \alpha_i & \text{if } d = 1, \\ D_x^{(d)} x_i - xD_x^{(d)} \alpha_i & \text{if } d > 1. \end{cases} \tag{7.53}$$

On the other hand,

$$D_x^{(d)}\gamma_{ij} = \sum_{k=0}^{d} (D_x^{(k)}x_i D_x^{(d-k)}\alpha_j - D_x^{(k)}x_j D_x^{(d-k)}\alpha_i). \quad (7.54)$$

From (7.53) and (7.54), if $D_x^{(d)}\alpha_i = 0$ for $2 \le i \le r$ and $1 \le d \le q'$, then

$$D_x^{(d)}\gamma_{ij} = \alpha_j D_x^{(d)}x_i - \alpha_i D_x^{(d)}x_j; \quad (7.55)$$

$$D_x^{(d)}\beta_i = \begin{cases} 0 & \text{if } d = 1, \\ D_x^{(d)}x_i & \text{if } d > 1. \end{cases} \quad (7.56)$$

By (7.55)and (7.56), it follows that (7.51) is equivalent to the condition

$$D_x^{(d)}\alpha_i = D_x^{(d)}D_x^{(1)}x_j = 0,$$

for all $2 \le i \le r$ and $d \ge 1$, while

$$D_x^{(d)}x_i = 0,$$

when $d > 1$. This condition is equivalent to (7.52). □

7.10 COMPLETE LINEAR SERIES OF SMALL ORDER

It is a difficult problem to determine the smallest n such that a curve has a complete linear series of order n and positive dimension, that is, a g_n^r with $r > 0$; see also Definition 9.49.

In this section, those curves are investigated for which this minimum order is either 1 or 2. By Theorem 6.81 the former possibility only occurs for rational curves. Also, by Theorems 6.81 and 7.42, if a curve has more than one g_2^1, it is either rational or elliptic, and in the latter but not in the former case, g_2^1 is complete. Also, if a curve has genus 2 then it is hyperelliptic since its canonical series is a g_2^1.

THEOREM 7.89 *An irreducible plane curve $\mathcal{F} = \mathbf{v}(f(X,Y))$ has genus 1 if and only if $\mathcal{F}$ is birationally equivalent to a plane non-singular cubic curve. A canonical form for $f(X,Y)$ is one of the following:*

(i) $Y^2 - X^3 - uX - v$ *with* $4u^3 + 27v^2 \neq 0$, *for* $p \neq 2, 3$;

(ii) $Y^2 - X^3 - uX^2 - vX - w$ *with* $v^3 = u^3 w$, *for* $p = 3$;

(iii) $Y^2 + uY + X^3 + vX + w$ *with* $u \neq 0$, *for* $p = 2$;

(iv) $Y^2 + XY + X^3 + uX + v$ *with* $v \neq 0$, *for* $p = 2$.

Proof. By Theorem 5.57, any non-singular cubic curve has genus 1. Conversely, let $\mathcal{C}$ be an elliptic curve, and take a place $\mathcal{P}$ of $\Sigma = K(\mathcal{C})$. By Theorem 7.40, the cubic curve derived from the rational transformation associated to the complete linear series $|3\mathcal{P}|$ of Σ is a non-singular plane curve $\mathcal{F}$. Take the coordinate system in the plane such that the line at infinity is a tangent to $\mathcal{F}$ at the point Y_∞ corresponding to $\mathcal{P}$. Then $\mathcal{F} = \mathbf{v}(f(X,Y))$ with $f(X,Y)$ containing only one term of degree 3, namely X^3. Replacing X by cX and Y by dX for suitable non-zero elements $c, d \in K$, the polynomial $f(X,Y)$ becomes the following:

$$Y^2 + a_1XY + a_3Y - X^3 + a_2X^2 + a_4X + a_5.$$

If $p \neq 2$, it is possible to get rid of both terms XY and Y by replacing Y by $aX + bY + c$ for suitable $a, b, c \in K$. So, without loss of generality, take $a_1 = a_3 = 0$. For $p = 3$, this leads to (ii), the condition $v^3 = u^3 w$ being necessary and sufficient for $\mathcal{F}$ to be non-singular.

If $p \neq 2, 3$, the X^2-term can also be removed by replacing X by $X - \frac{1}{3}a_2$, giving the canonical form (i). The condition $4u^3 + 27v^2 \neq 0$ says that $\mathcal{F}$ is non-singular.

Suppose $p = 2$. If $a_1 \neq 0$, then linear transformations of X and Y can be made to give $a_3 = a_4 = 0$ and $a_1 = 1$. If $a_1 = 0$, then $a_3 \neq 0$ and $a_2 = 0$ can be obtained by a linear transformation of X. Also, a linear transformation in Y gives $a_3 = 1$. The resulting canonical forms are those in (iii) and (iv); the conditions are necessary and sufficient for $\mathcal{F}$ to be non-singular. □

For $p \neq 2, 3$, the following result states when two plane non-singular cubics of $\mathrm{PG}(2, K)$ are birationally equivalent.

THEOREM 7.90 (Salmon) *For $p \neq 2, 3$, let $\mathcal{F}$ be a plane non-singular cubic.*

(i) *There are four tangents to $\mathcal{F}$ from a point $P \in \mathcal{F}$, other than the tangent at P, and the cross-ratio of the four tangents or, more precisely, the set of the six cross-ratios is constant.*

(ii) *Two plane non-singular cubics are birationally equivalent if and only if they have the same cross-ratio.*

(iii) *Two plane non-singular cubics have the same cross-ratio if and only if they have the same j-invariant, with*

$$j = j(\mathcal{F}) = \frac{u^3}{4u^3 + 27v^2}.$$

THEOREM 7.91 (Salmon) *Two non-singular cubics are birationally equivalent if and only if they are projectively equivalent.*

Proof. Let P be any non-inflexion point of a non-singular plane cubic $\mathcal{F}$. If ℓ is the tangent to $\mathcal{F}$ at P, then ℓ meets $\mathcal{F}$ at a second point, say P'. Let $\mathcal{L}$ be the simple, base-point-free, linear series cut out on $\mathcal{F}$, up to the fixed divisor $2P' + P$, by the conics containing P and tangent to $\mathcal{F}$ at P'; then $\mathcal{L} = |3P|$. Since $\deg \mathcal{L} = 3$, the Riemann–Roch Theorem 6.61 gives $\dim \mathcal{L} = 2$; hence $\mathcal{L} = g_3^2$. From Theorem 7.40, the curve Γ arising from $\mathcal{L}$ is non-singular. By Theorem 7.33, $\mathcal{L}$ is cut out on Γ by lines. Hence P is an inflexion point of Γ. This shows that, if $\mathcal{P}$ is any place of $K(\mathcal{F})$, there is a non-singular model Γ of $K(\mathcal{F})$ such that the unique branch centred at P is an inflexion of Γ.

Now assume that ω is birational transformation of $\mathcal{F}$ to another non-singular plane cubic $\mathcal{F}'$. Let O' be an inflexion point of $\mathcal{F}'$. Choose $P \in \mathcal{F}$ such a way that O' is the image of P under ω. If P is not an inflexion point, let τ be the birational transformation of $K(\mathcal{F})$ onto itself associated to the linear series $\mathcal{L}$. Then $\rho = \omega\tau^{-1}$ is a birational transformation of $\mathcal{F}$ to $\mathcal{F}'$ which takes the inflexion point P of Γ to O'. Let $\mathcal{L}'$ be the linear series cut out on $\mathcal{F}'$ by lines. By Corollary 6.105 (ii), ρ takes $\mathcal{L}$ to $\mathcal{L}'$. Therefore ρ sends every one-dimensional linear subseries of

$\mathcal{L}$ to another contained in $\mathcal{L}'$. Since such a one-dimensional linear subseries is cut out by lines through a point, this shows that ρ is a collineation taking Γ to Φ'. As $\deg \Gamma = \deg \mathcal{F}'$, such a collineation is a projectivity of $\mathrm{PG}(2, K)$. □

For more on elliptic cubics, see Sections 9.9, 9.10.

Now hyperelliptic curves are investigated.

THEOREM 7.92 *Every hyperelliptic curve of genus g is birationally equivalent to a curve Γ_0 of degree $g+2$ with only one singular point. This singular point is a g-fold point, and the reference system can be chosen so that*

$$\Gamma_0 = \mathbf{v}(a_0(X)Y^2 + a_1(X)Y + a_2(X)). \tag{7.57}$$

The unique g_2^1 is cut out by the vertical lines, apart from a fixed divisor $g\mathcal{P}$, where $\mathcal{P}$ is the place corresponding to the g-fold point Y_∞ of Γ_0.

Proof. Let Γ be irreducible of genus g. There exist distinct places $\mathcal{P}_1, \ldots, \mathcal{P}_g$ of $\Sigma = K(\Gamma)$ which arise from branches of Γ centred at non-singular points $P_1, \ldots, P_g$ such that the linear series $|E|$, with $E = \sum_{i=1}^{g} \mathcal{P}_i$, is non-special.

Now assume Γ to be hyperelliptic, and let $D = \mathcal{Q}_1 + \mathcal{Q}_2$ be a divisor in the unique g_2^1 of Σ such that the branches of Γ corresponding to $\mathcal{Q}_1$ and $\mathcal{Q}_2$ have different centres Q_1 and Q_2. Since the linear series $|D+E|$ is non-special, from the Riemann–Roch Theorem 6.61, $\dim |D+E| = 2$. Then, since $i(|E|) = -1$, so $i(|E - \mathcal{P}_i|) = 0$ for each $i = 1, \ldots, g$. Therefore there is a unique canonical divisor C_i containing $E - \mathcal{P}_i$.

If $|D+E|$ has a fixed place $\mathcal{P}$, then $\mathcal{P}$ is one of the places in E because g_2^1 has no fixed place. Also, $i(|D+E-\mathcal{P}_i|) = 1$; that is, there is a unique canonical divisor C_i containing $D+E-\mathcal{P}_i$. Therefore, to ensure that $D+E$ has no fixed place, it suffices to choose Q_1 different from the finitely many places appearing in the canonical divisors C_i. So suppose that $D+E = g_{g+2}^2$ has no fixed place. Let

$$g_2^1 = \{\mathrm{div}(c_0 + c_1 x) + D \mid \mathbf{c} = (c_0, c_1) \in \mathrm{PG}(1, K)\}.$$

Then there exists $y \in K(\Gamma)$ such that

$$g_{g+2}^2 = \{\mathrm{div}(c_0 + c_1 x + c_2 y) + D + E \mid \mathbf{c} = (c_0, c_1, c_2) \in \mathrm{PG}(2, K)\}.$$

Since g_{g+2}^2 has no fixed place, from Theorem 6.82 it follows that $\mathrm{div}(z)_\infty = D+E$ for some $z \in \Sigma$. Hence

$$z = c_0 + c_1 x + c_2 y$$

with $[\Sigma : K(z)] = g+2$. In particular, z cannot be written as $d_0 + d_1 x$ with $d_0, d_1 \in K$. Therefore y may be replaced by z in the above representation of $|D+E|$. By Theorem 7.44, $z \notin K(x)$, whence it follows that $\Sigma = K(x, z)$. In other words, $|D+E|$ is simple, that is, not composed of an involution. Hence, up to birationally equivalence, Γ may be supposed to be an irreducible plane curve of degree $g+2$, and g_{g+2}^2 is cut out on Γ by lines.

If $D' = \mathcal{Q}_1' + \mathcal{Q}_2'$ and $D' = D + \mathrm{div}(\xi)$ for some $\xi \in \Sigma$, then

$$D' + E = D + E' + \mathrm{div}(\xi).$$

In particular, if the branches of Γ corresponding to $\mathcal{Q}'_1$ and $\mathcal{Q}'_2$ have different centres Q'_1 and Q'_2, then the common point of the lines Q_1Q_2 and $Q'_1Q'_2$ is the common centre U of the branches corresponding to $\mathcal{P}_1, \ldots, \mathcal{P}_g$. This is possible only when U is a g-fold point of Γ. This implies that Γ has no more singular points. In fact, its virtual genus g^* is equal to g, since

$$g^* \leq \tfrac{1}{2}((g+1)g - g(g-1)) = g,$$

while $g^* \geq g$ by the remark after Definition 5.55. Without loss of generality, U may be assumed to be Y_∞. Then $\Gamma = \mathbf{v}(g(X,Y))$ with

$$g(X,Y) = a_0(X)Y^2 + a_1(X)Y + a_2(X),$$

where $\deg g(X,Y) \leq 2g+2$. The unique g_2^1 is cut out on Γ by the vertical lines, minus the fixed divisor E. □

COROLLARY 7.93 *A hyperelliptic curve of genus g is birationally equivalent to a curve*

$$\Gamma_1 = \mathbf{v}(Y^2 + h(X)Y + g(X)), \tag{7.58}$$

with $h(X)$, $g(X) \in K[X]$ and $\deg h(X) \leq g+1$, $\deg g(X) \leq 2g+2$.

Proof. Let $\Gamma = \mathbf{v}(a_0(X)Y^2 + a_1(X)Y + a_2(X))$. For a generic point $P = (x,y)$ of Γ, the map

$$\tau : x' = x,\ y' = a_0(x)$$

is a rational transformation of the function field $K(\Gamma) = K(x,y)$ of Γ. Then τ is birational, and the image curve $\Gamma' = \mathbf{v}(g'(X,Y))$, with

$$g'(X,Y) = Y^2 + h(X)Y + g(X),$$

where $h(X) = a_1(X)$ and $g(X) = a_0(X)a_1(X)$. This proves the corollary. □

Two cases are distinguished according as K has even characteristic or not.

THEOREM 7.94 *Let K to be of zero or odd characteristic. Then*

(i) *a curve of genus g is hyperelliptic if and only if it is birationally equivalent to a curve $\mathcal{X} = \mathbf{v}(Y^2 - f(X))$, where the polynomial $f(X) \in K[X]$ has degree $2g+1$ and no square factors*;

(ii) *the unique g_2^1 is cut out by the vertical lines apart from a fixed divisor*;

(iii) *there are finitely many places $\mathcal{P}$ of $K(\mathcal{X})$ such that $2\mathcal{P} \in g_2^1$.*

Proof. In zero or odd characteristic, the form (7.57) can be simplified further by means of the birational transformation $x' = x$, $y' = y + \frac{1}{2}h(x)$ to the form

$$\mathbf{v}(Y^2 - f(X)), \quad f(X) \in K[X],\ \deg f(X) \leq 2g+2. \tag{7.59}$$

Suppose that $f(X) = g(X)^2k(X)$, with $g(X), k(X) \in K[X]$, where $k(X)$ has no square factor. Then the birational transformation, $x' = x$, $y' = g(x)^{-1}y$, reduces it to the same form (7.59) but so that $f(X)$ has no square factors.

Suppose $\deg h = 2m$, and let α be a root of $f(X)$. The birational transformation, $x' = (x - \alpha)^{-1}$, $y' = yx^m$, reduces the curve to the form (7.59) with m odd. Example 5.59 shows that $\deg h = 2g + 1$.

An alternative proof depending on the Hurwitz formula is also possible. For a generic point $P = (x, y)$ of Γ, let $\Sigma = K(x, y)$ be the associated function field and $\Sigma' = K(x)$ its rational subfield. The field extension Σ/Σ' has degree 2, and hence it is separable, as the characteristic of K is distinct from 2. From the calculation in Example 5.59, $\omega : \Sigma \to \Sigma'$ only ramifies at the places corresponding to branches centred at the points $P = (a, 0)$ with $f(a) = 0$ and at Y_∞. The ramification index is equal to 2 for each of these places. Hence the degree of the different divisor $D(\Sigma/\Sigma')$ is equal to $2g + 2$. If $g(\Sigma)$ denotes the genus of Σ, then (7.6) becomes

$$2g(\Sigma) - 2 = 2(0 - 2) + 2g + 2,$$

whence $g(\Sigma) = g$. The places $\mathcal{P}$ such that $2\mathcal{P} \in g_2^1$ are those arising from the branches centred at Y_∞ and at the points $(a, 0)$ with $f(a) = 0$. So they are finitely many. □

THEOREM 7.95 *Suppose the following conditions hold:*

(a) *K is of zero or odd characteristic;*

(b) *the polynomials $g(X), h(X) \in K[X]$ have both degree $2g+1$ and no square factors;*

(c) *$\alpha_0, \ldots, \alpha_{2g}$ and $\beta_0, \ldots, \beta_{2g}$ are the roots of $g(X)$ and $h(X)$.*

Then the hyperelliptic curves

$$\Gamma = \mathbf{v}(Y^2 - f(X)), \quad \Gamma' = \mathbf{v}(Y^2 - g(X))$$

are birationally equivalent if and only if there is rational function

$$w(X) = (aX + b)/(cX + d), \quad ad - bc \neq 0$$

which maps the set $\{\alpha_0, \ldots, \alpha_{2g}, \infty\}$ onto the set $\{\beta_0, \ldots, \beta_{2g}, \infty\}$.

Proof. For a generic point $P = (x, y)$ of Γ, let $\Sigma = K(x, y)$ be the function field of Γ. The unique g_2^1 of Σ consists of all divisors

$$\operatorname{div}(c_0 + c_1 x) + B, \quad c_0, c_1 \in K, \ (c_0, c_1) \neq 0,$$

where $B = 2\mathcal{P}_\infty$ and $\mathcal{P}_\infty$ is the place corresponding to the unique branch of Γ centred at Y_∞; see Example 5.59.

Now assume that Γ' is a birationally equivalent to Γ. Then it may also be assumed that $\Sigma = K(\xi, \eta)$ for a generic point $Q = (\xi, \eta)$ of Γ'. Taking into account the uniqueness of g_2^1, Theorem 6.17 ensures the existence of $\zeta \in \Sigma$, for which

$$1 = \zeta(cx + d), \quad \xi = \zeta(ax + b)$$

with $ad - bc \neq 0$. Eliminating ζ gives $\xi = (ax + b)/(cx + d)$.

For $0 \leq i \leq 2g$, let γ_i be the branch of Γ centred at the point $P_i = (\alpha_i, 0)$. The corresponding place $\mathcal{P}_i$ of Σ is such that the divisor $2\mathcal{P}_i$ belongs to g_2^1. In the model $(\Gamma'; (\xi, \eta))$ of Σ, the place $2\mathcal{P}_i$ comes from a branch γ of Γ', centred at a point P, possible coincident with Y_∞. If σ is a primitive representation of $\mathcal{P}_i$, then

$$\sigma(\xi) = (a\sigma(x) + b)/(c\sigma(x) + d),$$

whence either P is affine and its X-coordinate is $(a\alpha_i + b)/(c\alpha_i + d)$, or $P = Y_\infty$ and $c\alpha_i + d = 0$. Hence $w(\alpha_i) \in \{\beta_0, \ldots, \beta_{2g}, \infty\}$. The previous argument still works if γ_i is replaced by the branch of Γ centred at Y_∞, showing that

$$w(\infty) = b/d \in \{\beta_0, \ldots, \beta_{2g}, \infty\}.$$

To show the converse, put

$$\begin{aligned} \Sigma &= K(x, y) \quad \text{with } y^2 - f(x) = 0, \\ \Sigma' &= K(x', y') \quad \text{with } y'^2 - g(x') = 0. \end{aligned}$$

Assume first that $w(\infty) = \infty$. Then $w(X) = aX + b$. Relabel the roots of $g(X)$ such that $w(\alpha_i) = \beta_i$ for $i = 0, \ldots, 2g$, and write

$$\begin{aligned} f(X) &= u \textstyle\prod_{i=0}^{2g} (X - \alpha_i), \\ g(X) &= v \textstyle\prod_{i=0}^{2g} (X - \beta_i). \end{aligned}$$

Then the birational transformation

$$\omega : \xi = ax + b, \quad \eta = \epsilon y \quad \text{with} \quad \epsilon^2 = a^{2g+1} v/u$$

maps Γ onto Γ'. In fact, since

$$\begin{aligned} g(\xi) &= v \textstyle\prod_{i=0}^{2g} (\xi - \beta_i) \\ &= v \textstyle\prod_{i=0}^{2g} (ax + b - \beta_i) \\ &= v \textstyle\prod_{i=0}^{2g} (ax + b - (a\alpha_i + b) \\ &= va^{2g+1} \textstyle\prod_{i=0}^{2g} (x - a_i) \\ &= a^{2g+1} (v/u) f(x), \end{aligned}$$

so

$$\eta^2 - g(\xi) = a^{2g+1} (v/u)(y^2 - f(x)),$$

whence $\eta^2 - g(\xi) = 0$. Therefore not only $P = (x', y')$ but also $Q = (\xi, \eta)$ is a generic point of the curve $\mathbf{v}(Y^2 - g(X))$. By Theorem 5.7, $K(\xi, \eta)$ and $K(x', y')$ are isomorphic over K. On the other hand, as ω is birational, $K(x, y)$ and $K(\xi, \eta)$ are isomorphic over K. The proof in the case $w(\infty) = \infty$ is complete.

If $w(\infty) \neq \infty$, relabel the indices of the roots of $f(X)$ and $g(X)$ such that

$$w(\infty) = \beta_0, \; u(\alpha_0) = \infty, \; w(\alpha_i) = \beta_i,$$

for $i = 1, \ldots, 2g$. Then

$$w(X) = (\beta_0 X + b)/(X - \alpha_0), \quad \beta_0 \alpha_i + b = \alpha_i \beta_i - \alpha_0 \beta_i,$$

for $1 \leq i \leq 2g$. It follows that

$$\omega : \xi = \frac{\beta_0 x + b}{x - \alpha_0}, \quad \eta = \epsilon \frac{y}{(x - \alpha_0)^{g+1}},$$

with

$$\epsilon^2 = vu^{-1}(b + \alpha_0\beta_0)\prod_{i=1}^{2g}(\beta_0 - \beta_i),$$

is a birational transformation of Γ onto Γ'. In fact,

$$\begin{aligned} g(\xi) &= v\prod_{i=0}^{2g}(x - \beta_i) \\ &= v\prod_{i=0}^{2g}\left(\frac{\beta_0 x + b}{x - \alpha_0} - \beta_i\right) \\ &= v(b + \alpha_0\beta_0)\frac{1}{x - \alpha_0}\prod_{i=1}^{2g}(\beta_0 - \beta_i)\frac{x - \alpha_i}{x - \alpha_0} \\ &= \frac{v}{u}(b + \alpha_0\beta_0)\left(\prod_{i=1}^{2g}(\beta_0 - \beta_i)\right)\frac{1}{(x - \alpha_0)^{2g+2}}f(x). \end{aligned}$$

This shows that $y^2 - f(x) = 0$ implies $\eta^2 - g(\xi) = 0$. Now the previous argument depending on Theorem 5.7 can be repeated to finish the proof. □

Analogous results hold in the case of even characteristic.

THEOREM 7.96 *When $p = 2$, a hyperelliptic curve is birationally equivalent to a curve*

$$\Gamma = \mathbf{v}(Y^2 + h(X)Y + g(X)) \tag{7.60}$$

as in (7.58), with no affine singular points and the leading coefficient of $g(X)$ equal to 1.

Proof. From Theorem 7.92, consider a hyperelliptic curve Γ in the form 7.60, with $h(X), g(X) \in K[X]$, $\deg h(X) \le g + 1$, $\deg g(X) \le 2g + 2$. If $P = (a, b)$ is an affine singular point of Γ, then, by a transformation $(X, Y) \mapsto (X, Y + b)$, the point $P = (a, 0)$; hence $g(a) = 0$. Since P is a singular point of Γ, so

$$\partial(Y^2 + h(X)Y + g(X))/\partial Y = 0$$

at $(a, 0)$; hence $h(a) = 0$. Also,

$$\partial(Y^2 + h(X)Y + g(X))/\partial X = 0$$

at $(a, 0)$. Putting $g'(X) = dg(X)/dX$, it follows that $g'(a) = 0$. Thus

$$h(X) = (X - a)h_1(X), \quad g(X) = (X - a)^2 g_1(X),$$

with $h_1(X), g_1(X) \in K[X]$, $\deg h_1(X) < \deg h(X)$, $\deg g_1(X) < \deg g(X)$. The birational transformation $x' = x, y' = y/(x - a)$ maps Γ onto the curve $\Gamma' = \mathbf{v}(Y^2 + h_1(X)Y + g_1(X))$. By repeated application of this argument, a hyperelliptic curve with no affine singular point is obtained. □

LEMMA 7.97 *For $p = 2$, let $\Gamma = \mathbf{v}(Y^2 + h(X)Y + g(X))$ be a hyperelliptic curve with no affine singular point. Then, under the birational transformation*

$$x' = x, \quad y' = y + r(X),$$

with $r(X) \in K[X]$, the image curve Γ' of Γ has no affine singular points.

Proof. It is shown that, if Γ' had an affine singular point, then Γ would also have an affine singular point. Note that

$$\Gamma' = \mathbf{v}(Y^2 + h(X)Y + f(X)),$$

with $f(X) = r(X)^2 + h(X)r(X) + g(X)$. If $P = (a, b)$ is a singular point of Γ', then both partial derivatives are zero at P. Putting

$$r'(X) = dr(X)/dX, \quad h'(X) = dh(X)/dX, \quad g'(X) = dg(X)/dX$$

gives

$$\begin{aligned} b^2 + h(a)b + r(a)^2 + h(a)r(a) + g(a) &= 0, \\ h(a) &= 0, \\ bh'(a) + r'(a)h(a) + r(a)h'(a) + g'(a) &= 0. \end{aligned}$$

Note that the third equation can also be written as $(r(a) + b)h'(a) + g'(a) = 0$. Since

$$\begin{aligned} \partial(Y^2 + h(X)Y + g(X))/\partial Y &= h(X), \\ \partial(Y^2 + h(X)Y + g(X))/\partial X &= Yh'(X) + g'(X), \end{aligned}$$

it follows that the point $Q = (a, c)$, with $c = r(a) + b$, is a singular point of Γ. □

LEMMA 7.98 *In Theorem 7.96, the polynomials $h(X)$, $g(X)$ can be chosen such that*

$$\deg g(X) \geq 2 \deg h(X) + 1. \tag{7.61}$$

Proof. Let Γ be as in Theorem 7.96. Put $k = \deg h(X)$, $n = \deg g(X)$. If n is even, then the birational transformation, $x' = x$, $y' = y + x^{n/2}$, reduces it to the same form (7.58) with $\deg h(X) = k$, $\deg g(X) < n$. Lemma 7.97 ensures that no affine point becomes singular. If $\deg g(X)$ is still even, the argument can be repeated. Thus $\deg g(X)$ can be taken to have odd degree $2m + 1$. If $k > m$, then the birational transformation, $x' = x$, $y' = y + x^{k-1}$, maps Γ onto the curve (7.58) with $\deg g(X) \geq 2k - 1$. Hence $m \geq k$; that is, (7.61) holds. □

LEMMA 7.99 *In Lemma 7.96, the polynomials $h(X)$, $g(X)$ can be chosen so that*

$$\deg g(X) = 2 \deg h(X) + 1. \tag{7.62}$$

Proof. Assume that $h(X)$ and $g(X)$ satisfy (7.61); let $\deg g(X) = 2m+1$. Choose an element $a \in K$ for which $h(a) \neq 0 \neq g(a)$. Replacing X by $X + a$ ensures that $h(0) \neq 0 \neq g(a)$. Now $x' = x^{-1}$, $y' = yx^{m+1}$ is a birational transformation of Γ onto the curve $\Gamma_1 = \mathbf{v}(Y^2 + h_1(X)Y + g_1(X))$ with $h_1(X), g_1(X) \in K[X]$ defined as follows:

$$h_1(X) = h(X^{-1})X^{m+1}, \; g_1(X) = g(X^{-1})X^{2m+2}.$$

Note that $\deg h_1(X) = m + 1$, and $\deg g_1(X) = 2m + 2$. If g_{2m+2} is the leading coefficient of $g_1(X)$, let $u \in K$ be a root of the polynomial

$$X^2 + g_{2m+2}.$$

Then the birational transformation, $x' = x$, $y' = y + ux^{m+1}$, maps Γ_1 onto a hyperelliptic curve Γ_2 as in (7.58) satisfying (7.62). If Γ_2 has some affine singular points, use the argument employed in the proof of Theorem 7.96 to obtain a curve with no affine singular point. Such a curve is still of the form (7.58) with $h(X)$, $g(X)$ satisfying (7.62). □

THEOREM 7.100 *For $p = 2$, a hyperelliptic curve of genus g is birationally equivalent to a curve*

$$\Gamma = \mathbf{v}(Y^2 + h(X)Y + g(X)), \quad \deg g(X) = 2g + 1, \deg h(X) \leq g$$

with the following properties:

(i) *Γ has no affine singular point;*

(ii) *Γ has only one branch centred at its unique point at infinity Y_∞;*

(iii) *if $\mathcal{P}_\infty$ is the place corresponding to Y_∞, then the g_2^1 is cut out on Γ by the vertical lines minus the fixed divisor $(2g-1)\mathcal{P}_\infty$;*

(iv) *there are finitely many places $\mathcal{P}$ of Σ such that $2\mathcal{P} \in g_2^1$.*

Proof. With the notation used in the preceding two proofs, $k + 1 < n = 2m + 1$ and hence Γ has a unique point at infinity, namely Y_∞. More precisely, Y_∞ is a $(2m-1)$-fold point of Γ.

Now it is shown that Γ has exactly one branch centred at Y_∞. Write

$$h(X) = h_0 + \cdots + h_k X^k,$$

with $h_k \neq 0$, and let $h(X, Y)$ be the associated homogeneous polynomial; that is,

$$h(X, Y) = h_0 Y^k + \cdots + h_k X^k.$$

Similarly, write

$$g(X) = f_0 + \cdots + X^{2m+1},$$

and let

$$g(X, Y) = f_0 Y^{2m+1} + \cdots + X^{2m+1}.$$

Then, in homogeneous coordinates,

$$\Gamma = \mathbf{v}(Y^2 Z^{2m-1} + YZ^{2m-k} h(X, Z) + g(X, Z)).$$

Take Y_∞ to the origin O by the projectivity $\mathfrak{T} : (X, Y, Z) \mapsto (X, Z, Y)$. The curve Γ is mapped to the curve

$$\Gamma' = \mathbf{v}(Y^{2m-1} Z^2 + Y^{2m-k} h(X, Y) Z + g(X, Y))$$

or, in inhomogeneous coordinates,

$$\Gamma' = \mathbf{v}(Y^{2m-1} + Y^{2m-k} h(X, Y) + g(X, Y)).$$

The mapping $\mathfrak{T}$ takes the line at infinity to the X-axis, which is then a tangent to Γ' at the origin.

Now it is shown that there is only one branch of Γ' centred at O. Let $(x(t), y(t))$ be a primitive representation of a branch of Γ' centred at the origin and tangent to the X-axis. Write

$$x(t) = at^i + \cdots, \quad y(t) = bt^j + \cdots, \quad \text{with } i < j.$$

Then

$$y(t)^{2m-1} + y(t)^{2m-k}h(x(t), y(t)) + g(x(t), y(t)) = 0. \tag{7.63}$$

Substitution gives

$$\begin{aligned}(bt^j + \cdots)^{2m-1} &+ (bt^j + \cdots)^{2m-k}h(at^i + \cdots, bt^j + \cdots)\\ &+g(at^i + \cdots, bt^j + \cdots) = 0,\end{aligned}$$

which shows that the terms of possible minimum degree in t are

$$(2m-1)j,\ (2m-k)j + ik,\ (2m+1)i.$$

Therefore one of the following cases occurs:

(i) $(2m-1)j = (2m-k)j + ki < (2m+1)i$;

(ii) $(2m+1)i = (2m-k)j + ki < (2m-1)j$;

(iii) $(2m-1)j = (2m-k)j + ki = (2m+1)i$;

(iv) $(2m-1)j = (2m+1)i < (2m-k)j + ki$.

In fact, none of the first three cases occurs. To show that (i) is impossible, note that $(2m-1)j = (2m-k)j+ki$ implies that $(k-1)j = ki$; hence $i = k-1,\ j = k$, since $k-1$ and k are relatively prime. Then

$$\begin{aligned}(2m-1)j &< (2m+1)i,\\ (2m-1)k &< (2m+1)(k-1),\\ m &< k,\end{aligned}$$

contradicting that $k \le m$.

Similarly for (ii), $(2m+1)i = (2m-k)j+ki$ implies $i = 2m-k, j = 2m-k+1$. Thus

$$\begin{aligned}(2m-k)j + ki &< (2m-1)j,\\ (2m-k)(2m-k+1) + k(2m-k) &< (2m-1)(2m-k+1),\\ m &< k,\end{aligned}$$

again a contradiction.

To rule out (iii), $(2m-1)j = (2m+1)i$ implies that $i = 2m-1,\ j = 2m+1$, as in (i), while $(2m+1)i = (2m-k)j+ki$ implies that $i = 2m-k,\ j = 2m-k+1$, as in (ii).

This leaves (iv). The only positive integers i, j which can occur in (IV) are $i = 2m-1$ and $j = 2m+1$. Hence Γ' has a branch γ centred at the origin such that $I(O, \mathbf{v}(Y) \cap \gamma) = 2m+1$. But then no other branch of Γ' is centred at the origin. This follows from Theorem 4.50, since $I(O, \Gamma' \cap \mathbf{v}(Y)) = 2m+1$.

It remains to show that $g = m$. As $i = 2m - 1$ is odd, Theorem 4.27 gives a substitution of type $t \mapsto \bar{t} = t + \cdots$, for which

$$x(t) = \bar{x}(\bar{t}) = \bar{t}^{2m-1}, \quad y(t) = \bar{y}(\bar{t}) = \alpha\bar{t}^{2m+1} + \cdots,$$

with $\alpha \in K\backslash\{0\}$. Replace $\bar{t}$ by t and omit the bar over x and y. If $y(t)$ contains some terms of odd degree, then βt^{2m+1+n} denotes the lowest degree for which this happens. Otherwise, let $\beta = 0$. Write

$$y(t) = \alpha t^{2m+1} + \cdots + \beta t^{2m+1+n} + \cdots$$

with $\beta \in K$. Now the left-hand side of (7.63) can be written as $t^{(2m-1)(2m+1)}u(t)$, with

$$\begin{aligned} u(t) = {} & (\alpha + \cdots + \beta t^n + \cdots)^{2m-1} + t^{2m-2k+1}(\alpha + \cdots + \beta t^n + \cdots)^{2m-k} \\ & [h_k + \cdots + h_{k-r}t^{2r}(\alpha + \cdots + \beta t^n + \cdots)^r \\ & \quad + \cdots + h_0 t^{2k}(\alpha + \cdots + \beta t^n + \cdots)^k] \\ & +1 + g_{2m}t^2(\alpha + \cdots + \beta t^n + \cdots) \\ & \quad + g_e t^{4m-2e+2}(\alpha + \cdots + \beta t^n + \cdots)^{2m+1-e} \\ & \quad + \cdots + g_0 t^{2(2m+1)}(\alpha + \cdots + \beta t^n + \cdots)^{2m+1}. \end{aligned}$$

If $g(X)$ contains some non-zero terms with positive index, then s denotes the smallest for which this happens. Otherwise, let $g_s = 0$. From the definition of n,

$$\begin{aligned} du/dt = {} & (\alpha + \cdots + \beta t^n + \cdots)^{2m-2}(\beta t^{n-1} + t^{n+1}G_1(t)) \\ & + t^{2m-2k}(\alpha + \cdots + \beta t^n + \cdots)^{2m-k}(h_k + t^2 G_2(t)) \\ & + t^{2m-2k+1}(\alpha + \cdots + \beta t^n + \cdots)^{2m-k-1}(\beta t^{n-1} + \cdots)(h_k + t^2 G_2(t)) \\ & + t^{2m-2k+1}(\alpha + \cdots + \beta t^n + \cdots)^{2m-k}(t^2 G_4(t)) \\ & + (g_s t^{2(2m-s)+2}(\alpha + \cdots \beta t^n + \cdots)^{2m-s}(\beta t^{n-1} + t^{n+1}G_1(t)). \end{aligned}$$

It should be noted that, if $\beta = 0$, then the term, $\beta t^{n-1} + t^{n+1}G_1(t)$, vanishes by definition. In this case, du/dt has only one term with lowest degree, namely $\alpha^{2m-k}h_k t^{2m-2k}$, but for $\beta \neq 0$ there is one more, namely $\alpha^{2m-2}\beta t^{n-1}$. By (7.63), $u(t) = 0$, and hence $du(t)/dt = 0$. This rules out the former possibility. Thus the latter possibility occurs, and hence $2m - 2k = n - 1$.

Going back to Γ, its unique branch centred at Y_∞ has the following primitive representation:

$$\xi(t) = \frac{t^{2m-1}}{\alpha t^{2m+1} + \cdots + \beta t^{4m-2k+2} + \cdots} = t^{-2}\frac{1}{\alpha + \cdots + \beta t^{2m-2k+3} + \cdots}$$

$$\eta(t) = \frac{1}{\alpha t^{2m+1} + \cdots \beta t^{4m-2k+2} + \cdots},$$

with $\alpha, \beta \neq 0$.

Since K has even characteristic,

$$\frac{d\xi(t)}{dt} = t^{-2}\frac{\beta t^{2m-2k} + \cdots}{(\alpha + \cdots + \beta t^{2m-2k+1} + \cdots)^2}.$$

Hence, if $\mathcal{P}$ is the corresponding place of Σ, then

$$\text{ord}_{\mathcal{P}}(d\xi) = \text{ord}_t(d\xi(t)/dt) = 2m - 2k - 2. \tag{7.64}$$

Note that $d\xi(t)/dt \neq 0$, and this shows that ξ is a separable variable of $K(\Gamma)$.

To calculate the genus g of Γ, take all branches γ of Γ with $\text{ord}_t\, dx(t)/dt \neq 0$, where $(x(t), y(t))$ is a primitive representation of γ. Apart from the branch centred at Y_∞, all branches are centred at affine points, and hence $\text{ord}_t\, dx(t)/dt \geq 0$. So attention is focused on branches with $\text{ord}_t\, dx(t)/dt > 0$. Since Γ has no affine singular point, a necessary and sufficient condition for $\text{ord}_t\, dx(t)/dt$ to be positive is that the tangent line to Γ at the centre $P = (a, b)$ of γ is the vertical line $\mathbf{v}(X-a)$. If this happens, then

$$\text{ord}_t\, dx(t)/dt = \text{ord}_t \left(h(t)\frac{dy(t)}{dt} \right), \quad \text{with } h(t) = h(x(t)),$$

by Theorem 5.56. Since K has even characteristic,

$$\text{ord}_t \left(\frac{dx(t)}{dt} \right) = \text{ord}_t \left(h(t)\frac{dy(t)}{dt} \right) = \text{ord}_t \left(\frac{dy(t)}{dt} \right) + \text{ord}_t\, h(t).$$

Let j denote the multiplicity of a root a of $h(X)$; that is,

$$h(X) = (X-a)^j h_1(X),$$

with $h_1(X) \in K[X]$ and $h_1(a) \neq 0$. A primitive representation of the unique branch centred at $P = (a, b)$ is

$$x(t) = a + t^i + \cdots, \quad y(t) = b + t,$$

where $i \geq 2$. Actually, $i = 2$ since Γ has degree $2m+1$ and Y_∞ is a $(2m-1)$-fold point of Γ; then $\text{ord}_t\, h(t) = 2j$.

Let $a_1, \ldots, a_s$ with multiplicities $j_1, \ldots, j_s$ be the roots of $h(X)$. Then

$$j_1 + \cdots + j_s = k.$$

This, together with (7.64), shows that

$$\sum \text{ord}_t\, dx(t)/dt = 2k + 2m - 2k - 2 = 2m - 2.$$

By Definition 5.55, this sum is equal to the genus of Γ, and hence $m = g$. The places $\mathcal{P}$ such that $2\mathcal{P} \in g_2^1$ are those arising from the branches centred at Y_∞ and at the points $(a, \sqrt{g(a)})$ with $h(a) = 0$. So they are finitely many. □

THEOREM 7.101 *For $p = 2$, let Γ and Γ' be two hyperelliptic curves of genus g given in their canonical form as in Theorem 7.100; that is,*

$$\Gamma = \mathbf{v}(Y^2 + h(X)Y + g(X)), \ \deg g(X) = 2g+1, \ n = \deg h(X) \leq g;$$
$$\Gamma_1 = \mathbf{v}(Y^2 + h_1(X)Y + g_1(X)), \ \deg g_1(X) = 2g+1, \ m = \deg h_1(X) \leq g.$$

Let $\alpha_0, \ldots, \alpha_n$ and $\beta_0, \ldots, \beta_m$ be the roots of $h(X)$ and $h_1(X)$, with multiple roots counted. Then Γ and Γ_1 are birationally equivalent if and only if each of the following conditions are satisfied:

(i) *$n = m$, and there is rational function $w(X) = (aX+b)/(cX+d)$, with $ad - bc \neq 0$, which maps the multi-set $\{\alpha_1, \ldots, \alpha_n, \infty\}$ onto the multi-set $\{\beta_1, \ldots, \beta_n, \infty\}$;*

(ii) *there is a rational function* $v(X) \in K(X)$ *such that*

$$v(X)^2 + v(X) = \frac{g(X)}{h(X)^2} + \frac{g_1(w(X))}{h_1(w(X))^2}.$$

Proof. For a generic point $P = (x, y)$ of Γ, let $K(\Gamma) = K(x, y)$. Assume that both (i) and (ii) hold. Then the birational transformation

$$\omega: \quad \xi = w(x), \quad \eta = v(x)h_1(w(x)) + \frac{h_1(w(x))}{h(x)} y$$

maps Γ onto Γ_1. In fact, a direct computation shows that $\eta^2 + h_1(\xi)\eta + g_1(\xi) = 0$ follows from $y^2 + h(x)y + g(x) = 0$.

To prove the converse, assume that $\Sigma = K(\xi, \eta)$ with $\eta^2 + h_1(\xi)\eta + g_1(\xi) = 0$. Let γ_i be a branch of Γ centred at the point $P_i = (\alpha_i, \sqrt{g(\alpha_i)})$; let $\mathcal{P}_i$ be the corresponding place of $K(\Gamma)$.

As in the proof of Theorem 7.95, the uniqueness of g_2^1 implies that $\xi = w(x)$ with $w(X) = (aX + b)/(cX + d)$ and $ad + bc \neq 0$. The argument used in that proof also shows that the set $\{\beta_1, \ldots, \beta_m, \infty\}$ contains $w(\alpha_i)$, for $i = 1, \ldots, n$, and $w(\infty)$. Interchanging the roles of Γ and Γ_1, it follows that the set $\{\alpha_1, \ldots, \alpha_n, \infty\}$ contains $w(\beta_i)$, for $i = 1, \ldots, m$, and $w(\infty)$. Therefore $n = m$ and (i) holds.

To show (ii), note that

$$\eta = a(x) + b(x)y,$$

with $a(X), b(X) \in K(X)$ since $[K(\Gamma) : K(x)] = 2$. Since $y^2 = h(x)y + g(x)$, so $\eta^2 + h_1(\xi)\eta + g_1(\xi) = 0$ can be written as follows:

$$(b(x)^2 h(x) + b(x)h_1(w(x))y + a(x)^2 + b(x)^2 g(x) + a(x)h_1(w(x)) + g_1(w(x)) = 0.$$

Since $y \notin K(x)$, this is only possible when $b(x) = h_1(w(x))/h(x)$ and

$$\left(\frac{a(x)}{h_1(w(x))}\right)^2 + \frac{a(x)}{h_1(w(x))} = \frac{g(x)}{h(x)^2} + \frac{g_1(w(x))}{h_1(w(x))^2}.$$

Putting $v(X) = a(X)/h_1(w(X))$, (ii) follows. □

THEOREM 7.102 *Let* $p = 2$. *For* $u, u_1 \in K(X)$, *put*

$$\Sigma = K(x, y) \text{ with } y^2 + y + u(x) = 0,$$
$$\Sigma_1 = K(\xi, \eta) \text{ with } \eta^2 + \eta + u_1(\xi) = 0.$$

Suppose that both Σ *and* Σ_1 *are hyperelliptic, and that* $\Sigma_1 \subset \Sigma$. *Then* $\Sigma = \Sigma_1$ *if and only if there exist* $w(X) = (aX+b)/(cX+d)$, $ad-bc \neq 0$, *and* $v(X) \in K(X)$ *such that*

$$\xi = w(x), \quad v(x)^2 + v(x) = u(x) + u_1(w(x)). \tag{7.65}$$

Proof. The hypothesis that Σ is hyperelliptic means that $y \notin K(x)$; that is, no element $f(x) \in K(x)$ exists for which $u(x) = f(x)^2 + f(x)$, and similarly for Σ_1. A direct calculation shows that, if (7.65) holds, then $y^2 + y + u(x) = 0$ implies that

$$(y + v(x))^2 + y + v(x) + u_1(\xi) = 0.$$

Hence either $\eta = y + v(x)$ or $\eta = y + v(x) + 1$. In both cases, $y \in K(\xi, \eta)$. Since $K(x) = K(w(x)) = K(\xi)$, so $\Sigma = \Sigma_1$.

Conversely, if $\Sigma = \Sigma_1$, the uniqueness of g_2^1 in any hyperelliptic function field implies that $\xi = w(x)$, where $w(x) = (ax + b)/(cx + d)$, $ad - bc \neq 0$ and $a, b, c, d \in K$. Also, $\eta = a(x) + b(x)y$ with $a(x), b(x) \in K(x)$ and $b(x) \neq 0$. Then $\eta^2 + \eta + u_1(w(x)) = 0$ implies that

$$(b(x)^2 + b(x))y + a(x)^2 + b(x)^2u(x) + a(x) + u_1(w(x)) = 0.$$

Since $y \notin K(x)$, this is only possible when $b(x) = 1$ and

$$(a(x))^2 + a(x) = u(x) + u_1(w(x)).$$

Letting $v(x) = a(x)$, the second equation in (7.65) follows. □

THEOREM 7.103 *Let $\mathcal{F}$ be a hyperelliptic curve of genus g. If w is the number of distinct Weierstrass points of $\mathcal{F}$, then*

$$\begin{array}{ll} w = 2g + 2, & \textit{when either } p = 0 \textit{ or } p > 2; \\ 1 \leq w \leq g + 1, & \textit{when } p = 2. \end{array}$$

Proof. For $p \neq 2$, the assertion follows from Theorem 7.94 and Example 6.90. For $p = 2$, if the canonical form in Theorem 7.100 is adopted to represent $\mathcal{F}$, then the arguments in Example 6.90 can be used to prove that $w - 1$ is equal to the number of distinct roots of $h(X)$. Hence $0 \leq w \leq g$. □

7.11 EXAMPLES OF CURVES

In this section, q denotes a power of p such that $q \equiv -1 \pmod 3$. Two examples are worked out to illustrate some points of the theory of space curves.

In the first example, $\mathcal{F}$ is the irreducible plane curve considered in Example 1.41. Then $\mathcal{F} = \mathbf{v}(g(X, Y))$, with

$$g(X, Y) = Y + X^q + XY^q - 3(XY)^{(q+1)/3}.$$

Also, $\mathcal{H}_q = \mathbf{v}(Y + X^q + Y^qX)$ is the Hermitian curve, $P = (x, y)$ is a generic point of $\mathcal{H}_q$, and $\Sigma = K(x, y)$ is the associated function field. Note that

$$\begin{aligned} &Y^3 + X^{q3} + (XY^q)^3 - 3(XY)^{q+1} \\ &\qquad = (Y + X^q + Y^qX)(Y + \epsilon X^q + \epsilon^2 Y^q X)(Y + \epsilon^2 X^q + \epsilon Y^q X), \end{aligned}$$

where $\epsilon^3 = 1$ but $\epsilon \neq 1$. Putting $\xi = x^3$, $\eta = y^3$, it follows that

$$\eta + \xi^q + \xi\eta^q - 3(\xi\eta)^{(q+1)/3} = 0.$$

Therefore $Q = (\xi, \eta)$ is a generic point of $\mathcal{F}$, and the function field $\Sigma' = K(\xi, \eta)$ of $\mathcal{F}$ is a subfield of Σ.

From Example 1.41, $\mathcal{F}$ has exactly $\frac{1}{3}(q^2 - q + 1)$ singular points, each of which is an ordinary double point. By Theorem 5.57, the genus of $\mathcal{F}$ is

$$\tfrac{1}{2}\left[\tfrac{1}{3}(q^2 - q + 1) - 1\right].$$

Let L denote the linear series cut out on $\mathcal{H}_q$ by the linear system of plane cubics $\mathcal{F}' = \mathbf{v}(c_0 + c_1X^3 + c_2Y^3 + XY)$. Then L has no fixed place and $L = g^3_{3(q+1)}$. The associated subfield $\Sigma'' = K(\xi, \eta, xy)$ of Σ contains Σ'. It is shown below that $\Sigma' = \Sigma''$; see Lemma 7.106.

LEMMA 7.104 *The series $g^3_{3(q+1)}$ is composed of an involution of order three.*

Proof. Let $\mathcal{P}_1$ be the place of Σ arising from the unique branch γ of $\mathcal{H}_q$ centred at the fundamental point $P_1 = (1,0,0)$. To compute the ramification index $e_{\mathcal{P}_1}$, a representation of the place $\mathcal{P}'_1$ of Σ'' lying under $\mathcal{P}_1$ is needed. As shown in Example 7.29, $\mathcal{P}_1$ has a primitive representation σ such that $\sigma(x) = t$ and

$$\sigma(y) = -(t^q + t^{q^2+1} + \cdots + a_j t^{q+\alpha_j(q^2-q+1)} + \cdots).$$

Note that

$$\begin{aligned}
-\sigma(y)^3 &= (t^q + t^{q^2+1} + \cdots)^3 = \left[t^q(1 + t^{q^2-q+1} + \cdots)\right]^3 \\
&= (t^3)^q(1 + t^{q^2-q+1} + \cdots)^3 = (t^3)^q\left[1 + (t^3)^{(q^2-q+1)/3} + \cdots\right]^3,
\end{aligned}$$

$$\begin{aligned}
-\sigma(xy) &= t(t^q + t^{q^2+1} + \cdots) = t\left[t^q(1 + t^{q^2-q+1} + \cdots)\right] \\
&= t^{q+1}(1 + t^{q^2-q+1} + \cdots) = (t^3)^{(q+1)/3}\left[1 + (t^3)^{(q^2-q+1)/3} + \cdots\right].
\end{aligned}$$

So the substitution $s = t^3$ provides a primitive representation of $\mathcal{P}'_1$, and hence $e_{\mathcal{P}_1} = 3$.

The point $P = (1, x^3, y^3, xy)$ determines an irreducible curve Γ' in $\mathrm{PG}(3,K)$. The branch γ'_1 of Γ' associated with $\mathcal{P}'_1$ has a primitive representation

$$(x_0(s) = 1,\ x_1(s) = s,\ x_2(s) = s^q u(s)^3,\ x_3(s) = s^{(q+1)/3}u(s)), \tag{7.66}$$

with

$$u(s) = 1 + s^{(q^2-q+1)/3} + \cdots + a_j s^{\alpha_j(q^2-q+1)/3} + \cdots.$$

Its centre is the point $P'_1 = (1,0,0,0)$.

If $\mathcal{P}_2$ is the place of Σ arising from the branch with centre at $P_2 = (0,1,0)$, the argument may proceed as above or use the fact that Γ is invariant under the following cyclic change of the coordinate system:

$$(X_0, X_1, X_2) \to (X_2, X_0, X_1).$$

The ramification index $e_{\mathcal{P}_2}$ is 3. Also,

$$x_0(s) = (s^q u(s)^3,\ x_1(s) = 1,\ x_2(s) = s,\ x_3(s) = s^{(q+1)/3}u(s))$$

is a primitive representation of the branch γ'_2 of Γ' associated with $\mathcal{P}_2$. Its centre is the point $P_2 = (0,1,0,0)$.

Similarly, the place $\mathcal{P}_3$ of Σ arising from the branch centred at $P_3 = (0,0,1)$ has ramification index 3, and

$$(x_0(s) = s,\ x_1(s) = s^q u(s)^3,\ x_2(s) = 1,\ x_3(s) = s^{q+1)/3}u(s))$$

is a primitive representation of the branch γ_3' associated with the place $\mathcal{P}_3'$ of Σ' lying under $\mathcal{P}_3$. The centre of γ_3' is the point $P_3' = (0, 0, 1, 0)$.

Every other point P of $\mathcal{F}$ is affine: $P = (1, a, b)$. Let $\mathcal{P}$ be the unique place of Σ centred at P. The corresponding branch γ centred at P has a primitive representation

$$(x(t) = a + a_1 t + a_2 t^q + \cdots, \quad y(t) = b + t)$$

with $a_1 = -b^{-q}$ and $a_2 = b^{-(q^2+q)} - ab^{-q}$. Hence the branch γ' corresponding to $\mathcal{P}'$ has a representation $(x_0(t) = 1, x_1(t), x_2(t), x_3(t))$ with

$$\begin{aligned} x_1(t) &= a^3 + 3a^2 a_1 t + 3aa_1^2 t^2 + a_1^3 t^3 + 3a^2 a_2 t^q + 6aa_1 a_2 t^{q+1} + \cdots, \\ x_2(t) &= b^3 + 3b^2 t + 3bt^2 + t^3, \\ x_3(t) &= ab + (a + a_1 b)t + a_1 t^2 + a_2 b t^q + a_2 t^{q+1} + \cdots. \end{aligned}$$

Its centre is the point $P' = (1, a^3, b^3, ab)$. For a primitive cube root of unity ϵ, the point $P_\epsilon = (1, \epsilon a, \epsilon^2 b) \in \mathcal{H}_q$ if and only if $P = (1, a, b) \in \mathcal{H}_q$.

Let $\mathcal{P}_\epsilon$ be the place of Σ arising from the branch γ_ϵ of $\mathcal{H}_q$ centred at P_ϵ. Since $q \equiv -1 \pmod 3$ implies that $\epsilon^{q+1} = 1$, it follows that the branch γ_ϵ' is centred at the point $P' = (1, a^3, b^3, ab)$. Therefore the places lying over $\mathcal{P}'$ are $\mathcal{P}, \mathcal{P}_\epsilon$ and $\mathcal{P}_{\epsilon^2}'$. In particular, the ramification index $e_\mathcal{P}$ is 1. □

As a corollary, the following lemma is obtained.

Lemma 7.105 *The curve Γ' has degree $q + 1$.*

From the results above, the different $D(\Sigma/\Sigma')$ is $2(P_1' + P_2' + P_3')$. By Hurwitz's Theorem 7.6, $2g - 2 = 3(2g' - 2) + 6$, where $g = \frac{1}{2}(q^2 - q)$ is the genus of Γ. Therefore Γ' and $\mathcal{F}$ have the same genus. Since Σ' is a subfield of Σ'', this gives the following result.

Lemma 7.106 *The fields Σ' and Σ'' are the same.*

In particular, Γ' is birationally equivalent to $\mathcal{F}$. A geometric explanation of this is that $\mathcal{F}$ can be viewed as the projection of Γ' from the point $P_4 = (0, 0, 0, 1)$ to the plane $\mathbf{v}(X_3)$, where $P_4 \notin \Gamma'$.

Now the L-orders are calculated. First, the branch point P_1' is considered. For $i = 0, 1, 2, 3$, let $H_i = \mathbf{v}(X_i)$. From (7.66),

$$I(P_1', H_i \cap \Gamma') = \begin{cases} 0 & \text{for } i = 0, \\ 1 & \text{for } i = 1, \\ \frac{1}{3}(q+1) & \text{for } i = 2, \\ q & \text{for } i = 3. \end{cases}$$

Hence the (L, P_1')-order sequence is $(0, 1, \frac{1}{3}(q+1), q)$. This holds true for both P_2' and P_3'.

If $P = (1, a^3, b^3, ab)$ is another point of Γ', the first three (L, P)-orders are $0, 1, 2$, and the last one is either q or $q+1$. To decide this, $I(P, H \cap \Gamma')$ is calculated for any plane

$$H = \mathbf{v}(u_0 X_0 + u_1 X_1 + u_2 X_2 + u_3 X_3)$$

with $u_0 = -(u_1a^3 + u_2b^3 + u_3ab)$. Then $I(P, H \cap \Gamma')$ takes both values 1 and 2 for some choice of u_1, u_2, u_3. However, $I(P, H \cap \Gamma') = 3$ cannot occur because the matrix

$$\begin{bmatrix} 3a^2a_1 & 3b^2 & a + a_1b \\ 3aa_1^2 & 3b & a_1 \\ a_1^3 & 1 & 0 \end{bmatrix}$$

has zero determinant. Also, $I(P, H \cap \Gamma') > 3$ implies that $I(P, H \cap \Gamma') \geq q$. On the other hand, from Lemma 7.66, $I(P, H \cap \Gamma') \leq q + 1$. Since $\binom{q+1}{q} \not\equiv 0 \pmod p$, Lemma 7.62 shows that the fourth L-order is q. In particular, there are finitely many Weierstrass points with L-orders $(0, 1, 2, q + 1)$.

The precise number M of the Weierstrass points is calculated from the ramification divisor R. Note that

$$\deg R = (1 + 2 + q)(\tfrac{1}{3}(q^2 - q - 8)) + 4(q + 1) = \tfrac{1}{3}(q^3 + 2q^2 + q - 12).$$

The contribution of each of the three fundamental points P_1', P_2', P_3' is $\frac{1}{3}(q+1)-2$. Therefore

$$M = \deg R - (q - 5) = \tfrac{1}{3}(q^3 + 2q^2 - 2q + 3).$$

It may be noted that $M = q^2 + 1 + 2g'q$, where g' is the genus of Γ'.

7.12 THE LINEAR GENERAL POSITION PRINCIPLE

The Riemann–Roch Theorem 6.61 shows how the dimension and the degree of a complete linear series are related. Some more results in this direction are stated in this section using the *Linear General Position Principle*.

DEFINITION 7.107 Let Γ be a non-degenerate irreducible curve in $\mathrm{PG}(r, K)$. A hyperplane H of $\mathrm{PG}(r, K)$ is in *linear general position* if there is no subspace of $\mathrm{PG}(r, K)$ of dimension less than $r - 1$ that contains r common points of Γ and H.

The main result on the Linear General Position Principle is the following.

THEOREM 7.108 *Let Γ be a non-degenerate irreducible curve in* $\mathrm{PG}(r, K)$. *If Γ is not strange, then some hyperplane is in linear general position.*

In particular, if K has zero characteristic, then the Linear General Position Principle is valid. Since the irreducible conic in characteristic 2 is the only non-singular strange curve, see Section 7.19, the Linear General Position Principle is valid for every non-singular curve.

7.13 CASTELNUOVO'S BOUND

If $\mathcal{F}$ is an irreducible plane curve of degree n and genus g, then $g \leq \frac{1}{2}(n-1)(n-2)$ by Definition 5.58. Castelnuovo's bound is a generalisation of this for space curves. The following two lemmas are the main ingredients in the proof of Castelnuovo's Bound.

LEMMA 7.109 *Let A and B be two positive divisors such that both* $\dim |A|$ *and* $\dim |B|$ *are positive. Then*

$$\dim |A| + \dim |B| \leq \dim |A+B| \leq \deg A + \dim |B|.$$

Proof. It may be that $|A|$ has a fixed divisor F; if this is the case, replace A by $D = A - F$. Since $\dim |A| = \dim |D|$ and $\dim |D+B| \leq \dim |A+B|$, if suffices to prove the assertion for D in place of A. Put $r = \dim A = \dim D$. Since $|D|$ has no fixed divisor, $\dim(D - \mathcal{P}) = r - 1$ for every place. Geometrically speaking, the divisors of $|D|$ correspond to the points of $\mathrm{PG}(r, K)$, and those in $|D - \mathcal{P}|$ to the points of a hyperplane $H_{\mathcal{P}}$ of $\mathrm{PG}(r, K)$. Since finitely many hyperplanes cannot cover the whole $\mathrm{PG}(r, K)$, there is a divisor $D_0 \in |D|$ which does not contain any place $\mathcal{P}$ from B. Replace D_0 by D, and put

$$\begin{aligned} |D| &= \{\mathrm{div}(c_0 + \textstyle\sum_{i=1}^{r} c_i x_i) + D \mid \mathbf{c} = (c_0, \ldots, c_r) \in \mathrm{PG}(r, K)\}, \\ |B| &= \{\mathrm{div}(c_0 + \textstyle\sum_{i=1}^{s} c_{r+i} x_{r+i}) + B \mid \mathbf{c} = (c_0, c_{r+1}, \ldots, c_{r+s}) \in \mathrm{PG}(s, K)\}, \\ L &= \{\mathrm{div}(c_0 + \textstyle\sum_{i=1}^{r+s} c_i x_i) + D_0 + B \mid \mathbf{c} = (c_0, \ldots, c_{r+s}) \in \mathrm{PG}(r+s, K)\}. \end{aligned}$$

First, $\mathrm{div}(x_i) \succ -(D+B)$ for every $1 \leq i \leq r+s$. Therefore L consists of effective divisors. Also, $\dim L = r + s$. To show this, it is enough to show that $1, x_1, \ldots, x_r, \ldots, x_{r+s}$ are linearly independent over K. Taking into account that both sets $\{1, x_1, \ldots, x_r\}$ and $\{1, x_{r+1}, \ldots, x_{r+s}\}$ are linearly independent over K, suppose by contradiction that

$$\xi = u_0 + u_1 x_1 + \cdots + u_r x_r = v_1 x_{r+1} + \cdots + v_s x_{r+s} = \eta,$$

with coefficients not all zero. Let $\mathcal{P}$ be a pole of ξ; then $\mathcal{P}$ is also a pole of η. It follows, on the one hand, that $\mathcal{P}$ is in D and, on the other hand, that $\mathcal{P}$ is in B. But this contradicts the choice of D_0. Since $D + B$ is a common divisor of L and $|D + B|$, Theorem 6.23 shows that L is a subseries of $|D + B|$. Hence $r + s = \dim L \leq \dim |D + B|$.

To show the other bound, put $C = D + B$. Then $C \succ B$. By Theorem 6.40,

$$|B| = |C - D| = ||C| - |D||$$

is the subseries of $|C|$ consisting of all the divisors containing D, minus its fixed divisor D. From Theorem 6.26,

$$\dim |B| \geq \dim |C| - \mathrm{ord}\, D,$$

whence it follows that $\dim |D + B| \leq \deg D + \dim B$. □

LEMMA 7.110 *Let Γ be a non-degenerate, irreducible curve in* $\mathrm{PG}(r, K)$ *of degree n that is not strange. If D is the divisor cut out on Γ by a general hyperplane, then*

(i)
$$\dim |kD| - \dim |(k-1)D| \geq k(r-1) + 1, \tag{7.67}$$

for every positive integer $k \leq m$, where $m = \lfloor \frac{n-1}{r-1} \rfloor$;

(ii)
$$\dim |kD| - \dim |(k-1)D| = n, \tag{7.68}$$

for $k > m$.

Proof. Let D be the divisor cut out on Γ by a general hyperplane. Then D can be written as the sum of places corresponding to branches of Γ centred at distinct points. Assume that $k \leq m$, and choose $k(n-1)+1$ places from D. Let Ω be the set of the centres of the corresponding branches of Γ. For each of the $k(n-1)+1$ points $Q_j \in \Omega$, a hypersurface of degree k exists which passes through $\Omega \backslash \{Q_j\}$ but Q_j. To show this, partition the remaining $k(n-1)$ points of Ω into k sets

$$S_1 = \{P_{1,1}, P_{1,2}, \ldots, P_{1,n-1}\}, \ \ldots, \ S_k = \{P_{k,1}, P_{k,2}, \ldots, P_{k,n-1}\}$$

of $n-1$ points each. By the Linear General Position Principle, each is linearly independent over K. Therefore, for every $1 \leq u \leq k$, there is a hyperplane H_u through every point in S_u that does not contain Q_j. The reducible hypersurface $\mathcal{K}_j$ of degree k whose components are the hyperplanes H_u, $u = 1, \ldots, k$, is the required hypersurface of degree k. The resulting $k(n-1)+1$ hypersurfaces $\mathcal{K}_j$ with Q_j ranging over Ω are linearly independent over K. From this, (7.67) follows.

In fact, let Γ be the curve arising from Q with $Q = (1, x_1, \ldots, x_n)$, and let

$$|(k-1)D| = \text{div}(c_0y_0 + c_1y_1 + \cdots + c_ry_r) + (k-1)D$$

with $r = \dim|(k-1)D|$. If the hyperplane $H = \mathbf{v}(F(X_0, X_1, \ldots, X_n))$ cuts out D, put $\xi = F(1, x_1, \ldots, x_n)$. Similarly, if $\mathcal{K}_j = \mathbf{v}(F_j(X_0, X_1, \ldots, X_n))$, put $\eta_j = F_j(1, x_1, \ldots, x_n)$. Let L be the linear series

$$\begin{aligned}&\{\text{div}\,(\textstyle\sum_{i=0}^{r} c_iy_i\xi + \sum_{j=1}^{k(n-1)+1} d_j\eta_j) + kD \mid \\ &\qquad (c_0, \ldots, c_r, d_1, \ldots, d_{k(n-1)+1}) \in \text{PG}(r + k(n-1)+1, K)\}.\end{aligned}$$

The $r+1$ elements $y_i\xi$ together with the $k(n-1)+1$ elements η_j form a linearly independent set over K. Hence (7.67) holds.

For $k > m$, the same argument shows how to find k hyperplanes in $\text{PG}(n, K)$ containing all but any one of the points of Ω. From this, (7.68) follows. □

THEOREM 7.111 (Castelnuovo's Bound) *Let Γ be a non-degenerate irreducible curve in $\text{PG}(n, K)$ of degree d and genus g. If Γ is not strange, then*

$$g \leq \tfrac{1}{2}m(m-1)(n-1) + m\epsilon,$$

where $m = \lfloor \frac{d-1}{n-1} \rfloor$ and $d - 1 = m(n-1) + \epsilon$.

Proof. Let L be the linear series cut out on Γ by hyperplanes. Since L may not be complete, consider the complete linear series L' containing L. Then $L' = |D|$, where D is a divisor cut out on Γ by a hyperplane. As $\dim|D| \geq n$, from Lemma 7.110, it follows by induction that

$$\begin{aligned}\dim|kD| &\geq \tfrac{1}{2}k(k+1)(n-1) + k, \quad k = 1, \ldots, m;\\ \dim|(m+j)D| &\geq \tfrac{1}{2}k(k+1)(n-1) + k + jd, \quad j \geq 1.\end{aligned}$$

However, for sufficiently large j, the divisor $|(m+j)D|$ is non-special. So, by the Riemann–Roch Theorem 6.61, $\dim|(m+j)D| = (m+j)d - g$. Hence

$$\begin{aligned}g &\leq (j+m)d - \frac{1}{2}m(m+1)(n-1) - m - jd\\ &= \tfrac{1}{2}m(m-1)(n-1) + m(d - m(n-1) - 1),\end{aligned}$$

whence the assertion follows. □

A slightly different wording of Castelnuovo's Bound is stated in the following theorem.

THEOREM 7.112 *Let Γ be as in Theorem 7.111. If ϵ is the unique integer with $0 \le \epsilon \le r-2$ and $d-1 \equiv \epsilon \pmod{(r-1)}$, then*

$$g \le c_0(d,r) = \frac{d-1-\epsilon}{2(r-1)}(d-r+\epsilon). \tag{7.69}$$

LEMMA 7.113 *Let D be as in Lemma 7.110, let $\epsilon' = d - m(r-1)$ for ϵ' in $\{2,\ldots,r\}$, and let $g = c_0(d,r)$. If $m \ge 2$, then*

(i) *the dimension of the linear series $|2D|$ is $3r-1$;*

(ii) *there exists a base-point-free $(\epsilon'-2)$-dimensional complete linear series $\mathcal{D}'$ of order $(\epsilon'-2)(m+1)$ such that $(m-1)|D| + \mathcal{D}'$ is the canonical linear series.*

7.14 A GENERALISATION OF CLIFFORD'S THEOREM

Clifford's Theorem 6.79 states that $n \ge 2r$ for any special g_n^r. Here, an alternative proof depending on the Linear General Position Principle is given, which also shows that $n = 2r$ with $n > 0$ only possible for canonical or hyperelliptic curves.

THEOREM 7.114 *A non-canonical special g_n^r, with $n > 0$, on a non-hyperelliptic curve $\mathcal{C}$ satisfies $n > 2r$.*

Proof. Suppose g_n^r is not complete. Choose an effective special divisor D such that $|D| = g_n^r$. Then $|W - D|$ is a $g_{n'}^{r'}$ with $r' \ge 0$. Note that $n' > 0$, since $|D|$ is not the canonical series. From the Reciprocity Theorem 6.78 of Brill and Noether, $n - 2r = n' - 2r'$. Since $n + n' = 2g - 2$, this gives $r - r' = n - g + 1$. Also, $r + r' \le g - 1$ by Lemma 7.109. Therefore $n \ge 2r$ and equality holds if and only if $n' = 2r'$.

Since $n + n' = 2g - 2$, either n or n' is less than or equal to $g - 1$, and suppose that this occurs for n; so $0 < n \le g - 1$. Note as well that $n = 2r$ only occurs when $r + r' = g - 1$; that is, every canonical divisor is the sum of a divisor from $|D|$ and a divisor from $|W - D|$.

Now assume that $n = 2r$ for a non-hyperelliptic curve and consider the canonical curve Γ. Let H be a hyperplane in $\mathrm{PG}(g-1, K)$ such that $\Gamma \cap H$ consists of $2g - 2$ distinct points. Then $\Gamma \cap H$ contains n points, say $P_1, \ldots, P_n$, such that the corresponding places define a divisor $\mathcal{P}_1 + \cdots + \mathcal{P}_n$ in $|D|$. By Theorem 7.45 (iii), these points span a $(\frac{1}{2}n - 1)$-dimensional subspace of $\mathrm{PG}(g-1, K)$. Since $n \le g - 1$, either $\frac{1}{2}n - 1 = g - 2$, or H is not in a linear general position. But Theorem 7.108, together with Theorem 7.45 (i), ensures the existence of a linear general hyperplane section, and this completes the proof. □

This generalisation of Clifford's Theorem 6.79 gives a refinement of the lower bound in Theorem 7.64.

THEOREM 7.115 *Let N be the number of Weierstrass points of a function field Σ which is not rational, elliptic, or hyperelliptic. If its canonical series is classical, then $N \geq 2g+6$.*

Proof. It suffices to note in the second part of the proof of Theorem 7.64 that the bound (7.16) may be replaced by $j_i \leq i-2$, by Theorem 7.114. □

7.15 THE UNIFORM POSITION PRINCIPLE

Another position principle on hyperplane sections of an algebraic curve, which is valid in zero characteristic but not for all curves in positive characteristic, is treated briefly.

DEFINITION 7.116 A finite set S of points in $\mathrm{PG}(n, K)$ is in *uniform position* if, for any two positive integers k, m, all subsets of S of a size k impose the same number of conditions of hypersurfaces of degree m.

In other words, the the dimension of the linear system of all hypersurfaces of degree m containing a subset of S of size k depends only on k but is independent of the choice of the subset.

With this definition, the *Uniform Position Principle* states that the points of a general hyperplane section of an algebraic curve are in uniform position.

Like the General Linear Position Principle, the Uniform Position Principle has been an important tool in investigating relationships between the genus g and the degree d of a space curves of $\mathrm{PG}(n, K)$. Here a few results are quoted without proofs.

A classical theorem of Halphen improves Castelnuovo's genus bound for certain curves in $\mathrm{PG}(3, \mathbf{C})$. Halphen's Theorem holds true in positive characteristic for non-singular curves of sufficiently large degree.

THEOREM 7.117 (Halphen) *In $\mathrm{PG}(3, K)$ let $\mathcal{X}$ be an irreducible non-singular curve of degree d. If either*

$$d \geq \begin{cases} 7, & \text{for } p = 0, \\ 7, & \text{for } p > 2 \text{ and } \mathcal{X} \text{ reflexive}, \\ 25, & \text{for } p > 0 \text{ and } \mathcal{X} \text{ non-reflexive} \end{cases}$$

or $d = 17$ for $p > 0$, then $\mathcal{X}$ lies on a quadric surface provided that

$$g > c_1(d, 3) = \lfloor \tfrac{1}{6}(d^2 - 3d + 6) \rfloor .$$

For the definition of reflexivity of Γ, see Section 7.19.

THEOREM 7.118 *Let $p \neq 2$. Then Γ is not reflexive if and only if $\epsilon_2 > 2$.*

Halphen's Theorem 7.117 extends to certain curves in $\mathrm{PG}(r, K)$ for $r \geq 4$, and it appears to be very useful when seeking a bound $c_\alpha(d, r)$ for the genus of a curve of degree d in $\mathrm{PG}(r, K)$ not lying on any irreducible surface of degree less than $r + \alpha - 1$. The result for the smallest case $\alpha = 1$ is stated below since it plays a role in Section 10.5.

LEMMA 7.119 *Let Γ be an irreducible curve of $\mathrm{PG}(r, K)$ of degree d and genus g, and let*

$$d \geq \begin{cases} 36r, & \textit{for } r \leq 6, \\ 288, & \textit{for } r = 7, \\ 2^{r+1}, & \textit{for } r \geq 8. \end{cases}$$

Then $\mathcal{X}$ lies on a surface of degree at most $r - 1$, provided that

$$g > c_1(d, r) = \frac{d - 1 - \epsilon_1}{2r}(d - r + \epsilon_1 + 1) + \begin{cases} 0 & \textit{if } \epsilon_1 \leq r - 2, \\ 1 & \textit{if } \epsilon_1 = r - 1, \end{cases} \tag{7.70}$$

where ϵ_1 is the unique integer such that $0 \leq \epsilon_1 \leq r - 1$ and $d - 1 \equiv \epsilon_1 \pmod{r}$.

These numbers $c_1(d, r)$ are *Halphen's numbers.* Note that (7.70) for $r = 3$ coincides with the formula in Theorem 7.117.

7.16 VALUATION RINGS

Valuations of a field Σ over a base field F occur frequently in algebraic geometry. Although, for the theory of algebraic curves as developed in this book, this concept is not really needed, it is useful to establish the relation between valuations and places. In the case that $F = K$ and Σ is a field of transcendence degree 1 over K, any place $\mathcal{P}$ of Σ gives rise to a valuation. In fact, all valuations of Σ/K are obtained in this way.

DEFINITION 7.120 The group G is *ordered* if it is an abelian group with a transitive relation $<$ satisfying the following properties:

(i) for all $\alpha, \beta, \gamma \in G$,

$$\alpha < \beta \Rightarrow \alpha + \gamma < \beta + \gamma; \tag{7.71}$$

(ii) for every element $\alpha \in G$, precisely one of the three relations

$$\alpha = 0, \quad \alpha < 0, \quad 0 < \alpha$$

holds.

It may be observed that the additive group $\mathbf{Z}$ with the natural order is the classical example. Another ordered abelian group consists of all ordered pairs (α, β) of integers with respect to the usual sum, $(\alpha, \beta) + (\gamma, \delta) = (\alpha + \gamma, \beta + \delta)$, where $(\alpha, \beta) < (\gamma, \delta)$ if $\alpha < \gamma$ or $\alpha = \gamma$ and $\beta < \delta$.

From (7.71) it follows that, if $\alpha < \beta$ and $\gamma < \delta$, then $\alpha + \gamma < \beta + \delta$.

DEFINITION 7.121 A *valuation of a field* Σ is a surjective mapping

$$v : \Sigma\backslash\{0\} \to \Gamma,$$

where Γ is an ordered abelian group, satisfying the following properties for all $a, b \in \Sigma\backslash\{0\}$:

(i) $v(ab) = v(a) + v(b)$;

(ii) $v(a+b) \geq \min\{v(a), v(b)\}$ when $a + b \neq 0$.

If the symbol ∞ is added to Σ with the natural rules

$$\alpha < \infty, \quad \alpha + \infty = \infty + \alpha = \infty,$$

then both (i) and (ii) hold for all elements $a, b \in \Sigma \cup \{\infty\}$. A valuation is *trivial* if $v(a) = 0$ for any $a \neq 0$. If $\Gamma = \mathbf{Z}$, the valuation is *normalised*.

LEMMA 7.122 *A valuation v satisfies the following properties:*

(i) $v(1) = v(-1) = 0$;

(ii) $v(a) = v(-a)$ *for all* $a \in \Sigma$;

(iii) *if* $v(a) \neq v(b)$ *then* $v(a+b) = \min\{v(a), v(b)\}$.

If F is a subfield of Σ, the valuation v is *over* F provided that $v(a) = 0$ for all $a \in F\backslash\{0\}$.

In the case that $F = K$ and Σ is a field of transcendence degree 1 over K, every place $\mathcal{P}$ of Σ defines a valuation by putting $v(f) = \text{ord}_{\mathcal{P}}\, f$ for every $f \in \Sigma$.

EXAMPLE 7.123 Given a prime number d, any non-zero rational number c can be written in the form

$$c = d^k \cdot \frac{m}{n}$$

with k, m, n integers and n positive such that $d \nmid mn$. Here the integer k, indicating how often d divides c, is uniquely determined by c. Then $v(c) = k$ defines a normalised valuation of $\mathbf{Q}$, the *d-adic valuation of* $\mathbf{Q}$. Any valuation on $\mathbf{Q}$ is either trivial or equivalent to a d-adic valuation for some prime number d.

A major result result on valuations states that, if Σ' is a finite extension of Σ, then any normalised valuation of Σ has an extension to Σ', again a normalised valuation.

THEOREM 7.124 *Let v be a valuation of Σ. Then*

(i) $\mathcal{O} = \{a \mid v(a) \geq 0\}$ *is a ring, the* valuation ring of v;

(ii) *for every non-zero element a of Σ, either $a \in \mathcal{O}$ or $a^{-1} \in \mathcal{O}$*;

(iii) *conversely, if a ring $\mathcal{O}$ has this property, there is a valuation v of Σ for which $\mathcal{O}$ is the ring.*

Proof. First, $\mathcal{O}$ is a ring by (i) and (ii) of Definition 7.121. Since

$$0 = v(1) = v(a \cdot a^{-1}) = v(a) + v(a^{-1}),$$

both $v(a) < 0$ and $v(a^{-1}) < 0$ cannot hold, because together these imply $0 < 0$. This proves (ii).

To show the converse, a relation $\sim$ is defined for $a, b \in \Sigma\backslash\{0\}$: $a \sim b$ if both a/b and b/a are in $\mathcal{O}$; that is, a/b is a unit in $\mathcal{O}$. This is an equivalence relation. For every $a \in \Sigma\backslash\{0\}$, put

$$L_a = \{b \mid a \sim b;\ b \in \Sigma\backslash\{0\}\};$$
$$\Gamma' = \{L_a \mid a \in \Sigma\backslash\{0\}\}.$$

Then Γ' is an ordered group with addition defined by $L_a + L_b = L_{ab}$ and an ordering given by $L_a < L_b$ when a/b is a non-unit in $\mathcal{O}$. Both addition and order do not depend on the representatives chosen. Note that L_1 is the zero element of Γ'. Also, the mapping $a \mapsto L_a$ is a valuation, and the set $\{a \mid L_a \geq 0\}$ coincides with $\mathcal{O}$. □

Let v be a valuation of Σ over K and let Γ be its group. If $\tau : \Gamma \to \Gamma'$ is an order-preserving monomorphism from Γ to an ordered group Γ', then $v' : a \mapsto \tau(v(a))$ is also a valuation over K, and v and v' are regarded as the same valuation. Then Theorem 7.124 shows that a valuation is uniquely determined by its ring.

To give a complete description of all valuations, the concept of morphism of projective fields is used.

DEFINITION 7.125 For a field F, the associated *projective field* is $F \cup \{\infty\}$ together with the following rules:

$$a + \infty = \infty + a = \infty, \quad \text{for } a \in F;$$
$$a \cdot \infty = \infty \cdot a = \infty, \quad \text{for } a \in F \cup \{\infty\}.$$

Here, $\infty + \infty,\ \infty \cdot 0,\ 0 \cdot \infty$ are undefined.

DEFINITION 7.126 Let $F \cup \{\infty\}$ and $F' \cup \{\infty\}$ be two projective fields. A *morphism*

$$\tau : F \cup \{\infty\} \to F' \cup \{\infty\},$$
$$a \mapsto a'$$

has the following properties:

(i) $0 \mapsto 0, \quad 1 \mapsto 1, \quad \infty \mapsto \infty;$

(ii) $a + b \mapsto a' + b'$, when both sides are defined;

(iii) $a \cdot b \mapsto a' \cdot b'$, when both sides are defined.

Let F'' be the set of images under τ. If $\tau(a), \tau(b) \in F''$, then

$$\tau(a + b) = \tau(a) + \tau(b) \in F'$$

and hence $\tau(a) + \tau(b) \in F''$. Similarly, if $\tau(a) \in F''$ and $\tau(a) \neq 0$, then

$$1 = \tau(a \cdot 1/a) = \tau(a) \cdot \tau(1/a);$$

so $\tau(1/a) = 1/\tau(a) \in F'$ and hence $1/\tau(a) \in F''$. From this it follows that F'' is a field. This shows that every morphism $F \cup \{\infty\} \to F' \cup \{\infty\}$ can be assumed to be surjective. A morphism is *trivial* if it fixes ∞ and maps Σ isomorphically onto Σ'. If K is a common subfield of Σ and Σ', the morphism is a *K-morphism* if $\tau(a) = a$ for every $a \in K$.

REMARK 7.127 In number theory, morphisms $F \cup \{\infty\} \to F' \cup \{\infty\}$ are usually called *places* of F on F'.

EXAMPLE 7.128 Let t be an indeterminate over K, and let F be the rational function field $K(t)$ over K. Any element in $K(t)$ is represented by a fraction $f(t)/g(t)$ with $f(X), g(X) \in K[X]$ having no common root. For $a \in K$ with $g(a) \neq 0$, let $f(t)/g(t) \mapsto f(a)/g(a)$; if $g(a) = 0$, and hence $f(a) \neq 0$, let $f(t)/g(t) \mapsto \infty$. This mapping is a K-morphism $F \cup \{\infty\} \to K \cup \{\infty\}$.

EXAMPLE 7.129 Let $u = 1/t$, so that $K(u) = K(t) = F$. Putting $u \mapsto 0$ in the previous example, another morphism $F \cup \{\infty\} \to K \cup \{\infty\}$ is obtained. It can be shown that these are all the K-morphisms of $F \cup \{\infty\}$ onto $K \cup \{\infty\}$ over K.

EXAMPLE 7.130 Let Σ be a field of transcendence degree 1 over K. For every place $\mathcal{P}$ of Σ, put

$$\operatorname{res}_{\mathcal{P}} x = \begin{cases} a & \text{if } \operatorname{ord}_{\mathcal{P}}(x-a) \geq 0, \\ \infty & \text{if } \operatorname{ord}_{\mathcal{P}} x < 0. \end{cases}$$

Then the *residue map* $x \mapsto \operatorname{res}_{\mathcal{P}}$ *associated to* $\mathcal{P}$ is a K-morphism of $\Sigma \cup \{\infty\}$ onto $K \cup \{\infty\}$. In other words, the following result is established.

THEOREM 7.131 *Every residue map is a K-morphism.*

Another result on valuations is that the converse of Theorem 7.131 also holds. The proof needs some preliminary results on a K-morphism $a \mapsto a'$.

LEMMA 7.132 *If $x \mapsto \infty$, then $x \neq 0$ and $1/x \mapsto 0$.*

Proof. As $0 \mapsto 0$ and hence $0 \mapsto \infty$ cannot occur, $x \neq 0$. Since $x \cdot (1/x) = 1$, it follows that $x' \cdot (1/x') = 1$. Therefore $\infty \cdot (1/x)' = 1$, unless $\infty \cdot (1/x)'$ is undefined. Now, whenever $a \neq 0$, then $\infty \cdot a$ is defined as ∞. Since $\infty \cdot (1/x)'$ is undefined, so $(1/x)' = 0$. □

LEMMA 7.133 *If the K-morphism is non-trivial, then $\Sigma' = K$.*

Proof. Let $x \in \Sigma$ be transcendental over K. It suffices to show that if its image x' neither is ∞ nor is in K, then the K-morphism is an isomorphism.

Let $y \neq 0$ be another element in Σ. Since Σ has transcendence degree 1,

$$a_0(x)y^m + a_1(x)y^{m-1} + \cdots + a_m(x) = 0,$$

where $a_0(X), a_1(X), \cdots, a_m(X) \in K[X]$ and $a_0(X) \neq 0$. Hence

$$a_0(x) + a_1(x)y^{-1} + \cdots + a_m(x)(y^{-1})^m = 0.$$

This shows that $y^{-1} \mapsto 0$ is impossible, since $a_0(x)' = a_0(x') \neq 0$. Hence $y \mapsto \infty$ does not occur. Thus no element of Σ is mapped into ∞. Hence

$$(a+b)' = a' + b', \quad (ab)' = a'b'.$$

Thus the K-morphism is an isomorphism of Σ onto Σ' unless $a' = 0$ for some $a \neq 0$.

Now, if $a \neq 0$, then $a \cdot (1/a) = 1$; hence $a' \cdot (1/a)'$ is defined and $a' \cdot (1/a)' = 1$, whence $a' \neq 0$. This completes the proof. □

THEOREM 7.134 *Every non-trivial K-morphism of Σ is the residue map of some place of Σ.*

Proof. By Theorem 7.40, there is an irreducible non- singular curve Γ in $\mathrm{PG}(r, K)$ whose function field $K(\Gamma)$ is Σ. Let Γ be given by the point $P = (x_0, x_1, \ldots, x_r)$ and, as usual, suppose that $x_0 = 1$. Then $\Sigma = K(x_1, \ldots, x_r)$.

Let τ be a non-trivial K-morphism of Σ. By Lemma 7.133, $\tau : \Sigma \to K$. Not all $\tau(x_j) = 0$ or ∞ since $\tau(x_0) = \tau(1) = 1$, but it may happen that $\tau(x_i) = 0$ for s indices with $s \geq 1$. Nevertheless, by changing the coordinate functions in the following way, it may be assumed that this does not occur. If $s \geq 1$ and $\tau(x_1) = \infty$, then Γ is also defined by the point $P_i = (x_1^{-1}, 1, x_2 x_1^{-1}, \ldots, x_r x_1^{-1})$. If $\tau(x_j) \neq \infty$, then

$$\tau(x_j x_1^{-1}) = \tau(x_j \cdot (1/x_1)) = \tau(x_j)\tau(1/x_1) = 0.$$

Thus more finite quantities appear for the new form of the coordinate function than for the old; 0 and 1 are in the first two coordinates. There are now at most $s - 1$ indices at which τ is infinite. In this way s can be reduced to 0.

Suppose henceforth that $s = 0$; that is, $\tau(x_1), \ldots, \tau(x_r)$ are all finite. Then

$$(x_1, \ldots, x_r) \mapsto (\tau(x_1), \ldots, \tau(x_r))$$

is a specialisation; see Section 7.17. Hence $Q = (\tau(x_1), \ldots, \tau(x_r))$ is a point of Γ, the centre of τ. Since Q is the centre of a branch γ of Γ, there is a place $\mathcal{P}$ of Σ arising from γ. Then the K-morphism $\mathrm{res}_{\mathcal{P}}$ is centred at Q.

It remains to prove that there is only one K-morphism centred at Q, as this shows that $\tau = \mathrm{res}_{\mathcal{P}}$.

Assume on the contrary that τ' is another K-morphism centred at Q. Choose $y \in \Sigma$ satisfying $\tau(y) \neq \tau'(y)$. Actually, y may be chosen such that $\tau(y)$, $\tau'(y)$ are both finite. To show this, suppose that $\tau'(y) = \infty$. Two cases occur according as either $\tau(y) \neq 0$ or $\tau(y) = 0$. In the former case, $\tau(y^{-1}) \neq \infty$ as well; also, $\tau(y \cdot y^{-1}) = \tau(y) \cdot \tau(y^{-1})$ is defined and is equal to 1. Then $\tau(y^{-1}) = \tau'(y^{-1}) = 0$, and replacing y by y^{-1} gives the result. In the latter case, $\tau(y) = 0$ and $\tau'(y) = \infty$. Replacing y by $1 + y$, the former case occurs again. Therefore, in each case, there exists y with $\tau(y)$, $\tau'(y)$ finite and unequal.

Now consider the irreducible curve Γ' given by the point $P' = (x_1, \ldots, x_r, y)$. Then $Q = (\tau(x_1), \ldots, \tau(x_r), \tau(y))$ and $Q' = (\tau(x_1), \ldots, \tau(x_r), \tau(y))$ are two distinct points of Γ'. Let δ and δ' be branches of Γ' centred at Q and Q'. Then $\delta \neq \delta'$ as $Q \neq Q'$. If $\mathcal{P}$ and $\mathcal{P}'$ are the corresponding places of Σ, then $\mathcal{P} \neq \mathcal{P}'$. On the other hand, these places correspond to branches of Γ having the same centre Q. But this is impossible since Q is a non-singular point of Γ. □

The relation between morphisms and valuations is described in the following theorem. Recall that morphisms are tacitly assumed to be surjective.

THEOREM 7.135 *Every morphism $\tau : \Sigma \cup \{\infty\} \to \Sigma' \cup \{\infty\}$ defines a valuation in a standard way; namely, the set of elements in Σ mapping to Σ' forms a valuation ring, and this yields the valuation. Conversely, every valuation ring $\mathcal{O}$ yields a morphism in a standard way, and from the morphism the valuation ring can be recovered.*

Proof. First, the elements in Σ mapping to Σ' form a ring $\mathcal{O}$. If $a \in \Sigma$ but $a \notin \mathcal{O}$, that is, $\tau(a) = \infty$, then $\tau(a^{-1}) = 0$; hence $a^{-1} \in \mathcal{O}$. Therefore $\mathcal{O}$ is a valuation ring.

Before proving the converse, it is shown how Σ' can be described up to an isomorphism in terms of $\mathcal{O}$. For $a, b \in \mathcal{O}$, write $a \sim b$ when $\tau(a) = \tau(b)$; this is an equivalence relation. Note that $\tau(a) = \tau(b)$ if and only if $\tau(a - b) = 0$. In fact, $\tau(-b) = -\tau(b)$ follows from the definition of a morphism, and this implies that $\tau(a - b) = \tau(a) - \tau(b)$. Further, $\tau(x) = 0$ if and only if x is a non-unit in $\mathcal{O}$. In fact, if x is a non-unit, then $\tau(x^{-1}) = \infty$, as otherwise $1 = \tau(x \cdot x^{-1}) = \tau(x) \cdot \tau(x^{-1}) = 0$, that is, $x \notin \mathcal{O}$. In this way, τ is eliminated from the definition of $a \sim b$. Hence $a \sim b$ if and only if $a - b$ is a non-unit in $\mathcal{O}$.

Let $L_a = \{x \mid x \sim a;\ x \in \mathcal{O}\}$. Then $L_a \mapsto \tau(a)$ is a one-to-one correspondence between the set L consisting of all L_a and the field Σ'. Defining

$$L_a + L_b = L_{a+b}, \quad L_a \cdot L_b = L_{ab},$$

the correspondence becomes an isomorphism.

One more observation is needed to prove the converse. If $\mathcal{O}$ is the valuation ring arising from a morphism τ, then the non-units in $\mathcal{O}$ form an ideal. This can be shown by observing that the non-units are those elements $x \in \mathcal{O}$ for which $\tau(x) = 0$ holds. Hence, if $\tau(x_1) = \tau(x_2 = 0$ and $a \in \mathcal{O}$, then $\tau(x_1 + x_2) = 0$ and $\tau(ax_1) = 0$. Actually, a standard argument shows that this holds without assuming that $\mathcal{O}$ arises from τ. Also, in any valuation ring $\mathcal{O}$, the non-units of $\mathcal{O}$ form the unique maximal ideal of $\mathcal{O}$. a non-unit, the trivial case

Now, starting from the valuation ring $\mathcal{O}$, define $a \sim b$ if $a - b$ is a non-unit in $\mathcal{O}$; this is an equivalence relation. Put $L_a = \{x \mid x \sim a; x \in \mathcal{O}\}$, and define $L_a + L_b = L_{a+b}$ and $L_a \cdot L_b = L_{ab}$. The set L consisting of all L_a with these two operations is a field Σ'. Also, putting $\tau(a) = L_a$ for $a \in \mathcal{O}$ and $\tau(a) = \infty$ for $a \notin \mathcal{O}$, the map $\tau : \Sigma \cup \{\infty\} \to \Sigma' \cup \{\infty\}$ is a morphism such that $\mathcal{O}$ is the associated valuation ring. □

REMARK 7.136 In the theory of function fields, the maximal ideal of a valuation ring is usually called a *place*.

Theorem 7.124 shows that there is a standard one-to-one correspondence between the valuations of a field Σ and its morphisms or, more precisely, the morphisms of $\Sigma \cup \{\infty\}$. Also, if there is a base field F of Σ, then a valuation over F gives rise to an F-morphism, and vice versa. Thus, to find all valuations of Σ over

F, it is sufficient to find all F-morphisms of Σ. When $F = K$ and Σ is a field of transcendence degree 1 over K, all K-morphisms are determined, and hence all valuations over K are known. The K-morphisms all arise as residue maps of places, by Theorem 7.134. Let $\mathcal{P}$ be a place of Σ. Then its valuation ring $\mathcal{O}_{\mathcal{P}}$ consists of all elements with finite residue or, in other words, of the elements of non-negative order. Thus the ring of the valuation $\text{ord}_{\mathcal{P}}$ is the same as the ring of the residue map $\text{res}_{\mathcal{P}}$. From this discussion the following result is obtained.

THEOREM 7.137 *If Σ is a field of transcendence degree 1 over K, then every valuation of Σ is of the form* $\text{ord}_{\mathcal{P}}$ *for some place $\mathcal{P}$ of Σ.*

REMARK 7.138 The concept of a place of Σ is usually defined either via morphisms or via valuations. The account in this section justifies the use of the term *place* for a formally different, although equivalent, concept.

7.17 CURVES AS ALGEBRAIC VARIETIES OF DIMENSION ONE

It is shown that the concept of an irreducible variety of dimension 1 is equivalent to the concept of an irreducible space curve as defined and investigated in this book.

Let $\mathcal{F}$ be an irreducible plane curve, and let $K(\mathcal{F})$ be its function field. Given a point $Q = (x_0, x_1, \ldots, x_r)$ in $\text{PG}(r, K(\mathcal{F}))$ such that $K(Q) = K(\mathcal{F})$, the associated irreducible space curve Γ of $\text{PG}(r, K)$ consists of the centres P of the branches $(\sigma(x_0, \sigma(x_1), \ldots, \sigma(x_r))$ with the corresponding place $\mathcal{P}$ ranging over all places of $K(\mathcal{F})$ with σ a primitive representation of $\mathcal{P}$. Since not all coordinates of Q are zero, take $x_0 \neq 0$, and write $y_i = x_i/x_0$. Here, and in the rest of the section, i ranges from 1 to r. Let $\text{AG}(r, K)$ be the r-dimensional affine space whose hyperplane at infinity is $\mathbf{v}(X_0)$. Then the affine points of Γ are centres of the branches of Γ that arise from the places $\mathcal{P}$ for which $\text{ord}_{\mathcal{P}}(y_i) \geq 0$. The set of these points may be considered as an affine part or as an affine curve, Γ_0, coming from Γ.

It is shown in this section that Γ_0 is an irreducible affine variety of dimension 1 and conversely. From this, the equivalence of the concepts of an irreducible curve and of a projective algebraic variety of dimension 1 can be deduced by standard arguments.

For the present purpose, the most convenient wording of the definition of an affine variety depends on the concept of a specialisation.

DEFINITION 7.139 If $x_1, \ldots, x_r \in \Sigma$, with Σ an extension field of K, then a point $A = (a_1, \ldots, a_r)$ of $\text{AG}(r, K)$ is a *specialisation* of the point $Q = (x_1, \ldots, x_r)$ of $\text{AG}(r, \Sigma)$ when every $f \in K[X_1, \ldots, X_r]$ that vanishes at $Q = (x_1, \ldots, x_r)$ also vanishes at $A = (a_1, \ldots, a_r)$; write $(x_1, \ldots, x_r)\underset{K}{\rightarrow}(a_1, \ldots, a_r)$ or, briefly, $Q\underset{K}{\rightarrow}A$.

DEFINITION 7.140 A non-empty set points in $\text{AG}(r, K)$ is an *affine irreducible variety* if it consists of all specialisations in $\text{AG}(r, K)$ of a point $Q = (x_1, \ldots, x_r)$ of $\text{AG}(r, \Sigma)$, where Σ is a field extension of K. The point Q is a *generic point* of the variety. If the transcendence degree $[\Sigma : K]$ is equal to d, then the variety has *dimension d*.

This can be carried over to projective space by using homogeneous polynomials in such a way that the following holds. Let $\mathcal{V}$ be an irreducible projective variety of $\mathrm{PG}(r, K)$. If H is any hyperplane in $\mathrm{PG}(r, K)$, than $\mathcal{V}_1 = \mathcal{V} \cap H$ is either empty or an irreducible affine variety.

Let Γ be an irreducible curve of $\mathrm{PG}(r, K)$ given by a point $Q = (x_1, \ldots, x_r)$. It is first shown that the affine points of Γ are points of an one-dimensional affine variety V_1 associated to the same point Q. Let $\mathcal{P}$ be a place of $K(\Gamma) = K(Q)$ such that $\mathrm{ord}_{\mathcal{P}}\, x_i \geq 0$. Choose a primitive representation of $\mathcal{P}$ and write

$$\sigma(x_i) = a_i + b_i t + \cdots .$$

Let $f \in K[X_1, \ldots, X_r]$ be a polynomial that vanishes at $(x_1, \ldots, x_r)$. Since σ is a K-monomorphism from $K(\Gamma)$ to K, this implies that f also vanishes at

$$(a_1 + \cdots,\ \ldots,\ a_r + \cdots).$$

Putting $t = 0$ gives $f(a_1, \ldots, a_r) = 0$. Therefore $(x_1, \ldots, x_r) \mapsto (a_1, \ldots, a_r)$, and the assertion follows.

To show the converse, the following result must be proved.

Lemma 7.141 *Every point of V_1 is the centre of a branch of Γ.*

Proof. At least one of the coordinates x_i of Q, say x_1, is a separable variable. Thus the other coordinates are separable variables of $K(x_1, \ldots, x_r)/K(x_1)$. Theorem A.1, that is, the Theorem of the Primitive Element ensures the existence of a primitive element of $K(x_1, \ldots, x_r)/K(x_1)$, which can be written as a linear combination $c_2x_2 + \cdots + c_rx_r$ with $c_j \in K$. Here, and in the rest of the section, j ranges from 2 to r. Also, replacing $\alpha + \Lambda\beta$ by $\Lambda_2x_2 + \cdots + \Lambda_rx_r$ in the proof of Theorem A.1, a polynomial $g(\Lambda_2, \ldots, \Lambda_r)$ exists such that $c_2x_2 + \cdots + c_rx_r$ is a primitive element of $K(x_1, \ldots, x_r)/K(x_1)$ provided that $g(c_2, \ldots, c_r) \neq 0$. Now $P = (x_1, c_2x_2 + \cdots + c_rx_r)$ is viewed as the generic point of an irreducible plane curve $\mathcal{F} = \mathbf{v}(f)$. By Definition 5.1, $\mathcal{F}$ is the locus of the specialisations of its generic point. Since $(x_1, \ldots, x_r) \mapsto (0, \ldots, 0)$, every polynomial in two indeterminates which vanishes at $(x_1, c_2x_2 + \cdots + c_rx_r)$ also vanishes at $(0, 0)$. Therefore

$$(x_1, c_2x_2 + \cdots + c_rx_t) \mapsto (0, 0).$$

In particular, $f(0, 0) = 0$; that is, $\mathcal{F}$ contains the point $O = (0, 0)$. Hence at least one of the branches of $\mathcal{F}$ arising from the places $\mathcal{P}_i$ is centred at O. Such a place is a zero of $c_2x_2 + \cdots + c_rx_r$.

It may be supposed that $x_2, \ldots, x_r$ are linearly independent over K. Otherwise, one of them, say x_r, is a linear combination of $x_2, \ldots, x_{r-1}$. Then an inductive argument on r shows the existence of a place $\mathcal{P}$ of

$$K(x_1, \ldots, x_{r-1}) = K(x_1, \ldots, x_r)$$

such that $x_2, \ldots, x_{r-1}$ are zeros of $\mathcal{P}$. Then x_r, being a linear combination of $x_2, \ldots, x_{r-1}$, is also a zero of $\mathcal{P}$. Hence the branch of Γ associated to $\mathcal{P}$ is centred at $P = (0, \ldots, 0)$, and the proof of the lemma is finished.

Choose elements $c_{jk} \in K$ with k ranging over $1, \ldots, r-1$ such that the matrix

$$\begin{bmatrix} c_{21} & \cdots & c_{r1} \\ \vdots & \vdots & \vdots \\ c_{2,r-1} & \cdots & c_{r,r-1} \end{bmatrix}$$

has non-zero determinant. In this way, $r-1$ elements of the form $c_2 x_2 + \cdots + c_r x_r$ are obtained which are linearly independent over K.

Generalising this, for any positive integer $N \geq r-1$, there are $N-1$ elements of the same form $c_2 x_2 + \cdots + c_r x_r$, no $n-1$ of them linearly dependent; it is enough to choose the entries c_{jk} by requiring certain determinants not to vanish. It is also possible to require the conditions $g(c_{2k}, \ldots, c_{r-1,k}) \neq 0$, so that all $c_{2k} x_2 + \cdots + c_{r-1,k} x_r$ are primitive elements of $K(x_1, \ldots, x_r)/K(x_1)$.

Now assume $N = (k+1)(r-1)$. Then, for some $r-1$ values of the index k, the elements $c_{2,k} x_2 + \cdots + c_{r,k} x_r$ are zeros of the same place $\mathcal{P}$. Solving these linear homogeneous equations shows that $x_2, \ldots, x_r$, as x_1, all are zeros of $\mathcal{P}$. Thus the corresponding branch of Γ is centred at $P = (0, \ldots, 0)$. □

REMARK 7.142 In the literature, an irreducible projective variety of dimension 1 is usually called a *projective, geometrically irreducible, algebraic curve*. It has been shown here that this term is equivalent, even if formally different, to the term used here of an irreducible algebraic curve.

7.18 EXERCISES

1. (Lüroth's Theorem) Show that every subfield of a rational field properly containing K is also rational.

2. For $p > 2$, let $\mathcal{F} = \mathbf{v}(f(X,Y))$, with
$$f(X,Y) = V_0^p + V_1^p X + V_2^p Y + V_3^p X^2 + V_4^p XY + V_5^p Y^2,$$
where $V_0, V_1, \ldots, V_5 \in K[X,Y]$ are polynomials satisfying the equation
$$V_0(4V_3V_5 - V_4^2) = V_1^2 V_5 + V_2^2 V_3 - V_1 V_2 V_4.$$
Show that $\mathcal{F}$ is non-classical with respect to lines, as its Hessian curve is $\mathbf{v}(2(4V_2V_5 - V_4^2)^p f(X,Y))$.

3. For $p = 2$, let $\mathcal{F} = \mathbf{v}(Y + X^3 + Y^8)$. Show that the orders of $\mathcal{F}$ with respect to the linear system of conics are $0, 1, 2, 3, 4, 6$.

4. For $p > 5$, let $\mathcal{G} = \mathbf{v}(G(X,Y))$ be the non-singular curve with
$$G(X,Y) = 1 + X^p + XY^p + Y^{p+1} + X^{p-1}.$$
Show that $\mathcal{G}$ is non-classical with respect to the linear system of conics, and that (7.22) holds with $h(X,Y) = X$ but not with $h(X,Y) = 1$.

5. Let $\mathcal{F}$ be the curve in Example 1.42. Show that, if $p \geq 5$ and $k^2 \not\equiv 1 \pmod{p}$, then $\mathcal{F}$ is classical with respect to the linear system of conics.

6. Given $g+1$ distinct elements $a_i \in K$, let $\Sigma = K(x,y)$ with
$$y^2 + y = \sum_{i=1}^{g+1} (x+a_i)^{-1}.$$
Show that, if $p=2$, then Σ has $g+1$ distinct Weierstrass points.

7. Show that the Klein quartic, as in Example 7.18, is not hyperelliptic.

8. Let $\mathcal{F}$ be a hyperelliptic curve. In characteristic $p \neq 2$, deduce from Example 6.90 and Theorem 7.94 that the smallest non-gap number $m_1(\mathcal{P})$ at an ordinary place of $K(\mathcal{F})$ is $g+1$. Extend the result to characteristic $p=2$.

9. Show that two distinct irreducible curves of $\mathrm{PG}(r,K)$ have only finitely many common points.

10. Show that $\epsilon_m + j_i(P) \leq j_{i+m}(P)$ for $i+m \leq r$.

11. Let $p \geq 5$. For a non-hyperelliptic curve $\mathcal{F}$ of genus 4, show the existence of a place $\mathcal{P}$ of $K(\mathcal{F})$ such that 4 is the first non-gap at $\mathcal{P}$.

12. Let $(0,1,p^m)$ be the order sequence of an irreducible plane curve $\mathcal{F}$ which is non-classical with respect to lines, and let $\mathcal{P}$ be any place of $K(\mathcal{F})$ with order sequence $(0,j_1,j_2)$. Adapting the techniques of Section 7.8, show that one of the following holds:

    (a) $j_1 \equiv 0 \pmod{p^n}$;

    (b) $j_2 \equiv 0 \pmod{p^n}$;

    (c) $j_2 \equiv j_1 \pmod{p^m}$.

13. Show that a plane non-singular cubic $\mathbf{v}(f(X,Y))$ is supersingular over $\overline{\mathbf{F}}_q$, $q=p^h$, if and only if the coefficient of $(XY)^{p-1}$ in $F(X,Y)^{p-1}$ is zero.

7.19 NOTES

The first four sections follow Seidenberg [400, Chapter 20]. For studies of the Klein quartic, see Elkies [117] and Duursma [109].

For coverings of curves, see also [7].

Section 7.6 follows Stöhr and Voloch [432]; see also Schmidt [387] and Laksov [289].

For rational strange curves of degree $q+1$, there are three types; see Ballico and Hefez [38]. Bayer and Hefez [40] described equations and singularities of plane strange curves and gave an upper bound on the genus of a strange curve of $\mathrm{PG}(r,K)$ in terms of its degree. For the General Order of Contact Theorem, see Hefez and Kleiman [204].

Proposition 7.70 is due to Pardini [357] for $m=1$ and to Homma [232] for arbitrary m. If singular points are allowed in Proposition 7.70, then $h(X,Y)=1$ can still be ensured under some natural conditions; see [200]. However, it is not possible in general to get rid of $h(X,Y)$, the curve in Exercise 2 being such an example; here, $\deg \mathcal{F} \leq q$ with $q = p^m$. Hefez [200] and Homma [229] show that equality holds for just one type of curve.

Fermat curves which are not classical with respect to L_2 were investigated by Garcia and Voloch [161]; if $p > 5$, then the Fermat curve $\mathcal{F} = \mathbf{v}(X^n + Y^n + 1)$ is non-classical with respect to L_2 if and only if

$$p \mid (n-2)(n-1)(n+1)(2n-1).$$

The classical proof of Salmon's Theorem 7.90 works for $p \neq 2, 3$; see [497, Section 9.3], [353, Chapter 15] and [216, Chapter 11].

Related to Theorem 7.103 for $p = 2$, Schmidt [387] exhibited a family of hyperelliptic curve with only one Weierstrass point. Henn [207] exhibited non-hyperelliptic curves with few Weierstrass points.

The Linear Position Principle in positive characteristic and, more generally, the general hyperplane section of a curve, were investigated by Ballico [27, 28, 29, 30, 31, 32, 34]; see also Ballico and Migliore [39], and Ballico and Cossidente [35].

The original paper of Castelnuovo is [72]. Curves hitting Castelnuovo's Bound, that is, those with genus equal to Castelnuovo's number $c_0(d, r)$ are *extremal* and they have several remarkable properties; see [11], [191, Chapter 3], [19, Chapter 3, Section 2]. One of these properties is stated in Lemma 7.113 since it is used in Section 10.5 where certain curves with $g = c_0(d, r)$ are considered. The proof can be deduced from that of Theorem 7.111. For more information, see [11, p. 361 and Lemma 3.5].

The main reference for the Gaussian duality of space curves is [163], in which the authors determine when Γ coincides with its Gaussian dual Γ', up to a linear transformation. A third generalisation of the concept of the dual curve of an irreducible plane curve $\mathcal{C}$ is the dual hypersurface of an irreducible curve Γ of $\mathrm{PG}(r, K)$. The idea is to consider all hyperplanes of $\mathrm{PG}(r, K)$ containing at least one tangent line to Γ and regard them as points in the dual projective space $\mathrm{PG}(r, K)^* \cong \mathrm{PG}(r, K)$. Such points, but finitely many of them, lie on an irreducible hypersurface of Π which is the *dual hypersurface* Δ *of* Γ. To find a generalisation of the notion of a reflexive curve, more is needed, namely the concept of the *conormal variety* $C(\mathcal{Z})$ of an irreducible algebraic variety $\mathcal{Z}$ defined over K; see [200].

The variety $C(\mathcal{Z})$ is defined to be the closure in $\mathrm{PG}(r, K) \times \mathrm{PG}(r, K)^*$ of the set of all pairs (P, H), where P is a non-singular point of $\mathcal{Z}$ and H is a hyperplane of $\mathrm{PG}(r, K)$ containing the tangent space of $\mathcal{Z}$ at P. The dimension of $C(\mathcal{Z})$ is $r - 1$. The *dual variety* of $\mathcal{Z}'$ of $\mathcal{Z}$ is the projection of $C(\mathcal{Z})$ onto $\mathrm{PG}(r, K)^*$. A variety is *reflexive* when $C(\mathcal{Z}) = C(\mathcal{Z}')$. From the definition, if $\mathcal{Z}$ is reflexive then $(\mathcal{Z}')' = \mathcal{Z}$ and $\mathcal{Z}'$ is also reflexive.

To identify the inequivalent elliptic curves in Theorem 7.89, see [216, Chapter 11].

A characteristic-free proof of Theorem 7.108 is found in [365].

A proof of the Uniform Position Principle in characteristic zero, due to Eisenbud and Harris, is found in [191]. Rathmann [365] extended the result to positive characteristic provided that $n \geq 4$. He also showed the validity of the Uniform Position Principle for a large family of curves of $\mathrm{PG}(3, K)$ including all reflexive curves. It should be noted that the validity of both position principles depends on the monodromy group of the curve, which is a highly transitive permutation group. In characteristic p, the classification of highly-transitive permutation groups depends on the classification of all finite simple groups. For more results, see [191].

The general theory of algebraic varieties is found in several textbooks, notably in [410] and [503]. Good references for Section 7.17 are [301] and [19].

The result that the irreducible conic in characteristic 2 is the only non-singular strange curve is due to Samuel; see [192, Theorem IV.3.9].

The characterisation of reflexive curves in Theorem 7.118 is due to Hefez and Kleiman [200]; see also [204].

For a proof of Lemma 7.119, see [191, Theorem 3.22] and [365, Corollary 2.8].

A full account of results related to Halphen's Theorem is found in the books, [191], [182], as well as in the papers, [79], [78].

For a historical account of Halphen's Theorem, see [192, p. 349], [182] and [190]. Halphen's Theorem remains essentially valid in any characteristic; see [191, Theorem 3.13] for $p = 0$ and [33] for $p > 0$.

For places in the function field setting, see [428, I.1.8. Definition].

For Exercise 5, see [1].

For Exercise 10, see [123], [234], [233], and [140].

For Exercise 11 and more results on Weierstrass points of curves with low genus, see [261].

PART 2
Curves over a finite field

Chapter Eight

Rational points and places over a finite field

In this chapter, $K = \overline{\mathbf{F}}_q$, the algebraic closure of the finite field $\mathbf{F}_q$ of order q. For any r, the space $\mathrm{PG}(r, K)$ contains the finite projective spaces $\mathrm{PG}(r, q^i)$ with $i \geq 1$.

If Γ is an algebraic curve embedded in $\mathrm{PG}(r, K)$, the set of points of Γ lying in $\mathrm{PG}(r, q)$ is a natural geometric object to investigate. To do this in a birational spirit, the concept of an $\mathbf{F}_q$-rational place, that is, a place defined over $\mathbf{F}_q$, is required. As in Section 5.3, the idea is to derive this concept from that of an $\mathbf{F}_q$-rational branch of Γ, even though the approach works properly if any non-singular point of Γ is the centre of a unique $\mathbf{F}_q$-rational branch. The latter hypothesis is satisfied by all curves which are defined over $\mathbf{F}_q$; for an irreducible plane curve $\mathcal{F} = \mathbf{v}(f(X, Y))$, this just means that $f(X, Y) \in \mathbf{F}_q[X, Y]$. It should be noted, however, that a singular point of $\mathcal{F}$ lying in $\mathrm{PG}(2, q)$ may happen not to be the centre of any $\mathbf{F}_q$-rational branch of $\mathcal{F}$. Therefore a bijection between $\mathbf{F}_q$-rational places of $K(\mathcal{F})$ and points of $\mathcal{F}$ lying in $\mathrm{PG}(2, q)$ can only be ensured for non-singular curves.

The foundation of the theory of the algebraic curves defined over a finite field is developed in the first four sections. The central problem on algebraic curves over finite fields is to estimate the number of their $\mathbf{F}_q$-rational branches. The study starts with the Stöhr–Voloch Theorem 8.65 which is obtained in a way that parallels Section 7.6. An interesting consequence of the Stöhr–Voloch Theorem 8.65 is that curves with a large number of $\mathbf{F}_q$-rational branches are mostly non-classical. This will be apparent in Sections 8.6 and 8.7, where $\mathbf{F}_q$-rational branches of irreducible plane curves that are non-classical with respect to lines and conics are thoroughly investigated.

The study of this problem is continued in Chapters 9 and 10.

8.1 PLANE CURVES DEFINED OVER A FINITE FIELD

Definition 8.1 A plane curve $\mathcal{F} = \mathbf{v}(f(X, Y))$ is *defined over* $\mathbf{F}_q$ if there is a non-zero constant $c \in K$ such that $cf(X, Y) \in \mathbf{F}_q[X, Y]$.

An irreducible plane curve $\mathcal{F}$ defined over $\mathbf{F}_q$ is $\mathbf{F}_q$-*rational.*

Lemma 8.2 *An irreducible plane curve $\mathcal{F}$ is defined over $\mathbf{F}_q$ if and only if, for every point $P = (a_0, a_1, a_2) \in \mathcal{F}$, its image point $P^q = (a_0^q, a_1^q, a_2^q)$ under the Frobenius collineation Φ is also in $\mathcal{F}$.*

Proof. Let $\mathcal{F} = \mathbf{v}(F)$ where $F \in K[X_0, X_1, X_2]$ is a homogeneous polynomial of degree d. If

$$F = \sum a_{ij} X_0^i X_1^j X_2^{d-(i+j)},$$

then the homogeneous polynomial

$$G = \sum a_{ij}^q X_0^i X_1^j X_2^{d-(i+j)}$$

defines the irreducible plane curve $\mathcal{G} = \mathbf{v}(G)$. Thus $P = (a_0, a_1, a_2) \in \mathcal{F}$ if and only if $P^q = (a_0^q, a_1^q, a_2^q) \in \mathcal{G}$. Now, if $\mathcal{F}$ is defined over $\mathbf{F}_q$, then $\mathcal{F} = \mathcal{G}$ and hence $P^q \in \mathcal{F}$.

Conversely, if $P^q \in \mathcal{F}$ then P^q is a common point of $\mathcal{F}$ and $\mathcal{G}$. If this occurs for every point $P \in \mathcal{F}$, the irreducible curves $\mathcal{F}$ and $\mathcal{G}$ have infinitely many common points. By Bézout's Theorem, $\mathcal{F} = \mathcal{G}$ and hence $\mathcal{F}$ is defined over $\mathbf{F}_q$. □

LEMMA 8.3 *The irreducible plane curve $\mathcal{F} = \mathbf{v}(F(X,Y))$ is defined over $\mathbf{F}_q$ if and only if, for every generic point $P = (x,y)$ of $\mathcal{F}$, the image point $P^q = (x^q, y^q)$ of $P = (x,y)$ under the Frobenius collineation is also a generic point of $\mathcal{F}$.*

Proof. If $F(X,Y) \in \mathbf{F}_q[X,Y]$, then $F(x,y) = 0$ implies $F(x^q, y^q) = 0$. Therefore $P^q = (x^q, y^q)$ is a point of $\mathcal{F}$, and hence a generic point of it. To show the converse, assume without loss of generality that some coefficient of $F(X,Y)$ is equal to 1. Now $F(x,y) = 0$ implies that $G(x^q, y^q) = 0$, where

$$F = \sum a_{ij} X^i Y^j, \quad G = \sum a_{ij}^q X^i Y^j.$$

If $P^q = (x^q, y^q) \in \mathcal{F}$, then it is a common generic point of $\mathcal{F}$ and the irreducible plane curve $\mathcal{G} = \mathbf{v}(G)$. By Remark 5.2, $F(X,Y) = c\,G(X,Y)$. Actually, $c = 1$ by the assumption. Therefore $F(X,Y) \in \mathbf{F}_q[X,Y]$. □

Let $\mathcal{F} = \mathbf{v}(f(X,Y))$ be an irreducible plane curve defined over $\mathbf{F}_q$.

DEFINITION 8.4 For a generic point $P = (x,y)$ of $\mathcal{F}$, the set of all elements $u(x,y)/v(x,y)$, with $u(X,Y), v(X,Y) \in \mathbf{F}_q[X,Y]$ and $F(X,Y) \nmid v(X,Y)$, form a subfield $\mathbf{F}_q(\mathcal{F})$ of $K(\mathcal{F})$, the *$\mathbf{F}_q$-rational function field of $\mathcal{F}$.*

By Theorem 5.7, this definition is independent of the choice of the generic point.

DEFINITION 8.5 (i) A rational transformation ω of $\Sigma = K(\mathcal{F})$ which maps $\mathbf{F}_q(\mathcal{F})$ into itself is an *$\mathbf{F}_q$-rational transformation*, and $\Sigma' = \omega(\Sigma)$ is an $\mathbf{F}_q$-rational function field, as well.

(ii) If $\mathcal{F}'$ is the image of $\mathcal{F}$ under an $\mathbf{F}_q$-rational transformation, then $\mathcal{F} \to \mathcal{F}'$ is an *$\mathbf{F}_q$-rational covering*. This term is also used if $\mathcal{F}$ and $\mathcal{F}'$ are replaced by $\mathbf{F}_q$-rational non-singular models $\mathcal{X}$ and $\mathcal{X}'$ of Σ and $\Sigma' = K(\mathcal{F}')$.

8.2 $\mathbf{F}_q$-RATIONAL BRANCHES OF A CURVE

Let $f = \sum a_i t^i \in K((t))$. For any integer n, the *n-th conjugate of f* is

$$f^{(n)} = \sum a_i^{q^n} t^i.$$

For any $f, g \in K((t))$ and integers n, m, the following properties hold:

$$(f+g)^{(n)} = f^{(n)} + g^{(n)}; \quad (fg)^{(n)} = f^{(n)}g^{(n)}; \quad (f^{(n)})^{(m)} = f^{(n+m)}. \quad (8.1)$$

Further, every $f \in K((t))$ can be written as $f = g^{(n)}$ with $g \in K((t))$. Hence the n-th conjugate map $\kappa : f \mapsto f^{(n)}$ is an automorphism of $K((t))$. Note that κ is not a K-automorphism as $\kappa(a) \neq a$ for any constant $a \notin \mathbf{F}_{q^n}$. Now $f = f^{(n)}$ if and only if $f \in \mathbf{F}_{q^n}((t))$. Also, if $f_2 = \alpha(f_1)$ for a K-automorphism α of $K((t))$, then $f_2^{(n)} = \kappa\alpha\kappa^{-1}(f_1^{(n)})$. Note that $\kappa\alpha\kappa^{-1}$ is a K-automorphism of $K((t))$. Thus, given $\xi, \xi_1, \eta, \eta_1 \in K[[t]]$, if the branch representations (ξ, η) and (ξ_1, η_1) are K-equivalent, then $(\xi^{(n)}, \eta^{(n)})$ and $(\xi_1^{(n)}, \eta_1^{(n)})$ are K-equivalent branch representation as well. From this, if (ξ, η) is a primitive branch representation, then $(\xi^{(n)}, \eta^{(n)})$ is also a primitive branch representation.

Let γ be a branch with a primitive representation (ξ, η). The branch $\Phi^n(\gamma)$ given by the primitive representation $(\xi^{(n)}, \eta^{(n)})$ is the *n-th Frobenius image of* γ.

DEFINITION 8.6 A branch γ is $\mathbf{F}_q$*-rational* if γ has a primitive representation (ξ, η) with $\xi, \eta \in \mathbf{F}_q[[t]]$, that is, if $\xi^{(1)} = \xi$, $\eta^{(1)} = \eta$.

It should be observed that, if γ is $\mathbf{F}_q$-rational having a primitive representation (ξ, η) with $\xi, \eta \in \mathbf{F}_q[[t]]$, then another primitive representation (ξ_1, η_1) of γ does not have the property that $\xi_1^{(1)}, \eta_1^{(1)} \in \mathbf{F}_q[[t]]$ unless there exists $\alpha \in \mathrm{Aut}_{\mathbf{F}_q} K[[t]]$ such that $\xi_1 = \alpha(\xi)$, $\eta_1 = \alpha(\eta)$. Here $\mathrm{Aut}_{\mathbf{F}_q} K[[t]]$, the $\mathbf{F}_q$*-automorphism group of* $K[[t]]$, consists of all K-automorphisms α such that $\alpha(f) \in \mathbf{F}_q[[t]]$ for every $f \in \mathbf{F}_q[[t]]$. The observation follows from the following two lemmas. The first is a refinement of Theorem 4.21.

LEMMA 8.7 *For an $\mathbf{F}_q$-rational primitive branch representation $(x(t), y(t))$, the subfield $\mathbf{F}_q(x(t), y(t))$ of $K(x(t), y(t))$ contains an element ξ of order* 1.

Proof. Let τ be an element in $\mathbf{F}_q(x(t), y(t))$ of minimal positive order. Arguing as in the proof of Theorem 4.21, both $x(t)$ and $y(t)$ can be expanded into power series of τ. Since they are the components of a primitive branch representation, this is only possible for $\mathrm{ord}_t\, \tau = 1$. □

LEMMA 8.8 *Let $\xi \in \mathbf{F}_q[[t]]$ with* $\mathrm{ord}_t\, \xi = 1$, *and let* $\alpha \in \mathrm{Aut}(K[[t]])$ *with $\alpha(\xi)$ in $\mathbf{F}_q[[t]]$. Then* $\alpha \in \mathrm{Aut}_{\mathbf{F}_q} K[[t]]$.

Proof. It suffices to prove that $\alpha(t) \in \mathbf{F}_q[[t]]$. Write

$$\xi = u_1 t + \cdots + u_k t^k + \cdots,$$

with $u_1, \ldots, u_k, \ldots \in \mathbf{F}_q$, and $\alpha(t) = a_1 t + \cdots + a_m t^m + \cdots$, where both u_1 and a_1 are non-zero. Then

$$\alpha(\xi) = u_1 \alpha(t) + \cdots + u_k \alpha(t)^k + \cdots = a_1 u_1 t + t^2\, G(t).$$

Since $\alpha(\xi) \in \mathbf{F}_q[[t]]$, so $a_1 u_1 \in \mathbf{F}_q$, whence $a_1 \in \mathbf{F}_q$.

To prove that this holds true for every a_m, induction on m is used. Suppose that $a_1, \ldots, a_{m-1} \in \mathbf{F}_q$. Then

$$\alpha(\xi) = b_1 t + \cdots + b_{m-1} t^m + (a_m + c_m) t^m + \cdots$$

with $b_1, \ldots, b_{m-1}, c_m \in \mathbf{F}_q$. Since $\alpha(\xi) \in \mathbf{F}_q[[t]]$, so $a_m + c_m \in \mathbf{F}_q$, implying that $a_m \in \mathbf{F}_q$. □

Now branches of an irreducible plane curve $\mathcal{F}$ defined over $\mathbf{F}_q$ are considered. From the definition, it may appear that a branch of $\mathcal{F}$ is $\mathbf{F}_q$-rational if and only if its centre is a point in $\mathrm{PG}(2, q)$. This is false in general, even if it holds for non-singular points.

THEOREM 8.9 *Let P be a non-singular point of an irreducible plane curve $\mathcal{F}$ defined over $\mathbf{F}_q$. If $P \in \mathrm{PG}(2, q)$, then the unique branch of $\mathcal{F}$ centred at P is $\mathbf{F}_q$-rational.*

Proof. After a change of coordinates over $\mathbf{F}_q$, suppose that $P = (1, 0, 0)$, and that the tangent line to $\mathcal{F}$ at P is not the Y-axis. By Theorem 4.6, the branch of $\mathcal{F}$ centred at P has a primitive representation $(x(t), y(t))$ with

$$x(t) = t, \quad y(t) = c_1 t + \cdots + c_k t^k + \cdots .$$

From the proof of that theorem, each coefficient c_i is in the subfield of K generated by the coefficients of $F(X, Y)$. Since $\mathcal{F}$ is defined over $\mathbf{F}_q$, this subfield is contained in $\mathbf{F}_q$. □

THEOREM 8.10 *Let $P \in \mathrm{PG}(2, q)$ be a singular point of an irreducible plane curve $\mathcal{F}$ defined over $\mathbf{F}_q$. If γ is a linear branch of $\mathcal{F}$ with centre P and ℓ is the common tangent line to $\mathcal{F}$ and γ at P, then γ is $\mathbf{F}_q$-rational if and only if ℓ is defined over $\mathbf{F}_q$.*

Proof. Again, assume that $P = (1, 0, 0)$, and that ℓ is not the Y-axis. The proof of Theorem 8.9 still works for linear branches, and gives the 'if' part of the assertion.

Conversely, any primitive representation $(x(t), y(t))$ of γ is of type

$$x(t) = at + \cdots, \; a \neq 0, \quad y(t) = bt + \cdots,$$

such that b/a is the slope of ℓ. If γ is $\mathbf{F}_q$-rational, then there is a representation with $a, b \in \mathbf{F}_q$. Hence ℓ is defined over $\mathbf{F}_q$, as it passes through a point of $\mathrm{PG}(2, q)$ and its slope is in $\mathbf{F}_q$. □

The notation for the set of all $\mathbf{F}_q$-rational branches of $\mathcal{F}$ is $\mathcal{F}(\mathbf{F}_q)$, and S_q for $|\mathcal{F}(\mathbf{F}_q)|$. Theorems 8.9 and 8.10 show that $\mathcal{F}(\mathbf{F}_q)$ is a contained in the set $\mathcal{F}(\mathbf{F}_q)^*$ consisting of all branches of $\mathcal{F}$ centred at points of $\mathrm{PG}(2, q)$. Hence $S_q \leq B_q$ where $B_q = |\mathcal{F}(\mathbf{F}_q)^*|$ and, if $\mathcal{F}$ is non-singular, then equality holds. However, it may happen for singular curves that $S_q < B_q$. This possibility is investigated in Section 9.6. Here an example is given.

EXAMPLE 8.11 Let q be odd, and choose a non-square element $c \in \mathbf{F}_q$. The irreducible cubic curve $\mathcal{F} = \mathbf{v}(cX^2 - Y^2 + X^3)$ defined over $\mathbf{F}_q$ contains the point $P = (1, 0, 0)$, which is a double point. Two branches γ^+ and γ^- of $\mathcal{F}$ are centred at P, namely those with primitive representation $(x(t), y(t))$, where

$$x(t) = t, \quad y(t) = \pm\nu t + \cdots,$$

with $\nu^2 = c$. Here, $\nu \in \mathbf{F}_{q^2} \backslash \mathbf{F}_q$ and hence neither branch is $\mathbf{F}_q$-rational.

On the other hand, the following result holds.

Lemma 8.12 *Let γ be a branch of an irreducible plane curve $\mathcal{F}$ defined over $\mathbf{F}_q$. Then there is an overfield $\mathbf{F}_{q^n}$ of $\mathbf{F}_q$ depending on γ such that γ has a primitive representation $(x(t), y(t))$ with $x(t), y(t) \in \mathbf{F}_{q^n}[[t]]$.*

Proof. Let $P' = (x', y')$ be a generic point of $\mathcal{F}$. By Theorem 3.27, there is an irreducible plane curve $\mathcal{G} = \mathbf{v}(g(X, Y))$ with only ordinary singularities which is birationally equivalent to $\mathcal{F}$. Without loss of generality, suppose that no point of $\mathcal{G}$ at infinity is singular and that no vertical line is tangent to $\mathcal{G}$ at a singular point. Let

$$\omega : x' = u(x, y)/v(x, y),\ y' = w(x, y)/z(x, y)$$

be a birational transformation that takes $\mathcal{G}$ to $\mathcal{F}$, where $P = (x, y)$ is a generic point of $\mathcal{G}$. The subfield $\mathbf{F}_{q^k}$ of K generated by $\mathbf{F}_q$, together with the coefficients of the polynomials $u, v, w, z \in K[X, Y]$, contains all coefficients of $g(X, Y)$. A possible larger subfield $\mathbf{F}_{q^s}$ of K also contains the slopes of the tangents to $\mathcal{G}$ at singular points.

Now let γ be any branch of $\mathcal{F}$. By Theorems 8.9 and 8.10, $\omega^{-1}(\gamma)$ is $\mathbf{F}_{q^n}$-rational if $\mathbf{F}_{q^n}$ is the subfield generated by $\mathbf{F}_{q^s}$ and the coordinates of the centre of $\omega^{-1}(\gamma)$. Since $\omega(\omega^{-1}(\gamma)) = \gamma$, the subfield $\mathbf{F}_{q^n}$ of K has the desired property. □

A corollary to Lemma 8.12 is the following result.

Theorem 8.13 *Let γ be a branch of an irreducible plane curve $\mathcal{F}$ defined over $\mathbf{F}_q$. Then there is a positive integer n depending on γ such that $\Phi^n(\gamma) = \gamma$.*

8.3 $\mathbf{F}_q$-RATIONAL PLACES, DIVISORS AND LINEAR SERIES

Let $f = \sum a_{ij} X^i Y^j$ be any polynomial in $K[X, Y]$. For any integer n, the *n-th conjugate of f* is the polynomial

$$f^{(n)} = \sum a_{ij}^{q^n} X^i Y^j.$$

The *n-th conjugate map* $f \mapsto f^{(n)}$ is an automorphism of $K[X, Y]$. Here f is irreducible if and only if $f^{(n)}$ is irreducible; also, (8.1) holds for any f, g in $K[X, Y]$. Then $f = f^{(n)}$ only when $f \in \mathbf{F}_{q^n}[X, Y]$, and

$$f \cdot f^{(1)} \cdots f^{(n-1)} \in \mathbf{F}_q[X, Y] \quad \text{if and only if } f \in \mathbf{F}_{q^n}[X, Y]; \tag{8.2}$$

$$f + f^{(1)} + \cdots + f^{(n-1)} \in \mathbf{F}_q[X, Y] \quad \text{if and only if } f \in \mathbf{F}_{q^n}[X, Y]. \tag{8.3}$$

Now these results together with those obtained in Section 8.2 are applied to the function field $K(\mathcal{F})$ of an irreducible plane curve $\mathcal{F} = \mathbf{v}(F(X, Y))$ defined over $\mathbf{F}_{q^n}$. Let $P = (x, y)$ be a generic point of $\mathcal{F}$. Those elements in $K(\mathcal{F})$ having a representation $u(x, y)/v(x, y)$, with $u, v \in \mathbf{F}_{q^n}[X, Y]$ constitute a subfield, the *$\mathbf{F}_{q^n}$-rational function field $\mathbf{F}_{q^n}(\mathcal{F})$ of $\mathcal{F}$*. Since any element $f \in K(\mathcal{F})$ can be written as

$$f = \frac{u(x, y)}{v(x, y)}, \quad \text{with } u[X, Y], v[X, Y] \in K[X, Y],\ F(X, Y) \nmid v(X, Y), \tag{8.4}$$

the *n-th conjugate of f* can be defined:

$$f^{(n)} = \frac{u^{(n)}(x,y)}{v^{(n)}(x,y)}.$$

Note that $v^{(n)}(x,y) \neq 0$ because $F(X,Y) \nmid v^{(n)}(X,Y)$.

LEMMA 8.14 *The definition of $f^{(n)}$ does not depend on the representation.*

Proof. It must be shown that

$$\frac{u(x,y)}{v(x,y)} = \frac{w(x,y)}{z(x,y)} \Longrightarrow \frac{u^{(n)}(x,y)}{v^{(n)}(x,y)} = \frac{w^{(n)}(x,y)}{z^{(n)}(x,y)}.$$

To do this, write the first equation as

$$u(x,y)z(x,y) - v(x,y)w(x,y) = 0.$$

Raising this to the q^n-th power gives

$$u^{(n)}(x^{q^n}, y^{q^n})z^{(n)}(x^{q^n}, y^{q^n}) - v^{(n)}(x^{q^n}, y^{q^n})w^{(n)}(x^{q^n}, y^{q^n}) = 0.$$

By Lemma 8.3, $P = (x^{(n)}, y^{(n)})$ is a generic point of $\mathcal{F}$. Thus

$$F(X,Y)G(X,Y) = u^{(n)}(X,Y)z^{(n)}(X,Y) - v^{(n)}(X,Y)w^{(n)}(X,Y),$$

for some polynomial $G \in K[X,Y]$, whence it follows that

$$u^{(n)}(x,y)z^{(n)}(x,y) - v^{(n)}(x,y)w^{(n)}(x,y) = 0,$$

as required. □

LEMMA 8.15 *If $f = f^{(1)}$ then $f \in \mathbf{F}_q(\mathcal{F})$.*

Proof. Assume first that $f \in K(x)$ and write $f = c \cdot u(x)/v(x)$ with $c \in K$ and $u(X), v(X) \in K[X]$ monic polynomials for which $\gcd(u(X), v(X)) = 1$. By Lemma 8.14, $f^{(1)} = c^q \cdot u^{(1)}(x)/v^{(1)}(x)$. From $f = f^{(1)}$,

$$u(X)v^{(1)}(X) = u^{(1)}(X)v(X)$$

and $c \in \mathbf{F}_q$. Since $\gcd(u(X), v(X)) = 1$ and $\deg u(X) = \deg u^{(1)}(X)$, this implies that $u(X) = u^{(1)}(X)$, whence $u(X) \in \mathbf{F}_q[X]$; similarly, $v(X) \in \mathbf{F}_q[X]$. Therefore the assertion holds when $f \in K(x)$.

Let $n = [K(x,y) : K(x)]$. Since $K(\mathcal{F}) = K(x,y)$, every element $f \in K(\mathcal{F})$ can be uniquely written in the form

$$f = c_0(x) + c_1(x)y + \cdots + c_{n-1}(x)y^{n-1},$$

with $c_0(x), \cdots, c_{n-1}(x) \in K(x)$. Then

$$f^{(1)} = c_0^{(1)}(x) + c_1^{(1)}(x)y + \cdots + c_{n-1}^{(1)}(x)y^{n-1}.$$

From $f = f^{(1)}$ it follows that $c_i(x) = c_i^{(1)}(x)$ for $i = 0, \ldots, n-1$. By the preceding assertion, this implies that $c_i(x) \in \mathbf{F}_q(x)$. Then $c_i(X) \in \mathbf{F}_q[X]$, and the result follows. □

Note that (8.1) holds for any $f, g \in K(\mathcal{F})$. Also, for every $f \in K(\mathcal{F})$ there is a $g \in K(\mathcal{F})$ such that $f = g^{(n)}$. Hence the n-th conjugate map is an automorphism

ι of $K(\mathcal{F})$ that fixes every element in $\mathbf{F}_{q^n}(\mathcal{F})$. The constants fixed by ι are those in $\mathbf{F}_{q^n}$.

Let $\mathcal{F} = \mathbf{v}(F(X,Y))$ be an irreducible plane curve defined over $\mathbf{F}_q$. For a generic point $P = (x,y)$ of $\mathcal{F}$, let $K(\mathcal{F}) = K(x,y)$ be the associated function field of $\mathcal{F}$. Given a place $\mathcal{P}$ of $K(\mathcal{F})$, let σ be any primitive place representation σ of $\mathcal{P}$. As usual, put $x(t) = \sigma(x), y(t) = \sigma(y)$. Then $F(x(t), y(t)) = 0$ and, by Lemma 8.3,

$$F(x^{(n)}(t), y^{(n)}(t)) = 0.$$

This means that the pair $(x^{(n)}(t), y^{(n)}(t))$ is a primitive representation of a branch of $\mathcal{F} = \mathbf{v}(F(X,Y))$. The corresponding place $\Phi^n(\mathcal{P})$ is the *n-th Frobenius image of* $\mathcal{P}$. This definition is independent of the choice of σ.

Now a primitive representation of $\Phi^n(\mathcal{P})$ is given. Next, $\sigma\iota^{-1}$ is a monomorphism from $K(\mathcal{F})$ into $K((t))$. This does not fix every constant, but the monomorphism

$$\tau = \kappa\sigma\iota^{-1} \tag{8.5}$$

does. To check that τ is a primitive place representation of the $\Phi^n(\mathcal{P})$ of $\mathcal{P}$, let $\iota^{-1}(x) = u(x,y)/v(x,y)$, that is, $x = u^{(n)}(x,y)/v^{(n)}(x,y)$. Then

$$\tau(x) = \kappa\left(\frac{u(x(t), y(t))}{v(x(t), y(t))}\right) = \frac{u^{(n)}(x^{(n)}(t), y^{(n)}(t))}{v^{(n)}(x^{(n)}(t), y^{(n)}(t))} = x^{(n)}(t).$$

Similarly, $\tau(y) = y^{(n)}(t)$, and the result is established.

DEFINITION 8.16 For an irreducible plane curve $\mathcal{F}$ defined over $\mathbf{F}_q$, a place $\mathcal{P}$ of the function field $K(\mathcal{F})$ is an $\mathbf{F}_{q^n}$*-rational place* if $\Phi^n(\mathcal{P}) = \mathcal{P}$.

Note that $\Phi^n(\mathcal{P}) = \mathcal{P}$ if and only if, for any primitive representation σ, also (8.5) is a primitive representation of $\mathcal{P}$. Further, from Theorem 8.13, for every place $\mathcal{P}$ of $K(\mathcal{F})$, there is a positive integer n depending on $\mathcal{P}$ such that $\Phi^n(\mathcal{P}) = \mathcal{P}$.

LEMMA 8.17 *Let $\mathcal{F}$ be an irreducible plane curve defined over $\mathbf{F}_q$. If the branch γ of $\mathcal{F}$ is $\mathbf{F}_q$-rational, then the corresponding place $\mathcal{P}$ is also $\mathbf{F}_q$-rational.*

Proof. With the notation above applied to the case $n = 1$, if $(x(t), y(t))$ is an $\mathbf{F}_q$-rational branch representation of γ, then $x^{(1)}(t) = x(t)$, $y^{(1)}(t) = y(t)$. Therefore $\tau = \sigma$, showing that $\mathcal{P}$ is $\mathbf{F}_q$-rational. □

It is shown later that the converse of Lemma 8.17 also holds; see Theorem 8.31.

THEOREM 8.18 *Let P be a non-singular point of an irreducible plane curve $\mathcal{F}$ defined over $\mathbf{F}_q$. Then $P \in \mathrm{PG}(2,q)$ if and only if the place $\mathcal{P}$ of $K(\mathcal{F})$ corresponding to the unique branch γ of $\mathcal{F}$ centred at P is $\mathbf{F}_q$-rational.*

Proof. Suppose that $P = (1, a, b)$ is the centre of γ, and that the tangent to $\mathcal{F}$ at P is not the vertical line through P. Assume first that $P \in \mathrm{PG}(2,q)$. By Theorem 8.9, the branch γ has a primitive representation $(x(t), y(t))$ such that $x(t) = x^{(1)}(t)$, $y(t) = y^{(1)}(t)$. Therefore $\mathcal{P}$ is an $\mathbf{F}_q$-rational place.

To show the converse, let σ be a primitive representation of $\mathcal{P}$ such that

$$\sigma(x) = a + t, \quad \sigma(y) = b + \varphi,$$

with $\mathrm{ord}_t\, \varphi \geq 1$. Since $\Phi^{(1)}(\mathcal{P}) = \mathcal{P}$ means that $\sigma = \lambda\tau$ for a K-automorphism λ of $K[[t]]$, it follows that

$$a + t = a^q + \lambda(t), \quad b + \varphi = b^q + \lambda(\varphi^{-1}).$$

Since both $\mathrm{ord}_t\, \lambda$ and $\mathrm{ord}_t\, \lambda(\varphi^{-1})$ are positive, this shows that $P \in \mathrm{PG}(2, q)$. □

REMARK 8.19 Let $\mathcal{F}$ and $\mathcal{G}$ be irreducible curves defined over $\mathbf{F}_q$, and suppose that $\mathcal{F}$ and $\mathcal{G}$ are birationally equivalent over K. It is not true generally that they must be birationally equivalent over $\mathbf{F}_q$, as well.

Consider the following counter-example. For a square q, let $\mathcal{H}_{\sqrt{q}} = \mathbf{v}(H)$ be the non-singular Hermitian curve defined over $\mathbf{F}_q$ with

$$H = X_0^{\sqrt{q}+1} + X_1^{\sqrt{q}+1} + X_2^{\sqrt{q}+1}.$$

For a generic point $P = (x_0, x_1, x_2)$ of $\mathcal{H}_{\sqrt{q}}$, let $K(\mathcal{H}_{\sqrt{q}})$ be the associated function field. Now the non-singular plane curve $\mathcal{F} = \mathbf{v}(F(X_0, X_1, X_2))$ is defined to be the image of $\mathcal{H}_{\sqrt{q}}$ under the projectivity ω of $\mathrm{PG}(3, K)$ associated to the matrix

$$M = \begin{bmatrix} t & 1 & t^{q+1} \\ t^{q+1} & t & 1 \\ 1 & t^{q+1} & t \end{bmatrix},$$

where t is a suitable element of order $q^2 + q + 1$ in the cubic extension $\mathbf{F}_{q^3}$ of $\mathbf{F}_q$. More precisely, the element t is chosen such that

$$\begin{aligned} t^{q\sqrt{q}+\sqrt{q}} + t^{q+\sqrt{q}+1} + t &= 0, \\ t^{q\sqrt{q}+q+\sqrt{q}+1} + t^{\sqrt{q}+1} + 1 &= 0, \\ t^{q\sqrt{q}+\sqrt{q}+1} + t^{q+1} + t^{\sqrt{q}} &\neq 0, \end{aligned}$$

and $\det M \neq 0$. Then

$$F(X_0, X_1, X_2) = X_0^{\sqrt{q}} X_1 + X_1^{\sqrt{q}} X_2 + X_2^{\sqrt{q}} X_0.$$

Hence $\mathcal{F}$ can also be regarded as a curve over $\mathbf{F}_q$. If $\mathcal{H}_{\sqrt{q}}$ and $\mathcal{F}$ were birationally equivalent over $\mathbf{F}_q$, then they would have the same number of $\mathbf{F}_q$-rational branches. But, as in Section 8.5, the number of $\mathbf{F}_q$-rational branches of $\mathcal{H}_{\sqrt{q}}$ is $q\sqrt{q}+1$, while $\mathcal{F}$ has at most $\frac{1}{2}(q\sqrt{q} + 2q + \sqrt{q})$ such points.

The map Φ^n can be extended to divisors. If $D = \sum n_{\mathcal{P}} \mathcal{P}$, then

$$\Phi^n(D) = \sum n_{\mathcal{P}} \Phi^n(\mathcal{P}).$$

DEFINITION 8.20 Let D be a divisor of $K(\mathcal{F})$, where $\mathcal{F}$ is an irreducible plane curve defined over $\mathbf{F}_q$.

(i) If $\Phi^n(D) = D$, then D is an $\mathbf{F}_{q^n}$*-rational divisor.*

(ii) The $\mathbf{F}_{q^n}$-rational divisors form the $\mathbf{F}_{q^n}$*-rational divisor group* $\mathrm{Div}(\mathbf{F}_{q^n}(\mathcal{F}))$, a subgroup of $\mathrm{Div}(\Sigma)$.

(iii) The subgroup of $\mathrm{Div}(\mathbf{F}_{q^n}(\mathcal{F}))$ consisting of all divisors of degree zero is the *zero-degree* $\mathbf{F}_{q^n}$*-rational divisor group* $\mathrm{Div}_0(\mathbf{F}_{q^n}(\mathcal{F}))$.

If $D = \sum n_\mathcal{P}\mathcal{P}$ where all $\mathcal{P}$ are $\mathbf{F}_{q^n}$-rational places, then D is an $\mathbf{F}_{q^n}$-rational divisor. The converse does not hold in general. To understand why this can happen, for instance when $n = 1$, choose any place $\mathcal{P}$ of $K(\mathcal{F})$, and define k to be the smallest positive integer for which $\Phi^k(\mathcal{P}) = \mathcal{P}$ occurs. Then the divisor

$$D = \mathcal{P} + \Phi(\mathcal{P}) + \cdots + \Phi^{k-1}(\mathcal{P})$$

is $\mathbf{F}_q$-rational even if $\mathcal{P}$ is not, that is, if $k > 1$. The divisor D is the *closed place* of $\mathcal{P}$, and $N(\mathcal{P}) = \deg D$ is the *degree of the closed place* $\mathcal{P}$.

Analogous to the trace of an element in a finite field, this observation leads to a similar concept for divisors.

Definition 8.21 For any $\mathbf{F}_{q^n}$-rational divisor D, the divisor

$$\mathfrak{T}(D) = D + \Phi(D) + \cdots + \Phi^{n-1}(D)$$

is an $\mathbf{F}_q$-rational divisor, the *trace of the* $\mathbf{F}_{q^n}$*-rational divisor* D *over* $\mathbf{F}_q$.

The *trace map* defines a homomorphism

$$\mathfrak{T} : \mathrm{Div}(\mathbf{F}_{q^n}(\mathcal{F})) \to \mathrm{Div}(\mathbf{F}_q(\mathcal{F})),$$
$$D \mapsto \mathfrak{T}(D).$$

It may be noted that $\mathrm{ord}_\mathcal{P}\, \xi = \mathrm{ord}_{\Phi(\mathcal{P})}\, \xi^{(1)}$ for $\mathcal{P} \in \mathfrak{P}(\Sigma)$ and $\xi \in K(\mathcal{F})$, $\xi \neq 0$. Therefore

$$\mathfrak{T}(\mathrm{div}\, \xi) = \mathrm{div}\, \xi + \mathrm{div}\, \xi^{(1)} + \cdots + \mathrm{div}\, \xi^{(n-1)}, \ \xi \in \mathbf{F}_{q^n}(\mathcal{F}),\ \xi \neq 0,$$

which justifies the terminology; see Section A.4.

Proposition 8.22 *If n is not divisible by p, then the trace map is surjective.*

Proof. Let D' be any $\mathbf{F}_q$-rational divisor of $K(\mathcal{F})$. Then $\Phi^k(D') = D'$ for every k. Choose $m \in \mathbf{N}$ such that $mn \equiv 1 \pmod p$, and let $D = m\,D'$. Then D is an $\mathbf{F}_q$-rational divisor, and hence an $\mathbf{F}_{q^n}$-rational divisor of $K(\mathcal{F})$ as well. Thus

$$\mathfrak{T}(D) = \textstyle\sum_{k=0}^{n-1} \Phi^k(D) = \sum_{k=0}^{n-1} m\Phi^k(D') = \sum_{k=0}^{n-1} m\,D' = D'.$$

This proves the assertion. □

Lemma 8.23 *Let $\mathcal{F}$ be an irreducible plane curve defined over $\mathbf{F}_q$. If $\xi \in \mathbf{F}_q(\mathcal{F})$, then both* $\mathrm{div}(\xi)$ *and $d\xi$ are $\mathbf{F}_q$-rational divisor.*

Proof. With the notation above, as in (8.5), if $\sigma(\xi) = \sum a_i t^i$, then $\tau(\xi) = \sum a_i^q t^i$. Also,

$$d\sigma(\xi)/dt = \textstyle\sum i a_i t^{i-1}, \quad d\tau(\xi)/dt = \sum i a_i^q t^{i-1}.$$

Thus

$$\mathrm{ord}_\mathcal{P}\, \xi = \mathrm{ord}_{\Phi(\mathcal{P})}\, \xi, \quad \mathrm{ord}_\mathcal{P}\, d\xi = \mathrm{ord}_{\Phi(\mathcal{P})}\, d\xi.$$

□

LEMMA 8.24 *Let $\mathcal{F}$ be an irreducible plane curve defined over $\mathbf{F}_q$. If ξ is a non-zero element of $K(\mathcal{F})$ and $\operatorname{div}\xi$ is an $\mathbf{F}_q$-rational divisor, then $\xi \in \mathbf{F}_q(\mathcal{F})$ up to a constant factor.*

Proof. If $\operatorname{div}\xi$ is an $\mathbf{F}_q$-rational divisor, then $\operatorname{div}\xi = \operatorname{div}\xi^{(1)}$. By Corollary 5.35, there is a a non-zero constant $a \in K$ such that $a\xi^{(1)} = \xi$. Choose $b \in K$ such that $b^{q-1} = a$, and replace ξ by $\eta = b\xi$. Then $\eta^{(1)} = \eta$, as

$$\eta^{(1)} = b^q\xi^{(1)} = b^q a^{-1}\xi = b^q a^{-1} b^{-1}\eta = \eta.$$

Hence $\eta \in \mathbf{F}_q(\mathcal{F})$, and the result follows from Lemma 8.15. □

DEFINITION 8.25 For $\mathcal{F}$ an irreducible curve defined over $\mathbf{F}_q$, a virtual linear series L of $K(\mathcal{F})$ is $\mathbf{F}_q$-*rational* if there exist $x_0, \ldots, x_r$ in $\mathbf{F}_q(\mathcal{F})$ and an $\mathbf{F}_q$-rational divisor B such that L consists of all divisors

$$A_{\mathbf{c}} = \operatorname{div}\left(\textstyle\sum_{i=0}^r c_i x_i\right) + B, \quad (c_0, \ldots, c_r) \in \mathrm{PG}(r, K).$$

THEOREM 8.26 *If $\mathcal{F}$ is an $\mathbf{F}_q$-rational curve and B an $\mathbf{F}_q$-rational divisor of $\mathbf{F}_q(\mathcal{F})$, then $|B|$ is an $\mathbf{F}_q$-rational linear series.*

Proof. As usual, $P = (x, y)$ is a generic point of $\mathcal{F}$, and $K(\mathcal{F})$ is represented by $K(x, y)$. Choose $g, f_0, \ldots, f_r$ in $K[X, Y]$ for which $x_i = f_i(x, y)/g(x, y)$, for $i = 0, \ldots, r$. It may be assumed that $g(X, Y) \in \mathbf{F}_q[X, Y]$ by the following argument.

The coefficients of $g(X, Y)$ generate a finite subfield $\mathbf{F}_{q^n}$ of K. Put

$$h(X, Y) = g(X, Y)^{(1)} \cdots g(X, Y)^{(n-1)}.$$

From (8.2), $g(X, Y)h(X, Y) \in \mathbf{F}_{q^n}[X, Y]$. For brevity, write

$$h = h(x, y), \quad h_i = hf_i(x, y),$$

for $i = 0, \ldots, r$, and put $D = B - \operatorname{div}(gh)$. By Lemma 8.23, D is still an $\mathbf{F}_q$-rational divisor. Then $|B| = |D|$ and $|B|$ consists of all divisors

$$A_{\mathbf{c}} = \operatorname{div}\left(\textstyle\sum_{i=0}^r c_i h_i\right) + D, \quad (c_0, \ldots, c_r) \in \mathrm{PG}(r, K).$$

The coefficients of the polynomials $h_i(X, Y)$, $0 \le i \le r$, together with $\mathbf{F}_q$ generate a finite subfield $\mathbf{F}_{q^k}$ of K. The simple field extension $\mathbf{F}_{q^k}/\mathbf{F}_q$ has degree k. So $\mathbf{F}_{q^k} = \mathbf{F}_q(\theta)$, with θ a root of a separable polynomial $s(X) \in \mathbf{F}_q[X]$ of degree k, irreducible over $\mathbf{F}_q$. The roots of $s(X)$ are $\theta_j = \theta^j$ where $j = 0, 1, \ldots, k-1$. Let $d \in \mathbf{F}_{q^k}[X, Y]$. For $j = 0, 1, \ldots, k-1$, put

$$d_j(X, Y) = \theta_j d(X, Y) + {\theta_j}^q d^{(1)}(X, Y) + \cdots + {\theta_j}^{q^{k-1}} d^{(k-1)}(X, Y).$$

Then $d_j(X, Y) = d_j^{(1)}(X, Y)$, and hence $d_j \in \mathbf{F}_q(X, Y)$. Also,

$$\begin{aligned} d_0(x, y) &= d(x, y) + d(x, y)^{(1)} + \cdots + d(x, y)^{(k-1)}, \\ d_1(x, y) &= \theta d(x, y) + \theta^q d(x, y)^{(1)} + \cdots + \theta^{q^{k-1}} d(x, y)^{(k-1)}, \\ &\ \ \vdots \\ d_{k-1}(x, y) &= \theta_{k-1} d(x, y) + \theta_{k-1}^q d(x, y)^{(1)} + \cdots + \theta_{k-1}^{q^{k-1}} d(x, y)^{(k-1)}. \end{aligned}$$

Note that the determinant of the matrix

$$\begin{bmatrix} 1 & 1 & \dots & 1 \\ \theta & \theta^q & \dots & \theta^{q^{k-1}} \\ \vdots & \vdots & \dots & \vdots \\ \theta_{k-1} & \theta_{k-1}^q & \dots & \theta_{k-1}^{q^{k-1}} \end{bmatrix}$$

does not vanish as $\{1, \theta_1, \dots, \theta_{k-1}\}$ is a basis of $\mathbf{F}_{q^k}$ over $\mathbf{F}_q$. Therefore there exist $e_0, e_1, \dots, e_{k-1} \in \mathbf{F}_{q^k}$ such that

$$d(x, y) = e_0 d_0(x, y) + e_1 d_1(x, y) + \dots + e_{k-1} d_{k-1}(x, y).$$

Applying this to $d(X, Y) = h_i(X, Y)$ for every $i = 0, 1, \dots, r$, it turns out that $\sum c_i h_i$ can be written as a K-linear combination of rk elements from $\mathbf{F}_q(x, y)$. Since these elements are chosen independently of $\mathbf{c} = (c_0, \dots, c_r)$, the assertion follows. □

REMARK 8.27 Theorem 8.26, together with Lemma 8.23, ensures the existence of $\mathbf{F}_q$-rational linear series. In particular, the canonical series of $\mathbf{F}_q(\mathcal{F})$ is an $\mathbf{F}_q$-rational linear series. Also, if L is an $\mathbf{F}_q$-rational linear series, and $\mathcal{P}$ is an $\mathbf{F}_q$-rational place, then the sub-series of L consisting of all divisors containing $\mathcal{P}$ is also an $\mathbf{F}_q$-rational linear series.

8.4 SPACE CURVES OVER $\mathbf{F}_q$

In this section, $\mathcal{F}$ is an irreducible plane curve defined over $\mathbf{F}_q$. Also, for a generic point $P = (x, y)$ of $\mathcal{F}$, the field $\mathbf{F}_q(\mathcal{F})$ is the $\mathbf{F}_q$-rational subfield of the function field $K(\mathcal{F})$ of $\mathcal{F}$. Consistent with the concept of an $\mathbf{F}_q$-rational linear series, some but not all space curves arising from $\mathcal{F}$ are considered to be $\mathbf{F}_q$-rational curves. The main goal is to define and characterise the $\mathbf{F}_q$-rational branches of an $\mathbf{F}_q$-rational space curve in terms of $\mathbf{F}_q$-rational places of $K(\mathcal{F})$.

DEFINITION 8.28 (i) An irreducible curve Γ of $\mathrm{PG}(r, K)$ is *defined over* $\mathbf{F}_q$ or is $\mathbf{F}_q$*-rational* if the curve is given by a point $Q = (x_0, x_1, \dots, x_r)$, with $x_i \in \mathbf{F}_q(\mathcal{F})$ for $i = 0, 1, \dots, r$.

(ii) If Γ is defined over $\mathbf{F}_q$, then a branch γ of Γ is $\mathbf{F}_q$*-rational* if it has a branch representation

$$(x_0(t), x_1(t), \dots, x_r(t)),$$

with $x_i(t) \in \mathbf{F}_q[[t]]$ for $i = 0, 1, \dots, r$.

THEOREM 8.29 *Let Γ be an irreducible curve Γ of $\mathrm{PG}(r, K)$ defined over $\mathbf{F}_q$. For a non-singular point of Γ the following conditions are equivalent:*

(i) $P \in \mathrm{PG}(r, q)$;

(ii) *the unique branch γ of Γ with centre at P is $\mathbf{F}_q$-rational;*

(iii) *the place $\mathcal{P}$ of $K(\mathcal{F})$ associated to γ is $\mathbf{F}_q$-rational.*

Proof. As usual, let $x_0 = 1$ and let σ be a primitive representation of $\mathcal{P}$. Put $\sigma(x_i) = x_i(t)$ for $i = 1, \ldots, r$. Then $(1, x_1(t), \ldots, x_r(t))$ is a primitive representation of γ. Since Γ is defined over $\mathbf{F}_q$, so $(1, x_1^{(1)}(t), \ldots, x_r^{(1)}(t))$ is a primitive representation of the branch γ' of Γ associated to $\Phi(\mathcal{P})$.

(i) $\Rightarrow$ (ii) As P is a non-singular point of $\mathcal{F}$, take $x_1(t) = a_1 + t$. Write

$$x_i(t) = a_i + \varphi_i(t),$$

with $\varphi_i \in K[[t]]$ and $\text{ord}_t\, \varphi_i(t) \geq 1$. Then

$$x_i^{(1)}(t) = a_i^q + \varphi_i^{(1)}(t),$$

for $i = 1, \ldots, r$. As $P \in \text{PG}(r, q)$, so P is the centre of γ' as well. Since P is the centre of exactly one branch, this implies that $\gamma = \gamma'$. So there exists $\tau \in K[[t]]$ such that $x_i(t) = x_i^{(1)}(\tau)$. For $i = 1$, this gives $t = \tau$. But then $x_i(t) = x_i^{(1)}(t)$ for every $i = 1, \ldots, r$; that is, γ is an $\mathbf{F}_q$-rational branch of Γ.

(ii) $\Rightarrow$ (iii) By hypothesis, γ has a primitive representation $(1, x_1(t), \ldots, x_r(t))$ with $x_i(t) = x_i^{(1)}(t)$ for $i = 1, \ldots, r$. If σ is the corresponding primitive representation of $\mathcal{P}$, then $\sigma(x_i) = x_i(t) = x_i^{(1)}(t)$ for $i = 1, \ldots, r$. As in (8.5), let $\tau = \kappa\sigma\iota^{-1}$. Then $\tau(x_i) = x_i^{(1)}(t)$, for $1 \leq i \leq r$. This means that $\tau = \sigma$, and hence $\Phi(\mathcal{P}) = \mathcal{P}$.

(iii) $\Rightarrow$ (i) If $\tau = \sigma$, then $x_i(t) = x_i^{(1)}(t)$, implying that $a_i = a_i^q$ for each $i = 1, \ldots, r$. □

REMARK 8.30 Theorem 8.26 shows how to obtain non-singular models of $\mathcal{F}$ defined over $\mathbf{F}_q$; in particular, the canonical curve is such a curve provided that $K(\mathcal{F})$ is neither rational, nor elliptic, nor hyperelliptic. For each non-singular model, the above three conditions are equivalent and P is an $\mathbf{F}_q$-*rational* point.

THEOREM 8.31 *If Γ is an irreducible curve of* $\text{PG}(r, K)$ *defined over $\mathbf{F}_q$, then a branch γ of Γ is $\mathbf{F}_q$-rational if and only if the corresponding place $\mathcal{P}$ is $\mathbf{F}_q$-rational.*

Proof. By Theorems 7.41 and 8.26, Γ is birationally equivalent over $\mathbf{F}_q$ to a non-singular model $\mathcal{X}$ of $K(\mathcal{F})$ defined over $\mathbf{F}_q$. So, if $\Gamma = K(Q)$ and $\mathcal{X} = K(R)$ with

$$R = (1, y_1, \ldots, y_r), \quad y_i \in \mathbf{F}_q(\mathcal{F}),$$

then the rational transformation

$$\omega : y_i = \frac{u_i(x_1, \ldots, x_r)}{v_i(x_1, \ldots, x_r)}, \quad i = 1, \ldots, r,$$

is $\mathbf{F}_q$-rational; that is, $u_i, v_i \in \mathbf{F}_q[X_1, \ldots, X_r]$. Note that ω^{-1} is also $\mathbf{F}_q$-rational.

Now assume that γ is $\mathbf{F}_q$-rational. Then the image γ' of γ under ω is also an $\mathbf{F}_q$-rational branch of $\mathcal{X}$. By Theorem 8.29, $\mathcal{P}$ is $\mathbf{F}_q$-rational. To prove the converse, choose a primitive representation of γ',

$$(1, y_1(t) = a_1 + \varphi_1(t), \ldots, y_r(t) = a_r + \varphi_r(t)), \ \text{ord}_t\, \varphi_i(t) \geq 1, \ i = 1, \ldots, r,$$

such that $\tau(y_i(t)) = \sigma(y_i(\alpha(t))$ with an automorphism α of $K[[t]]$. Then $a_i^q = a_i$, that is, $P \in \mathrm{PG}(r, q)$, Again, Theorem 8.29 applies, whence γ' is $\mathbf{F}_q$-rational. Since $\gamma = \omega^{-1}(\gamma')$, the result follows. □

The notation for the set of all $\mathbf{F}_q$-rational places of a function field Σ defined over $\mathbf{F}_q$ of transcendence degree 1 is $\Sigma(\mathbf{F}_q)$. By Theorem 8.31, there is a one-to-one correspondence between $\Sigma(\mathbf{F}_q)$ and $\Gamma(\mathbf{F}_q)$ for each birational model Γ of Σ; in particular, $|\Sigma(\mathbf{F}_q)| = S_q$.

REMARK 8.32 If a non-singular model $\mathcal{X}$ of $\mathcal{F}$ is defined over $\mathbf{F}_q$, and places of $\mathcal{F}$ are identified with points of $\mathcal{X}$, any closed place of $K(\mathcal{F})$ is viewed as a *closed point*. For a point $A = (a_0, \ldots, a_r) \in \mathrm{PG}(r, K)$ belonging to $\mathcal{X}$ and having a coordinate equal to 1, let $\mathbf{F}_{q^k}$ be the subfield of K generated by $\mathbf{F}_q$ and the coordinates of A. Theorem 8.31 shows that the closed point of A is the set consisting of the k points

$$A = \{A_0, A_1, \ldots, A_{k-1}\},$$

where $A_j = (a_0^{q^j}, \ldots, a_r^{q^j})$ for $j = 0, \ldots, k-1$.

LEMMA 8.33 *For every n, the field Σ has only finitely many effective $\mathbf{F}_q$-rational divisors of degree n.*

Proof. With the notation in Remark 8.32, if $D = P_1 + \cdots + P_n$ is an $\mathbf{F}_q$-rational divisor, and P is a point in D, then either $P \in \mathcal{X}(\mathbf{F}_q)$ or $P \in \mathcal{X}(\mathbf{F}_{q^i}) \backslash \mathcal{X}(\mathbf{F}_q)$ for some i with $1 < i \leq n$. If the latter case occurs then the Frobenius image $\Phi(P)$ of P is another point in D, and this holds true for $\Phi^j(P)$ with $j = 2, \ldots, i-1$. Therefore every point appearing in D with positive weight belongs to $\mathrm{PG}(r, q^m)$, where m is the least common multiple of all integers j with $1 < j \leq n$. As D ranges over all $\mathbf{F}_q$-rational divisors of degree n, such points form a finite set $\{P_1, \ldots, P_s\}$ as $\mathrm{PG}(r, q^n)$ has only finitely many points. Put $\deg P_i = v_i$. It suffices to show that the equation

$$\textstyle\sum_{i=1}^{s} n_i v_i = n$$

has finitely many solutions $(n_1, \ldots, n_s)$ in non-negative integers n_i. Actually, the number of the solutions is bounded above by $(n+1)^s$. □

Using Lemmas 8.24 and 8.15, it is straightforward to show that the $\mathbf{F}_q$-rational principal divisors form a subgroup $\mathrm{Prin}(\mathbf{F}_q(\mathcal{F}))$ of $\mathrm{Prin}(\Sigma)$. A divisor class $[A]$, that is, an element $[A]$ of the divisor class group $\mathrm{Pic}(\Sigma)$, is an *$\mathbf{F}_q$-rational divisor class* if $[A]$ contains at least one $\mathbf{F}_q$-rational divisor. By Theorem 8.26, if $[A]$ is an $\mathbf{F}_q$-rational divisor class, then the linear series $|A|$ is $\mathbf{F}_q$-rational. The set of all $\mathbf{F}_q$-rational divisors classes form a subgroup $\mathrm{Pic}(\mathbf{F}_q(\mathcal{F}))$ of $\mathrm{Pic}(\Sigma)$, the *group of the $\mathbf{F}_q$-rational divisor classes of the $\mathbf{F}_q$-rational function field* $\mathbf{F}_q(\mathcal{F})$. Its subgroup $\mathrm{Pic}_0(\mathbf{F}_q(\mathcal{F}))$ consisting of all $\mathbf{F}_q$-rational divisor classes of degree zero is the *group of the zero-degree $\mathbf{F}_q$-rational divisor classes of the $\mathbf{F}_q$-rational function field* $\mathbf{F}_q(\mathcal{F})$.

THEOREM 8.34 *The group* $\mathrm{Pic}_0(\mathbf{F}_q(\mathcal{F}))$ *is finite.*

Proof. Choose a non-constant element $f \in \mathbf{F}_q(\mathcal{F})$ and let $g = f^m$ for a sufficiently large positive integer m. From Theorem 5.33, $\deg(\mathrm{div}(f)_0) > 0$. So

$$\deg(\mathrm{div}(g)_0) = m \deg(\mathrm{div}(f)_0) = n > 0$$

and m is chosen so that $n \geq 2g$. By Lemma 8.33, it suffices to show that any $\mathbf{F}_q$-rational divisor of degree zero is the difference of the effective $\mathbf{F}_q$-rational divisors of degree n. Let $D \in \mathrm{div}_0(\mathbf{F}_q(\mathcal{F}))$. By the Riemann–Roch Theorem 6.61,

$$\dim(\mathrm{div}(g)_0 - D)) \geq \deg(\mathrm{div}(g)_0) - g \geq g \geq 0.$$

Since $\mathrm{div}(g)_0 - D$ is an $\mathbf{F}_q$-rational divisor, there exists a non-zero element h in $\mathbf{F}_q(\mathcal{F})$ such that $\mathrm{div}(h) + \mathrm{div}(g)_0 - D \succ 0$. Put

$$D_1 = \mathrm{div}(g)_0, \quad D_2 = \mathrm{div}(h) + \mathrm{div}(g)_0 - D.$$

Both divisors D_1 and D_2 are $\mathbf{F}_q$-rational of the same degree n; therefore the equivalence $D \equiv D_1 - D_2$ follows from the fact that $D = D_1 - D_2 + \mathrm{div}(h)$ □

This theorem gives rise to following definition.

DEFINITION 8.35 Let $\mathcal{F}$ be an irreducible algebraic curve defined over a finite field $\mathbf{F}_q$. The order of the group $\mathrm{Pic}_0(\mathbf{F}_q(\mathcal{F}))$ is the *class number.*

EXAMPLE 8.36 Assume that $\mathcal{F}$ is rational. Let A and B be any two $\mathbf{F}_q$-rational divisors of the same degree n. From Theorem 6.81, the linear series $|B|$ has dimension n, and hence $A \in |B|$. Since $|B|$ is also $\mathbf{F}_q$-rational by Theorem 8.26, there exists $f \in \mathbf{F}_q(\mathcal{F})$ such that $A = B + \mathrm{div} f$. Therefore the class number is 1.

Let $\mathcal{P}_0 \in \mathcal{F}(\mathbf{F}_q)$. For any $\mathcal{P} \in \mathcal{F}(\mathbf{F}_q)$, the divisor class $[\mathcal{P} - \mathcal{P}_0]$ is contained in $\mathrm{Pic}_0(\mathbf{F}_q(\mathcal{F}))$. If $g \geq 1$, distinct places $\mathcal{P}$ of $\mathcal{F}(\mathbf{F}_q)$ give rise to distinct elements of $\mathrm{Pic}_0(\mathbf{F}_q(\mathcal{F}))$. Such elements generate a subgroup G of $\mathrm{Pic}_0(\mathbf{F}_q(\mathcal{F}))$. The following result shows that, if q is large enough compared to g, then G is the whole group.

THEOREM 8.37 *Let $\mathcal{F}$ be an $\mathbf{F}_q$-rational curve of genus $g \geq 2$ containing an $\mathbf{F}_q$-rational place $\mathcal{P}_0$, and let $G = \langle[\mathcal{P} - \mathcal{P}_0]\rangle$ with $\mathcal{P}$ ranging over $\mathcal{F}(\mathbf{F}_q)$. If $q \geq (8g-2)^2$, then $G = \mathrm{Pic}_0(\mathbf{F}_q(\mathcal{F}))$.*

Let e denote the smallest degree of all positive degree $\mathbf{F}_q$-rational divisors of $\mathbf{F}_q(\mathcal{F})$. It follows that the degree of every $\mathbf{F}_q$-rational divisor D is divisible by e, that is, $\deg D = m \cdot e$ with $m \in \mathbf{Z}$. Actually, $e = 1$, but this is shown later; see Proposition 9.4. Meanwhile no use is made of it.

PROPOSITION 8.38 *Let*

$$[C_1], \ldots, [C_h]$$

be all zero-degree $\mathbf{F}_q$-rational divisor classes of $\mathbf{F}_q(\mathcal{F})$, and let C_0 be a fixed $\mathbf{F}_q$-rational divisor of $\mathbf{F}_q(\mathcal{F})$ of degree e. Then

(i) *any $\mathbf{F}_q$-rational divisor class $[C]$ can be written as $[C] = [vC_0] + [C_i]$, where $\deg C = v \cdot e$ and $1 \leq i \leq h$;*

(ii) *for any $v \geq 0$, the number of $\mathbf{F}_q$-rational divisor classes of degree $v \cdot e$ is equal to the class number.*

Proof. Let $[C]$ be an $\mathbf{F}_q$-rational divisor class of $\mathbf{F}_q(\mathcal{F})$; then $\deg C = v \cdot e$. The class $v[C_0] = [vC_0]$ has the same degree, and hence $[C - vC_0]$ is a degree zero $\mathbf{F}_q$-rational divisor class of $\mathbf{F}_q(\mathcal{F})$. Therefore $[C - vC_0] = [C_i]$ for some i with $1 \leq i \leq h$. □

THEOREM 8.39 *The number $n_q(|C|)$ of distinct effective $\mathbf{F}_q$-rational divisors in an $\mathbf{F}_q$-rational divisor class $[C]$ is*

$$n_q(|C|) = \frac{q^{\dim|C|+1} - 1}{q - 1}.$$

Proof. Let $r = \dim|C|$. By Theorem 8.26 there are linearly independent elements $x_i \in \mathbf{F}_q(\mathcal{F})$ such that the complete linear series $|C|$ consists of the divisors

$$A_{\mathbf{c}} = \operatorname{div}(c_0x_0 + c_1x_1 + \cdots + c_rx_r) + C,$$

where $\mathbf{c} = (c_0, \ldots, c_r)$ runs over all points of $\operatorname{PG}(r, K)$. The divisor $A_{\mathbf{c}}$ is $\mathbf{F}_q$-rational if and only if the corresponding point $\mathbf{c}$ belongs to $\operatorname{PG}(r, q)$. The result follows from the fact that $|\operatorname{PG}(r, q)| = (q^{r+1} - 1)/(q - 1)$. □

An infinite sequence associated to $\operatorname{Pic}(\mathbf{F}_q(\mathcal{F}))$ is $A_0, A_1, \ldots, A_n, \ldots$, where

$$A_n = |\{A \in \operatorname{Div}(\mathbf{F}_q(\mathcal{F})) \mid A \succ 0;\ \deg A = n\}|.$$

Note that $A_0 = 1$, and $A_1 = N_1$, the number of all $\mathbf{F}_q$-rational places of $\mathbf{F}_q(\mathcal{F})$. For $n > 2g - 2$ the number A_n can be calculated from the class number h of $\mathbf{F}_q(\mathcal{F})$.

PROPOSITION 8.40 *For any integer $n > 2g - 2$ divisible by e,*

$$A_n = \frac{h}{q-1}\left(q^{n+1-g} - 1\right).$$

Proof. There are h divisor classes of degree n, say $[C_1], \ldots, [C_h]$. By Theorem 8.39,

$$|\{A \in [C_j] \cap \operatorname{div}(\mathbf{F}_q(\mathcal{F}));\ A \succ 0\}| = \frac{1}{q-1}\left(q^{\dim|C_j|+1} - 1\right).$$

This, together with the Riemann–Roch Theorem 6.61, implies that

$$|\{A \in [C_j] \cap \operatorname{Div}(\mathbf{F}_q(\mathcal{F})); A \succ 0\}| = \frac{1}{q-1}\left(q^{n+1-g} - 1\right).$$

Any $\mathbf{F}_q$-rational divisor of degree n of $\mathbf{F}_q(\mathcal{F})$ lies in exactly one of the above divisor classes $[C_1], \ldots, [C_h]$. Therefore

$$A_n = \textstyle\sum_{j=1}^{h} |\{A \in [C_j] \cap \operatorname{div}(\mathbf{F}_q(\mathcal{F})); A \succ 0\}| = \dfrac{h}{q-1}\left(q^{n+1-g} - 1\right). \quad \square$$

Write $\ell_1 = \mathbf{F}_{q^n}$, $\ell = \mathbf{F}_q$. Then the trace map $\mathrm{T}_{\ell_1|\ell}$ for fields also defines a homomorphism $\mathfrak{T}_0 = \mathfrak{T}_{\ell_1|\ell}$:

$$\begin{aligned} \mathfrak{T}_0 : \operatorname{Pic}_0(\mathbf{F}_{q^n}(\mathcal{F})) &\to \operatorname{Pic}_0(\mathbf{F}_q(\mathcal{F})), \\ [D] &\mapsto [\mathfrak{T}(D)]. \end{aligned} \tag{8.6}$$

Like the trace map for fields, such a homomorphism is surjective provided that n is not divisible by p. Therefore Proposition 8.22 holds true for $\mathfrak{T}_0$.

8.5 THE STÖHR–VOLOCH THEOREM

In this section, the fundamental theorem of Stöhr and Voloch on the number S_q of $\mathbf{F}_q$-rational points on a non-singular curve $\mathcal{F}$ is obtained. The essential idea is to count those points P on a non-singular model $\mathcal{X}$ for which the osculating hyperplane at P contains the image P^q of P under the Frobenius collineation Φ. Since every $\mathbf{F}_q$-rational point has this property, an upper bound for S_q is obtained.

If an explicit presentation of a non-singular model $\mathcal{X}$ is not available, or too complicated, it is convenient to use some simpler singular model Γ defined over $\mathbf{F}_q$ and count the $\mathbf{F}_q$-rational branches of Γ. This can be done by using the same idea but points on the curve must be regarded as branch points, in the same way as Section 7.6.

An elementary proof of a simple case of the Stöhr–Voloch Theorem is given first.

THEOREM 8.41 *Let $p > 2$, and let $\mathcal{F}$ be an irreducible plane curve of degree d defined over $\mathbf{F}_q$. If $\mathcal{F}$ has only finitely many points of inflexion, then the number S_q of $\mathbf{F}_q$-rational points of $\mathcal{F}$ satisfies the inequality*

$$2S_q + N' \leq d(q + d - 1), \tag{8.7}$$

where N' counts the non-$\mathbf{F}_q$-rational points $Q \in \mathcal{F}$ such that the tangent line at Q contains the image Q^q of Q under the Frobenius collineation.

Proof. If $\mathcal{F} = \mathbf{v}(F(X,Y))$, the tangent to $\mathcal{F}$ at a non-singular point $P = (1, a, b)$ is

$$\mathbf{v}((X - a)F_X(a, b) + (Y - b)F_Y(a, b)).$$

Let

$$G(X, Y) = (X - X^q)F_X + (Y - Y^q)F_Y \tag{8.8}$$

and put $\mathcal{G} = \mathbf{v}(G)$. Suppose that $F(X, Y)$ does not divide $G(X, Y)$. Then the curves $\mathcal{G}$ and $\mathcal{F}$ have no common components. Since

$$\begin{aligned} G_X &= F_X + (X - X^q)F_{XX} + (Y - Y^q)F_{XY}, \\ G_Y &= F_Y + (X - X^q)F_{XY} + (Y - Y^q)F_{YY}, \end{aligned}$$

the tangent to $\mathcal{F}$ at $P \in \mathrm{PG}(2, q)$ is also tangent to $\mathcal{G}$ at the same point. Hence the intersection number $I(P, \mathcal{F} \cap \mathcal{G}) \geq 2$; this remains valid when P is a singular point of $\mathcal{F}$. Also, a direct calculation shows that $I(P, \mathcal{F} \cap \mathcal{G}) \geq 2$ for a point $P = (0, a, b)$ of $\mathcal{F}$ with $a, b \in \mathbf{F}_q$.

For any other common point Q of $\mathcal{F}$ and $\mathcal{G}$, the tangent to $\mathcal{F}$ at Q contains the Frobenius image Q^q. Hence, if there are N' points of the latter type, then Bézout's Theorem implies (8.7).

In the case that $F(X, Y)$ divides $G(X, Y)$, write $G(X, Y) = H(X, Y)F(X, Y)$. Then

$$\begin{aligned} H_X F + H F_X &= G_X = F_X + (X - X^q)F_{XX} + (Y - Y^q)F_{XY}, \\ H_Y F + H F_Y &= G_Y = F_Y + (X - X^q)F_{XY} + (Y - Y^q)F_{YY}. \end{aligned} \tag{8.9}$$

Let $P = (x, y)$ be a generic point of $\mathcal{F}$. Either x or y is a separable variable of $K(\mathcal{F})$. Without loss of generality, suppose that this is x. Then $F_Y(x, y) \neq 0$. Since $G(x, y) = 0$, from (8.8),

$$\frac{F_X(x,y)}{F_Y(x,y)} = -\frac{y^q - y}{x^q - x}. \tag{8.10}$$

Also, from (8.9),

$$\frac{F_X(x,y)}{F_Y(x,y)} = \frac{F_X(x,y) + (x - x^q)F_{XX}(x,y) + (y - y^q)F_{XY}(x,y)}{F_Y(x,y) + (x - x^q)F_{XY}(x,y) + (y - y^q)F_{YY}(x,y)}. \tag{8.11}$$

Eliminating $x - x^q$ and $y - y^q$ from (8.10) and (8.11) shows that (x, y) is a zero of the polynomial

$$(F_Y)^2 F_{XX} - 2F_X F_Y F_{XY} + (F_X)^2 F_{YY}. \tag{8.12}$$

Since $P = (x, y)$ is a generic point of $\mathcal{F}$, it follows that $F(X, Y)$ divides (8.12). By Theorem 1.37, all non-singular points of $\mathcal{F}$ are inflexions. □

Now the general case is considered. The notation and terminology from Section 7.6 are used. In addition, Γ is defined over $\mathbf{F}_q$, and the simple base-point-free linear series L is assumed to be $\mathbf{F}_q$-rational. The symbol $\Gamma(\mathbf{F}_q)$ stands for the set of all $\mathbf{F}_q$-rational branches of Γ. As in Section 7.6, the term 'point P' is adopted to denote the branch of Γ associated to the place $\mathcal{P}$ of $K(\Gamma)$. In doing so, $\Phi(P)$ denotes the Frobenius image of P, and the point $(x_0(P)^q, \ldots, x_r(P)^q)$ is the centre of $\Phi(P)$. Further, $P \in \Gamma(\mathbf{F}_q)$ means that the point P, viewed as a branch of Γ, is $\mathbf{F}_q$-rational. Bearing this in mind, $P \in \Gamma(\mathbf{F}_q)$ does not merely mean that the centre of P is in $\mathrm{PG}(r, q)$, even if this is true when the centre is a non-singular point of Γ.

Let ζ be a local parameter at $\mathcal{P}$. The osculating hyperplane to Γ at P contains the centre of $\Phi(P)$ if and only if $\Delta = 0$, where

$$\Delta = \begin{vmatrix} x_0(P)^q & x_1(P)^q & \cdots & x_r(P)^q \\ D_\zeta^{(j_0)} x_0(P) & D_\zeta^{(j_0)} x_1(P) & \cdots & D_\zeta^{(j_0)} x_r(P) \\ D_\zeta^{(j_1)} x_0(P) & D_\zeta^{(j_1)} x_1(P) & \cdots & D_\zeta^{(j_1)} x_r(P) \\ \vdots & \vdots & \cdots & \vdots \\ D_\zeta^{(j_{r-1})} x_0(P) & D_\zeta^{(j_{r-1})} x_1(P) & \cdots & D_\zeta^{(j_{r-1})} x_r(P) \end{vmatrix}. \tag{8.13}$$

As in Section 7.6, this motivates the study of the Wronskian-type determinant

$$W_\zeta^{\nu_0,\ldots,\nu_{r-1}}(x_0, \ldots, x_r) = \det M(\nu_0, \ldots, \nu_{r-1}),$$

where ζ denotes a separable variable, $\nu_0, \ldots, \nu_{r-1}$ are non-negative integers, and

$$M(\nu_0, \ldots, \nu_{r-1}) = \begin{bmatrix} x_0^q & x_1^q & \cdots & x_r^q \\ D_\zeta^{(\nu_0)} x_0 & D_\zeta^{(\nu_0)} x_1 & \cdots & D_\zeta^{(\nu_0)} x_r \\ D_\zeta^{(\nu_1)} x_0 & D_\zeta^{(\nu_1)} x_1 & \cdots & D_\zeta^{(\nu_1)} x_r \\ \vdots & \vdots & \cdots & \vdots \\ D_\zeta^{(\nu_{r-1})} x_0 & D_\zeta^{(\nu_{r-1})} x_1 & \cdots & D_\zeta^{(\nu_{r-1})} x_r \end{bmatrix}.$$

PROPOSITION 8.42 *There exists a sequence of increasing non-negative integers* $\nu_0, \ldots, \nu_{r-1}$ *with* $\nu_0 \geq 0$ *such that*

$$W_\zeta^{\nu_0,\ldots,\nu_{r-1}}(x_0, \ldots, x_r) \neq 0.$$

If the ν_i *are chosen minimally in lexicographical order, then there exists an integer* I *with* $0 < I \leq r$ *such that*

$$\nu_i = \begin{cases} \epsilon_i & \textit{for } i < I, \\ \epsilon_{i+1} & \textit{for } i \geq I. \end{cases}$$

Proof. Let I be the smallest integer such that the row $(x_0^q, x_1^q, \ldots, x_r^q)$ is a linear combination of the vectors $(D_\zeta^{(\epsilon_i)} x_0, D_\zeta^{(\epsilon_i)} x_1, \ldots, D_\zeta^{(\epsilon_i)} x_r)$ with $i = 0, \ldots, I$. Then $\{\nu_0, \ldots, \nu_{r-1}\} = \{\epsilon_0, \ldots, \epsilon_n\} \backslash \{\epsilon_I\}$. Since $\nu_0 = 0$, so $I > 0$. □

EXAMPLE 8.43 Let $p = 5$. Take $\mathcal{F} = \mathbf{v}(Y^5 + Y - X^3)$ and $\Gamma = \{(1, x, y, y^2)\}$ as in Examples 7.50 (ii) and 7.53. Let $q = p^2$. Then

$$\det M(0,1,2) = \det M(0,1,3) = \det M(0,1,4) = 0,$$

$$\det M(0,1,5) = \begin{vmatrix} 1 & x^{25} & y^{25} & y^{50} \\ 1 & x & y & y^2 \\ 0 & 1 & 3x^2 & x^2 y \\ 0 & 0 & 2x^{10} & x - x^{10} y \end{vmatrix} = 2x^4(1 - x^{12})^3.$$

The minimal choice of the ν_i has a stronger significance.

COROLLARY 8.44 *If* $m_0, \ldots, m_k$ *are any increasing integers with* $m_0 \geq 0$ *such that the row vectors of the matrix* $M(m_0, \ldots, m_k)$ *are linearly independent, then* $\nu_i \leq m_i$ *for* $i = 0, \ldots, k$.

Proof. The matrix $M(\nu_0, \ldots, \nu_{k-1})$ has rank $k + 1$, whence $\nu_{k-1} < m_k$; that is, $\nu_k \leq m_k$. □

DEFINITION 8.45 The *Stöhr–Voloch divisor of* L is

$$S = \text{div}\,(W_\zeta^{\nu_0,\ldots,\nu_{r-1}}(x_0, \ldots, x_r)) + (\nu_1 + \cdots + \nu_{r-1})\,\text{div}(d\zeta) + (q + r)E,$$

where $E = \sum e_\mathcal{P} \mathcal{P}$ and $e_\mathcal{P} = -\min\{\text{ord}_\mathcal{P}\, x_0, \ldots, \text{ord}_\mathcal{P}\, x_r\}$.

By the lemma below, the integers $\nu_0, \ldots, \nu_{r-1}$ and the Stöhr–Voloch divisor depend only on the linear series L.

DEFINITION 8.46 The integers $\nu_0, \ldots, \nu_{r-1}$ are the *Frobenius orders* of the linear series L, or, equivalently, of the curve Γ, and $(\nu_0, \ldots, \nu_{r-1})$ is the *Frobenius order sequence.*

Since Frobenius orders of L are also L-orders, $\nu_i \leq \deg \Gamma$ for $i = 0, 1, \ldots, r - 1$.

LEMMA 8.47 (i) *If* $y_i = \sum a_{ij} x_j$ *with* $(a_{ij}) \in \text{GL}(r + 1, q)$, *then*

$$W_\zeta^{\nu_0,\ldots,\nu_{r-1}}(y_0, \ldots, y_r) = \det(a_{ij})\, W_\zeta^{\nu_0,\ldots,\nu_{r-1}}(x_0, \ldots, x_r).$$

(ii) *If* $h \in \mathbf{F}_q(\Gamma)$, *then*

$$W_\zeta^{\nu_0,\ldots,\nu_{r-1}}(hx_0,\ldots,hx_r) = h^{q+r}\, W_\zeta^{\nu_0,\ldots,\nu_{r-1}}(x_0,\ldots,x_r).$$

(iii) *If* η *is another separable variable, then*

$$W_\eta^{\nu_0,\ldots,\nu_{r-1}}(x_0,\ldots,x_r) = (d\zeta/d\eta)^{\nu_1+\cdots+\nu_{r-1}}\, W_\zeta^{\nu_0,\ldots,\nu_{r-1}}(x_0,\ldots,x_r).$$

Proof. This is similar to that of Lemma 7.51. □

Corollary 5.35 together with the definition of the genus (5.55) shows that

$$\deg S = (\nu_1 + \cdots + \nu_{r-1})(2g-2) + (q+r)n. \tag{8.14}$$

PROPOSITION 8.48 *The coordinate functions* $x_0,\ldots,x_r$ *may be chosen so that*

$$W_\zeta^{\nu_0,\ldots,\nu_{r-1}}(x_0,\ldots,x_r) = \begin{vmatrix} x_1 - x_1^q & \cdots & x_r - x_r^q \\ D_\zeta^{(\nu_1)} x_1 & \cdots & D_\zeta^{(\nu_1)} x_r \\ D_\zeta^{(\nu_2)} x_1 & \cdots & D_\zeta^{(\nu_2)} x_r \\ \vdots & \cdots & \vdots \\ D_\zeta^{(\nu_{r-1})} x_1 & \cdots & D_\zeta^{(\nu_{r-1})} x_r \end{vmatrix}. \tag{8.15}$$

Proof. Dividing $x_0,\ldots,x_r$ by x_0 is equivalent to taking $x_0 = 1$. Now the result is obtained by putting $\nu_0 = 0$. □

REMARK 8.49 By (5.20) and Remark 5.80, $D_\zeta^{(\nu_i)} x_j^q = 0$ if ν_i is not a multiple of q. Hence, if $\nu_i < q$, then $\nu_0,\ldots,\nu_i$ are the first $i+1$ orders of the irreducible curve of $\mathrm{PG}(r-1,K)$ given by the point $R = (x_1 - x_1^q,\ldots,x_r - x_r^q)$. Such a curve is defined over $\mathbf{F}_q$.

PROPOSITION 8.50 (i) *If* ν *is a Frobenius order of* L *less than* q, *then every integer* p*-adically less than* ν *is also a Frobenius order of* L.

(ii) *If* $\nu_i < p$, *then* $(\nu_0,\ldots,\nu_i) = (0,\ldots,i)$.

Proof. This follows from Corollary 7.62. □

Now the weight $v_\mathcal{P}(S)$ of the place $\mathcal{P}$ in the Stöhr–Voloch divisor S is examined. Given a local parameter $\zeta \in K(\Gamma)$ at $\mathcal{P}$, if the components $x_0,\ldots,x_r$ are divided by $\zeta^{e_\mathcal{P}}$, then $e_\mathcal{P}$ becomes zero. Therefore

$$v_\mathcal{P}(S) = v_\mathcal{P}(W_\zeta^{\nu_0,\ldots,\nu_{r-1}}(x_0,\ldots,x_r)) \geq 0;$$

in particular, S is an effective divisor. Further, $\mathcal{P}$ is in $\mathrm{Supp}(S)$ if and only if

$$\begin{vmatrix} x_0^q(P) & x_1^q(P) & \cdots & x_r^q(P) \\ x_0(P) & x_1(P) & \cdots & x_r(P) \\ D_\zeta^{(\nu_1)} x_0(P) & D_\zeta^{(\nu_1)} x_1(P) & \cdots & D_\zeta^{(\nu_1)} x_r(P) \\ \vdots & \vdots & \cdots & \vdots \\ D_\zeta^{(\nu_{r-1})} x_0(P) & D_\zeta^{(\nu_{r-1})} x_1(P) & \cdots & D_\zeta^{(\nu_{r-1})} x_r(P) \end{vmatrix} = 0. \tag{8.16}$$

Let $(j_0, \ldots, j_{r-1})$ be the $(L, \mathcal{P})$-order sequence. In the case that

$$(\nu_0, \ldots, \nu_{r-1}) = (j_0, \ldots, j_{r-1}),$$

$\mathcal{P} \in \text{Supp}(S)$ if and only if $\Phi(P)$ is centred at a point lying in the osculating hyperplane of Γ at P. If $\nu_i < j_i$ for some i, then $\mathcal{P} \in \text{Supp}(S)$.

Also, if the centre of P is in $\text{PG}(r, q)$, the determinant in (8.16) vanishes as its first two rows coincide. Hence $\text{Supp}(S)$ contains every place that corresponds to a branch centred at a point in $\text{PG}(r, q)$. In particular, all $\mathbf{F}_q$-rational places of $K(\Gamma)$ are in $\text{Supp}(S)$.

DEFINITION 8.51 The linear series L or, equivalently, the curve Γ is *Frobenius classical* if the Frobenius order sequence $(\nu_0, \ldots, \nu_{r-1})$ is $(0, \ldots, r-1)$.

REMARK 8.52 If Γ is Frobenius non-classical, then Γ is also non-classical in general. Exceptions only occur when the integer I in Definition 8.42 satisfies both conditions: $I + 1 \equiv 0 \pmod{p}$, and $I < r$. This follows from Propositions 8.42 and 8.50. An exception for $\nu_1 = \epsilon_2 = q = 2$ is the linear series g_4^2 cut out by lines on the irreducible plane quartic

$$\mathcal{F} = \mathbf{v}((X^2 + X)(Y^2 + Y) + 1)$$

defined over $\mathbf{F}_2$.

THEOREM 8.53 *The curve Γ is Frobenius non-classical if and only if, for infinitely many places $\mathcal{P}$ of $K(\Gamma)$, the osculating hyperplane at P contains the centre of $\Phi(P)$.*

Proof. Let

$$\xi = \begin{vmatrix} x_0^q & x_1^q & \cdots & x_r^q \\ x_0 & x_1 & \cdots & x_r \\ D_\zeta^{(\epsilon_1)} x_0 & D_\zeta^{(\epsilon_1)} x_1 & \cdots & D_\zeta^{(\epsilon_1)} x_r \\ \vdots & \vdots & \cdots & \vdots \\ D_\zeta^{(\epsilon_{r-1})} x_0 & D_\zeta^{(\epsilon_{r-1})} x_1 & \cdots & D_\zeta^{(\epsilon_{r-1})} x_r \end{vmatrix}.$$

If Γ is Frobenius classical, then $\xi = 0$. Hence $\Delta = 0$, with Δ as in (8.13), for every P of Γ. Conversely, $\Delta = 0$ implies that ξ has infinitely many zeros. Then ξ itself is zero, and hence $(\nu_0, \ldots, \nu_{r-1}) \neq (0, \ldots, r-1)$. □

As in Remark 8.52, Frobenius non-classical curves are not classical in general. When this is the case, they have properties established in Theorem 7.65. The following result shows how that theorem can be strengthened for Frobenius non-classical curves.

THEOREM 8.54 *Let Γ be a non-classical curve with $(\epsilon_0, \epsilon_1, \ldots, \epsilon_r)$ its order sequence. Assume that $\epsilon_r \leq p^m \leq q = p^h$ and that (7.17) holds for some $z_0, \ldots, z_r$ in $K(\Gamma)$. Then Γ is Frobenius non-classical if and only if*

$$z_0 x_0^{p^{h-m}} + z_1 x_1^{p^{h-m}} + \cdots + z_r x_r^{p^{h-m}} = 0. \tag{8.17}$$

Proof. For a primitive representation σ of $\mathcal{P}$, let

$$\sigma(x_i) = x_i(P) + a_i t + \cdots,$$
$$\sigma(z_i) = z_i(P) + b_i t + \cdots,$$

for all i. Also, let

$$\eta = z_0^{p^m} x_0^q + z_1^{p^m} x_1^q + \cdots + z_r^{p^m} x_r^q.$$

Then

$$\eta = z_0(P)^{p^m} X_0(P)^q + \cdots + z_r(P)^{p^m} X_r(P)^q + tG(t).$$

This shows that the osculating plane at P contains the centre of $\Phi(P)$ if and only if P is a zero of η.

Suppose that Γ is Frobenius non-classical. Then $\eta = 0$ as it has infinitely many zeros. Therefore (8.17) holds, because η is the p^m-th power of the left-hand side in (8.17). Conversely, (8.17) implies $\eta = 0$. Hence, for every point P of Γ, the osculating plane at P contains the centre of $\Phi(P)$. By Theorem 8.53, Γ is a Frobenius non-classical curve. □

EXAMPLE 8.55 Let $p = 5$ and $q = p^2$. Take $\mathcal{F}$ and Γ as in Examples 7.50 (ii) and 8.43. Then

$$z_0 + z_1 X^5 + z_2 Y^5 + z_3 Y^{10} = Y^2 - X^6 + 2Y^6 + Y^{10} = (Y^5 + Y - X^3)^2 = 0.$$

Therefore Γ is Frobenius non-classical.

Now some numerical results follow. For $P \in \Gamma$, let $(j_0, \ldots, j_r)$ be the $(L, \mathcal{P})$-order sequence. By convention in this chapter and Section 7.6, $\mathcal{P}$ is the place of $K(\Gamma)$ corresponding to the branch point $P \in \Gamma$.

PROPOSITION 8.56 (i) *If $P \in \Gamma(\mathbf{F}_q)$, then*

$$v_{\mathcal{P}}(S) \geq \sum_{i=1}^{r} (j_i - \nu_{i-1}),$$

with equality if and only if

$$\det\left(\binom{j_i}{\nu_k}\right) \not\equiv 0 \pmod{p},$$

$0 \leq k \leq r-1,\ 1 \leq i < r.$

(ii) *If $P \notin \Gamma(\mathbf{F}_q)$, then*

$$v_{\mathcal{P}}(S) \geq \sum_{i=1}^{r-1} (j_i - \nu_i).$$

If

$$\det\left(\binom{j_i}{\nu_k}\right) \equiv 0 \pmod{p},$$

with $0 \leq k \leq r-1,\ 0 \leq i \leq r-1$, then strict equality holds.

Proof. The argument is similar to the proof of Theorem 7.55 with τ a primitive representation of $\mathcal{P}$.

(i) Since $\mathcal{P}$ is $\mathbf{F}_q$-rational, the osculating plane at P is over $\mathbf{F}_q$; that is, it is

$$H_P = \mathbf{v}(a_0X_0 + \cdots + a_rX_r),$$

with $a_0, \ldots, a_r \in \mathbf{F}_q$. As in the proof of Theorem 7.47, it is possible to replace x_i by $y_i \in \mathbf{F}_q(\Gamma)$ such that

$$\begin{aligned} y_0 &= 1, \\ \tau(y_i) &= t^{j_i} + \cdots, \end{aligned}$$

with $j_i \geq 1$ for $i = 1, \ldots, r$. Then $\mathrm{ord}_{\mathcal{P}}\, W_\zeta^{\nu_0,\ldots,\nu_{r-1}}(y_0, \ldots, y_r)$, as given in (8.15), may be computed as follows. Since $\nu_0 = 0$ and $j_i > 0$ for $i = 1, \ldots, r$,

$$\begin{aligned} &\mathrm{ord}_t\,(W_t^{\nu_0,\ldots,\nu_{r-1}}(1, \sigma(y_1), \ldots, \sigma(y_r)) \\ &\quad= \mathrm{ord}_t\ \det\left(\binom{j_i}{\nu_k} t^{j_i-\nu_k} + \cdots\right) \\ &\quad= \mathrm{ord}_t\ \det\left(\binom{j_i}{\nu_k}\right) t^{j_1+\cdots+j_r-\nu_0-\cdots-\nu_{r-1}} + \cdots. \end{aligned}$$

The result follows.

(ii) There is a non-singular $(r+1) \times (r+1)$ matrix (a_{ij}) with entries in K, but not necessarily in $\mathbf{F}_q$, such that

$$y_i = \textstyle\sum_{j=0}^r a_{ij}x_j, \quad \tau(y_i) = t^{j_i} + \cdots, \quad i = 0, \ldots, r.$$

Let $h_i = \sum_{j=0}^r a_{ij}x_j^q$. In contrast to (i), it may not be true that $h_i = y_i^q$. Nevertheless, $\mathrm{ord}_{\mathcal{P}}\, h_i \geq 0$ for each i. As

$$\begin{aligned} W_\zeta^{\nu_0,\ldots,\nu_{r-1}}(x_0, \ldots, x_r)\det(a_{ij}) &= \begin{vmatrix} h_0 & \cdots & h_r \\ D_t^{(\nu_0)}y_0 & \cdots & D_t^{(\nu_0)}y_r \\ D_t^{(\nu_1)}y_0 & \cdots & D_t^{(\nu_1)}y_r \\ \vdots & \cdots & \vdots \\ D_t^{(\nu_{r-1})}y_0 & \cdots & D_t^{(\nu_{n-1})}y_r \end{vmatrix} \\ &= \textstyle\sum_0^r (-1)^i h_i d_i, \end{aligned}$$

where d_i is the minor of h_i in the matrix, $i = 0, \ldots, n$, so

$$v_{\mathcal{P}}(S) \geq \min\{\mathrm{ord}_t\, \tau(d_0), \ldots, \mathrm{ord}_t\, \tau(d_r)\}.$$

Now, by the same computations as above,

$$\mathrm{ord}_t\, \tau(d_i) \geq j_0 + \cdots + j_r - j_i - \nu_0 - \cdots - \nu_{r-1}.$$

If

$$\det\left(\binom{j_i}{\nu_k}\right)_{i,k=0,\ldots,r-1} \equiv 0 \pmod p,$$

then

$$\mathrm{ord}_t\, \tau(d_r) > j_0 + \cdots + j_{r-1} - \nu_0 - \cdots - \nu_{r-1}.$$

This proves (ii). □

Now numerical relations between the Frobenius orders $\nu_0, \ldots, \nu_{n-1}$ and the Hermitian orders $j_0, \ldots, j_n$ of an $\mathbf{F}_q$-rational place $\mathcal{P}$ are considered.

PROPOSITION 8.57 *Let $P \in \Gamma(\mathbf{F}_q)$. If $m_0, \ldots, m_{r-1}$ are integers satisfying both the conditions that $0 \le m_0 < \cdots < m_{r-1}$ and*

$$\det\left(\binom{j_i - j_1}{m_k}\right)_{i=1,\ldots,r;\ k=0,\ldots,r-1} \not\equiv 0 \pmod{p},$$

then $\nu_i \le m_i$ for each i.

Proof. The best choices for the integers m_i are the orders of the rational map $\mathrm{PG}(1,K) \to \mathrm{PG}(r-1,K)$ defined as follows:

$$(1,x) \mapsto (1, x^{j_2-j_1}, \ldots, x^{j_r-j_1}) = (x^{j_1}, x^{j_2}, \ldots, x^{j_r}).$$

It may therefore be supposed that $m_0 = 0$ and that

$$\det\left(\binom{j_i}{m_k}\right)_{i=1,\ldots,r;\ k=0,\ldots,r-1} \not\equiv 0 \pmod{p}.$$

Assume again that $x_0 = 1$ and $\sigma(x_i) = t^{j_i} + \cdots$ for each i. Thus, as in the proof of Proposition 8.56 (i),

$$W_\zeta^{m_0,\ldots,m_{r-1}}(x_0,\ldots,x_r) = \det\left(\binom{j_i}{m_k}\right) t^{j_1+\cdots+j_r-m_0-\cdots-m_{r-1}} + \cdots \ne 0.$$

Hence $\nu_i \le m_i$ for all i, by the minimality of the ν_i. □

Some useful corollaries to Propositions 8.56 and 8.57 are now stated.

COROLLARY 8.58 *If $P \in \Gamma(\mathbf{F}_q)$, then*

$$\nu_i \le j_{i+1} - j_1 \text{ for each } i,$$
$$v_{\mathcal{P}}(S) \ge r\, j_1.$$

In the case that $m_i = i$, this gives the next result.

COROLLARY 8.59 *Let $P \in \Gamma(\mathbf{F}_q)$. If the integer $\prod_{r \ge k > s \ge 0}(j_k - j_s)/(k-s)$ is not divisible by p, then $\nu_i = i$ for every i, and*

$$v_{\mathcal{P}}(S) = n + \textstyle\sum_{i=0}^{r}(j_i - i).$$

When $m_i = \epsilon_i$, Proposition 8.57 reads as follows.

COROLLARY 8.60 *Suppose $P \in \Gamma(\mathbf{F}_q)$. If*

$$\det\left(\binom{j_i - j_1}{\epsilon_k}\right)_{i=1,\ldots,r;\ k=0,\ldots,r-1} \not\equiv 0 \pmod{p},$$

then $\nu_i \le \epsilon_i$ for each i.

COROLLARY 8.61 *If the Frobenius order sequence $(\nu_0, \ldots, \nu_{r-1})$ of Γ is not classical, then every $P \in \Gamma(\mathbf{F}_q)$ is an L-osculating point.*

Proof. If there exists $P \in \Gamma(\mathbf{F}_q)$ with $(L, \mathcal{P})$-order sequence $(0, \ldots, r)$, then Corollary 8.58 would imply that $\nu_i = i$ for all i. □

COROLLARY 8.62 *If the Frobenius order sequence $(\nu_0, \ldots, \nu_{r-1})$ is other than the sequence $(\epsilon_0, \ldots, \epsilon_{r-1})$, then every $P \in \Gamma(\mathbf{F}_q)$ is an L-Weierstrass point.*

Proof. If there is an L-ordinary point, then $\nu_i \leq \epsilon_{i+1} - \epsilon_1$ by Corollary 8.58; hence $\nu_i = \epsilon_i$ by Proposition 8.42. □

COROLLARY 8.63 *If L is complete and Γ has some $\mathbf{F}_q$-rational point, then $\nu_i = i$ for each $i < n - 2g$.*

Proof. Let $P \in \Gamma(\mathbf{F}_q)$. Since L is complete, so $j_i = i$ for $i \leq d - 2g$. Hence, by Corollary 8.58, $\nu_i = i$ for each $i < n - 2g$. □

COROLLARY 8.64 *If some branch point P is $\mathbf{F}_q$-rational, then $\nu_i \leq i + n - r$.*

Proof. Since $j_r \leq n$, so $j_i \leq i + n - r$ for each i, and Corollary 8.58 gives the result. □

THEOREM 8.65 (Stöhr–Voloch) *Let Γ be a curve with the following properties:*

(a) *Γ is an irreducible curve in* $\mathrm{PG}(r, K)$ *defined over* $\mathbf{F}_q$;

(b) *n is the degree and g is the genus of Γ;*

(c) *S_q is the number of $\mathbf{F}_q$-rational points of Γ;*

(d) *$(\nu_0, \ldots, \nu_{r-1})$ is the Frobenius order sequence of the linear series $L = g_n^r$ cut out by hyperplanes.*

Then the following results hold:

(i)
$$S_q \leq r^{-1}\{(\nu_1 + \cdots + \nu_{r-1})(2g - 2) + (q + r)n\}; \tag{8.18}$$

(ii) *in stronger form,*
$$S_q \leq r^{-1}\{(\nu_1 + \cdots + \nu_{r-1})(2g - 2) + (q + r)n - \textstyle\sum A(P)\}, \tag{8.19}$$

with
$$A(P) = \begin{cases} \sum_{k=1}^{r} (j_k - \nu_{k-1}) - r, & \text{for } P \in \Gamma(\mathbf{F}_q); \\ \sum_{k=0}^{r-1} (j_k - \nu_k), & \text{otherwise.} \end{cases}$$

Proof. By Corollary 8.58, $v_P(S) \geq r$ for each $P \in \Gamma(\mathbf{F}_q)$. Since S is effective, so $S_q \leq \deg(S)/r$. Now, from (8.14), the first result (8.18) follows. Using Proposition 8.56 gives (8.19). □

COROLLARY 8.66 *If $S_q > r^{-1}\{(r-1)(g-1)+(q+r)n\}$, then each $\mathbf{F}_q$-rational point is an L-osculating point.*

EXAMPLE 8.67 (i) Let C be an irreducible plane curve of degree n. By Lemma 3.24, the genus
$$g \leq \tfrac{1}{2}(n-1)(n-2).$$

(a) If $L_1 = g^2_n$ is the linear series cut out by on $\mathcal{C}$ by lines in $\mathrm{PG}(2, K)$, then

$$S_q \leq \tfrac{1}{2}\{\nu_1(2g-2) + n(q+2)\}. \tag{8.20}$$

Here either $(\nu_0, \nu_1) = (0, \epsilon_1)$ with $\epsilon_1 = 1$ or $(\nu_0, \nu_1) = (0, \epsilon_2)$, where ϵ_2 is the order of contact of $\mathcal{C}$ with the tangent at a generic point; so $\epsilon_2 \leq n$. Also, $\epsilon_2 = 2$ or $\epsilon_2 = p^v$ for some integer v.

If not every point of $\mathcal{C}$ is an inflexion and $p \neq 2$, then $(\nu_0, \nu_1) = (0, 1)$ and

$$S_q \leq \tfrac{1}{2}\{n(n-3) + (q+2)n\} = \tfrac{1}{2}n(n-1+q). \tag{8.21}$$

If, under the same condition that $\mathcal{C}$ is classical and $p = 2$, then

$$S_q \leq \tfrac{1}{2}\{2n(n-3) + (q+2)n\} = \tfrac{1}{2}n(2n-4+q). \tag{8.22}$$

(b) Let $L_2 = g^5_{2n}$ be the linear series cut out on $\mathcal{C}$ by conics in $\mathrm{PG}(2, K)$. Then

$$S_q \leq \tfrac{1}{5}\{(\nu_1 + \nu_2 + \nu_3 + \nu_4)(2g-2) + 2n(q+5)\}. \tag{8.23}$$

Suppose that $\mathcal{C}$ is classical with respect to g^2_n; then, with respect to g^5_{2n}, the order sequence is $(0, 1, 2, 3, 4, \epsilon_5)$ for some $\epsilon_5 \leq 2n$. If $p \neq 2, 3$, then $\epsilon_5 = 5$ or $\epsilon_5 = p^v$.

If $\mathcal{C}$ is classical with respect to g^5_{2n}, then

$$(\nu_0, \nu_1, \nu_2, \nu_3, \nu_4) = (0, 1, 2, 3, 4),$$
$$S_q \leq \tfrac{1}{5}\{10n(n-3) + (q+5)2n\} = \tfrac{2}{5}n\{5(n-2) + q\}. \tag{8.24}$$

(c) Now let $\mathcal{C}$ be defined over $\mathbf{F}_q$ and suppose that $n \geq 3$. If $n \leq \frac{1}{2}p$ and $\mathcal{C}$ has s double points, then

$$S_q \leq \tfrac{2}{5}n\{5(n-2) + p\} - 4s. \tag{8.25}$$

This is obtained from (8.23). By Corollary 7.61, the order sequence is classical as $d = 2n \leq p$. Also, the genus of the curve is

$$g \leq \tfrac{1}{2}(n-1)(n-2) - s.$$

Then (8.23) gives (8.25).

(ii) Now take the Hermitian curve

$$\mathcal{H}_{\sqrt{q}} = \mathbf{v}(X_0^{\sqrt{q}+1} + X_1^{\sqrt{q}+1} + X_2^{\sqrt{q}+1}),$$

and the complete linear series $L_1 = g^2_{\sqrt{q}+1}$ cut out by the lines of the plane. The tangent at $P = (a_0, a_1, a_2)$ to $\mathcal{H}_{\sqrt{q}}$ is

$$\mathbf{v}(a_0^{\sqrt{q}} X_0 + a_1^{\sqrt{q}} X_1 + a_2^{\sqrt{q}} X_2).$$

This meets $\mathcal{H}_{\sqrt{q}}$ at $Q = (b_0, b_1, b_2)$, where

$$b_0^{\sqrt{q}+1} + b_1^{\sqrt{q}+1} + b_2^{\sqrt{q}+1} = 0,$$

with

$$a_0^{\sqrt{q}+1} + a_1^{\sqrt{q}+1} + a_2^{\sqrt{q}+1} = 0,$$
$$a_0^{\sqrt{q}} b_0 + a_1^{\sqrt{q}} b_1 + a_2^{\sqrt{q}} b_2 = 0.$$

Substitution for b_2 gives

$$(a_1^{\sqrt{q}} b_0^{\sqrt{q}} - a_0^{\sqrt{q}} b_1^{\sqrt{q}})(a_1^q b_0 - a_0^q b_1) = 0.$$

Thus the tangent at P meets $\mathcal{H}_{\sqrt{q}}$ with multiplicity $\sqrt{q}$ at P and with multiplicity one at $P^q = (a_0^q, a_1^q, a_2^q)$. In particular, $Q = P^q$. So, with respect to L_1, at a generic point, $j_2 = \sqrt{q}$, whereas at an $\mathbf{F}_q$-rational point, $j_2 = \sqrt{q}+1$. Therefore the order sequence for $\mathcal{H}_{\sqrt{q}}$ is

$$(\epsilon_0, \epsilon_1, \epsilon_2) = (0, 1, \sqrt{q})$$

and the Frobenius order sequence is $(\nu_0, \nu_1) = (0, \sqrt{q})$. For a non-singular curve $\mathcal{C}$ of degree $\sqrt{q}+1$ with a linear series giving these ν_i, the genus

$$g = \tfrac{1}{2}\sqrt{q}(\sqrt{q}-1),$$

and Theorem 8.65 shows the following:

$$\begin{aligned} S_q &\le r^{-1}\{(2g-2)(\nu_0+\nu_1) + (q+r)n\} \\ &= \tfrac{1}{2}\{(q-\sqrt{q}-2)\sqrt{q} + (q+2)(\sqrt{q}+1)\} \\ &= q\sqrt{q}+1. \end{aligned}$$

The upper bound is achieved by $\mathcal{H}_{\sqrt{q}}$,

To prove this, change the coordinate system over $\mathbf{F}_q$ such that $\mathcal{H}_{\sqrt{q}}$ becomes

$$\mathbf{v}(Y^{\sqrt{q}} + Y + X^{\sqrt{q}+1}).$$

Now Y_∞ is an $\mathbf{F}_q$-rational point of $\mathcal{H}_{\sqrt{q}}$. Each vertical line $\mathbf{v}(X-c)$ with $c \in \mathbf{F}_q$ meets $\mathcal{H}_{\sqrt{q}}$ in q affine points, each an $\mathbf{F}_q$-rational point of $\mathcal{H}_{\sqrt{q}}$. In fact, if $c \in \mathbf{F}_q$, then $u = c^{\sqrt{q}+1} \in \mathbf{F}_{\sqrt{q}}$, and the polynomial $Y^{\sqrt{q}} + Y - u$ has exactly $\sqrt{q}$ roots in $\mathbf{F}_q$.

There are only a few curves over $\mathbf{F}_q$ whose number of $\mathbf{F}_q$-rational places can be calculated by using a simple formula. However, certain Frobenius non-classical curves are of this kind.

THEOREM 8.68 (Hefez–Voloch) *Let $\mathcal{X}$ be an irreducible non-singular curve of degree n defined over $\mathbf{F}_q$. If $\nu_1 \ge 2$, then $S_q = n(q-n+2)$.*

Proof. The idea is to compute $v_{\mathcal{P}}(S) - v_{\mathcal{P}}(R)$ for any place $\mathcal{P}$ of Γ. Up to a change of the coordinate functions within $\mathbf{F}_q(\mathcal{X})$, the centre P is an affine point, and the hyperplane $\mathbf{v}(X_1)$ does not contain the tangent line to $\mathcal{X}$ at P. Then a local parameter at $\mathcal{P}$ is $\zeta = x_1 - x_1(P)$. Since $\nu_1 > 1$,

$$v_{\mathcal{P}}(S) = \begin{vmatrix} 1 & x_1^q & x_2^q & \cdots & x_r^q \\ 1 & x_1 & x_2 & \cdots & x_r \\ 0 & 0 & & & \\ \vdots & \vdots & & V & \\ 0 & 0 & & & \end{vmatrix} = \operatorname{ord}_{\mathcal{P}}((x_1 - x_1^q)\det V),$$

where V denotes the $(r-2)\times(r-2)$ matrix with general entry $D_\zeta^{(i)} x_j$. On the other hand, since $\nu_1 > 1$, from Proposition 8.42, $\nu_i = \epsilon_{i+1}$ for $i = 1, \ldots, r-1$. Thus (7.14) becomes the following:

$$v_{\mathcal{P}}(R) = \begin{vmatrix} 1 & x_1 & x_2 & \cdots & x_r \\ 0 & 1 & D_\zeta^{(1)} x_2 & \cdots & D_\zeta^{(1)} x_r \\ 0 & 0 & & & \\ \vdots & \vdots & & V & \\ 0 & 0 & & & \end{vmatrix} = \operatorname{ord}_{\mathcal{P}}(\det V).$$

Hence

$$v_{\mathcal{P}}(S) - v_{\mathcal{P}}(R) = \mathrm{ord}_{\mathcal{P}}(\det V). \tag{8.26}$$

Now it is shown that

$$\mathrm{ord}_{\mathcal{P}}(x_1 - x_1^q) = \begin{cases} 1 & \text{for } P \in \mathbf{F}_q(\mathcal{X}), \\ 0 & \text{for } P \notin \mathbf{F}_q(\mathcal{X}). \end{cases} \tag{8.27}$$

Since $\nu_1 > 1$, the matrix

$$\begin{bmatrix} 1 & x_1^q & x_2^q & \dots & x_r^q \\ 1 & x_1 & x_2 & \dots & x_r \\ 0 & 1 & D_\zeta x_2 & \dots & D_\zeta x_r \end{bmatrix}$$

has rank 2. In particular,

$$(x_1 - x_1^q) D_\zeta x_i = (x_i - x_i^q)$$

for $i = 2, \dots, r$. So, if $\mathrm{ord}_{\mathcal{P}}(x_1 - x_1^q) > 0$, then $\mathrm{ord}_{\mathcal{P}}(x_i - x_i^q) > 0$ for $i = 2, \dots r$, and hence $P \in \mathbf{F}_q(\mathcal{X})$. If $D_\zeta(x_1 - x_1^q) = 1$, then $P \notin \mathbf{F}_q(\mathcal{X})$. Therefore (8.27) holds. From (8.26) and (8.27), $S_q = \mathrm{ord}(S) - \mathrm{ord}(R)$. By (7.13) and (8.14), the result follows. □

REMARK 8.69 Let $\mathcal{F}$ be the DLS curve, that is, the irreducible plane curve considered in Examples 5.24 and 7.73; then $\mathcal{F}$ is defined over $\mathbf{F}_q$. More precisely, $\mathcal{F}$ is a Frobenius non-classical plane curve with $\epsilon = \nu = 2q_0$. This follows from Theorem 8.54 for

$$z_0(X, Y) = X^{q_0+1} + Y^{q_0},\ z_1(X, Y) = X,\ z_2(X, Y) = 1.$$

As noted before, Y_∞ is the unique singular point of $\mathcal{F}$, and it is the centre of only one branch of $\mathcal{F}$. Theorem 8.68 does not hold true for $\mathcal{F}$, since $\mathcal{F}$ has exactly q^2+1 $\mathbf{F}_q$-rational branch points, but

$$n(q - n + 2) = (q + 2q_0)(-2q_0 + 2) \neq q^2 + 1.$$

EXAMPLE 8.70 Let $q = p^h$ be a power of $p > 3$. For a power $d = p^m$ with $m < h$ and for $a, b \in K$ with $b \neq a^2$, the plane curve $\mathcal{F} = \mathbf{v}(f(X, Y))$, with

$$f(X, Y) = X^d Y^d - (Y + a)X^d - (X + a)Y^d + XY + a(X + Y) + b,$$

has the following properties:

(i) $\mathcal{F}$ has two singular points, namely X_∞ and Y_∞, both ordinary d-fold points;

(ii) the tangent lines to $\mathcal{F}$ at X_∞ are $\ell = \mathbf{v}(Y - \lambda)$ and at Y_∞ are $\ell' = \mathbf{v}(X - \lambda)$, with λ ranging over the solutions of $\lambda^d - \lambda - a = 0$;

(iii) $I(X_\infty, \ell \cap \mathcal{F}) = I(Y_\infty, \ell' \cap \mathcal{F}) = 2d$;

(iv) $\mathcal{F}$ has no linear components at either X_∞ or Y_∞;

(v) the genus of $\mathcal{F}$ is $(d-1)^2$.

From (ii), (iii), (iv), Exercise 4 in Chapter 1 implies the irreducibility of $\mathcal{F}$. For a generic point $P=(x,y)$ of $\mathcal{F}$, the birational transformation

$$\omega: \quad x_1 = x, \quad x_2 = y, \quad x_3 = xy$$

defines an irreducible curve Γ of $\mathrm{PG}(3,K)$. Since $f_Y = -(X^d - X - a)$ is not the zero polynomial, x is a separable variable in the function field $K(\mathcal{F})$ of $\mathcal{F}$. Also,

$$D_x^{(1)}y = -\frac{y^d-y-a}{x^d-x-a},\ D_x^{(2)}y = \frac{D_x^{(1)}y}{x^d-x-a},\ D_x^{(3)}y = \frac{D_x^{(2)}y}{x^d-x-a}. \tag{8.28}$$

The generalised Wronskian $W=\det(D_x^{(\epsilon_j)})$ arising from ω is

$$W = \begin{vmatrix} 1 & x & y & xy \\ 0 & 1 & D_x^{(1)}y & y + xD_x^{(1)}y \\ 0 & 0 & D_x^{(2)}y & D_x^{(1)}y + xD_x^{(2)}y \\ 0 & 0 & D_x^{(3)}y & D_x^{(2)}y + xD_x^{(3)}y \end{vmatrix} = \begin{vmatrix} 1 & x & y & 0 \\ 0 & 1 & D_x^{(1)}y & y \\ 0 & 0 & D_x^{(2)}y & D_x^{(1)}y \\ 0 & 0 & D_x^{(3)}y & D_x^{(2)}y \end{vmatrix}.$$

From (8.28), $W=0$; hence Γ is a non-classical curve of $\mathrm{PG}(3,K)$. The orders are $0,1,2,\epsilon_3$, with $\epsilon_3 = p^v$ for some $v \geq 1$ by Theorem 7.65(i). Note that $f(X,Y)$ can be written in the form

$$f(X,Y) = z_0(X,Y)^d + z_1(X,Y)^dX + z_2(X,Y)^dY + z_3(X,Y)^dXY, \tag{8.29}$$

where

$$z_0(X,Y) = XY - c(X+Y) + e, \quad z_1(X,Y) = -Y + c,$$
$$z_2(X,Y) = -X + c, \quad z_3(X,Y) = 1,$$

with $c^d = a$ and $e^d = b$. Since $z_2(x,y) = -x + c$ is a separable variable of $K(\mathcal{F})$, so $\epsilon_3 = d$ by Theorem 7.65 (i).

Now take $a, b \in \mathbf{F}_q$, and view $\mathcal{F}$ as a curve defined over $\mathbf{F}_q$. Then Γ is Frobenius non-classical if and only if

$$q = d^2,\ a^q + a = 0,\ b = b^d.$$

For this, the determinant (8.48) with $\nu_i = i$, for $i = 0,1,2,3$, is calculated. By (8.28), this is

$$W' = \begin{vmatrix} x^q - x & y^q - y & x^qy^q - xy^q \\ 1 & D_x^{(1)}y & y \\ 0 & D_x^{(2)}y & D_x^{(1)}y \end{vmatrix}.$$

Since

$$x^dy^d - (y+a)x^d - (x+a)y^d + xy + a(x+y) + b = 0,$$

so

$$\begin{aligned} W' &= (xy)^q - (y^d - a)x^q - (x^d - a)y^q + x^dy^d - a(x^d + y^d) + b \\ &= [(xy)^{q/d} - (y-c)x^{q/d} - (x-c)y^{q/d} + xy - c(x+y) + e]^d. \end{aligned}$$

Thus a necessary and sufficient condition for Γ to be Frobenius non-classical is $g(x,y)=0$, where

$$g(X,Y)=(XY)^{q/d}-(Y-c)X^{q/d}-(X-c)Y^{q/d}+XY-c(X+Y)+e.$$

Let $\mathcal{G}=\mathbf{v}(g(X,Y))$. Note that, replacing q/d by d and (c,e) by (a,b) as well, $g(X,Y)$ becomes $f(X,Y)$. Hence $\mathcal{G}$ is also irreducible. From this discussion, $\mathcal{F}$ is Frobenius non-classical if and only if $P=(x,y)$ is a common generic point of $\mathcal{F}$ and $\mathcal{G}$. From Remark 5.2, $\mathcal{F}=\mathcal{G}$. Lemma 1.7 implies that $f(X,Y)=kg(X,Y)$ with $k\in K\backslash\{0\}$. Finally, a straightforward calculation shows that this occurs if and only if

$$q=d^2,\ b=b^d,\ a^d+a=0.$$

The Stöhr–Voloch Theorem applied to ω gives the inequality

$$S_q\le\begin{cases}\frac{2}{3}(d+1)((d-1)^2-1), & \text{for } q=d^2,\ b=b^d,\ a^d+a=0;\\ 2(d-1)^2-2+\frac{2}{3}d(q+3), & \text{otherwise.}\end{cases}\tag{8.30}$$

If $\mathcal{F}$ is defined over $\mathbf{F}_q$, then the following result can be added to Theorem 7.68.

THEOREM 8.71 *With the notation in Theorem 7.68, the curve $\mathcal{F}$ is Frobenius non-classical if and only if there exists $t(X,Y)$ in $K[X,Y]$ such that*

$$t(X,Y)f(X,Y)=z_0(X,Y)^{p^m}+z_1(X,Y)^{p^m}X^q+\cdots+z_r(X,Y)^{p^m}Y^{qs}.\tag{8.31}$$

Also, Remark 7.69 remains valid. Despite the characterisation given in Theorems 7.68 and 8.71 it is hard to find Frobenius non-classical curves. What emerges is that they are rare but important curves that can have many $\mathbf{F}_q$-rational points. In the next two sections, the two simplest cases, $s=1,2$, are considered. The case $s=1$ refers to lines and the case $s=2$ to conics.

8.6 FROBENIUS CLASSICALITY WITH RESPECT TO LINES

In this section, classicality and Frobenius classicality are intended with respect to the linear series $L_1=g_n^2$ cut out by lines on an irreducible curve $\mathcal{F}$ of degree n defined over $\mathbf{F}_q$. Also, if a branch of $\mathcal{F}$ centred at the point P and $\mathcal{P}$ is the associated place of $\Sigma=K(\mathcal{F})$, then $j_0=0<j_1<j_2$ denote the L_1-orders of $\mathcal{P}$.

For $s=1$, Theorem 8.71 reads as follows.

THEOREM 8.72 *Let $\mathcal{F}=\mathbf{v}(f(X,Y))$ be an irreducible plane curve defined over $\mathbf{F}_q$ that is non-classical and so satisfies (7.21). Then $\mathcal{F}$ is Frobenius non-classical if and only if there exists $t(X,Y)$ in $K[X,Y]$ such that*

$$t(X,Y)f(X,Y)=z_0(X,Y)^{p^m}+z_1(X,Y)^{p^m}X^q+z_2(X,Y)^{p^m}Y^q.\tag{8.32}$$

REMARK 8.73 For every square q, the Hermitian curve

$$\mathcal{H}_{\sqrt{q}}=\mathbf{v}(X^{\sqrt{q}+1}+Y^{\sqrt{q}+1}+1)$$

is Frobenius non-classical. By Theorem 8.68, the number of its $\mathbf{F}_{q^2}$-rational points is equal to $q\sqrt{q}+1$.

The following result comes from the discussion in Example 8.67.

THEOREM 8.74 *If S_q is the number of $\mathbf{F}_q$-rational points of $\mathcal{F}$ and $q' = \nu_1$, then*

$$S_q \le \tfrac{1}{2}\{q'(2g-2)+(q+2)n] \le \tfrac{1}{2}n[(n-3)q'+q+2\}. \tag{8.33}$$

Non-classical plane curves with respect to lines are investigated in Section 7.7. Now some more material is added on Frobenius non-classical plane curves.

PROPOSITION 8.75 *Let $\mathcal{F}$ be a non-singular plane curve of degree $n \ge 3$ defined over $\mathbf{F}_q$ such that $\mathcal{F}$ is Frobenius non-classical with $\nu_1 = q'$.*

(i) *If $q' > 2$, then $\mathcal{F}$ is a non-classical, and $\nu_1 = \epsilon_2 \le q$.*

(i) *If $\mathcal{F}$ is a strange curve, then it has genus 0.*

Proof. Let $(0, 1, \epsilon)$ be the order sequence of $\mathcal{F}$. From Proposition 8.42, $1 \le \nu_1 \le \epsilon$. If $\nu_1 > 2$ then $\nu_1 = \epsilon_2$ and hence $\epsilon > 2$. Therefore $\mathcal{F}$ is non-classical. In particular, $\epsilon = p^m$ with $m \ge 2$ when $p = 2$. Let $P = (x, y)$ be a generic point of $\mathcal{F}$. It may be assumed that either x or y is a separable variable of $K(\mathcal{F})$. From (7.21) and (8.32),

$$z_0^{p^m} + z_1^{p^m} x + z_2^{p^m} y = 0, \tag{8.34}$$

$$z_0^{p^m} + z_1^{p^m} x^q + z_2^{p^m} y^q = 0. \tag{8.35}$$

Assume on the contrary that $p^m > q$. Let $p^k = p^m/q$. Then, if $k \ge 1$, (8.35) is equivalent to the equation

$$z_0^{p^k} + z_1^{p^k} x + z_2^{p^k} y = 0. \tag{8.36}$$

This and (8.34) define a system of linear equations with solution (x, y).

If (8.34) and (8.36) are dependent, then $z_i^{q-1} = z_j^{q-1}$ when $0 \le i, j \le 2$ and $z_i \ne 0$, $z_j \ne 0$. The latter condition means that $z_i = \lambda z_j$ with $(q-1)$-st root of unity $\lambda_i \in K$. If this occurs, (8.34) can be written as $\alpha_0 + \alpha_1 x + \alpha_2 y = 0$ with $\alpha_0, \alpha_1, \alpha_2 \in K$ not all zero. But then $\mathcal{F}$ must be a line, a contradiction.

If (8.34) and (8.36) are independent, then $x = \xi^p$, $y = \eta^p$, with $\xi, \eta \in K(\mathcal{F})$. But this is impossible since at least one of the elements x and y is a separable variable of $K(\mathcal{F})$.

To show the last assertion in the proposition, assume that $\mathcal{F}$ is a strange curve. Then z_0, z_1 and z_3 are linearly dependent over K, which implies that

$$z_0^{p^m}\alpha_0 + z_1^{p^m}\alpha_1 + z_2^{p^m}\alpha_2 = 0$$

with $\alpha_0, \alpha_1, \alpha_2 \in K$ not all zero. Therefore

$$\begin{vmatrix} 1 & x & y \\ 1 & x^q & y^q \\ \alpha_0 & \alpha_1 & \alpha_2 \end{vmatrix} = 0,$$

whence $\alpha(x - x^q) = y - y^q$ with $\alpha \in K\backslash\{0\}$. Therefore $\mathcal{F}$ is a component of the plane curve $\mathcal{G} = \mathbf{v}(\alpha(X - X^q) - (Y - Y^q))$. If $\alpha \in \mathbf{F}_q$, then $\mathcal{G}$ splits into q lines, and hence $\mathcal{F}$ itself a line. If $\alpha \notin \mathbf{F}_q$, let $\beta^q = \alpha$ with $\beta \in K$. Then $\alpha \ne \beta$ and $\mathcal{G}$ is an irreducible curve of degree q. This implies that $\mathcal{F} = \mathcal{G}$. Since the point $(0, 1, \beta)$ is an $(q-1)$-fold point of $\mathcal{G}$, Exercise 6 of Chapter 1 shows that $\mathcal{F}$ has genus 0. □

For Frobenius non-classical curves of degree $n < q$ with $\nu_1 > 2$, Corollary 8.58 extends to any branch centred at a point of $\mathrm{PG}(2, q)$, that is, for any branch in $\mathcal{F}(\mathbf{F}_q)^*$.

PROPOSITION 8.76 *Let $\mathcal{F}$ be an irreducible plane curve of degree n defined over $\mathbf{F}_q$, and suppose that $\mathcal{F}$ is Frobenius non-classical. If $\nu_1 > 2$, then*

$$\nu_1 \le j_2 - j_1,$$

for every branch in $\mathcal{F}(\mathbf{F}_q)^*$.

Proof. If the tangent to γ lies in $\mathrm{PG}(2, q)$ then γ is an $\mathbf{F}_q$-rational branch by Theorem 8.10 and the assertion follows from Corollary 8.58.

From Proposition 8.75, $\mathcal{F}$ is non-classical with order sequence $(0, 1, \epsilon_2)$ with $\epsilon_2 = \nu_1$ and $\epsilon_2 = p^m$. Here $\nu_1 = p^m < q$ by Proposition 8.75.

Let $(x(t), y(t))$ be a primitive representation of a branch γ of $\mathcal{F}$ centred at a point P of $\mathrm{PG}(2, q)$; it may be assumed that P is the origin. Then take

$$\mathrm{ord}_t\, x(t) = \mathrm{ord}_t\, y(t) = j_1;$$

otherwise the tangent to γ is either $\mathbf{v}(X)$ or $\mathbf{v}(Y)$ and hence it lies in $\mathrm{PG}(2, q)$.

From equations (7.21) and (8.32),

$$m_0(t)^{p^m} + m_1(t)^{p^m} x(t) + m_2(t)^{p^m} y(t) = 0, \tag{8.37}$$

$$m_0(t) + m_1(t)x(t)^{q/p^m} + m_2(t)y(t)^{q/p^m} = 0, \tag{8.38}$$

where $z_i(t) = z_i(x(t), y(t))$ and $m_i(t) = t^{-u_P} z_i(t)$, with $u_P = \min\{\mathrm{ord}_t\, z_i(t)\}$ for $i = 0, 1, 2$.

Let $\mu_i = m_i(0)$. From (8.38), $\mu_0 = 0$; more precisely, (8.38) implies that $\mathrm{ord}_t\, m_0(t) > j_1$. Then either μ_1 or μ_2 is different from zero, and $\mu_1 \ne 0$ without loss of generality. If μ_2 is zero, then (8.37) implies that $j_1 = p^m \mathrm{ord}_t\, m_0(t)$, while (8.37) would give $j_1 q = p^m \mathrm{ord}_t\, m_0(t)$. This contradiction shows that $\mu_2 \ne 0$. Now write (8.37) in the form

$$m_0(t)^{p^m} + (m_1(t) - \mu_1)^{p^m} x(t) = -m_2(t)^{p^m} y(t) - \mu_1^{p^m} x(t).$$

As $\mathrm{ord}_t\, m_0(t)^{p^m} \ge (j_1 + 1)p^m$ and $\mathrm{ord}_t[(m_1(t) - \mu_1)^{p^m} x(t)] \ge p^m + j_1$, this gives the relation

$$\mathrm{ord}_t(-m_2(t)^{p^m} y(t) - \mu_1^{p^m} x(t)) \ge p^m + j_1.$$

Since $\mu_2 \ne 0$, it follows that

$$\mathrm{ord}_t(-\mu_2^{p^m} y(t) - \mu_1^{p^m} x(t)) \ge p^m + j_1.$$

Therefore the intersection number of γ with the line $\ell = \mathbf{v}(-\mu_2^{p^m} Y - \mu_1^{p^m} X)$ is at least $p^m + j_1$. Hence $j_2 \ge p^m + j_1 = \nu_1 + j_1$. □

THEOREM 8.77 *Let $\mathcal{F}$ be a non-singular plane curve of degree n defined over $\mathbf{F}_q$ such that $\mathcal{F}$ is Frobenius non-classical with $\nu = q'$.*

(i) *If* $q' > 2$, *then*

$$\sqrt{q}+1 \leq n \leq \frac{q-1}{q'-1} \quad \textit{and} \quad q' \leq \sqrt{q}.$$

(ii) *If* $q' = \sqrt{q}$, *then* $\mathcal{F}$ *is projectively equivalent to the Hermitian curve*

$$\mathcal{H}_{\sqrt{q}} = \mathbf{v}(X_0^{\sqrt{q}+1} + X_1^{\sqrt{q}+1} + X_2^{\sqrt{q}+1}).$$

Proof. Since $\mathcal{F}$ is non-singular, its genus $g = (n-1)(n-2)/2$ by Theorem 5.57. From (9.46), which is the Hasse–Weil Bound for non-singular plane curves, as in Section 9.6,

$$S_q \leq q+1+(n-1)(n-2)\sqrt{q}.$$

On the other hand, from Theorem 8.68, $S_q = n(q-n+2)$, whence the lower bound on n in (i) follows.

Put $q = p^h$. Since $q' > 1$, from Theorem 8.111, $\mathcal{F}$ is non-classical and

$$1 < \nu_1 \leq \epsilon_2 = p^m \leq p^h.$$

Then Theorem 7.65, together with Remark 7.70, shows that

$$F(X_0, X_1, X_2) = z_0^{p^m} X_0 + z_1^{p^m} X_1 + z_2^{p^m} X_2,$$

with homogeneous polynomials $z_0, z_1, z_2 \in \mathbf{F}_q[X_0, X_1, X_2]$ of the same degree d and $n = dp^m + 1$. On the other hand, as $\mathcal{F}$ is Frobenius non-classical, from the proof of Theorem 8.54 there is a homogeneous polynomial $G \in K[X_0, X_1, X_2]$ such that

$$GF = z_0 X_0^{p^{h-m}} + z_1 X_1^{p^{h-m}} + z_2 X_2^{p^{h-m}}.$$

So $\deg F \leq \deg(GF)$. Therefore $n \leq d + p^{h-m}$, whence the upper bound on n in (i) follows.

Eliminating n from the first inequality in (i) gives $q' \leq \sqrt{q}$. If equality holds, then $d = 1$, and a direct computation shows that this implies (ii). □

Now it is shown that (ii) remains true for singular curves of low degree.

THEOREM 8.78 *If* $\mathcal{F}$ *is an irreducible plane curve of degree* n *defined over* $\mathbf{F}_q$, *with* q *square, such that*

(a) $\mathcal{F}$ *is Frobenius non-classical with* $\nu = \sqrt{q}$,

(b) $n < 2\sqrt{q}$ *and* $\mathcal{F}$ *is not rational*,

then (ii) *in Theorem 8.77 holds.*

Proof. The idea is to consider the dual curve $\mathcal{D}$ of $\mathcal{F}$. First it is shown that $\mathcal{D}$ and $\mathcal{F}$ are birationally equivalent over $\mathbf{F}_q$. The second step consists in proving that $\mathcal{D}$ is a non-singular plane curve which is still Frobenius non-classical with the same $\nu_1 = q'$. After that, the assertion follows from Theorem 8.77 (ii).

From Proposition 8.75, either $q = 4$ and $\mathcal{F}$ is classical, or $\mathcal{F}$ is not only Frobenius non-classical but also non-classical with order sequence $(0, 1, \sqrt{q})$. In the former case, $\mathcal{F}$ is a non-singular plane cubic curve, and the assertion follows from Theorem 8.77 (ii).

Therefore $\mathcal{F}$ may be taken as non-classical with order sequence $(0, 1, \sqrt{q})$. If $Q = (x, y)$ is a generic point of $\mathcal{F}$, from (7.17) and (8.17),

$$z_0^{\sqrt{q}} + z_1^{\sqrt{q}}x + z_2^{\sqrt{q}}y = 0, \tag{8.39}$$

$$z_0 + z_1 x^{\sqrt{q}} + z_2 y^{\sqrt{q}} = 0, \tag{8.40}$$

where at least one of the elements z_i is a separable variable. Let γ be a branch of $\mathcal{F}$ centred at the point $A = (a, b) \in \mathcal{F}$, corresponding to the place $\mathcal{P}$ of $K(\mathcal{F})$. If $\text{ord}_{\mathcal{P}}\, z_j = \min \text{ord}_{\mathcal{P}}\, z_i$ with $i \in \{0, 1, 2\}$, then, by (8.39), the tangent to γ is

$$\mathbf{v}(m_0(a,b)X_0 + m_1(a,b)X_1 + m_2(a,b)X_2).$$

By Proposition 8.75 is not a strange curve. Hence z_0, z_1, z_2 are linearly independent over K.

The dual curve $\mathcal{D}$ of $\mathcal{F}$ is the image curve of $\mathcal{F}$ under the rational transformation

$$\omega: \quad x_0' = z_0,\ x_1' = z_1,\ x_2' = z_2$$

defined over $\mathbf{F}_q$. Let n' denote the degree of $\mathcal{D}$, and suppose that ω is not birational. Then the linear series $g_2^{n'}$ consisting of all divisors

$$A_{\mathbf{c}} = \text{div}\,(c_0 z_0 + c_1 z_1 + c_2 z_2) + B,$$

where $B = \sum e_{\mathcal{P}}\mathcal{P}$ and $\mathbf{c} = (c_0, c_1, c_2) \in \text{PG}(2, K)$, is composed of an involution. Therefore, for infinitely many points $A \in \mathcal{F}$, the tangent ℓ to $\mathcal{F}$ at A is tangent to $\mathcal{F}$ at another point A', as well. Since both $I(A, \mathcal{F} \cap \ell)$ and $I(A', \mathcal{F} \cap \ell)$ are at least $\sqrt{q}$, this implies that $n \geq 2\sqrt{q}$, a contradiction.

Now it is shown that every branch γ' of $\mathcal{D}$ is linear. To do this, take a primitive representation σ of $\mathcal{P}$ of $K(\mathcal{F})$, and let $\sigma(x) = x(t)$, $\sigma(y) = y(t)$ where $x(t), y(t) \in K[[t]]$. Then the branch γ has the primitive representation

$$\begin{aligned} x(t) &= a + a_1 t + \cdots + a_k t^k + \cdots, \\ y(t) &= b + b_1 t + \cdots + b_k t^k + \cdots. \end{aligned} \tag{8.41}$$

Put $z_i(t) = \sigma(z_i)$ and $\sigma(m_i) = m_i(t)$. Then a homogeneous primitive representation of the branch γ' of $\mathcal{D}$ associated to $\mathcal{P}$ is $(m_0(t), m_1(t), m_2(t))$. From (8.39),

$$z_0(t)^{\sqrt{q}} + z_1(t)^{\sqrt{q}}x(t) + z_2(t)^{\sqrt{q}}y(t) = 0,$$

whence

$$m_0(t)^{\sqrt{q}} + m_1(t)^{\sqrt{q}}x(t) + m_2(t)^{\sqrt{q}}y(t) = 0. \tag{8.42}$$

When the formal power series $m_i(t)$ is written in the form

$$m_i(t) = \mu_i^{(0)} + \mu_i^{(1)}t + \cdots + \mu_i^{(k)}t^k + \cdots,$$

for $i = 0, 1, 2$, then (8.42) gives the following:

$$\begin{aligned} &[(\mu_0^{(0)})^{\sqrt{q}} + (\mu_1^{(0)})^{\sqrt{q}}x(t) + (\mu_2^{(0)})^{\sqrt{q}}y(t)] \\ &+ t^{\sqrt{q}}[(\mu_0^{(1)})^{\sqrt{q}} + (\mu_1^{(1)})^{\sqrt{q}}x(t) + (\mu_2^{(1)})^{\sqrt{q}}y(t)] \\ &+ \cdots \\ &+ t^{k\sqrt{q}}[(\mu_0^{(k)})^{\sqrt{q}} + (\mu_1^{(k)})^{\sqrt{q}}x(t) + (\mu_2^{(k)})^{\sqrt{q}}y(t)] \\ &+ \cdots = 0; \end{aligned} \tag{8.43}$$

here $\left((\mu_0^{(0)})^{\sqrt{q}}, (\mu_1^{(0)})^{\sqrt{q}}, (\mu_2^{(0)})^{\sqrt{q}}\right)$ is the centre of the branch γ'. This shows that the line ℓ, where

$$\ell = \mathbf{v}((\mu_0^{(0)})^{\sqrt{q}} X_0 + (\mu_1^{(0)})^{\sqrt{q}} X_1 + (\mu_2^{(0)})^{\sqrt{q}} X_2), \tag{8.44}$$

meets $\mathcal{F}$ in P with multiplicity at least $\sqrt{q}$; that is, $I(P, \ell \cap \mathcal{F}) \geq \sqrt{q}$.

If there were another branch of $\mathcal{D}$ with the same centre, then $I(R, \ell \cap \mathcal{F}) \geq \sqrt{q}$ would hold for another branch R of $\mathcal{F}$, contradicting the hypothesis that $n < 2\sqrt{q}$.

Now suppose that γ' is a non-linear branch of $\mathcal{D}$. Then $\mu_i^{(1)} = 0$ for $i = 0, 1, 2$, and equation (8.43) implies that

$$\begin{aligned}
&[(\mu_0^{(0)})^{\sqrt{q}} + (\mu_1^{(0)})^{\sqrt{q}} x(t) + (\mu_2^{(0)})^{\sqrt{q}} y(t)] \\
&+ t^{2\sqrt{q}}[(\mu_0^{(2)})^{\sqrt{q}} + (\mu_1^{(2)})^{\sqrt{q}} x(t) + (\mu_2^{(2)})^{\sqrt{q}} y(t)] \\
&+ \cdots \\
&+ t^{k\sqrt{q}}[(\mu_0^{(k)})^{\sqrt{q}} + (\mu_1^{(k)})^{\sqrt{q}} x(t) + (\mu_2^{(k)})^{\sqrt{q}} y(t)] \\
&+ \cdots = 0.
\end{aligned} \tag{8.45}$$

This shows that the line ℓ given by (8.44) meets $\mathcal{F}$ at P with multiplicity at least $2\sqrt{q}$. But this is impossible, again by the hypothesis $n < 2\sqrt{q}$. Hence $\mathcal{D}$ has only linear branches and just one is centred at each point. Therefore $\mathcal{D}$ is a non-singular plane curve.

To prove that $\mathcal{D}$ is non-classical, (8.40) is used. From (8.40),

$$m_0(t) + m_1(t)x(t)^{\sqrt{q}} + m_2(t)y(t)^{\sqrt{q}} = 0, \tag{8.46}$$

whence

$$m_0(t) + m_1(t)x_0^{\sqrt{q}} + m_2(t)y_0^{\sqrt{q}} + t^{\sqrt{q}}G(t) = 0. \tag{8.47}$$

This shows that the line $\ell = \mathbf{v}(X_0 + a^{\sqrt{q}}X_1 + b^{\sqrt{q}}X_2)$ is the tangent to $\mathcal{D}$ at the centre of γ', and that $j_2 \geq \sqrt{q}$. Thus $\mathcal{D}$ is non-classical. Since either x or y is a separable variable of $K(\mathcal{F})$, from Theorem 7.65 (iii) it follows that the orders of $\mathcal{D}$ are $0, 1, \sqrt{q}$. Also, ℓ passes through the point which is the image of the centre of γ' under the Frobenius collineation. This follows from (8.39).

By Theorem 8.77 (ii), $\mathcal{D}$ is the Hermitian curve $\mathcal{H}_{\sqrt{q}}$ as in (ii), up to a change of the coordinate system over $\mathbf{F}_q$. Since the dual curve of the Hermitian curve coincides with itself, $\mathcal{F}$ itself is birationally equivalent to the Hermitian curve, by means of a birational transformation defined over $\mathbf{F}_q$.

To finish the proof, it suffices to show that $\mathcal{F}$ is a non-singular plane curve. From (8.47) it follows that $\mathcal{F}$ has only linear branches. In fact, if the branch γ were not linear, then (8.47) would imply for the corresponding place $\mathcal{P}$ of $K(\mathcal{F})$ that

$$\operatorname{ord}_{\mathcal{P}}[m_0(t) + \xi_0 m_1(t) + \eta_0 m_2(t)] \geq 2\sqrt{q};$$

that is, the line $\ell = \mathbf{v}(X_0 + \xi_0 X_1 + \eta_0 X_2)$ would have intersection multiplicity at least $2\sqrt{q}$ with the branch γ' of $\mathcal{D}$, contradicting that $\deg \mathcal{D} = \sqrt{q} + 1$.

If the point $(1, a^q, b^q)$ were also the centre of another branch of $\mathcal{F}$, then ℓ would be a bitangent of $\mathcal{D}$, which would therefore have degree at least $2\sqrt{q}$, again contradicting that $\deg \mathcal{D} = \sqrt{q} + 1$. Therefore each point of $\mathcal{F}$ is the centre of a single linear branch. Thus $\mathcal{F}$ is non-singular, and this completes the proof. □

Since a non-singular point P of a classical curve $\mathcal{F}$ is an inflexion if and only if $j_2(\mathcal{P}) \neq 2j_1(\mathcal{P})$, the classical idea to distinguish inflexions from other non-singular points can be extended to branches. For branches centred at points of $\mathrm{PG}(2, q)$, this produces a partition of $\mathcal{F}(\mathbf{F}_q)^*$ into two subsets

$$\begin{array}{rcl} S_1 & = & \{\gamma \in \mathcal{F}(\mathbf{F}_q)^* \mid j_2 = 2j_1\}, \\ S_2 & = & \{\gamma \in \mathcal{F}(\mathbf{F}_q)^* \mid j_2 \neq 2j_1\}, \end{array} \tag{8.48}$$

where $(j_0 = 0, j_1, j_2)$ is the order sequence at the place arising from the branch γ.

This gives rise to important constants, M_q and M_q'.

NOTATION 8.79 (i) M_q is the number of branches $\gamma \in S_1$, each one counted $j_1(\mathcal{P})$ times;

(ii) M_q' is the number of branches $\gamma \in S_2$, each one counted $j_1(\mathcal{P})$ times.

Here $B_q = M_q + M_q'$. As stated in Theorems 8.9, 8.10 and Remark 8.11, $\mathcal{F}$ may have branches centred at points in $\mathrm{PG}(2, q)$ that are not $\mathbf{F}_q$-rational branches of $\mathcal{F}$. In other words, $B_q \geq S_q$ but it is not necessary that $B_q = S_q$. Estimates for B_q are obtained in Section 9.6.

For any Frobenius classical curve, Proposition 8.75 implies that $M_q = 0$, and hence $S_q = M_q'$. On the other hand, if $\mathcal{F}$ is classical and has degree n, then

$$M_q' \leq 3n(n-2). \tag{8.49}$$

As shown in Section 8.7, for Frobenius non-classical curves with respect to conics, it is easier to estimate $2M_q + M_q'$ rather than S_q. Then S_q can be estimated from that bound and some bound on M_q'. Also, if $\mathcal{F}$ is classical, then the inequality (8.49) can be used. The latter occurs when $q \geq 25$, $n \leq \sqrt{q}$ and

$$2M_q + M_q' \geq n(q - \sqrt{q} + 1). \tag{8.50}$$

This is a consequence of the following result.

THEOREM 8.80 *Let $\mathcal{F}$ be an irreducible plane curve of degree n defined over $\mathbf{F}_q$ with q odd and $q \geq 25$. If $n \leq \sqrt{q} + 1$ and (8.50) holds, then $\mathcal{F}$ is either classical or projectively equivalent to the Hermitian curve*

$$\mathcal{H}_{\sqrt{q}} = \mathbf{v}(X_0^{\sqrt{q}+1} + X_1^{\sqrt{q}+1} + X_2^{\sqrt{q}+1}).$$

Proof. Suppose that $\mathcal{F}$ is non-classical. Then Corollary 7.60 implies that

$$j_1\, j_2\, (j_2 - j_1) \equiv 0 \pmod{p}, \tag{8.51}$$

for any branch γ of $\mathcal{F}$ centred at a point P of $\mathrm{PG}(2, K)$. First, the case where $\mathcal{F}$ is Frobenius classical is investigated. From Theorem 8.56,

$$v_{\mathcal{P}}(S) \geq (j_2 - 1) + j_1.$$

If $\gamma \in S_2$, this implies that $v_{\mathcal{P}}(S) \geq 2j_1$. If $\gamma \in S_1$, it follows from (8.51) that $j_1 \equiv 0 \pmod{p}$. From this and Theorem 8.56 (ii), $v_{\mathcal{P}}(S) > j_2 + j_1 - 1$, whence $v_{\mathcal{P}} \geq 3j_1$.

Therefore

$$(2g-2)+(q+2)n=\deg S\geq 3M_q+2M'_q\geq \tfrac{3}{2}\left(2M_q+M'_q\right). \qquad (8.52)$$

Since $n(n-3)\geq 2g-2$, it follows from (8.52) and (8.50) that

$$(n-3)+(q+2)\geq \tfrac{3}{2}(q-\sqrt{q}+1);$$

that is, $2n\geq (q-3\sqrt{q}+5)$. But this cannot actually occur when both $n\leq\sqrt{q}+1$ and $q\geq 25$.

Suppose that $\mathcal{F}$ is $\mathbf{F}_q$-Frobenius non-classical. It is shown first that

$$n(2g-2)+3n\geq \deg(R)=(1+\epsilon)(2g-2)+3n\geq 2M_q+M'_q. \qquad (8.53)$$

To prove the first inequality in (8.53), it must be shown that $n\geq\epsilon+1$. If $\mathcal{F}$ has an $\mathbf{F}_q$-rational point, this follows from Corollary 8.58 as $\nu_1=\epsilon$. Otherwise $\mathcal{F}(\mathbf{F}_q)^*$ consists of singular points each being the centre of at least two branches. Take two branches γ and γ' of $\mathcal{F}$ centred at $P\in\mathcal{F}$, and let ℓ be the tangent to γ. Then

$$n\geq I(P,\ell\cap\gamma)+I(P,\ell\cap\gamma')\geq\epsilon+1.$$

To prove the second equality in (8.53), note that, from Theorem 7.55,

$$v_{\mathcal{P}}(R)\geq (j_2-\epsilon_2)+(j_1-1)=j_2-j_1-\epsilon_2+2j_1-1.$$

If $\gamma\in S_1$ then $j_2\equiv 0 \pmod p$ by (8.51), and hence the strict inequality holds. This, together with Proposition 8.76 and $\epsilon_2=\nu_1$, implies that $v_{\mathcal{P}}(R)\geq 2j_1$. If $\gamma\in S_2$, the above argument shows that $v_{\mathcal{P}}(R)\geq 2j_1-1\geq j_1$. Therefore (8.53) holds.

Since $n(n-3)\geq 2g-2$ by Lemma 3.24, (8.53) shows that $n\geq\sqrt{q}+1$. This and the hypothesis $n\leq\sqrt{q}+1$ imply that $n=\sqrt{q}+1$. Hence $n(n-3)=2g-2$ and $\epsilon_2=\nu_1=\sqrt{q}$. By Lemma 3.24, $\mathcal{F}$ is non-singular and the assertion follows from Theorem 8.77 (ii). □

THEOREM 8.81 *For a polynomial $f(X)\in\mathbf{F}_q[X]$ having some simple root, let $\mathcal{F}=\mathbf{v}(Y^n-f(X))$ with $p\nmid n$ be an irreducible plane curve. If $p>2$, then the following hold.*

(i) *$\mathcal{F}$ is non-classical if and only if $n\equiv 1 \pmod p$ and there exist polynomials $g(X),h(X)\in K[X]$ with $\deg g(X)\geq\deg h(X)$ such that*

$$f(X)=g(X)^pX+h(X)^p. \qquad (8.54)$$

(ii) *$\mathcal{F}$ is Frobenius non-classical if and only if*

$$g(X)^pX^q+h(X)^p=(g(X)^pX+h(X)^p)^{p^sm'}, \qquad (8.55)$$

where m' is given by the system

$$\begin{cases} n=p^sm+1, & m\not\equiv 0 \pmod p,\\ q-1=(p^sm'-1)(p^s+m), & m'\not\equiv 0 \pmod p.\end{cases} \qquad (8.56)$$

(iii) *If $\mathcal{F}$ is Frobenius non-classical, then $\deg f(X)=n$. If, in addition, the multiplicities of the roots of $f(X)$ are relatively prime to p, then all roots of $f(X)$ belong to $\mathbf{F}_q$.*

Proof. First, $K(\mathcal{F}) = K(x, y)$ with $y^n - f(x) = 0$. Here x is a separable variable of $K(\mathcal{F})$ by the hypothesis $p \nmid n$. The non-classicality of $\mathcal{F}$ is expressed by the condition $D_x^{(2)} y = 0$. By (5.24) and Lemma 5.80, differentiating $y^n - f(x)$ twice with respect to x gives the equation

$$2ny^{n-1}D_x^{(2)}y + \frac{(n-1)f'(x)^2}{nf(x)} = f''(x),$$

where $f'(x)$ is $f'(X)$ evaluated at x, and similarly for $f''(x)$. So $D_x^{(2)}y = 0$ is equivalent to that $(n-1)f'(x)^2 = nf''(x)f(x)$. Calculating both sides at a simple root of $f(X)$, this equation holds if and only if $n \equiv 1 \pmod p$ and $f''(x) = 0$. Since x is not a constant, $f''(x) = 0$ implies $f''(X) = 0$. Write $f(X) = \sum a_i X^i$.

Now $f''(X) = 0$ occurs if and only if every a_i with $i \not\equiv 0, 1 \pmod p$ vanishes. Hence $f(X) = g(X)^p X + h(X)^p$ with $g(X), h(X) \in \mathbf{F}_q[X]$. Further, $\deg f(X)$ is not divisible by p if and only if $\deg g(X) \geq \deg h(X)$. Therefore (i) is proved.

To prove (ii), the following hypotheses may be assumed:

$$\begin{aligned} &Y^n = g(X)^p X + h(X)^p;\\ &n \equiv 1 \pmod p);\\ &\deg g(X) \geq \deg h(X). \end{aligned} \tag{8.57}$$

Using the first equation, the Frobenius classicality of $\mathcal{F}$ expressed by the relation

$$y^q - y = (x^q - x)D_x^{(1)}y$$

becomes the following:

$$y^{q-1+n} = g(x)^p x^p + h(x)^p. \tag{8.58}$$

So, if $\mathcal{F}$ is Frobenius non-classical, then $y^{q-1+n} \in K(x)$. No power of y with exponent smaller than n is in $K(x)$, since $[K(\mathcal{F}) : K(x)] = n$. Hence n divides $q - 1 + n$, and so $q - 1 = \lambda n$ for a positive integer λ. Since $n \equiv 1 \pmod p$, there is a positive integer m such that $n = p^s m + 1$ with m a positive integer not divisible by p. Thus $\lambda \equiv -1 \pmod p$; so $\lambda = p^r m' - 1$ with m' a positive integer not divisible by p. Hence $q = p^{sr} mm' + p^r m' - p^s m$. From this, it follows that $r = s$; otherwise either m or m' would be divisible by p. Then $m \equiv m' \pmod p$. Therefore

$$y^{q-1+n} = (y^n)^{p^s m'} = (g(x)^p x + h(x)^p)^{p^s m'}, \tag{8.59}$$

whence

$$g(x)^p x^q + h(x)^p = (g(x)^p x + h(x)p)^{p^s m'}.$$

Since x is not a constant, this shows (8.55).

Conversely, if (8.55) holds, then $\mathcal{F}$ is Frobenius non-classical. This completes the proof of (ii).

Comparison of degrees in (8.55) gives the equation

$$p \deg g(X) + q = (p \deg g(X) + 1)p^s m'.$$

Since $q - 1 = (p^s m' - 1)(p^s m + 1)$, it follows that $\deg g(X) = p^{s-1} m$. Thus

$$\deg f(X) = p \deg g(X) + 1 = p^s m + 1 = n,$$

which proves (iii). □

Note that, writing $f(X) = a_0X^n + a_1X^{n-1} + \cdots + a_n$, a change of the indeterminate X of type $X' = aX + b$ can be made with $a, b \in \mathbf{F}_q$ such that $f'(X) = f(aX + b)$ has no term of degree $n - 1$. Therefore it may be assumed in Theorem 8.81 that $\deg g(X) > \deg h(X)$ without loss of generality.

Now the cases $q = p^2$ or $q = p^3$ in Theorem 8.81 are considered more closely. Assume that both (8.54) and (8.55) hold. If $q = p^2$, then $s = m = m' = 1$ can only occur in (8.56). Also, $\deg g(X) = p^{s-1}m = 1$. Hence $h(X)$ is a constant polynomial. It turns out that $\mathcal{F}$ is projectively equivalent over $\mathbf{F}_q$ to the Hermitian curve $\mathcal{H}_{\sqrt{q}}$.

If $q = p^3$, there are examples which are not Fermat curves. Since

$$p^3 - 1 = (p - 1)(p(p + 1) + 1),$$

from (8.56), $s = m' = 1$, $m = p + 1$, and $\deg g(X) = p + 1$. Since it is assumed that $\deg g(X) > \deg h(X)$, so (8.55) gives the equation

$$g(X)X^{p^2} + h(X) = g(X)^pX + h(X)^p.$$

Comparing the coefficients gives that

$$g(X) = \alpha X^{p+1} + \beta^{p^2}, \quad h(X) = \beta^pX^p + \beta X + \delta \tag{8.60}$$

with $\alpha, \delta \in \mathbf{F}_p$, $\alpha \neq 0$, and $\beta \in \mathbf{F}_{p^3}$. For instance, $\alpha = \beta = 1$ and $\delta = 0$ is a good choice, giving rise to the Frobenius non-classical curve

$$\mathcal{F} = \mathbf{v}(Y^{p^2+p+1} - ((X^{p+1} + 1)^pX + (X^p + X)^p)). \tag{8.61}$$

It may be noted that this curve is non-singular. By Theorem 8.68, the number S_q of its $\mathbf{F}_q$-rational places is $(p^2 + p + 1)(p^3 - p^2 - p + 1)$.

REMARK 8.82 Let $q = p^3$. Up to birational equivalence over $\mathbf{F}_q$, the curve $\mathcal{F}$ from (8.61) is the only Frobenius non-classical curve of the form $\mathbf{v}(Y^n - f(X))$ over $\mathbf{F}_q$. This follows from the fact that (8.56) is only possible when $s = m' = 1$ and $m = p + 1$.

8.7 FROBENIUS CLASSICALITY WITH RESPECT TO CONICS

This section continues the study of irreducible curves $\mathcal{F}$ which are non-classical with respect to the linear series L_2 cut out by conics. The L_2-orders are assumed to be $0, 1, 2, 3, 4, p^m$ with $p > 2$. In particular, $\mathcal{F}$ is classical with respect to L_1. The notation is that of Section 7.8. In addition, $\mathcal{F}$ is defined over $\mathbf{F}_q$ and it is assumed that $p^m < q = p^h$.

The hypothesis on $\mathcal{F}$ to be Frobenius non-classical is expressed by the existence of a polynomial $t(X, Y) \in K[X, Y]$ such that

$$\begin{aligned} &t(X,Y)f(X,Y) = \\ &\quad z_0(X,Y)^{p^m} + z_1(X,Y)^{p^m}X^q + z_2(X,Y)^{p^m}Y^q \\ &\quad +z_3(X,Y)^{p^m}X^{2q} + z_4(X,Y)^{p^m}X^qY^q + z_5(X,Y)^{p^m}Y^{2q}. \end{aligned} \tag{8.62}$$

Equivalently,

$$v_0^{p^m} + v_1^{p^m} x^q + v_2^{p^m} y^q + v_3^{p^m} x^{2q} + v_4^{p^m} x^q y^q + v_5^{p^m} y^{2q} = 0. \qquad (8.63)$$

As in the proof of Theorem 8.77, in (8.63), $h \geq m$. Then

$$v_0 + v_1 x^{p^{h-m}} + v_2 y^{p^{h-m}} + v_3 x^{2(p^{h-m})} + v_4 (xy)^{p^{h-m}} + v_5 y^{2(p^{h-m})} = 0. \quad (8.64)$$

In terms of osculating conics, as in Lemma 7.74, Frobenius non-classicality is expressed in the following theorem.

THEOREM 8.83 *If L_2 is Frobenius non-classical, then the conic $\mathcal{C}_0$ passes through the image $A^q = (1, a^q, b^q)$ of the centre $A = (1, a, b)$ of γ under the Frobenius collineation. In addition, if $A \in \mathrm{PG}(2, q)$, then $\mathcal{C}_0$ meets γ with multiplicity greater than p^m.*

Proof. Take ξ, η in $K[[t]]$ such that $\xi(t^q) = x(t)^q$ and $\eta(t^q) = y(t)^q$. From (8.63),

$$\begin{aligned}
&[(\mu_0^{(0)})^{p^m} + (\mu_1^{(0)})^{p^m}\xi(t^q) + (\mu_2^{(0)})^{p^m}\eta(t^q) + (\mu_3^{(0)})^{p^m}\xi(t^q)^2 \\
&\quad +(\mu_4^{(0)})^{p^m}\xi(t^q)\eta(t^q) + (\mu_5^{(0)})^{p^m}\eta(t^q)^2]+ \\
&t^{p^m}[(\mu_0^{(1)})^{p^m} + (\mu_1^{(1)})^{p^m}\xi(t^q) + (\mu_2^{(1)})^{p^m}\eta(t^q) + (\mu_3^{(1)})^{p^m}\xi(t^q)^2+ \\
&\quad +(\mu_4^{(1)})^{p^m}\xi(t^q)\eta(t^q) + (\mu_5^{(1)})^{p^m}\eta(t^q)^2] + \quad \cdots \\
+ \quad &t^{kp^m}[(\mu_0^{(k)})^{p^m} + (\mu_1^{(k)})^{p^m}\xi(t^q) + (\mu_2^{(k)})^{p^m}\eta(t^q) + (\mu_3^{(k)})^{p^m}\xi(t^q)^2 \\
&\quad +(\mu_4^{(k)})^{p^m}\xi(t^q)\eta(t^q) + (\mu_5^{(k)})^{p^m}\eta(t^q)^2] + \cdots = 0.
\end{aligned}$$

Since $x(t) = a + tG(t)$ and $y(t) = b + tH(t)$, it follows that

$$\xi(t^q) = a^q + t^q G_1(t), \quad \eta(t^q) = b^q + t^q H_1(t).$$

Hence

$$\begin{aligned}
&(\mu_0^{(0)})^{p^m} + (\mu_1^{(0)})^{p^m} a^q + (\mu_2^{(0)})^{p^m} b^q + (\mu_3^{(0)})^{p^m} (a^q)^2 \\
&\qquad +(\mu_4^{(0)})^{p^m} a^q b^q + (\mu_5^{(0)})^{p^m} (b^q)^2 = 0.
\end{aligned}$$

This shows that that $\mathcal{C}_0$ passes through the point $A^q = (1, a^q, b^q)$. Also, since $p^m < q$,

$$\begin{aligned}
&(\mu_0^{(1)})^{p^m} + (\mu_1^{(1)})^{p^m} a^q + (\mu_2^{(1)})^{p^m} b^q) + (\mu_3^{(1)})^{p^m} (a^q)^2 \\
&\qquad +(\mu_4^{(1)})^{p^m} a^q b^q + (\mu_5^{(1)})^{p^m} (b^q)^2 = 0.
\end{aligned}$$

If $a^q = a$, $b^q = b$, then this equation is $s_1(a, b) = 0$. By Theorem 7.74 (ii), the second assertion follows. □

For the rest of this section, $\mathcal{F}$ is a Frobenius non-classical curve with Frobenius orders $0, 1, 2, 3, p^m$.

The Stöhr–Voloch Theorem applied to L_2 reads as follows:

$$S_q \leq \tfrac{1}{5}\{(6 + p^m)(2g - 2) + (q + 5)2n\}. \qquad (8.65)$$

The aim is to find other good upper bounds on the number S_q of $\mathbf{F}_q$-rational branches of such a curve. The idea is to exploit Frobenius classicality with respect to conics to compute the possible $(L_1, \mathcal{P})$-orders, rather than counting $\mathbf{F}_q$-rational branches only. If there are only a few possibilities for the $(L_1, \mathcal{P})$-orders, and the weights $v_{\mathcal{P}}(S)$ can be calculated using the previous numerical formulas, this idea can work effectively, and the Stöhr–Voloch divisor arising from the linear series L_1 cut out by lines can bring about an improvement to (8.65).

By Theorem 7.76, the hypothesis $n < p^m$ ensures that only three cases can occur for $(L_1, \mathcal{P})$-orders, and in each case j_2 is calculated from j_1. Now these three cases are investigated further.

PROPOSITION 8.84 *Let $n < p^m$. If (7.33) holds and the centre $A = (1, a, b)$ of γ is not in $\mathrm{PG}(2, q)$, then the Frobenius image $A^q = (1, a^q, b^q)$ does not lie on the tangent to γ; also, $(a - a^q)v_{21} - (b - b^q)v_{11} \neq 0$.*

Proof. By Theorem 8.83, $\mathcal{C}_0$ passes through $A^q = (1, a^q, b^q)$. On the other hand, when (7.33) holds, $\mathcal{C}_0$ is a non-degenerate conic whose tangent line at $A = (1, a, b)$ coincides with the tangent of γ. □

PROPOSITION 8.85 *Let $n < p^m$. If (7.34) holds and the centre $A = (1, a, b)$ of γ is not in $\mathrm{PG}(2, q)$, then the Frobenius image $A^q = (1, a^q, b^q)$ lies on the tangent to γ; also, $(a - a^q)v_{21} - (b - b^q)v_{11} = 0$.*

Proof. If (7.34) holds, then $\mathcal{C}_0$ is a degenerate conic and consists of the tangent line of γ counted twice. Thus the assertion is an immediate consequence of Theorem 8.83. □

PROPOSITION 8.86 *If (7.35) holds, then the centre of γ is not in $\mathrm{PG}(2, q)$.*

Proof. In this case, $\mathcal{C}_0$ splits into two distinct lines, one of which coincides with the tangent of γ. This shows that $\mathcal{C}_0$ meets γ with multiplicity $j_1 + j_2 = p^m$. Hence the result follows from the second assertion in Theorem 8.83. □

Now it is shown that Frobenius non-classicality sharply limits the possible values of j_1.

In the following theorem, $\mathcal{Z}$ is the curve introduced in Section 7.8, P is the branch of $\mathcal{Z}$ associated to a branch γ of $\mathcal{F}$, and $\mathcal{P}$ is the corresponding place.

THEOREM 8.87 *There is a hyperplane H in $\mathrm{PG}(5, K)$ such that*

$$I(P, \mathcal{Z} \cap H) \geq j_1 p^{h-m}.$$

Proof. Without loss of generality, let γ be centred at an affine point $A = (1, a, b)$. Let $\alpha, \beta \in K$ such that $\alpha = a^{p^{h-m}}$ and $\beta = b^{p^{h-m}}$. The intersection multiplicity of $\mathcal{Z}$ with the hyperplane

$$H = \mathbf{v}(X_0 + \alpha X_1 + \beta X_2 + \alpha^2 X_3 + \alpha\beta X_4 + \beta^2 X_5) \tag{8.66}$$

at P is

$$I(P, \mathcal{Z} \cap \mathcal{H}) = \mathrm{ord}_{\mathcal{P}}(v_0 + \alpha v_1 + \beta v_2 + \alpha^2 v_3 + \alpha\beta v_4 + \beta^2 v_5).$$

Putting $\xi = x - a, \eta = y - b$ gives $j_1 = \min\{\mathrm{ord}_{\mathcal{P}}\, \xi, \mathrm{ord}_{\mathcal{P}}\, \eta\}$. From (8.64),

$$\begin{aligned} &v_0 + \alpha v_1 + \beta v_2 + \alpha^2 v_3 + \alpha\beta v_4 + \beta^2 v_5 \\ &\quad + \xi^{p^{h-m}} v_1 + \eta^{p^{h-m}} v_2 + \xi^{2(p^{h-m})} v_3 + (\xi\eta)^{p^{h-m}} v_4 + \eta^{2(p^{h-m})} v_5 = 0, \end{aligned}$$

whence

$$\mathrm{ord}_{\mathcal{P}}(v_0 + \alpha v_1 + \beta v_2 + \alpha^2 v_3 + \alpha\beta v_4 + \beta^2 v_5) \geq j_1 p^{h-m},$$

which proves the result. □

It is worth mentioning the following result that is valid for singular curves.

PROPOSITION 8.88 *If $\mathcal{F}$ is singular, then* $\deg \mathcal{Z} \geq 2p^{h-m}$.

Proof. Let $A = (1, a, b)$ be a singular point of $\mathcal{F}$, and let H be the hyperplane (8.66). If A is the centre of only one branch of $\mathcal{F}$, it must have order greater than 1, and the assertion follows from Proposition 8.87.

Suppose that A is the centre of two branches γ_1 and γ_2 of $\mathcal{F}$. The corresponding branches P_1 and P_2 of $\mathcal{Z}$ are distinct. Therefore

$$\deg \mathcal{Z} \geq I(P_1, \mathcal{Z} \cap H) + I(P_2, \mathcal{Z} \cap H).$$

From Proposition 8.87, $I(P_i, \mathcal{Z} \cap H) \geq p^{h-m}$ for $i = 1, 2$, and the assertion follows. □

If $n < p^{h-m}$ also holds, then Proposition 8.87 together with Theorem 7.80 gives the following sharp bound for j_1.

PROPOSITION 8.89 *If $n \leq \min\{p^m, p^{h-m}\}$, then $j_1 \leq 3$.*

From this, an upper bound on S_q is obtained after giving an explicit formula for $v_{\mathcal{P}}(S)$ that actually holds without any limitation on n with respect to p^m and p^{h-m}.

Since $v_{\mathcal{P}}(S)$ remains invariant under the change of coordinates (7.27) only when all coefficients are from $\mathbf{F}_q$, it is necessary to know how $v_{\mathcal{P}}(S)$ changes in general under (7.27). For this, take a local parameter ζ at $\mathcal{P}$, and adopt the usual simpler notation u' in place of $D_\zeta(u)$ for $u \in K(\mathcal{F})$. Since

$$\begin{aligned}
&(x - x^q)y' - (y - y^q)x' \\
&= [a - a^q + v_{11}\bar{x} + v_{12}\bar{y} - (v_{11}\bar{x} + v_{12}\bar{y})^q][v_{21}\bar{x}' + v_{22}\bar{y}'] \\
&\quad -[b - b^q + v_{21}\bar{x} + v_{22}\bar{y} - (v_{21}\bar{x} + v_{22}\bar{y})^q][v_{11}\bar{x}' + v_{12}\bar{y}'] \\
&= [(a - a^q)v_{21} - (b - b^q)v_{11}]\bar{x}' + [(a - a^q)v_{22} - (b - b^q)v_{12}]\bar{y}' \\
&\quad +(v_{11}v_{22} - v_{12}v_{21})(\bar{x}\bar{y}' - \bar{y}\bar{x}') - (v_{11}\bar{x} + v_{12}\bar{y})^q(v_{21}\bar{x}' + v_{22}\bar{y}') \\
&\quad +(v_{21}\bar{x} + v_{22}\bar{y})^q(v_{11}\bar{x}' + v_{12}\bar{y}'),
\end{aligned}$$

it follows that

$$\begin{aligned}
v_{\mathcal{P}}(S) \quad = \quad & \mathrm{ord}_{\mathcal{P}}\,\bar{x}' \\
& + \mathrm{ord}_{\mathcal{P}}\{[(a - a^q)v_{21} - (b - b^q)v_{11}] \\
& \qquad + [(a - a^q)v_{22} - (b - b^q)v_{12}]\bar{y}'/\bar{x}' \\
& \qquad + (v_{11}v_{22} - v_{12}v_{21})(\bar{x}\bar{y}'/\bar{x}' - \bar{y}) \\
& \qquad - (v_{11}\bar{x} + v_{12}\bar{y})^q(v_{21} + v_{22}\bar{y}'/\bar{x}') \\
& \qquad + (v_{21}\bar{x} + v_{22}\bar{y})^q(v_{11} + v_{12}\bar{y}/\bar{x}')\}.
\end{aligned} \tag{8.67}$$

THEOREM 8.90 *Let $n \leq \min\{p^m - 4, p^{h-m}\}$.*

(i) *If $j_1 \neq 3$ for every branch of $\mathcal{F}$, then*

$$S_q \leq \frac{(q-1)n - 2(2g-2)}{2}. \tag{8.68}$$

(ii) *If $p > 3$, then*

$$S_q < \frac{(q-1)n - 2(2g-2) + 12n}{2}. \tag{8.69}$$

Proof. The idea is to calculate $v_{\mathcal{P}}(S) - v_{\mathcal{P}}(R)$. By Proposition 8.89,

$$v_{\mathcal{P}} = j_1 + j_2 - 3$$

with just one exception: $p = j_1 = 3$. Apart from this exception, in each of the three cases in Theorem 7.76, $j_2 \not\equiv j_1 \pmod{p}$. From (8.67), there are three possibilities:

(a) if $a, b \in \mathbf{F}_q$, then $v_{\mathcal{P}}(S) \geq j_1 + j_2 - 1$;

(b) if $(a - a^q)v_{21} - (b - b^q)v_{11} = 0$, but either $a \notin \mathbf{F}_q$ or $b \notin \mathbf{F}_q$, then $v_{\mathcal{P}}(S) = j_2 - 1$;

(c) if $(a - a^q)v_{21} - (b - b^q)v_{11} \neq 0$, then $v_{\mathcal{P}}(S) = j_1 - 1$.

The hypothesis, $n \leq p^m - 4$, rules out case (c). Hence, if $p = j_1 = 3$ does not occur, then

$$v_{\mathcal{P}}(S) - v_{\mathcal{P}}(R) \geq \begin{cases} 2 & \text{if (a) holds;} \\ -2 & \text{if (b) holds and } j_1 = 3; \\ 0 & \text{if (b) holds and } j_1 \leq 2; \end{cases} \tag{8.70}$$

Now sum this inequality as γ ranges over all branches of $\mathcal{F}$. Since

$$\begin{aligned} \deg S &= (2g-2) + (q+2)n, \\ \deg R &= 3(2g-2) + 3n, \end{aligned}$$

this gives

$$2S_q \leq (q-1)n - 2(2g-2) + 2N,$$

where N denotes the number branches γ satisfying (7.34) with $j_1 = 3$.

Assume that $N \neq 0$. Since, by Theorem 7.76, $v_{\mathcal{P}}(R) = j_2 = \frac{1}{2}(p^m + 3)$ for any γ counted in N, an upper bound on N is

$$N \leq \frac{6(2g-2) + 6n}{p^m + 3} < 6n.$$

This completes the proof. □

THEOREM 8.91 *Let $\mathcal{F}$ be non-singular. If $n < p^m - 1$, then*

$$S_q = \tfrac{1}{2}\{n(q - 2n + 5) - I\}, \tag{8.71}$$

where I is the number of all non-$\mathbf{F}_q$-rational inflexion points of $\mathcal{F}$.

Proof. The arguments in the proof of Theorem 8.90 can be used, but two more pieces are needed.

First, $v_{\mathcal{P}}(S) = j_2$ when (a) holds. To prove this, show that

$$v_{\mathcal{P}}(S) = \mathrm{ord}_{\mathcal{P}}(\bar{x}\bar{y}'/\bar{x}' - \bar{y}) = j_2.$$

Since $j_1 = 1$, so $\bar{x}$ is a local parameter at $\mathcal{P}$. Hence there is a primitive representation σ of $\mathcal{P}$ such that

$$\sigma(\bar{x}) = t, \quad \sigma(\bar{y}) = ct_2^j + \cdots$$

with $c \neq 0$. Then

$$\sigma(\bar{x}\bar{y}'/\bar{x}' - \bar{y}) = c(j_2 - 1)t^{j_2} + \cdots .$$

Since $j_2 = \frac{1}{2}(p^m + 1)$ when (a) holds, the coefficient $c(j_2 - 1) \neq 0$. Hence $\mathrm{ord}_t(\sigma(\bar{x}\bar{y}'/\bar{x}' - \bar{y})) = j_2$, and the assertion is proved.

Secondly, (c) does not occur since $n < p^m - 1$. Therefore

$$v_{\mathcal{P}}(S) - v_{\mathcal{P}}(R) = \begin{cases} 2 & \text{if (a) holds;} \\ 1 & \text{if (b) holds;} \\ 0 & \text{otherwise.} \end{cases}$$

Since $\deg S - \deg R = qn - n - 2(2g - 2)$ and $\mathcal{F}$ is non singular, so

$$\deg S - \deg R = n(q - 2n + 5).$$

Further, case (b) occurs if and only if P is an inflexion that is not an $\mathbf{F}_q$-rational point. From this the assertion follows. □

EXAMPLE 8.92 The Fermat curve $\mathcal{F}$ in Example 7.81 is Frobenius non-classical with respect to conics and its inflexion points are in $\mathrm{PG}(2, q)$. Hence (8.71) applies with $n = \frac{1}{2}(\sqrt{q} + 1)$, $I = 0$, giving

$$S_q = q + 1 + \tfrac{1}{4}(\sqrt{q} - 1)(\sqrt{q} - 3)\sqrt{q}.$$

Theorem 8.90 together with (8.65) gives the following result.

THEOREM 8.93 *Let* $n \leq \min\{p^m - 4, p^{h-m}\}$.

(i) *If* $j_1 \neq 3$ *for every branch of* $\mathcal{F}$, *then*

$$S_q \leq \frac{(6 + p^m)(q - 1)n + (q + 5)4n}{22 + 2p^m}. \tag{8.72}$$

(ii) *If* $p > 3$, *then*

$$S_q \leq \frac{(6 + p^m)(q - 1)n + (q + 3p^m + 23)4n}{22 + 2p^m}. \tag{8.73}$$

REMARK 8.94 Let $q \geq 49$ be a square power of p with $p > 2$. The irreducible plane curve $\mathcal{G} = \mathbf{v}(g(X_0, X_1, X_2))$, with

$$\begin{aligned} g(X_0, X_1, X_2) = {} & X_2^2 X_0^{2\sqrt{q}} + X_1^2 X_2^{2\sqrt{q}} + X_0^2 X_1^{2\sqrt{q}} \\ & -2(X_0 X_1^{\sqrt{q}+1} X_2^{\sqrt{q}} + X_0^{\sqrt{q}+1} X_1^{\sqrt{q}} X_2 + X_0^{\sqrt{q}} X_1 X_2^{\sqrt{q}+1}), \end{aligned}$$

is classical with respect to lines, but non-classical with respect to conics, since the L_2-orders are $0, 1, 2, 3, 4, \sqrt{q}$. In fact, (7.22) holds for $f(X, Y) = G(1, X, Y)$ when

$$h = 1,\ z_0 = X^2,\ z_1 = -2XY,\ z_2 = -2X,\ z_3 = Y^2,\ z_4 = -2Y,\ z_5 = 1.$$

Since (8.62) does not hold, $\mathcal{G}$ is Frobenius classical with respect to conics. Also, $\mathcal{G}$ has three singular points, namely the vertices of the fundamental triangle. Each of them is a double point and is the centre of only one branch. The $(L_1, \mathcal{P})$-orders of the corresponding place $\mathcal{P}$ of such a branch are $0, 2, 2\sqrt{q}$. The genus of $\mathcal{G}$ is $\frac{1}{2}(q - \sqrt{q})$. Now the irreducible plane curve $\mathcal{F} = \mathbf{v}(f(X_0, X_1, X_2))$ is defined to be the image of $\mathcal{F}$ under the projectivity of $\mathrm{PG}(3, K)$ associated to the matrix

$$\begin{bmatrix} \alpha & 1 & \alpha^{q+1} \\ \alpha^{q+1} & \alpha & 1 \\ 1 & \alpha^{q+1} & \alpha \end{bmatrix}, \tag{8.74}$$

where α is an element of order $q^2 + q + 1$ in the cubic extension $\mathbf{F}_{q^3}$ of $\mathbf{F}_q$. Equivalently,

$$\begin{aligned} &f(\alpha X_0 + X_1 + \alpha^{q+1} X_2, \alpha^{q+1} X_0 + \alpha X_1 + X_2, X_0 + \alpha^{q+1} X_1 + \alpha X_2) \\ &\quad = g(X_0, X_1, X_2). \end{aligned}$$

Now $\mathcal{G}$ and $\mathcal{F}$ have the same L_1- and L_2-orders. In particular, $\mathcal{F}$ is classical with respect to lines, but non-classical with respect to conics. Although the projectivity (8.74) is defined over $\mathbf{F}_{q^3}$, the curve $\mathcal{F}$ is defined over $\mathbf{F}_q$; this is not trivial.

It might be thought that $\mathcal{F}$ is also Frobenius classical with respect to conics. Actually, this is false. In fact, $\mathcal{F}$ is the image of the Hermitian curve

$$\mathcal{H}_{\sqrt{q}} = \mathbf{v}(X_0^{\sqrt{q}+1} + X_1^{\sqrt{q}+1} + X_2^{\sqrt{q}+1})$$

under a birational transformation defined over $\mathbf{F}_q$. In particular, $S_q = q\sqrt{q} + 1$. This is consistent with (8.65) only for $p^m = \sqrt{q}$, yielding that the last Frobenius order is $\sqrt{q}$. This example also shows that the hypothesis $n \le \sqrt{q} - 4$ in Theorem 8.90 cannot be weakened to $n \le 2\sqrt{q} + 2$.

To obtain an upper bound on S_q, begin with (7.43), evaluate $\mathrm{ord}_{\mathcal{P}}\, B(\bar{x}, \bar{y})$ using $v_{\mathcal{P}}(S)$, and calculate $\deg S$.

PROPOSITION 8.95 *Let $n < p^m$.*

(i) *If $a, b \in \mathbf{F}_q$, then*

$$\mathrm{ord}_{\mathcal{P}}\, B \le \begin{cases} v_{\mathcal{P}}(S) - (3j_1 - 1) & \text{if (7.33) holds,} \\ v_{\mathcal{P}}(S) - (j_1 - 1) & \text{if (7.34) holds.} \end{cases} \tag{8.75}$$

(ii) *If either $a \notin \mathbf{F}_q$ or $b \notin \mathbf{F}_q$, then*

$$\mathrm{ord}_{\mathcal{P}}\, B \le \begin{cases} v_{\mathcal{P}}(S) - (j_1 - 1) & \text{if (7.33) holds,} \\ v_{\mathcal{P}}(S) + 1 & \text{if (7.34) or (7.35) holds.} \end{cases} \tag{8.76}$$

Proof. Let $a, b \in \mathbf{F}_q$. From (8.67), $v_\mathcal{P}(S) \geq j_1 + j_2 - 1$. By Proposition 8.86, either (7.33) or (7.34) occurs. In the former case, $\text{ord}_\mathcal{P}\, B = 0$ and $j_2 = 2j_1$; in the latter case, $\text{ord}_\mathcal{P}\, B = j_2$ and $2j_2 - j_1 = p^m$. From this, (i) is obtained. The argument is similar for (ii). □

Summing up the inequalities in Proposition 8.95 gives the inequality

$$\sum \text{ord}_\mathcal{P}\, B \leq \sum v_\mathcal{P}(S) - 2M_q - M'_q - \sum{}' \,(j_1 - 1) + N, \tag{8.77}$$

where M_q and M'_q are as in Definition 8.79, while the summation $\sum'$ is over all branches of $\mathcal{F}$ of type (7.33), and N is the number branches of type (7.34) or (7.35).

An upper bound on N is

$$N \leq 6(2g - 2 + d)/(p^m - 3). \tag{8.78}$$

To show (8.78), from the inequality $v_\mathcal{P} \geq j_2 + j_1 - 3$, it follows that

$$\sum (j_2 + j_1 - 3) \leq \sum v_\mathcal{P}(R) = 3(2g - 2) + 3n.$$

On the other hand,

$$\sum (j_2 + j_1 - 3) = N_0(p^m - 3) + \sum{}'' \,(p^m + 3j_1 - 6)/2,$$

where N_0 is the number of branches of type (7.33) while the summation $\sum''$ is over the $N - N_0$ branches of type (7.34). Therefore

$$\sum (j_2 + j_1 - 3) \geq N_0(p^m - 3) + (N - N_0)(p^m - 3)/2.$$

Hence

$$\sum (j_2 + j_1 - 3) \geq \tfrac{1}{2}N(p^m - 3),$$

and (8.78) follows.

It should be noted that, since $2g - 2 \leq n(n - 3)$ and $n \leq p^m - 1$, the expression on the right of (8.78) is at most $6n$. Hence (8.77) can be put in a more manageable but somewhat weaker form as follows:

$$\sum \text{ord}_\mathcal{P} B \leq \sum v_\mathcal{P}(S) - 2M_q - M'_q + 6n. \tag{8.79}$$

This together with (7.44) gives the inequality

$$p^m \deg \mathcal{Z} < c + (2g - 2) + [(q + 1)n - 2M_q - M'_q] + 6n. \tag{8.80}$$

Also, from Theorem 8.87,

$$j_1 \leq \frac{c + (2g - 2) + [(q + 1)n - 2M_q - M'_q] + 6n}{q}. \tag{8.81}$$

Since $c \leq n(n - 1)$ and $2g - 2 \leq n(n - 3)$, from (8.80) and (8.81), the following result is obtained.

PROPOSITION 8.96 *Let $n < p^m$. If (8.50) holds, then*

$$p^m \deg \mathcal{Z} \leq n(2n + \sqrt{q} + 2), \tag{8.82}$$

$$j_1 \leq n(2n + \sqrt{q} + 2)/q. \tag{8.83}$$

Now the case that q is a square and $p^m = \sqrt{q}$, that is,

$$\begin{cases} \text{(i)} & \text{the order sequence is } (0,1,2,3,4,\sqrt{q}), \\ \text{(ii)} & \text{the Frobenius order sequence is } (1,2,3,\sqrt{q}), \end{cases} \tag{8.84}$$

is investigated in detail. Since $n \le \sqrt{q}-1$,

$$n(2n+\sqrt{q}+2)/q \le 3 - 3/\sqrt{q} < 3.$$

By (8.83), this gives the following result.

PROPOSITION 8.97 *If $n < \sqrt{q}$, and both (8.50) and (8.84) hold, then*

(i) $j_1 = 1$ *for branches of type (7.34) while* $j_1 \le 2$ *for those of type (7.33) and (7.35);*

(ii) *for branches of type (7.35), there are only two cases:*

(a) $n = \sqrt{q} - 2$ *and* $j_1 = 2$;

(b) $n = \sqrt{q} - 1$ *and* $j_1 = 1$.

It should be noted that the linearity of the branches of type (7.34) depends on the condition that $p > 2$. A first consequence of Proposition 8.97 is the following result.

PROPOSITION 8.98 *Let τ_1 and τ_2 be the numbers of branches of type (7.34) and (7.35). Then*

$$3 \deg \mathcal{Z} = 2\tau_1 + \tau_2; \tag{8.85}$$

$$3 \deg \mathcal{Z} = 2\tau_1, \quad \textit{when } n < \sqrt{q} - 2. \tag{8.86}$$

Proof. From Proposition 8.97, $j_1 \le 2 < p$. So, by Corollary 7.60, the classical formula $v_{\mathcal{P}}(R) = j_1 + j_2 - 3$ holds. Also, $\mathrm{ord}_{\mathcal{P}}\, d\bar{x} = j_1 - 1$. From the discussion after (7.43), the exact value of $\mathrm{ord}_{\mathcal{P}}\, B$ is known, namely

$$\mathrm{ord}_{\mathcal{P}}\, B = \begin{cases} 0 & \text{if the branch is of type (7.33),} \\ j_2 & \text{if the branch is of type (7.34),} \\ j_1 & \text{if the branch is of type (7.35).} \end{cases}$$

This, together with (7.42), (7.43) and Lemma 7.79, gives (8.85). The second assertion is a consequence of (8.85) and Proposition 8.97 (ii). □

PROPOSITION 8.99 *If each of the conditions (8.86), (8.50), (8.84) is satisfied, then*

$$\deg \mathcal{Z} \le 2n < 2(\sqrt{q} - 2).$$

Proof. Since $n < \sqrt{q} - 2$, it follows from Proposition 8.97 that no branch of type (7.35) exists. Hence, for a branch γ,

$$3v_{\mathcal{P}}(S) - v_{\mathcal{P}}(R)$$

$$= \begin{cases} 6j_1 & \text{if } \gamma \text{ is of type (7.33) centred at a point in } \mathrm{PG}(2,q), \\ 0 & \text{if } \gamma \text{ is of type (7.33) centred at a point not in } \mathrm{PG}(2,q), \\ \sqrt{q} + 3j_1 & \text{if } \gamma \text{ is of type (7.34) centred at a point in } \mathrm{PG}(2,q), \\ \sqrt{q} & \text{if } \gamma \text{ is of type (7.34) centred at a point not in } \mathrm{PG}(2,q). \end{cases}$$

Since

$$\sum 3v_{\mathcal{P}}(S) - v_{\mathcal{P}}(R) = 3(2g-2) + 3(q+2)n - 3(2g-2) - 3n = 3nq + 3n,$$

it follows that

$$3nq + 3n = 6M_q + 3M'_q + \tau_1\sqrt{q},$$

whence

$$2(nq - 2M_q - M'_q + n) = \sqrt{q}\deg \mathcal{Z},$$

by Proposition 8.98. Now, from (8.50), the assertion follows. □

THEOREM 8.100 *If each of the conditions (8.86), (8.50), (8.84) is satisfied, then*

(i) *$\mathcal{F}$ is non-singular*;

(ii) $n = \frac{1}{2}(\sqrt{q}+1)$;

(iii) *the number of $\mathbf{F}_q$-rational points of $\mathcal{F}$ is*

$$S_q = q + 1 + \tfrac{1}{4}(\sqrt{q}-1)(\sqrt{q}-3)\sqrt{q}.$$

Proof. Suppose that $\mathcal{F}$ has a singular point. Since $p^{h-m} = \sqrt{q}$, Proposition 8.88 implies that $\deg \mathcal{Z} \geq 2\sqrt{q}$; this contradicts Proposition 8.99.

As $\mathcal{F}$ is non-singular and classical with respect to lines, (7.20) applies. Since the inflexions of $\mathcal{F}$ are exactly the points of type (7.34), it follows that

$$\tau_1 = 6n(n-2)/(\sqrt{q}-3).$$

Hence, by Proposition 8.98, $\deg \mathcal{Z} = n(n-2)/(\sqrt{q}-3)$. Therefore the first assertion follows from Proposition 7.80. To prove the second it suffices to show that every inflexion point of $\mathcal{F}$ lies in $\mathrm{PG}(2,q)$ and apply Theorem 8.91. Assume on the contrary that $\mathcal{F}$ has an inflexion point P outside $\mathrm{PG}(2,q)$. If ℓ is the tangent to $\mathcal{F}$ at P, then Theorem 7.78 shows that ℓ contains the Frobenius image P^q of P. Since $P \neq P^q$, Bézout's Theorem gives the inequality

$$n = |\mathcal{F} \cap \ell| \geq \tfrac{1}{2}(\sqrt{q}+1) + 1,$$

contradicting $n = \frac{1}{2}(\sqrt{q}+1)$. □

For the Fermat curve $\mathcal{F} = \mathbf{v}(X^n + Y^n + 1)$ with $n = \frac{1}{2}(\sqrt{q}+1)$, described in Example 7.81,

$$M_q = \tfrac{1}{4}(\sqrt{q}+1)(q-\sqrt{q}-2), \qquad M'_q = \tfrac{3}{2}(\sqrt{q}+1).$$

This shows that the conditions in Proposition 8.100 can be fulfilled. In fact, this Fermat curve is, up to projectivities in $\mathrm{PG}(2,q)$, the only non-singular plane curve of degree n defined over $\mathbf{F}_q$ that satisfies both conditions in Proposition 8.100. Therefore the following result holds.

THEOREM 8.101 *If (8.86), (8.50) and (8.84) hold, then $n = \frac{1}{2}(\sqrt{q}+1)$ and $\mathcal{F}$ is the Fermat curve.*

Theorem 8.101 shows that an irreducible plane curve defined over $\mathbf{F}_q$ is fully determined by three conditions, namely (8.86) on the degree, (8.50) on the number of $\mathbf{F}_q$-rational points, and (8.84) on the Frobenius order sequence. Actually, the following proposition shows that the latter condition is independent of the other two only for small values of q.

PROPOSITION 8.102 *Let $q \geq 25$ be a power of an odd prime p such that if $p = 3$ then q is a square. Let $\mathcal{G}$ be an irreducible plane curve of degree $n \geq 3$ and defined over $\mathbf{F}_q$. If condition (8.50) is satisfied and*

$$n < \begin{cases} \sqrt{q}+1 & \text{for } q > 23^2, q \neq 3^6, 5^5, \\ 23 & \text{for } q = 3^6, \\ 49 & \text{for } q = 5^5, \\ \min\{(q-5\sqrt{q}+45)/20\,, & \\ \qquad (q-5\sqrt{q}+57)/24\} & \text{for } q \leq 23^2, \end{cases} \tag{8.87}$$

then q is square and (8.84) holds.

Proof. By Theorem 8.80, $\mathcal{G}$ is classical with respect to lines. Consequently, the L_2-order sequence of $\mathcal{G}$ is $(0,1,2,3,4,\epsilon)$ and the Frobenius L_2-order sequence is $(0,1,2,3,\nu)$ with either $\nu = 4$, or $\nu = \epsilon$.

To show that $\nu = \epsilon$, suppose on the contrary that $\nu = 4$. Since

$$5S_q = 5(M_q + M_q') \geq \tfrac{5}{2}(2M_q + M_q'),$$

from (8.65),

$$10(n-3) + (q+5)n \geq \tfrac{5}{2}(q - \sqrt{q} + 1),$$

whence $n \geq (q - 5\sqrt{q} + 45)/20$, a contradiction.

Hence $\epsilon = \nu \geq 5$. If equality holds, then either $p = 3$, or $p = 5$. From Remark 8.52, ϵ is a power of p for $p \geq 5$, apart from just one possibility for $p = 3$, namely when $\epsilon = \nu = 6$. The latter case cannot actually occur under the hypothesis (8.87). In fact, for $\nu = 6$, (8.65) implies that

$$12(n-3) + (q+5)2 \geq \tfrac{5}{2}(q - \sqrt{q} + 1),$$

so that $n \geq (q - 5\sqrt{q} + 57)/24$ and hence $q > 23^2$ by (8.87). From the inequality

$$(q - 5\sqrt{q} + 57)/24 \leq n \leq \sqrt{q},$$

it follows that $q - 29\sqrt{q} + 57 \leq 0$, whence $529 < q < 722$. This is a contradiction as no power of 3 lies in this interval.

To complete the proof it must be shown that neither $\nu > \sqrt{q}$ nor $n < \sqrt{q}$.

Suppose that $\nu > \sqrt{q}$, that is, $\nu^2 = p^e q$ with $e \geq 1$. Since $p \geq 3$ and q is a square for $p = 3$, this implies that $\nu > 2\sqrt{q}$. On the other hand, $\epsilon \leq 2n < 2\sqrt{q}$, a contradiction.

Suppose that $\nu < \sqrt{q}$; that is, $q = p^e\nu^2$, with $e \geq 1$. Since $n(n-3) \geq 2g - 2$, $n < \sqrt{q}$, the hypothesis (8.50), together with (8.65), gives the following:

$$(6+\nu)(\sqrt{q}-3)+(q+5)2 \geq (6+\nu)(n-3)+(q+5)2 \geq \tfrac{5}{2}(q-\sqrt{q}+1). \tag{8.88}$$

In particular,

$$n - 3 \geq \frac{q - 5\sqrt{q} - 15}{2(6 + \nu)}. \tag{8.89}$$

From (8.88) it follows that $2\nu\sqrt{q} - 6\nu \geq q - 17\sqrt{q} + 21$, whence

$$2\nu - 6/\sqrt{p^e} \geq \sqrt{q} - 17 + 21/\sqrt{q}.$$

Therefore

$$17 - \frac{21}{\sqrt{q}} - \frac{6}{\sqrt{p^e}} \geq \nu(\sqrt{p^e} - 2). \tag{8.90}$$

Since $\nu \geq 5$, this gives $p^e < 36$. As q is a square for $p = 3$ and $\nu \geq 5$ is a power of p, only five possibilities are left; namely,

$$(p^e, \nu, q) \in \{(9, 9, 729), (5, 5, 125), (5, 25, 3125), (7, 7, 343), 11, 11, 1331)\}.$$

But, for these, (8.89) implies that

$$n \geq 23, 6, 49, 13, 37,$$

contradicting the hypothesis (8.87). □

REMARK 8.103 If $\mathcal{G}$ is a non-singular plane curve defined over $\mathbf{F}_q$ with q an odd square such that $\mathcal{G}$ has degree $n = (\sqrt{q} + 1)/2$ and $S_q = q + 1 + n(n-3)\sqrt{q}$ $\mathbf{F}_q$-rational points, then $\mathcal{G}$ is the Fermat curve, up to a projectivity over $\mathbf{F}_q$.

This result, together with Theorems 8.101 and 8.102, has the following corollary.

THEOREM 8.104 *Let $q \geq 25$ be a power of an odd prime p such that, if $p = 3$, then q is a square. Let $\mathcal{G}$ be an irreducible plane curve of degree $n < \sqrt{q} - 2$ defined over $\mathbf{F}_q$. If both (8.50) and (8.87) hold, then $n = \frac{1}{2}(\sqrt{q} + 1)$ and $\mathcal{G}$ is the Fermat curve, up to a projectivity over $\mathbf{F}_q$.*

REMARK 8.105 For the Fermat curve

$$\mathcal{F}_n = \mathbf{v}(X^n + Y^n + 1)$$

of degree n with $n \not\equiv 0 \pmod{p}$, viewed as an irreducible curve over $\mathbf{F}_q$, the values of n for which $\mathcal{F}_n$ is Frobenius non-classical with respect to L_2 are known.

THEOREM 8.106 *Let $q = p^h$ with $p > 5$. Then $\mathcal{F}_n$ is Frobenius non-classical with respect to conics if and only if one of the following holds:*

(i) $n - 1 \equiv 0 \pmod{p}$, *and $\mathcal{F}_n$ is Frobenius classical with respect to lines;*

(ii) $n - 2 \equiv 0 \pmod{p}$, *and $n = 2(p^h - 1)/(p^r - 1)$ with r a divisor of h and $r < h$;*

(iii) $2n - 1 \equiv 0 \pmod{p}$, *and $n = (p^h - 1)/[2(p^r - 1)]$ with r a divisor of h such that h/r is even.*

The Stöhr–Voloch Theorem in its stronger form (8.19) can be applied to $\mathcal{F}_n$ for $n < p$ to show that

$$S_q \leq \tfrac{1}{2}(M-1)n(n-3) + M^{-1}\{sn(q+M) - 3n(A - dB)\},$$

where

$$\begin{aligned} A &= \tfrac{1}{6}\{(n-s-1)s(s-1)(s+4) + \tfrac{1}{4}s(s-1)(s-2)(s+5)\},\\ M &= \tfrac{1}{2}(s+2)(s+1) - 1,\\ B &= sn - M,\\ d &= \text{number of } \mathbf{F}_q\text{-rational points of } \mathcal{F}_n \text{ lying on the coordinate axes,} \end{aligned}$$

with $1 \leq s \leq n-3,\ sn \leq p$.

Remark 8.107 For the number S_p of $\mathbf{F}_p$-rational points of the Fermat curve, $\mathcal{F} = \mathbf{v}(X^n + Y^n + c)$ with $c \in \mathbf{F}_p$ and $n \mid p-1$, the Stöhr–Voloch Theorem in its stronger form (8.19) gives the following result.

Theorem 8.108 *If* $(n - \frac{1}{2})^4 \geq p - 1$, *then* $S_p \leq 4n^{4/3}(p-1)^{2/3}$.

The constant 4 in the bound can be improved to $3 \cdot 2^{-(2/3)}$.

Remark 8.109 The number N_{q^k} of $\mathbf{F}_{q^k}$-rational points of $\mathcal{F}_n$, when

$$N = (q^{k-1} + \cdots + q + 1)/n,$$

is an integer and p is sufficiently large with respect to N is as follows:

$$N_{q^k} = n^2[(q-2) + (d-1)(d-2)] + 3n;$$

here $d = (N/n, t+1),\ q \equiv t \pmod{N/n}$ with $0 < t < N/n$, and

$$p > \left(\frac{2}{\sqrt[t+1]{\sin(n\pi/2N))}} + 1\right)^{(t-1)(N/n-d)}.$$

8.8 THE DUAL OF A FROBENIUS NON-CLASSICAL CURVE

The notation is the same as Section 7.9. Also, Γ is defined over $\mathbf{F}_q$; then the Gaussian dual Δ is also defined over $\mathbf{F}_q$.

Theorem 8.110 *If* $\nu_1 > 1$, *then* $K(\Gamma)/K(\Delta)$ *is a purely inseparable extension with inseparability degree* q' *and with* $q' \leq q$.

Proof. From the definition of Frobenius orders, $\nu_1 > 1$ if and only if the matrix

$$\begin{bmatrix} 1 & x^q & x_2^q & \cdots & x_r^q \\ 1 & x & x_2 & \cdots & x_r \\ 0 & 1 & D_x^{(1)}x_2 & \cdots & D_x^{(1)}x_r \end{bmatrix}$$

has rank 2. Hence $\nu_1 > 1$ yields the equations

$$(x - x^q)D_x^{(1)}x_i = x_i - x_i^q, \quad i = 2, \ldots, r. \tag{8.91}$$

Since $x_i = x\alpha_i + \beta_i$, it follows that

$$(x - x^q)\alpha_i = x\alpha_i + \beta_i - (x\alpha_i + \beta_i)^q,$$

whence $x^q(\alpha_i^q - \alpha_i) = \beta_i - \beta_i^q$. Thus

$$x^q = \frac{\beta_i^q - \beta_i}{\alpha_i^q - \alpha_i}.$$

As $K(\Gamma) = K(\Delta)(x) = K(\alpha_2, \ldots, \alpha_r, \beta_2, \ldots, \beta_r, x)$, the assertion follows. □

This, together with Theorem 7.88, has the following corollary.

THEOREM 8.111 *If $\nu_1 > 1$, then $\epsilon_2 = q'$ with q' a power of p not exceeding q.*

Frobenius non-classicality has implications for the geometry of dual curves of plane curves.

THEOREM 8.112 *Let $\mathcal{F} = \mathbf{v}(f(X,Y))$ be an irreducible plane curve defined over $\mathbf{F}_q$ which is Frobenius non-classical. If $\epsilon_2 < q$, the dual curve $\mathcal{D}$ of $\mathcal{F}$ is non-reflexive.*

Proof. By Theorem 5.91, it is assumed that $p > 2$. From Remark 8.52, $\mathcal{F}$ is not only Frobenius non-classical but also non-classical with order sequence $(0, 1, q')$ with $\epsilon_2 = \nu_1 = q'$. Since $q > q'$, the argument at the beginning of the proof of Theorem 8.78 can be used. If $Q = (x, y)$ is a generic point of $\mathcal{F}$, from (7.65) and (8.17),

$$z_0 + z_1 x^{q/q'} + z_2 y^{q/q'} = 0, \tag{8.92}$$

where at least one of the elements z_i is a separable variable. Since $\mathcal{F}$ is neither a line nor a strange curve by Exercise 3 of this chapter, so $K(\mathcal{F}) = K(\mathcal{D})$. Without loss of generality, suppose that $z_2 \neq 0$, and let $u = z_0/z_2$, $v = z_1/z_2$ with $d = q/q'$. Then (8.92) reads as follows:

$$y^d + ux^d + v = 0.$$

Therefore $\mathcal{D}$ is non-classical, and hence is non-reflexive by Theorem 5.90. □

8.9 EXERCISES

1. For any two positive integers m, n, both prime to $p > 2$, the irreducible plane curve $\mathcal{F} = \mathbf{v}(X^n + Y^m + 1)$ is viewed as a curve defined over $\mathbf{F}_q$. Show that $\mathcal{F}$ is Frobenius classical if and only if $m = n = (p^k - 1)/(p^s - 1)$, where s divides k.

2. Show by direct computation that, if $\nu_1 > 1$ and $p > 2$, then $\epsilon_2 > 2$.

3. Prove that, if $\mathcal{F}$ is a strange curve defined over $\mathbf{F}_q$, then $\nu_1 = 1$.

4. Let $p > 2$. Assume that $f(X) \in \mathbf{F}_q[X]$ has only simple roots and that its degree n is not divisible by p. Show that, if the irreducible plane curve $\mathcal{F} = \mathbf{v}(Y^n - f(X))$ defined over $\mathbf{F}_q$ is Frobenius non-classical, then the roots of $f(X)$ belong to $\mathbf{F}_q$.

5. Let $\mathcal{F}$ be the curve considered in Example 1.42 and Exercise 3 in Chapter 7 regarded as a curve defined over $\mathbf{F}_q$ with $q = p^h$ and $p \geq 5$.

   (a) If $k \leq \frac{1}{2}(q-1)$ and $k \equiv 1 \pmod{p}$, show that $\mathcal{F}$ is Frobenius classical with respect to the linear system of conics.

   (b) If $q \equiv 1 \pmod{k}$ and $k \equiv -1 \pmod{p}$, show that $\mathcal{F}$ is Frobenius classical with respect to linear system of lines but it can be Frobenius non-classical with respect to the linear system of conics.

   The second case in (b) occurs if and only if $(q-1)/k = p^n + 1$ with $2n \mid h$ and $u, v \in \mathbf{F}_{q^n}$. For instance, this happens for $h = 2$, $k = p+1$, $n = 1$, and also for $h = 6$, $k = (p^2 - p + 1)(p^3 - 1)$, $n = 1$.

6. Let $\mathcal{F} = \mathbf{v}(f(X,Y))$ be a Frobenius non-classical plane curve defined over $\mathbf{F}_q$, let $Q = (x, y)$ be a generic point of $K(\mathcal{F})$ such that x is a separable variable of $K(\mathcal{F})$, and let $\epsilon_2 = q'$.

   (a) If $q > q'$, show that
   $$D_x^{(q')}y\, D_x^{(2q'+1)}y - D_x^{(2q')}y - D_x^{(q'+1)}y = 0.$$

   (b) If $q = q'$, show that
   $$(x^q - x)(D_x^{(q)}y\, D_x^{(2q+1)}y - D_x^{(2q)}y\, D_x^{(q+1)}y) = (D_x^{(1)}y - (D_x^{(1)}y)^q) D_x^{(2q)}y.$$

   (c) For $q = q'$ and $p > 2$, prove that the dual curve $\mathcal{D}$ of $\mathcal{F}$ is non-reflexive if and only if $D_x^{(2q)}y = 0$.

7. Let $\mathcal{C}$ be a plane curve, possibly reducible, defined over $\mathbf{F}_p$ of degree $d < p$, and suppose that $\mathcal{C}$ has no linear component defined over $\mathbf{F}_p$. Prove the following properties of the number R_p of points of $\mathcal{C}$ lying in $\mathrm{PG}(2, p)$:

   (a) $R_p \leq \frac{1}{2}d(d + p - 1)$;

   (b) If $R_p \geq \frac{1}{2}d(d + p - 1) - (d - 1)$ then $\mathcal{C}$ is irreducible.

8. Let $\mathcal{C} = \mathbf{v}(f(X,Y))$, where, with $c \in \mathbf{F}_p$ a non-square,
   $$f(X,Y) = (Y + cX)^{(p-1)/2} + Y^{(p-1)/2} - X^{(p-1)/2} - 1.$$
   Show that, if $p \equiv 1 \pmod{4}$, then $R_p = 3(p-1)^2/8$.

9. Let $g(X) = (X + c)^{(p-1)/2} + X^{(p-1)/2}$, with $c \in \mathbf{F}_p$ a non-square, and let $\mathcal{C} = \mathbf{v}(f(X,Y))$, with
   $$f(X,Y) = \frac{g(X) - g(Y)}{X - Y}.$$
   Show that, if $p \equiv 3 \pmod{4}$, then $R_p = (p-3)(3p-5)/8$.

10. Show that the number of points N_1 on an elliptic cubic, for which the number n of rational inflexions is $n = 0, 1, 3, 9$, satisfies the following:

    (a) if $n = 0$, then $N_1 \equiv 0 \pmod{3}$;

(b) if $n = 1$, then $N_1 \equiv \pm 1 \pmod 3$;

(c) if $n = 3$, then $N_1 \equiv 0 \pmod 3$;

(d) if $n = 9$, then $N_1 \equiv 0 \pmod 9$.

11. Show that, for the curve in Exercise 6 of Chapter 4, the number of $\mathbf{F}_3$-rational points is 13: three are non-singular points and ten arise from branches centred at singular points.

12. Let $\mathcal{F}$ be a plane curve defined over $\mathbf{F}_q$ containing every point in PG$(2, q)$, and let $\varphi_{ij} = X_i^q X_j - X_j^q X_i$ for $0 \leq i, j \leq 2$. Prove the following.

    (a) deg $\mathcal{F} \geq q + 1$, and equality holds if and only if $\mathcal{F}$ splits into the $q + 1$ lines of PG$(2, q)$ through a point P.

    (b) Each of the homogeneous polynomials over $\mathbf{F}_q$ below defines an irreducible plane curve $\mathcal{F} = \mathbf{v}(F)$ of degree $q + 2$ containing every point of PG$(2, q)$:

    (i) $F = c(X_0 + X_1)\varphi_{01} + X_0\varphi_{20} + X_2\varphi_{12}$, for any q, where $T^3 - cT - c \in \mathbf{F}_q[T]$ is irreducible;

    (ii) $F = cX_1\varphi_{01} + X_0\varphi_{20} + X_2\varphi_{12}$, for $q \equiv 1 \pmod 3$, where $T^3 - c \in \mathbf{F}_q[T]$ is irreducible;

    (iii) $F = (X_1 + cX_2)\varphi_{01} - X_0\varphi_{20} - X_2\varphi_{12} = 0$, for $p = 3$, where $T^3 + cT + 1 \in \mathbf{F}_q[T]$ is irreducible.

    Conversely, if $\mathcal{F} = \mathbf{v}(F(X_0, X_1, X_2))$ is an irreducible plane curve of degree $q + 2$ defined over $\mathbf{F}_q$ containing every point of PG$(2, q)$, then there exists a homogeneous coordinate system (X_0, X_1, X_2) over $\mathbf{F}_q$ such that $\mathcal{F}$ is as in (i), (ii) or (iii). If irreducible plane curves defined over an algebraic extension of $\mathbf{F}_q$ are allowed, then there exist curves of degree $q+1$ containing all points of PG$(2, q)$.

8.10 NOTES

The counter-example in Remark 8.19 comes from [91].

Theorem 8.31 comes from [415].

Theorem 8.37 is due to Voloch, see [496], who also pointed out the existence of curves with $G \neq \text{Pic}_0(\mathbf{F}_q(\mathcal{F}))$. The Hermitian curve $\mathcal{F} = (Y^{q'} + Y - X^{q'+1})$ of genus $g = \frac{1}{2}q'(q' - 1)$ regarded as a curve over $q = q'^4$ is such an example. For any divisor n of $(q - 1)^{2g}$, this allows the existence of $\mathbf{F}_q$-rational curves of genus $n(g - 1) + 1$ containing at least $n(q^3 + 1)$ $\mathbf{F}_q$-rational points. A related result of Özbudak [355] is the following bound valid for any $\mathbf{F}_q$-rational curve $\mathcal{F}$ of genus g:

$$S_q \leq \max\left\{2 + (q - 1)\lfloor \tfrac{1}{2}\,\text{ex}(G)\rfloor,\ \tfrac{1}{2}N_q(g)(\text{ex}(G) - 2)(\text{ex}(G) - 1)\right\},$$

where ex(G) is the exponent of G and $N_q(g)$ is as in Section 9.4. Explicit computations for Pic$_0(\mathbf{F}_q(\mathcal{F}))$ of particular families of curves are found in Stichtenoth and Xing [430] and Blache, Cherdieu, and Sarlabous [53].

Lachaud and Martin-Deschamps [287] proved that, if the irreducible curve $\mathcal{F}$ of genus g defined over $\mathbf{F}_q$ has an $\mathbf{F}_q$-rational point, then

$$|\mathrm{Pic}_0(\mathbf{F}_q(\mathcal{F}))| \geq (g^q - 1)(q-1)/(q+g+gq).$$

The Stöhr–Voloch Theorem 8.65 is the central result in the book and comes from [432].

A geometric interpretation of S_q in Example 8.70 is found in [270].

Theorems 8.68 and 8.77 are due to Hefez and Voloch; see [205].

Theorem 8.81 comes from Garcia [137], where some particular families of Frobenius non-classical curves are investigated.

Related to Proposition 8.42, further properties of the integer I are found in Garcia and Homma [144]. These authors proved that $I > m$ if and only if Γ is *extremely strange of order* m; that is, the intersection of the m-osculating spaces of Γ at all but finitely many points is an $(m-1)$-dimensional projective subspace.

Theorem 8.78 comes from [219].

Section 8.7 is based on [218]; see also [217], [12].

For a similar but weaker result to Theorem 8.104 for curves of degree one less, namely, $n = \sqrt{q} - 2$, see [12].

Theorem 8.91 comes from Giulietti [168].

The result that the curve $\mathcal{F}$ in Remark 8.94 is the image of $\mathcal{H}_q$ is shown in [89] and [90].

The characterisation of the Fermat curve in Theorem 8.100 is found in [88].

The result in Remark 8.103 is shown in [88]; that in Remark 8.109 is shown in [280].

Theorems 8.106 and 8.108 are due to Garcia and Voloch [162].

Remark 8.107 is related to Waring's problem for finite fields on the least integer m for which every element of $\mathbf{F}_p$ is a sum of this number of n-th powers; see [162]. The improvement on Theorem 8.108 is due to Matterei [325]; here, recent results on Waring's problem obtained by using character sums are also surveyed.

The number of $\mathbf{F}_q$-rational points of Fermat curves $\mathcal{F} = \mathbf{v}(aX^n + bY^n + c)$ was investigated by Anuradha and Katre [18]; see also [16] and [333].

Section 8.8 is based on Hefez and Voloch [205].

Motivated also by applications of curves to public-key cryptography, as described in Koblitz's book [274], and to certain linear codes, much work has been done on low-genus curves defined over $\mathbf{F}_q$ and, more generally, over any field of positive characteristic. Projective classifications, the number of $\mathbf{F}_q$-rational points, automorphism groups and zero-divisor class groups (Jacobians) of hyperelliptic and other curves of low genera, including computational aspects, have been investigated in great detail in a series of papers by Anuradha, Auer, Barcelo, Barreiero, Blache, Cardona, Cherdieu, Choie, Deng, Enge, Estrada, Gebhardt, Gutierrez, Hernandez, Ibukiyama, Katsura, Keller, Lomont, Lopez, Luengo, Manin, Menezes, Moreno, Mulan, Muñoz, Nart, Oort, Pujolás, Ritzenhaler, Sadorni, Sevilla, Shaska, Takizawa, van der Geer, van der Vlugt, Völklein, Yasin, D. Zinoviev, V. Zinoviev, and others. For this, see [23], [53], [76], [80], [81], [99], [100], [184], [185], [120], [124], [125], [164], [210], [211], [312], [245], [246], [266], [315], [322], [337], [68], [344], [345], [411], [412], [413], [445], [475], [487], [498], [510]; see also

Hurt's book [242].

For Exercise 6, see [205]. For Exercise 12, see [447, 448]. For Exercises 8 and 9 see [370]. For Exercise 7, see [70]. For Exercise 5, see [1].

The result in Remark 8.103 is shown in [88], that in Remark 8.105 is shown in [162], and that in Remark 8.109 is shown in [280].

For a similar but weaker result to Theorem 8.104 for curves of degree one less, namely, $n = \sqrt{q} - 2$, see [12].

Section 8.8 is based on [205].

For Exercise 12, see Tallini [447, 448].

Chapter Nine

Zeta functions and curves with many rational points

In this chapter as in the last, $K = \overline{\mathbf{F}}_q$, the algebraic closure of $\mathbf{F}_q$. Let Γ be an irreducible algebraic curve of $\mathrm{PG}(r, K)$ defined over $\mathbf{F}_q$ and equipped with the places of its function field Σ. By definition, as in Section 7.1, Γ arises from an irreducible plane curve $\mathcal{F}$ defined over $\mathbf{F}_q$. If $\mathcal{F} = \mathbf{v}(f(X,Y))$ with $f(X,Y)$ in $\mathbf{F}_q[X,Y]$, then $\Sigma = K(x,y)$ with $f(x,y) = 0$.

The aim here is to investigate birational properties of Γ using its $\mathbf{F}_q$-rational points and, in particular, the number of its $\mathbf{F}_q$-rational places. For this purpose, a non-singular model $\mathcal{X}$ of Σ is chosen which is birationally equivalent over $\mathbf{F}_q$ to $\mathcal{F}$. So $\mathcal{X}$ is an irreducible non-singular curve in an appropriate projective space $\mathrm{PG}(r, K)$, and places of Σ are identified with points of $\mathcal{X}$. Consequently, divisors are finite sums of points of $\mathcal{X}$.

Since $\mathcal{X}$ can also be viewed as an irreducible algebraic curve over $\mathbf{F}_{q^i}$ for every finite extension $\mathbf{F}_{q^i}$ of $\mathbf{F}_q$, the number N_{q^i} of the $\mathbf{F}_{q^i}$-rational points of $\mathcal{X}$ is defined, and every point of $\mathcal{X}$ is an $\mathbf{F}_{q^i}$-rational point for some i. It is conceivable, at least intuitively, that from the infinite sequence

$$N_1, N_2, \ldots, N_i, \ldots, \text{ with } N_i = N_{q^i}, \tag{9.1}$$

it is possible to obtain information about $\mathcal{X}$, including properties apparently independent of the behaviour of the sets $\mathcal{F}(\mathbf{F}_{q^i})$ of the $\mathbf{F}_{q^i}$-rational points of $\mathcal{X}$, such as the class number and the fundamental equation. To move in this direction, the technical tool is the *zeta function*.

9.1 THE ZETA FUNCTION OF A CURVE OVER A FINITE FIELD

The *Riemann zeta function*,

$$\zeta(s) = \textstyle\sum_{n=1}^{\infty} n^{-s},$$

is important in algebraic number theory. It is convergent for $\mathrm{Re}(s) > 1$ and extends to a meromorphic function of s with a simple pole $s = 1$. Equivalently, the Riemann zeta function can also be written as an infinite product, the *Euler product*

$$\zeta(s) = \textstyle\prod_p (1 - p^{-s})^{-1},$$

where p ranges over all primes of $\mathbf{Q}$.

A similar function for the curve $\mathcal{X}$ is the formal product

$$\textstyle\prod_P \left(1 - N(P)^{-s}\right)^{-1}, \tag{9.2}$$

where P ranges over all closed points of $\mathcal{X}$ and $N(P) = q^{\deg P}$. The aim is to give an expression for this product in terms of the $\mathbf{F}_q$-rational divisors of $\mathcal{X}$.

DEFINITION 9.1 For a complex variable s, the *zeta function* of $\mathcal{X}$, viewed as a curve over $\mathbf{F}_q$, is

$$\zeta(\mathcal{X}, s) = \textstyle\sum_D N(D)^{-s}, \tag{9.3}$$

where D runs over all effective $\mathbf{F}_q$-rational divisors of $\mathcal{X}$, and $N(D) = q^{\deg D}$ is the *norm* of D.

Note that, if $D = \sum n_P\, P$, then

$$N(D) = \textstyle\prod_P N(P)^{n_P}.$$

Also, $N(P) = q^i$ if $P \in \mathrm{PG}(r, q^i)\backslash\mathrm{PG}(r, q^{i-1})$.

THEOREM 9.2 *For* $\mathrm{Re}(s) > 1$, *the series (9.3) is convergent to the function*

$$F_0(q) = F(q^{-s}) + \frac{hq^{1-g}q^{(1-s)m}}{(q-1)(1-q^{e(1-s)})} - \frac{h}{(q-1)(1-q^{-es})},$$

where

(i) $F(q^{-s})$ *is a polynomial in* q^{-s} *of degree at most* $2g-2$;

(ii) h *is the class number of* $\mathcal{X}$;

(iii) g *is the genus of* $\mathcal{X}$;

(iv) e *is the smallest positive degree of divisors in* $\mathrm{Div}(\mathbf{F}_q(\mathcal{X}))$;

(v)
$$m = \begin{cases} 0, & \textit{for } g = 0, \\ 2g-2+e, & \textit{for } g \geq 1. \end{cases}$$

Proof. The minimality of e implies that the degree of any $\mathbf{F}_q$-rational divisor of $\mathcal{X}$ is divisible by e, but it does not imply that every multiple of e is the degree of some $\mathbf{F}_q$-rational divisor of $\mathcal{X}$. In fact, $e = 1$, but this is shown later, in Proposition 9.4; meanwhile no use is made of it.

The $\mathbf{F}_q$-rational effective divisors of $\mathcal{X}$ are partitioned into divisor classes, two such divisors D_1 and D_2 being equivalent if and only if there exists $u \in \mathbf{F}_q(\mathcal{F})$ such that $D_1 = D_2 + \mathrm{div}\, u$. Each class consists of all $\mathbf{F}_q$-rational divisors of a complete linear series $|C|$ for an $\mathbf{F}_q$-divisor C of $\mathcal{X}$; that is, it coincides with the set $|C| \cap \mathrm{Div}(\mathbf{F}_q(\mathcal{X}))$. By Theorem 8.39, the number $n_q(|C|)$ of such $\mathbf{F}_q$-rational divisors is equal to $(q^{\ell(C)}-1)/(q-1)$ with the usual notation $\ell(C) = \dim|C| + 1$.

From the Riemann–Roch Theorem 6.61,

$$\ell(C) = \begin{cases} 0, & \text{if } \deg C < 0; \\ 1, & \text{if } C \text{ is the zero divisor;} \\ 0, & \text{if } C \text{ is a non-zero divisor and } \deg C = 0\,; \\ g-1, & \text{if } C \text{ is a non-canonical divisor, } \deg C = 2g-2; \\ g, & \text{if } C \text{ is a canonical divisor;} \\ \deg C - g + 1, & \text{if } \deg C > 2g-2. \end{cases}$$

By Remark 8.27, $\mathcal{X}$ has some $\mathbf{F}_q$-rational canonical divisor. Hence e divides $2g-2$.

In the summations below, $|C|$ runs over all $\mathbf{F}_q$-rational linear series of $\mathcal{X}$ containing some effective $\mathbf{F}_q$-rational divisor. Now

$$\begin{aligned}\zeta(\mathcal{X},s) &= \textstyle\sum_D N(D)^{-s} = \sum_{|C|}\sum_{D\in|C|} N(D)^{-s}\\ &= \textstyle\sum_{|C|}\sum_{D\in|C|} q^{-s\deg D} = \sum_{|C|} n(|C|)q^{-s\deg C}.\end{aligned}$$

Hence

$$\begin{aligned}\zeta(\mathcal{X},s) &= \textstyle\sum_{|C|} q^{-s\deg C}\,(q^{\ell(C)}-1)/(q-1)\\ &= (q-1)^{-1}\textstyle\sum_{\deg C\geq 0} q^{\ell(C)-s\deg C} - (q-1)^{-1}\sum_{\deg C\geq 0} q^{-s\deg C}.\end{aligned}$$

If $\mathrm{Re}(s) > 1$, then

$$\textstyle\sum_{\deg C\geq 0} q^{-s\deg C} = \sum_{\deg C=ve}\sum_{v=0}^{\infty} q^{-s\deg C} = h\sum_{v=0}^{\infty} q^{-esv} = \dfrac{h}{1-q^{-es}},$$

whence

$$\zeta(\mathcal{X},s) = (q-1)^{-1}\textstyle\sum_{\deg C\geq 0} q^{\ell(C)-s\deg C} - \dfrac{h}{(q-1)(1-q^{-es})}.$$

If $g=0$, then $\ell(C) = \deg C + 1$, and $h=1$ by Example 8.36. So, for $g=0$,

$$\begin{aligned}\zeta(\mathcal{X},s) &= (q-1)^{-1}\textstyle\sum_{v=0}^{\infty}\sum_{\deg|C|=ve} q^{\deg C+1-s\deg C} - \dfrac{1}{(q-1)(1-q^{-es})}\\ &= q(q-1)^{-1}\textstyle\sum_{v=0}^{\infty} q^{e(1-s)v} - \dfrac{1}{(q-1)(1-q^{-es})}\\ &= \dfrac{q}{(q-1)(1-q^{e(1-s)})} - \dfrac{1}{(q-1)(1-q^{-es})}.\end{aligned}$$

For $g\geq 1$,

$$\begin{aligned}\zeta(\mathcal{X},s) &= (q-1)^{-1}\textstyle\sum_{0\leq\deg C\leq 2g-2} q^{\ell(C)-s\deg C}\\ &\quad +(q-1)^{-1}\textstyle\sum_{\deg C>2g-2} q^{\ell(C)-s\deg C} - \dfrac{h}{(q-1)(1-q^{-es})}\\ &= (q-1)^{-1}\textstyle\sum_{v=0}^{(2g-2/e} q^{-esv}\sum_{i=1}^{h} q^{\ell(C_i^{ve})}\\ &\quad + h(q-1)^{-1}\textstyle\sum_{v=(2g-2)/e+1}^{\infty} q^{e(s-1)v-g+1} - \dfrac{h}{(q-1)(1-q^{-es})},\end{aligned}$$

where $|C_i^{ve}|$ denotes an $\mathbf{F}_q$-rational divisor class of degree $v\cdot e$. Hence $\zeta(\mathcal{X},s)$ converges to

$$F_0(q) = F(q^{-s}) + \frac{hq^{1-g}q^{(1-s)m}}{(q-1)(1-q^{e(1-s)})} - \frac{h}{(q-1)(1-q^{-es})},$$

with

$$F(q^{-s}) = (q-1)^{-1}\textstyle\sum_{v=0}^{(2g-2/e} q^{-esv} - \sum_{i=1}^{h} q^{\ell(C^{ve})},$$

which is a polynomial in q^{-s} of degree at most $2g-2$. □

Theorem 9.2 shows that $\zeta(\mathcal{X},s)$ has an analytic continuation to the whole complex plane. Its poles of first order are of two types, namely

$$s_u = \frac{2\pi i}{e\log q}u, \qquad s_v = 1 - \frac{2\pi i}{e\log q}v,$$

where u, v range over $\mathbf{Z}$.

THEOREM 9.3 *For* $\mathrm{Re}(s) > 1$, *the product (9.2) is absolutely convergent to the function* $\zeta(\mathcal{X}, s)$. *In particular, the product is independent of the order of its factors.*

Proof. For any integer $N \geq 1$,

$$\prod_{N(P)\leq N} (1 - N(P)^{-s})^{-1} = \prod_{N(P)\leq N} \left(\sum_{n=0}^{\infty} N(P)^{-ns}\right).$$

Since the product on the right-hand side has finitely many factors and each factor is an absolutely convergent series, it follows by multiplication that

$$\prod_{N(P)\leq N} (1 - N(P)^{-s})^{-1} = \sum_{N(D)\leq N} N(D)^{-s} + \sum_{N(D)>N} N(D)^{-s},$$

where the first summation on the right-hand side is over all effective $\mathbf{F}_q$-rational divisors with $N(D) \leq N$ while the second is over all effective $\mathbf{F}_q$-rational divisors D with $N(D) > N$ which contain no closed point with $N(P) > N$. Then

$$\left|\prod_{N(P)\leq N} (1 - N(P)^{-s})^{-1} - \sum_{N(D)\leq N} N(D)^{-s}\right| \leq \sum_{N(D)>N} N(D)^{-\mathrm{Re}(s)}.$$

The sum on the right-hand side can be viewed as the error term of the convergent series $\zeta(\mathcal{X}, s)$, and hence it tends to zero. This proves the convergence of (9.2). The absolute convergence follows from the inequality

$$\left|\sum_{n=1}^{\infty} N(D)^{-ns}\right| \leq \sum_{n=1}^{\infty} N(D)^{-n(\mathrm{Re}(s))}.$$

□

Theorem 9.3 implies that $\zeta(\mathcal{X}, s)$ has no zeros for $\mathrm{Re}(s) > 1$. Also, it is the key in the proof of the following proposition which has already been enunciated in Section 8.3 as well as in the proof of Theorem 9.2.

PROPOSITION 9.4 *The curve* $\mathcal{X}$ *has an* $\mathbf{F}_q$*-rational divisor of degree* 1.

Proof. With the notation of Theorem 9.2, it suffices to prove that $e = 1$. By the minimality of e, the divisor D is $\mathbf{F}_q$-rational of degree e if and only if D is a closed point of degree e; that is,

$$D = P + \Phi(P) + \cdots + \Phi^{e-1}(P)$$

for some $\mathbf{F}_{q^e}$-rational point P of $\mathcal{X}$. Note that the Frobenius images $\bar{P}_i = \Phi^i(P)$ of P, for $i = 1, \ldots, e-1$, are also $\mathbf{F}_{q^e}$-rational points of $\mathcal{X}$. Let

$$\bar{\zeta}(\mathcal{X}, s) = \prod_{\bar{P}} (1 - N(\bar{P})^{-s})^{-1}$$

be the zeta function of $\mathcal{X}$ regarded as a curve over $\mathbf{F}_{q^e}$. Then

$$\bar{\zeta}(\mathcal{X}, s) = \prod_{\bar{P}} (1 - q^{-se\deg \bar{P}})^{-1} = \prod_{i=1}^{e} \left(\prod_{\bar{P}_i} (1 - q^{-se\deg \bar{P}_i})^{-1}\right).$$

Since $\deg P = e \deg \bar{P}_i$, it follows that

$$\bar{\zeta}(\mathcal{X}, s) = \prod_{i=1}^{e} \left(\prod_{P} (1 - q^{-s\deg P})^{-1}\right) = \zeta(\mathcal{X}, s)^e.$$

As both $\zeta(\mathcal{X}; s)$ and $\bar{\zeta}(\mathcal{X}, s)$ have a pole of the same order 1 at $s = 1$, this implies that $e = 1$. □

Proposition 9.4 makes it possible to simplify Theorem 9.2.

THEOREM 9.5 *The zeta function of $\mathcal{X}$ can be written as*

$$\zeta(\mathcal{X}, s) = \frac{L(q^{-s})}{(1-q^{-s})(1-q^{1-s})},$$

where

$$L(q^{-s}) = \textstyle\sum_{j=0}^{2g} a_j q^{-js}, \quad \textit{with } a_0 = 1,\ a_{2g} = q^g.$$

Proof. The proof is carried out in three parts according as g is $0, 1, \geq 2$.

For $g = 0$,

$$\zeta(\mathcal{X}, s) = \frac{q}{(q-1)(1-q^{1-s})} - \frac{1}{(q-1)(1-q^{-s})} = \frac{1}{(1-q^{-s})(1-q^{1-s})}.$$

For $g = 1$,

$$\begin{aligned}\zeta(\mathcal{X}, s) &= \frac{1}{q-1} \sum_{\deg C=0} q^{\ell(C)-s\deg C} + \frac{hq^{1-s}}{(q-1)(1-q^{1-s})} - \frac{h}{(q-1)(1-q^{-s})} \\ &= \frac{h-1}{q-1} + \frac{q}{q-1} + \frac{hq^{1-s}}{(q-1)(1-q^{1-s})} - \frac{h}{(q-1)(1-q^{-s})} \\ &= \frac{1+(h-q-1)q^{-s}+q\cdot q^{-2s}}{(1-q^{1-s})(1-q^{-s})}.\end{aligned}$$

For $g \geq 2$,

$$\begin{aligned}\zeta(\mathcal{X}, s) &= \frac{1}{q-1} \sum_{\deg C=0} q^{\ell(C)-s\deg C} + \frac{1}{q-1} \sum_{1\leq \deg C<2g-2} q^{\ell(C)-s\deg C} \\ &+ \frac{1}{q-1} \sum_{\deg C\geq 2g-2} q^{\ell(C)-s\deg C} + \frac{hq^{1-g}q^{(1-s)(2g-1)}}{(q-1)(1-q^{1-s})} \\ &- \frac{h}{(q-1)(1-q^{-s})} = \frac{h+q-1}{q-1} + \frac{1}{q-1} \sum_{1\leq \deg C<2g-2} q^{\ell(C)-s\deg C} \\ &+ \frac{(h-1)q^{g-1-s(2g-2)} + q^{g-s(2g-2)}}{q-1} \\ &+ \frac{hq^{1-g}q^{(1-s)(2g-1)}}{(q-1)(1-q^{1-s})} - \frac{h}{(q-1)(1-q^{-s})} \\ &= \frac{1}{q-1} \sum_{j=1}^{2g-3} \alpha_j q^{-js} + \frac{(h+q-1)(q^{g-1}q^{-2(g-1)s}+1)}{q-1} \\ &+ \frac{hq^g q^{-(2g-1)s}}{(q-1)(1-q^{1-s})} - \frac{h}{(q-1)(1-q^{-s})} \\ &= \frac{1+\alpha_1 q^{-s} + \cdots + \alpha_{2g-1}q^{-(2g-1)s} + q^g q^{-2gs}}{(1-q^{-s})(1-q^{1-s}))}.\end{aligned}$$

□

The following result is the *functional equation* of the zeta function.

THEOREM 9.6 *The zeta function of $\mathcal{X}$ satisfies the equation*

$$q^{(g-1)(2s-1)}\zeta(\mathcal{X},s) = \zeta(\mathcal{X},1-s). \tag{9.4}$$

Proof. Again, three cases are treated separately according as g is $0, 1, \geq 2$.
For $g = 0$,

$$\zeta(\mathcal{X},1-s) = \frac{1}{(1-q^s)(1-q^{s-1})} = g^{1-2s}\zeta(\mathcal{X},s).$$

For $g = 1$,

$$\zeta(\mathcal{X},1-s) = \frac{1+(h-q-1)q^{s-1}+q\cdot q^{2s-2}}{(1-q^s)(1-q^{s-1})} = \zeta(\mathcal{X},s).$$

For $g \geq 2$, let

$$\zeta(\mathcal{X},s) = \zeta_1(\mathcal{X},s) + \zeta_2(\mathcal{X},s),$$

with

$$\begin{aligned}\zeta_1(\mathcal{X},s) &= (q-1)^{-1}\textstyle\sum_{1\leq \deg C<2g-2} q^{\ell(C)+1-s\deg C},\\ \zeta_2(\mathcal{X},s) &= 1+q^{-(g-1)(2s-1)}\\ &\quad+\frac{h}{q-1}\left(1+q^{(g-1)(2s-1)}+\frac{q^g q^{-(2g-1)s}}{1-q^{1-s}}-\frac{1}{1-q^{s-1}}\right).\end{aligned}$$

Then

$$\begin{aligned}\zeta_2(\mathcal{X},1-s) &= 1+q^{(g-1)(2s-1)}\\ &\quad+\frac{h}{q-1}\left(1+q^{(g-1)(2s-1)}+\frac{q^g q^{(2g-1)(s-1)}}{1-q^s}-\frac{1}{1-q^{s-1}}\right)\\ &= q^{(g-1)(2s-1)}\zeta_2(\mathcal{X},s).\end{aligned}$$

It remains to show that

$$q^{(g-1)(2s-1)}\zeta_1(\mathcal{X},s) = \zeta_1(\mathcal{X},1-s).$$

Let $\rho(C) = \ell(C) - \frac{1}{2}\deg C$. Then

$$\begin{aligned}\zeta_1(\mathcal{X},s) &= (q-1)^{-1}\textstyle\sum_{1\leq \deg C<2g-2} q^{\ell(C)-s\deg C}\\ &\quad+(q-1)^{-1}\textstyle\sum_{1\leq \deg C<2g-2} q^{\rho(C)-(2s-1)(\deg C)/2}.\end{aligned}$$

From the Riemann–Roch Theorem 6.61,

$$\rho(C) = \rho(W-C)$$

for a $\mathbf{F}_q$-rational canonical divisor W of $\mathcal{X}$. Hence, if C runs over the set of all divisors classes satisfying the condition

$$1 \leq \deg C < 2g-2,$$

then $W - C$ does the same. Therefore

$$\begin{aligned}\zeta_1(\mathcal{X},s) &= (q-1)^{-1}\textstyle\sum_{1\leq \deg(W-C)<2g-2} q^{\rho(W-C)-(2s-1)(2g-2-\deg C)/2}\\ &= (q-1)^{-1}\textstyle\sum_{1\leq \deg C<2g-2} q^{\rho(C)-(2s-1)(2g-2-\deg C)/2},\end{aligned}$$

whence

$$\begin{aligned}&\zeta_1(\mathcal{X},1-s)\\&=(q-1)^{-1}\textstyle\sum_{1\le\deg C<2g-2} q^{\rho(C)+(2s-1)(\deg C)/2}\\&=q^{(g-1)(2s-1)}(q-1)^{-1}\textstyle\sum_{1\le\deg C<2g-2} q^{\rho(C)-(2s-1)(2g-2-\deg C)/2}\\&=q^{(g-1)(2s-1)}\zeta_1(\mathcal{X},s).\end{aligned}$$

□

Now the connection between the zeta function $\zeta(\mathcal{X},s)$ and the infinite sequence (9.1) is established. For this purpose, replace q^{-s} with t in $\zeta(\mathcal{X},s)$. The resulting formal power series $Z(\mathcal{X},t)$ satisfies the equation $\zeta(\mathcal{X},s)=Z(\mathcal{X},t)$.

THEOREM 9.7 *If $|t|<q^{-1}$, then*

$$Z(\mathcal{X},t)=\exp\left(\textstyle\sum_{i=1}^{\infty} N_i t^i/i\right).$$

Proof. The hypothesis, $|t|<q^{-1}$, holds if and only if $\mathrm{Re}(s)>1$. For $\mathrm{Re}(s)>1$,

$$\begin{aligned}\zeta(\mathcal{X},s)&=\textstyle\prod_P (1-N(P)^{-s})^{-1}=\prod_P (1-q^{-s\deg P})^{-1}\\&=\textstyle\prod_P (1-t^{\deg P})^{-1}=Z(\mathcal{X},t).\end{aligned}$$

Hence

$$\begin{aligned}\log Z(\mathcal{X},t)&=-\textstyle\sum_P \log(1-t^{\deg P})=\sum_P\sum_{m=1}^{\infty} t^{m\deg P}/m\\&=\textstyle\sum_{n=1}^{\infty}\left(\sum_{m\deg P=n} m^{-1}\right)t^n=\sum_{n=1}^{\infty}\left(\sum_{\deg P\mid n}\deg P\right)t^n/n.\end{aligned}$$

Let $N_n^*=\sum_{\deg P\mid n}\deg P$. Then

$$Z(\mathcal{X},t)=\exp\left(\textstyle\sum_{n=1}^{\infty} N_n^*\, t^n/n\right).$$

This shows that it suffices to prove the following equality:

$$\textstyle\sum_{\deg P\mid n}\deg P=N_n. \tag{9.5}$$

Every $\mathbf{F}_{q^n}$-rational point Q of $\mathcal{X}$ gives rise to the $\mathbf{F}_q$-rational closed point

$$P=Q+\Phi(Q)+\cdots+\Phi^{n-1}(Q)$$

of $\mathcal{X}$. In this context, two $\mathbf{F}_{q^n}$-rational points of $\mathcal{X}$ are equivalent if they define the same $\mathbf{F}_q$-rational closed point of $\mathcal{X}$. So the set of $\mathbf{F}_{q^n}$-rational points of $\mathcal{X}$ is partitioned into equivalence classes. If Δ is a representative system of such classes, then

$$N_q^*=\textstyle\sum_{P\in\Delta}\deg P.$$

On the other hand, if the closed point P arises from a point Q of $\mathcal{X}$, that is,

$$P=Q+\Phi(Q)+\cdots+\Phi^{m-1}(Q),$$

then the condition $\deg P\mid n$ in this summation means that $\Phi^m(Q)=Q$ with $m\mid n$. Therefore Q is an $\mathbf{F}_{q^m}$-rational point of $\mathcal{X}$, and the divisibility condition $m\mid n$ implies that $\mathbf{F}_{q^m}$ is a subfield of $\mathbf{F}_{q^n}$. From this argument, the desired equation (9.5) follows. □

As in Section 8.3, let A_n denote the number of all effective $\mathbf{F}_{q^n}$-rational divisors of $\mathcal{X}$ of degree n. Then

$$Z(\mathcal{X},t) = 1 + \textstyle\sum_{n=1}^{\infty} A_n t^n. \tag{9.6}$$

This follows from Theorem 9.3, since

$$\textstyle\prod_P (1 - N(P)^{-s})^{-1} = \zeta(\mathcal{X}, s) = Z(\mathcal{X},t) = \prod_P (1 - t^{\deg P})^{-1},$$

where P ranges over all $\mathbf{F}_q$-rational closed points of $\mathcal{X}$, and each factor can be written as a geometric series, giving the equation

$$\textstyle\prod_P (1 - t^{\deg P})^{-1} = \prod_P \sum_{n=0}^{\infty} t^{\deg(nP)} = \sum_D t^{\deg D} = 1 + \sum_{n=1}^{\infty} A_n t^n,$$

where D ranges over all effective $\mathbf{F}_q$-rational divisors of $\mathcal{X}$.

Proposition 9.8 *If $g \geq 1$, then $Z(\mathcal{X},t) = G(t) + H(t)$, with*

$$G(t) = (q-1)^{-1} \textstyle\sum_{0 \leq \deg C \leq 2g-2} q^{\ell(C)} t^{\deg C},$$
$$H(t) = h(q-1)^{-1} \left(q^{1-g}(qt)^{2g-1}(1-qt)^{-1} - (1-t)^{-1}\right).$$

Proof. A straightforward computation depending on Theorem 8.39 and Proposition 8.40 with $e = 1$ gives the following:

$$\begin{aligned} Z(\mathcal{X},t) &= \textstyle\sum_{n=0}^{\infty} A_n t^n \\ &= \textstyle\sum_{\deg C \geq 0} n_q(|C|)\, t^{\deg C} + (q-1)^{-1} \sum_{\deg C \geq 0} (q^{\ell(C)} - 1) t^{\deg C} \\ &= (q-1)^{-1} \textstyle\sum_{0 \leq \deg C \leq 2g-2} q^{\ell(C)} t^{\deg C} \\ &\qquad + (q-1)^{-1} (\textstyle\sum_{\deg C > 2g-2} q^{\ell(C)+1-g}\, t^{\deg C} - \sum_{\deg C \geq 0} t^{\deg C}) \\ &= G(t) + H(t). \end{aligned}$$

□

Theorem 9.5 shows that the zeta function $Z(\mathcal{X},t)$ has the form

$$Z(\mathcal{X},t) = \frac{L(t)}{(1-t)(1-qt)},$$

where

$$L(t) = L_q(t) = \begin{cases} 1, & \text{for } g = 0, \\ 1 + \sum_{i=1}^{2g-1} a_i t^i + q^g t^{2g}, & \text{for } g \geq 1, \end{cases} \tag{9.7}$$

is the *L-polynomial* of $\mathcal{X}$, viewed as a curve defined over $\mathbf{F}_q$. Note that $L(t) \in \mathbf{Z}[t]$. For $g \geq 1$, from Proposition 9.8,

$$L(t) = (1-t)(1-qt)G(t) + h(q-1)^{-1}\{q^g t^{2g-1}(1-t) - (1-qt)\}. \tag{9.8}$$

Proposition 9.9 *The L-polynomial has the following properties:*

(i) $L(t) = q^g t^{2g} L((qt)^{-1}) = 1 + a_{2g-1} q^{g-1} t + \cdots + q^g t^{2g}$;

(ii) $a_{2g-i} = q^{g-i} a_i$ *for* $i = 0, \ldots, g$.

Proof. The first assertion is a consequence of the functional equation (9.4). Comparing the coefficients of t^i, the second assertion follows. □

Let

$$L(t) = \prod_{i=1}^{2g} (1 - \omega_i t) \tag{9.9}$$

be a factorisation of $L(t)$ into linear factors in some finite extension of $\mathbf{Q}$. The following result plays an essential role in the study of $\mathbf{F}_q$-rational points of $\mathcal{X}$.

THEOREM 9.10 *Let N_n be the number of $\mathbf{F}_{q^n}$-rational points of an irreducible, non-singular, algebraic curve $\mathcal{X}$ of genus g defined over $\mathbf{F}_q$. Then*

$$N_n = q^n + 1 - \sum_{i=1}^{2g} \omega_i^n, \tag{9.10}$$

where $\omega_1, \ldots, \omega_{2g}$ are the reciprocals of the roots of the L-polynomial of $\mathcal{X}$.

Proof. From the relation

$$Z(\mathcal{X}, t) = \exp\left(\sum_{n=1}^{\infty} \frac{N_n}{n} t^n\right) = \frac{\prod_{i=1}^{2g}(1 - \omega_i t)}{(1-t)(1-qt)},$$

it follows that

$$\begin{aligned}\sum_{n=1}^{\infty} N_n t^n / n &= \sum_{i=1}^{2g} \log(1 - \omega_i t) - \log(1-t) - \log(1-qt)\\ &= \sum_{n=1}^{\infty} n^{-1}\left(q^n + 1 - \sum_{i=1}^{2g} \omega_i^n\right) t^n.\end{aligned}$$

Now the result follows by comparing coefficients. □

Note that $\omega_1, \ldots, \omega_{2g}$ are algebraic integers as $L(t) \in \mathbf{Z}[t]$.

REMARK 9.11 If $g = 1$, that is, $\mathcal{X}$ is a non-singular plane cubic curve defined over $\mathbf{F}_q$, then

$$N_1 = q + 1 - 2\sqrt{q} \cos \varphi,$$

where $\omega_1 = \sqrt{q}(\cos \varphi + i \sin \varphi)$. From this, the *Hasse Bound*,

$$q + 1 - 2\sqrt{q} \le N_1 \le q + 1 + 2\sqrt{q},$$

follows. There exists a non-singular elliptic cubic over $\mathbf{F}_q$, $q = p^h$, whose number of $\mathbf{F}_q$-rational points is precisely $N_1 = q + 1 - t$, where $|\,t\,| \le 2\sqrt{q}$, for the values of t shown in Table 9.3.

If the numbers N_n are known for $n = 1, \ldots, r$ for sufficiently large r, then the coefficients of the L-polynomial can be calculated as follows.

PROPOSITION 9.12 *Let $L(t) = \sum_{i=0}^{2g} a_i t^i$ be the L-polynomial of $\mathcal{X}$ viewed as a curve over $\mathbf{F}_q$, and let $s_n = N_n - (q^n + 1)$.*

(i) *The logarithmic derivative is the following*:

$$\frac{dL(t)/dt}{L(t)} = \sum_{n=1}^{\infty} s_n t^{n-1}.$$

(ii) *For* $n = 1, \ldots, g$,

$$na_n = s_n + s_{n-1}a_1 + \cdots + s_1 a_{n-1}. \tag{9.11}$$

In particular, from $N_1, \ldots, N_g$, *the coefficients of* $L(t)$ *can be calculated using (9.11) and the equations* $a_0 = 1$, $a_{2g-n} = q^{g-n}a_n$, *for* $n = 1, \ldots, g$.

Proof. (i) From Theorem 9.10,

$$\begin{aligned}\frac{dL(t)/dt}{L(t)} &= \textstyle\sum_{n=1}^{2g} -\omega_n/(1-\omega_n t) = \sum_{n=1}^{2g}(-\omega_n)\cdot\sum_{k=0}^{\infty}(\omega_n t)^k \\ &= \textstyle\sum_{k=1}^{\infty}\left(\sum_{n=1}^{2g} -\omega_n^k\right)t^{k-1} = \sum_{k=1}^{\infty} s_k t^{k-1}.\end{aligned}$$

(ii) From (i),

$$\begin{aligned}&a_1 + 2a_2 t + \cdots + (2g-1)a_{2g-1}t^{2g-2} + 2gqt^{2g-1} \\ &= (1 + a_1 t + \cdots + a_{2g-1}t^{2g-2} + 2gqt^{2g})\cdot\textstyle\sum_{k=1}^{\infty} s_k t^{k-1}.\end{aligned}$$

So (9.11) follows by comparing the coefficients of t^i for $i = 0, \ldots, g-1$. □

EXAMPLE 9.13 Let $q = 2$ and let $\mathcal{X} = \mathbf{v}(X_0^3 + X_1^3 + X_2^3)$. The set

$$\mathcal{X}(\mathbf{F}_2) = \{(0,1,1),\ (1,0,1),\ (1,1,0)\}).$$

Then

$$a_1 = s_1 = N_1 - 3 = 0.$$

So

$$Z(\mathcal{X}, t) = \frac{1+2t^2}{(1-t)(1-2t)} \tag{9.12}$$

and

$$\textstyle\sum N_i t^i/i = \log Z(\mathcal{X}, t) = \sum t^i/i + \sum (2t)^i/i + \sum (-1)^{j-1}(2t^2)^j/j.$$

Hence

$$N_h = \begin{cases} 1 + 2^h & \text{for } h \text{ odd}, \\ 1 + 2^h + 2.2^{h/2} & \text{for } h \equiv 2 \pmod 4, \\ 1 + 2^h - 2.2^{h/2} & \text{for } h \equiv 0 \pmod 4. \end{cases}$$

Also, from (9.12),

$$Z(\mathcal{X}, t) = 1 + 3\textstyle\sum_{i>0}(2^i - 1)t^i,$$

whence $A_i = 3(2^i - 1)$ for $i > 0$.

EXAMPLE 9.14 Let $\mathcal{X} = \mathbf{v}(X_0X_1^3 + X_1X_2^3 + X_2X_0^3)$ be the Klein quartic, as in Example 7.18, viewed as a curve defined over $\mathbf{F}_2$. The vertices $\mathbf{U}_i$, $i = 0, 1, 2$, of the fundamental triangle are points of PG(2, 2) lying on $\mathcal{X}$. Since no other point of PG(2, 2) has this property, so $N_1 = 3$. To count the $\mathbf{F}_2$-rational closed points of degree 2, consider the quadratic extension $\mathbf{F}_4 = \mathbf{F}_2(\omega)$ with $\omega^2 + \omega + 1 = 0$. The points of $\mathcal{X}$ over $\mathbf{F}_4$ distinct from the points $\mathbf{U}_i$ are two, namely $V_1 = (\omega, \omega^2, 1)$

and $V_2 = (\omega^2, \omega, 1)$. There are 21 points of $\mathcal{X}$ over the cubic extension $\mathbf{F}_8 = \mathbf{F}_2(\epsilon)$ with $\epsilon^3 + \epsilon + 1 = 0$ distinct from the points $\mathbf{U}_i$, namely

$$\begin{array}{lllllllll} W_1 & = & (1,\epsilon,1), & W_2 & = & (1,\epsilon^2,1), & W_3 & = & (1,\epsilon^4,1), \\ W_4 & = & (\epsilon^6,\epsilon^6,1), & W_5 & = & (\epsilon^5,\epsilon^5,1), & W_6 & = & (\epsilon^3,\epsilon^3,1), \\ W_7 & = & (\epsilon,1,1), & W_8 & = & (\epsilon^2,1,1), & W_9 & = & (\epsilon^4,1,1), \\ W_{10} & = & (\epsilon,\epsilon^4,1), & W_{11} & = & (\epsilon^2,\epsilon,1), & W_{12} & = & (\epsilon^4,\epsilon^2,1), \\ W_{13} & = & (\epsilon^3,\epsilon^4,1), & W_{14} & = & (\epsilon^6,\epsilon,1), & W_{15} & = & (\epsilon^5,\epsilon^2,1), \\ W_{16} & = & (\epsilon^3,\epsilon^6,1), & W_{17} & = & (\epsilon^6,\epsilon^5,1), & W_{18} & = & (\epsilon^5,\epsilon^3,1), \\ W_{19} & = & (\epsilon,\epsilon^5,1), & W_{20} & = & (\epsilon^2,\epsilon^3,1), & W_{21} & = & (\epsilon^4,\epsilon^6,1). \end{array}$$

Thus

$$\begin{aligned} & s_1 = s_2 = 0,\ s_3 = 24 - (8+1) = 15; \\ & a_1 = a_2 = 0,\ a_3 = 5,\ a_4 = a_5 = 0; \\ & L_8(t) = 1 + 5t^3 + 8t^6. \end{aligned}$$

Any curve $\mathcal{X}$ defined over $\mathbf{F}_q$ can also be regarded as a curve over $\mathbf{F}_{q^n}$. So, for every n, the curve $\mathcal{X}$ viewed as a curve over $\mathbf{F}_{q^n}$ has its zeta function $Z_{q^n}(\mathcal{X}, t)$ and L-polynomial $L_{q^n}(t)$. The relationship between $Z(\mathcal{X}, t)$ and $Z_{q^n}(\mathcal{X}, t)$, as well as that between $L(t) = L_q(t)$ and $L_{q^n}(t)$, is described in the following results.

Proposition 9.15 *The zeta function*

$$Z_{q^n}(\mathcal{X}, t^n) = \textstyle\prod_\epsilon Z(\mathcal{X}, \epsilon t),$$

where ϵ ranges over the n-th roots of unity.

Proof. It suffices to prove the assertion for $|t| < q^{-1}$. In this region,

$$Z_{q^n}(\mathcal{X}, t^n) = \textstyle\prod_{P'} (1 - t^{n \deg P'})^{-1}, \tag{9.13}$$

where P' ranges over all closed points of $\mathcal{X}$, viewed as a curve over $\mathbf{F}_{q^n}$. Write $P' = Q + \Phi^n(Q) + \cdots + \Phi^{(k-1)n}(Q)$ with $Q \in \mathcal{X}$ where $k = \deg P'$. If

$$P = Q + \Phi(Q) + \cdots + \Phi^n(Q) + \cdots + \Phi^{m-1}(Q)$$

with $\Phi^m(Q) = Q$, then write $P' \mid P$. Note that P is a closed point of $\mathcal{X}$ of degree m, and that

$$\textstyle\prod_{P'} (1 - t^{n \deg P'})^{-1} = \prod_P \prod_{P'|P} (1 - t^{n \deg P'})^{-1}.$$

With $d = \gcd(n, m)$,

$$\begin{aligned} \textstyle\prod_{P'|P} (1 - t^{n \deg P'}) &= (1 - t^{nm/d})^d \\ &= \textstyle\prod_\epsilon (1 - (\epsilon t)^m) = \prod_\epsilon (1 - (\epsilon t)^{\deg P}), \end{aligned}$$

where ϵ ranges over the n-th roots of unity. From (9.13), the result follows. □

Proposition 9.16 *The L-function*

$$L_{\mathbf{F}_{q^n}}(t) = \textstyle\prod_{i=1}^{2g} (1 - \omega_i^n t).$$

Proof. By Proposition 9.15, if ϵ ranges over the n-th roots of unity, then

$$\begin{aligned} L_{\mathbf{F}_{q^n}}(t^n) &= (1-t^n)(1-q^nt^n)\, Z_{q^n}(\mathcal{X},t^n) \\ &= (1-t^n)(1-q^nt^n)\textstyle\prod_\epsilon Z(\mathcal{X},\epsilon t) \\ &= (1-t^n)(1-q^nt^n)\textstyle\prod_\epsilon \dfrac{L(\epsilon t)}{(1-\epsilon t)(1-q\epsilon t)} \\ &= \textstyle\prod_\epsilon L(\epsilon t) \;=\; \prod_{i=1}^{2g}\prod_\epsilon (1-\omega\epsilon t) \;=\; \prod_{i=1}^{2g}(1-\omega_i^n t^n). \end{aligned}$$ □

The next result relates L-polynomials of two curves.

THEOREM 9.17 (Serre) *If $\mathcal{X} \to \mathcal{X}'$ is an $\mathbf{F}_q$-rational covering, then the L-polynomial of $\mathcal{X}'$ divides the L-polynomial of $\mathcal{X}$ in $\mathbf{Z}[t]$.*

Exercises 9 and 10 illustrate Theorem 9.17.

9.2 THE HASSE–WEIL THEOREM

The fundamental result on curves over finite fields is the Hasse–Weil Bound (9.14), which was originally proved by Hasse for $g = 1$ and then by Weil for $g > 1$.

THEOREM 9.18 (Hasse–Weil Bound) *With N_1 the number of $\mathbf{F}_q$-rational points of an irreducible non-singular algebraic curve defined over $\mathbf{F}_q$,*

$$q + 1 - 2g\sqrt{q} \le N_1 \le q + 1 + 2g\sqrt{q}. \tag{9.14}$$

Here, (9.14) is proved by using the Stöhr–Voloch Theorem 8.65 for the upper bound, and Galois Theory for the lower bound. From the Hasse–Weil Bound, via zeta functions, the Hasse–Weil Theorem is deduced, which is actually the Riemann hypothesis for function fields over finite fields.

THEOREM 9.19 (Hasse–Weil Theorem) *The reciprocals of the roots $\omega_1, \ldots, \omega_{2g}$ of the L-polynomial of an irreducible non-singular curve defined over $\mathbf{F}_q$ satisfy*

$$|\omega_i| = \sqrt{q}, \qquad i = 1, \ldots, 2g.$$

The Hasse–Weil Bound (9.14) is stated for irreducible non-singular curves defined over $\mathbf{F}_q$, but it also implies bounds for irreducible singular plane curves. Refinements and relevant improvements of (9.14) are also possible; all these are treated in Section 9.3.

The first step is the following result on the inverse roots.

LEMMA 9.20 *The Hasse–Weil Theorem holds over $\mathbf{F}_q$ if and only if it holds over every $\mathbf{F}_{q^i}$ with $i \ge 1$.*

Proof. Let $\omega_1, \ldots, \omega_{2g}$ be the reciprocals of the roots of the L-polynomial $L_q(t)$ of $\mathcal{X}$. If $L_{q^n}(t)$ is the L-polynomial of $\mathcal{X}$, then its roots are $\omega_1^n, \ldots, \omega_{2g}^n$, by Proposition 9.16. The result follows since $|\omega_i| = \sqrt{q}$ if and only if $|\omega_i^n| = \sqrt{q^n}$. □

The next step is to show that a somewhat weaker version of the bound (9.14) implies the Hasse–Weil Theorem 9.19.

LEMMA 9.21 *If there is a constant $c \in \mathbf{R}$ such that, for every $n \geq 1$,*

$$|N_n - (q^n + 1)| \leq c\sqrt{q^n}, \tag{9.15}$$

then the Hasse–Weil Theorem 9.19 holds.

Proof. Theorem 9.10 and (9.15) imply, for every $n \geq 1$, that

$$\left|\sum_{i=1}^{2g} \omega_i^n\right| \leq c\sqrt{q^n}. \tag{9.16}$$

Denote by ρ the minimum of $|\omega_i^{-1}|$ for $i \in \{1, \ldots, 2g\}$. Then the radius of convergence of the power series expansion of the meromorphic function

$$H(t) = \sum_{i=1}^{2g} \frac{\omega_i t}{1 - \omega_i t}$$

is ρ, the only singularities of $H(t)$ being $\omega_1^{-1}, \ldots, \omega_{2g}^{-1}$. For $|t| < \rho$,

$$H(t) = \sum_{i=1}^{2g} \sum_{n=1}^{\infty} (\omega_i t)^n = \sum_{n=1}^{\infty} \left(\sum_{i=1}^{2g} \omega_i^n\right) t^n.$$

By (9.16), this is a convergent series for $|t| < q^{-1/2}$, whence $q^{-1/2} \leq \rho$. Therefore $\sqrt{q} \geq |\omega_i|$ for $i = 1, \ldots, 2g$. Finally, $|\omega_i| = \sqrt{q}$ since $\prod_{i=1}^{2g} \omega_i = q^g$, which is a consequence of (9.7) and (9.9). □

In view of Lemma 9.21, the Hasse–Weil Theorem 9.19 can be established in two further steps by showing the existence of constants $c_1 > 0$ and $c_2 > 0$ satisfying the inequalities

$$N_n \leq q^n + 1 + c_1\sqrt{q^n}, \tag{9.17}$$

$$N_n \geq q^n + 1 - c_2\sqrt{q^n} \tag{9.18}$$

for any $n \geq 1$. The latter condition on n can be weakened to require only $n \geq n_0$ for some $n_0 \geq 1$; see Lemma 9.20. This is useful in proving (9.18).

The upper bound (9.17) is achieved in the following result, which is essentially a corollary to the Stöhr–Voloch Theorem 8.65.

PROPOSITION 9.22 *If q^n is a square greater than $4g^4(g-1)^2$, then*

$$N_n \leq q^n + 1 + 2g\sqrt{q^n}.$$

Proof. It may be supposed that $\mathcal{X}$ has an $\mathbf{F}_q$-rational point P. For any integer $d \geq 2g$, let $L = |d\mathcal{P}|$. Then $L = g_r^d$ is a non-special base-point-free $\mathbf{F}_q$-rational linear series. In particular, $r = d - g \geq g$. By Corollaries 8.63 and 8.64, $\nu_i = i$ for $i < r - g$ and $\nu_i \leq i + g$ for $r - g \leq i \leq r - 1$. Hence the Stöhr–Voloch Theorem 8.65 shows that

$$N_1 \leq q + 1 + (r + q/r)g + 2g^2(g-1)/r$$

for any $r \geq g$. If q is a square and $q > 4g^4(g-1)^2$, then take $r = \sqrt{q}$ to give the result. □

To obtain the lower bound (9.18), Galois theory is also needed.

PROPOSITION 9.23 *If q is a square, then*

$$N_n = q^n + O(\sqrt{q^n}),$$

where, for a constant c,

$$\lim_{n\to\infty} \frac{O(\sqrt{q^n})}{\sqrt{q^n}} = c.$$

Proof. Choose a separable variable $u \in K(\mathcal{X})$. From (VII) in Section A.2, $K(u)$ has a normal extension Σ' that is also a normal extension of $K(\mathcal{X})$. Let G and H be the corresponding Galois groups with H a subgroup of G. In geometric terms, if $\mathcal{X}'$ is an irreducible non-singular curve of $\mathrm{PG}(r', K')$ which is a model of Σ', then both $\mathcal{X}$ and the projective line $\ell = \mathbf{v}(X)$ with generic point $P = (u)$ are quotient curves of $\mathcal{X}'$; more precisely, $\mathcal{X} = \mathcal{X}'/H$ and $\ell = \mathcal{X}'/G$. Here, G and H are linear collineation groups of $\mathrm{PG}(r', K')$ which preserve $\mathcal{X}'$. Although the curve $\mathcal{X}'$ and the group G need not be defined over $\mathbf{F}_{q^n}$, but only over some finite extension of $\mathbf{F}_{q^n}$, here $\mathbf{F}_{q^n}$ may be replaced by any of its finite extensions. So $\mathcal{X}'$ is defined over $\mathbf{F}_{q^n}$, and G is a subgroup of $\mathrm{PGL}(r'+1, q^n)$.

Let Φ^n be the Frobenius collineation of $\mathrm{PG}(r', K)$. For any $\gamma \in G$, the product

$$\gamma' = \Phi^n \gamma^{-1}$$

is also a collineation of $\mathrm{PG}(r', K')$ which preserves $\mathcal{X}'$. Let $N_n(\mathcal{X}', \gamma)$ be the number of all fixed points of γ' on $\mathcal{X}'$. Such points of $\mathcal{X}'$ lie in an s-dimensional subspace Π_γ of $\mathrm{PG}(r', K)$ consisting of all fixed points of γ' in $\mathrm{PG}(r', K)$. Note that Π_γ need not consist of points with coordinates in $\mathbf{F}_{q^n}$ since Π_γ is the image of $\mathrm{PG}(s, q^n)$ by a linear collineation belonging to $\mathrm{PGL}(r, q^{in})$ for some $i \geq 1$. Again, for the present purpose, the finite extension $\mathbf{F}_{q^{in}}$ may be replaced by $\mathbf{F}_{q^n}$. Then the fixed points of γ' are $\mathbf{F}_{q^n}$-rational points of the $\mathbf{F}_{q^n}$-rational curve $\mathcal{X}'$.

To apply the Stöhr–Voloch Theorem 8.65 in the same way as in the proof of Proposition 9.22, a further replacement of $\mathbf{F}_{q^n}$ by a larger finite field may be needed in order to ensure that $q^n > 4g'^4(g'-1)^2$. With this, $N_n \leq q^n + 1 + 2g\sqrt{q^n}$, and so

$$N_n(\mathcal{X}', \gamma) \leq q^n + 1 + 2g'\sqrt{q^n}. \tag{9.19}$$

If Q is a point of ℓ, and $P_1, \ldots, P_N$ are the distinct points of $\mathcal{X}'$ lying over Q in the covering $\mathcal{X}' \to \ell$, then Q is an $\mathbf{F}_q$-rational point of ℓ if and only if $\Phi^n(P_1)$ coincides with $\gamma(P)$ for some $\gamma \in G$. Since the points $P_1, \ldots, P_N$ form an orbit of G, either $N = |G|$ and the orbit is long, or $|N|$ is a divisor of $|G|$, with $|N| < |G|$, and the orbit is short. The latter case only occurs for at most $g - 1 + |G|$ points Q, by Theorem 11.57 and Remark 11.61. Since the number of $\mathbf{F}_q$-rational points on ℓ is $q^n + 1$, it follows that

$$\textstyle\sum_{\gamma' \in G} N_n(\mathcal{X}', \gamma') = |G|(q^n + 1) - c, \tag{9.20}$$

where $c > 0$ and c only depends on g and G but is independent of q^n.

From (9.19) and (9.20),

$$N_n(\mathcal{X}', \gamma) = q^n + O(\sqrt{q^n}).$$

The above argument still works if ℓ is replaced by $\mathcal{X}$, and it shows that

$$\textstyle\sum_{\gamma'\in H} N_n(\mathcal{X}',\gamma') = |H|N_n + O(\sqrt{q^n}).$$

Therefore

$$N_n = q^n + O(\sqrt{q^n}). \qquad \square$$

The Hasse–Weil Bound also provides an estimate for the number B_n of closed $\mathbf{F}_q$-rational points of a given degree n. The idea is to use the equation

$$N_n = \textstyle\sum_{d|n} d \cdot B_d, \tag{9.21}$$

where the sum is extended over all positive divisors d of n. The Möbius Inversion Formula inverts (9.21) to the equation

$$n \cdot B_n = \textstyle\sum_{d|n} \mu(n/d) \cdot N_d, \tag{9.22}$$

where $\mu : \mathbf{N} \to \{0, 1, -1\}$ is the Möbius function. Since $n > 1$,

$$\textstyle\sum_{d|n} \mu(n/d) = 0.$$

Substituting (9.10) into (9.22) gives the following result.

PROPOSITION 9.24 *For any $n \geq 2$,*

$$B_n = n^{-1}\textstyle\sum_{d|n} \mu(n/d) \left(q^d - \sum_{i=1}^{2g} \omega_i^n\right),$$

where $\omega_i, \ldots, \omega_{2g}$ are the reciprocals of the roots of the L-polynomial $L_q(t)$ of $\mathcal{X}$.

From Proposition 9.24, an upper bound on B_n is obtained which only depends on q^n and g.

THEOREM 9.25 (i) *For every $n \geq 2$,*

$$\left|B_n - \frac{q^n}{n}\right| \leq \left(\frac{g}{q-1} + 2g\frac{\sqrt{q}}{\sqrt{q}-1}\right) \cdot \frac{\sqrt{q^n}-1}{n} < (2+7g) \cdot \frac{\sqrt{q^n}}{n}.$$

(ii) *For any n satisfying the condition*

$$2g + 1 \leq \sqrt{q^{n-1}}(\sqrt{q} - 1), \tag{9.23}$$

the curve $\mathcal{X}$ has at least one closed point of degree n.

(iii) *If $n \geq 4g + 3$ then $B_n \geq 1$.*

Proof. (i) The assertion for $n = 1$ follows readily from the Hasse–Weil Bound (9.14). For $n \geq 2$, Proposition 9.25 gives the following:

$$B_n - q^n/n = n^{-1}\textstyle\sum_{d|n,d<n} \mu(n/d)q^n - n^{-1}\sum_{d|n} \sum_{i=1}^{2g} \omega_i^d.$$

Let $m = \lfloor \frac{1}{2}n \rfloor$. Then, from the Hasse–Weil Theorem 9.19, $|\sum_{i=1}^{2g} \omega_i^d| \leq 2g\sqrt{q^d}$. Therefore

$$\begin{aligned}\left|B_n - \frac{q^n}{n}\right| &\leq \frac{1}{n}\sum_{d=1}^{m} q^n + \frac{2g}{n}\sum_{d=1}^{n} \sqrt{q^d} = \frac{q}{n} \cdot \frac{q^m-1}{q-1} + \frac{2g\sqrt{q}}{n} \cdot \frac{\sqrt{q^n}-1}{\sqrt{q}-1}\\ &\leq \left(\frac{q}{q-1} + 2g\frac{\sqrt{q}}{\sqrt{q}-1}\right) \cdot \frac{\sqrt{q^n}-1}{n} \leq (2+7g) \cdot \frac{\sqrt{q^n}}{n}.\end{aligned}$$

(ii) By (i), the inequality $B_n > 0$ holds, provided that

$$\frac{q^n}{n} > \left(\frac{q}{q-1} + 2g\frac{\sqrt{q}}{\sqrt{q}-1}\right) \cdot \frac{\sqrt{q^n}-1}{n}. \tag{9.24}$$

If $g = 0$, then (9.24) holds for all $n \geq 1$. So it may be assumed that $g \geq 1$. A direct calculation shows that (9.24) can be written as follows:

$$2g + \frac{\sqrt{q}}{\sqrt{q}+1} < \frac{q^n(\sqrt{q}-1)}{\sqrt{q}(\sqrt{q^n}-1)}. \tag{9.25}$$

Taking into account the inequalities

$$2g + \frac{\sqrt{q}}{\sqrt{q}+1} < 2g+1, \qquad \sqrt{q^{n-1}}(\sqrt{q}-1) < \frac{q^n(\sqrt{q}-1)}{\sqrt{q}(\sqrt{q^n}-1)},$$

the assumption (9.23) implies (9.25). Therefore $B_n > 0$.

(iii) If $n \geq 4g+3$, then

$$2g+1 < 2^{2g+1}(\sqrt{q}-1) \leq \sqrt{2}^{n-1}(\sqrt{2}-1) \leq \sqrt{q^{n-1}}(\sqrt{q}-1),$$

whence it follows that $B_n > 0$. □

It is of interest to compare the bounds (8.18) and (9.14), at least for the series g_n^2 cut out by lines in the case of a plane non-singular curve of degree n.

LEMMA 9.26 *When $n \geq \frac{1}{2}\sqrt{q}+3$, the bound given by Theorem 8.65 is better than that given by Theorem 9.18.*

Proof. For $g = \frac{1}{2}(n-1)(n-2)$, the bounds given by (8.18) and (9.14) are as follows:

$$\begin{aligned} &\tfrac{1}{2}n(n+q-1), && \text{(SV)} \\ &q+1+(n-1)(n-2)\sqrt{q}. && \text{(HW)} \end{aligned}$$

Then

$$\begin{aligned} &\tfrac{1}{2}n(n+q-1) \leq q+1+(n-1)(n-2)\sqrt{q} \\ &\iff n^2 + n(q-1) \leq 2q+2+2(n^2-3n+2)\sqrt{q} \\ &\iff n^2(2\sqrt{q}-1) - n(q+6\sqrt{q}-1) + 2(q+2\sqrt{q}+1) \geq 0. \end{aligned} \tag{9.26}$$

The discriminant of (9.26) is

$$\begin{aligned} \delta &= (q+6\sqrt{q}-1)^2 - 8(2\sqrt{q}-1)(q+2\sqrt{q}+1) \\ &= (q-2\sqrt{q}+3)^2. \end{aligned}$$

Hence (9.26) holds when

$$n \leq \frac{(q+6\sqrt{q}-1)-(q-2\sqrt{q}+3)}{2(2\sqrt{q}-1)} = 2$$

or

$$\begin{aligned} n &\geq \frac{(q+6\sqrt{q}-1)+(q-2\sqrt{q}+3)}{2(2\sqrt{q}-1)} = \frac{(\sqrt{q}+1)^2}{2\sqrt{q}-1} \\ &= \frac{1}{2}\sqrt{q} + \frac{5}{4} + \frac{9}{4(2\sqrt{q}-1)}, \end{aligned}$$

which holds for $n \geq \frac{1}{2}\sqrt{q}+3$. □

9.3 REFINEMENTS OF THE HASSE–WEIL THEOREM

The Hasse–Weil Bound 9.18 is a powerful result even though it only depends on two parameters, the genus g of the curve $\mathcal{X}$ and the order q of the finite field over which $\mathcal{X}$ is defined. Even though the bound is sharp for an infinite number of values of q and g, see Chapter 10, some refinements of the bound are possible, and this is the aim of this section.

An immediate refinement derives from the fact that the term $2g\sqrt{q}$ in the Hasse–Weil Bound (9.18) may be replaced by its integer part. Therefore

$$|N_1 - (q+1)| \leq \lfloor 2g\sqrt{q} \rfloor. \tag{9.27}$$

A non-trivial improvement of (9.27) occurs when q is a non-square.

THEOREM 9.27 (Serre Bound)

$$|N_1 - (q+1)| \leq g\lfloor 2\sqrt{q} \rfloor. \tag{9.28}$$

Proof. For $g = 0$, equality holds in (9.28). So $g \geq 1$ is assumed and the L-polynomial (9.7) is considered. Its roots $\omega_1, \ldots, \gamma_{2g}$ are algebraic integers, that is, they belong to the subring $\mathbf{A}$ of $\mathbf{C}$ consisting of all complex numbers which are roots of polynomials with integer coefficients. All integers are in $\mathbf{A}$, and

$$\mathbf{A} \cap \mathbf{Q} = \mathbf{Z}. \tag{9.29}$$

By the Hasse–Weil Theorem 9.19, $|\omega_i| = \sqrt{q}$ for $i = 1, \ldots, 2g$. Also, by the last statement in Proposition 9.9, these roots may be ordered so that $\omega_i\,\omega_{g+i} = q$ for $i = 1, \ldots, g$. So, with $\overline{z}$ the complex conjugate of z,

$$\overline{\omega_i} = \omega_{g+i} = q/\omega_i, \qquad \text{for} \quad 1 \leq i \leq g.$$

Let

$$\alpha_i = \omega_i + \overline{\omega_i} + \lfloor 2\sqrt{q} \rfloor + 1;$$
$$\beta_i = -(\omega_i + \overline{\omega_i}) + \lfloor 2\sqrt{q} \rfloor + 1.$$

Then α_i, β_i are real algebraic integers. Also, since $|\omega_i| = \sqrt{q}$,

$$\alpha_i > 0, \quad \beta_i > 0, \qquad \text{for} \quad i = 1, \ldots, g. \tag{9.30}$$

The monic minimal polynomial of the field extension $\mathbf{Q}(\omega_1, \ldots, \omega_g)$ of $\mathbf{Q}$ is

$$L^{\perp}(t) = \prod_{i=1}^{2g} (t - \omega_i),$$

whose roots are the reciprocals of the roots of $L(t)$. Since $L(t) \in \mathbf{Z}[t]$, the same holds true for $L^{\perp}(t)$.

Let σ be any embedding of $\mathbf{Q}(\omega_1, \ldots, \omega_g)$ in $\mathbf{C}$. If $\sigma(\omega_i) = \omega_j$, then

$$\sigma(\overline{\omega_i}) = \sigma(q/\omega_i) = q/\sigma(\omega_i) = \overline{\sigma(\omega_i)} = \overline{\omega_j}.$$

Therefore σ induces permutations on both the sets $\{\alpha_1, \ldots, \alpha_g\}$ and $\{\beta_1, \ldots, \beta_g\}$. Hence both algebraic integers

$$\alpha = \prod_{i=1}^{g} \alpha_i, \qquad \beta = \prod_{i=1}^{g} \beta_i$$

are left invariant by all embeddings of $\mathbf{Q}(\omega_1, \ldots, \omega_g)$ into $\mathbf{C}$. This implies that $\alpha, \beta \in \mathbf{Q} \cap \mathbf{A}$. Hence, by (9.29), $\alpha, \beta \in \mathbf{Z}$. From (9.30), $\alpha \geq 1$ and $\beta \geq 1$.

Since the arithmetic mean of positive real numbers is greater than or equal to the geometric mean of the same numbers,

$$g^{-1}\sum_{i=1}^{g} \alpha_i \geq \left(\prod_{i=1}^{g} \alpha_i\right)^{1/g} \geq 1,$$

whence

$$g \leq \sum_{i=1}^{g} (\omega_i + \overline{\omega_i}) + g\lfloor 2\sqrt{q} \rfloor + g = \sum_{i=1}^{2g} \omega_i + g\lfloor 2\sqrt{q} \rfloor + g.$$

From (9.10), $\sum_{i=1}^{2g} \omega_i = (q+1) - N_1$. Hence

$$N_1 \leq q + 1 + g\lfloor 2\sqrt{q} \rfloor.$$

Similarly, the inequality

$$g^{-1}\sum_{i=1}^{g} \beta_i \geq \left(\prod_{i=1}^{g} \beta_i\right)^{1/g} \geq 1$$

implies that

$$N_1 \geq q + 1 - g\lfloor 2\sqrt{q} \rfloor.$$

□

THEOREM 9.28 *In the Serre Bound, equality is attained if and only if the L-polynomial of $\mathcal{X}$ is*

$$L_q(t) = (1 \pm \lfloor 2\sqrt{q} \rfloor t + qt^2)^g, \tag{9.31}$$

with $+$ or $-$ according as the bound is upper or lower.

Proof. The arithmetic mean of positive real numbers x_i equals to their geometric mean if and only if all x_i are equal. From the proof of Theorem 9.27, if equality holds in the upper bound (9.28) then $\alpha_1 = \cdots = \alpha_g$ and hence $\omega_1 = \cdots = \omega_g$. Let $\omega = \omega_1$. Then

$$g(\omega + q/\omega) = q + 1 - N_1 = -\lfloor 2\sqrt{q} \rfloor,$$

whence (9.31) follows with a plus sign.

For the converse, (9.31) with $+$ implies that

$$-\lfloor 2\sqrt{q} \rfloor = \omega_1 + q/\omega_1 = \cdots = \omega_g + q/\omega_g.$$

From (9.10), $N_1 - (q+1) = g\lfloor 2\sqrt{q} \rfloor$.

A similar argument using β shows the assertion for the lower bound in (9.28) with a minus sign. □

REMARK 9.29 The Klein quartic over $\mathbf{F}_8$, as in Example 9.14, shows that the upper Serre Bound is sharp. For other curves attaining the Serre Bound, see Section 12.5.

Now another general formula for N_1 is given. It may appear complicated but can provide good estimates for appropriate choices of the polynomials.

THEOREM 9.30 (Serre's Explicit Formula) *Take any sequence $c_1, c_2, \ldots$ of real numbers and, for any $m \leq 1$, let*

$$\lambda_m(t) = \textstyle\sum_{i=1}^m c_i t^i, \qquad f_m(t) = 1 + \lambda_m(t) + \lambda_m(t^{-1}). \tag{9.32}$$

Then

$$\lambda_m\left(\frac{1}{\sqrt{q}}\right) N_1 = \lambda_m(\sqrt{q}) + \lambda_m\left(\frac{1}{\sqrt{q}}\right) + g - \sum_{i=1}^{g} f_m(\alpha_i) - \sum_{n=1}^{m}(N_n - N_1)c_n \frac{1}{\sqrt{q^n}}. \tag{9.33}$$

Proof. Let

$$\alpha_i = \omega_i/\sqrt{q}, \qquad i = 1, \ldots, 2g, \tag{9.34}$$

and observe that $|\alpha_i| = 1$ by the Hasse–Weil Theorem 9.19. As in the proof of Theorem 9.27, these elements may be ordered in such a way that

$$\alpha_{g+i} = \overline{\alpha_i} = \alpha_i^{-1}, \qquad \text{for} \quad i = 1, \ldots, g.$$

From Theorem 9.10,

$$\frac{N_n}{\sqrt{q^n}} = \sqrt{q^n} + \frac{1}{\sqrt{q^n}} - \sum_{i=1}^{g}\left(\alpha_i^n + \frac{1}{\alpha_i^n}\right). \tag{9.35}$$

Note that $f_m(t) \in \mathbf{R}$ for all $t \in \mathbf{C}$ with $|t| = 1$. Multiply (9.35) by c_n and write the result as follows:

$$c_n \frac{N_1}{\sqrt{q^n}} = c_n\sqrt{q^n} + c_n\frac{1}{\sqrt{q^n}} - \sum_{i=1}^{g} c_n\left(\alpha_i^n + \frac{1}{\alpha_i^n}\right) - (N_n - N_1)c_n\frac{1}{\sqrt{q^n}}. \tag{9.36}$$

Summing the equations (9.36) for $n = 1, \ldots, m$ gives the result. □

A useful corollary is the following result.

PROPOSITION 9.31 (i) *If $c_1, \ldots, c_m$ is a sequence of non-negative real numbers not all zero, and the polynomial $f_m(t)$ defined in (9.32) is non-negative for all $t \in \mathbf{C}$ with $|t| = 1$, then*

$$N_1 \leq \frac{g}{\lambda_m(\sqrt{q^{-1}})} + \frac{\lambda_m(\sqrt{q})}{\lambda_m(\sqrt{q^{-1}})} + 1, \tag{9.37}$$

with $\lambda_m(t)$ as defined in (9.32).

(ii) *Equality holds in (9.37) if and only if $N_n = N_1$, for $n = 1, \ldots, m$, and $\sum_{i=1}^{g} f_m(\alpha_i) = 0$.*

Proof. Note that $\lambda_m(\sqrt{q^{-1}}) \neq 0$, as the sequence $c_1, \ldots, c_m$ contains a non-zero element. Since $N_1 \leq N_n$, Serre's Explicit Formula implies that

$$\lambda_m(\sqrt{q^{-1}})N_1 \leq \lambda_m(\sqrt{q}) + \lambda_m(\sqrt{q^{-1}}) + g.$$

Division by $\lambda_m(\sqrt{q^{-1}})$ gives the result. □

Before showing two important consequences of Proposition 9.31, an equation similar to (9.33) is stated.

With the notation of (9.9) and (9.34), let

$$\alpha_j = \cos\varphi_j + i\sin\varphi_j$$

for $j = 1, \dots, 2g$. Then (9.10) can also be written as follows:

$$N_n = q^n + 1 - \sqrt{q^n}\,\textstyle\sum_{j=1}^{g} 2\cos n\alpha_j. \tag{9.38}$$

If B_d denotes the number of $\mathbf{F}_q$-rational closed points of $\mathcal{X}$ of degree d, then

$$N_n = \textstyle\sum_{d|n} d\,B_d.$$

For a sequence $c_1, c_2, \dots$ of real numbers, define the trigonometric series

$$f(\alpha) = 1 + 2\textstyle\sum_{j=1}^{\infty} c_j \cos j\alpha,$$

and, for every positive integer n, the power series

$$\Psi_n(t) = \textstyle\sum_{j=1}^{\infty} c_{jn} t^{jn}.$$

From this,

$$\textstyle\sum_{j=1}^{g} f(\alpha_j) + \sum_{d=1}^{\infty} d\,B_d\,\Psi_d(t) = g + \Psi_1\,(1/\sqrt{q})\,\Psi_1(\sqrt{q}), \tag{9.39}$$

which is *Weil's Explicit Formula.*

Now two cases are examined, related to the maximal curves having the Suzuki and Ree simple groups as their automorphism groups.

PROPOSITION 9.32 *Let $q = 2q_0^2$ with $q_0 = 2^s$ and $s \geq 1$.*

(i) *If $g = q_0(q-1)$, then $N_1 \leq q^2 + 1$.*

(ii) *If $N_1 = q^2 + 1$, then*

$$Z(\mathcal{X}, t) = \frac{(1 + 2q_0 t + qt^2)^g}{(1-t)(1-qt)};$$
$$N_n = 1 + q^n - g\sqrt{q^n}\kappa,$$

where

$$\kappa = \begin{cases} -2 & \textit{if}\ n \equiv 4 \pmod 8, \\ -\sqrt{2} & \textit{if}\ n \equiv 1,7 \pmod 8, \\ 0 & \textit{if}\ n \equiv 2 \pmod 4, \\ \sqrt{2} & \textit{if}\ n \equiv 3,5 \pmod 8, \\ 2 & \textit{if}\ n \equiv 0 \pmod 8. \end{cases}$$

Proof. Let $m = 2,\ c_1 = \frac{1}{2}\sqrt{2},\ c_2 = \frac{1}{4}$. Then

$$\lambda_2(t) = \tfrac{1}{\sqrt{2}}t + \tfrac{1}{4}qt^2.$$

If $|t| = 1$, write $t = \cos\varphi + i\sin\varphi$, and calculate $f_2(t)$. The result is that

$$f_2(t) = (\tfrac{1}{\sqrt{2}} + \cos\varphi)^2,$$

and hence $f_2(t) \geq 0$ for $|t| = 1$. Applying (9.37) gives $N_1 \leq q^2 + 1$.

Suppose that equality holds and write $\alpha_j = \cos\phi_j + i\sin\phi_j$. Then

$$\sum_{j=1}^{g} f_2(\alpha_j) = \sum_{j=1}^{g} \left(\tfrac{1}{\sqrt{2}} + \cos\phi_j\right)^2 = 0,$$

which implies that $\cos\phi_1 = \cdots = \cos\phi_g = -\frac{1}{\sqrt{2}}$. Therefore

$$\alpha_1 = \cdots = \alpha_g = -\tfrac{1}{\sqrt{2}} + i\tfrac{1}{\sqrt{2}}, \qquad \alpha_{g+1} = \cdots = \alpha_{2g} = -\tfrac{1}{\sqrt{2}} - i\tfrac{1}{\sqrt{2}}.$$

Direct computation shows the remaining parts. □

REMARK 9.33 The DLS curve satisfies the hypothesis of Proposition 9.32.

PROPOSITION 9.34 *Let $q = 3q_0^2$ with $q_0 = 3^s$ and $s \geq 1$.*

(i) *If*

$$g = \tfrac{3}{2}\, q_0(q-1)(q+q_0+1),$$

then $N_1 \leq q^3 + 1$.

(ii) *If $N_1 = q^3 + 1$, then*

$$Z(\mathcal{X}, t) = \frac{(1 + 3q_0t + qt^2)^{q_0(q^2-1)}(1 + qt^2)^{q_0(q-1)(q+3q_0+1)/2}}{(1-t)(1-qt)},$$

$$N_n = 1 + q^n - q_0\sqrt{q}(q-1)\{(q+3q_0+1)\cos\tfrac{1}{2}\pi n + 2(1+q)\cos\tfrac{5}{6}\pi n\}.$$

Proof. The same argument in the proof of Proposition 9.32 is used, but with a different choice of the sequence $c_1, c_2, \ldots$, namely,

$$c_1 = \frac{\sqrt{3}}{2},\ c_2 = \frac{7}{12},\ c_3 = \frac{\sqrt{3}}{6},\ c_4 = \frac{1}{12}.$$

The calculations give the following result: for $t = \cos\phi + i\sin\phi$,

$$f_2(t) = \tfrac{1}{3}(\sqrt{3} + 2\cos\varphi)^2\cos^2\varphi.$$

This and $\sum_{j=1}^{g} f_2(\alpha_j) = 0$ imply that either $\alpha_j = \pm\frac{1}{2}i$ or $\alpha_j = -\frac{\sqrt{3}}{2} \pm \frac{1}{2}i$. Therefore $\omega_j + \overline{\omega_j}$ can only assume two values, namely, 0 and $-\sqrt{3q}$.

Now let k denote the number of j with $1 \leq j \leq g$ such that $\omega_j + \overline{\omega_j} = -\sqrt{3q}$. Then (9.10) applied to N_1 gives $q^3 - q = m_2\sqrt{3q}$, whence $k = q_0(q^2-1)$. This also shows that $\omega_j + \overline{\omega_j} = 0$ occurs $g - k = \frac{1}{2}q_0(q-1)(q+3q_0+1)$ times. Therefore, up to re-labelling the indices,

$$\omega_1 = \cdots = \omega_k = \tfrac{1}{2}\sqrt{q}\,i, \qquad \omega_{g+1} = \cdots = \omega_{g+k} = -\tfrac{1}{2}\sqrt{q}\,i,$$

$$\omega_{k+1} = \cdots \omega_g = (\tfrac{-\sqrt{3}}{2} + \tfrac{1}{2}i)\sqrt{q}, \qquad \omega_{g+k+1} = \cdots = \omega_{2g} = (\tfrac{-\sqrt{3}}{2} - \tfrac{1}{2}i)\sqrt{q}.$$

From this and equation (9.10), $Z(\mathcal{X}, t)$ and N_n can be obtained. □

REMARK 9.35 The DLR curve satisfies the hypotheses of Proposition 9.32.

9.4 ASYMPTOTIC BOUNDS

Proposition 9.31 is also useful for investigating the asymptotic behaviour of N_1 as the genus goes to infinity.

DEFINITION 9.36 (i) Let $N_g(q) = \max N_1$, taken over all irreducible non-singular algebraic curves of genus g defined over $\mathbf{F}_q$.

(ii) Let $A^+(q) = \limsup_{g\to\infty} N_g(q)/g$.

(iii) Let $A^-(q) = \liminf_{g\to\infty} N_g(q)/g$.

The Hasse–Weil Bound implies that $A^+(q) \leq 2\sqrt{q}$. This bound was improved using the Hasse–Weil Bound for N_1 and N_2 and the inequality $N_1 \leq N_2$.

THEOREM 9.37 (Ihara Bound)

$$A^+(q) \leq \tfrac{1}{2}(\sqrt{8q+1}-1).$$

Refining the ideas in this, a better bound can be obtained.

THEOREM 9.38 (Drinfeld–Vlăduţ Bound)

$$A^+(q) \leq \sqrt{q}-1.$$

Proof. For a fixed positive integer m, consider the sequence $c_1, \ldots, c_m$, where

$$c_n = 1 - \frac{n}{m}, \qquad \text{for} \quad n = 1, \ldots, m.$$

Then the functions in (9.32) are as follows:

$$\lambda_m(t) = \sum_{n=1}^{m}\left(1-\frac{n}{m}\right)t^n = \frac{t}{(1-t)^2}\left(\frac{t^m-1}{m}+1-t\right); \tag{9.40}$$

$$f_m(t) = 1 + \lambda_m(t) + \lambda_m(t^{-1}) = \frac{2-(t^m+t^{-m})}{m(t-1)(t^{-1}-1)}. \tag{9.41}$$

Since $t^{-1} = \bar{t}$ for $|t| = 1$, equation (9.41) implies that $f_m(t) \geq 0$ for all $t \in \mathbf{C}$. From Proposition 9.31 applied to the sequence $c_1, \ldots, c_m$,

$$\frac{N_1}{g} \leq \frac{1}{\lambda_m(\sqrt{q^{-1}})} + \frac{1}{g}\left(1+\frac{\lambda_m(\sqrt{q})}{\lambda_m(\sqrt{q^{-1}})}\right). \tag{9.42}$$

From (9.40),

$$\lim_{m\to\infty}\lambda_m(\sqrt{q^{-1}}) = \frac{\sqrt{q^{-1}}}{(1-\sqrt{q^{-1}})^2}(1-\sqrt{q^{-1}}) = \frac{1}{\sqrt{q}-1};$$

so, for any $\epsilon > 0$, there exists m_0 such that

$$\lambda_{m_0}(q^{-1/2}) < (\sqrt{q}-1)^{-1} + \tfrac{1}{2}\epsilon.$$

Now choosing g_0 such that

$$\frac{1}{g_0}\left(1+\frac{\lambda_m(\sqrt{q})}{\lambda_m(\sqrt{q^{-1}})}\right) < \tfrac{1}{2}\,\epsilon,$$

(9.42) implies that

$$N_1/g < \sqrt{q} - 1 + \epsilon$$

for every $g \geq g_0$. Therefore $A^+(q) \leq \sqrt{q} - 1$. □

On the other hand,

$$A^+(q) \geq \sqrt{q} - 1 \text{ for } q \text{ square.} \tag{9.43}$$

Therefore the Drinfeld–Vlăduţ Bound is sharp for q square.

DEFINITION 9.39 (i) A *tower* $\mathcal{T}$ over $\mathbf{F}_q$, or an $\mathbf{F}_q$-tower, is an infinite sequence of irreducible non-singular curves together with surjective and separable maps defined over $\mathbf{F}_q$,

$$\cdots \to \mathcal{X}_{k+1} \to \mathcal{X}_k \to \mathcal{X}_2 \to \mathcal{X}_1,$$

such that

$$\lim_{k\to\infty} g(\mathcal{X}_k) = \infty,$$

where $g(\mathcal{X}_k)$ is the genus of $\mathcal{X}_k$.

(ii) The *limit of the tower* $\mathcal{T}$ is

$$\lambda(\mathcal{T}) = \lim_{k\to\infty} \frac{|\mathcal{X}_k(\mathbf{F}_q)|}{g(\mathcal{X}_k)}.$$

(iii) A tower is *asymptotically good* if $\lambda(\mathcal{T}) > 0$, and is *recursive* if it uses the same equation to construct all curves in the tower.

Note that $A^+(q) \geq \lambda(\mathcal{T})$ for any $\mathbf{F}_q$-tower $\mathcal{T}$. Therefore, to show that equality holds over fields of square order in the Drinfeld–Vlăduţ Bound, it suffices to construct an $\mathbf{F}_{q^2}$-tower $\mathcal{T}$ satisfying $\lambda(\mathcal{T}) = q - 1$.

As illustrative examples, two such $\mathbf{F}_{q^2}$-towers are presented by explicit algebraic equations.

EXAMPLE 9.40 $\lambda(\mathcal{T}) = q - 1$ for the $\mathbf{F}_{q^2}$-tower $\mathcal{T}$ defined as follows:

(i) $\mathcal{X}_1$ is the line $\mathbf{v}(X_1)$, that is, an irreducible non-singular model of the rational function field $K(x_1)$;

(ii) $\mathcal{X}_2$ is the non-singular plane curve $\mathbf{v}(Y^2 + Y - X^{q+1})$, that is, a non-singular model over $\mathbf{F}_q$ of the function field $K(x_1, z_2)$ with $z_2^q + z_2 = x_1^q$;

(iii) $\mathcal{X}_3$ is a non-singular model over $\mathbf{F}_q$ of the function field $K(x_1, x_2, z_2, z_3)$ with

$$z_2^q + z_2 = x_1^q, \quad z_3^q + z_3 = x_2^{q+1}, \quad x_2 = \frac{z_2}{x_1};$$

(iv) $\mathcal{X}_4$ is a non-singular model over $\mathbf{F}_q$ of the field $K(x_1, x_2, x_3, z_2, z_3, z_4)$ with

$$z_2^q + z_2 = x_1^{q+1}, \quad z_3^q + z_3 = x_2^{q+1}, \quad z_4^q + z_4 = x_3^{q+1}, \quad x_3 = \frac{z_3}{x_2};$$

(v) $\mathcal{X}_5, \mathcal{X}_6, \ldots$ continue in this way.

EXAMPLE 9.41 The limit $\lambda(\mathcal{T}) = q - 1$ for the $\mathbf{F}_{q^2}$-tower $\mathcal{T}$ defined as follows:

(i) $\mathcal{X}_1$ is the line $\mathbf{v}(X_1)$, that is, an irreducible non-singular model of the rational function field $K(x_1)$;

(ii) $\mathcal{X}_2$ is a non-singular model over $\mathbf{F}_{q^2}$ of the field $K(x_1, x_2)$ with

$$x_2^q + x_2 = x_1^q/(x_1^{q-1} + 1);$$

(iii) $\mathcal{X}_3$ is a non-singular model over $\mathbf{F}_{q^2}$ of the field $K(x_1, x_2, x_3, x_4)$ with

$$x_2^q + x_2 = \frac{x_1^q}{x_1^{q-1} + 1}, \quad x_3^q + x_3 = \frac{x_2^q}{x_2^{q-1} + 1}, \quad x_4^q + x_4 = \frac{x_3^q}{x_3^{q-1} + 1};$$

(iv) the curves $\mathcal{X}_4, \mathcal{X}_5, \ldots$ continue in this way.

The map $\mathcal{X}_{k+1} \to \mathcal{X}_k$, that is, the rational transformation

$$K(\mathcal{X}_{k+1}) = K(x_1, \ldots, x_k, x_{k+1}) \to K(\mathcal{X}_k) = K(x_1, \ldots, x_k)$$

is given by

$$(x_1, \ldots, x_k, x_{k+1}) \mapsto (x_1, \ldots, x_k).$$

The tower is recursively defined by the following equation:

$$y^q + y = x^q/(x^{q-1} + 1).$$

THEOREM 9.42 *For arbitrary q, there exists a positive constant c such that*

$$A^+(q) > c \log q. \tag{9.44}$$

As a consequence, $A^+(q) > 0$ for all q.

EXAMPLE 9.43 Let $q > p$, and let $m = (q - 1)/(p - 1)$. Then

$$A^+(q) \geq \lambda(\mathcal{T}) \geq \frac{2}{q - 2}$$

for the $\mathbf{F}_q$-tower $\mathcal{T}$ defined as follows:

(i) $\mathcal{X}_1$ is the line $\mathbf{v}(X_1)$, that is, an irreducible non-singular model of the rational function field $K(x_1)$;

(ii) $\mathcal{X}_2$ is the non-singular plane curve $\mathbf{v}(Y^m + (X + 1)^m - 1)$, that is, a non-singular model over $\mathbf{F}_q$ of the field $K(x_1, x_2)$ with $x_2^m + (x_1 + 1)^m = 1$;

(iii) $\mathcal{X}_3$ is a non-singular model over $\mathbf{F}_q$ of the field $K(x_1, x_2, x_3)$ with

$$x_3^m + (x_2 + 1)^m = 1, \qquad x_2^m + (x_1 + 1)^m = 1;$$

(iv) the curves $\mathcal{X}_4, \mathcal{X}_5, \ldots$ continue in this way.

REMARK 9.44 (i) For small q, the following lower bounds are known:

$$A^+(2) \geq \tfrac{2}{9}; \quad A^+(3) \geq \tfrac{1}{3}; \quad A^+(5) \geq \tfrac{1}{2}.$$

(ii)
$$A^+(q^3) \geq \frac{2(q^2-1)}{q+2}.$$

(iii) If l is a prime number such that $q > 4l + 1$ and $q \equiv 1 \pmod{l}$, then

$$A^+(q^l) \geq (\sqrt{l(q-1)} - 2l)/(l-1).$$

The significance of $A^-(q)$ is that for it to be greater than some constant c it is necessary that, for every sufficiently large g, there is a curve of genus g defined over $\mathbf{F}_q$ with at least cg points.

THEOREM 9.45 *For every q, the constant $A^-(q)$ is positive.*

THEOREM 9.46 *There exists a constant d such that, for all q,*

$$A^-(q) \geq d \log q.$$

For q square, this can be improved.

THEOREM 9.47

$$A^-(q) \geq \begin{cases} \dfrac{\sqrt{q}-1}{2 + (\log 2)/(\log q)}, & \text{if } q \text{ is an even square;} \\[2ex] \dfrac{\sqrt{q}-1}{2 + (\log 4)/(\log q)}, & \text{if } q \text{ is an odd square.} \end{cases}$$

9.5 OTHER ESTIMATES

Ad hoc methods can produce estimates on N_1 that are better than both the Hasse–Weil and the Stöhr–Voloch Bounds in some special cases. Two such methods for producing upper bounds on N_1 derive from the gonality of the curve $\mathcal{X}$ and the first non-gap at an $\mathbf{F}_q$-rational point.

For any $\xi \in \mathbf{F}_q(\mathcal{X})$, the $\mathbf{F}_q$-rational transformation $x' = x$, $y' = 0$ defines an $\mathbf{F}_q$-rational covering of degree $d = [K(\mathcal{X}) : K(x)]$.

PROPOSITION 9.48 *For an $\mathbf{F}_q$-rational covering of degree d,*

$$N_1 \leq d(q+1). \tag{9.45}$$

Proof. The projective line ℓ, which is a non-singular model of $K(x)$, has $q+1$ rational points. If P is an $\mathbf{F}_q$-rational point of $\mathcal{X}$ then the point of ℓ lying under P is an $\mathbf{F}_q$-rational point of ℓ. Although the points of $\mathcal{X}$ lying over $Q' \in \ell$ need not be $\mathbf{F}_q$-rational, the bound is obtained by counting the points of $\mathcal{X}$ lying over the $\mathbf{F}_q$-rational points of ℓ. □

DEFINITION 9.49 (i) The *Lüroth semigroup* S of an irreducible curve $\mathcal{F}$ is the additive integer semigroup containing all integers n for which $K(\mathcal{F})$ has a base-point-free linear series of degree n.

(i) The *gonality* of $\mathcal{F}$ is the least positive integer t such that $\mathcal{F}$ has a linear series of degree t and dimension 1, necessarily without base point. Equivalently, the gonality is the smallest degree of a non-trivial covering of a rational subfield of $K(\mathcal{F})$ by $K(\mathcal{F})$.

(iii) If $\mathcal{F}$ is defined over $\mathbf{F}_q$, the $\mathbf{F}_q$-*gonality* of $\mathcal{F}$ is the least positive integer $t_q(\mathcal{F})$ such that $\mathcal{F}$ has an $\mathbf{F}_q$-rational linear series of order $t_q(\mathcal{F})$ and dimension 1.

REMARK 9.50 Knowledge of the Lüroth semigroup gives worthwhile information about the curve; see Section 7.10.

(i) The gonality of $\mathcal{X}$ is 1 if and only if $\mathcal{X}$ is rational.

(ii) The gonality of $\mathcal{X}$ is 2 if and only if $\mathcal{X}$ is either elliptic or hyperelliptic.

(iii) If $\mathcal{F}$ is birationally equivalent to an irreducible plane curve of degree m, then the gonality is $m-1$.

Proposition 9.48 can be stated in terms of $\mathbf{F}_q$-gonality.

COROLLARY 9.51 $t_q(\mathcal{X}) \geq N_1/(q+1)$.

On the other hand, the following results hold.

PROPOSITION 9.52 (i) $t_q(\mathcal{X}) \leq g+1$.

(ii) *If* $N_1 > 0$ *and* $\deg E \geq t_q(\mathcal{X})-1$, *then* $\ell(E) \leq \deg E + 2 - t_q(\mathcal{X})$.

The result below shows that, as long as $N_1 > 0$, the parameter d may be chosen so that $d \leq g$ for $g \geq 2$, and this choice can be made in such a way that the number $d(P'_\infty)$ of $\mathbf{F}_q$-rational points lying over the infinite point P'_∞ of ℓ is very small.

LEMMA 9.53 *Let $\mathcal{X}$ have genus $g \geq 2$. If $\mathcal{X}$ has an $\mathbf{F}_q$-rational point P, then there exists $x \in \mathbf{F}_q(\mathcal{X})$ for which $[K(\mathcal{X}) : K(x)] \leq g$ and $d(P'_\infty) \leq 2$.*

Proof. From the Riemann–Roch Theorem 6.61, there is a canonical divisor E of $\mathcal{X}$ such that $E \succ (g-1)P$. By Remark 8.27, take E to be $\mathbf{F}_q$-rational.

Write $E = (g-1)P + A$ and choose a point $Q \in \operatorname{Supp} A$. Since A is an $\mathbf{F}_q$-rational divisor and D is the closed point $D = Q + \Phi(Q) + \cdots + \Phi^{k-1}(Q)$ of degree k, then $A \succ D$.

Let $B = (g-d)P + D$. Since $E \succ B$, the index of speciality $i(B)$ is positive. From the Riemann–Roch Theorem 6.61, $\dim B \geq 1$. Since B is an $\mathbf{F}_q$-rational divisor, there exists $x \in \mathbf{F}_q(\mathcal{X})$ such that $\operatorname{div}(x)_\infty = \deg B = g$; this proves the first assertion.

Now P'_∞ is the point of ℓ lying under P. Therefore $d(P'_\infty) \leq 2$, and equality holds if and only if $k=1$, that is, $Q \in \mathcal{X}(\mathbf{F}_q)$. □

As a corollary of Lemma 9.53, the following upper bound on N_1 is obtained.

PROPOSITION 9.54 *If $g \geq 2$, the following hold:*

(i) $N_1 \leq qd + 2$;

(ii) *if* n *is a non-gap at an* $\mathbf{F}_q$*-rational point of* $\mathcal{X}$, *then* $N_1 \leq qn + 1$.

9.6 COUNTING POINTS ON A PLANE CURVE

If $\mathcal{F}$ is a non-singular plane curve of degree n defined over $\mathbf{F}_q$, then $\mathcal{X}$ can be identified with $\mathcal{F}$, and the Hasse–Weil Bound (9.14) reads as follows:

$$q + 1 - (n-1)(n-2)\sqrt{q} \leq S_q \leq q + 1 + (n-1)(n-2)\sqrt{q}, \tag{9.46}$$

where n is the degree of $\mathcal{F}$ and $S_q = N_1$ is the number of all $\mathbf{F}_q$-rational points of $\mathcal{F}$. Now (9.46) is extended to irreducible singular plane curves defined over $\mathbf{F}_q$.

If a curve is not identified with its non-singular model, then there is some ambiguity in the definition of an $\mathbf{F}_q$-rational point. For a non-singular plane curve $\mathcal{F}$ defined over $\mathbf{F}_q$, there is a one-to-one correspondence between the $\mathbf{F}_q$-rational points of $\mathcal{F}$ and the $\mathbf{F}_q$-rational places of the associated function field $K(\mathcal{F})$; see Theorem 8.18. If $\mathcal{F}$ is singular and $\mathcal{X}$ is a non-singular model of $\mathcal{F}$, defined over $\mathbf{F}_q$ as well, then $\mathcal{F}$ and $\mathcal{X}$ have the same function field and hence they have the same number N_1 of $\mathbf{F}_q$-rational points, but N_1 is, in general, different from the number R_q of points of $\mathcal{F}$ which lie in $\mathrm{PG}(2, q)$.

To compare R_q with $S_q = N_1$, certain other parameters for $\mathcal{F}$ must be defined, some of which have appeared previously:

R_q = number of points $P \in \mathcal{F}$ that lie in $\mathrm{PG}(2, q)$;

$S_q = N_1$ = number of $\mathbf{F}_q$-rational points of $\mathcal{F}$;

R_q^* = number of points $P \in \mathrm{PG}(2, q)$ which are centres of $\mathbf{F}_q$-rational branches;

$\widetilde{R}_q$ = number of points $P \in \mathcal{F}$ in $\mathrm{PG}(2, q)$ with each r-fold point counted with multiplicity r;

B_q = number of branches of $\mathcal{F}$ centred at points of $\mathrm{PG}(2, q)$;

E_q = number of singular points of $\mathcal{F}$;

b_P = number of $\mathbf{F}_q$-rational branches of $\mathcal{F}$ with centre at $P \in \mathrm{PG}(2, q)$;

c_P = number of all branches of $\mathcal{F}$ with centre at $P \in \mathrm{PG}(2, q)$;

m_P = multiplicity of a point $P \in \mathcal{F}$.

To see, in a simple case, the differences between $R_q, S_q, B_q, \widetilde{R}_q$, consider the three singular plane cubics, $\mathcal{F}_2, \mathcal{F}_1, \mathcal{F}_0$, with respectively two, one, and zero tangents over $\mathbf{F}_q$ at the singular point P; these are cubics with a node, a cusp, and an isolated double point at P. Then Table 9.1 is straightforward to verify.

From now on, $\mathcal{F}$ is any irreducible plane curve of degree n and genus g defined over $\mathbf{F}_q$. To prove the desired result

$$q + 1 - (n-1)(n-2) \leq R_q \leq q + 1 + (n-1)(n-2), \tag{9.47}$$

it must be shown that (9.46) holds true when N_1 is replaced by R_q.

Table 9.1 Numbers of points on singular cubics

	R_q	S_q	B_q	$\widetilde{R}_q$
$\mathcal{F}_2$	q	$q+1$	$q+1$	$q+1$
$\mathcal{F}_1$	$q+1$	$q+1$	$q+1$	$q+2$
$\mathcal{F}_0$	$q+2$	$q+1$	$q+3$	$q+3$

Let $\mathcal{F}(\mathbf{F}_q)'$ be the set of all points $P \in \mathrm{PG}(2,q)$ lying on $\mathcal{F}$. With this notation,

$$S_q = \textstyle\sum_{P\in\mathcal{F}(\mathbf{F}_q)'} b_P, \qquad R_q = |\mathcal{F}(\mathbf{F}_q)'|.$$

Also, $R_q^* \leq S_q$ and equality holds if and only if no two distinct $\mathbf{F}_q$-rational branches of $\mathcal{F}$ have the same centre.

The starting point are upper and lower bounds for $S_q - R_q$.

LEMMA 9.55 *An upper bound for $S_q - R_q$ is the following:*

$$S_q - R_q \leq \tfrac{1}{2}(n-1)(n-2) - g. \tag{9.48}$$

Proof.
$$\begin{aligned} S_q - R_q &= \textstyle\sum_{P\in\mathcal{F}(\mathbf{F}_q)'} (b_P - 1) \leq \sum_{P\in\mathcal{F}(\mathbf{F}_q)'} (c_P - 1) \\ &\leq \textstyle\sum_{P\in\mathcal{F}(\mathbf{F}_q)'} (m_P - 1) \leq \sum_{P\in\mathcal{F}(\mathbf{F}_q)'} \frac{1}{2}m_P(m_P-1) \\ &\leq \textstyle\sum_{P\in\mathcal{F}} \frac{1}{2}m_P(m_P-1) \leq \frac{1}{2}(n-1)(n-2) - g. \end{aligned}$$

The first equality follows from the definitions. The first two inequalities follow from the fact that $b_P \leq c_P \leq m_P$; see (iii) of Theorem 4.36. The last inequality follows from Lemma 3.24 and Definition 5.55. □

LEMMA 9.56 *A lower bound for $S_q - R_q$ is the following:*

$$R_q - S_q \leq \tfrac{1}{2}(n-1)(n-2) - g. \tag{9.49}$$

Proof. Since $R_q^* \leq S_q$ it follows that $R_q - S_q \leq R_q - R_q^*$. On the other hand, $R_q - R_q^* \leq E_q$, as any point $P \in \mathcal{F}(\mathbf{F}_q)'$ for which $b_P = 0$ and $c_P > 0$ is a singular point of $\mathcal{F}$. Therefore

$$R_q - S_q \leq R_q - R_q^* \leq E_q \leq \tfrac{1}{2}(n-1)(n-2) - g. \tag{9.50}$$

□

THEOREM 9.57 *Let $\mathcal{F}$ be an irreducible plane curve defined over $\mathbf{F}_q$ of degree n and genus g. If R_q is the number of points of $\mathrm{PG}(2,q)$ lying on $\mathcal{F}$, then*

(i) $|R_q - (q+1)| \leq g\lfloor 2\sqrt{q}\rfloor + \frac{1}{2}(n-1)(n-2) - g$;

(ii) $|R_q - (q+1)| \leq \frac{1}{2}(n-1)(n-2)\lfloor 2\sqrt{q}\rfloor$; *(Serre Bound)*

(iii) $|R_q - (q+1)| \leq (n-1)(n-2)\sqrt{q}$. *(Hasse–Weil Bound)*

Proof. (i) The Serre Bound (9.28), together with Lemmas 9.55 and 9.56, implies the following:

$$\begin{aligned}|R_q-(q+1)| &\leq |R_q-S_q)|+|S_q-(q+1)|\\ &\leq \tfrac{1}{2}(n-1)(n-2)-g+g\lfloor 2\sqrt{q}\rfloor.\end{aligned}$$

(ii) From (i) and the inequality $g\leq \frac{1}{2}(n-1)(n-2)$ of Lemma 3.24,

$$\begin{aligned}|R_q-(q+1)| &\leq g(\lfloor 2\sqrt{q}\rfloor-1)+\tfrac{1}{2}(n-1)(n-2)\\ &\leq \tfrac{1}{2}(n-1)(n-2)(\lfloor 2\sqrt{q}\rfloor-1)+\tfrac{1}{2}(n-1)(n-2)\\ &= \tfrac{1}{2}(n-1)(n-2)\lfloor 2\sqrt{q}\rfloor.\end{aligned}$$

(iii) This is a consequence of (ii). □

Under some additional hypotheses, equality can be attained in the above bounds.

THEOREM 9.58 (i) *Equality occurs in Lemma 9.56 if and only if every singular point of $\mathcal{F}$ is an isolated double point lying in* PG$(2,q)$.

(ii) *Equality occurs in Lemma 9.55 if and only if every singular point of $\mathcal{F}$ is a node.*

Proof. (i) Assume that equality holds in (9.49). From (9.50), $S_q=R_q^*$. Hence every point $P\in\mathcal{F}(\mathbf{F}_q)'$ is the centre of at most one $\mathbf{F}_q$-rational branch of $\mathcal{F}$. Therefore R_q-S_q gives the number of singular points $P\in\mathcal{F}(\mathbf{F}_q)'$ which are not the centre of an $\mathbf{F}_q$-rational branch of $\mathcal{F}$.

Equality in (9.50) implies that $R_q-S_q=E_q$. Hence every singular point P of $\mathcal{F}$ lies in PG$(2,q)$, and P is not the centre of any $\mathbf{F}_q$-rational branch of $\mathcal{F}$. In particular, at least two branches of $\mathcal{F}$ are centred at P.

Now

$$\textstyle\sum_{P\in\mathcal{F}}\frac{1}{2}m_P(m_P-1)\geq g.$$

Since equality in (9.50) also implies that $E_q=\frac{1}{2}(n-1)(n-2)-g$, every singular point P of $\mathcal{F}$ is an ordinary double point. Therefore every singular point of $\mathcal{F}$ is an isolated double point lying in PG$(2,q)$.

To prove the converse, suppose that every singular point of $\mathcal{F}$ is an isolated double point. In particular, $\mathcal{F}$ has only ordinary singularities and hence, by Theorem 5.57, $E_q=\frac{1}{2}(n-1)(n-2)-g$. As $R_q-R_q^*$ counts the points in $\mathcal{F}(\mathbf{F}_q)'$ that are not the centre of any $\mathbf{F}_q$-rational branch, it follows that $E_q=R_q-R_q^*$. Thus only the non-singular points in $\mathcal{F}(\mathbf{F}_q)'$ are the centres of $\mathbf{F}_q$-rational branches of $\mathcal{F}$. Hence $S_q=R_q^*$. Therefore

$$R_q-S_q=R_q-R_q^*=E_q=\tfrac{1}{2}(n-1)(n-2)-g.$$

(ii) Assume that equality holds in (9.48). Then, from the proof of Lemma 9.55,

$$\textstyle\sum_{P\in\mathcal{F}(\mathbf{F}_q)'}(b_P-1)=\sum_{P\in\mathcal{F}(\mathbf{F}_q)'}(c_P-1).$$

Since $b_P\leq c_P$ for all $P\in\mathcal{F}(\mathbf{F}_q)'$, this implies that $b_P=c_P$; that is, every branch of $\mathcal{F}$ centred at a point in $\mathcal{F}(\mathbf{F}_q)'$ is $\mathbf{F}_q$-rational. Also, from the proof of Lemma 9.55,

$$\textstyle\sum_{P\in\mathcal{F}(\mathbf{F}_q)}\frac{1}{2}m_P(m_P-1)=\sum_{P\in\mathcal{F}}\frac{1}{2}m_P(m_P-1),$$

which shows that every singular point of $\mathcal{F}$ is in $\mathcal{F}(\mathbf{F}_q)'$. From the proof of Lemma 9.55, every singular point of $\mathcal{F}$ is a double point. Again from the proof of Lemma 9.55, $b_P = c_P = m_P$ and hence $v_P = 2$ for every singular point P of $\mathcal{F}$. This shows that each such point P is a node.

Conversely, if the singularities of $\mathcal{F}$ are nodes, then $E_q = \frac{1}{2}(n-1)(n-2) - g$, every singular point of $\mathcal{F}$ belongs to $\mathcal{F}(\mathbf{F}_q)'$, and $b_P = c_P = m_P$ for every point $P \in \mathcal{F}(\mathbf{F}_q)'$. Therefore $S_q - R_q = E_q$ and hence

$$S_q - R_q = \tfrac{1}{2}(n-1)(n-2) - g. \qquad \square$$

The following corollaries show that when equality occurs either in Lemma 9.55 or in Lemma 9.56, q cannot be too small compared to n.

COROLLARY 9.59 *If equality holds in Lemma 9.55, then*

(i) $(n-1)(n-2) - 2g \leq S_q$;

(ii) $(n-1)(n-2) - 2g \leq q + 1 + g\lfloor 2\sqrt{q} \rfloor$.

Proof. From Theorem 9.58 (ii), every singular point of $\mathcal{F}$ is a node, and hence $E_q = \frac{1}{2}(n-1)(n-2) - g$. Therefore $(n-1)(n-2) - 2g = 2E_q \leq S_q$. This, together with Theorem 9.57 (ii), gives part (ii). $\square$

COROLLARY 9.60 *Let $\mathcal{F}$ be an irreducible singular plane curve of degree $n \geq 3$ defined over $\mathbf{F}_q$ for which equality holds in Lemma 9.56.*

(i) $q \geq n - 2 + (S_q - 2g - 2)/(n-2)$.

(ii) *If* $g = 0$, *then* $q \geq n - 1$.

(iii) *If n is odd and* $n \geq 5$, *then* $q \geq n - 1 - 2g/(n-3)$.

Proof. (i) If P is any point in $\mathrm{PG}(2,q)$, the pencil Π_P of lines of $\mathrm{PG}(2,q)$ through P covers every other point in $\mathrm{PG}(2,q)$ just once.

Now suppose P is a singular point of $\mathcal{F}$. From Theorem 9.58(ii), P is an isolated double point of $\mathcal{F}$. Also, from the proof of Lemma 9.55, $S_q = R_q^*$.

For any line ℓ through P, Bézout's Theorem 3.14 becomes the following:

$$\textstyle\sum_{Q \in \ell \cap \mathcal{F}(\mathbf{F}_q)'} I(P, \ell \cap \mathcal{F}) \leq n.$$

Summing over all lines in the pencil Π_P, this gives the inequality

$$\textstyle\sum_{\ell \in \Pi_P} \sum_{Q \in \ell \cap \mathcal{F}(\mathbf{F}_q)'} I(P, \ell \cap \mathcal{F}) \leq n(q+1).$$

Counting this double sum in a different way,

$$\textstyle\sum_{\ell \in \Pi_P} \sum_{Q \in \ell \cap \mathcal{F}(\mathbf{F}_q)'} I(P, \ell \cap \mathcal{F}) = \sum_{\ell \in \Pi_P} I(P, \ell \cap \mathcal{F}) + \sum_{Q \in \mathrm{Sing}(\mathcal{F}) \setminus P} I((Q, \ell \cap \mathcal{F}) + \sum_{Q \in \mathcal{F}(\mathbf{F}_q)' \setminus \mathrm{Sing}(\mathcal{F})} I(Q, \ell \cap \mathcal{F}),$$

where $\mathrm{Sing}(\mathcal{F})$ is the set of all singular points of $\mathcal{F}$. Also, since no $\mathbf{F}_q$-rational branch of $\mathcal{F}$ is centred at a singular point of $\mathcal{F}$ and the singularities of $\mathcal{F}$ are double points,

$$\textstyle\sum_{\ell \in \Pi_P} I(P, \ell \cap \mathcal{F}) = 2(q+1),$$

$$\textstyle\sum_{Q \in \mathrm{Sing}(\mathcal{F}), Q \neq P} I((Q, \ell \cap \mathcal{F}) = 2(E_q - 1).$$

Also, since $S_q = R_q^*$,

$$\sum_{Q\in\mathcal{F}(\mathbf{F}_q)'\setminus\mathrm{Sing}(\mathcal{F})} I(Q,\ell\cap\mathcal{F}) = S_q.$$

Therefore

$$2(q+1)+2(E_q-1)+S_q \leq n(q+1).$$

As $2E_q=(n-1)(b-2)-2g$ when equality holds in Lemma 9.56, so (i) follows.

(ii) If $g=0$, then $S_q=q+1$ and (i) implies that

$$q \geq n-2+(q-1)/(n-2).$$

Since $q\geq 2$ and $n>2$, part (ii) follows.

(iii) Since $\mathcal{F}$ is an irreducible plane curve, Bézout's Theorem 3.14 implies that no line ℓ contains more than $\lfloor\frac{1}{2}n\rfloor$ singular points of $\mathcal{F}$. As before, fix a singular point P and consider the pencil Π_P of lines of $\mathrm{PG}(2,q)$ through P. Each line contains at most $\lfloor\frac{1}{2}n\rfloor-1$ singular points of $\mathcal{F}$ other than P. As already noted, $E_q=\frac{1}{2}(n-1)(n-2)-g$. Therefore the inequality

$$\left(\lfloor\tfrac{1}{2}n\rfloor-1\right)(q+1)+1\geq E_q=\tfrac{1}{2}(n-1(n-2)-g$$

follows.

If n is odd and $n\geq 5$, then $\lfloor\frac{1}{2}n\rfloor=\frac{1}{2}(n-1)$, which implies that

$$q\geq n-1-2g/(n-3).$$

Note that, if n is even and $n\geq 4$, then $\lfloor\frac{1}{2}n\rfloor=\frac{1}{2}n$, whence

$$q\geq n-2-(2g-2)/(n-2),$$

which is weaker than (i). □

Remark 9.61 The above results are sharp for curves of low degree and genus. For $g=2$ this is illustrated by two examples. Let

$$\begin{aligned} &f(X_0,X_1,X_2)\\ &\quad = X_0^2(X_1^2+2X_2^2)+X_0(X_1^3+2X_1X_2^2+X_2^3)+3X_1^3X_2+3X_1X_2^3,\\ &g(X_0,X_1,X_2)\\ &\quad = X_0^2(X_1^2-X_2^2)+X_0(X_1^2X_2-X_2^3)+(6X_1^4+6X_1^2X_2^2-4X_2^4). \end{aligned}$$

Then the parameters are given in Table 9.2.

Table 9.2 Two curves

Curve	q	n	g	R_q	S_q	Bound on S_q	Bound on R_q
$\mathcal{F}=\mathbf{v}(f)$	5	4	2	13	12	$0\leq S_q\leq 12$	$0\leq R_q\leq 13$
$\mathcal{F}=\mathbf{v}(g)$	13	4	2	1	2	$2\leq S_q\leq 28$	$1\leq R_q\leq 27$

The following Theorems 9.62 and 9.63 provide bounds on B_q.

THEOREM 9.62 *Let $\mathcal{F}$ be an irreducible plane curve of degree n and genus g defined over $\mathbf{F}_q$.*

(i) *If $\mathcal{F}$ is either classical, or non-classical but Frobenius classical, then*

$$B_q \leq \tfrac{1}{2}\{(2g-2)+(q+2)n\}, \tag{9.51}$$

except for classical but Frobenius non-classical curves with $\nu_1 = 2$, in which case

$$B_q \leq \tfrac{1}{2}\{2(2g-2)+(q+2)n\}. \tag{9.52}$$

(ii) *If $\mathcal{F}$ is Frobenius non-classical,*

$$B_q \leq \tfrac{1}{2}\{\nu_1(2g-2)+(q+2)n\}. \tag{9.53}$$

Proof. From (8.56)(i), $v_{\mathcal{P}}(S) \geq 2$ for every $\mathcal{P} \in \mathcal{F}(\mathbf{F}_q)$, that is, for every $\mathbf{F}_q$-rational branch γ of $\mathcal{F}$. By (8.14), it suffices to extend this result to every branch γ of $\mathcal{F}$ centred at a point in $\mathrm{PG}(2,q)$. This is done after stating a useful formula for $v_{\mathcal{P}}(S)$.

Let $Q=(x,y)$ be a generic point of $\mathcal{F}$. For any place $\mathcal{P}$ of $K(\mathcal{F})$, let $(0,j_1,j_2)$ the order sequence at $\mathcal{P}$ with respect to lines. Then $j_1 = \min\{\mathrm{ord}_{\mathcal{P}}\, x, \mathrm{ord}_{\mathcal{P}}\, y\}$. Let γ be the branch of $\mathcal{F}$ corresponding to $\mathcal{P}$. Without loss of generality, suppose that its centre A is affine. Assume that $A=(1,a,b)$ lies in $\mathrm{PG}(2,q)$, and denote by ℓ the tangent to γ. As $v_{\mathcal{P}}(S)$ is invariant under $\mathbf{F}_q$-linear transformations, A may be taken to be the origin, and it may be assumed that $\mathrm{ord}_{\mathcal{P}}\, x \leq \mathrm{ord}_{\mathcal{P}}\, y$, where $\mathcal{P}$ is the place of $K(\mathcal{F})$ arising from γ. Then

$$\ell = \mathbf{v}(m_{21}X - m_{11}Y),$$

with $m_{11}(\neq 0), m_{21} \in K$. Now introduce a new coordinate system $(\bar{X},\bar{Y})$ so that the tangent ℓ becomes the X-axis. This change of coordinates from (X,Y) to $(\bar{X},\bar{Y})$ is given by the linear substitutions

$$X = m_{11}\bar{X} + m_{12}\bar{Y}, \quad Y = m_{21}\bar{X} + m_{22}\bar{Y}.$$

Define $\bar{x},\bar{y}$ by the equations,

$$x = m_{11}\bar{x} + m_{12}\bar{y}, \quad y = m_{21}\bar{x} + m_{22}\bar{y}.$$

Then $\mathcal{F}(K) = K(\bar{x},\bar{y})$.

Choose a local parameter ζ at $\mathcal{P}$, and write ν for ν_1. Now

$$\begin{aligned} D_\zeta^{(\nu)}x &= m_{11}D_\zeta^{(\nu)}\bar{x} + m_{12}D_\zeta^{(\nu)}\bar{y},\\ D_\zeta^{(\nu)}y &= m_{21}D_\zeta^{(\nu)}\bar{x} + m_{22}D_\zeta^{(\nu)}\bar{y}, \end{aligned}$$

whence

$$\begin{aligned} (x-x^q)D_\zeta^{(\nu)}y &= (x-x^q)(m_{21}D_\zeta^{(\nu)}\bar{x} + m_{22}D_\zeta^{(\nu)}\bar{y}),\\ (y-y^q)\,D_\zeta^{(\nu)}x &= (y-y^q)(m_{11}D_\zeta^{(\nu)}\bar{x} + m_{12}D_\zeta^{(\nu)}\bar{y}). \end{aligned}$$

Therefore

$$\begin{aligned}
&(x-x^q)D_\zeta^{(\nu)}y-(y-y^q)D_\zeta^{(\nu)}x\\
&=(x-x^q)\,(m_{21}D_\zeta^{(\nu)}\bar{x}+m_{22}D_\zeta^{(\nu)}\bar{y})-(y-y^q)(m_{11}D_\zeta^{(\nu)}\bar{x}+m_{12}D_\zeta^{(\nu)}\bar{y})\\
&=[m_{11}\bar{x}+m_{12}\bar{y}-(m_{11}\bar{x}+m_{12}\bar{y})^q]\,(m_{21}D_\zeta^{(\nu)}\bar{x}+m_{22}D_\zeta^{(\nu)}\bar{y})\\
&\qquad-[m_{21}\bar{x}+m_{22}\bar{y}-(m_{21}\bar{x}+m_{22}\bar{y})^q]\,(m_{11}D_\zeta^{(\nu)}\bar{x}+m_{12}D_\zeta^{(\nu)}\bar{y})\\
&=\bar{x}[(m_{11}m_{21}-m_{21}m_{11})D_\zeta^{(\nu)}\bar{x}+(m_{11}m_{22}-m_{21}m_{12})D_\zeta^{(\nu)}\bar{y}]\\
&\qquad+\bar{y}[(m_{12}m_{21}-m_{22}m_{11})D_\zeta^{(\nu)}\bar{x}+(m_{12}m_{22}-m_{22}m_{12})D_\zeta^{(\nu)}\bar{y}]+\cdots\\
&=(m_{11}m_{22}-m_{21}m_{12})[\bar{x}D_\zeta^{(\nu)}\bar{y}-\bar{y}D_\zeta^{(\nu)}\bar{x}]+\cdots,
\end{aligned}$$

where each omitted expression has order at least j_1q at $\mathcal{P}$. Hence, if $v_{\mathcal{P}}(S)<j_1q$, then

$$v_{\mathcal{P}}(S)\geq \mathrm{ord}_t(\bar{x}D_\zeta^{(\nu)}\bar{y}-\bar{y}D_\zeta^{(\nu)}\bar{x}). \tag{9.54}$$

Now it is shown that, if the branch γ of $\mathcal{F}$ is centred at a point in $\mathrm{PG}(2,q)$, then

$$v_{\mathcal{P}}(S)\geq 2j_1. \tag{9.55}$$

If $\nu_1=\nu=1$, then $D_\zeta^{(\nu)}\bar{y}=D_\zeta\bar{y}$, and hence $\mathrm{ord}_{\mathcal{P}}\,D_\zeta^{(\nu)}\bar{y}\geq j_2-1$. From (9.54),

$$v_{\mathcal{P}}(S)\geq \min\{2j_1,j_2+j_1-1\}.$$

Since $j_2>j_1$, the assertion follows.

If $\nu_1=\nu=2$, the above argument shows that $v_{\mathcal{P}}(S)\geq\min\{2j_1,j_2+j_1-2\}$, whence the assertion follows for $j_2>j_1+1$. The case $j_2=j_1+1$ cannot actually occur since $\bar{x}D_\zeta\,\bar{y}=\bar{y}D_\zeta\,\bar{x}$.

Suppose now that $\mathcal{F}$ is Frobenius non-classical with $\nu_1=\epsilon=p^m>2$. By (9.54), to prove the result, it suffices to prove that $\mathrm{ord}_{\mathcal{P}}\,\bar{y}\geq p^m+j_1$. However, this has already been done in the proof of Proposition 8.75. □

THEOREM 9.63 *Let $\mathcal{F}$ be a non-classical irreducible plane curve of degree n and genus g defined over $\mathbf{F}_q$. If $\mathcal{F}$ is Frobenius non-classical, and has only tame branches, then*

$$B_q\geq (q-1)n-(2g-2). \tag{9.56}$$

Also, equality holds if and only if every non-linear branch of $\mathcal{F}$ is centred at a point in $\mathrm{PG}(2,q)$.

Proof. As in the proof of Theorem 8.68, the idea is to compute $v_{\mathcal{P}}(S)-v_{\mathcal{P}}(R)$ and apply (7.13) together with (8.14). Put $\nu=\epsilon=p^m>2$. From (7.17) and (8.17),

$$\begin{aligned}
z_1^{p^m}(x^q-x)+z_2^{p^m}(y^q-y)&=0,\\
z_1^{p^m}D_\zeta^{(1)}x+z_2^{p^m}D_\zeta^{(1)}y&=0.
\end{aligned}$$

These relations allow the simplification of both the Wronskian determinants

$$\det(D_\zeta^{(\epsilon_i)}x_j) = \begin{vmatrix} 1 & x & y \\ 0 & D_\zeta^{(1)}x & D_\zeta^{(1)}y \\ 0 & D_\zeta^{(p^m)}x & D_\zeta^{(p^m)}y \end{vmatrix} = \begin{vmatrix} D_\zeta^{(1)}x & D_\zeta^{(1)}y \\ D_\zeta^{(p^m)}x & D_\zeta^{(p^m)}y \end{vmatrix},$$

$$\det(D_\zeta^{(\nu_i)}x_j) = \begin{vmatrix} 1 & x & y \\ 1 & x^q & y^q \\ 0 & D_\zeta^{(p^m)}x & D_\zeta^{(p^m)}y \end{vmatrix} = \begin{vmatrix} x^q - x & y^q - y \\ D_\zeta^{(p^m)}x & D_\zeta^{(p^m)}y \end{vmatrix}.$$

Multiplying the second column by $z_2^{p^m}$ and adding to it $z_1^{p^m}$ times the first column, the second column becomes

$$[0, z_1^{p^m} D_\zeta^{(p^m)}x + z_2^{p^m} D_\zeta^{(p^m)}y]^T$$

in both cases. Putting $u = z_1^{p^m} D_\zeta^{(p^m)}x + z_2^{p^m} D_\zeta^{(p^m)}y$, it follows that

$$v_{\mathcal{P}}(S) = \text{ord}_{\mathcal{P}}(x^q - x) + \text{ord}_{\mathcal{P}}\, u - p^m \text{ord}_{\mathcal{P}}\, z_2, \tag{9.57}$$

$$v_{\mathcal{P}}(R) = \text{ord}_{\mathcal{P}}(D_\zeta^{(1)}x) + \text{ord}_{\mathcal{P}}\, u - p^m \text{ord}_{\mathcal{P}}\, z_2. \tag{9.58}$$

Hence

$$v_{\mathcal{P}}(S) - v_{\mathcal{P}}(R) = \text{ord}_{\mathcal{P}}(x^q - x) - \text{ord}_{\mathcal{P}}(D_\zeta^{(1)}x). \tag{9.59}$$

Up to a non-degenerate linear transformation

$$(X, Y) \mapsto (uX + vY, wX + zY),$$

with $u, v, w, z \in \mathbf{F}_q$, it may be supposed that the tangent to the branch γ is not the vertical line through the centre. Then

$$x(t) = a + t^{j_1} + \cdots, \quad y(t) = b + bt^s + \cdots,$$

with $s \geq j_1$. Since γ is tame, p does not divide j_1. Hence

$$j_1 - 1 = \text{ord}_{\mathcal{P}}(D_\zeta^{(1)}x) \leq \text{ord}_{\mathcal{P}}(D_\zeta^{(1)}y).$$

Now $\text{ord}_{\mathcal{P}}(x^q - x) > 0$ if and only if $a \in \mathbf{F}_q$. If this is the case, then

$$\text{ord}_{\mathcal{P}}(x^q - x) = j_1,$$

and $(1, a, b) \in \text{PG}(2, q)$.

To prove this last statement, it must be checked that $a \in \mathbf{F}_q$ implies $b \in \mathbf{F}_q$. Since $\mathcal{F}$ is Frobenius non-classical,

$$(x^q - x)D_\zeta^{(1)}y - (y^q - y)D_\zeta^{(1)}x = 0.$$

If $b \notin \mathbf{F}_q$, then $\text{ord}_{\mathcal{P}}(y^q - y) = 0$ and hence $\text{ord}_{\mathcal{P}}(D_\zeta^{(1)}x) > \text{ord}_{\mathcal{P}}(D_\zeta^{(1)}y)$; but this contradicts that $\text{ord}_{\mathcal{P}}(D_\zeta^{(1)}x) \leq \text{ord}_{\mathcal{P}}(D_\zeta^{(1)}y)$. Since $\text{ord}_{\mathcal{P}}(x^q - x) = 0$ only if $a \notin \mathbf{F}_q$, the following formula is obtained and can be viewed as an extension of (8.27) to singular branches:

$$\text{ord}_{\mathcal{P}}(x^q - x) = \begin{cases} j_1, & \text{if the centre of } \gamma \text{ is in PG}(2, q); \\ 0, & \text{otherwise.} \end{cases} \tag{9.60}$$

From (9.59) and ((9.60),

$$v_{\mathcal{P}}(S) - v_{\mathcal{P}}(R) = \begin{cases} 1, & \text{if the centre of } \gamma \text{ is in PG}(2, q); \\ -(j_1 - 1), & \text{otherwise.} \end{cases}$$

Since $\deg S - \deg R = (q-1)n - (2g-2)$, this implies that

$$B_q = (q-1)n - (2g-2) + \sum (r-1),$$

where the sum is over all branches γ of $\mathcal{F}$ whose centre $(1, a, b)$ is not in PG$(2, q)$. This completes the proof of Theorem 9.63. □

EXAMPLE 9.64 (1) The Hermitian curve $\mathcal{H}_{\sqrt{q}} = \mathbf{v}(X^{\sqrt{q}+1} + Y^{\sqrt{q}+1} + 1)$ attains the upper bound (9.53) and the lower bound (9.56).

(2) Another example which illustrates Theorems 9.62 and 9.63 for $q = p^3$ and p odd, is the dual curve $\mathcal{C}$ of the plane curve $\mathcal{F}$ in (8.61). The main properties of $\mathcal{C}$ are as follows:

(i) $\mathcal{C}$ is a singular plane curve defined over $\mathbf{F}_q$ birationally equivalent to $\mathcal{C}$;

(ii) $\mathcal{C}$ has degree $(p^2+p+1)(p+1)$, and genus $g = (p^2+p)(p^2+p-1)/2$;

(iii) $\mathcal{C}$ is a Frobenius non-classical plane curve with $\epsilon_2 = \nu_1 = p^2$;

(iv) $\mathcal{C}$ has only one non-linear branch; it is centred at a point in PG$(2, q)$ and has order $p+1$.

Apply Theorem 9.63:

$$\begin{aligned} B_{p^3} &= (p^2+p+1)(p+1)(p^3-1) - (p^2+p+1)(p^2+p-2) \\ &= (p^2+p+1)(p-1)(p^3+2p^2+p-1). \end{aligned}$$

When $p = 3$, this gives $B_{27} = 1222,\ N_{27} = 208$.

(3) Now let $q = p = 2$, and $\mathcal{F} = \mathbf{v}((X^2+X)(Y^2+Y)+1)$. As noted in Remark 8.52, $\mathcal{F}$ is classical but Frobenius non-classical with $\epsilon_2 = \nu_1 = 2$. Here, $\mathcal{F}$ has two points in PG$(2, q)$, namely X_∞ and Y_∞, both singular. Also, X_∞ is a double point and both branches of $\mathcal{F}$ centred at X_∞ are $\mathbf{F}_q$-rational; the similar property holds for Y_∞. Therefore $\mathcal{F}$ has only linear and hence tame branches. Since $\mathcal{F}$ has genus $g = 1$, so $B_2 = 4$, as in (9.51).

(4) To illustrate Theorem 9.63, put $q = 2^{2e+1}$, $q_0 = 2^e$, with $e \geq 1$, and consider the Deligne–Lusztig–Suzuki irreducible plane curve, or DLS curve for short,

$$\mathcal{F} = \mathbf{v}(Y^q + Y + X^{q_0}(X^q + X)),$$

of genus $q_0(q-1)$. It has several interesting properties; see also Section 12.2. First, $\mathcal{F}$ is Frobenius non-classical with $\epsilon = \nu = q_0$; in fact, both (7.65) and (8.54) hold for

$$z_0(x, y) = x^{q/q_0+1} + y^{q/q_0},\ z_1(x, y) = x,\ z_2(x, y) = 1.$$

Also, $\mathcal{F}$ has only one singular point, namely Y_∞. More precisely, Y_∞ is the centre of exactly one branch of $\mathcal{F}$, and hence $B_q = q^2 + 1$. This branch P has order $r = q_0$ and class $s = q$; in particular, P is a non-tame branch. Theorem 9.63 fails in the sense that here equality does not hold in (9.56); the unique singular branch of $\mathcal{F}$ is centred at a point in PG$(2, q)$, but

$$q^2 + 1 > q^2 - qq_0 - q + q_0 + 2 = (q-1)(q+q_0) - (2g-2).$$

An improvement on the bound in Proposition 9.45 is the following result.

PROPOSITION 9.65 *If $\mathcal{C}$ is an irreducible $\mathbf{F}_q$-rational curve of degree $n > 1$, then* $R_q \leq (n-1)q + \lfloor n/2 \rfloor$.

Proof. For each point $P \in \mathcal{C}$ lying in $\mathrm{PG}(2,q)$ choose an $\mathbf{F}_q$-rational line t_P such that $I(P, \mathcal{C} \cap t_P) > 1$. If P is non-singular, then t_P is the tangent line to $\mathcal{C}$ at P. If P is singular, then any $\mathbf{F}_q$-rational line through P has the required property. The number u_P of common points of $\mathcal{C}$ and t_P is at most $n-1$. Define k to be the smallest value of u_P as P ranges over all points of $\mathcal{C}$ lying in $\mathrm{PG}(2,q)$. Then $R_q \leq (n-1)q + k$, as through P there are exactly $q+1$ lines and one of these lines is t_P.

A point P is said to *see a tangency* at $Q \in \mathcal{C}$ if t_Q contains P; here $P = Q$ is allowed. The points of $\mathcal{C}$ lying in $\mathrm{PG}(2,q)$ can see at least $k\,R_q$ tangencies; so there must be some point $P \in \mathcal{C}$ lying in $\mathrm{PG}(2,q)$ which sees at least k tangencies. Counting the points of $\mathcal{C}$ lying in $\mathrm{PG}(2,q)$ by viewing from P shows that

$$R_q \leq q(n-1) + n - k.$$

Finally, since

$$\min\{(n-1)q + k, (n-1)q + n - k\} \leq (n-1)q + \lfloor n/2 \rfloor,$$

the assertion follows. □

Some applications require information on the number of points of $\mathrm{PG}(2,q)$ lying on a plane curve not defined over $\mathbf{F}_q$. An upper bound on this number is as follows.

LEMMA 9.66 *If an irreducible plane curve of degree n is defined over $\mathbf{F}_{q^k}$ but not over $\mathbf{F}_q$, then the number N of its points lying in $\mathrm{PG}(2,q)$ does not exceed n^2.*

Proof. By Bézout's Theorem 3.14, two distinct irreducible curves of degree n may intersect in up to n^2 points. So there cannot be two such curves through n^2+1 points. Thus an irreducible plane curve, defined over $\mathbf{F}_{q^k}$ and containing n^2+1 points in $\mathrm{PG}(2,q)$, is unique and is defined over $\mathbf{F}_q$. Hence $N \leq n^2$. □

In conclusion, two more bounds are established; these are applied in Chapter 13.

THEOREM 9.67 *Let $\mathcal{C}_n$ be a plane curve of degree n defined over $\mathbf{F}_q$ with no $\mathbf{F}_q$-linear components, and let T_q be the number of rational simple points of $\mathcal{C}_n$. If*

$$\sqrt{q} > n - 1, \tag{9.61}$$

then

$$T_q < n(q + 2 - n). \tag{9.62}$$

Proof. (i) *$\mathcal{C}_n$ irreducible.*

Suppose (9.62) does not hold. Then, from (9.46), which is the Hasse–Weil Bound for plane curves,

$$n(q+2-n) \leq T_q \leq S_q \leq q + 1 + (n-1)(n-2)\sqrt{q}.$$

Hence

$$(n-1)(\sqrt{q}+1)(\sqrt{q}+1-n) \leq 0,$$

and so $\sqrt{q} \leq n-1$, contradicting (9.61).

(ii) $\mathcal{C}_n$ *reducible.*

Suppose that the components of $\mathcal{C}_n$ are $\mathcal{F}_1, \ldots, \mathcal{F}_m$, and let $\mathcal{F}_i$ have degree n_i and $T_q^{(i)}$ simple points in $\mathrm{PG}(2,q)$ for $i \in \mathbf{N}_m$. Since $T_q \leq \sum T_q^{(i)}$, it suffices to show that, for each i,

$$T_q^{(i)} < n_i(q+2-n). \tag{9.63}$$

Then $T_q \leq \sum T_q^{(i)} < \sum n_i(q+2-n) = n(q+2-n)$.

Two cases must be distinguished, according as a component is $\mathbf{F}_q$-rational or not.

(a) $\mathcal{F}_i$ *is an* $\mathbf{F}_q$*-rational component.*

By hypothesis, $n_i \geq 2$ and, since $\mathcal{C}_n$ could not have just one non-$\mathbf{F}_q$-rational linear component, $n \geq n_i + 2 \geq 4$.

Suppose that (9.63) is false. Then, again by (9.46),

$$\begin{aligned}
&n_i(q+2-n) \leq T_q^{(i)} \leq q+1+(n_i-1)(n_i-2)\sqrt{q}\\
&\Longleftrightarrow q(n_i-1) - \sqrt{q}(n_i-1)(n_i-2) \leq n_i(n-2)+1\\
&\Longleftrightarrow \sqrt{q}(n_i-1)(\sqrt{q}-n_i+2) \leq n_i(n-2)+1\\
&\Longrightarrow (n-1)(n_i-1)(n-n_i+1) \leq n_i(n-2)+1\\
&\Longrightarrow (n-1)(n_i-1)(n-n_i+1) < n_i(n-1)\\
&\Longrightarrow (n_i-1)(n-n_i+1) < n_i\\
&\Longrightarrow 3(n_i-1) < n_i\\
&\Longleftrightarrow 2n_i < 3,
\end{aligned}$$

a contradiction. So $T_q^{(i)} < n_i(q+2-n)$.

(b) $\mathcal{F}_i$ *is a non-*$\mathbf{F}_q$*-rational component.*

By Lemma 9.66, $T_q^{(i)} \leq n_i^2$. Now

$$\begin{aligned}
&(\sqrt{q}-1)^2 \geq 0\\
\Longleftrightarrow\quad & 2\sqrt{q}+1 \leq q+2\\
\Longrightarrow\quad & 2n-1 < q+2\\
\Longrightarrow\quad & n+n_i < q+2\\
\Longleftrightarrow\quad & n_i < q+2-n\\
\Longleftrightarrow\quad & n_i^2 < n_i(q+2-n)\\
\Longrightarrow\quad & T_q^{(i)} < n_i(q+2-n).
\end{aligned}$$

□

REMARK 9.68 It is not true that $n^2 \leq q+1+(n-1)(n-2)\sqrt{q}$ for all values of n and q.

THEOREM 9.69 *Let $\mathcal{D}_{2t}$ be a plane curve of degree $2t$ defined over $\mathbf{F}_q$ and let $\mathcal{C}_n$ be a component of $\mathcal{D}_{2t}$, with degree $n \geq 3$ if $\mathcal{C}_n$ is $\mathbf{F}_q$-rational. Let T_q be the number of rational simple points of $\mathcal{C}_n$ and let s be the number of its rational singular points, of which d are double points. If*

$$\sqrt{q} > 2t+n-1, \tag{9.64}$$

then

$$T_q + d < \tfrac{1}{2}n(q+2-t). \tag{9.65}$$

Proof. (i) $\mathcal{C}_n$ *an* $\mathbf{F}_q$*-rational component with* $n \geq 3$.

Suppose that (9.65) is false. Then, from (9.46),

$$\begin{aligned}
&\tfrac{1}{2}n(q-t+2) \leq T_q + d \leq T_q + s \leq q+1+(n-1)(n-2)\sqrt{q}\\
&\Longrightarrow n(q-t+2) \leq 2(q+1)+2(n-1)(n-2)\sqrt{q}\\
&\Longleftrightarrow (n-2)q - 2(n-1)(n-2)\sqrt{q} \leq n(t-2)+2\\
&\Longleftrightarrow (n-2)\sqrt{q}[\sqrt{q}-2(n-1)] \leq n(t-2)+2\\
&\Longrightarrow (n-2)(2t+n-1)(2t-n+1) < n(t-2)+2\\
&\Longleftrightarrow (n-2)[4t^2-(n-1)^2] < n(t-2)+2\\
&\Longrightarrow (n-2)[4t^2-(2t-1)^2] < n(t-2)+2 \text{ (since } 2t \geq n)\\
&\Longleftrightarrow (n-2)(4t-1) < n(t-2)+2\\
&\Longleftrightarrow (3n-8)t+n < 0,
\end{aligned}$$

which is impossible for $n \geq 3$.

(ii) $\mathcal{C}_n$ *not* $\mathbf{F}_q$*-rational.*

Suppose the result is false. Then, by Lemma 9.66,

$$\begin{aligned}
&\tfrac{1}{2}n(q-t+2) \leq T_q + d \leq T_q + s \leq n^2\\
&\Longrightarrow q-t+2 \leq 2n\\
&\Longrightarrow (2t+n-1)^2-(t-2) < 2n\\
&\Longrightarrow (2t+n-1)^2-(2t-1)^2 < 2n\\
&\Longleftrightarrow n(4t+n-2) < 2n\\
&\Longleftrightarrow 4t+n-2 < 2\\
&\Longleftrightarrow 4(t-1)+n < 0,
\end{aligned}$$

which is impossible as $t \geq 1,\ n \geq 1$. □

9.7 FURTHER APPLICATIONS OF THE ZETA FUNCTION

The zeta function is a useful tool also in the study of the group $\mathrm{Pic}_0(\mathbf{F}_q(\mathcal{X}))$ of all zero-degree $\mathbf{F}_q$-rational divisors classes of $\mathcal{X}$. This becomes apparent below in the fundamental role of the L-polynomial in the proof of two major results on $\mathrm{Pic}_0(\mathbf{F}_q(\mathcal{X}))$, previously stated without proof in Section 6.7.

Here, the notation of Sections 8.3 and 8.4 is changed as follows. If $l = \mathbf{F}_{q^n}$ with $n \geq 1$, then

$$\begin{aligned}
\mathrm{Div}^{(l)}(\mathcal{X}) &= \text{the } l\text{-divisor group } \mathrm{Div}(l(\mathcal{F}));\\
\mathrm{Div}_0^{(l)}(\mathcal{X}) &= \text{the zero-degree } l\text{-divisor group } \mathrm{Div}_0(l(\mathcal{F}));\\
\mathrm{Pic}^{(l)}(\mathcal{X}) &= \text{the } l\text{-divisor class group } \mathrm{Pic}(l(\mathcal{F}));\\
\mathrm{Pic}_0^{(l)}(\mathcal{X}) &= \text{the zero-degree } l\text{-divisor class group } \mathrm{Pic}_0(l(\mathcal{F})).
\end{aligned}$$

By Theorem 8.34, $\mathrm{Pic}_0^{(l)}(\mathcal{X})$ is a finite abelian group. Its order, that is, the class number h of $\mathcal{X}$, is closely related to the zeta function of $\mathcal{X}$ via the L-polynomial $L_q(t)$.

THEOREM 9.70 $h = L_q(1)$.

Proof. If $g = 0$, then $L_q(1) = 1$ and $h = 1$ by Example 8.36. If $g \geq 1$, from (9.8) $L(1) = h$. □

Theorem 6.94 was stated without proof. Now, using the zeta function, a proof is given in the case that K is the algebraic closure of $\mathbf{F}_q$. An essential tool in the proof is the algebraic number field $\mathbf{Q}(\omega, \ldots, \omega_{2g})$ generated by the inverse roots of the L-polynomial of $\mathcal{X}$. As suggested in Section 7.16, see especially Example 7.123, function fields of transcendence degree 1 are closely analogous to algebraic number fields, although number-theoretic methods have no direct counterpart in function field theory. A detailed discussion of these methods is beyond the scope of this book. The concepts and results used in the proofs below mostly concern the theory of divisibility.

Two more pieces of notation are needed:

$$\mathrm{Pic}_0^{(l)}(\mathcal{X}, d) = \text{the subgroup of } \mathrm{Pic}_0^{(l)}(\mathcal{X}) \text{ of all elements of order } d;$$

$$\mathrm{Pic}_0^{(l)}(\mathcal{X}, d_\infty) = \text{the subgroup of } \mathrm{Pic}_0^{(l)}(\mathcal{X}) \text{ of all elements of order a power of } d.$$

The following lemma is the main ingredient in the proof.

LEMMA 9.71 *Let d be a prime different from p. Then, for the curve $\mathcal{X}$ of genus g, there exists a finite extension k of $\mathbf{F}_q$ such that the following condition is satisfied by all finite extensions l of k, where l_1 is an extension of l of degree d^r:*

$$[\mathrm{Pic}_0^{(l_1)}(\mathcal{X}, d_\infty) : \mathrm{Pic}_0^{(l)}(\mathcal{X}, d_\infty)] = d^{2gr}. \tag{9.66}$$

Proof. The group $\mathrm{Pic}_0^{(l)}(\mathcal{X}, d)$ is finite and abelian, in which every element has order a power of d. Thus $|\mathrm{Pic}_0^{(l)}(\mathcal{X}, d)| = d^{s(l)}$. By Theorem 9.70, the order of the group $\mathrm{Pic}_0^{(l)}(\mathcal{X})$ containing $\mathrm{Pic}_0^{(l)}(\mathcal{X}, d)$ is

$$L_{q^m}(1) = \prod_{i=1}^{2g}(1 - \omega_i^m),$$

where $l = \mathbf{F}_{q^m}$; that is, m is the degree of the extension $l/\mathbf{F}_q$. Therefore $q^{s(l)}$ is the largest power of q which divides

$$\prod_{i=1}^{2g}(1 - \omega_i^m).$$

Replacing l by l_1 in this argument shows that $s(l_1)$ is the largest power of q dividing

$$\prod_{i=1}^{2g}(1 - \omega_i^{md^r}) = \prod_{i=1}^{2g}(1 - \omega_i^m)(1 + \omega_i^m + \cdots + (\omega_i^m)^{d^r - 1}).$$

Choose n such that $\omega_i^n \equiv 1 \pmod{d}$ for every $i = 1, \ldots, 2g$, and let $k = q^n$. Then n divides m, and the condition is satisfied. □

Lemma 9.71 can be interpreted in terms of the trace map (8.6). Let ϕ be the restriction of $\mathfrak{T}_0$ to $\mathrm{Pic}_0^{(l_1)}(\mathcal{X}, d_\infty)$. Since $\mathfrak{T}_0$ is surjective, ϕ is a surjective homomorphism. Then (9.66) reads as follows:

$$|\ker \phi| = d^{2gr}. \tag{9.67}$$

Now (9.67) implies that $\ker \phi \leq \mathrm{Pic}_0^{(l_1)}(\mathcal{X}, d^{2gr})$. On the other hand, $\ker \phi$ contains $\mathrm{Pic}_0^{(l)}(\mathcal{X}, d^r)$.

To show this, let $l = \mathbf{F}_{q^n}$, $l_1 = \mathbf{F}_{(q^n)^{d^r}}$, and choose any $[D] \in \mathrm{Pic}_0^{(l)}(\mathcal{X}, d^r)$. Then $\Phi^n([D]) = ([D])$ for the n-th Frobenius map on divisor classes, and hence

$$[0] = d^r([D]) = [D] + \Phi^n([D]) + \cdots + (\Phi^n)^{d^r-1}([D]) = \mathfrak{T}_0([D]),$$

which shows that $[D] \in \ker\phi$. Therefore

$$\mathrm{Pic}_0^{(l)}(\mathcal{X}, d^r) \le \ker\phi \le \mathrm{Pic}_0^{(l_1)}(\mathcal{X}, d^{2gr}) \tag{9.68}$$

To prove Theorem 6.94, it suffices to show the following result.

THEOREM 9.72 *For all positive integers r and primes d distinct from p,*

$$\mathrm{Pic}_0(K(\mathcal{X}), d^r) \cong (\mathbf{Z}_{d^r})^{2g}.$$

Proof. Choose k, l, l_1 as in Lemma 9.71. Then

$$|\mathrm{Pic}_0^{(l)}(\mathcal{X}, d^r)| \le |\ker\phi| = d^{2rg}.$$

This shows that, for sufficiently large k, the following condition is satisfied by all finite extensions $\mathbf{F}_l$ of $\mathbf{F}_k$. If l_1' is a finite extension of l', then

$$\mathrm{Pic}_0^{(l')}(\mathcal{X}, d^r) = \mathrm{Pic}_0^{(l_1')}(\mathcal{X}, d^r). \tag{9.69}$$

Replacing d^r by d^{2gr}, this becomes the following:

$$\mathrm{Pic}_0^{(l')}(\mathcal{X}, d^{2gr}) = \mathrm{Pic}_0^{(l_1')}(\mathcal{X}, d^{2gr}). \tag{9.70}$$

Now choose a sufficiently large k such that both (9.69) and (9.70) hold whenever $\mathbf{F}_{l'}$ is a finite extension of $\mathbf{F}_k$ and $\mathbf{F}_{l_1'}$ is a finite extension of $\mathbf{F}_{l'}$. Now, from (9.68),

$$\mathrm{Pic}_0^{(l')}(\mathcal{X}, d^r) \le \ker\phi \le \mathrm{Pic}_0^{(l')}(\mathcal{X}, d^{2gr}).$$

Let $[D] \in \ker\phi$, and $l' = \mathbf{F}_{q^{n'}}$. Then

$$r^d[D] = [D] + \Phi^{n'}([D]) + \cdots + (\Phi^{n'})^{d^r-1}([D]) = \mathfrak{T}_{0\,l_1'|l'}([D]) = 0.$$

This implies that $[D] \in \mathrm{Pic}_0^{(l')}(\mathcal{X}, d^r)$ showing that $\ker\phi \le \mathrm{Pic}_0^{(l')}(\mathcal{X}, d^r)$. Therefore $\ker\phi = \mathrm{Pic}_0^{(l')}(\mathcal{X}, d^r)$. From (9.67),

$$\mathrm{Pic}_0^{(l')}(\mathcal{X}, d^r) \cong (\mathbf{Z}_{d^r})^{2g}.$$

This, together with (9.69), implies the result. □

A similar result enunciated in Chapter 6 is Theorem 6.96. Here, a proof is given in the case that $K = \overline{\mathbf{F}}_q$.

Write $L_q(t) = 1 + a_1 t + \cdots + a_i t^i + \cdots + t^{2g}$, and denote by $\gamma(\mathcal{X})$ the highest index i with $1 \le i \le g$ for which $a_i \not\equiv 0 \pmod{p}$. To prove Theorem 6.96, it suffices to show that $\gamma(\mathcal{X})$ is equal to the p-rank γ of $\mathcal{X}$. For this, two lemmas are required.

As before, $\omega_1, \ldots, \omega_{2g}$ are the inverse roots of the L-polynomial $L_q(t)$ of $\mathcal{X}$. The idea is to use properties of the algebraic number field $\Omega = \mathbf{Q}(\omega_1, \ldots, \omega_{2g})$. Let τ be a prime divisor of Ω lying over p.

LEMMA 9.73 *The index $\gamma(\mathcal{X})$ is the number of i for which $\omega_i \not\equiv 0 \pmod{\tau}$.*

Proof. Order $\omega_1, \ldots, \omega_{2g}$ so that

$$\omega_j \not\equiv 0 \pmod{\tau}, \qquad j = 1, \ldots, s;$$
$$\omega_j \equiv 0 \pmod{\tau}, \qquad j = s+1, \ldots, 2g.$$

From (9.9),

$$1 + a_1 t + \cdots + a_i t^i + \cdots + q t^{2g} = \prod_{i=1}^{2g} (1 - \omega_i t);$$

it follows that

$$a_j = (-1)^j \prod_{i_1 < \cdots < i_j} \omega_{i_1} \cdots \omega_{i_j}.$$

For $j \geq s+1$, each term in the sum contains a factor from τ while, for $j = s$, the product $\omega_1 \cdots \omega_s$ is the only term not lying in τ. □

The second preliminary result is analogous to Lemma 9.71.

Lemma 9.74 *There exists a finite extension k of $\mathbf{F}_q$ such that the following condition is satisfied by all finite extensions l of k. If l_1 is an extension of l of degree p, then*

$$[\mathrm{Pic}_0^{(l_1)}(\mathcal{X}, p_\infty) : \mathrm{Pic}_0^{(l)}(\mathcal{X}, p_\infty)] = p^{\gamma(\mathcal{X})}. \tag{9.71}$$

Proof. To τ there is associated an additive normalised valuation v_τ of Ω. Choose a positive integer n so that

$$\omega_i^n \equiv 1 \pmod{\tau^{v_\tau(p)+1}}$$

for all i with $\omega_i \not\equiv 0 \pmod{\tau}$. By Proposition 9.16, $L_k(t) = \prod_{i=1}^{2g}(1 - \omega_i^n t)$ for an extension k of $\mathbf{F}_q$ of degree n. Therefore, for any extension l of k,

$$L_l(t) = \prod_{i=1}^{2g} (1 - \eta_i t),$$

with

$$\eta_i \equiv 0 \text{ or } 1 \pmod{\tau^{v_\tau(p)+1}}.$$

By Lemma 9.74, the latter case occurs as many times as $\gamma(\mathcal{X})$. So (9.71) is equivalent to the equation

$$v_\tau\, L_{l_1}(1)/L_l(1) = \gamma(\mathcal{X}) \cdot v_\tau(p), \tag{9.72}$$

where l_1 is an extension of l of degree p. Now (9.72) is shown as follows:

$$\begin{aligned} v_\tau\, L_{l_1}(1)/L_l(1) &= \gamma(\mathcal{X}) \cdot v_\tau(p) = v_\tau \prod_{i=1}^{2g} (1 - \eta_i^p)/(1 - \eta_i) \\ &= \sum_{\eta_i \equiv 0 \pmod{\tau}} (1 + \eta_i + \cdots + \eta_i^{p-1}). \end{aligned}$$

Note that, if $\eta_i \not\equiv 0 \pmod{\tau}$, then

$$1 + \eta_i + \cdots + \eta_i^{p-1} \equiv p \pmod{\tau^{v_\tau(p)+1}}.$$

Therefore

$$v_\tau(1 + \eta_i + \cdots + \eta_i^{p-1}) = v_\tau(p),$$

whence the result follows. □

THEOREM 9.75 *The index $\gamma(\mathcal{X})$ equals the p-rank γ of $\mathcal{X}$.*

Proof. The arguments in the proof of Theorem 9.72 remain valid when Lemma 9.71 is replaced by Lemma 9.74. □

A further characterisation of the p-rank γ of $\mathcal{X}$ is given by the following result.

THEOREM 9.76 (i) *$\gamma = 0$ if and only if $A_i \equiv 1 \pmod{p}$.*

(ii) *If $\gamma \neq 0$, then $\gamma = \max\{i \mid 1 \leq i \leq g;\ A_{i-1} \not\equiv A_i \pmod{p}\}$.*

Proof. Comparing the coefficients in $L_q(t) = (1 - t(1 - qt))Z(\mathcal{X}, t)$ modulo p shows that $a_i \equiv A_i - A_{i-1} \pmod{p}$ for $i \geq 1$. Thus Theorem 9.76 is a corollary to Theorem 9.75. □

9.8 THE FUNDAMENTAL EQUATION

The study of $\mathrm{Pic}_0(K(\mathcal{X}))$ and its endomorphisms benefits from both group theory and algebraic number theory, but it is best achieved by using the powerful tools of algebraic geometry.

To explain how to associate an algebraic variety with $\mathrm{Pic}_0(K(\mathcal{X}))$, choose a point $P_0 \in \mathcal{X}$, and embed $\mathcal{X}$ in $\mathrm{Pic}_0(K(\mathcal{X}))$ by the map sending $P \in \mathcal{X}$ to the degree-zero divisor class $[P - P_0]$. If $\mathcal{X}$ is a plane cubic, then such a map is bijective, thus allowing in this case the identification of the group $\mathrm{Pic}_0(K(\mathcal{X}))$ with the points of $\mathcal{X}$, which is a particular case of an abelian variety. As a matter of fact, the group $\mathrm{Pic}_0(K(\mathcal{X}))$ can always be identified in a functorial way with the set of points of an abelian variety $\mathcal{A}$, which is an irreducible non-singular projective algebraic variety endowed with an abelian group compatible with the structure of the variety. Compatibility means that the group law

$$\mu : \mathcal{A} \times \mathcal{A} \to \mathcal{A}$$

and the inverse map

$$\iota : \mathcal{A} \to \mathcal{A}$$

are morphisms of varieties. Such an abelian variety is the *Jacobian variety* $\mathcal{J}_{\mathcal{X}}$ of $\mathcal{X}$. It has dimension g and is defined over $\mathbf{F}_q$. If P_0 is an $\mathbf{F}_q$-rational point of $\mathcal{X}$, then the identification is carried out by a morphism of algebraic varieties

$$\mathcal{X} \to \mathcal{J}_{\mathcal{X}},$$

defined over $\mathbf{F}_q$, that sends P_0 to the identity element of $\mathcal{J}_{\mathcal{X}}$. This means that the $\mathbf{F}_{q^n}$-rational points of $\mathcal{J}_{\mathcal{X}}$ are the elements of $\mathrm{Pic}_0^{(q^n)}(\mathcal{X})$, that is, the $\mathbf{F}_{q^n}$-rational degree-zero divisor classes of $\mathcal{X}$.

The action of the Frobenius map Φ on the points of $\mathcal{X}$ has a natural extension on the divisors of $\mathcal{X}$: if $D = \sum n_{\mathcal{P}}\mathcal{P}$, then $\Phi(D) = \sum n_{\mathcal{P}}\Phi(\mathcal{P})$. This defines an endomorphism of $\mathrm{Div}(K(\mathcal{X}))$, and hence gives rise to an endomorphism of $\mathrm{Pic}_0(K(\mathcal{X}))$, the *Frobenius endomorphism*, that fixes each element of $\mathrm{Pic}_0^{(q)}(\mathcal{X})$.

On the Jacobian variety $\mathcal{J}_{\mathcal{X}}$ embedded in a projective space over $\mathbf{F}_q$, the Frobenius map Φ can be thought of as the map induced by the Frobenius collineation whose fixed points are exactly the $\mathbf{F}_q$-rational points of $\mathcal{J}_{\mathcal{X}}$. So Φ is also the *Frobenius endomorphism of* $\mathcal{J}_{\mathcal{X}}$.

The main results on the Frobenius endomorphism are summarised. The characteristic polynomial $P(t)$ of the Frobenius endomorphism Φ is a monic polynomial of degree $2g$ with coefficients in $\mathbf{Z}$; it is related to the L-polynomial of $\mathcal{X}$.

THEOREM 9.77 *The characteristic polynomial* $P(t) = t^{2g}L_q(1/t)$.

For any prime d distinct from p, Theorem 9.72 shows that $\mathrm{Pic}_0(K(\mathcal{X}), d)$ can be viewed as a $2g$-dimensional vector space $V(2g, d)$ over the finite field $\mathbf{F}_d$. Then the Frobenius transformation Φ acts faithfully on $V(2g, d)$ as an endomorphism whose characteristic polynomial is described in the following result.

THEOREM 9.78 *Let d be a prime distinct from p. The Frobenius endomorphism acts linearly on* $\mathrm{Pic}_0(K(\mathcal{X}), d)$ *and the characteristic polynomial of the induced endomorphism is* $P(t)$ *with its coefficients reduced modulo d.*

THEOREM 9.79 *If* $P(t) = \prod P_i^{r_i}(t)$ *is the factorisation of the characteristic polynomial* $P(t)$ *of the Frobenius endomorphism of* $\mathcal{J}_{\mathcal{X}}$*, then*

$$\prod P_i(\Phi) = 0 \quad \text{on } \mathcal{J}_{\mathcal{X}}. \tag{9.73}$$

Assume that $\mathcal{X}(\mathbf{F}_q)$ is non-empty, and choose an $\mathbf{F}_q$-rational point P_0 of $\mathcal{X}$. Let $m = |\prod P_i(1)|$, and write

$$\prod P_i(t) = \textstyle\sum_{i=0}^{u} a_i t^{u-i}.$$

Then (9.73) is equivalent to the *Fundamental Equation*

$$\textstyle\sum_{i=1}^{u} a_i \Phi^{u-i}(P) + \Phi^u(P) \equiv mP_0. \tag{9.74}$$

EXAMPLE 9.80 Let $q = 2q_0^2$ with $q_0 = 2^s$ and $s \geq 1$, as in Proposition 9.32. If $\mathcal{X}$ has genus $g = q_0(q-1)$ and the number of $\mathbf{F}_q$-rational points is q^2+1, then the Fundamental Equation reads as follows:

$$qP + 2q_0\Phi(P) + \Phi^2(P) \equiv (q + 2q_0 + 1)P_0.$$

The effective $\mathbf{F}_q$-rational divisor mP_0 of $\mathcal{X}$ is the *Frobenius divisor* of $\mathcal{X}$, and the complete linear series $|mP_0|$ is the *Frobenius linear series* of $\mathcal{X}$. Some properties of the Frobenius linear series are established in the following propositions.

PROPOSITION 9.81 (i) *If* P, Q *are* $\mathbf{F}_q$*-rational points, then* $mP_0 \equiv mQ$*. In particular,* m *is a non-gap at each* $P \in \mathcal{X}(\mathbf{F}_q)$ *provided that* $N_1 \geq 2$.

(ii) *If* $p \nmid m$ *and* $N_1 \geq 2g+3$*, then there exists* $P_1 \in \mathcal{X}(\mathbf{F}_q)$ *such that both* m *and* $m-1$ *are non-gaps at* P_1.

Proof. Part (i) follows from (9.74). To show (ii), note that Lemma 8.24 ensures the existence of an element $x \in \mathbf{F}_q(\mathcal{X})$ such that $\operatorname{div} x = mP_0 - mQ$. Since $p \nmid m$, so x is a separable variable. Then Hurwitz's Theorem 7.27 applied to the rational transformation

$$\omega : x' = x,\ y' = 0$$

gives $2g - 2 = -2m + \deg D$, where D is the different. Note that

$$\deg D \geq 2(m-1) + (N_1 - 2 - k),$$

where k counts the unramified $\mathbf{F}_q$-rational points of $\mathcal{X}$. Therefore

$$k \geq N_1 - (2g+2).$$

So the hypothesis implies that $k > 0$. Let $P_1 \in \mathcal{X}(\mathbf{F}_q)$ be such an unramified point. Then P_1 is distinct from P and Q and there exist $c \in \mathbf{F}_q$ and an effective divisor D with $P_1, Q \notin \operatorname{supp}(D)$ such that $\operatorname{div}(x-c) = P_1 + D - mQ$. Choose $y \in \mathbf{F}_q(\mathcal{X})$ such that $\operatorname{div}(y) = mQ - mP_1$. Now (ii) follows using the fact that $(x-c)y \in \mathbf{F}_q(\mathcal{X})$. □

The first assertion of Proposition 9.81 shows that the Frobenius linear series is independent of the choice of P_0. Two more consequences are stated in the following proposition.

PROPOSITION 9.82 *Let $\mathcal{D}$ be the Frobenius linear series of $\mathcal{X}$.*

(i) *If $N_1 \geq 2$, then $\mathcal{D}$ is base-point-free.*

(ii) *If $p \nmid m$ and $N_1 \geq 2g+3$, then $\mathcal{D}$ is simple.*

There is a link between the $(\mathcal{D}, P)$-orders and the Fundamental Equation. Notation from Section 8.5 is used and three conditions are assumed:

$$\begin{array}{l} |\mathcal{X}(\mathbf{F}_q)| \geq 2; \quad \prod P_i(1) > 0; \\ a_1 \geq 1, \quad a_u \geq 1, \quad a_i \geq a_{i-1} \text{ for } i = 1, \ldots, u-1. \end{array} \tag{9.75}$$

Let $r = \dim \mathcal{D}$. Then $m_r(P) = m$. From Section 6.6, the Weierstrass semigroup $H(P)$ at $P \in \mathcal{X}$ contains the elements of the increasing sequence

$$m_0(P) = 0,\ m_1(P), \ldots,\ m_i(P), \ldots,$$

where $m_i(P)$ is the i-th non-gap at P.

PROPOSITION 9.83 *The Frobenius linear series $\mathcal{D}$ of $\mathcal{X}$ has the following properties:*

(i) *the $(\mathcal{D}, P)$-orders at an $\mathbf{F}_q$-rational point P are $j_i(P) = m - m_{r-i}(P)$ for $i = 0, \ldots, r$;*

(ii) *if $\Phi^i(P) \neq P$ for $i = 1, \ldots, u$, then $j_1(P) = 1$ and, in particular, $\epsilon_1 = 1$;*

(iii) *the integers $1, a_1, \ldots, a_u$ are $\mathcal{D}$-orders, so that $r \geq u+1$;*

(iv) *if* $\Phi^i(P) \neq P$ *for* $i = 1, \ldots, u+1$, *then* a_u *is a non-gap at* P;

(v) *if* $\Phi^i(P) \neq P$ *for* $i = 1, \ldots, u$, *but* $\Phi^{u+1}(P) = P$, *then* a_{u-1} *is a non-gap at* P.

Proof. (i) Let $m(P)$ be a non-gap at an $\mathbf{F}_q$-rational point P of $\mathcal{X}$. There exists an effective divisor E for which $E \equiv m(P)P$ and $P \notin \operatorname{supp}(E)$. From this, $E + (m - m(P))P \equiv mP$. If $m(P) \leq m$, then $m - m(P)$ is a $(P, \mathcal{D})$-order.

(ii) Since Φ, and hence Φ^u, is surjective on $\mathcal{X}$, there is a unique point $P' \in \mathcal{X}$ whose image under Φ^u is P. From (9.74) applied to P',

$$\sum_{i=1}^{u} a_i \Phi^{u-i}(P') + P \equiv mP_0.$$

For $1 \leq j \leq u$, the hypothesis that $\Phi^j(P) \neq P$ implies that $\Phi^{u-j}(P') \neq P$. Therefore 1 is a $(P, \mathcal{D})$-order.

(iii) Take a point $P \in \mathcal{X}$ such that $\Phi^j(P) \neq P$ for $j = 1, \ldots, u$. Then the assertion follow from (9.74).

(iv),(v) Again, the Fundamental Equation (9.74) applies, this time to P and $\Phi(P)$, giving the equivalence

$$\begin{aligned} mP_0 &\equiv \sum_{i=1}^{u} a_i \Phi^{u-i}(P) + \Phi^u(P) \\ &\equiv a_1 \Phi^u(P) + \sum_{i=1}^{u-1} a_{i+1} \Phi^{u-i}(P) + \Phi^{u+1}(P), \end{aligned}$$

whence

$$a_u P \equiv \Phi^{u+1}(P) + (a_1 - 1)\Phi^u(P) + \sum_{i=1}^{u-1} (a_{i+1} - a_i)\Phi^{u-i}(P).$$

Taking (9.75) into account, the results follow. □

PROPOSITION 9.84 *If* $N_1 > q(m - a_u) + 1$, *then* $j_{r-1}(P) < a_u$ *for every* $\mathbf{F}_q$-*rational point* P *of* $\mathcal{X}$.

Proof. By (i) of Lemma 9.83, $j_{r-1}(P) = m - m_1(P)$. On the other hand, Proposition 9.54 shows that $N_1 \leq 1 + m_1(P)q$. Taking the hypothesis into account, this gives the inequality

$$j_{r-1}(P) \leq m - (N_1 - 1)/q < a_u.$$ □

PROPOSITION 9.85 (i) *The order* $\epsilon_r = \nu_{r-1} = a_u$, *and every* $\mathbf{F}_q$-*rational point is in the support of the ramification divisor* R *of* $\mathcal{D}$.

(ii) *If* $N_1 \geq qa_u + 1$, *then* $m_1(P) = q$ *for every* $\mathbf{F}_q$-*rational point* P *of* $\mathcal{X}$.

Proof. (i) First, $\epsilon_{r-1} \leq j_{r-1}(P)$ for any $P \in \mathcal{X}$. Since a_u is a $\mathcal{D}$-order by Proposition 9.83, this implies that $\epsilon_r = a_u$. From (9.74), $\Phi(P)$ lies in the $(r-1)$-th osculating hyperplane. Thus $\nu_{r-1} = \epsilon_r$. By Lemma 9.83 (i), $j_r(P) = m$ for each $\mathbf{F}_q$-rational point of $\mathcal{X}$. Since $m > a_u$, the result follows from Theorem 7.55.

(ii) Let $P \in \mathcal{X}(\mathbf{F}_q)$. Since $m_1(P) \leq m_1(Q)$ for all but finitely many points Q of $\mathcal{X}$, Lemma 9.83 (iv) shows that $m_1(P) \leq a_u$. On the other hand, $m_1(P) \geq a_u$ by Proposition 9.54 and the hypothesis on the size of $\mathcal{X}(\mathbf{F}_q)$. □

Now return to $\operatorname{Pic}_0(K(\mathcal{X}))$. The machinery applied to any endomorphism α of $\mathcal{J}_{\mathcal{X}}$ or, equivalently, of $\operatorname{Pic}_0(K(\mathcal{X}))$ can give an equation of type (9.74) depending on the characteristic polynomial $P_\alpha(X)$ of α. To gain information on $P_\alpha(X)$, it is appropriate to use a representation of the ring End$(\mathcal{J}_\mathcal{X})$ on a module of dimension $2g$ over the l-adic integers $\mathcal{Z}_l$ with l a prime not equal to p. Such a representation depends on the Tate group.

Let A be an abelian group, written additively. Assume that A is infinitely divisible by powers of a fixed prime l, that is, $lA = A$. The *Tate group* $T_l(A)$ of A consists of vectors

$$(a_1, a_2, \ldots, a_n, \ldots)$$

with denumerably many components $a_i \in A$ satisfying $la_{i+1} = a_i$ and $la_1 = 0$. Addition in $T_l(A)$ is carried out component-wise. It is possible to define on $T_l(A)$ the structure of a module over the l-adic integers $\mathcal{Z}_l$ as follows. For $z \in \mathcal{Z}_l$, let m be an integer such that $m \equiv z \pmod{l^n}$. If $l^n a = 0$, define $za = ma$, this being independent of the choice of m. Then an operation of z on $T_l(A)$ is defined:

$$z \cdot (a_1, a_2, \ldots) = (za_1, za_2, \ldots).$$

From now on A is assumed to be $\mathrm{Pic}_0(K(\mathcal{X}), l_\infty)$. This is possible since the group $\mathrm{Pic}_0(K(\mathcal{X}), l_\infty)$ is a divisible group.

Lemma 9.86 *The Tate group $T_l(A)$ is a module over $\mathcal{Z}_l$ of dimension $2g$.*

Proof. Let $x_1, \ldots, x_{2g}$ be vectors of $T_l(A)$ whose first components $a_{1,1}, \ldots, a_{2g,1}$ are linearly independent over $\mathbf{F}_l$. Then these vectors are linearly independent over $\mathcal{Z}_l$. In fact, if a relation of linear dependence existed, it would be possible to assume that not all the coefficients are divisible by l, and hence this relation considered on the first component would contradict the hypothesis made on the $a_{i,j}$.

Now it is shown by an inductive argument that these vectors x_i form a basis of $T_l(A)$ over $\mathcal{Z}_l$. Suppose that every element $w \in T_l(A)$ can be written as a linear combination

$$w = z_1x_1 + \cdots + z_{2g}x_{2g} \qquad \pmod{l^nT_l(A)}, \tag{9.76}$$

with integers $z_j \in \mathbf{Z}$. Let $w = (b_1, \ldots, b_n, b_{n+1}, \ldots)$. By definition, for the first $n+1$ components,

$$\begin{aligned} z_1(a_{1,1,}, \ldots, a_{1,n+1}) + \cdots + z_{2g}(a_{2g,1}, \ldots, a_{2g,n+1}) \\ = (b_1, \ldots, b_n, b_{n+1}) + (0, \ldots, 0, c_{n+1}), \end{aligned}$$

for some c_{n+1} with $l^{n+1}c_{n+1} = 0$. By the choice of the vectors, there exist integers $d_1, \ldots, d_{2g}$ such that

$$c_{n+1} = d_1l^na_{1,n+1} + \cdots + d_{2g}l^na_{2g,n+1}.$$

Replacing $z_1, \ldots, z_{2g}$ by $z_1 + d_1l^n, \ldots, z_{2g} + d_{2g}l^n$ shows that the congruence (9.76) extends from n to $n+1$. This gives the result. □

If $\alpha \in \mathrm{End}(\mathcal{J}_\mathcal{X})$ then α induces a linear map

$$\begin{aligned} \alpha_T : T_l(A) &\to T_l(A), \\ (a_1, a_2, \ldots) &\mapsto (\alpha a_1, \alpha a_2, \ldots). \end{aligned}$$

The $\mathbf{Z}$-module End$(\mathcal{J}_X)$ becomes a module over the l-adic integers $\mathcal{Z}_l$ by defining $(z\alpha)x = \alpha(zx)$ for $x \in T_l(A)$ and $z \in \mathbf{Z}$. This gives the desired representation of End$(\mathcal{J}_\mathcal{X})$. The main results in this area are summarised in the following theorems.

THEOREM 9.87 (i) *If $\alpha_1, \dots, \alpha_m$ are elements of* $\mathrm{End}(\mathcal{J}_{\mathcal{X}})$ *linearly independent over* $\mathbf{Z}$, *then they are independent over* $\mathcal{Z}_l$.

(ii) *The map $\alpha \mapsto \alpha_T$ of* $\mathrm{End}(\mathcal{J}_{\mathcal{X}})$ *into the ring of $\mathcal{Z}_l$-linear transformations of* $T_l(\mathrm{Pic}_0(K(\mathcal{X}), l)$ *is a ring isomorphism.*

(iii) $\mathrm{End}(\mathcal{J}_{\mathcal{X}})$ *is a module of finite type of dimension at most* $16g^2$.

THEOREM 9.88 *The characteristic polynomial of an endomorphism α of $\mathcal{J}_{\mathcal{X}}$ is equal to the characteristic polynomial of the linear transformation α_T induced by α on* $T_l(\mathrm{Pic}_0(\mathcal{K}(\mathcal{X}), l))$.

THEOREM 9.89 *The coefficients of the characteristic polynomial of an endomorphism α of $\mathcal{J}_{\mathcal{X}}$ are integers.*

9.9 ELLIPTIC CURVES OVER $\mathbf{F}_q$

Let $\mathcal{F}$ be a non-singular plane cubic curve defined over $\mathbf{F}_q$. The points of $\mathcal{F}$ with respect to the operation described in the proof of Theorem 6.104 is an abelian group $G_{\mathcal{F}}$.

THEOREM 9.90 *The $\mathbf{F}_q$-rational points of $\mathcal{F}$ form a subgroup $G_{\mathcal{F}(\mathbf{F}_q)}$ of $G_{\mathcal{F}}$.*

Proof. The proof is a consequence of the following facts: (i) any line containing two distinct $\mathbf{F}_q$-rational points of the cubic $\mathcal{F}$ contains a third $\mathbf{F}_q$-rational point of $\mathcal{F}$; (ii) the tangent to $\mathcal{F}$ at an $\mathbf{F}_q$-rational point which is not an inflexion contains another $\mathbf{F}_q$-rational point of $\mathcal{F}$. □

From Theorem 6.104, $G_{\mathcal{F}(\mathbf{F}_q)} \cong \mathrm{Pic}_0^{(q^n)}(\mathcal{X})$.

Some further properties of the group structure are now summarised.

THEOREM 9.91 (i) *The characteristic polynomial is* $\psi(t) = t^2 - mt + 1$;

(ii) $|\,m\,| \le 2\sqrt{q}$;

(iii) $N_1 = q + 1 - m$;

(iv) $p \mid m$ *if and only if $\mathcal{F}$ has p-rank* 0, *that is, $\mathcal{F}$ is supersingular.*

From Theorem 9.18,

$$(\sqrt{q} - 1)^2 \le N_1 \le (\sqrt{q} + 1)^2.$$

In fact, the precise values that N_1 can take are given by the next result.

THEOREM 9.92 (Waterhouse) *There exists an elliptic cubic over $\mathbf{F}_q$, $q = p^h$, with precisely $N_1 = q + 1 - m$ rational points, where $|\,m\,| \le 2\sqrt{q}$, for only the values of m in Table 9.3.*

Table 9.3 Values of m

	m	p	h
(i)	$m \not\equiv 0 \pmod p$		
(ii)	$m = 0$		odd
(iii)	$m = 0$	$p \not\equiv 1 \pmod 4$	even
(iv)	$m = \pm\sqrt{q}$	$p \not\equiv 1 \pmod 3$	even
(v)	$m = \pm 2\sqrt{q}$		even
(vi)	$m = \pm\sqrt{2q}$	$p = 2$	odd
(vii)	$m = \pm\sqrt{3q}$	$p = 3$	odd

Let $N_q(1)$ denote the maximum number of rational points on any non-singular cubic over $\mathbf{F}_q$ and $M_q(1)$ the minimum number. The prime power $q = p^h$ is *exceptional* if h is odd, $h \geq 3$, and p divides $\lfloor 2\sqrt{q} \rfloor$.

REMARK 9.93 *The only exceptional* $q < 1000$ *is* $q = 128$.

COROLLARY 9.94 *The bounds* $N_q(1)$ *and* $M_q(1)$ *are as follows*:

(i) $N_q(1) = \begin{cases} q + \lfloor 2\sqrt{q} \rfloor, & \textit{if } q \textit{ is exceptional}, \\ q + 1 + \lfloor 2\sqrt{q} \rfloor, & \textit{if } q \textit{ is non-exceptional}; \end{cases}$

(ii) $M_q(1) = \begin{cases} q + 2 - \lfloor 2\sqrt{q} \rfloor, & \textit{if } q \textit{ is exceptional}, \\ q + 1 - \lfloor 2\sqrt{q} \rfloor, & \textit{if } q \textit{ is non-exceptional}. \end{cases}$

Proof. This is an immediate consequence of Theorem 9.92. □

COROLLARY 9.95 *The number* N_1 *takes every value between* $q + 1 - \lfloor 2\sqrt{q} \rfloor$ *and* $q + 1 + \lfloor 2\sqrt{q} \rfloor$ *if and only if* (a) $q = p$ *or* (b) $q = p^2$ *with* $p = 2$ *or* $p = 3$ *or* $p \equiv 11 \pmod{12}$.

COROLLARY 9.96 *For* $q \leq 128$, *the values that* N_1 *cannot take between* $M_q(1)$ *and* $N_q(1)$ *are given in Table 9.4. The values of* $M_q(1)$ *and* $N_q(1)$ *for* $q \leq 128$ *are given in Table 9.6.*

For any integer n and any prime divisor ℓ, let $v_\ell(n)$ be the highest power of ℓ dividing n; that is, $\prod_\ell \ell^{v_\ell(n)}$ is the prime decomposition of n. Also, write $\mathbf{Z}/(r)$ for the additive group of the ring $\mathbf{Z}_r$.

THEOREM 9.97 (Voloch, Rück) *The groups* $G_{\mathcal{F}(\mathbf{F}_q)}$ *of the cubics* $\mathcal{F}$ *occurring in Theorem 9.92 in cases* (i) – (vii) *are as follows*:

(i) $G_{\mathcal{F}(\mathbf{F}_q)} = \mathbf{Z}/(p^{v_p(N_1)}) \times \prod_{\ell \neq p} \left(\mathbf{Z}/(\ell^{r_\ell}) \times \mathbf{Z}/(\ell^{s_\ell})\right)$,
with $r_\ell + s_\ell = v_\ell(N_1)$ *and* $\min(r_\ell, s_\ell) \leq v_\ell(q-1)$;

Table 9.4 The values that N_1 cannot take between $M_q(1)$ and $N_q(1)$ for $q \leq 128$

q	Forbidden values
8	$7, 11$
16	$11, 15, 19, 23$
25	26
27	$22, 25, 31, 34$
32	$23, 27, 29, 31, 35, 37, 39, 43$
49	$43, 57$
64	$51, 53, 55, 59, 61, 63, 67, 69, 71, 75, 77, 79$
81	$67, 70, 76, 79, 85, 88, 94, 97$
	$106, 111, 116, 121, 131, 136, 141, 146$
128	$109, 111, 115, 117, 119, 121, 123, 125, 127,$
	$131, 133, 135, 137, 139, 141, 143, 147, 149$

(ii) ,(iii) $$G_{\mathcal{F}(\mathbf{F}_q)} = \begin{cases} \mathbf{Z}/(q+1) & \textit{when } q \not\equiv -1 \pmod 4, \\ \mathbf{Z}/(q+1) \textit{ or} & \\ \mathbf{Z}/(2) \times \mathbf{Z}/((q+1)/2) & \textit{when } q \equiv -1 \pmod 4; \end{cases}$$

(iv) $G_{\mathcal{F}(\mathbf{F}_q)} = \mathbf{Z}/(N_1)$;

(v) $G_{\mathcal{F}(\mathbf{F}_q)} = \mathbf{Z}/(\sqrt{N_1}) \times \mathbf{Z}/(\sqrt{N_1}),\ N_1 = (\sqrt{q} \pm 1)^2$;

(vi) $G_{\mathcal{F}(\mathbf{F}_q)} = \mathbf{Z}/(N_1)$;

(vii) $G_{\mathcal{F}(\mathbf{F}_q)} = \mathbf{Z}/(N_1)$.

EXAMPLE 9.98 For $F = X_0^3 + X_1^3 + X_2^3$, Table 9.6 gives the number N_1 of rational points of $\mathbf{v}(F)$ for $2 \leq q \leq 179$ with also the values of $N_q(1)$ and $M_q(1)$.

To calculate N_1, note the following.

1. When $q \equiv 0, -1 \pmod 3$, then
$$N_1 = q + 1.$$

2. When $q = p$ and $p \equiv 1 \pmod 3$, then
$$N_1 = q + 1 + A,$$
where $4q = A^2 + 27B^2$ with $A, B \in \mathbf{N}$ and $A \equiv 1 \pmod 3$.

3. $N_2 = N_1\{2(q+1) - N_1\}$. In particular, if $N_1 = q + 1$, then
$$N_2 = (q+1)^2.$$

4. When $q \equiv 1 \pmod 3$, then 9 divides N_1, as the nine inflexions form a subgroup.

Table 9.5 The number A when $q = p$ and $q \equiv 1 \pmod 3$

q	$4q$	$=$	$A^2 + 27B^2$	A
7	28	$=$	$1 + 27$	1
13	52	$=$	$25 + 27$	-5
19	76	$=$	$49 + 27$	7
31	124	$=$	$16 + 108$	4
37	148	$=$	$121 + 27$	-11
43	172	$=$	$64 + 108$	-8
61	244	$=$	$1 + 243$	1
67	268	$=$	$25 + 243$	-5
73	292	$=$	$49 + 243$	7
79	316	$=$	$289 + 27$	-17
97	388	$=$	$361 + 27$	19
103	412	$=$	$169 + 243$	13
109	436	$=$	$4 + 432$	-2
127	508	$=$	$400 + 108$	-20
139	556	$=$	$529 + 27$	-23
151	604	$=$	$361 + 243$	19
157	628	$=$	$196 + 432$	-14
163	652	$=$	$625 + 27$	25

9.10 CLASSIFICATION OF NON-SINGULAR CUBICS OVER $\mathbf{F}_q$

For plane non-singular cubics defined over $\mathbf{F}_q$, Salmon's Theorem 7.91 states that birational and projective equivalence over $\overline{\mathbf{F}}_q$ are the same. This is not true when equivalences over $\mathbf{F}_q$ are considered.

Nevertheless, the proof of Salmon's Theorem 7.91 gives some ideas for the case where $\mathcal{F}$ and $\mathcal{F}'$ are two non-singular plane cubic curves each having at least one point in $\mathrm{PG}(2, q)$. The first part of that proof shows indeed how to find a non-singular plane cubic Γ, birationally equivalent over $\mathbf{F}_q$ to $\mathcal{F}$, that has an inflexion point O in $\mathrm{PG}(2, q)$. From the second part, if there is a birational transformation $\mathfrak{T}$ over $\mathbf{F}_q$ taking $\mathcal{F}$ to $\mathcal{F}'$ and O to O', then the non-singular models Γ and Γ' are projectively equivalent over $\mathbf{F}_q$. This motivates the following definition: if $\mathcal{F}$, $\mathcal{F}'$ and a birational transformation $\mathfrak{T} : K(\mathcal{F}) \to K(\mathcal{F}')$ are defined over $\mathbf{F}_q$, then $\mathcal{F}$ and $\mathcal{F}'$ are $\mathbf{F}_q$*-isomorphic*.

Let A_q be the total number of $\mathbf{F}_q$- isomorphism classes and P_q the total number of projective equivalence classes over $\mathbf{F}_q$ Here, n_i, for $i = 0, 1, 3, 9$, is the number of projective equivalence classes over $\mathbf{F}_q$ with exactly i rational inflexions. Hence

$$A_q = n_9 + n_3 + n_1, \qquad P_q = n_9 + n_3 + n_1 + n_0.$$

The number of inequivalent types of plane non-singular cubic with a fixed number of $\mathbf{F}_q$- rational points can also be given. Let $A_q(m)$ and $P_q(m)$ be the number

Table 9.6 Number N_1 of points on the cubic $\mathbf{v}(X_0^3 + X_1^3 + X_2^3)$

q	2	3	4	5	7	8	9	11	13	16	17
N_1	3	4	9	6	9	9	10	12	9	9	18
$N_q(1)$	5	7	9	10	13	14	16	18	21	25	26
$M_q(1)$	1	1	1	2	3	4	4	6	7	9	10
q	19	23	25	27	29	31	32	37	41	43	47
N_1	27	24	36	28	30	36	33	27	42	36	48
$N_q(1)$	28	33	36	38	40	43	44	50	54	57	61
$M_q(1)$	12	15	16	18	20	21	22	26	30	31	35
q	49	53	59	61	64	67	71	73	79	81	83
N_1	63	54	60	63	81	63	72	81	63	82	84
$N_q(1)$	64	68	75	77	81	84	88	91	97	100	102
$M_q(1)$	36	40	45	47	49	52	56	57	63	64	66
q	89	97	101	103	107	109	113	121	125	127	128
N_1	90	117	102	1117	108	108	114	144	126	108	129
$N_q(1)$	108	117	122	124	128	130	135	144	148	150	150
$M_q(1)$	72	79	82	84	88	90	93	100	104	106	108
q	131	137	139	149	151	157	163	167	169	173	179
N_1	132	138	117	150	171	144	189	168	171	174	180
$N_q(1)$	154	161	163	176	176	183	189	193	196	200	206
$M_q(1)$	110	115	117	126	128	133	139	143	144	148	154

of inequivalent non-singular cubics, under $\mathbf{F}_q$-isomorphism and projective equivalence over $\mathbf{F}_q$, whose number of $\mathbf{F}_q$-rational points is exactly $q+1-m$. So

$$A_q = \textstyle\sum_m A_q(m), \qquad P_q = \textstyle\sum_m P_q(m).$$

In the formulas below, both certain Legendre and Legendre–Jacobi symbols are used:

$$\left(\frac{x}{3}\right) = \begin{cases} 1 & \text{if } x \equiv 1 \pmod 3, \\ 0 & \text{if } x \equiv 0 \pmod 3, \\ -1 & \text{if } x \equiv -1 \pmod 3; \end{cases}$$

$$\left(\frac{-4}{c}\right) = \begin{cases} 1 & \text{if } c \equiv 1 \pmod 4, \\ 0 & \text{if } c \equiv 0 \pmod 2, \\ -1 & \text{if } c \equiv -1 \pmod 4; \end{cases}$$

$$\left(\frac{-3}{c}\right) = \begin{cases} 1 & \text{if } c \equiv 1 \pmod 3, \\ 0 & \text{if } c \equiv 0 \pmod 3, \\ -1 & \text{if } c \equiv -1 \pmod 3. \end{cases}$$

THEOREM 9.99 (i) $A_q = 2q + 3 + \left(\frac{-4}{q}\right) + 2\left(\frac{-3}{q}\right)$;

(ii) $P_q = 3q + 2 + \left(\frac{-4}{q}\right) + \left(\frac{-3}{q}\right)^2 + 3\left(\frac{-3}{q}\right).$

It is necessary to define the *class number of an integral quadratic form.* Let

$$\mathcal{S} = \{f(X,Y) = aX^2 + bXY + cY^2 \mid a, b, c \in \mathbf{Z};\ a > 0\}.$$

Consider $G = SL(2, \mathbf{Z})$ acting on $\mathcal{S}$. For $\mathfrak{T}$ in G with $\mathfrak{T} = \begin{bmatrix} A & B \\ C & D \end{bmatrix}$ and $AD - BC = 1$, let

$$\mathfrak{T}(f) = a(AX + CY)^2 + b(AX + CY)(BX + DY) + c(BX + DY)^2.$$

With $\Delta(f) = b^2 - 4ac$, also $\Delta(\mathfrak{T}(f)) = \Delta(f)$. So all quadratic forms in the same orbit have the the same discriminant.

The *class number* $H(\Delta)$ is the number of orbits of G on quadratic forms in $\mathcal{S}$ with discriminant Δ.

THEOREM 9.100 *Let* $(C_1), (C_2)$ *be the following two conditions:*
$(C_1):\ c > a$ *and* $-a < b \leq a$;
$(C_2):\ c = a$ *and* $0 \leq b \leq a$.
The class number,

$$H(\Delta) = |\{(a, b, c) \in \mathbf{Z}^3 \mid b^2 - 4ac = \Delta;\ a > 0;\ (C_1) \text{ or } (C_2) \text{ is satisfied}\}|.$$

For $0 < -\Delta \leq 103$, the class number $H(-\Delta)$ is given in Table 9.7.

LEMMA 9.101 *Let* $q = p^h$.

(i) *For* $p \equiv 1 \pmod 3$, *there exists a unique solution for* s *such that*

$$(s, p) = 1,\ s \equiv q + 1 \pmod 9,\ s^2 - 4q = -3x^2 \text{ for some } x \in \mathbf{Z}.$$

(ii) *For* $p \equiv 1 \pmod 4$, *there exists a unique solution for* s *such that*

$$(s, p) = 1,\ s \equiv q + 1 \pmod 9,\ s^2 - 4q = -4x^2 \text{ for some } x \in \mathbf{Z}.$$

Define the integers a_0, a_1 as follows. When $q \equiv 1 \pmod 3$,

$$a_0 = \begin{cases} \text{the unique solution for } s \text{ in Lemma 9.101 (i)} & \text{if } p \equiv 1 \pmod 3, \\ 2\left(\frac{\sqrt{q}}{3}\right)\sqrt{q} & \text{if } p \not\equiv 1 \pmod 3; \end{cases}$$

when $q \equiv 1$ or $4 \pmod{12}$,

$$a_1 = \begin{cases} \text{the unique solution for } s \text{ in Lemma 9.101 (ii)} & \text{if } p \equiv 1 \pmod 4, \\ 2\left(\frac{\sqrt{q}}{3}\right)\sqrt{q} & \text{if } p \not\equiv 1 \pmod 4. \end{cases}$$

THEOREM 9.102 *The number* $A_q(m)$ *of* $\mathbf{F}_q$*- isomorphism classes of non-singular plane cubics over* $\mathbf{F}_q$, $q = p^h$, *with* $q + 1 - m$ *points, where* $|m| \leq 2\sqrt{q}$, *is given in Table 9.8. In all other cases,* $A_q(m) = 0$.

Table 9.7 Values of the class number

$-\Delta$	$H(\Delta)$	$-\Delta$	$H(\Delta)$	$-\Delta$	$H(\Delta)$
3	1	36	3	71	7
4	1	39	4	72	3
7	1	40	2	75	3
8	1	43	3	76	4
11	1	44	4	79	5
12	2	47	5	80	6
15	2	48	4	83	3
16	2	51	2	84	4
19	1	52	2	87	6
20	2	55	4	88	2
23	3	56	4	91	2
24	2	59	3	92	6
27	2	60	4	95	8
28	2	63	5	96	6
31	3	64	4	99	3
32	3	67	1	100	3
35	2	68	4	103	5

Table 9.8 Number of $\mathbf{F}_q$- isomorphism classes of cubics

			m	$A_q(m)$
(i)			$(m,p)=1$	$H(m^2-4q)$
(ii)	h odd			
		(a)	0	$H(-4p)$
		(b) $p=2$	$\pm\sqrt{2q}$	1
		(c) $p=3$	$\pm\sqrt{3q}$	1
(iii)	h even			
		(a)	0	$1-\left(\frac{-4}{p}\right)$
		(b)	$\pm\sqrt{q}$	$1-\left(\frac{-3}{p}\right)$
		(c)	$\pm 2\sqrt{q}$	D_p

where $D_p = \frac{1}{12}\left[p+6-4\left(\frac{-3}{p}\right)-3\left(\frac{-4}{p}\right)\right]$

For the projective classification over $\mathbf{F}_q$, define the functions $B(m), C(m)$ as follows. Let $B(m)$ be the number of isomorphism classes of elliptic curves $\mathcal{F}$ with 3 dividing $N_1(\mathcal{F}) = q+1-m$; also, let $C(m)$ be the number of isomorphism classes of elliptic curves $\mathcal{F}$ with $N_1(\mathcal{F}) = q+1-m$. The group $G_{\mathcal{F}}[3] \cong \mathbf{Z}_3 \times \mathbf{Z}_3$, where $G_{\mathcal{F}}[3]$ is the subgroup of elements of order a power of 3 in the group $G_{\mathcal{F}(\mathbf{F}_q)}$ of the curve.

Theorem 9.103 *The number $P_q(m)$ of projectively distinct non-singular plane cubics over $\mathbf{F}_q$, $q = p^h$, with $q+1-m$ points, where $|\, m \,| \leq 2\sqrt{q}$, is as follows:*

$$P_q(m) = A_q(m) + B(m) + 3C(m) - \epsilon(m),$$

where

$$\epsilon(m) = \begin{cases} 2, & \text{if } m = a_0 \text{ or } a_1 \text{ with } a_0 \neq a_1; \\ 3, & \text{if } m = a_0 = a_1 \text{ and } p = 2; \\ 4, & \text{if } m = a_0 = a_1 \text{ and } p \neq 2; \\ 0, & \text{otherwise.} \end{cases}$$

9.11 EXERCISES

1. Show the following corollary of Weil's Explicit Formula (9.39). If $f(\alpha) \geq 0$ for all $\alpha \in \mathbf{R}$ and $c_n \geq 0$ for all $n \geq 1$, then

$$N_1 \leq \frac{g}{\Psi_1(\sqrt{q^{-1}})} + \frac{\Psi_1(\sqrt{q})}{\Psi_1(\sqrt{q^{-1}})} + 1;$$

equality holds if and only if

$$\textstyle\sum_{j=1}^{g} f(\alpha_j) = 0, \qquad \sum_{d=2}^{\infty} dB_d \Psi_d(t) = 0.$$

2. Let $\mathcal{F}$ be a plane curve of degree n over $\mathbf{F}_q$ which splits into n rational lines. Prove the following upper and lower bounds on the number R_q of points of $\mathrm{PG}(2,q)$ lying on $\mathcal{F}$:

$$nq - \tfrac{1}{2}n(n-3) \leq R_q \leq nq + 1.$$

Show also that the lower limit is achieved if and only if no three of the n lines are concurrent while the upper limit is achieved if and only if the n lines are concurrent.

3. For $n \geq 1$, let $h_{\mathbf{F}_{q^n}}(\mathcal{X})$ be the class number of $\mathcal{X}$. Also, for a prime d, let

$$\delta(d) = \begin{cases} \operatorname{lcm}(d-1, \ldots, d^i - 1, \ldots, d^{2g} - 1) & \text{when } d \neq p, \\ \operatorname{lcm}(d-1, \ldots, d^i - 1, \ldots, d^{g} - 1) & \text{when } d = p. \end{cases}$$

Prove the following.

(a) If d is a prime dividing $h_{\mathbf{F}_q}(\mathcal{X})$ and if $d \mid n(q^n-1)/(q-1)$, then also $d \mid h_{\mathbf{F}_{q^n}}(\mathcal{X})/h_{\mathbf{F}_q}(\mathcal{X})$.

(b) If $d \nmid h_{\mathbf{F}_q}(\mathcal{X})$ and n is relatively prime to $\delta(d)$, then $d \nmid h_{\mathbf{F}_{q^n}}(\mathcal{X})$.

(iii) If n is the smallest integer for which $d \mid h_{\mathbf{F}_{q^n}}(\mathcal{X})$, then $n \mid \delta(d)$.

4. Let $p \equiv 3 \pmod 4$ and $q = p^{2n}$ for n odd. For $\lambda \neq -3, 1, 0$, let $\mathcal{F}_\lambda$ be the irreducible $\mathbf{F}_q$-rational quartic of genus 3 with

$$\mathcal{F}_\lambda = \mathbf{v}(X_0^4 + X_1^4 + X_2^4 - (\lambda + 1)(X_0^2 X_1^2 + X_1^2 X_2^2 + X_2^2 X_0^2)).$$

Show that $\mathcal{F}_\lambda$ attains the Hasse–Weil Bound; that is, $S_q = q + 1 + 6p^n$. Compare Chapter 10, Exercise 1.

5. Let $\mathcal{F}$ be an irreducible $\mathbf{F}_q$-rational curve of genus 3 with more than $2q + 6$ $\mathbf{F}_q$-rational points. Show that one of the following holds:

(a) $q = 8,\ N_1 = 24$, and $\mathcal{F}$ is birationally equivalent over $\mathbf{F}_8$ to the curve

$$\begin{aligned}\mathbf{v}(\mathbf{X}_0^4 &+ X_1^4 + X_2^4 + X_0^2 X_1^2 + X_1^2 X_2^2 \\ &+ X_0^2 X_2^2 + X_0^2 X_1 X_2 + X_1^2 X_0 X_2 + X_2^2 X_0 X_1);\end{aligned}$$

(b) $q = 9,\ N_1 = 28$, and $\mathcal{F}$ is birationally equivalent over $\mathbf{F}_q$ to the Hermitian curve

$$\mathbf{v}(X_0^4 + X_1^4 + X_2^4).$$

6. Show that an irreducible $\mathbf{F}_3$-rational curve of genus 5 has $N_1 \leq 13$.

7. Show that the irreducible $\mathbf{F}_5$-rational curve

$$\mathbf{v}(X^4 + X^3 Y^3 - X^2 - XY^5 + Y^5 + 2Y)$$

of genus 3 has $N_1 = 13$.

8. Show that an irreducible $\mathbf{F}_8$-rational curve of genus 4 has $N_1 \leq 25$.

9. Let $\mathcal{X} = \mathbf{v}(Y^3 + Y + 1 - X^3 Y)$ be defined over $\mathbf{F}_2$. Show that

(a) $\mathcal{X}$ is non-singular and has genus 3;

(b) $N_1 = 3,\ N_2 = 11,\ N_3 = 9$;

(c) $L_2(t) = 1 + 3t^2 + 6t^4 + 8t^6 = (1 + 2t^2)(1 + t^2 + 4t^4)$;

(d) the 2-rank of $\mathcal{X}$ is 2.

10. Over $\mathbf{F}_2$, let

$$\mathcal{F} = \mathbf{v}(Y^6 + Y^5 + Y^4 + Y^3 + Y^2 + Y + 1 - X^3(Y^2 + Y)).$$

Show that

(a) $\mathcal{F}$ has only one singular point X_∞;

(b) X_∞ is an ordinary triple point, and each of the branches of $\mathcal{F}$ centred at X_∞ is $\mathbf{F}_2$-rational;

(c) $\mathcal{F}$ has genus $g = 7$;

(d) $N_1 = 3,\ N_2 = 9,\ N_3 = 9,\ N_4 = 45,\ N_5 = 33, \mathrm{N}_6 = 69,\ N_7 = 129$;

(e) $$\begin{aligned} L_2(t) &= 1 + 2t^2 + 9t^4 + 16t^6 + 32t^8 + 72t^{10} + 64t^{12} + 128t^{14} \\ &= (1 + 2t^2)(1 - 2t^2 + 4t^4)(1 + t^2 + 4t^4)^2; \end{aligned}$$

(f) the 2-rank of $\mathcal{F}$ is 4;

(g) $\mathcal{X}$, as given in Example 9, is covered over $\mathbf{F}_2$ by $\mathcal{F}$.

11. Let $\mathcal{F} = \mathbf{v}(Y^q - Y - f(X))$ be an Artin–Schreier curve defined over $\mathbf{F}_{q^m}$, where

$$f(X) = \lambda_1 X^{i_1} + \lambda_2 X^{i_2} + \cdots + \lambda_s X^{i_s}, \qquad \lambda_j \in \mathbf{F}_{q^m}.$$

Show that $\mathcal{F}$ has at most q^{m+1} points lying in $\mathrm{AG}(2, q^m)$.

12. Define the symmetric polynomial $\sigma_2 \in \mathbf{F}_q[X_0, \ldots, X_{r-1}]$ as follows:

$$\sigma_2(X_0, \ldots, X_{r-1}) = \sum X_i X_j,$$

summed for $0 \leq i < j \leq r - 1$, and let

$$S(X) = \sigma_2(X, X^q, \ldots, X^{q^{r-1}}).$$

For $r \geq 2$, let $\mathcal{F} = \mathbf{v}(f(X, Y))$ be the irreducible plane curve defined over $\mathbf{F}_q$, where

$$f(X, Y) = Y^{q^{r-1}} + \cdots + Y^q + Y - S(X).$$

Show that $\mathcal{F}$ has genus $g = \frac{1}{2}q^{r-1}(q^{r-1} - 1)$ and that the number N_r of its $\mathbf{F}_{q^r}$-rational points is $1 + q^{2r-1}$.

13. Let $\mathcal{F} = \mathbf{v}(f(X, Y))$ be the irreducible plane curve defined over $\mathbf{F}_q$, where

$$f(X, Y) = Y^{q^3+q^2+q+1} - (X^{q^3+q} + X^{q^2+1}).$$

Show that $\mathcal{F}$ has genus $g = \frac{1}{2}\, q(q^5 + q^3 - q^2 - 2q - 1)$ and that the number N_4 of its $\mathbf{F}_{q^4}$-rational points is $(q^2 + 1)(q^5 + 1)$.

14. For $q = p^m$ with $p > 3$, a Picard curve over $\mathbf{F}_q$ is a non-singular planar curve, $\mathcal{F} = \mathbf{v}(Y^3 - p_4(X))$, where $p_4(X)$ is a polynomial of degree 4 with distinct roots. For $j \in \mathbf{N}$, let

$$c_{ij} = \begin{cases} \text{the coefficient of } X^i \text{ in } p_4(X)^{q-1-j/3} & \text{for } 3 \mid j, \\ 0 & \text{for } 3 \nmid j. \end{cases}$$

Show that

$$S_q \equiv \begin{cases} 1 \pmod{p} & \text{for } 3 \mid (p-2), \\ 1 - c_{q-1,q-1} - c_{2q-1,q-1} - c_{q-1,2q-1} \pmod{p} & \text{for } 3 \nmid (p-2). \end{cases}$$

15. For $p > 2$ and $c \in \mathbf{F}_q \backslash \{0\}$, let $\mathcal{F} = \mathbf{v}(Y^2 - X^{2g+1} + c)$ be the irreducible hyperelliptic curve of genus g with $\gcd(2g + 1, p) = 1$. Show that the p-rank of $\mathcal{F}$ is zero if and only if there exists $n \in \mathbf{N}$ such that $p^n \equiv -1 \pmod{4g + 2}$.

16. Let $\mathcal{F}$ be a non-singular plane cubic defined over $\mathbf{F}_q$ that has an inflexion point in $\mathrm{PG}(2,q)$. Prove the following.

    (a) A canonical form for $\mathcal{F}$ over $\mathbf{F}_q$ is

    $$\mathbf{v}(Y^2Z + a_1XYZ + a_3YZ^2 - X^3 - a_2X^2Z - a_4XZ^2 - a_6Z^3).$$

    (b) Let

    $$\begin{aligned}&b_2 = a_1^2 + 4a_2,\ b_4 = a_1a_3 + 2a_4,\ b_6 = a_3^2 + 4a_6,\\ &b_8 = a_1^2a_6 - a_1a_3a_4 + 4a_2a_6 + a_2a_3^2 - a_4^2,\\ &c_4 = b_2^2 - 24b_4,\ c_6 = -b_2^3 + 36b_2b_4 - 216b_6,\\ &\Delta = -b_2^2b_8 - 8b_4^3 - 27b_6^2 + 9b_2b_4b_6.\end{aligned}$$

    Then the *discriminant* Δ does not vanish and, for $p \neq 2, 3$, $1728\Delta = c_4^3 - c_6^2$.

    (c) The j-invariant of $\mathcal{F}$ is equal to c_4^3/Δ.

    (d) For $p \neq 2$, the cubic $\mathcal{F}$ becomes the curve

    $$\mathbf{v}(Y^2Z - 4X^3 - b_2X^2Z - 2b_4XZ^2 - b_6Z^3)$$

    with $4b_8 = b_2b_6 - b_4^2$.

    (v) If μ is given by $4\mu = 27j/(j - 1728)$, then $\mathcal{F}$ is *equianharmonic* for $\mu = 0, j = 0$ and *harmonic* for $\mu = \infty, j = 1728$.

9.12 NOTES

The general theory of zeta functions of an arbitrary fields of functions was developed by Schmidt in [384]; see also [374]. The presentation in Section 9.1, as in several other books, follows Hasse [195]. An introduction to algebraic varieties over a finite field is found in Joly [253].

Theorem 9.17 is proved in [21, Proposition 5].

The proof of Proposition 9.22 comes from [432]. Proposition 9.23 was proved by Bombieri [54], [55] via the approach initiated by Stepanov [418] and Schmidt [388]; see also [389] and [467]. The original proof of Theorem 9.18 is due to Weil [504].

For the Möbius Inversion Formula, see, for example, [310, Theorem 3.24],

Curves with many $\mathbf{F}_q$-rational points have important applications to coding theory, and especially to algebraic-geometry codes. The methods for constructing and investigating such curves vary widely. Linear equations and quadratic forms were used by van der Geer and van der Vlugt [478, 480] and by Kawakita and Miura [265]. An elementary approach to the construction of curves of Kummer type with many $\mathbf{F}_q$-rational points is due to van der Geer and van der Vlugt [482]; see also Garcia and Quoos [146], Garcia and Garzon [143] and Kawakita [262]. On the other hand, sophisticated methods from algebraic geometry are often effective. Inspired by Serre's work, Lauter [296, 297, 298, 299] and Howe and

Lauter [237] used Galois descent and Honda–Tate theory, endomorphisms of Jacobians of curves, Hermitian modules, and ray class fields. For constructions involving fibre products of Artin–Schreier curves, see van der Geer and van der Vlugt [476, 477, 479], Stepanov [420], Stepanov and Shalalfekh [422]. For cyclotomic function fields and Drinfeld modules of rank 1, see Neiderreiter and Xing [347] and [350, Chapter 3].

Hurt's monograph [242] provides an overview of these methods and related results, including applications to coding theory. For survey papers, see van der Geer and van der Vlugt [479, 480], van der Geer [470, 471, 472], Garcia [141], Hajir and Maire [186], Garcia and Stichtenoth [153], and Voight [490].

Related to Remark 9.29, Fuhrmann and Torres [133] showed that, if q is square, then no curve of genus g, with

$$\tfrac{1}{4}(\sqrt{q}-1)^2 < g < \tfrac{1}{2}(q-\sqrt{q}),$$

attains the Serre Bound. Also, Serre showed the non-existence of curves of genus $g > 2$ that miss his bound by just one. In other terms, if a curve of *defect* k is an $\mathbf{F}_q$-rational curve for which $N_1 = q+1+\lfloor 2g \rfloor q - k$ with $k \geq 1$, then no curve of genus $g > 2$ is a curve of defect 1. Serre also provided a list of possible zeta functions for curves of defect $k \leq 6$ and genus g. Using this list, together with results on abelian varieties, Lauter [298] found that the Serre Bound can only be obtained if

$$g \leq \frac{q^2-q}{\lfloor 2\sqrt{q} \rfloor + \lfloor 2\sqrt{q} \rfloor^2 - 2q}.$$

The non-existence of various curves of defect 2 and genus $g > 2$ was also shown; for instance, no such curve exists when $q = 2^{2s}, s > 1$.

On the positive side, Lauter [299] showed that, for all finite fields $\mathbf{F}_q$, either there exists a curve of genus 3 and defect k with $k \leq 3$, or a curve is within three of Serre's lower bound. In [294] a method is given for improving the Serre Bound for a wide range of small genera, namely,

$$g \leq (q^2-q)/(\lfloor 2\sqrt{q} \rfloor + \lfloor 2\sqrt{q} \rfloor^2 - 2q),$$

for $q = 8, 32, 2^{13}, 27, 243, 125$ and 2^{2s}, $s > 1$. For more results, see [237] and, especially for a survey on curves of low defect and genus $g \leq 5$, see [242].

Proposition 9.37 implies that

$$g \geq \sup\{(N_1-1)\lambda_m(\sqrt{q^{-1}}) - \lambda_m(\sqrt{q})\}, \tag{9.77}$$

taken over all $\lambda_m(t)$ with $m \in \mathbf{N}$. This bound was optimised by Oesterlé; see [391]: the right-hand side of (9.77) equals

$$\frac{(N_1-2)\sqrt{q}\cos\alpha_0 + q - (N_1-1)}{q+1-2\sqrt{q}\cos\alpha_0},$$

where α_0 is defined as follows. If k be the unique integer for which

$$\sqrt{q}^{\,k} + 1 \leq N_1 \leq \sqrt{q}^{\,k+1} + 1$$

and

$$u = \frac{\sqrt{q}^{\,k+1} - (N_1-1)}{(N_1-1)\sqrt{q} - \sqrt{q}^{\,k}}.$$

Then $\alpha_0 \in [0,1)$ and is the unique solution of the equation

$$\cos\tfrac{1}{2}(k+1)X + u\cos\tfrac{1}{2}(k-1)X = 0$$

in the interval $[\lfloor\pi/(k+1)\rfloor, \pi/k)$.

For irreducible singular space curves, Zúñiga-Galindo [520] refined the Hasse-Weil Bound 9.14 in terms of their singular points.

Using certain hyperplane configurations in $\mathrm{AG}(r,q)$, Özbudak and Stichtenoth [356] constructed curves whose number of $\mathbf{F}_q$-rational points is close to Oesterlé's bound.

The problem of determining the exact value of $N_q(g)$ for small g was initiated by Serre [405, 407], giving $N_q(1)$ and $N_q(2)$.

A computer-assisted search by Luengo and Lopez [315] provided irreducible plane curves defined over $\mathbf{F}_3$ of genus $4 \leq g \leq 8$ that have many $\mathbf{F}_3$-rational points, giving the following lower bounds:

$$N_3(g) \geq \begin{cases} 12 & \text{for } g=4,\\ 13 & \text{for } g=5,\\ 14 & \text{for } g=6,\\ 16 & \text{for } g=7,\\ 17 & \text{for } g=8. \end{cases}$$

On the other hand, $N_3(5) \leq 14$ and $N_3(7) \leq 17$. For lower bounds on $N_3(g)$ for $g \geq 51$, see also [349].

The lower bound (9.43) was shown by Ihara [247], and by Tsfasman and Vlăduţ [460, 462]. The constructive but not explicit proofs of these results are based on modular curves and class field theory; see [247], [462], [336], [405], [519], [392], [508], [350], [118], [283], [119].

In 1995, Garcia and Stichtenoth [148] gave the first explicit example of a sequence of curves over $\mathbf{F}_q$ with square q attaining the Drinfeld-Vlăduţ Bound. The tower in Example 9.40 comes from that paper; see also Elkies [116]. The $\mathbf{F}_{q^2}$-tower in Example 9.41 is recursive; see Garcia and Stichtenoth [150].

Theorem 9.42 is due to Serre [405]. In the case that q is a proper power of p, that is, $q > p$, this also follows from Example 9.43, due to Garcia, Stichtenoth and Thomas [157].

In Remark 9.44, the lower bounds in (i) for $q = 2,3,5$ are due to Schoof [392] and Xing [508]. For improvements, see Niederreiter and Xing [348], Angles and Maire [15] and Temkine [452]. The result in (ii) is due to Zink [519]; that in (iii) is due to Perret [363]. For improvements to (iii), see Temkine [452] and [308].

Following the paper by Garcia and Stichtenoth [148], much work has been done on towers, especially on those given explicitly, by Beelen, Bezera, Elkies, Garcia, Li, Ling, Maharaj, Rück, Stichtenoth, van der Geer and van der Vlugt, Wulftange, Zaytsev and Zieve; see [142], [483], [156], [41], [42], [43], [44], [50], [51], [110], [309], [320], [507], [514] [153], [311]. For a survey, see [155].

Surveys on $A^+(q)$ are found in [142], [350, Chapter 5], [421, Chapters 6, 9], and in [460, Chapter 4].

Theorems 9.45, 9.46 and 9.47 on $A^-(q)$ come from [119].

The Lüroth semigroup S was introduced by Heinzer and Moh [331]. Greco and Raciti [181] considered irreducible plane curves $\mathcal{F}$ of degree $d \geq 4$ with only ordinary singularities. They showed that if $(a-1)d+1 \leq n \leq ad-(a^2+1)$ for $a \geq 1$ then $n \notin S$.

For the gonality of a curve of degree m, see Homma [229]. The connection between Clifford's Theorem 6.79 and gonality was investigated by Hartshorne [193].

For Proposition 9.52, see [360].

Proposition 9.54 is due to Lewittes [307].

Theorems 9.57, 9.58 and Corollaries 9.59, 9.60 are from [300]; see also [20] and [24].

Theorems 9.62 and 9.63 come from [220]. Proposition 9.65 is due to Sziklai [442]. For an upper bound on the difference between the number of points on a curve over a $\mathbf{F}_q$ and on a certain covering of the curve, see [22]. For more results similar to Theorems 9.67 and 9.69, see [216]. For curves with no rational points, see [511].

Bounds on the number of $\mathbf{F}_q$-rational points have widely been used in the study of polynomials in one variable over finite fields. The main reference is [310]. Some applications on the Hasse–Weil Bound 9.18 to graph theory and coding theory are treated in Cohen [83].

The proof of Theorem 9.72 comes from Frey [130]. Theorems 9.75 and 9.76 and related proofs are due to Stichtenoth; see [425].

For proofs of the results on the theory of divisibility in Section 9.7, see [56]; see also [352].

A proof of Theorem 9.79 via Castelnuovo's inequality can be found in [114, Chapter V. 5.4].

The propositions on the Frobenius linear series in Section 9.8 come from [134].

The presentation of the general theory of the abelian varieties is beyond the scope of this book. For background on this topic, see the books by Mumford [338, 339], Lang [292], and Milne [330]; see also the papers by Tate [449], Mazur [326], and Waterhouse and Milne [501]. Theorems 6.94 and 6.96 proved in Section 9.7, describing the structure of $\mathrm{Pic}_0(K(\mathcal{X}))$, extend to any abelian variety.

For the fact that $\mathrm{Pic}_0(K(\mathcal{X}), l_\infty)$ is a divisible group, see [313, Theorem XI.3.2].

Proofs, for abelian varieties, of the last results in Section 9.8 can be found in [292, Chapter VII].

For computations in $\mathrm{Pic}_0^{(\mathbf{F}_p)}(\mathcal{F})$ of curves $\mathcal{F}$ of type $\mathbf{v}(Y^3-\gamma X^5+\delta)$, see [53].

Theorem 9.92 giving the possible numbers of rational points on an elliptic curve is due to Waterhouse [500]; see Ughi [466] for another proof. For the bounds on N_1, see Zimmer [517, 518]. Theorem 9.97 giving the group structure of the set of rational points is due independently to Rück [377] and to Voloch [491].

The formulas for the numbers of non-isomorphic and projectively distinct cubics with a certain number of rational points are due to Schoof [390]; for the prime case, see also Deuring [101]. For Theorem 9.91, see Schoof [390] and Tate [450].

For a summary of non-isomorphic cubics, see also Menezes [328, Chapter 7].

For the theory of singular plane cubics, see [216, §11.4].

For more general accounts of elliptic curves, see Cassels [71], Husemöller [243], Koblitz [273], Silverman [416], Tate [450]. For classical accounts of plane cubics, see Enriques [122], Hilton [212], Salmon [380], Seidenberg [400], Walker [497].

For the projective classification of plane cubics, including the cases of characteristic 2 and 3, see [216, Chapter 11]; the values of n_0, n_1, n_3, n_9 are given in Tables 11.28 to 11.30.

For Exercises 4 and 5, see [23]; for Exercise 6, see [295]; for Exercise 7, see [242, Chapter 2, Section 6]; for Exercise 8, see [381]; for Exercises 12 and 13, see [152]; for Exercise 11, see [183]; for Exercise 14, see [124]; for Exercise 15, see [276].

PART 3
Further developments

Chapter Ten

Maximal and optimal curves

In the study of curves with many $\mathbf{F}_q$-rational points the main problem is to determine $N_q(g)$, the largest number of $\mathbf{F}_q$-rational points that an irreducible $\mathbf{F}_q$-rational curve of genus g can have. The asymptotic behaviour of $N_q(g)$ with respect to g together with the Drinfeld–Vlăduţ Bound (9.38) are discussed in Section 9.3, where some examples with $N_q(g) \approx (\sqrt{q}-1)g$ for a fixed square q and for g large enough are also constructed. In general, $N_q(g)$ is known for only a few pairs (q, g).

The aim of this chapter is threefold:

(1) to describe the theory of $\mathbf{F}_q$-*maximal curves*, that is, curves of genus g defined over $\mathbf{F}_q$ whose number S_q of $\mathbf{F}_q$-rational points is $q + 1 + 2g\sqrt{q}$, the Hasse–Weil Bound of Theorem 9.18;

(2) to give the known examples of maximal curves and some of their characterisations;

(3) to describe some examples of $\mathbf{F}_q$-*optimal curves*, that is, curves of genus g defined over $\mathbf{F}_q$ whose number of $\mathbf{F}_q$-rational points equals $N_q(g)$.

The Hasse–Weil Bound implies that $N_g(q) \le q + 1 + 2g\sqrt{q}$. If equality holds, then the $\mathbf{F}_q$-optimal curve is $\mathbf{F}_q$-maximal, and $\mathbf{F}_q$-maximal curves of genus $g > 0$ can only exist for q square. An $\mathbf{F}_q$-maximal curve is, by definition, $\mathbf{F}_q$-optimal.

For $g = 0$, every curve is $\mathbf{F}_q$-optimal, as the number of $\mathbf{F}_q$-rational points of the projective line over $\mathbf{F}_q$ is $q + 1$, while $N_q(0) \le q + 1$ by the Hasse–Weil Bound 9.18.

Non-trivial examples of maximal and optimal curves appear in other chapters:

(1) the Hermitian curve, Example 1.40, is both $\mathbf{F}_q$-optimal and $\mathbf{F}_q$-maximal for $q = p^{2e}$;

(2) the DLS curve, Example 5.24, is both $\mathbf{F}_q$-optimal and also $\mathbf{F}_{q^4}$-maximal for $q = 2^{2e+1}$ with $e \ge 1$;

(3) the DLR curve, Section 12.4, is $\mathbf{F}_q$-optimal and $\mathbf{F}_{q^6}$-maximal for $q = 3^{2e+1}$ with $e \ge 1$.

Notation and terminology are the same as in Chapter 9. To avoid trivial cases, it is assumed that $g > 0$, unless otherwise stated.

First, maximal curves are investigated. So the underlying field is $\mathbf{F}_{q^2}$.

10.1 BACKGROUND ON MAXIMAL CURVES

A characterisation of maximal curves can be deduced from the Hasse–Weil Theorem 9.19 and (9.10) for $n = 1$.

THEOREM 10.1 *When $\mathcal{X}$ be an irreducible non-singular curve of genus g defined over $\mathbf{F}_{q^2}$, then the following conditions are equivalent:*

(i) *$\mathcal{X}$ is $\mathbf{F}_{q^2}$-maximal, that is, $N_1 = q^2 + 1 + 2qg$;*

(ii) *$\omega_i = -q$ for $i = 1, 2, \ldots, 2g$;*

(iii) *$L_{\mathbf{F}_{q^2}}(t) = (t + q)^{2g}$.*

Also, if $\mathcal{X}$ is $\mathbf{F}_{q^2}$-maximal and N_m is the number of its $\mathbf{F}_{q^{2m}}$-rational points, then

$$N_m = q^{2m} + 1 + (-1)^{m-1} 2gq^m, \qquad \text{for } m = 1, \ldots. \tag{10.1}$$

Theorem 10.1 shows that, if the curve is maximal then its zeta function has particularly simple form. Consequently, the zeta function via its enumerator $L_{q^2}(t)$, the corresponding fundamental equation, the Frobenius linear series, and the automorphisms combine to produce significant results.

Beginning with L-polynomials, Theorem 9.17 together with Theorem 10.1 has the following corollary.

THEOREM 10.2 *If $\mathcal{X}'$ is an $\mathbf{F}_{q^2}$-rational curve covered by an $\mathbf{F}_{q^2}$-maximal curve $\mathcal{X}$, with an $\mathbf{F}_{q^2}$-rational covering, then $\mathcal{X}'$ is also $\mathbf{F}_{q^2}$-maximal.*

There are $\mathbf{F}_{q^2}$-maximal curves with simple equations such as those of Kummer type

$$\mathcal{K}_n = \mathbf{v}(Y^{q+1} - f(X)), \tag{10.2}$$

and those of Artin–Schreier type

$$\mathcal{A} = \mathbf{v}(Y^q - Y - f(X)), \tag{10.3}$$

as well as the Hermitian curve

$$\mathcal{H}_q = \mathbf{v}(Y^q + Y - X^{q+1}). \tag{10.4}$$

EXAMPLE 10.3 (i) For q odd, the irreducible plane curve

$$\mathcal{E}^1_{(q+1)/2} = \mathbf{v}(Y^q + Y - X^{(q+1)/2}) \tag{10.5}$$

is covered by the Hermitian curve $\mathcal{H}_q$, since $K(\mathcal{E}^1_{(q+1)/2})$ is the subfield $K(x^2, y)$ of $K(\mathcal{H}_q) = K(x, y)$. Note that $K(x^2, y)$ is a proper subfield of $K(\mathcal{H}_q)$ with $y^q + y - x^{q+1} = 0$, as the genus g' of $\mathcal{E}^1_{(q+1)/2}$ is smaller than $\frac{1}{2}(q^2 - q)$, the genus of $\mathcal{H}_q$. Therefore $[K(\mathcal{H}_q) : K(\mathcal{E}^1_{(q+1)/2})] = 2$, and the rational transformation $\omega : (x, y) \mapsto (x^2, y)$ provides a two-fold covering of $\mathcal{E}^1_{(q+1)/2}$ by $\mathcal{H}_q$. From Lemma 12.1, $g' = \frac{1}{4}(q - 1)^2$. Since $x^2, y \in \mathbf{F}_{q^2}(\mathcal{H}_q)$, then $\mathcal{E}^1_{(q+1)/2}$, ω and

Table 10.1 Families of curves $\mathcal{F} = \mathbf{v}(F)$ containing $\mathbf{F}_{q^2}$-maximal curves

$\mathcal{F}$	F
$\mathcal{D}_r$	$X^r + Y^r + 1$
$\mathcal{E}^1_m$	$Y^q + Y - X^m$
$\mathcal{T}_p$	$X^{q+1} - (Y + Y^p + Y^{p^2} + \cdots + Y^{q/p})$
$\mathcal{T}'_3$	$(Y + Y^3 + \cdots + Y^{q/3})^2 - X^q - X$
$\mathcal{T}''_3$	$Y + Y^3 + \cdots + Y^{q/3} + cX^{q+1},\ c^{q-1} = -1$
$\mathcal{K}$	$Y^{q+1} - f(X)$
$\mathcal{A}$	$Y^q - Y - f(X)$
$\mathcal{F}_0$	$X^{(q+1)/3} + X^{2(q+1)/3} + Y^{q+1}$
$\mathcal{F}'_0$	$YX^{(q-2)/3} + Y^q + X^{(2q-1)/3}$
$\mathcal{G}$	$Y^q - YX^{2(q-1)/3} + \omega X^{(q-1)/3},\ \omega^{q+1} = -1$
$\mathcal{C}_n$	$X^nY + Y^n + X$
$\mathcal{C}_{n,k}$	$X^nY^k + Y^n + X^k$
$\mathcal{C}^m_i$	$X^{mi+m} + X^{mi} + Y^{q+1}$
$\mathcal{X}_r$	$Y^{2^r} + a_1Y^{2^{r-1}} + \cdots + a_{r-1}Y^2 + Y + X^{q+1}$

the associated two-fold covering are $\mathbf{F}_{q^2}$-rational, as well. From Theorem 10.2, $\mathcal{E}^1_{(q+1)/2}$ is an $\mathbf{F}_{q^2}$-maximal curve. See also Example 10.35.

(ii) The corresponding example for q even is the irreducible plane curve

$$\mathcal{T}_2 = \mathbf{v}(X^{q+1} + Y + Y^2 + Y^4 + \cdots + Y^{q/2}), \qquad (10.6)$$

whose genus is $\frac{1}{4}q(q-2)$ by Lemma 12.1. Since $K(\mathcal{T}_2)$ is the subfield $K(x, y^2+y)$ of $K(\mathcal{H}_q)$, the same argument shows that $\mathcal{T}_2$ is two-fold covered by the Hermitian curve $\mathcal{H}_q$, and hence it is an $\mathbf{F}_{q^2}$-maximal curve. See also Example 10.36.

(iii) See Example 10.33 for the curve $\mathcal{F}_0 = \mathbf{v}(F_0)$ with

$$F_0 = X^{(q+1)/3} + X^{2(q+1)/3} + Y^{q+1},$$

when $q \equiv 2 \pmod 3$; its genus is $\frac{1}{6}(q^2 - q + 4)$.

(iv) See Example 10.34 for the curve $\mathcal{F}'_0 = \mathbf{v}(F'_0)$ with

$$F'_0 = YX^{(q-2)/3} + Y^q + X^{(2q-1)/3},$$

when $q \equiv 2 \pmod 3$; its genus is $\frac{1}{6}(q^2 - q - 2)$.

(v) For curves of genus $\frac{1}{6}(q^2 - q)$, examples are $\mathcal{T}'_3 = \mathbf{v}(T'_3)$ with

$$T'_3 = \mathrm{T}(Y)^2 - X^q - X, \quad \mathrm{T}(Y) = Y + Y^3 + \cdots + Y^{q/3},$$

when $q \equiv 0 \pmod 3$ as in Example 10.37, and $\mathcal{G} = \mathbf{v}(G)$ with

$$G = Y^q - YX^{2(q-1)/3} + \omega X^{(q-1)/3}, \quad \omega^{q+1} = -1,$$

when $q \equiv 1 \pmod 3$ as in Exercise 11.

As in Section 11.5, every non-trivial automorphism group G of $\mathcal{X}$ gives rise to a covering of $\mathcal{X}$. In Example 10.3 (i), $\mathcal{F} = \mathcal{H}_q/\langle\alpha\rangle$ with $\alpha : (X,Y) \mapsto (-X,Y)$. Such a covering and the corresponding quotient curve $\mathcal{X}' = \mathcal{X}/G$ are $\mathbf{F}_{q^2}$-rational if G is an $\mathbf{F}_{q^2}$-automorphism group; that is, G is the restriction to $\mathcal{X}$ of a subgroup of $\mathrm{PGL}(r, q^2)$. Each of the initial examples, (a), (b), (c), of $\mathbf{F}_{q^2}$-maximal curves has a large $\mathbf{F}_{q^2}$-automorphism group with many non-conjugate subgroups; see Chapter 12. From this, the existence of numerous $\mathbf{F}_{q^2}$-rational maximal curves is deduced. As in Sections 12.2, 12.3 and 12.4, the genera of such quotient curves can often be computed using the techniques developed in Chapter 11; see especially Theorems 11.57 and 11.72, but the problem of finding an explicit equation has been solved so far only in a few cases; see, for instance, Theorem 12.28.

EXAMPLE 10.4 For a power ℓ of p, let $q = \ell^3$. Then the irreducible plane curve $\mathcal{F} = \mathbf{v}(f(X,Y))$, with

$$f(X,Y) = Y^{\ell^2} - Y - X^{\ell^2-\ell+1}, \tag{10.7}$$

is an $\mathbf{F}_{q^2}$-maximal curve of genus $g = \frac{1}{2}(\ell^2-1)(\ell^2-\ell)$. Observe first that $\mathcal{F}$ belongs to the family of curves studied in Section 12.1. From Lemma 12.1, $\mathcal{F}$ has only one singular point, namely $X_\infty = (0,1,0)$. More precisely, X_∞ is an $(\ell-1)$-fold point and is the centre of a unique branch. In particular, this branch is $\mathbf{F}_{q^2}$-rational. Therefore the number of all $\mathbf{F}_{q^2}$-rational points of $\mathcal{F}$ is equal to the number of points of $\mathcal{F}$ lying in $\mathrm{PG}(2,q)$.

To count such points, note first that X_∞ is the unique point of $\mathcal{F}$ at infinity and that the Y-axis meets $\mathcal{F}$ in exactly ℓ^2 points; the latter are the points $P = (0,b)$ with $b \in \mathbf{F}_{\ell^2}$ and hence they lie in $\mathrm{AG}(2,q^2)$. Also, every other point of $\mathcal{F}$ is of type $Q = (u,v)$ with $u \neq 0$. Let $\xi = u^{\ell^2-\ell+1}$. Since $u^{q^2-1} = 1$ if and only if $u \in \mathbf{F}_{q^2}\backslash\{0\}$, from the identity

$$q^2 - 1 = (\ell^3-1)(\ell+1)(\ell^2-\ell+1),$$

it follows that $u \in \mathbf{F}_{q^2}\backslash\{0\}$ and $\xi^{\ell+1} \in \mathbf{F}_q\backslash\{0\}$ are equivalent conditions. For each such ξ, the equation $Y^{\ell^2} - Y = \xi$ has ℓ^2 distinct solutions in $\mathbf{F}_{q^2}$. Counting, therefore, gives $\ell^2(\ell^2+1)(\ell+1)(\ell^2-\ell+1)$ points.

Since

$$q^2 + 1 + 2gq = 1 + \ell^2 + \ell^2(\ell^2+1)(\ell+1)(\ell^2-\ell+1),$$

so $\mathcal{F}$ is $\mathbf{F}_{q^2}$-maximal.

THEOREM 10.5 (i) *For $\ell = 2$, the curve $\mathcal{F}$ in Example 10.4 is a quotient curve of the Hermitian curve $\mathcal{H}_8$.*

(ii) *For $\ell = 3$, the curve $\mathcal{F}$ in Example 10.4 is not a quotient curve of the Hermitian curve $\mathcal{H}_{27}$; by Theorem 11.36, $\mathcal{H}_{27}$ is not even a Galois covering of $\mathcal{F}$.*

Table 10.2 Known $\mathbf{F}_{q^2}$-maximal curves of large genera

	Genus g	Condition on q	Curves
1	$\frac{1}{2}q(q-1)$		$\mathcal{H}_q = \mathcal{D}_{q+1}$
2	$\frac{1}{4}(q-1)^2$	$q \equiv 1 \pmod 2$	$\mathcal{E}^1_{(q+1)/2}$
3	$\frac{1}{4}q(q-2)$	$q \equiv 0 \pmod 2$	$\mathcal{T}_2$
4	$\frac{1}{6}(q^2-q+4)$	$q \equiv 2 \pmod 3$	$\mathcal{F}_0$
5	$\frac{1}{6}(q^2-q)$	$q \equiv 1 \pmod 3$	$\mathcal{G}$
6	$\frac{1}{6}(q^2-q)$	$q \equiv 0 \pmod 3$	$\mathcal{T}_3'$
7	$\frac{1}{6}(q^2-q-2)$	$q \equiv 2 \pmod 3$	$\mathcal{F}_0'$
8	$\frac{1}{6}(q-1)(q-2)$	$q \equiv 2 \pmod 3$	$\mathcal{E}^1_{(q+1)/3}$
9	$\frac{1}{6}q(q-3)$	$q \equiv 0 \pmod 3$	$\mathcal{T}_3''$
10	$\frac{1}{8}(q^2-2q+5)$	$q \equiv 3 \pmod 4$	$(*)$
11	$\frac{1}{8}(q-1)^2$	$q \equiv 1 \pmod 4$	$(*)$
12	$\frac{1}{8}q(q-2)$	$q \equiv 0 \pmod 4$	$(*)$
13	$\frac{1}{8}(q-1)(q-3)$	$q \equiv 1 \pmod 4$	$\mathcal{E}^1_{(q+1)/4}$
14	$\frac{1}{8}(q-1)(q-3)$	$q \equiv 3 \pmod 4$	$\mathcal{E}^1_{(q+1)/4}, \mathcal{D}_{(q+1)/2}$
15	$\frac{1}{8}q(q-4)$	$q \equiv 0 \pmod 2$	$\mathcal{X}_2$

For $(*)$ see Remark 10.64.

The classification of maximal curves is currently out of reach. However, for larger values of g for which there exists an $\mathbf{F}_{q^2}$-maximal curve, it seems that there are few curves: see Table 10.2. This has been shown so far for $g \geq \lfloor\frac{1}{6}(q^2-q+4)\rfloor$. The proof is postponed to Section 10.5.

10.2 THE FROBENIUS LINEAR SERIES OF A MAXIMAL CURVE

The starting point is Theorem 10.1, showing that the Fundamental Equation (9.74) for an irreducible, non-singular $\mathbf{F}_{q^2}$-maximal curve $\mathcal{X}$ is as follows:

$$qP + \Phi(P) \equiv (q+1)P_0, \tag{10.8}$$

where $P_0 \in \mathcal{X}(\mathbf{F}_{q^2})$ and Φ is the Frobenius map. Let $\mathcal{D} = |(q+1)P_0|$ be the Frobenius linear series of $\mathcal{X}$, and put $r = \dim \mathcal{D}$. From (10.8), $\dim |qP| = r-1$ for every $P \in \mathcal{X}$. By the Weierstrass Gap Theorem 6.89 and Corollary 6.75, an immediate consequence of (10.8) for the non-gap sequence of $\mathcal{X}$ at a point $P \in \mathcal{X}$ is the following:

$$0 < m_1(P) < \cdots < m_{r-1}(P) \le q < m_r(P). \tag{10.9}$$

Further, Propositions 9.81 to 9.85 applied to an $\mathbf{F}_{q^2}$-maximal curve, that is, for $u = 1$, $m = q+1$, $a_0 = 1$, $a_1 = q$, provide a number of basic facts that are collected in the next proposition.

PROPOSITION 10.6 *With the notation above, the following hold:*

(I) *if P and Q are $\mathbf{F}_{q^2}$-rational points, then $(q+1)P \equiv (q+1)Q$, and $q+1$ is a non-gap at each $P \in \mathcal{X}(\mathbf{F}_{q^2})$;*

(II) *there exists $P_1 \in \mathcal{X}(\mathbf{F}_{q^2})$ such that both $q+1$ and q are non-gaps at P_1;*

(III) *the linear series $\mathcal{D}$ is complete, base-point-free, simple and defined over $\mathbf{F}_{q^2}$; it gives rise to an $\mathbf{F}_{q^2}$-rational curve Γ of $\mathrm{PG}(r,q)$ that is $\mathbf{F}_{q^2}$-birationally equivalent to $\mathcal{X}$;*

(IV) *the $(\mathcal{D},P)$-orders j_i at an $\mathbf{F}_{q^2}$-rational point P are the terms of the sequence*

$$0 < q+1-m_{r-1}(P) < \cdots < q+1-m_1(P) < q+1;$$

that is, $j_{r-i}(P) + m_i(P) = q+1$ for $i = 0,\ldots,r-1$;

(V) *if $P \notin \mathcal{X}(\mathbf{F}_{q^2})$, then $j_1(P) = 1$ and so $\epsilon_1 = 1$;*

(VI) *the integer q is a $\mathcal{D}$-order, and so $r \ge 2$;*

(VII) *if $P \in \mathcal{X}(\mathbf{F}_{q^4})\backslash\mathcal{X}(\mathbf{F}_{q^2})$ then $q-1$ is a non-gap at P; if $P \notin \mathcal{X}(\mathbf{F}_{q^4})$ then q is a non-gap at P;*

(VIII) *if P is an $\mathbf{F}_{q^2}$-rational point of $\mathcal{X}$, then $j_{r-1}(P) < q$;*

(IX) *$\epsilon_r = \nu_{r-1} = q$, so Γ is Frobenius non-classical, and every $\mathbf{F}_{q^2}$-rational point of $\mathcal{X}$ is in the support of the ramification divisor R of Γ;*

(X) *if $N_1 \ge q^3+1$, then $m_1(P) = q$ for every $\mathbf{F}_{q^2}$-rational point P of $\mathcal{X}$.*

Some refinements are worthwhile.

(XI) *If $P \in \mathcal{X}$ is not an $\mathbf{F}_{q^2}$-rational point, then*

$$0 \le q-m_{r-1}(P) < \cdots < q-m_1(P) < q$$

are $(\mathcal{D},P)$-orders at P; in particular, $j_r(P) = q$.

Arguing as in the proof of Proposition 9.83 (i) shows that, if $m(P)$ is a non-gap at a point $P \notin \mathcal{X}(\mathbf{F}_{q^2})$, then there exists an effective divisor E for which $E \equiv m(P)P$ and $P \notin \mathrm{Supp}(E)$. From this,

$$E + (q-m(P))P + \Phi(P) \equiv qP + \Phi(P) \equiv (q+1)P_0.$$

If $m(P) \leq q$, then $q - m(P)$ is a $(\mathcal{D}, P)$-order. Assume on the contrary that $j_r(P) \neq q$. Then $j_r(P) = q + 1$, and hence $(q+1)P \equiv qP + \Phi(P)$ by (10.8). But then $P \equiv \Phi(P)$ which is not possible since the genus of $\mathcal{X}$ is positive.

A similar argument shows that (II) is true for any $P_1 \in \mathcal{X}(\mathbf{F}_{q^2})$.

(XII) *If P is an $\mathbf{F}_{q^2}$-rational point then both q and $q+1$ are non-gaps at P; in particular, $j_1(P) = 1$ for every $\mathbf{F}_{q^2}$-rational point P.*

As (IX) implies that $r \leq q$, this may be refined as follows.

(XIII) *Either $r = q - (g-1)$ or $r \leq \frac{1}{2}(q+1)$.*

In fact, from the Riemann–Roch Theorem 6.61 it follows that either $\mathcal{D}$ is non-special and $r = q + 1 - g$ or it is special and Clifford's Theorem 6.79 implies that $r \leq \frac{1}{2}(q+1)$.

By (IV) and (XII), $j_1(P) = 1$ for every $P \in \Gamma$; that is, the curve Γ has only linear branches. This does not imply *a priori* that Γ is non-singular. So the following result is an improvement.

THEOREM 10.7 (i) *The irreducible $\mathbf{F}_{q^2}$-rational curve Γ associated to the Frobenius linear series $\mathcal{D}$ of an $\mathbf{F}_{q^2}$-maximal curve $\mathcal{X}$ is non-singular.*

(ii) *The curves $\mathcal{X}$ and Γ are isomorphic over $\mathbf{F}_{q^2}$.*

Proof. Choose $f_0, \ldots, f_r \in \mathbf{F}_{q^2}(\mathcal{X})$ such that $\mathcal{D}$ consists of all divisors

$$A_{\mathbf{c}} = \text{div}\,(c_0 f_0 + \cdots + c_r f_r) + (q+1)P_0$$

with $\mathbf{c} = (c_0, \ldots, c_r) \in \mathrm{PG}(r, K)$. By (III), let Γ be the irreducible curve arising from the point $(f_0, \ldots, f_r)$; in other words, the $\mathbf{F}_{q^2}$-rational transformation

$$\omega : x_0 = f_0, \ldots, x_r = f_r$$

defines an $\mathbf{F}_{q^2}$-rational morphism $\pi : \mathcal{X} \to \Gamma$. Since every branch of Γ is linear, it must be shown that no point of Γ is the centre of two or more branches.

Let $P \in \mathcal{X}$ and $\pi(P)$ be a branch point of Γ. After multiplying $(f_0, \ldots, f_r)$ by a suitable element of $\mathbf{F}_{q^2}(\mathcal{X})$, it may be assumed that P is neither a pole of any f_k nor a zero of all the f_k. Since $\epsilon_r = q$ by (XI), Theorem 7.65 applied to $p^m = q$ ensures the existence of $z_0, \ldots, z_r$ in $\mathbf{F}_{q^2}(\mathcal{X})$ satisfying the following three properties:

(i) P is neither a pole of any z_k nor a zero of all the z_k;

(ii) $$z_0^q f_0 + \cdots + z_r^q f_r = 0; \qquad (10.10)$$

(iii) the osculating hyperplane L_P at the branch point $\pi(P)$ of Γ is

$$\mathbf{v}(z_0(P)^q X_0 + \cdots + z_r(P)^q X_r). \qquad (10.11)$$

Since Γ is Frobenius non-classical, Theorem 8.54 gives for $q^2 = p^h$ that

$$z_0 f_0^q + \cdots + z_r f_r^q = 0. \qquad (10.12)$$

Suppose, on the contrary, that $\pi(P)$ and $\pi(Q)$ have the same centre $A \in \Gamma$. Then the osculating hyperplane to Γ at $\pi(P)$ contains the centre of $\pi(Q)$. From (10.8),

$\Phi(P) = Q$, and $A \in \mathrm{PG}(r, q^2)$. After a suitable change of the coordinate system of $\mathrm{PG}(r, q^2)$, let $A = (1, 0, \ldots, 0)$. The coordinate functions f_k also change, but it is still true that P is not a pole of any f_k, and that P is not a zero of some coordinate function, say f_0; the same holds for z_k. So take $f_0 = 1$. If τ is a primitive place representation of the branch $\pi(P)$, then $\pi(P)$ has a primitive branch representation

$$x_0(t) = 1,\ x_1(t) = \tau(f_1) = a_1 t + \cdots,\ \ldots,\ x_r(t) = \tau(f_r) = a_r t + \cdots,$$

with some $a_k \neq 0$ as $\pi(P)$ is linear. For $k = 0, 1 \ldots, r$, write

$$z_k(t) = \tau(z_k) = z_k(P) + u_k(t)$$

with $\mathrm{ord}\, u_k(t) \geq 1$. Apply τ to (10.10):

$$\mathrm{ord}(z_0(P)^q + \cdots + z_r(P)^q x_r(t)) = \mathrm{ord}(u_0(t)^q + \cdots + u_r(t)^q x_r(t)).$$

Hence $z_0(P) = 0$. From (10.12), $u_0(t) \geq q > 1$. This together with the preceding equation, implies that

$$\mathrm{ord}(z_0(P)^q + \cdots + z_r(P)^q x_r(t)) \geq q + 1.$$

Therefore $j_r(P) = I(P, \Gamma \cap L_P) = q + 1$. But then $I(Q, \Gamma \cap L_P) = 0$, which contradicts the hypothesis that the centre of $\pi(Q)$ lies in L_P. □

COROLLARY 10.8 *Let $\mathcal{X}$ be an $\mathbf{F}_{q^2}$-maximal curve of genus g and, for some point $P_0 \in \mathcal{X}(\mathbf{F}_{q^2})$, suppose that there exist positive integers $a, b \in H(P_0)$ such that all non-gaps at P which are at most $q+1$ can be written as $\lambda a + \mu b$ for non-negative integers λ, μ. Then $H(P_0) = \langle a, b\rangle$ and $g = \frac{1}{2}(a-1)(b-1)$.*

Proof. Choose $x, y \in \mathbf{F}_{q^2}(\mathcal{X})$ such that $\mathrm{div}(x)_\infty = aP_0$ and $\mathrm{div}(y)_\infty = bP_0$. Since $q, q+1 \in H(P_0)$ by (XII), the integers a, b are coprime, and so let $a < b$. Therefore $K(\mathcal{X}) = K(x, y)$ and $\mathbf{F}_{q^2}(\mathcal{X}) = \mathbf{F}_{q^2}(x, y)$. Let $f(X, Y) \in \mathbf{F}_{q^2}[X, Y]$ be an irreducible polynomial such that $f(x, y) = 0$. Its Weierstrass normal form, see Theorem 6.92, is of type

$$f(X, Y) = X^a + \beta Y^b + \textstyle\sum_{ai+bj<ab} \alpha_{ij} X^i Y^j,$$

with $\beta, \alpha_{ij} \in \mathbf{F}_{q^2}$. Note that $\mathcal{F}$ has a unique point at infinity, namely X_∞.

As $K(\mathcal{X}) = K(x, y)$, the $\mathbf{F}_{q^2}$-rational irreducible plane curve $\mathcal{F} = \mathbf{v}(f(X, Y))$ is a birational model of $\mathcal{X}$ over $\mathbf{F}_{q^2}$. By Theorem 10.7 this holds true if $\mathcal{X}$ is replaced by Γ. It must be shown that X_∞ is the centre of a unique branch of $\mathcal{F}$ and that every other point of $\mathcal{F}$ is non-singular. From this and Proposition 6.93 (ii) it follows that $\mathcal{F}$ has genus $\frac{1}{2}(a-1)(b-1)$.

Choose the coordinate functions $f_0, f_1, \ldots, f_r$ of Γ such that

$$f_0 = 1, \quad \mathrm{div}(f_i)_\infty = m_i P_0, \quad \text{for } i = 1, \ldots, r.$$

Note that two of them are x and y. There exist polynomials $g_i(X, Y) \in \mathbf{F}_{q^2}[X, Y]$ for $i = 1, \ldots, r$ such that

$$f_1 = g_1(x, y),\ \ldots,\ f_r = g_r(x, y).$$

Therefore, if $P' = (u, v) \in \mathcal{F}$ and $x = x(t), y = y(t)$ is a primitive representation of a branch γ' of $\mathcal{F}$ centred at P, then the point $P = (1, g_1(u, v), \ldots, g_r(u, v))$ is the centre of the branch γ of Γ with primitive representation

$$x_1(t) = g_1(x(t), y(t)), \ldots, x_r(t) = g_r(x(t), y(t)).$$

So, if P' were the centre of two distinct branches of $\mathcal{F}$, then the corresponding point P would be the centre of two distinct branches of Γ, in contradiction with Theorem 10.7. Similarly, if γ' were non-linear, then γ would also be non-linear, again contradicting Theorem 10.7.

Finally, if X_∞ is the centre of a branch γ' of $\mathcal{F}$, then the corresponding place of $K(\mathcal{X})$, that is, a point $P \in \Gamma$ is a pole of the coordinate function x of Γ. But then $P = P_0$, and hence γ' is uniquely determined. □

From (XII), $m_{r-1}(P) = q$ for every $\mathbf{F}_{q^2}$-rational point P of $\mathcal{X}$. This remains true for any point P of $\mathcal{X}$, but the analogous result $m_{r-2} = q-1$ can fail for some points.

Proposition 10.9 *Let $\mathcal{X}$ be an $\mathbf{F}_{q^2}$-maximal curve.*

(XIV) *$m_{r-1}(P) = q$ for every $P \in \mathcal{X}$.*

(XV) *If $r \geq 3$ and $m_{r-2}(P) < q-1$, then P is a Weierstrass point of $\mathcal{X}$.*

Proof. (XIV) Assume that $P \notin \mathcal{X}$. As $\mathcal{D}$ is base-point-free by (III), Theorems 10.7 and 6.26 imply that $\dim |(q-1)P| = r-2$; so $m_{r-2}(P) \leq q-1 < m_{r-1}(P)$. This together with (10.9) shows (XIV).

(XV) From the definition of an ordinary place given in Section 6.6, any two ordinary points of $\mathcal{X}$ have the same non-gap numbers; that is, $H(P) = H(P')$ for any two ordinary points $P, P' \in \mathcal{X}$. Also, for every $0 \leq i \leq r$, the i-th non-gap $m_i(P)$ at an ordinary point $P \in \mathcal{X}$ is at least the i-th non-gap $m_i(Q)$ at any point $Q \in \mathcal{X}$.

Let $r \geq 3$. If P is a non-$\mathbf{F}_{q^2}$-rational point of $\mathcal{X}$, then $r-1$ of the $(\mathcal{D}, P)$-orders smaller than q are known from (XI). If $m_{r-2}(P) < q-1$, the missing number in (XI) is 1; in fact, 1 is also a $(\mathcal{D}, P)$-order by (V). Therefore the $(\mathcal{D}, P)$-orders at such a point P are completely determined. Since $\mathcal{X}$ has only finitely many Weierstrass points and $\mathbf{F}_{q^2}$-rational points, so, if $m_{r-2}(P) < q-1$ for some point $P \in \mathcal{X}$ which is neither a Weierstrass point nor an $\mathbf{F}_{q^2}$-rational point, then the $\mathcal{D}$-order sequence is the following:

$$(\epsilon_0 = 0,\ \epsilon_1 = 1, \ldots,\ \epsilon_2 = q - m_{r-2}(P), \ldots,\ \epsilon_{r-1} = q - m_1(P), q).$$

Now take an $\mathbf{F}_{q^2}$-rational point $Q \in \mathcal{X}$. From (IV) and Theorem 7.55,

$$v_Q(R) \geq \textstyle\sum_{i=1}^{r} (j_i(Q) - \epsilon_i) = \sum_{i=1}^{r} (q + 1 - m_{r-i}(Q) - \epsilon_i).$$

Hence

$$v_Q(R) \geq 1 + \textstyle\sum_{i=1}^{r-2} (m_i(P) - m_i(Q) + 1) \geq r - 1.$$

Since

$$|\mathcal{X}(\mathbf{F}_{q^2})| = q^2 + 1 + 2gq = (q+1)^2 - (2g-2)q,$$

it follows that $\deg R \geq (r-1)\{(q+1)^2 - (2g-2)q\}$. From (7.13),

$$\left(\textstyle\sum_i \epsilon_i\right)(2g-2) + (r+1)(q+1) \geq (r-1)(q+1)^2 + (r-1)q(2g-2).$$

Since $\epsilon_i \leq \epsilon_r = q$, this implies that

$$(r+1)\epsilon_r(g-1) + (r+1)(q+1) \geq (r-1)q(2g-2) + (r-1)(q+1)^2.$$

Hence

$$(q+1)(q(r-1)-2)+q(g-1)(r-3)\leq 0.$$

But this inequality does not hold for $r \geq 3$. □

From (XIV), a refinement of (XIII) can be deduced.

(XVI) *If $\mathcal{X}$ has classical gap sequence at a general point, then* $g = q-(r-1)$.

In fact, if $\mathcal{X}$ has classical gap sequence at a general point P, then the positive non-gaps at P are the numbers $g+i$ for $i = 1, \ldots$.

If $\mathcal{X}$ is a non-hyperelliptic curve of genus $g > 2$, it has classical gap sequence at a general point if and only if the canonical curve of $\mathcal{X}$ is classical; see Remark 7.56. By Corollary 7.61, this is the case whenever $p > 2g-2$. The following example shows that (XVI) can occur.

EXAMPLE 10.10 The Klein quartic $\mathcal{F} = \mathbf{v}(X_0X_1^3 + X_1X_2^3 + X_2X_0^3)$, as in Example 7.18, is a non-hyperelliptic curve of genus 3 and is a canonical curve. Also, for $p > 2g-2 = 4$, $\mathcal{F}$ has classical gap sequence at a general point. Now let $\mathcal{F}$ be viewed as an $\mathbf{F}_{p^2}$-rational curve. By Theorem 10.78, $\mathcal{F}$ is $\mathbf{F}_{q^2}$-maximal if and only if $q \equiv 6 \pmod 7$; see also Corollary 10.82 (ii). Note that the classical Dirichlet Theorem states that there exist infinitely many primes satisfying $p \equiv 6 \pmod 7$. In particular, $\mathcal{F}$ is a classical $\mathbf{F}_{p^2}$-maximal curve for an infinite number of primes p.

It was noted in the proof of Theorem 10.9 that all $\mathcal{D}$-orders but one can be deduced from (XI) whenever the non-gaps at an ordinary point are known. Let J denote the missing $\mathcal{D}$-order; that is, at an ordinary point $P \in \mathcal{X}$,

$$\epsilon_{J-1} = q - m_{r-J}(P) < \epsilon_J < \epsilon_{J+1} = q - m_{r-(J+1)}(P)\,. \tag{10.13}$$

From Proposition 8.42, the Frobenius $\mathcal{D}$-orders are obtained from the $\mathcal{D}$-orders by removing one of them, say ϵ_I. In fact, the integers I and J coincide.

PROPOSITION 10.11 *The Frobenius orders of $\mathcal{D}$ are the $\mathcal{D}$-orders apart from ϵ_J with J as in (10.13).*

Proof. Let P be an ordinary point of $\mathcal{X}$ and suppose that $P \notin \mathcal{X}(\mathbf{F}_{q^2})$. Choose $\xi_t \in K(\mathcal{X})$, for $t = 1, \ldots, r-1$, such that $\operatorname{div} \xi_t = D_t - m_t(P)P$ with $D_t \succ 0$ and $P \notin \operatorname{Supp} D_t$. Also, choose $\xi, v \in K(\mathcal{X})$ for which

$$\operatorname{div} \xi = qP + \Phi(P) - (q+1)P_0, \quad \operatorname{div} v = \epsilon_J P + D - (q+1)P_0,$$

with $D \succ 0$ and $P \notin \operatorname{Supp} D$. From (XI),

$$\operatorname{div}(\xi\,\xi_{r-i}) = \begin{cases} -(q+1)P_0 + \epsilon_{i-1}P + \Phi(P) + D_{r-i}, & i \leq J, \\ -(q+1)P_0 + \epsilon_i P + \Phi(P) + D_{r-i}, & i > J. \end{cases} \tag{10.14}$$

Let $x_{i-1} = \xi\,\xi_{r-i}$ for $i = 1, \ldots, r-1$. From (10.14), $\operatorname{div} x_{i-1} + (q+1)P_0 \in \mathcal{D}$. More precisely, $\mathcal{D}$ consists of all divisors

$$A_{\mathbf{c}} = \operatorname{div}(\textstyle\sum_{i=0}^{j-1} c_i x_i + c_j v_j + \sum_{i=j}^{r-1} c_{i+1} x_i) + (q+1)P_0$$

with $\mathbf{c} = (c_0, \ldots, c_r) \in \mathrm{PG}(r, K)$. Since these divisors are cut out on Γ by hyperplanes, it also follows from (10.14) that every hyperplane Δ of $\mathrm{PG}(r, K)$ with $I(P, \Delta \cap \Gamma) > J$ passes through $\Phi(P)$. On other hand, the hyperplane Δ' arising from v does not have this property. Indeed, if $\Phi(P) \in \Delta'$, then $\Phi(P) \in \mathrm{Supp}\, D$ and hence

$$\mathrm{div}\, v = \epsilon_J P + \Phi(P) - (q+1)P_0 + D',$$

with $D' \succ 0$ and $P \notin \mathrm{Supp}\, D'$. Thus there exists $w \in K(\mathcal{X})$ such that

$$\mathrm{div}\, w = \epsilon_J P + \Phi(P) - (qP + \Phi(P)) + D' = (q - \epsilon_j)P + D'.$$

But this implies that $q - \epsilon_J$ is a non-gap at $P \notin \mathcal{X}(\mathbf{F}_{q^2})$, contradicting the definition of J. So the assertion is proved. Note that, in terms of the Frenet frame of $\mathcal{X}$ at P, see Section 7.6, this assertion can be reworded as saying that $\Phi(P) \in (\Pi_J \backslash \Pi_{J-1})$ for any ordinary non-$\mathbf{F}_q$-rational point P of $\mathcal{X}$. Using Proposition 8.42, the result is established. □

PROPOSITION 10.12 *Let $\mathcal{W}$ be the set of all Weierstrass points of an $\mathbf{F}_{q^2}$-maximal curve $\mathcal{X}$.*

(i) *If $g > q - (r-1)$, then $\mathcal{X}(\mathbf{F}_{q^2}) \subset \mathcal{W}$.*

(ii) *If S is the Stöhr–Voloch divisor of $\mathcal{X}$, then*

$$\mathrm{Supp}\, S \subset \mathcal{W} \cup \mathcal{X}(\mathbf{F}_{q^2}).$$

Proof. (i) By (XVI) and Exercise 8 in Chapter 7, the curve $\mathcal{X}$ is neither elliptic nor hyperelliptic. Let $\mathcal{K}$ be its canonical curve, and take an ordinary point P of $\mathcal{X}$. Then the orders $\epsilon_0 = 0 < \cdots < \epsilon_j < \cdots < \epsilon_{g-1}$ of $\mathcal{K}$ are also the orders of $\mathcal{K}$ at P. From Section 7.6, the gaps at P are exactly the integers $\epsilon_j + 1$, and hence $H(P)$ consists of all non-negative integers distinct from

$$1, \ldots, \epsilon_j + 1, \ldots, \epsilon_{g-1} + 1.$$

Also, for any integer i with $1 \le i \le q$, Lemma A.6 implies that

$$\binom{q+i-1}{q} \not\equiv 0 \pmod{p}.$$

Suppose that $q + i \notin H(P)$ for an integer i with $1 \le i \le q$. Then $q + i - 1$ is an order of $\mathcal{K}$ at P. From Lemma 7.62, q is also an order of $\mathcal{K}$ at P. Hence $q + 1$ is a gap at P. By (I), $P \notin \mathcal{X}(\mathbf{F}_{q^2})$. Therefore, if $P \in \mathcal{X}(\mathbf{F}_{q^2})$, then $q+1, \ldots, 2q \in H(P)$.

Assume that $P \in \mathcal{X}(\mathbf{F}_{q^2})$. Then (XII) shows that $\{q, q+1, \ldots\} \subset H(P)$. Since P is an ordinary point, from (10.9) it follows that the number of gaps at P is equal to $q - (r-1)$. By the Weierstrass Gap Theorem 6.89, $g = q - (r-1)$. But by the hypothesis, this contradicts the assumption that $P \in \mathcal{X}(\mathbf{F}_{q^2})$.

(ii) Let D_i denote the i-th row vector in the generalised Wronskian determinant (7.12) of the coordinate functions $f_0, \ldots, f_r$ of Γ; that is,

$$D_i = (D^{(\epsilon_i)} f_0, D^{(\epsilon_i)} f_1, \ldots, D^{(\epsilon_i)} f_r), \quad i = 0, 1, 2, \ldots, r.$$

Also, let E be the first row vector of the Wronskian, $\det M(\nu_0, \ldots, \nu_{r-1})$, from which the Stöhr–Voloch divisor S is derived, as in Section 8.5; that is,

$$E = (f_0^{q^2}, f_1^{q^2}, f_2^{q^2}, \ldots, f_r^{q^2}).$$

If J is defined as in (10.13), Proposition 10.11 shows that

$$\nu_0 < \nu_1 < \cdots < \nu_{J-1} < \epsilon_J < \nu_J < \cdots < \nu_{r-1}$$

is the order sequence of $\mathcal{D}$. Let $P \notin \mathcal{W} \cup \mathcal{X}(\mathbf{F}_{q^2})$. By the last remark in the proof of Proposition 10.11, $\Phi(P) \in \Pi_J \backslash \Pi_{J-1}$. Therefore E is a linear combination of $D_0, \ldots, D_J$ but not of $D_0, \ldots, D_{J-1}$. Hence $\det M(\nu_0, \ldots, \nu_{r-1})$ does not alter if its first row E is replaced by D_J. In other words,

$$\det M(\nu_0, \ldots, \nu_{r-1}) = \det(D_0, \ldots, D_r).$$

Now, on the contrary, let $P \in \operatorname{Supp} S$. This assumption means that P is a zero of $\det M(\nu_0, \ldots, \nu_{r-1})$. Then P is also a zero of $\det(D_0, \ldots, D_r)$; that is, $(j_0(P), \ldots, j_r(P)) \neq (\epsilon_0, \ldots, \epsilon_r)$. But then $P \in \mathcal{W} \cup \mathcal{X}(\mathbf{F}_{q^2})$, by (XI). □

REMARK 10.13 Let $\mathcal{X}$ have classical gap sequence at a general point. Then, from (XVI), the first $r-1$ positive non-gaps at any ordinary point are

$$g+1, \ldots, g+r-1 = q.$$

Thus the orders of $\mathcal{D}$ are $0, \ldots, r-2, \epsilon_{r-1}, q$ with $r-1 \le \epsilon_{r-1} < q$. By Proposition 10.11, the $\mathbf{F}_{q^2}$-Frobenius orders are $0, \ldots, r-2, q$. A direct computation shows that $\deg S = N_1 + \deg R$. It may be that the stronger equation $S = \sum_{P \in \mathcal{X}(\mathbf{F}_{q^2})} P + R$ also holds, but no proof has yet been found.

The following result regards hyperelliptic curves which are $\mathbf{F}_{q^2}$-maximal.

PROPOSITION 10.14 *Let $\mathcal{X}$ be an $\mathbf{F}_{q^2}$-maximal curve of genus $g \ge 2$.*

(i) *If $\mathcal{X}$ is hyperelliptic, then $q \le 2r-2$.*

(ii) *If there exists $P \in \mathcal{X}$ such that $j_{r-1}(P) = q-1$, then $\mathcal{X}$ is hyperelliptic.*

Proof. If $\mathcal{X}$ is hyperelliptic, then $m_1(P) = g+1$ for any ordinary point P; see Exercise 8 in Chapter 7. Then $m_{r-1}(P) = g+r-1$, by the Weierstrass Gap Theorem 6.89. From (XI), it follows that $m_{r-1}(P) = q$; hence $g = q-r+1$. On the other hand, $N_1 \le 2(q^2+1)$ and the maximality of $\mathcal{X}$ shows that $2g \le q$. Therefore (i) holds.

Let $P \in \mathcal{X}(\mathbf{F}_{q^2})$ with $j_{r-1}(P) = q-1$. Since $j_r(P) = q-1$, this and (IV) imply that $m_1(P) = 2$. As $g \ge 2$, this proves that $\mathcal{X}$ is hyperelliptic.

Finally, let $P \notin \mathcal{X}(\mathbf{F}_{q^2})$ with $j_{r-1}(P) = q-1$. Then (10.8) implies the existence of a divisor E for which $(q-1)P + E \equiv qP + \Phi(P)$ with $P \notin \operatorname{Supp}(E)$. Hence $E \equiv P + \Phi(P)$. This shows that the linear series $|P + \Phi(P)|$ has positive dimension. Since $g \ge 2$, Theorem 6.81 (i) implies that $\dim |P + \Phi(P)| = 1$. Therefore, by Theorem 7.42, $\mathcal{X}$ is hyperelliptic. □

PROPOSITION 10.15 *Suppose that $j_{r-1}(P) = r-1$ for every point $P \in \mathcal{X}$. Then*

$$(r-1)r(g-1) = (q+1)(q-r).$$

Proof. Under the hypothesis, a point $P \in \mathcal{X}$ is a $\mathcal{D}$-Weierstrass point of $\mathcal{X}$ if and only if P is an $\mathbf{F}_{q^2}$-rational point of $\mathcal{X}$. Also, if $R_\mathcal{D}$ is the ramification divisor of $\mathcal{X}$, then $v_P(R_\mathcal{D}) = 1$ for $P \in \mathcal{X}(\mathbf{F}_{q^2})$; see Theorem 7.55. Since $\mathcal{X}$ is $\mathbf{F}_{q^2}$-maximal, the assertion follows from Definition 7.52. □

10.3 EMBEDDING IN A HERMITIAN VARIETY

Although it should not matter which model of a curve is considered, some models are better than others in revealing birationally invariant properties.

For $\mathbf{F}_{q^2}$-maximal curves, the model arising from the Frobenius linear series $\mathcal{D}$ appears to be a natural candidate, even though a slight change is necessary when the smallest linear subseries L of $\mathcal{D}$ containing all divisors $qP + \Phi(P)$ is properly contained in $\mathcal{D}$, that is, when $m = \dim L < \dim \mathcal{D} = r$. As explained later, this replacement prevents the model from having higher-dimensional strangeness; that is, it ensures that there is no point lying in an infinite number of osculating hyperplanes.

Therefore, throughout this section, let $\mathcal{X}$ be the curve given by Theorem 10.7. The idea is to use the geometric properties of the results of Section 10.2 to find geometric characterisations of $\mathcal{X}$ as an irreducible non-singular curve lying on certain hypersurfaces.

Such a characterisation is the main aim in this section. Up to an isomorphism over $\mathbf{F}_{q^2}$, Theorems 10.21 and 10.31 establish that $\mathbf{F}_{q^2}$-maximal curves are the irreducible $\mathbf{F}_{q^2}$-rational curves of degree $q+1$ lying on a non-degenerate Hermitian variety of $\mathrm{PG}(m, K)$ with $2 \le m \le r$.

The Hermitian variety of $\mathrm{PG}(n, K)$, with $K = \overline{\mathbf{F}}_q$, is a hypersurface which is the natural generalisation of the Hermitian curve to higher dimensions; canonically, it is

$$\mathcal{H}_{n,q} = \mathbf{v}(X_0^{q+1} + \cdots + X_n^{q+1}).$$

By Theorem 10.7, the curve $\mathcal{X}$ satisfies the Linear General Position Principle.

If $m \ge 3$ and m is small, then Castelnuovo's Bound provides an upper bound on the genus g, which has applications for larger g in the case of maximal curves; see Sections 10.5, 10.6.

If $m = 3$ and g is large, such a curve $\mathcal{X}$ also lies on a low degree surface, by Halphen's Theorem and its generalisations; see Section 7.15. In the known examples, as at the end of this section, $\mathcal{X}$ is a component of degree $q+1$ of the intersection of the Hermitian surface with an $\mathbf{F}_{q^2}$-rational quadric or cubic surface in $\mathrm{PG}(3, K)$. The characterisation, together with Castelnuovo's Bound and Halphen's Theorem, is the main ingredient in the classification of $\mathbf{F}_{q^2}$-maximal curves with $\dim \mathcal{D} = 3$. For more details, see Section 10.6.

In the characterisation, the idea is to show that an $\mathbf{F}_{q^2}$-maximal curve $\mathcal{X}$ coming from its Frobenius divisor lies on a non-degenerate Hermitian variety of $\mathrm{PG}(r, K)$. This is demonstrated under the hypothesis that $\mathcal{X}$ coincides with its dual curve up to the Frobenius collineation $(X_0, \ldots, X_r) \mapsto (X_0^q, \ldots, X_r^q)$ of $\mathrm{PG}(r, K)$. When this hypothesis fails, a weaker result is obtained, namely that the dual curve lies on a degenerate Hermitian variety $\mathcal{H}_{m,q}$ of PG(r, K). However, it is still possible to recover from the dual curve a non-singular model $\mathcal{X}'$ of $\mathcal{X}$ lying on a non-degenerate Hermitian variety in some PG(m, K) with $2 \leq m < r$.

The starting point is the rational transformation

$$\omega : x'_0 = z_0, \ldots, x'_r = z_r,$$

with $z_0, z_1, \ldots, z_r$ satisfying (10.10), (10.11), (10.12), which defines a morphism $\pi^\star$ since the osculating hyperplanes of $\mathcal{X}$ at distinct points are distinct.

By (10.11), $\rho \circ \pi^\star$ is the Gauss map $P \mapsto L_P^{(r-1)}$, where ρ is the Frobenius collineation $(X_0, \ldots, X_r) \mapsto (X_0^q, \ldots, X_r^q)$. So consider the irreducible $\mathbf{F}_{q^2}$-rational curve $\mathcal{X}' = \pi^\star(\mathcal{X})$ of PG(r, K). Note that $\mathcal{X}'$ is the dual curve of $\mathcal{X}$ up to ρ; see Section 7.9. Also, $\mathcal{X}'$ may be degenerate in the sense that it may be contained in a proper subspace $\Pi_m = $ PG(m, K) of PG(r, K). If this is the case and m is chosen to be minimal, then there is an $(r - m)$-dimensional subspace Π'_{r-m} of PG(r, K) which is the intersection of the osculating hyperplanes to $\mathcal{X}$ at general points $P \in \mathcal{X}$, except for finitely many points P.

Here, no point of $\mathcal{X}$ lies in Π'_{r-m}. To show this, let $R \in \Pi'_{r-m}$ and assume on the contrary that $R \in \mathcal{X}$. Choose a point $Q \in \mathcal{X}$ such that $Q \neq R$ but the osculating hyperplane L_Q to $\mathcal{X}$ at Q contains Π'_{r-m}. Since the common points of L_Q and $\mathcal{X}$ are only two, namely Q and $\Phi(Q)$, it follows that $\Phi(Q) = R$. Hence Q is uniquely determined by R. But this is a contradiction, as Q can be chosen in many different ways.

The Frobenius collineation $(X_0, \ldots, X_r) \mapsto (X_0^{q^2}, \ldots, X_r^{q^2})$ fixes Π_m. Take a projective frame in PG(r, q^2) in such a way that

$$\Pi_m = \mathbf{v}(X_{m+1}, \ldots, X_r).$$

Then $z_{m+1} = 0, \ldots, z_r = 0$ and the coordinate functions of $\pi^\star : \mathcal{X} \to \Pi_m$ are $(z_0, \ldots, z_m)$. Hence, by (10.11), the osculating hyperplane to $\mathcal{X}$ at Q is

$$L_Q = \mathbf{v}(\gamma_0^q X_0 + \cdots + \gamma_m^q X_m),$$

where $C = (\gamma_0, \ldots, \gamma_m)$ is the centre of the branch $\pi^\star(Q)$. The linear series $\mathcal{D}'$ cut out on $\mathcal{X}'$ by hyperplanes of Π_m consists of all divisors

$$A_{\mathbf{c}} = \text{div}(c_0 z_0 + \cdots + c_m z_m) + E,$$

where $\mathbf{c} = (c_0, \ldots, c_m) \in \Pi_m$ and, as usual,

$$E = \sum e_P P, \quad e_P = -\min\{\text{ord}_P\, z_0, \ldots, \text{ord}_P\, z_m\}.$$

Lemma 10.16 (i) *The irreducible $\mathbf{F}_{q^2}$-rational curve $\mathcal{X}' = \pi^\star(\mathcal{X})$ of Π_m has degree $q + 1$.*

(ii) *The $\mathbf{F}_{q^2}$-rational linear series $\mathcal{D}'$ cut out on $\mathcal{X}'$ by hyperplanes of Π_m contains all divisors $qP + \Phi(P)$ with $P \in \mathcal{X}$.*

Proof. Choose a point $P_0 = (\alpha_0, \ldots, \alpha_r) \in \mathcal{X}(\mathbf{F}_{q^2})$. If $\alpha_i = 0$ for $i = 0, \ldots, m$, then P_0 lies in the osculating hyperplane at a general point of $\mathcal{X}$; with the above notation, $P_0 \in \Pi'_{r-m}$. But this has already been shown to be impossible. Therefore $\alpha_i \neq 0$ for some i with $0 \leq i \leq m$.

Now consider the hyperplane

$$H = \mathbf{v}(\alpha_0^q X_0 + \cdots + \alpha_m^q X_m)$$

of $\mathrm{PG}(r, K)$ which can also be regarded as a hyperplane of Π_m. For a point $P \in \mathcal{X}$, let $C = (\gamma_0, \ldots, \gamma_m)$ be the centre of the branch $\pi^\star(P)$ of $\mathcal{X}'$. If $C \in H$ then $\alpha_0^q\gamma_0 + \cdots + \alpha_m^q\gamma_m = 0$, and so $\gamma_0^q\alpha_0 + \cdots + \gamma_m^q\alpha_m = 0$. By (10.11), the osculating hyperplane to $\mathcal{X}$ at P passes through P_0. Since $P_0 \in \mathcal{X}(\mathbf{F}_{q^2})$, this is only possible when $P = P_0$. Thus $H \cap \mathcal{X}'$ contains no point other than the centre of $\pi^\star(P_0)$.

The next step is to show that $I(\pi^\star(P_0), \mathcal{X} \cap H) = q + 1$; that is,

$$\mathrm{ord}_{P_0}(\alpha_0^q w_0 + \cdots + \alpha_r^q w_r) = q + 1\,,$$

where

$$w_i = z_i/z_k, \quad \mathrm{ord}_{P_0}(z_k) = \min\{\mathrm{ord}_{P_0}(z_0), \ldots, \mathrm{ord}_{P_0}(z_r)\}. \tag{10.15}$$

After an $\mathbf{F}_{q^2}$-linear transformation of $\mathrm{PG}(r, K)$, let

$$P_0 = (1, 0, \ldots, 0).$$

Then it must be shown that $\mathrm{ord}_{P_0}(w_0) = q + 1$. An $\mathbf{F}_{q^2}$-linear transformation may be chosen so that

$$f_0 = 1, \quad \mathrm{ord}_{P_0} f_i < \mathrm{ord}_{P_0} f_{i+1}, \quad \text{for } i = 0, \ldots, r-1. \tag{10.16}$$

Then the unique branch γ of $\mathcal{X}$ centred at P_0 has a primitive representation

$$x_0(t) = 1,\ x_1(t) = a_1 t + \cdots,\ x_k(t) = a_k t^{j_k} + \cdots,\ x_r(t) = a_r t^{q+1} + \cdots,$$

where $(0, 1, \ldots, j_k, \ldots, q + 1)$ is the $(\mathcal{D}, P_0)$-order sequence of $\mathcal{X}$. If ρ is a primitive representation of the place arising from γ, then (10.12) gives the following:

$$w_0(t) = -\{w_1(t)(a_1 t + \cdots)^q + \cdots + w_r(t)(a_r t^{q+1} + \cdots)^q\}\,. \tag{10.17}$$

Therefore it must be shown that both $a_1 \neq 0$ and $\mathrm{ord}_t\, w_1(t) = 1$. From (10.10),

$$\begin{aligned} w_0(t)^q + w_1(t)^q(a_1 t + \cdots) + \cdots + w_k(t)^q(a_k t^{j_k} + \cdots)& \\ + \cdots + w_r(t)^q(a_r t^{q+1} + \cdots) = 0.& \end{aligned}$$

Also, from the definition of w_i, it follows that $\mathrm{ord}_t\, w_i(t) = 0$ for at most one index i. Since $1 = j_1 < j_2 < \cdots < j_r = q + 1$ and $j_{r-1} < q$, the only possibility is $i = r$, whence the assertion follows.

Thus $(q+1)\pi^\star(P_0)$ is obtained as the intersection divisor of $\mathcal{X}'$ with a hyperplane of Π_m. Therefore $\deg \mathcal{X}' = q + 1$. Also, $a_1 \neq 0$ implies that the branch $\pi^\star(P)$ of $\mathcal{X}'$ is linear.

This argument applied to a point $P \in \mathcal{X} \backslash \mathcal{X}(\mathbf{F}_{q^2})$ in place of P_0 shows that the hyperplane H cuts out the divisor $q\pi^\star(P) + \Phi(\pi^\star(P))$ on $\mathcal{X}'$. □

By Lemma 10.16, the invariants j_m, ϵ_m of $\mathcal{X}'$ are equal to the corresponding invariants j_r, ϵ_r of $\mathcal{X}$. In the proof of Lemma 10.16, every branch of $\mathcal{X}'$ is linear. From that proof, $z_{m+1} = 0, \ldots, z_r = 0$, since $\Pi_m = \mathbf{v}(X_{m+1}, \ldots, X_r)$. In particular, (10.10) and (10.12) become the following:

$$z_0^q f_0 + \cdots + z_m^q f_m = 0, \qquad z_0 f_0^q + \cdots + z_m f_m^q = 0. \tag{10.18}$$

The osculating hyperplane at the branch point $\pi^\star(P)$ of $\mathcal{X}'$ is

$$\mathbf{v}(f_0(P)^q X_0 + \cdots + f_r(P)^q X_r). \tag{10.19}$$

Therefore the proof of Theorem 10.7 may be used without alteration to prove that $\mathcal{X}'$ has no singular points. This, together with Theorem 10.7, gives the following result.

THEOREM 10.17 *Every $\mathbf{F}_{q^2}$-maximal curve is isomorphic over $\mathbf{F}_{q^2}$ to both the following curves:*

(i) *the curve $\mathcal{X}$ arising from its Frobenius divisor;*

(ii) *the dual curve $\mathcal{X}'$ of $\mathcal{X}$.*

REMARK 10.18 All proofs in this section remain valid for $g = 0$, that is, for the projective line $\ell = \mathbf{v}(Y)$ over $\mathbf{F}_{q^2}$. The Frobenius divisor $\mathcal{D}$ of ℓ has dimension $r = q + 1$ and consists of all effective divisors of degree $q + 1$. The corresponding curve $\mathcal{X}$ is the normal rational curve of $\mathrm{PG}(q + 1, K)$ defined by the coordinate functions $f_i = x^i$ with $i = 0, \ldots, q + 1$ and $K(\mathcal{F}) = K(x)$. Given any point $P(a) = (1, a, \ldots, a^{q+1}) \in \mathcal{X}$, the hyperplane

$$\mathbf{v}(a^{q^2+q} X_0 - a^q X_1 - a^{q^2} X_q + X_{q+1})$$

of $\mathrm{PG}(q + 1, K)$ cuts out the divisor $qP + \Phi(P)$ on $\mathcal{X}$. To show this, note that $\Phi(P(a)) = P(a^{q^2})$, and that

$$x_0(t) = 1, \ \ldots, \ x_i(t) = (a + t)^i, \ \ldots, \ x_{q+1}(t) = (a + t)^{q+1}$$

is a primitive representation of the unique branch centred at $P(a)$. Then the result follows from the identities

$$a^{q^2+q} - a^q(a + t) - a^{q^2}(a + t)^q + (a + t)^{q+1} = t^q(a - a^{q^2} + t),$$
$$a^{q^2+q} - a^q a^{q^2} - a^{q^2}(a^q)^{q^2} + (a^{q+1})^{q^2} = 0.$$

Therefore the smallest linear series containing all divisors $qP + \Phi(P)$ with $P \in \mathcal{X}$ is cut out by the three-dimensional linear system of all hyperplanes

$$\mathbf{v}(u_0 X_0 + u_1 X_1 + u_q X_q + u_{q+1} X_{q+1});$$

that is, $m = 3$ in this case. For $q > 2$, it follows that $m < r$. No other such example is known to exist.

The next step is to show that $\mathcal{X}'$ lies on a non-degenerate Hermitian variety of Π_m. Since $\mathcal{D}'$ is a linear subseries of the complete linear series $\mathcal{D}$, Theorem 6.17 implies the following result.

LEMMA 10.19 *There exists a non-zero element $\eta \in K(\mathcal{X})$ and a non-singular matrix $C = (c_{ij})$ with entries in $K(\mathcal{X})$ such that*

$$\eta z_i = \textstyle\sum_{j=0}^r c_{ij} f_j, \quad \textit{for } i = 0, 1, 2, \ldots, r. \tag{10.20}$$

Since both $\mathcal{D}$ and $\mathcal{D}'$ are $\mathbf{F}_{q^2}$-rational linear series, from the proof of Theorem 6.17 it also follows that $\eta \in \mathbf{F}_{q^2}(\mathcal{X})$ and that the entries in C are elements of $\mathbf{F}_{q^2}$. Note that $c_{ij} = 0$ for $m+1 \leq i \leq r$ and the matrix C has rank $m+1$. Without loss of generality, let

$$f_0 = 1, \quad z_0 = 1. \tag{10.21}$$

Then, from (10.20) with $i = 0$,

$$\eta = 1, \quad c_{00} = 1, \quad c_{0j} = 0 \quad \text{for } j = 1, \ldots, r.$$

Lemma 10.20 *If (10.21) holds, then the matrix C in Lemma 10.19 is Hermitian; that is,* $c_{ij}^q = c_{ji}$ *for all* i, j.

Proof. The method is to substitute z_i in (10.12) by (10.20) and re-write the result in the form

$$\begin{aligned} 1 + \textstyle\sum_{i=1}^r (\sum_{j=0}^r c_{ij} f_j) f_i^q &= 1 + \textstyle\sum_{j=0}^r (\sum_{i=1}^r c_{ij} f_i^q) f_j \\ &= 1 + \textstyle\sum_{i=1}^r (\sum_{j=0}^r c_{ji} f_j^q) f_i \\ &= 0. \end{aligned}$$

This shows that the osculating hyperplane H_P of $\mathcal{X}$ at a general point P is

$$\mathbf{v}(X_0 + \textstyle\sum_{j=0}^r (c_{j1} f_j^q(P)) X_1 + \cdots + \sum_{j=0}^r (c_{jr} f_j^q(P)) X_r).$$

But then $z_i^q(P) = \sum_{j=0}^r (c_{ji} f_j^q(P))$ for $i = 1, \ldots, r$. Hence $z_i^q - \sum_{j=0}^r c_{ji} f_j^q$ has an infinite number of zeros. By Theorem 5.33, for $i = 1, \ldots, r$,

$$z_i^q = \textstyle\sum_{j=0}^r c_{ji} f_j^q.$$

On the other hand, (10.20) implies that $z_i^q = \sum_{j=0}^r c_{ij} f_j^q$. Since $f_1, \ldots, f_r$ are linearly independent over K, the same holds for $f_1^q, \ldots, f_r^q$. Therefore $c_{ij} = c_{ji}^q$ for all i, j with $0 \leq i, j \leq r$. □

Lemma 10.21 *Up to a projectivity over* $\mathbf{F}_{q^2}$*, the curve* $\mathcal{X}'$ *lies on the non-degenerate Hermitian variety* $\mathcal{H}_{m,q}$ *of* Π_m :

$$\mathcal{H}_{m,q} = \mathbf{v}(X_0^{q+1} + \cdots + X_m^{q+1}), \tag{10.22}$$

Proof. From Lemma 10.20, $c_{ij} = 0$ for $i = m+1, \ldots, r$ implies $c_{ij} = 0$ for $j = m+1, \ldots, r$. Therefore (10.20) becomes the following:

$$z_i = \textstyle\sum_{j=0}^r c_{ij} f_j, \qquad i = 0, 1, 2, \ldots, m, \quad c_{ij} = c_{ji}^q. \tag{10.23}$$

Here, the $(m+1) \times (m+1)$ matrix $C = (c_{ij})$ is non-singular and Hermitian. Its inverse matrix D is also Hermitian; hence

$$f_i = \textstyle\sum_{j=0}^r d_{ij} z_j, \qquad i = 0, 1, 2, \ldots, m, \quad d_{ij} = d_{ji}^q. \tag{10.24}$$

Substitution of these f_i in (10.12) gives

$$\textstyle\sum_{i,j=0}^m d_{ij} z_i^q z_j = 0,$$

which shows that $\mathcal{X}'$ lies on the hypersurface

$$\mathcal{H}' = \mathbf{v}(\textstyle\sum_{i,j=0}^m d_{ij} X_i^q X_j)$$

of Π_m, with $d_{ij} = d_{ji}^q$ and $d_{ij} \in \mathbf{F}_{q^2}$. The result follows since $\mathcal{H}'$ is a non-degenerate Hermitian variety projectively equivalent over $\mathbf{F}_{q^2}$ to $\mathcal{H}_{m,q}$. □

Therefore the following result is obtained.

THEOREM 10.22 (Natural Embedding Theorem) *Every $\mathbf{F}_{q^2}$-maximal curve $\mathcal{X}$ of genus $g \geq 0$ is isomorphic over $\mathbf{F}_{q^2}$ to a curve of* $\mathrm{PG}(m,K)$ *of degree $q+1$ lying on a non-degenerate Hermitian variety $\mathcal{H}_{m,q}$ defined over $\mathbf{F}_{q^2}$.*

REMARK 10.23 The dimension m is less then or équal to the dimension r of the Frobenius linear series $\mathcal{D}$ of $\mathcal{X}$. In particular, if $r=2$, then also $m=2$ and $\mathcal{X}$ is isomorphic over $\mathbf{F}_{q^2}$ to the Hermitian curve $\mathcal{H}_q$.

REMARK 10.24 From the proof of Theorem 10.22, the osculating hyperplane of $\mathcal{X}'$ at any point $P \in \mathcal{X}'$ coincides with the tangent hyperplane at P to the non-degenerate Hermitian variety $\mathcal{H}_{m,q}$ in which $\mathcal{X}'$ lies. As any point of $\mathrm{PG}(m,K)$ is the common point of only finitely many tangent hyperplanes of $\mathcal{H}_{m,q}$, the same holds for the osculating hyperplanes at $\mathcal{X}'$. Therefore the $\mathbf{F}_{q^2}$-maximal curve $\mathcal{X}$ in the Natural Embedding Theorem 10.22 has no higher-dimensional strangeness.

The Natural Embedding Theorem 10.22 and Castelnuovo's Bound (7.69) give the following result.

COROLLARY 10.25 *Let $\mathcal{X}$ be an $\mathbf{F}_{q^2}$-maximal curve of genus g. If the dimension of the Frobenius linear series of $\mathcal{X}$ is equal to r, then*

$$g \leq c_0(q+1,r) \leq F(r) = \begin{cases} \dfrac{(2q-(r-1))^2-1}{8(r-1)}, & \text{if } r \text{ is even;} \\ \dfrac{(2q-(r-1))^2}{8(r-1)}, & \text{if } r \text{ is odd.} \end{cases} \tag{10.25}$$

It may be noted that $F(r) \leq F(s)$ for $s \leq r$. Thus $F(r) \leq F(3) \leq \frac{1}{4}(q-1)^2$.

Next it is shown that the property given in Theorem 10.22 characterises $\mathbf{F}_{q^2}$-maximal curves. For this purpose, assume from now on that $\mathcal{Y}$ is an $\mathbf{F}_{q^2}$-rational irreducible, possibly singular, non-degenerate curve of $\mathrm{PG}(m,K)$ having the following property:

$$\deg \mathcal{Y} = q+1 \quad \text{and} \quad \mathcal{Y} \subset \mathcal{H}_{m,q}. \tag{10.26}$$

Then the aim is to prove that $\mathcal{Y}$ is an $\mathbf{F}_{q^2}$-maximal curve. It is also shown that $\mathcal{Y}$ is non-singular and that the linear series cut out on $\mathcal{Y}$ by hyperplanes is contained in the Frobenius linear series of $\mathcal{Y}$.

Since $\mathcal{Y}$ is not necessarily non-singular, points of $\mathcal{Y}$ are considered as branch points; that is, a point P of $\mathcal{Y}$ indicates the branch of $\mathcal{Y}$ associated to the place $\mathcal{P}$ of $K(\mathcal{Y})$. This convention comes from Section 7.6. Alternatively, choose a non-singular model of $\mathcal{Y}$ given by an $\mathbf{F}_{q^2}$-rational irreducible non-singular curve $\mathcal{Z}$ that is birationally equivalent to $\mathcal{Y}$ over $\mathbf{F}_{q^2}$. Then there exists a morphism defined over $\mathbf{F}_{q^2}$, and the branches of $\mathcal{Y}$ arise from points of $\mathcal{Z}$.

If $f_0,\ldots,f_M$ are the coordinate functions of $\mathcal{Y}$ with $f_0,\ldots,f_M \in \mathbf{F}_{q^2}(\mathcal{Y})$, the hypothesis (10.26) means that

$$f_0^{q+1} + \cdots + f_m^{q+1} = 0. \tag{10.27}$$

For $P \in \mathcal{Y}$, a primitive branch representation of P is

$$x_0 = f_0(t),\ \ldots,\ x_m = f_m(t).$$

Let $Q = (a_0 \dots, a_m)$ be the centre of P, Then

$$f_i(t) = \sum_{j=0}^{\infty} a_{ij}t^j \in K[[t]], \qquad a_{00} = a_0, \dots, a_{m0} = a_m.$$

Hence (10.27) implies that

$$\sum_{i=0}^{m} (\sum_{j=0}^{\infty} a_{ij})^{q+1} = 0.$$

The tangent hyperplane H_Q to $\mathcal{H}_{m,q}$ at $Q = (a_0, \dots, a_m)$ is

$$\mathbf{v}(a_0^q X_0 + \dots + a_m^q X_m).$$

The first step is to show that Lemma 10.7 (ii) holds for $\mathcal{Y}$.

LEMMA 10.26 *The linear series $\mathcal{R}$ cut out on $\mathcal{Y}$ by hyperplanes contains the divisor $qP + \Phi(P)$ for every $P \in \mathcal{Y}$.*

Proof. It is shown that H_Q cuts out the divisor $qP + \Phi(P)$ on $\mathcal{Y}$. From (10.27),

$$(\sum_{j=0}^{\infty} a_{0j}t^j)^q f_0 + \dots + (\sum_{j=0}^{\infty} a_{mj}t^j)^q f_m = 0\,. \tag{10.28}$$

Writing out the terms of lower order in t,

$$\sum_{i=0}^{m} a_{i0}^q\, f_i + t^q \sum_{i=0}^{m} a_{i0}\, a_{i1}^q + t^{q+1} \sum_{i=0}^{m} a_{i1}^{q+1} + t^{q+2} g(t) = 0\,.$$

Hence the intersection number $I(P, H_Q \cap \mathcal{Y})$ is at least q and equality holds if and only if $\sum_{i=0}^{m} a_{i1}^q a_{i0} \neq 0$.

Now it is shown that, if $P \in \mathcal{Y}(\mathbf{F}_{q^2})$, then $\sum_{i=0}^{m} a_{i1}^q a_{i0} = 0$. From (10.28),

$$\sum_{i=0}^{m} a_{i0}^{q+1} + t\sum_{j=0}^{m} a_{i0}^q\, a_{i1} + t^q\, h(t) = 0\,.$$

Thus $\sum_{i=0}^{m} a_{i0}^q a_{i1} = 0$. Since

$$(\sum_{i=0}^{m} a_{i0}^q\, a_{i1})^q = \sum_{i=0}^{m} a_{i0}\, a_{i1}^q$$

when $P \in \mathcal{Y}(\mathbf{F}_{q^2})$, the assertion follows.

Since $\deg(\mathcal{Y}) = q + 1$, this implies that H_Q cuts out the divisor $(q + 1)P$ on $\mathcal{Y}$. Therefore the lemma is established for every $P \in \mathcal{Y}(\mathbf{F}_{q^2})$.

For the case $P \notin \mathcal{Y}(\mathbf{F}_{q^2})$, it must also be shown that the image point of Q under the Frobenius collineation $(X_0, \dots, X_m) \mapsto (X_0^q, \dots, X_m^q)$ lies in H_Q. This occurs when

$$\sum_{i=0}^{m} a_i^{q^2+q} = 0.$$

Since this relation is a consequence of (10.28), the result follows. Therefore H_Q cuts out the divisor $qP + \Phi(P)$ on $\mathcal{Y}$ for every $P \notin \mathcal{Y}(\mathbf{F}_{q^2})$. □

Now Lemma 10.26 makes it possible to use the proof of Theorem 10.7 to show that $\mathcal{Y}$ is an $\mathbf{F}_{q^2}$-rational irreducible non-singular curve.

If $m = 2$, then $\mathcal{Y}$ is the Hermitian curve and so $\mathcal{Y}$ is $\mathbf{F}_{q^2}$-maximal. Therefore let $m \geq 3$.

The approach used for the proof is based on the relationship between the Wronskian determinant of $\mathcal{Y}$ and the projection $\bar{\mathcal{Y}}$ of $\mathcal{Y}$ to an $(m-1)$-dimensional subspace of $\mathrm{PG}(m, K)$. More precisely, let

$$\begin{array}{rl} \bar{\pi}: & \mathrm{PG}(m, K) \to \Pi_{m-1}, \\ & (f_0, \dots, f_m) \mapsto (f_0, \dots, f_{m-1}), \end{array} \tag{10.29}$$

be the projection $\bar{\mathcal{Y}}$ of $\mathcal{Y}$ from the point $\mathbf{U}_m = (0, \dots, 0, 1)$ to the hyperplane $\mathbf{v}(X_m)$. It may happen that $\mathcal{Y}$ and $\bar{\mathcal{Y}}$ are not birationally equivalent over $\mathbf{F}_{q^2}$. However, it is always possible to avoid this situation by changing the coordinate system in $\mathrm{PG}(m, K)$.

LEMMA 10.27 (i) *In* $\mathrm{PG}(m,K)$ *there is a point* P *with coordinates in* $\mathbf{F}_{q^2}$ *that satisfies each of the following conditions*:

(a) $P \notin \mathcal{H}_{m,q}$;

(b) *no tangent line to* $\mathcal{Y}$ *at an* $\mathbf{F}_{q^2}$*-rational point passes through* P;

(c) *no chord through two* $\mathbf{F}_{q^2}$*-rational points of* $\mathcal{Y}$ *passes through* P.

(ii) *If* ℓ *is an* $\mathbf{F}_{q^2}$*-rational line through an* $\mathbf{F}_{q^2}$*-rational point* R *of* $\mathcal{Y}$, *then each point of* $\ell \cap \mathcal{Y}$ *is an* $\mathbf{F}_{q^2}$*-rational point of* ℓ.

Proof. (i) Take an $\mathbf{F}_{q^2}$-rational point $Q \in \mathcal{Y}$. Since the number of $\mathbf{F}_{q^2}$-rational points of $\mathcal{Y}$ is at most $q^2+1+2gq \le q^3+1$, there are at most q^3 chords joining Q to another $\mathbf{F}_{q^2}$-rational point of $\mathcal{Y}$. But, since $m \ge 3$, the number of $\mathbf{F}_{q^2}$-rational lines through Q is at least q^4+q^2+1 and hence one of these lines is neither a line contained in $\mathcal{H}_{m,q}$, nor a tangent line to $\mathcal{Y}$ at Q, nor a chord through Q and another $\mathbf{F}_{q^2}$-rational point of $\mathcal{Y}$. Now any $\mathbf{F}_{q^2}$-rational point P outside $\mathcal{H}_{m,q}$ is a suitable choice for P.

(ii) Assume on the contrary that ℓ meets $\mathcal{Y}$ in a non-$\mathbf{F}_{q^2}$-rational point S of ℓ. Then ℓ is the line joining S and the image point S' of S under the Frobenius collineation $(X_0,\ldots,X_m) \mapsto (X_0^{q^2},\ldots,X_m^{q^2})$. Therefore ℓ is contained in the osculating hyperplane of $\mathcal{Y}$ at S. Hence the common points of ℓ and $\mathcal{Y}$ are only two, namely S and S'. But this contradicts the hypothesis that $R \in \ell \cap \mathcal{Y}$. □

Take a point P as in (i) of this lemma. Since the projective group $\mathrm{PGU}(m+1,q)$ preserving $\mathcal{H}_{m,q}$ acts transitively on the set of all points of $\mathrm{PG}(m,q^2)\backslash\mathcal{H}_{m,q}$, it follows that a projectivity of $\mathrm{PG}(m,q^2)$ exists which preserves $\mathcal{H}_{m,q}$ and takes P to the point $\mathbf{U}_m=(0,\ldots,0,1)$. Lemma 10.27 now ensures that $\mathcal{Y}$ and $\bar{\mathcal{Y}}$ are birationally equivalent over $\mathbf{F}_{q^2}$.

Choose a separable variable ζ of $K(\mathcal{Y})$, and consider the Wronskian matrices of type investigated in Section 7.6:

$$\mathcal{W}(f_0,\ldots,f_{m-1}) = \det\begin{bmatrix} D_\zeta^{\epsilon_0} f_0 & D_\zeta^{\epsilon_0} f_1 & \cdots & D_\zeta^{\epsilon_0} f_{m-1} \\ \vdots & \vdots & \cdots & \vdots \\ D_\zeta^{\epsilon_{m-1}} f_0 & D_\zeta^{\epsilon_{m-1}} f_1 & \cdots & D_\zeta^{\epsilon_{m-1}} f_{m-1} \end{bmatrix},$$

$$\mathcal{W}(f_0,\ldots,f_m) = \det\begin{bmatrix} D_\zeta^{\epsilon_0} f_0 & D_\zeta^{\epsilon_0} f_1 & \cdots & D_\zeta^{\epsilon_0} f_m \\ \vdots & \vdots & \cdots & \vdots \\ D_\zeta^{\epsilon_m} f_0 & D_\zeta^{\epsilon_m} f_1 & \cdots & D_\zeta^{\epsilon_m} f_m \end{bmatrix}; \tag{10.30}$$

here $D_\zeta^{(i)}$ is the i-th Hasse derivative with respect to a separable variable ζ in $K(\mathcal{Y})$. From Lemma 10.26, $\epsilon_0=0$, $\epsilon_1=1$, $\epsilon_m=q$. So (10.30) becomes the following:

$$\mathcal{W}(f_0,\ldots,f_m) = \det\begin{bmatrix} f_0 & f_1 & \cdots & f_m \\ D_\zeta f_0 & D_\zeta f_1 & \cdots & D_\zeta f_m \\ D_\zeta^{\epsilon_2} f_0 & D_\zeta^{\epsilon_2} f_1 & \cdots & D_\zeta^{\epsilon_2} f_m \\ \vdots & \vdots & \cdots & \vdots \\ D_\zeta^{\epsilon_{m-1}} f_0 & D_\zeta^{\epsilon_{m-1}} f_1 & \cdots & D_\zeta^{\epsilon_{m-1}} f_m \\ D_\zeta^q f_0 & D_\zeta^q f_1 & \cdots & D_\zeta^q f_m \end{bmatrix}. \tag{10.31}$$

LEMMA 10.28

$$\begin{aligned}&\operatorname{div}(\mathcal{W}(f_0,\ldots,f_m))\\&=\operatorname{div}(\mathcal{W}(f_0,\ldots,f_{m-1}))-q\operatorname{div}(f_m)+\operatorname{div}(f_0D_\zeta^q f_0^q+\cdots+f_mD_\zeta^q f_m^q).\end{aligned}$$

Proof. If the columns of the matrix in (10.31) are $c_0, c_1, \ldots, c_m$, for c_m put

$$f_0^q c_0 + f_1^q c_1 + \cdots + f_{m-1}^q c_{m-1} + f_m^q c_m.$$

Then

$$\begin{aligned}&f_m^q\mathcal{W}(f_0,\ldots f_m)\\&=\det\begin{bmatrix} f_0 & f_1 & \ldots & f_{m-1} & f_0^{q+1}+\cdots+f_m^{q+1}\\ D_\zeta f_0 & D_\zeta f_1 & \ldots & D_\zeta f_{m-1} & f_0^q D_\zeta f_0+\cdots+f_m^q D_\zeta f_m\\ \vdots & \vdots & \ldots & \vdots & \vdots\\ D_\zeta^q f_0 & D_\zeta^q f_1 & \ldots & D_\zeta^q f_{m-1} & f_0^q D_\zeta^q f_0+\cdots+f_m^q D_\zeta^q f_m\end{bmatrix}.\end{aligned}$$

Each element in the last column is 0 apart from the last: this follows from (10.27) by derivation. Also, the q-th Hasse derivative of the same relation gives the relation

$$f_0^q D_\zeta f_0 + \cdots + f_m^q D_\zeta f_m + f_0 D_\zeta^q f_0^q + \cdots + f_m D_\zeta^q f_m^q = 0,$$

and this completes the proof. □

Let $R_m = \sum v_P(R_m)P$ be the ramification divisor of the linear series cut out on $\mathcal{Y}$ by hyperplanes of $\mathrm{PG}(m, K)$. Let $P \in \mathcal{Y}$ and choose a local parameter ζ at P as a separable variable. Then

$$v_P(R_m) = \mathrm{ord}_P(\mathcal{W}(f_0, \ldots, f_m)). \tag{10.32}$$

Similarly, let R_{m-1} be the ramification divisor of the linear series cut out on $\overline{\mathcal{Y}}$ by hyperplanes of $\mathrm{PG}(m, K)$.

LEMMA 10.29 *If $P \in \mathcal{Y}$ and ζ is a local parameter of $\mathcal{Y}$ at P, then*

$$v_P(R_{m-1}) = \mathrm{ord}_P(\mathcal{W}(f_0, \ldots, f_{m-1})).$$

Proof. By Definition 7.52,

$$v_P(R_{m-1}) = \mathrm{ord}_P(\mathcal{W}(f_0,\ldots,f_{m-1}))+(\epsilon_0+\epsilon_1+\cdots+\epsilon_{m-1})\mathrm{ord}_P(d\zeta)+m\bar{e}_P,$$

where

$$\bar{e}_P = -\min\{\mathrm{ord}_P(f_0), \ldots, \mathrm{ord}_P(f_{m-1})\}.$$

Here, $\bar{e}_P = 0$. In fact, if $\bar{e}_P > 0$ and $e_P = 0$, then $\mathbf{U}_m = (0, \ldots, 0, 1)$ lies on $\mathcal{Y}$, contradicting (10.27). Since ζ is a local parameter at P, so $\mathrm{ord}_P(d\zeta) = 0$, and the result follows. □

LEMMA 10.30

$$\mathrm{ord}_P(f_0 D_\zeta^q f_0^q + \cdots + f_M D_\zeta^q f_M^q) = \begin{cases} 1 & \textit{when } P \in \mathcal{Y}(\mathbf{F}_{q^2}),\\ 0 & \textit{when } P \notin \mathcal{Y}(\mathbf{F}_{q^2}).\end{cases}$$

Proof. The proof of Lemma 10.26 also provides a proof of the following result.

(i) For any point $P \in \mathcal{Y}$,

(a) $P \notin \mathcal{Y}(\mathbf{F}_{q^2})$ if and only if $\sum_{i=0}^{m} a_{i1}^q a_{m1} \neq 0$;

(b) $P \in \mathcal{Y}(\mathbf{F}_{q^2})$ if and only if $\sum_{i=0}^{m} a_{i1}^q a_{m1} = 0$ and $\sum_{i=0}^{m} a_{i1}^{q+1} \neq 0$.

On the other hand,

$$\begin{aligned} & f_0 D_\zeta^q f_0^q + \cdots + f_m D_\zeta^q f_m^q \\ &= (\textstyle\sum_{j=0}^{\infty} a_{0j} t^j)(a_{01}^q + t^q g_1(t)) + \cdots + (\textstyle\sum_{j=0}^{\infty} a_{mj} t^j)(a_{m1}^q + t^q g_2(t)) . \end{aligned}$$

This implies the following.

(ii)

(a) $\mathrm{ord}_P(f_0 D_\zeta^q f_0^q + \cdots + f_m D_\zeta^q f_m^q) = 0$ if and only if $\sum a_{i0} a_{i1}^q \neq 0$;

(b) $\mathrm{ord}_P(f_0 D_\zeta^q f_0^q + \cdots + f_m D_\zeta^q f_m^q) = 1$ if and only if $\sum a_{i0} a_{i1}^q = 0$ and $\sum a_{i,1}^{q+1} \neq 0$.

Now comparison of (i) with (ii) proves Lemma 10.30. □

THEOREM 10.31 *Every irreducible curve $\mathcal{Y}$ defined over $\mathbf{F}_{q^2}$ satisfying (10.26) is a non-singular $\mathbf{F}_{q^2}$-maximal curve.*

Proof. By (7.13),

$$\begin{aligned} \textstyle\sum v_P(R_m) &= (\epsilon_0 + \epsilon_1 + \cdots + \epsilon_m)(2g-2) + (m+1)(q+1) , \\ \textstyle\sum v_P(R_{m-1}) &= (\epsilon_0 + \epsilon_1 + \cdots + \epsilon_{m-1})(2g-2) + m(q+1) . \end{aligned}$$

Hence

$$\textstyle\sum (v_P(R_m) - v_P(R_{m-1})) = q(2g-2) + q + 1.$$

Lemmas 10.28, 10.29, 10.30, (10.32) and the relation, $\sum \mathrm{ord}_P(f_M) = q+1$, finish the proof. □

REMARK 10.32 In (10.26) the condition 'degree $q+1$' may be replaced by 'minimum degree with respect to the property of lying on a non-degenerate Hermitian variety of $\mathrm{PG}(m, K)$ defined over $\mathbf{F}_{q^2}$'.

To show this, let $f_0, f_1, \ldots, f_M$ be the coordinate functions of an irreducible non-degenerate curve Γ of $\mathrm{PG}(m, K)$ lying on the Hermitian variety $\mathcal{H}_{m,q}$ in $\mathrm{PG}(m, K)$; then (10.27) holds. If

$$(x_0 = f_0(t) = a_0 + h_0(t), \ldots, x_m = f_m(t) = a_m + h_m(t)), \quad \mathrm{ord}_P\, h_i(t) \geq 1,$$

is a primitive representation of a branch of Γ centred at the point $Q = (a_0, \ldots, a_m)$, then (10.27) implies that

$$\mathrm{ord}_t(a_0^q f_0(t) + \cdots + a_m^q f_m(t)) = \mathrm{ord}_t(h_0(t)^q f_0(t) + \cdots + h_m(t)^q f_m(t)),$$

whence

$$\mathrm{ord}_t(a_0^q f_0(t) + \cdots + a_m^q f_m(t)) \geq q.$$

This shows that the tangent hyperplane $H_Q = \mathbf{v}(a_0^q X_0 + \cdots + a_m^q X_m)$ to $\mathcal{H}_{m,q}$ at Q meets $\mathcal{Y}$ at the branch γ with multiplicity at least q. If $P \notin \mathrm{PG}(r, q^2)$, then its image

$$P' = (a_0^{q^2}, \ldots, a_m^{q^2})$$

under the Frobenius collineation $(X_0, \ldots, X_m) \mapsto (X_0^{q^2}, \ldots, X_m^{q^2})$ is another common point of Γ and H_P. In fact, $a_0^{q+1} + \cdots + a_m^{q+1} = 0$ implies that

$$a_0^q a_0^{q^2} + \cdots + a_m^q a_m^{q^2} = 0.$$

Since $\mathcal{Y}$ is a non-degenerate curve, it is not contained in H_Q. Thus $\mathcal{Y}$ must have degree at least $q+1$.

As an illustration of this material, it is shown how the known examples of $\mathbf{F}_{q^2}$-maximal curves with $r = m = 3$ are embedded in a non-degenerate Hermitian surface of PG$(3, K)$. In this way, an independent proof of the maximality of these curves is obtained.

EXAMPLE 10.33 Let $q \equiv 2 \pmod 3$, and fix a primitive cube root of unity ϵ in $\mathbf{F}_{q^2}$. For $i = 0, 1, 2$, let $\mathcal{F}_i = \mathbf{v}(F_i)$ be the $\mathbf{F}_{q^2}$-rational irreducible plane curve with

$$F_i(X, Y) = \epsilon^i X^{(q+1)/3} + \epsilon^{2i} X^{2(q+1)/3} + Y^{q+1},$$

and let $K(x, y)$, with $F_i(x, y) = 0$, be its function field. Let Γ_i be the $\mathbf{F}_{q^2}$-rational irreducible curve of PG$(3, K)$, birationally equivalent over $\mathbf{F}_{q^2}$ to $\mathcal{F}_i$, defined by the coordinate functions,

$$f_0 = x,\ f_1 = x^2,\ f_2 = y^3,\ f_3 = xy.$$

Note that the three curves Γ_i are projectively equivalent in $PG(3, K)$. In fact, the projectivity over $\mathbf{F}_{q^2}$ induced by the matrix

$$T_4^{(i)} = \begin{bmatrix} \epsilon^i & 0 & 0 & 0 \\ 0 & \epsilon^{2i} & 0 & 0 \\ 0 & 0 & 1 & 0 \\ 0 & 0 & 0 & \epsilon^i \end{bmatrix}$$

maps Γ_0 to Γ_j with $j \equiv 2i \pmod 3$.

Now it is shown that Γ_i is an $\mathbf{F}_{q^2}$-rational irreducible non-singular curve of degree $q+1$ lying, up to a projectivity, on the Hermitian surface

$$\mathcal{H}_{3,q} = \mathbf{v}(X_0^{q+1} + X_1^{q+1} + X_2^{q+1} + X_3^{q+1}). \tag{10.33}$$

To do this, note that the classical identity

$$a^3 + b^3 + c^3 - 3abc = (a + b + c)(a + \epsilon b + \epsilon^2 c)(a + \epsilon^2 b + \epsilon c), \tag{10.34}$$

with

$$a = Y^{q+1}, \quad b = X^{(q+1)/3}, \quad c = X^{2(q+1)/3},$$

becomes the identity

$$\begin{aligned}
&(X^{(q+1)/3} + X^{2(q+1)/3} + Y^{q+1}) \\
&\quad \times (\epsilon X^{(q+1)/3} + \epsilon^2 X^{2(q+1)/3} + Y^{q+1}) \\
&\quad\quad \times (\epsilon^2 X^{(q+1)/3} + \epsilon X^{2(q+1)/3} + Y^{q+1}) \\
&\quad\quad\quad = X^{q+1} + X^{2(q+1)} + Y^{3(q+1)} - 3X^{q+1}Y^{q+1}.
\end{aligned}$$

This yields the equation

$$x^{q+1} + x^{2(q+1)} + y^{3(q+1)} - 3x^{q+1}y^{q+1} = 0,$$

whence $f_0^{q+1} + f_1^{q+1} + f_2^{q+1} - 3f_3^{q+1} = 0$. Therefore Γ_i lies on $\mathcal{H}_{3,q}$, up to the projectivity

$$(x_0, x_1, x_2, x_3) \mapsto (x_0, x_1, x_2, wx_3), \qquad w^{q+1} = -3.$$

Also, Γ_i is contained in the cubic surface

$$\mathcal{S}_3 = \mathbf{v}(X_3^3 + w^3 X_0 X_1 X_2) \subset \mathrm{PG}(3, K).$$

More precisely, the intersection of $\mathcal{H}_{3,q}$ and $\mathcal{S}_3$ splits into the three curves, $\Gamma_0, \Gamma_1, \Gamma_2$, each of degree $q+1$. By Theorem 10.31, each Γ_i is a non-singular maximal curve defined over $\mathbf{F}_{q^2}$. From Theorem 12.28 (IV)(b), its genus is $\frac{1}{6}(q^2 - q + 4)$.

EXAMPLE 10.34 For a similar but non-isomorphic example, again let $q \equiv 2 \pmod 3$, and fix a primitive cube root of unity $\epsilon \in \mathbf{F}_{q^2}$. For $i = 0, 1, 2$, let $\mathcal{F}'_i = \mathbf{v}(F'_i)$ be the $\mathbf{F}_{q^2}$-rational irreducible plane curve with

$$F'_i(X, Y) = \epsilon^i Y X^{(q-2)/3} + Y^q + \epsilon^{2i} X^{(2q-1)/3},$$

and $K(x, y)$ with $F'_i(x, y) = 0$ be its function field. Let Γ'_i be the $\mathbf{F}_{q^2}$-rational irreducible curve of $\mathrm{PG}(3, K)$, birationally equivalent to $\mathcal{F}'_i$ over $\mathbf{F}_{q^2}$, with coordinate functions

$$f_0 = x,\ f_1 = x^2,\ f_2 = y^3,\ f_3 = -3xy.$$

The three curves Γ'_i are projectively equivalent in $\mathrm{PG}(3, K)$. In fact, the projectivity induced by the matrix $T_4^{(i)}$ in Example 10.33 maps Γ'_0 to Γ'_i.

Arguing as in Example 10.33, it is shown now that Γ'_i is an $\mathbf{F}_{q^2}$-rational irreducible non-singular curve of degree $q+1$ lying on $\mathcal{H}_{3,q}$ as in (10.33). From the identity (10.34), with

$$a = Y^q, \quad b = Y X^{(q-2)/3}, \quad c = X^{(2q-1)/3},$$

it follows that

$$y^3 x^{q-2} + y^{3q} + x^{2q-1} - 3x^{q-1}y^{q+1} = 0,$$

whence

$$y^3 x^q + y^{2q} x^2 + x^{2q+1} - 3x^{q+1}y^{q+1} = 0.$$

Hence $f_2 f_0^q + f_2^q f_1 + f_1^q f_0 - 3f_3^{q+1} = 0$ and this shows that Γ'_i lies on the surface

$$\mathcal{S}'_{q+1} = \mathbf{v}(X_0^q X_1 + X_1^q X_2 + X_2^q X_0 - \tfrac{1}{3} X_3^{q+1}).$$

Also, Γ'_i lies on the cubic surface

$$\mathcal{S}_3 = \mathbf{v}(X_3^3 + 27 X_0 X_1 X_2).$$

Hence the intersection of $\mathcal{S}'_{q+1}$ and $\mathcal{S}_3$ splits into the three curves, Γ'_0, Γ'_1, Γ'_2, each of degree $q+1$.

To prove that $\mathcal{S}'_{q+1}$ is projectively equivalent to $\mathcal{H}_{3,q}$, choose a root α of the polynomial $X^{q+1}+X+1$. Then $\alpha^{q^2+q+1}=1$, and hence $\alpha \in \mathbf{F}_{q^3}$. From this,

$$\alpha^{q+1}+\alpha^{q^2+q+1}+\alpha=0, \quad \alpha^{q^2+q+2}+\alpha^{q+1}+1=0;$$

but $\alpha^{q+2}+\alpha^{q^2+1}+\alpha^q \neq 0$ as $(\alpha^{q+2}+\alpha^{q^2+1}+\alpha^q)^{q-1}=\alpha^{-1}$. Also, the matrix

$$M_3=\begin{bmatrix} \alpha & 1 & \alpha^{q^2+1} \\ \alpha^{q^2+1} & \alpha & 1 \\ 1 & \alpha^{q^2+1} & \alpha \end{bmatrix}$$

is non-singular. Choose an element $\mu \in \mathbf{F}_{q^2}$ satisfying the equation

$$-3\mu^{q+1}=\alpha^{q^3+q+1}+\alpha^{q^2+1}+\alpha^q,$$

and define $\mathfrak{T}$ to be the projectivity of $\mathrm{PG}(3,K)$ associated to the non-singular matrix

$$M_4=\begin{bmatrix} \alpha & 1 & \alpha^{q^2+1} & 0 \\ 1 & \alpha^{q^2+1} & \alpha & 0 \\ \alpha^{q^2+1} & \alpha & 1 & 0 \\ 0 & 0 & 0 & -\mu \end{bmatrix}.$$

Then $\mathfrak{T}^{-1}$ maps $\mathcal{S}'_{q+1}$ to $\mathcal{H}_{3,q}$, and $\mathcal{S}_3$ to the cubic surface $\mathcal{S}_3^*=\mathbf{v}(F_3)$, with

$$\begin{aligned} F_3= & (X_0^3+X_1^3+X_2^3)+\mathrm{T}(a^{q+1})(X_0^2X_1+X_1^2X_2+X_2^2X_0) \\ & +\mathrm{T}(\alpha)(X_0^2X_2+X_1^2X_0+X_2^2X_1) \\ & +(3+\mathrm{T}(\alpha^{q-1}))X_0X_1X_2-\alpha^{q-1}\mu^3X_3^3, \end{aligned}$$

where $\mathrm{T}(u)=u+u^q+u^{q^2}$ is the trace of $u \in \mathbf{F}_{q^3}$ over $\mathbf{F}_q$. Also, $\alpha^{q-1}\mu^3 \in \mathbf{F}_{q^2}$, and this shows that $\mathcal{S}_3^*$ is defined over $\mathbf{F}_{q^2}$. Now Γ'_i is mapped by $\mathfrak{T}^{-1}$ to an $\mathbf{F}_{q^2}$-rational irreducible curve $\mathcal{C}'_i$ of degree $q+1$ lying on $\mathcal{H}_{3,q}$. By Theorem 10.31, $\mathcal{C}'_i$ is a non-singular $\mathbf{F}_{q^2}$-maximal curve. From Theorem 12.28 (V) with $d=3$, its genus is $\frac{1}{6}(q^2-q-2)$; see also Section 7.11.

EXAMPLE 10.35 Let q be odd. For $i=0,1$ and $m=\frac{1}{2}(q+1)$, let $\mathcal{E}_m^i=\mathbf{v}(E_i)$ be the $\mathbf{F}_{q^2}$-rational irreducible plane curve, with

$$E_i(X,Y)=Y^q+Y+(-1)^iX^{(q+1)/2}$$

and $K(x,y)$ with $E_i(x,y)=0$ its function field.

Let Γ''_i be the $\mathbf{F}_{q^2}$-rational irreducible curve of $\mathrm{PG}(3,K)$, birationally equivalent to $\mathcal{E}_i$ over $\mathbf{F}_{q^2}$, with coordinate functions

$$f_0=1,\ f_1=x,\ f_2=y,\ f_3=y^2.$$

The curves Γ''_1 and Γ''_0 are projectively equivalent, since the projectivity induced by the matrix

$$T_4=\begin{bmatrix} 1 & 0 & 0 & 0 \\ 0 & \epsilon & 0 & 0 \\ 0 & 0 & 1 & 0 \\ 0 & 0 & 0 & 1 \end{bmatrix},$$

with $\epsilon^{(q+1)/2} = -1$, maps Γ_0'' to Γ_1''. The polynomial identity

$$(Y^q + Y - X^{(q+1)/2})(Y^q + Y + X^{(q+1)/2}) = Y^{2q} + 2Y^{q+1} + Y^2 - X^{q+1}$$

implies that $y^{2q} + 2y^{q+1} + y^2 - x^{q+1} = 0$ and so $f_3^q + f_3 + 2f_2^{q+1} - f_1^{q+1} = 0$. This shows that Γ_i'' lies on the surface

$$\mathcal{H}' = \mathbf{v}(X_3^q X_0 + X_3 X_0^q + 2X_2^{q+1} - X_1^{q+1}),$$

which is a non-degenerate Hermitian surface of $\mathrm{PG}(3, K)$. Also, Γ_i'' lies on the quadric cone

$$\mathcal{Q} = \mathbf{v}(X_2^2 - X_0 X_3),$$

and hence the intersection of $\mathcal{H}'$ and $\mathcal{Q}$ splits into the curves Γ_0'' and Γ_1''. By Theorem 10.31, Γ_i'' is a non-singular $\mathbf{F}_{q^2}$-maximal curve. From Theorem 10.41, its genus is $\frac{1}{4}(q-1)^2$; see also Theorem 12.28 (I).

EXAMPLE 10.36 Let $q = 2^t$, and put $\mathrm{T}(Y) = Y + Y^2 + \cdots + Y^{q/4} + Y^{q/2}$. For each $i = 0, 1 \in \mathbf{F}_2 \subset \mathbf{F}_{q^2}$, let $\mathcal{C}_i = \mathbf{v}(T_i)$ be the $\mathbf{F}_{q^2}$-rational irreducible plane curve with

$$T_i(X, Y) = \mathrm{T}(Y) + X^{q+1} + i,$$

and $K(x, y)$ with $T_i(x, y) = 0$ its function field. It may be noted that $\mathcal{C}_0 = \mathcal{T}_2$, as in Example 10.3 (ii).

Let Φ_i be the $\mathbf{F}_{q^2}$-rational irreducible curve of $\mathrm{PG}(3, K)$, birationally equivalent to $\mathcal{C}_i$ over $\mathbf{F}_{q^2}$, with coordinate functions

$$f_0 = 1,\ f_1 = x,\ f_2 = y,\ f_3 = x^2 .$$

Since

$$(\mathrm{T}(Y) + X^{q+1})(\mathrm{T}(Y) + X^{q+1} + 1) = Y^q + Y + X^{q+1} + X^{2q+2},$$

so $y^q + y + x^{q+1} + x^{2q+2} = 0$. Therefore Φ_i lies on the non-degenerate Hermitian variety

$$\mathcal{H}'' = \mathbf{v}(X_2^q X_0 + X_2 X_0^q + X_1^{q+1} + X_3^{q+1}).$$

Also, the quadric cone $\mathcal{Q} = \mathbf{v}(X_0 X_3 - X_1^2)$ contains Φ_i. Hence $\mathcal{H}'' \cap \mathcal{Q}$ splits into Φ_0 and Φ_1. Note that the curves Φ_0 and Φ_1 are projectively equivalent over $\mathbf{F}_{q^2}$ in $\mathrm{PG}(3, K)$, and hence both have degree $q + 1$. Again, by Theorem 10.31, they are non-singular $\mathbf{F}_{q^2}$-maximal curves and, from Theorem 10.45, of genus $\frac{1}{4}q(q-2)$; see also Theorem 12.28 (II) (a) for $p = 2$.

EXAMPLE 10.37 Let $q = 3^t$, and put $\mathrm{T}(Y) = Y + Y^3 + \cdots + Y^{q/3}$. For each $i = 0, 1, 2 \in \mathbf{F}_3 \subset \mathbf{F}_{q^2}$, let $\mathcal{C}_i = \mathbf{v}(T_i')$ be the $\mathbf{F}_{q^2}$-rational irreducible plane curve, with

$$T_i'(X, Y) = \mathrm{T}(Y)^2 - X^q - X + i(\mathrm{T}(Y) + i)\,,$$

and let $K(x, y)$ with $T_i'(x, y) = 0$ be its function field. Also, $\mathcal{C}_0 = \mathcal{T}_3'$, as in Example 10.3. From the identity

$$\begin{aligned}(\mathrm{T}(Y)^2 &- X^q - X)\\ &\times(\mathrm{T}(Y)^2 + \mathrm{T}(Y) + 1 - X^q - X)\\ &\quad\times(\mathrm{T}(Y)^2 - \mathrm{T}(Y) + 1 - X^q - X)\\ &\quad= (Y^q - Y)^2 - (X^q + X)(X^q + X - 1)^2,\end{aligned}$$

it follows that

$$(x^3 + x^2 - y^2 + x)^q + (x^3 + x^2 - y^2 + x) - x^{q+1} - y^{q+1} = 0\,.$$

Let Φ'_i be the $\mathbf{F}_{q^2}$-rational irreducible curve of $\mathrm{PG}(3, K)$, birationally equivalent to $\mathcal{C}_i$ over $\mathbf{F}_{q^2}$, with coordinate functions

$$f_0 = 1,\ f_1 = x,\ f_2 = y,\ f_3 = x^3 + x^2 - y^2 + x\,.$$

As in the preceding examples, the three curves, Φ'_0, Φ'_1, Φ'_2, are $\mathbf{F}_{q^2}$-projectively equivalent in $\mathrm{PG}(3, K)$. Also, Φ'_i lies on the non-degenerate Hermitian variety

$$\mathcal{H}^* = \mathbf{v}(X_0X_3^q + X_0^qX_3 - X_1^{q+1} - X_2^{q+1}).$$

Finally, the cubic surface

$$\mathcal{S}'_3 = \mathbf{v}(X_3X_0^2 - X_1^3 + X_1^2X_0 + X_2^2X_0 - X_1X_0^2)$$

also contains Φ'_i. Therefore Φ'_i has degree $q + 1$, and Theorem 10.31 ensures that each Φ'_i is a non-singular $\mathbf{F}_{q^2}$-maximal curve. From Theorem 12.28 (II)(b), their genus is $\frac{1}{6}q(q-1)$.

10.4 MAXIMAL CURVES LYING ON A QUADRIC SURFACE

The Natural Embedding Theorem 10.22 offers a geometric approach to the study of $\mathbf{F}_{q^2}$-maximal curves. The idea is to use the specific properties of the model of $\mathbf{F}_{q^2}$-maximal curves described in that theorem and the successive three remarks. To do this, consider the following.

CONDITION 10.38 *An $\mathbf{F}_{q^2}$-maximal curve is an $\mathbf{F}_{q^2}$-rational, irreducible, non-singular curve of degree $q + 1$ lying on a non-degenerate Hermitian variety of $\mathrm{PG}(m, K)$, where $2 \le m \le r = \dim \mathcal{D}$ with $\mathcal{D}$ the Frobenius linear series of $\mathcal{X}$.*

As in many geometric problems, low dimensions can be tackled successfully. Nevertheless, it is not apparent which of the low-dimensional results are of general type; this is the main difficulty in deciding what happens in higher dimensions.

Since $m = 2$ only occurs when $\mathcal{X}$ is a Hermitian curve, the smallest dimension to consider is $m = 3$. Examples 10.35 and 10.36 show that there are $\mathbf{F}_{q^2}$-maximal curves of $\mathrm{PG}(3, K)$ lying on a quadric. The aim of this section is to show that no other $\mathbf{F}_{q^2}$-maximal curve of $\mathrm{PG}(3, K)$ has this property.

Therefore let $\dim \mathcal{D} = m = 3$. Then the Frobenius linear series $\mathcal{D}$ of $\mathcal{X}$ is the linear series cut out on $\mathcal{X}$ by hyperplanes. For a point $P \in \mathcal{X}$, the positive $\mathcal{D}$-orders are

$$j_1(P) = 1 < j_2(P) < j_3(P),$$

where

$$j_3(P) = \begin{cases} q+1, & \text{when } P \text{ is } \mathbf{F}_{q^2}\text{-rational,} \\ q, & \text{when } P \text{ is not } \mathbf{F}_{q^2}\text{-rational.} \end{cases} \tag{10.35}$$

Lemma 10.39 *Let* $\dim \mathcal{D} = 3$ *and suppose that* $\mathcal{X}$ *lies on a quadric surface* $\mathcal{Q}$ *of* $\mathrm{PG}(3, K)$. *If* $q \geq 4$, *then the following properties hold for any point* $P \in \mathcal{X}$:

(i) $j_2(P) \in \{2, \frac{1}{2}j_3(P), \frac{1}{2}(j_3(P)+1)\}$;

(ii) $j_2(P) > 2$ *if and only if the tangent line* ℓ_P *of* $\mathcal{X}$ *at* P *lies on* $\mathcal{Q}$;

(iii) *if* $j_2 > 2$ *and* P *is a non-singular point of* $\mathcal{Q}$, *then*

$$j_2(P) = \begin{cases} \frac{1}{2}q, & \text{when } q \text{ is even and } P \text{ is not } \mathbf{F}_{q^2}\text{-rational}, \\ \frac{1}{2}(q+1), & \text{when } q \text{ is odd and } P \text{ is } \mathbf{F}_{q^2}\text{-rational}; \end{cases}$$

(iv) *if there exists* $P \in \mathcal{X}$ *with* $j_2(P) > 2$, *then* $\mathcal{Q}$ *is a cone.*

Proof. Let the quadric $\mathcal{Q} = \mathbf{v}(F)$, with

$$\begin{aligned} &F(X_0, X_1, X_2, X_3) \\ &\quad = a_{00}X_0^2 + a_{01}X_0X_1 + a_{02}X_0X_2 + a_{03}X_0X_3 + a_{11}X_1^2 \\ &\quad\quad + a_{12}X_1X_2 + a_{13}X_1X_3 + a_{22}X_2^2 + a_{23}X_2X_3 + a_{33}X_3^2 . \end{aligned}$$

First it is shown that $\mathcal{Q}$ is defined over $\mathbf{F}_{q^2}$. Otherwise $\mathcal{Q}$ is distinct from its image $\mathcal{Q}'$ under the Frobenius collineation

$$(X_0, X_1, X_2, X_3) \mapsto (X_0^{q^2}, X_1^{q^2}, X_2^{q^2}, X_3^{q^2}). \tag{10.36}$$

Hence $\mathcal{X}$ would be in the intersection of two distinct quadrics, and $\deg \mathcal{X} = q+1$ would be less than 5 by Corollary 7.9, contradicting the hypothesis that $q \geq 4$.

The coordinate functions $f_0, f_1, f_2, f_3 \in \mathbf{F}_{q^2}(\mathcal{X})$ of $\mathcal{X}$ may be chosen such that $\mathrm{ord}_P(f_i) = j_i(P)$ for $i = 0, 1, 2, 3$. The hypothesis on $\mathcal{X}$ to lie on $\mathcal{Q}$ implies that $F(f_0, f_1, f_2, f_3) = 0$, by Theorem 7.5.

After a change of coordinates in $\mathrm{PG}(3, K)$, let the point P be a vertex of the projective frame, say $P = (1, 0, 0, 0)$, so that the osculating plane at P is $\mathbf{v}(X_3)$ and the tangent line to $\mathcal{X}$ at P is $\mathbf{v}(X_2, X_3)$; see the proof of Theorem 7.47. Also, let $f_0 = 1$.

To use the equation

$$F(1, f_1, f_2, f_3) = 0, \tag{10.37}$$

the orders at P of some other elements from $\mathbf{F}_{q^2}(\mathcal{X})$, namely,

$$f_1^2,\ f_1f_2,\ f_1f_3,\ f_2^2,\ f_2f_3,\ f_3^2,$$

are required. By direct computation, these orders are

$$2j_1,\ j_1 + j_2,\ j_1 + j_3,\ 2j_2,\ j_2 + j_3, 2j_3. \tag{10.38}$$

Hence (10.37) together with $P \in \mathcal{Q}$ implies that $a_{00} = a_{01} = 0$.

To show (i), note that $j_2 + 1 < j_3$ by (VIII) and (XI). So, from (10.38) and the inequalities

$$2 \leq j_2 < j_2 + 1 < j_3 < j_3 + 1 < j_3 + j_2 < 2j_3,$$

(i) follows. Also, (10.38) shows that $j_2 > 2$ if and only if $a_{11} = 0$. Now, as $F(X_0, X_1, 0, 0) = a_{11}X_1^2$, the last condition is satisfied if and only if the tangent line ℓ_P to $\mathcal{X}$ at P is contained in the quadric $\mathcal{Q}$. From this, (ii) follows.

Also, if $j_2 > 2$, from (i),

$$a_{11} = a_{02} = a_{12} = 0.$$

Now P is a non-singular point of $\mathcal{Q}$ if and only if $a_{03} \neq 0$. Therefore $2j_2 = j_3$, and (iii) follows from the fact that $j_1(P) = 1$ and (10.35).

To show (iv), note that if P is a non-singular point of $\mathcal{Q}$ then $\mathcal{Q}$ is a cone whose vertex is the point $V = (-a_{13}, a_{03}, 0, 0)$. Otherwise, P is singular and hence $\mathcal{Q}$ is a cone with vertex P. □

LEMMA 10.40 *Let $q \geq 4$ and $\dim \mathcal{D} = 3$. Then*

(i) $\dim(2\mathcal{D}) \geq 8$;

(ii) *if* $\dim(2\mathcal{D}) = 8$, *the curve $\mathcal{X}$ lies on a quadric surface in* $\mathrm{PG}(3, K)$;

(iii) *the quadric surface in* (ii) *is uniquely determined and is defined over* $\mathbf{F}_{q^2}$.

Proof. (i) Let $P \in \mathcal{X}(\mathbf{F}_{q^2})$. Then $m_2(P) = q$ and $m_3(P) = q + 1$ by (IV) and (XIV). Now, as $2m_1(P) \geq m_2(P) = q$ and $q \geq 4$, it follows that there are at least eight non-gaps in the interval $[m_1(P), 2m_3(P)]$. Hence $\dim 2\mathcal{D} \geq 8$.

(ii) Geometrically, the linear system of all quadrics has dimension 9 and the linear series L cut out on $\mathcal{X}$ by quadrics is contained in $2\mathcal{D}$. So if $\dim(2\mathcal{D}) \geq 8$ then $L = 2\mathcal{D}$ and some quadrics must contain $\mathcal{X}$.

(iii) See the first part in the proof of Lemma 10.39. □

THEOREM 10.41 *Let q be odd with $q \geq 5$, and let $\mathcal{X}$ be an $\mathbf{F}_{q^2}$-maximal curve of genus $g = \frac{1}{4}(q-1)^2$. Then $\mathcal{X}$ is birationally equivalent over $\mathbf{F}_{q^2}$ to the $\mathbf{F}_{q^2}$-rational irreducible plane curve $\mathcal{E}^1_{(q+1)/2} = \mathbf{v}(E_1)$ in Example 10.35, with*

$$E_1(X, Y) = Y^q + Y - X^{(q+1)/2}. \tag{10.39}$$

Proof. From Lemma 10.39, $j_2(P) = 2$ at each point $P \in \mathcal{X}$ which is not $\mathbf{F}_{q^2}$-rational while the $\mathbf{F}_{q^2}$-rational points P of $\mathcal{X}$ are of two types:

(I) $j_2(P) = 2$;

(II) $j_2(P) = \frac{1}{2}(q+1)$.

In particular, the set of all $\mathcal{D}$-Weierstrass points of $\mathcal{X}$ coincides with $\mathcal{X}(\mathbf{F}_{q^2})$.

Since the $\mathcal{D}$-orders are $0, 1, 2, q$, the weight of P in the ramification divisor R can be calculated from Theorem 7.55:

$$v_P(R) = \begin{cases} \frac{1}{2}(q-1), & \text{when } P \in \mathcal{X}(\mathbf{F}_{q^2}) \text{ and } j_2(P) = \frac{1}{2}(q+1), \\ 1, & \text{when } P \in \mathcal{X}(\mathbf{F}_{q^2}) \text{ and } j_2(P) = 2, \\ 0, & \text{when } P \notin \mathcal{X}(\mathbf{F}_{q^2}). \end{cases}$$

Let n_1 and n_2 denote the numbers of $\mathbf{F}_{q^2}$-rational points of type (I) and (II). From (7.13),

$$\tfrac{1}{2}(q+1)n_2 + n_1 = (q+3)(\tfrac{1}{2}(q-1)^2 - 2) + 4(q+1).$$

Since $\mathcal{X}$ is $\mathbf{F}_{q^2}$-maximal,

$$n_1 + n_2 = q^2 + 1 + \tfrac{1}{2}(q-1)^2 q.$$

From these two equations, $n_1 = q+1$ and, in particular, $n_1 \geq 4$.

Now the configuration of the points of type (I) is investigated. Such points are not collinear, since no non-degenerate irreducible curve of $\mathrm{PG}(3,K)$ of degree k contains k collinear points. By Lemma 10.39 (iv), these points lie on a cone $\mathcal{Q}$ with vertex T. From the proof of that lemma, the tangent line ℓ_P to $\mathcal{X}$ at a point P of type (I) is contained in $\mathcal{Q}$. Therefore every line joining T to a point of type (I) is a generator of $\mathcal{Q}$.

If P and P' are distinct points of type (I), then ℓ_P does not contain $\ell_{P'}$. Hence T is not a point of type (I). Also, T lies in every plane which osculates $\mathcal{X}$ at a point of type (I).

Another property of a point P of type (I), established implicitly in the proof of Lemma 10.39, is as follows.

LEMMA 10.42 *The following three planes coincide:*

(a) *the osculating plane π_P of $\mathcal{X}$ at P;*

(b) *the tangent plane α_P to the Hermitian surface $\mathcal{H}_{3,q}$ at P;*

(c) *the tangent plane β_P to the cone $\mathcal{Q}$ at P.*

A first consequence is that $T \notin \mathcal{H}_{3,q}$, as an $\mathbf{F}_{q^2}$-rational line ℓ of $\alpha(P)$ which passes through P either lies in $\mathcal{H}_{3,q}$ or meets $\mathcal{H}_{3,q}$ only in P. The group $\mathrm{PGU}(4,q)$ that fixes $\mathcal{H}_{3,q}$ acts transitively on the points of $\mathrm{PG}(3,q^2)\backslash\mathcal{H}_{3,q}$. Therefore a projective frame over $\mathbf{F}_{q^2}$ may be chosen so that

$$\mathcal{H}_{3,q} = \mathbf{v}(X_0^q X_2 + X_0 X_2^q + 2X_1^{q+1} + X_3^{q+1}), \qquad T = (0,0,0,1). \tag{10.40}$$

The polar plane Π of T under the unitary polarity of $\mathcal{H}_{3,q}$ is $\mathbf{v}(X_3)$. So Π cuts out on $\mathcal{H}_{3,q}$ the non-degenerate Hermitian curve

$$\mathcal{H}_{2,q} = \mathbf{v}(X_0^q X_2 + X_0 X_2^q + 2X_1^{q+1}),$$

and on $\mathcal{Q}$ an $\mathbf{F}_{q^2}$-rational irreducible conic $\mathcal{C}$. Another consequence of Lemma 10.42 is that $\mathcal{C}$ and $\mathcal{H}_{2,q}$ have the same tangent at every point P of type (I). From Bézout's Theorem 3.14, the common points of $\mathcal{C}$ with $\mathcal{H}_{2,q}$ are exactly the points of type (I).

This shows that $\mathcal{C}$ and $\mathcal{H}_{2,q}$ are in *permutable position* with respect to the two polarities arising from $\mathcal{C}$ and $\mathcal{H}_{2,q}$. Given a non-degenerate Hermitian curve, the irreducible conics in permutable position with it form a unique orbit under the action of the group $\mathrm{PGU}(3,q^2)$ fixing the Hermitian curve.

The $\mathbf{F}_{q^2}$-rational irreducible conic $\mathcal{C}' = \mathbf{v}(X_0X_2 + X_1^2)$ provides an example of such a conic. There exists a projectivity (S) of Π preserving $\mathcal{H}_{2,q}$ that takes $\mathcal{C}$ to $\mathcal{C}'$. Also, δ can be lifted to an element $(S)'$ of $\mathrm{PGU}(4,q^2)$; then $(S)'$ fixes T. Hence the cone $\mathcal{Q}$ may be taken as

$$\mathcal{Q} = \mathbf{v}(X_0X_2 + X_1^2). \tag{10.41}$$

Now let $A = (a_0, a_1, a_2, a_3)$ be any point of $\mathcal{X}$. Then

$$a_0^q a_2 + a_0 a_2^q + 2a_1^{q+1} + a_3^{q+1} = 0, \quad a_0 a_2 = a_1^2.$$

If $a_0 = 0$ then $A = (0, 0, 1, 0)$. Otherwise, let $a_0 = 1$; then these equations imply that

$$(a_2^q + a_2)^2 + a_3^{q+1} = 0,$$

whence either $a_2^q + a_2 + a_3^{(q+1)/2} = 0$ or $a_2^q + a_2 - a_3^{(q+1)/2} = 0$. Comparison with Example 10.35 shows that $\mathcal{H}_{3,q} \cap \mathcal{Q}$ is $\Gamma_1 \cup \Gamma_2$. From Exercise 9 in Chapter 7, either $\mathcal{X} = \Gamma_1$ or $\mathcal{X} = \Gamma_2$. Since Γ_1 and Γ_2 are isomorphic over $\mathbf{F}_{q^2}$, the result follows. □

Theorem 10.41 together with Corollary 10.8 gives the case that $m = \frac{1}{2}(q+1)$ of the following general result.

THEOREM 10.43 *Let $\mathcal{X}$ be an $\mathbf{F}_{q^2}$-maximal curve of genus g, and suppose that there is a point $P \in \mathcal{X}(\mathbf{F}_{q^2})$ with a non-gap m dividing $q+1$. Then $\mathcal{X}$ is birationally equivalent over $\mathbf{F}_{q^2}$ to the curve $\mathcal{E}_m^1 = \mathbf{v}(E_1)$, with*

$$E_1(X, Y) = Y^q + Y - X^m. \qquad (10.42)$$

PROPOSITION 10.44 *Let q be even with $q > 4$, and let* $\dim \mathcal{D} = 3$. *If $\mathcal{X}$ lies on a quadric $\mathcal{Q}$ in* PG$(3, K)$, *then*

(i) *$\mathcal{Q}$ is a cone;*

(ii) *the vertex V of $\mathcal{Q}$ belongs to $\mathcal{X}$;*

(iii) *$V \in \mathcal{X}(\mathbf{F}_{q^2})$ and $j_2(V) = \frac{1}{2}(q+2)$.*

Proof. The following properties of quadrics of PG$(3, K)$ are used, where, for a non-singular point P of $\mathcal{Q}$, let β_P be the tangent plane to $\mathcal{Q}$ at P:

(a) if $P \in \mathcal{X}$, then $\beta_P \supset \ell_P$;

(b) if ℓ and ℓ_1 are distinct lines such that $P \in \ell \subset \mathcal{Q}$ and $\ell_1 \subset \beta_P$, then β_P is spanned by ℓ and ℓ_1;

(c) there exist lines ℓ and ℓ_1 such that $P \in \ell \cap \ell_1$ and $\mathcal{Q} \cap \beta_P = \ell \cup \ell_1$;

(d) if $\mathcal{Q}$ is non-singular, then no two tangent hyperplanes of $\mathcal{Q}$ at different points coincide.

(i) Since $\mathcal{X}$ is non-degenerate, $\mathcal{Q}$ is irreducible. Also, $\mathcal{Q}$ is a cone if and only if $\mathcal{Q}$ is singular. Therefore let $\mathcal{Q}$ be non-singular, that is, a hyperbolic quadric of PG$(3, K)$.

From Lemma 10.39 (iii), $j_2(Q) = 2$ for each $Q \in \mathcal{X}(\mathbf{F}_{q^2})$. Note that there exists $P \notin \mathcal{X}(\mathbf{F}_{q^2})$ such that $j_2(P) > 2$. Otherwise, by Lemma 10.15,

$$6(g-1) = (q+1)(q-3);$$

but then q would be odd, a contradiction. Hence $j_2(P) = \frac{1}{2}q$ again by Lemma 10.39 (iii).

Let $Q_1 \in \mathcal{X}(\mathbf{F}_{q^2})$. Then $Q_1 \notin \ell_P$ since

$$\mathcal{X} \cap \ell_P \subset \mathcal{X} \cap \pi_P = \{P, \Phi(P)\},$$

and hence the plane $H = H_{Q_1}$ spanned by ℓ_P and Q_1 is well defined. Thus $H \neq \pi_P$, and the intersection divisor $\mathcal{X} \cdot H$ is given by

$$\mathcal{X} \cdot H = \tfrac{1}{2}P + D, \tag{10.43}$$

where $D = D_{Q_1}$ is a divisor on $\mathcal{X}$ of degree $\frac{1}{2}(q+2)$ with $Q_1 \in \text{Supp}(D)$ and $P \notin \text{Supp}(D)$. In addition, Lemma 10.39 (ii) ensures the existence of a line $\ell = \ell_{Q_1}$ such that

$$\mathcal{Q} \cap H = \ell_P \cup \ell. \tag{10.44}$$

The line ℓ is defined over $\mathbf{F}_{q^2}$ as is $\mathcal{Q}$ by Lemma 10.40 (iii); the fact that Q_1 is on $\mathcal{X}(\mathbf{F}_{q^2})$ but not on ℓ_P implies that $Q_1 \in \ell$.

Now it is shown that

$$\mathcal{X} \cap \ell \subset \mathcal{X}(\mathbf{F}_{q^2}). \tag{10.45}$$

If there exists $Q \in \mathcal{X} \cap \ell \backslash \mathcal{X}(\mathbf{F}_{q^2})$, then $\Phi(Q) \in \ell$ as ℓ is defined over $\mathbf{F}_{q^2}$. Thus $\ell \subset \pi_Q$, and hence $\ell \cap \mathcal{X} \subset \{Q, \Phi(Q)\}$. Therefore $Q_1 \notin \ell$; but this is a contradiction.

(A) If $Q \in \text{Supp}(D)\backslash\{\Phi(P)\}$, then $Q \in \mathcal{X}(\mathbf{F}_{q^2})$ and $n_Q(D) = 1$.

To prove (A) note that from (10.45) and the relation

$$\text{Supp}(D)\backslash\{\Phi(P)\} \subset \ell \cap \mathcal{X},$$

it follows that $Q \in \mathcal{X}(\mathbf{F}_{q^2})$. Now, if $n_Q(D) \geq 2$, then $H \supset \ell_Q$ since $j_2(Q) = 2$ and by (IV). Also, $\ell \neq \ell_Q$ because $\ell_Q \not\subset \mathcal{Q}$ by Lemma 10.39 (ii). Therefore the plane H is spanned by the lines ℓ and ℓ_Q, and hence H is the tangent plane π_{Q_1} to $\mathcal{Q}$ at Q_1. If ℓ_1 is the line defined by the property that $Q_1 \in \ell_1$, then $\mathcal{Q} \cap \pi_{Q_1} = \ell \cap \ell_1$. From (10.44), $\ell_P = \ell_1$ and so $Q_1 \in \ell_P$; but this is a contradiction.

(B) $\Phi(P) \notin \text{Supp}(D)$.

To prove (B) suppose, on the contrary, that $\Phi(P) \in \text{Supp}(D)$. By (10.45), this assumption is equivalent to $\Phi(P) \in \ell_P$. The argument in the proof of (A) can be used to deduce that $H = \pi_{\Phi(P)}$ since $n_{\Phi(P)}(D) \neq 1$ and $\ell_P \neq \ell_{\Phi(P)}$. But this leads to a contradiction with (10.8). Therefore $n_{\Phi(P)}(D) = 1$. Hence, for every $Q \in \mathcal{X}(\mathbf{F}_{q^2})$, the divisor D in (10.43) may also be written as

$$D = D_Q = \Phi(P) + D'_Q,$$

and this may be done so that not only (10.44) remains valid but also that

$$\text{Supp}(D'_Q) \subset \mathcal{X}(\mathbf{F}_{q^2}), \quad \deg(D'_Q) = \tfrac{1}{2}q.$$

This gives the following:

(C) H_Q is spanned by ℓ_P and Q', where Q' is any point of $\text{Supp}(D'_Q)$.

Now let $Q_1, Q_2 \in \mathcal{X}(\mathbf{F}_{q^2})$ such that $Q_2 \notin \text{Supp}(D'_{Q_1})$. Then

$$\text{Supp}(D'_{Q_1}) \cap \text{Supp}(D'_{Q_2}) = \emptyset;$$

otherwise $H_{Q_1} = H_{Q_2}$ by (C). So $\frac{1}{2}q$ must divide the number of $\mathbf{F}_{q^2}$-rational points of $\mathcal{X}$, which is a contradiction since $|\mathcal{X}(\mathbf{F}_{q^2})| = q^2 + 1 + 2gq$ is odd.

So far it has been shown that each $Q_1 \in \mathcal{X}(\mathbf{F}_{q^2})$ gives rise to a plane H_{Q_1} together with a line $\ell = \ell_{Q_1}$ and a divisor $D = D_{Q_1}$ such that (10.43) and (10.44) hold, with

$$D = Q_1 + Q_2 + \cdots + Q_{(q+2)/2}$$

being the sum of $\frac{1}{2}(q+2)$ points all $\mathbf{F}_{q^2}$-rational. Note that $\mathrm{Supp}(D) = \mathcal{X} \cap \ell$.

Let ℓ_1 be chosen in such a way that $Q_1 \in \ell_1$ and that

$$\mathcal{Q} \cap \pi_{Q_1} = \ell \cup \ell_1 . \tag{10.46}$$

Here ℓ_1 is $\mathbf{F}_{q^2}$-rational, and thus $\mathcal{X} \cap \ell_1 \subset \mathcal{X}(\mathbf{F}_{q^2})$ as in the proof of (10.45). Therefore

$$\mathcal{X} \cdot \pi_{Q_1} = 2Q_1 + Q_2 + \cdots + Q_{(q+2)/2} + D' , \tag{10.47}$$

where D' is a divisor of $\mathcal{X}$ of degree $\frac{1}{2}(q-2)$ such that $Q_1 \notin \mathrm{Supp}(D') \subset \mathcal{X}(\mathbf{F}_{q^2})$.

Consider the following statement:

(D) $\mathrm{Supp}(D) \cap \mathrm{Supp}(D') = \emptyset$ and $n_S(D') = 1$ for each $S \in \mathrm{Supp}(D')$.

To show (D), let $S \in \mathrm{Supp}(D')$. Suppose on the contrary that $S = Q_i$ for some i. Then π_{Q_1} contains ℓ_{Q_i} which is different from ℓ as $j_2(Q_i) = 2$. Hence π_{Q_1} is generated by ℓ_{Q_i} and ℓ. These lines also span π_{Q_i} and so $i = 1$, contradicting that $Q_1 \notin \mathrm{Supp}(D')$.

Finally suppose, on the contrary, that $n_S(D_2) \geq 2$. Replacing ℓ by ℓ_1, the above argument shows that $\pi_S = \pi_{Q_1}$, whence $S = Q_1$, again a contradiction. Thus, to each Q_1, there is associated two lines ℓ and ℓ_1 such that both (10.46) and (10.47) hold, where D' is a divisor of degree $\frac{1}{2}(q-2)$, where $\mathrm{Supp}(D') \subset \mathcal{X}(\mathbf{F}_{q^2})$, and where $\mathrm{Supp}(D) \cap \mathrm{Supp}(D') = \{Q_1\}$. A hyperbolic quadric $\mathcal{Q}$ contains exactly two families of lines and any two lines of the same family are disjoint. This implies again that $|\mathcal{X}(\mathbf{F}_{q^2})|$ must be a multiple of $\frac{1}{2}q$, contradicting the $\mathbf{F}_{q^2}$-maximality of $\mathcal{X}$.

(ii) As q is even, Lemma 10.15 ensures the existence of a point $P \in \mathcal{X}$ with $j_2(P) > 2$. Suppose that $P \notin \mathcal{X}(\mathbf{F}_{q^2})$; then $j_2(\Phi(P)) = j_2(P) > 2$. Since $j_2(P)P + D \equiv (q+1)P_0$,

$$j_2(P)\Phi(P) + \Phi(D) \equiv (q+1)P_0$$

and so $j_2(\Phi(P)) = j_2(P) > 2$. Therefore both the lines ℓ_P and $\ell_{\Phi(P)}$ are contained in $\mathcal{Q}$ by Lemma 10.39 (ii), and hence their common point is V. Now, since V is $\mathbf{F}_{q^2}$-rational by Lemma 10.40 (iii), so $\Phi(P) \neq V$, and hence $\ell_{\Phi(P)}$ is spanned by $\Phi(P)$ and V. In particular, $\ell_{\Phi(P)} \subset \pi_P$ and thus

$$1 = n_{\Phi(P)}(\mathcal{X} \cdot \pi_P) \geq j_2(\Phi(P)),$$

a contradiction.

Therefore P must be $\mathbf{F}_{q^2}$-rational and hence $\mathcal{Q}$ must have a singularity at P by Lemma 10.39 (iii). Then $P = V$ and $j_2(P) = \frac{1}{2}(q+2)$; these follow from Lemma 10.39 (i) and the assumption that q is even. □

THEOREM 10.45 *Let q be even with $q \geq 16$, and let $\mathcal{X}$ be an $\mathbf{F}_{q^2}$-maximal curve of genus $g = \frac{1}{4}q(q-2)$. Then $\mathcal{X}$ is birationally equivalent over $\mathbf{F}_{q^2}$ to the curve $\mathcal{T}_2 = \mathbf{v}(T_2)$ in Example 10.36, with*

$$T_2(X,Y) = Y^{q/2} + Y^{q/4} + \cdots + Y^2 + Y + X^{q+1}. \tag{10.48}$$

Proof. From Castelnuovo's Bound, Corollary 10.25, $\dim \mathcal{D} \leq 3$. As $\dim \mathcal{D} = 2$ only occurs when $\mathcal{X}$ is a Hermitian curve, whose genus is $\frac{1}{2}q(q-1)$, let $\dim \mathcal{D} = 3$. By Halphen's Theorem 7.117, $\mathcal{X}$ lies on a quadric of $\mathrm{PG}(3, K)$. From (IV) and Proposition 10.44 (ii), there exists $P_0 \in \mathcal{X}(\mathbf{F}_{q^2})$ such that $\frac{1}{2}q$ is a non-gap at P_0. By (XII), the same holds for $q+1$. Therefore the Weierstrass semigroup $H(P_0)$ is generated by $\frac{1}{2}q$ and $q+1$. Choose $x, y \in \mathbf{F}_{q^2}(\mathcal{X})$ such that $\mathrm{div}(x)_\infty = \frac{1}{2}qP_0$ and $\mathrm{div}(y)_\infty = (q+1)P_0$, and consider the $\mathbf{F}_{q^2}$-rational linear series $|\frac{1}{2}q(q+1)P_0|$ of order $\frac{1}{2}q(q+1)$. Since

$$\tfrac{1}{2}q(q+1) > 2g - 2 = \tfrac{1}{2}q(q-2) - 2,$$

the Riemann–Roch Theorem 6.61 implies that $\dim |\frac{1}{2}q(q+1)P_0| = \frac{1}{4}q^2 + q$. On the other hand, $\mathrm{div}(x^{q+1}) + \frac{1}{2}q(q+1)P_0$, as well as each of the following $\frac{1}{4}(q+2)^2$ divisors, belongs to $|\frac{1}{2}q(q+1)P_0|$:

$$\mathrm{div}(x^j y^i) + \tfrac{1}{2}q(q+1)P_0, \qquad 0 \leq i \leq \tfrac{1}{2}q,\ 0 \leq j \leq q - 2i.$$

Therefore x and y must satisfy a non-trivial polynomial relation; that is,

$$cy^{q/2} + \textstyle\sum_{i=0}^{q/2-1} A_i(x)y^i = x^{q+1}, \tag{10.49}$$

for $c \in \mathbf{F}_{q^2}\backslash\{0\}$ and for $A_i[X] \in \mathbf{F}_{q^2}[X]$ with $\deg A_i(x) \leq q - 2i$. A lengthy calculation reduces (10.49) to the equation

$$\textstyle\sum_{i=1}^{t} a_i y^{q/2^i} + b = x^{q+1}, \tag{10.50}$$

where $a_i \in \mathbf{F}_{q^2}\backslash\{0\}$ for all i, and the polynomial $\sum_{i=1}^{t} a_i T^{q/2^i} + b$ has a root $\alpha \in \mathbf{F}_{q^2}$. Replacing y by $y + \alpha$, (10.50) becomes the following:

$$\textstyle\sum_{i=1}^{t} a_i y^{q/2^i} = x^{q+1}. \tag{10.51}$$

Replacing x by $a_1^{-1}x$ and y by $a_{t-1}y$ give the desired equation:

$$\textstyle\sum_{i=1}^{t} y^{q/2^i} = x^{q+1}.$$

Therefore $\mathcal{X}$ is birationally equivalent to the $\mathbf{F}_{q^2}$-rational irreducible plane curve $\mathcal{T}_2$ of (10.48). □

10.5 MAXIMAL CURVES WITH HIGH GENUS

In this section, the aim is to show that the upper part of the spectrum of genera of $\mathbf{F}_{q^2}$-maximal curves consists of the values of g in Table 10.2. Although the whole spectrum depends on the nature of q, the highest three genera in the spectrum have the same positions independently of q.

Since the Hermitian curve $\mathcal{H}_q$ is $\mathbf{F}_{q^2}$-maximal, its genus $g = \frac{1}{2}q(q-1)$ belongs to the spectrum, and is the maximum.

PROPOSITION 10.46 *If $\mathcal{X}$ is an $\mathbf{F}_{q^2}$-maximal curve of genus g, then*

$$g \leq \tfrac{1}{2}q(q-1).$$

Proof. Since $N_2 \geq N_1$, from (10.1),

$$q^4 + 1 - 2gq^2 \geq q^2 + 1 + 2gq,$$
$$q^4 - q^2 \geq 2g(q^2 + q),$$

whence the result follows. □

To present other results on the spectrum, let

$$g_i = g_i(q)$$

denote the i-th largest genus in the spectrum; then

$$g_i - g_{i-1}$$

is the i-th *hole* or *gap* in the spectrum. Proposition 10.46 and the existence of $\mathcal{H}_q$ show that

$$g_1 = \tfrac{1}{2}q(q-1).$$

The following theorem implies the uniqueness of an $\mathbf{F}_{q^2}$-maximal curve of genus g_1. Also, it shows that $g_2 = \lfloor \frac{1}{4}(q-1)^2 \rfloor$, as the curves (i) and (ii) in Example 10.3 have this value.

THEOREM 10.47 *If $\mathcal{X}$ is an $\mathbf{F}_{q^2}$-maximal curve of genus g, the following statements are equivalent:*

(i) *$\mathcal{X}$ is isomorphic over $\mathbf{F}_{q^2}$ to the Hermitian curve $\mathcal{H}_q$;*

(ii) $g > \lfloor \frac{1}{4}(q-1)^2 \rfloor$;

(iii) $r = 2$;

(iv) $\nu_1 > 1$.

Proof. (i) $\Rightarrow$ (ii), (iii), (iv) This follows from properties of $\mathcal{H}_q$.

(iii) $\Rightarrow$ (iv) For $r = 2$, $\nu_1 = \nu_{r-1}$ and the assertion follows from (IX).

(iv) $\Rightarrow$ (ii) Theorem 8.68 together with the Natural Embedding Theorem 10.22 imply that if $\nu_1 > 1$ then $N_1 = q^3 + 1$, whence $g = \frac{1}{2}q(q-1)$.

(ii) $\Rightarrow$ (iii) From Corollary 10.25, $F(3) = (q-1)^2/4$, and hence $r = 2$.

(iii) $\Rightarrow$ (i) This follows from the Natural Embedding Theorem 10.22. □

The second hole in the spectrum is determined in the next result.

THEOREM 10.48 *For $q \geq 7$, let $\mathcal{X}$ be an $\mathbf{F}_{q^2}$-maximal curve of genus g satisfying (10.38). Then the following conditions are equivalent:*

(i) $\lfloor \frac{1}{6}(q^2 - q + 4) \rfloor < g \leq \lfloor \frac{1}{4}(q-1)^2 \rfloor$;

(ii) $\dim \mathcal{D} = 3$, *the curve $\mathcal{X}$ lies on a quadric surface in* PG$(3, K)$, *and*

$$g \neq \tfrac{1}{6}(q^2 - 2q + 3), \quad \textit{for } q \equiv 3, 5 \pmod 6;$$

(iii) $\dim \mathcal{D} = 3$, $\dim(2\mathcal{D}) = 8$, *and*

$$g \neq \tfrac{1}{6}(q^2 - 2q + 3), \quad \textit{for } q \equiv 3, 5 \pmod 6;$$

(iv) $\dim \mathcal{D} = 3$ *and there exists* $P \in \mathcal{X}(\mathbf{F}_{q^2})$ *such that*

$$j_2(P) = \begin{cases} \tfrac{1}{2}(q+1) & \textit{for } q \textit{ odd}, \\ \tfrac{1}{2}(q+2) & \textit{for } q \textit{ even}; \end{cases}$$

(v) $\mathcal{X}$ *is birationally equivalent over* $\mathbf{F}_{q^2}$ *to the curve* (i) *or* (ii) *in Example 10.3 according as* q *is odd or even*;

(vi) $g = \frac{1}{4}(q-1)^2$ *if* q *is odd and* $g = \frac{1}{4}q(q-2)$ *if* q *is even; in particular, the genus* $g = c_0(q+1, 3)$, *Castelnuovo's number.*

Proof. (i) $\Rightarrow$ (ii) From the hypothesis on g, Halphen's Theorem 7.117 and (10.8) imply that $\dim \mathcal{D} = 3$. Since, in Corollary 10.25, $c_0(q+1,3) = \lfloor \frac{1}{6}(q^2-q+4) \rfloor$, the Natural Embedding Theorem 10.22 together with Halphen's Theorem shows that $\mathcal{X}$ lies on a quadric provided that either $q \notin \{7, 8, 9, 11, 13, 17, 19, 23\}$ or $p > 2$ and $\mathcal{X}$ is reflexive. So, for q even, let $q = 8$. Then $g > \frac{1}{6}(q^2 - q + 4) = 10$. By Lemma 10.40 (i),(ii), it is enough to show that $\dim(2\mathcal{D}) \leq 8$. Suppose, however, that $\dim(2\mathcal{D}) \geq 9$. Then, by Lemma 7.113, $g \leq \frac{1}{4}(q-1)(q-2) = 21/2$, a contradiction.

Now let q be odd with $q \geq 7$, and suppose that $\mathcal{X}$ is not reflexive. By Theorem 7.118, $\epsilon_2 > 2$. If $P \in \mathcal{X}(\mathbf{F}_{q^2})$, then, from Propositions 10.47 and 8.56,

$$v_P(S) \geq \textstyle\sum_{i=1}^{3} (j_i(P) - \nu_{i-1}) \geq \epsilon_2 + 1 \geq 4.$$

So the Stöhr–Voloch Theorem applied to $\mathcal{X}$ implies that

$$(3q-1)(2g-2) \leq (q+1)(q^2 - 4q - 1).$$

On the other hand, $2g - 2 > \frac{1}{3}(q+1)(q-2)$ by hypothesis, and thus $5q + 5 < 0$, a contradiction.

(iii) $\Rightarrow$ (ii) This follows from Lemma 10.40 (ii).

(ii) $\Rightarrow$ (iv) Let q be odd. Then there exists $P \in \mathcal{X}$ such that $j_2(P) > 2$; otherwise $g = \frac{1}{6}(q^2 - 2q + 3)$ by Proposition 10.15. If such a point $P \in \mathcal{X}$ were not in $\mathcal{X}(\mathbf{F}_{q^2})$ then, by Lemma 10.39 (iii), both P and $\Phi(P)$ would be singular points of the quadric, a contradiction. Therefore $P \in \mathcal{X}(\mathbf{F}_{q^2})$ and hence $j_2(P) = \frac{1}{2}(q+1)$ by Lemma 10.39 (i).

If q is even, the result follows from Proposition 10.44 (ii).

(iv) $\Rightarrow$ (v) From (IV) and the hypothesis, $m_1(P) = \frac{1}{2}(q+1)$ for q odd and $m_1(P) = \frac{1}{2}q$ for q even. In the odd case, q is another non-gap at P, and hence $g = \frac{1}{4}(q-1)^2$ by Corollary 10.8. A similar argument works in the even case; since $g = \frac{1}{4}q(q-2)$ and $q+1$ is another non-gap at P, Corollary 10.8 applies.

Finally, the implications (v) $\Rightarrow$ (vi), (vi) $\Rightarrow$ (i) and (v) $\Rightarrow$ (iii) follow readily. □

REMARK 10.49 For $q = 2, 3, 4, 5$, the spectra of the genera of $\mathbf{F}_{q^2}$-maximal curves are $\{0, 1\}, \{0, 1, 3\}, \{0, 1, 2, 6\}, \{0, 1, 2, 3, 4, 10\}$.

10.6 CASTELNUOVO'S NUMBER

In this section, certain $\mathbf{F}_{q^2}$-maximal curves whose genus is Castelnuovo's number $c_0(q+1,r)$ for $4 \leq r \leq 5$ are considered. First the case

$$q \equiv 1,2 \pmod 3$$

is investigated. In this case, $c_0(q+1,4) = \frac{1}{6}(q-1)(q-2)$. The main result, see Theorem 10.54 below, provides a complete classification of $\mathbf{F}_{q^2}$-maximal curves of genus

$$g = \tfrac{1}{6}(q-1)(q-2), \quad q \equiv 1,2 \pmod 3, \quad q \geq 11.$$

Such $\mathbf{F}_{q^2}$-maximal curves can only exist for $q \equiv 2 \pmod 3$, and they are birationally equivalent over $\mathbf{F}_{q^2}$ to the $\mathbf{F}_{q^2}$-rational curve

$$\mathcal{E}^1_{(q+1)/3} = \mathbf{v}(Y^q + Y - X^{(q+1)/3}). \tag{10.52}$$

The proof of Theorem 10.54 requires three preliminary lemmas.

Lemma 10.50 *For $q \equiv 1,2 \pmod 3$, let $\mathcal{X}$ be an $\mathbf{F}_{q^2}$-maximal curve of genus $g = \frac{1}{6}(q-1)(q-2)$. Then $\dim \mathcal{D} = 4$ and $g = c_0(q+1,4)$.*

Proof. From (10.8) and Theorem 10.47, $3 \leq r \leq 4$. Suppose on the contrary that $r = 3$.

If $\epsilon_2 = 2$, then

$$\deg R = (3+q)(2g-2) + 4(q+1),$$

while the $\mathbf{F}_{q^2}$-maximality of $\mathcal{X}$ implies that

$$\deg R \geq q^2 + 1 + 2gq,$$

since $v_P(R) \geq 1$ for every $P \in \mathbf{F}_{q^2}(\mathcal{X})$. But then $g \geq \frac{1}{6}(q^2-2q+3)$, contradicting the hypothesis on g.

If $\epsilon_2 > 2$, then $\epsilon_2 \geq 5$ by Lemma 7.63 and since $q \not\equiv 0 \pmod 3$. Replacing the ramification divisor R by the Stöhr–Voloch divisor S, the previous argument again yields a contradiction. In fact,

$$\deg S = (1+q)(2g-2) + (q^2+3)(q+1),$$

while

$$\deg S \geq (q^2 + 1 + 2gq)(\epsilon_2 + 1),$$

by the $\mathbf{F}_{q^2}$-maximality of $\mathcal{X}$ and the lower bound $v_P(S) \geq \epsilon_2 + 1$ for $P \in \mathbf{F}_{q^2}(\mathcal{X})$, shown in the proof of Theorem 10.48. Since $\epsilon_2 \geq 5$, so

$$(5q-1)(2g-2) \leq (q+1)(q^2 - 6q - 3),$$

whence $2q^2 - 3q + 13 \leq 0$ for $g = \frac{1}{6}(q-1)(q-2)$, a contradiction. □

The assumption that the genus of $\mathcal{X}$ is Castelnuovo's number $c_0(q+1,4)$ is used via Lemma 7.113 (i). Indeed, from this lemma, $\dim(2\mathcal{D}) = 11$ and hence the possibilities for $(\mathcal{D}, P)$-orders can be determined. To show how to do this, let $j_i = j_i(P)$ and denote by Δ_P the set of $(2\mathcal{D}, P)$-orders. Then Δ_P contains both sets

$$\begin{aligned} \Delta_1 &= \{0, 1, 2, j_3, j_4, j_4+1, j_4+j_2, j_4+j_3, 2j_4\}, \\ \Delta_2 &= \{j_2, j_2+1, j_3+1, 2j_2, j_3+j_2, 2j_3\}, \end{aligned} \tag{10.53}$$

where $j_4 = q+1$ for $P \in \mathcal{X}(\mathbf{F}_{q^2})$, and $j_4 = q$ otherwise.

LEMMA 10.51 *Let $P \in \mathcal{X}$ with $j_2(P) = 2$. If $r = \dim \mathcal{D} = 4,\ \dim(2\mathcal{D}) = 11$, and $q \geq 9$, then $j_3(P) = 3$.*

Proof. The hypothesis on q together with Proposition 10.15 implies that

$$j_3 < \begin{cases} j_4 - 2 & \text{for } P \in \mathcal{X}(\mathbf{F}_{q^2}), \\ j_4 - 1 & \text{for } P \in \mathcal{X} \backslash \mathcal{X}(\mathbf{F}_{q^2}). \end{cases} \tag{10.54}$$

Suppose $j_3 > 3$. If $P \in \mathcal{X}(\mathbf{F}_{q^2})$, from (10.53) and (10.54),

$$\Delta_P = \Sigma_1 \cup \{3, j_3 + 1, j_3 + 2\},$$

and $2j_2, 2j_3 \in \Delta_P$. Hence $j_3 = 2j_2 = 4$, so that $2j_3 = 8 = j_4 = q + 1$; thus $q = 7$. If $P \notin \mathcal{X}(\mathbf{F}_{q^2})$ and $j_3 > 4$, from (10.53) and (10.54) it follows that

$$\Delta_P = \Delta_1 \cup \{3, 4, j_3 + 1\},$$
$$(j_3 + 2, 2j_3) \in \{(q, q + 1), (q, q + 2), (q + 1, q + 2)\}.$$

Then $j_3 \leq 4$, a contradiction. Finally, if $P \notin \mathcal{X}(\mathbf{F}_{q^2})$ and $j_3 = 4$, then (10.53) together with (10.54) show that

$$\Delta_P = \Delta_1 \cup \{3, 5, 6, 8\}.$$

Hence $j_4 = q = 8$, and this completes the proof. □

The previous lemma together with Lemma 10.15 gives the following result.

COROLLARY 10.52 *If $q \geq 9$, $\dim(\mathcal{D}) = 4$, $\dim(2\mathcal{D}) = 11$, $j_2(P) = 2$ for any $P \in \mathcal{X}$, then $q \equiv 1, 2 \pmod 3$ and $g = \frac{1}{12}(q^2 - 3q + 8)$.*

Next, the case that $j_2(P) > 2$ for some point $P \in \mathcal{X}$ is considered.

LEMMA 10.53 *Let $P \in \mathcal{X}$ with $j_2(P) > 2$, and suppose that*

$$\dim \mathcal{D} = 4,\ \dim(2\mathcal{D}) = 11,\ q \geq 7.$$

(i) *If $P \in \mathcal{X}(\mathbf{F}_{q^2})$ and $g > \frac{1}{8}q(q - 2)$ for q even, then one of the following holds:*

(a) $q \equiv 2 \pmod 3$, $\quad j_2(P) = \frac{1}{3}(q + 1)$, $j_3(P) = \frac{1}{3}(2q + 2)$;

(b) $q \equiv 0 \pmod 3$, $\quad j_2(P) = \frac{1}{3}(q + 3)$, $j_3(P) = \frac{1}{3}(2q + 3)$.

(ii) *If $P \notin \mathcal{X}(\mathbf{F}_{q^2})$, then one of the following holds:*

(a) $q \equiv 1 \pmod 3$, $\quad j_2(P) = \frac{1}{3}(q + 2)$, $j_3(P) = \frac{1}{3}(2q + 1)$;

(b) $q \equiv 0 \pmod 3$, $\quad j_2(P) = \frac{1}{3}q$, $j_3(P) = \frac{2}{3}q$;

(c) $q \equiv 1 \pmod 2$, $\quad j_2(P) = \frac{1}{2}(q - 1)$, $j_3(P) = \frac{1}{2}(q + 1)$;

(d) $q \equiv 0 \pmod 2$, $\quad j_2(P) = \frac{1}{2}q$, $j_3(P) = \frac{1}{2}(q + 2)$.

Proof. Suppose first that $j_3 > j_2 + 1$. From (10.53) and (10.54), there are only three possibilities, namely,

$$\Delta_P = \Delta_1 \cup \{j_2, j_2+1, j_3+1\},$$
$$(j_3+j_2, 2j_3) \in \{(j_4, j_4+1), (j_4, j_4+j_2), (j_4+1, j_4+j_2)\}.$$

The first cannot occur since $j_3 \neq j_2 + 1$; in the second case,

$$j_4 \equiv 0 \pmod 3,\ j_2 = \tfrac{1}{3} j_4,\ j_3 = \tfrac{2}{3} j_4;$$

in the third case,

$$j_4 \equiv 1 \pmod 3,\ j_2 = \tfrac{1}{3}(j_4+2),\ j_3 = \tfrac{1}{3}(2j_4+1).$$

Suppose next that $j_3 = j_2 + 1$. Then $2j_2 \notin \{j_3, j_3+1\}$ since $j_2 > 2$. Further, either $2j_2 \neq j_4 + 1$ or $j_2 = \frac{1}{2}(j_4+1)$, $j_3 = \frac{1}{2}(j_4+3)$. In the latter case, from (10.53) and (10.54),

$$\Delta_P = \Delta_1 \cup \{j_2, j_3+1, j_4+2, j_4+3\}.$$

But this implies that $j_4 + j_2 = j_4 + 3$, whence $j_4 = 5$ and so $q \leq 5$. If $2j_2 = j_4$, then either $P \notin \mathcal{X}(\mathbf{F}_{q^2})$ or $j_3 = \frac{1}{2}(q+3)$. Again, the latter case cannot actually occur because then $m_1(P) = \frac{1}{2}(q-1)$ by Proposition 10.6 (IV), and this would imply that $\dim(\mathcal{D}) \geq 5$.

Finally, assume that $2j_2 \notin \{j_3, j_3+1, j_4, j_4+1\}$. Then, from (10.53) and (10.54),

$$\Delta_P = \{j_2, j_3+1, 2j_2\},$$

and $j_3+j_2 \in \{j_4, j_4+1\}$. If $j_3+j_2 = j_4+1$, then $2j_2 = j_4$, whence $j_3+j_2 = j_4$. Then $j_2 = \frac{1}{2}(j_4-1)$ and $j_3 = \frac{1}{2}(j_4+1)$. Here $P \notin \mathcal{X}(\mathbf{F}_{q^2})$; otherwise, $j_2 = \frac{1}{2}q$, $j_3 = \frac{1}{2}(q+2)$ and hence $m_1(P) = \frac{1}{2}q$, $m_2(P) = \frac{1}{2}(q+2)$ again by (IV). But then $g \leq \frac{1}{8}q(q-2)$, a contradiction. □

Theorem 10.54 *Let* $q \geq 11$.

(i) *If* $q \equiv 1 \pmod 3$, *there is no* $\mathbf{F}_{q^2}$*-maximal curve of genus* $\frac{1}{6}(q-1)(q-2)$.

(ii) *If* $q \equiv 2 \pmod 3$, *the statements below are equivalent for an* $\mathbf{F}_{q^2}$*-maximal curve* $\mathcal{X}$ *of genus* g :

(a) $g = \frac{1}{6}(q-1)(q-2)$;

(b) *there exist* $P \in \mathcal{X}(\mathbf{F}_{q^2})$ *and* $m \in H(P)$ *such that* $3m = q+1$;

(c) $\mathcal{X}$ *is birationally equivalent over* $\mathbf{F}_{q^2}$ *to the curve in (10.52).*

Proof. (i) Suppose on the contrary that $\mathcal{X}$ is an $\mathbf{F}_{q^2}$-maximal curve of genus

$$g = \tfrac{1}{3}(q-1)(q-2), \qquad q \equiv 1 \pmod 3.$$

Since $q+1 = \frac{1}{3}(q-1)\cdot 3 + 2$, it follows from Lemma 10.50 that $g = c_0(q+1, 3)$. Hence Lemma 7.113 implies that $\dim(2\mathcal{D}) = 11$ and that $|\frac{1}{3}(q-4)\mathcal{D}|$ is the canonical linear series on $\mathcal{X}$. Then

$$1 + a_1 i_1 + \cdots + a_{(q-4)/3}\, i_{(q-4)/3} \notin H(P), \tag{10.55}$$

where the integers i_j are $(\mathcal{D}, P)$-orders, and the coefficients a_j are non-negative integers such that $\sum_j a_j \le (q-4)/3$.

Let $P \in \mathcal{X}$ with $j_2(P) > 2$ as in Corollary 10.52. By Lemma 10.53, P is not in $\mathcal{X}(\mathbf{F}_{q^2})$. Recall that $m_3(P) = q$ by (XIV). So three cases are considered.

Case (A): $\quad j_2(P) = \frac{1}{3}(q+2),\ j_3(P) = \frac{1}{3}(2q+1)$.

From (XI),

$$\{q - m_2(P), q - m_1(P)\} \subset \{1, \tfrac{1}{3}(q+2), \tfrac{1}{3}(2q+1)\}.$$

This implies that $q - m_1(P) = \frac{1}{3}(q+2)$, since otherwise $m_1(P) = \frac{1}{3}(q-1)$ and hence $q \ge m_4(P)$, a contradiction. Thus $m_1(P) = \frac{1}{3}(2q-2)$. But this again leads to a contradiction as (10.55) implies that $\frac{1}{3}(q-7) + \frac{1}{3}(q+2) + 1 = \frac{1}{3}(2q-2)$ does not belong to $H(P)$.

Case (B): $\quad q \equiv 1 \pmod 2,\ j_2(P) = \frac{1}{2}(q-1),\ j_3(P) = \frac{1}{2}(q+1)$.

From (10.55), $2j_2(P) + 1 = q$ does not belong to $H(P)$, a contradiction.

Case (C): $\quad q \equiv 0 \pmod 2,\ j_2(P) = \frac{1}{2}q,\ j_3(P) = \frac{1}{2}(q+2)$.

The argument used in Case (A) shows this time that either $m_1(P) = \frac{1}{2}q - 1$ or $m_1(P) = \frac{1}{2}q$. In the former case, $q - 2 \in H(P)$ and thus (XI) implies that $j_2(P) = 2$. Since this contradicts the hypothesis, only the latter case can occur. Then $m_1(P) = \frac{1}{2}q$ and $m_2(P) = q - 1$. Now, as $\dim(2\mathcal{D}) = 11$, so $m_9(P) = 2q$. From this, $2q - m_4(P) = \frac{1}{2}q + 2$. In fact, $2q - m_i(P)$ is a $(2\mathcal{D}, P)$-order for $i = 0, \ldots, 9$, and the set of $(2\mathcal{D}, P)$-orders is

$$\{0, 1, 2, \tfrac{1}{2}q, \tfrac{1}{2}(q+2), \tfrac{1}{2}(q+4), q, q+1, q+2, \tfrac{3}{2}q, 2q\}.$$

Hence $m_4(P) = \frac{3}{2}q - 2$.

Finally, let $d = \frac{1}{3}(q-4)(\frac{1}{2}q+1) + 1$. From (10.55), $d \notin H(P)$. On the other hand; $d = m_4(P) + \frac{1}{6}(q-10)m_2(P) \in H(P)$, a contradiction.

(ii) (a) $\Rightarrow$ (b) From Lemma 10.50, $g = c_0(q+1, 4)$. Since $q+1 = \frac{1}{3}(q-2)\cdot 3+3$, Lemma 7.113 shows that $\dim(2\mathcal{D}) = 11$ and that $\frac{1}{3}(q-5)\mathcal{D} + \mathcal{D}'$ is the canonical linear series, where $\mathcal{D}'$ is a base-point-free 1-dimensional linear series of order $\frac{1}{3}(q+1)$. In particular, there exists $x \in K(\mathcal{X})$ with $[K(\mathcal{X}) : K(x)] = \frac{1}{3}(q+1)$.

Let $P \in \mathcal{X}$ and assume by Corollary 10.52 that $j_2(P) > 2$. If $P \in \mathcal{X}(\mathbf{F}_{q^2})$, the result follows from Lemma 10.53 (i). Otherwise, $P \notin \mathcal{X}(\mathbf{F}_{q^2})$, and there are two possibilities according as q is odd or even; see Lemma 10.53 (ii).

Case (A): $\quad q \equiv 1 \pmod 2,\ j_2(P) = \frac{1}{2}(q-1),\ j_3(P) = \frac{1}{2}(q+1)$.

A property similar to (10.55) holds; namely,

$$\delta + 1 \notin H(P) \text{ for any } (\tfrac{1}{3}(q-5)\mathcal{D}, P)\text{-order } \delta.$$

Hence $2j_2(P) + 1 = q \notin H(P)$, a contradiction.

Case (B): $\quad q \equiv 0 \pmod 2,\ j_2(P) = \frac{1}{2}q,\ j_3(P) = \frac{1}{2}(q+2)$.

From Case (C) in the proof of (i), it also follows that $m_1(P) = \frac{1}{2}q$. Let y be an element of $K(\mathcal{X})$ such that $\mathrm{div}(x)_\infty = \frac{1}{2}q$. Since $\gcd(\frac{1}{3}(q+1), \frac{1}{2}q)$, so $K(\mathcal{X}) = K(x, y)$. From Theorem 5.61, $g \le \frac{1}{6}(q-2)^2$, a contradiction.

Finally, (b) $\Rightarrow$ (c) is Theorem 10.43 for $m = \frac{1}{3}(q+1)$, and (c) $\Rightarrow$ (a) follows immediately. $\square$

Now $\mathbf{F}_{q^2}$-maximal curves of genus $\frac{1}{6}q(q-3)$ are considered in the case

$$r = 4, \quad q = 3^h.$$

Note that $c_0(q+1,4) = \frac{1}{6}q(q-3)$. The main result, see Theorem 10.57 below, states that such $\mathbf{F}_{q^2}$-maximal curves are birationally equivalent over $\mathbf{F}_{q^2}$ to the $\mathbf{F}_{q^2}$-rational curve

$$\mathcal{T}_3 = \mathbf{v}(Y^{q/3} + Y^{q/9} + \cdots + Y^3 + Y - X^{q+1}). \tag{10.56}$$

The proof of Theorem 10.57 depends on two preliminary lemmas.

Lemma 10.55 *For $q = 3^h$, let $\mathcal{X}$ be an $\mathbf{F}_{q^2}$-maximal curve of genus $\frac{1}{6}q(q-3)$. Then $r = \dim \mathcal{D}$ is either 3 or 4. If $r = 4$, the following hold:*

(i) $\dim(2\mathcal{D}) = 11$;

(ii) *there exists a complete linear series $\mathcal{D}'$ of order $\frac{2}{3}q$ and dimension 2 such that $\frac{1}{3}(q-6)\mathcal{D} + \mathcal{D}'$ is the canonical linear series on $\mathcal{X}$;*

(iii) *if $j_2 = 2$, then $j_3 = 3$;*

(iv) *if $P \in \mathcal{X}(\mathbf{F}_{q^2})$ and $j_2 > 2$, then $j_2 = \frac{1}{3}(q+3)$, and $j_3 = \frac{1}{3}(2q+3)$; in particular, the Weierstrass semigroup at P is generated by $\frac{1}{3}q$ and $q+1$;*

(v) *if $q \geq 27$ and $P \notin \mathcal{X}(\mathbf{F}_{q^2})$, then $j_2 = 2$.*

Proof. This can be done using arguments and calculations similar to those in the preceding proofs for the case $q \equiv 1, 2 \pmod 3$. $\square$

Lemma 10.56 *For $q \geq 27$, there exists $P \in \mathcal{X}(\mathbf{F}_{q^2})$ such that $H(P)$ is generated by $\frac{1}{3}q$ and $q+1$.*

Proof. By Lemma 10.55 (ii), it is enough to show the existence of $P \in \mathcal{X}(\mathbf{F}_{q^2})$ with $j_2(P) > 2$. Suppose on the contrary that $j_2(P) = 2$ for every $P \in \mathcal{X}(\mathbf{F}_{q^2})$. If $P \in \mathcal{X}(\mathbf{F}_{q^2})$ then $j_4(P) = q+1$, and this, together with Lemma 10.55 (iii), implies that the $(\mathcal{D}, P)$-orders are $0, 1, 2, 3, q+1$. Likewise, if $P \notin \mathcal{X}(\mathbf{F}_{q^2})$, then $j_4(P) = q$ and, from this and Lemma 10.55 (v), the $(\mathcal{D}, P)$-orders are $0, 1, 2, 3, q$. Therefore, by Corollary 8.58,

$$\deg(R) = (6+q)(2g-2) + 5(q+1) = |\mathcal{X}(\mathbf{F}_{q^2})| = (q+1)^2 + q(2g-2),$$

whence $2g - 2 = \frac{1}{6}(q-4)(q+1)$; that is, $q^2 - 3q - 8 = 0$, a contradiction. $\square$

Theorem 10.57 *For $q = 3^h$, let $\mathcal{X}$ be an $\mathbf{F}_{q^2}$-maximal curve of genus $\frac{1}{6}(q-3)q$. If $\dim \mathcal{D} \neq 3$, then $\dim \mathcal{D} = 4$ and $\mathcal{X}$ is birationally equivalent over $\mathbf{F}_{q^2}$ to the $\mathbf{F}_{q^2}$-rational curve $\mathcal{T}_3$ in (10.56).*

Proof. From Lemma 10.55, $r = \dim \mathcal{D} = 4$. Let $P \in \mathcal{X}(\mathbf{F}_{q^2})$ be as in Corollary 10.56. Then the Weierstrass semigroup $H(P)$ is generated by $\frac{1}{3}q$ and $q+1$.

Choose $x, y \in \mathbf{F}_{q^2}(\mathcal{X})$ for which $\operatorname{div}(x)_\infty = \frac{1}{3}qP$ and $\operatorname{div}(y)_\infty = (q+1)P$. Then the Frobenius linear series $\mathcal{D}$ consists of all divisors

$$A_{\mathbf{c}} = \operatorname{div}(c_0 + c_1x + c_2x^2 + x_3x^3 + c_4y) + (q+1)P,$$

with $\mathbf{c} = (c_0, c_1, c_2, c_3, c_4) \in \mathrm{PG}(4, K)$. Since $\operatorname{ord}_P(x^jy^i) \geq -\frac{1}{3}q(q+1)+1$ unless either $(j,i) = (q+1, 0)$ or $(j,i) = (0, \frac{1}{3}q)$, the complete linear series $|\frac{1}{3}q(q+1)P)|$ comprises the divisors

$$B_{\mathbf{b}} = \operatorname{div}(b_0x^{q+1} + \textstyle\sum_{i=0}^{q/3}\sum_{j=0}^{q-3i} b_{ij}x^jy^i) + \frac{1}{3}q(q+1)P,$$

where $\mathbf{b} = (b_0, \ldots, b_{ij}, \ldots) \in \mathrm{PG}(m, K)$ with $m = \frac{1}{6}(q^2+5q)+1$. By the Riemann–Roch Theorem 6.61, $\dim|\frac{1}{3}q(q+1)P)| = \frac{1}{6}(q^2+5q)$. Therefore

$$x^{q+1} + \textstyle\sum_{i=0}^{q/3} A_i(x)y = 0, \tag{10.57}$$

where $A_i(X) \in \mathbf{F}_{q^2}[X]$ with $\deg A_i(X) \leq q - 3i$ and, in particular, $A_{q/3}(X)$ is a non-zero constant belonging to $\mathbf{F}_{q^2}$. Finally, some ad hoc arguments depending on higher Hasse derivatives and changes of coordinates reduce (10.57) to the simpler (10.56). □

Now consider the case

$$g = \tfrac{1}{8}(q-1)(q-3), \quad q \text{ odd}.$$

Up to a birational transformation over $\mathbf{F}_{q^2}$, Theorem 10.63 below states that, for $p \geq 5$ and q large enough, the two curves

$$\mathcal{F} = \mathcal{E}^1_{(q+1)/4} = \mathbf{v}(Y^q + Y - X^{(q+1)/4}), \quad q \equiv 3 \pmod 4, \tag{10.58}$$

$$\mathcal{F} = \mathcal{D}_{(q+1)/2} = \mathbf{v}(X^{(q+1)/2} + Y^{(q+1)/2} + 1) \tag{10.59}$$

are the only $\mathbf{F}_{q^2}$-maximal curves of genus $g = \frac{1}{8}(q-1)(q-3)$ when $\dim \mathcal{D} = 5$.

It should be noted that the condition on $\dim \mathcal{D}$ enters into play since, from the proof of Lemma 10.50, it can only be deduced that $4 \leq \dim \mathcal{D} \leq 5$, whereas $\frac{1}{8}(q-1)(q-3)$ is Castelnuovo's number for $\dim \mathcal{D} = 5$ but not for $\dim \mathcal{D} = 4$. So the condition, $\dim \mathcal{D} = 5$, is necessary to take advantage of Lemma 7.113. It may also be noted that the two curves are not birationally equivalent even over K since their Weierstrass semigroups are different; see Exercises 6 and 7 in Chapter 6.

THEOREM 10.58 *When $q \geq 11$, the Fermat curve in (10.59) is the unique $\mathbf{F}_{q^2}$-maximal non-singular plane curve of degree $\frac{1}{2}(q+1)$.*

As $\dim(2\mathcal{D}) = 14$ by Lemma 7.113 (i), the possibilities for the $(\mathcal{D}, P)$-order sequences at $P \in \mathcal{X}$ can be determined. The two relevant results are stated below but their proofs are omitted as being similar to those of Lemmas 10.51 and 10.53, and Corollary 10.52.

LEMMA 10.59 *For $P \in \mathcal{X}$,*

(i) $j_1(P) = 1$;

(ii) $j_5(P) = q+1$ *if* $P \in \mathcal{X}(\mathbf{F}_{q^2})$ *and* $j_5(P) = q$ *if* $P \in \mathcal{X}\backslash\mathcal{X}(\mathbf{F}_{q^2})$.

LEMMA 10.60 *For* $q \geq 11$ *and* $P \in \mathcal{X}$, *let* $\dim(\mathcal{D}) = 5,\ \dim(2\mathcal{D}) = 14$.

(i) *If* $j_3(P) = 3$, *then* $j_4(P) = 4$.

(ii) *Let* $j_2(P) = 2,\ j_3(P) > 3$;

(a) *if* $P \in \mathcal{X}(\mathbf{F}_{q^2})$, *then* q *is odd*, $j_3(P) = \frac{1}{2}(q+1),\ j_4(P) = \frac{1}{2}(q+3)$;
(b) *if* $P \notin \mathcal{X}(\mathbf{F}_{q^2})$, *then* q *is even*, $j_3(P) = \frac{1}{2}q,\ j_4(P) = \frac{1}{2}(q+2)$.

(iii) *If* $P \in \mathcal{X}(\mathbf{F}_{q^2}),\ j_2(P) > 2$, *and*

$$g > \begin{cases} \frac{1}{9}(q-2)^2, & \text{for } q \equiv 2 \pmod 3,\\ \frac{1}{9}q(q-3), & \text{for } q \equiv 0 \pmod 3,\end{cases}$$

then one of the following holds:

(a) $q \equiv 3 \pmod 4,\ j_2(P) = \frac{1}{4}(q+1),\ j_3(P) = \frac{1}{2}(q+1),$
$j_4(P) = \frac{3}{4}(q+1)$;
(b) $q \equiv 0 \pmod 4,\ j_2(P) = \frac{1}{4}(q+4),\ j_3(P) = \frac{1}{2}(q+2),$
$j_4(P) = \frac{1}{4}(3q+4)$.

(iv) *If* $P \notin \mathcal{X}(\mathbf{F}_{q^2}),\ j_2(P) > 2$, *then one of the following holds*:

(a) $q \equiv 1 \pmod 4,\ j_2(P) = \frac{1}{4}(q+3),\ j_3(P) = \frac{1}{2}(q+1),$
$j_4(P) = \frac{1}{4}(3q+1)$;
(b) $q \equiv 0 \pmod 4,\ j_2(P) = \frac{1}{4}q,\ j_3(P) = \frac{1}{2}q,\ j_4(P) = \frac{3}{4}q$;
(c) $q \equiv 1 \pmod 3,\ j_2(P) = \frac{1}{3}(q-1),\ j_3(P) = \frac{1}{3}(q+2),$
$j_4(P) = \frac{1}{3}(2q+1)$;
(d) $q \equiv 0 \pmod 3,\ j_2(P) = \frac{1}{3}q,\ j_3(P) = \frac{1}{3}(q+3),\ j_4(P) = \frac{2}{3}q$.

COROLLARY 10.61 *If* $q \geq 11,\ \dim(\mathcal{D}) = 5,\ \dim(2\mathcal{D}) = 14$, *and* $j_3(P) = 3$ *for every* $P \in \mathcal{X}$, *then* $q \equiv 0, 4 \pmod 5$ *and* $g = \frac{1}{20}(q^2 - 4q + 15)$.

COROLLARY 10.62 *If, for* $q \geq 11$ *with* q *odd*, $\mathcal{X}$ *is an* $\mathbf{F}_{q^2}$*-maximal curve of genus* $g = \frac{1}{8}(q-1)(q-3)$ *with* $\dim(\mathcal{D}) = 5$, *then the following hold*:

(i) $\mathcal{X}$ *is birationally equivalent over* $\mathbf{F}_{q^2}$ *to the curve* $\mathcal{E}^1_{(q+1)4}$ *in (10.58) if and only if there exists* $P \in \mathcal{X}(\mathbf{F}_{q^2})$ *with* $j_2(P) > 2$;

(ii) $\mathcal{X}$ *is birationally equivalent over* $\mathbf{F}_{q^2}$ *to the curve* $\mathcal{D}_{(q+1)/2}$ *in (10.59) if and only if there exists* $P \in \mathcal{X}(\mathbf{F}_{q^2})$ *with* $j_2(P) = 2,\ j_3(P) > 3$.

Proof. (i) From Lemma 12.1, the curve $\mathcal{F}$ in (10.58) has a unique branch centred at $Y_\infty = (0,0,1)$. This branch is $\mathbf{F}_{q^2}$-rational and hence it corresponds to a point $P \in \mathcal{X}(\mathbf{F}_{q^2})$. By a straightforward computation, $m_3(P) = \frac{3}{4}(q+1)$. Hence $j_2(P) = \frac{1}{4}(q+1)$ by (IV). Conversely, from Lemma 10.60 (iii), $j_4(P) = \frac{3}{4}(q+1)$

and so $m_1(P) = \frac{1}{4}(q+1)$ by (IV). Therefore the result follows from Theorem 10.43.

(ii) The Fermat curve $\mathcal{F}$ in (10.58) is non-singular. By Theorem 6.51, the linear system of conics cuts out on $\mathcal{F}$ a complete linear series g^5_{q+1}.

The point $P_0 = (1, \alpha, 0)$ of $\mathcal{F}$, with $\alpha^{(q+1)/2} = -1$, is an inflexion and the tangent $\mathcal{F}$ at P is $\ell = \mathbf{v}(\alpha X_0 - X_1)$. Therefore $I(P_0, \mathcal{F} \cap \ell) = \frac{1}{2}(q+1)$. Hence $I(P_0, \mathcal{F} \cap \mathcal{C}) = q+1$, where $\mathcal{C} = \mathbf{v}((\alpha X_0 - X_1)^2)$ is the conic consisting of the repeated line ℓ. Thus g^5_{q+1} contains the $\mathbf{F}_{q^2}$-rational divisor $(q+1)P$, and hence $\mathcal{D} = g^5_{q+1}$ by Theorem 6.34.

Choose a line ℓ_1 through P and a line ℓ_2 disjoint from P. Then

$$I(P_0, \mathcal{F} \cap \mathcal{C}') = \begin{cases} 2 & \text{when the conic } \mathcal{C}' = \ell_1^2\,, \\ \frac{1}{2}(q+1) & \text{when the conic } \mathcal{C}' = \ell\ell_2\,. \end{cases}$$

Therefore $j_2(P) = 2,\ j_3(P) > 3$.

Conversely, (IV) and Lemma 10.60 (ii) imply that $m_1(P) = \frac{1}{2}(q-1)$, and $m_2(P) = \frac{1}{2}(q+1)$. Choose $x, y \in K(\mathcal{X})$ such that $\operatorname{div}(x)_\infty = \frac{1}{2}(q-1)$ and $\operatorname{div}(y)_\infty = \frac{1}{2}(q+1)$. Since these numbers are coprime, $K(\mathcal{X}) = K(x,y)$ and there exists $f \in \mathbf{F}_{q^2}[X,Y]$ such that $\deg f \le \frac{1}{2}(q+1)$ and $f(x,y) = 0$. In other words, $\mathcal{X}$ is birationally equivalent over $\mathbf{F}_{q^2}$ to an irreducible plane curve $\mathcal{F}$ of degree at $\frac{1}{2}(q+1)$. If $\mathcal{F}$ were not singular, from Lemma 3.24 its virtual genus

$$g^* < \tfrac{1}{2}[(\tfrac{1}{2}(q+1) - 1)(\tfrac{1}{2}(q+1) - 2] = \tfrac{1}{8}(q-1)(q-3).$$

But then $g^* < g$, contradicting Lemma 3.28. Hence $\mathcal{F}$ is non-singular and the result follows from Theorem 10.58. □

THEOREM 10.63 *If q is odd, $p \ge 5$, and $\mathcal{X}$ is an $\mathbf{F}_{q^2}$-maximal curve with genus $g = \frac{1}{8}(q-1)(q-3)$ and $\dim(\mathcal{D}) = 5$, the following hold.*

(i) *If $q \ge 17$ and $q \equiv 1 \pmod 4$, then $\mathcal{X}$ is birationally equivalent over $\mathbf{F}_{q^2}$ to the curve $\mathcal{D}_{(q+1)/2}$ in (10.59).*

(ii) *If $q \ge 19$ and $q \equiv 3 \pmod 4$, then $\mathcal{X}$ is birationally equivalent over $\mathbf{F}_{q^2}$ either to the curve $\mathcal{E}^1_{(q+1)/4}$ in (10.58) or to the curve $\mathcal{D}_{(q+1)/2}$ in (10.59).*

Proof. As already observed, $g = c_0(q+1, 5)$ and thus $\dim(2\mathcal{D}) = 14$. In particular, by Corollary 10.61, there exists $P \in \mathcal{X}$ with $j_3(P) > 3$.

(i) Let $q \equiv 1 \pmod 4$. If $P \in \mathcal{X}(\mathbf{F}_{q^2})$, then Lemma 10.60 (ii), (iii) imply that $j_2(P) = 2$. Hence the result follows from Corollary 10.62 (ii). To show that this is the only possible case, let $P \notin \mathcal{X}(\mathbf{F}_{q^2})$. As $\mathcal{D}' = \frac{1}{4}(q-5)\mathcal{D}$ is the canonical linear series by Lemma 7.113 (ii), and hence $\delta + 1 \notin H(P)$ for any $(\mathcal{K}, P)$-order δ.

Since $m_4(P) = q$ by (XIV), Lemma 10.60 together with the hypothesis $p \ge 5$ leads to the following two cases (A) and (B).

Case (A): $j_2(P) = \frac{1}{4}(q+3),\ j_3(P) = \frac{1}{2}(q+1),\ j_4(P) = \frac{1}{4}(3q+1)$.

In this case,

$$\{q - m_3, q - m_2, q - m_1\} \subset \{1, \tfrac{1}{4}(q+3), \tfrac{1}{2}(q+1), \tfrac{1}{4}(3q+1)\}$$

by (XIV). Thus

$$m_1(P) = \tfrac{1}{2}(q-1),\ m_2(P) = 3(q-1),\ m_3(P) = q-1.$$

Now $\delta = \frac{1}{4}(q-9)+\frac{1}{4}(3q+1) = q-2$ is a $(\mathcal{D}', P)$-order and hence $q-1 \notin H(P)$, a contradiction.

Case (B): $q \equiv 1 \pmod 3,\ j_2(P) = \frac{1}{3}(q-1),\ j_3(P) = \frac{1}{3}(q+2),$
$j_4(P) = \frac{1}{3}(2q+1).$

Since $\frac{1}{4}(q-5) \geq 3$, in this case $\delta = 3j_2(P)$ must be a $(\mathcal{D}', P)$-order. Hence $q \notin H(P)$, a contradiction.

(ii) $q \equiv 3 \pmod 4$. Again, it suffices to show that $P \in \mathcal{X}(\mathbf{F}_{q^2})$, since the result then follows from Corollary 10.62. If $P \notin \mathcal{X}(\mathbf{F}_{q^2})$, Lemma 10.60 (ii) and (iv), together with the hypothesis $p \geq 5$, imply that $j_2(P) = \frac{1}{3}(q-1)$. From Lemma 7.113 (ii) , $\delta + 1 \notin H(P)$ for every $(\frac{1}{4}(q-7)\mathcal{D}, P)$-order δ. On the other hand, since $\frac{1}{4}(q-7) \geq 3$ it follows that $3j_2(P)+1 = q \in H(P)$, a contradiction. □

REMARK 10.64 There exist $\mathbf{F}_{q^2}$-maximal curves whose genus equals Halphen's number $c_1(4, q+1)$. The known examples are listed below, but in (ii) and (iii) no explicit equation is known; see Section 10.12:

(i) for $q \equiv 0 \pmod 4$, curves of genus $\frac{1}{8}(q^2-2q)$;

(ii) for $q \equiv 1 \pmod 4$, curves of genus $\frac{1}{8}(q-1)^2$ constructed as a quotients of the Hermitian curve $\mathcal{H}_q$ by a subgroup of the automorphism group;

(iii) for $q \equiv 3 \pmod 4$, curves of genus $\frac{1}{8}(q^2-2q+5)$ constructed similarly to those in (ii).

10.7 PLANE MAXIMAL CURVES

In this section $\mathbf{F}_{q^2}$-maximal non-singular plane curves are considered.

THEOREM 10.65 *For every proper divisor d of $q+1$, the Fermat curve*

$$\mathcal{F} = \mathcal{D}_{(q+1)/d} = \mathbf{v}(X^{(q+1)/d} + Y^{(q+1)/d} + 1)$$

is $\mathbf{F}_{q^2}$-maximal.

Proof. The rational transformation $x' = x^d,\ y' = y^d$ of the function field $K(\mathcal{H}_q)$ of the Hermitian curve $\mathcal{H}_q = \mathbf{v}(X^{q+1}+Y^{q+1}+1)$ is defined over $\mathbf{F}_{q^2}$ and provides a d-fold covering of $K(\mathcal{F})$. The result follows from Theorem 10.2. □

Let $\mathcal{F}$ be an $\mathbf{F}_{q^2}$-maximal non-singular plane curve of degree n with $n \geq 3$. From Theorem 5.57, the genus of $\mathcal{F}$ is $g = \frac{1}{2}(n-1)(n-2)$. From Proposition 10.46, $n \leq q+1$, and equality holds if and only if $\mathcal{F}$ is projectively equivalent over $\mathbf{F}_{q^2}$ to the Hermitian curve $\mathcal{H}_q$, as in Theorems 10.47 and 10.22.

From now on, let $n < q+1$. Upper bounds on n, and equivalently on g, are obtained from the complete linear series $L_1 = g_n^2$ cut out on $\mathcal{F}$ by lines. Notation and terminology from Sections 7.7 and 8.6 are used. In particular, $\epsilon_0 < \epsilon_1 < \epsilon_2$ with $\epsilon_0 = 0$, $\epsilon_1 = 1$ are the L_1-orders and $\nu_0 < \nu_1$ with $\nu_0 = 0$ are the Frobenius orders of L_1.

LEMMA 10.66 *Let $\mathcal{F}$ be an $\mathbf{F}_{q^2}$-maximal non-singular plane curve of degree n with $n < q+1$. Then the following hold:*

(i) $\epsilon_2 \leq q, \quad \nu_1 = 1$;

(ii) *if*

$$n_2(q) = \begin{cases} \lfloor (q+2)/2 \rfloor & \text{when } q \geq 4 \text{ and } q \neq 3,5, \\ 3 & \text{when } q = 3, \\ 4 & \text{when } q = 5, \end{cases}$$

then $n \leq n_2(q)$;

(iii) *if* $n = \frac{1}{2}(q+1)$, *then $\mathcal{F}$ is classical.*

Proof. (i) For any line ℓ and $P \in \mathcal{F}$,

$$I(P, \ell \cap \mathcal{F}) \leq n < q+1.$$

Hence $\epsilon_2 \leq j_2 \leq n \leq q$. If $\nu_1 > 1$, then, from Theorem 8.68,

$$q^2 + 1 + (n-1)(n-2)q = n(q^2 - n + 2)$$
$$n^2 - (q+2)n + q + 1 = 0,$$

whence $n = 1$ or $q + 1$, a contradiction. Hence $\nu_1 = 1$; that is, $\mathcal{F}$ is Frobenius classical.

(ii) From the Stöhr–Voloch Theorem 8.18,

$$2(q^2 + 1 + (n-1)(n-2)q) \leq n(n-3) + (q^2+2)n,$$

whence $n \leq n_2(q)$.

(iii) Suppose that $j_2(P) \geq 3$ for all P. From Corollary 8.58, $v_P(S) \geq 3$. Then, from the Stöhr–Voloch Theorem 8.18,

$$3(q^2 + 1 + (n-1)(n-2)q) \leq n(n-3) + (q^2+2)n.$$

But this is impossible for $n = \frac{1}{2}(q+1)$. □

COROLLARY 10.67 (i) *If $n \geq 3$ is the degree of an $\mathbf{F}_{q^2}$-maximal non-singular plane curve, then either $n = q+1$ or $n \leq n_2$.*

(ii) *For $q \neq 3, 5$, there is no $\mathbf{F}_{q^2}$-maximal non-singular plane curve with genus g in the interval I given as follows:*

$$\tfrac{1}{8}q(q-2) < g \leq \tfrac{1}{4}q(q-2), \quad \textit{for } q \textit{ even},$$
$$\tfrac{1}{8}(q-1)(q-3) < g \leq \tfrac{1}{4}(q-1)^2, \quad \textit{for } q \textit{ odd}.$$

Proof. To show (ii), assume that $n \leq n_2(d)$. By Lemma 10.66 (i), $\mathcal{F}$ is Frobenius classical. From the Stöhr–Voloch Theorem 8.18,

$$n \leq \frac{q^2 + 2q + 1}{2q - 1} = F(q).$$

It follows that $F(q) < \frac{1}{2}(q+3)$ for $q > 5$ and that $F(3) = 16/5$, $F(4) = 25/7$, $F(5) = 4$. Finally, a straightforward computation shows that $n \notin I$ for $q \neq 3, 5$. □

REMARK 10.68 Let $3 \le d \le n_2(q)$.

(i) If q is odd, then the $\mathbf{F}_{q^2}$-maximal curve in (10.59) shows that the upper bound $n_2(q) = \frac{1}{2}(q+1)$ in Corollary 10.67 is the best possible for $q \ne 3, 5$.

(ii) The value $n_2(q) = 3$ is sharp for $q = 3$.

(iii) There exists an $\mathbf{F}_{25}$-maximal non-singular plane quintic; so $n_2(q) = 4$ is sharp for $q = 5$.

(iv) From part (ii) of the lemma, $n_2(q) = 3$ is sharp for $q = 4$. For q even and $q \ge 8$, it is not known if the bound, $n_2(q) = \frac{1}{2}(q+2)$, is good or bad.

Refinements to the above upper bound can be obtained by similar computations depending on the Stöhr–Voloch Theorem 8.18 applied to L_1.

PROPOSITION 10.69 *If $\mathcal{F}$ is an $\mathbf{F}_{q^2}$-maximal non-singular plane curve of degree n such that $3 \le n \le n_2(q)$, then $\mathcal{X}$ is classical with respect to L_1 when one of the following conditions holds:*

(i) $p > n$ *or* $n \not\equiv 1 \pmod p$;

(ii) $q = 4, 8, 16, 32$;

(iii) $p \ge 3$ *and either* $q = p$ *or* $q = p^2$;

(iv) $p = 2,\ q \ge 64$, *and either* $d \le 4$ *or*

$$n \le n_3(q) = \begin{cases} \frac{1}{4}q - 1, & \textit{for } q = 64, 128, 256, \\ \frac{1}{4}q, & \textit{for } q \ge 512 \textit{ and } q \textit{ odd}; \end{cases}$$

(v) $p \ge 3,\ q = p^v$ *with* $v \ge 3$, *and* $n \ge n_3(q) = q/p - p + 2$.

REMARK 10.70 For $q = p^3$, $p \ge 3$, the bound $n_3(q)$ in Proposition 10.69 is sharp, as there exists a plane $\mathbf{F}_{p^6}$-maximal curve of degree $p^2 - p + 1$ which is non-classical with respect to L_1; see Corollary 10.76 and Remark 10.86.

COROLLARY 10.71 *If $\mathcal{F}$ is as in Proposition 10.69 and non-classical with respect to L_1, then*

(i) $q \ge 64$ *for* $p = 2$, *and* $q \ge p^3$ *for* $p \ge 3$;

(ii) $\epsilon_2^2 \le q/p$.

The example in Remark 10.70 shows that Corollary 10.71 (i) is sharp for $p \ge 3$.

DEFINITION 10.72 (i) For $q = 8$ and $q \ge 11$, let

$$n_4(q) = \frac{2q^2 + 15q - 20 + \sqrt{4q^4 - 40q^3 + 145q^2 - 300q + 600}}{10(q-2)}.$$

(ii) For $q = p^v$, $v \geq 2$, let

$$n_4(p,q) = \frac{2q^2 + 3(5-1/p)q - 8}{2(5-1/p)q - 12}$$
$$+ \frac{\sqrt{4q^4 - 8(5-1/p)q^3 + (113 - 50/p + 9/p^2)q^2 - 4(25 - 17/p)q + 184}}{2(5-1/p)q - 12}.$$

THEOREM 10.73 *If $q = 8$ or $q \geq 11$ and n is the degree of a plane $\mathbf{F}_{q^2}$-maximal curve $\mathcal{X}$, with $3 \leq n < q+1$, then one of the following holds:*

(i) $n = \lfloor (q+2)/2 \rfloor$;

(ii)
$$n \leq n_5(q) = \begin{cases} n_4(q) & \text{when } q = p, \\ n_4(p,q) & \text{when } q = p^v,\ v \geq 2. \end{cases}$$

10.8 MAXIMAL CURVES OF HURWITZ TYPE

DEFINITION 10.74 A *Hurwitz curve of degree* $n+1$ is a non-singular plane curve

$$\mathcal{C}_n = \mathbf{v}(X_0^n X_1 + X_1^n X_2 + X_2^n X_0), \tag{10.60}$$

where $n^2 - n + 1 \not\equiv 0 \pmod p$.

LEMMA 10.75 *The Hurwitz curve $\mathcal{C}_n$ is covered by the Fermat curve*

$$\mathcal{D}_{n^2-n+1} = \mathbf{v}(X_0^{n^2-n+1} + X_1^{n^2-n+1} + X_2^{n^2-n+1}).$$

Proof. Let $K(\mathcal{F}) = K(x,y)$, with $x^{n^2-n+1} + y^{n^2-n+1} + 1 = 0$, be the function field of $\mathcal{F}$. Put

$$\xi = x^{n-1}y^{-1}, \quad \eta = xy^{n-1};$$

then $\xi^n\eta + \eta^n + \xi = 0$. So the function field $K(\xi,\eta)$ of $\mathcal{C}_n$ is a subfield of $K(\mathcal{F})$, and the rational transformation ω, given by $x' = \xi$, $y' = \eta$, provides an $\mathbf{F}_{q^2}$-rational covering of $\mathcal{C}_n$ by $\mathcal{D}_{n^2-n+1}$. □

Consider the following condition:

$$q + 1 \equiv 0 \pmod{n^2 - n + 1}. \tag{10.61}$$

LEMMA 10.76 *If (10.61) holds, then the curves $\mathcal{C}_n$ and $\mathcal{D}_{n^2-n+1}$ are both covered over $\mathbf{F}_{q^2}$ by the Hermitian curve $\mathcal{H}_q$, and hence are $\mathbf{F}_{q^2}$-maximal.*

Proof. The argument of Lemma 10.75 can be adapted to show that $\mathcal{D}_{n^2-n+1}$ is $\mathbf{F}_{q^2}$-covered by the Hermitian curve $\mathcal{H}_q = \mathbf{v}(X_0^{q+1} + X_1^{q+1} + X_2^{q+1})$. Then Theorem 10.2 gives the result. □

LEMMA 10.77 *The Weierstrass semigroup of $\mathcal{C}_n$ at the point $P_1 = (0,1,0)$ is generated by the set $S = \{s(n-1)+1 \mid s = 1,\ldots,n\}$.*

Proof. Let $P_0 = (1,0,0)$ and $P_2 = (0,0,1)$. Then $\operatorname{div} x = nP_2 - (n-1)P_1 - P_0$ and $\operatorname{div} y = (n-1)P_0 + P_2 - nP_1$, so that

$$\operatorname{div}(x^{s-1}y) = ((n(s-1)+1)P_2 + (n-s)P_0 - (s(n-1)+1)P_1 .$$

This shows that S is contained in the Weierstrass semigroup $H(P_1)$ at P_1, and hence $H(P_1) \supset \langle S \rangle$. Therefore the result follows from a result on arithmetic progressions: $|\mathbf{N} \backslash \langle S \rangle| = \frac{1}{2}n(n-1)$. □

THEOREM 10.78 *The Hurwitz curve $\mathcal{C}_n$ is $\mathbf{F}_{q^2}$-maximal if and only if (10.61) holds.*

Proof. If (10.61) holds, then $\mathcal{C}_n$ is $\mathbf{F}_{q^2}$-maximal by Lemma 10.76.

Conversely, assume that $\mathcal{C}_n$ is $\mathbf{F}_{q^2}$-maximal. Then $(q+1)P_1 \equiv (q+1)P_2$ by (10.8), and the case $s = n$ in the proof of Lemma 10.77 gives that

$$(n^2 - n + 1)P_1 \equiv (n^2 - n + 1)P_2.$$

Therefore $u = \gcd(n^2 - n + 1, q + 1) \in H(P_1)$. From Lemma 10.77,

$$u = A(n-1) + B \text{ with } A \geq B \geq 1.$$

Now there exists $C \geq 1$ such that

$$(A(n-1) + B)C = n^2 - n + 1,$$

and so $BC = D(n-1) + 1$ for some $D \geq 0$. Therefore

$$AD(n-1) + A + BD = Bn.$$

Here $D = 0$, as otherwise the left-hand side would exceed Bn. So

$$B = C = 1, \quad A = n;$$

that is, $u = n^2 - n + 1$. □

COROLLARY 10.79 *The Fermat curve $\mathcal{D}_{n^2-n+1}$ is $\mathbf{F}_{q^2}$-maximal if and only if (10.61) holds.*

Proof. If (10.61) is satisfied, the result follows from Corollary 10.76. Now if $\mathcal{D}_{n^2-n+1}$ is $\mathbf{F}_{q^2}$-maximal, then $\mathcal{C}_n$ is also $\mathbf{F}_{q^2}$-maximal by Theorem 10.2. Then the corollary follows from Theorem 10.78. □

REMARK 10.80 For a given positive integer n, is there a power q of a prime p such that $q + 1 \equiv 0 \pmod{m}$ with $m = n^2 - n + 1$?

Since $m \not\equiv 0 \pmod{p}$, and $p \not\equiv 0 \pmod{m}$, a necessary and sufficient condition for q to have property (10.61) is that $p \equiv x \pmod{m}$, where x is a solution of the congruence $T^w + 1 \equiv 0 \pmod{m}$, and w is defined by $q = p^{\phi(m)v+w}$, where ϕ is the Euler function.

REMARK 10.81 The Hurwitz curve $\mathcal{C}_n$ is $\mathbf{F}_{q^2}$-maximal in the following cases:

(i) $n = 3,\ q = p^{6v+3}$ and $p \equiv 3, 5 \pmod{7}$;

(ii) $n = 4,\ q = p^{12v+6}$ and $p \equiv 2, 6, 7, 11 \pmod{13}$.

By using Theorem 10.78 and Remark 10.80, this is refined as follows.

COROLLARY 10.82 *The Hurwitz curves $\mathcal{C}_2$, $\mathcal{C}_3$, $\mathcal{C}_4$ are $\mathbf{F}_{q^2}$-maximal if and only if the one of the following conditions on p and q holds.*

(i) $\mathcal{C}_2 : q = p^{2v+1}$ *and* $p \equiv 2 \pmod 3$.

(ii) $\mathcal{C}_3$:
  (a) $q = p^{6v+1}$ *and* $p \equiv 6 \pmod 7$;
  (b) $q = p^{6v+3}$ *and* $p \equiv 3, 5, 6 \pmod 7$;
  (c) $q = p^{6v+5}$ *and* $p \equiv 6 \pmod 7$.

(iii) $\mathcal{C}_4$:
  (a) $q = p^{12v+1}$ *and* $p \equiv 12 \pmod{13}$;
  (b) $q = p^{12v+2}$ *and* $p \equiv 5, 8 \pmod{13}$;
  (c) $q = p^{12v+3}$ *and* $p \equiv 4, 10, 12 \pmod{13}$;
  (d) $q = p^{12v+5}$ *and* $p \equiv 12 \pmod{13}$;
  (e) $q = p^{12v+6}$ *and* $p \equiv 2, 5, 6, 7, 8, 11 \pmod{13}$;
  (f) $q = p^{12v+7}$ *and* $p \equiv 12 \pmod{13}$;
  (g) $q = p^{12v+9}$ *and* $p \equiv 4, 10, 12 \pmod{13}$;
  (h) $q = p^{12v+11}$ *and* $p \equiv 12 \pmod{13}$.

COROLLARY 10.83 *For a positive integer n, let $m = n^2 - n + 1$.*

(i) *If $n = p^e$ with $e \geq 1$, then the curve $\mathcal{C}_n$ is $\mathbf{F}_{q^2}$-maximal for $q = p^{\phi(m)v+3e}$.*

(ii) *If $p \equiv 3 \pmod 4$, $n \equiv 0, 1 \pmod p$, m is prime and $m \equiv 3 \pmod 4$, then the curve $\mathcal{C}_n$ is $\mathbf{F}_{q^2}$-maximal for $q = p^{(m-1)v+(m-1)/2}$.*

Proof. (i) This is a consequence of Theorem 10.78 and the identity

$$p^{3e} + 1 = (p^e + 1)(p^{2e} - p^e + 1).$$

(ii) It is enough to show that $p^{(m-1)/2} + 1 \equiv 0 \pmod m$. Recall the Legendre symbol (a/p):

$$(a/p) = \begin{cases} 1 & \text{if } x^2 \equiv a \pmod p \quad \text{has two solutions in } \mathbf{Z}_p, \\ -1 & \text{if } x^2 \equiv a \pmod p \quad \text{has no solution in } \mathbf{Z}_p. \end{cases}$$

Since $m \equiv 1 \pmod p$, here $(m/p) = 1$. By the Quadratic Reciprocity Law,

$$(m/p)(p/m) = (-1)^{((m-1)/2)((p-1)/2)},$$

and the condition that $m \equiv 3 \pmod 4$, it follows that $(p/m) = (-1)^{(p-1)/2}$. Now, as $p \equiv 3 \pmod 4$, this gives $(p/m) = -1$. In other words, p viewed as an element in $\mathbf{F}_m$ is a non-square. Since -1 is also a non-square in $\mathbf{F}_m$, this implies that $p \equiv (-1)u^2 \pmod m$ for an integer u with $u \not\equiv 0 \pmod m$. Hence $p^{(m-1)/2} \equiv -1 \pmod m$ as m is odd and $u^{m-1} \equiv 1 \pmod m$. □

REMARK 10.84 The condition that $m \equiv 3 \pmod 4$ in this corollary cannot be relaxed. If $n = 4$, then $m = 13$ but, from Corollary 10.82, $\mathcal{C}_4$ is not $\mathbf{F}_{3^6}$-maximal.

REMARK 10.85 Under the hypotheses in Corollary 10.83 (ii) with m not necessarily prime, the congruence (10.61) may be investigated by means of the group U_m of units in $\mathbf{Z}_m$. This group has order $\phi(m)$, and $p \in U_m$ since $m \equiv 1 \pmod p$. Now suppose that p, as an element of U_m, has even order $2i$. Then $p^{2i} \equiv 1 \pmod m$ and hence $(p^i + 1)(p^i - 1) \equiv 0 \pmod m$. Since p has order greater than i, so $p^i - 1 \not\equiv 0 \pmod m$ unless both $p^i + 1$ and $p^i - 1$ are zero divisors in $\mathbf{Z}_m$. When this does not happen, then (10.61) follows for $q = p^{\phi(m)v+i}$.

REMARK 10.86 Let p be a prime, $n = p^e u$ with $e \geq 1$ and $\gcd(p, u) = 1$. Assume that $e \geq 2$ when $p = 2$. Then the curves $\mathcal{C}_n$ and $\mathcal{D}_{n^2-n+1}$ are both non-classical with respect to L_1, and $0,\ 1,\ p^e$ are their L_1-orders.

DEFINITION 10.87 A *generalised Hurwitz curve of degree* $n+k$ is the irreducible plane curve

$$\mathcal{C}_{n,k} = \mathbf{v}(X_0^n X_1^k + X_1^n X_2^k + X_2^n X_0^k)\,,$$

where $n \geq k \geq 2$ and p does not divide $Q(n,k) = n^2 - nk + k^2$.

The singular points of $\mathcal{C}_{n,k}$ are $P_0 = (1,0,0)$, $P_1 = (0,1,0)$, and $P_2 = (0,0,1)$; each of them is the centre of a unique branch of $\mathcal{C}_{n,k}$. The genus of $\mathcal{C}_{n,k}$ is

$$g = \frac{n^2 - nk + k^2 + 2 - 3\gcd(n,k)}{2}\,,$$

The $\mathbf{F}_{q^2}$-maximality of these curves can be investigated by the same methods as for Hurwitz curves, with analogous results.

LEMMA 10.88 *The curve $\mathcal{C}_{n,k}$ is $\mathbf{F}_{q^2}$-covered by the Fermat curve $\mathcal{D}_{n^2-nk+k^2}$.*

Consider the following condition:

$$n^2 - nk + k^2 \equiv 0 \pmod{q+1}\,. \tag{10.62}$$

COROLLARY 10.89 *The curve $\mathcal{C}_{n,k}$ is $\mathbf{F}_{q^2}$-maximal provided that (10.62) holds.*

COROLLARY 10.90 *The curve $\mathcal{D}_{n^2-nk+k^2}$ is $\mathbf{F}_{q^2}$-maximal provided that (10.62) holds.*

Let $\mathcal{P}_1$ be the place arising from the branch of $\mathcal{C}_{n,k}$ centred at P_1.

LEMMA 10.91 *If* $\gcd(n,k) = 1$, *then the Weierstrass semigroup at $\mathcal{P}_1$ is*

$$H(\mathcal{P}_1) = \{(n-k)s + nt \mid s,t \in \mathbf{Z};\ t \geq 0;\ -kt/n \leq s \leq (n-k)/(kt)\}\,. \tag{10.63}$$

THEOREM 10.92 *If* $\gcd(n,k) = 1$ *and* $n^2 - nk + k^2$ *is prime, then the curve* $\mathcal{C}_{n,k}$ *is* $\mathbf{F}_{q^2}$*-maximal if and only if* (10.62) *holds.*

COROLLARY 10.93 *If n, k are as in Theorem 10.92, then the curve $\mathcal{D}_{n^2-nk+k^2}$ $\mathbf{F}_{q^2}$-maximal if and only if (10.62) holds.*

REMARK 10.94 There are infinitely many pairs n, k with $n > k \geq 1$ such that $n^2 - nk + k^2$ is prime. In fact, for a prime p' such that $p' \equiv 1 \pmod 6$, there exist such n, k with $p' = n^2 - nk + k^2$.

10.9 NON-ISOMORPHIC MAXIMAL CURVES

In this section, a 2-parameter family of curves $\mathcal{X}_i^m$ is presented; for each fixed m, there is a large number of non-isomorphic curves all with some identical properties.

With $K = \overline{\mathbf{F}}_q$, let $\mathcal{X}_i^m$ be a non-singular model over K of the plane curve

$$\mathcal{C}_i^m = \mathbf{v}(X^{mi+m} + X^{mi} + Y^{q+1}), \tag{10.64}$$

where m is a positive divisor of $q+1$ for which $d = (q+1)/m > 3$ is prime. The curve $\mathcal{X}_i^m$ is the quotient curve of the Hermitian curve $\mathcal{H}_q$ arising from an automorphism group of $\mathcal{H}_q$ of the same order d. Let $\mathcal{D} = |(q+1)P|$ denote the associated complete linear series at a point P of $\mathcal{X}_i^m$.

THEOREM 10.95 *Assume* $1 \leq i \leq d-2$.

(i) (a) *The curves $\mathcal{X}_i^m$ and $\mathcal{X}_j^m$ are K-equivalent if and only if one of the following equations holds modulo d:*

$$\begin{array}{lll} i \equiv j, & ij \equiv 1, & ij + i + j \equiv 0, \\ i + j + 1 \equiv 0, & ij + i + 1 \equiv 0, & ij + j + 1 \equiv 0. \end{array}$$

(b) *The number of K-isomorphism classes of curves $\mathcal{X}_i^m$ is given by*

$$n(d) = \begin{cases} \frac{1}{6}(d+1) & \text{if } d \equiv 2 \pmod 3, \\ \frac{1}{6}(d-1)+1 & \text{if } d \equiv 1 \pmod 3. \end{cases} \tag{10.65}$$

(c) *Each of these classes consists of six curves, apart from two exceptions of sizes 2 and 3. The corresponding indices i are as follows:*

(1) i_1, i_2, *where i_1 and i_2 are the solutions of $t^2+t+1 = 0 \pmod d$, with $d \equiv 1 \pmod 3$;*

(2) $1, \frac{1}{2}(d-1), d-2$.

(ii) *The genus of $\mathcal{X}_i^m$ is $g = \frac{1}{2}m(q-2)+1$.*

(iii) *The K-automorphism group of $\mathcal{X}_i^m$ is*

$$\mathrm{Aut}(\mathcal{X}_i^m) = \begin{cases} \mathbf{Z}_3 \rtimes (\mathbf{Z}_{q+1} \times \mathbf{Z}_m) & \text{in case (c)(1)}, \\ \mathbf{Z}_2 \rtimes (\mathbf{Z}_{q+1} \times \mathbf{Z}_m) & \text{in case (c)(2)}, \\ \mathbf{Z}_{q+1} \times \mathbf{Z}_m & \text{otherwise.} \end{cases}$$

(iv) *When $m = 2$, the series $\mathcal{D}$ has projective dimension $\frac{1}{2}(d+3)$. There are at least six K-rational points P such that, if*

$$(j_0 = 0,\ j_1 = 1, \ldots,\ j_{(d+1)/2},\ j_{(d+3)/2})$$

is the $\mathcal{D}$-order sequence at P, then $j_{(d+1)/2} = d$, $j_{(d+3)/2} = q+1$.

(v) *When $m = 2$ and q is prime, then the $\mathcal{D}$-order sequence at a generic point is $(0, 1, \ldots, \frac{1}{2}(d+1), q)$.*

THEOREM 10.96 (i) *The curves $\mathcal{X}_0^m$ and $\mathcal{X}_{d-1}^m$ are K-isomorphic.*

(ii) $\mathcal{X}_0^m$ *has genus* $g = \frac{1}{2}(m-1)(q-1)$, *and is hyperelliptic when* $m = 2$.

(iii) *The centre* Z *of the* K*-automorphism group* $\mathrm{Aut}_K(\mathcal{X}_0^m)$ *is a cyclic group of order* m, *and the group* $\mathrm{Aut}_K(\mathcal{X}_0^m)/Z$ *is isomorphic to* $\mathrm{PGL}(2, q)$.

(iv) *The complete linear series on* $\mathcal{X}_0^2$ *has projective dimension* $d+1$, *and the* $\mathcal{D}$*-order sequence at a Weierstrass point is one of*

$$(0, 1, 2, \ldots, d, q), \quad (0, 1, 2, 4, 6, \ldots, q-1, q+1).$$

(v) *The* $\mathcal{D}$*-order sequence of* $\mathcal{X}_0^2$ *is* $(0, 1, \ldots, d, q)$.

10.10 OPTIMAL CURVES

The previous sections contain examples of maximal curves such as the Hermitian curve $\mathcal{H}_q$ that is both $\mathbf{F}_{q^2}$-maximal and $\mathbf{F}_{q^2}$-optimal, as well as families of maximal curves.

Optimal curves that are not maximal have only been classified for $g = 1$; see Sections 9.9 and9.10. The value of $N_q(g)$ is known for $g = 1, 2$ and for various other pairs (g, q).

The aim of this section is a characterisation of the DLS curve as the unique optimal curve over $\mathbf{F}_q$ which has genus $g = q_0(q+1)$ when $q = 2q_0^2$, $q_0 = 2^s$ and $s \geq 1$. The proof of Theorem 10.102 requires some preliminary results.

From Proposition 9.32, the L-polynomial of $\mathcal{X}$ is $L_q(t) = (t^2 + 2q_0t + q)^g$. Hence, as already observed in Example 9.80, the Fundamental Equation (9.74) of $\mathcal{X}$ becomes, for $P, P_0 \in \mathcal{X}$ with $P_0 \in \mathcal{X}(\mathbf{F}_q)$:

$$qP + 2q_0\Phi(P) + \Phi^2(P) \equiv (q + 2q_0 + 1)P_0. \tag{10.66}$$

Some results from Section 9.8 to be applied to the corresponding linear series $\mathcal{D} = |(q + 2q_0 + 1)P_0|$ are summarised below.

From Section 9.8,

(1) $r = \dim \mathcal{D}$;

(2) $\epsilon_0 = 0 < \epsilon_1 = 1 < \cdots < \epsilon_r$ are the $\mathcal{D}$-orders of $\mathcal{X}$;

(3) $\nu_0 = 0 < \cdots < \nu_{r-1}$ are the Frobenius orders of $\mathcal{X}$;

(4) $m_1(P) < \cdots < m_r(P) < \cdots$ are the non-gaps of $\mathcal{X}$ at P.

Then

(A) $j_r(P) = m_r(P) = q + 2q_0 + 1$ for any $P \in \mathcal{X}(\mathbf{F}_q)$, and there exists $P_1 \in \mathcal{X}(\mathbf{F}_{q^2})$ such that $m_{r-1}(P_1) = q + 2q_0$;

(B) $j_i(P) = m_r(P) - m_{r-i}(P)$ for any $P \in \mathcal{X}(\mathbf{F}_q)$ and $i = 0, \ldots, r$;

(C) $\mathcal{D}$ is simple and base-point-free;

(D) $2q_0$ and q are $\mathcal{D}$-orders, and so $r \geq 3$;

(E) $\epsilon_r = \nu_{r-1} = q$;

(F) $m_1(P) = q$ for any $P \in \mathcal{X}(\mathbf{F}_q)$.

From (A), (C), and Proposition 9.83 (i), for any $P \in \mathcal{X}(\mathbf{F}_q)$,

$$j_{r-1}(P) = j_r(P) - m_1(P) = 2q_0 + 1,$$

whence

(G) $2q_0 \leq \epsilon_{r-1} \leq 2q_0 + 1$.

LEMMA 10.97 *The order* $\epsilon_{r-1} = 2q_0$.

Proof. Suppose on the contrary that $\epsilon_{r-1} > 2q_0$. Then

$$\epsilon_{r-2} = 2q_0, \quad \epsilon_{r-1} = 2q_0 + 1.$$

By Corollary 8.58,

$$\nu_{r-2} \leq j_{r-1}(P) - j_1(P) \leq 2q_0 = \epsilon_{r-2},$$

and thus the Frobenius orders of $\mathcal{D}$ are $\epsilon_0, \epsilon_1, \ldots, \epsilon_{r-2}$, and ϵ_r. By Proposition 8.56 (i), for $P \in \mathcal{X}(\mathbf{F}_q)$,

$$v_P(S) \geq \textstyle\sum_{i=1}^{r} (j_i(P) - \nu_{i-1}) \geq (r-1)j_1(P) + 1 + 2q_0 \geq r + 2q_0 \,. \quad (10.67)$$

Hence

$$\deg S = (\textstyle\sum_{i=0}^{r} \nu_i)(2g-2) + (q+r)(q+2q_0+1) \geq (r+2q_0)|\mathcal{X}(\mathbf{F}_q)|.$$

Since $2g-2 = (2q_0-2)(q+2q_0+1)$ and $|\mathcal{X}(\mathbf{F}_q)| = (q-2q_0+1)(q+2q_0+1)$,

$$\textstyle\sum_{i=1}^{r-2} \nu_i = \sum_{i=1}^{r-2} \epsilon_i \geq (r-1)q_0 \,.$$

Since $\epsilon_i + \epsilon_j \leq \epsilon_{i+j}$ for $i + j \leq r$ by Exercise 9 in Chapter 7, it follows that

$$(r-1)2q_0 = (r-1)\epsilon_{r-2} \geq 2\textstyle\sum_{i=0}^{r-2} \epsilon_i \geq 2(r-1)q_0 \,.$$

Hence $\epsilon_i + \epsilon_{r-2-i} = \epsilon_{r-2}$ for $i = 0, \ldots, r-2$. In particular, $\epsilon_{r-3} = 2q_0 - 1$ and, by Lemma 7.62, $\epsilon_i = i$ for $i = 0, 1, \ldots, r-3$. Then $r = 2q_0 + 2$. Now Castelnuovo's Bound, see Theorem 7.112, applied to $\mathcal{D}$ gives the following:

$$2g = 2q_0(q-1) \leq \frac{q + 2q_0 - (r-1)/2)^2}{r-1}.$$

Since $r = 2q_0 + 2$, this implies that

$$2q_0(q-1) < \tfrac{1}{2}(q+q_0)^2 q_0 = q_0 q + \tfrac{1}{2}q + \tfrac{1}{2}q_0,$$

a contradiction. □

COROLLARY 10.98 *There exists* $P_1 \in \mathcal{X}(\mathbf{F}_q)$ *such that*

$$j_1(P_1) = \begin{cases} 1 & \textit{for } i = 1, \\ \nu_{i-1} + 1 & \textit{for } i = 2, \ldots, r-1. \end{cases}$$

Proof. Taking (10.67) into account, it suffices to show the existence of a point $P_1 \in \mathcal{X}(\mathbf{F}_{q^2})$ for which $v_{P_1}(S) = r + 2q_0$. So suppose that $v_P(S) \geq r + 2q_0 + 1$ for every $P \in \mathcal{X}(\mathbf{F}_q)$. Then (8.14) implies that

$$\sum_{i=0}^{r-1} \nu_i \geq q + rq_0 + 1\,.$$

Therefore

$$\sum_{i=0}^{r-1} \epsilon_i \geq rq_0 + 2\,,$$

since $\epsilon_1 = 1$, $\nu_{r-1} = q$ and $\nu_i \leq \epsilon_{i+1}$. Then, from Exercise 9 in Chapter 7, $r\epsilon_{r-1} \geq 2rq_0 + 4$. But this means that $\epsilon_{r-1} > 2rq_0$, which contradicts Lemma 10.97. □

LEMMA 10.99 (i) *The Frobenius order* $\nu_1 > \epsilon_1 = 1$;

(ii) *the order* $\epsilon_2 = 2^e$.

Proof. (i) Suppose on the contrary that $\nu_1 = 1$. Then, from Corollary 10.98, there exists a point $P_1 \in \mathcal{X}(\mathbf{F}_q)$ such that $j_1(P_1) = 1, j_2(P_1) = 2$. Therefore

$$H(P_1) \subset H = \langle q, q + 2q_0 - 1, q + 2q_0, q + 2q_0 + 1\rangle\,,$$

by (A), (B) and (F). In particular, $g = q_0(q-1) \leq \tilde{g}$, where $\tilde{g}$ counts the non-negative integers which do not belong to H.

On the other hand, $g > \tilde{g}$. To show this, it suffices to prove the following:

$$\tilde{g} = g - \tfrac{1}{4}q_0^2. \tag{10.68}$$

The key idea is to observe that $\Lambda = \cup_{i=1}^{2q_0-1}\Lambda_i$ is a complete system of residues modulo q, where

$$\begin{aligned}
\Lambda_i &= \{iq + i(2q_0-1) + j \mid j = 0,\ldots,2i\} \quad \text{if } 1 \leq i \leq q_0 - 1;\\
\Lambda_{q_0} &= \{q_0q + q - q_0 + j \mid j = 0,\ldots,q_0-1\};\\
\Lambda_{q_0+1} &= \{(q_0+1)q + 1 + j \mid j = 0,\ldots,q_0-1\} \quad \text{for } 2 \leq i \leq \tfrac{1}{2}q_0;\\
\Lambda_{q_0+i} &= \{(q_0+i)q + (2i-3)q_0 + i - 1 + j \mid j = 0,\ldots,q_0 - 2i + 1\}\\
&\quad \cup\{(q_0+i)q + (2i-2)q_0 + i + j \mid j = 0,\ldots q_0 - 1\}\\
&\qquad \text{for } 1 \leq i \leq \tfrac{1}{2}q_0 - 1;\\
\Lambda_{3q_0/2+i} &= \{(3q_0/2+i)q + (q_0/2+i-1)(2q_0-1) + q_0 + 2i - 1 + j \mid\\
&\qquad j = 0,\ldots,q_0 - 2i - 1\}.
\end{aligned}$$

Also, $d \in \Lambda$, $d \in H$ and $d - q \notin H$ for each $d \in \Lambda$. Hence $\tilde{g}$ can be calculated by summing the coefficients of q in this list. The result is as follows:

$$\begin{aligned}
\tilde{g} &= \sum_{i=1}^{q_0-1} i(2i+1) + q_0^2 + (q_0+1)q_0 + \sum_{i=2}^{q_0/2} (q_0+i)(2q_0-2i+2)\\
&\quad + \sum_{i=1}^{q_0/2-1} (3q_0/2+i)(q_0-2i)\\
&= q_0(q-1) - q_0^2/4\,.
\end{aligned}$$

(ii) This follows from (i) and Lemma 7.62. □

Now let $P_0 = P_1$ be an $\mathbf{F}_q$-rational point satisfying Corollary 10.98, and let $m_i = m_i(P_1)$.

Lemma 10.99 (i) implies that $\nu_i = \epsilon_{i+1}$ for $i = 1, \ldots, r-1$. Therefore, from (A), (B) and Corollary 10.98,

$$m_i = 2q_0 + q - \epsilon_{r-i} \quad \text{for } i = 1, \ldots r-2,$$
$$m_{r-1} = 2q_0 + q, \quad m_r = 1 + 2q_0 + q. \tag{10.69}$$

Let $x, y_2, \ldots, y_r \in \mathbf{F}_q(\mathcal{X})$ be such that

$$\operatorname{div}(x)_\infty = m_1 P_1, \quad \operatorname{div}(y_i)_\infty = m_i P_1 \text{ for } i = 2, \ldots, r.$$

Note that x is a separable variable of $K(\mathcal{X})$. Since $\nu_1 > 1$, Proposition 8.42 shows that the matrix

$$\begin{bmatrix} 1 & x^q & y_2^q & \cdots & y_r^q \\ 1 & x & y_2 & \cdots & y_r \\ 0 & 1 & D_x y_2 & \cdots & D_x y_r \end{bmatrix}$$

has rank 2. Therefore

$$y_i^q - y_i = D_x y_i (x^q - x) \quad \text{for } i = 2, \ldots, r. \tag{10.70}$$

LEMMA 10.100 (i) *The divisor $(2g-2)P$ is canonical for any $P \in \mathcal{X}(\mathbf{F}_{q^2})$.*

(ii) *If $m \in H(P_1)$ satisfies $m < q + 2q_0$, then $m \leq q + q_0$.*

(iii) *There exists $g_i \in \mathbf{F}_q(\mathcal{X})$ such that $D_x y_i = g_i^{\epsilon_2}$ for $i = 2, \ldots, r$, and*

$$\operatorname{div}(g_i)_\infty = \frac{q m_i - q^2}{\epsilon_2} P_1.$$

Proof. (i) It may be assumed that $P = P_1$ from the identity (10.66) and the fact that $2g - 2 = (2q_0 - 2)(q + 2q_0 + 1)$. When $i = r$ in (10.70), $\operatorname{ord}_{P_1}(dx) = 2g - 2$ and the result follows since $\operatorname{ord}_Q(dx) \geq 0$ for $Q \neq P_1$.

(ii) From (10.69), q, $q + 2q_0$, $q + 2q_0 + 1 \in H(P_1)$. Then

$$(2q_0 - 2)q + q - 4q_0 + j, \quad j = 0, \ldots, q_0 - 2,$$

are also non-gaps at P_1. Since (i) implies that $H(P)$ is symmetric, it follows that

$$q + q_0 + 1 + j, \quad j = 0, \ldots, q_0 - 2,$$

are gaps at P_1, and this completes the proof.

(iii) Put $f_i = D_x y_i$. Then

$$D_x^{(j)} y_i = (x^q - x) D_x^{(j)} f_i + D_x^{(j-1)} f_i,$$

for $1 \leq j < q$, by Lemmas 5.72 and 5.80. Then $D_x^{(j)} f_i = 0$ for $1 \leq j < \epsilon_2$, since the matrix

$$\begin{bmatrix} 1 & x & y_2 & \cdots & y_r \\ 0 & 1 & D_x y_2 & \cdots & D_x y_r \\ 0 & 0 & D_x^{(j)} y_2 & \cdots & D_x^{(j)} y_r \end{bmatrix}$$

has rank 2 for each j in $2 \leq j < \epsilon_2$. Consequently, as ϵ_2 is a power of 2 by Lemma 10.99 (ii), by Lemma 5.84, there exists $g_i \in K(\mathcal{X})$ such that $f_i = g_i^{\epsilon_2}$. Since $f_i \in \mathbf{F}_q(\mathcal{X})$, also $g_i \in \mathbf{F}_q(\mathcal{X})$. Finally, from the proof of (i), $x - x(P)$ is a local parameter at P if $P \neq P_1$. Then, by the choice of the elements y_i, the function g_i has no pole other than P_1. From (10.70), $\operatorname{ord}_{P_1}(g_i) = -(q m_i - q^2)/\epsilon_2$. □

LEMMA 10.101 *The dimension $r = 4$ and the order $\epsilon_2 = q_0$.*

Proof. From Theorem 10.47, $r \geq 3$. Suppose that $r = 3$. Then

$$\epsilon_2 = 2q_0,\ n_1 = q,\ n_2 = q + 2q_0,\ n_3 = q + 2q_0 + 1,$$

and hence $\mathrm{ord}_{P_1} g_2 = -q$ for the same g_2 as in Lemma 10.100 (iii). Therefore, after a change over $\mathbf{F}_q$ of the coordinate system of PG$(3, K)$, the case $i = 2$ of (10.70) is as follows:

$$y_2^q - y_2 = x^{2q_0}(x^q - x).$$

Now the element $z = y_2^{q_0} - x^{q_0+1}$ satisfies $z^q - z = x^{q_0}(x^q - x)$. Therefore $q_0 + q$ is a non-gap at P_1. This contradiction implies that $r \neq 3$.

Let $r \geq 4$ and $2 \leq i \leq r$. By Lemma 10.100 (iii),

$$\frac{qn_i - q^2}{\epsilon_2} \in H(P_1).$$

Since

$$\frac{qm_i - q^2}{\epsilon_2} \geq mn_{i-1} \geq q,$$

from (10.69), $2q_0 \geq \epsilon_2 + \epsilon_{r-i}$ for $i = 2, \ldots, r-2$. In particular, $\epsilon_2 \leq q_0$. On the other hand, by Lemma 10.100 (ii), $m_{r-2} \leq q + q_0$ and hence, by (10.69), $\epsilon_2 \geq q_0$; that is, $\epsilon_2 = q_0$.

Note that $\epsilon_2 = q_0$ implies $\epsilon_{N-2} \leq q_0$. Since $n_2 \leq q + q_0$ by Lemma 10.100) (ii), it follows from (10.69) that $\epsilon_{r-2} \geq q_0$. Therefore $\epsilon_{r-2} = q_0 = \epsilon_2$ and hence $r = 4$. □

THEOREM 10.102 *If $q = 2q_0^2$, $q_0 = 2^s$ and $s \geq 1$, and $\mathcal{X}$ is an irreducible non-singular $\mathbf{F}_q$-rational curve of genus g such that*

$$g = q_0(q-1), \qquad |\mathcal{X}(\mathbf{F}_q)| = q^2 + 1, \tag{10.71}$$

then $\mathcal{X}$ is birationally equivalent over $\mathbf{F}_q$ to the DLS curve.

Proof. Let $P_1 \in \mathcal{X}(\mathbf{F}_q)$ be as above. From (10.70), Lemma 10.100 (iii) and Lemma 10.101,

$$y_2^q - y_2 = g_2^{q_0}(x^q - x)\,,$$

where g_2 has no pole other than P_1. Also, from (10.69), $m_2 = q_0 + q$ and so $\mathrm{ord}_{P_1} g_2 = -q$ by Lemma 10.100 (iii). Thus $g_2 = ax + b$ with $a, b \in \mathbf{F}_q$, $a \neq 0$. After a change over $\mathbf{F}_q$ of the coordinate system of PG$(4, K)$, the result follows. □

REMARK 10.103 The DLS curve $\mathcal{X}$ is also investigated in Section 12.2. From the proof of Theorem 10.102, some further properties of $\mathcal{X}$ related to its Frobenius linear series $\mathcal{D} = |(q + 2q_0 + 1)P_0|$, $P_0 \in \mathcal{X}(\mathbf{F}_q)$, can be deduced.

(i) Since $\mathcal{D} = g^4_{q+2q_0+1}$, the curve $\mathcal{X}$ has a complete, simple, base-point-free, $\mathbf{F}_q$-invariant linear series with orders $0, 1, q_0, 2q_0, q$ and with Frobenius orders

$0, q_0, 2q_0, q$. If $K(\mathcal{X})$ is given by $K(x,y)$ with $x^{2q_0}(x^q + x) = y^q + y$ as in Section 12.2, and $u(x,y) = x^{2q_0+1} + y^{2q_0}$, then $\mathcal{D}$ consists of all divisors

$$A_{\mathbf{c}} = \mathrm{div}(c_0 + c_1 x + c_2 y + c_3 u(x,y) + c_4(xu(x,y) + y^2)) + (q + 2q_0 + 1)P_0$$

with $\mathbf{c} = (c_0, c_1, c_2, c_3, c_4) \in \mathrm{PG}(4, K)$. Therefore the birational transformation $\pi : \mathcal{F} \mapsto \mathcal{X}$, where $\mathcal{F} = \mathbf{v}(X^{2q_0}(X^q+X) - (Y^q+Y))$, is given by the coordinate functions

$$x_0 = 1,\ x_1 = x,\ x_2 = y,\ x_3 = u(x,y),\ x_4 = xu(x,y) + y^2.$$

From this representation of $\mathcal{X}$ it can be verified that $\mathcal{X}$ is a non-singular curve of $\mathrm{PG}(4,K)$. To do this, note that the points of $\mathcal{X}$ coming from the points of $\mathcal{F}$ other than $Y_\infty = (0,0,1)$ are non-singular since $Y_\infty = (0,0,1)$ is the unique infinite point of $\mathcal{F}$ and all other points are non-singular; see Theorem 12.13. As Y_∞ is the centre of a unique branch of $\mathcal{F}$, the corresponding point $P = (0,0,0,0,1)$ of $\mathcal{X}$ is the only candidate to be a singular point of $\mathcal{X}$. But, if P were singular, then every automorphism of $\mathcal{X}$ would fix P and the automorphism group $\mathrm{Sz}(q)$ of $\mathcal{X}$ could not act transitively on the set of all $\mathbf{F}_q$-rational point of $\mathcal{X}$, contradicting Theorem 12.13 (iv).

(ii) There exists $P_1 \in \mathcal{X}(\mathbf{F}_q)$ whose $(\mathcal{D}, P_1)$-orders are

$$j_0(P_1) = 0,\ j_1(P_1) = 1,\ j_2(P_1) = q_0{+}1,\ j_3(P_1) = 2q_0{+}1,\ j_4(P_1) = q{+}2q_0{+}1.$$

They are also the $(\mathcal{D}, P)$-orders for each $P \in \mathcal{X}(\mathbf{F}_q)$. To see this, note that

$$\deg S = (3q_0 + q)(2g - 2) + (q+4)(q + 2q_0 + 1) = (4 + 2q_0)|\mathcal{X}(\mathbf{F}_q)|.$$

From (10.67),

$$v_P(S) = \textstyle\sum_{i=1}^{4} (j_i(P) - \nu_{i-1}) = 4 + 2q_0.$$

Now, using Corollary 8.58, it follows that $j_i(P) = j_i(P_1)$ for all i.

(iii) From (ii) and (B) it follows that $H(P)$ contains the semigroup

$$H = \langle q, q + q_0, q + 2q_0, q + 2q_0 + 1\rangle,$$

whenever $P \in \mathcal{X}(\mathbf{F}_q)$. Actually, $H(P) = H$ since the number of non-negative integers not in H is $g = q_0(q-1)$. This can be shown as in the proof of Lemma 10.99 (i).

(iv) From (i) and (ii),

$$\deg R = \textstyle\sum_{i=0}^{4} \epsilon_i(2g-2) + 5(q + 2q_0 + 1) = (2q_0 + 3)|\mathcal{X}(\mathbf{F}_q)|,$$

and $v_P(R) = 2q_0 + 3$ for $P \in \mathcal{X}(\mathbf{F}_q)$. Therefore the $\mathcal{D}$-Weierstrass points of $\mathcal{X}$ are precisely the $\mathbf{F}_q$-rational points of $\mathcal{X}$. In particular, the $(\mathcal{D}, P)$-orders for $P \notin \mathcal{X}(\mathbf{F}_q)$ are $0,\ 1,\ q_0,\ 2q_0,\ q$.

(v) Regarding the order sequence of the canonical curve of $\mathcal{X}$, from (iv) and Lemma 10.100 (i), it follows that every integer

$$m = a + q_0 b + 2q_0 c + qd \quad \text{with} \quad a + b + c + d \le 2q_0 - 2$$

is an order of the canonical curve at every non-$\mathbf{F}_q$-rational point.

(vi) From Proposition 9.83, the canonical curve of $\mathcal{X}$ is non-classical as $g \ge q$.

10.11 EXERCISES

1. Over $\mathbf{F}_7$, compare the Serre and Stöhr–Voloch Bounds for a plane quartic curve. Show that the curve $\mathcal{F} = \mathbf{v}(F)$ over $\mathbf{F}_7$, with
$$F = X^4 + Y^4 + Z^4 + 3(X^2Y^2 + X^2Z^2 + Y^2Z^2),$$
has 20 rational points, namely,
$$(\pm 1, \pm 3, 1),\ (\pm 3, \pm 1, 1),\ (\pm 3, \pm 2, 1),\ (\pm 2, \pm 3, 1),\ (\pm 2, \pm 2, 1).$$
Deduce that $\mathcal{F}$ is optimal; that is, $N_7(3) = 20$.

2. If $q = 3^t \geq 27$ and $\mathcal{X}$ is an $\mathbf{F}_{q^2}$-maximal curve over of genus $\frac{1}{6}q(q-3)$ with $\dim \mathcal{D} = 4$, show that $\mathcal{X}$ is birationally equivalent over $\mathbf{F}_{q^2}$ to the plane curve $\mathcal{T}_3'' = \mathbf{v}(T_3'')$, with
$$T_3'' = Y^{q/3} + Y^{q/9} + \cdots + Y^3 + Y + cX^{q+1},$$
where $c \in \mathbf{F}_q$ with $c^{q-1} = -1$.

3. Show that there is a maximal curve over $\mathbf{F}_{q^2}$ whose genus is Halphen's number $c_1(q+1, 4) = \lfloor \frac{1}{8}(q^2 - 2q + 5) \rfloor$.

4. Let $\mathcal{X}$ be a non-singular $\mathbf{F}_{q^2}$-rational plane curve and let L_1 be the linear series cut out on $\mathcal{X}$ by lines. Show that

   (a) L_1 is both non-classical and Frobenius non-classical if and only $\mathcal{X}$ is projectively equivalent over $\mathbf{F}_{q^2}$ to the Hermitian curve;

   (b) if $q = p$ or $q = p^2$, then L_1 is classical.

5. Let d be a positive integer such that $q \equiv -1 \pmod{d^2}$. Show that the curve $\mathcal{F} = \mathbf{v}(F)$, with
$$F = Y^{(q+1)/d} + X^{(d+1)(q+1)/d^2} + X^{(q+1)/d^2},$$
is an $\mathbf{F}_{q^2}$-maximal curve of genus $1 + (q+1)(q - d^2 - 1)/(2d^2)$.

6. Let d be a divisor of $q+1$. Use the Natural Embedding Theorem 10.22 to show that the Fermat curve $\mathcal{D}_m = \mathbf{v}(F)$, with $m = (q+1)/d$ and
$$F = X^m + Y^m + 1,$$
is $\mathbf{F}_{q^2}$-maximal. It has genus $g = \frac{1}{2}(m-1)(m-2)$ and the number of its $\mathbf{F}_{q^2}$-rational points is $q^2 + 1 + (m-1)(m-2)q$.

7. With d be a divisor of $q+1$ and $m = (q+1)/d$, show that the Artin–Schreier curve $\mathcal{A}_m = \mathbf{v}(F)$, with
$$F = Y^q + Y - X^m,$$
is $\mathbf{F}_{q^2}$-maximal. It has genus $g = \frac{1}{2}(m-1)(q-1)$, and the number of its $\mathbf{F}_{q^2}$-rational points is $mq^2 - (m-1)q + 1$.

8. Prove Theorem 10.47 without using the Natural Embedding Theorem 10.22.

9. Let $\mathcal{X}$ be a $\mathbf{F}_{q^2}$-maximal curve and P an ordinary point, that is, an ordinary point of $\mathcal{X}$. Show that, if $q+1 \in H(P)$, then $g = q - N_1 + 1$.

10. Let q be a prime power of and let $d > 2$, $i > 0$ be integers such that d^i divides $q+1$. Prove that there exists a sequence of curves
$$\mathcal{X}_0 = \mathcal{H} \to \mathcal{X}_1 \to \ldots \to \mathcal{X}_i,$$
such that each covering $\mathcal{X}_{j-1} \to \mathcal{X}_j$, $j = 1, \ldots, i$, is unramified, where $\mathcal{H}$ is the Hermitian curve.

11. For $q \equiv 1 \pmod 3$, let $\mathcal{G} = \mathbf{v}(G)$ with
$$G = Y^q - YX^{2(q-1)/3} + \omega X^{(q-1)/3},$$
where $\omega^{q+1} = -1$. Show that the genus of $\mathcal{G}$ is $g = \frac{1}{6}(q^2 - q)$.

12. Find all $\mathbf{F}_{64}$-maximal curves of genus 3 up to isomorphism.

13. Let $\mathcal{X}$ be an $\mathbf{F}_{q^2}$-maximal curve of genus g. Show that, if $\mathcal{X}$ is hyperelliptic, then $g \le \frac{1}{2}q$.

14. Show that no $\mathbf{F}_{64}$-maximal curve has genus 8.

15. Show that the curves
$$\mathcal{F}_1 = \mathbf{v}(X^4 + X + Y^3),$$
$$\mathcal{F}_2 = \mathbf{v}(X^4 + X^2 + X + Y^3)$$
both have 113 rational points over $\mathbf{F}_{64}$, and so are maximal but not isomorphic over $\mathbf{F}_{64}$.

16. Let $\mathcal{F}$ be an irreducible curve of genus g defined over $\mathbf{F}_2$. Show that, if $N_i = 2^i + 1$ for every $1 \le i \le g$, then $\mathcal{F}$ is a $\mathbf{F}_{2^{2g}}$-maximal curve.

10.12 NOTES

Surveys on maximal curves are found in papers by Garcia, Stichtenoth and van der Geer: [149], [141], [142], [472]. See also Fuhrmann [131]. Minimal curves have not received much attention so far, but fundamental results are treated in the survey paper by Viana and Rodriguez [488]; see also Lauter [299].

Theorem 10.2 was originally stated by Serre; see Lachaud [286]. An alternative proof of the $\mathbf{F}_{q^2}$-maximality of the curve in Example 10.3 is due to Ebin, Hor, Jatz and Shchogolev [113].

Maximal curves of Kummer type (10.2) over $\mathbf{F}_{q^2}$, with $f(X) \in \mathbf{F}_q[X]$ additive and separable, were studied by Garcia, Kawakita and Miura [145], Moisio [332], and Özbudak [354]. In particular, $\mathcal{K}_{q+1}$ is covered over $\mathbf{F}_{q^2}$ by the Hermitian curve; see [145].

Maximal Artin–Schreier curves over $\mathbf{F}_{p^{2m}}$ of type
$$\mathcal{F} = \mathbf{v}(Y^{p^n+1} - Y + cX^{p^a+1} + L(X)),$$

where $n \mid a$, $a \mid m$, and $L(X) = \sum_{i=0}^{(2m/n)-1} b_i X^{p^{ni}}$, were investigated by Coulter [95], Çakçak and Özbudak [62], van der Geer and van der Vlugt [474], and for $L = 0$ by Wolfmann [506].

Very recently, a family of $\mathbf{F}_{q^2}$-maximal curves not covered over $\mathbf{F}_{q^2}$ by the Hermitian curve $\mathcal{H}_q$ have been found; see Addendum. An $\mathbf{F}_{q^2}$-maximal curve other than the quotient curves of $\mathcal{H}_q$ for $q = 27$ is described in Example 10.4, and is due to Garcia and Stichtenoth. An unpublished result by Rains and Zieve is the existence of another example for $q = 27$; it was shown that, for $q_0 = 1$, the DLR curve $\mathcal{F} = \mathbf{v}(f(X,Y)$ with $f(X,Y)$ as in (12.34) is not a quotient of $\mathcal{H}_{27}$. The most likely candidate for a further example seems to be the DLS curve for $q = 8$.

Regarding Example 10.4, the case $\ell = 2$ and the first statement in Theorem 10.5 are due to Serre. For the extension to any ℓ and the proof of the second statement in Theorem 10.4, see Garcia and Stichtenoth [154]. The problem of whether the second statement holds true for any ℓ is still open.

In Table 10.2 the sources and characterisations are as follows:

1. $g = \frac{1}{2}q(q-1)$, [378], [223];
2. $g = \frac{1}{4}(q-1)^2$, [132];
3. $g = \frac{1}{4}q(q-2)$, [5];
4. $g = \frac{1}{6}(q^2-q+4)$, [92];

5, 6. $g = \frac{1}{6}(q^2-q)$, [158];

7. $g = \frac{1}{6}(q^2-q-2)$, [91], [158];
8. $g = \frac{1}{6}(q-1)(q-2)$, [282];
9. $g = \frac{1}{6}q(q-3)$, [91], [5];
10. $g = \frac{1}{8}(q^2-2q+5)$, [282];
11. $g = \frac{1}{8}(q-1)^2$, [282];
12. $g = \frac{1}{8}q(q-2)$, [282];

13, 14. $g = \frac{1}{8}(q-1)(q-3)$, [282];

15. $g = \frac{1}{8}q(q-4)$, [3].

The minimality of m in the preamble to Lemma 10.16 follows from a result due to Kaji [255, Prop. 1]; see also [163, Prop. 2],

General properties of quadrics in $\mathrm{PG}(3,q)$ can be found in [213, Chapter 16]. For detailed studies of Hermitian varieties, see [397], [213, Chapter 19], and [224, Chapter 23].

Section 10.2 is based on the survey paper [134]. The proof of Theorem 10.7 comes from [281]. A slight modification of the proof of Proposition 10.11 shows that each $P \in \mathrm{Supp}\, \mathcal{D} \backslash \mathcal{X}(\mathbf{F}_q)$ is a Weierstrass point of $\mathcal{X}$; see [159].

Related to Proposition 10.12, Garcia and Viana [160] showed that the Weierstrass points of the $\mathbf{F}_{q^2}$-maximal curves, $\mathcal{F} = \mathbf{v}(Y^q + Y - X^m)$ with $m \mid q+1$ and $m > 2$, are exactly the $\mathbf{F}_{q^2}$-rational points of $\mathcal{F}$.

Theorem 10.47 is due to Stichtenoth and Xing [509]. For a proof of this theorem independent of the the Stöhr–Voloch approach, see Fuhrmann [131].

For Remark 10.49, see Garcia, Stichtenoth and Xing [158, Remark 6.1].

The results in Section 10.3, in particular, the Natural Embedding Theorem 10.22 and Theorem 10.31, come from [281].

Sections 10.4, 10.5 and 10.6 follow [282].

The proof of Theorem 10.43 given in [132] is independent of the Natural Embedding Theorem 10.22 but it uses Galois theory. An alternative proof may be carried out by adapting the arguments in the proof of Theorem 10.43 and using results on Hermitian varieties and quadrics in permutable position.

For the lengthy calculation omitted from Theorem 10.45, see [5].

For Theorem 10.58, see [88].

Sections 10.7 and 10.8 are based on [13].

Theorem 10.102 and its proof come from [134].

For the examples of maximal curves in Remark 10.64, see the following:

(i) [478, Prop. 5.2(ii)], [158, Theorem 3.3];

(ii) [92, Proposition 3.3(3)];

(iii) [92, Proposition 3.3(3)(1)], [158, Example 5.10].

For the quintic in Remark 10.68 (iii), see [406, Section 4].

For Proposition 10.69, see [13].

For the result on arithmetic progressions in Lemma 10.77, see [180].

For Remark 10.81, see [65, Lemmes 3.3, 3.6] and [66].

For generalised Hurwitz curves, see [47, Section 4], [66], [45, Example 4.5], [13].

The Weierstrass semigroup $H(\mathcal{P}_1)$ in Lemma 10.91 is calculated for $k = n - 1$, and $(n, k) = (5, 2)$ in [47].

For Remark 10.94, see [47, Remarque 4].

The proofs of Theorems 10.95 and 10.96 are found in [171]. For a generalisation of these theorems to m for which $d = (q + 1)/m$ is not prime, see [172].

Section 10.10 follows [134].

Theorem 10.2 implies that the quotient curves of an $\mathbf{F}_{q^2}$-maximal curve are also $\mathbf{F}_{q^2}$-maximal. This makes it possible to obtain several infinite families of $\mathbf{F}_{q^2}$-maximal curves, especially from the Hermitian curve; see Section 12.3, and [4], [5], [3], [6], [91], [92], [158], [151], [171], [172]. Others can be obtained from the DLS and DLR curves; see Sections 12.2, 12.4, as well as [174] and [60]. However, Theorem 10.2 fails for optimal curves in general. But the curves attaining the Serre Bound 9.27 for a non-square q are optimal curves of this type, such as the DLS and DLR curves over $\mathbf{F}_q$. Sporadic examples are the Klein quartic over $\mathbf{F}_8$, as in Remark 9.29, and the curve $\mathbf{v}(Y^5 + Y^4 - X^{23})$ over $\mathbf{F}_{2^{11}}$; see Section 12.5. See also Remark 9.29.

The value of $N_q(g)$ was found by Waterhouse for $g = 1$ and by Serre for $g = 2$ and for $g = 3$ when $q \leq 25$. Also, $N_q(g)$ is known for a number of sporadic pairs (g, q); see the survey in [242, Chapter 2]. See also van der Geer and van der Vlugt [481], where tables of known values of $N_q(g)$ are maintained. See also Howe and Lauter [237].

For Remark 10.68, see [401].

For an explicit description of the action of $\mathrm{Sz}(q)$ on the DLS curve $\mathcal{X}$, see [174]. Also, $\mathcal{X}(\mathbf{F}_q)$ is an ovoid of the non-singular quadric $\mathbf{v}(X_0X_4 - X_1X_3 - X_2^2)$ of $\mathrm{PG}(4, q)$; see [134].

For other results on the orders of the canonical curve of $\mathcal{X}$, see [147, Section 4]. The non-classicality of the canonical curve is also shown in [147].

For the curve in Exercise 1, see [456]. Exercise 16 is related to *quasi-Hermitian* curves $\mathcal{F} = \mathbf{v}(Y^n + bY + cX^m)$; see [327] and [371].

ADDENDUM

Let $q = n^3$, with n a power of p, and $K = \overline{\mathbf{F}}_{q^2}$. In $\mathrm{PG}(3, K)$, let (X_0, X_1, X_2, X_3) be homogeneous coordinates, and define

$$h(X_0, X_3) = \sum_{i=0}^{n} (-1)^{i+1} X_0^{i(n-1)} X_3^{n^2-n-i(n-1)},$$

viewed as a polynomial over K. Then the complete intersection of the surface

$$\mathbf{v}(X_2^{n^2-n+1} - X_1 h(X_0, X_3)) \tag{10.72}$$

and the Hermitian cone

$$\mathbf{v}(X_0^n X_3 + X_0 X_3^n - X_1^{n+1}) \tag{10.73}$$

is an $\mathbf{F}_{q^2}$-maximal curve $\mathcal{X}$ with genus

$$g = \tfrac{1}{2}(n^3 + 1)(n^2 - 2) + 1.$$

This curve $\mathcal{X}$ is not covered over $\mathbf{F}_{q^2}$ either by the Hermitian curve or by the DLS curve or by the DLR curve; see the preliminary report:

M. Giulietti and G. Korchmáros, A new family of $\mathbf{F}_{q^2}$-maximal curves, available on line http://arxiv.org/archive/math : 07110445v1[math.A.G.], 2007.

Chapter Eleven

Automorphisms of an algebraic curve

The concept of a symmetry, or automorphism in modern terminology, plays a prominent role in all branches of geometry. The saying, *the larger its automorphism group the richer its geometry*, is especially appropriate for algebraic curves. For instance, if there exists an automorphism of an irreducible curve Γ, then another irreducible curve, the quotient curve, can be constructed, and it inherits important properties of Γ. Also, certain families of curves have characterisations obtained from the structure and action of their automorphism groups.

As in several concepts introduced previously, an automorphism of an algebraic curve is a birational invariant. In the function field $K(\Gamma)$, the corresponding term is a *K-automorphism*, that is, a field automorphism of $K(\Gamma)$ which fixes every element of K. So K-automorphisms are mostly investigated by using algebraic rather than geometric tools. However, since Γ is birationally equivalent to an irreducible non-singular curve $\mathcal{X}$ embedded in a projective space $\mathrm{PG}(r, K)$ such that every K-automorphism of $K(\Gamma)$ can be represented as a linear collineation of $\mathrm{PG}(r, K)$ preserving $\mathcal{X}$, so a K-automorphism can be viewed as a truly geometric concept. In this spirit, $\mathrm{Aut}_K(\Sigma)$ is the K-automorphism group of $\mathcal{X}$. More generally, the term, *K-automorphism group of $\mathcal{F}$*, is adopted for any curve $\mathcal{F}$ birationally equivalent to Γ, that is, for every model of $K(\Gamma)$.

Since the K-automorphisms of Γ form a group, their study greatly benefits from group theory. A fundamental result is that the K-automorphism groups of an irreducible curve is finite, apart from rational and elliptic curves. So finite group theory is relevant in this context, which is a further great advantage as it is in the finite case that group theory has been most developed. So algebraic geometry and group theory combine to make it possible to find an upper bound for the order of $\mathrm{Aut}_K(\Sigma)$ depending only on the genus g of Σ. The classical bound is $84(g-1)$. In positive characteristic, the result is achieved in Section 11.12 by showing that $|\mathrm{Aut}_K(\Sigma)| < 8g^3$, apart from four exceptions.

As throughout the book, Σ denotes a function field of transcendence degree 1, that is, the function field of an irreducible curve $\mathcal{F}$. Further, $\mathrm{Aut}_K(\Sigma)$ stands for the group of all K-automorphisms of Σ. For the special case that $\mathcal{F}$ is defined over $\mathbf{F}_q$, the concept of $\mathbf{F}_q$-rationality for K-automorphisms is introduced.

Let $\mathbf{F}_q(\mathcal{F})$ be the $\mathbf{F}_q$-rational function field of $\mathcal{F}$. A K-automorphism ω of $\mathcal{F}$ is *$\mathbf{F}_q$-rational* if ω preserves $\mathbf{F}_q(\mathcal{F})$; that is, the birational transformation

$$x' = \omega(x),\ y' = \omega(y)$$

of Σ is $\mathbf{F}_q$-rational. Such elements form a subgroup, the *$\mathbf{F}_q$-rational K-automorphism group of $\mathcal{F}$*, of $\mathrm{Aut}_K(\Sigma)$.

11.1 THE ACTION OF K-AUTOMORPHISMS ON PLACES

The first result is a characterisation of K-automorphisms in terms of birational transformations.

LEMMA 11.1 *Let α be a K-automorphism of Σ. If $(\mathcal{F};(x,y))$ is any model of Σ, then the birational transformation ω given by the equations*

$$x' = \alpha(x), \quad y' = \alpha(y)$$

preserves the curve $\mathcal{F}$.

Proof. Since $\alpha \in \mathrm{Aut}_K(\Sigma)$, from $f(x,y) = 0$ and $\alpha(f(x,y)) = f(x',y')$ it follows that $f(x',y') = 0$. Hence ω preserves $\mathcal{F}$. □

LEMMA 11.2 *Let $(\mathcal{F};(x,y))$ be a model of Σ. If a birational transformation ω of Σ, defined by*

$$(x,y) \mapsto (x',y'),$$

preserves the curve $\mathcal{F}$, then there is a K-automorphism α of Σ such that $x' = \alpha(x)$ and $y' = \alpha(y)$.

Proof. Since ω preserves $\mathcal{F}$, the point $P' = (x',y')$ is another generic point of $\mathcal{F}$. Now let $x' = \alpha(x)$, $y' = \alpha(y)$, and extend this definition to any element $\xi \in \Sigma$ in a natural way; namely,

$$\alpha(\xi) = \frac{u(x',y')}{v(x',y')} \quad \text{when} \quad \xi = \frac{u(x,y)}{v(x,y)},$$

with $u(X,Y), v(X,Y) \in K[X,Y]$ and $f(X,Y) \nmid v(X,Y)$. Note that the expression of ξ as a rational function of x, y is not unique; however, $\alpha(\xi)$ does not depend on the form of the expression. To verify this, let

$$\frac{u(x,y)}{v(x,y)} = \frac{u_1(x,y)}{v_1(x,y)}.$$

Then

$$u(X,Y)v_1(X,Y) - u_1(X,Y)v(X,Y)$$

is a polynomial that vanishes at the generic point $P = (x,y)$ of $\mathcal{F}$. By the definition of generic point, this polynomial is divisible by $f(X,Y)$. Since $f(x',y') = 0$, this yields the result:

$$\frac{u(x',y')}{v(x',y')} = \frac{u_1(x',y')}{v_1(x',y')}.$$

So α is correctly defined on Σ, and it is indeed a K-automorphism of Σ. □

It is intuitively apparent that K-automorphisms of Σ are an important tool in investigating places, divisors, and linear series by means of the action of $\mathrm{Aut}_K(\Sigma)$ on such objects. In this context, the fixed field Σ^G consisting of all elements in Σ which are fixed by every element in G plays an important role.

A geometric view of this is also possible, since there is a one-to-one correspondence between places of Σ and points of any non-singular model $\mathcal{X}$ of Σ, which gives rise to a natural action of $\mathrm{Aut}_K(\Sigma)$ on the points of $\mathcal{X}$. In some circumstances, such an action is *linear* in the sense that it arises from a linear collineation group of the projective space in which $\mathcal{X}$ is embedded.

The first step is to define the action of K-automorphisms on places.

Let $\alpha \in \mathrm{Aut}_K(\Sigma)$ and let σ be a primitive representation of a place $\mathcal{P}$ of Σ. Then the image $\sigma\alpha$ of σ under α is defined by the equation

$$(\sigma\alpha)(\xi) = \sigma(\alpha(\xi)), \quad \xi \in \Sigma.$$

As $\sigma\alpha$ is a primitive place representation, it defines a place of Σ: this place is independent of the choice of σ, since equivalent primitive place representations have equivalent images. Such a place is called the *image* of $\mathcal{P}$ by α, and denoted by $\mathcal{P}^\alpha$.

Once the action of α on places has been defined, this extends to $\mathrm{Div}(\Sigma)$ in a natural way. Let $D \in \mathrm{Div}(\Sigma)$ with $D = \sum n_\mathcal{P}\mathcal{P}$; then

$$D^\alpha = \textstyle\sum n_\mathcal{P}\, \mathcal{P}^\alpha.$$

In particular, α preserves the set of all effective divisors of Σ. For any α-invariant subfield $\Sigma' = K(\xi, \eta)$ of Σ, it is appropriate to keep the notation α to denote the induced K-automorphism of Σ'. Then α has an action, which may be trivial, on the set of all places $\mathcal{P}_{\Sigma'}$ of Σ'. It follows that, if $\mathcal{P} \in \mathcal{P}_\Sigma$ and $\mathcal{P}' \in \mathcal{P}_{\Sigma'}$ are places such that $\mathcal{P}$ lies over $\mathcal{P}'$, then $\mathcal{P}^\alpha$ lies over $\mathcal{P}'^\alpha$.

First, some elementary properties of K-automorphisms of Σ are collected.

Lemma 11.3 *Let $\alpha \in \mathrm{Aut}_K(\Sigma)$ and $\mathcal{P} \in \mathcal{P}_\Sigma$. Then, for any non-zero $z \in \Sigma$,*

$$\mathrm{ord}_\mathcal{P}\, z = \mathrm{ord}_{\mathcal{P}^\alpha}(\alpha^{-1}(z)). \tag{11.1}$$

Proof. Let σ be a primitive representation of $\mathcal{P}$. Then $\mathrm{ord}_\mathcal{P}\, z = \mathrm{ord}_t\, \sigma(z)$ for $z \in \Sigma$. Also, $\mathrm{ord}_t\, \sigma(\alpha(\alpha^{-1}(z))) = \mathrm{ord}_t(\sigma\alpha)\alpha^{-1}(z) = \mathrm{ord}_{\mathcal{P}^\alpha}(\alpha^{-1}(z))$. This proves the assertion. □

Note that (11.1) can also be written in the following way:

$$\mathrm{ord}_\mathcal{P}(\alpha(z)) = \mathrm{ord}_{\mathcal{P}^\alpha}(z). \tag{11.2}$$

Lemma 11.3 has some useful corollaries.

Lemma 11.4 *If $\alpha \in \mathrm{Aut}_K(\Sigma)$ and $z \in \Sigma\backslash\{0\}$, then $\mathrm{div}(z)^\alpha = \mathrm{div}(\alpha^{-1}(z))$; more particularly,*

$$(\mathrm{div}(z)_0)^\alpha = \mathrm{div}(\alpha^{-1}(z))_0, \quad (\mathrm{div}(z)_\infty)^\alpha = \mathrm{div}(\alpha^{-1}(z))_\infty. \tag{11.3}$$

Proof. First, $\mathrm{div}(z)_0 = \sum_\mathcal{P} \mathrm{ord}_\mathcal{P}(z)\mathcal{P}$, where the summation is over all places $\mathcal{P} \in \mathcal{P}_\Sigma$ with $\mathrm{ord}_\mathcal{P}(z) > 0$. Then $(\mathrm{div}(z)_0)^\alpha = \sum_\mathcal{P} \mathrm{ord}_\mathcal{P}(z)\mathcal{P}^\alpha$. By (11.1),

$$\textstyle\sum_\mathcal{P} \mathrm{ord}_\mathcal{P}(z)\mathcal{P}^\alpha = \sum_\mathcal{P} \mathrm{ord}_{\mathcal{P}^\alpha}(\alpha^{-1}(z))\mathcal{P}^\alpha.$$

By (11.1), $\mathrm{ord}_\mathcal{P}(z) > 0$ if and only if $\mathrm{ord}_{\mathcal{P}^\alpha}(\alpha^{-1}(z)) > 0$. Hence

$$\textstyle\sum_\mathcal{P} \mathrm{ord}_{\mathcal{P}^\alpha}(\alpha^{-1}(z))\mathcal{P}^\alpha = \mathrm{div}(\alpha^{-1}(z)),$$

which shows the second assertion. The equation for the divisor of poles is proved similarly. The first assertion is consequence of the second and third assertions. □

LEMMA 11.5 *If $A \equiv B$ with $A, B \in \mathrm{Div}(\Sigma)$, then $A^\alpha \equiv B^\alpha$ for $\alpha \in \mathrm{Aut}_K(\Sigma)$.*

Proof. $A - B = \mathrm{div}(z)$ for some $z \in \Sigma$. By (11.1), $(A - B)^\alpha = \mathrm{div}(\alpha^{-1}(z))$. Since $(A - B)^\alpha = A^\alpha - B^\alpha$, the result follows. □

A similar argument may be used to prove the following two results.

LEMMA 11.6 *Every K-automorphism of Σ preserves the set of all Weierstrass points of Σ.*

LEMMA 11.7 *Let F be a subfield of Σ of transcendence degree 1 such that the field extension Σ/F is separable. If $\alpha \in \mathrm{Aut}_K(\Sigma)$ and $F' = \alpha(F)$, then*

$$D(\Sigma/F)^\alpha = D(\Sigma/F'),$$

where $D(\Sigma/F)$ and $D(\Sigma/F')$ are the different divisors of Σ/F and Σ/F'.

LEMMA 11.8 *If $\alpha \in \mathrm{Aut}_K(\Sigma)$ and $\mathrm{div}(z)^\alpha = \mathrm{div}(z)$ for every non-zero element $z \in \Sigma$, then α is the identity automorphism of Σ.*

Proof. As α is a K-automorphism, it suffices to prove that $\alpha(z) = z$ for any z in $\Sigma \backslash K$. Since α^{-1} also satisfies the hypothesis, so, from Lemma 11.4,

$$\mathrm{div}(z) = \mathrm{div}(\alpha(z)),$$

whence $\mathrm{div}(z/\alpha(z)) = 0$. By Theorem 6.4, $\alpha(z) = cz$ for a non-zero constant $c \in K$. Replacing z by $z + 1$, there is $c' \in K$ such that $\alpha(z+1) = c'(z+1)$. Thus $\alpha(z) + \alpha(1) = c'z + c'$. These equations show that $cz + 1 = c'z + c'$, whence $(c - c')z = c' - 1$. Since $z \in \Sigma \backslash K$, this is only possible when $c - c' = 0$ and $c' = 1$. Hence $c = 1$, and the result follows. □

LEMMA 11.9 *If $\alpha \in \mathrm{Aut}_K(\Sigma)$ and $\mathcal{P}^\alpha = \mathcal{P}$ for every place of Σ, then α is the identity automorphism of Σ.*

Proof. From the hypothesis, $D^\alpha = D$ for every $D \in \mathrm{Div}(\Sigma)$. In particular, the hypothesis in the preceding lemma is satisfied. Hence the assertion is just a corollary of that lemma. □

In fact, Lemma 11.9 still holds if it is only required that α fixes infinitely many places of $\mathcal{P}_\Sigma$. To show this, the residue map $\mathrm{res}_\mathcal{P}$ associated to any place $\mathcal{P} \in \mathcal{P}_\Sigma$, as in Example 7.130, is needed.

LEMMA 11.10 *If $\alpha \in \mathrm{Aut}_K(\Sigma)$ and $\mathcal{P} \in \mathcal{P}_\Sigma$, then $\mathrm{res}_\mathcal{P}\, z = \mathrm{res}_{\mathcal{P}^\alpha}\, \alpha^{-1}(z)$ for any $z \in \Sigma$.*

Proof. Let σ be a primitive representation of $\mathcal{P}$. If $z \in \Sigma$, then

$$\sigma(z) = \sigma(\alpha\alpha^{-1}(z)) = (\sigma\alpha)(\alpha^{-1}(z)),$$

and hence

$$\mathrm{res}_\mathcal{P}\, z = \mathrm{res}(\sigma(z)) = \mathrm{res}(\sigma\alpha(\alpha^{-1}(z)) = \mathrm{res}_{\mathcal{P}^\alpha}\, \alpha^{-1}(z).$$ □

THEOREM 11.11 *If $\alpha \in \mathrm{Aut}_K(\Sigma)$ and $\mathcal{P}^\alpha = \mathcal{P}$ for infinitely many places $\mathcal{P}$ of Σ, then α is the identity automorphism of Σ.*

Proof. Choose an α-invariant place $\mathcal{P}$ of Σ. Lemma 11.10 applied to α^{-1} shows that $\mathrm{res}_{\mathcal{P}}\, \alpha(z) = \mathrm{res}_{\mathcal{P}}\, z$ holds for all $z \in \Sigma$. If neither z nor $\alpha(z)$ has a pole at $\mathcal{P}$, the additive property of $\mathrm{res}_{\mathcal{P}}$ yields that

$$\mathrm{res}_{\mathcal{P}}\, \alpha(z) - \mathrm{res}_{\mathcal{P}}\, z = \mathrm{res}_{\mathcal{P}}\, (\alpha(z) - z),$$

giving $\mathrm{res}_{\mathcal{P}}(\alpha(z) - z) = 0$. Now assume that α has infinitely many fixed places, and remove those places that are poles for either z or $\alpha(z)$. The remaining fixed places of α are still infinite, but none of them is a pole of any of the elements z, $\alpha(z)$ and $\alpha(z) - z$. For each such place $\mathcal{P}$, the equation $\mathrm{res}_{\mathcal{P}}(\alpha(z) - z) = 0$ implies that $\mathcal{P}$ is a zero of $\alpha(z) - z$. By Theorem 5.33, there exists $c \in K$ such that $\alpha(z) - z = c$. Since α is a K-automorphism, so $\alpha(z) - z = 0$. □

The number of fixed places that a K-automorphism can have is an important topic to which the whole Section 11.11 is devoted. Meanwhile, the following result is required.

LEMMA 11.12 *Every non-trivial K-automorphism of Σ has at most $2g + 2$ fixed places.*

Proof. Let α be a non-trivial K-automorphism of Σ. By Theorem 11.11, α has a finite number of fixed places. Take $g + 1$ distinct places $\mathcal{P}_1, \ldots, \mathcal{P}_{g+1}$ of Σ such that the linear series $|D|$ with $D = \mathcal{P}_1 + \cdots + \mathcal{P}_{g+1}$ has no fixed point and that D and its image D^α share no place. By Theorem 6.82, Σ contains an element $z \notin K$ such that $\mathrm{div}(z)_\infty = D$. Now consider $w = z - \alpha(z)$. Since z and $\alpha(z)$ have different poles, $w \neq 0$; more precisely, w has exactly $2g + 2$ poles, because each pole of w is either a pole of $\alpha(z)$ or a pole of z. By Theorem 5.35, α has exactly $2g + 2$ zeros. On the other hand, every fixed place of α is a zero of w. Hence the number of such places is at most $2g + 2$. □

LEMMA 11.13 *Let m be the smallest non-gap of Σ at a place $\mathcal{P}$, and suppose that $\xi \in \Sigma$ has a pole of order m at $\mathcal{P}$ and is regular elsewhere. If $\mathcal{P}^\alpha = \mathcal{P}$ for an element $\alpha \in \mathrm{Aut}_K(\Sigma)$, then $\alpha(\xi) = c\xi + d$ with $c, d \in K$, $c \neq 0$.*

Proof. Choose a primitive representation σ of $\mathcal{P}$. Then

$$\sigma(\xi) = c_1 t^{-m} + c_2 t^{-m+1} + \cdots,$$

with $c_1 \neq 0$. By Lemma 11.3,

$$\sigma(\alpha(\xi)) = d_1 t^{-m} + d_2 t^{-m+1} + \cdots,$$

with $d_1 \neq 0$. Let $\eta = d_1\xi - c_1\alpha(\xi)$. Then

$$\sigma(\eta) = e_2 t^k + \cdots, \text{ with } k \geq -m + 1.$$

The minimality of m implies that $k \geq 0$. This shows that $\mathcal{P}$ is not a pole of η. Since ξ, and by Lemma 11.3 also $\alpha(\xi)$, is regular outside $\mathcal{P}$, the same holds for η. Hence η has no poles at all. By Theorem 5.33, $\eta \in K$, whence the result follows. □

THEOREM 11.14 *Let $\Sigma = K(\xi)$ be rational.*

(i) *The group* $\mathrm{Aut}_K(\Sigma) \cong \mathrm{PGL}(2,K)$, *and* $\mathrm{Aut}_K(\Sigma)$ *acts on the set* $\mathcal{P}_\Sigma$ *of all places of* Σ *as* $\mathrm{PGL}(2,K)$ *naturally on* $\mathrm{PG}(1,K)$.

(ii) (a) $\alpha \in \mathrm{Aut}_K(\Sigma)$ *if and only if* $\alpha(\xi)$ *is bilinear; that is,*

$$\alpha(\xi) : \xi \mapsto \frac{a\xi + b}{c\xi + d}, \quad ad - bc \neq 0,\ a, b, c, d \in K;$$

(b) *the identity of* $\mathrm{Aut}_K(\Sigma)$ *is the only K-automorphism in* $\mathrm{Aut}_K(\Sigma)$ *fixing at least three places of* $\mathcal{X}$;

(c) *every K-automorphism of* $\mathrm{Aut}_K(\Sigma)$ *fixes a place* $\mathcal{P} \in \mathcal{P}_\Sigma$;

(d) *if* $p \nmid \mathrm{ord}\,\alpha$, *then α has exactly two fixed places*;

(e) *if* $p \mid \mathrm{ord}\,\alpha$, *then*

(1) $\mathrm{ord}\,\alpha = p$;

(2) *α has exactly one fixed place*;

(3) *when the fixed place is a pole of ξ, the map is $\alpha(\xi) = \xi + b$ with b in $K\backslash\{0\}$.*

Proof. The projective line $\ell = \mathbf{v}(Y)$ is a non-singular model of $\Sigma = K(x)$ where $P = (x, 0)$ is a generic point of ℓ. For every $a, b, c, d \in K$ such that $ad - bc \neq 0$, the rational transformation $\omega : x' = (ax + b)/cx + d)$ of $K(x)$ is birational, and preserves ℓ. By Lemma 11.2, ω defines a K-automorphism α of $K(x)$ such that

$$\alpha : \xi \mapsto \frac{a\xi + b}{c\xi + d}, \quad ad - bc \neq 0, \quad a, b, c, d \in K. \tag{11.4}$$

These constitute a subgroup L of $\mathrm{Aut}_K(\Sigma)$ isomorphic to the projective linear group $\mathrm{PGL}(2,K)$. Also, L acts on the points of ℓ, that is on the places of $K(x)$, as $\mathrm{PGL}(2,K)$ acts on $\mathrm{PG}(1,K)$ in its natural representation.

Now let β be any K-automorphism in $\mathrm{Aut}_K(\Sigma)$. Let $\mathcal{P}_\infty$ denote the unique place centred at the infinite point of ℓ. Since $\mathrm{PGL}(2,K)$ is transitive, there exists $\delta \in L$ such that the K-automorphism $\beta\delta$ of Σ fixes $\mathcal{P}_\infty$. Since $\Sigma = K(\xi)$ and since $\mathrm{div}(\xi)_\infty = \mathcal{P}_\infty$, so 1 is the smallest non-gap at $\mathcal{P}_\infty$. Thus Lemma 11.13 applies, showing that $\beta\delta \in L$. But this shows that β itself belongs to L. This completes the proof of (ii)(a) and (ii)(b).

To show (ii)(c), let α be given as (11.4). If either $c \neq 0$, or $c = 0$ but $a \neq d$, then the equation $aX + b = X(cX + d)$ has a solution u in K, and the unique place $\mathcal{P}_u$ centred at u is fixed by α. Otherwise, $\alpha(\xi) = \xi + b$ and hence α fixes $\mathcal{P}_\infty$. Conversely, if α fixes $\mathcal{P}$ and its order is divisible by p, then $\alpha(\xi) = \xi + b$ with $b \in K\backslash\{0\}$. In particular, $\mathrm{ord}\,\alpha = p$, which proves (ii)(e).

To show (ii)(d), a little more is needed. Since $\mathrm{PGL}(2,K)$ is transitive on ℓ, so assume that $\mathcal{P}_\infty$ is fixed by α. Then $\alpha(\xi) = a\xi + b$ with $a, b \in K$ and $a \neq 0$. Since $p \nmid \mathrm{ord}\,\alpha$, so $a \neq 1$. Therefore α fixes not only P_∞ but also the place centred at the point $P = (u, 0)$ with $u = -b/(a - 1)$. □

REMARK 11.15 All finite subgroups of the group $\mathrm{PGL}(2,K)$ are known; see Theorem 11.91.

Let $\mathcal{P}$ be any place of Σ. Choose a primitive place representation σ of $\mathcal{P}$. Assume that $\alpha \in \mathrm{Aut}_K(\Sigma)$ fixes $\mathcal{P}$. Then $\sigma\alpha$ is a primitive place representation σ' of $\mathcal{P}$, and $\lambda\sigma = \sigma'$ for a K-automorphism λ of $K[[t]]$. Here, λ depends on α, and this is stressed by writing λ_α and saying that λ_α is the *companion K-automorphism of α in* $\mathrm{Aut}(K[[t]])$. For any K-automorphism group G of Σ which fixes $\mathcal{P}$, this leads to the introduction of the following map:

$$\varphi : G \to \mathrm{Aut}(K[[t]]), \quad \alpha \mapsto \lambda_\alpha. \tag{11.5}$$

Lemma 11.16 *The mapping φ is an epimorphism.*

Proof. Define the image $\Lambda_G = \{\lambda_\alpha \mid \alpha \in G\}$ of φ to be the *companion group of* G. Let $\alpha, \beta \in G$. Then

$$\lambda_{\beta\alpha}\sigma = \sigma(\beta\alpha) = (\sigma\beta)\alpha = \lambda_\alpha(\sigma\beta) = \lambda_\alpha(\lambda_\beta\sigma) = (\lambda_\alpha\lambda_\beta)\sigma,$$

whence $\lambda_{\beta\alpha} = \lambda_\alpha\lambda_\beta$. Let $\alpha \in \ker\varphi$. Then $\sigma = \sigma\alpha$. As σ is an isomorphism, α is the the identity in $\mathrm{Aut}_K(\Sigma)$. This completes the proof. □

Lemma 11.17 *Let α be a finite K-automorphism of Σ that fixes a place $\mathcal{P}$. Then, α has order $p^r s$ with $r \geq 0$ and $p \nmid s$ if and only if $\lambda_\alpha(t) = ut + \cdots$, where s is the smallest integer satisfying $u^s = 1$.*

Proof. Let $G = \langle\alpha\rangle$. The elements $\beta \in G$ with $\lambda_\beta(t) = t + \cdots$ form a subgroup H of G. To show that H is a p-group, take a prime divisor r of $|H|$, and an element $\beta \in H$ of order r. If $\lambda_\beta(t) = t + vt^i + \cdots$, then $\lambda_{\beta^r}(t) = t + rvt^i + \cdots$, which gives $r = p$.

On the other hand, every element $\beta \in G$ whose order is a power of p is in H. This follows from the fact that, if $\lambda_\beta(t) = ut + \cdots$, then $\lambda_{\beta^n}(t) = u^n t + \cdots$, while $u^{p^m} = 1$ implies that $u = 1$. Since G is cyclic, $G = H \times N$ with N a cyclic group whose order s is prime to p. If $N = \langle\delta\rangle$, then $\lambda_\delta(t) = ut + \cdots$ and s is the smallest positive integer such that $u^s = 1$. From this, the assertion follows. □

It may be noted that Λ_G depends on the choice of σ, even though only formally. In fact, if σ is replaced by another primitive place representation $\bar{\sigma}$ of $\mathcal{P}$, then the resulting groups Λ_G and $\bar{\Lambda}_G$ are conjugate in $\mathrm{Aut}(K[[t]])$. To show this, write $\tau\sigma = \bar{\sigma}$ with $\tau \in \mathrm{Aut}(K[[t]])$. Then $\tau\sigma\alpha = \bar{\sigma}\alpha$; that is, $\tau\lambda_\alpha\sigma = \bar{\lambda}_\alpha\bar{\sigma}$, which implies that $\tau\lambda_\alpha\sigma = \bar{\lambda}_\alpha\tau\sigma$. Hence $\lambda_\alpha\sigma = (\tau^{-1}\bar{\lambda}_\alpha\tau)\sigma$. Since $\sigma \in \mathrm{Aut}(K[[t]])$ and α is independent of τ, the result follows.

11.2 LINEAR SERIES AND AUTOMORPHISMS

The following theorem shows how K-automorphisms of Σ act on linear series.

Theorem 11.18 *Let $\alpha \in \mathrm{Aut}_K(\Sigma)$ and let*

$$L = \{\mathrm{div}(\textstyle\sum c_i x_i) + B \mid \mathbf{c} = (c_0, \ldots, c_r) \in \mathrm{PG}(r, K)\}, \quad B \in \mathrm{Div}(\Sigma), \tag{11.6}$$

be a linear series of Σ. Then the divisors D^α with D ranging over L are the divisors of a linear series L^α of Σ :

$$L^\alpha = \{\mathrm{div}(\textstyle\sum c_i \alpha(x_i)) + B^\alpha \mid \mathbf{c} = (c_0, \ldots, c_r) \in \mathrm{PG}(r, K)\}.$$

Note that $\deg L = \deg L^\alpha$ and $\dim L = \dim L^\alpha$. Also, if both $L^\alpha = L$ and $B^\alpha = B$, then α can be viewed as an element of $\mathrm{End}_K(E)$, where E is the Riemann–Roch space, that is, the $(r+1)$-dimensional vector space over K with basis $\{x_0, \ldots, x_r\}$. By Theorem 6.72 (i), the canonical series of Σ is the only effective linear series of Σ of order $2g-2$ and dimension $g-1$. This gives the following result.

Lemma 11.19 *Every K-automorphism of Σ preserves the canonical series of Σ.*

Similarly, the following result holds.

Lemma 11.20 *If a K-automorphism of Σ fixes a place $\mathcal{P}$, then α preserves the complete linear series $|k\mathcal{P}|$ for every non-negative integer k.*

If Σ is the function field of a hyperelliptic curve, then every K-automorphism of Σ preserves g_2^1, the unique complete linear series of dimension 1 and order 2 of Σ. By Theorems 7.94 and 7.100, the set consisting of all places $\mathcal{P}$ of Σ such that $2\mathcal{P} \in g_2^1$ is finite. This has the following consequence.

Lemma 11.21 *For a hyperelliptic curve $\mathcal{F}$, the K-automorphism group of $\mathcal{F}$ preserves a finite set of places.*

From a geometric point of view, it is important to have a linear representation of α in the projective space in which a suitable irreducible model of Σ is embedded. This can be done by using the following result.

Lemma 11.22 *For a function field Σ, let*

(a) *(11.6) be a simple, base-point-free linear series of Σ;*

(b) *$\mathcal{Y}$ be the irreducible curve of $\mathrm{PG}(r, K)$ arising from L;*

(c) *$C(\mathcal{P}) \in \mathcal{Y}$ denote the centre of the corresponding branch of $\mathcal{Y}$ for any place $\mathcal{P} \in \mathbb{P}_\Sigma$;*

(d) *$\alpha \in \mathrm{Aut}_K(\Sigma)$ preserve L.*

Then there is a projectivity $\mathfrak{T}$ of $\mathrm{PG}(r, K)$ such that

(i) *$C(\mathcal{P}^\alpha) = \mathfrak{T}(C(\mathcal{P}))$;*

(ii) *$\mathfrak{T}$ preserves $\mathcal{Y}$;*

(iii) *if $\mathcal{Y}$ is non-singular, $\mathfrak{T}$ acts on the points of $\mathcal{Y}$ as α on $\mathbb{P}_\Sigma$.*

Proof. The series L may be taken to be normalised. Then $B \in L$ by Theorem 6.22. Hence there is a non-zero element $z = \sum c_i x_i$ such that $B^\alpha = \mathrm{div}(z) + B$. From this, for $i = 0, \ldots, r$,

$$\alpha(x_i)z = \sum a_{ij}x_j,$$

with $a_{ij} \in K$, and $\det(a_{ij}) \neq 0$. Let $\mathcal{P}$ be a place of Σ. If σ is a primitive representation of $\mathcal{P}$, it follows that

$$(\sigma\alpha)(x_i)\sigma(z) = \sum a_{ij}\sigma(x_j). \tag{11.7}$$

Let $P = (u_0, \ldots, u_r)$ and $P' = (u'_0, \ldots, u'_r)$ denote the centres of the branches arising from $\mathcal{P}$ and $\mathcal{P}^\alpha$, let ord $\sigma(x_i) \geq 0$ for all indices $i = 0, \ldots, r$, and suppose that ord $\sigma(x_i) = 0$ for some index i. Now, from (11.7), $u'_i = \sum a_{ij}u_j$. This shows that $P' = \mathfrak{T}(P)$, where $\mathfrak{T}$ is the projectivity of $\mathrm{PG}(r, K)$ associated to the matrix $T = (t_{ij})$. □

A corollary of Lemmas 11.19 and 11.22 is the following.

Theorem 11.23 *Let $\mathcal{F}$ be an irreducible curve which is not rational, elliptic or hyperelliptic. Then the K-automorphism group of $\mathcal{F}$ acts on the points of the canonical curve of $K(\mathcal{F})$ as a linear collineation group of* $\mathrm{PG}(g-1, K)$.

Theorem 11.23 makes possible the use of geometric methods in investigating automorphisms. Important results in this direction are the following theorems that extend to all characteristics p in Section 11.7.

Theorem 11.24 *Let $\mathcal{F}$ be an irreducible curve which is not rational, elliptic or hyperelliptic. If either $p = 0$ or $p > 2g-2$, then every K-automorphism of $\mathcal{F}$ is of finite order.*

Proof. From Theorem 7.64, the set of all Weierstrass points of $\Sigma = K(\mathcal{F})$ is finite. By Lemma 11.6, it suffices to prove the finiteness of any K-automorphism α fixing every Weierstrass point. By Corollary 7.61, the canonical curve Γ of Σ is classical. Let $P \in \Gamma$ be a Weierstrass point. The hyperplanes H through P with intersection multiplicity $I(P, H \cap \Gamma)$ at least j_{g-2} constitute a pencil through a common subspace Π of dimension $g-3$. Then Π is preserved by α. It may be that Π contains some more points from Γ, and let $k+1$, with $k \geq 0$, denote the number of distinct points of Γ lying in Π. Then every hyperplane in the pencil has at most $2g-2-(g-2)-k = g-k$ points outside Π. On the other hand, the lower bound in (7.15) shows the existence of at least $2g+2$ Weierstrass points. Therefore at least three, but finitely many, distinct hyperplanes in the pencil contain a Weierstrass point not lying in Π. Therefore a power β of α preserves at least three hyperplanes in the pencil.

Since, in the dual space of $\mathrm{PG}(g-1, K)$, the pencil corresponds to a line, it follows from Theorem 11.14 (ii)(b) that β preserves every hyperplane through Π. If H is such a hyperplane, then a sufficiently large power of β fixes every common point of H and Γ. Repeating this argument for another hyperplanes through Π, it turns out that, if α is infinite, then a power of α has more than $2g+2$ fixed points in Γ. Therefore, to prove the assertion, it suffices to show that no non-trivial K-automorphism can have more than $2g+2$ fixed points in Γ. This has already been shown; see Lemma 11.12. Here a different proof is given.

To do this, let δ be a non-trivial K-automorphism, and take an ordinary point $Q \in \Gamma$ such that $Q^\delta \neq Q$. Note that such a point exists by Theorem 11.11. Since Γ is classical, $g+1$ is a non-gap at Q, and hence there is a function $x \in \Sigma$ such that

$$\mathrm{div}(x)_\infty = (g+1)\mathcal{Q},$$

where $\mathcal{Q}$ is the place arising from Q. Letting $y = x - \delta(x)$, then

$$\mathrm{div}(y)_\infty = (q+1)\mathcal{Q} + (g+1)\mathcal{Q}^\delta.$$

Hence y has $2g+2$ zeros. On the other hand, every fixed place of δ is a zero of y. This proves the assertion and completes the proof of the theorem. □

Theorem 11.24 extends to the hyperelliptic case provided that either $p = 0$ or p is large enough with respect to g.

Lemma 11.25 *If $p = 0$ and $g \geq 2$, then every K-automorphism of an irreducible curve $\mathcal{F}$ of genus g is finite.*

Proof. Let W be a canonical divisor of $\Sigma = K(\mathcal{F})$. Then the tri-canonical series $L = |3W|$ of Σ is a non-special linear series of order $n = 3(2g-2) > 2g$ and dimension $r = 5g - 6$. By Theorem 7.40, the associated irreducible curve Γ in $\mathrm{PG}(r, K)$ is non-singular. As $p = 0$, so Γ is a classical curve. By Corollary 7.61, $v_\mathcal{P}(R) \leq \frac{1}{2}g(g+1)$ for every $\mathcal{P} \in \mathcal{P}_\Sigma$, and the number of L-Weierstrass points, each counted according to its weight, is $\sum v_\mathcal{P}(R) = 25(g-1)^2g$. Thus Γ has at least

$$\frac{25(g-1)^2g}{\frac{1}{2}g(g+1)}$$

L-Weierstrass points. For $g \geq 2$, this number is larger than $2g+2$. Since every K-automorphism α of Σ preserves the set $\mathcal{W}$ of all L-Weierstrass points, Lemma 11.12 shows that α cannot fix $\mathcal{W}$ element-wise. Hence α acts faithfully on $\mathcal{W}$, and α can be viewed as a permutation in the symmetric group on $\mathcal{W}$, which is finite. Therefore ord α is finite. □

In the elliptic case, Theorem 11.24 holds for K-automorphisms fixing a place. This is shown here for $p \neq 2, 3$. For the general case, see Section 11.4.

Lemma 11.26 *Let $p \neq 2, 3$. If $\mathcal{F}$ is an elliptic curve, then every K-automorphism $\mathcal{F}$ fixing a place is finite, and its order divides 4 or 6.*

Proof. Since $g = 1$, the smallest non-gap at $\mathcal{P}$ is equal to 2. Choose $\xi \in K(\mathcal{F})$ with $\mathrm{div}(\xi)_\infty = 2\mathcal{P}$. By Lemma 11.13, $\alpha(\xi) = a\xi + b$ with $a, b \in K$ and $a \neq 0$. Since $p \neq 2, 3$, there exists $\eta \in \Sigma$ such that

$$\eta^2 = 4\xi^3 - g_2\xi - g_3. \tag{11.8}$$

In other words, Σ has a model $(\mathcal{F}; (\xi, \eta))$ with a generic point $P = (\xi, \eta)$ which satisfies (11.8). Then η has a pole of order 3 in $\mathcal{P}$. Hence there exists $c \in K$ such that $\alpha(\eta) - c\eta$ has a pole of order at most 2. Therefore $\alpha(\eta) - c\eta = d\xi + e$ with $d, e \in K$. Hence $\alpha(\eta)$ is a K-linear combination of ξ, η and 1. Applying α to (11.8), this yields that

$$(c\eta + d\xi + e)^2 = 4(a\xi + b)^3 - g_2(\xi + b) - g_3,$$

which can also be written as

$$\begin{aligned} &c^2\xi^2 + d^2\eta^2 + e^2 + 2ce\xi + 2de\eta + 2cd\xi\eta \\ &= 4(a^3\xi^3 + 3a^2b\xi^2 + 3ab^2\xi + b^3) - g_2(a\xi + b) - g_3. \end{aligned}$$

By Lemma 11.1, this differs from (11.8) only by a constant factor. Comparing coefficients of $\xi\eta, \eta, \xi^2$ shows that $d \neq 0$ and that $c = e = b = 0$. Hence

$$\alpha(\xi) = a\xi, \quad \alpha(\eta) = d\eta, \quad with \quad d^2 = a^3 = 1.$$

If $g_3 = 0,\ g_2 \neq 0$ then either $a = 1,\ d = \pm 1$ or $a = -1,\ d = \pm j$ with $j^2 = -1$, and hence $\operatorname{ord}\alpha \in \{1, 2, 4\}$. If $g_2 = 0,\ g_3 \neq 0$, then $a = \pm 1,\ d^3 = 1$, and hence $\operatorname{ord}\alpha \in \{1, 2, 3, 6\}$. Otherwise, $a = 1,\ d = \pm 1$, and hence $\operatorname{ord}\alpha \in \{1, 2\}$. □

11.3 AUTOMORPHISM GROUPS OF PLANE CURVES

The K-automorphism groups of rational, elliptic, hyperelliptic, and many other irreducible curves are known; see Section 11.10 and Chapter 12. On the other hand, every finite group is isomorphic to the K-automorphism group of some irreducible curve.

For any irreducible curve $\mathcal{F}$ of genus g which is not rational, elliptic, or hyperelliptic, Theorem 11.19 shows that its K-automorphism group is isomorphic to a subgroup of $\mathrm{PGL}(g, K)$ which preserves the canonical curve Γ of $K(\mathcal{F})$ embedded in $\mathrm{PG}(g-1, K)$. This provides information about K-automorphisms, especially when the dimension $g - 1$ is low. In fact, $\Sigma = K(\mathcal{F})$ may have other, possibly singular, models in $\mathrm{PG}(r, K)$ such that $\mathrm{Aut}_K(\Sigma)$ is a subgroup of $\mathrm{PG}(r+1, K)$. If this is the case for an irreducible plane curve $\mathcal{C}$, a plane model of Σ, the structure and the action of $\mathrm{Aut}_K(\Sigma)$ can be examined by using the classification of all subgroups of $\mathrm{PGL}(3, K)$. However, finding out whether a given function field Σ has such a plane model may be very involved and may depend on special features of Σ.

An important, general result in this direction is that this problem can be solved for those function fields that have a non-singular plane model. The elementary proof depends on two lemmas of independent interest.

Lemma 11.27 *Let $\{P_1, \ldots, P_n\}$ be a set of $n > 3$ distinct points in the plane. If there are at least $\frac{1}{2}(n-2)(n-3)$ linearly independent curves $\mathcal{C}^{n-3}$ of order $n-3$ passing through all the points $P_1, \ldots, P_n$, then these points are collinear.*

Proof. The result is true for $n = 4$; so proceed by induction on n.

Let $n > 4$. On a line ℓ through one of these points, say P_1, disjoint from the remaining points $P_2, \ldots, P_n$, choose $n-3$ distinct points $Q_1, \ldots, Q_{n-3}$ which are also distinct from P_1. The number of linearly independent curves C^{n-3} of order $n-3$ passing all the points $P_1, \ldots, P_n, Q_1, \ldots, Q_{n-3}$ is at least

$$\tfrac{1}{2}(n-2)(n-3) - n - 3 = \tfrac{1}{2}(n-3)(n-4).$$

Such curves are reducible as each of them contains $n-2$ points from ℓ, namely the points $P_1, Q_1, \ldots, Q_{n-3}$. It follows that they split into ℓ and curves $\mathcal{C}^{n-4}$ of order $n-4$ passing through all the points $P_2, \ldots, P_n$. The number of such linearly independent curves is at least $\frac{1}{2}(n-3)(n-4)$. By induction, $P_2, \ldots, P_n$ are collinear. Therefore any $n-1$ of the points $P_1, \ldots, P_n$ are collinear. This can only happen when all these points lie on a line. □

LEMMA 11.28 *Let $\mathcal{C}^n$ be a non-singular plane curve of order $n > 3$. The linear series cut out on $\mathcal{C}^n$ by lines is the unique linear series g^2_n of $\mathcal{C}^n$.*

Proof. Choose a divisor $D = \mathcal{P}_1 + \cdots + \mathcal{P}_n$ in g^2_n such that the places $\mathcal{P}_i$ are centred at distinct points, say $P_1, \ldots, P_n$. Since $\mathcal{C}^n$ is non-singular, its genus is $g = \frac{1}{2}(n-1)(n-2)$. From Theorem 6.61 applied to g^2_n, at least $i = g - n - 2$ linearly independent curves $\mathcal{C}^{n-2}$ of order $n-2$ pass through $P_1, \ldots, P_n$. By Lemma 11.27, these points are collinear, and the assertion follows. □

So the following theorem is a corollary of Lemmas 11.28 and 11.22.

THEOREM 11.29 *The K-automorphism group of any non-singular plane curve $\mathcal{X}$ of order $n > 3$ is linear; that is, it is isomorphic to the subgroup of* $\mathrm{PGL}(3, K)$ *that fixes $\mathcal{X}$.*

It should be noted that Theorem 11.29 holds true for $n = 2$ but it fails for both $n = 1$ and $n = 3$; see Theorem 11.14 and the proof of Theorem 11.94. An application of Theorem 11.29 is the following result.

PROPOSITION 11.30 *If K is the algebraic closure of $\mathbf{F}_{q^2}$ with $q > 2$, then the K-automorphism group of the Hermitian curve*

$$\mathcal{H}_q = \mathbf{v}(X^q + X + Y^{q+1})$$

is the projective unitary group $\mathrm{PGU}(3, q)$.

Proof. Since $\mathcal{H}_q$ is a non-singular plane curve, Theorem 11.29 implies that its K-automorphism group G consists of all projectivities fixing $\mathcal{H}_q$. A straightforward computation shows that every $\alpha \in G$ is defined over $\mathbf{F}_{q^2}$; that is, a matrix associated to α has all entries from $\mathbf{F}_{q^2}$. Therefore G preserves the set $\mathcal{H}_q(\mathbf{F}_{q^2})$ of all $\mathbf{F}_{q^2}$-rational points of $\mathcal{H}_q$. By Example A.9, $\mathcal{H}_q(\mathbf{F}_{q^2})$ coincides with the set $\mathcal{U}$ of all self-conjugate points of a unitary polarity of $\mathrm{PG}(2, q^2)$, and $\mathrm{PGU}(3, q)$ is a subgroup of G.

It remains to show that every $\alpha \in G$ belongs to $\mathrm{PGU}(3, q)$. As $\mathrm{PGU}(3, q)$ acts 2-transitively on $\mathcal{U}$, with the notation used in Example A.9 it may be assumed that α fixes both the points X_∞ and O. After a change of coordinates, $\mathcal{U}$ consists of Y_∞ plus all points $U = (1, u, cu^{q+1} + m)$ as u ranges over $\mathbf{F}_{q^2}$ and m ranges over M. By another computation, the hypothesis that α preserves $\mathcal{U}$ implies that α is associated to a matrix

$$\begin{bmatrix} t^{q+1} & 0 & 0 \\ 0 & t & 0 \\ 0 & 0 & 1 \end{bmatrix}$$

for a non-zero element t in $\mathbf{F}_{q^2}$. Therefore $\alpha \in \mathrm{PGU}(3, q)$. □

THEOREM 11.31 *If $p \nmid n$ and $n \neq q + 1$, then the K-automorphism group G of the Fermat curve $\mathcal{F}_n = \mathbf{v}(X^n + Y^n + 1)$ has order $6n^2$ and has a normal abelian subgroup N of order n^2 such that $G/N \cong \mathbf{S}_3$.*

Proof. For any two n-th roots of unity, $\lambda, \mu \in K$, the projectivity

$$(X_0, X_1, X_2) \mapsto (X_0, \lambda X_1, \mu X_2)$$

preserves $\mathcal{F}_n$. These projectivities form an abelian group N of order n^2 fixing each vertex of the fundamental triangle. Both the projectivities

$$(X_0, X_1, X_2) \mapsto (X_1, X_2, X_0), \quad (X_0, X_1, X_2) \mapsto (X_0, X_2, X_1)$$

also preserve $\mathcal{F}_n$ and permute the vertices of the fundamental triangle. The projective group H generated by all these projectivities has order $6n^2$, contains N as a normal subgroup, and induces $\mathbf{S}_3$ on the vertices of the fundamental triangle. It remains to shows that H coincides with G.

Since $\mathcal{F}_n$ is a non-singular plane curve, Theorem 11.29 implies that G is a subgroup of PGL$(3, q)$. A straightforward computation shows that, if $\alpha \in$ PGL$(3, q)$ preserves $\mathcal{F}_n$, then α also preserves the fundamental triangle, except for $n = q+1$. Hence $\alpha \in H$, and the result follows. □

11.4 A BOUND ON THE ORDER OF A K-AUTOMORPHISM

A major result on automorphisms of algebraic curves is that, apart from rational curves, every K-automorphism fixing a place is finite. In characteristic zero, this follows from Lemmas 11.25 and 11.26. For $p > 0$, the proof is much more involved and requires some lemmas of independent interest on the invariant subfields of K-automorphisms.

In this section it is assumed that $p > 0$ although almost all results, but not all proofs, remain valid in zero characteristic.

LEMMA 11.32 *If $\alpha \in \mathrm{Aut}_K(\Sigma)$ fixes every element of a subfield F of Σ of transcendence degree* 1, *then α is finite and its order does not exceed* $[\Sigma : F]$.

Proof. Let $n = [\Sigma : F]$. As the field extension Σ/F is algebraic, it is also simple. There exists $z \in \Sigma$ such that $\Sigma = F(z)$. Let $f(X) = X^n + \cdots + a_{n-1}X + a_n$ be the minimal polynomial of z over F. Since α fixes every coefficient a_i of $P(X)$, it follows that $\alpha(z)$ is also a root of $f(X)$. So the set $S = \{\alpha^i(z) \mid i \geq 0\}$ has cardinality at most n. Since $\alpha^i(z) = \alpha^j(z)$ if and only if $\alpha^{i-j}(z) = z$ and since $\Sigma = F(z)$, this last condition can only occur when α^{i-j} is the identity automorphism of Σ. Hence ord $\alpha \leq |S| \leq n$. □

LEMMA 11.33 *Assume that Σ is non-rational and that $\alpha \in \mathrm{Aut}_K(\Sigma)$ preserves a rational subfield $K(\xi)$ of Σ. If $p \nmid \rho$, where $\rho = [\Sigma : K(\xi)]$, then α is finite and*

$$\mathrm{ord}\,\alpha \leq 2p(g+1)\rho^2.$$

Proof. By Lemma 11.32, the idea is to find an element $\eta \in \Sigma \backslash K$ which is fixed by α or a power of α.

By Theorem 11.14 (ii) and Lemma 11.13, it may be assumed that, after a K-linear change of ξ, either $\alpha(\xi) = \xi + v$ or $\alpha(\xi) = c\xi$ for some $v, c \in K$, and $c \neq 0$. In the former case, as $p \neq 0$, so $\alpha^p(\xi) = \xi$, and hence ord α cannot exceed $p\,[\Sigma : F]$, by Lemma 11.32.

Henceforth, let $\alpha(\xi) = c\xi$, where $c \in K\backslash\{0\}$. By Theorem A.5, there exists $\eta \in \Sigma$ such that $\Sigma = K(\xi, \eta)$. Let $\mathcal{F} = \mathbf{v}(f(X,Y))$ be the irreducible plane curve with generic point $P = (\xi, \eta)$. The projection of $\mathcal{F}$ onto the X-axis, that is, the rational transformation

$$\tau : \qquad x' = \xi, \quad y' = 0$$

is not birational, as Σ is not rational. Let $D = D(\Sigma/K(\xi))$ be the associated different divisor. Note that, if ξ is a separating variable of Σ, then Hurwitz's Theorem 7.27 implies that

$$\deg D = 2g - 2 + 2\rho, \tag{11.9}$$

where $\rho = [\Sigma : K(\xi)]$. By Lemma 11.7, $D^\alpha = D$ since α preserves $K(\xi)$. Also, α viewed as a K-automorphism of $K(\xi)$ preserves the set of places $\mathcal{Q} \in \mathcal{P}_{K(\xi)}$ which lie under the ramified places of $\mathcal{P}_{K(\mathcal{X})}$.

First, consider the case where $\mathrm{div}(\xi) = \rho\mathcal{P}_1 - \rho\mathcal{P}_2$. As in the proof of Theorem 5.33, after replacing η by $c_0(\xi)\eta$ with $c_0(\xi) \in K(\xi)$, it may be assumed that

$$f(X,Y) = Y^\rho + g(X,Y),$$

where $g(X,Y) \in K[X,Y]$ has degree in Y less than ρ. Since

$$f_Y(X,Y) = \rho Y^{\rho-1} + g_Y(X,Y)$$

does not vanish, the hypothesis $p \nmid \rho$ ensures that ξ is indeed a separating variable of Σ. Let σ_1 and σ_2 be primitive place representations of $\mathcal{P}_1$ and $\mathcal{P}_2$. Then

$$\sigma_1(\xi) = ct^\rho + \cdots, \quad c \in K\backslash\{0\},$$

and the contribution of $\mathcal{P}_1$ in D is $\rho - 1$ since $p \nmid \rho$. Similarly,

$$\sigma_2(\xi) = ct^{-\rho} + \cdots, \quad c \in K\backslash\{0\},$$

and the contribution of $\mathcal{P}_2$ in D is again $\rho - 1$. Thus $\deg D = 2(\rho - 1) + m$, with $m \geq 0$. Comparison with (11.9) gives $m = 2g$. This yields that, apart from $\mathcal{P}_1$, $\mathcal{P}_2$, at least one but at most $2g$ places contribute to D. Therefore a power $\beta = \alpha^\ell$, with $\ell \leq 2g$, of α fixes a place $\mathcal{P}_3$ different from both $\mathcal{P}_1, \mathcal{P}_2$.

If $\mathcal{Q}_1$, $\mathcal{Q}_2$, $\mathcal{Q}_3$ are the places of $K(\xi)$ under $\mathcal{P}_1, \mathcal{P}_2$, $\mathcal{P}_3$, then β fixes them. By Theorem 11.14 (ii)(b), it follows that β fixes every place $\mathcal{Q} \in \mathcal{P}_{K(\xi)}$. By Lemma 11.9, the subfield $\Sigma^\beta = \{z \in \Sigma \mid \beta(z) = z\}$ contains $K(\xi)$. Hence $\mathrm{ord}\,\beta \leq \rho$, yielding that $\mathrm{ord}\,\alpha \leq 2g\rho$.

Therefore it may be assumed that either $\mathrm{div}(\xi)_0$ or $\mathrm{div}(\xi)_\infty$, say the latter, contains two different places $\mathcal{P}_1$ and $\mathcal{P}_2$. Note that $\mathcal{P}_1^\beta = \mathcal{P}_1$ for a power $\beta = \alpha^\ell$ with $\ell \leq \rho$. Put $\mathrm{ord}_{\mathcal{P}_1}(\xi) = -k_1$ and $\mathrm{ord}_{\mathcal{P}_2}(\xi) = -k_2$, where k_1, k_2 are positive integers. Let n_1 be the smallest non-gap of Σ at $\mathcal{P}_1$. Then there exists $\zeta \in \Sigma$ such that $\mathrm{div}(\zeta)_\infty = n_1\mathcal{P}_1$. Since ξ has at least two poles, $\zeta \neq k\xi$ for every $k \in K$.

After changing ζ to $u\zeta + v$, with suitable $u, v \in K$, Lemma 11.13 allows that $\beta(\zeta) = b\zeta$, $b \in K\backslash\{0\}$.

Let $f_1(X,Y) = \sum a_{ij}X^iY^j$ be an irreducible polynomial with coefficients in K such that $f_1(\xi, \zeta) = 0$. From $\beta \in \mathrm{Aut}_K(\Sigma)$, it follows that $f_1(\beta(\xi), \beta(\zeta)) = 0$. Since $\beta(\xi) = c^\ell\xi$, $\beta(\zeta) = b\zeta$, this implies that

$$f_1(c^l X, bY) = ef_1(X,Y), \quad e \in K\backslash\{0\}.$$

Now choose two distinct pairs (i, j) and (m, n) with $n > j$ such that both a_{ij} and a_{mn} are distinct from zero. Then $c^{\ell i} b^j = e$ and $c^{\ell m} b^n = e$. Thus $\eta = \xi^{i-m} \zeta^{j-n}$ is fixed by β.

It must be shown that $\eta \notin K$. Assume by contradiction that $\zeta^{n-j} = k\xi^{i-m}$ with $k \in K$. Then $\text{div}(\xi^{i-m})_\infty = \text{div}(\zeta^{n-j})_\infty$, but

$$\text{div}(\xi^{i-m})_\infty = (i-m)k_1\mathcal{P}_1 + (i-m)k_2\mathcal{P}_2 + \cdots,$$
$$\text{div}(\zeta^{n-j})_\infty = (n-j)n_1\mathcal{P}_1.$$

Hence $(i-m)k_2 = 0$; but then $i = m$ whence $n - j = 0$, a contradiction.

Finally, it is shown that $[\Sigma : K(\eta)] \le 2\rho(g+1)$. Since the degree of $f_1(X, Y)$ in X is at most $[\Sigma : K(\zeta)] = n_1 \le g+1$, while in Y it is at most $[\Sigma : K(\xi)] = \rho$, so $|i-m| \le g+1$ and $|j-n| \le \rho$. Then

$$[\Sigma : K(\xi^{i-m})] = [\Sigma : K(\xi)] \cdot [K(\xi) : K(\xi^{i-m})] \le \rho(g+1),$$
$$[\Sigma : K(\zeta^{j-n})] = [\Sigma : K(\zeta)] \cdot [K(\zeta) : K(\zeta^{j-n})] \le (g+1)\rho.$$

Since $\text{div}\,\xi^{i-m}_\infty + \text{div}\,\zeta^{j-n}_\infty \succeq \text{div}\,\eta$, the assertion follows. From Lemma 11.32, $\text{ord}\,\beta \le 2\rho(g+1)$. Therefore $\text{ord}\,\alpha \le 2p\rho^2(g+1)$, and Lemma 11.33 is completely proved. □

Now the finiteness of every $\alpha \in \text{Aut}_K(\Sigma)$ fixing a place $\mathcal{P}$ of Σ is shown under the hypothesis that $g \ge 1$. By the Weierstrass Gap Theorem 6.89, there are exactly g gaps of Σ at $\mathcal{P}$, and each of them is at most $2g$. In particular, $2g+1$ is a non-gap of Σ at $\mathcal{P}$. By Corollary 6.75, $|(2g+1)\mathcal{P}|$ is a simple, base-point-free, complete linear series. Its dimension r is $g+1 \ge 2$, by the Riemann–Roch Theorem 6.61.

Let $m_0 = 0 < m_1 < \cdots < m_r$ denote the first $r+1 = g+2$ non-gaps of Σ at $\mathcal{P}$. In particular, $m_{r-1} = 2g$, $m_r = 2g+1$. Choose $x_0 = 1, x_1, \ldots, x_r \in \Sigma$ such that $\text{div}(x_i)_\infty = m_i$, for $i = 0, 1, \ldots, r$. Since $x_0, x_1, \ldots, x_r$ are linearly independent over K, the linear series,

$$L = \{\text{div}(\textstyle\sum c_i x_i) + (2g+1)\mathcal{P} \mid \mathbf{c} = (c_0, \ldots, c_r) \in \text{PG}(r, K)\},$$

coincides with $|(2g+1)\mathcal{P}|$. By Lemma 11.20, α preserves $|(2g+1)\mathcal{P}|$. By Lemma 11.3, α preserves each subseries,

$$L_j = \{\text{div}(\textstyle\sum_{i=j}^r c_i x_i) + (2g+1)P \mid \mathbf{c} = (c_0, \ldots, c_r) \in \text{PG}(r, K)\}.$$

Consider the Riemann–Roch space $\mathcal{L}((2g+1)\mathcal{P})$, that is, the $(r+1)$-dimensional vector space E over K with basis $\{x_0, \ldots, x_r\}$. Then $\alpha \in \text{End}(E_j)$, and

$$\alpha(x_j) = \textstyle\sum_{i=0}^j a_{ji} x_i$$

for $j = 0, \ldots, r$. In terms of E, the matrix A associated to α is lower-triangular:

$$A = \begin{bmatrix} a_0 & 0 & \ldots & 0 \\ * & a_1 & & \vdots \\ \vdots & & \ddots & 0 \\ * & \ldots & * & a_r \end{bmatrix}.$$

THEOREM 11.34 *Every K-automorphism of a non-rational irreducible curve that fixes a place is finite and its order is bounded above by $2p(g+1)(2g+1)^2$.*

Proof. Let ξ_i and ξ_j be eigenvectors of α associated to the eigenvalues a_i and a_j. It is shown first that, if $a_i \neq a_j$, then $\text{ord}_{\mathcal{P}}\, \xi_i \neq \text{ord}_{\mathcal{P}}\, \xi_j$. Since $\mathcal{P}$ cannot be a pole of $\xi_i\xi_j^{-1}$ and its inverse $\xi_i^{-1}\xi_j$, it may be assumed that $\mathcal{P}$ is not a pole of $\xi_i\xi_j^{-1}$. Then, for the residue map $\text{res}_{\mathcal{P}}$ associated to $\mathcal{P}$,

$$\text{res}_{\mathcal{P}}\, \xi_i\xi_j^{-1} = \text{res}_{\mathcal{P}}\, \alpha(\xi_i\xi_j^{-1}) = \text{res}_{\mathcal{P}}\, a_i a_j^{-1}\xi_i\xi_j^{-1} = a_i a_j^{-1}\text{res}_{\mathcal{P}}\, \xi_i\xi_j^{-1};$$

see Example 7.130. As $a_i a_j^{-1} \neq 1$, this shows that $\text{res}_{\mathcal{P}}\, \xi_i\xi_j^{-1} = 0$. Therefore $\mathcal{P}$ is a zero of $\xi_i\xi_j^{-1}$, and the assertion is proved.

Now suppose that all eigenvalues are different from one other. Then, for every non-gap m_i of Σ at $\mathcal{P}$ with $0 \leq i \leq r = g+1$, there is an eigenvector ξ_i of α such that $\text{ord}_{\mathcal{P}}\, \xi_i = m_i$. The subfields $K(\xi_i)$ are left invariant by α. Here

$$[\Sigma : K(\xi_{r-1})] = 2g, \quad \text{and} \quad [\Sigma : K(\xi_r)] = 2g+1.$$

Since p cannot divide two consecutive integers, Lemma 11.33 applies. This immediately implies that $\text{ord}\, \alpha \leq 2p(g+1)(2g+1)^2$.

Next, assume that two eigenvalues, say a_i and a_j, coincide. Then the Jordan form of α is either diagonal or contains a non-trivial block. So there are independent elements $\xi_1, \xi_2 \in E$, such that $\alpha(\xi_1) = a_i\xi_1$ and $\alpha(\xi_2) = a_j(\lambda\xi_1 + \xi_2)$ with $\lambda \in K$. Put $z = \xi_2\xi_1^{-1}$. Then $\alpha(z) = z + \lambda$, and the subfield $K(z)$ is left invariant by α. Since $p \neq 0$, this implies that $\alpha^p(z) = z$.

Also, $\rho = [\Sigma : K(z)]$ is at most $2(2g+1)$ since both $\text{div}(\xi_1)_\infty$ and $\text{div}(\xi_2)_\infty$ have degree less than or equal to $2g+1$. From Lemma 11.32, $\text{ord}\, \alpha \leq 2p(g+1)$. This completes the proof. □

11.5 AUTOMORPHISM GROUPS AND THEIR FIXED FIELDS

For any subgroup G of $\text{Aut}_K(\Sigma)$, the set

$$\Sigma^G = \{z \in \Sigma \mid \sigma(z) = z \text{ for all } \sigma \in G\}$$

is a subfield of Σ, the *fixed subfield of* G. The most important result on such subfields states that the field extension Σ/Σ^G is finite if and only if G is finite; also, if this is the case, then $[\Sigma : \Sigma^G] = |G|$. Actually, as shown later, this occurs for any $g > 1$, $\text{Aut}_K(\Sigma)$ being infinite only for rational and elliptic function fields. For a finite subgroup G of $\text{Aut}_K(\Sigma)$, this yields that Σ/Σ^G is a finite Galois extension with Galois group $\text{Gal}(\Sigma/\Sigma^G)$ isomorphic to G.

Geometrically speaking, if Γ is a model of Σ, the irreducible curve whose function field is Σ^G is the *quotient curve of* Γ *with respect to* G and denoted by Γ/G. So, in the study of quotient curves using function field theory, the principal technical tool is Galois theory. Here, the main goal is to prove that $\text{Aut}_K(\Sigma)$ is finite for $g > 1$; this is achieved after some preliminary results established in Section 11.7.

In this section, some properties of G that depend on Σ^G are discussed. First, a straightforward generalisation of Lemma 11.35 to any subgroup G of the K-automorphism group $\text{Aut}_K(\Sigma)$ is stated.

LEMMA 11.35 *Let G be a subgroup of* $\mathrm{Aut}_K(\Sigma)$ *which fixes every element of a subfield F of Σ of transcendence degree* 1. *Then G is finite and*

$$|G| \leq [\Sigma : F].$$

For a finite subgroup of $\mathrm{Aut}_K(\Sigma)$, the fundamental result on Σ^G is the following.

THEOREM 11.36 *Let G be a finite subgroup of* $\mathrm{Aut}_K(\Sigma)$. *Then Σ/Σ^G is a finite Galois extension, and $G \cong \mathrm{Gal}(\Sigma/\Sigma^G)$.*

Proof. Let $\xi \in \Sigma \backslash K$, and let $\{\alpha_1, \ldots, \alpha_r\}$ be a maximal set of elements of G such that $\alpha_1(\xi), \ldots, \alpha_r(\xi)$ are distinct. If $\beta \in G$, then $\beta\alpha_1(\xi), \ldots, \beta\alpha_r(\xi)$ is just a permutation of $\alpha_1(\xi), \ldots, \alpha_r(\xi)$; otherwise, the latter set is not maximal. Let $f(X) = \prod_{i=1}^{r}(X - \alpha_i(\xi))$. Since the coefficients of a monic polynomial are symmetric polynomials of its roots, every coefficient of $f(X)$ is in Σ^G. Further, $f(X)$ is separable. Therefore every element $\xi \in \Sigma$ is a root of a separable polynomial $f(X)$ of degree at most n with coefficients in Σ^G. Also, $f(X)$ splits into linear factors in Σ. Thus the field extension Σ/Σ^G is normal and separable, and hence a Galois extension. By Lemma 11.35, it remains to show that $[\Sigma : \Sigma^G] \geq n$.

Let $G = \{1, \alpha_1, \ldots, \alpha_{n-1}\}$. For every $1 \leq i \leq n-1$, let F_i be the subfield of Σ fixed by α_i. Choose an element $\eta \in \Sigma$ such that no F_i contains η. Then $\eta, \alpha_1(\eta), \ldots, \alpha_{n-1}(\eta)$ are distinct and hence the above polynomial $f(X)$ has degree n. Also, the notation F is used for Σ^G, and an element $\zeta \in \Sigma$ is chosen such that $[F(\zeta) : F]$ is as large as possible, say m. Then $m \geq n$.

To prove that $[\Sigma : F] \geq n$, it must be shown that $F(\zeta) = \Sigma$. If this is not true, then there exists an element $\delta \in \Sigma$ such that $\delta \notin F(\zeta)$. By Theorem A.5, there is an element $\omega \in F(\zeta, \delta)$ such that $F(\omega) = F(\zeta, \delta)$. But, from the tower $F \subset F(\zeta) \subset F(\zeta, \delta)$, it follows that $[F(\zeta, \delta) : F] > m$, whence $[F(\omega) : F] > m$; this contradicts the choice of m. □

LEMMA 11.37 *Let G be a finite subgroup of* $\mathrm{Aut}_K(\Sigma)$. *Then two places $\mathcal{P}_1$ and $\mathcal{P}_2$ of Σ lie over the same place $\mathcal{P}'$ of Σ^G if and only if there is a K-automorphism $\alpha \in G$ such that $\mathcal{P}_2 = \mathcal{P}_1^{\alpha}$.*

Proof. Let $\tau_1, \tau_2 \in \Sigma$ be primitive representations of the places $\mathcal{P}_1, \mathcal{P}_2$. Suppose first that $\mathcal{P}_2 = \mathcal{P}_1^{\alpha}$. Then $\tau_1\alpha = \lambda\tau_2$ with λ a K-automorphism of $K[[t]]$. Since α fixes Σ^G element-wise, it follows that $\tau_1(u') = (\lambda\tau_2)(u')$ for every $u' \in \Sigma^G$. This shows that the place of Σ^G represented by τ_1 coincides with that represented by τ_2. If $\mathcal{P}'$ is such a place, both places $\mathcal{P}_1$ and $\mathcal{P}_2$ are over $\mathcal{P}'$.

Conversely, suppose that $\mathcal{P}_1$ and $\mathcal{P}_2$ lie over the same place $\mathcal{P}'$ of Σ^G. There is a K-automorphism λ of $K[[t]]$ such that $\tau_1(u') = (\lambda\tau_2)(u')$ for every $u' \in \Sigma^G$. Hence $(\tau_1^{-1}\lambda\tau_2)(u') = u'$ for each $u' \in \Sigma^G$. This shows that $\beta = \tau_1^{-1}\lambda\tau_2$ is a K-automorphism of Σ which fixes the subfield Σ^G element-wise. Hence $\beta \in G$, and $\alpha = \beta^{-1}$ takes $\mathcal{P}_1$ to $\mathcal{P}_2$. □

As a by-product, there is also the following result.

LEMMA 11.38 *Let G be a finite subgroup of* $\mathrm{Aut}_K(\Sigma)$. *If $\mathcal{P}_1$ and $\mathcal{P}_2$ are two places of Σ lying over the same place $\mathcal{P}'$ of Σ^G, then $\mathcal{P}_1$ and $\mathcal{P}_2$ have the same ramification index ρ.*

The next step is to investigate subgroups G of $\mathrm{Aut}_K(\Sigma)$ by means of properties depending on the action of G on $\mathcal{P}_\Sigma$. Now the definition of a group acting on a set is particularised.

DEFINITION 11.39 For a group G,

(i) the *orbit* of a place $\mathcal{P}$ under G is the set

$$\mathcal{P}^G = \{\mathcal{P}^\alpha \mid \alpha \in G\};$$

(ii) the *stabiliser of* $\mathcal{P}$ in G is the subgroup

$$G_\mathcal{P} = \{\alpha \in G \mid \mathcal{P}^\alpha = \mathcal{P}\};$$

(iii) an orbit $\mathcal{P}^G$ is either *long* or *short* according as $G_\mathcal{P}$ is trivial or not.

REMARK 11.40 For G finite, the orbit $\mathcal{P}^G$ is short if and only if $|\mathcal{P}^G| < |G|$.

With this terminology, Lemma 11.37 can be reworded as follows.

LEMMA 11.41 *Let G be a finite subgroup of* $\mathrm{Aut}_K(\Sigma)$. *Then two places of Σ lie over the same place of Σ^G if and only if they are in the same orbit under the action of G. In other words, there is a one-to-one correspondence between places of Σ^G and G-orbits of places of Σ.*

It may be noted that the number of the short orbits under a finite subgroup G of $\mathrm{Aut}_K(\Sigma)$ is finite. If this were not true, then the stabiliser $G_\mathcal{P}$ for infinitely many $\mathcal{P} \in \mathcal{P}_\Sigma$ would contain some non-trivial K-automorphisms. Since G is finite, this could only occur when at least one non-trivial K-automorphism in G fixes an infinite number of places of Σ, a contradiction to Lemma 11.9.

A numerical result on short orbits is given in the following theorem.

THEOREM 11.42 *Let $\mathcal{P}$ be a place of Σ lying over a place $\mathcal{P}'$ of Σ^G. If $n = |G|$ and $m = |G_\mathcal{P}|$, then the number of distinct places lying over $\mathcal{P}'$ is n/m, and the ramification index of each of them is $e_\mathcal{P} = m$.*

Proof. If $\mathcal{P} = \mathcal{P}_1, \ldots, \mathcal{P}_s$ are the places of Σ over $\mathcal{P}'$, then $s = n/m$. In fact,

$$|G| = |G_\mathcal{P}| \cdot |\mathcal{P}^G| = |G_\mathcal{P}| \cdot \left[\Sigma : \Sigma^G\right],$$

whence $s = \left[\Sigma : \Sigma^G\right] = |G|/|G_\mathcal{P}| = n/m$. Finally, from Lemma 11.38, it follows that $\rho_1 = \cdots = \rho_s = m$. □

To end this section, a result applied in later sections is given.

LEMMA 11.43 *Let α be a non-trivial K-automorphism of Σ. If α fixes an element $\xi \in \Sigma$ and a place $\mathcal{P}$ of Σ, then ξ is not a uniformising element of Σ at $\mathcal{P}$.*

Proof. By (7.1), $\mathrm{ord}_\mathcal{P}\,\xi = e_\mathcal{P}\,\mathrm{ord}_{\mathcal{P}'}\xi$, where $\mathcal{P}'$ is the place of Σ^α under $\mathcal{P}$, and $e_\mathcal{P}$ is the ramification index of $\mathcal{P}$ with respect to the extension Σ/Σ^α. Since α fixes $\mathcal{P}$, so $e_\mathcal{P} = \mathrm{ord}\,\alpha$, whence $\mathrm{ord}_\mathcal{P}\,\xi \neq 1$. □

11.6 THE STABILISER OF A PLACE

In this section, the genus g of Σ is assumed to be positive.

LEMMA 11.44 *Let z be a uniformising element of Σ at $\mathcal{P} \in \mathcal{P}_\Sigma$, and let $G = G_\mathcal{P}$.*

(i) *For any $\alpha \in G$, there is just one $c_\alpha \in K$ such that* $\mathrm{ord}_\mathcal{P}(\alpha(z) - c_\alpha z) > 1$.

(ii) (a) *c_α is independent of the choice of the uniformising element;*

(b) *the mapping*

$$\begin{aligned} \varphi: \quad & G \to K \\ & \alpha \mapsto c_\alpha, \end{aligned} \tag{11.10}$$

is a group homomorphism with kernel N;

(c) *the group $G/N \cong \mathbf{Z}_m$ with* $\gcd(m, p) = 1$;

(d) *if $p = 0$, then $N = \{1\}$;*

(e) *if $p > 0$, then N is a nilpotent group consisting of all elements in $G_\mathcal{P}$ whose order is a power of p, and $G = N \rtimes H$, with H a subgroup of G whose order is prime to p.*

Proof. Choose a primitive place representation $\sigma : \Sigma \mapsto K((t))$ of $\mathcal{P}$ such that $\sigma(z) = t$. By (11.3), $\sigma(\alpha(z)) = kt + \cdots$. Thus $\mathrm{ord}_\mathcal{P}(\sigma(\alpha(z)) - c\sigma(z)) > 1$ if and only if $c = k$. It is straightforward to check that c is independent of the choice of σ. For another uniformising element z' at $\mathcal{P}$, there exists $u \in K\backslash\{0\}$ such that $\sigma(z') = ut + \cdots$. Hence

$$\mathrm{ord}_\mathcal{P}(\alpha(z') - u\alpha(z)) = \mathrm{ord}_\mathcal{P}(z' - uz) > 1. \tag{11.11}$$

Then

$$\mathrm{ord}_\mathcal{P}(\alpha(z') - kz') = \mathrm{ord}_\mathcal{P}((\alpha(z') - u\alpha(z)) + u(\alpha(z) - kz) + k(uz - z')),$$

which is greater than 1 by (11.11). Also, $\mathrm{ord}_\mathcal{P}(\alpha(z) - kz) > 1$. So (ii)(a) holds.

Since k depends on both $\mathcal{P}$ and α, but is independent of the choice of the place representation σ of $\mathcal{P}$, so φ is well defined. Let $\alpha, \beta \in G$. Then

$$\begin{aligned} \mathrm{ord}_\mathcal{P}((\beta\alpha)(z) - c_\beta c_\alpha z) &= \mathrm{ord}_\mathcal{P}((\beta\alpha)(z) - c_\beta\alpha(z)) + c_\beta(\alpha(z) - c_\alpha z)) \\ &\geq \min\{\mathrm{ord}_\mathcal{P}((\beta\alpha)(z) - c_\beta\alpha(z)), \mathrm{ord}_\mathcal{P}(c_\beta(\alpha(z) - c_\alpha z))\} \\ &= \min\{\mathrm{ord}_\mathcal{P}(\beta(\alpha(z)) - c_\beta\alpha(z)), \mathrm{ord}_\mathcal{P}(c_\beta(\alpha(z) - c_\alpha z))\}. \end{aligned}$$

Since both weights are greater than 1, so

$$\mathrm{ord}_\mathcal{P}((\beta\alpha)(z) - c_\beta c_\alpha(z)) > 1.$$

By the uniqueness, shown previously, $c_{\beta\alpha} = c_\beta c_\alpha$ which shows that φ is indeed a homomorphism. Hence $G/\ker\varphi$ is isomorphic to a subgroup of the multiplicative group of K. By Theorem 11.34, there is an integer M which is larger than the order of each element in G. This holds true for any factor group of G, and in particular for $G/\ker\varphi$. On the other hand, every finite multiplicative subgroup of K is the group of m-th roots of unity in K for a suitable integer m prime to p.

As H ranges over the finite subgroups of $G/\ker\varphi$, only a finite number of distinct images $\varphi(H)$ are obtained, as $|H|$, and hence $|\varphi(H)|$, is bounded by M. So all such subgroups $\varphi(H)$ generate a multiplicative cyclic subgroup Γ of K whose order is prime to p and bounded above by the product of all integers smaller than M. Since $\Gamma \cong G/\ker\varphi$, so (ii) (c) follows.

To prove the last two parts, it is first shown that every eigenvalue of α is a power of c_α. With the notation in the proof of Theorem 11.34, every eigenvalue a_j can be extracted from $\alpha(x_j) = \sum_{i=0}^{j} a_{ji}x_i$ by taking the last coefficient $a_j = a_{jj}$. Since $\text{ord}_{\mathcal{P}} x_j = -m_j$, so $x_j z^{m_j+1}$ is a uniformising element of Σ at $\mathcal{P}$. Write $\sigma(x_j) = b_j t^{-m_j} + \cdots$, with $b_j \in K^*$. Then

$$\begin{aligned}\sigma(\alpha(x_j z^{m_j+1})) - c_\alpha x_j z^{m_j+1}) &= \sigma(\alpha(x_j)\sigma(\alpha(z^{m_j+1})) - c_\alpha\sigma(x_j)\sigma(z^{m_j+1}) \\ &= (a_j b_j c_\alpha^{m_j+1} t + \cdots) - (c_\alpha b_j t + \cdots),\end{aligned}$$

whence $a_j = c_\alpha^{-m_j}$; this proves the assertion.

In particular, if $\alpha \in \ker\varphi$, then every eigenvalue of α is equal to 1. Hence $\ker\varphi$ consists of all those $\alpha \in G$ for which the associated lower-triangular matrix A has the form

$$A = \begin{bmatrix} 1 & 0 & \dots & 0 \\ * & 1 & & \vdots \\ \vdots & & \ddots & 0 \\ * & \dots & * & 1 \end{bmatrix}.$$

However, if $p = 0$, such an α cannot have finite order unless A is the identity matrix, and hence α is the identity automorphism, see Lemmas 11.25 and 11.26 This proves (ii) (d).

On the other hand, if p is positive, $\ker\varphi$ is a nilpotent group, and any non-trivial element in $\ker\varphi$ has order a power of p. In fact, $\ker\varphi$ is isomorphic to a group of matrices of the above form. This completes the proof of (ii) (e). □

LEMMA 11.45 *Let $p > 0$ and $g > 0$. Assume that H is a non-trivial abelian subgroup of the stabiliser of $\mathcal{P}$ satisfying the following properties:*

(a) *the order of each element in $H\backslash\{1\}$ is a power of p;*

(b) *the fixed field of any non-trivial finite subgroup of H is rational.*

Then H is a cyclic group of order p or p^2.

Proof. The result holds trivially when H has order p.

Let U be a subgroup of H of order p generated by $\alpha \in H$. Then $\Sigma^U = \Sigma^\alpha$ is a rational subfield $K(\xi)$ of Σ. From Theorem 11.36, $[\Sigma : K(\xi)] = p$. An element in $K(\xi)$ with a pole of multiplicity p at $\mathcal{P}$ may be found by a suitable substitution $\xi \to (a\xi + b)/(c\xi + d)$, $ad - bc \neq 0$, $a, b, c, d \in K$. Actually, $\mathcal{P}$ is the unique pole of ξ by (7.1) and Theorem 11.42.

So there is $\xi \in \Sigma$ with $\Sigma^U = K(\xi)$, $\text{div}(\xi)_\infty = p\mathcal{P}$, and $[\Sigma : \Sigma^U] = p$.

Now, to prove the assertion, assume by contradiction that there is an element τ of order p such that $\tau \in H$ but $\tau \notin U$. Since H is abelian, τ and α generate an elementary abelian subgroup V of H of order p^2. By Lemma 11.32, $\Sigma^V \neq \Sigma^U$, showing that τ does not fix all elements in Σ^U. On the other hand, $\alpha\tau = \tau\alpha$ implies that τ preserves Σ^U. Since $\mathcal{P}^\tau = \mathcal{P}$, from Theorem 11.14 (ii)(e) it follows that $\tau(\xi) = \xi + a$ with $a \in K\backslash\{0\}$. Replacing ξ by ξ/a gives the relations

$$\alpha(\xi) = \xi, \quad \tau(\xi) = \xi + 1, \quad \mathrm{div}(\xi)_\infty = p\mathcal{P}. \tag{11.12}$$

Interchanging the roles of α and τ in the above argument shows the existence of $\eta \in \Sigma$ with the following properties:

$$\alpha(\eta) = \eta + 1, \quad \tau(\eta) = \eta, \quad \mathrm{div}(\eta)_\infty = p\mathcal{P}. \tag{11.13}$$

Since $[\Sigma : K(\xi,\eta)] \cdot [K(\xi,\eta) : K(\xi)] = [\Sigma : K(\xi)] = p$, and $K(\xi,\eta) \neq K(\xi)$, so $\Sigma = K(\xi,\eta)$. Then the curve $\mathcal{F}$ with generic point $P = (\xi,\eta)$ has genus g, and both α and τ act on the plane as translations.

In particular, the centre of the branch of $\mathcal{F}$ associated to $\mathcal{P}$ lies on the line at infinity. To find an equation $f(X,Y) = 0$ of $\mathcal{F}$, deduce from the first part of (11.13) that $\eta^p - \eta \in K(\xi)$. Since both $\eta^p - \eta$ and ξ have no pole centred at an affine point, it follows that $\eta^p - \eta = f_1(\xi)$ with $f_1(X) \in K[X]$. As $\mathrm{div}(\eta^p - \eta)_\infty = p^2\mathcal{P}$ while $\mathrm{div}(\xi)_\infty = p$, so $\deg f_1(X) = p$. Also, since $(\eta - i)^p - (\eta - i) = \eta^p - \eta$, so

$$f_1(\xi - i) = f_1(\xi)$$

for every $i = 0, 1, \ldots, p-1$. Given a root a of $f_1(X)$, it follows that the roots of the polynomial $f_2(X) = f_1(X - a)$ are $0, 1, \ldots, p-1$. Hence $f_2(X) = c(X^p - X)$. Thus

$$\eta^p - \eta = c(\xi^p - \xi) + d, \quad \text{with } c, d \in K. \tag{11.14}$$

Therefore $\mathcal{F} = \mathbf{v}(d - cX + Y + cX^p - Y^p)$. This shows that the point $E = (0, 1, e)$, with $e^p = c$, is a $(p-1)$-fold point of $\mathcal{F}$. But then $\mathcal{F}$ is a rational curve by Exercise 6 in Chapter 1, contradicting the hypothesis $g > 0$. □

Lemma 11.46 *Let $p > 0$ and $g > 0$. Assume that H is a non-trivial abelian subgroup of the stabiliser of $\mathcal{P}$. If H has property* (a) *of Lemma 11.45, then H is finite and its order does not exceed $p^2(g-1)$.*

Proof. From Hurwitz's Theorem 7.27,

$$2(g-1) = n(2g' - 2) + \deg(\Sigma/\Sigma^U),$$

where U is a subgroup of H of order n, and g' is the genus of Σ^U. Further, $\deg(\Sigma/\Sigma^U) > n - 1$, because $\mathcal{P}$ completely ramifies in the extension Σ/Σ^U. For $g' > 0$, it follows that

$$n \leq 2(g-1)/(2g'-1) < 2g.$$

Therefore either the fixed field of every non-trivial finite subgroup of H has genus 0 or there is a subgroup U of H of order $n < 2g$ such that Σ^U has genus $g' \geq 1$. In the former case, Lemma 11.45 applies and the assertion follows. In the latter case, replace U by a finite maximal subgroup V of H containing U. Both properties hold true for V; that is, $|V| < 2g$ and Σ^V has genus $g' > 0$.

Now it is shown that Lemma 11.45 applies to Σ^V in so far as the factor group $\overline{H} = H/V$ is regarded as a K-automorphism group of Σ^V, and the place under

$\mathcal{P}$ is taken as the fixed place of $\overline{H}$. Since both commutativity and property (a) are inherited by $\overline{H}$, it is only necessary to deal with property (a). Under the natural homomorphism $H \mapsto \overline{H}$, any subgroup $\overline{W}$ of $\overline{H}$ corresponds to a subgroup W of H containing V. By the maximality of V, the genus of Σ^W is 0. Then Σ^W viewed as a subfield of Σ^V has genus 0, showing that $\overline{H}$ has property (a) of Lemma 11.45. Consequently, the order of $\overline{H}$ is at most p^2 and hence the order of H itself does not exceed $p^2(2g-1)$. □

At this point an elementary result from group theory is needed.

LEMMA 11.47 *Let G be a finite or infinite group of order at least n, containing its centre Z of order p, such that $\overline{G} = G/Z$ is an elementary abelian p-group. Then G contains an abelian subgroup of order at least $\sqrt{pn}$.*

Lemma 11.44 is improved by showing the finiteness of N.

LEMMA 11.48 *Let $p > 0$ and $g > 0$. Then N is finite and its order does not exceed $p^2(g+1)(2g-1)^2$.*

Proof. Choose an element $x_1 \in \Sigma$ such that $\text{div}(x_1)_\infty = m_1\mathcal{P}$, where m_1 denotes the smallest non-gap of Σ at $\mathcal{P}$. The subfield $K(x_1)$ of Σ is invariant under any K-automorphism fixing $\mathcal{P}$. From Theorem 11.14 (ii)(e), if $\alpha \in N$, then

$$\alpha(x_1) = x_1 + a_\alpha,$$

with $a_\alpha \in K$. Hence the map $\alpha \mapsto a_\alpha$ is a homomorphism of N into the additive group of K; let N_1 be its kernel. If N_1 is trivial, then $|N| \le p^2(2g-1)$, from Lemma 11.46.

So, for the rest of the proof, N_1 may be assumed non-trivial. Then the factor group N/N_1 is an elementary abelian p-group, possibly trivial. Also, since any $\alpha \in N_1$ fixes $K(x_1)$ element-wise, the order of N_1 is at most $m_1 = [\Sigma : K(x_1)]$ by Lemma 11.32. As N is nilpotent, N_1 has a subgroup N_2 of index p such that N_2 is a normal subgroup of N and N_1/N_2 is contained in the centre of the factor group N/N_2. By Theorem 11.36,

$$p \cdot [\Sigma : \Sigma^{N_2}] = p \cdot |N_2| = |N_1| \le [\Sigma : K(x_1)].$$

As $[\Sigma : K(x_1)] = [\Sigma : \Sigma^{N_2}] \cdot [\Sigma^{N_2} : K(x_1)]$, it follows that $p \le [\Sigma^{N_2} : K(x_1)]$. From this, Σ^{N_2} has positive genus g', for, otherwise, $\Sigma^{N_2} = K(\xi)$ with $\xi \in \Sigma$ and hence div $(\xi)_\infty = m'\mathcal{P}$ with $m' < m$.

Now, as the factor group N/N_2 can be regarded as a group of K-automorphisms of Σ^{N_2}, from Lemma 11.46 it follows that any abelian subgroup of N/N_2 has order at most $p^2(2g'-1)$.

On the other hand, for $Z = N_1/N_2$, the quotient group N/N_2 has the structure described in Lemma 11.47. Therefore, if n' is chosen such that the order of N/N_2 is not less than n', then N/N_2 contains an abelian subgroup whose order is at least $\sqrt{pn'}$. It follows then that $\sqrt{pn'} \le p^2(2g'-1)$. Hence N/N_2 has order at most $p^3(2g'-1)^2$, and the order of N does not exceed

$$p^3(2g'-1)^2 m_1 p^{-1} = p^2 m_1 (2g'-1)^2.$$

As $m_1 \le g+1$ and $g' \le g$, the order $|N| \le p^2(g+1)(2g-1)^2$. □

To sum up, a complete description of the structure of the stabiliser of a place $\mathcal{P}$ of Σ is obtained.

THEOREM 11.49 *Let Σ be a function field of genus $g > 0$. If $\mathcal{P}$ is a place of Σ, then the subgroup of $\text{Aut}_K(\Sigma)$ fixing $\mathcal{P}$ is finite and its structure is described as follows.*

(i) *If $p = 0$, then $G_{\mathcal{P}}$ is a cyclic group of order at most $8(2g+1)^3$.*

(ii) (a) *If $p > 0$, then a p-Sylow subgroup N of $G_{\mathcal{P}}$ is a normal subgroup with*

$$|N| \le p^2(g+1)(2g-1)^2, \tag{11.15}$$

and the quotient group $G_{\mathcal{P}}/N$ is a cyclic group with

$$|G_{\mathcal{P}}/N| \le 2p(2g+1)^2(g+1). \tag{11.16}$$

(b) *$G_{\mathcal{P}}$ contains a cyclic subgroup $H \cong G_{\mathcal{P}}/N$ such that $G_{\mathcal{P}} = N \rtimes H$. All such subgroups H are conjugate in $G_{\mathcal{P}}$.*

(c) *$|G_{\mathcal{P}}|$ has an upper bound depending only on p and g.*

Both (11.15) and (11.16) are approximate bounds. As is shown below, (11.16) can be significantly improved to $4g+2$, and (11.15) can be sharpened at least to $|N| \le 4pg^2/(p-1)^2$.

11.7 FINITENESS OF THE K-AUTOMORPHISM GROUP

By Theorems 11.14 and 11.94, the K-automorphism group of any rational or elliptic curve is infinite, and these are the only irreducible curves with this property. This important result has already been discussed before. Now a proof is given.

THEOREM 11.50 *The K-automorphism group of an irreducible curve $\mathcal{F}$ of genus $g \ge 2$ is finite.*

Proof. Let $\Sigma = K(\mathcal{F})$. By Theorem 11.49, it suffices to show that $\text{Aut}_K(\Sigma)$ preserves a finite set of places of Σ. In fact, if an $\text{Aut}_K(\Sigma)$-invariant set of places has size at most m, and the stabiliser of a place in the set has order n, then mn is an upper bound on $\text{Aut}_K(\Sigma)$. If Σ is hyperelliptic, Lemma 11.21 provides such a set of places of size at most $2g+2$. In the non-hyperelliptic case, the canonical series is used. By Lemma 11.23, $\text{Aut}_K(\Sigma)$ is a linear collineation group of $\text{PG}(g-1, K)$ which preserves the canonical curve Γ of Σ. In particular, $\text{Aut}_K(\Sigma)$ preserves the set of all Weierstrass points of Γ. This set is finite as its size does not exceed

$$\left(\textstyle\sum_{i=0}^{g-1} \epsilon_i\right)(2g-2) + g(2g-2),$$

by (7.13). □

REMARK 11.51 Let $\mathcal{F}$ be an irreducible, curve of genus $g > 2$. Assume that $\mathcal{F}$ is not hyperelliptic, and consider its canonical curve $\mathcal{K}$. Then the order sequence of $\mathcal{K}$ is

$$\epsilon_0 = 0 < \epsilon_1 = 1 < \cdots < \epsilon_{g-1} \leq 2g-2,$$

whence the number of its Weierstrass points is at most $(g-1)(2g-1)(2g-2)$. Hence $|\mathrm{Aut}_K(\Sigma)| < 4g^3\,|G_{\mathcal{P}}|$ for $\mathcal{P}$ a Weierstrass point. Thus an upper bound for $|\mathrm{Aut}_K(\Sigma)|$ follows from the upper bound for $|\mathrm{Aut}_K(\Sigma)_{\mathcal{P}}|$ given in Theorem 11.49.

An important consequence of Theorem 11.50 is that, if Σ has genus $g \geq 2$, then Hurwitz's Theorem 7.27 applies to every extension Σ/Σ^G such that Σ^G is the fixed field of a non-trivial subgroup G of $\mathrm{Aut}_K(\Sigma)$. Using the geometric term of quotient curve, the following result is obtained.

COROLLARY 11.52 *Let $\mathcal{F}$ be an irreducible curve of genus $g > 1$. For a non-trivial K-automorphism group G of $\mathcal{F}$, let g' be the genus of the quotient curve $\mathcal{F}/G$. Then*

$$2g-2 = |G|(2g'-2) + d, \tag{11.17}$$

where $d = \deg D(\Sigma/\Sigma^G)$ with $\Sigma = K(\mathcal{F})$ and $\Sigma^G = K(\mathcal{F}/G)$.

REMARK 11.53 Corollary 11.52 holds true for rational and elliptic curves when G is finite.

Although the aim of the proof of Theorem 11.50 was only to prove the finiteness of $\mathrm{Aut}_K(\Sigma)$, an upper bound for the order of $\mathrm{Aut}_K(\Sigma)$ in terms of p and g can be extracted from it; see Remark 11.51. A further effort shows that such an upper bound can be improved significantly. In this direction, the first step is Theorem 11.56, which makes evident the importance of short orbits in investigating curves with large K-automorphism groups. Two types of short orbits are distinguished.

DEFINITION 11.54 A short orbit of a K-automorphism group G is *tame* or *non-tame* (*wild*) according as the order of $G_{\mathcal{P}}$ for one and hence every place $\mathcal{P}$ in the orbit is prime or not to the characteristic p of K.

REMARK 11.55 If $p = 0$, then every short orbit is tame.

THEOREM 11.56 *Let $\mathcal{F}$ be an irreducible curve of genus $g \geq 2$.*

(i) *If G is a K-automorphism group of $\mathcal{F}$, then Hurwitz's upper bound*

$$|G| \leq 84(g-1)$$

holds in general with exceptions occurring only in positive characteristic.

(ii) *In terms of the function field Σ of $\mathcal{F}$, exceptions can only occur when $p > 0$, the fixed field Σ^G is rational, and G has at most three short orbits, as follows:*

  (a) *exactly three short orbits, two tame and one non-tame, with $p \geq 3$;*

(b) *exactly two short orbits, both non-tame*;

(c) *only one short orbit which is non-tame*;

(d) *exactly two short orbits, one tame and one non-tame.*

Proof. The proof uses Corollary 11.52 following the original idea of Hurwitz. By Theorem 11.42, equation (11.17) may be rewritten in the form

$$2g - 2 = |G| \cdot (2g' - 2 + d'), \tag{11.18}$$

where

$$d'_{\mathcal{P}} = \frac{d_{\mathcal{P}}}{e_{\mathcal{P}}} = \frac{d_{\mathcal{P}}}{|G_{\mathcal{P}}|}, \quad \text{and} \quad d' = \sum d'_{\mathcal{P}};$$

here, the summation is only over a set of representatives of places in $\mathcal{P}_\Sigma$, exactly one from each short orbit. So it is necessary to investigate the possibilities for $|G|$ according to the number r of short orbits of $|G|$ on $\mathcal{P}_\Sigma$. Now take (7.7) into account; this gives $d_{\mathcal{P}} \geq e_{\mathcal{P}} - 1$, with equality holding if and only if either $p = 0$ or $e_{\mathcal{P}}$ is prime to p. Therefore, if $d_{\mathcal{P}} > 0$, then $d'_{\mathcal{P}} \geq \frac{1}{2}$. Also, if $d > 0$, then $d' \geq \frac{1}{2}$.

If $g' \geq 2$, then $|G| \leq g - 1$. For $g' = 1$, it follows that $d' > 0$ since $g \geq 2$, and hence $|G| \leq 4(g-1)$. So assume that $g' = 0$; that is, Σ^G is rational. Then

$$2g - 2 = |G| \cdot (d' - 2). \tag{11.19}$$

In particular, $d' > 2$. Therefore G has some, say $r \geq 1$, short orbits on $\mathcal{P}_\Sigma$.

Take representatives, $\mathcal{Q}_1, \ldots, \mathcal{Q}_r$, from each short orbit, and let $d'_i = d'_{\mathcal{Q}_i}$. After a change of indices, it may be assumed that $d'_i \leq d'_j$ for $i \leq j$.

(I) When $r \geq 5$, then $d' \geq \frac{5}{2}$, and hence $|G| \leq 4(g-1)$.

(II) When $r = 4$, then $d' > 2$ and $d'_i > \frac{1}{2}$ for at least one place $\mathcal{P}$. As $d'_i > \frac{1}{2}$ implies $d'_i \geq \frac{2}{3}$, so $d' - 2 \geq \frac{1}{6}$, whence $|G| \leq 12(g-1)$.

(III) When $r = 3$, then again use $d' - 2 > 0$. If $d'_1 = \frac{2}{3}$ then $d'_3 \geq \frac{3}{4}$ and hence $|G| \leq 24(g-1)$. If $d'_1 = \frac{1}{2}$, $d'_2 \geq \frac{3}{4}$, then $|G| \leq 40(g-1)$. Also, $d'_1 = \frac{1}{2}$ and $d'_2 = \frac{2}{3}$ imply that $d'_3 \geq \frac{6}{7}$, whence $|G| \leq 84(g-1)$. Finally, if $d'_1 = d'_2 = \frac{1}{2}$, then $d'_3 > 1$, which can only occur when $p > 2$ and $e_{\mathcal{Q}_1} = e_{\mathcal{Q}_2} = 2$ while $e_{\mathcal{Q}_3}$ is divisible by p. This leads to (a).

(IV) When $r = 2$, then $d' = d'_1 + d'_2 > 2$. This can only occur when either d'_1, or d'_2, or both are greater than 1. Hence one of (b) and (d) occurs.

(V) When $r = 1$, then $d' = d'_1 > 2$, and case (c) occurs.

□

For $p \neq 0$, there are so many exceptions to Hurwitz's bound that their classification appears out of range. However, those with $|G|$ greater than g^3 seem not too numerous, at least for g large enough, and their determination might be possible. So far, this has been done for $|G| \geq 8g^3$, and the classification is stated in Theorem 11.127; it requires some deeper preliminary results that are proved in the next two sections.

11.8 TAME AUTOMORPHISM GROUPS

Let G be a non-trivial K-automorphism group of an irreducible curve $\mathcal{F}$. In this and the next section, the important equation (11.17) linking the genus g of the curve $\mathcal{F}$ to the genus g' of its quotient curve $\mathcal{F}' = \mathcal{F}/G$ is given a more explicit form. The idea is to examine the action of G on the set of places of the function field Σ of $\mathcal{F}$. Here, the tame case is considered; that is, it is supposed that the orders of the stabilisers are all prime to p.

THEOREM 11.57 *Let G be a finite K-automorphism group of an irreducible curve $\mathcal{F}$. If $|G_{\mathcal{P}}|$ is prime to p for every place $\mathcal{P}$ of $\Sigma = K(\mathcal{F})$, then*

$$2g - 2 = n(2g' - 2) + \textstyle\sum_{i=1}^{s} (n - l_i), \tag{11.20}$$

where $n = |G|$, and $l_1, \ldots, l_s$ denote the sizes of the short orbits of G.

Proof. Choose a place $\mathcal{P}$ of Σ, and let $\mathcal{P}'$ be a place of Σ^G lying under $\mathcal{P}$. By Theorem 11.42, the ramification index ρ is equal to m, with $m = |G_{\mathcal{P}}|$. Since by hypothesis $p \nmid n$, by the remark made after the proof of Hurwitz's Theorem 7.27,

$$\mathrm{ord}_{\mathcal{P}}\, d\xi = (\rho - 1) + \mathrm{ord}_{\mathcal{P}'}\, d\xi.$$

Hence

$$D(\Sigma/\Sigma^G) = \textstyle\sum (m-1)\,\mathcal{P}, \tag{11.21}$$

with $\mathcal{P}$ ranging over all places of Σ. Therefore $\deg D(\Sigma/\Sigma^G) = \sum(m-1)$.

To evaluate $\sum(m-1)$, consider the finite sub-sum $\sum(m_i - 1)$ restricted to those places $\mathcal{P}$ of Σ which lie over a given place $\mathcal{P}_i'$ of Σ^G. By Theorem 11.42, there are precisely n/m_i distinct places over $\mathcal{P}_i'$, each of them contributing $m_i - 1$. Hence, from every place $\mathcal{P}_i'$, a contribution $n(m_i - 1)/m_i$ is obtained. Therefore

$$\textstyle\sum (m-1) = \sum (n/m_i)(m_i - 1) = \sum (n - n/m_i).$$

As $\ell_i = n/m_i$, so

$$\deg(\Sigma/\Sigma^G) = \textstyle\sum (m-1) = \sum_{i=1}^{s} (n - l_i).$$

From Corollary 11.52,

$$\begin{aligned} 2g - 2 &= \left[\Sigma : \Sigma^G\right] \cdot (2g' - 2) + \deg D(\Sigma/\Sigma^G) \\ &= n(2g' - 2) + \textstyle\sum_{i=1}^{s} (n - l_i). \end{aligned}$$

□

EXAMPLE 11.58 Over $\mathbf{F}_{13}$, let

$$\begin{aligned} \mathcal{F} &= \mathbf{v}(X_0^6 + X_1^6 + X_2^6 + 3X_0^2X_1^2X_2^2), \\ \mathcal{F}' &= \mathbf{v}(X_0^3 + X_1^3 + X_2^3 + 3X_0X_1X_2). \end{aligned}$$

Then $\mathcal{F} \to \mathcal{F}'$ by the map $(x_0, x_1, x_2) \mapsto (x_0^2, x_1^2, x_2^2)$, and

$$G = \{(X_0, X_1, X_2) \mapsto (\pm X_0, \pm X_1, X_2)\} \cong \mathbf{Z}_2 \times \mathbf{Z}_2.$$

Then $n = 4$, $g' = 1$, $g = 10$, $s = 9$, $l_i = 2$ for all i; so (11.20) is satisfied for these values. The short orbits consist of the pairs of points of $\mathcal{F}$ mapping to an inflexion of $\mathcal{F}'$; see Table 11.1.

Table 11.1 A sextic curve 4-covers a cubic

$(\pm 1, \pm 2, 1)$	$\mapsto$	$(1, 4, 1)$	$(\pm 5, \pm 2, 1)$	$\mapsto$	$(-1, 4, 1)$
$(\pm 2, \pm 1, 1)$	$\mapsto$	$(4, 1, 1)$	$(\pm 3, \pm 5, 1)$	$\mapsto$	$(-4, -1, 1)$
$(\pm 2, \pm 5, 1)$	$\mapsto$	$(4, -1, 1)$	$(\pm 5, \pm 3, 1)$	$\mapsto$	$(-1, -4, 1)$
$(\pm 4, \pm 6, 1)$	$\mapsto$	$(3, -3, 1) = (-1, 1, 4)$			
$(\pm 6, \pm 4, 1)$	$\mapsto$	$(-3, 3, 1) = (1, -1, 4)$			
$(\pm 6, \pm 6, 1)$	$\mapsto$	$(-3, -3, 1) = (1, 1, 4)$			
$(0, \pm 2, 1)$	$\mapsto$	$(0, 4, 1)$	$(0, \pm 5, 1)$	$\mapsto$	$(0, -1, 1)$
$(\pm 2, 0, 1)$	$\mapsto$	$(4, 0, 1)$	$(\pm 5, 0, 1)$	$\mapsto$	$(-1, 0, 1)$
$(\pm 2, 1, 0)$	$\mapsto$	$(4, 1, 0)$	$(\pm 5, 1, 0)$	$\mapsto$	$(-1, 1, 0)$
$(0, \pm 6, 1)$	$\mapsto$	$(0, -3, 1) = (0, 1, 4)$			
$(\pm 6, 0, 1)$	$\mapsto$	$(-3, 0, 1) = (1, 0, 4)$			
$(\pm 6, 1, 0)$	$\mapsto$	$(-3, 1, 0) = (1, 4, 0)$			

In the special case that G has prime order, Theorem 11.57 reads as follows.

THEOREM 11.59 *Let α be a finite non-trivial K-automorphism of an irreducible curve whose order is prime to p. If $\rho(\alpha)$ is the number of fixed places of α and $n = \text{ord}\,\alpha$ with n prime, then*

$$2g - 2 = n(2g' - 2) + \rho(\alpha)(n - 1). \tag{11.22}$$

For tame K-automorphism groups, a significant improvement on the bounds in (ii) of Theorem 11.49 is the following result.

THEOREM 11.60 *Let $\mathcal{F}$ be an irreducible curve of genus $g > 0$, and let $G_\mathcal{P}$ be a K-automorphism group of $\mathcal{F}$ fixing a place $\mathcal{P}$. If the order n of $G_\mathcal{P}$ is prime to p, then*

$$n \le 4g + 2. \tag{11.23}$$

Proof. The place $\mathcal{P}$ itself constitutes a short orbit of size 1. With $\ell_s = 1$ in (11.20),

$$2g - 1 = n(2g' - 1) + n \textstyle\sum_{i=1}^{s-1} (1 - \ell_i/n), \tag{11.24}$$

where $\ell_1, \ldots, \ell_{s-1}$ denote the sizes of the remaining short orbits of $G_\mathcal{P}$. If $g' > 0$, the assertion follows.

Hence assume that $g' = 0$. Then (11.24) becomes the following:

$$2g - 1 = \textstyle\sum_{i=1}^{s-1} (n - \ell_i) - n. \tag{11.25}$$

Since ℓ_i divides n, so $n - \ell_i \ge \frac{1}{2}n$. Hence, if $s \ge 4$, then $2g - 1 \ge \frac{3}{2}n - n$, whence $n \le 2(2g - 1) \le 4g + 2$. The same argument works for $s = 3$ as long as $\ell_1 + \ell_2 \le \frac{1}{2}n$. Further, neither $s = 2$ nor $s = 1$ can actually occur.

So it remains to investigate the case that $2g - 1 = n - (\ell_1 + \ell_2)$ with $\ell_1 + \ell_2 > \frac{1}{2}n$. Up to an index change, it may be assumed that $\ell_1 \ge \ell_2$. Then the possibilities for (ℓ_1, ℓ_2) are the following:

$$(\tfrac{1}{3}n, \tfrac{1}{5}n), \quad (\tfrac{1}{3}n, \tfrac{1}{4}n), \quad (\tfrac{1}{3}n, \tfrac{1}{3}n), \quad (\tfrac{1}{2}n, \le \tfrac{1}{3}n). \tag{11.26}$$

Write $G = G_{\mathcal{P}}$ for brevity, and denote by H_i the subgroup of G which fixes one, and hence every, place in the short orbit of size ℓ_i for $i = 1, 2$. Let H be the subgroup of G generated by H_1 and H_2. Since G is cyclic by Theorem 11.49, so

$$|H| = n/\gcd(\ell_1, \ell_2).$$

It will be shown that $G = H$.

By way of contradiction, Σ^G is assumed to be a proper subfield of Σ^H. Then $[\Sigma^H : \Sigma^G] = |G|/|H|$. Let $\mathcal{P}'$ be the place of Σ^H lying under $\mathcal{P}$ in the extension Σ/Σ^H. Then $\mathcal{P}'$ completely ramifies in Σ^H/Σ^G.

From the definition of H, it can be deduced that no further place of Σ^H ramifies in Σ^H/Σ^G. To do this, note that, for any place $\mathcal{Q}$ from the orbit of length ℓ_1, $G_{\mathcal{Q}} = H_{\mathcal{Q}}$ and

$$\frac{|G|}{|H|} = \frac{|G_{\mathcal{Q}}|}{|H_{\mathcal{Q}}|} \cdot \frac{\ell_1}{|\mathcal{Q}^H|} = \frac{\ell_1}{|\mathcal{Q}^H|} = \frac{|\mathcal{Q}^G|}{|\mathcal{Q}^H|}.$$

Then the same is true for ℓ_2. If $\mathcal{Q}$ is a place not in a short orbit, then $G_{\mathcal{Q}}$ is the identity subgroup, and hence $G_{\mathcal{Q}} = H_{\mathcal{Q}}$. Therefore $|G|/|H| = |\mathcal{Q}^G|/|\mathcal{Q}^H|$ for every place $\mathcal{Q} \neq \mathcal{P}$ of Σ. This proves the assertion.

Corollary 11.52 gives the equation

$$2\overline{g} - 2 = -2m' + m' - 1,$$

where $\overline{g}$ is the genus of Σ^H, and $m' = |G|/|H|$. But this can only occur for $m' = 1$, giving the desired result that $H = G$. Hence ℓ_1 and ℓ_2 are prime to each other. From (11.26), the following cases can occur:

$$\begin{aligned} (\ell_1, \ell_2) &= (\tfrac{1}{3}n, \tfrac{1}{5}n), \quad |G_{\mathcal{P}}| = 15, \quad g = 4; \\ (\ell_1, \ell_2) &= (\tfrac{1}{3}n, \tfrac{1}{4}n), \quad |G_{\mathcal{P}}| = 12, \quad g = 3; \\ (\ell_1, \ell_2) &= (\tfrac{1}{2}n, \tfrac{1}{k}n),\ k \text{ odd}, \quad |G_{\mathcal{P}}| = 2k, \quad g = \tfrac{1}{2}(k-1). \end{aligned}$$

In each of these cases, $n \leq 4g + 2$. This completes the proof. □

REMARK 11.61 If the hypothesis that $p \nmid n$ is dropped, then the proof of Theorem 11.57 only provides a lower bound for $2g - 2$; that is,

$$2g - 2 \geq n(2g' - 2) + \textstyle\sum_{i=1}^{s} (n - \ell_i). \tag{11.27}$$

However, for a p-group G of order n, if g is replaced by the p-rank γ, a relationship similar to (11.20) holds.

THEOREM 11.62 (Deuring–Shafarevich) *Let G be a K-automorphism group of an irreducible curve $\mathcal{F}$ whose order n is a power of p. Then*

$$\gamma - 1 = n(\gamma' - 1) + \textstyle\sum_{i=1}^{s} (n - \ell_i), \tag{11.28}$$

where γ and γ' are the p-ranks of $\mathcal{F}$ and its quotient curve $\mathcal{F}/G$, while $\ell_1, \ldots, \ell_s$ are the sizes of the short orbits of G on the places of Σ.

If G has no short orbits, the possibilities for the structure of G is described in the following theorem.

THEOREM 11.63 *Let G be a K-automorphism group of an irreducible curve $\mathcal{F}$ whose order n is a power of p. If G has no short orbits, then G is isomorphic to a quotient group of the free group with γ generators.*

Since G has no short orbits means that the extension Σ/Σ^G is unramified, the question of determining the maximum number of unramified independent extensions of a given degree p^h arises. Interestingly, the answer for $h = 1$ gives a characterisation of the p-rank of Σ.

THEOREM 11.64 *The p-rank of Σ is equal to the maximum number of cyclic unramified independent extensions of degree p of Σ.*

Theorem 11.63 may be reworded in terms of such an extension.

11.9 NON-TAME AUTOMORPHISM GROUPS

As pointed out in Remark 11.61, the exact formula (11.20) depending only upon the action of G on places does not hold true for non-tame subgroups. However, a similar, although more complicated, formula, relying on more concepts and arguments exists for non-tame subgroups; this is the *Hilbert Different Formula* stated in Theorem 11.70. The goal in this section is to give a proof of this. Further, for non-tame K-automorphism groups, a significant improvement to Theorem 11.49 (ii) is obtained by using Theorem 11.70.

Given a place $\mathcal{P}$ of Σ, choose a primitive representation τ of $\mathcal{P}$ together with a uniformising element $x \in \Sigma$ at $\mathcal{P}$, and assume that $\tau(x) = t$. By Lemma 11.43, $\alpha(x) \neq x$ for any non-trivial K-automorphism $\alpha \in G_{\mathcal{P}}$. Also, $n = |G_{\mathcal{P}}|$ divides $\text{ord}_t\, \tau(\xi)$ for any $\xi \in \Sigma^{G_{\mathcal{P}}}$; see (7.1) and Theorem 11.36. More precisely, there exists $s(t) \in K[[t]]$ such that $\tau(\xi) = \sigma(s(t))$ with $\sigma \in K((s))$. Such an element $s(t)$ is independent of the choice of ξ, and σ is a primitive representation of the place $\mathcal{P}'$ of $\Sigma^{G_{\mathcal{P}}}$ lying under $\mathcal{P}$; see Section 7.2.

LEMMA 11.65 *The set $\{1, x, x^2, \ldots, x^{n-1}\}$ is a basis for $\Sigma/\Sigma^{G_{\mathcal{P}}}$.*

Proof. By Theorem 11.36, it suffices to prove that $1, x, x^2, \ldots, x^{n-1}$ are linearly independent over $\Sigma^{G_{\mathcal{P}}}$. Suppose, on the contrary, that

$$\xi_0 + \xi_1 x + \cdots + \xi_{n-1} x^{n-1} = 0,$$

where $\xi_0, \xi_1, \ldots, \xi_{n-1} \in \Sigma^{G_{\mathcal{P}}}$ but not all $\xi_i = 0$. Then

$$\tau(\xi_0) + \tau(\xi_1)\tau(x) + \cdots + \tau(\xi_{n-1})(\tau(x))^{n-1} = 0,$$

showing that two or more of the n terms on the left-hand side have the same order. This cannot actually occur by the following argument. If $0 \leq i \leq n-1$, then

$$\text{ord}_t\, \tau(\xi_i)\tau(x^i) = \text{ord}_t\, \tau(\xi_i) + i\, \text{ord}_t\, \tau(x) = k_i\, n + i$$

for an integer k_i; the same holds for every j with $0 \leq j \leq n-1$. This implies that $\text{ord}_t\, \tau(\xi_i)\tau(x^i) = \text{ord}_t\, \tau(\xi_j)\tau(x^j)$, and hence $n(k_j - k_i) = i - j$. But this is impossible since $0 \leq i, j < n$ and $i \neq j$. □

LEMMA 11.66 *There exist $\xi_0, \dots, \xi_{n-1} \in \Sigma^{G_\mathcal{P}}$ satisfying*

$$\xi_0 + \xi_1 x + \cdots + \xi_{n-1} x^{n-1} + x^n = 0, \tag{11.29}$$

with $\text{ord}_t\, \tau(\xi_0) = n$, *and* $\text{ord}_t\, \tau(\xi_i) \geq n$ *for* $i = 1, \dots, n-1$.

Proof. By Lemma 11.65, there exist $\eta_0, \dots, \eta_n \in \Sigma^{G_\mathcal{P}}$, $\eta_n \neq 0$, satisfying

$$\eta_0 + \eta_1 x + \cdots + \eta_{n-1} x^{n-1} + \eta_n x^n = 0. \tag{11.30}$$

It follows that

$$\tau(\eta_0) + \tau(\eta_1)\tau(x) + \cdots + \tau(\eta_n)(\tau(x))^n = 0.$$

By the argument used in the proof of Lemma 11.65, the equation

$$\text{ord}_t(\tau(\eta_i)\tau(x^i)) = \text{ord}_t(\tau(\eta_j)\tau(x^j))$$

implies that $n(k_j - k_i) = i - j$. Here, this is only possible for $i = n, j = 0$ and

$$\text{ord}_t\, \tau(\eta_0) = n + \text{ord}_t\, \tau(\eta_n).$$

It also follows that $\text{ord}_t\, \tau(\eta_i) \geq n + \text{ord}_t\, \tau(\eta_n)$.

Let $\xi_i = \eta_i/\eta_n$; then $\text{ord}_t\, \tau(\xi_0) = n$, while $\text{ord}_t\, \tau(\xi_i) \geq n$ for $1 \leq i \leq n-1$. Dividing both sides in equation (11.30) by η_n, the result follows. □

For $i = 0, \dots, n-1$, let $\tau(\xi_i) = \sigma_i$ where $\sigma_i \in K((s))$ and $s \in K[[t]]$ with $\text{ord}_t\, s = n$. By Lemma 11.66, $\text{ord}_s\, \sigma_0 = 1$ and $\text{ord}_s\, \sigma_i \geq 1$ for $i = 1, \dots, n-1$. From equation (11.29),

$$\sigma_0 + \sigma_1 t + \cdots + \sigma_{n-1} t^{n-1} + t^n = 0, \tag{11.31}$$

Derivation of (11.31) gives the equation

$$\left(\frac{d\sigma_0}{ds} + \frac{d\sigma_1}{ds}t + \cdots + \frac{d\sigma_{n-1}}{ds}t^{n-1}\right)\frac{ds}{dt}$$
$$+\sigma_1 + 2\sigma_2 t + \cdots + (n-1)\sigma_{n-1}t^{n-2} + nt^{n-1} = 0.$$

Define the polynomial $\Phi(T) \in K(s)[T]$:

$$\Phi(T) = \sigma_0 + \sigma_1 T + \cdots + \sigma_{n-1}T^{n-1} + T^n.$$

Then

$$\left(\frac{d\sigma_0}{ds} + \frac{d\sigma_1}{ds}t + \cdots + \frac{d\sigma_{n-1}}{ds}t^{n-1}\right)\frac{ds}{dt} = -\left.\frac{d\Phi(T)}{dT}\right|_{T=t}. \tag{11.32}$$

LEMMA 11.67

$$\Phi(T) = \textstyle\prod_{\alpha \in G_\mathcal{P}}(T - (\tau\alpha)(x)). \tag{11.33}$$

Proof. From (11.31), $\Phi(t) = 0$. To find the other roots of $\Phi(T)$, first note in (11.29) that $\alpha(\xi_i) = \xi_i$ for every $\alpha \in G_\mathcal{P}$. From (11.29), it follows that

$$\xi_0 + \xi_1\alpha(x) + \cdots + \xi_{n-1}\alpha(x)^{n-1} + \alpha(x)^n = 0.$$

Let $\delta(t) = (\tau\alpha)(x)$; then, since $\alpha(x)$ is also a uniformising element of Σ, so $\text{ord}_t\, \delta(t) = 1$. Applying τ to the last equation shows that

$$\sigma_0 + \sigma_1\delta(t) + \cdots + \sigma_{n-1}\delta(t)^{n-1} + \delta(t)^n = 0.$$

As α ranges over $G_{\mathcal{P}}$, the number of distinct $\delta(t)$ obtained is n. This depends on Lemma 11.43 taking into account that τ is a K-isomorphism from Σ to $K((t))$. Since $\Phi(T)$ has degree n, it follows that the roots of $\Phi(T)$ in Σ are the n distinct elements $\delta(t) = (\tau\alpha)(x)$ with α ranging over $G_{\mathcal{P}}$. □

From (11.33),

$$\left.\frac{d\Phi(T)}{dT}\right|_{T=t} = \prod\nolimits_{\alpha\in G^*_{\mathcal{P}}}(t-(\tau\alpha)(x)), \tag{11.34}$$

with α ranging over the set $G^*_{\mathcal{P}}$ of non-trivial K-automorphisms in $G_{\mathcal{P}}$. From (11.32) and (11.34),

$$\left(\frac{d\sigma_0}{ds}+\frac{d\sigma_1}{ds}t+\cdots+\frac{d\sigma_{n-1}}{ds}t^{n-1}\right)\frac{ds}{dt} = -\prod\nolimits_{\alpha\in G^*_{\mathcal{P}}}(t-(\tau\alpha)(x)).$$

Since $\mathrm{ord}_s\,\sigma_0 = 0$ and $\mathrm{ord}_s\,\delta_i \geq 1$ for $i = 1,\ldots,n-1$, this gives the equation

$$\mathrm{ord}_t\,\frac{ds}{dt} = \sum\nolimits_{\alpha\in G^*_{\mathcal{P}}}\mathrm{ord}_t((\tau\alpha)(x)-t). \tag{11.35}$$

Therefore the following result is obtained.

THEOREM 11.68 *For a subgroup G of $\mathrm{Aut}_K(\Sigma)$ and a place $\mathcal{P}$ of Σ, let $G^*_{\mathcal{P}}$ be the set of all non-trivial elements of $G_{\mathcal{P}}$. Then the different $D(\Sigma/\Sigma^G)$ of the separable extension Σ/Σ^G is given by the formula*

$$D(\Sigma/\Sigma^G) = \sum\nolimits_{\mathcal{P}\in\mathcal{P}_\Sigma}\left(\sum\nolimits_{\alpha\in G^*_{\mathcal{P}}}\mathrm{ord}_{\mathcal{P}}(\alpha(x)-x)\right)\mathcal{P} \tag{11.36}$$

for any uniformising element $x\in\Sigma$ at $\mathcal{P}$.

For $i = 0,1,\ldots$, define $G^{(i)}_{\mathcal{P}}$ to be the set of all elements $\alpha\in G_{\mathcal{P}}$ such that

$$\mathrm{ord}_{\mathcal{P}}(\alpha(x)-x)) \geq i+1.$$

For $\alpha,\beta\in G^{(i)}_{\mathcal{P}}$,

$$\begin{aligned}\mathrm{ord}_{\mathcal{P}}(\beta\alpha(x)-x) &= \mathrm{ord}_{\mathcal{P}}(\beta\alpha(x)-\alpha(x)+\alpha(x)-x)\\ &\geq \min\{\mathrm{ord}_{\mathcal{P}}(\beta(\alpha(x))-\alpha(x)),\mathrm{ord}_{\mathcal{P}}(\alpha(x)-x)\}.\end{aligned}$$

Since $\alpha(x)$ is still a uniformising element at $\mathcal{P}$, so

$$\mathrm{ord}_{\mathcal{P}}(\beta(\alpha(x))-\alpha(x)) \geq i+1.$$

Hence $\mathrm{ord}_{\mathcal{P}}(\beta\alpha(x)-x) \geq i+1$. This leads to the following definition.

DEFINITION 11.69 The *i-th ramification group* $G^{(i)}_{\mathcal{P}}$ at $\mathcal{P}$ is

$$G^{(i)}_{\mathcal{P}} = \{\alpha \mid \mathrm{ord}_{\mathcal{P}}(\alpha(x)-x) \geq i+1,\ \alpha\in G_{\mathcal{P}}\},$$

where $x\in\Sigma$ is a uniformising element of Σ at $\mathcal{P}$.

From Theorem 11.68, the following important result is deduced.

THEOREM 11.70 (Hilbert Different Formula)

$$D(\Sigma/\Sigma^G) = \sum\nolimits_{\mathcal{P}\in\mathcal{P}_\Sigma} d_{\mathcal{P}}\mathcal{P}, \quad \textit{with} \quad d_{\mathcal{P}} = \sum\nolimits_{i\geq 0}(|G^{(i)}_{\mathcal{P}}|-1). \tag{11.37}$$

EXAMPLE 11.71 As in Example 4.33, $\mathcal{H}_q = \mathbf{v}(Y^q + Y - X^{q+1})$ is the Hermitian curve; it is non-singular and contains the origin. If $\Sigma = K(x, y)$ with the relation $y^q + y = x^{q+1}$, then $\Sigma = K(\mathcal{H})$. For any $a, b \in \mathbf{F}_{q^2}$ with $b^q + b = a^{q+1}$, the birational transformation

$$\alpha_{(a,b)} : x' = \frac{x + ay}{a^q x + by + 1}, \quad y' = \frac{y}{a^q x + by + 1}$$

defines a K-automorphism of Σ. Such a K-automorphism fixes the place $\mathcal{P}$ arising from the unique branch of $\mathcal{H}_q$ centred at the origin. Further, these automorphisms form a subgroup G of $\mathrm{Aut}_K(\Sigma)$ of order q^3, and those with $a = 0$ form a subgroup M of order q. Since the tangent to $\mathcal{H}$ at the origin is $\mathbf{v}(Y)$, so x is a uniformising element of Σ at $\mathcal{P}$, and

$$\alpha_{(a,b)}(x) - x = \frac{-a^q x^2 + ay - bxy}{1 + a^q x + by}.$$

From Example 4.33, $\mathrm{ord}_{\mathcal{P}}\, x = 1$, $\mathrm{ord}_{\mathcal{P}}\, y = q + 1$. Hence

$$\mathrm{ord}_{\mathcal{P}}(\alpha_{(a,b)}(x) - x) = \begin{cases} 2 & \text{for} \quad a \neq 0, \\ q+2 & \text{for} \quad a = 0. \end{cases}$$

Therefore $G_{\mathcal{P}}^{(0)} = G_{\mathcal{P}}^{(1)} = G$ and $G_{\mathcal{P}}^{(2)} = \cdots G_{\mathcal{P}}^{(q+1)} = M$, but $G_{\mathcal{P}}^{(q+2)}$ is trivial. From this, $d_{\mathcal{P}} = 2(q^3 - 1) + (q-1)q = 2q^3 + q^2 - q - 2$.

Theorem 11.70 and Hurwitz's Theorem 7.27, see also Corollary 11.52, have the following corollary.

THEOREM 11.72 *For a finite K-automorphism group G of an irreducible curve $\mathcal{F}$, let $\mathcal{F}/G$ be the corresponding quotient curve; equivalently, in terms of function fields, $\Sigma = K(\mathcal{F})$ and $\Sigma^G = K(\mathcal{F}/G)$. Then*

$$2g - 2 = |G|(2g' - 2) + \textstyle\sum_{\mathcal{P} \in \mathcal{P}_\Sigma} \sum_{i \geq 0} (|G_{\mathcal{P}}^{(i)}| - 1), \tag{11.38}$$

where g and g' are the genera of $\mathcal{F}$ and $\mathcal{F}/G$.

EXAMPLE 11.73 With the notation as in Example 11.71, no non-trivial element in G fixes a place of Σ other than $\mathcal{P}$. Further, Σ^G is rational. Now (11.38) becomes $2g - 2 = -2q^3 + 2q^3 + q^2 - q - 2$, verifying that $\mathcal{H}_q$ has genus $g = \frac{1}{2}q(q-1)$.

The subgroup chain $G_{\mathcal{P}}^{(0)} \geq G_{\mathcal{P}}^{(1)} \geq \cdots$ is described in the following theorem.

THEOREM 11.74 (i) *$G_{\mathcal{P}}^{(0)}$ is the stabiliser of $\mathcal{P}$.*

(ii) *$G_{\mathcal{P}}^{(1)}$ is the unique maximal normal p-subgroup of $G_{\mathcal{P}}$.*

(iii) *For $i \geq 1$, the group $G_{\mathcal{P}}^{(i)}$ is normal in $G_{\mathcal{P}}$ and the group $G_{\mathcal{P}}^{(i)}/G_{\mathcal{P}}^{(i+1)}$ is an elementary abelian p-group.*

Proof. Parts (i) and (ii) follow from definitions; see also Lemma 11.44.

To prove (iii), let $i \geq 1$. From (11.34),

$$(\tau\alpha)(x) - t = c_\alpha t^i + a(t), \qquad a(t) \in K[[t]],\ \mathrm{ord}_t\, a(t) > i,$$

for every K-automorphism $\alpha \in G_{\mathcal{P}}^{(i)}$. The mapping

$$\phi : G_{\mathcal{P}}^{(i)} \to (K, +),$$
$$\alpha \mapsto c_\alpha \tag{11.39}$$

is a group homomorphism. show this, let $\alpha, \beta \in G_{\mathcal{P}}^{(i)}$ and

$$a(x) = \alpha(x) - x - c_\alpha x^i, \quad b(x) = \beta(x) - x - c_\beta x^i.$$

Then both $\text{ord}_{\mathcal{P}}\, b(x)$ and $\text{ord}_{\mathcal{P}}\, c(x)$ are greater than i. Now

$$\begin{aligned}(\alpha\beta)(x) = \alpha(\beta(x)) &= \alpha(x + c_\beta x^i + b(x)) = \alpha(x) + c_\beta \alpha(x)^i + \alpha(b(x)) \\ &= x + c_\alpha x^i + a(x) + c_\beta(x + c_\alpha x^i + a(x))^i + \alpha(b(x)) \\ &= x + (c_\alpha + c_\beta)x^i + d(x).\end{aligned}$$

Since $\text{ord}_{\mathcal{P}}\, d(x) > i$, the assertion follows. On the other hand, $\ker \phi$ consists of all K-automorphisms belonging $G_{\mathcal{P}}^{(i+1)}$. Hence $G_{\mathcal{P}}^{(i+1)}$ is a normal subgroup of $G_{\mathcal{P}}^{(i)}$, and the factor group $G_{\mathcal{P}}^{(i)}/G_{\mathcal{P}}^{(i+1)}$ is isomorphic to an additive subgroup of K. Part (iii) now follows from the fact that finite additive subgroups of K are abelian groups in which every non-trivial element has order p. □

In the subgroup chain, $G_{\mathcal{P}} = G_{\mathcal{P}}^{(0)} > G_{\mathcal{P}}^{(1)} > \cdots$, consecutive non-trivial subgroups may coincide. Sometimes, repetitions in the chain are deleted so that the Hilbert Different Formula (11.37) only involves distinct ramification groups. To do this, let

$$v_{-1} = 0 < v_0 < v_1 < \cdots$$

be the strictly increasing sequence of all non-negative integers which are of the form $\text{ord}_{\mathcal{P}}(\alpha(x) - x)$ for some $\alpha \in G_{\mathcal{P}}$. Then v_i is the i-th *ramification number* of $G_{\mathcal{P}}$. Note that

$$v_0 = \min\{\text{ord}_{\mathcal{P}}(\alpha(x) - x) \mid \alpha \in G_{\mathcal{P}}\},\ v_i = \min\{\text{ord}_{\mathcal{P}}(\alpha(x) - x) \mid \alpha \in U_{\mathcal{P}}^{(i)}\},$$

where $U_{\mathcal{P}}^{(0)} = G_{\mathcal{P}}$ and

$$U_{\mathcal{P}}^{(i+1)} = \{\alpha \in U_{\mathcal{P}}^{(i)} \mid \text{ord}_{\mathcal{P}}(\alpha(x) - x) \geq v_i + 1\}.$$

Also, $v_0 = 1$ if and only if $G_{\mathcal{P}} > G_{\mathcal{P}}^{(1)}$; that is, $G_{\mathcal{P}} = G_{\mathcal{P}}^{(1)} \rtimes H$ with H a non-trivial tame subgroup.

The subgroup chain $G_{\mathcal{P}} = U_{\mathcal{P}}^{(0)} > U_{\mathcal{P}}^{(1)} > \cdots$ is strictly decreasing, and

$$d_{\mathcal{P}} = \textstyle\sum_{i \geq 0} (v_i - v_{i-1})(|U_{\mathcal{P}}^{(i)}| - 1). \tag{11.40}$$

Therefore (11.38) can also be written as follows:

$$2g - 2 = |G|(2g' - 2) + \textstyle\sum_{\mathcal{P} \in \mathcal{P}_\Sigma} \sum_{i \geq 0} (v_i - v_{i-1})(|U_{\mathcal{P}}^{(i)}| - 1). \tag{11.41}$$

A useful fact is that $\alpha \in U_{\mathcal{P}}^{(i)} \setminus U_{\mathcal{P}}^{(i+1)}$ if and only if $v_i = \text{ord}_{\mathcal{P}}(\alpha(\xi) - \xi)$ for a uniformising element ξ of Σ.

Some further properties of the ramification subgroups may be proved by using companion K-automorphisms, introduced in Section 11.1. This essentially depends on the fact that, in terms of the companion K-automorphism λ_α of α, the equation, $\text{ord}_{\mathcal{P}}(\alpha(x) - x) = k$, means that

$$\lambda_\alpha(t) = t + ct^k + \cdots,\ c \neq 0.$$

LEMMA 11.75 *Let $\alpha \in G_{\mathcal{P}}$ and $\beta \in G_{\mathcal{P}}^{(k)}$, $k \geq 1$.*

(i) *If $\alpha \notin G_{\mathcal{P}}^{(1)}$, then the commutator $[\alpha, \beta] = \alpha\beta\alpha^{-1}\beta^{-1}$ belongs to $G_{\mathcal{P}}^{(k+1)}$ if and only if either $\alpha^k \in G_{\mathcal{P}}^{(1)}$ or $\beta \in G_{\mathcal{P}}^{(k+1)}$.*

(ii) *If $\alpha \in G_{\mathcal{P}}^{(l)}$ with $l \geq 1$, then the commutator $[\alpha, \beta]$ belongs to $G_{\mathcal{P}}^{(k+l+1)}$.*

(iii) *Let* $\operatorname{ord} \alpha = p^r s$ *with $p \nmid s$. If $\beta \notin G_{\mathcal{P}}^{(k+1)}$ and $\alpha\beta = \beta\alpha$, then s divides k.*

(iv) *Let k be an integer not divisible by $|G_{\mathcal{P}}|/|G_{\mathcal{P}}^{(1)}|$. If $G_{\mathcal{P}}$ is abelian, then $G_{\mathcal{P}}^{(k)} = G_{\mathcal{P}}^{(k+1)}$.*

(v) *The integers $k \geq 1$ satisfying $G_{\mathcal{P}}^{(k)} \neq G_{\mathcal{P}}^{(k+1)}$ are all congruent modulo p; in terms of ramification indices, $v_i \equiv v_j \pmod{p}$, for all $i, j \geq 1$.*

Proof. Since $G_{\mathcal{P}}^{(k)}$ is a normal subgroup of $G_{\mathcal{P}}$, the commutator $\gamma = [\alpha, \beta]$ belongs to $G_{\mathcal{P}}^{(k)}$. Write

$$\begin{aligned}\lambda_\alpha(t) &= ut + \textstyle\sum_{i \geq i_0} u_i t^i, \text{ with } u_{i_0} \neq 0,\\ \lambda_\beta(t) &= t + \textstyle\sum_{j \geq j_0} w_j t^j, \text{ with } w_{j_0} \neq 0,\\ \lambda_\gamma(t) &= t + \textstyle\sum_{m \geq m_0} v_m t^m.\end{aligned}$$

By Lemma 11.17, $\operatorname{ord} \alpha = p^r s$, where s is the smallest positive integer satisfying $u^s = 1$; the hypothesis that $\beta \in G_{\mathcal{P}}^{(k)}$ with $k \geq 1$ means that $k \leq j_0 - 1$. Also, it may be that $\lambda_\gamma(t) = t$, and this occurs if and only if $\alpha\beta = \beta\alpha$; otherwise, γ is a non-trivial K-automorphism belonging to $\in G_{\mathcal{P}}^{(m_0-1)}$. Then

$$\begin{aligned}\lambda_\beta\lambda_\alpha(t) &= ut + \textstyle\sum_{i \geq i_0} u_i t^i + \sum_{j \geq j_0} w_j (ut + \sum_{i \geq i_0} u_i t^i)^j,\\ \lambda_\gamma\lambda_\alpha\lambda_\beta(t) &= ut + u\textstyle\sum_{j \geq j_0} w_j t^j + \sum_{i \geq i_0} u_i (t + \sum_{j \geq j_0} w_j t^j)^i\\ &\quad + \textstyle\sum_{m \geq m_0} v_m [ut + u\sum_{j \geq j_0} w_j t^j + \sum_{i \geq i_0} u_i (t + \sum_{j \geq j_0} w_j t^j)^i]^m.\end{aligned}$$

If $u \neq 1$, then $\beta\alpha = \gamma\alpha\beta$ implies that

$$0 = w_{j_0}(u^{j_0} - u)t^{j_0} + v_{m_0} u^{m_0} t^{m_0} + ct^r + \cdots,$$

where $r \geq j_0 + 1$. Hence two cases can occur:

(a) $u^{j_0-1} = 1$ and either $\gamma = 1$, or $m_0 \geq j_0 + 1$;

(b) $u^{j_0-1} \neq 1$ and $j_0 = m_0$.

In the former case, α^{j_0-1} has order a power of p, while in the latter case $\gamma \neq 1$. Now a straightforward argument proves (i).

If $u = 1$, then $\alpha\beta = \gamma\beta\alpha$ implies $m_0 \geq i_0 + j_0 - 1$. Hence

$$m_0 - 1 \geq (i_0 - 1) + (j_0 - 1) + 1,$$

whence (ii) follows.

Let $u \neq 1$. If $\alpha\beta = \beta\alpha$, then $\lambda_\gamma = t$, and hence $u^{j_0-1} = 1$. Thus $\alpha^{j_0-1} \in G_{\mathcal{P}}^{(1)}$. This implies that $s \mid (j_0 - 1)$. If $\beta \notin G_{\mathcal{P}}^{(k+1)}$, then $k = j_0 - 1$, and hence $s \mid k$, as stated in (iii).

Since $G_{\mathcal{P}}$ is the semidirect product of its normal subgroup $G_{\mathcal{P}}^{(1)}$ by a cyclic subgroup of order $s = |G_{\mathcal{P}}|/|G_{\mathcal{P}}^{(1)}|$, some $\alpha \in G_{\mathcal{P}}$ has order s. If k does not divide s, then $\alpha^k \neq 1$. By (i), $\beta \in G_{\mathcal{P}}^{(k+1)}$. This proves (iv).

Choose an element α from the last non-trivial ramification group, say $G_{\mathcal{P}}^{(l)}$; then $|G_{\mathcal{P}}^{(l+1)}| = 1$. Since $\beta \in G_{\mathcal{P}}^{(k)}$ with $k \geq 1$, from (ii) $\alpha\beta = \beta\alpha$ follows. To show that $i_0 - j_0$ is divisible by p, assume that p does not divide both i_0 and j_0. Then $\alpha\beta = \beta\alpha$ implies that

$$0 = (j_0 - i_0)u_{j_0}w_{j_0}t^{j_0-1+i_0} + \cdots .$$

Therefore $j_0 \equiv i_0 \pmod{p}$. However, $\beta \notin G_{\mathcal{P}}^{(k+1)}$ implies that $k = j_0 - 1$; similarly, $l = i_0 - 1$. Hence $k \equiv l \pmod{p}$. From this, (v) follows. □

The following two results are corollaries of Lemma 11.75 (i).

LEMMA 11.76 *Let A be an abelian subgroup of* $\mathrm{Aut}_K(\Sigma)$ *that fixes $m \geq 1$ places.*

(i) *If $\mathcal{P}$ is one of the fixed places and g' is the genus of Σ^A, then*

$$2g - 2 \geq |A|(2g' - 2) + m\Big(|A| - 1 + \frac{|A|}{|A_{\mathcal{P}}^{(1)}|}(|A_{\mathcal{P}}^{(1)}| - 1)\Big). \tag{11.42}$$

(ii) *If G is non-tame, then*

$$2g - 2 \geq |A|(2g' + \tfrac{3}{2}m - 2) - m. \tag{11.43}$$

Proof. For a fixed place $\mathcal{P} \in \mathcal{P}_\Sigma$ of A, let $A = A_{\mathcal{P}} = A_{\mathcal{P}}^{(1)} \rtimes H$ with $p \nmid |H|$; see Theorem 11.49. Since A is abelian, from Lemma 11.75 (i), it follows that $A_{\mathcal{P}}^{(1)} = \cdots = A_{\mathcal{P}}^{(h)}$ with $h = |H|$. Now (11.42) follows from Theorem 11.72 applied to A. If $A_{\mathcal{P}}^{(1)}$ is non-trivial, then $|A_{\mathcal{P}}^{(1)}| - 1 \geq \frac{1}{2}|A_{\mathcal{P}}^{(1)}|$. Hence (11.42) implies (11.43). □

LEMMA 11.77 *Let G be an abelian non-tame subgroup of* $\mathrm{Aut}_K(\Sigma)$ *fixing a place $\mathcal{P}$. If $G = G_{\mathcal{P}}^{(1)} \rtimes H$ with $p \nmid |H|$, then*

$$2g \geq 2g'|G_{\mathcal{P}}^{(1)}| + (|H| - 1)(|G_{\mathcal{P}}^{(1)}| - 1). \tag{11.44}$$

Proof. Since G is abelian, from (i) of Lemma 11.75, $G_{\mathcal{P}}^{(1)} = \cdots = G_{\mathcal{P}}^{(h)}$ where $h = |H|$. Then (11.44) follows from Theorem 11.72 applied to $G_{\mathcal{P}}^{(1)}$. □

From Theorem 11.49, if $g > 0$, then $G_{\mathcal{P}}$ is finite and its order has an upper bound depending only on p and g. In the notation of this section, Theorem 11.49 states that $|G_{\mathcal{P}}^{(1)}| \leq p^2(g+1)(2g-1)^2$. The aim is to improve this bound.

THEOREM 11.78 *Let $\mathcal{F}$ be an irreducible curve of positive genus g, and let $G_\mathcal{P}$ be a K-automorphism group of $\mathcal{F}$ fixing a place $\mathcal{P}$. Then*

$$|G_\mathcal{P}^{(1)}| \leq \frac{4p}{p-1}g^2.$$

More precisely, if F_i is the fixed subfield of $G_\mathcal{P}^{(i)}$, then one of the following cases occurs:

(i) *F_1 is not rational, and $|G_\mathcal{P}^{(1)}| \leq g$;*

(ii) *F_1 is rational, $G_\mathcal{P}^{(1)}$ has a short orbit on $\mathcal{P}_\Sigma$ other than $\{\mathcal{P}\}$, and*

$$|G_\mathcal{P}^{(1)}| \leq \frac{p}{p-1}g;$$

(iii) *F_1 and F_2 are rational, $\{\mathcal{P}\}$ is the unique short orbit of $G_\mathcal{P}^{(1)}$ on $\mathcal{P}_\Sigma$, and*

$$|G_\mathcal{P}^{(1)}| \leq \frac{4|G_\mathcal{P}^{(2)}|}{(|G_\mathcal{P}^{(2)}|-1)^2}g^2 \leq \frac{4p}{(p-1)^2}g^2.$$

Proof. From Theorem 11.72 applied to $G_\mathcal{P}^{(1)}$,

$$2g-2 = |G_\mathcal{P}^{(1)}|(2g_1-2)+d, \tag{11.45}$$

where g_1 is the genus of F_1 and $d = \deg D(\Sigma/F_1)$. Since $\mathcal{P}$ completely ramifies in Σ/F_1, so $e_\mathcal{P} = |G_\mathcal{P}^{(1)}|$. Since $|G_\mathcal{P}^{(1)}|$ is a power of p, hence $|G_\mathcal{P}^{(1)}|$ is equal to the order of the first ramification group. Write $D(\Sigma/F_1) = \sum d_\mathcal{P}\mathcal{P}$. By the Hilbert Different Formula (11.37),

$$d_\mathcal{P} \geq 2|G_\mathcal{P}^{(1)}|-2. \tag{11.46}$$

When $g_1 > 0$, (11.45) and (11.46) imply that

$$2g-2 \geq d \geq d_\mathcal{P} \geq 2|G_\mathcal{P}^{(1)}|-2,$$

whence $|G_\mathcal{P}^{(1)}| \leq g$.

From now on, assume that $g_1 = 0$. Then (11.45) becomes the following:

$$2g-2 = -2|G_\mathcal{P}^{(1)}|+d. \tag{11.47}$$

In case (ii), there is one more place $\mathcal{Q}$ with the above properties. As usual, $e_\mathcal{Q}$ is its ramification index, and $d_\mathcal{Q}$ is its weight in the different divisor $D(\Sigma/F_1)$. In particular, $e_\mathcal{Q} \leq |G_\mathcal{P}^{(1)}|$. Since $e_\mathcal{Q}$ is a power of p and F_1 is rational, again from the Hilbert Different Formula (11.37), $d_\mathcal{Q} \geq 2e_\mathcal{Q}-2$. Taking (11.46) into account, it follows that

$$\begin{aligned} 2g-2 &\geq -2|G_\mathcal{P}^{(1)}| + d_\mathcal{P} + \frac{|G_\mathcal{P}^{(1)}|}{e_\mathcal{Q}}d_\mathcal{Q} \\ &\geq -2|G_\mathcal{P}^{(1)}| + 2|G_\mathcal{P}^{(1)}| - 2 + 2|G_\mathcal{P}^{(1)}|\frac{e_\mathcal{Q}-1}{e_\mathcal{Q}}; \end{aligned}$$

hence

$$g \geq \frac{e_Q - 1}{e_Q}|G_{\mathcal{P}}^{(1)}| \geq \frac{p-1}{p}|G_{\mathcal{P}}^{(1)}|,$$

which is the desired result.

Now consider the case that F_1 is rational and $\mathcal{P}$ is the unique ramified place. First, it is shown that F_2 is rational as well. Put

$$d_1 = \deg D(\Sigma/F_1), \quad d_2 = \deg D(\Sigma/F_2), \quad d^* = \deg D(F_2/F_1).$$

From (7.8),

$$d_1 = d_2 + [\Sigma : F_2]d^* = d_2 + |G_{\mathcal{P}}^{(2)}|d^*. \tag{11.48}$$

Also,

$$\begin{aligned} d_1 &= 2(|G_{\mathcal{P}}^{(1)}| - 1) + \textstyle\sum_{i\geq 2}(|G_{\mathcal{P}}^{(i)}| - 1), \\ d_2 &= 2(|G_{\mathcal{P}}^{(2)}| - 1) + \textstyle\sum_{i\geq 2}(|T^{(i)}| - 1), \end{aligned} \tag{11.49}$$

where $T^{(i)}$ denotes the i-th ramification group of Σ/F_2 at $\mathcal{P}$. As F_2 is the fixed field of $G_{\mathcal{P}}^{(2)}$, the groups $T^{(i)}$ and $G_{\mathcal{P}}^{(i)}$ coincide for $i \geq 2$. Hence

$$d_1 - 2(|G_{\mathcal{P}}^{(1)}| - 1) = d_2 - 2(|G_{\mathcal{P}}^{(2)}| - 1),$$

which, together with (11.48), gives the relation

$$d^* = 2\frac{|G_{\mathcal{P}}^{(1)}|}{|G_{\mathcal{P}}^{(2)}|} - 2.$$

Then Theorem 11.72 shows that F_2 has genus 0; that is, F_2 is rational.

Let k denote the smallest integer such that $k \geq 3$ and $G_{\mathcal{P}}^{(k)} \neq G_{\mathcal{P}}^{(2)}$. All ramification groups, and thus both $G_{\mathcal{P}}^{(2)}$ and $G_{\mathcal{P}}^{(k)}$, are normal subgroups of $G_{\mathcal{P}}$. Hence there is a normal subgroup G' containing $G_{\mathcal{P}}^{(k)}$, which is a subgroup of $G_{\mathcal{P}}^{(2)}$ of index p.

The next step is to compute the genus g' of the fixed field F' of G'. Let

$$d' = \deg D(\Sigma/F'), \quad \overline{d} = \deg D(F'/F_2).$$

As before, the following equations hold:

$$d_2 = d' + p^{-1}|G_{\mathcal{P}}^{(2)}|\overline{d}, \quad d_2 = k(|G_{\mathcal{P}}^{(2)}| - 1) + \textstyle\sum_{i\geq k}(|G_{\mathcal{P}}^{(i)}| - 1).$$

Also,

$$d' = k(p^{-1}|G_{\mathcal{P}}^{(p)}| - 1) + \textstyle\sum_{i\geq k}(|G_{\mathcal{P}}^{(i)}| - 1),$$

as G' lies between $G_{\mathcal{P}}^{(2)}$ and $G_{\mathcal{P}}^{(k)}$. It follows that $\overline{d} = k(p-1)$. Now Theorem 11.72 shows that

$$g' = \tfrac{1}{2}(k-2)(p-1). \tag{11.50}$$

Since $k \geq 3$, so F' is not rational. Let $\mathcal{P}'$ the place of F' lying under $\mathcal{P}$ in the extension Σ/F', and let $G_{\mathcal{P}'}^{(1)}$ be the stabiliser of $\mathcal{P}'$ in the K-automorphism group of F'. As G' is a normal subgroup of $G_{\mathcal{P}}^{(1)}$, the factor group $G_{\mathcal{P}}^{(1)}/G'$ can be viewed as a subgroup of $G^{(1)}_{\mathcal{P}'}$. Since $[G^{(2)}_{\mathcal{P}} : G'] = p$, so $[F' : F_2] = p$. Also, $\mathcal{P}'$ is the only place which ramifies in $F'|F_2$. Choose $x \in F_2$ having $\mathcal{P}'$ as a pole. Then F'/F_2 is an Artin–Schreier extension, and Theorem 12.5 ensures the existence of $y \in F'$ such that $y^p - y = B(x)$ with $B(X) \in K[X]$. From Theorem 12.7,

$$|G^{(1)}_{\mathcal{P}'}| \leq \frac{4p}{(p-1)^2}\, g'^2.$$

As the quotient group $G^{(1)}_{\mathcal{P}}/G'$ is isomorphic to a subgroup of $G^{(1)}_{\mathcal{P}'}$, from (11.50),

$$|G^{(1)}_{\mathcal{P}}| \leq |G'|\frac{4p}{(p-1)^2}g'^2 = \frac{4|G^{(2)}_{\mathcal{P}}|}{(p-1)^2}g'^2 = |G^{(2)}_{\mathcal{P}}|(k-2)^2. \tag{11.51}$$

From Theorem 11.72 applied to $G^{(2)}_{\mathcal{P}}$,

$$2g - 2 \geq -2|G^{(2)}_{\mathcal{P}}| + k(|G^{(2)}_{\mathcal{P}}| - 1),$$

whence

$$k - 2 \leq \frac{2g}{|G^{(2)}_{\mathcal{P}}| - 1}.$$

By substituting this in (11.51) the final result is obtained:

$$|G^{(1)}_{\mathcal{P}}| \leq \frac{4|G^{(2)}_{\mathcal{P}}|}{(|G^{(2)}_{\mathcal{P}}| - 1)^2}\, g^2.$$

□

The next objective is to obtain sharper estimates on the orders of abelian K-automorphism groups of Σ.

THEOREM 11.79 *Let G be an abelian K-automorphism group of an irreducible curve $\mathcal{F}$ of genus $g \geq 2$. Then*

$$|G| \leq \begin{cases} 4g+4 & \text{for} \quad p \neq 2, \\ 4g+2 & \text{for} \quad p = 2. \end{cases}$$

Proof. Put $\Sigma = K(\mathcal{F})$. If Σ/Σ^G does not ramify, then $|G| \leq g - 1$ by Theorem 11.72. Another case in which the proof is a direct consequence of Theorem 11.72 when $|G| = p$, and Σ^G is rational. In fact, if k denotes the number of non-trivial ramification groups of G at $\mathcal{P}$, then $2g-2 = -2p+k(p-1)$ follows from Theorem 11.72. This implies that $k \geq 3$. whence $|G| \leq 2g+1 < 4g+2$.

Let $\mathcal{P}$ be a place of Σ which ramifies in Σ/Σ^G. Since G is abelian, $G_{\mathcal{P}}$ fixes every place in the orbit of $\mathcal{P}$ under G. Let $A = G_{\mathcal{P}}$ and $a = |A|$. Then $|G|/a$ is the size of the orbit of $\mathcal{P}$ under G. Let g' denote the genus of Σ^A. From Theorem 11.72 applied to A,

$$2g - 2 \geq a(2g'-2) + |G|(a-1)/a \geq a(2g'-2) + \tfrac{1}{2}|G|.$$

If $g' \geq 1$, this gives $|G| \leq 4g - 4$. Therefore Σ^A is assumed to be rational. Then Theorem 11.72 applied to A shows that

$$2g - 2 \geq -2a + |G|(a-1)/a, \tag{11.52}$$

and the inequality is strict when A is non-tame.

Suppose that $a = 2$. Then (11.52) implies that $|G| \leq 4g + 4$ giving the assertion for $p > 2$. If $p = 2$, then A is non-tame, and hence $|G| < 4g + 4$. To prove the assertion for $p = 2$, the possibility of $|G| = 4g + 3$ must be ruled out. In such a case, G is tame, and the strict inequality occurs in (11.52). From Theorem 11.57, some place $\mathcal{Q}$ outside the G-orbit of $\mathcal{P}$ must be ramified in Σ/Σ^G. In other words, $G_{\mathcal{Q}}$ contains a non-trivial element α. Since α has order at least 3, this implies that (11.52) remains valid when 2 is subtracted from the right-hand side. But then $|G| \leq 4g + 2$, a contradiction. Therefore (11.79) holds for $a = 2$.

Suppose that $a = 3$. From (11.52), $|G| \leq 3g+6$ which implies $|G| \leq 4g+2$ for $g > 3$. The preceding argument can be used to show that if $p = 2$ then $|G| = 14$ is impossible. Therefore (11.79) holds for $a = 3$.

From now on, let $a > 3$. Theorem 11.79 holds if the size ℓ of the G-orbit of $\mathcal{P}$ is at least 4. In fact, from (11.52),

$$2g + 2 \geq \tfrac{1}{2}|G| + \tfrac{1}{2}\ell(a-1) - 2a + 4 \geq \tfrac{1}{2}|G| + (\tfrac{1}{2}a - 1)(\ell - 4), \qquad (11.53)$$

which implies that $|G| \leq 4g + 4$ for $\ell \geq 4$ (and $|G| \leq 4g + 2$ for $\ell \geq 5$). The possibility of $4g + 3 \leq |G| \leq 4g + 4$ for $p = 2$ can be ruled out by the argument previously used for $a = 2$.

Let $\ell_1, \ldots, \ell_s$ be the sizes of the short G-orbits. From the above result, $\ell_i \leq 3$.

Suppose that A is tame. Since Σ^G is rational, Theorem 11.57 applied to G gives the relation

$$2g + 1 = -2|G| + s|G| - (\ell_1 + \cdots + \ell_s) + 3.$$

In particular, $s \geq 3$. If $\ell_i = 3$ for some $i = 1$, then $|G| \geq 12$, and

$$2g + 1 \geq \tfrac{1}{2}|G| + (s - \tfrac{5}{2})|G| - 3s + 3 \geq \tfrac{1}{2}|G|,$$

whence $|G| \leq 4g + 2$. Otherwise, $\ell_1 = \cdots = \ell_s = 2$, $|G| \geq 8$, and, in this case,

$$2g + 1 = -2|G| + s|G| - 2s + 3 = \tfrac{1}{2}|G| + (s - \tfrac{5}{2})|G| - 2s + 3 > \tfrac{1}{2}|G|,$$

which gives the result.

Suppose that A is non-tame. If a short G-orbit has length 3, then A has at least three fixed places. Now, for $m = 3$, (11.43) gives the inequality

$$2g + 1 \geq \tfrac{5}{2}|A| = \tfrac{5}{6}|G|,$$

whence $|G| \leq \frac{1}{5}(12g + 6) < 4g + 2$. The same argument with $m = 2$ shows that, if G has a short orbit of length 2, then $2g \geq |A| = \frac{1}{2}|G|$, whence $|G| \leq 4g$. Also, if G fixes at least two places, then (11.43) applied to G gives $|G| \leq 2g$.

It remains to consider the case that $G = A = G_{\mathcal{P}}$, that $G_{\mathcal{P}}^{(1)}$ is non-trivial and that the identity is the only K-automorphism in $G_{\mathcal{P}}$ that fixes a place distinct from $\mathcal{P}$. Write $G_{\mathcal{P}} = G_{\mathcal{P}}^{(1)} \rtimes H$ with $p \nmid |H|$.

Suppose that H is non-trivial. From Lemma 11.77,

$$2g \geq \tfrac{1}{2}(|H| - 1)(|G_{\mathcal{P}}^{(1)}| - 1).$$

Since $|H| = |G_{\mathcal{P}}^{(1)}| = 2$ is not possible, so

$$(|H| - 1)(|G_{\mathcal{P}}| - 1) \geq \tfrac{1}{2}|H||G_{\mathcal{P}}| - 1$$

Therefore $|G| = |G_{\mathcal{P}}| = |G_{\mathcal{P}}||H| \leq 4g + 2$.

Suppose that $G = G_{\mathcal{P}} = G_{\mathcal{P}}^{(1)}$. A slightly sharper bound is shown, namely that $|G| \leq 4g$. By Theorem 11.99, this assertion holds for $g = 2$, but assume on the contrary that it does not extend to every $g > 2$. Let g be denote the minimum value of such exceptional genera g. Then there is an abelian subgroup $G = G_{\mathcal{P}}^{(1)}$ of $\mathrm{Aut}_K(\Sigma)$ that fixes no place distinct from $\mathcal{P}$ and that its order $|G|$ is at least $4g+1$.

Since $|G| \neq p$, it has a non-trivial subgroup M. If Σ^M has genus $g' \geq 2$, Theorem 11.72 applied to M implies that $g \geq g'|M|$. Therefore

$$4g + 1 \geq 4g'|M| + 1,$$

whence $|G| \geq 4g'|M| + 1$. Since $|G|/|M|$ is an integer, it follows that that $|G|/|M| \geq 4g' + 1$. Note that $\bar{G} = G/M$ can be viewed as a K-automorphism group of Σ^M whose genus g' is smaller than g. From what has been shown so far, $\bar{G} = \bar{G}_{\mathcal{P}'}$, where $\mathcal{P}'$ is the place of Σ^M lying under $\mathcal{P}$ in the extension Σ/Σ^M, and $\bar{G}$ fixes no place of Σ^M distinct from $\mathcal{P}'$. But this contradicts the minimal choice of g.

If Σ^M is elliptic, Theorem 11.72 applied to M implies that $g \geq |M|$. Since $\bar{G} = G/M$ can be viewed as a K-automorphism group of Σ^M, Theorem 11.94 shows that either $p = 2$ and $|G|/|M| = 2, 4$, or $p = 3$ and $|G/M| = 3$. Therefore $|G| \leq 4g$ for $p = 2$ and $|G| \leq 3g$ for $p = 3$.

If Σ^M is rational for every non-trivial subgroup M of G, Lemma 11.45 implies that either $|G| = p$, or $|G| = p^2$. Therefore $|G| = p^2$ is assumed. If $p = 2$ then $|G| = 4 \leq 4g + 1$. Let $p \geq 3$. Since $g > 0$, Theorem 11.72 implies that $G_{\mathcal{P}}^{(2)}$ is not trivial. From this, $|G| \leq 2g - 1$ when $G = G_{\mathcal{P}}^{(1)} = G_{\mathcal{P}}^{(2)}$. In the case where $|G_{\mathcal{P}}^{(1)}| = p^2$ and $|G_{\mathcal{P}}^{(2)}| = p$, Lemma 11.75 (v) gives the equalities

$$G_{\mathcal{P}}^{(2)} = G_{\mathcal{P}}^{(3)} = \cdots = G_{\mathcal{P}}^{(p+1)}.$$

Therefore Theorem 11.72 gives $2g \geq p(p-1)$, whence $4g > p^2 = |G|$. □

The following lemmas are related to the Hilbert Different Formula (11.37) and Theorem 11.72; they are used in later sections.

LEMMA 11.80 *Let $D(\Sigma/\Sigma^G) = \sum d_{\mathcal{P}}\mathcal{P}$. Then*

$$d_{\mathcal{P}} \geq |G_{\mathcal{P}}| + |G_{\mathcal{P}}^{(1)}| - 2.$$

LEMMA 11.81 *Suppose that a subgroup M of $G_{\mathcal{P}}^{(1)}$ contains no non-trivial element from the second ramification group $G_{\mathcal{P}}^{(2)}$. If $H \leq G_{\mathcal{P}}$ is in the normaliser of M and $|H|$ is prime to p, then $|H| \leq |M| - 1$.*

Proof. Let $h \in H$ commute with a non-trivial element of M. By Theorem 11.75 (i), this can only occur when $h = 1$. So $|M| > |H|$. □

LEMMA 11.82 *Let $G = G_{\mathcal{P}}$ be a K-automorphism group of an irreducible curve $\mathcal{F}$ that fixes a place $\mathcal{P}$, and let $\nu = d_{\mathcal{P}} - |G_{\mathcal{P}}|$ with $d_{\mathcal{P}} = d_{\mathcal{P}}(\Sigma/\Sigma^{G_{\mathcal{P}}})$ and $q_i = [U_{\mathcal{P}}^{(i)} : U_{\mathcal{P}}^{(i+1)}]$. Then, with v_i the i-th ramification number of G at $\mathcal{P}$,*

(i) $\nu = (\mu + 1)|H| - 1$ *with* μ *a non-negative integer*;

(ii) $|H| \mid (v_i - 1)(q_i - 1)$;

(iii) $q_i \mid (v_{i+1} - v_i)^2$, *for every* $i \geq 1$.

Proof. (i) Since $d_{\mathcal{P}}(\Sigma/\Sigma^H) = |H| - 1$, and $d_{\mathcal{P}}(\Sigma/\Sigma^{G_{\mathcal{P}}}) = |H||G_{\mathcal{P}}^{(1)}| + v$, equation (7.8) reads as follows:

$$|H|\,|G_{\mathcal{P}}^{(1)}| + \nu = |H| - 1 + |H|\,d_{\mathcal{P}'}(\Sigma^H/\Sigma^{G_{\mathcal{P}}}),$$

whence (i) follows for $\mu = d_{\mathcal{P}'}(\Sigma^H/\Sigma^{G_{\mathcal{P}}}) - |G_{\mathcal{P}}^{(1)}|$. Since $[\Sigma^H : \Sigma^{G_{\mathcal{P}}}] = |G_{\mathcal{P}}^{(1)}|$ and p divides $|G_{\mathcal{P}}^{(1)}$, from the remark after the Hurwitz Theorem 7.27 it follows that $d_{\mathcal{P}'}(\Sigma^H/\Sigma^{G_{\mathcal{P}}}) \geq |G_{\mathcal{P}}^{(1)}|$, whence $\mu \geq 0$.

(ii) For $i \geq 1$, put $U_i = U_{\mathcal{P}}^{(i)} \rtimes H$. Applying (i) to U_i gives that

$$v_i \equiv -1 \pmod{|H|}.$$

Hence $|H|$ divides $v_i - v_{i+1}$. This, together with the equations

$$v_i - v_{i+1} = v_i(|U_{\mathcal{P}}^{(i)}| - |U_{\mathcal{P}}^{(i+1)}|) - |U_{\mathcal{P}}^{(i)}| + |U_{\mathcal{P}}^{(i+1)}| = |U_{\mathcal{P}}^{(i+1)}|(v_i - 1)(q_i - 1),$$

proves (ii).

(iii) It may assumed that $i = 1$ after replacing $G_{\mathcal{P}}$ by $U_{\mathcal{P}}^{(i)} \rtimes H$. Also, take $|U_{\mathcal{P}}^{(2)}| = p$, after replacing Σ by Σ^N, where N is a normal subgroup of $U_{\mathcal{P}}^{(2)}$ of index p.

First, consider the case that $G_{\mathcal{P}}^{(1)}$ is not abelian. Note that a non-abelian group has non–trivial commutator subgroup. Hence the commutator subgroup C of $G_{\mathcal{P}}^{(1)}$ is non–trivial. By Theorem 11.74 (iii), the quotient group $G_{\mathcal{P}}^{(1)}/U_{\mathcal{P}}^{(2)}$ is abelian, and hence $U_{\mathcal{P}}^{(2)}$ contains C. Since $U_{\mathcal{P}}^{(2)}$ has no non-trivial subgroup, this implies that $C = U_{\mathcal{P}}^{(2)}$. Also, $Z(G_{\mathcal{P}}^{(1)}) \cap U_{\mathcal{P}}^{(2)}$ is non-trivial, as $U_{\mathcal{P}}^{(2)}$ is a normal subgroup of $G_{\mathcal{P}}^{(1)}$; see Theorem 11.74 (ii). Thus $Z(G_{\mathcal{P}}^{(1)})$ contains C, and hence $G_{\mathcal{P}}^{(1)}/Z(G_{\mathcal{P}}^{(1)})$ is an elementary abelian p–group.

From this,

$$[G_{\mathcal{P}}^{(1)} : Z(G_{\mathcal{P}}^{(1)})] = p^{2m},$$

and, if A is a maximal abelian normal subgroup of $G_{\mathcal{P}}^{(1)}$, then

$$Z(G_{\mathcal{P}}^{(1)}) \leq A \quad \text{and} \quad |A| = p^m|Z(G_{\mathcal{P}}^{(1)})|.$$

Since every non-trivial element in $A/G_{\mathcal{P}}^{(2)}$ has order p, the possible orders of elements in A are $1, p, p^2$. Therefore the Main Theorem for Finite Abelian Groups implies that A is a direct product of cyclic groups

$$A = A_1 \times \langle \tau_1 \rangle \times \cdots \times \langle \tau_{m+n-2} \rangle,$$

where

$$|A_1| = p^2, \quad Z(G_{\mathcal{P}}^{(1)}) = p^n, \quad \tau_1^p = \cdots = \tau_{m+n-2}^p = 1,$$

and A_1 contains $U_{\mathcal{P}}^{(2)}$. In particular, $|G_{\mathcal{P}}^{(1)}| = p^{2m+n}$ and $q_1 = p^{2m+n-1}$.

Let $A_2 = \langle \tau_1 \rangle \times \cdots \times \langle \tau_{m+n-2} \rangle$ and $B = A_2 U_{\mathcal{P}}^{(2)}$. Then $|A_2| = p^{m+n-2}$ and $|B| = p^{m+n-1}$. Also, $A_2 \triangleleft B \triangleleft A$.

Now the contribution of $\mathcal{P}$ to $d = \deg(D(\Sigma/\Sigma^B)$ is calculated in two different ways by applying (7.8) to the towers $\Sigma \supset \Sigma^{U_{\mathcal{P}}^{(2)}} \supset \Sigma^B$ and $\Sigma \supset \Sigma^{A_2} \supset \Sigma^B$. This gives the equation

$$v_1(p^{m+n-2} - 1)p + v_2(p-1) = \bar{v}_2(p-1)p^{m+n-2} + v_1(p^{m+n-2} - 1),$$

where $\bar{v}_2$ is the ramification number in the extension Σ^{A_2}/Σ^B. It follows that

$$v_2 = \bar{v}_2 p^{m+n-2} - v_1(p^{m+n-2} - 1).$$

In particular, $\bar{v}_2 > v_1$. Since the group A/A_2 is abelian and v_1 is also the ramification number in the extension Σ^{A_2}/Σ^A, Lemma 11.75 (v) implies that $\bar{v}_2 \equiv v_1 \pmod{p}$; that is, $\bar{v}_2 = v_1 + kp$ for a positive integer k. Therefore

$$v_2 = v_1 p^{m+n-2} + kp^{m+n-1} - v_1 p^{m+n-2} + v_1 = v_1 + kp^{m+n-1},$$

whence $v_2 - v_1 = kp^{m+n-1}$. This implies that q_1 divides $(v_2 - v_1)^2$.

If $G_{\mathcal{P}}^{(1)}$ is abelian, then the Main Theorem for Finite Abelian Groups applies directly to $G_{\mathcal{P}}^{(1)}$. Therefore $G_{\mathcal{P}}^{(1)}$ contains an elementary abelian subgroup A such that

$$A \cap U_{\mathcal{P}}^{(2)} = \{1\} \quad \text{and} \quad |A| = |U_{\mathcal{P}}^{(2)}|/p^2.$$

Using the above argument, the sharper assertion that $q_1 \mid (v_2 - v_1)$ follows. □

LEMMA 11.83 *Let $z \in \Sigma$ be an element such that $p \nmid \operatorname{ord}_{\mathcal{P}} z$. If $\alpha \in U_{\mathcal{P}}^{(i)} \setminus U_{\mathcal{P}}^{(i+1)}$, then $\alpha(z) = z + z'$ with $z' \neq 0$, and $v_i = 1 - \operatorname{ord}_{\mathcal{P}} z + \operatorname{ord}_{\mathcal{P}} z'$.*

Proof. There exists $c \in K\backslash\{0\}$ such that $z' = \alpha(z) - cz$ with $\operatorname{ord}_{\mathcal{P}} z' > a$ and $a = \operatorname{ord}_{\mathcal{P}} z$. Equivalently, $\alpha(z) = cz + z'$. If $\operatorname{ord} \alpha = p^m$, then $c^{p^m} = 1$. Therefore $c = 1$. Note that $z' \neq 0$, otherwise $z \in \Sigma^\alpha$ which contradicts $p \mid a$ by (7.1).

Now let $x \in \Sigma^\alpha$ with $\operatorname{ord}_{\mathcal{P}'} x = 1$ where $\mathcal{P}'$ is the place of Σ^α lying under $\mathcal{P}$. By (7.1), $b = \operatorname{ord}_{\mathcal{P}} x = p^m$ and hence $\gcd(a, b) = 1$. Further, choose integers n_1, n_2 such that $n_1 a + n_2 b = 1$. Now $p \mid b$ implies that $p \nmid n_1$ and that $w = z^{n_1} x^{n_2}$ is a uniformising element at $\mathcal{P}$. Therefore

$$\begin{aligned} v_i &= \operatorname{ord}_{\mathcal{P}}(\alpha(w) - w) \\ &= \operatorname{ord}_{\mathcal{P}}((z + z')^{n_1} x^{n_2} - z^{n_1} x^{n_2}) \\ &= \operatorname{ord}_{\mathcal{P}}(n_1 z^{n_1 - 1} z' x^{n_2}) \\ &= (n_1 - 1)a + n_2 b + \operatorname{ord}_{\mathcal{P}} z' \\ &= 1 - \operatorname{ord}_{\mathcal{P}} z + \operatorname{ord}_{\mathcal{P}} z'. \end{aligned}$$

□

The following result is a refinement of Theorem 11.78.

THEOREM 11.84 *If γ is the p-rank of Σ, then every p-subgroup G of* $\operatorname{Aut}_K(\Sigma)$ *has the following properties.*

(i) *If $\gamma \geq 2$, then $|G| \leq \kappa_p(\gamma - 1)$ with $\kappa_p = p/(p-2)$ for $p \geq 3$, and $\kappa_2 = 4$.*

(ii) *If* $\gamma = 1$ *and* $p \geq 3$, *then* G *is a cyclic group and* $|G|$ *divides* $\gamma - 1$; *also,* G *has no fixed place.*

(iii) *If* $\gamma = 1$ *and* $p = 2$, *then* $|G| \leq 4(g-1)$.

(iv) *If* $\gamma = 0$, *then* $|G| \leq \max\{g, 4pg^2/(p-1)^2\}$.

Proof. Let $\Sigma' = \Sigma^G$, $n = |G|$ and $\lambda = (\gamma' - 1)/n$ where γ' denotes the p-rank of Σ'. If G has s short orbits with lengths $\ell_1, \ldots, \ell_s$, then (11.28) reads as follows:

$$\lambda = \gamma' - 1 + \textstyle\sum_{i=1}^{s} (1 - \ell_i/n). \tag{11.54}$$

Since G is a p-group, $n/\ell_i \geq p$ for $i = 1, \ldots, s$. Assume that $\gamma \geq 2$; that is, $\lambda > 0$. Then (i) can be reworded as $\lambda \geq \kappa_p^{-1}$. If $\gamma' \geq 2$, (11.54) shows that $\lambda \geq 1 > \kappa_p^{-1}$. Also, it shows that if $\gamma' = 1$ then $s \geq 1$, and hence

$$\lambda \geq 1 - \ell_i/n \geq 1 - p > \kappa_p^{-1}.$$

If $\gamma' = 0$ and $p \geq 3$, then $s \geq 2$, and hence

$$\lambda \geq -1 + 2(1 - 1/p) = \kappa_p^{-1}.$$

If $\gamma' = 0$ and $p = 2$, two cases occur: either there are at least two short orbits, one of length ≥ 4, or there at least three short orbits. From (11.54), $\lambda \geq \frac{1}{4}$ in the former case and $\lambda \geq \frac{1}{2}$ in the latter. Thus (i) is established.

Assume next that $\gamma = 1$; that is, $\lambda = 0$. From (11.54), either $\gamma' = 0$, or $\gamma' = 1$. If $\gamma' = 1$, no short orbit exists, and G must be cyclic by Theorem 11.63. Further, from Theorem 11.72, $2g - 2 = |G|(2g' - 2)$, which shows that $|G|$ divides $g - 1$. In particular, $|G| \leq g - 1$. The case $g' = 0$ cannot occur when $p \geq 3$, since

$$\tfrac{2}{3} \leq 1 - 1/p \leq 1 - \ell/n < 1$$

for each short orbit of length ℓ. Hence, if $p \geq 3$, then $\sum(1 - \ell_i/n) \neq 1$. This proves (ii).

If $\gamma' = 0$ and $p = 2$, there are exactly two short orbits both of length $\frac{1}{2}|G|$, and (11.18) reads as follows:

$$\begin{aligned} 2g - 2 = {} & |G|\,(2g' - 2) + \tfrac{1}{2}|G|\,[2(|G_{\mathcal{P}}^{(1)} - 1) + \textstyle\sum_{i\geq 2} (|G_{\mathcal{P}}^{(i)} - 1)] \\ & + \tfrac{1}{2}|G|\,[2(|G_{\mathcal{Q}}^{(1)} - 1) + \textstyle\sum_{i\geq 2} |G_{\mathcal{Q}}^{(i)} - 1)], \end{aligned} \tag{11.55}$$

where $\mathcal{P}$ and $\mathcal{Q}$ come from the two distinct short orbits. Since $|G_{\mathcal{P}}| = |G_{\mathcal{Q}}| = 2$, this gives $|G| \leq g - 1$ when either $g' \geq 1$, or $g' = 0$ and $|G_{\mathcal{P}}^{(1)}| > 2$ or $|G_{\mathcal{Q}}^{(1)}| > 2$. In the remaining case, $g' = 0$ and $|G_{\mathcal{P}}^{(1)}| = |G_{\mathcal{Q}}^{(1)}| = 2$. As $g \geq 2$, in (11.55) either $|G_{\mathcal{P}}^{(2)}| \leq 2$ or $|G_{\mathcal{Q}}^{(1)}| \geq 2$. Therefore (11.55) becomes the equation

$$2g - 2 \geq -2|G| + 2|G| + \tfrac{1}{2}|G|,$$

whence (iii) follows.

Finally, assume that $\gamma = 0$; that is, $\lambda < 0$. Then $\gamma' = 0$ from (11.54). If there were at least two short orbits, the right-hand side in (11.54) would be at least $-1 + 2(1 + 1/p) \geq 0$, contradicting that $\lambda < 0$. So there is just one short orbit and, in this case, $\lambda = -\ell/n$. But then $\ell = 1$; that is, G fixes a place $\mathcal{P}$. From Theorem 11.78, part (iv) follows. □

REMARK 11.85 Theorem 11.84 shows that the order of a K-automorphism group of an irreducible curve, whose order is a power of p, is bounded above by a linear polynomial in g except when the p-rank γ is zero. Some of the exceptions are characterised in Theorem 11.140.

THEOREM 11.86 *Let $g \geq 2$.*

(i) *If $\gamma = 1$ and $p \geq 3$, then $\mathrm{Aut}_K(\Sigma)$ is tame.*

(ii) *If $p \geq 5$ and $2 \leq \gamma \leq p-2$, then $|\mathrm{Aut}_K(\Sigma)|$ is not divisible by p.*

Proof. Part (i) follows from Theorem 11.84 (ii). Also, Theorem 11.84 (i) shows that $\gamma \leq p-2$, whence $|G| \leq p(p-3)/(p-2) < p$. Therefore $|G| = 1$ since G has order a power of p. This proves that $\mathrm{Aut}_K(\Sigma)|$ has no p-subgroups. □

DEFINITION 11.87 For $p \neq 0$, an irreducible curve $\mathcal{F}$ is *ordinary* if the p-rank γ of $K(\mathcal{F})$ is equal to the genus g of $\mathcal{F}$.

THEOREM 11.88 *For any ordinary curve $\mathcal{F}$ of genus $g \geq 2$, the K-automorphism group of $\mathcal{F}$ has order at most $84(g-1)g$.*

EXAMPLE 11.89 An example of an ordinary curve is $\mathcal{F} = \mathbf{v}(F)$ with

$$F(X,Y) = (X^q - X)(Y^q - Y) - 1,$$

where $q = p^{2^n}$. It has genus $g = (q-1)^2$ and its K-automorphism group has order at least $2q^2(q-1)$. An interesting property of $\mathcal{F}$, valid for $n = 0$, is established in Theorem 11.93 (i).

THEOREM 11.90 *Let $p = 2$. If $G_{\mathcal{P}}^{(1)}$ contains a cyclic characteristic subgroup U, then $G_{\mathcal{P}}$ has cyclic subgroup V such that $[G_{\mathcal{P}} : V] = [G_{\mathcal{P}}^{(1)} : U]$.*

Proof. By Theorem 11.44, $G_{\mathcal{P}} = G_{\mathcal{P}}^{(1)} \rtimes H$, where H is a cyclic group of odd order, and $G_{\mathcal{P}}^{(1)}$ is a normal 2-subgroup of $G_{\mathcal{P}}$. From the latter assertion, since U is characteristic, it follows that U is a normal subgroup of $G_{\mathcal{P}}$. Therefore the set $V = UH$ is a subgroup of $G_{\mathcal{P}}$. More precisely, $V = U \rtimes H$, and U is a cyclic Sylow 2-subgroup of V. Therefore $V = O(V) \rtimes U$. So both U and $O(V)$ are normal subgroups, and $V = O(V) \times U$. Since both $O(V)$ and U are cyclic, V itself is cyclic. Also

$$|G_{\mathcal{P}}| = |G_{\mathcal{P}}^{(1)}||H| = [G_{\mathcal{P}}^{(1)} : U]|U||H| = [G_{\mathcal{P}}^{(1)} : U]|V|,$$

whence the assertion follows. □

11.10 K-AUTOMORPHISM GROUPS OF PARTICULAR CURVES

This section describes the groups of curves of low genus, namely rational and elliptic curves, as well as those of hyperelliptic and Artin–Schreier curves. Theorem 11.14 states that $\mathrm{Aut}(K(x)) \cong \mathrm{PGL}(2,K)$. The finite subgroups of $\mathrm{PGL}(2,K)$ can be determined by the techniques used in Section 11.9, although a deeper result, namely Theorem A.16, is also required. Here the two main results are stated without proofs.

THEOREM 11.91 *Let G be a non-trivial finite group of K-automorphisms of a rational function field $K(x)$. Let s be the number of short orbits of G on the set of all places of $K(x)$, and let $\ell_1, \ldots, \ell_s$ be their lengths. Then G is a group of one of the following types:*

(i) *the cyclic group $\mathbf{Z}_n$ of order n, with $p \nmid n$, $s = 2$, $\ell_1 = \ell_2 = 1$;*

(ii) *an elementary abelian p-group, with $s = 1$, $\ell_1 = 1$;*

(iii) *the dihedral group $\mathbf{D}_n$ of order $2n$, with $p \nmid n$, $s = 2$, $\ell_1 = 2$, $\ell_2 = n$, or $p \neq 2$, $s = 3$, $\ell_1 = \ell_2 = n$, $\ell_3 = 2$;*

(iv) *the alternating group $\mathbf{A}_4$, with $p \neq 2, 3$, and $s = 3$, $\ell_1 = 6$, $\ell_2 = \ell_3 = 4$;*

(v) *the symmetric group $\mathbf{S}_4$, with $p \neq 2, 3$, $s = 3$, $\ell_1 = 12$, $\ell_2 = 8$, $\ell_3 = 6$;*

(vi) *the alternating group $\mathbf{A}_5$, with $p = 3$, $s = 2$, $\ell_1 = 10$, $\ell_2 = 12$, or $p \neq 2, 3, 5$, $s = 3$, $\ell_1 = 30$, $\ell_2 = 20$, $\ell_3 = 12$;*

(vii) *the semidirect product of an elementary abelian p-group of order q with a cyclic group of order n, with $n \mid (q-1)$, $s = 2$, $\ell_1 = 1$, $\ell_2 = q$;*

(viii) $\mathrm{PSL}(2, q)$, *with $p \neq 2$, $q = p^m$, $s = 2$, $\ell_1 = q(q-1)$, $\ell_2 = q+1$;*

(ix) $\mathrm{PGL}(2, q)$, *with $q = p^m$, $s = 2$, $\ell_1 = q(q-1)$, $\ell_2 = q+1$.*

THEOREM 11.92 *Let $K = \overline{\mathbf{F}}_q$ and let G be a subgroup of $\mathrm{Aut}(K(x))$ preserving the set of all $\mathbf{F}_q$-rational places of $K(x)$. If $G \cong \mathrm{PGL}(2, q)$, then the extension $K(x)/K(x)^G$ has the following properties.*

(i) *Every $\mathbf{F}_q$-rational $\mathcal{P}$ place wildly ramifies; $G_\mathcal{P}$ is the semidirect product of an elementary abelian p-group of order q by a cyclic group of order $q-1$; $G_\mathcal{P}^{(1)}$ is an elementary abelian p-group of order q; $G_\mathcal{P}^{(2)}$ is trivial; all $\mathbf{F}_q$-rational places form a single orbit of size $q+1$ under G.*

(ii) *Every $\mathbf{F}_{q^2}$-rational place $\mathcal{P}$ which is not $\mathbf{F}_q$-rational tamely ramifies; $G_\mathcal{P}$ is a cyclic group of order $q+1$; such places form a single orbit of size $q^2 - q$ under G.*

(iii) *No other place of $K(x)$ ramifies.*

The main result on Artin–Schreier curves in terms of the associated function field is the following.

THEOREM 11.93 *Let $\Sigma/K(x)$ be an Artin–Schreier extension such that the genus of Σ is greater than 1. Then the K-automorphism group G of Σ is an extension of a cyclic group of order p by a finite group of K-automorphisms of $K(x)$ except in the following cases for $p > 2$:*

(i) *$\Sigma = K(x, y)$, $(x^p - x)(y^p - y) = c$ with $c \in K \backslash \{0\}$, and G is the semidirect product of an elementary abelian p-group of order p^2 with $\mathbf{D}_{p-1}$;*

(ii) $\Sigma = K(x,y),\ y^3 - y = \nu/(x(x-1))$ *where* $p = 3,\ \nu^2 = 2$, *and* G *is an extension of a cyclic group of order* 2 *by* $\mathbf{S}_4$;

(iii) $\Sigma = K(x,y),\ y^p - y = x^m$ *with* $m \mid (p+1),\ m < p+1$, *and* G *is an extension of* $\mathbf{Z}_m$ *by* $\mathrm{PGL}(2,p)$;

(iv) $\Sigma = K(x,y),\ y^p - y = x^{p+1}$; *that is,* Σ *is the Hermitian function field over* $\mathbf{F}_{p^2}$, *and* $G \cong \mathrm{PGU}(3,p)$.

The K-automorphism groups of elliptic curves are known. Here the following result, which is an extension of Lemma 11.26, is shown.

THEOREM 11.94 *he automorphism group of an elliptic curve* $\mathcal{F}$ *has the following properties.*

(i) *The* K*-automorphism group of* $\mathcal{F}$ *is infinite, and it acts on the places as a transitive permutation group.*

(ii) *For every place* $\mathcal{P}$*, the stabiliser of* $\mathcal{P}$ *is finite. More precisely, if* G *is a non-trivial* K*-automorphism group of* $\mathcal{F}$*, then*

$$|G_{\mathcal{P}}| = \begin{cases} 2,4,6 & \text{when } p \neq 2,3, \\ 2,4,6,12 & \text{when } p = 3, \\ 2,4,6,8,12,24 & \text{when } p = 2. \end{cases} \tag{11.56}$$

(iii) *For* $p = 2$*, the stabiliser* $G_{\mathcal{P}}^{(1)}$ *is cyclic when* $|G_{\mathcal{P}}^{(1)}| \leq 4$*, and it is the quaternion group* $\mathbf{Q}_8$ *when* $|G_{\mathcal{P}}^{(1)}| = 8$.

Proof. Let $p \neq 2,3$. By Theorem 7.89, $\mathcal{F}$ is taken to be the plane cubic curve

$$\mathbf{v}(Y^2 - X^3 - uX - v),$$

with $4u^3 + 27v^2 \neq 0$. Let $\Sigma = K(\mathcal{F})$, and let $\mathcal{P}_\infty$ be the place of $\Sigma = K(\mathcal{F})$ arising from the branch of $\mathcal{F}$ with centre at Y_∞. For another place $\mathcal{P}$ of Σ, let $P = (a,b)$ be the centre of the corresponding branch of $\mathcal{F}$. It is straightforward to show that the birational transformation $\omega(a,b)$, given by the equations

$$x' = \left(\frac{y-b}{x-a}\right)^2 - x - a, \quad y' = \frac{y-b}{x-a}(x'-a) + b$$

extends to a K-automorphism of Σ; see Lemma 11.2. Let $\mathcal{Q}$ be the place of Σ corresponding to the branch centred at the the point $Q = (a,-b)$. Then $\omega(a,b)$ sends $\mathcal{Q}$ to $\mathcal{P}_\infty$; hence $\omega(a,b)^{-1}$ sends $\mathcal{P}_\infty$ to $\mathcal{Q}$. Therefore the K-automorphism group generated by $\omega(a,b)$ acts transitively on the set of all places of Σ. The structure of the stabiliser of a place $\mathcal{P}$ can be deduced from Lemma 11.26. In fact, Lemma 11.26 shows that $\mathrm{Aut}_K(\Sigma)_{\mathcal{P}}$ contains no p-element, and hence it is tame. By Theorem 11.49, $\mathrm{Aut}_K(\Sigma)_{\mathcal{P}}$ is cyclic. Therefore Lemma 11.26 shows that the possibilities for $|\mathrm{Aut}_K(\Sigma)_{\mathcal{P}}|$ are $2,3,4,6$.

Now let $p = 3$. By Theorem 7.89, $\Sigma = K(\mathcal{F})$, where

$$\mathcal{F} = \mathbf{v}(Y^2 - X^3 - uX^2 - vX + w),$$

with $v^3 \neq u^3w$. To show (i), the previous argument still works, the only change being in the definition of $\omega(a,b)$ where $-u$ is added in x'. Also, the stabiliser G of $\mathcal{P}_\infty$ contains $\omega(a,b)$. If G is tame, then G is a cyclic group of order 2 or 4 by Theorem 11.60. If G is non-tame, then G has a normal subgroup N of order 3, and either $|G| = 6$ or $|G| = 12$.

Finally, let $p = 2$. Theorem 7.89 gives two possibilities. If $\mathcal{F}$ is supersingular, then

$$\mathcal{G} = \mathbf{v}(Y^2 + uY + X^3 + vX + w),$$

with $u \neq 0$, is a non-singular model of $\Sigma = K(\mathcal{F})$. It is possible to use the previous argument, provided that $\omega(a,b)$ is defined as follows:

$$x' = \left(\frac{y+b}{x+a}\right)^2 + x + a, \quad y' = \frac{y-b}{x-a}(x'+a) + b + u.$$

The stabiliser $G_{\mathcal{P}_\infty}$ is non-tame as it contains the involutory K-automorphism

$$\omega_0 : \quad x' = x, \quad y' = y + u.$$

By Theorem 11.60, a Sylow 2-subgroup of $G_{\mathcal{P}_\infty}$ has order at most 8. From (11.20), $G_{\mathcal{P}_\infty}$ has no element of order 5. Hence, by Theorem 11.60, 3 is the only possibility for the order of a non-trivial K-automorphism of Σ of odd order. Hence $|G_{\mathcal{P}_\infty}|$ is an even divisor of 24. From the proof of Theorem 12.7, $G_{\mathcal{P}_\infty}$ consists of linear transformations of type

$$\omega' : \quad x' = x + d, \quad y' = y + ax + b,$$

with $a, b, d \in K$. Then ω' is involutory if and only if $ad = 0$. As $\omega \in G_{\mathcal{P}_\infty}$, the condition $ad = 0$ implies that $a = d = 0$ and $b = u$. Therefore ω_0 is the unique involutory automorphism in $G_{\mathcal{P}_\infty}$. In particular, if $|G^{(1)}_{\mathcal{P}_\infty}| \leq 4$, then $G^{(1)}_{\mathcal{P}_\infty}$ is cyclic. If $|G^{(1)}_{\mathcal{P}_\infty}| = 8$, Theorem 11.72 applied to $G^{(1)}_{\mathcal{P}_\infty}$ implies that $|G^{(2)}_{\mathcal{P}_\infty}| = 2$. From this, the group $G^{(1)}_{\mathcal{P}_\infty}/G^{(2)}_{\mathcal{P}_\infty}$ is elementary abelian of order 4. Therefore $G^{(1)}_{\mathcal{P}_\infty}$ is not cyclic, and hence is a quaternion group of order 8.

Finally, if $\mathcal{F}$ is not supersingular, then

$$\mathcal{G} = \mathbf{v}(Y^2 + XY + X^3 + uX^2 + v),$$

with $v \neq 0$, is a non-singular model of $\Sigma = K(\mathcal{F})$. Again, the previous argument can be used with

$$\omega(a,b) : x' = \left(\frac{y+b}{x+a}\right)^2 + \frac{y+b}{x+a} + x + a + u, \; y' = \frac{y+b}{x+a}(x'+a) + x' + b,$$

$$\omega_0 : x' = x, \quad y' = y + x,$$

to obtain the results. □

REMARK 11.95 Theorem 11.94 implies that, for every place $\mathcal{P}$ of an elliptic curve,

$$|G_\mathcal{P}| \leq \begin{cases} 6 & \text{when } p \neq 2,3, \\ 12 & \text{when } p = 3, \\ 24 & \text{when } p = 2. \end{cases} \tag{11.57}$$

In each case, this upper bound is achieved by some elliptic curves. For $p \neq 2, 3$, see the proof of Lemma 11.26. Here, a few examples are given for $p = 2, 3$.

EXAMPLE 11.96 Let $p = 3$. Let $\mathcal{F} = \mathbf{v}(Y^2 - X^3 + X)$. If δ is a primitive fourth root of unity, then both birational transformations

$$\begin{aligned} \omega: \quad & x' = -x + 1, \quad y' = \delta y, \\ \omega': \quad & x' = x + 1, \quad\;\; y' = y \end{aligned}$$

extend to K-automorphisms of Σ whose orders are 4 and 3. The group generated by ω and ω' has order 12.

EXAMPLE 11.97 Let $p = 2$, let $\mathcal{F} = \mathbf{v}(Y^2 + Y + X^3)$, and choose a cube root of unity ϵ in Σ. Then the birational transformation

$$\omega(i,k): \quad x' = x + \epsilon^i, \quad y' = y + \epsilon^{2i}x + \epsilon^k,$$

with $i = 0, 1, 2$ and $k = 1, 2$, extends to a K-automorphism of Σ. These, together with the K-automorphism

$$\omega: \quad x' = x, \quad y' = y + 1$$

and the identity form a quaternion group $\mathbf{Q}_8$ of order 8. Also, the birational transformation

$$\omega': \quad x' = \epsilon x, \quad y' = y$$

extends to a K-automorphism of Σ of order 3; this and $\mathbf{Q}_8$ generate a group of order 24.

Now the K-automorphism groups of hyperelliptic curves are investigated. By Theorem 11.50, such a group is finite. As in Section 7.10, two cases are distinguished.

First, suppose that K has zero or odd characteristic. By Theorem 7.94, $\mathcal{F}$ is an irreducible plane curve $\mathbf{v}(Y^2 - f(X))$, where $f(X) \in K[X]$ is monic, has degree $2g+1$, and its roots $\alpha_0, \ldots, \alpha_{2g}$ are distinct. Let $\Sigma = K(x,y)$ with $y^2 = f(x)$ be the function field Σ of $\mathcal{F}$. The Weierstrass points are the places associated to the branches of $\mathcal{F}$ centred at the non-singular points $U_i = (\alpha_i, 0)$ for each i in $\{0, \ldots, 2g\}$, together with the place associated to the unique branch centred at Y_∞. As in Theorem 7.95,

$$w(X) = \frac{aX + b}{cX + d}, \quad \text{with } ad - bc \neq 0,$$

is a rational function which maps the set $\{\alpha_0, \ldots, \alpha_{2g}, \infty\}$ onto itself. Such rational functions constitute a subgroup W of $\mathrm{PGL}(2, K)$. Hence, if $w \in W$ fixes three distinct elements from the set $\{\alpha_0, \ldots, \alpha_{2g}, \infty\}$, then $w(X) = X$. Since $g \geq 2$, this implies that W is a finite group.

From the proof of Theorem 7.95, the K-automorphisms α of $\mathcal{F}$ are as follows:

(i) $$\begin{aligned} &\alpha(x) = ax + b, \text{ with } aX + b \in W, \\ &\alpha(y) = \epsilon y \text{ with } \epsilon^2 = a^{2g+1}; \end{aligned} \tag{11.58}$$

(ii) $$\begin{aligned} &\alpha(x) = \frac{ax+b}{x-\alpha_0}, \text{ with } \frac{aX+b}{X-\alpha_0} \in W, \\ &\alpha(y) = \epsilon \frac{y}{(x-\alpha_0)^{g+1}}, \text{ with } \epsilon^2 = (b + a\alpha_0)\textstyle\prod_{i=1}^{2g} (\alpha_0 - \alpha_i). \end{aligned} \tag{11.59}$$

Hence $|\mathrm{Aut}_K(\Sigma)| = 2|W|$; more particularly, the hyperelliptic involution of $\mathcal{F}$, that is, the birational transformation $(x, y) \mapsto (x, -y)$, extends to a K-automorphism ω of $\mathcal{F}$ such that $\omega \in Z(\mathrm{Aut}_K(\Sigma))$ and $W \cong \mathrm{Aut}_K(\Sigma)/\langle\omega\rangle$.

In even characteristic, let

$$\mathcal{F} = \mathbf{v}(Y^2 + h(X)Y + g(X)),$$

where $\deg h(X) \leq g + 1$ and $\deg g(X) = 2g + 1$. Then the function field of $\mathcal{F}$ is $\Sigma = K(x, y)$, with $y^2 + h(x)y + g(x) = 0$.

If $n = \deg h(X) \geq 1$, let $\alpha_1, \ldots, \alpha_n$ denote the roots of $h(X)$, each counted with multiplicity. As before, W is the subgroup of $\mathrm{PGL}(2, K)$ that preserves the multi-set $\{\infty, \alpha_1, \ldots, \alpha_n\}$. The Weierstrass points are the places arising from the branches centred at the non-singular points $U_i = (\alpha_i, \sqrt{\alpha_i})$, with $i = 1, \ldots, n$, together with the place associated to Y_∞. From the proof of Theorem 7.101, the K-automorphisms α of $\mathcal{F}$ are as follows:

(i) $$\alpha(x) = w(x), \quad \alpha(y) = v(x)h(w(x)) + \frac{h(w(x))}{h(x)}y,$$

$$\text{with } w(X) \in W \text{ and } v(x)^2 + v(x) = \frac{g(x)}{h(x)^2} + \frac{g(w(x))}{h(w(x))^2};$$

(ii) $$\alpha(x) = w(x), \quad \alpha(y) = h(w(x)) + \frac{h(w(x))}{h(x)}y,$$

$$\text{with } w(X) \in W \text{ and } \frac{h(w(x))^2}{h(x)^2}g(x) + g(w(x)) = 0.$$

The centre $Z(\mathrm{Aut}_K(\Sigma))$ of $\mathrm{Aut}_K(\Sigma)$ is non-trivial as the hyperelliptic involution of $\mathcal{F}$, that is, the birational transformation $(x, y) \mapsto (x, y + 1)$ extends to a K-automorphism ω of $\mathcal{F}$ lying in $Z(\mathrm{Aut}_K(\Sigma))$. Therefore $W \cong \mathrm{Aut}_K(\Sigma)/\langle\omega\rangle$ and $|\mathrm{Aut}_K(\Sigma)| = 2|W|$.

If $h(X)$ is constant, then $\mathrm{Aut}_K(\Sigma)$ fixes the unique Weierstrass point. Therefore the following result is obtained.

THEOREM 11.98 *Let $\mathcal{F}$ be a hyperelliptic curve. Then*

(i) *its hyperelliptic involution ω is in the centre of the K-automorphism group G of $\mathcal{F}$;*

(ii) *the quotient group $G/\langle\omega\rangle$ is isomorphic to a subgroup W of* $\mathrm{PGL}(2, K)$ *that preserves the set or, for $p = 2$, the multi-set consisting of all Weierstrass points of $\mathcal{F}$;*

(iii) *if $\mathcal{F}$ has at least three Weierstrass points, the identity is the only K-automorphism of $\mathcal{F}$ that fixes each Weierstrass point.*

By Remark 11.15, all possibilities for W are known.

Now some particular classes of hyperelliptic curves are considered. The following proposition is a special case of a more general result that is established later; see Theorem 11.138.

PROPOSITION 11.99 *Let $\mathcal{F}$ be an irreducible hyperelliptic curve of genus* 2.

(i) *Assume that either $p \neq 2$, or $p = 2$ and $\mathcal{F}$ has more than one Weierstrass point, and let G be a K-automorphism group of $\mathcal{F}$. Then*

$$|G_{\mathcal{P}}^{(1)}| \leq \begin{cases} 1 & \textit{for } p = 0 \textit{ and } p \geq 7, \\ 5 & \textit{for } p = 5, \\ 3 & \textit{for } p = 3, \\ 8 & \textit{for } p = 2. \end{cases}$$

(ii) *If $\mathcal{F}$ has only one Weierstrass point and G is an abelian p-group, then the order $|G| \leq 8$.*

(iii) *If* (i) *holds and G is a soluble group, then $|G| \leq 48$.*

Proof. By Theorem 11.78, (i) holds for $p \neq 2, 3$.

To prove (i) for $p = 3$, Theorem 11.98 is relevant, as Σ has six Weierstrass points. If G has a subgroup of order at least 9, then some non-trivial elements of H fixes three Weierstrass points, in contradiction to Theorem 11.98 (iii).

To prove (i) for $p = 2$, let $\omega \in G$; Theorem 11.98 and Remark 11.15 are used. If Σ has three Weierstrass points, then $\mathrm{Aut}_K(\Sigma)/\langle\omega\rangle$ is a subgroup of $\mathbf{S}_3$. In particular, the order of a Sylow 2-subgroup S_2 of G is at most 4. Similarly, $|G/\langle\omega\rangle| \leq 2$, and hence $|S_2| \leq 4$ when Σ has two Weierstrass points.

To prove (ii), let $p = 2$. The essential tools are Theorem 11.94 and Lemma 11.46. Let M be a non-trivial subgroup of G. Suppose first that Σ^M is elliptic. Since M fixes the unique Weierstrass point $\mathcal{P}$, Theorem 11.72 applied to M implies that $|M| = 2$. Therefore G/M is an abelian K-automorphism group of an elliptic curve. Theorem 11.94 (II) implies that $|G/M| \leq 4$, whence it follows that $|G| \leq 8$. Finally, in the case that the fixed field of every non-trivial subgroup of G is rational, Lemma 11.46 implies that $|G| \leq 4$.

In proving (iii), Theorem 11.98 and Remark 11.15 play a role since they, together with the solubility condition, imply that, if Σ has more than two Weierstrass points, then $|G|/\langle\omega\rangle| \leq 24$. Therefore (iii) holds for $p \neq 2$,

To complete the proof for $p = 2$ it remains to examine the case that Σ has exactly two Weierstrass points, say $\mathcal{P}_1$ and $\mathcal{P}_2$. Then G has a subgroup U of index 2 fixing both $\mathcal{P}_1$ and $\mathcal{P}_2$. Since G contains ω, the group $\bar{U} = U/\langle\omega\rangle$ is a subgroup of $\mathrm{PGL}(2, K)$ fixing two objects, namely the places $\mathcal{P}_1'$ and $\mathcal{P}_2'$ of Σ^ω lying under $\mathcal{P}_1$ and $\mathcal{P}_2$ in the extension Σ/Σ^ω. From Theorem 11.14, $\bar{U}$ is a cyclic group of odd order. By Theorem 11.60, $|\bar{U}| \leq 4g + 2 = 10$. Therefore $|G| \leq 18$, and (iii) follows. □

EXAMPLE 11.100 Let $p > 2$. For a power q of p, the curve

$$\mathcal{F} = \mathbf{v}(Y^2 - g(X))$$

with $g(X) = X^q - X$ is a special case of Example 5.59. So $\mathcal{F}$ has genus $\frac{1}{2}(q-1)$. The field $\mathbf{F}_q$ is viewed as a simple extension of the prime field $\mathbf{F}_p$ of K, and $\mathbf{F}_q$ is assumed to be a subfield of K. The distinct roots of $g(X)$ are elements in $\mathbf{F}_q$. Hence the group W consists of all rational functions

$$w(X) = \frac{aX + b}{cX + d}, \quad \text{with } a, b, c, d \in \mathbf{F}_q,\ ad - bc \neq 0.$$

Therefore $W \cong \mathrm{PGL}(2,q)$ and W acts on $\mathbf{F}_q \cup \{\infty\}$ as $\mathrm{PGL}(2,q)$ acts naturally on $\mathrm{PG}(1,q)$. If $\omega : x' = x,\ y' = -y$ is the central involution in $\mathrm{Aut}_K(\Sigma)$, then $\mathrm{Aut}_K(\Sigma)/\langle\omega\rangle \cong \mathrm{PGL}(2,q)$. It may be noted that if $q = 5$ then $g = 2$ and $|\mathrm{Aut}_K(\Sigma)| = 240$. This shows that the solubility condition in Proposition 11.99 cannot be relaxed.

The following result is a characterisation of the curve in Example 11.100.

PROPOSITION 11.101 *Let $p > 2$ and $q = p^h = 2g + 1$. If $W \cong \mathrm{PSL}(2,q)$ in Theorem 11.98, then $\mathcal{F}$ is birationally equivalent to the curve in Example 11.100.*

Proof. The notation is the same as in the proof of Theorem 11.98. Take $\alpha_0 = 0$, $\alpha_1 = 1$ and $f(X)$ monic, after a suitable substitution of type $X' = uX + v$, with $u, v \in K$ and $u \neq 0$.

Since $W \cong \mathrm{PSL}(2,q)$, the polynomial $f(X)$ is invariant under the substitution $X \mapsto cX$ for every $\frac{1}{2}(q-1)$-st root of unity $c \in K$; it should be noted that $c \in \mathbf{F}_q$. Since $\alpha_1 = 1$, and $\{c\,\alpha_1, \ldots, c\,\alpha_{2g}\} = \{\alpha_1, \ldots, \alpha_{2g}\}$, this shows that every c with $c^{(q-1)/2} = 1$ is a root of $f(X)$. Also, since $W \cong \mathrm{PSL}(2,q)$, so $f(X)$ is invariant under the substitution $X \mapsto X + 1$. Choose a $\frac{1}{2}(q-1)$-st root of unity $u \in K$ in such a way that $v = u + 1$ is not also such a root of unity. Then the elements cv with c ranging over all $\frac{1}{2}(q-1)$-st roots of unity provide the remaining $\frac{1}{2}(q-1)$ non-zero roots of $f(X)$. Hence the roots of $f(X)$ are precisely the elements of $\mathbf{F}_q$. Thus $f(X) = X^q - X$. □

EXAMPLE 11.102 Let $p = 2$, $\mathcal{F} = \mathbf{v}(Y^2 + Y + X^5)$ and $\Sigma = K(\mathcal{F})$. The field $\mathbf{F}_{16}$ is viewed as a subfield of K and as an extension of the prime field $\mathbf{F}_2$ of K. Since $h(X) = 1$, the place $\mathcal{P}_\infty$ of Σ arising from the unique branch of $\mathcal{F}$ centred at Y_∞ is the only Weierstrass point. Therefore $\mathrm{Aut}_K(\Sigma)$ fixes $\mathcal{P}_\infty$. A straightforward computation shows the following results.

For every $b, c \in \mathbf{F}_{16}$ with $c^2 + c + b^5 = 0$, the birational transformation

$$\alpha_{b,c}(x) = x + b, \quad \alpha_{b,c}(y) = b^8x^2 + b^4x + c + y$$

of Σ is a K-automorphism fixing $\mathcal{P}_\infty$. For $b = c = 0$, the identity is obtained; otherwise the order of $\alpha_{b,c}$ is either 2 or 4 according as $b^{10} + b^5$ is 0 or 1. The set

$$\{\alpha_{b,c} \mid b, c \in \mathbf{F}_{16};\ c^2 + c + b^5 = 0\}$$

is a group U of order 32 containing 11 elements of order 2 and 20 elements of order 4. Such a group is uniquely determined up to isomorphism, and hence it is isomorphic to $\mathbf{Z}_4 \times \mathbf{D}_4$. Since U fixes $\mathcal{P}_\infty$, Theorem 11.78 implies that $G_{\mathcal{P}}^{(1)} = U$. Further, for every $d \in \mathbf{F}_{16}$ with $d^5 = 1$, the birational transformation

$$\alpha_d(x) = dx, \quad \alpha_d(y) = y$$

is a K-automorphism of Σ fixing $\mathcal{P}_\infty$. These form a subgroup C_5 of $\mathrm{Aut}_K(\Sigma)$ of order 5. By Theorem 11.60, C_5 is a maximal subgroup of odd order in the stabiliser of $\mathcal{P}$. Therefore $\mathrm{Aut}_K(\Sigma) = U \rtimes C_5$; so $\mathrm{Aut}_K(\Sigma) \cong (\mathbf{Z}_4 \rtimes \mathbf{D}_4) \rtimes \mathbf{Z}_5$, and $|\mathrm{Aut}_K(\Sigma)| = 160$. This shows that Proposition 11.99 (iii) can fail when Σ has only one Weierstrass point.

EXAMPLE 11.103 Let $p > 2$ and $a \in \mathbf{F}_q\setminus\{0\}$. The hyperelliptic curve

$$\mathcal{G}_0 = \mathbf{v}(X(Y^p - Y) - (aX^2 + 1))$$

of genus $p - 1$ has the following properties:

(i) the $\mathbf{F}_q$-rational K-automorphism group of $\mathcal{G}_0$ is the direct product of the hyperelliptic involution by a dihedral group $\mathbf{D}_p$ of order $2p$;

(ii) the fixed places of $\mathbf{D}_p$ are $\mathbf{F}_q$-rational.

Conversely, up to birational equivalence over $\mathbf{F}_q$, the curve $\mathcal{G}_0$ is the unique hyperelliptic curve defined over $\mathbf{F}_q$ of genus $g = p - 1$ which satisfies both (i) and (ii).

If $\mathbf{D}_p$ is replaced by the cyclic group $\mathbf{Z}_p$, similar properties and characterisations hold for hyperelliptic curves $\mathcal{G}_b$ of genus $p - 1$ with

$$\mathcal{G}_b = \mathbf{v}(X(Y^p - Y) - (aX^2 + bX + 1)), \tag{11.60}$$

where $a, b \in \mathbf{F}_q$, $a \neq 0$ and $\mathrm{T}_{\mathbf{F}_q/\mathbf{F}_p}(b) \neq 0$.

EXAMPLE 11.104 Let $\mathcal{F}_{n,m} = \mathbf{v}(X^n + Y^m + 1)$ with $n > m$. Assume that p does not divide n and m, and that $\mathcal{F}_{n,m}$ has genus $g > 1$. Let G be the K-automorphism group of $\mathcal{F}_{n,m}$. Then G has a normal cyclic subgroup N order m, and one of the following holds:

(i) G/N is a cyclic group of order n and $m \nmid n$;

(ii) G/N is a dihedral group of order $2n$ and $m \mid n$ but $n \neq q + 1$;

(iii) $G/N \cong \mathrm{PGL}(2, q)$ with $m \mid n$ and $n = q + 1$.

11.11 FIXED PLACES OF AUTOMORPHISMS

Some simplification in the statements of the results in this section is made by using the following notation. For a non-trivial K-automorphism α, the integer $\rho(\alpha)$ denotes the number of fixed places of α in the action of α on the set $\mathcal{P}_\Sigma$ of all places of Σ. More generally, for a non-trivial subgroup G of $\mathrm{Aut}_K(\Sigma)$, let $\rho(G)$ be the number of places of Σ fixed by every K-automorphism in G. If G has prime order and $G = \langle\alpha\rangle$, then $\rho(G) = \rho(\alpha)$, but this is not always true for subgroups G of composite order.

LEMMA 11.105 *Let $\alpha \in \mathrm{Aut}_K(\Sigma)$ be a non-trivial K-automorphism such that $\rho(\alpha) > 2n$, where $n = [\Sigma : K(z)]$ for a non-constant element z of Σ. Then $\alpha(z) = z$, and n is divisible by* ord α. *If, in addition, n is prime, then* ord α *is also a prime and Σ^α is a rational function field.*

Proof. Let $\mathcal{P}$ be a place of Σ and take a primitive place representation σ of α. If $\sigma(z) = ct^i + \cdots$ and $\mathcal{P}^\alpha = \mathcal{P}$, then there is a K-automorphism λ of $K[[t]]$ such that

$$\sigma(\alpha(z)) = c\lambda(t)^i + \cdots = ct^i + \cdots .$$

Then $\text{ord}_{\mathcal{P}}(\alpha(z) - z) > \text{ord}_{\mathcal{P}}\, z$ provided that $\alpha(z) - z \notin K$ and $\mathcal{P} = \mathcal{P}^{\alpha}$.

Now let $h = \alpha(z) - z$, and assume that $h \notin K$. Then every pole of h is either a pole of z or a pole of its image $\alpha(z)$. Since $n = [\Sigma : K(\alpha(z))]$, this implies that $[\Sigma : K(h)] \leq 2n$. Using this observation, this can be improved to $[\Sigma : K(h)] \leq 2n - r$, where r is the number of poles of z fixed by α. In particular, $r \leq 2n$. From the above observation it also follows that each fixed place of α that is not a pole of z is a zero of h. Thus h has $\rho(\alpha) - r > 2n - r$ zeros. But this is in contradiction with Corollary 5.35; hence $h \in K$. More precisely, $h = 0$; that is, $\alpha(z) = z$, because α fixes some places which are not poles of z.

To show the other two assertions, choose a place $\mathcal{P}$ fixed by α, and let $\mathcal{P}'$ be a place lying under $\mathcal{P}$ in the extension Σ/Σ^{α}. Then $\mathcal{P}$ completely ramifies. Hence $[\Sigma : \Sigma^{\alpha}] = \text{ord}\, \alpha$ since $e_{\mathcal{P}} = \text{ord}\, \alpha$ by Theorem 11.42. From the equation,

$$[\Sigma : K(z)] = [\Sigma : \Sigma^{\alpha}] \cdot [\Sigma^{\alpha} : K(z)],$$

n is the product of ord α and $[\Sigma^{\alpha} : K(z)]$. If n is prime, this gives $n = \text{ord}\, \alpha$ and $\Sigma^{\alpha} = K(z)$. □

LEMMA 11.106 *Let $\alpha \in \text{Aut}_K(\Sigma)$ be a non-trivial K-automorphism of order n, where Σ has genus g. Then*

$$\rho(\alpha) \leq 2 + \frac{2g}{n-1}. \tag{11.61}$$

Proof. Since $\rho(\alpha) \leq \rho(\alpha^j)$ for $j \geq 1$, Remark 11.61 applied to $G = \langle \alpha \rangle$ gives that

$$2g - 2 \geq n(2g' - 2) + \rho(\alpha)(n-1),$$

where g' is the genus of the function field Σ^{α}. Since $g' \geq 0$, the result follows. □

LEMMA 11.107 *If the order n of a K-automorphism α is a prime different from p, then*

$$\rho(\alpha) = 2 + \frac{2g - 2g'n}{n-1}, \tag{11.62}$$

where g and g' are the genera of Σ and Σ^{α}. In this case, equality holds in (11.61) if and only if $g' = 0$.

Proof. If n is prime, then $\rho(\alpha) = \rho(\alpha^j)$ for $j \geq 1$. Then the proof of Lemma 11.106 gives the result. □

THEOREM 11.108 *If a prime number n other than p is the order of a K-automorphism of Σ, then one of the following holds:*

(i) $n \leq g + 1$;

(ii) $n = 2g - 1$ *and* $g > 2$;

(iii) $n = 2g + 1$.

Proof. Using the formula (11.27) several times, an upper bound for n can be obtained.

If $g' \geq 2$, then $2g-2 \geq 2n+\rho(\alpha)(n-1) \geq 2n$; hence $n \leq g-1$, yielding that $n < g+1$. Again, by (11.22), $g' = 1$ and $\rho(\alpha) = 0$ cannot occur simultaneously. If $g' = 1$ and $\rho(\alpha) = 1$, then $n = 2g-1$. If $g' = 1$ and $\rho(\alpha) \geq 2$, then $2g-2 \geq 2(n-1)$, and hence $n \leq g$, yielding $n < g+1$. Similarly, by (11.22), $g' = 0$ and $\rho(\alpha) \leq 2$ cannot occur simultaneously. If $g' = 0$ and $\rho(\alpha) = 3$, then $n = 2g+1$. If $g' = 0$ and $\rho(\alpha) \geq 4$, then $2g-2 \geq -2n+4(n-1)$, which implies that $n \leq g+1$. This completes the proof. □

REMARK 11.109 In the extremal case that $n = 2g+1$, all possibilities are known. Let n be as in Lemma 11.108. Then $n = 2g+1$ if and only if Σ is the function field of the curve $\mathcal{F} = \mathbf{v}(F(X,Y))$, where F is one of the following:

(i) $F(X,Y) = Y^{m-r}(Y-1)^r - X^n$, with $1 \leq r < m \leq g+1,\ n \neq p$;

(ii) $F(X,Y) = Y^2 - X^n + X$ with $n = p$.

Theorem 11.108 is a slight improvement on Theorem 11.79 for groups of prime order. The following result is another refinement for particular cyclic groups.

THEOREM 11.110 *Let $\mathcal{F}$ be a non-singular plane curve of degree d defined over $\mathbf{F}_q$, and suppose that its K-automorphism group contains a Singer group, that is, a cyclic subgroup G of* $\mathrm{PGL}(3,q)$ *of order q^2+q+1 acting transitively on the points of* $\mathrm{PG}(2,q)$. *Then one of the following holds*:

(i) (a) $d = q+2$;

(b) *$\mathcal{F}$ is projectively equivalent over $\mathbf{F}_{q^3}$ to the curve*
$$\mathcal{G} = \mathbf{v}(X^{q+1}Y + Y^{q+1} + X);$$

(c) $\mathrm{Aut}_K(\mathcal{F})$ *is the normaliser of G in* $\mathrm{PGL}(3,q)$;

(ii) $d \geq q^2+q+1$.

Theorem 11.56 shows that large K-automorphism groups may have only one short orbit. To improve Theorem 11.116, the following lemma is used.

LEMMA 11.111 *Let G be a K-automorphism group of Σ with only one short orbit. If Σ has genus $g \geq 2$, then the size of the short orbit divides $2g-2$.*

Proof. Let g' be the genus of Σ^G. If the stabiliser $G_{\mathcal{P}}$ of a place $\mathcal{P}$ in the short orbit o is tame, then Theorem 11.57 reads as follows:
$$2g-2 = |G|(2g'-2) + |G| - |o|.$$
As $|G| = |G_{\mathcal{P}}|\,|o|$, the result follows.

In the non-tame case, this equation may fail. However, in the Hilbert Different Formula (11.37), $d_{\mathcal{P}} = d_{\mathcal{Q}}$ for $\mathcal{P}, \mathcal{Q} \in o$ and $d_{\mathcal{Q}} = 0$ for $\mathcal{Q} \notin o$. Hence $|o|$ divides d, and the result follows as in the tame case. □

The next two theorems are classical results linking Weierstrass points and fixed places of K-automorphisms. For their validity in positive characteristic, some hypotheses are needed.

THEOREM 11.112 *Let $\mathcal{F}$ be an ordinary curve. If*

(a) *p does not divide the order of a non-trivial K-automorphism α of $\mathcal{F}$,*

(b) *α has more than four fixed places,*

then every fixed place of α is a Weierstrass point.

Proof. It suffices to deal with the case that $n = \operatorname{ord}\alpha$ is prime, since a suitable power α^j of α has prime order, and then α may be replaced by α^j, as every fixed place of α is fixed by α^j.

Let $\mathcal{P}$ be a fixed place of α, and take a place $\mathcal{P}'$ of Σ^α under $\mathcal{P}$. By the Weierstrass Gap Theorem 6.89, there exists $z \in \Sigma^\alpha$ such that $\operatorname{div}(z)_\infty = m'\mathcal{P}'$ with $m' \leq g' + 1$ where g' is the genus of Σ^α. Theorem 11.36, together with (7.1), shows that

$$\operatorname{ord}_{\mathcal{P}} z = n \cdot \operatorname{ord}_{\mathcal{P}'} z = n \cdot m' \leq (g'+1)n.$$

On the other hand, Theorem 11.72 applied to the subgroup G generated by α, together with the hypothesis $\rho(\alpha) > 4$, gives the inequalities

$$2g - 2 \geq n(2g' - 2) + (n-1)\rho(\alpha) > n(2g'-2) + 4(n-1),$$

showing that $g + 1 > n(g'+1)$. Hence $\operatorname{ord}_{\mathcal{P}} z < g + 1$. Since $\mathcal{F}$ is ordinary, $\mathcal{P}$ is a Weierstrass point. □

THEOREM 11.113 *Let G be a tame K–automorphism group of Σ. Suppose that $\alpha, \beta \in G\backslash\{1\}$ satisfy the following conditions:*

(a) *$g > n^2g' + (n-1)^2$, where $n = \operatorname{ord}\alpha$ is a prime, and g' is the genus of Σ^α;*

(b) *$\rho(\beta) > 2n(g'+1)$.*

Then the following hold:

(i) *each fixed place of α is a fixed by β, and $\operatorname{ord}\beta \leq n$;*

(ii) *if $\operatorname{ord}\beta = n$, then $\langle\alpha\rangle = \langle\beta\rangle$;*

(iii) *if $\operatorname{ord}\beta = n$, then $\langle\alpha\rangle$ is a normal subgroup of G;*

(iv) *if $\operatorname{ord}\beta = n = 2$, then $\alpha \in Z(G)$.*

Proof. Let $\mathcal{P}_1, \ldots, \mathcal{P}_{\rho(\alpha)}$ denote the fixed places of α. From the proof of Theorem 11.112, for each j there is a non-constant $z_j \in \Sigma^\alpha$ such that $\operatorname{div}(z_j)_\infty = r_j\mathcal{P}_j$ with $r_j \leq n(g'+1)$. From Lemma 11.105, $z_j = \beta(z_j)$ for each j, and thus $\mathcal{P}_j = \mathcal{P}_j^\beta$. So the first part of (i) holds. Also, $\rho(\beta) \geq \rho(\alpha)$.

Rewrite $g > n^2g' + (n-1)^2$ in the form

$$\frac{g - ng'}{n-1} > \frac{g+n-1}{n}. \tag{11.63}$$

Now

$$\begin{aligned}\rho(\beta) \geq \rho(\alpha) &= 2 + \frac{2(g - ng')}{n-1} > 2 + 2\frac{(g+n-1)}{n} \\ &= 2 + \frac{2g}{n} + 2\frac{(n-1)}{n} > 2 + \frac{2g}{n}.\end{aligned}$$

This, together with (11.61), gives ord $\beta \leq n$, completing the proof of (i).

Both α and β are in the stabiliser of $\mathcal{P}_1$. Since G is tame, $G_{\mathcal{P}_1}$ is a cyclic group, by Theorem 11.49. So ord $\alpha =$ ord β yields (ii).

Next, suppose $\delta \in G$ is conjugate to β under G. Then

$$\rho(\delta) = \rho(\beta) > 2n(g'+1).$$

Thus, by (ii), $\langle\delta\rangle = \langle\beta\rangle$ showing that $\langle\beta\rangle$ is a normal subgroup of G. This and (ii) imply (iii).

Finally, (iv) is a consequence of (iii). □

THEOREM 11.114 *Let G be a subgroup of* $\mathrm{Aut}_K(\Sigma)$ *of order n which has a partition with components $G_1, \ldots, G_k$, with $n_i = |G_i|$ for $i = 1, \ldots, k$, and let g, g', g'_i be the genera of Σ, Σ^G, Σ^{G_i} for $i = 1, \ldots, k$. Then*

$$(k-1)g + ng' = \textstyle\sum_{i=1}^{k} n_i g'_i.$$

Proof. Let $G_0 = G$ and $n_0 = n$. Let $D_i = D(\Sigma/\Sigma^{G_i})$ denote the different divisor of the extension Σ/Σ^{G_i}, for $i = 0, 1, \ldots, k$. Then $D_0 = \sum_{i=1}^{k} D_i$. From Corollary 11.52,

$$2g - 2 = n_i(2g'_i - 2) + \deg D_i,$$

for $i = 0, 1, \ldots, k$. Summing over $i = 1, \ldots, k$ and subtracting for $i = 0$, since $\deg D_0 = \sum_{1=1}^{k} \deg D_i$,

$$(k-1)(2g-2) = \textstyle\sum_{i=1}^{k} n_i(2g'_i - 2) - n(2g' - 2).$$

Hence

$$(k-1)g = \textstyle\sum_{i=1}^{k} n_i g'_i - ng' - n + (k-1) - \sum_{i=1}^{k} n_i.$$

Since $n = \sum_{i=1}^{k} n_i - (k-1)$, the result follows. □

REMARK 11.115 Three important examples of finite groups with a partition are $\mathrm{PGL}(2, q)$, $\mathrm{PSL}(2, q)$, $\mathrm{Sz}(q)$.

11.12 LARGE AUTOMORPHISM GROUPS OF FUNCTION FIELDS

In this section, Theorem 11.56 is refined.

THEOREM 11.116 *In Theorem 11.56, the upper bounds can be sharpened for the cases* (a), (b), (c) *as follows*:

(a′) $|G| < 24g^2$;

(b′) $|G| < 16g^2$;

(c′) $|G| < 8g^3$.

Proof. The notation used in the proof of Theorem 11.56 is maintained.

Assume first that (a) occurs. Then $d'_1 = d'_2 = \frac{1}{2}$, $d'_3 = d_{Q_3}/e_{Q_3}$, and hence

$$d' - 2 = \frac{d_{Q_3} - e_{Q_3}}{e_{Q_3}}. \tag{11.64}$$

Write $e_{Q_3} = e = e_1 e'_{Q_3}$ with $e_1 = |G^{(1)}_{Q_3}| \geq p$ and $p \nmid e'_Q$. In other words,

$$e_{Q_3} = |G_{Q_3}|, \quad e_1 = |G^{(1)}_{Q_3}|, \quad e'_{Q_3} = |H|,$$

where $G_{Q_3} = G^{(1)}_{Q_3} \rtimes H$.

By Lemma 11.80, $d_{Q_3} \geq e + e_1 - 2$. Since $p \geq 3$, from (11.64),

$$d' - 2 \geq \frac{e - 1 + e_1 - 1 - e}{e} = \frac{1}{e'_{Q_3}}\left(1 - \frac{2}{e_1}\right) \geq \frac{1}{e'_{Q_3}}\left(1 - \frac{2}{p}\right) \geq \frac{1}{3e'_{Q_3}}.$$

From (11.19) and (11.23),

$$|G| \leq 3e'_{Q_3}(2g-2) \leq 3(4g+2)(2g-2) < 24g^2.$$

Now assume that (b) occurs. There are two short orbits, both non-tame. As above, for $p \geq 3$,

$$d' - 2 = \frac{d_{Q_1} - e_{Q_1}}{e_{Q_1}} + \frac{d_{Q_2} - e_{Q_2}}{e_{Q_2}} \geq \frac{1}{3e'_{Q_1}} + \frac{1}{3e'_{Q_2}} \geq \frac{2}{3(4g+2)},$$

whence $|G| \leq 3(2g+1)(2g-2) < 12g^2$.

For $p = 2$, a further argument is required. Since $|G_{Q_1}| = e_{Q_1}$, the Hilbert Different Formula (11.37) implies that

$$\begin{aligned} d_{Q_1} &= |G_{Q_1}| - 1 + |G^{(1)}_{Q_1}| - 1 + \textstyle\sum_{i\geq 2} (|G^{(i)}_{Q_1}| - 1) \\ &= e_{Q_1} + G^{(1)}_{Q_1} - 2 + \textstyle\sum_{i\geq 2} (|G^{(i)}_{Q_1}| - 1). \end{aligned}$$

Since the same holds for Q_2, so

$$\begin{aligned} d' - 2 &= \textstyle\sum_{i=1}^{2} \dfrac{d_{Q_i} - e_{Q_i}}{e_{Q_i}} \\ &= \textstyle\sum_{j=1}^{2} \dfrac{1}{e_{Q_j}} \left(|G^{(1)}_{Q_j}| - 2 + \textstyle\sum_{i\geq 2} (|G^{(i)}_{Q_j}| - 1)\right). \end{aligned} \tag{11.65}$$

Three cases (I), (II), (III) are investigated separately.

(I) For $|G^{(1)}_{Q_1}| \geq 4$ and, similarly for $|G^{(1)}_{Q_2}| \geq 4$, (11.65) shows that

$$d' - 2 \geq \frac{|G^{(1)}_{Q_1}| - 2}{e_{Q_1}} = \frac{|G^{(1)}_{Q_1}|}{e_{Q_1}}\left(1 - \frac{2}{|G^{(1)}_{Q_1}|}\right) \geq \frac{|G^{(1)}_{Q_1}|}{2\,e_{Q_1}}.$$

Then (11.23) together with (11.19) gives the bound

$$|G| \leq 2(4g+2)(2g-2) < 16g^2.$$

(II) The two equations $|G^{(1)}_{\mathcal{Q}_1}| = |G^{(1)}_{\mathcal{Q}_2}| = 2$ and $|G^{(2)}_{\mathcal{Q}_1}| = |G^{(2)}_{\mathcal{Q}_2}| = 1$ cannot hold simultaneously because of (11.65) and $d' - 2 > 0$.

(III) If $|G^{(1)}_{\mathcal{Q}_1}| = |G^{(1)}_{\mathcal{Q}_2}| = 2$, and either $|G^{(2)}_{\mathcal{Q}_1}| = 2$ or $|G^{(2)}_{\mathcal{Q}_2}| = 2$, from (11.65),

$$d' - 2 \geq \frac{1}{e_{\mathcal{Q}_1}} = \frac{|G^{(1)}_{\mathcal{Q}_1}|}{2\, e_{\mathcal{Q}_1}},$$

whence again $|G| < 16g^2$, by (11.23) and (11.19). This completes the proof of (b$'$).

To investigate (c), two cases are distinguished according as the unique short orbit consists of just one place or more.

In the former case, $G = G_{\mathcal{P}}$. If $F_1 = \Sigma^{G^{(1)}_{\mathcal{P}}}$ is not rational, then $|G| \leq g(4g+2)$ by Theorems 11.60 and 11.78. So assume that F_1 is rational. It is shown that $G = G^{(1)}_{\mathcal{P}}$. Assume on the contrary that G contains a tame subgroup H whose order is prime to p. Choose a non-trivial K-automorphism α in H. Since α is in the normaliser of $G^{(1)}_{\mathcal{P}}$, it induces a K-automorphism α' of F_1. By Theorem 11.14 (ii)(d), α' has at least two fixed places, and hence one of them lies under a place $\mathcal{Q}$ distinct from $\mathcal{P}$. So the group generated by α together with $G^{(1)}_{\mathcal{P}}$ has a non-trivial stabiliser at $\mathcal{Q}$. Then the orbit of such a place $\mathcal{Q}$ under the action of G is short. But then G has at least two short orbits, a contradiction. Hence $G = G_{\mathcal{P}}$. Now, from Theorem 11.78, $|G| = |G^{(1)}_{\mathcal{P}}| \leq 8g^2$.

In the latter case, G does not fix $\mathcal{P}$. Let o denote the G-orbit of $\mathcal{P}$, that is, the unique short orbit of G. Then $|o| > 1$, and $\deg D(\Sigma/\Sigma^G) = |o| \deg D(\Sigma/\Sigma^{G_{\mathcal{P}}})$. Since $|G| = |G_{\mathcal{P}}||o|$, this gives the equation

$$2g - 2 = |o|(\deg D(\Sigma/\Sigma^{G_{\mathcal{P}}}) - 2|G_{\mathcal{P}}|). \tag{11.66}$$

Therefore $|o| \leq 2g - 2$. If F_1 is not rational, then $|G^{(1)}_{\mathcal{P}}| \leq g$, by Theorem 11.78 (i). This together with Theorem 11.60 give $|G_{\mathcal{P}}| \leq g(4g+2)$, whence

$$|G| \leq g(2g-2)(4g+2) < 8g^3.$$

Assume that F_1 is rational. Then

$$\deg D(\Sigma/\Sigma^{G_{\mathcal{P}}}) = |G_{\mathcal{P}}| - |G^{(1)}_{\mathcal{P}}| + \deg D(\Sigma/\Sigma^{G^{(1)}_{\mathcal{P}}}) = |G_{\mathcal{P}}| + |G^{(1)}_{\mathcal{P}}| + 2g - 2.$$

Thus (11.66) reads:

$$2g - 2 = |o|(2g - 2 + |G^{(1)}_{\mathcal{P}}| - |G_{\mathcal{P}}|).$$

If $|o| = 2g - 2$, then $2g - 2 + |G^{(1)}_{\mathcal{P}}| - |G_{\mathcal{P}}| = 1$ with $G_{\mathcal{P}} = G^{(1)}_{\mathcal{P}} \rtimes H$, whence

$$2g - 2 = (|H| - 1)|G^{(1)}_{\mathcal{P}}| + 1.$$

This implies that H is non-trivial. Also, $2g - 2 > \frac{1}{2}|H||G^{(1)}_{\mathcal{P}}| = \frac{1}{2}|G_{\mathcal{P}}|$, whence

$$|G| = |o||G_{\mathcal{P}}| < 2(2g-2)^2 < 8g^3.$$

So let $|o| \leq g - 1$. If F_1 is rational and $G^{(1)}_{\mathcal{P}}$ has a short orbit disjoint from $\mathcal{P}$, then Theorem 11.78 (ii) again gives the desired upper bound, $|G| < 8g^3$. Otherwise, o contains a long orbit of $G^{(1)}_{\mathcal{P}}$, and $|o| \geq 1 + |G^{(1)}_{\mathcal{P}}|$. As $|o| \leq g - 1$, this shows that $|G_{\mathcal{P}}^{(1)}| < g$. Using the above argument, it follows that $|G| < 4g^3$, and this completes the proof of (c′). □

The methods in this study of $G_{\mathcal{P}}^{(1)}$ may also be used to investigate case (d) of Theorem 11.56. But the upper bound on $|G|$ which may be shown in this way is only about g^5. Therefore a substantial improvement is needed. This requires a careful analysis using deeper group-theoretical results. The final result is stated in Theorem 11.127.

In Theorem 11.56 (d), G has two short orbits, one tame and one non-tame. Choose a place $\mathcal{P}$ in the non-tame orbit. By Lemma 11.44 (e), $G_{\mathcal{P}} = G_{\mathcal{P}}^{(1)} \rtimes H$ with $|H|$ prime to p. As before, write F_1 for $\Sigma^{G_{\mathcal{P}}^{(1)}}$, and F_2 for $\Sigma^{G_{\mathcal{P}}^{(2)}}$. The following five cases are treated separately.

(iv.1) F_1 is not rational.

(iv.2) F_1 is rational, and there is a place $\mathcal{R}$ distinct from $\mathcal{P}$ such that the stabiliser of $\mathcal{R}$ in $G_{\mathcal{P}}$ has order p^t with $t \geq 1$.

(iv.3) F_1 is rational, and there is a place $\mathcal{R}$ distinct from $\mathcal{P}$ such that the stabiliser of $\mathcal{R}$ in $G_{\mathcal{P}}$ has order hp^t with $t \geq 1$, $h > 1$ and h prime to p.

(iv.4) F_1 is rational, no non-trivial element of $G_{\mathcal{P}}^{(1)}$ fixes a place distinct from $\mathcal{P}$, and there is a place $\mathcal{R}$ distinct from $\mathcal{P}$ but lying in the orbit of $\mathcal{P}$ in G such that the stabiliser of $\mathcal{R}$ in $G_{\mathcal{P}}$ is trivial.

(iv.5) F_1 is rational, no non-trivial element of $G_{\mathcal{P}}^{(1)}$ fixes a place distinct from $\mathcal{P}$, and, for every place $\mathcal{R}$ distinct from $\mathcal{P}$ but lying in the orbit of $\mathcal{P}$ in G, the stabiliser of $\mathcal{R}$ in $G_{\mathcal{P}}$ is non-trivial.

Before starting with (iv.1), another closed formula for the order of G is deduced. Choose a place $\mathcal{Q}$ from the tame orbit of G. Then $d_{\mathcal{Q}} = e_{\mathcal{Q}} - 1 = |G_{\mathcal{Q}}| - 1$. From Theorem 11.72,

$$|G| = 2(g-1)\,\frac{|G_{\mathcal{P}}^{(1)}|\,|H|\,|G_{\mathcal{Q}}|}{N}, \tag{11.67}$$

where

$$N = 2g_1\,|G_{\mathcal{P}}|\,|G_{\mathcal{Q}}| + d_{\mathcal{P}}\,|G_{\mathcal{Q}}| - |G_{\mathcal{P}}|\,|G_{\mathcal{Q}}| - |G_{\mathcal{P}}|. \tag{11.68}$$

If Σ^G is rational, which is the relevant case, then $N = |G_{\mathcal{Q}}|(d_{\mathcal{P}} - |G_{\mathcal{P}}|) - |G_{\mathcal{P}}|$.

Case (iv.1). If F_1 is not rational, then its genus $g_1 \geq 1$. Since $\mathcal{P}$ is fixed by $G_{\mathcal{P}}^{(1)}$ and $|G_{\mathcal{P}}^{(1)}|$ is a power of p,

$$d_{\mathcal{P}} \geq 2(e_{\mathcal{P}} - 1) = 2(|G_{\mathcal{P}}^{(1)}| - 1).$$

Hence, by Theorem 11.72 applied to $G_{\mathcal{P}}^{(1)}$,

$$2g - 2 \geq |G_{\mathcal{P}}^{(1)}|(2g_1 - 2) + 2(|G_{\mathcal{P}}^{(1)}| - 1).$$

This implies that $|G_{\mathcal{P}}^{(1)}| \leq g/g_1$. As the group $H \cong G_{\mathcal{P}}/G_{\mathcal{P}}^{(1)}$ can be viewed as a K-automorphism group of F_1, Theorem 11.60 shows that $|H| \leq 4g_1 + 2$. The same theorem also shows that $|G_{\mathcal{Q}}| \le 4g+2$ as $\mathcal{Q}$ belongs to the tame orbit. Also, $N \ge 6$, since

$$N = |G_{\mathcal{P}}|2g_1(|G_{\mathcal{Q}}|-1) + |G_{\mathcal{Q}}|(d_{\mathcal{P}} - |G_{\mathcal{P}}|) + (2g_1-1)|G_{\mathcal{P}}^{(1)}|$$

and $d_{\mathcal{P}} \ge |G_{\mathcal{P}}| \ge 2$. By (11.67),

$$\begin{aligned} |G| &\le \tfrac{2}{6}(g-1)(4g+2)(4g_1+2)g/g_1 \\ &< \tfrac{8}{6}g(g-1)(g+1)(4+2/g_1) < 8g^3. \end{aligned} \tag{11.69}$$

Case (iv.2). Here, F_1 is rational, and the orbit of $\mathcal{R}$ under H has size $|H|$. Let M be the subgroup of $G_{\mathcal{P}}^{(1)}$ fixing $\mathcal{R}$, that is $M = G_{\mathcal{P},\mathcal{R}}$. Then

$$d(\Sigma/F_1) \ge 2(|G_{\mathcal{P}}^{(1)}|-1) + \frac{|G_{\mathcal{P}}^{(1)}|}{|M|}(2|M|-2)\,|H|.$$

By Theorem 11.72, it follows that

$$g \ge |G_{\mathcal{P}}^{(1)}|\,(1-1/|M|)\,|H| \ge \tfrac{1}{2}\,|G_{\mathcal{P}}^{(1)}|\,|H| = \tfrac{1}{2}|G_{\mathcal{P}}|.$$

If $|G_Q| \le 2g+2$, this and (11.67) imply that

$$|G| \le 2(g-1)(2g+2)2g/N \le 8(g^3-g) < 8g^3.$$

If $|G_{\mathcal{Q}}| > 2g+2$, then a sharper bound is obtained from (11.67), namely,

$$|G| < (2g-2)2g < 4g^3.$$

This depends on the fact that $|G_{\mathcal{Q}}| < N$ when $|G_{\mathcal{Q}}| > 2g+2$.

That $|G_{\mathcal{Q}}| > 2g+2$ implies $|G_{\mathcal{Q}}| < N$ is shown in the following way. Since $|G_{\mathcal{P}}|/|G_{\mathcal{Q}}| \le 2g/(2g+2) < 1$,

$$N > |G_{\mathcal{Q}}|(d_{\mathcal{P}} - |G_{\mathcal{P}}| - 1).$$

Therefore it may be assumed that $d_{\mathcal{P}} - |G_{\mathcal{P}}| = 1$.

On the other hand, (11.38) can be re-written in the form

$$d_{\mathcal{P}} - |G_{\mathcal{P}}| - 1 = -2 + \textstyle\sum_{i\ge 1}(|G_{\mathcal{P}}^{(i)}|-1).$$

Therefore

$$\frac{N}{|G_{\mathcal{Q}}|} > |G_{\mathcal{P}}^{(1)}| - 3 + |G_{\mathcal{P}}^{(2)}| - 1 + |G_{\mathcal{P}}^{(3)}| - 1.$$

If $|G_{\mathcal{P}}^{(1)}| \ge 4$, in particular for $p \ge 5$, the assertion follows.

To show that this holds true for $|G_{\mathcal{P}}^{(1)}| \le 3$, a little more is needed. For $|G_{\mathcal{P}}| \le 3$, (11.67) implies that

$$|G| \le 3(2g-2)(4g+2),$$

which is smaller than $8g^3$ for $g \ge 2$. So this case can be dismissed.

Let $|G_{\mathcal{P}}| \ge 4$ and $|G_{\mathcal{P}}^{(1)}| \le 3$. If $|G_{\mathcal{P}}^{(1)}| = 3$ and $|G_{\mathcal{P}}| = 6$, then $N = |G_{\mathcal{Q}}| - 6$. From (11.67),

$$|G| = 12(g-1)\,\frac{|G_{\mathcal{Q}}|}{|G_{\mathcal{Q}}|-6} < 24(g-1) < 8g^3;$$

so this case can also be dismissed. If either $|G_{\mathcal{P}}^{(1)}| = 3$ and $|G_{\mathcal{P}}| > 6$, or $|G_{\mathcal{P}}^{(1)}| = 2$ and $|G_{\mathcal{P}}| > 4$, then

$$|H| > |G_{\mathcal{P}}^{(1)}| - 1 \text{ with } G_{\mathcal{P}} = G_{\mathcal{P}}^{(1)} \rtimes H.$$

Since $G_{\mathcal{P}}^{(1)}$ is a normal subgroup of $G_{\mathcal{P}}$, this implies the existence of a non-trivial element $\alpha \in H$ in the centraliser of $G_{\mathcal{P}}^{(1)}$. From Theorem 11.75 (i), $G_{\mathcal{P}}^{(1)} = G_{\mathcal{P}}^{(2)}$ when $p = 3$, and $G_{\mathcal{P}}^{(1)} = G_{\mathcal{P}}^{(2)} = G_{\mathcal{P}}^{(3)}$ when $p = 2$. Therefore $|G_{\mathcal{Q}}| < N$.

This completes the proof for (iv.2).

Case (iv.3). The stabiliser of $\mathcal{R}$ in $G_{\mathcal{P}}$ contains a non-trivial subgroup T of $G_{\mathcal{P},\mathcal{R}}$ of prime order m with $p \neq m$. By Sylow's theorem, a subgroup H_1 conjugate to H in $G_{\mathcal{P}}$ contains T. Now consider the action of T on F_1 or, equivalently, on the set of all orbits of $G_{\mathcal{P}}^{(1)}$. By Theorem 11.14 (ii)(d), T preserves exactly two such orbits: one is the trivial orbit consisting of $\mathcal{P}$ and the other is the orbit $o(\mathcal{R})$ containing $\mathcal{R}$. Note that either $o(\mathcal{R}) = \{\mathcal{R}\}$ or $|o(\mathcal{R})|$ is a power of p.

Since H_1 is cyclic and its order is prime to p, a generator of H_1 has exactly two invariant orbits of $G_{\mathcal{P}}^{(1)}$, again by Theorem 11.14 (iv). Therefore $\{\mathcal{P}\}$ and $o(\mathcal{R})$ are the invariant orbits of H_1 as well. Since $o(\mathcal{R})$ is either trivial or its size is a power of p, so H_1 fixes a place in $o(\mathcal{R})$. Since H and H' are conjugate in $G_{\mathcal{P}}$, this implies that H fixes a place $\mathcal{R}'$ in $o(\mathcal{R})$.

After replacing this place by $\mathcal{R}$, it may be assumed that H fixes $\mathcal{R}$. From the above discussion, a non-trivial element of H may fix some other places of Σ, but such places must be in $o(\mathcal{R})$. So, if some non-trivial element of $G_{\mathcal{P}}^{(1)}$ fixes a place distinct from those in $\mathcal{P} \cup o(\mathcal{R})$, then (iv.2) occurs. Therefore it is assumed that this does not happen, since $|G| < 8g^3$ has already been shown in Case (iv.2). In particular, if $o(\mathcal{R}) = \{\mathcal{R}\}$, then $\mathcal{P}$ and $\mathcal{R}$ are the only fixed places of any non-trivial element in H, and $G_{\mathcal{P}}^{(1)}$ fixes both $\mathcal{P}$ and $\mathcal{R}$; but no non-trivial element of $G_{\mathcal{P}}^{(1)}$ fixes another place of Σ. Since $G_{\mathcal{P}} = G_{\mathcal{P}}^{(1)} \rtimes H$, this extends to $G_{\mathcal{P}}$.

Let $M = G_{\mathcal{P}}^{(1)} \cap G_{\mathcal{R}}^{(1)}$. Then M is non-trivial p-group. A first consequence is the following result.

Lemma 11.117 *Assume that* (iv.3) *holds. If $o(\mathcal{R})$ is larger than $\mathcal{R}$, then $|G_{\mathcal{P}}^{(1)}|$ is divisible by p^2.*

Proof. If $|G_{\mathcal{P}}^{(1)}| = p$, then $G_{\mathcal{P}}^{(1)} = M$, and hence $G_{\mathcal{P}}^{(1)} = G_{\mathcal{R}}^{(1)}$. In other words, $G_{\mathcal{P}}^{(1)}$ fixes $\mathcal{R}$. □

Now upper bounds on $|H|$ and $|G_{\mathcal{P}}^{(1)}|$, depending only on g, are given.

Lemma 11.118 *If* (iv.3) *holds, then $|G_{\mathcal{P}}| \leq \frac{2}{3}(2g+1)g$.*

Proof. By Theorem 11.72 applied to $G_{\mathcal{P}}$, and Lemma 11.82 (i),

$$\begin{aligned} 2g-2 &\geq -2|G_{\mathcal{P}}| + d_{\mathcal{P}} + |o(\mathcal{R})|\, d_{\mathcal{R}} \\ &= v - |G_{\mathcal{P}}| + |o(\mathcal{R})|\, d_{\mathcal{R}} \geq |H| - 1 - |G_{\mathcal{P}}| + |o(\mathcal{R})|\, d_{\mathcal{R}}. \end{aligned}$$

From Lemma 11.82 (i), applied this time to $G_{\mathcal{P}} \cap G_{\mathcal{R}} = M \rtimes H$, it follows that $d_{\mathcal{R}} \geq |M||H| + |H| - 1$. Therefore

$$2g-2 \geq |H| - 1 - |G_{\mathcal{P}}| + |o(\mathcal{R})|(|M||H| + |H| - 1) \geq (|o(\mathcal{R})| + 1)(|H| - 1),$$

whence

$$|H| \le \begin{cases} (2g+1)/3 & \text{for } |o(\mathcal{R})| \ge 2, \\ g & \text{for } o(\mathcal{R}) = \{\mathcal{R}\}. \end{cases}$$

For $|o(\mathcal{R})| \ge 2$, this gives $|H| \le (2g+1)/3$.

Also, by Theorem 11.72 applied to $G_{\mathcal{P}}^{(1)}$ and Lemma 11.80,

$$2g-2 \ge -2|G_{\mathcal{P}}^{(1)}| + (2|G_{\mathcal{P}}^{(1)}| - 2) + |o(\mathcal{R})|(2|M| - 2),$$

whence

$$g \ge |o(\mathcal{R})|(|M|-1) \ge \tfrac{1}{2}|o(\mathcal{R})||M| = \tfrac{1}{2}|G_{\mathcal{P}}^{(1)}|.$$

If $|o(\mathcal{R})| \ge 2$, the result follows.

Therefore let $o(\mathcal{R}) = \{\mathcal{R}\}$. Then Theorem 11.72 applied to $G_{\mathcal{P}}^{(1)}$ gives the inequality,

$$2g-2 \ge -2\,|G_{\mathcal{P}}^{(1)}| + 2(2\,|G_{\mathcal{P}}^{(1)}| - 2) = 2|G_{\mathcal{P}}^{(1)}| - 4,$$

whence $g+1 \ge |G_{\mathcal{P}}^{(1)}|$. Thus $|G_{\mathcal{P}}| = |G_{\mathcal{P}}^{(1)}||H| \le g(g+1) < \frac{2}{3}(2g+1)g$. □

The desired upper bound $|G| < 8g^3$ can be deduced from Lemma 11.118 if the size of the G-orbit of $\mathcal{P}$ does not exceed $6(g-1)$. The next two lemmas show that this hypothesis is fulfilled in many cases, for instance, when $G_{\mathcal{P}}^{(2)}$ is trivial, apart from a few exceptions described in the following lemma. However, such exceptional curves do not provide counter-examples for the bound, $|G| < 8g^3$.

LEMMA 11.119 *Assume that* (iv.3) *holds. If the second ramification group $G_{\mathcal{P}}^{(2)}$ is trivial and if the orbit of $\mathcal{P}$ under G has size $r \ge 6(g-1)$, then $|G| < 8g^3$.*

Proof. Since $|G_{\mathcal{P}}^{(2)}| = 1$,

$$N = |G_{\mathcal{Q}}|(|G_{\mathcal{P}}^{(1)}| - 2) - |G_{\mathcal{P}}^{(1)}||H|.$$

Thus $N > 0$ implies that $|G_{\mathcal{Q}}| = |H| + \epsilon$ with $\epsilon > 0$. So $N = \epsilon(|G_{\mathcal{P}}^{(1)}| - 2) - 2|H|$, which, together with

$$|G_{\mathcal{P}}^{(1)}| - 2 = v = (\mu+1)|H| - 1,$$

given in Lemma 11.82 (i), shows the following equations for $\mu \ge 1$ and $\epsilon \ge 2$:

$$\frac{r}{2g-2} = \frac{|G_{\mathcal{Q}}|}{N} = \frac{|H| + \epsilon}{|H|[(\mu+1)\epsilon - 2] - \epsilon} \le \frac{|H| + \epsilon}{|H|(2\epsilon - 2) - \epsilon} \le \frac{|H| + 2}{2|H| - 2} \le 2.$$

If $\epsilon = 1$, then $\mu \ge 2$ and

$$\frac{r}{2g-2} = \frac{|G_{\mathcal{Q}}|}{N} = \frac{|H| + 1}{|H|(\mu - 1) - 1} \le \frac{|H| + 1}{|H| - 1} \le 3.$$

If $\mu = 0$, then $\epsilon \ge 3$ and $|H| = |G_{\mathcal{P}}^{(1)}| - 1$. For $|H| \ge 6$,

$$\frac{r}{2g-2} = \frac{|G_{\mathcal{Q}}|}{N} = \frac{|H| + \epsilon}{|H|(\epsilon - 2) - \epsilon} \le 3,$$

Table 11.2 Order of automorphism group

$\lvert G_{\mathcal{P}}^{(1)}\rvert$	$\lvert H\rvert$	$\lvert G_{\mathcal{Q}}\rvert$	$\frac{r}{2g-2}$	$\lvert G\rvert$	$o(\mathcal{R}) \neq \{R\}$	$o(\mathcal{R}) = \{R\}$
5	4	7	7	$280(g-1)$	impossible	$g = 4$
4	3	7	7/2	$84(g-1)$	$g \geq 3$	$g = 3$
3	2	7	7	$84(g-1)$	impossible	$g = 2$
3	2	8	4	$48(g-1)$	impossible	$g = 2$

while, for $|H| = |G_{\mathcal{P}}^{(1)}| - 1 < 6$,

$$2 \leq |H| \leq 4.$$

In the last case, $r \leq 6(g-1)$ provided that either $|G_{\mathcal{Q}}| < 7$ and $3 \leq |H| \leq 4$, or $|G_{\mathcal{Q}}| > 8$ and $|H| = 2$. Otherwise, by Lemmas 11.117 and 11.118, one of the cases in Table 11.2 occurs.

In the second and fourth cases, $|G| < 8g^3$. In the third case, Σ is hyperelliptic and has more than one Weierstrass point as G fixes no place. Since no non-soluble group has order 84, Proposition 11.99 shows that again $|G| < 8g^3$.

Some more is needed to rule out the first case. Here, $|G| = 840 = 2^3 \cdot 3 \cdot 5 \cdot 7$, and the orbit $o(\mathcal{Q})$ of $\mathcal{Q}$, that is, the tame orbit of G, has size 120. The key idea is to prove that $G_{\mathcal{Q}}$ must have another fixed place in $o(\mathcal{Q})$.

Let T be minimal normal subgroup of G. If T is an elementary abelian group of prime power order w^k, then $w^k \in \{2, 2^2, 2^3, 3, 5, 7\}$. If $|T| < 7 = |G_{\mathcal{Q}}|$, then a non-trivial element $\alpha \in T$ is contained in the centraliser of $G_{\mathcal{Q}}$. This implies that $\mathcal{Q}^\alpha$ is also fixed by $G_{\mathcal{Q}}$. Since $\alpha \notin G_{\mathcal{Q}}$, it follows that $\mathcal{Q} \neq \mathcal{Q}^\alpha$, and the assertion is proved.

If $|T| = 7$, then $G_{\mathcal{Q}} \cong T$, and hence $G_{\mathcal{Q}}$ is a normal subgroup of G. Then G preserves the set of all fixed places of $G_{\mathcal{Q}}$, showing that $G_{\mathcal{Q}}$ fixes each place in $o(\mathcal{Q})$.

If $|T| = 8$, then T is a Sylow 2-subgroup of G. Since T is a normal subgroup of G, H is a subgroup of T. Therefore T contains an element of order 4, and hence T has at most 5 involutions. So at least one involution of T is contained in the centraliser of $G_{\mathcal{Q}}$. As before, this implies that $G_{\mathcal{Q}}$ has at least two fixed places.

Otherwise, T is either a non-abelian simple group or the direct product of isomorphic non-abelian simple groups. The latter possibility does not occur in this situation, since no factor of $|G|$ distinct from 2 is a power of an integer with exponent greater than 1. Also there are only two simple groups whose order divides 840, namely $\mathbf{A}_5$ and $\mathrm{PSL}(2,7)$. If $T \cong \mathbf{A}_5$, then $G_{\mathcal{Q}}$ is contained in the centraliser of T in G. This follows from the fact that 7 is not a factor of 120, and that $|\mathrm{Aut}(\mathbf{A}_5)| = |\mathrm{PGL}(2,5)|$. Therefore $G_{\mathcal{Q}}$ must have at least two fixed places in $o(\mathcal{Q})$.

As 7 is not a factor of 120 and $|\mathrm{Aut}(\mathrm{PSL}(2,7))| = |\mathrm{PGL}(2,7)|$, a similar argument may be used for the case $T \cong \mathrm{PSL}(2,7)$ showing that $G_{\mathcal{Q}}$ coincides with one of the eight Sylow 7-subgroups of T. Then an element of order 3 is in the normaliser of $G_{\mathcal{Q}}$. As before, this implies that $G_{\mathcal{Q}}$ cannot have just one fixed place.

Once it has been shown that $G_{\mathcal{Q}}$ fixes not only $\mathcal{Q}$ but some more places in $o(\mathcal{Q})$, it follows that $G_{\mathcal{Q}}$ must have at least 7 fixed places. In fact, 120 minus the number of fixed places of $G_{\mathcal{Q}}$ must be divisible by 7. By Theorem 11.72 applied to $G_{\mathcal{Q}}$,

$$2g-2 \geq 7(2g'-2)+7\cdot 6 \geq 42-14=28.$$

From this, $g \geq 15$, a contradiction. □

LEMMA 11.120 *Assume that* (iv.3) *holds. Then* $|G| < 8g^3$ *provided that at least one of the following conditions is satisfied*:

(a) $v < |G_{\mathcal{P}}^{(1)}|$, *where* $v = d_{\mathcal{P}} - |G_{\mathcal{P}}|$;

(b) $|G_{\mathcal{Q}}| > |H|$;

(c) $(\mu+1)|G_{\mathcal{Q}}| \neq |G_{\mathcal{P}}^{(1)}|+1$, *where* $(\mu+1)|H| = v+1$, *as in Lemma 11.82*;

(d) $v \geq 2|G_{\mathcal{P}}^{(1)}|-3$.

Proof. By Lemma 11.118, it suffices to show that $r \leq 6(g-1)$ if some of the above conditions are satisfied. By (11.67), the bound to be established is the following:

$$\frac{r}{2g-2} = \frac{|G_{\mathcal{Q}}|}{N} = \frac{|G_{\mathcal{Q}}|}{|G_{\mathcal{Q}}|(d_{\mathcal{P}}-|G_{\mathcal{P}}|)-|G_{\mathcal{P}}|} \leq 3.$$

By the Hilbert Different Formula (11.37) and Lemma 11.75 (iv), Case (a) can only occur in one of two circumstances:

(i) $v = |G_{\mathcal{P}}^{(1)}|-1, \quad v_1 = 3, \quad |G_{\mathcal{P}}^{(2)}| = 2;$

(ii) $v = |G_{\mathcal{P}}^{(1)}|-2, \quad v_1 = 2, \quad |G_{\mathcal{P}}^{(2)}| = 1.$

In case (i), Lemma 11.82 (i) implies that $|G_{\mathcal{P}}^{(1)}| = (\mu+1)|H|$, a contradiction, as both $\gcd(|H|, |G_{\mathcal{P}}^{(1)}|) = 1$ and $|H| > 1$. In case (ii), either $|G| < 8g^3$ or $r < 6(q-1)$; see Lemma 11.119.

Assume that (b) holds. From (a), take $v \geq |G_{\mathcal{P}}^{(1)}|$. Put $|G_{\mathcal{Q}}| = |H|+\epsilon$ with $\epsilon > 0$. Then

$$0 < \frac{r}{2g-2} = \frac{|G_{\mathcal{Q}}|}{N} = \frac{|H|+\epsilon}{|H|(v-|G_{\mathcal{P}}^{(1)}|)+v\epsilon} \leq \frac{|H|+\epsilon}{v\epsilon} \leq \frac{|H|+1}{v}.$$

On the other hand, by Lemma 11.82 (i), $|H|$ divides $(v+1)$. Thus $|H| \leq 3v$, and hence $r \leq 6(g-1)$.

Now assume that (c) holds. From (b), take $|H| \geq |G_{\mathcal{Q}}|$. From Lemma 11.82 (i),

$$\begin{aligned} 0 < \frac{r}{2g-2} &= \frac{|G_{\mathcal{Q}}|}{|G_{\mathcal{Q}}|v-|G_{\mathcal{P}}|} = \frac{|G_{\mathcal{Q}}|}{|G_{\mathcal{Q}}|(\mu+1)|H|-1)-|G_{\mathcal{P}}|} \\ &= \frac{|G_{\mathcal{Q}}|}{|H|(|G_{\mathcal{Q}}|(\mu+1)-|G_{\mathcal{P}}^{(1)}|)-|G_{\mathcal{Q}}|} \leq \frac{|G_{\mathcal{Q}}|}{2|H|-|G_{\mathcal{Q}}|} \leq \frac{|G_{\mathcal{Q}}|}{|H|}. \end{aligned}$$

Since $|H| \geq |G_{\mathcal{Q}}|$ by hypothesis, so $r \leq 2(g-1)$.

Finally, assume that (d) holds. From (c), let $|G_{\mathcal{Q}}|(\mu+1) = |G_{\mathcal{P}}^{(1)}| + 1$. Again, from Lemma 11.82 (i),

$$0 < \frac{r}{2g-2} = \frac{|G_{\mathcal{P}}|}{|G_{\mathcal{Q}}|v - |G_{\mathcal{P}}|} = \frac{|G_{\mathcal{Q}}|}{|H| - |G_{\mathcal{Q}}|}$$
$$= \frac{|G_{\mathcal{Q}}|(\mu+1)}{(|H| - |G_{\mathcal{Q}}|)(\mu+1)} = \frac{|G_{\mathcal{P}}^{(1)}|+1}{v - |G_{\mathcal{P}}^{(1)}|} \leq \frac{|G_{\mathcal{P}}^{(1)}|+1}{|G_{\mathcal{P}}^{(1)}|-3}.$$

Therefore $|G_{\mathcal{P}}^{(1)}| \neq 2$ and $r \leq 6(g-1)$ for $|G_{\mathcal{P}}^{(1)}| \geq 5$. Let $|G_{\mathcal{P}}^{(1)}| = 3$. Then

$$v - |G_{\mathcal{P}}^{(1)}| = 2(v_1 - 1) - 1 - 3 = 2(v_1 - 3).$$

If $v_1 = 3$, then $v = |G_{\mathcal{P}}^{(1)}|$, and hence $v - |G_{\mathcal{P}}| < 0$ which is impossible. Thus $v_1 \geq 4$. Therefore $v - |G_{\mathcal{P}}^{(1)}| \geq 2$, whence $r \leq 6(g-1)$. Finally, if $|G_{\mathcal{P}}^{(1)}| = 4$, then $r \leq 5(g-1)$. In fact, if $v - |G_{\mathcal{P}}^{(1)}| = 1$, then (b) holds, since from the above equations for $v = 5$, it follows that $|H| = 3$ and $|G_{\mathcal{Q}}| = 5$. □

Two more technical results are needed to finish the proof for Case (iv.3).

Lemma 11.121 *Assume that* (iv.3) *holds, and let* $v' = d_{\mathcal{R}}(\Sigma/\Sigma^M) - |M|$. *For the place* $\bar{\mathcal{R}}$ *under* $\mathcal{R}$ *in* Σ/Σ^M, *let* $\bar{v} = d_{\bar{\mathcal{R}}}(\Sigma^M/F_1) - |G_{\mathcal{P}}^{(1)}|/|M|$. *If* $o(\mathcal{R})$ *is larger than* $\mathcal{R}$, *then*

(i)
$$2(g-1) = v + |o(\mathcal{R})|v'; \tag{11.70}$$

(ii)
$$|G_{\mathcal{P}}| + v = (|o(\mathcal{R})| + \bar{v})|M| + |M| + v'; \tag{11.71}$$

(iii)
$$|H| - 1 \leq \min\{v, \bar{v}\}. \tag{11.72}$$

Proof. The identity automorphism is the only element in $G_{\mathcal{P}}^{(1)}$ fixing a place of Σ distinct from both $\mathcal{P}$ and the places in $o(\mathcal{R})$. As these two places are in the same orbit of G, the unique non-tame orbit of G, some element $\alpha \in G$ takes $\mathcal{P}$ to $\mathcal{R}$. Therefore $|G_{\mathcal{P}}| = |G_{\mathcal{R}}|$. Let $M^{(i)}$ be i-th ramification group of M at $\mathcal{R}$. If $|M^{(s)}| > 1$ but $|M^{(s+1)}| = 1$, then

$$v' = -1 + \textstyle\sum_{i=1}^{s} (|M^{(i)}| - 1).$$

From Theorem 11.72 applied to $G_{\mathcal{P}}^{(1)}$,

$$2g - 2 = -2|G_{\mathcal{P}}^{(1)}| + d_{\mathcal{P}}(\Sigma/F_1) + |o(\mathcal{R})|(|M| + v').$$

Since $|G_{\mathcal{P}}^{(1)}| = |o(\mathcal{R})||M|$, this gives (11.70). Then (11.71) follows from (7.8) applied to the tower $\Sigma \supset \Sigma^M \supset F_1$.

Since $v' = d_{\mathcal{R}}(\Sigma/\Sigma^{M \rtimes H}) - |M \rtimes H|$, the first inequality in (11.72) is Lemma 11.82 (i) with $M \rtimes H$ in place of $G_{\mathcal{P}}$. Similarly, the second can be deduced from Lemma 11.82 (i) if this is applied to Σ^M and $\bar{\mathcal{R}}$ taking into account that $\bar{G}_{\mathcal{R}} \cong (G_{\mathcal{R}}^{(1)}/M) \rtimes H$. □

These preliminary results provide enough information for a proof of the bound $|G| < 8g^3$, under the hypothesis that $o(\mathcal{R}) \neq \{\mathcal{R}\}$. Suppose $2|o(\mathcal{R})| \geq \bar{v} + 1$. Then (11.72) and (11.70) imply that

$$2(g-1) > |o(\mathcal{R})|v' \geq \tfrac{1}{2}|H|(|H|-1) \geq \tfrac{1}{2}|H||G_{\mathcal{Q}}|.$$

From (11.67),

$$|G| \leq 2(g-1)|H|\,|G_{\mathcal{P}}^{(1)}|\,|G_{\mathcal{Q}}| \leq 4(g-1)^2|G_{\mathcal{P}}^{(1)}| \leq 4(g-1)^2 g < 8g^3.$$

Suppose $2|o(\mathcal{R})| < \bar{v} + 1$. Then (11.71) gives the inequalities

$$|G_{\mathcal{P}}^{(1)}| + v \geq (3|o(\mathcal{R})| - 1)|M| + |M| + v' \geq 3|G_{\mathcal{P}}^{(1)}|.$$

Therefore (d) in Lemma 11.120 occurs. This completes the proof for the case where $o(R)$ is larger than $\mathcal{R}$.

If $o(\mathcal{R}) = \{\mathcal{R}\}$, the bound $|G| < 8g^3$ can be established by showing that (b), (c) or (d) of Lemma 11.120 holds. To do this, suppose (b) does not hold. By Lemma 11.119, take $G_{\mathcal{P}}^{(2)}$ to be non-trivial; that is, $v_1 \geq 2$. Actually, it may be assumed that $v_1 = 2$, as $v_1 \geq 3$ implies that case (d) of Lemma 11.120 holds.

LEMMA 11.122 *Assume that* (iv.3) *holds with* $o(\mathcal{R}) = \{R\}$. *If* $G_{\mathcal{P}}^{(2)}$ *is not trivial, then* $g_2 = q_1 - 1$ *where* g_2 *is the genus of* $F_2 = \Sigma^{G_{\mathcal{P}}^{(2)}}$ *and* $q_1 = |G_{\mathcal{P}}^{(1)}|/|G_{\mathcal{P}}^{(2)}|$.

Proof. No non-trivial element of $G_{\mathcal{P}}^{(1)}$ fixes a place distinct from $\mathcal{P}$ and $\mathcal{R}$. Also, $G_{\mathcal{P}}$ is a subgroup of $G_{\mathcal{R}}$. However, $|G_{\mathcal{P}}| = |G_{\mathcal{R}}|$, as some element of G sends $\mathcal{P}$ to $\mathcal{R}$; therefore $G_{\mathcal{P}} = G_{\mathcal{R}}$. From (11.17) applied to $G_{\mathcal{P}}^{(1)}$ and $G_{\mathcal{P}}^{(2)}$,

$$2g-2 = -2|G_{\mathcal{P}}^{(1)}| + 2[2(|G_{\mathcal{P}}^{(1)}|-1) + \textstyle\sum_{i\geq 2}(|G_{\mathcal{P}}^{(i)}|-1)],$$
$$2g-2 = (2g_2-2)|G_{\mathcal{P}}^{(2)}| + 2[2|G_{\mathcal{P}}^{(2)}|-1) + \textstyle\sum_{i\geq 2}(|G_{\mathcal{P}}^{(i)}|-1)].$$

Subtracting the two equations gives the result. □

Let $\mathcal{P}'$ be the place of F_2 lying under $\mathcal{P}$ in the covering $\Sigma \to F_2$. Also, let $\bar{\mathcal{P}}$ be the place of F_1 lying under $\mathcal{P}$ in the covering $\Sigma \to F_1$. Then $\bar{\mathcal{P}}$ is the place lying under $\mathcal{P}'$ in the covering $F_2 \to F_1$, and similarly for $\mathcal{R}$, $\mathcal{R}'$ and $\bar{\mathcal{R}}$.

Since F_1 is rational, there exists $y \in F_1$ such that $\operatorname{div} y = \bar{\mathcal{R}} - \bar{\mathcal{P}}$ in F_1. As F_1 is a subfield of F_2, so y may be viewed as an element of F_1. From (7.1), $\operatorname{div} y = q_1\mathcal{R}' - q_1\mathcal{P}'$ in F_2. Let m be the smallest non-gap at $\mathcal{P}'$, and choose $x \in F_2$ such that $\operatorname{div}(x)_\infty = m\mathcal{P}'$ in F_2.

If $m < q_1$, then $x \notin F_1$ by (7.1). Hence there exists $\alpha \in G_{\mathcal{P}}^{(1)} \backslash G_{\mathcal{P}}^{(2)}$ which induces a non-trivial K-automorphism α' of F_2 such that $\alpha'(x) \neq x$. From Lemma 11.13, $\alpha'(x) = x + u$ with $u \in K$ and $u \neq 0$. Since $\mathcal{R}'$ is a zero of x, this implies that α' does not fix $\mathcal{R}'$, a contradiction. Therefore $m = q_1$.

From the Weierstrass Gap Theorem 6.6 and Lemma 11.122, $m + 1 = q_1 + 1$ is also a non-gap. Choose $z \in F_2$ such that

$$\operatorname{div} z = k\mathcal{R}' + D - (q_1+1)\mathcal{P}',$$

where $k \geq 0$ and D is an effective divisor of F_2 of degree $q_1 + 1 - k$. Since $z \notin F_1$, there exists $\alpha \in G_{\mathcal{P}}^{(1)} \backslash G_{\mathcal{P}}^{(2)}$ that induces a non-trivial K-automorphism α' of F_2

such that $\alpha'(z) \neq z$. Since $q_1 + 1$ is the second smallest non-gap at $\mathcal{R}'$, it follows that $\alpha'(z) = z + ux + v$ with $u, v \in K$. Since α' also fixes $\mathcal{R}'$, so $v = 0$ and $u \neq 0$. From Lemma 11.83,

$$2 = v_1 = q_1 + 1 - k,$$

showing that $\deg D = 2$.

Now take a generator h of H. Then h is in the normaliser of $G_{\mathcal{P}}^{(2)}$, and hence h may be viewed as a K-automorphism of F_2 of order $|H|$ such that

$$h(z) = uz + vx + w$$

with $u, v, w \in K$ and $u \neq 0$. Since h fixes both $\mathcal{P}'$ and $\mathcal{R}'$, so $v = w = 0$ up to a substitution of z by $z + u'y$ with $u' \in K$. Thus $h(z) = uz$, whence it follows that $\text{div}\, h(z) = \text{div}\, z$ and hence h preserves D. Since $\deg D = 2$, so h^2 fixes both places in D. Therefore h^2 as a K-automorphism of F_1 fixes at least four places, namely, $\bar{\mathcal{P}}, \bar{\mathcal{R}}$ and the places lying under those of D in the covering $F_2 \to F_1$.

From Theorem 11.14 (b), $\text{ord}\, h = 2$, and hence $|H| = 2$.

Since case (b) of Lemma 11.120 does not hold by hypothesis, $|H| = 2$ implies that $|G_{\mathcal{Q}}| = 2$. Hence $N = 2(v - |G_{\mathcal{P}}^{(1)}|)$. But then case (d) of Lemma 11.120 holds, as $N \neq 0$.

Therefore $|G| < 8g^3$ holds true for case (iv.3) when $o(\mathcal{R}) = \{R\}$.

Case (iv.4). Here, F_1 is rational and no non-trivial element in $G_{\mathcal{P}}$ fixes $\mathcal{R}$. The latter hypothesis means that the orbit $o(\mathcal{R})$ of $\mathcal{R}$ under $G_{\mathcal{P}}$ is long. Let $o'(\mathcal{R})$ denote the orbit of $\mathcal{R}$ under G. Then $|o'(\mathcal{R})| \cdot |G_{\mathcal{R}}| = |G|$. Since $\mathcal{P}$ and $\mathcal{R}$ lie in the same orbit of G, so $G_{\mathcal{R}} \cong G_{\mathcal{P}}$. Also, $o(\mathcal{R})$ is contained in $o'(\mathcal{R})$. From (11.67),

$$|G_{\mathcal{P}}| \leq \frac{|G|}{|G_{\mathcal{P}}|} = 2(g-1) \cdot \frac{|G_{\mathcal{Q}}|}{N} \leq 2(g-1)|G_{\mathcal{Q}}|.$$

Now a lower bound on N, as in (11.68), is given. As

$$N = d_{\mathcal{P}}|G_{\mathcal{Q}}| - |G_{\mathcal{P}}||G_{\mathcal{Q}}| - |G_{\mathcal{P}}| \geq d_{\mathcal{P}}|G_{\mathcal{Q}}| - |G_{\mathcal{P}}||G_{\mathcal{Q}}| - 2(g-1)|G_{\mathcal{Q}}|,$$

so

$$N \geq |G_{\mathcal{Q}}|(d_{\mathcal{P}} - |G_{\mathcal{P}}| - 2(g-1)) \tag{11.73}$$

From Theorem 11.72 applied to $G_{\mathcal{P}}$,

$$2(g-1) = -2|G_{\mathcal{P}}| + (d_{\mathcal{P}} - |G_{\mathcal{P}}^{(1)}|(|H| - 1)) = d_{\mathcal{P}} - |G_{\mathcal{P}}| - |G_{\mathcal{P}}^{(1)}|.$$

Hence

$$d_{\mathcal{P}} - |G_{\mathcal{P}}| - 2(g-1) = |G_{\mathcal{P}}^{(1)}|.$$

Taking (11.73) into account, $N \geq |G_{\mathcal{P}}^{(1)}||G_{\mathcal{Q}}|$. Then (11.67), together with Theorem 11.60, gives the desired result:

$$|G| \leq 2(g-1) \cdot \frac{|G_{\mathcal{P}}^{(1)}||H||G_{\mathcal{Q}}|}{|G_{\mathcal{P}}^{(1)}||G_{\mathcal{Q}}|} \leq 8(g+1)(g-1) < 8g^3.$$

Case (iv.5). The type of argument used for (iv.4) only gives an estimate that $|G| < cg^5$ in this case. So further investigation is needed, which also requires a few technical results, such as the inverse formula of (11.67):

$$2g - 2 = \frac{|G|\,(|G_{\mathcal{P}}| - |G_{\mathcal{P}}^{(1)}|\,|G_{\mathcal{Q}}|)}{|G_{\mathcal{Q}}|(|G| - |G_{\mathcal{P}}|)}; \tag{11.74}$$

this holds whenever $|G_{\mathcal{P}} \cap G_{\mathcal{Q}}| = 1$.

First, the possibility that the unique non-tame orbit consists of a single place $\mathcal{P}$ is considered. From Theorem 11.72 applied to $G_{\mathcal{P}}^{(1)}$,

$$2g - 2 = -2 + \textstyle\sum_{i\geq 2} (|G_{\mathcal{P}}^{(i)}| - 1).$$

Arguing as in the proof of (11.72), it follows that $2g \geq |H|$. By Theorem 11.78,

$$|G| = |H||G_{\mathcal{P}}^{(1)}| \leq 2g \frac{4|G_{\mathcal{P}}^{(2)}|}{(|G_{\mathcal{P}}^{(2)}| - 1)^2} g^2,$$

which is greater than or equal to $8g^3$ only if $p = 2$, $|G_{\mathcal{P}}^{(2)}| = 2$, and F_2 is rational.

Theorem 12.5 implies that $\Sigma = K(x, y)$ with $y^2 - y = B(x)$, where $B(X)$ is in $K[X]$ and has odd degree m. Now Theorem 12.7 provides all the necessary information:

$$p^n = 2, \quad m - 1 = 2g, \quad |H| \mid m.$$

From Theorems 12.8 and 12.9, $|G| \geq 8g^3$ only occurs when $m - 1 = 2^k \geq 4$, $|H| = m$ and $|G_{\mathcal{P}}^{(1)}| = 2(m-1)^2$. Therefore $|G| < 8g^3$ apart from just one exception occurring for every $k > 1$ and $p = 2$, namely, the hyperelliptic curve

$$\mathbf{v}(Y^2 + Y + X^{2^k+1}),$$

which has genus 2^{k-1}. In fact, if $\mathcal{P}$ is the unique place centred at $Y_\infty = (0, 0, 1)$, then $G_{\mathcal{P}}$ has order $2^{2k+1}(2^k + 1)$, which is greater than $8g^3 = 2^{3k}$.

The possibility that the non-tame orbit is non-trivial is now investigated. Permutation group theory plays an important role, as in the next result.

LEMMA 11.123 *If case* (iv.5) *occurs, then G acts on its non-tame short orbit o as a 2-transitive permutation group. In particular, o has size $q + 1$ with $q = p^t$, and the possibilities for the permutation group $\bar{G}$ induced by G on o are as follows*:

(I) $\bar{G} \cong \mathrm{PSL}(2, q)$ *or* $\mathrm{PGL}(2, q)$;

(II) $\bar{G} \cong \mathrm{PSU}(3, n)$ *or* $\mathrm{PGU}(3, n)$, *with* $q = n^3$;

(III) $\bar{G} \cong \mathrm{Sz}(n)$, *with* $p = 2$, $n = 2n_0^2$, $n_0 = 2^k$, *with k odd, and* $q = n^2$;

(IV) $\bar{G} \cong \mathrm{Ree}(n)$ *with* $p = 3$, $n = 3n_0^2$, $n_0 = 3^k$, *and* $q = n^3$;

(V) *a minimal normal subgroup of $\bar{G}$ is soluble, and the size of o is a prime power.*

Proof. For a place $\mathcal{P} \in o$, let $o_0 = \{\mathcal{P}\}, o_1, \ldots, o_k$ denote the orbits of $G_{\mathcal{P}}^{(1)}$ contained in o; then $o = \bigcup_{i=0}^{k} o_i$. To prove that G acts 2-transitively on o, it must be shown that $k = 1$.

For any i with $1 \leq i \leq k$, take a place $\mathcal{R} \in o_i$. In case (iv.5), $\mathcal{R}$ is fixed by an element $\alpha \in G_{\mathcal{P}}$ whose order m is a prime different from p, and hence divides $|H|$. By Sylow's theorem, there is a subgroup H' conjugate to H in $G_{\mathcal{P}}$ which contains α; here, α preserves o_k. As noted previously, F_1 being rational implies that α fixes at most two orbits of $G_{\mathcal{P}}^{(1)}$. Therefore o_0 and o_i are the orbits preserved by α. As H' is abelian and it fixes o_0, so the orbits o_0 and o_i are also the only orbits of $G_{\mathcal{P}}^{(1)}$ which are fixed by H'. Since $G_{\mathcal{P}} = G_{\mathcal{P}}^{(1)} \rtimes H'$, this implies that the whole group $G_{\mathcal{P}}$ fixes o_i. As i can be any integer between 1 and k, it follows that $G_{\mathcal{P}}$ fixes each of the orbits $o_0, o_1, \ldots, o_k$. Therefore either $k = 1$ or $G_{\mathcal{P}}$ fixes at least three orbits of $G_{\mathcal{P}}^{(1)}$. The latter case cannot actually occur, by Theorem 11.14 applied to F_1. Also, the size of o is of the form $q + 1$ with $q = |G_{\mathcal{P}}^{(1)}|$; in particular, q is a power of p.

Let $\bar{G}$ denote the 2-transitive permutation group induced by G on o. The 2-point stabiliser of $\bar{G}$ is cyclic, as the subgroup of G fixing two distinct places is cyclic when (iv.5) occurs; see Theorem 11.49. Two-transitivity together with this special behaviour of the 2-point stabiliser is enough to determine completely both the abstract structure and the action of $\bar{G}$. Theorem A.17 states indeed that, up to isomorphism, $\bar{G}$ is one of the groups in the list, with $\bar{G}$ acting in each of the first four cases in its natural 2-transitive permutation representation. □

A consequence of Lemma 11.123 is the following result.

LEMMA 11.124 *In case* (iv.5), *$G_{\mathcal{P}}$ and $G_{\mathcal{Q}}$ have trivial intersection.*

Proof. Let $\alpha \in G_{\mathcal{P}} \cap G_{\mathcal{Q}}$ be non-trivial. Then $p \nmid \operatorname{ord} \alpha$, and hence $\alpha \in H$. This shows that α fixes not only $\mathcal{P}$ but another place in o, say $\mathcal{R}$. Since $\mathcal{Q} \notin o$, this shows that α has at least three fixed places. These places are in three different orbits of $G_{\mathcal{P}}^{(1)}$. Since F_1 is rational, Theorem 11.14 implies that α fixes every orbit of $G_{\mathcal{P}}^{(1)}$, a contradiction. □

Now the cases in Lemma 11.123 are investigated further. Let M be the kernel of the above permutation representation of G over o. Then $\bar{G} = G/M$. Since M is a normal subgroup of G, the group $\bar{G}$ may also be viewed as an automorphism group of Σ^M. When M is not trivial, then it influences the size of G.

THEOREM 11.125 *Assume that case* (iv.5) *occurs. If the subgroup M fixing all places in o is non-trivial, then the following assertions hold.*

(i) *If Σ^M is rational, then*

(a) *the genus* $$g = \tfrac{1}{2}(q-1)(|M|-1); \tag{11.75}$$

(b) $|G| \geq 8g^3$ *only occurs when* $\Sigma = K(x, y)$ *with* $y^q - y = x^2$, *and* $q = p^k$ *is odd,* $g = \frac{1}{2}(q-1)$, $|M| = 2$, $\bar{G} = G/M \cong \mathrm{PGL}(2, q)$.

(ii) *If* Σ^M *is elliptic, then* $g = \frac{1}{2}(q+1)(|M|-1)+1$, *and* $|G| \le 8g^3$.

(iii) *If* Σ^M *has genus* $\bar{g} \ge 2$, *then* $g = \bar{g}|M| + \frac{1}{2}(q-1)(|M|-1)$, *and*

$$\frac{|G|}{g^3} < \frac{|\bar{G}|}{\bar{g}^3}\frac{1}{|M|^2} < \frac{|\bar{G}|}{\bar{g}^3}.$$

Proof. Since F_1 is rational, M has no fixed place outside o, by Theorem 11.14 (iv). From Theorem 11.72 applied to M,

$$2g-2 \ge |M|(2\bar{g}-2) + (q+1)(|M|-1). \tag{11.76}$$

For $\bar{g} \le 1$, this shows the first statement in both (i) and (ii). If $\bar{g} > 1$, write (11.76) as follows:

$$g = \bar{g}|M| + \tfrac{1}{2}(q-1)(|M|-1).$$

Since $|M| > 1$ and $q > 1$, it follows that $g > \bar{g}|M|$. This together with the relation $|G| = |\bar{G}||M|$ proves (iii).

To prove (i)(b) and (ii)(b), take a non-trivial element μ in M. From Theorem 11.49, μ is in the normaliser of $G_{\mathcal{P}}^{(1)}$. Also, μ fixes two orbits of $G_{\mathcal{P}}^{(1)}$, namely $\{\mathcal{P}\}$ and $o\backslash\{\mathcal{P}\}$. Since F_1 is rational, they are all the orbits of $G_{\mathcal{P}}^{(1)}$ fixed by μ. This implies that μ fixes no place outside o. Hence $\mu \notin G_{\mathcal{Q}}$, showing that M and $G_{\mathcal{Q}}$ have trivial intersection; that is, $|G_{\mathcal{Q}}| = |\bar{G}_{\mathcal{Q}}|$. So (11.74) reads as follows:

$$2g-2 = \frac{|\bar{G}|(|\bar{G}_{\mathcal{P}}||M| - |\bar{G}_{\mathcal{P}}^{(1)}||\bar{G}_{\mathcal{Q}})|}{|\bar{G}_{\mathcal{Q}}|(|\bar{G}| - |\bar{G}_{\mathcal{P}}|)} \tag{11.77}$$

For the proof of (i)(b), let $\bar{g} = 0$. Then no non-trivial K-automorphism in $\bar{G}$ fixes three distinct places of Σ^M in o. From the list in Theorem 11.123, the groups of type (II) and (IV) do not comply with this requirement.

If $\bar{G} \cong \mathrm{PSL}(2,q)$ as in Lemma 11.123 (I), then

$$|\bar{G}| = \tfrac{1}{2}(q^3-q), \quad |\bar{G}_{\mathcal{P}}| = \tfrac{1}{2}(q^2-q), \quad |G_{\mathcal{P}}^{(1)}| = q.$$

From (11.77),

$$2g-2 = \tfrac{1}{2}(q+1)\,\frac{(q-1)|M| - 2|\bar{G}_{\mathcal{Q}}|}{|\bar{G}_{\mathcal{Q}}|}.$$

This together with $2g = (q-1)(|M|-1)$ gives $|\bar{G}_{\mathcal{Q}}| = \frac{1}{2}(q+1)$.

Now let $|G| \ge 8g^3$. Since $|G| = \frac{1}{2}(q^3-q)|M|$, from the above computation,

$$\frac{q(q+1)}{(q-1)^2} > 2\frac{(|M|-1)^3}{|M|}. \tag{11.78}$$

Since $g > 1$, (11.78) implies that $|M| = 2$. Hence $g = \frac{1}{2}(q-1)$. In particular, $p > 2$. Since $|M| = [\Sigma : \Sigma^M] = 2$ by Theorem 11.36, Σ is hyperelliptic. From Proposition 11.101, (ii)(b) follows for $\bar{G} = \mathrm{PSL}(2,q)$.

The above argument also works for $\bar{G} = \mathrm{PGL}(2,q)$. If $\bar{G} \cong \mathrm{Sz}(n)$ as in Lemma 11.123 (III), then

$$|\bar{G}| = (n^2+1)n^2(n-1), \ |\bar{G}_{\mathcal{P}}| = n^2(n-1), \ |\bar{G}_{\mathcal{P}}^{(1)}| = n^2.$$

From (11.77),

$$2g-2=(n^2+1)\frac{(n-1)|M|-|\bar{G}_Q|}{|\bar{G}_Q|}.$$

Comparison with (i)(a) shows that $|G_Q| = n-1+2/(n+1)$ which is impossible for $n>1$. Therefore the case $\bar{G}\cong \mathrm{Sz}(n)$ cannot actually occur.

If T is a minimal normal subgroup of $\bar{G}$ regular on o, then T is an elementary abelian w-group acting on o as a sharply transitive permutation group. Therefore $|o|=|T|=w^k$ for a prime $w\neq p$ and $k\geq 1$. Thus

$$w^k = q+1.$$

Therefore either $q=8,\ w^k=9$ or $k=1,\ q=2^r$ or $q=p$ and k is prime.

Up to an isomorphism, there is only one soluble 2-transitive permutation group of prime degree w, namely, $\mathrm{AGL}(1,w)$. Hence, when $k=1$, Lemma 11.124 implies that $|G_Q|=w$, whence $2g-2=|M|-w$ by (11.77). Comparison with (i)(a) shows that $|M|-w+2$ is equal to $(w-2)(|T|-1)$, whence

$$w=3,\ q=2,\ g=\tfrac{1}{2}(|M|-1),\ |\bar{G}|=6,\ |G|=6|M|.$$

Therefore $|G|\geq 8g^3$ only occurs when $|M|=3$. But then $g=1$ contradicting the hypothesis.

In the other case, $w^k=9$. From Lemma 11.124, either $|G_Q|=9$ or 3. In the former case, (i)(a) together with (11.77) implies that $|\bar{G}_{\mathcal{P}}|=40$. But this is impossible, because $\mathrm{GL}(2,3)$, the normaliser of T in $\bar{G}$, has order 48 and hence contains no element of order 5. If $|G_Q|=3$, then (i)(a) and (11.77) cannot hold simultaneously.

If $q=p$ and k is prime, from Huppert's classification either $|G|=(p+1)p|M|$ or $|G|=(p+1)pk|M|$. In both cases, (ii)(a) implies that $8g^3>|G|$.

Finally, let $\bar{g}=1$. From Theorem 11.94, $|\bar{G}_{\mathcal{P}}|$ divides 24. This is consistent with the 2-transitive action of $\bar{G}$ on o only if $|o|$ is one of $3,4,5,7,9,13,25$. Since $|o|$ divides $|\bar{G}|$ and $\bar{G}$ is a K-automorphism group of Σ^M, Lemmas 11.60 and 11.60 together with (11.59) also imply that the possibilities for $|o|$ are actually four, namely, $3,4,5,9$. The last one has already been ruled out.

If $|o|=3$, then $q=p=2$ and $|\bar{G}|=6$, as the 2-transitivity of $\bar{G}$ on o implies that $\bar{G}\cong S_3$. Hence $|G|=6|M|$ and $g=\frac{3}{2}(|T|-1)+1$. Therefore $|G|<8g^3$.

If $|o|=4$, then $q=p=3$, and $g=2(|T|-1)+1$. From the 2-transitive action of $\bar{G}$ on o, either $|\bar{G}|=12$ or $|\bar{G}|=24$. In both cases, $|G|<8g^3$.

If $|o|=5$, then $q=4$, and $g=\frac{5}{2}(|T|-1)+1$. Since a 2-transitive group on 5 letters with a regular normal subgroup is sharply 2-transitive, $|\bar{G}|=20$. Hence, again, $|G|<8g^3$. □

THEOREM 11.126 *If* $|G|\geq 8g^3$, *then* (iv.5) *only occurs in the following cases*:

(I) $p=2$, *the field* $\Sigma=K(x,y)$ *with* $y^2+y=x^{2^k+1}$ *and genus* $g=2^{k-1}$, *where* G *fixes a place* $\mathcal{P}$, *and* $|G|=2^{2k+1}(2^k+1)$;

(II) $p>2$, *the field* $\Sigma=K(x,y)$ *with* $y^2=x^q-x$ *and genus* $g=\frac{1}{2}(q-1)$, *with* $G/M\cong \mathrm{PSL}(2,q)$ *or* $G/M\cong \mathrm{PGL}(2,q)$, *where* $q=p^r$ *and* $|M|=2$;

(III) *$p \geq 2$, the field Σ is the Hermitian function field $K(x,y)$ with $y^q + y = x^{q+1}$ and genus $g = \frac{1}{2}(q^2 - q)$, with $G \cong \mathrm{PSU}(3,q)$ or $G \cong \mathrm{PGU}(3,q)$;*

(IV) *$p = 2$, $q_0 = 2^r$, $q = 2q_0^2$, the field $\Sigma = K(x,y)$ with $x^{q_0}(x^q + x) = y^q + y$ and genus $g = q_0(q-1)$, with $G \cong \mathrm{Sz}(q)$.*

Proof. If the non-tame orbit consists of a single place, then (I) holds. This was the first result in case (iv.5).

Therefore the non-tame orbit o is assumed non-trivial. The few possibilities for the 2-transitive permutation group $\bar{G}$ induced by G on o are described in Lemma 11.123.

First, the faithful actions are investigated. In doing so, Lemma 11.123 together with classification theorems of subgroups of the groups in that lemma are the main ingredients. Also, by Lemma 11.124, no non-trivial element in G_Q fixes a place in o. In particular, G_Q, which is cyclic by Theorem 11.49 (ii)(b), contains no element of order 2 or p.

Take $G \cong \mathrm{PGL}(2,q)$. Then

$$|G| = q^3 - q, \quad |G_\mathcal{P}| = q^2 - q, \quad |G_\mathcal{P}^{(1)}| = q.$$

From (11.74),

$$2g - 2 = (q+1)\left(\frac{q-1}{|G_Q|} - 1\right).$$

If $|G| \geq 8g^3$ then

$$1 < \frac{q-1}{|G_Q|} < 2. \tag{11.79}$$

On the other hand, (11.79) and Theorem A.8 imply that $|G_Q| = \frac{1}{2}(q+1)$ with q odd. Then $g = \frac{1}{2}(q-1)$. In particular, $q \geq 5$.

If $q \equiv 1 \pmod 4$, then $2g-2 \equiv 2 \pmod 4$. Also, G_Q has odd order and hence no involutory element in $\mathrm{PGL}(2,q)\backslash\mathrm{PSL}(2,q)$ has a fixed place. From Theorem 11.72, $2g-2 \equiv 0 \pmod 4$, a contradiction.

If $q \equiv 3 \pmod 4$, then $2g-2 \equiv 0 \pmod 4$. Since every involutory element in $\mathrm{PGL}(2,q)\backslash\mathrm{PSL}(2,q)$ has exactly two fixed places, so $2g-2 \equiv 2 \pmod 4$, a contradiction. Therefore $G \cong \mathrm{PGL}(2,q)$ does not occur.

For $G \cong \mathrm{PSL}(2,q)$, the above computation shows that

$$2g - 2 = \tfrac{1}{2}(q+1)\left(\frac{q-1}{|G_Q|} - 2\right).$$

If $|G| \geq 8g^3$ holds then $|G_Q| = \frac{1}{3}(q+1)$ and $g = \frac{1}{4}(q-1)$; in particular, $q \geq 5$. From Theorem A.8, $|G_Q|$ must divide $\frac{1}{2}(q+1)$, a contradiction. Therefore $G \cong \mathrm{PSL}(2,q)$ cannot happen.

Take $G \cong \mathrm{PSU}(3,q)$ with $q \geq 3$, and let $d = \gcd(3, q+1)$. Then

$$|G| = \frac{(q^3+1)q^3(q^2-1)}{d}, \quad |G_\mathcal{P}| = \frac{q^3(q^2-1)}{d}, \quad |G_\mathcal{P}^{(1)}| = q^3.$$

From (11.74),

$$2g = \frac{(q^3+1)(q^2-1)}{d\,|G_Q|} - (q^3 - 1).$$

Assume that $|G| \geq 8g^3$. Since $|G| < q^8$, this implies that

$$q^{8/3} > \frac{(q^3+1)(q^2-1)}{d\,|G_Q|} - (q^3-1).$$

whence $|G_Q| > q+1$. Further, the order of G_Q divides $(q^3+1)(q+1)(q^2-1)$. Therefore G_Q must be contained in one of the subgroups (ii), (iii), (iv) in Theorem A.10. Since $|G_Q| > q+1$, (ii) and (iii) cannot occur; hence

$$|G_Q| = \frac{q^2-q+1}{td}$$

for an odd integer t, and

$$2g = (q+1)(q^2-1)t - (q^3-1) = (q-1)(t(q+1)^2 - (q^2+q+1)).$$

As $|G| \geq 8g^3$, this implies that $t=1$ and $2g = q^2 - q$. From Theorem 12.25 case (III) follows.

The same argument works for $G \cong \mathrm{PGU}(3,q)$.

In the case that $G \cong \mathrm{Sz}(q)$, $q_0 = 2^r$ and $q = 2q_0^2$,

$$|G| = (q^2+1)q^2(q-1), \quad |G_{\mathcal{P}}| = q^3 - q, \quad |G_{\mathcal{P}}^{(1)}| = q^2.$$

From (11.74),

$$2g = \frac{(q+2q_0+1)(q-2q_0+1)(q-1)}{|G_Q|} - q^2 + 1.$$

By Lemma 11.124 and Theorem A.12, $|G_Q|$ must be prime to $q-1$. Hence $|G_Q|$ divides either $q+2q_0+1$ or $q-2q_0+1$.

If $|G_Q| = (q+2q_0+1)/t$, then

$$2g = (t-1)(q^2-1) - 2tq_0(q-1),$$

and hence $t>1$. On the other hand, $|G| < 8g^3$ when $t > 1$.

If $|G_Q| = (q-2q_0+1)/t$, then

$$2g = (t-1)(q^2-1) + 2tq_0(q-1).$$

If, in addition, $|G| \geq 8g^3$, then $t=1$ and hence $g = q_0(q-1)$. From Theorem 12.14, case (IV) follows.

Now assume that $G \cong \mathrm{Ree}(q)$, $q = 3q_0^2$ and $q_0 = 3^h$. Then

$$|G| = (q^3+1)q^3(q-1), \quad |G_{\mathcal{P}}| = q^3(q-1), \quad |G_{\mathcal{P}}^{(1)}| = q^3.$$

From (11.74),

$$2g = (q-1)\left(\frac{(q+3q_0+1)(q-3q_0+1)(q+1)}{|G_Q|} - (q^2+q+1)\right).$$

From Theorem A.14, no prime divisor of $|G_Q|$ divides $q-1$.

Suppose that $|G_Q| = (q+1)/t$. Then

$$2g = (q-1)((t-1)(q^2+1) - q);$$

hence $t \geq 2$ and $2g \geq (q-1)(q^2-q+1)$. Thus $|G| < 8g^3$.

Suppose that $|G_Q| = (q + 3q_0 + 1)/t$. Then

$$2g = (q-1)((t-1)q^2 - t(3q_0q - 2q + 3q_0 - 1) - q - 1);$$

hence $t \geq 2$ and

$$2g \geq (q-1)(q^2 - 3(2qq_0 - q + 2q_0) + 1).$$

Thus $|G| < 8g^3$.

Suppose that $|G_Q| = (q - 3q_0 + 1)/t$. Then

$$2g = (q-1)((t-1)q^2 + t(3q_0q + 2q + 3q_0 + 1) - q - 1);$$

hence $t \geq 2$ and

$$2g \geq (q-1)(q^2 + 3(2q_0q + q + 2q_0) + 1).$$

Again, $|G| < 8g^3$.

Now assume that a minimal normal subgroup T of G is regular on o. Then the size $|o| = q + 1$ is a power of a prime. It is shown in the proof of Theorem 11.125 that this is only possible when $q = 8$ or $q+1 = w$ is prime or $q = p$ and k is prime. By (11.74), G is not sharply 2-transitive on Ω. This leaves only one case, namely, $|G| = wpk$ with $q = p$ and k prime. For $p = 3$, $|G| = 24$, which is less than $8g^3$. So it may be assumed that $p \geq 7$. From (11.38) applied to $G_P^{(1)}$,

$$2g - 2 = -2p + \lambda(p-1).$$

As $g \geq 2$, this implies that $\lambda \geq 3$. Hence $g \geq \frac{1}{2}(p-1)$. But then $8g^3 > |G|$ as $(p-1)^3 > p(p+1)k$ for $p \geq 7$.

Finally, the possibility of a 2-transitive permutation representation $\bar{G}$ of G on o with non-trivial kernel M is considered. Suppose that $|G| \geq 8g^3$. Theorem 11.125 states that, if Σ^M is rational, then (I) occurs while no example exists when Σ^M is elliptic. Also, if the genus $\bar{g}$ of Σ^M is at least 2, then $|\bar{G}| > 8\bar{g}^3$. Hence case (iv.5) occurs, where $\bar{G}$ is viewed as a K-automorphism group of Σ^M. As $\bar{G}$ acts faithfully on its 2-transitive non-tame orbit, from the first part of this proof there are three possibilities for $\bar{G}$. But each can be ruled out as follows.

If $\bar{G} \cong \mathrm{PSU}(3, q)$, then

$$2\bar{g} = q^2 - q, \quad 2g = (q^2 - q)|M| + (q^3 - 1)(|M| - 1)$$

by Theorem 11.125 (iii). Since $|G| = |\bar{G}||M|$, it follows that

$$|G| = (q^3+1)q^3(q^2-1)|M| < (q^3+1)q^3(q^2-1)(\tfrac{1}{2}|M|)^3(q-1)^3(q^2+3q+1)^3,$$

which is smaller than $8g^3$. The same argument works for $\bar{G} \cong \mathrm{PGU}(3, q)$.

If $\bar{G} = \mathrm{Sz}(q)$, then

$$2\bar{g} = 2q_0(q-1), \quad 2g = 2q_0(q-1)|M| + (q^2+1)(|M| - 1).$$

Thus

$$\begin{aligned}|G| = |\bar{G}||M| &= (q^2+1)q^2(q-1)|M| \\ &< (\tfrac{1}{2}|M|)^3(q^2 + 4q_0q - 4q_0 + 1)^3 < 8g^3.\end{aligned}$$

□

In geometric terms, the following result is obtained.

THEOREM 11.127 *Let $\mathcal{F}$ be an irreducible curve of genus $g \geq 2$. If $|G| \geq 8g^3$, then $\mathcal{F}$ is birationally equivalent to one of the following:*

(I) *the hyperelliptic curve* $\mathbf{v}(Y^2+Y+X^{2^k+1})$ *with* $p=2,\ g=2^{k-1}$; *also,* G *fixes a place* $\mathcal{P}$ *and* $|G|=2^{2k+1}(2^k+1)$;

(II) *the hyperelliptic curve* $\mathbf{v}(Y^2-(X^q-X))$ *with* $p>2,\ g=\frac{1}{2}(q-1)$; *also,* $G/M \cong \mathrm{PSL}(2,q)$ *or* $G/M \cong \mathrm{PGL}(2,q)$, *where* $q=p^r$ *and* $|M|=2$;

(III) *the Hermitian curve* $\mathbf{v}(Y^q+Y-X^{q+1})$ *with* $p \geq 2,\ g=\frac{1}{2}(q^2-q)$; *also,* $G \cong \mathrm{PSU}(3,q)$ *or* $G \cong \mathrm{PGU}(3,q)$;

(IV) *the DLS curve* $\mathbf{v}(X^{q_0}(X^q+X)-(Y^q+Y))$ *with* $p=2,\ q_0=2^r,\ q=2q_0^2$, $g=q_0(q-1)$; *also,* $G \cong \mathrm{Sz}(q)$.

11.13 K-AUTOMORPHISM GROUPS FIXING A PLACE

In this section, K-automorphism groups of Σ whose order is divisible by p and in which

$$\textit{every element of order } p \textit{ has exactly one fixed place} \tag{11.80}$$

are investigated.

REMARK 11.128 A suitable power β of any non-trivial element α of a p-group has order p. Also, every fixed place of α is fixed by β as well. Since every p-subgroup has non-trivial centre, condition (11.80) is satisfied by a subgroup G of $\mathrm{Aut}_K(\Sigma)$ if and only if every p-element in G has exactly one fixed place. Then the non-trivial elements of a Sylow p-subgroup S_p of a K-automorphism G of $\mathrm{Aut}_K(\Sigma)$ satisfying (11.80) have the same fixed place.

Condition (11.80) occurs in several circumstances, for instance when G has property (iv.5) in the proof of Theorem 11.116, or when G is the K-automorphism group of certain curves $\mathbf{v}(A(X)-B(Y)=0)$ investigated in Section 12.1. Another such circumstance is described in the following two lemmas.

LEMMA 11.129 *If Σ has p-rank 0, then (11.80) holds in* $\mathrm{Aut}_K(\Sigma)$.

Proof. Let $\alpha \in \mathrm{Aut}_K(\Sigma)$ have order p. Applying the Deuring–Shafarevich formula (11.28) to the group generated by α gives $-1=p(\gamma'-1)+m(p-1)$, where γ' is the p-rank of Σ^α and m is the number of fixed places of α. This is only possible when $\gamma'=0$ and $m=1$. □

REMARK 11.130 The converse of Lemma 11.129 is not true, a counter-example being the curve $\mathcal{F}$ of 2-rank equal to 4 given in Exercise 10 in Chapter 9. To show this, let

$$\Sigma=K(x,y), \quad y^6+y^5+y^4+y^3+y^2+y+1=x^3(y^2+y),$$

be the function field of $\mathcal{F}$. Then the birational transformation $x'=x,\ y'=y+1$ is a K-automorphism of Σ which has only one fixed place, namely that arising from the unique branch of $\mathcal{F}$ tangent to the infinite line.

LEMMA 11.131 *Assume that* $\mathrm{Aut}_K(\Sigma)$ *contains a p-subgroup G of order* p^r. *If* Σ^G *has p-rank zero, and (11.80) holds, then* Σ *has p-rank zero.*

Proof. Let $\mathcal{P}$ be the fixed place of G. Since $\gamma' = 0$, (11.28) applied to G gives $\gamma - 1 = -|G| + |G| - 1$, whence $\gamma = 0$. □

REMARK 11.132 Exercises 9 and 10 in Chapter 9 show that the hypothesis on the p-rank of Σ^G is essential.

Suppose that a subgroup G of $\mathrm{Aut}_K(\Sigma)$ has a Sylow p-subgroup with property (11.80). Then (11.80) is satisfied by all Sylow p-subgroups of G. Suppose further that G has no normal Sylow p-subgroup. Let S_p and S'_p two distinct Sylow p-subgroups of G. By Remark 11.128, both S_p and S'_p each have exactly one fixed place, say $\mathcal{P}$ and $\mathcal{P}'$. Also, S_p is the unique Sylow p-subgroup of the stabiliser of $\mathcal{P}$ under G, and $G_\mathcal{P} = G_\mathcal{P}^{(1)} \rtimes H$; see Theorem 11.49. Further, $\mathcal{P} \neq \mathcal{P}'$; for, if $\mathcal{P} = \mathcal{P}'$, then S_p and S'_p are two distinct Sylow p-subgroups in $G_\mathcal{P}$. However, this is impossible as $G_\mathcal{P}^{(1)}$ is a normal subgroup of $G_\mathcal{P}$. Therefore any two distinct Sylow p-subgroups in G have trivial intersection.

A subgroup H of a group G is a *trivial intersection set* if, for every $g \in G$, either $H = g^{-1}Hg$ or $H \cap g^{-1}Hg = \{1\}$. A Sylow p-subgroup T of G is a trivial intersection set in G if and only if that T meets every other Sylow p-subgroup of G trivially. Hence Lemma 11.129 has the following corollary.

THEOREM 11.133 *If* Σ *has p-rank* 0, *then every Sylow p-subgroup* S_p *in* $\mathrm{Aut}_K(\Sigma)$ *is a trivial intersection set.*

A classical result on trivial intersection sets is the following.

THEOREM 11.134 (Burnside) *If a Sylow p-subgroup* S_p *of a finite soluble group is a trivial intersection set, then either* S_p *is normal or cyclic, or* $p = 2$ *and* S_2 *is a generalised quaternion group.*

Finite groups whose Sylow 2-subgroups are trivial intersection sets have been classified. This has been refined to groups containing a subgroup of even order which intersects each of its distinct conjugates trivially.

THEOREM 11.135 (Hering) *Let Q be a subgroup of a finite group G with trivial normaliser intersection; that is, the normaliser* $N_G(Q)$ *of Q in G has the following two properties:*

(i) $Q \cap x^{-1}Qx = \{1\}$ *for all* $x \in G \backslash N_G(Q)$;

(ii) $N_G(Q) \neq G$.

If $|Q|$ *is even and S is the normal closure of Q in G, then* $S = O(S) \rtimes Q$, *the semidirect product of* $O(S)$ *by Q, with Q a Frobenius complement, unless S isomorphic to one of the groups*

$$\mathrm{PSL}(2,n),\ \mathrm{Sz}(n),\ \mathrm{PSU}(3,n),\ \mathrm{SU}(3,n), \tag{11.81}$$

where n is a power of 2. *Let* $\bar{G}$ *be the permutation group induced by G on the set* Ω *of all conjugates of Q under G. Then*

(iii) $C_G(S)$ *is the kernel of this representation;*

(iv) S *is transitive on* Ω;

(v) $|\Omega|$ *is odd;*

(vi) *in the exceptional cases,*

(a) $|\Delta| = n+1, n^2+1, n^3+1, n^3+1$;

(b) $\bar{G} = G/C_G(S)$ *contains a normal subgroup isomorphic to one of the first three groups in (11.81);*

(c) $\bar{G}$ *is isomorphic to an automorphism group of* $\bar{S} = SC_G(S)/C_G(S)$ *containing* $\bar{S}$.

Hering's result does not extend to subgroups of odd order with trivial normal intersection. A major result which can be viewed as a generalisation of Theorem 11.134 is the following.

THEOREM 11.136 *Let* S_d, *with* $d > 11$, *be a Sylow d-subgroup of a finite group G that is a trivial intersection set but not a normal subgroup. Then* S_d *is cyclic if and only if G has no composition factors isomorphic to either* $\mathrm{PSL}(2, d^n)$ *with* $n > 1$ *or* $\mathrm{PSU}(3, d^m)$ *with* $m \geq 1$.

Now some consequences of the above results are stated. Two cases are distinguished according as $p = 2$ or $p > 2$.

When $p = 2$, if (11.80) is satisfied by a subgroup G of $\mathrm{Aut}_K(\Sigma)$, then Theorem 11.135 applies to G with Q a 2-subgroup of a Sylow 2-subgroup S_2 of G. In fact, (i) is true, while (ii) holds provided that G is larger than $G_{\mathcal{P}}$, where $\mathcal{P}$ is the fixed place of S_2.

When $G_{\mathcal{P}} = G$, the structure of G is fully described in Theorem 11.49. Otherwise, G has no fixed place, and Hering's Theorem determines the abstract structure of the normal subgroup S of G generated by all elements whose order is a power of 2. Note that, since S_2 has only one fixed point, $N_G(S_2) = G_{\mathcal{P}}$ holds. If $S = O(S) \rtimes S_2$, then S_2 is a Frobenius complement, and hence it has a unique involutory element. Note that it is not claimed that S is a Frobenius group. In the exceptional cases, if $C_G(S)$ is the centraliser of S in G, then $G/C_G(S)$ is the K-automorphism group of a group listed in Hering's Theorem.

Therefore the following result is obtained.

THEOREM 11.137 *Let* $p = 2$ *and assume that (11.80) holds. If S is the subgroup of* $\mathrm{Aut}_K(\Sigma)$ *of even order generated by all elements whose order is a power of* 2, *then one of the following alternatives holds.*

(i) *S fixes no place and either* (a) *or* (b) *is satisfied:*

(a) *S is isomorphic to one of the groups*

$$\mathrm{PSL}(2,n),\ \mathrm{PSU}(3,n),\ \mathrm{SU}(3,n),\ \mathrm{Sz}(n),\ \textit{with } n = 2^r; \qquad (11.82)$$

(b) $S = O(S) \rtimes S_2$ *with* S_2 *a 2-Sylow subgroup of* G; *here,* S_2 *is either a cyclic group or a generalised quaternion group.*

(ii) $\mathrm{Aut}(\mathcal{X})$ *fixes a place, and* $\mathrm{Aut}_K(\Sigma)(\mathcal{X}) = S \rtimes H$, *with* H *a subgroup of odd order.*

Using Theorem 11.135, the action of S on the set Ω of all places of Σ fixed by some involution or, equivalently, on the set of all involutions in G can be investigated.

Assume that $\mathrm{Aut}(\mathcal{X})$ fixes no place, and let $\bar{S}$ be the permutation group induced by S on Ω. If M is the kernel of the permutation representation of S, then M is the centre of $\mathrm{Aut}(\mathcal{X})$, the group $\bar{S} = S/M$, and $\bar{S}_2 = S_2M/M$ is a 2-Sylow subgroup of $\bar{S}$. Here, $\bar{S}$ is transitive on Ω; also, $\bar{S}_2$ fixes $\mathcal{P}$ and acts on $\Omega\backslash\{P\}$ as a semiregular permutation group. Further, $M = C_G(S)$. If S is isomorphic to one of the groups in (11.82), then $\bar{S}$ is 2-transitive on Ω, and $|\Omega| = n+1,\ n^2+1, n^3+1$ according as $\bar{S} \cong \mathrm{PSL}(2,n),\ \mathrm{Sz}(n),\ \mathrm{PSU}(3,n)$.

Theorem 11.134 is the main ingredient in the proof of the following result.

THEOREM 11.138 *Let* G *be a soluble* K*-automorphism group of* Σ *satisfying condition (11.80). Assume that* G *fixes no place, and that*

$$|G| \text{ is divisible by } \begin{cases} p^2 & \text{when } p > 2, \\ 16 & \text{when } p = 2. \end{cases} \tag{11.83}$$

If Σ *has genus* $g \geq 2$, *then* $|G| \leq 24g(g-1)$.

Proof. For $g = 2$, Theorem 11.138 follows from Theorem 11.99.

Let N be a minimal normal subgroup of G. Since G is soluble, N is an elementary abelian group of order d^r with a prime d. If $d = p$, then (11.80) implies that N fixes a unique place. But then G itself must fix $\mathcal{P}$, contradicting one of the hypotheses. Therefore $d \neq p$. In particular, the Sylow p-subgroups of G and G/N are isomorphic, and hence Theorem 11.134 applies to G/N, while (11.83) remains valid for G/N.

This suggests that $\bar{G} = G/N$, viewed as a K-automorphism group of Σ^N, should be investigated. The aim is to show that $\bar{G}$ also satisfies (11.80).

The place $\mathcal{P}'$ lying below $\mathcal{P}$ in the extension Σ/Σ^N is fixed by a Sylow p-subgroup of $\bar{G}$. Since G/N can be viewed as a K-automorphism group of Σ^N, this rules out the possibility that Σ^N is elliptic, because (11.83) is inconsistent with Theorem 11.94. Similarly, Σ^N cannot be rational. In fact, if Σ^N were rational, then no p-subgroup of G/N and hence of G would be either cyclic or a quaternion group, by Theorem 11.14.

Therefore Σ^N has genus $\bar{g} \geq 2$. Assume that Theorem 11.138 does not hold, and choose a minimal counter-example with genus g as small as possible. Then $|G| > 24g(g-1)$, but Theorem 11.138 holds for all genera g' with $2 \leq g' < g$. Theorem 11.72 applied to N gives $2g-2 \geq |N|(2\bar{g}-2)$, whence

$$|G| > 12g(2g-2) > |N|\,12g(2\bar{g}-2).$$

Therefore $|G/N| > 24\bar{g}(\bar{g}-1)$. By the divisibility condition (11.83) on $|G|$, the order of $\bar{G} = G/N$ is not a prime. Hence $\bar{G}$ is a soluble group. Since $d \neq p$, so

$\bar{G}$ satisfies (11.83). On the other hand, $\bar{G}$ can be regarded as a K-automorphism group of Σ^N.

To show that $\bar{G}$ also satisfies condition (11.80), let $\bar{\alpha}$ be any element of $\bar{G}$ of order p. Choose an element α in G whose image is $\bar{\alpha}$ under the natural homomorphism $G \to G/N$. By (11.80), Σ has a unique place $\mathcal{P}$ fixed by α. Then $\bar{\alpha}$ fixes the place $\bar{\mathcal{P}}$ lying under $\mathcal{P}$ in the extension Σ/Σ^N. Assume on the contrary that $\bar{\alpha}$ fixes another place of Σ^N, say $\bar{\mathcal{Q}}$. Then the set of places of Σ lying over $\bar{\mathcal{Q}}$ is preserved by α. The number of such places is prime to p since their number divides $|N|$. But then α must fix one of these places, contradicting (11.80).

Since G is a minimal counter-example, $\bar{G}$ fixes a place of Σ^N. Equivalently, an orbit of N is preserved by G. Such an orbit consists of places of Σ fixed by p-elements of G.Since S_p has only one fixed place, so $|S_p| \leq |N| - 1$. This together with (11.83) implies that $|N| \geq 10$. As $g - 1 \geq |N|(\bar{g} - 1)$, it follows that $g \geq 11$.

From Theorem 11.49 (ii)(b), $|\bar{G}| = |\bar{G}_{\mathcal{P}}| = |\bar{G}^{(1)}_{\mathcal{P}}|\,|\bar{H}|$, with $\bar{H}$ a cyclic group of order prime to p. By Theorem 11.60, $|\bar{H}| \leq 4\bar{g} + 2$. Let $\bar{S}_p$ be the Sylow p-subgroup of $\bar{G}$; then $\bar{S}_p = \bar{G}^{(1)}_{\bar{\mathcal{P}}}$. Since $p \nmid |N|$, so $\bar{S}_p$ is isomorphic to S_p.

Assume that S_p is cyclic. From Theorem 11.79, $|\bar{G}^{(1)}_{\bar{\mathcal{P}}}| \leq 4\bar{g} + 4$. Therefore

$$|\bar{G}| \leq (4\bar{g} + 4)(4\bar{g} + 2),$$

whence $|G| \leq (4\bar{g}+4)(4\bar{g}+2)|N|$. Since $g - 1 \geq |N|(\bar{g} - 1)$, $\bar{g} \geq 2$ and $|N| \geq 10$, it follows that

$$\begin{aligned}|G| &\leq 16(\bar{g} + 1)(\bar{g} + \tfrac{1}{2})|N| \leq \frac{16}{10} \cdot \frac{\bar{g} + 1}{\bar{g} - 1}(\bar{g} - 1)|N| \cdot \frac{\bar{g} + \frac{1}{2}}{\bar{g} - 1}(\bar{g} - 1)|N| \\ &\leq \frac{8}{5} \cdot \frac{2\bar{g}^2 + 3\bar{g} + 1}{2(\bar{g} - 1)^2}(g - 1)^2 \leq 12(g - 1)^2 < 24g(g - 1).\end{aligned}$$

But then G is not a counter-example.

If S_p is not cyclic, Theorem 11.134 implies that $p = 2$ and that S_2 is a quaternion group. Since a quaternion group has a cyclic subgroup of index 2, Theorem 11.79 shows that $|\bar{G}^{(1)}_{\mathcal{P}}| \leq 2(4\bar{g} + 4)$. Now the argument above can be used to show that $\frac{1}{2}|G| \leq 12(g - 1)^2$, whence $|G| \leq 24(g - 1)^2 < 24g(g - 1)$. Again, G is not a counter-example. □

Finally, an alternative proof for $p = 2$ which does not require the hypothesis (11.83) is given.

THEOREM 11.139 *Let $p = 2$, $g \geq 2$, and assume that (11.80) holds. If G is a soluble K-automorphism group of Σ fixing no place of Σ, then $|G| \leq 24g^2$.*

Proof. Since $84(g - 1) < 24g^2$, the group G may be supposed to be one of the exceptions in Theorem 11.56. Since G has no fixed place, from Theorem 11.137, $S = O(S) \rtimes S_2$, and $O(S)$ acts transitively on the set o consisting of all places fixed by involutory elements in G. In particular, o is the unique non-tame orbit of G. Let $\mathcal{P} \in o$. By assumption (11.80), $G^{(1)}_{\mathcal{P}}$ is a Sylow 2-subgroup S_2 of G. From $|O(S)| = |o||O(S)_{\mathcal{P}}|$,

$$|G| = \frac{|G_{\mathcal{P}}||O(S)|}{|O(S)_{\mathcal{P}}|}. \tag{11.84}$$

In particular, o has odd size. As $O(S)$ is transitive on o and $N_G(S_2) = G_{\mathcal{P}}$, each element in G is the product of an element fixing $\mathcal{P}$ and an element from $O(S)$, that is, $G = G_{\mathcal{P}}O(S)$. Also, as $O(S)$ is a characteristic subgroup of S and $S \triangleleft G$, so $O(S) \triangleleft G$. Hence $\bar{G} = G/O(S)$ is a factor group which can be viewed as a K-automorphism group of $\Sigma^{O(S)}$.

Assume that $\Sigma^{O(S)}$ is rational. By Theorem 11.14, every 2-subgroup of $\bar{G}$ is elementary abelian. On the other hand, as $S_2 \cong S_2O(S)/O(S)$, from Theorem 11.135 the factor group $S_2O(S)/O(S)$ has only one involutory element. Thus $|S_2O(S)/O(S)| = 2$, whence $|S_2| = 2$. This implies that $O(S) = O(G)$ and $G = O(G) \rtimes S_2$. Hence

$$|G| = 2|O(G)| \leq 168(g-1) \leq 24g^2$$

for $g \geq 6$.

To show that this holds true for $2 \leq g \leq 5$, note that, if $O(G)$ has more than three short orbits, then $|G| \leq 12(g-1) < 24g^2$, as observed in the proof of Theorem 11.56 (I). Since S_2 has exactly one fixed place, it preserves only one orbit of $O(G)$. Hence the number of short orbits of $O(G)$ is odd, and thus it is one or three. Since $p = 2$, in the latter case, $|G| \leq 84(g-1)$, as in the proof of Theorem 11.56 (III). Hence $|G| < 24g^2$.

If o is the unique short orbit of $O(G)$, then Lemma 11.111 implies that $|o| = 3$, $g = 4$. Hence $O(G)$ acts on o as a group of order 3. Let H be the subgroup of $O(G)$ fixing every place in o. Then $|O(G)| = 3|H|$ and, from Theorem 11.57, $|H|(2\tilde{g}+1) = 9$, where $\tilde{g}$ is the genus of Σ^H, a contradiction.

Assume that $\Sigma^{O(S)}$ is elliptic. Then $O(S)$ has at least one short orbit o_1. Since $|O(S)| - |o_1| > \frac{1}{2}|O(S)|$, Corollary 11.52 implies that $2g - 2 \geq |o_1| > \frac{1}{2}|O(S)|$, whence $|O(S)| < 4(g-1)$. From Theorem 11.94, $|G| < 96(g-1) \leq 24g^2$.

It remains to consider the case that the genus of $\Sigma^{O(S)}$ is greater than 1. From Corollary 11.52, $|O(S)| \leq g-1$. Hence, by (11.84),

$$|G| \leq (g-1)|G_{\mathcal{P}}|.$$

Assume that $F_1 = \Sigma^{S_2}$ has genus $g' > 1$. Then $|S_2|(g'-1) < g-1$ by the same Corollary 11.52. Since $G_{\mathcal{P}}/S_2$ is a tame automorphism group of F_1 fixing the place of F_1 lying under $\mathcal{P}$, from Theorem 11.60, $|G_{\mathcal{P}}/S_2| \leq 4g'+2$. Hence

$$|G_{\mathcal{P}}| < \frac{4g'+2}{g'-1}(g'-1)|S_2| \leq 10(g-1).$$

Since $|o| \leq |O(S)| \leq g-1$, this implies that $|G| < 10(g-1)^2 < 24g^2$.

If F_1 is elliptic, then $|G_{\mathcal{P}}/S_{\mathcal{P}}| \leq 24$ by Theorem 11.94 while $|S_{\mathcal{P}}| \leq g$ by Theorem 11.78. Therefore $|G| \leq 24g(g-1) < 24g^2$.

So take F_1 to be rational. Let $\mathcal{R} \in o$ be a place distinct from $\mathcal{P}$. If the stabiliser $G_{\mathcal{P},\mathcal{R}}$ of $\mathcal{R}$ in $G_{\mathcal{P}}$ is trivial, then $|G_{\mathcal{P}}| < |o|$, and hence

$$|G| = |G_{\mathcal{P}}||o| < |o|^2 \leq |O(S)|^2 \leq (g-1)^2.$$

Therefore let $|G_{\mathcal{P},\mathcal{R}}| > 1$ for every $\mathcal{R} \in o$. Then, from the proof of Theorem 11.123, G acts on o as a 2-transitive permutation group $\bar{G}$, and o comprises $\mathcal{P}$ together with all places in the unique non-trivial orbit of S_2. The latter assertion implies that $|o| = |S_2| + 1 = q + 1$, $q = 2^r$. On the other hand, as $\bar{G}$ is soluble, $|o| = w^k$ with an odd prime w. From $q + 1 = w^k$, either $k = 1$, or $k = 2$, $q = 8$, $w = 3$. Let M be the kernel of the permutation representation of G on o.

It may be that M is trivial, that is $G = \bar{G}$, and this possibility is investigated first. Assume that $k = 1$. Then $G = \mathrm{AGL}(1, w)$ and G is sharply 2-transitive on o. Let N be the normal subgroup of G of order w. If Σ^N has genus $g' \geq 2$, then

$$2g - 2 \geq w(2g' - 2) \geq 2w,$$

and hence

$$|G| = |o|(|o| - 1) = w(w - 1) \leq (g - 1)(g - 2) < 24g^2.$$

If Σ^N is elliptic, then $|S_2|$ is either 2, or 4, or 8 by Theorem 11.94. The last case cannot occur because 9 is not a prime number. If $|S_2| = 2$, then $w = 3$, and hence $|G| = 6$, while $|S_2| = 4$ occurs when $w = 5$ and hence $|G| = 20$.

If Σ^N is rational, then S_2, viewed as a K-automorphism group of Σ^N, must be an elementary abelian group by Theorem 11.14. On the other hand, S_2 is the 1-point stabiliser of $\mathrm{AGL}(1, w)$ which is cyclic. It follows that $|S_2| = 2$. Therefore $|o| = 3$, whence $|G| = 6$.

If $k = 2, w = 3, q = 8$, then G has a normal subgroup N of order 9, and S_2 is the quaternion group. Since G is a subgroup of the normaliser of N, it follows that $G_{\mathcal{P}}$ is a subgroup of $\mathrm{GL}(2, 3)$ whose order is divisible by eight but not by sixteen. Since $|\mathrm{GL}(2, 3)| = 48$, either $|G_{\mathcal{P}}| = 8$, or $|G_{\mathcal{P}}| = 24$. In the former case, $|G| = 72$ and hence $|G| < 24g^2$. In the latter case, $|G| = 216$, which is greater than $24g^2$ only for $g = 2$. But this does not actually occur, as the assertion holds for $g = 2$.

Suppose M is non-trivial. By Theorem 11.49, M is cyclic of odd order, and $\bar{G} = G/M$ can be viewed as a K-automorphism group of Σ^M. The Sylow 2-subgroup $\bar{S}_2 = S_2M/M$ of $\bar{G}$ is isomorphic to S_2. Now it is shown that $\bar{G}$ has property (11.80). If this did not hold, then some non-trivial K-automorphism $\bar{\alpha}$ of Σ^M in $\bar{S}_2$ would fix not only the place $\mathcal{P}'$ under $\mathcal{P}$ but at least one more place $\mathcal{Q}'$. Equivalently, some non-trivial element $\alpha \in S_2$ would fix $\mathcal{P}$ and the orbit θ of M consisting of all places $\mathcal{Q}$ of Σ lying over $\mathcal{Q}'$. Since θ has odd size, then α would fix at least one place in θ, contradicting (11.80).

From above, if the genus $\bar{g}$ of Σ^M is at least 2, then $|\bar{G}| \leq 24\bar{g}^2$; in consequence, $|G| \leq 24|M|\bar{g}^2$. Since $|o| \geq 3$, from Theorem 11.57,

$$2g - 2 \geq |M|(2\bar{g} - 2) + 3(|M| - 1) \geq 2|M|\bar{g} - 2.$$

This shows that $|G| \leq 24g\bar{g} < 24g^2$.

If Σ^M is elliptic, then S_2 as a subgroup of $\mathrm{Aut}_K(\Sigma^M)$ fixes a place and hence has order at most 8, by Theorem 11.94, whence

$$|G| = |S_2||\Omega||H| = |S_2|(|S_2| + 1)|H| \leq 72(4g + 2).$$

For $g \geq 4$, this gives the desired result. If $2 \leq g \leq 3$, then

$$2g - 2 \geq |\Omega|(|M| - 1) \geq |\Omega| = |S_2| + 1,$$

a contradiction.

Finally, if Σ^M is rational, then $|S_2| \leq 2$, and hence

$$|G| = |S_2||\Omega||H| = |S_2|(|S_2| + 1)|H| \leq 6(4g + 2).$$

Therefore $|G| < 6(4g + 2) < 24g^2$ by Theorem 11.60. □

11.14 LARGE p-SUBGROUPS FIXING A PLACE

THEOREM 11.140 *Let $\mathcal{F}$ be a curve of genus $g \geq 2$ with function field Σ, and let G be a K-automorphism group of Σ such that Σ has a place $\mathcal{P}$ satisfying the condition,*

$$|G_{\mathcal{P}}^{(1)}| > 2g + 1. \tag{11.85}$$

Then one of the following occurs:

(i) $G = G_{\mathcal{P}}$;

(ii) *one of the cases* (III), (IV) *in Theorem 11.127 holds*;

(iii) (V) $p = 3,\ g = n_0(n-1)$, *with* $n = 3n_0^2,\ n_0 = 3^r,\ r \geq 0$, *and $\mathcal{F}$ is the DLR curve with*

$$\mathcal{F} = \mathbf{v}(Y^{n^2} - (1+(X^n-X)^{n-1})Y^n + (X^n-X)^{n-1}Y - X^n(X^n-X)^{n+3n_0});$$

here, $G \cong \mathrm{Ree}(n)$.

Proof. By Theorem 11.78, it may be assumed that the fixed subfield F_1 of $G_{\mathcal{P}}^{(1)}$ is rational.

Let Ω be the set of all places $\mathcal{R}$ of Σ with non-trivial first ramification group $G_{\mathcal{R}}^{(1)}$. Since $\mathcal{P} \in \Omega$ and $G \neq G_{\mathcal{P}}$, so Ω contains at least two places. From (XXVIII) of Section A.5, Ω is a full G-orbit, and hence Ω is the only non-tame G-orbit.

If there are at least two more short G-orbits, say Ω_1 and Ω_2, then, from Theorem 11.72,

$$\begin{aligned} 2g-2 \geq & -2|G| + (|G_{\mathcal{P}}| + |G_{\mathcal{P}}^{(1)}| + |G_{\mathcal{P}}^{(2)}| - 3)\,|\Omega| \\ & + (|G_{\mathcal{Q}_1}| - 1)\,|\Omega_1| + (|G_{\mathcal{Q}_2}| - 1)\,|\Omega_2|, \end{aligned}$$

where $\mathcal{Q}_i \in \Omega_i$ for $i = 1, 2$. Note that

$$(|G_{\mathcal{Q}_1}| - 1)|\Omega_1| + (|G_{\mathcal{Q}_2}| - 1)|\Omega_2| \geq |G|$$

since $|G_{\mathcal{Q}_i}| - 1 \geq \frac{1}{2}|G_{\mathcal{Q}_i}|$, and $|G| = |G_{\mathcal{Q}_i}|\,|\Omega_i|$. Also, $|G| = |G_{\mathcal{P}}|\,|\Omega|$, and $|G_{\mathcal{P}}^{(2)}| > 1$ by Theorem 11.78. Therefore

$$2g-2 \geq (|G_{\mathcal{P}}^{(1)}| - 1)\,|\Omega|.$$

Since $|\Omega| \geq 2$, this implies that $g \geq |G_{\mathcal{P}}^{(1)}|$.

If Σ^G has genus $g' \geq 2$, then Theorem 11.72 gives $|G| \leq g-1$. If Σ^G is elliptic, then

$$2g-2 \geq 2(|G_{\mathcal{P}}| - 1 + |G_{\mathcal{P}}^{(1)}| - 1) \geq 2|G_{\mathcal{P}}|,$$

whence $|G| \leq g-1$. From what has been shown so far, it may be assumed that Σ^G is rational and that either Theorem 11.56 (iii) holds, or case (iv.4) or (iv.5) in Section 11.11 occurs.

In case (iii), Ω is the unique short G-orbit. By Theorem 11.72 applied to G,

$$2g-2 = -|G| + \deg D(\Sigma/\Sigma^G) = |\Omega|\,(\deg D(\Sigma/\Sigma^{G_{\mathcal{P}}}) - 2|G_{\mathcal{P}}|). \tag{11.86}$$

Therefore $|\Omega| \le 2g-2$. On the other hand, $|\Omega| > g-1$ as $|\Omega| \ge |G_{\mathcal{P}}^{(1)}|+1$ is inconsistent with (11.85) for $|\Omega| \le g-1$.

Since F_1 is rational,

$$\begin{aligned}\deg D(\Sigma/\Sigma^{G_{\mathcal{P}}}) &= |G_{\mathcal{P}}| - |G_{\mathcal{P}}^{(1)}| + \deg D(\Sigma/F_1)\\ &= |G_{\mathcal{P}}| + |G_{\mathcal{P}}^{(1)}| + 2g-2.\end{aligned}$$

Thus (11.86) reads as follows:

$$2(g-1) = |\Omega|(2g-2) + |G_{\mathcal{P}}^{(1)}| - |G_{\mathcal{P}}|).$$

Therefore $|\Omega| = 2g-2$. Here, $2g-2+|G_{\mathcal{P}}^{(1)}|-|G_{\mathcal{P}}| = 1$. Since $G_{\mathcal{P}} = G_{\mathcal{P}}^{(1)} \rtimes H$, hence $2g-2 = (|H|-1)|G_{\mathcal{P}}^{(1)}|+1$. This and (11.85) imply that $|H| = 2$.

Therefore $p \ne 2$ and $|G_{\mathcal{P}}^{(1)}| = 2g-3$. Since $|\Omega| = 2g-2$ and $G_{\mathcal{P}}$ only ramifies at $\mathcal{P}$, this implies that $G_{\mathcal{P}}^{(1)}$ acts on $\Omega\backslash\{\mathcal{P}\}$ as a transitive permutation group. Hence $|\Omega| = q+1$ with $q = |G_{\mathcal{P}}^{(1)}|$. Since Ω is a G-orbit, it follows that G induces on Ω a 2-transitive permutation group $\bar{G}$ whose one-point stabiliser has order either q or $2q$ according as $|G| = 2|\bar{G}|$ or $G = \bar{G}$.

If $|G| = 2|\bar{G}|$, the subgroup H is the kernel of the permutation representation of G on Ω; that is, H fixes every place in Ω. In particular, H is a normal subgroup of G. Therefore $\bar{G}$ can be viewed as a K-automorphism group of Σ^H. Let $\mathcal{P}'$ be the place of Σ^H lying under $\mathcal{P}$. Since $2 \nmid |G_{\mathcal{P}}^{(1)}|$ it follows that $G_{\mathcal{P}}^{(1)} \cong \bar{G}_{\mathcal{P}'}^{(1)}$. Also, the places of Σ^H lying under the places in Ω form the unique short $\bar{G}$-orbit. Therefore case (iii) occurs for Σ^H and $\bar{G}$. From what was shown before, this implies that $\bar{G}_{\mathcal{P}'} = \bar{G}_{\mathcal{P}'} \rtimes \bar{H}$ with $|\bar{H}| = 2$ But this is impossible as the stabiliser $\bar{G}_{\mathcal{P}'}$ has order q.

If $G = \bar{G}$, two cases are distinguished according as G is soluble or not.

In the former case, from (XXVI) of Section A.4, $q+1 = d^k$ with d prime. Since $|\Omega| = 2g-2$ is even, so $d = 2$. From Theorem A.18, $G_{\mathcal{P}}$ is a subgroup of the 1-point stabiliser of $\mathrm{A\Gamma L}(1, q+1)$, and hence $|G_{\mathcal{P}}|$ divides kq. On the other hand, $q+1 = 2^k$ can only occur when k and q are both primes. Since $|G_{\mathcal{P}}| = 2q$, this implies that $k = 2$. Hence $q = g = 3$; that is, $p = 2$ and $|G_{\mathcal{P}}^{(1)}| = g = 3$.

Suppose that G is not soluble. If G has a soluble regular normal subgroup M, then G/M is not soluble. On the other hand, $|G_{\mathcal{P}}| = 2q$, and hence $|G/M| = 2q$. But the latter is not possible for a non-soluble group. From Theorem A.17, $|\Omega| = 6$ and $G \cong \mathrm{PSL}(2,5)$, as Ree(3) with $q = 27$ does not occur. Therefore $|G_{\mathcal{P}}^{(1)}| = 5$ and $g = 4$. Thus (11.85) does not hold in case (iii).

In case (iv), G has not only a short non-tame orbit Ω but also a short tame orbit Δ. Let $\mathcal{P} \in \Omega$ and $\mathcal{Q} \in \Delta$.

If (iv.4) occurs, from the proof of Theorem 11.116, $|G_{\mathcal{P}}| \le 2(g-1)|G_{\mathcal{Q}}|/N$ and $N \ge |G_{\mathcal{P}}^{(1)}||G_{\mathcal{Q}}|$, whence

$$|G_{\mathcal{P}}||G_{\mathcal{P}}^{(1)}| \le 2(g-1).$$

Since $|G_{\mathcal{P}}| \ge 2$, this implies that $|G_{\mathcal{P}}| \le g-1$ contradicting (11.85).

Assume that case (iv.5) holds. From the proof of Theorem 11.116, the possible actions of $G = \bar{G}$ on Ω are those in Lemma 11.123. First the case when this action is faithful is considered.

If $G \cong \mathrm{PGL}(2,q)$, then

$$|G| = q^3 - q,\ |G_{\mathcal{P}}| = q^2 - q,\ |G_{\mathcal{P}}^{(1)}| = q.$$

From Theorem 11.116, the second ramification group $G_{\mathcal{P}}^{(2)}$ of $G_{\mathcal{P}}$ is non-trivial. Since $G_{\mathcal{P}}$ has a unique conjugacy class of elements of order p,

$$G_{\mathcal{P}}^{(1)} = G_{\mathcal{P}}^{(2)} = \cdots = G_{\mathcal{P}}^{(k)},\ |G_{\mathcal{P}}^{(k+1)}| = 1.$$

Since F_1 is rational, from the Hilbert Different Formula (11.37),

$$2g = (q-1)(k-1).$$

But then (11.85) does not hold.

If $G \cong \mathrm{PSL}(2,q)$, then

$$|G| = \tfrac{1}{2}(q^3 - q),\ |G_{\mathcal{P}}| = \tfrac{1}{2}(q^2 - q),\ |G_{\mathcal{P}}^{(1)}| = q.$$

The above argument showing that 2g=(q-1)(k-1) still holds, as $G_{\mathcal{P}}$ has two conjugacy classes of elements of order p, and each of them generates $G_{\mathcal{P}}^{(1)}$. Again, this contradicts (11.85).

If $G \cong \mathrm{PSU}(3,n)$ with $q = n^3$, then

$$|G| = (n^3+1)n^3(n^2-1)/\mu,\ |G_{\mathcal{P}}| = n^3(n^2-1)/\mu,\ |G_{\mathcal{P}}^{(1)}| = n^3,$$

where $\mu = \gcd(3, n+1)$. From the proof of Theorem 11.126, there is a divisor t of $(n^2 - n + 1)/\mu$ such that

$$2g = (n-1)(t(n+1)^2 - (n^2+n+1)), \quad |G_{\mathcal{Q}}| = (n^2 - n + 1)/(t\mu).$$

Since m is odd, this and (11.85) imply that $t = 1$. Then $g = \frac{1}{2}n(n-1)$, and hence $8g^3 < |G|$. Therefore (III) in Theorem 11.126 holds.

The same argument works for $G \cong \mathrm{PGU}(3,n)$.

If $G \cong \mathrm{Sz}(n)$ with $q = n^2$ and $n = 2n_0^2$ for a power $n_0 \geq 2$ of 2, then

$$|G| = (n^2+1)n(n-1),\ |G_{\mathcal{P}}| = n^2(n-1),\ |G_{\mathcal{P}}^{(1)}| = n^2.$$

From the proof of Theorem 11.126, there is an odd integer t such that either (A) or (B) holds, where

$$\text{(A)} \qquad 2g = (t-1)(n^2-1) - 2tn_0(n-1), \quad |G_{\mathcal{Q}}| = (n + 2n_0 + 1)/t;$$
$$\text{(B)} \qquad 2g = (t-1)(n^2-1) + 2tn_0(n-1), \quad |G_{\mathcal{Q}}| = (n - 2n_0 + 1)/t.$$

In the former case, t must be at least 3. But then (11.85) does not hold. In the latter case, (11.85) implies that $t = 1$. Then $g = n_0(n-1)$, and hence $8g^3 < |G|$. Therefore (IV) in Theorem 11.126 holds.

If $G \cong \mathrm{Ree}(n)$ with $q = n^3$ and $n = 3n_0^2$ for a power of 3, then

$$|G| = (n^3+1)n^3(n-1), \quad |G_{\mathcal{P}}| = n^3(n-1), \quad |G_{\mathcal{P}}^{(1)}| = n^3.$$

As was shown in the proof of Theorem 11.126, three cases may occur.

If $|G_{\mathcal{Q}}| = (n+1)/t$, then $2g = (n-1)((t-1)(n^2+1) - n)$, with $t \geq 2$. Let $t = 2$. Then $|G_{\mathcal{Q}}| = \frac{1}{2}(n+1)$. Since $\frac{1}{2}(n+1)$ is even, $G_{\mathcal{Q}}$ must contain an involution. But this is impossible by Theorem A.15 and Lemma 11.124. When $t > 2$, (11.85) does not hold.

If $|G_Q| = (n + 3n_0 + 1)/t$, then

$$2g = (n-1)((t-1)n^2 - t(3n_0n - 2n + 3n_0 - 1) - n - 1).$$

Hence $t \geq 2$. Since t is odd and $t \neq 3$, t must be at least 5. But then (11.85) does not hold.

If $|G_Q| = (n - 3n_0 + 1)/t$, then

$$2g = (n-1)((t-1)n^2 + t(3n_0n + 2n + 3n_0 + 1) - n - 1).$$

When $t = 1$, this becomes $2g = 3n_0(n-1)(n+n_0+1)$. From Theorem 12.31, case (b) follows. Otherwise, $t \geq 5$. But then (11.85) does not hold.

It remains to show the impossibility of the permutation representation $\bar{G}$ of G on Ω having non-trivial kernel. Such a kernel M is a cyclic normal subgroup of G whose order is relatively prime to p. Note that no place outside Ω is fixed by a non-trivial element in M, by Lemma 11.124.

Let $\tilde{g}$ be the genus of Σ^M. From Theorem 11.57 applied to M,

$$2g - 2 = |M|(2g\tilde{g} - 2) + (|M| - 1)(q + 1).$$

Therefore $2g \geq (|M| - 1)(q - 1)$. But this is inconsistent with (11.85). □

11.15 NOTES

The proof of the finiteness of the K-automorphism group of a complex irreducible curve of genus $g \geq 2$ was a major achievement of the 19th century, due to Schwartz, Klein, Noether, Weierstrass, Poincaré, and Hurwitz. A detailed account of the group $\mathrm{Aut}_K(\Sigma)$ in zero characteristic is found in [11, 127]; see also [306], [284].

The classical result was extended to curves over any field by Schmid [383]. The proof given in Sections 11.4 and 11.6 is based on [249, 250, 251]; see also [372]. An alternative proof using Jacobian varieties is found in [376].

For the result that every finite group is isomorphic to the K-automorphism group of some irreducible curve, see [319], [379], [426]. Related to this is Abhyankar's project to find families of 'nice' polynomials whose Galois groups are linear groups over $\mathbf{F}_q$. For a survey, see [10]; for symplectic groups, see also [9].

In connection with Section 11.2, the uniqueness of certain linear series on curves is important. So far, this has been investigated in zero characteristic; see [82].

Lemmas 11.27, 11.28 and Theorem 11.29 are due to Segre [393].

Theorem 11.29 can be generalised to show that any birational transformation between two plane non-singular curves of the same degree $n \geq 4$ is induced by a projectivity of $\mathrm{PG}(2, K)$; see [73].

For Theorem 11.31, see [303] and [414].

For the proof of Lemma 11.47, see [249, Lemma 5].

Related to Section 11.5, several good textbooks on the topic of quotient curves using function field theory are available in the literature, such as [428] and [335].

Theorems 11.56 11.60 11.78 are due to Stichtenoth [423].

The proof of Theorem 11.56 follows [244].

The plane sextic $\mathcal{F}$ in Example 11.58 arises in the study of arcs in $\mathrm{PG}(2,13)$; see Example 13.46. Since $\mathcal{F}$ over $\mathbf{F}_{13}$ achieves the Stöhr–Voloch Bound of 54 for a plane curve of this degree over $\mathbf{F}_{13}$, so $\mathcal{F}$ is a candidate to be an optimal curve for the genus $g = 10$. For more on this curve, see [170].

The study of plane curves of low degree with large automorphism groups was initiated by Klein and Wiman. The Klein quartic $\mathcal{F}$, regarded as a complex curve, has long been known to attain Hurwitz's upper bound, the K-automorphism group G of $\mathcal{F}$ being isomorphic to $\mathrm{PSL}(2,7)$; see [117]. Tuffery [463, 464] showed that this holds true for $p \neq 3, 7$. In both exceptional characteristics, $|G| = 6048$ and G contains a subgroup isomorphic to $\mathrm{PSL}(2,7)$. The Wiman sextic

$$\mathcal{F} = \mathbf{v}(10X^3Y^3 + 9(X^5 + Y^5) - 45X^2Y^2 - 135XY + 27),$$

is defined for $p \neq 2, 3, 5$ and its K-automorphism group G of order 360 is isomorphic to the alternating group $\mathbf{A}_6$. In particular, $\mathcal{F}$ possesses the maximum number of K-automorphisms of all curves of genus 10, and $\mathcal{F}$ is the unique non-singular plane sextic, up to projectivity, whose K-automorphism group contains a subgroup isomorphic to $\mathbf{A}_6$: this was shown by Kaneta, Marcugini and Pambianco [256].

Theorem 11.62 was stated by Deuring for $s \geq 1$ and then by Shafarevich for $s = 0$. Later Rosen and, independently, Sullivan gave proofs for special cases. Subrao [435] proved the theorem in full generality. Madan [318] revised Deuring's original proof fixing some errors and giving a proof for every s using Deuring's approach. A short proof valid for $n = p$ and $K = \overline{\mathbf{F}}_q$ is found in [375]; see also [369].

Theorem 11.63 is due to Witt [505] and independently to Shafarevich [409].

The group-theoretic result required in Lemma 11.82 is [239, Theorem 13.7] in which the hypothesis that $Z(G_{\mathcal{P}}^{(1)})$ is cyclic is unnecessary, since $G_{\mathcal{P}}^{(2)}$ has order p.

Related to Section 11.9, a detailed account of ramification groups is found in [404, Chapter IV]; see also [428].

Theorems 11.79, 11.84, 11.86 and 11.88 are due to Nakajima, see [342], [341] and the survey paper [343]. Bounds on the order of automorphism groups in terms of the p-rank of the curve are also found in [465].

Example 11.89 comes from [435]. For another example of an ordinary curve with large K-automorphism group, see [341].

Theorem 11.79 for complex curves goes back to Wiman; a proof valid for $p = 0$ is found in [194]. For a tame cyclic group G with at least two fixed places, Theorem 11.79 can be sharpened to $|G| \leq 4g$; see [441]. Also, if G is cyclic of order at least $2g + 1$ and $p = 0$, then either $\mathcal{F}$ is a quotient curve of a Fermat curve or $\mathcal{F}$ is hyperelliptic and $|G| = 2g + 2$. This was shown by Irokawa and Sasaki [248]. For improvements to Theorem 11.79, also valid in positive characteristic but under some further conditions, see [458].

Proofs of Theorems 11.91, 11.92, and 11.93, are found in [319]; see also [279]. For further results on the K-automorphism group of an Artin–Schreier curve

$$\mathbf{v}(Y^{p^h} - Y - f(X)),$$

with $f(X) \in K[X]$, see Section 12.1.

There is a large literature on automorphism groups of complex hyperelliptic curves, whose study was initiated by Clebsch, Hurwitz, and Bolza in the late nineteenth century; see [11] and [127]. Many of the results hold true in zero characteristic, see for instance [59], but this does not happen in positive characteristic when non-tame automorphisms come into play.

Automorphism groups of curves of genus 2 of $\mathrm{PG}(2, K)$ with $p = 2$ were investigated by Ancochea [14] in 1943. As mentioned in Section 8.10, important applications in cryptography have stimulated research on hyperelliptic curves over finite fields, especially of genus 2, with particular attention to their automorphism groups; see [274]. Recent results improving Proposition 11.99 are found in a series of papers by Cardona, Duursma, Gonzalez, Gutierrez, Kiyavash, Lario, Quer, Rio, Shaska, and Völklein; see [111], [67], [69], [413]. A relevant result is that, when $p \neq 2$, the possible K-automorphism groups of curves of genus 2 are the following:

$$\mathbf{Z}_2,\ \mathbf{Z}_{10},\ \mathbf{Z}_2 \times \mathbf{Z}_2,\ \mathbf{D}_8,\ \mathbf{D}_{12},\ \mathbf{Z}_3 \rtimes \mathbf{D}_8,\ \mathrm{GL}(2,3),$$

and an extension of $\mathbf{S}_5$ by the hyperelliptic involution.

The classification of curves of genus 2 by their K-automorphism groups is being investigated. For $p \neq 2$, the curves of genus 2 whose K-automorphism group G is isomorphic to $\mathbf{Z}_3 \rtimes \mathbf{D}_8, \mathrm{GL}(2,3)$, or $\mathbf{Z}_{10}$ are determined. Such curves are birationally equivalent to the irreducible plane curves $\mathbf{v}(Y^2 - f(X))$, with

$$f(X) = \begin{cases} X^6 - 1 & G \cong \mathbf{Z}_3 \rtimes \mathbf{D}_{12}, \\ X^5 - X & G \cong \mathrm{GL}(2,3), \\ X^5 - 1 & G \cong \mathbf{Z}_{10}. \end{cases}$$

Results on K-automorphisms of hyperelliptic curves of genus 3 are found in [184].

In [516] and [411], hyperelliptic curves whose hyperelliptic involution is the only K-automorphism are considered. In [185] and [411] hyperelliptic curves with a K-automorphism group isomorphic to $\mathbf{Z}_2 \times \mathbf{Z}_2$ are considered.

Duma's paper [108] is a survey of results contained in the unpublished thesis of Brandt, who gave a complete classification of K-automorphism groups of Kummer extensions of the rational field $K(x)$, by exhibiting explicit equations for the curves to which they correspond. The main ingredients in Brandt's analysis are a refinement of Dickson's classification and the computation of all central extensions of cyclic groups with these groups.

Example 11.103 comes from [473]. Example 11.104 is due to Kontogeorgis [278]; see also [334].

Section 11.11 comes from [11]. For the result in Remark 11.109, see [228].

For a classification of finite groups with partitions, see [438].

The curve in case (ii) has large K-automorphism group; see Theorem 11.127(II).

Theorem 11.110 is due to Cossidente and Siciliano [93, 94]. The set of $\mathbf{F}_q$-rational points of the curve $\mathcal{G}$ has interesting configurational properties; this was pointed out by Pellikaan [361].

Theorem 11.112 was generalised by Garcia in [138, 139] to higher-order Weierstrass points; such a generalisation in zero characteristic was previously given by Farkas and Kra [127]. The number of fixed places of an involutory K-automorphism was investigated by Towse [459] and, independently, in [457] using the earlier work of Garcia [136] and Kato [258, 259] on double coverings and Weierstrass points.

Levcovitz [305] gave a bound on the number of fixed points of an automorphism α acting on a non-singular model $\mathcal{X}$ embedded in $\mathrm{PG}(r, K)$. Ideas from the Stöhr-Voloch approach, as in Section 8.5, are used to count the number of points $P \in \mathcal{X}$ for which the m-osculating space to $\mathcal{X}$ at P meets the $(r-m-1)$-osculating space to $\mathcal{X}$ at P^α. The result is an inequality involving the number of fixed points of α, the order sequence of $\mathcal{X}$ and certain orders arising from the counting argument.

Section 11.12 is based on [206] and [423]. Theorem 11.127, relying on previous results of Stichtenoth, is due to Henn. As a consequence, Hurwitz's upper bound, $|G| \leq 84(g-1)$, remains valid in positive characteristic p provided that p is large with respect to the genus, namely, $p > g+1$. This was first pointed out by Roquette in [373], who also showed that, if $p = g+1$ and $|G|$ exceeds Hurwitz's upper bound, then (II) of Theorem 11.127 holds.

Sing's bound [417] valid for $p \leq 2g+1$,

$$|G| \leq \left(\frac{4pg^2}{p-1} \cdot \frac{2g}{p-1} + 1\right)\left(\frac{4pg^2}{(p-1)^2} + 1\right),$$

is weaker than the bound in Theorem 11.127.

For a proof of Theorem 11.134, due to Burnside, see [179].

In finite group theory, trivial intersection sets play an important role. Suzuki [440] classified all finite groups whose Sylow 2-subgroups are trivial intersection sets. Hering [209] refined and extended Suzuki's classification to finite groups containing a subgroup of even order which intersects each of its distinct conjugates trivially. Here, Hering's result Theorem 11.135 is quoted. In the special case that Q is a Sylow 2-subgroup of G, this is essentially Suzuki's classification. Theorem 11.136 is stated in [515].

For more on the curves given in Theorems 11.127 and 11.140, see Sections 12.2, 12.3 and 12.4.

For Huppert's classification referred to in the proof of Theorem 11.140, see [241, Theorem 7.3]

Theorem 11.140 holds true for $|G_{\mathcal{P}}| > pg/(p-1)$ provided that the DLR curve and case (II) from Theorem 11.127 are added to (ii); see [173].

In connection with Example 11.103 and Kloosterman sums, see van der Geer and van der Vlugt [473], where it is shown that the irreducible plane curves

$$\mathbf{v}(A(Y) - R(X)),$$

with $A(Y) \in \mathbf{F}_q[Y]$ and $R(X) \in \mathbf{F}_q(X)$, are precisely the curves Γ over $\mathbf{F}_q$ of genus $p-1$ with the three properties listed.

If $\mathcal{C}$ is an irreducible curve of genus $g \geq 2$ covering the projective line $\mathrm{PG}(1, K)$ with a cyclic p-covering ramified in exactly one point, Lehr and Matignon [302] showed that p-subgroups of $\mathrm{Aut}_K(\Sigma)$ providing such coverings are exactly the extensions of a cyclic p-group by an elementary abelian p-group. They also characterised curves $\mathcal{C}$ of genus $g \geq 2$ and p-subgroups of $\mathrm{Aut}_K(\Sigma)$ such that

$$2p/(p-1) < |G|/g \quad \text{and} \quad 4/(p-1)^2 \leq |G|/g^2.$$

Automorphism groups of modular curves were investigated by Bending, Camina, and Guralnick [46].

Chapter Twelve

Some families of algebraic curves

The aim of this chapter is to provide families of algebraic curves in positive characteristic with properties that a complex algebraic curve cannot have. As mentioned in the earlier chapters, these curves illustrate concepts and results unknown in the classical literature on curves.

12.1 PLANE CURVES GIVEN BY SEPARATED POLYNOMIALS

In this section, the curve with

$$\mathcal{C} = \mathbf{v}(A(Y) - B(X)), \tag{12.1}$$

is investigated in detail under the following conditions:

(I) $\deg \mathcal{C} \geq 4$ and $\gcd(p, \deg B(X)) = 1$;

(II) $A(Y) = a_n Y^{p^n} + a_{n-1} Y^{p^{n-1}} + \cdots + a_0 Y, \quad a_j \in K,\ a_0,\ a_n \neq 0$;

(III) $B(X) = b_m X^m + b_{m-1} X^{m-1} + \cdots + b_1 X + b_0, \quad b_j \in K,\ b_m \neq 0$;

(IV) $m \not\equiv 0 \pmod{p}$;

(V) $n \geq 1,\ m \geq 2$.

Note that (II) occurs if and only if $A(Y + a) = A(Y) + A(a)$ for every $a \in K$, that is, the polynomial $A(Y)$ is *additive*. The basic properties of $\mathcal{C}$ are collected in the following lemma.

LEMMA 12.1 *The curve $\mathcal{C}$ is an irreducible plane curve with at most one singular point.*

(i) *If $|m - p^n| = 1$, then $\mathcal{C}$ is non-singular.*

(ii) (a) *If $m > p^n + 1$, then $P_\infty = (0, 0, 1)$ is an $(m - p^n)$-fold point of $\mathcal{C}$.*

(b) *If $p^n > m + 1$, then $P_\infty = (0, 1, 0)$ is a $(p^n - m)$-fold point of $\mathcal{C}$.*

(c) *In both cases, P_∞ is the centre of only one branch of $\mathcal{C}$; also, P_∞ is the unique infinite point of $\mathcal{C}$.*

(iii) *If $\Sigma = K(x, y)$, with $A(y) = B(x)$, is the function field of $\mathcal{C}$, and $\mathcal{P}_\infty$ is the place associated to the branch centred at P_∞, then*

(a) $\operatorname{div}(dx) = ((p^n - 1)(m - 1) - 2)\mathcal{P}_\infty$;

(b) *$\mathcal{X}$ has genus* $g = \frac{1}{2}(p^n - 1)(m - 1)$;

(c) *a translation* $(x, y) \mapsto (x, y + a)$ *preserves* $\mathcal{C}$ *if and only if* $A(a) = 0$;

(d) *these translations form an elementary abelian group of order* p^n, *and* $\mathrm{Aut}_K(\Sigma)$ *contains an elementary abelian* p*-group* G *of order* p^n *that fixes a unique place* $\mathcal{P}_\infty$ *and acts transitively on the zeros of* x;

(e) *the sequence of ramification groups of* G *at* $\mathcal{P}_\infty$ *is*

$$G = G^{(1)}_{\mathcal{P}_\infty} = G^{(2)}_{\mathcal{P}_\infty} = \cdots = G^{(m)}_{\mathcal{P}_\infty}, \qquad G^{(m+1)}_{\mathcal{P}_\infty} = 1;$$

(f) $\{\mathcal{P}_\infty\}$ *is the unique short orbit of* G, *and*

$$\mathrm{div}(\Sigma/\Sigma^G) = (p^n - 1)(m + 1)\mathcal{P}_\infty;$$

(g) Σ^G *is rational, and* Σ *has* p*-rank zero.*

Proof. Put $F(X, Y) = A(Y) - B(X)$. Since $\partial F/\partial Y = a_0 \neq 0$, no affine point is a singular point of $\mathcal{C}$.

The case $m > p^n$ is worked out in detail; the arguments can be adapted to the case $p^n \geq m$.

First, $P_\infty = (0, 0, 1)$ is the unique point of $\mathcal{C}$ at infinity and $\mathcal{P}_\infty$ is, in fact, an $(m - p^n)$-fold point, with a unique tangent line $\mathbf{v}(X_0)$.

It is now shown that P_∞ is the centre of only one branch of $\mathcal{C}$. To do this, P_∞ is taken to the origin by interchanging X_0 and X_2. Then $\mathcal{C} = \mathbf{v}(A(Y) - C(X, Y))$, where

$$C(X, Y) = b_m X^m + b_{m-1} X^{m-1} Y + \cdots + b_1 X Y^{m-1} + b_0 Y^m.$$

Let $x(t) = ut^i + \cdots, y(t) = vt^j + \cdots$, with $u, v \neq 0$, $i < j$, be a primitive representation of a branch of $\mathcal{C}$ centred at O. Since $A(y(t)) = C(x(t), y(t))$, it follows that $j(m - p^n) = im$. As $\gcd(m, p^n) = 1$, this leaves only one case, namely $i = m - p^n$, $j = m$. Since O is an $(m - p^n)$-fold point, the assertion follows. As a consequence, $\mathcal{C}$ is irreducible.

Going back to the original frame,

$$(x(t), y(t)) = \left(\frac{ut^{m-p^n} + \cdots}{vt^m + \cdots}, \frac{1}{vt^m + \cdots} \right)$$
$$= \left(t^{-p^n}(u_1 + \cdots), t^{-m}(v_1 + \cdots) \right)$$

is a primitive representation of the unique branch of $\mathcal{C}$ centred P_∞. As $\mathrm{ord}_t\, x(t)$ is divisible by p, the calculation of $\mathrm{ord}_t(dx(t)/dt)$ is not simple. To overcome the difficulty, take derivatives in $A(y(t)) = B(x(t))$. This gives the equation

$$a_0 \frac{dy}{dt} = \left((m - 1) b_m x(t)^{m-1} + \cdots + b_1 \right) \frac{dx}{dt}.$$

Since $\mathrm{ord}_t\, y(t)$ is not divisible by p and $\mathrm{ord}_t\, x(t) < 0$, so

$$-(m + 1) = -p^n(m - 1) + \mathrm{ord}_t(dx/dt).$$

Therefore $\mathrm{ord}_t(dx/dt) = (p^n - 1)(m - 1) - 2$. By assumption, $\mathcal{C}$ has degree at least 4 and this implies that $(p^n - 1)(m - 1) > 2$. In particular, x is a separable variable of Σ. Also, as no vertical line is tangent to $\mathcal{C}$, so $\mathrm{ord}_\gamma(dx(t)/dt) = 0$ for every branch γ of $\mathcal{C}$ centred at an affine point. Since $\mathcal{C}$ has degree m, so (iii)(b) follows from Definition 5.55. Let $\mathcal{P}_\infty$ be the place associated with the branch γ. So (iii) (a) holds.

The roots of the polynomial $A(Y)$ are the elements of an additive subgroup A of K of order p^n. Also, for every $a \in A$, the linear collineation $(x, y) \mapsto (x, y + a)$ preserves $\mathcal{C}$; that is, $\alpha_a : (x, y) \mapsto (x, y + a)$ is a K-automorphism of the function field $\Sigma = K(x, y)$ of $\mathcal{C}$ which fixes the place $\mathcal{P}_\infty$. Let G be the group consisting of these α_a with a ranging over A. For $a \neq 0$, the automorphism α_a fixes no more place of Σ showing (iii) (c) and the first part of (f).

To prove (e), take $\mathcal{P}_\infty$ again to the origin by interchanging the coordinates X_0 and X_2. Then Σ is assumed to be $K(x, y)$ with $A(x) = C(x, y)$ and, for every $a \in A$,

$$\alpha_a : (x, y) \mapsto \left(\frac{x}{1 + ay}, \frac{y}{1 + ay}\right).$$

A uniformising element at O can be found by using the following argument. Since $\gcd(m, p^n) = 1$, there exist integers k, ℓ such that $k(m - p^n) + \ell m = 1$. Let $\xi = x^k y^\ell$. As shown before, $x(t) = ut^{m-p^n} + \cdots$, $y(t) = vt^m + \cdots$ is a primitive representation of the unique branch centred at O. If $\mathcal{O}$ is the corresponding place of Σ, then

$$\mathrm{ord}_{\mathcal{O}}\, \xi = \mathrm{ord}_t(t^{k(m-p^n)+\ell m} + \cdots) = 1,$$

showing that $\xi = x^k y^\ell$ is a uniformising element of Σ at O.

As $\gcd(p^n, m) = 1$), for every $a \in A$, $a \neq 0$,

$$\begin{aligned}\mathrm{ord}_t(\alpha(\xi(t)) - \xi(t)) &= \mathrm{ord}_t\left(\frac{x(t)^k y(t)^\ell}{(1 + ay(t))^{\ell+k}} - x(t)^k y(t)^\ell\right) \\ &= \mathrm{ord}_t(aty(t) + \cdots) = m + 1,\end{aligned}$$

whence (e) follows. Hilbert's Different Formula (11.37), together with (e), gives the second part of (f). From this, by Theorem 11.72, the first part in (g) follows. Finally, Lemma 11.131 shows the last assertion.

□

LEMMA 12.2 (i) *For every $k > 1$,*

$$|k\mathcal{P}_\infty| = \mathrm{div}(\textstyle\sum c_{ij} x^i y^j) + k\mathcal{P}_\infty,$$

summed over all $i, j \geq 0$ satisfying both $j \leq p^n - 1$ and $ip^n + jm \leq k$.

(ii)

$$g_{2g-2}^{g-1} = \mathrm{div}(\textstyle\sum c_{ij} x^i y^j) + (2g - 2)\mathcal{P}_\infty,$$

summed over all $i, j \geq 0$ satisfying $ip^n + jm \leq (p^n - 1)(m - 1) - 2$.

Proof. With the notation in the previous proof, $\mathrm{ord}_t(x(t)^i y(t)^j) = -(ip^n + jm)$. This shows that $x^i y^j$ has a pole at $\mathcal{P}_\infty$ of order at most k when $ip^n + jm \leq k$. Since every place distinct from $\mathcal{P}_\infty$ arises from a branch of $\mathcal{C}$ centred at an affine point, $x^i y^j$ has no more poles; hence $\mathrm{div}(x^i y^j) + k\mathcal{P} \in |k\mathcal{P}|$. This implies that $\mathrm{div}(\sum c_{ij} x^i y^j) + k\mathcal{P} \in |k\mathcal{P}|$ provided that

$$ip^n + jm \leq k \text{ when } c_{ij} \neq 0. \tag{12.2}$$

Those $x^i y^j$ satisfying not only (12.2) but also $j < p^n$ provide a linearly independent set over K.

Every element in Σ has a representation $a(x,y)/b(x)$ where $a(X,Y) \in K[X,Y]$ and $b(X) \in K[X]$. As $\mathcal{C} = \mathbf{v}(A(Y) - B(X))$, with $\deg A(Y) = p^n$, it may be assumed that $\deg_Y a(X,Y) < p^n$. Now suppose that

$$\operatorname{div}\left(\frac{a(x,y)}{b(x)}\right) \succ k\mathcal{P}. \tag{12.3}$$

Then $a(x,y)/b(x)$ has no pole at an affine point.

To show that $b(X)$ is a constant polynomial, assume on the contrary that $u \in K$ be a root of $b(X)$. Then every point $P = (u,v)$ of $\mathcal{C}$ is a zero of $a(x,y) \in \Sigma$; that is, every root of the polynomial $A(Y) - B(u)$ is a root of the polynomial $a(u,Y)$. The former polynomial has p^n distinct roots since $dA(y)/dY = 1$, but the latter one has fewer roots, as its degree does not exceed $p^n - 1$. This contradiction proves the assertion. So it may be assumed that $b(X) = 1$.

Let $a(X,Y) = \sum c_{ij} X^i Y^j$. If some (i,j) satisfies (12.2), replace $a(X,Y)$ by $a(X,Y) - c_{ij} X^i Y^j$. In doing so, $ip^n + jm > k$ with $c_{ij} \neq 0$. To show that this leads to a contradiction, let a primitive representation of the unique branch centred at $\mathcal{P}_\infty$ be

$$(x(t), y(t)) = \left(t^{-p^n}(u_1 + \cdots), t^{-m}(v_1 + \cdots)\right).$$

Therefore, if $c_{ij} \neq 0$, then $\operatorname{ord}_t\,(x(t)^i y(t)^j) = -(ip^n + jm)$. As $k < ip^n + jm$ for $c_{ij} \neq 0$, it turns out that either $\operatorname{ord}_t a(x(t), y(t)) < -k$, or there are two terms in $a(X,Y)$, say $c_{ij} X^i Y^j$ and $c_{\bar{\imath},\bar{\jmath}} X^{\bar{\imath}} Y^{\bar{\jmath}}$ such that

$$\operatorname{ord}_t\,(x(t)^i y(t)^j) = \operatorname{ord}_t\,(x(t)^{\bar{\imath}} y(t)^{\bar{\jmath}}).$$

In the former case, the contradiction is with (12.3) while, in the latter case,

$$(\bar{\imath} - i)p^n = (\bar{\jmath} - j)m,$$

which gives $j \equiv \bar{\jmath} \pmod{p^n}$. Since $\deg_Y a(X,Y) < p^n$ this implies that $j = \bar{\jmath}$ and hence $i = \bar{\imath}$, a contradiction.

The second assertion is a corollary to the first when applied to $k = 2g - 2$, as $(2g-2)\mathcal{P}_\infty$ is the canonical series on $\mathcal{X}$. □

From Lemma 12.2, it also means that the Weierstrass semigroup of Σ at $\mathcal{P}$ is generated by p^n and m. Hence the following result holds.

LEMMA 12.3 *There are exactly $\frac{1}{2}(p^n - 1)(m-1)$ gaps, every other positive integer being a linear combination of m and p^n with positive integer coefficients.*

A characterisation of the curve $\mathcal{C}$ of (12.1) is given in the following theorem.

THEOREM 12.4 *Let $\mathcal{F}$ be an irreducible plane curve such that a place $\mathcal{P}$ of its function field $\Sigma = K(\mathcal{F})$ satisfies the following conditions:*

(a) *Σ has a K-automorphism group N of order p^n whose non-trivial elements fix $\mathcal{P}$ but no other place of Σ;*

(b) Σ^N *is rational*;

(c) Σ *contains an element* y *such that* $\operatorname{div}(y)_\infty = m\mathcal{P}$ *with* $\gcd(m,p) = 1$;

(d) Σ *has genus* $g \geq \frac{1}{2}(p^n - 1)(m - 1)$.

Then $\mathcal{F}$ *is birationally equivalent to the irreducible plane curve* $\mathcal{C}$ *of (12.1) with* $A(Y)$, $B(X)$ *satisfying* (I)–(V).

Proof. Choose $x \in \Sigma$ such that $\Sigma^N = K(x)$. Let σ be a primitive representation of the place $\mathcal{P}$. If $\sigma(x) = c_i t^i + \cdots$ with $c_i \neq 0$, $i \geq 0$, replace x by x^{-1} or $(x - c_0)^{-1}$ depending on whether $i > 0$ or $i = 0$. Then $\mathcal{P}$ may be taken to be a pole of x and is the unique pole of x. If $\mathcal{Q}$ were another pole of x, then every place in the orbit of $\mathcal{Q}$ under N would be a pole of x as well. So x would have more than p^n poles. But this contradicts Theorem 5.34 since

$$p^n = |N| = [\Sigma : \Sigma^N] = [\Sigma : K(x)].$$

Thus $\operatorname{div}(x)_\infty = p^n\mathcal{P}$. From this two facts follow. First, the Weierstrass semigroup at $\mathcal{P}$ contains the semigroup generated by p^n and m, where m is defined as in (c). From the Weierstrass Gap Theorem 6.89, $2g \leq (p^n - 1)(m - 1)$. This and (d) imply that $g = \frac{1}{2}(p^n - 1)(m - 1)$. Secondly, $\operatorname{div}(x)_\infty = p^n\mathcal{P}$, together with (c), implies that $\Sigma = K(x, y)$. In fact, $[\Sigma : K(x, y)]$ divides $\gcd(m, p^n) = 1$, as

$$[\Sigma : K(x,y)][K(x,y) : K(x)] = [\Sigma : K(x)] = p^n,$$
$$[\Sigma : K(x,y)][K(x,y) : K(y)] = [\Sigma : K(y)] = m.$$

It remains to show that x and y satisfy an equation of type $A(y) = B(x)$, with $A(Y)$, $B(X)$ as in (I)–(V). To do this, two cases are distinguished.

Assume first that $m < p^n$, and choose $\alpha \in N$. Since m is the smallest non-gap at $\mathcal{P}$, there exists $d_\alpha \in K$ such that $\alpha(y) = y + d_\alpha$, by Lemma 11.13. Then $d_\alpha + d_\beta = d_{\alpha\beta}$ for any two $\alpha, \beta \in N$. As N is a group, $\{d_\alpha \mid \alpha \in N\}$ is an additive subgroup of K of order p^n. Then the polynomial $A(Y) = \prod(Y + d_\alpha)$ with α ranging of all elements in N, is of type (II). Also, $A(y) \in K(x)$, because $A(y)$ is fixed by N. Therefore $A(y) = u(x)/w(x)$, where $u(X), w(X) \in K[X]$ are polynomials without common roots.

In fact, $w(X)$ must be constant. If c is a root of $w(X)$, replace x by $x - c$. Let $\mathcal{Q}$ be a zero of x; then $\mathcal{Q}$ is not a zero of $u(x)$. Also, $\mathcal{Q} \neq \mathcal{P}$ because $\mathcal{P}$ is a pole of x. Hence $\mathcal{Q}$ is a pole of $A(y)$. As $A(Y) \in K[Y]$, this is only possible when y itself has a pole at $\mathcal{Q}$, contradicting that $\operatorname{div}(y)_\infty = m\mathcal{P}$. So there exists $B(X) \in K[X]$ such that $A(y) = B(x)$. Then

$$\operatorname{div}(A(y))_\infty = \operatorname{div}(B(x))_\infty.$$

As $\operatorname{div}(A(y))_\infty = p^n m\mathcal{P}$, and $\operatorname{div}(x)_\infty = p^n\mathcal{P}$, this implies that $\deg B(X) = m$. So the proof is complete for $m < p^n$.

Now assume that $m > p^n$, and take a non-trivial K-automorphism $\alpha \in N$. If σ is a primitive representation of $\mathcal{P}$, then $\sigma(y) = ct^{-m} + \cdots$. If $\lambda_\alpha : t \mapsto t + ut^i + \cdots$ is the companion automorphism of α, then

$$\sigma(\alpha(y)) = c(t + ut^i + \cdots)^{-m} + \cdots.$$

Thus $\sigma(\alpha(y)-y)=dt^{-j}+\cdots$ with $j<m$. As $m>p^n$, this implies that

$$\mathrm{div}(\alpha(y)-y)_\infty=\ell p^n\mathcal{P}. \tag{12.4}$$

Since every non-gap at $\mathcal{P}$ smaller than m is a multiple of p^n,

$$|\ell p^n\mathcal{P}|=\mathrm{div}(\textstyle\sum_{i=0}^{\ell}c_ix^i)+\ell p^n\mathcal{P}.$$

Hence there is a polynomial $P_\alpha(X)$ such that

$$\alpha(y)-y=P_\alpha(x),\quad s=\deg\, P_\alpha(X)\leq\ell. \tag{12.5}$$

Change y to a uniformising element η by putting $\eta=y^ux^{-v}$ with $-um+vp^n=1$. Since $\alpha(x)=x$, so (12.4) becomes the equation

$$\mathrm{div}(\alpha(\eta)-\eta)_\infty=(m+1-sp^n)\mathcal{P}. \tag{12.6}$$

In fact,

$$\alpha(\eta)-\eta=x^{-v}(\alpha(y)^u-y^u)=x^{-v}(uy^{u-1}P_\alpha(x)+\cdots+P_\alpha(x)^u),$$

which implies that

$$\mathrm{ord}_t\,\sigma(\alpha(\eta)-\eta)=p^nv-m(u-1)-p^ns=m+1-p^ns.$$

On the other hand, the Hilbert Different formula (11.37) applied to N gives the relation

$$(m+1)(p^n-1)=\textstyle\sum_{i\geq0}(|N_{\mathcal{P}}^{(i)}|-1),$$

where $N_{\mathcal{P}}^{(i)}$ is the i-th ramification group of N at $\mathcal{P}$. From (12.6) and $\ell p^n<m$ it follows that $N_{\mathcal{P}}^{(i)}$ is trivial for any $i>m$. Hence N must coincide with $N_{\mathcal{P}}^{(i)}$ for $0\leq i\leq m$. Again by (12.6), $P_\alpha(X)$ is a constant, say $d_\alpha\in K$, for every $\alpha\in N$. Now the proof can be completed in the same way as in the case $m<p^n$. □

For $n=1$, Theorem 12.4 becomes the following.

THEOREM 12.5 *Let $\mathcal{F}$ be an irreducible plane curve such that its function field $\Sigma=K(\mathcal{F})$ satisfies the following conditions:*

(a) *Σ has an automorphism α of order p fixing only one place;*

(b) *Σ^α is rational;*

If $\mathcal{P}$ is the fixed place of α and m is the smallest pole number at $\mathcal{P}$ prime to p, then $\mathcal{F}$ is birationally equivalent to the curve

$$\mathcal{C}=\mathbf{v}(Y^p-Y-B(X))$$

with $B(X)\in K[X]$ of degree m.

Proof. The background and arguments from the preceding proof are used. Let $\Sigma=K(x,y)$, $\Sigma^\alpha=K(x)$, $\mathrm{div}(x)_\infty=p\mathcal{P}$, $\mathrm{div}(y)_\infty=m\mathcal{P}$.

If $m<p$, then $\alpha(y)=y+d$ with $d\in K$. Therefore $\alpha^i(y)=y+id$, for $i=0,1,\ldots,p-1$. Since $\prod_{i=0}^{p-1}\alpha^i(y)$ is fixed by α,

$$y^p-d^{p-1}y=\textstyle\prod_{i=0}^{p-1}(y+id)\in K(x).$$

Replacing y by yd^{-1} gives $y^d - y \in K(x)$. This implies that $y^p - y = B(x)$ with $B(X) \in K[X]$. In particular, $\deg B(X) = m$. Therefore $\mathcal{C}$ has equation of type (12.1) with $A(Y) = Y^p - Y$.

If $m > p$, then $\alpha(y) = y + P(x)$ with $P(X) \in K[X]$ and $p \cdot \deg P(X) < m$. To prove that $P(X)$ is a constant polynomial, suppose on the contrary that $P(X)$ has a root, say a. Replacing x by $x - a$, it may be assumed without loss of generality that $P(X) = XQ(X)$ with $Q(X) \in K[X]$. This implies that

$$y^p - Q(x)^{p-1}x^{p-1}y = B(x)$$

for a polynomial $B(X) \in K[X]$, whose degree is m, since

$$\operatorname{ord}_{\mathcal{P}} y = -m, \quad \operatorname{ord}_{\mathcal{P}} x = p, \quad p \cdot \deg Q(X) < m - p.$$

Also, $B(0) = 0$ after replacing y by $y - u$ where $u^p = B(0)$. Therefore the following equations are valid:

$$\begin{aligned} y^p - Q(x)^{p-1}x^{p-1}y &= a_x^m + \cdots + a_1 x; \\ \alpha(x) &= x; \\ \alpha(y) &= y + xQ(x), \quad p \cdot \deg Q(x) < m - p. \end{aligned}$$

The plane curve

$$\mathcal{G} = \mathbf{v}(Y^p - Q(X)^{p-1}X^{p-1}Y - (a_m X^m + \cdots + a_1 X))$$

is a birational model of Σ. If the place $\mathcal{Q}$ is a zero of x, then $\mathcal{Q}$ is a zero of y, as well. Hence the branch of $\mathcal{G}$ corresponding to $\mathcal{Q}$ is centred at the origin O. Since the set of zeros of x is preserved by α, but none of these is fixed by α, the origin is the centre of at least p distinct branches of $\mathcal{G}$. On the other hand, O is at most a p-fold point of $\mathcal{G}$. Hence the places corresponding to branches centred at O form a single orbit under α, and hence

$$\operatorname{div}(x)_0 = \textstyle\sum_{i=0}^{p-1} \mathcal{Q}^{\alpha^i}.$$

As $\operatorname{div}(y)_0 \succ \sum_{i=0}^{p-1} \mathcal{Q}^{\alpha^i}$, it follows that $\mathcal{P}$ is the unique pole of y/x. Further, $\operatorname{ord}_{\mathcal{P}}(y/x) = -(m-p)$. Since $\gcd(p, m-p) = 1$, this contradicts the choice of y, and proves that $P(X)$ is a constant polynomial. Now the proof finishes in the same way as in the case $m < p$. □

REMARK 12.6 The following two results are refinements of Theorem 12.5.

(i) If $m - 1 = p^r s$, $r > 0, s > 1$, $\gcd(p, s) = 1$, then

$$|G_{\mathcal{P}}^{(1)}| \le \begin{cases} p^{r+1} & \text{for } p > 2, \\ 2^r & \text{for } p = 2. \end{cases}$$

(ii) If $|G_{\mathcal{P}}^{(1)}| \le pg/(p-1)$, then $B(X) = XR(X)$ with $R(X) \in K[X]$.

THEOREM 12.7 *Let G be the K-automorphism group of a plane curve $\mathcal{C}$ as in (12.1). Let $G_{\mathcal{P}_\infty} = G_{\mathcal{P}_\infty}^{(1)} \rtimes H$ with $p \nmid |H|$. Then*

(i) *$|H|$ divides $m(p^n - 1)$;*

(ii) $|G^{(1)}_{\mathcal{P}_\infty}| \leq p^n(m-1)^2 = \dfrac{4p^n}{(p^n-1)^2}\, g^2;$

(iii) $|G^{(1)}_{\mathcal{P}_\infty}| = p^n$ *when* $m \not\equiv 1 \pmod{p^n}$, *and so* $g \not\equiv 0 \pmod{p^n}$;

(iv) $|G^{(2)}_{\mathcal{P}_\infty}| = p^n$ *when* $m \equiv 1 \pmod{p^n}$, *and so* $g \equiv 0 \pmod{p^n}$.

Proof. Let $\alpha \in G_{\mathcal{P}_\infty}$. By Lemma 11.20, α preserves both $|m\mathcal{P}_\infty|$ and $|p^m\mathcal{P}_\infty|$.

First, consider the case $m < p^n$. By Lemma 12.2,

$$\begin{aligned}
|m\mathcal{P}| &= \{\operatorname{div}(c_0 + c_1 y) + m\mathcal{P} \mid \mathbf{c} = (c_0, c_1) \in \mathrm{PG}(1,K)\},\\
|p^n\mathcal{P}| &= \{\operatorname{div}(cx + \textstyle\sum_{i=0}^r d_i y^i) + p^n\mathcal{P} \mid \mathbf{c} = (d_0,\ldots,d_r,c) \in \mathrm{PG}(r+1,K)\},
\end{aligned}$$

with r the integer defined by $rm < p^n < (r+1)m$. Hence

$$\alpha(y) = ay + b, \quad \alpha(x) = cx + P(y),$$

where $P(Y) \in K[Y]$ such that $m \cdot \deg P(Y) < p^n$, and $a, b, c \in K$ with $ac \neq 0$. As $\alpha(A(y)) = \alpha(B(x))$, the polynomial,

$$\begin{aligned}
H(X,Y) &= A(aY+b) - B(cX + P(Y))\\
&= a_n a^{p^n} Y^{p^n} + a_{n-1} a^{p^{n-1}} Y^{p^{n-1}} + \cdots + a_0 aY + A(b) - b_m c^m X^m\\
&\quad - c^{m-1}(b_m m P(Y) + b_{m-1}) X^{m-1} - \cdots - B(P(Y)),
\end{aligned}$$

is divisible by $A(Y) - B(X)$. Also,

$$\deg H(X,Y) = p^n = \deg(A(Y) - B(X)),$$

which implies that $H(X,Y)$ coincides with $A(Y) - B(X)$ up to a constant factor; that is, $H(X,Y) = k(A(Y) - B(X))$ with $k \in K\backslash\{0\}$. Since

$$k = a^{p^n} = a = c^m, \qquad c^{m-1}(b_m m P(Y) + b_{m-1}) = k b_{m-1},$$

so $P(Y)$ is a constant; namely, $P(Y) = (c-1)b_{m-1}/(m b_m)$. Hence

$$\alpha(x) = cx + \frac{(c-1)b_{m-1}}{m b_m}, \quad \alpha(y) = ay + b \quad \text{with} \quad a^{p^n} = a = c^m. \tag{12.7}$$

The map $\alpha \mapsto c$ is a homomorphism of $G_{\mathcal{P}_\infty}$ into the multiplicative group of K, whose kernel is $G^{(1)}_{\mathcal{P}_\infty}$ as $c = 1$ implies that $a = 1$.

To calculate the order of $G^{(1)}_{\mathcal{P}_\infty}$, the number of elements b when α varies in $G^{(1)}_{\mathcal{P}_\infty}$ needs to be found. These are given by the roots of the polynomial

$$A(Y) = kB\left(\frac{(c-1)b_{m-1}}{m b_m}\right),$$

whose number is p^n. Further, H is isomorphic to a subgroup of the group of all $m(p^n-1)$-st roots of unity, as $c^{m(p^n-1)} = a^{p^n-1} = 1$. This completes the proof for $m < p^n$.

The proof for the case $m > p^n$ is similar. Again, by Lemma 12.2,

$$\begin{aligned}
|p^n\mathcal{P}| &= \{\operatorname{div}(c_0 + c_1 x) + p^n\mathcal{P} \mid \mathbf{c} = (c_0, c_1) \in \mathrm{PG}(1,K)\},\\
|m\mathcal{P}| &= \{\operatorname{div}(cy + \textstyle\sum_{i=0}^r d_i x^i) + m\mathcal{P} \mid \mathbf{c} = (d_0,\ldots,d_r,c) \in \mathrm{PG}(r+1,K)\},
\end{aligned}$$

with r defined by $rp^n < m < (r+1)p^n$. Therefore

$$\alpha(x) = cx + d, \qquad \alpha(y) = ay + Q(x), \tag{12.8}$$

where $Q(X) \in K[X]$ such that $p^n \cdot \deg Q(X) < m$, and $a, c, d \in K$ with $ac \neq 0$. The polynomial

$$\begin{aligned} H(X,Y) &= A(aY + Q(X)) - B(cX + d) \\ &= a_n a^{p^n} Y^{p^n} + a_{n-1} a^{p^{n-1}} Y^{p^{n-1}} + \cdots + a_0 aY + A(Q(X)) \\ &\quad -b_m c^m X^m - c^{m-1}(b_m md + b_{m-1}) X^{m-1} - \cdots - B(d) \end{aligned}$$

is $A(Y) - B(X)$ up to a constant factor; that is, $H(X,Y) = k(A(Y) - B(X))$ with $k \in K\backslash\{0\}$. As before, this gives $a^{p^n} = a = c^m$. Again, the map $\alpha \mapsto c$ is a homomorphism of $G_{\mathcal{P}_\infty}$ into the multiplicative group of K. Since

$$c^{m(p^n-1)} = a^{p^n-1} = 1,$$

it follows that H is isomorphic to a subgroup of the group of all $m(p^n-1)$-st roots of unity. As $m < p^n$, the kernel of this homomorphism coincides with $G^{(1)}_{\mathcal{P}_\infty}$ since $c = 1$ implies that $a = 1$.

To calculate the order of $G^{(1)}_{\mathcal{P}_\infty}$, first consider the case $m \not\equiv 1 \pmod{p^n}$. Here, no monomial in $A(Q(X))$ has degree $m-1$. Therefore, from the equation

$$H(X,Y) = k(A(Y) - B(X)), \tag{12.9}$$

it follows that

$$c^{m-1}(b_m md + b_{m-1}) = 0;$$

that is, $d = -b_{m-1}/(mb_m)$. Then $d = 0$ after substituting $x - b_{m-1}/(mb_m)$ for x. As before, (12.9) implies that $A(Q(X)) = 0$ and this only occurs when $Q(X)$ is constant. More precisely, $Q(X) = u$ with $A(u) = 0$. Hence $\alpha \in G^{(1)}_{\mathcal{P}_\infty}$ if and only if

$$\alpha(x) = x, \qquad \alpha(y) = y + u, \quad u \in K, \quad A(u) = 0.$$

As $A(Y)$ has p^n distinct roots, so $|G^{(1)}_{\mathcal{P}_\infty}| = p^n$. This completes the proof of (iii).

For the case $m \equiv 1 \pmod{p^n}$, more is needed. Let $m = vp^k + 1$ with $k \geq n$ and $\gcd(p, v) = 1$. By (12.8), any $\alpha \in G^{(1)}_{\mathcal{P}_\infty}$ has an equation of type

$$\alpha(x) = x + d, \quad \alpha(y) = y + Q(x),$$

with

$$Q(x) = f_1 x + \cdots + f_v x^v + \cdots + f_{vp^{k-n}} x^{vp^{k-n}} + b.$$

Since

$$\begin{aligned} 0 = A(\alpha(y)) - B(\alpha(x)) &= A(y) + A(Q(x)) - B(x+d) \\ &= A(Q(x)) + B(x) - B(x+d), \end{aligned}$$

it follows that

$$A(f_1 X + \cdots + f_v X^v + \cdots + f_{vp^{k-n}} X^{vp^{k-n}} + b) + B(X) - B(X+d)$$

is the zero polynomial, and hence

$$a_n f_{vp^{k-n}}^{p^n} - m b_m d = 0.$$

To calculate the second ramification group $G_{\mathcal{P}_\infty}^{(2)}$, note that $|G_{\mathcal{P}_\infty}^{(2)}| \geq p^n$ by Lemma 12.1 (iii) (f). To see that equality holds, let $\alpha \in G_{\mathcal{P}_\infty}^{(2)}$, that is,

$$\operatorname{ord}_t \sigma(\alpha(z) - z) \geq 3,$$

where $z = x^{vp^{k-n}}/y$ is a local parameter at $\mathcal{P}_\infty$. As

$$\alpha(z) - z = \frac{y(x^{p^{k-n}} + d^{p^{k-n}})^v - yx^{vp^{k-n}} - Q(x)x^{vp^{k-n}}}{(y + Q(x))y},$$

straightforward computation shows that $f_{vp^{k-n}} = 0$, whence $d = 0$. Therefore $\alpha \in G_{\mathcal{P}_\infty}^{(2)}$ if and only if

$$\alpha(x) = x, \qquad \alpha(y) = y + u, \quad u \in K, \quad A(u) = 0.$$

Since $A(Y)$ has p^n distinct roots, so $|G_{\mathcal{P}_\infty}^{(2)}| = p^n$. By Theorem 11.78, the proof is complete. □

THEOREM 12.8 *If equality holds in* (i) *of Theorem 12.7, and* $|G_{\mathcal{P}_\infty}^{(1)}| > (p-1)g/p$ *then*

$$B(X) = b_m X^m.$$

Proof. Let $\alpha \in H$. From the proof of Theorem 12.7,

$$\alpha(x) = cx + d, \qquad \alpha(y) = ay + Q(x),$$

with $a^{p^n} = a = c^m$ and $Q(X) \in K[X]$. Also, $Q(X)$ is linear for $m < p^n$ while its degree is smaller than m/p^n for $m > p^n$. Assume that α has order $m(p^n - 1)$. Then $\beta = \alpha^{p^n - 1}$ has order m. After replacing x by $x + u$ with a suitable $u \in K$,

$$\beta(x) = cx, \quad \beta(y) = y + Q(x), \quad \text{with} \quad c^m = 1, \quad Q(X) \in K[X]. \tag{12.10}$$

The key idea in the proof is to show that $Q(X)$ is the zero polynomial. To do this, it is first shown that β fixes every place arising from a branch centred on the Y-axis. As, by Theorem 11.78, $\Sigma^{G_{\mathcal{P}_\infty}^{(2)}}$ is rational, Lemma 11.14 (ii)(d) implies that β fixes two places of $\Sigma^{G_{\mathcal{P}_\infty}^{(2)}}$, one of them is the place under $\mathcal{P}_\infty$, the other lies under p^n distinct places of Σ, say $\mathcal{R}_1, \ldots, \mathcal{R}_{p^n}$. Since the K-automorphisms in $G_{\mathcal{P}_\infty}^{(2)}$ act as

$$(x, y) \mapsto (x, y + u), \quad \text{with } u \in K \text{ and } A(u) = 0,$$

they commute with β. Thus $G_{\mathcal{P}_\infty}^{(2)} \rtimes H$ is an abelian group. As p does not divide the order of β, this can only occur when β fixes each of the places $\mathcal{R}_1, \ldots, \mathcal{R}_{p^n}$. In particular, $Q(0) = 0$. By $\beta(x) = cx$, the centres of the corresponding branches lie on the Y-axis, and they are indeed all the places of Σ centred at an affine point on the Y-axis. Now, by Theorem 11.72 applied to the K-automorphism group generated by β,

$$2g - 2 = (2g_1 - 2)m + (p^n + 1)(m - 1),$$

where g_1 denotes the genus of Σ^β. This implies that $g_1 = 0$. Hence Σ^β is rational.

Let $\Sigma^\beta = K(\eta)$. Then $[\Sigma : K(\eta)] = m$ is prime to p, and hence $\Sigma = K(x, \eta)$. After replacing η by y in the proof of Theorem 12.7, the argument used there shows that $A(x) = B(\eta)$ with $A(X)$ and $B(Y)$ as in (12.1). The advantage is that the action of β on Σ has become as simple as possible; that is,

$$\beta(x) = cx, \qquad \beta(\eta) = \eta.$$

As β is a K-automorphism of Σ, so $B(cx) = B(x)$. Since $B(X)$ has degree m and c is an element of order m in the multiplicative group of K, this only happens when $B(X) = b_m X^m$. □

THEOREM 12.9 *Suppose that*

(a) *$\mathcal{C}$ is an irreducible plane curve of degree ≥ 4 as in (12.1) with*

$$A(Y) = Y^{p^n} + Y, \quad B(X) = X^m, \quad \gcd(p, m) = 1;$$

(b) *G is the K-automorphism group of $\mathcal{C}$;*

(c) *$\mathcal{P}_\infty$ is the place arising from the unique branch centred at the unique infinite point of $\mathcal{C}$;*

(d) *$G_{\mathcal{P}_\infty} = G^{(1)}_{\mathcal{P}_\infty} \rtimes H$ with $p \nmid |H|$.*

Then

(i) $|H| = m(p^n - 1)$;

(ii)
$$|G^{(1)}_{P_\infty}| = \begin{cases} p^n, & \text{if } m - 1 \text{ is not a power of } p^n, \\ p^{n(2k+1)}, & \text{if } m - 1 = p^{nk}. \end{cases}$$

Proof. For a primitive $m(p^n - 1)$-st root of unity, the birational transformation α of $\Sigma = K(\mathcal{C})$ defined by

$$\alpha(x) = cx, \qquad \alpha(y) = c^m y,$$

is a K-automorphism generating a cyclic group of order $m(p^n - 1)$. Theorem 12.7 (i) implies that $H = \langle \alpha \rangle$, showing (i).

By Theorem 12.7 (iii), part (ii) holds for $m \not\equiv 1 \pmod{p^n}$. So consider the case $m \equiv 1 \pmod{p^n}$ and $s = (m - 1)/p^n$. As in the proof of Theorem 12.7, if $\alpha \in G_{\mathcal{P}_\infty}$, then

$$\alpha(x) = x + d, \quad \alpha(y) = y + Q(x), \quad p^n \cdot \deg Q(X) < m. \tag{12.11}$$

Hence $\alpha \in G_{\mathcal{P}_\infty}$ if and only if $Q(x)^{p^n} + Q(x) = (x+d)^m - x^m$. So the polynomial equation

$$Q(X)^{p^n} + Q(X) = (X + d)^m - X^m \tag{12.12}$$

is to be solved. For $d = 0$, only the trivial solution exists; that is, $Q(X) = c$ with $c^{p^n} + c = 0$.

Now let $d \neq 0$. To find which coefficients of $(X+d)^m - X^m$ do not vanish, consider the p-adic representation of s:

$$s = s_i p^i + s_{i+1} p^{s+1} + \cdots, \quad s_i \neq 0.$$

Suppose first that s is not a power of p; then $s - p^n \geq \frac{1}{2}s$. Put $r = m - p^{i+n}$. The coefficients in the p-adic representation of r are smaller than or equal to those of m. Hence $\binom{m}{r}$ does not vanish, by Lemma A.6. Therefore X^r has no zero coefficient on the right-hand side in (12.12). So the same holds for the left-hand side. As $\gcd(r,p) = 1$, $Q(X)$ itself contains X^r which implies that $r \leq \deg Q(X)$. On the other hand, $\deg Q(X) \leq s$ by $p^n \cdot \deg Q(X) < m$. However, this leads to a contradiction, since

$$r = m - p^{i+n} = (s - p^i)p^m + 1 > \tfrac{1}{2}sp^n \geq s.$$

This shows that $d = 0$ provided that s is not a power of p.

It remains to investigate the case that $s = p^i$; that is, $m = p^{n+i} + 1$. This time,

$$(X+d)^m - X^m = dX^{p^{n+i}} + d^{p^n}X + d^{p^{n+i}}.$$

With

$$Q(X) = q_0 + q_1 X + \cdots + q_r X^r, \qquad q_r \neq 0,$$

it follows from (12.12) that

$$q_0^{p^n} + q_0 = d^{p^{i+n}+1}, \quad q_1 = d^{p^{n+i}+1}, \quad q_2 = \cdots = q_{p^n-1} = 0.$$

In fact, the non-constant terms in $Q(X)$ have degree of the form p^{jn}, as a direct argument shows. Therefore $m = p^{kn} + 1$.

From Theorem 12.7 (ii),

$$|G^{(1)}_{P_{infty}}| \leq p^n p^{2kn} = p^{(2k+1)n}.$$

To finish the proof, it suffices to exhibit $p^{(2k+1)n}$ K-automorphisms of type (12.11). To do this, choose elements $d, e \in K$ such that

$$d^{p^{2kn}} + (-1)^k d = 0, \quad e^{p^n} + e = d^{p^n+1}, \tag{12.13}$$

and let

$$\begin{aligned} \alpha(x) &= x + d, \\ \alpha(y) &= y + e + d^{p^{kn}}x - d^{p^{(k-1)n}} + \cdots + (-1)^{k-1} d^{p^{(2k-1)n}} x^{p^{(k-1)n}}. \end{aligned} \tag{12.14}$$

It is straightforward to verify that α is a K-automorphism of $\mathcal{C}$:

$$\begin{aligned} \alpha(y)^{p^n} + \alpha(y) &= y^{p^n} + y + e^{p^n} + e + d^{p^{kn}}x + (-1)^{k-1} d^{p^{2kn}} x^{p^{kn}} \\ &= x^{p^{kn}+1} + d^{p^{kn+1}} + d^{p^{kn}}x + dx^{kn} \\ &= (x+d)^{p^{kn+1}} = \alpha(x)^{p^{kn}+1}. \end{aligned}$$

For each of the p^{kn} possible values of d, there exist exactly p^n values of e that satisfy (12.13). This shows that (12.14) provides $p^{(2k+1)n}$ distinct K-automorphisms of $\mathcal{C}$. □

Let $\mathcal{C}$ again be the curve (12.1), and let $\Sigma = K(\mathcal{C})$. Also, define the following notation for Weierstrass semigroups:

$$\begin{aligned} H(\mathcal{P}) &= \text{the Weierstrass semigroup at } \mathcal{P};\\ H^0(\mathcal{P}) &= \text{the Weierstrass semigroup at } \mathcal{P} \text{ generated by } m \text{ and } p^n;\\ \mathcal{S}^0 &= \{\mathcal{P} \mid H(\mathcal{P}) = H^0(\mathcal{P})\}. \end{aligned}$$

Note that $\mathcal{P}_\infty \in \mathcal{S}^0$ by Lemma 12.3.

THEOREM 12.10 *Let Σ have a place $\mathcal{P} \neq \mathcal{P}_\infty$ with $\mathcal{P} \in \mathcal{S}^0$. Then, up to a linear substitution on X and Y, one of the following holds:*

(i) $B(X) = X^m$ *with* $m < p^n$, $p^n \equiv -1 \pmod{m}$, *and $\mathcal{S}^0$ consists of the p^n distinct zeros of x, together with $\mathcal{P}_\infty$;*

(ii) $A(Y) = Y^{p^n} + Y$, $B(X) = X^{p^n+1}$, *that is, $\mathcal{C}$ is the Hermitian curve over $\mathbf{F}_{p^{2n}}$, and $\mathcal{S}^0$ consists of the places centred at the $p^{3n}+1$ rational points of $\mathcal{C}$;*

(iii) $A(Y) = Y^{p^n} + a_j Y^{p^j} + \cdots + Y$, $B(X) = X^{p^n+1}$, *with $1 \le j \le n-1$ and $a_j \neq 0$, and $\mathcal{S}^0$ consists of the p^n distinct zeros of x, together with $\mathcal{P}_\infty$.*

Proof. Take the centre of the branch associated with $\mathcal{P}$ to the origin $P = (0,0)$ by a translation. Since $\mathcal{C}$ remains of type (12.1), all previous results hold true. Also, $\mathrm{ord}_{\mathcal{P}}\, x = 1$ because P is a simple point of $\mathcal{C}$ and the tangent to $\mathcal{C}$ is not the vertical line.

By Lemma 12.2, the canonical series of Σ is $|W|$, where

$$W = ((p^n - 1)(m-1) - 2)\mathcal{P}_\infty.$$

Hence

$$\mathrm{div}(x^i y^j) + W \succ 0 \quad \text{whenever} \quad ip^n + jm \le (p^n - 1)(m-1) - 2. \tag{12.15}$$

First, take the case $m > p^n + 1$. If $p^n > 2$, then (12.15) implies that

$$\mathrm{div}\, x^{p^n - 1} + W \succ 0.$$

This is also true for $p^n = 2$, since $\gcd(m, p) = 1$ together with $\deg C \ge 4$ implies that $m \ge 5$.

From the beginning of Section 7.6, the Weierstrass gaps are characterised by means of the canonical series as follows: a positive integer i is not in the Weierstrass semigroup $H(\mathcal{P})$ at $\mathcal{P}$ if and only if there exists $z \in \Sigma$ such that $\mathrm{div}(z) + W \succ 0$ and $\mathrm{ord}_{\mathcal{P}}\, z = i$. This characterisation is used several times in this proof. Hence p^n does not belong to $H(\mathcal{P})$, which gives a contradiction.

Assume that $m \le p^n - 1$ with $p^n > 4$. From (12.15),

$$\begin{aligned} \mathrm{div}\, x^i + W \succ 0 \quad &\text{for} \quad 0 \le i \le m-2,\\ \mathrm{div}\, y^j + W \succ 0 \quad &\text{for} \quad 0 \le j \le p^{n-1}. \end{aligned}$$

Since $A(y) = B(x)$,

$$z = a_n y^{p^n} - b_m x^m - b_{m-1} x^{m-1}$$

satisfies $\operatorname{div} z + W \succ 0$. If $b_{m-1} \neq 0$, then $m < p^n$ implies that $\operatorname{ord}_{\mathcal{P}} z = m - 1$, which is a contradiction by the above characterisation of Weierstrass gaps. Hence either $B(X) = b_m X^m$ or

$$B(X) = b_m X^m + b_{m-2} X^{m-2} + \cdots + b_k X^k \text{ with } 1 \leq k \leq m-2,\ b_k \neq 0.$$

The latter case cannot in fact occur. For $m = p^n - 1$, this is a consequence of the relation

$$\operatorname{ord}_{\mathcal{P}} z = m = p^n - 1$$

and the above characterisation of Weierstrass gaps. Similarly, for $m \leq p^n - 2$, there exists a rational function η such that $\operatorname{div} \eta + W \succ 0$ and $\operatorname{ord}_{\mathcal{P}} \eta = m - 1$. Letting

$$\eta = \begin{cases} y^{m-1}, & \text{if } k = 1, \\ yx^{m-k+1}, & \text{if } k \geq 2, \end{cases}$$

this again gives a contradiction to the characterisation of Weierstrass gaps. Thus $B(X) = b_m X^m$. In particular, $I(P, \mathcal{C} \cap \ell) = m$, where P is the centre of the branch associated to $\mathcal{P}$ and $\ell = \mathbf{v}(Y)$.

Now choose positive integers $s < m$ and t such that $p^n + s \equiv 0 \pmod{m}$; so $lm = p^n + s$. Note that

$$\operatorname{div}(xy^{-l}) = \mathcal{P}_2 + \cdots + \mathcal{P}_n + s\mathcal{P}_\infty - (p^n + s - 1)\mathcal{P},$$

where $\mathcal{P}_2, \ldots, \mathcal{P}_n$ are the zeros of x other than $\mathcal{P}$. Hence $p^n + s - 1 \in H(\mathcal{P})$ and this gives a contradiction for $s > 1$. If $s = 1$, then

$$\operatorname{ord}_{\mathcal{P}}(xy^{-t}) = -p^n, \quad \operatorname{ord}_{\mathcal{P}}(y^{-1}) = -m.$$

Thus $H(\mathcal{P})$ is generated by m and p^n. This result holds true for every zero of x, that is, for every place of Σ arising from a branch centred at a common point of $\mathcal{C}$ and the line $X = 0$. This depends on the fact that each translation $(x, y) \mapsto (x, y + c)$, for a root c of $A(Y)$, preserves $\mathcal{C}$.

The possibility of having another place with the same Weierstrass semigroup is ruled out by using the invariance of the intersection multiplicity under translations. If such a place exists and it corresponds to a branch centred at the point $P = (u, v)$, the translation $(x, y) \mapsto (x - u, y - v)$ takes P to the origin O and the image of $\mathcal{C}$ under any translation is a curve $\mathcal{C}'$ of the same type. Hence what has already been shown for $\mathcal{C}$ applies to $\mathcal{C}'$. Therefore, if $\ell' = \mathbf{v}(Y)$ and $\ell = \mathbf{v}(Y - v)$, then

$$I(O, \mathcal{C}' \cap \ell') = I(P, \mathcal{C} \cap \ell), \qquad I(O, \mathcal{C}' \cap \ell') = m;$$

so $I(P, \mathcal{C} \cap \ell) = m$. But a straightforward computation shows that this only occurs when $u = 0$.

Let $m \leq p^n - 1$ with $p^n \leq 4$. The case $p^n \leq 3$ cannot occur because $\deg \mathcal{C} \geq 4$. Since $\gcd(m, p) = 1$, if $p^n = 4$ then $m = 3$, and

$$A(Y) = a_2 Y^4 + a_1 Y^2 + a_0 Y, \quad B(X) = b_3 X^3 + b_2 X^2 + b_1 X.$$

By (12.15), $\operatorname{div} z + W \succ 0$, where $z = a_0 y + b_1 x$. On the other hand, a calculation shows that

$$\operatorname{ord}_{\mathcal{P}} z = \begin{cases} 2, & \text{if } a_1 a_0^{-1} b_1^2 + b_2 \neq 0, \\ 3, & \text{if } a_1 a_0^{-1} b_1^2 + b_2 = 0. \end{cases}$$

But again the characterisation of Weierstrass gaps leads to a contradiction since both $p^n = 4$ and $m = 3$ are in $H(\mathcal{P})$.

Let $m = p^n + 1$. Without loss of generality, let $a_0 = 1$. With $\eta = y + b_1 x$,

$$\begin{aligned} A(\eta) - B(x) &= A(y) + A(b_1 x) - B(x) &= A(b_1 x), \\ A(\eta) &= B(x) + A(b_1 x) &= b_m x^m + \cdots + b_2' x^2, \end{aligned}$$

with $b_2' = b_2$ for $p > 2$ and $b_2' = b_2 + a_1 b_1^2$ for $p = 2$. After the linear transformation $(x, y) \mapsto (x, a_0 y + b_1 x)$, it is possible to put $b_1 = 0$, or equivalently, to assume that

$$p^n + 1 \geq \mathrm{ord}_{\mathcal{P}}\, y \geq 2.$$

In fact, $\mathrm{ord}_{\mathcal{P}}\, y = p^n + 1$. To show this, suppose on the contrary that

$$\mathrm{ord}_{\mathcal{P}}\, y < p^n + 1.$$

If $\mathrm{ord}_{\mathcal{P}}\, y = p^n = m - 1$, there is a contradiction to the characterisation of Weierstrass gaps since $\mathrm{div}\, y + W \succ 0$ by (12.15).

If $\mathrm{ord}_{\mathcal{P}} y < p^n$, let

$$z = y x^{p^n - k - 1}, \quad \text{with } k = \mathrm{ord}_{\mathcal{P}}\, y.$$

As $\mathrm{ord}_{\mathcal{P}}\, z = p^n - 1$ and $\mathrm{div}\, z + W \succ 0$, once again there is a contradiction to the characterisation of Weierstrass gaps.

Therefore $B(X) = X^m$. In other words, the point $P = (0, 0)$ of $\mathcal{C}$ has the property that $I(P, \mathcal{C} \cap \ell) = m$ where ℓ is the tangent line to $\mathcal{C}$ at P.

By a preceding argument depending on the invariance of the intersection multiplicity under translations, this holds true for any point of $\mathcal{C}$ which is the centre of a branch corresponding to a place of Σ with Weierstrass semigroup $H^0(\mathcal{P})$

Now, if $A(y) = a_n y^{p^n} + a_0 y$, then $\mathcal{C}$ is the Hermitian curve over $\mathbf{F}_{p^{2n}}$. Otherwise, there exists j with $0 < j < n$ such that

$$A(y) = a_n y^{p^n} + a_j y^{p^j} \cdots + a_0 y, \qquad a_j \neq 0.$$

Since $\mathrm{ord}_{\mathcal{P}}(y^{-1}) = -m$, so $H(\mathcal{P}) = H^0(\mathcal{P})$. As before, this holds true for every zero of x, again by the fact that every translation $(X, Y) \mapsto (X, Y + c)$, for a root c of $A(Y)$, preserves $\mathcal{C}$.

It remains to show that no other place of Σ has Weierstrass semigroup $H^0(\mathcal{P})$. Again, if $P = (u, v)$ is the centre of a branch corresponding to such a place, then $I(P, \mathcal{C} \cap \ell) = m$, where ℓ is tangent line to $\mathcal{C}$ at P. But another calculation shows that this is not the case as soon as $u \neq 0$. □

THEOREM 12.11 *If $\mathcal{C}$ is a curve of type (12.1), then the K-automorphism group $\mathrm{Aut}_K(\Sigma)$ of $\mathcal{C}$ fixes the place $\mathcal{P}_\infty$ except in the following two cases.*

(i) (a) $\mathcal{C} = \mathbf{v}(Y^{p^n} + Y - X^m)$, *with* $m < p^n$, $p^n \equiv -1 \pmod{m}$;

(b) *the group* $\mathrm{Aut}_K(\Sigma)$ *contains a cyclic normal subgroup* C_m *of order* m *such that* $\mathrm{Aut}_K(\Sigma)/C_m \cong \mathrm{PGL}(2, p^n)$;

(c) C_m *fixes each of the* $p^n + 1$ *places with the same Weierstrass semigroup as* $\mathcal{P}_\infty$;

(d) $\mathrm{Aut}_K(\Sigma)/C_m$ *acts on the set of such* $p^n + 1$ *places as* $\mathrm{PGL}(2, p^n)$.

(ii) (a) $\mathcal{C} = \mathbf{v}(Y^{p^n} + Y - X^{p^n+1})$, *the Hermitian curve*;

(b) $\mathrm{Aut}_K(\Sigma) \cong \mathrm{PGU}(3, q)$;

(c) $\mathrm{Aut}_K(\Sigma)$ *acts on the set of all places with the same Weierstrass semigroup as* $\mathcal{P}_\infty$;

(d) $\mathrm{Aut}_K(\Sigma)$ *acts on the set of such places as* $\mathrm{PGU}(3, q)$ *on a Hermitian curve.*

Proof. Assume that $G = \mathrm{Aut}_K(\Sigma)$ does not fix $\mathcal{P}_\infty$. The invariance of the Weierstrass semigroup under the action of $\mathrm{Aut}_K(\Sigma)$ implies that the orbit of $\mathcal{P}_\infty$ is a subset of $\mathcal{S}^0$. From this, Theorem 12.10 shows that there are only a few cases to consider.

First, let

$$B(X) = X^m,\ m < p^n, \quad p^n \equiv -1 \pmod m), \quad b_0 = 0;$$
$$A(Y) = Y^{p^n} + a_j Y^{p^j} + \cdots + Y,\ 1 \le j \le n-1, \quad a_j \ne 0.$$

The K-automorphisms

$$(x, y) \mapsto (x, y + d), \quad A(d) = 0$$

of Σ form a group acting transitively on the set of all places which are zeros of x. If $G \ne G_{\mathcal{P}_\infty}$, then the orbit of $\mathcal{P}_\infty$ consists of all zeros of x together with $\mathcal{P}_\infty$, and one of them, say $\mathcal{P}_O$, corresponds to a branch centred at the origin O. The Weierstrass semigroup $H(\mathcal{P}_O)$ is generated by m and p^n. Also, $\dim |m\mathcal{P}_O| = 1$ and $\dim |p^n \mathcal{P}_O| = s$ where $s = (p^n + 1)/m$. Since the tangent to $\mathcal{C}$ at O is not the the line $\mathbf{v}(X)$, a primitive representation of the unique branch centred at O is $(x = t, y(t) = vt^i + \cdots)$. Since $A(y(t)) = B(x(t))$,

$$y(t) = a_O^{-1} t^m + \cdots ;$$

so $\mathrm{ord}_{\mathcal{P}_O}(y) = m$. From the hypothesis that $m < p^n$,

$$|m\mathcal{P}_O| = \{\mathrm{div}(c_0 + c_1 y^{-1}) + m\mathcal{P}_O \mid \mathbf{c} = (c_0, c_1) \in \mathrm{PG}(1, K)\},$$
$$|p^n\mathcal{P}_O| = \{\mathrm{div}(c_0 + c_1 y^{-1} + \cdots + c_{s-1} y^{-(s-1)} + c_s x y^{-s}) + m\mathcal{P}_O \mid \mathbf{c} = (c_0, \ldots, c_s) \in \mathrm{PG}(s, K)\},$$

Assume that $\alpha \in \mathrm{Aut}_K(\Sigma)$ takes $\mathcal{P}_\infty$ to $\mathcal{P}_O$. Then α takes $|m\mathcal{P}_\infty|$ to $|m\mathcal{P}|$, and $|p^n\mathcal{P}_\infty|$ to $|p^n\mathcal{P}|$. Hence

$$\alpha(x) = y^{-1}(cx + d_1 y + \cdots + d_s y^s), \quad \alpha(y) = y^{-1}(ay + b). \tag{12.16}$$

Here, $b \ne 0$, $c \ne 0$. Then

$$A(\alpha(y)) = A(a) + b^{p^n} y^{-p^n} + a_j b^{p^j} y^{-p^j} + \cdots + by^{-1},$$
$$A(\alpha(x)) = y^{-ms}(cx + d_1 y + \cdots + d_s y^s)^m.$$

On the other hand,

$$A(\alpha(y)) = \alpha(A(y)) = \alpha(B(x)) = B(\alpha(x))$$

implies that $y^{ms} A(\alpha(y)) = y^{p^n+1} B(\alpha(x))$. Hence

$$\begin{aligned} A(a) y^{p^n+1} + b^{p^n} y + a_j b^{p^j} y^{p^n - p^j + 1} + \cdots + b y^{p^n} &= c^m x^m + m c^{m-1} x^{m-1}(d_1 y + \cdots + d_s y^s) + \cdots \\ &= c^m A(y) + m c^{m-1} x^{m-1}(d_1 y + \cdots + d_s y^s) + \cdots . \end{aligned}$$

Since $P = (x, y)$ is a generic point of $\mathcal{C}$, the polynomial

$$H(X,Y) = A(a)Y^{p^n+1} + b^{p^n}Y + a_j b^{p^j} Y^{p^n-p^j+1} + \cdots + bY^{p^n} \qquad (12.17)$$
$$-(c^m A(Y) + mc^{m-1}X^{m-1}(d_1 y + \cdots + d_s Y^s) + \cdots)$$

is divisible by $A(Y) - B(X)$. Note that

$$\deg_X H(X,Y) < m, \quad \deg_X = A(Y) - B(X) = m.$$

Hence no term in X can actually occur in $H(X, Y)$. Therefore

$$A(a)Y^{p^n+1} + b^{p^n}Y + a_j b^{p^j} Y^{p^n-p^j+1} + \cdots + bY^{p^n} = c^m A(Y).$$

Since $A(Y)$ has only terms of type Y^{p^i}, either $b = 0$ or $a_j = 0$, a contradiction.

The next case is the following:

$$b(X) = X^m, \ m = p^n + 1, \ b_0 = 0,$$
$$A(Y) = Y^{p^n} + a_j Y^{p^j} + \cdots + Y, \ 1 \le i \le n-1, \ a_j \ne 0.$$

This can be ruled out in a similar way, since

$$|p^n \mathcal{P}_O| = \{\text{div}(c_0 + c_1 x y^{-1}) + p^n \mathcal{P}_O \mid \mathbf{c} = (c_0, c_1) \in \text{PG}(1, K)\},$$
$$|(p^n+1)\mathcal{P}_O| = \{\text{div}(c_0 + c_1 x y^{-1} + c_2 y^{-1}) + (p^n+1)\mathcal{P}_O \mid$$
$$\mathbf{c} = (c_0, c_1, c_2) \in \text{PG}(2, K)\}.$$

In the case

$$\mathcal{C} = \mathbf{v}(Y^{p^n} + Y - X^m), \quad p^n \equiv -1 \pmod{m},$$

the rational transformation

$$\alpha(x) = xy^{-s}, \quad \alpha(y) = y^{-1}, \qquad (12.18)$$

with $s = (p^n + 1)/m$, is a K-automorphism of Σ. In fact,

$$A(\alpha(y)) = \frac{1}{y^{p^n}} + \frac{1}{y} = \frac{y^{p^n} + y}{y^{p^n+1}} = \frac{x^m}{y^{sm}} = B(\alpha(x)).$$

This shows that G is larger than $G_{\mathcal{P}_\infty}$. Also, the orbit of $\mathcal{P}_\infty$ consists of all zeros of x together with $\mathcal{P}_\infty$. To give an explicit permutation representation for G, the zeros of x are identified with the elements of $\mathbf{F}_{p^n}$ in the following way. Choose an element $u \in K$ satisfying $u^{p^n-1} = -1$, and change (X, Y) to (X, uY). Then

$$\mathcal{C} = \mathbf{v}(Y^{p^n} - Y - X^m).$$

The places arising from the branch centred at the point $P_v = (0, v)$ of $\mathcal{C}$ and the Y-coordinates v of P_v are in one-to-one correspondence; this can be extended to $\mathcal{P}_\infty$ by associating the symbol ∞ to it. Note that $v^{p^n} = v$; that is, $v \in \mathbf{F}_{p^n}$.

To prove that G acts on $\mathbf{F}_{p^n} \cup \{\infty\}$ as $\text{PGL}(2, p^n)$ on $\text{PG}(1, p^n)$, let $\bar{G}$ be the permutation group induced by G on $\mathbf{F}_{p^n} \cup \{\infty\}$. By Lemma 12.1 (iii)(c), every permutation of $\mathbf{F}_{p^n} \cup \{\infty\}$ of type

$$v = v \mapsto \begin{cases} v + b & \text{for } v \in \mathbf{F}_{p^n}, \\ \infty & \text{for } v = \infty, \end{cases} \qquad (12.19)$$

with $b \in \mathbf{F}_{p^n}$ is in $\bar{G}$. Also, every permutation of $\mathbf{F}_{p^n} \cup \{\infty\}$ of type

$$v \mapsto \begin{cases} cv & \text{for } v \in \mathbf{F}_{p^n}, \\ \infty & \text{for } v = \infty, \end{cases} \tag{12.20}$$

with $c \in \mathbf{F}_{p^n} \backslash \{0\}$ is in $\bar{G}$. Further, α defined as in (12.18) acts on $\mathbf{F}_{p^n} \cup \{\infty\}$ as the permutation

$$v \mapsto \begin{cases} v^{-1} & \text{for } v \in \mathbf{F}_{p^n}, \\ \infty & \text{for } v = 0, \\ 0 & \text{for } v = \infty. \end{cases} \tag{12.21}$$

The group generated by the permutations (12.19), (12.20), (12.21) is $\mathrm{PGL}(2, p^n)$. In fact, $\bar{G} = \mathrm{PGL}(2, p^n)$. This is a consequence of Theorem A.17.

An elementary proof is also possible, using Theorem 12.7 in the following way. The K-automorphisms $(x, y) \mapsto (dx, y)$ with d ranging over the m-th roots of unity form a cyclic group C_m of order m. Also, the K-automorphisms in (12.20) form a cyclic subgroup C_{p^n-1}. These two groups generate an automorphism group of order $m(p^n - 1)$. On the other hand, from Theorem 12.7, $|G_\infty| \leq m(p^n - 1)p^n$, which implies that $|G_{\mathcal{P}_\infty, \mathcal{P}_O}| \leq m(p^n - 1)$. Therefore $|G_{\mathcal{P}_\infty, \mathcal{P}_O}| = m(p^n - 1)$. Since G acts 2-transitively on the orbit of $\mathcal{P}_\infty$, it follows that

$$|G| = m(p^n + 1)p^n(p^n - 1).$$

As every place in this orbit is fixed by C_m, this shows that $G/C_m \cong \mathrm{PGL}(2, p^n)$.

Finally, if

$$\mathcal{C} = \mathbf{v}(Y^{p^n} + Y - X^{p^n+1}),$$

then $\mathcal{C}$ is the Hermitian curve, and (ii) follows from Theorem 11.30. □

REMARK 12.12 Let $A(Y)$ be a separable additive polynomial of degree p^n, and $B(X)$, $C(X)$ polynomials with no common roots such that the rational function $R(X) = B(X)/C(X)$ is admissible; that is, no pole of $R(x)$ regarded as an element of the rational function field $K(x)$ has multiplicity divisible by p. If

$$R(X) = f_0(X) + \textstyle\sum_{i=1}^{r-1} f_i(X)(X - e_i)^{-1},$$

with $f_i \in K[X]$ for $0 \leq i \leq r - 1$, then $R(X)$ is admissible if and only if the degree d_i of each f_i is relatively prime to p. If, also, $d_i \geq 1$ for $0 \leq i \leq r - 1$, then $R(x)$ has exactly r distinct poles in $K(x)$. In this situation, the curve

$$\mathcal{C} = \mathbf{v}(C(X)A(Y) - B(X))$$

is irreducible and has genus

$$g = \tfrac{1}{2}(p^n - 1)(d_0 - 1 + \textstyle\sum_{i=1}^{r-1} (d_i + 1)).$$

Also, if γ is the p-rank of the function field $\Sigma = K(\mathcal{C})$, then

$$\gamma = (p^n - 1)(r - 1).$$

Therefore $g = \gamma$ if and only if $d_i = 1$ for $0 \leq i \leq r - 1$; that is, every pole of $R(x)$ is simple.

In the spirit of Theorem 12.5, the particular case

$$R(X) = aX + X^{-1},$$

where $q = p$, is characterised, up to an $\mathbf{F}_q$-birational equivalence, as an irreducible hyperelliptic curve $\mathcal{F}$ defined over $\mathbf{F}_q$ of genus $p - 1$ whose $\mathbf{F}_q$-automorphism group G has the following two properties:

(i) $G = \langle\omega\rangle \times D$, where ω is the hyperelliptic involution and D is a dihedral group of order $2p$;

(ii) the set of fixed places of D is exactly $\mathcal{F}(\mathbf{F}_q)$.

12.2 CURVES WITH SUZUKI AUTOMORPHISM GROUP

Let $p = 2$, $q_0 = 2^s$, with $s \geq 1$ and $q = 2q_0^2 = 2^{2s+1}$.

Let $\mathcal{F}$ be the irreducible plane curve defined over $\mathbf{F}_q$ with

$$\mathcal{F} = \mathbf{v}(X^{2q_0}(X^q + X) + (Y^q + Y)). \tag{12.22}$$

The *Deligne–Lusztig curve associated with the Suzuki group* or DLS curve, for short, is any algebraic curve birationally isomorphic to $\mathcal{F}$.

So the function field of a DLS curve is $K(\mathcal{F}) = K(x, y)$ with

$$x^{2q_0}(x^q + x) = y^q + y.$$

Note that $\mathcal{F}$ is birationally isomorphic to the curve

$$\mathbf{v}(X^{q_0}(X^q + X) + (Y^q + Y)).$$

To show this, let $\Sigma = K(x, y)$ with $x^{q_0}(x^q + x) = y^q + y$. If $z = x^{2q_0+1} + y^{2q_0}$, that is, $y^q = z^{q_0} + x^{q+q_0}$, then $\Sigma = K(x, z)$. Also,

$$z^{q_0} = x^{q+q_0} + y^q = x^{q_0+1} + y,$$

and hence $z^q = x^{q+2q_0} + y^{2q_0}$. Now, since

$$z^q + z = x^{q+2q_0} + y^{2q_0} + x^{2q_0+1} + y^{2q_0} = x^{2q_0}(x^q + x),$$

the assertion follows. So the definition here is equivalent to the usual definition of a DLS curve.

THEOREM 12.13 *If $\mathcal{F}$ is the curve (12.22), $P_\infty = (0, 0, 1)$ and $\mathcal{P}_\infty$ is the place associated to the branch centred at P_∞, then the following properties hold:*

(i) *P_∞ is a q_0-fold point, it is the unique infinite point of $\mathcal{F}$, and is the centre of just one branch of $\mathcal{F}$;*

(ii) $\mathrm{div}(dx) = (2q_0(q-1) - 2)P_\infty$;

(iii) *$\mathcal{F}$ has genus $g = q_0(q-1)$;*

(iv) *the set $\mathcal{F}(\mathbf{F}_q)$ of all $\mathbf{F}_q$-rational places of $\mathcal{F}$ has size $q^2 + 1$;*

(v) *the K-automorphism group G of $\mathcal{F}$ is $\mathbf{F}_q$-rational and is isomorphic to the Suzuki group $\mathrm{Sz}(q)$;*

(vi) *G acts on $\mathcal{F}(\mathbf{F}_q)$ as $\mathrm{Sz}(q)$ on the Tits ovoid in $\mathrm{PG}(3, q)$;*

(vii) $|G| = (q^2 + 1)q^2(q-1)$, *and* $|G_{\mathcal{P}_\infty}| = q^2(q-1)$;

(viii) $|G^{(1)}_{\mathcal{P}_\infty}| = q^2$, *and $G^{(2)}_{\mathcal{P}_\infty}$ is an elementary abelian group of order q;*

(ix) $G^{(2)}_{\mathcal{P}_\infty} = \cdots = G^{(2q_0-1)}_{\mathcal{P}_\infty} = G^{(2q_0)}_{\mathcal{P}_\infty}$ *and* $G^{(2q_0+1)}_{\mathcal{P}_\infty} = \{1\}$.

Proof. It is immediate that P_∞ is the only singular point of $\mathcal{F}$ and that P_∞ is a $2q_0$-fold point of $\mathcal{F}$. First, it is shown that just one branch of $\mathcal{F}$ is centred at P_∞. As a consequence, the irreducibility of $\mathcal{F}$ follows.

With $h = xy + x^{2q_0+2} + y^{2q_0}$, the function h has a zero at the place $\mathcal{O}$ of the field $\Sigma = K(\mathcal{F})$ arising from the unique branch γ of $\mathcal{F}$ centred at the origin $O = (1,0,0)$. Actually, h has no more zeros corresponding to branches centred at an affine point. Equivalently, the system

$$x^{2q_0}(x^q + x) - (y^q + y) = 0, \tag{12.23}$$
$$xy + x^{2q_0+2} + y^{2q_0} = 0 \tag{12.24}$$

has only one solution over K, namely $(x,y) = (0,0)$, as follows From these equations,

$$x^{q_0}y^{q_0} + x^{q+2q_0} + y^q = x^{q_0}y^{q_0} + x^{2q_0+1} + y = 0,$$

whence

$$y^2 = x^{2q_0}y^{2q_0} + x^{4q_0+2} = x^{2q_0}(xy + x^{2q_0+2}) + x^{4q_0+2} = yx^{2q_0+1}.$$

This shows that either $x = y = 0$, or $y = x^{2q_0+1}$. If $y = x^{2q_0+1}$, from (12.22) $x^q = x$. But then (12.24) gives $x = 0$, and again $x = y = 0$.

The utility of h depends on the fact that the rational transformation φ, given by

$$(x,y) \mapsto (y/h, x/h) \tag{12.25}$$

is a K-automorphism of Σ. From above, φ takes any branch centred at an affine point to a branch centred at an affine point, except for γ. Since φ has order 2 and hence coincides with its inverse, it takes γ to a branch centred at P_∞.

Let $(x'(t), y'(t))$ be a primitive representation of γ. Then

$$x'(t) = t,\ y'(t) = t^{2q_0+1} + t^{q+2q_0} + \cdots .$$

Therefore

$$x(t) = \frac{t^{2q_0+1} + \cdots}{t^{q+2q_0+1} + t^{2q+2q_0} + \cdots} = t^{-q}\frac{1}{1 + t^{q-1} + \cdots} = t^{-q}(1 + \cdots),$$
$$y(t) = \frac{t}{t^{q+2q_0+1} + t^{2q+2q_0} + \cdots} = t^{-q-2q_0}\frac{1}{1 + t^{q-1} + \cdots} = t^{-q-2q_0}(1 + ...).$$

The resulting branch has order $2q_0$. Since P_∞ is a $2q_0$-fold point, it is the unique branch of $\mathcal{F}$ centred at P_∞. So, in terms of places, φ interchanges $\mathcal{P}_\infty$ and $\mathcal{O}$. In particular,

$$\mathrm{ord}_{\mathcal{P}_\infty} x = -q,\ \mathrm{ord}_{\mathcal{P}_\infty} y = -(q + 2q_0),\ \mathrm{ord}_{\mathcal{P}_\infty} h = -(q + 2q_0 + 1).$$

To calculate $\mathrm{div}(dx)$, begin from the equation

$$x(t)^{2q_0}(x(t)^q + x(t)) = y(t)^q + y(t).$$

Taking derivatives,

$$\frac{dy(t)}{dt} = x(t)^{2q_0+1}\frac{dx(y)}{dt}.$$

Since $\mathrm{ord}_t(dy(t)/dt) = (-2q_0 - 2)$, this gives

$$\mathrm{ord}_t \frac{dx(t)}{dt} = 2q_0(q-1) - 2.$$

Therefore $\mathcal{P}_\infty$ contributes weight $2q_0(q-1) - 2$ to $\mathrm{div}(dx)$.

To prove (ii), it must be shown that $\mathrm{ord}_{\mathcal{P}}\, dx = 0$ for every place $\mathcal{P}$ whose corresponding branch is centred at an affine point. By Lemma 5.56, this only requires that no vertical line is tangent to $\mathcal{F}$ or, equivalently, that the polynomial $Y^q + Y + c$ has q distinct roots.

By Definition 5.55, (ii) implies (iii). Further, $\mathcal{F}(\mathbf{F}_q)$ comprises $\mathcal{P}_\infty$ and every place which arises from a branch centred at a point in $\mathrm{AG}(2,q)$. This proves (iv).

It can be shown that, for any $a, c, d \in \mathbf{F}_q$ with $d \neq 0$, the following transformations are K-automorphisms of $K(\mathcal{F})$:

$$\psi_{a,c}: \quad x \mapsto x + a, \quad y \mapsto a^{2q_0}x + y + c; \tag{12.26}$$

$$\gamma_d: \quad x \mapsto dx, \quad y \mapsto d^{2q_0+1}y. \tag{12.27}$$

Let

$$\Psi = \{\psi_{a,c} \mid a, c \in \mathbf{F}_q\}.$$

It is straightforward to verify that Ψ is a group of order q^2 which preserves $\mathcal{F}(\mathbf{F}_q)$. Also, Ψ fixes $\mathcal{P}_\infty$ and acts on the other places in $\mathcal{F}(\mathbf{F}_q)$ as a sharply transitive permutation group. The involutory elements in Ψ together with the identity form an elementary abelian subgroup Ψ' of order q. Also,

$$H = \{\gamma_d \mid d \in \mathbf{F}_q \backslash \{0\}\}$$

is a cyclic group of order $q-1$ which fixes $\mathcal{P}_\infty$ and preserves $\mathcal{F}(\mathbf{F}_q)$. The group generated by Ψ and H has order $q^2(q-1)$ and is the semidirect product $\Psi \rtimes H$. This group together with the K-automorphism φ of (12.25) generate a larger group G which acts on $\mathcal{F}(\mathbf{F}_q)$ as a 2-transitive permutation group. From Theorem A.17, $G \cong \mathrm{Sz}(q)$, and G acts on $\mathcal{C}(\mathbf{F}_q)$ as $\mathrm{Sz}(q)$ in its natural 2-transitive representation.

Note that G is $\mathbf{F}_q$-rational as it is generated by $\mathbf{F}_q$-rational K-automorphisms. Also, $|G| = (q^2+1)q^2(q-1)$. Hence $\Psi = G^{(1)}_{\mathcal{P}_\infty}$, and $G_{\mathcal{P}_\infty} = G^{(1)}_{\mathcal{P}_\infty} \rtimes H$.

Now the ramification groups at $\mathcal{P}_\infty$ can be determined by using Hilbert's Different Formula 11.37. As y/h is a uniformising element, it suffices to calculate

$$\mathrm{ord}_{\mathcal{P}_\infty}(u/h + \psi_{a,c}(y/h)).$$

By a straightforward calculation,

$$\begin{aligned} &u/h + \psi_{a,c}(y/h) \\ &= \frac{ay^2 + a^{2q_0}xh + bh + (a^{2q_0+1} + b)xy + (a^{2q_0+2} + ab + b^{2q_0})y}{h^2 + ahy + (a^{2q_0+1} + b)hx + (a^{2q_0+2} + ab + b^{2q_0})h}, \end{aligned}$$

which shows that $\mathrm{ord}_{\mathcal{P}_\infty}(u/h + \psi_{a,c}(y/h))$ is either 2 or $q_0 + 2$ according as $a \neq 0$ or $a = 0$. Therefore $\Psi' = G^{(2)}_{\mathcal{P}_\infty} = \cdots = G^{(2q_0+1)}_{\mathcal{P}_\infty}$. □

A group-theoretic characterisation of the DLS curve is established in the following theorem.

THEOREM 12.14 *With $q_0 = 2^s, q = 2q_0^2$, let $\mathcal{F}$ be an irreducible curve of genus $q_0(q-1)$. If a K-automorphism group of $\mathcal{F}$ contains a subgroup G isomorphic to $\mathrm{Sz}(q)$, then $\mathcal{F}$ is the DLS curve.*

The proof is organised in a series of lemmas and uses some results obtained in Section 11.12. The same notation is employed here, and $F_i = \Sigma^{G_{\mathcal{P}}^{(i)}}$. Since $|G| > 8g^3$, the relevant case is (iv.5). Here, G has one non-tame orbit o of places, and its faithful action on o is the natural 2-transitive permutation representation of $\mathrm{Sz}(q)$.

For $\mathcal{P} \in o$, the following properties hold:

(i) $|G_{\mathcal{P}}| = q^2(q-1)$;

(ii) $G_{\mathcal{P}}^{(1)}$ is a group of order q^2 acting on $o \setminus \{\mathcal{P}\}$ as a sharply transitive permutation group;

(iii) $G_{\mathcal{P}} = G_{\mathcal{P}}^{(1)} \rtimes H$, where $H \cong C_{q-1}$;

(iv) $F_2 = \Sigma^{G_{\mathcal{P}}^{(2)}}$ is rational.

LEMMA 12.15 *The group $G_{\mathcal{P}}^{(2)}$ has order q and consists of all involutory elements in $G_{\mathcal{P}}$ together with the identity. In particular, $[G_{\mathcal{P}}^{(1)} : G_{\mathcal{P}}^{(2)}] = q$.*

Proof. Each involutory element in $\mathrm{Sz}(q)$ is the square of an element of order 4 in $\mathrm{Sz}(q)$. Since the factor group $G_{\mathcal{P}}^{(1)}/U_{\mathcal{P}}^{(2)}$ is an elementary abelian 2-group, every involutory element in $G_{\mathcal{P}}^{(1)}$ belongs to $U_{\mathcal{P}}^{(2)}$. In particular, $|U_{\mathcal{P}}^{(2)}| \geq q$.

Choose any non-involutory element $\alpha \in G_{\mathcal{P}}^{(1)}$, and look at the structure of the 1-point stabiliser of $\mathrm{Sz}(q)$. Every non-involutory element in $G_{\mathcal{P}}^{(1)}$ can be written as $\beta\alpha\beta^{-1}\delta$ with $\beta \in H$ and $\delta^2 = 1$. Now assume on the contrary that α is in $U_{\mathcal{P}}^{(2)}$. Since H is contained in the normaliser $U_{\mathcal{P}}^{(2)}$, it follows that $\beta\alpha\beta^{-1}\delta$ is also in $U_{\mathcal{P}}^{(2)}$, contradicting that $G_{\mathcal{P}}^{(1)} \neq U_{\mathcal{P}}^{(2)}$. If $\rho_1 > 2$, that is, if $G_{\mathcal{P}}^{(1)} = G_{\mathcal{P}}^{(2)}$, then Theorem 11.72 applied to $G_{\mathcal{P}}^{(1)}$ would imply that

$$2q_0(q-1) - 2 = 2g - 2 \geq -2q^2 + 3(q^2-1) = q^2 - 1,$$

which is impossible since $q = 2q_0^2$. Therefore $\rho_1 = 2$. □

LEMMA 12.16 $\rho_2 = 2q_0 + 2$.

Proof. In the 1-point stabiliser of $\mathrm{Sz}(q)$, any two involutory elements are conjugate. Thus, if $i \geq 3$, then either $G_{\mathcal{P}}^{(i)} = G_{\mathcal{P}}^{(2)}$, or $G_{\mathcal{P}}^{(i)} = \{1\}$. By Theorem 11.72 applied to $G_{\mathcal{P}}^{(2)}$, it follows that

$$2q_0(q-1) - 2 = -2q + 2q - 2 + (\rho_2 - 2)(q-1),$$

whence $\rho_2 = 2q_0 + 2$. □

LEMMA 12.17 *The smallest non-gap at $\mathcal{P}$ is q.*

Proof. Let m denote the smallest non-gap at $\mathcal{P}$. Take an element $x \in \Sigma$ for which $\text{div}(x)_\infty = m\mathcal{P}$. The group $G_\mathcal{P}^{(1)}$ induces an automorphism group on $K(x)$. The kernel of such a permutation representation contains $G_\mathcal{P}^{(2)}$ as every non-trivial K-automorphism of $K(x)$ whose order is a power of p has order p. Hence $K(x) \subset F_2$. Since

$$m = [\Sigma : K(x)] = [\Sigma : F_2][F_2 : K(x)],$$

and $|G_\mathcal{P}^{(2)} = [\Sigma : F_2]$ by Theorem 11.36, it follows that $m \geq q$. On the other hand, every $G_\mathcal{P}^{(2)}$-orbit distinct to $\mathcal{P}$ has length q. Therefore $m \geq q$, whence $m = q$. □

This result together with Theorem 11.14 also shows the following lemma.

LEMMA 12.18 *There is an element $x \in \Sigma$ such that* $\text{div}(x)_\infty = q\mathcal{P},\ F_2 = K(x)$, *and, for every $\alpha \in G_\mathcal{P}$,*

$$\alpha(x) = \begin{cases} x + a,\ a \in K,\ a \neq 0, & \textit{for } \alpha \in G_\mathcal{P}^{(1)} \backslash G_\mathcal{P}^{(2)}, \\ bx,\ b^{q-1} = 1,\ b \neq 1, & \textit{for } \alpha \in H, \\ x, & \textit{for } \alpha \in G_\mathcal{P}^{(2)}. \end{cases} \tag{12.28}$$

LEMMA 12.19 *The smallest non-gap m at $\mathcal{P}$ which is not divisible by q does not exceed $q + 2q_0$.*

Proof. Since F_1 is rational, $F_1 = K(x_1)$ with $x_1 \in \Sigma$. After a change of x_1 in $K(x_1)$, let $\mathcal{P}$ be a pole of x_1. Apart from the trivial orbit $\{\mathcal{P}\}$, all orbits of places under $G_\mathcal{P}^{(1)}$ have maximum size q^2. Since $[\Sigma : F_1] = q^2$, so x_1 has at most q^2 distinct poles, each counted with multiplicity. Hence $\mathcal{P}$ is the unique pole of x_1. Since $G_\mathcal{P} = G_\mathcal{P}^{(1)} \rtimes H$, an automorphism $\delta \in H$ of order $q-1$ preserves $K(x_1)$. By Theorem 11.14, $\delta(x_1) = ax_1$ with $a \in K,\ a \neq 0$. Hence δ preserves the set of zeros of x_1. Since the order of δ is prime to q^2, a power of δ fixes at least one such zero, say $\mathcal{R}$. As F_1 is rational, and o is the only $G_\mathcal{P}^{(1)}$-orbit preserved by δ, it follows that $\mathcal{R} \in o$. Hence o consists of all zeros of x_1; that is,

$$\text{div}(x_1) = \textstyle\sum_{\mathcal{R} \in o \backslash \{\mathcal{P}\}} \mathcal{R} - q^2\mathcal{P}.$$

Now take an element $\beta \in G$ which moves $\mathcal{P}$, and put $\mathcal{Q} = \mathcal{P}^\beta$. Note that $\mathcal{Q} \in o$. Replacing $\mathcal{P}$ by $\mathcal{Q}$ in the preceding argument gives the equation,

$$\text{div}(y_1) = \textstyle\sum_{\mathcal{R} \in o \backslash \{\mathcal{Q}\}} \mathcal{R} - q^2\mathcal{Q},$$

for an element $y_1 \in \Sigma$. Let $z_1 = x_1/y_1$; then $\text{div}(z_1) = q^2\mathcal{Q} - q^2\mathcal{P}$.

Similarly, there exists $z_2 \in \Sigma$ such that

$$\text{div}(z_2) = (2g-2)\mathcal{Q} - (2g-2)\mathcal{P}.$$

To show this, note the following two properties of the differential dx_1:

(1) dx_1 can have at most one pole, namely $\mathcal{P}$;

(2) the weight of $\mathcal{P}$ in $\text{div}(dx_1)$ does not exceed q^2, by $\text{div}(x_1)_\infty = q^2\mathcal{P}$.

As a consequence, dx_1 has at most $2g-2+q^2 < 2q^2$ zeros. Since $x_1 \in F_1$ and $|G_{\mathcal{P}}^{(1)}| = n^2$ this implies that the zeros of dx_1 are q^2. Also,

$$(\mathrm{div}(dx_1))^{\delta} = \mathrm{div}(d(\delta(x_1))) = \mathrm{div}(d(ax_1)) = \mathrm{div}(dx_1);$$

so δ fixes $\mathrm{div}(dx_1)$. Now the above arguments can be used to prove that $o\backslash\{\mathcal{P}\}$ is the set of all zeros of dx_1 each with the same weight. Again, moving $\mathcal{P}$ to $\mathcal{Q}$ by an element $\beta \in G$, it can also be shown that there exists y_1 such that

$$\mathrm{div}(dx_1) - \mathrm{div}(dy_1) = (2g-2)\mathcal{Q} - (2g-2)\mathcal{P}.$$

Since $z_2 = dy_1 - dx_1 \in \Sigma$, it follows that

$$\mathrm{div}(z_2) = (2q_0(q-1)-2)\mathcal{Q} - (2q_0(q-1)-2)\mathcal{P}.$$

Since $q^2+1 = (q+2q_0+1)(q-2q_0+1)$,

$$\gcd(q^2+1, 2(q(q_0-1)-1)) = q+2q_0+1.$$

Hence

$$\mathrm{div}(z_3) = (q+2q_0+1)\mathcal{Q} - (q+2q_0+1)\mathcal{P}$$

for some element $z_3 \in \Sigma$.

Finally, if $\alpha \in G_{\mathcal{P}}^{(1)}\backslash G_{\mathcal{P}}^{(2)}$, then $\mathrm{div}(\alpha(z_3) - z_3)_\infty = (q+2q_0)\mathcal{P}$. This follows from Lemma 11.83 and the fact that $\rho_1 = 2$. □

Choose an element $z \in \Sigma$ such that $\mathrm{div}(z)_\infty = m\mathcal{P}$ with m as in Lemma 12.19.

LEMMA 12.20 *There exists a polynomial $f(X) \in K[X]$ with $\deg f = m$ such that $\Sigma = K(x, z)$, where $z^q + z = f(x)$.*

Proof. For any element $u \in F_2$, the equation $[\Sigma : K(u)] = [\Sigma : F_2][F_2 : K(u)]$ shows that if $\mathrm{div}(u)_\infty = k\mathcal{P}$ then $q \mid k$. Hence $G_{\mathcal{P}}^{(2)}$ must contain some element δ such that $\delta(z) \neq z$. Choose $\alpha \in G_{\mathcal{P}}^{(1)}\backslash G_{\mathcal{P}}^{(2)}$ such that $\alpha^2 = \delta$. Then $\alpha(z) \neq z$, and the minimality of m implies that $\mathrm{ord}_{\mathcal{P}}(\alpha(z) - z) > -m$. As $2q > q + 2q_0$, this is only possible when $\mathrm{ord}_{\mathcal{P}}(\alpha(z) - z) = -q$. Therefore

$$\alpha(z) = z + cx + d \quad \text{with} \quad c, d \in K,\ c \neq 0.$$

Hence

$$\delta(z) = z + e_\delta \quad \text{with} \quad e_\delta \in K\backslash\{0\}. \tag{12.29}$$

Replacing z with $e_\delta^{-1}z$ ensures that $\delta(z) = z+1$.

Now take a generator β of H. From Lemma 11.17, either $\beta(z) = \epsilon^{-m}z$ or $\mathrm{div}(\beta(z) - \epsilon^{-m}z)_\infty < m\mathcal{P}$, where ϵ denotes a primitive $(q-1)$-st root of unity. As before, this implies that $\beta(z) = \epsilon^{-m}z + dx + e$ with $d, e \in K$. Replacing z by $z + d'$, where $d' = e/(c-1)$, gives $\beta(z) = \epsilon^{-m}z + dx$, and replacing β^2 by β, this becomes

$$\beta(z) = \epsilon^{-m}z. \tag{12.30}$$

From this, e_γ can be calculated for all non-trivial elements $\gamma \in G_{\mathcal{P}}^{(2)}$ using two facts:

(1) $\beta^{-k}\delta\beta^k(z) = z + \epsilon^{-mk}e_\delta$;

(2) any non-trivial element in $G_{\mathcal{P}}^{(2)}$ is conjugate to δ by an element in H.

So $e_\gamma = \epsilon^{-mk}e_\delta$, showing that $e_\gamma \neq 0$. If two of these values, say e_γ and $e_{g'}$, were equal, then $\delta' = \gamma\gamma'$ would fix x; that is, $e_{\delta'}$ would be zero. Since $\delta' \in G_{\mathcal{P}}^{(2)}$, this is impossible. As a consequence, ϵ^{-m} is a $(q-1)$-st of unity; that is, all non-zero elements of a subfield $\mathbf{F}_q$ of K are obtained. Also, as each $\alpha \in G_{\mathcal{P}}^{(1)}\backslash G_{\mathcal{P}}^{(2)}$ is a product of elements from H and $G_{\mathcal{P}}^{(1)}$, the coefficients of the equation of α are all in $\mathbf{F}_q$. Therefore $G_{\mathcal{P}}$ is defined over $\mathbf{F}_q$. Now, for every $\gamma \in G_{\mathcal{P}}^{(2)}$,

$$\gamma(z^q + z) = (z + e_\gamma)^q + z + e_\gamma = z^q + z.$$

Since $K(x) = F_2$, this implies that $z^q + z \in K(x)$. Hence $z^q + z = f(x)$ with $f(X) \in K[X]$ and $\deg f = m$, since both z and x have only one pole, namely $\mathcal{P}$, and $\mathrm{ord}_{\mathcal{P}}(z)_\infty = m$, $\mathrm{ord}_{\mathcal{P}}(x)_\infty = q$. □

Note that m is even by Lemma 11.83.

LEMMA 12.21 $f(X) = X^{q+q_0} + X^{q_0+1}$.

Proof. By (12.30) and $m < 2q$, applying β to $z^q + z + f(x) = 0$ gives

$$f(X) = uX^m + vX^{m+1-q}, \quad u \neq 0.$$

Note that $v \neq 0$, because an equation of type $z^q + z = x^m$ can be written as $z = (z^{q/2} + \sqrt{u}x^{m/2})^2$, showing that

$$\mathrm{div}(z^{m/2} + \sqrt{u}y^{m/2})_\infty = \tfrac{1}{2}m\mathcal{P},$$

in contradiction to the definition of m. Now take $\alpha \in G_{\mathcal{P}}^{(1)}\backslash G_{\mathcal{P}}^{(2)}$. Then

$$\alpha(x) = x + a,\ \alpha(z) = z + bx + c, \quad a, b, c \in \mathbf{F}_q,\ ab \neq 0.$$

Applying α to $z^q + z + ux^m + vx^{m+1-q}$ gives $\binom{m}{q} \neq 0$ but $\binom{m}{q+i} = 0$ for $1 \leq i \leq m - q$. Hence $m - q$ is a power of 2. Taking Lemma 12.19 into account, it follows that

$$f(X) = X^{2^k}(rX^q + sX), \quad 2^k \leq q_0,\ r, s \in K\backslash\{0\}.$$

Also, $ra^q + sa = 0$ for every $a \in \mathbf{F}_q$. Hence $r = s$. Further, $ca^{2^k} = b^q$, whence $r, s \in \mathbf{F}_q$. Replacing z by rz gives the equation

$$f(X) = X^{2^k}(X^q + X), \quad \text{with } 2^k \leq q_0.$$

Applying $\beta \in H$ to $z^q + z + f(X)$, where $\beta(x) = \eta x$, $\beta(z) = \epsilon z$ with ϵ, η primitive $(q-1)$-st roots of unity, gives $\epsilon = \eta^{2^k+1}$.

For brevity, put $q_1 = 2^k$, $q_2 = q/q_1$, and $q_3 = q_2/q_1 = q/q_1^2$. Note that each of these numbers q_1, q_2, q_3 is an integer. Let $v = z^{q_2} + x^{q_2+1}$; then

$$\begin{aligned} v^q + v &= z^{qq_2} + x^{(q_2+1)q} + z^{q_2} + x^{q_2+1} \\ &= (z^q + z)^{q_2} + x^{(q_2+1)} \\ &= [x^{q_1}(x^q + x)]^{q_2} + x^{(q_2+1)q} + x^{q_2+1} \\ &= x^q(x^{qq_2} + x^{q_2}) + x^{qq_2+q} + x^{q_2+1} \\ &= x^{q+q_2} + x^{q_2+1}, \end{aligned}$$

and $v^{q_1} = z^q + x^{q+q_1} = z + y^{q_1+1}$. Also, in succession,

$$\begin{aligned}(z^q + z)x^{q_1(q_3-1)} &= (x^q + x)x^{q_1}x^{q_1(q_3-1)},\\ z^q x^{q_1(q_3-1)} + zx^{q_1(q_3-1)} &= x^{q+q_1q_3} + x^{q_1q_3+1},\\ (z^q x^{(q_3-1)} + x^{q_2+q_3})^{q_1} &= zx^{q_1(q_3-1)} + x^{q_1q_3+1},\\ (z^{q_2}x^{(q_3-1)} + v^{q_2})^{q_1} &= zx^{q_1(q_3-1)} + v.\end{aligned}$$

Let $w = z^{q_2}x^{(q_3-1)} + v^{q_2}$. Since

$$\begin{aligned}&\text{ord}_{\mathcal{P}}\, z = -m = -(b+q_1), \quad \text{ord}_{\mathcal{P}}\, x = -q, \quad \text{ord}_{\mathcal{P}}\, v \le -q(q_2-1),\\ &w^{q_1} = zx^{q_1(q_3-1)} + v^{q_2},\end{aligned}$$

it follows that

$$\text{ord}_{\mathcal{P}}\, w = {q_2}^{-1}(-q - q_2 - q_1q_3q - q_1q) = -(1 + q_1 + q(q_3-1)).$$

Also, if $\delta \in G_{\mathcal{P}}^{(2)}$, then $\delta(w) = w + e_\delta^{q_2}x^{q_3-1} + e_\delta^{q_2^2}$. From Lemmas 12.16 and 11.83,

$$2 + 2q_0 = 1 + 1 + +q_2 + q(q_3-1) - q(q_3-1) = 2 + q_2,$$

whence $q_0 = 2q_2$. Also, $q_1 = q_0$, which gives the result. □

THEOREM 12.22 *If $\mathcal{F}' = \mathcal{F}/G$ is a tame quotient curve of the DLS curve, then it is one of the following types, where g' is the genus of $\mathcal{F}'$ and $r = |G|$:*

(i) *with $r \mid (q-1)$ and $g' = q_0(q-1)/r$,*
$$\mathcal{F}' = \mathbf{v}(Y^{(q-1)/r}f(X) - (1 + X^{q_0})(X^{q-1} + Y^{2(q-1)/r}));$$

(ii) *with $r \mid (q + 2q_0 + 1)$ and $g' = (q + 2q_0 + 1)(q_0 - 1)/r + 1$,*
$$\mathcal{F}' = \mathbf{v}(Y^{(q+2q_0+1)/r}g(X) - X^{q+2q_0+1} + Y^{2(q+2q_0+1)/r});$$

(iii) *with $r \mid (q - 2q_0 + 1)$ and $g' = (q - 2q_0 + 1)(q_0 + 1)/r - 1$,*
$$\begin{aligned}\mathcal{F}' = \mathbf{v}(&bY^{(q-2q_0+1)/r}h(X)\\ &-(X^{q_0-1} + X^{2q_0-1})(X^{q-2q_0+1} + Y^{2(q-2q_0+1)/r})).\end{aligned}$$

Here

$$\begin{aligned}f(X) &= 1 + \textstyle\sum_{i=0}^{s-1} X^{2^i(2q_0+1)-(q_0+1)}(1+X)^{2^i},\\ g(X) &= 1 + \textstyle\sum_{i=0}^{s-1} X^{2^i q_0}(1+X)^{2^i(q_0+1)-q_0} + X^{q/2},\\ h(X) &= 1 + \textstyle\sum_{i=0}^{s-1} X^{2^i(2q_0+1)-(q_0+1)}(1+X)^{2^i},\end{aligned}$$

and $b = \lambda^{q_0} + \lambda^{q_0-1} + \lambda^{-q_0} + \lambda^{-(q_0-1)}$ with $\lambda \in \mathbf{F}_{q^4}$ of order $q - 2q_0 + 1$.

Since $\text{Sz}(q)$ contains a large number of non-isomorphic subgroups of even order, especially 2-subgroups, a complete classification for non-tame quotient curves is out of reach.

THEOREM 12.23 *There exist non-tame quotient curves $\mathcal{F}' = \mathcal{F}/G$ of the DLS curve $\mathcal{F}$ given by the following examples, where g' is the genus of $\mathcal{F}'$ and $r = |G|$.*

(i) *With* $r = q_2$ *and* $g' = q_0(q/q_2 - 1)$,

$$\mathcal{F}' = \mathbf{v}(\textstyle\sum_{i=0}^{q-1} Y^{q_2^i} - X^{2q_0}(X^q + X)),$$

where $\mathbf{F}_{q_2}$ *is a proper subfield of* $\mathbf{F}_q$. *The* K*-automorphisms* $\psi_{0,c}$ *defined in (12.26) form an elementary abelian group* G *of order* $q_2 = 2^{s'+1}$ *with* $(2s'+1) \mid (2s+1)$.

(ii) *With* $r = 4$ *and* $g' = \frac{1}{4}q_0(q-2)$,

$$\mathcal{F}' = \mathbf{v}(\textstyle\sum_{i=0}^{2s} Y^{2^i} - (f_1(X) + f_2(X) + f_3(X))),$$

where

$$f_1(X) = \textstyle\sum_{i=0}^{s} \left(\sum_{j=i}^{s} X^{2^j}\right) X^{2^i},$$
$$f_2(X) = \textstyle\sum_{i=0}^{2s} X^{2^i},$$
$$f_3(X) = \textstyle\sum_{i=s+2}^{2s} \left(\sum_{j=0}^{i-s-2} X^{2^j}\right)^{2q_0} X^{2^i}.$$

Here $G \cong \mathbf{Z}_4$.

12.3 CURVES WITH UNITARY AUTOMORPHISM GROUP

Let $q = p^h > 2$. In this section, $\mathcal{H}_q$ is the irreducible plane curve defined over $\mathbf{F}_{q^2}$ with

$$\mathcal{H}_q = \mathbf{v}(Y^q + Y - X^{q+1}). \tag{12.31}$$

A *Hermitian curve* defined over $\mathbf{F}_{q^2}$ is any algebraic curve birationally isomorphic to $\mathcal{H}_q$. In projective coordinates, equivalent forms of (12.31) are the $\mathbf{v}(F_i)$, for $i = 1, 2, 3, 4$, with the following F_i:

(M1) $F_1 = X_0^{q+1} + X_1^{q+1} + X_2^{q+1}$;

(M2) $F_2 = X_2^q X_0 - X_2 X_0^q + \omega X_1^{q+1}$, where $\omega^{q-1} = -1$;

(M3) $F_3 = X_1 X_2^q - X_1^q X_2 + \omega X_0^{q+1}$, where $\omega^{q-1} = -1$;

(M4) $F_4 = X_0^q X_1 + X_1^q X_2 + X_2^q X_0$.

Each of the first three is obtained from (12.31) by a linear substitution defined over $\mathbf{F}_{q^2}$, but for (M4) this can be done only over $\mathbf{F}_{q^3}$; see Remark 8.19. The function field of $\mathcal{H}_q$ is $K(\mathcal{H}_q) = K(x, y)$ with $y^q + y + x^{q+1} = 0$.

The relevant properties of $\mathcal{H}_q$ are collected in the following theorem.

THEOREM 12.24 *The curve* $\mathcal{H}_q$ *is non-singular, with* $\mathcal{P}_\infty$ *the place associated to the branch centred at* $P_\infty = (0, 0, 1)$, *and has the following properties*:

(i) $\mathrm{div}(dx) = (q+1)(q-2)\mathcal{P}_\infty$;

(ii) $\mathcal{H}_q$ *has genus* $g = \frac{1}{2}q(q-1)$;

(iii) *the set* $\mathcal{F}(\mathbf{F}_{q^2})$ *of all* $\mathbf{F}_{q^2}$*-rational places of* $\mathcal{F}$ *has size* q^3+1;

(iv) *the* K*-automorphism group* $\mathrm{Aut}_K(\Sigma)$ *of* $\mathcal{H}_q$ *is* $\mathbf{F}_{q^2}$*-rational and isomorphic to* $\mathrm{PGU}(3,q)$, *acting projectively on* $\mathcal{H}_q(\mathbf{F}_{q^2})$.

Proof. The curve $\mathcal{H}_q$ is of type (12.1) with $A(Y) = Y^q + Y$ and with the degree $q+1$ of $B(X)$ not divisible by p. So the results obtained in Section 12.1 apply. Parts (i) and (ii) follow from Lemma 12.1, while (iii) is Theorem 12.10 (ii), and (iv) is Proposition 11.30. □

THEOREM 12.25 *For* $q = p^h > 2$, *let* $\mathcal{F}$ *be an irreducible algebraic curve of genus* $\frac{1}{2}q(q-1)$. *If a* K*-automorphism group of* $\mathcal{F}$ *contains a subgroup* G *isomorphic to* $\mathrm{PSU}(3,q)$, *then* $\mathcal{F}$ *is the Hermitian curve.*

The proof is similar to that of Theorem 12.14. Since $|G| > 8g^3$, case (iv.5) of Section 11.12 occurs. From Theorems 11.123 and 11.125, G has just one non-tame orbit o of places, and its action on o is faithful and 2-transitive, the same as $\mathrm{PSU}(3,q)$ on its permutation representation of the classical unital. For a place $\mathcal{P} \in o$, the following properties hold:

(i) $|G_{\mathcal{P}}| = q^3(q-1)$;

(ii) $G_{\mathcal{P}}^{(1)}$ is a group of order q^3 acting on $o\backslash\{\mathcal{P}\}$ as a sharply transitive permutation group;

(iii) $G_{\mathcal{P}} = G_{\mathcal{P}}^{(1)} \rtimes H$, where $H \cong C_{q-1}$;

(iv) $F_2 = \Sigma^{G_{\mathcal{P}}^{(2)}}$ is rational by Theorem 11.78.

The following lemmas can be proved using the same arguments as in the proofs of the analogous Lemmas 12.15 and 12.17. The necessary changes are as follows:

$$\begin{aligned} \text{involutory element} &\longmapsto \text{element of order } p;\\ n^2 &\longmapsto q^3;\\ n &\longmapsto q^2;\\ 2g-2 = 2n_0(n-1)-2 &\longmapsto 2g-2 = q(q-1)-2. \end{aligned}$$

LEMMA 12.26 $G_{\mathcal{P}}^{(2)}$ *has order* q *and consists of all elements of order* p *in* $G_{\mathcal{P}}$ *together with the identity.*

LEMMA 12.27 *The smallest non-gap at* $\mathcal{P}$ *is* q.

Theorem 12.25 now follows from these two lemmas and Theorems 12.4 and 12.11.

Since $\mathrm{PGU}(3,q)$ is particularly rich in subgroups, the corresponding quotient curves of $\mathcal{H}_q$ provide a large family of $\mathbf{F}_{q^2}$-maximal curves. However, their explicit determination appears to be difficult, especially for non-tame subgroups. For the simplest case, where G has prime order, the following result is a complete description of the quotient curves.

THEOREM 12.28 *Let $\mathcal{H}_q/G$ be the quotient curve of $\mathcal{H}_q$ with respect to a subgroup G of* PGU(3, q) *of prime order d.*

(i) *One of the following holds:*

  (a) $d = 2 \neq p$;

  (b) $d = p$;

  (c) $d \geq 3$ *and* $(q^2-1)(q^2-q+1) \equiv 0 \pmod{d}$.

(ii) *The curve $\mathcal{F}' = \mathcal{H}_q/G$ is birationally $\mathbf{F}_{q^2}$-equivalent to one the following cases of irreducible plane curves of genus g' defined over $\mathbf{F}_{q^2}$.*

  (I) $\mathcal{F}' = \mathbf{v}(Y^q + Y - X^{(q+1)/2})$,
  with $d = 2,\ p > 2,\ g' = \frac{1}{4}(q-1)^2$.

  (II) (a) $\mathcal{F}' = \mathbf{v}(\sum_{i=1}^{h} Y^{q/p^i} + \omega X^{q+1})$,
  with $d = p,\ q = p^h,\ g' = \frac{1}{2}q(q/p-1)$, *where* $\omega^{q-1} = -1$;

  (b) $\mathcal{F}' = \mathbf{v}(Y^q + Y - (\sum_{i=1}^{h} X^{q/p^i})^2)$,
  with $d = p,\ q = p^h,\ p \geq 3,\ g' = q(q-1)/(2p)$.

  (III) $\mathcal{F}' = \mathbf{v}(Y^q - YX^{2(q-1)/d} + \omega X^{(q-1)/d})$,
  with $d \geq 3,\ q \equiv 1 \pmod{d},\ g' = q(q-1)/(2d)$, *where* $\omega^{q-1} = -1$.

  (IV) (a) $\mathcal{F}' = \mathbf{v}(Y^q + Y - X^{(q+1)/d})$,
  with $d \geq 3,\ q \equiv -1 \pmod{d},\ g' = \frac{1}{2}(q-1)((q+1)/d-1)$;

  (b) $\mathcal{F}'_i = \mathbf{v}(X^{i(q+1)/d} + X^{(i+1)(q+1)/d} + Y^{q+1})$,
  with $d \geq 3,\ q \equiv -1 \pmod{d},\ g' = (q+1)(q-2)/(2d) + 1$,
  $1 \leq i \leq d-1$.

  (V) $\mathcal{F}'_i = \mathbf{v}(S(X^{q/d}, Y^{1/d}, X^{1/d}y^{q/d}))$,
  where
  $S(X,Y,Z) = \prod(\beta X + \beta^q Y + Z)$,
  with β ranging over all d-th roots of unity, $d \geq 3,\ (q^2-q+1) \equiv 0 \pmod{d},\ g' = \frac{1}{2}((q^2-q+1)/d-1)$.

(iii) *In* (IV) (b) *two such curves, $\mathcal{F}'_i$ and $\mathcal{F}'_j$, are birationally $\mathbf{F}_{q^2}$-equivalent if and only if one of the following relations holds modulo d* :

$$\begin{array}{lll} i \equiv j, & ij \equiv 1, & ij+i+j \equiv 0, \\ i+j+1 \equiv 0, & ij+i+1 \equiv 0, & ij+j+1 \equiv 0. \end{array} \tag{12.32}$$

The number of birationally $\mathbf{F}_{q^2}$-equivalent classes of curves $\mathcal{F}_i$ is

$$n(d) = \begin{cases} \frac{1}{6}(d+1) & \text{for } d \equiv 2 \pmod{3}, \\ \frac{1}{6}(d-1)+1 & \text{for } d \equiv 1 \pmod{3}. \end{cases} \tag{12.33}$$

12.4 CURVES WITH REE AUTOMORPHISM GROUP

In this section, $p = 3$, $q = 3q_0^2$, with $q_0 = 3^s$, $s \geq 0$.

Let $\mathcal{F} = \mathbf{v}(F)$ be the irreducible plane curve defined over $\mathbf{F}_q$ with

$$F(X,Y) = Y^{q^2} - [1+(X^q - X)^{q-1}]Y^q + (X^q - X)^{q-1}Y - X^q(X^q - X)^{q+3q_0}. \tag{12.34}$$

The *Deligne–Lusztig curve associated with the Ree group* or DLR curve, for short, is any algebraic curve birationally isomorphic to $\mathcal{F}$. The function field of $\mathcal{F}$ is $\Sigma = K(x,y)$ with $F(x,y) = 0$.

To obtain a simpler description of Σ, let

$$y_1 = \left(\frac{y^q - y}{x^q - x}\right)^{q_0} - x^{q_0}(x^q - x), \tag{12.35}$$

$$y_2 = x^{q_0}\left(\frac{y^q - y}{x^q - x}\right)^{q_0} - x^{2q_0}(x^q - x) - y^{q_0}. \tag{12.36}$$

Hence

$$y_1^q - y_1 = x^{q_0}(x^q - x), \tag{12.37}$$

$$y_2^q - y_2 = x^{q_0}(y_1^q - y_1). \tag{12.38}$$

Also, $y^q = x^q y_1^{3q_0} - y_2^{3q_0}$. Therefore $\Sigma = K(x, y_1, y_2)$. The relevant properties of $\mathcal{F}$ are collected in the following two theorems.

THEOREM 12.29 *If $\mathcal{F}$ is the curve (12.34), $P_\infty = (0,0,1)$, and $\mathcal{P}_\infty$ is the place associated to the branch centred at P_∞, then the following properties hold:*

(i) *P_∞ is a q_0-fold point, it is the unique infinite point of $\mathcal{F}$, and is the centre of just one branch of $\mathcal{F}$;*

(ii) $\mathrm{div}(dx) = (3q_0(q-1)(q+q_0+1) - 2)\mathcal{P}_\infty$;

(iii) *$\mathcal{F}$ has genus $g = \frac{3}{2}q_0(q-1)(q+q_0+1)$;*

(iv) *the set $\mathcal{F}(\mathbf{F}_q)$ of all $\mathbf{F}_q$-rational places of $\mathcal{F}$ has size q^3+1;*

(v) *the K-automorphism group G of $\mathcal{F}$ is $\mathbf{F}_q$-rational and is isomorphic to the Ree group* $\mathrm{Ree}(q)$;

(vi) *G acts on $\mathcal{F}(\mathbf{F}_q)$ as* $\mathrm{Ree}(q)$ *in its unique 2-transitive permutation representation on the Kantor ovoid;*

(vii) $|G| = (q^3+1)q^3(q-1)$ *and* $|G_{\mathcal{P}_\infty}| = q^3(q-1)$.

Proof. (ii) The rational subfield $K(x)$ of $\Sigma = K(\mathcal{F})$ is regarded as the function field of the line $\ell = \mathbf{v}(X)$. Then $[\Sigma : K(x)] = q^2$. Let $\mathcal{P}'_\infty$ be the unique place of $K(x)$ centred at the infinite point of ℓ. By Lemma 7.20, some places of Σ lie over $\mathcal{P}'$. If $\mathcal{P}_\infty$ is one of them, by (7.1), $e_\infty = -e_\mathcal{P}$, where e_∞ denotes the ramification index of $\mathcal{P}$. From (7.3), $e_\infty \leq q^2$. On the other hand, from Lemma 12.30 (ii) (I),

$$\mathrm{ord}_{\mathcal{P}_\infty}(w_8) = -(1 + {q_0}^{-1} + 2q^{-1} + (q_0 q)^{-1} + q^{-2})e_\infty.$$

Since $\mathrm{ord}_{\mathcal{P}_\infty}(w_8)$ is an integer, so $e_\infty = q^2$. This together with (7.3) implies that $\mathcal{P}_\infty$ is the unique place of Σ over $\mathcal{P}'_\infty$. Also,

$$\mathrm{ord}_{\mathcal{P}_\infty} x = -q^2, \ \mathrm{ord}_{\mathcal{P}_\infty} y_1 = -q(q+q_0), \ \mathrm{ord}_{\mathcal{P}_\infty} y_2 = -q(q+2q_0). \tag{12.39}$$

Now let $\mathcal{P}$ be any place of Σ other than $\mathcal{P}_\infty$. Denote by $\mathcal{P}'$ the place of $K(x)$ lying under $\mathcal{P}$. Let $P' = (u, 0)$ be the centre of $\mathcal{P}'$. Choose a primitive representation σ of $\mathcal{P}$. For every $a, b \in \mathbf{F}_q$, let $\sigma_{(a,b)}$ be the K-isomorphism from Σ into $K((t))$ defined by the relations

$$\sigma_{(a,b)}(x) = \sigma(x), \quad \sigma_{(a,b)}(y_1) = \sigma(y_1) + a, \quad \sigma_{(a,b)}(y_2) = \sigma(y_2) + b.$$

Then $\sigma_{(a,b)}$ is also a primitive place representation. By (12.37), (12.38), the corresponding place $\mathcal{P}_{(a,b)}$ of Σ lies over $\mathcal{P}'$. Note that $\mathcal{P} = \mathcal{P}_{(0,0)}$. Since $[\Sigma : \mathbf{F}_q] = q^2$, this shows that there are q^2 places of Σ over every place of $K(x)$. Therefore no place of Σ distinct from $\mathcal{P}_\infty$ ramifies. From (7.1) applied to $\xi = x - u$, it follows that $\mathrm{ord}_{\mathcal{P}}(x-u) = 1$. Hence $x - u$ is a local parameter of Σ at $\mathcal{P}$. For $u \neq 0$,

$$\begin{array}{rcl} \sigma(x) & = & u + t, \\ \sigma(y_1) & = & v + u^{q_0}t - (u^q - u)t^{q_0} + t^{q_0+1} + \cdots, \\ \sigma(y_2) & = & w + u^{2q_0}t + u^{q_0}(u^q - u)t^{q_0} - u^{q_0}t^{q_0+1} + \cdots, \end{array} \tag{12.40}$$

with

$$v^q - v = u^{q_0}(u^q - u), \quad w^q - v = u^{2q_0}(u^q - u), \tag{12.41}$$

while, for $u = 0$,

$$\begin{array}{rcl} \sigma(x) & = & t, \\ \sigma(y_1) & = & v + t^{q_0+1} + \cdots, \\ \sigma(y_2) & = & w + t^{2q_0+1} + \cdots, \end{array} \tag{12.42}$$

with $v, w \in \mathbf{F}_q$.

From parts (G) and (I) of Lemma 12.30, $\zeta = w_6/w_8$ is a local parameter at $\mathcal{P}_\infty$. Hence ζ is a separable variable, and it defines a differential $d\xi$ for every $\xi \in \Sigma$. Again from (G) and (I) of Lemma 12.30,

$$dw_6 = w_4^{3q_0}\,dx, \quad dw_8 = w_7^{3q_0}\,dx.$$

Since

$$1 = d(\zeta) = d\left(\frac{w_6}{w_8}\right) = \frac{w_6 dw_8 - w_8 dw_6}{w_8^2} = \frac{w_6 w_7^{3q_0} - w_8 w_4^{3q_0}}{w_8^2}\,dx,$$

it follows that

$$dx = \frac{w_8^2}{w_6 w_7^{3q_0} - w_8 w_4^{3q_0}}.$$

From parts (D) and (G) of Lemma 12.30,

$$\begin{aligned} \mathrm{ord}_{\mathcal{P}_\infty}(w_6 w_7^{3q_0} - w_8 w_4^{3q_0}) &= \mathrm{ord}_{\mathcal{P}_\infty}(w_6 w_7^{3q_0}) \\ &= -3(q_0 q^2 + q^2 + 2q_0 q + q + q_0), \\ \mathrm{ord}_{\mathcal{P}_\infty} w_8 &= -(q^2 + 3q_0 q + 2q + 3q_0 + 1). \end{aligned}$$

Hence

$$\mathrm{ord}_{\mathcal{P}_\infty}(dx) = 3q_0q^2 + q^2 - q - 3q_0 - 2.$$

By (12.40) and (12.42), $\mathrm{ord}_{\mathcal{P}}(dx) = 0$ for every $\mathcal{P} \neq \mathcal{P}_\infty$. Therefore (ii) holds. From this and Definition 5.55, (iii) follows.

As $\Sigma = K(x, y_1, y_2)$, the point $Q = (1, x, y_1, y_2)$ defines a model of Σ given by an irreducible curve Γ of PG$(4, K)$. Note that Γ is defined over $\mathbf{F}_q$. From (12.37) to (12.42), Γ consists of the point $P_\infty = (0, 0, 0, 1)$ together with all points $P_{(u,v,w)} = (1, u, v, w)$ where $u, v, w \in K$ satisfy (12.41). The point P_∞ is the unique singular point of Γ; more precisely, P_∞ is a (q_0q)-fold point, and it is the centre of a unique branch of Γ, namely that arising from the place $\mathcal{P}_\infty$ of Σ. Also, the $\mathbf{F}_q$-rational places of Γ are those centred at points with coordinates in $\mathbf{F}_q$, that is, either at P_∞ or at $P_{(u,v,w)}$ with $u, v, w \in \mathbf{F}_q$. This proves (iv).

A computation shows that, if $a, b, c, d \in \mathbf{F}_q$ with $a \neq 0$, then the birational transformation $\phi_{(a,b,c,d)}$ given by

$$\begin{aligned} &x \mapsto ax + b, \quad y_1 \mapsto ab^{q_0}x + a^{q_0+1}y_1 + c,\\ &y_2 \mapsto ab^{3q_0}x - a^{q_0+1}b^{q_0}y_1 + a^{3q_0+1}y_2 + d, \end{aligned} \tag{12.43}$$

of Σ is a K-automorphism of Σ. Also,

$$\phi_{(1,b,c,d)}\phi_{(1,b',c',d')} = \phi_{(1,b'',c'',d'')}$$

with

$$b'' = b + b', \quad c'' = c + c', \quad d'' = b'^{3q_0}b - b^{q_0}c + d' + d.$$

These K-automorphisms form a subgroup Φ of $\mathrm{Aut}_K(\Sigma)$ of order $q^3(q-1)$. Also,

$$\begin{aligned} \Phi_1 &= \{\phi_{(1,b,c,d)} \mid b, c, d \in \mathbf{F}_q\}, & \Phi_2 &= \{\phi_{(1,0,c,d)} \mid c, d \in \mathbf{F}_q\},\\ \Phi_3 &= \{\phi_{(1,0,0,d)} \mid d \in \mathbf{F}_q\}, & \Psi &= \{\phi_{(a,0,0,0)} \mid a \in \mathbf{F}_q\} \end{aligned}$$

are subgroups of Φ; their orders are

$$|\Phi_1| = q^3, \quad |\Phi_2| = q^2, \quad |\Phi_3| = q, \quad |\Psi| = q - 1,$$

and Φ_2 is the commutator subgroup of Φ_1. In addition, the set of cubes in Φ_1 is its centre Φ_3. Also, $\Phi = \Phi_1 \rtimes \Psi$.

A further K-automorphism of Σ is the birational transformation θ as follows:

$$x \mapsto w_6/w_8, \quad y_1 \mapsto w_{10}/w_8, \quad y_2 \mapsto w_9/w_8. \tag{12.44}$$

Also, θ is an involution. As a consequence, a primitive representation τ of the place $\mathcal{P}_\infty$ is

$$\tau(x) = \sigma(w_6/w_8), \quad \tau(y_1) = \sigma(w_{10}/w_8), \quad \tau(y_2) = \sigma(w_9/w_8), \tag{12.45}$$

where σ is given by (12.42).

These K-automorphisms of Σ all fix $\mathcal{F}(\mathbf{F}_q)$. Also, every $\phi_{a,b,c,d}$ fixes $\mathcal{P}_\infty$, and Φ_1 acts on the remaining q^3 places in $\mathcal{F}(\mathbf{F}_q)$ as a sharply transitive permutation group. The map θ interchanges $\mathcal{P}_\infty$ and the place $\mathcal{P}_{(0,0,0)}$ corresponding to the branch centred at $P_{(0,0,0)}$. Therefore the group G generated by Φ and θ acts on $\mathcal{F}(\mathbf{F}_q)$ as a 2-transitive permutation group.

To prove (v), the first step is to show that this permutation representation of G on $\mathcal{F}(\mathbf{F}_q)$ is faithful. Since $|\Phi_1| = q^3$ and Φ is a subgroup of $G^{(0)}_{\mathcal{P}_\infty}$, Theorem 11.78 together with (iii) imply that no non-trivial element of $G^{(1)}_{\mathcal{P}}$ has more than one fixed place. Hence the kernel of the representation is a tame subgroup of $G_{\mathcal{P}_\infty}$. By Theorem 11.60, this subgroup is cyclic, that is, generated by an element α. From Lemma 11.106, if α were non-trivial, then it would have at most $2g+2$ fixed places while (iii) implies that $2g+2 < q^3$. This contradiction shows that the kernel is trivial.

The possibilities for the structure of G are given by Theorem A.17, and it remains to show that only the case $G = \text{Ree}(q)$ can occur. If G had a regular normal subgroup, then $q^3+1 = 2^m$ for some m. But this equation has no solution, as q is a power of 3. Also, $\text{PSL}(2,q)$ has an abelian Sylow 3-subgroup, and the same holds for $\text{PGL}(2,q)$. So the final step in the proof of (v) consists in showing that G cannot contain $\text{PSU}(3,q)$. Since $|\text{PSU}(3,q)| = q^3(q^3+1)(q^2-1)/\mu$ with $\mu = \gcd(3, q+1)$, it follows from (iii) that $|G| > 8g^3$. From Theorem 11.127, Σ is the Hermitian function field of genus $\frac{1}{2}q(q-1)$. But this is a contradiction, as this value is different from the genus of Σ given by (iii). Finally, G is $\mathbf{F}_q$-rational as it is generated by $\mathbf{F}_q$-rational K-automorphisms. □

LEMMA 12.30 (i) *With $x, y_1, y_2 \in \Sigma$ as in (12.35), (12.36),*

$$\text{ord}_{\mathcal{P}_\infty}(y_1) = -(1+(3q_0)^{-1})e_\infty, \quad \text{ord}_{\mathcal{P}_\infty}(y_2) = -(1+2(3q_0)^{-1})e_\infty.$$

(ii) *Define the further functions on Σ:*

$$\begin{aligned}
&w_1 = x^{3q_0+1} - y_1^{3q_0},\ w_2 = xy_1^{3q_0+1} - y_2^{3q_0},\ w_3 = xy_2^{3q_0} - w_1^{3q_0},\\
&w_4 = xw_2^{3q_0} - y_1w_1^{3q_0},\ v = xw_3^{q_0} - y_2w_1^{q_0},\ w_5 = y_1w_3^{q_0} - y_2w_2^{q_0},\\
&w_6 = v^{3q_0} - w_2^{3q_0} + w_4^{3q_0},\\
&w_7 = y_1w_2^{q_0} - xw_3^{q_0} - w_6^{3q_0},\ w_8 = w_5^{3q_0} - xw_7^{3q_0},\\
&w_9 = w_2^{q_0}w_4 - y_1w_6^{q_0},\ w_{10} = y_2w_6^{q_0} - w_3^{q_0}w_4.
\end{aligned}$$

Then

(a) $w_1^{q_0} = x^{q_0+1} - y_1,\quad w_1^q - w_1 = x^{3q_0}(x^q - x),$
$\text{ord}_{\mathcal{P}_\infty}(w_1) = -(1+q_0^{-1})e_\infty;$

(b) $w_2^{q_0} = x^{q_0}y_1 - y_2,\quad w_2^q - w_2 = y_1^{3q_0}(x^q - x),$
$\text{ord}_{\mathcal{P}_\infty}(w_2) = -(1+q_0^{-1}+q^{-1})e_\infty;$

(c) $w_3^{q_0} = x^{q_0}y_2 - w_1,\quad w_3^q - w_3 = y_2^{3q_0}(x^q - x),$
$\text{ord}_{\mathcal{P}_\infty}(w_3) = -(1+q_0^{-1}+2q^{-1})e_\infty;$

(d) $w_4^{3q_0} = x^{3q_0}w_2 - y_1^{3q_0}w_1,\quad w_4^q - w_4 = w_2^{q_0}(x^q - x) - w_1^{q_0}(y_1^q - y_1),$
$\text{ord}_{\mathcal{P}_\infty}(w_4) = -(1+2(3q_0)^{-1}+q^{-1})e_\infty;$

(e) $v^{3q_0} = x^{3q_0}w_3 - y_2^{3q_0}w_1,\ v^q - v = w_3^{q_0}(x^q - x) - w_1^{q_0}(y_1^q - y_1),$
$\text{ord}_{\mathcal{P}_\infty}(v) = -(1+q_0^{-1}+q^{-1})e_\infty;$

(f) $w_5^{3q_0} = y_1^{3q_0}w_3 - y_2^{3q_0}w_2,\ w_5^q - w_5 = w_3^{q_0}(y_1^q - y_1) - w_2^{q_0}(y_2^q - y_2),$
$\text{ord}_{\mathcal{P}_\infty}(w_5) = -(1+q_0^{-1}+q^{-1}+(3q_0q)^{-1})e_\infty;$

(g) $w_6^{q_0} = -w_2 + x^{q_0}w_4, \quad w_6^q - w_6 = w_4^{3q_0}(x^q - x),$
$\mathrm{ord}_{\mathcal{P}_\infty}(w_6) = -(1 + q_0{}^{-1} + 2q^{-1} + (q_0q)^{-1})e_\infty;$

(h) $w_7^{3q_0} = y_1^{3q_0}w_2 - x^{3q_0}w_3 - w_6,\ w_7^q - w_7 = w_2^{q_0}(y_1^q - y_1) - w_3^{q_0}(x^q - x),$
$\mathrm{ord}_{\mathcal{P}_\infty}(w_7) = -(1 + 2(3q_0)^{-1} + q^{-1} + (3q_0q)^{-1})e_\infty;$

(i) $w_8^{q_0} = w_5 + x^{q_0}w_7, \quad w_8^q - w_8 = w_7^{3q_0}(x^q - x),$
$\mathrm{ord}_{\mathcal{P}_\infty}(w_8) = -(1 + q_0{}^{-1} + 2q^{-1} + (q_0q)^{-1} + q^{-2})e_\infty;$

(j) $w_9^{3q_0} = w_2w_4^{3q_0} - y_1^{3q_0}w_6,\ w_9^q - w_9 = w_2^{q_0}(w_4^q - w_4) - w_6^{q_0}(y_1^q - y_1),$
$\mathrm{ord}_{\mathcal{P}_\infty}(w_7) = -(1 + q_0{}^{-1} + 2q^{-1} + (3q_0q)^{-1})e_\infty;$

(k) $w_{10}^{3q_0} = y_2^{3q_0}w_6 - w_3w_4^{3q_0},\ w_{10}^q - w_{10} = w_6^{q_0}(y_2^q - y_2) - w_3^{q_0}(w_4^q - w_4),$
$\mathrm{ord}_{\mathcal{P}_\infty}(w_{10}) = -(1 + q_0{}^{-1} + 2q^{-1} + (3q_0q)^{-1})e_\infty.$

Proof. This is by straightforward calculation. □

A major result on the DLR curve is the following characterisation, analogous to Theorem 12.14.

THEOREM 12.31 *Let $p = 3$, $q = 3q_0^2$ with $q_0 = 3^s$, $s \geq 0$. If $\mathcal{F}$ is a curve of genus $g = \frac{3}{2}q_0(q-1)(q+q_0+1)$ such that the K-automorphism group of $\mathcal{F}$ contains a subgroup G isomorphic to* Ree(q), *then $\mathcal{F}$ is the DLR curve.*

As in the proof of Theorem 12.14, a series of preliminary results are first established.

LEMMA 12.32 *There is a place $\mathcal{P}$ of Σ such that the ramification groups of $G_\mathcal{P}$ have the following properties:*

(i) $|G_\mathcal{P}^{(1)}| = q^3$, *and* $\rho_1 = 2$, *that is,* $G_\mathcal{P}^{(1)} \neq G_\mathcal{P}^{(2)}$;

(ii) $G_\mathcal{P}^{(2)}$ *is the commutator group of* $G_\mathcal{P}^{(1)}$, *is an elementary abelian group of order* q^2, *and* $\rho_2 = 3q_0 + 2$, *that is,*
$$U_\mathcal{P}^{(2)} = G_\mathcal{P}^{(2)} = G_\mathcal{P}^{(3)} = \cdots = G_\mathcal{P}^{(3q_0+2)};$$

(iii) $G_\mathcal{P}^{(3q_0+3)}$ *is the centre of* $G_\mathcal{P}^{(1)}$, *is an elementary abelian group of order* q, *and* $\rho_3 = q + 3q_0 + 2$, *that is,*
$$U_\mathcal{P}^{(3)} = G_\mathcal{P}^{(3q_0+3)} = G_\mathcal{P}^{(3q_0+4)} = \cdots = G_\mathcal{P}^{(q+3q_0+2)};$$

(iv) $U_\mathcal{P}^{(4)} = G_\mathcal{P}^{(q+3q_0+3)} = \{1\}$;

(v) $\Sigma^{G_\mathcal{P}^{(i)}}$ *is rational for* $0 \leq i \leq 3q_0 + 2$, *and* $\Sigma^{G_\mathcal{P}^{(3q_0+3)}}$ *has genus* $\frac{3}{2}q_0(q-1)$.

Proof. A Sylow 3-subgroup S_3 of G has order q^3. Since $3g < q^3$ and the 3-rank of $\mathcal{F}$ is at most g, Theorem 11.84 implies that the 3-rank of $\mathcal{F}$ is zero. From Lemma 11.129, S_3 fixes a place $\mathcal{P}$ of Σ and no non-trivial element in S_3 fixes a place distinct from $\mathcal{P}$. In particular, $S_3 = G_\mathcal{P}^{(1)}$ for some place $\mathcal{P}$. Since G has $q^3 + 1$ Sylow 3-subgroups, the set Ω of all places fixed by some 3-element of G

has size q^3+1, and G acts on Ω as $\mathrm{Ree}(q)$ in its natural 2-transitive permutation representation.

From Theorem 11.78, $\Sigma^{G_{\mathcal{P}}^{(1)}}$ is rational. This implies that $\Sigma^{G_{\mathcal{P}}}$ also rational. The normaliser of S_3 in G coincides with $G_{\mathcal{P}}$, and hence $G_{\mathcal{P}} = S_3 \rtimes H$ with $H \cong \mathbf{Z}_{q-1}$.

Theorem 11.72 applied to $G_{\mathcal{P}}$ gives

$$2(g-1) \geq q^3(2g_1-2) + 2q^3 - 2 + |G_{\mathcal{P}}^{(2)}| - 1,$$

where g_1 denotes the genus of $\Sigma^{G_{\mathcal{P}}^{(1)}}$. Since $g < q^3$, this implies that $g_1 = 0$. Hence $G_{\mathcal{P}}^{(1)} \neq G_{\mathcal{P}}^{(2)}$, that is, $\rho_1 = 2$. From Lemma 11.82,

$$(q-1) \mid (\rho_i - 1)(n_i - 1). \tag{12.46}$$

where $n_i = |U_{\mathcal{P}}^{(i)}|/|U_{\mathcal{P}}^{(i+1)}|$. For $i=1$, this gives $n_1 = q$; that is, $|U_{\mathcal{P}}^{(2)}| = q^2$.

By Lemma 11.75 (ii), $G_{\mathcal{P}}^{(2)}$ contains the commutator subgroup of $G_{\mathcal{P}}^{(1)}$. On the other hand, the commutator subgroup of a 3-subgroup of Sylow of $\mathrm{Ree}(q)$ is elementary abelian of order q^2. Therefore $G_{\mathcal{P}}^{(2)}$ is the commutator subgroup of $G_{\mathcal{P}}^{(1)}$.

The genus g_2 of $G_{\mathcal{P}}^{(2)}$ is calculated from Theorem 11.72 applied to $G_{\mathcal{P}}^{(1)}$ regarded as a K-automorphism group of $\Sigma^{G_{\mathcal{P}}^{(2)}}$:

$$2g_2 - 2 = -2q^2 + 2(q^2-1) + \rho_1(q-1),$$

which gives $g_2 = 0$.

Further, $\rho_2 \equiv 2 \pmod 3$. This follows from Lemma 11.75 (iv) and the fact that $\rho_1 = 2$. The congruence $\rho_2 \equiv 2 \pmod 3$ and the equation

$$2g = (\rho_2 - 2)(q^2-1) + \textstyle\sum_{i\geq 3} (\rho_i - \rho_{i-1})(|U_{\mathcal{P}}^{(i)}| - 1), \tag{12.47}$$

which is actually a consequence of (11.40), provide the bound $\rho_2 \leq 3q_0 + 2$. On the other hand, Lemma 11.82 (iii) shows that $q \mid (\rho_2 - 2)^2$, whence $3q_0 \mid (\rho_2 - 2)$. Thus $\rho_2 = 3q_0 + 2$. From Lemma 11.82 (ii), $(q-1) \mid (3q_0+1)(n_2-1)$, which implies that $n_2 = q$. Therefore $|U_{\mathcal{P}}^{(3)}| = q$.

Since the centraliser of H in $G_{\mathcal{P}}$ intersects $G_{\mathcal{P}}^{(1)}$ trivially, every normal subgroup of $G_{\mathcal{P}}$ contains at least $q-1$ non-trivial elements. In particular, the last non-trivial ramification group of $G_{\mathcal{P}}$ has order at least q. Therefore $|U_{\mathcal{P}}^{(4)}| = \{1\}$, and $U_{\mathcal{P}}^{(3)}$ coincides with $Z(G_{\mathcal{P}}^{(1)})$.

Now (12.47) reads as follows:

$$2g = (\rho_2 - 2)(q^2-1) + (\rho_3 - \rho_2)(q-1). \tag{12.48}$$

It follows that $\rho_3 = q + 3q_0 + 2$.

Finally, the genus g_3 of $\Sigma^{U_{\mathcal{P}}^{(3)}}$ can be calculated by applying Theorem 11.72 to $U_{\mathcal{P}}^{(2)}$ regarded as a K-automorphism group of $\Sigma^{U_{\mathcal{P}}^{(3)}}$:

$$2(g_3 - 1) = -2q + \rho_2(q-1).$$

The result is that $g_3 = \frac{3}{2}q_0(q-1)$. □

To find generators of Σ with the desired polynomial relations, the first step consists in showing the existence of elements $x, y_1 \in \Sigma$ satisfying (12.37) such that

$\mathbf{F}_q(x, y_1)$ is K-isomorphic to $\Sigma^{U_{\mathcal{P}}^{(3)}} = \Sigma^{G_{\mathcal{P}}^{(3n_0+3)}}$. Next, an element $y_2 \in \Sigma$ is found satisfying both $\Sigma = K(x, y_1, y_2)$ and (12.38). Finally, it is shown that G preserves $\mathbf{F}_q(x, y_1, y_2)$.

Let $\mathcal{P}_\infty$ denote a place of Σ satisfying the properties listed in Lemma 12.32. The arguments used in the proofs of Lemmas 12.17 12.18 and 12.19 can be adapted to prove analogous results on Σ. The necessary changes are as follows:

$$q^2 \mapsto q^3; \quad q \mapsto q^2;$$
$$\begin{aligned} 2g-2 = 2q_0(q-1) - 2 \mapsto 2g-2 &= 3q_0(q-1)(q+q_0+1) - 2 \\ &= (q+1)(q+3q_0+1)(3q_0-2); \end{aligned}$$
$$\begin{aligned} q^2+1 &= (q+2q_0+1)(q-2q_0+1) \\ &\mapsto q^3+1 = (q+1)(q+3q_0+1)(q-3q_0+1); \end{aligned}$$
$$q+2q_0+1 \mapsto (q+1)(q+3q_0+1).$$

Lemma 12.33 (i) *The smallest non-gap at $\mathcal{P}_\infty$ is q^2.*

(ii) *There exists $x \in \Sigma$ such that*

(a) $\operatorname{div} x = -q^2\mathcal{P}_\infty + \sum_{i=1}^{q^2} \mathcal{P}_i$;

(b) $K(x) = U_{\mathcal{P}_\infty}^{(2)}$.

Lemma 12.34 *For $\alpha \in G_{\mathcal{P}_\infty}$,*

$$\alpha(x) = \begin{cases} x + a, & a \in K\backslash\{0\}, \quad \text{for } \alpha \in G_{\mathcal{P}_\infty}^{(1)} \backslash U_{\mathcal{P}_\infty}^{(2)}, \\ bx, & b^{q-1} = 1,\ b \neq 1, \quad \text{for } \alpha \in H, \\ x, & \quad \text{for } \alpha \in U_{\mathcal{P}_\infty}^{(2)}. \end{cases} \tag{12.49}$$

Lemma 12.35 *There exists $w \in \Sigma$ such that*

$$\operatorname{div}(w)_\infty = (q+1)(q+3q_0+1)\mathcal{P}_\infty.$$

By Lemma 12.35, the smallest non-gap m at $\mathcal{P}$ which is not divisible by q does not exceed $(q+1)(q+3q_0+1)$.

From Lemma 12.32, $\rho_1 = 2$, $\rho_2 = 3q_0 + 2$, $\rho_3 = q + 3q_0 + 3$. This together with Lemma 11.82 gives the following result.

Lemma 12.36 *Let w be as in Lemma 12.35. If α is a non-trivial element in $U_{\mathcal{P}_\infty}^{(3)}$ and $w' = \alpha(w) - w$, then*

$$\operatorname{div}(w')_\infty = q(q+3q_0+1)\mathcal{P}_\infty.$$

Lemma 12.37 *Let $y_1 \in \Sigma\backslash K(x_1)$ with $\operatorname{div}(y_1) = k\mathcal{P}_\infty$, such that k is minimal and $k \geq 1$.*

(i) *If β is a generator of H, then $\beta(y_1) = b_\beta y_1$ for a primitive $(q-1)$-st root of unity b_β.*

(ii) $y_1 \in \Sigma^{U_{\mathcal{P}_\infty}^{(3)}}$.

Proof. (i) If $z \in \Sigma\backslash K(x)$ with $\operatorname{div}(z) = k\mathcal{P}_\infty$, then, under β,

$$\begin{aligned} &x \mapsto bx, \quad \text{for a primitive } (q-1)\text{-st root of unity } b \in K; \\ &z \mapsto u_\beta x + v_\beta y_1 + w_\beta, \quad u_\beta, v_\beta, w_\beta \in K,\ u_\beta \neq 0. \end{aligned} \tag{12.50}$$

A calculation shows that there exist $v'_\beta, w'_\beta \in K$ for which $y_1 = z + v'_\beta x + w'_\beta$ satisfies (i).

(ii) Since $y_1 \notin K(x_1)$, from Lemma 12.36, $q^2 < k \leq (q+1)(q+3q_0+1)$. For $\alpha \in G^{(1)}_{\mathcal{P}_\infty}\backslash G^{(2)}_{\mathcal{P}_\infty}$,

$$\begin{aligned} &x \mapsto x + b_\alpha, \quad b_\alpha \in K\backslash\{0\}; \\ &y_1 \mapsto y_1 + d_\alpha x + e_\alpha, \quad d_\alpha, e_\alpha \in K,\ d_\alpha \neq 0, \end{aligned} \tag{12.51}$$

whence $\alpha^3(y_1) = y_1$. Since α^3, with α ranging over $G^{(1)}_{\mathcal{P}_\infty}\backslash G^{(2)}_{\mathcal{P}_\infty}$, produces all elements of $Z(G^{(1)}_{\mathcal{P}}) = U^{(3)}_{\mathcal{P}_\infty}$, so (ii) follows. □

From now on, let x and y_1 have the properties stated in Lemmas 12.33, 12.34 and 12.37. Choose an element α from $U^{(2)}_{\mathcal{P}_\infty}$. Since $U^{(3)}_{\mathcal{P}_\infty}$ is a normal subgroup of $U^{(2)}_{\mathcal{P}_\infty}$, so α induces a K-automorphism of $\Sigma^{U^{(3)}_{\mathcal{P}_\infty}}$ fixing every element in $\Sigma^{U^{(2)}_{\mathcal{P}_\infty}}$. By Lemma 12.32 (i), the latter function field is rational. From Lemma 11.14, it follows that $\alpha(y_1) = y_1 + a_\alpha$ with $a_\alpha \in K\backslash\{0\}$. Replacing y_1 by $a_\alpha^{-1}y_1$, it may be supposed that $\alpha(y_1) = y_1 + 1$.

PROPOSITION 12.38 *The field* $\Sigma^{U^{(3)}_{\mathcal{P}}} = K(x, y_1)$, *and* $y_1^q - y_1 = x^{q_0}(x^q - x)$.

Proof. Let $\alpha_j = \beta^j\alpha\beta^{-j}$ for $j = 1, \ldots, q-1$. From (i) of the previous lemma, $\alpha_j(y_1) = y_1 + b^j$. Since $b^{q-1} = 1$, this shows that $\alpha_j(y_1) = y_1 + b_j$ with $b_j \in \mathbf{F}_q$. Since $|U^{(2)}_{\mathcal{P}_\infty}/U^{(3)}_{\mathcal{P}_\infty}| = q$, every non-zero element in $\mathbf{F}_q$ is obtained exactly once in this way. Putting $b_0 = 0$, this implies that

$$\textstyle\prod_{j=0}^{q-1}(y_1 - b_j) = y_1^q - y_1 \in \mathbf{F}_q(x).$$

From the definition of x and Lemma 12.37 (ii), $K(x, y_1)$ is a subfield of $\Sigma^{U^{(3)}_{\mathcal{P}_\infty}}$. On the other hand,

$$[\Sigma^{U^{(2)}_{\mathcal{P}_\infty}} : K(x, y_1)] = q, \quad |U^{(2)}_{\mathcal{P}_\infty}/U^{(3)}_{\mathcal{P}_\infty}| = q.$$

Therefore $K(x, y_1) = \Sigma^{U^{(3)}_{\mathcal{P}_\infty}}$. Since y_1 is assumed to have no pole other than $\mathcal{P}_\infty$, this implies that

$$y_1^q - y_1 = f(x), \quad f(X) \in \mathbf{F}_q[X].$$

Put $r = \deg f(X)$; then $r = m/q$ as $q\,\mathrm{ord}_{\mathcal{P}_\infty}(y_1) = -r\,e_{\mathcal{P}_\infty}$. Now, since

$$q^2 < m \leq (q+1)(q+3q_0+1),$$

it follows that $q < r \leq q + 3q_0 + 2$. Therefore

$$y_1^q - y_1 = \textstyle\sum_{i=0}^{r} a_i x^i, \qquad q < r \leq q + 3q_0 + 2. \tag{12.52}$$

Without loss of generality, let $a_r = 1$. Let b and b_β be as in Lemma 12.37. Applying the map β, in which $x \mapsto bx$, $y_1 \mapsto b_\beta y_1$, to (12.52) gives $b_\beta a_i = b^i b_i$

for $i = 0, 1, \ldots, r$. Note that $b_\beta = b^r$ as $a_r = 1$. Thus $(b^r - b^i)a_i = 0$ for $i = 0, 1, \ldots, r$. Since $a_i = 0$ for i neither $r - q + 1$ nor r, (12.52) can be written as follows:

$$y_1^q - y_1 = x^r + cx^{r-q+1}, \qquad r \in \mathbf{F}_q. \tag{12.53}$$

Choose an element α from $G_{\mathcal{P}_\infty}^{(1)} \backslash G_{\mathcal{P}_\infty}^{(2)}$; then (12.51) holds. Applying α to (12.53) gives the following:

$$y_1^q - y_1 + d_\alpha(x^q - x) = (x + e_\alpha)^r + c(x + e_\alpha)^{r-q+1}. \tag{12.54}$$

From this, $c = -1$. Now replace $y_1^q - y_1$ in (12.54) by $x^r + cx^{r-q-1}$ and use the binomial expansion. Then (12.54) becomes the following:

$$x^{r-q} + d_\alpha = x^r - q + e_\alpha^{r-1} + \textstyle\sum_{i=1}^{r-q-1} \binom{r-q}{i} x^i e_\alpha^{r-1-i}.$$

This implies that

$$\binom{r-q}{i} \equiv 0 \pmod 3, \quad i = 1, \ldots, r - q - 1;$$

that is, $r - q$ is a power of 3. From (12.52), $1 \le r - q \le 3q_0$. From (12.53), the only possibility is $r - q = q_0$ as m is chosen to be minimal. Also, the genus of $K(x, y_1)$ must be $\frac{3}{2}q_0(q-1)$, the genus of $\Sigma^{U_{\mathcal{P}_\infty}^{(3)}}$. □

Using the arguments in the proof of Lemma 12.37 the following result can be obtained.

Lemma 12.39 *Let $y_2 \in \Sigma \backslash \Sigma^{U_{\mathcal{P}_\infty}^{(3)}}$ for which* $\operatorname{div} y_2 = k\mathcal{P}_\infty$ *with $k \ge 1$ and minimal. Then y_2 can be chosen such that*

$$\beta(y_2) = c_\beta y_2, \qquad c_\beta \in \mathbf{F}_q \backslash \{0\}.$$

In this lemma, it follows from Lemma 12.36 that

$$q(q + q_0) < k \le q(q + 3q_0 + 1),$$

as $w' \notin \Sigma^{U_{\mathcal{P}_\infty}^{(3)}}$.

Proof of Theorem 12.31. So far, the following results have been shown. The cyclic group H is generated by b and, for a primitive element b of $\mathbf{F}_q$, the map β acts as follows:

$$x \mapsto bx, \quad y_1 \mapsto b^{q_0+1}y_1, \quad y_2 \mapsto b^s y_2, \quad \text{for } 1 \le s \le q - 1, \tag{12.55}$$

while $G_{\mathcal{P}_\infty}^{(1)}$ consists of q^3 elements α that act as follows:

$$x \mapsto x + b_\alpha, \quad y_1 \mapsto y_1 + b_\alpha^q x + e_\alpha \quad y_2 \mapsto y_2 + f_\alpha y_1 + g_\alpha x + h_\alpha, \tag{12.56}$$

where $b_\alpha, e_\alpha, f_\alpha, g_\alpha, h_\alpha \in \mathbf{F}_q$. Note that $\alpha \in G_{\mathcal{P}_\infty}$ if and only if $b_\alpha = 0$. As the cube of any element in $G_{\mathcal{P}_\infty}^{(1)} \backslash G_{\mathcal{P}_\infty}^{(2)}$ belongs to $U_{\mathcal{P}_\infty}^{(3)}$, the necessary and sufficient condition for α to be in $U_{\mathcal{P}_\infty}^{(3)}$ is that it acts as follows:

$$x \mapsto x, \quad y_1 \mapsto y_1, \quad y_2 \mapsto y_2 + u, \quad \text{for } u \in \mathbf{F}_q. \tag{12.57}$$

Now the proof proceeds. By (12.57),

$$\textstyle\prod_{u \in \mathbf{F}_q} (y_2 - u) = y_2^q - y_2 \in \Sigma^{U_{\mathcal{P}_\infty}^{(3)}}. \tag{12.58}$$

Since $K(x, y_1, y_2)$ is a subfield of Σ, this implies that $\Sigma = K(x, y_1, y_2)$ where x, y_1, y_2 satisfy (12.37),(12.38) and (12.58).

The unique involutory element in H is $\beta' = \beta^{(q-1)/2}$. By (12.55), its action is the following:

$$x \mapsto -x, \quad y_1 \mapsto y_1, \quad y_2 \mapsto v y_2;$$

here either $v = 1$ or $v = -1$. Using (12.56), a calculation shows that β' commutes with $\psi \in G^{(1)}_{\mathcal{P}_\infty}$ if and only if ψ has the following action:

$$x \mapsto x, \quad y_1 \mapsto y_1 + u \quad \text{for } u \in \mathbf{F}_q, \quad y_2 \mapsto y_2.$$

In particular, the centraliser C of ψ in $G^{(1)}_{\mathcal{P}_\infty}$ has order q and is a subgroup of $G^{(2)}_{\mathcal{P}_\infty}$. Since $C \cap U^{(3)}_{\mathcal{P}_\infty} = \{1\}$, this implies that $G^{(2)}_{\mathcal{P}_\infty} = C \times U^{(3)}_{\mathcal{P}_\infty}$. Hence $G^{(2)}_{\mathcal{P}_\infty}$ consists of all $\tau \in G^{(1)}_{\mathcal{P}_\infty}$ acting as follows:

$$x \mapsto x, \quad y_1 \mapsto y_1 + u, \quad y_2 \mapsto y_2 + w, \quad \text{for } u, w \in \mathbf{F}_q.$$

Therefore

$$\prod_{w \in \mathbf{F}_q} (y_2 - w) = y_2^q - y_2 \in \mathbf{F}_q(x).$$

The assumption that y_2 has no pole other than from $\mathcal{P}_\infty$ implies that

$$y_2^q - y_2 = f(x), \quad \text{with } f(X) \in \mathbf{F}_q[X].$$

Arguing as before, put $t = \deg f(X)$. Then $t = k/q$ as $q \operatorname{ord}_{\mathcal{P}_\infty}(y_1) = -t\, e_{\mathcal{P}_\infty}$. Since $q^2 < k \le (q+1)(q+3q_0+1)$, so $q < t \le q + 3q_0 + 2$. Therefore

$$y_2^q - y_2 = \sum_{i=0}^t b_i x^i, \quad b_i \in \mathbf{F}_q, \quad q < t \le q + 3q_0 + 2. \tag{12.59}$$

Without loss of generality, let $b_t = 1$. Also, let b and b_β be as in Lemma 12.37. Applying the map β, with $x \mapsto bx$, $y_2 \mapsto b_\beta^s y_2$, to (12.52) gives $b_\beta b_i = b^i b_i$ for $i = 0, 1, \ldots, t$. Note that $b_\beta = b^t$ as $b_r = 1$. Thus $(b^r - b^i) b_i = 0$ for $i = 0, 1, \ldots, t$. As $b_i = 0$ for $i \ne t - q + 1$, t, so (12.59) can be written as follows:

$$y_2^q - y_2 = x^t + c x^{t-q+1}, \qquad c \in \mathbf{F}_q. \tag{12.60}$$

As above, choose an element α from $G^{(1)}_{\mathcal{P}_\infty} \backslash G^{(2)}_{\mathcal{P}_\infty}$; then (12.51) holds. Applying α to (12.60) gives the equation

$$y_2^q - y_2 + d_\alpha (y_1^q - y_1) + e_\alpha (x^q - x) = (x + f_\alpha)^{s-q}(x^q - x). \tag{12.61}$$

From this, $c = -1$.

In (12.61), replace $y_2^q - y_2$ by $x^s + cx^{s-q+1}$ and $y_1^q - y_1$ by $x^{q_0}(x^q - x)$, and use the binomial expansion. Then (12.54) becomes the following:

$$x^{t-q} + d_\alpha x^{q_0} + e_\alpha = x^{t-q} + f_\alpha^{t-1} + \sum_{i=1}^{t-q-1} \binom{t-q}{i} x^i e_\alpha^{t-1-i}. \tag{12.62}$$

In particular, $e_\alpha = f_\alpha^{t-q}$. Suppose that $d_\alpha = 0$. Then (12.62) implies that

$$\binom{s-q}{i} \equiv 0 \pmod 3, \quad i = 1, \ldots, r - q - 1;$$

that is, $r - q$ is a power of 3. Since, $q_0 \le s - q \le 3q_0 + 1$, this yields that $r - q = 3q_0$. However, in this case, $y_2 \in U^{(3)}_{\mathcal{P}_\infty}$, a contradiction. Therefore $d_\alpha \ne 0$.

Now, from (12.62), modulo 3,

$$\binom{t-q}{q_0} \not\equiv 0, \qquad \binom{t-q}{i} \equiv 0 \quad \text{for } i \ne q_0,\ 1 \le t - q - 1,$$

showing that $t - q = sq_0$ with s not divisible by 3. Since $q_0 < t - q \le 3q_0 + 1$, and $t - q$ is distinct from both q_0 and $3q_0$, the desired result, that $t - q = 2q_0$, follows. □

12.5 A CURVE ATTAINING THE SERRE BOUND

In this section, $K = \overline{\mathbf{F}}_2$. Consider the curve

$$\mathcal{C} = \mathbf{v}(X_0^{18}X_2^5 + X_0^{19}X_2 + X_1^{23}) \tag{12.63}$$

together with its function field $\Sigma = K(x, y)$, where

$$y^5 + y^4 = x^{23}. \tag{12.64}$$

THEOREM 12.40 *The curve $\mathcal{C}$ has the following properties:*

(i) *it is an irreducible plane curve of genus 11 with exactly two singular points, the 18-fold point $Y_\infty = (0, 0, 1)$ and the 4-fold point $O = (1, 0, 0)$, each the centre of a unique branch of $\mathcal{C}$;*

(ii) *if $\mathcal{P}_\infty$ and $\mathcal{P}_0$ are the associated places of Σ, then*

$$\mathrm{div}(dx) = 26\mathcal{P}_0 - 6\mathcal{P}_\infty; \tag{12.65}$$

(iii) *with*

$$\begin{array}{llll} z = y^{-1}, & x_0 = z, & x_1 = xz, & x_2 = x^2z, \\ x_3 = x^3z, & x_4 = x^5z^2, & x_5 = x^6z^2, & x_6 = x^7z^2, \\ x_7 = x^8z^2, & x_8 = x^{11}z^3, & x_9 = x^{12}z^3, & x_{10} = x^{17}z^4, \end{array}$$

the canonical series of Σ is

$$|dx| = \{\mathrm{div}(\textstyle\sum_{i=0}^{10} c_ix_i) + 26\mathcal{P}_0 - 6\mathcal{P}_\infty \mid \mathbf{c} = (c_0, \ldots, c_{11}) \in \mathrm{PG}(11, K)\}; \tag{12.66}$$

(iv) *the K-automorphism group of $\mathcal{C}$ contains a cyclic group of order 23.*

Proof. The arguments and calculations are similar to those in the proof of Lemma 12.1. The point Y_∞ is the centre of a unique branch of $\mathcal{C}$ with a primitive representation,

$$(x(t) = t^{-5} + \cdots, \, y(t) = t^{-23} + \cdots).$$

From this,

$$\mathrm{ord}_{\mathcal{P}_\infty}(dx) = \mathrm{ord}_t(dx/dt) = -6. \tag{12.67}$$

The other singular point of $\mathcal{C}$ is the origin O; the unique branch of $\mathcal{C}$ centred here has a primitive representation

$$(x(t) = t^4 + \cdots, \, y(t) = t^{23} + \cdots).$$

From (12.64), $y^4dy = x^{22}dx$. Since $dy/dt = t^{22} + \cdots$, so

$$\mathrm{ord}_{\mathcal{P}_0}(dx) = \mathrm{ord}_t(dx/dt) = 4\,\mathrm{ord}_t\, y(t) - 22\,\mathrm{ord}_t\, x(t) + \mathrm{ord}_t(dy/dt) = 26. \tag{12.68}$$

Since $\mathcal{C}$ has no more singular points, a first consequence is that every point of $\mathcal{C}$ is the centre of just one branch of $\mathcal{C}$. Hence $\mathcal{C}$ is an irreducible plane curve and, for a power q of 2, a place of Σ is $\mathbf{F}_q$-rational if and only if the corresponding branch of $\mathcal{C}$ is centred at a point with coordinates in $\mathbf{F}_q$.

Also, $\text{ord}_{\mathcal{P}}(dx) \neq 0$ only at these two places, $\mathcal{P}_\infty$ and $\mathcal{P}_0$. Therefore (12.67) and (12.68) give (12.65). From Definition 5.55, $2g - 2 = 26 - 6 = 20$, whence $g = 11$.

Let $\mathcal{P}_1$ be the place of Σ associated to the unique branch of $\mathcal{C}$ whose centre is the point $A = (1, 1, 0)$. Then

$$\text{div}\, x = 4\mathcal{P}_0 + \mathcal{P}_1 - 5\mathcal{P}_\infty, \quad \text{div}\, y = 23\mathcal{P}_0 - 23\mathcal{P}_\infty.$$

From this, (12.66) follows.

Finally, for a primitive twenty-third root of unity $\lambda \in K$, the birational transformation

$$\omega: \quad x' = \lambda x, \quad y' = \lambda y$$

of Σ is a K-automorphism of order 23. □

From (12.66), the gap sequences of Σ at the places $\mathcal{P}_\infty, \mathcal{P}_1$ and $\mathcal{P}_0$ are the following:

$$\begin{aligned} \mathcal{P}_\infty: &\quad (1, 2, 3, 4, 6, 8, 9, 11, 13, 16, 18); \\ \mathcal{P}_1: &\quad (1, 2, 3, 4, 6, 7, 8, 9, 12, 13, 18); \\ \mathcal{P}_0: &\quad (1, 2, 3, 4, 5, 6, 8, 9, 12, 13, 16). \end{aligned}$$

Next, the Weierstrass points of Σ are determined. From (12.66), the canonical curve Γ of Σ is given by the point $P = (x_0, x_1, \ldots, x_{11})$ of $\text{PG}(10, K)$ with the x_i as defined above. The 11×11 Wronskian matrix $W = (D_x^{(i)} x_j)$ with $0 \le i, j \le 10$ in Figure 12.1 can be calculated using the method illustrated in Example 5.81. In it, let

$$\begin{aligned} &f_i = y^i + y^{i-1} + \cdots + y + 1, \quad g_4 = y^4 + y + 1, \\ &h_3 = y^8 + y^3 + 1, \quad h_4 = y^8 + y^4 + 1. \end{aligned}$$

From the matrix,

$$\begin{aligned} \det W = \,& x^{17}(y^3 + y^2 + 1)(y^8 + y^7 + y^6 + y^5 + y^4 + y + 1) \\ & \times (y^{35} + y^{34} + y^{32} + y^{30} + y^{29} + y^{28} + y^{27} + y^{26} + y^{19} \\ & + y^{18} + y^{15} + y^{11} + y^8 + y^5 + y^4 + y^3 + 1)(y + 1)^3 y^{-76}. \end{aligned}$$

Since W is not identically zero, the canonical series is classical; that is, $\epsilon_i = i$ for $i = 0, 1, \ldots, 10$. Further, the zeros of W are exactly the Weierstrass points of Σ; their total number is $3 + (3 + 8 + 35) \cdot 23 = 1061$.

The weights of the places $\mathcal{P}_\infty, \mathcal{P}_0, \mathcal{P}_1$ in the ramification divisor R of Σ are as follows:

$$\begin{aligned} & \text{ord}_{\mathcal{P}_\infty} W = 536, \quad \text{ord}_{\mathcal{P}_\infty}(dx) = -6, \quad e_{\mathcal{P}_\infty} = -6, \\ & v_{\mathcal{P}_\infty}(R) = 536 - 330 - 66 = 140; \end{aligned}$$

$$\begin{aligned} & \text{ord}_{\mathcal{P}_0} W = -1680, \quad \text{ord}_{\mathcal{P}_0}(dx) = 26, \quad e_{\mathcal{P}_0} = 26, \\ & v_{\mathcal{P}_0}(R) = -1680 + 1430 + 286 = 36; \end{aligned}$$

$$\begin{aligned} & \text{ord}_{\mathcal{P}_1} W = 86, \quad \text{ord}_{\mathcal{P}_1}(dx) = 0, \quad e_{\mathcal{P}_1} = 0, \\ & v_{\mathcal{P}_1}(R) = 86 + 0 + 0 = 86. \end{aligned}$$

From (7.13), $\sum v_{\mathcal{P}}(R) = 1320$. Hence

$$\textstyle\sum_{i=1}^{1058} v_{\mathcal{P}}(R) = 1320 - (140 + 36 + 86) = 1058.$$

Since Σ has 1058 Weierstrass points other than $\mathcal{P}_\infty, \mathcal{P}_0, \mathcal{P}_1$, this is only possible when $v_{\mathcal{P}}(R) = 1$ for each of these 1058 places. Also, since the canonical series of Σ is classical, Theorem 7.55 shows that $v_{\mathcal{P}}(R) = 1$ if and only if the order sequence of Γ at the branch point associated to $\mathcal{P}$ is

$$(0, 1, 2, 3, 4, 5, 6, 7, 8, 9, 11).$$

Therefore the gap sequence of Σ at any Weierstrass point $\mathcal{P}$ other than $\mathcal{P}_\infty, \mathcal{P}_0, \mathcal{P}_1$ is

$$(1, 2, 3, 4, 5, 6, 7, 8, 9, 10, 12).$$

This shows that each of the three places $\mathcal{P}_\infty, \mathcal{P}_0, \mathcal{P}_1$ differs from any other place of Σ in its Weierstrass gap sequence.

As a consequence, $\mathrm{Aut}_K(\Sigma)$ has three fixed places. From Theorem 11.72 applied to $\mathrm{Aut}_K(\Sigma)$,

$$20 \geq -2|\mathrm{Aut}_K(\Sigma)| + 3(|\mathrm{Aut}_K(\Sigma)| - 1) = |\mathrm{Aut}_K(\Sigma)| - 3,$$

whence $|\mathrm{Aut}_K(\Sigma)| \leq 23$. By Proposition 12.40 (iv), $|\mathrm{Aut}_K(\Sigma)| = 23$. It may be noted that $|\mathrm{Aut}_K(\Sigma)| = 23 = 2g + 1$, which agrees with Remark 11.109 (ii) for $m = 5, r = 1, n = 23$.

Calculating the number of solutions of the equation $Y^4(Y+1) = X^{23}$ over $\mathbf{F}_{2^{11}}$, it is found that the number of $\mathbf{F}_{2^{11}}$-rational places on $\mathcal{C}$ is exactly 3039. Since

$$2^{11} + 1 + 11\lfloor 2\sqrt{2^{11}} \rfloor = 3039,$$

so Σ attains the Serre Bound. Now Theorem 9.28 can be applied to calculate the L-polynomial of Σ over $\mathbf{F}_{2^{11}}$.

These results are summarised in the following theorem.

THEOREM 12.41 (i) *The number of Weierstrass points of* Σ *is* 1061.

(ii) $|\mathrm{Aut}_K(\Sigma)| = 23$.

(iii) *The number of* $\mathbf{F}_{2^{11}}$*-rational points of* $\mathcal{C}$ *attains the Serre Bound*, 3039.

(iv) *The L-polynomial of* $\mathcal{C}$ *over* $\mathbf{F}_{2^{11}}$ *is* $(1 + 90t + 2048t^2)^{11}$.

12.6 NOTES

Section 12.1 is based on Stichtenoth [424].

The properties of the curve $\mathcal{C}$ in Remark 12.12 are due to Sullivan [436]. The characterisation of the particular case is due to van deer Geer and van der Vlugt [473].

The Weierstrass points of curves of type (12.1) were investigated by Garcia and Voloch [160] in the following cases:

(i) $A(Y) = Y^q + Y,\ B(X) = X^m,\ m \mid q+1,\ m > 2;$

(ii) $A(Y) = Y^q + Y,\ B(X) = X^{q^i+1},\ i \geq 1;$

(iii) $A(Y) = Y^4 + Y^2 + Y,\ B(X) = X^7 + \lambda X,\ \lambda \in K,\ p = 2.$

For a proof of Remark 12.6, see Lehr and Matignon [302].

Section 12.2 is based on [189], [134], [174]; see also Section 10.10 and [358].

For Theorems 12.22 and 12.23, see [174].

The explicit determination of quotient curves of $\mathcal{H}_q$ appears to be difficult, especially for non-tame subgroups; Garcia, Stichtenoth, and Xing [158] calculated the genera of many such quotient curves. For the simplest case, where G has prime order, the complete description of the quotient curves in Theorem 12.28 comes from [92].

Using the method developed by Garcia, Stichtenoth, and Xing [158] for the Hermitian curve, Çakçak and Özbudak determined many quotient curves of the DLR curve and their genera. Their papers [60], [61] also contains numerous results on the arithmetic and automorphism groups of these curves.

Section 12.4 is based on [359] and [188]; see also [187].

Section 12.5 is based on [264]. A similar example for a curve of degree 12 is found in [263]. For more curves which are quotients of Fermat curves, see [485].

$$W = \begin{bmatrix}
\frac{1}{y} & \frac{x}{y} & \frac{x^2}{y} & \frac{x^2}{y} & \frac{x^3}{y^2} & \frac{x^6}{y^2} & \frac{x^7}{y^2} & \frac{x^8}{y^2} & \frac{x^{11}}{y^3} & \frac{x^{12}}{y^3} & \frac{x^{17}}{y^4} \\
\frac{x^{22}}{y^6} & \frac{1}{y^2} & \frac{(y+1)x}{y^2} & \frac{x^2}{y^2} & \frac{x^4}{y^2} & 0 & \frac{x^8}{y^2} & 0 & \frac{x^{10}}{y^4} & \frac{(y+1)x^{11}}{y^4} & \frac{x^{16}}{y^4} \\
\frac{x^{21}}{y^7} & \frac{(y+1)x^{22}}{y^7} & \frac{y^2+y+1}{y^3} & \frac{x}{y^3} & \frac{(y^2+1)x^3}{y^4} & \frac{x^4}{y^4} & \frac{x^5}{y^4} & \frac{(y^2+1)x^6}{y^4} & \frac{(y+1)x^{10}}{y^4} & \frac{(y+1)x^{10}}{y^4} & 0 \\
\frac{x^{20}}{y^8} & \frac{(y+1)x^{21}}{y^8} & \frac{(y^2+1)x^{27}}{y^8} & \frac{1}{y^4} & \frac{(y^2+1)x^2}{y^4} & 0 & \frac{x^4}{y^4} & 0 & \frac{x^8}{y^4} & \frac{(y+1)x^9}{y^4} & 0 \\
\frac{g_4x^{19}}{y^{10}} & \frac{(y^4+f_2)x^{20}}{y^{10}} & \frac{(y^4+y^3+y+1)x^{21}}{y^{10}} & \frac{f_3x^{22}}{y^{10}} & \frac{(y^4+y^2+1)x}{y^6} & \frac{x^2}{y^6} & \frac{x^3}{y^6} & \frac{(y^2+1)x^4}{y^6} & \frac{(y^4+1)x^7}{y^8} & \frac{x^8}{y^8} & \frac{(y^4+1)x^{13}}{y^8} \\
\frac{x^{18}}{y^6} & \frac{(y+1)x^{19}}{y^{10}} & \frac{(y^2+1)x^{20}}{y^8} & \frac{f_3x^{21}}{y^{10}} & \frac{(y^4+y^2+1)}{y^6} & 0 & \frac{x^2}{y^6} & 0 & \frac{(y^4+1)x^6}{y^8} & 0 & \frac{(y^4+1)x^{12}}{y^8} \\
\frac{g_4x^{17}}{y^{12}} & \frac{(y^8+g_4)x^{18}}{y^{12}} & \frac{(y^6+f_4)x^{10}}{y^{12}} & \frac{f_3x^{20}}{y^{12}} & \frac{(y^5+g_4)x^{22}}{y^{12}} & \frac{1}{y^8} & \frac{x}{y^8} & \frac{(y^2+1)x^2}{y^8} & \frac{(y^4+1)x^5}{y^8} & 0 & 0 \\
\frac{x^{16}}{y^8} & \frac{(y+1)x^{17}}{y^{12}} & \frac{(y^2+1)x^{18}}{y^8} & \frac{f_3x^{19}}{y^{12}} & \frac{(y^5+g_4)x^{21}}{y^{12}} & 0 & \frac{1}{y^8} & 0 & \frac{(y^4+1)x^4}{y^8} & 0 & 0 \\
\frac{(y^4+f_2)x^{15}}{y^{15}} & \frac{(y^8+y^7+f_2)x^{16}}{y^{15}} & \frac{(y^8+y^7+f_4)x^{17}}{y^{15}} & \frac{(y^9+y^8+f_4)x^{18}}{y^{15}} & \frac{(y^6+1)f_3x^{20}}{y^{16}} & \frac{(y^8+1)x^{21}}{y^{16}(y+1)} & \frac{(y^8+1)x^{22}}{y^{16}(y+1)} & \frac{h_4}{y^{12}} & \frac{h_3x^3}{y^{11}} & \frac{(y^8+y^7+1)x^4}{y^{11}} & \frac{(y^4+1)x^9}{y^{12}} \\
\frac{(y^8+1)x^{14}}{y^{16}} & \frac{(y^9+y^8+f_3)x^{15}}{y^{16}} & \frac{x^{16}}{y^{16}} & \frac{(y^9+y^8+f_5)x^{17}}{y^{16}} & \frac{(y^6+1)f_3x^{19}}{y^{16}} & 0 & \frac{(y^8+1)x^{21}}{y^{16}(y+1)} & 0 & \frac{h_4x^2}{y^{12}} & \frac{(y^8+1)(y+1)x^3}{y^{12}} & \frac{(y^4+1)x^8}{y^{12}} \\
\frac{(y^9+1)x^{13}}{y^{16}} & \frac{(y+1)x^{14}}{y^8} & \frac{f_3x^{15}}{y^{16}} & \frac{f_2x^{16}}{y^{15}} & \frac{(y^5+g_4)x^{18}}{y^{12}} & \frac{(y^8+1)x^{19}}{y^{16}(y+1)} & \frac{(y^8+1)x^{20}}{y^{16}(y+1)} & 0 & \frac{h_3x}{y^{11}} & \frac{(y^8+1)(y+1)x^2}{y^{11}} & 0
\end{bmatrix}$$

Figure 12.1 The Wronskian for the curve C

Chapter Thirteen

Applications: codes and arcs

Algebraic curves over a finite field are the basic structure in the theory of algebraic-geometry codes, which combines algebraic geometry and error-correcting codes.

Here, only a brief exposition of the main construction, due to Goppa, and a few illustrative examples are presented in Section 13.1.

Coding theory is also connected with algebraic curves via finite geometry, since complete arcs in $\mathrm{PG}(r, q)$ are the geometric counterpart of maximum distance separable (MDS) codes, which are linear codes correcting the greatest number of errors with respect to their length and dimension. In the other sections, an account of this topic, especially on the Main Conjecture for MDS codes, is given.

The key definitions using the terminology of linear algebra are the following.

DEFINITION 13.1 (i) Let $V(n, q)$ be the vector space on $(\mathbf{F}_q)^n$ with the usual addition and scalar multiplication:

$$(x_1, \ldots, x_n) + (y_1, \ldots, y_n) = (x_1 + y_1, \ldots, x_n + y_n),$$
$$t(x_1, \ldots, x_n) = (tx_1, \ldots, tx_n),$$

for $t, x_i, y_i \in \mathbf{F}_q$.

(ii) An $[n, k, d]_q$ *code* C is a k-dimensional subspace of $V(n, q)$, such that the number of non-zero co-ordinates of any non-zero element of C is at least d, with precisely d for some element of C.

(iii) A *generator matrix* for C is a $k \times n$ matrix whose rows form a basis for C.

13.1 ALGEBRAIC-GEOMETRY CODES

In this section, K is the algebraic closure of $\mathbf{F}_q$, and $\mathcal{F}$ is an algebraic curve defined over $\mathbf{F}_q$.

Assume that $\mathcal{F}$ has some $\mathbf{F}_q$-rational points, that is, the set $\mathcal{F}(\mathbf{F}_q)$ is not empty. For an ordered set of distinct $\mathbf{F}_q$-rational places $\mathcal{P}_1, \ldots, \mathcal{P}_n$ of $K(\mathcal{F})$ associated to the $\mathbf{F}_q$-rational points $P_1, \ldots, P_n$, let

$$D = \mathcal{P}_1 + \cdots + \mathcal{P}_n$$

be the associated $\mathbf{F}_q$-rational divisor. Further, for $\mathcal{Q}_1, \ldots, \mathcal{Q}_s$ not-necessarily-distinct $\mathbf{F}_q$-rational places of $K(\mathcal{F})$, let

$$E = \textstyle\sum_{j=1}^{s} m_j \mathcal{Q}_j,$$

with $m_j \geq 0$ and $\sum m_j = m$. Suppose that the associated $\mathbf{F}_q$-rational divisor E is disjoint from D; that is,

$$\text{Supp}(D) \cap \text{Supp}(E) = \emptyset.$$

As in Section 6.4, $\mathcal{L}(E)$ is the Riemann–Roch space of E. Then the *evaluation map* θ at $\mathcal{P}$ is

$$\theta : \mathcal{L}(E) \longrightarrow V(n, q),$$

given by

$$f \longmapsto (f(P_1), \dots, f(P_n)).$$

Now

$$\text{im}\,\theta = C = C(D, E) = (\mathcal{F}, D, E)_{\mathcal{L}}$$

is an *algebraic-geometry* code.

If $f_1, \dots, f_k$ is a basis for $\mathcal{L}(E)$, then a generator matrix for C is

$$G = \begin{bmatrix} f_1(P_1) & \dots & f_1(P_n) \\ \vdots & & \vdots \\ f_k(P_1) & \dots & f_k(P_n) \end{bmatrix}.$$

THEOREM 13.2 *If $n > m > 2g - 2$, then*

(i) $k = m - g + 1$;

(ii) $d \geq n - m$.

Proof. (i) This is precisely part (iii) of Theorem 6.61.

(ii) If $f \in \mathcal{L}(E)$ and $w(\theta(f)) = d$, then $n - d$ places $\mathcal{P}_{i_1}, \dots, \mathcal{P}_{i_{n-d}}$ are zeros of f and their sum defines the divisor D', where $\deg D' = n - d$ and $D' \prec D$. So $\text{div}(f) \succ D' - E$, whence

$$\deg \text{div}(f) \geq \deg D' - \deg E;$$

that is, $0 \geq n - d - m$. □

COROLLARY 13.3 *With $R = k/n$, $\delta = d/n$,*

(i) $n - k + 1 - g \leq d \leq n - k + 1$;

(ii) $R + \delta \geq 1 - (g - 1)/n$.

Proof. The upper bound in (i) is the Singleton bound for any linear code. The rest follows from the theorem. □

Let $C^\perp$, the dual of $C(D, E)$, be an $[n, k^\perp, d^\perp]$ code. It turns out that $C^\perp$ is also an algebraic-geometry code.

COROLLARY 13.4 *If $n > m > 2g - 2$, then*

(i) $k^\perp = n - m + g - 1$;

(ii) $d^{\perp} \geq m - 2g + 2$;

(iii) $n - k^{\perp} + 1 - g \leq d^{\perp} \leq n - k^{\perp} + 1$.

EXAMPLE 13.5 Consider the case that the curve $\mathcal{F}$ is the line $\mathbf{v}(X_2)$, and write $\mathbf{F}_q = \{t_1, \ldots, t_q\}$. Then the $\mathbf{F}_q$-rational points P, the centres of the $\mathbf{F}_q$-rational places $\mathcal{P}$ of $\mathcal{F}$, are

$$P_{t_i} = (1, t_i, 0),\ i = 1, \ldots, q, \text{ and } P_\infty = (0, 1, 0).$$

In particular, $P_0 = (1, 0, 0)$. Let $E = m\mathcal{P}_\infty$. For a generic point $Q = (1, x, 0)$ of $\mathcal{F}$, the divisor div $x^r = r\mathcal{P}_0 - r\mathcal{P}_\infty$ and $\mathcal{L}(E)$ contains the functions

$$1, x, \ldots, x^m, \tag{13.1}$$

but not the function x^{m+1}.

Also,

$$D = \mathcal{P}_{t_1} + \cdots + \mathcal{P}_{t_q}.$$

So

$$n = q, \quad k = m + 1, \quad d = n - k + 1 = n - m,$$

and

$$G = \begin{bmatrix} 1 & 1 & \ldots & 1 \\ t_1 & t_2 & \ldots & t_q \\ \vdots & \vdots & & \vdots \\ t_1^m & t_2^m & \ldots & t_q^m \end{bmatrix} \quad \begin{matrix} 1 \\ x \\ \vdots \\ x^m \end{matrix}.$$

This is a Reed–Solomon code and is MDS; see Section 13.2.

EXAMPLE 13.6 As in Example 6.71, take $q = 4$ and $\mathcal{F} = \mathbf{v}(X_0^3 + X_1^3 + X_2^3)$. Let $Q = (1, x, y)$ be a generic point of $\mathcal{F}$. The $\mathbf{F}_q$-rational points of $\mathcal{F}$ are given in Table 13.1. Let $\mathcal{S}_i$ be the place associated with S_i.

Table 13.1 $\mathbf{F}_4$-rational points of $\mathbf{v}(X_0^3 + X_1^3 + X_2^3)$

0	0	0	1	1	1	1	1	1
1	1	1	0	0	0	1	ω	ω^2
1	ω	ω^2	1	ω	ω^2	0	0	0
S_0	S_1	S_2	S_3	S_4	S_5	S_6	S_7	S_8

Consider $f_1 = 1/(x + y)$ and $f_2 = x/(x + y)$. Then

$$\begin{aligned} \text{div } f_1 &= (\mathcal{S}_0 + \mathcal{S}_1 + \mathcal{S}_2) - 3\mathcal{S}_0 = (\mathcal{S}_1 + \mathcal{S}_2) - 2\mathcal{S}_0, \\ \text{div } f_2 &= (\mathcal{S}_3 + \mathcal{S}_4 + \mathcal{S}_5) - 3\mathcal{S}_0. \end{aligned} \tag{13.2}$$

With $S_1, \ldots, S_8$ as in Table 13.1, let $E = 3\mathcal{S}_0$ and $D = \mathcal{S}_1 + \cdots + \mathcal{S}_8$. The genus $g = 1$ and $C(D, E)$ has parameters

$$n = 8, \quad k = 3, \quad 5 \leq d \leq 6.$$

Also,

$$G = \begin{bmatrix} 1 & 1 & 1 & 1 & 1 & 1 & 1 & 1 \\ 0 & 0 & 1 & \omega^2 & \omega & 1 & \omega^2 & \omega \\ \omega & \omega^2 & 0 & 0 & 0 & 1 & 1 & 1 \end{bmatrix} \begin{matrix} 1 \\ 1/(x+y) \\ x/(x+y) \end{matrix}$$

As the last row of G reveals a word of weight 5, so $d = 5$. Hence C is an $[8, 3, 5]_4$ code.

EXAMPLE 13.7 As in the previous example, let $q = 4$ and $\mathcal{F} = \mathbf{v}(X_0^3+X_1^3+X_2^3)$. Here, let $D = \mathcal{S}_0 + \mathcal{S}_1 + \cdots + \mathcal{S}_8$. Also, with $\mathcal{G} = \mathbf{v}(X_0X_1 + X_0X_2 + X_1X_2)$, let $E = \mathcal{F} \cdot \mathcal{G}$; that is, E is the intersection divisor of the curves $\mathcal{F}$ and $\mathcal{G}$. So E is an $\mathbf{F}_4$-rational divisor, even though no point in its support is $\mathbf{F}_4$-rational. In fact, the six points of the divisor E are, with τ a primitive element of $\mathbf{F}_{64}$,

$$(\tau^{9i} + 1, \tau^{-9i} + 1, 1),$$

for $i = 1, \ldots, 6$. As $\deg E = 6$, so $k = 6$. Let $v = x + y + xy$. Then a basis for $\mathcal{L}(E)$ is the following:

$$\frac{1}{v}, \frac{x^2}{v}, \frac{y^2}{v}, \frac{x}{v}, \frac{y}{v}, \frac{xy}{v}.$$

Projectively, with $v^* = x_0x_1 + x_0x_2 + x_1x_2$, this basis is as follows:

$$\frac{x_0^2}{v^*}, \frac{x_1^2}{v^*}, \frac{x_2^2}{v^*}, \frac{x_0x_1}{v^*}, \frac{x_0x_2}{v^*}, \frac{x_1x_2}{v^*}.$$

Evaluating the places $\mathcal{S}_i$ at these functions of $\mathcal{L}(E)$ gives the generator matrix G below:

$$\begin{matrix} \mathcal{S}_0 & \mathcal{S}_1 & \mathcal{S}_2 & \mathcal{S}_3 & \mathcal{S}_4 & \mathcal{S}_5 & \mathcal{S}_6 & \mathcal{S}_7 & \mathcal{S}_8 \\ \\ 0 & 0 & 0 & 1 & 1 & 1 & 1 & 1 & 1 \\ 1 & 1 & 1 & 0 & 0 & 0 & 1 & \omega & \omega^2 \\ 1 & \omega & \omega^2 & 1 & \omega & \omega^2 & 0 & 0 & 0 \end{matrix}$$

$$G = \begin{bmatrix} 0 & 0 & 0 & 1 & \omega^2 & \omega & 1 & \omega^2 & \omega \\ 1 & \omega^2 & \omega & 0 & 0 & 0 & 1 & \omega & \omega^2 \\ 1 & \omega & \omega^2 & 1 & \omega & \omega^2 & 0 & 0 & 0 \\ 0 & 0 & 0 & 0 & 0 & 0 & 1 & 1 & 1 \\ 0 & 0 & 0 & 1 & 1 & 1 & 0 & 0 & 0 \\ 1 & 1 & 1 & 0 & 0 & 0 & 0 & 0 & 0 \end{bmatrix} \begin{matrix} 1/v \\ x^2/v \\ y^2/v \\ x/v \\ y/v \\ xy/v \end{matrix}$$

Here, C is a $[9, 6, 3]_4$ code.

The dual $C^\perp$ is a $[9, 3, 6]_4$ code. This follows from the geometry of the curve $\mathcal{F}$. A generator matrix for $C^\perp$ is

$$H = \begin{bmatrix} 0 & 0 & 0 & 1 & \omega & \omega^2 & 1 & \omega & \omega^2 \\ 1 & \omega & \omega^2 & 0 & 0 & 0 & 1 & \omega^2 & \omega \\ 1 & \omega^2 & \omega & 1 & \omega^2 & \omega & 0 & 0 & 0 \end{bmatrix}.$$

Here, the nine columns of H are the nine $\mathbf{F}_4$-rational points of the cubic curve $\mathcal{F}$. If the rows of H are h_1, h_2, h_3, then an element of $C^\perp$ is $a_1h_1 + a_2h_2 + a_3h_3$. So a coordinate of this codeword is zero if $a_1x_1 + a_2x_2 + a_3x_3 = 0$ for the corresponding column ${}^t(x_1, x_2, x_3)$. Any line of $\mathrm{PG}(2, 4)$ meets $\mathcal{F}$ in at most 3 points and some, namely 12, in precisely 3 points. Hence any line misses at least 6 points of $\mathcal{F}$ and 12 miss precisely 6; so $d^\perp = 6$.

13.2 MAXIMUM DISTANCE SEPARABLE CODES

Linear codes in $V(n, q)$ and subsets of $\mathrm{PG}(k-1, q)$ are connected as follows.

DEFINITION 13.8 An $[n, k]$ system is a multi-set of n points in $\mathrm{PG}(k-1, q)$ with at most $n-d$ in a hyperplane.

REMARK 13.9 Given a generator matrix G of an $[n, k, d]_q$ code, the columns form an $[n, k]$ system of n points in $\mathrm{PG}(k-1, q)$.

The following four concepts are equivalent for $n \geq k$:

(1) (Coding theory) a *maximum distance separable* (MDS) linear code C of length n, dimension k and hence minimum distance $d = n - k + 1$, that is, an $[n, k, n-k+1]$ code over $\mathbf{F}_q$;

(2) (Matrix theory) a $k \times (n-k)$ matrix A with entries in $\mathbf{F}_q$ such that every minor is non-zero;

(3) (Vector space) a set $\mathcal{K}'$ of n vectors in $V(k, q)$, with any k linearly independent;

(4) (Projective space) an n-*arc* in $\mathrm{PG}(k-1, q)$, that is, a set $\mathcal{K}$ of n points with at most $k-1$ in any hyperplane of the projective space of $k-1$ dimensions over $\mathbf{F}_q$.

To show the equivalence, consider a generator matrix G for such a code C in canonical form:

$$\overset{n}{k \begin{bmatrix} 1 & 0 & \dots & 0 & a_{11} & \dots & a_{1,n-k} \\ 0 & 1 & \dots & 0 & a_{21} & \dots & a_{2,n-k} \\ \vdots & \vdots & & \vdots & \vdots & & \vdots \\ 0 & 0 & \dots & 1 & a_{k1} & \dots & a_{k,n-k} \end{bmatrix}} = G.$$

Since C has minimum distance $n-k+1$, any linear combination of the rows of G has at most $k-1$ zeros; that is, considering the columns of G as a set $\mathcal{K}'$ of n vectors in $V(k, q)$, any k are linearly independent. Regarding the columns of G as a set $\mathcal{K}$ of points of $\mathrm{PG}(n, q)$ means that no k lie in a hyperplane; equivalently, any k points of $\mathcal{K}$ are linearly independent. This, in turn, implies that every minor of A is non-zero.

For given k and q, let $M(k, q)$ be the maximum value of n for such a code. Then

$$M(k, q) = k + 1 \text{ for } q \leq k.$$

A suitable set of vectors in $V(k, q)$ is

$$(1, 0, \dots, 0),\ (0, 1, 0, \dots, 0),\ \dots, (0, \dots, 0, 1), (1, 1, \dots, 1);$$

that is, for $q \leq k$, every element of $V(k, q)$ is a linear combination of at most $k-1$ of these $k+1$ vectors.

REMARK 13.10 Example 13.5 gives q points in $\mathrm{PG}(m,q)$ on a normal rational curve. The system can be extended to include the point $(0,0,\ldots,0,1)$, the remaining rational point on the normal rational curve

$$\mathcal{T} = \{(s^m, s^{m-1}t, \ldots, st^{m-1}, t^m) \mid s, t \in \mathbf{F}_q\}.$$

The code can be extended to the MDS code C' by adding the transpose of this vector as an extra column of G. For C',

$$n = q+1, \quad k = m+1, \quad d = n - m = q + 2 - k = n - k + 1.$$

The Main Conjecture MC_k for MDS Codes, always taking $q > k$, is the following:

$$M(k,q) = \begin{cases} q+2 & \text{for } k = 3 \text{ and } k = q-1 \text{ both with } q \text{ even}, \\ q+1 & \text{in all other cases.} \end{cases}$$

It is convenient to have the notation $m(k-1,q) = M(k,q)$.

In the rest of this section, the main problems and related results on MDS codes are presented in terms of projective geometry.

Three problems may be enunciated.

(I) For given k and q, what is the maximum value of n such that an n-arc exists in $\mathrm{PG}(k-1,q)$? What are the n-arcs corresponding to this value of n?

(II) For what values of k and q with $q > k$ is every $(q+1)$-arc of $\mathrm{PG}(k-1,q)$ is a normal rational curve?

(III) For given k and q with $q > k$, what are the values of $n(\leq q)$ such that each n-arc is contained in a normal rational curve of $\mathrm{PG}(k-1,q)$? In how many such curves is the n-arc contained?

An n-arc is *complete* if it is maximal with respect to inclusion; that is, it is not contained in an $(n+1)$-arc. Implicit in Problem III is Problem IV, which may be enunciated as follows:

(IV) What are the values of n for which a complete n-arc exists in $\mathrm{PG}(k-1,q)$? In particular, what is the size of the second largest complete arc in this space?

From above, $m(r,q)$ is the maximum size of an arc in $\mathrm{PG}(r,q)$; also, let $m'(r,q)$ denote the size of the second largest complete arc in $\mathrm{PG}(r,q)$. Then an n-arc in $\mathrm{PG}(r,q)$ with $n > m'(r,q)$ is contained in an $m(r,q)$-arc. This is an important inductive tool.

EXAMPLE 13.11 Let Γ be the normal rational curve of $\mathrm{PG}(r,K)$ viewed as a curve over $\mathbf{F}_q$. Then the set $\Gamma(\mathbf{F}_q)$ of all $\mathbf{F}_q$-rational points is a $(q+1)$-arc of $\mathrm{PG}(r,K)$, the *classical* arc of $\mathrm{PG}(r,K)$.

If $q \geq r+2$, then any $(r+3)$-arc of $\mathrm{PG}(r,q)$ is contained in a unique classical arc.

The next result gives simultaneous information on pairs of dimensions.

THEOREM 13.12 (i) *The dual code of an MDS code is also MDS.*

(ii) *An n-arc exists in* $\mathrm{PG}(k-1,q)$ *if and only if an n-arc exists in the space* $\mathrm{PG}(n-k-1,q)$.

Proof. If $G = [I_k \mid A]$ is a generator matrix for the code C, then a generator matrix for the dual code $C^\perp$ is $H = [-\,{}^tA \mid I_{n-k}]$, where ${}^tA$ is the transpose of A. □

COROLLARY 13.13 (i) *A $(q+1)$-arc exists in* $\mathrm{PG}(k-1,q)$ *if and only if a $(q+1)$-arc exists in* $\mathrm{PG}(q-k,q)$.

(ii) *A $(q+2)$-arc exists in* $\mathrm{PG}(k-1,q)$ *if and only if a $(q+2)$-arc exists in* $\mathrm{PG}(q-k+1,q)$.

(iii) *A $(q+3)$-arc exists in* $\mathrm{PG}(k-1,q)$ *if and only if a $(q+3)$-arc exists in* $\mathrm{PG}(q-k+2,q)$.

THEOREM 13.14 *Let $a_{n,r}$ be the number of n-arcs in* $\mathrm{PG}(r,q)$ *and let ν_r be the number of normal rational curves in* $\mathrm{PG}(r,q)$. *Then*

$$a_{n,n-2-r}/a_{n,r} = \nu_{n-2-r}/\nu_r.$$

An approach to the Main Conjecture is made through the fact that, in a certain dimension, a $(q+1)$-arc is classical. In the case of q odd, this occurs in the plane, see Theorem 13.32, but for q even, it is necessary to go to a higher dimension.

THEOREM 13.15 *In* $\mathrm{PG}(4,q)$, *q even, a $(q+1)$-arc is a normal rational curve.*

The following result is discussed in Section 13.3.

$$m(2,q) = M(3,q) = \begin{cases} q+1 & \text{for } q \text{ odd,} \\ q+2 & \text{for } q \text{ even.} \end{cases}$$

Theorems 13.32 and 13.15 imply a dependency of $m(r,q)$ and $m'(2,q)$ for q odd and of $m(r,q)$ and $m'(4,q)$ for q even. Define the integer functions F and G by the purely arithmetic conditions:

$$q+1 > m'(2,q)+r-2 \Longleftrightarrow q > F(r);$$
$$q+1 > m'(4,q)+r-4 \Longleftrightarrow q > G(r).$$

There are two results that produce an inductive argument on dimension.

THEOREM 13.16 *Let $\mathcal{K}$ be an n-arc in* $\mathrm{PG}(r,q)$ *with $q+1 \geq n \geq r+3 \geq 6$ and suppose there exist $P_0, P_1 \in \mathcal{K}$ and a hyperplane π containing neither P_0 nor P_1 such that, for $i = 0,1$, the projection $\mathcal{K}_i$ of $\mathcal{K}$ onto π is $\mathbf{F}_q$-rational in π. Then the arc $\mathcal{K}$ is contained in one and only one classical arc of* $\mathrm{PG}(r,q)$.

THEOREM 13.17 *Let $\mathcal{K}$ be a $(q+2)$-arc in* $\mathrm{PG}(r,q)$ *with $q+1 \geq r+3 \geq 6$. If a hyperplane π of* $\mathrm{PG}(r,q)$ *contains neither of the points P_0, P_1 of $\mathcal{K}$, then it cannot happen that both projections $\mathcal{K}_i$ of $\mathcal{K}$ from P_i, $i = 0,1$, onto π are $\mathbf{F}_q$-rational in π. In particular, if every $(q+1)$-arc in* $\mathrm{PG}(r-1,q)$ *is classical, then* $m(r,q) = q+1$.

These two results have as corollaries the following way of approaching solutions to Problems I, II, and III.

THEOREM 13.18 *In* $\mathrm{PG}(r,q)$, *q odd*, $r \geq 3$,

(i) *if $\mathcal{K}$ is an n-arc with $n > m'(2,q) + r - 2$, then $\mathcal{K}$ lies on a unique classical arc;*

(ii) *if $q > F(r)$, then every $(q+1)$-arc is classical;*

(iii) *if $q > F(r-1)$, then $m(r,q) = q+1$.*

THEOREM 13.19 *In* $\mathrm{PG}(r,q)$, *q even*, $q > 2$, $r \geq 4$,

(i) *if $\mathcal{K}$ is an n-arc with $n > m'(4,q) + r - 4$, then $\mathcal{K}$ lies on a unique classical arc;*

(ii) *if $q > G(r)$, then every $(q+1)$-arc is classical;*

(iii) *if $q > G(r-1)$, then $m(r,q) = q+1$.*

THEOREM 13.20 *If q is even and $q > 2$, then* $m'(4,q) \leq q - \frac{1}{2}\sqrt{q} + \frac{13}{4}$.

Theorems 13.16 and 13.17 with the upper bounds for $m'(2,q)$ in Sections 13.6, 13.7, 13.8 can be applied to give the following results for q odd.

THEOREM 13.21 *For q odd*, $r \geq 3$,

(i) $F(r) \leq (4r - \frac{23}{4})^2$;

(ii) $F(r) \leq 5(9r - 19)$ *for q prime;*

(iii) $F(r) \leq 4(r+2)^2$ *when* $p \geq 5$;

(iv) $F(r) \leq 4r^2$ *when* $p \geq 5$, $q \geq 23^2$, $q \neq 3^6, 5^5$;

(v) $F(r) \leq 4r^2$ *when* $q = 3^{2e}$, $q > 3^6$;

(vi) $F(r) \leq \frac{1}{3}(4r + \frac{55}{4})^2$ *when* $q = 3^{2e+1}$.

For q even, the results depend on Theorems 13.15 and 13.20.

THEOREM 13.22 *For q even*, $q > 2$, $r \geq 4$,

$$G(r) \leq (2r - \tfrac{7}{2})^2.$$

However, it should be noted that the classification of $(q+1)$-arcs for $r \geq 4$ will not produce a simple result.

THEOREM 13.23 (Glynn) *In* $\mathrm{PG}(4,9)$, *there are precisely two projectively distinct* 10*-arcs, a normal rational curve and a non-classical arc.*

A result on the Main Conjecture is the following.

THEOREM 13.24 *For q odd and $q \leq 27$, the largest size of square matrix over $\mathbf{F}_q$ with every minor non-zero is $\frac{1}{2}(q+1)$.*

As an illustration,

$$A = \begin{bmatrix} 1 & 1 & 1 & 1 & 1 & 1 & 1 & 1 & 1 & 1 \\ 1 & 10 & 13 & 5 & 4 & 16 & 11 & 12 & 17 & 2 \\ 10 & 13 & 5 & 4 & 16 & 11 & 12 & 17 & 2 & 7 \\ 13 & 5 & 4 & 16 & 11 & 12 & 17 & 2 & 7 & 8 \\ 5 & 4 & 16 & 11 & 12 & 17 & 2 & 7 & 8 & 3 \\ 4 & 16 & 11 & 12 & 17 & 2 & 7 & 8 & 3 & 15 \\ 16 & 11 & 12 & 17 & 2 & 7 & 8 & 3 & 15 & 14 \\ 11 & 12 & 17 & 2 & 7 & 8 & 3 & 15 & 14 & 6 \\ 12 & 17 & 2 & 7 & 8 & 3 & 15 & 14 & 6 & 9 \\ 17 & 2 & 7 & 8 & 3 & 15 & 14 & 6 & 9 & 18 \end{bmatrix}$$

is a 10×10 matrix with the property that every minor is non-zero modulo 19; further, there is no square matrix of larger size over $\mathbf{F}_{19}$ with this property.

Near-MDS codes are linear codes whose parameters differ only slightly from those of MDS codes.

DEFINITION 13.25 A linear $[n,k]_q$-code is a *near-MDS code* if $d(C) = n-k$ and $d(C^{\perp}) = k$.

Like MDS codes, near-MDS codes can be investigated within finite projective geometry, since a near-MDS code $[n,k]_q$ can be viewed as a point set C of size n in $\mathrm{PG}(k-1,q)$ satisfying the following conditions:

(i) every $k-1$ points in C generate a hyperplane of $\mathrm{PG}(k-1,q)$;

(ii) there exist k points in C lying in a hyperplane of $\mathrm{PG}(k-1,q)$;

(iii) any $k+1$ points in C generate $\mathrm{PG}(k-1,q)$.

In the special case of near-MDS codes of dimension 4, this geometric representation shows that a near-MDS $[n,4]_q$ code is a point set C of size n in $\mathrm{PG}(3,q)$ with the following properties:

(i) C is a cap, that is, no three points in C are collinear;

(ii) C is not an arc, that is, C contains four coplanar points;

(iii) no five points in C are coplanar.

Let $K = \overline{\mathbf{F}}_q$. The set of $\mathbf{F}_q$-rational points of an elliptic quartic Γ_4 of $\mathrm{PG}(3,K)$ defined over $\mathbf{F}_q$ is a natural example of such a point set C with n equal to the number N_1 of all $\mathbf{F}_q$-rational points of Γ_4. Since such an elliptic quartic is birationally equivalent over $\mathbf{F}_q$ to a non-singular plane cubic defined over $\mathbf{F}_q$, an $[n,4]_q$ near-MDS code exists provided that there exists a non-singular plane cubic over $\mathbf{F}_q$ whose number of $\mathbf{F}_q$-rational points is exactly n.

13.3 ARCS AND OVALS

As explained in Section 13.2, MDS codes and arcs in finite projective spaces are equivalent objects. In this section, arcs in planes are investigated by using previous results on curves with many $\mathbf{F}_q$-rational points.

Some standard notation from finite geometry used in this chapter is as follows.

NOTATION 13.26 (i) In $\mathrm{PG}(2,q)$, the coordinates are x_0, x_1, x_2 with corresponding indeterminates X_0, X_1, X_2.

(ii) In $\mathrm{PG}(2,q)$, let

$$\mathbf{U}_0 = (1,0,0),\ \mathbf{U}_1 = (0,1,0),\ \mathbf{U}_2 = (0,0,1),\ \mathbf{U} = (1,1,1);$$
$$\mathbf{u}_0 = \mathbf{v}(X_0),\ \mathbf{u}_1 = \mathbf{v}(X_1),\ \mathbf{u}_2 = \mathbf{v}(X_2),\ \mathbf{u} = \mathbf{v}(X_0 + X_1 + X_2).$$

(iii) Write $\mathbf{N}_s = \{1,2,\ldots,s\}$ for $s > 0$.

(iv) For binomial coefficients, write

$$\mathbf{c}(n,r) = \binom{n}{r}.$$

The concept of an oval in finite geometry arises from two combinatorial properties of a closed convex curve in the real plane.

DEFINITION 13.27 A set Ω of points in $\mathrm{PG}(2,q)$ is an *oval* if

(i) no three points in Ω are collinear;

(ii) at every point $P \in \Omega$ there is exactly one *unisecant*, that is, a line of $\mathrm{PG}(2,q)$ that meets Ω only in P.

In this setting, a k-arc in $\mathrm{PG}(2,q)$ is a set Ω of k-points that has property (i). In particular, there are three types of lines with respect to an arc: (a) bisecants or chords, (b) unisecants and (c) external lines or lines disjoint from the arc. A combinatorial argument shows that

$$k \leq \begin{cases} q+1 & \text{when } q \text{ odd,} \\ q+2 & \text{when } q \text{ even,} \end{cases}$$

for every k-arc in $\mathrm{PG}(2,q)$.

A *dual k-arc in* $\mathrm{PG}(2,q)$ is a k-arc in the dual plane, that is, a set of k lines in $\mathrm{PG}(2,q)$ with no three concurrent. In the study of arcs, the set of points of $\mathrm{PG}(2,q)$ lying on exactly one of the lines of a dual k-arc $\mathcal{K}$ plays an important role. Such a set $\Delta(\mathcal{K})$ has size kt with $t = q - k + 2$, each line of $\mathcal{K}$ containing exactly t points from $\Delta(\mathcal{K})$.

Ovals are exactly the $(q+1)$-arcs in $\mathrm{PG}(2,q)$. An example of an oval is the set of all $\mathbf{F}_q$-rational points of an irreducible conic defined over $\mathbf{F}_q$. For q odd, every oval in $\mathrm{PG}(2,q)$ is a complete arc. This does not hold true for q even because an irreducible conic in even characteristic is a strange curve, and hence, adding the nucleus to the $\mathbf{F}_q$-rational points of the conic, a $(q+2)$-arc in $\mathrm{PG}(2,q)$, necessarily complete, is obtained. More generally, every oval Ω in $\mathrm{PG}(2,q)$ with q even has a nucleus; that is, its unisecants have a common point N; the oval Ω, together with N, forms a complete $(q+2)$-arc of $\mathrm{PG}(2,q)$, called a *hyperoval*. Therefore

$$m(2,q) = \begin{cases} q+1 & \text{when } q \text{ is odd,} \\ q+2 & \text{when } q \text{ is even,} \end{cases}$$

Note that, if $\mathcal{K}$ is a k-arc with $k > m'(2,q)$, then $\mathcal{K}$ is contained in an oval of $\mathrm{PG}(2,q)$.

It is shown below that every oval in $\mathrm{PG}(2,q)$ with q odd consists of the $\mathbf{F}_q$-rational points of an irreducible conic defined over $\mathbf{F}_q$. Although ovals for q even have been classified for $q \leq 32$, the classification of ovals for $q \geq 64$ appears difficult and is a major challenge in finite geometry.

Menelaus' classical theorem that gives a condition for three points one on each side of a triangle to be collinear is generalised to show the existence of a plane curve passing through each point in $\Delta(\mathcal{K})$, the point set arising from a dual arc $\mathcal{K}$ of $\mathrm{PG}(2,q)$. Such a curve $\mathcal{F}$, which may be singular and even reducible, is defined over $\mathbf{F}_q$; it has degree t, as small as possible, for q even, and $2t$, larger but still low, for q odd. If k is large, then $\mathcal{F}$ is irreducible and an interesting case occurs; namely, $\mathcal{F}$ has low degree, and hence low genus, compared to its large number of $\mathbf{F}_q$-rational points.

Referring to the results in Chapter 9 on the maximum number of $\mathbf{F}_q$-rational points that a curve defined over $\mathbf{F}_q$ can have, it is conceivable that this is an unusual occurrence. Large complete arcs are rare, but they can be investigated by studying particular curves defined over $\mathbf{F}_q$ which have many $\mathbf{F}_q$-rational points. This is done in Section 13.4.

The starting point of the study of ovals in odd characteristic is a series of combinatorial results.

Lemma 13.28 *In* $\mathrm{PG}(2,q)$ *for q odd, every point off an oval* Ω *lies on exactly two or no unisecants of* Ω.

Proof. Since Ω consists of $q+1$ points, there is exactly one unisecant through each point of Ω. Let ℓ be the unisecant at P to Ω and let $Q \in \ell\backslash\{P\}$. If $\sigma_i(Q)$, with $i = 1,2$, denotes the number of i-secants to Ω through Q, then

$$\sigma_1(Q) + 2\sigma_2(Q) = q+1$$

and, since q is odd, $\sigma_1(Q)$ is even. As ℓ itself is a unisecant through Q, it follows that $\sigma_1(Q) \geq 2$. Since this is true for each of the q points Q in $\ell\backslash\{P\}$ and since Ω has exactly $q+1$ unisecants, $\sigma_1(Q) = 2$ for all such points Q; that is, through every point of $\ell\backslash\{P\}$, there is exactly one other unisecant. □

A point of $\mathrm{PG}(2,q)$ is *external* or *internal* to the oval Ω according as it lies on two or no unisecants of Ω. Hence, with respect to Ω, the q^2+q+1 points of $\mathrm{PG}(2,q)$ are partitioned into three classes:

(a) $q+1$ points on Ω;

(b) $\frac{1}{2}q(q+1)$ external points;

(c) $\frac{1}{2}q(q-1)$ internal points.

Since a k-arc in the dual plane is a dual k-arc, the q^2+q+1 lines of $\mathrm{PG}(2,q)$ are partitioned similarly into three classes with respect to Ω:

(a′) $q+1$ unisecants;

(b′) $\frac{1}{2}q(q+1)$ bisecants;

(c′) $\frac{1}{2}q(q-1)$ external lines.

For any k-arc $\mathcal{K}$ with $3 \leq k \leq q+1$, choose three of its points as the triangle of reference $\mathbf{U}_0\mathbf{U}_1\mathbf{U}_2$ of the coordinate system. A unisecant to $\mathcal{K}$ through one of $\mathbf{U}_0, \mathbf{U}_1, \mathbf{U}_2$ has the respective form

$$\mathbf{v}(X_1 - dX_2), \quad \mathbf{v}(X_2 - dX_0), \quad \mathbf{v}(X_0 - dX_1),$$

with $d \neq 0$. Here, d is the *coordinate* of such a line.

Suppose the $t = q+2-k$ unisecants to $\mathcal{K}$ at each of $\mathbf{U}_0, \mathbf{U}_1, \mathbf{U}_2$ are

$$\mathbf{v}(X_1 - a_iX_2), \quad \mathbf{v}(X_2 - b_iX_0), \quad \mathbf{v}(X_0 - c_iX_1),$$

$i \in \mathbf{N}_t$.

LEMMA 13.29 (Segre's Lemma of Tangents) *The coordinates a_i, b_i, c_i of the unisecants at $\mathbf{U}_0, \mathbf{U}_1, \mathbf{U}_2$ to a k-arc $\mathcal{K}$ through these points satisfy the identity*

$$\textstyle\prod_{i=1}^{t} a_ib_ic_i = -1.$$

Proof. If $P = (d_0, d_1, d_2)$ is any point of $\mathcal{K}$ other than $\mathbf{U}_0, \mathbf{U}_1, \mathbf{U}_2$, then the lines $P\mathbf{U}_0, P\mathbf{U}_1, P\mathbf{U}_2$ are, respectively,

$$\mathbf{v}(X_1 - eX_2), \quad \mathbf{v}(X_2 - fX_0), \quad \mathbf{v}(X_0 - gX_1),$$

where $e = d_1/d_2, f = d_2/d_0, g = d_0/d_1$ and so

$$efg = 1. \tag{13.3}$$

Through $\mathbf{U}_0$ there are $q-1$ lines other than $\mathbf{U}_0\mathbf{U}_1, \mathbf{U}_0\mathbf{U}_2$; they consist of t unisecants

$$\mathbf{v}(X_1 - a_iX_2),\ i \in \mathbf{N}_t,$$

and $k-3$ bisecants

$$\mathbf{v}(X_1 - e_jX_2),\ j \in \mathbf{N}_{k-3}.$$

Since the product of the non-zero elements of $\mathbf{F}_q$ is -1, so

$$\textstyle\prod a_i \prod e_j = -1.$$

Similarly, for the $q-1$ lines through $\mathbf{U}_1$ and through $\mathbf{U}_2$ other than the sides of the triangle of reference,

$$\textstyle\prod b_i \prod f_j = -1, \quad \prod c_i \prod g_j = -1.$$

Hence

$$\textstyle\prod (a_ib_ic_i) \prod (e_jf_jg_j) = (-1)^3 = -1.$$

However, as $e_jf_jg_j = 1$ for each j by (13.3), so $\prod(a_ib_ic_i) = -1$. □

COROLLARY 13.30 *If the unisecants to a $(q+1)$-arc $\mathcal{K}$ at the points $\mathbf{U}_0, \mathbf{U}_1, \mathbf{U}_2$ of $\mathcal{K}$ are $\mathbf{v}(X_1 - aX_2), \mathbf{v}(X_2 - bX_0), \mathbf{v}(X_0 - cX_1)$, then $abc = -1$.*

LEMMA 13.31 *For q odd, the triangles formed by three points of a $(q+1)$-arc $\mathcal{K}$ and the unisecants at these points are in perspective.*

Proof. Choose the three points as $\mathbf{U}_0, \mathbf{U}_1, \mathbf{U}_2$; then the unisecants are the lines $\mathbf{v}(X_1 - aX_2), \mathbf{v}(X_2 - bX_0), \mathbf{v}(X_0 - cX_1)$, with $abc = -1$. The vertices V_0, V_1, V_2 of the triangle with the unisecants as sides are

$$(c, 1, bc), \quad (ca, a, 1), \quad (1, ab, b).$$

Hence the lines $\mathbf{U}_0 V_0, \mathbf{U}_1 V_1, \mathbf{U}_2 V_2$ are

$$\mathbf{v}(X_2 - bcX_1), \quad \mathbf{v}(X_0 - caX_2), \quad \mathbf{v}(X_1 - abX_0).$$

These three lines are concurrent if and only if $a^2b^2c^2 = 1$, which is satisfied since $abc = -1$. It may be noted that, since the three unisecants are not concurrent, by Lemma 13.28, so $abc \neq 1$. □

THEOREM 13.32 (Segre) *In* PG$(2, q)$, *with q odd, every oval consists of all $\mathbf{F}_q$-rational points of an irreducible conic defined over $\mathbf{F}_q$.*

Proof. Let $\mathcal{K}$ be the oval and choose any three of its points as $\mathbf{U}_0, \mathbf{U}_1, \mathbf{U}_2$. Also, let the unit point $\mathbf{U}$ be the point of perspectivity of the triangle $\mathbf{U}_0\mathbf{U}_1\mathbf{U}_2$ and its circumscribed triangle as in the previous lemma. As $\mathbf{U}$ is the intersection of $\mathbf{v}(X_0 - caX_2)$ and $\mathbf{v}(X_1 - abX_0)$, which is $(ca, -a, 1)$, so $a = b = c = -1$. Thus the unisecants to $\mathcal{K}$ at $\mathbf{U}_0, \mathbf{U}_1, \mathbf{U}_2$ are

$$\mathbf{v}(X_1 + X_2), \quad \mathbf{v}(X_2 + X_0), \quad \mathbf{v}(X_0 + X_1).$$

Let $P = (y_0, y_1, y_2)$ be any other point of $\mathcal{K}$ and let the unisecant to $\mathcal{K}$ at P be $\ell = \mathbf{v}(d_0X_0 + d_1X_1 + d_2X_2)$. Lemma 13.29 applied to the triangle $P\mathbf{U}_1\mathbf{U}_2$ gives the condition

$$(d_0 - d_1 - d_2)[d_1(y_0 + y_1) - d_2(y_0 + y_2)] = 0.$$

However, by Lemma 13.28, the line ℓ cannot pass through the meet of the unisecants at $\mathbf{U}_1$ and $\mathbf{U}_2$; this point is $(1, -1, -1)$ and so $d_0 - d_1 - d_2 \neq 0$. Thus

$$d_1(y_0 + y_1) = d_2(y_0 + y_2);$$

similarly, from the triangles $P\mathbf{U}_2\mathbf{U}_0$ and $P\mathbf{U}_0\mathbf{U}_1$,

$$d_2(y_1 + y_2) = d_0(y_1 + y_0),$$
$$d_0(y_2 + y_0) = d_1(y_2 + y_1).$$

Hence

$$d_0 : d_1 : d_2 = y_1 + y_2 : y_2 + y_0 : y_0 + y_1.$$

As $d_0y_0 + d_1y_1 + d_2y_2 = 0$ and $2 \neq 0$, so

$$y_1y_2 + y_2y_0 + y_0y_1 = 0.$$

Hence all points of $\mathcal{K}$ lie on the irreducible conic $\mathcal{C} = \mathbf{v}(X_1X_2 + X_2X_0 + X_0X_1)$ defined over $\mathbf{F}_q$. As the set $\mathcal{C}(\mathbf{F}_q)$ of all $\mathbf{F}_q$-rational points of $\mathcal{C}$ has size $q + 1$, it coincides with $\mathcal{K}$. □

The preceding proof is typical in this subject. A known property of conics is proved for ovals and this property is used to show that an oval is a conic. In other words, a property which characterises conics among algebraic curves also characterises ovals among k-arcs.

13.4 SEGRE'S GENERALISATION OF MENELAUS' THEOREM

An *n-point* in $\mathrm{PG}(2, K)$ is a set of n points, no three of which are collinear, together with the $\frac{1}{2}n(n-1)$ lines that are joins of pairs of the points. The points and lines are called *vertices* and *sides* of the n-point. The vertices form an *n-arc*.

Dually, an *n-line* in $\mathrm{PG}(2, K)$ is a set of n lines, no three of which are concurrent, together with the $\frac{1}{2}n(n-1)$ points that are intersections of pairs of the lines. The points and lines are again called *vertices* and *sides* of the n-line. The sides form a *dual n-arc*.

An *n-gon* is an ordered set $(P_1, P_2, \ldots, P_n)$ of n points, no three of which are collinear: the P_i are the *vertices* and the n lines $P_1P_2, P_2P_3, \ldots, P_{n-1}P_n, P_nP_1$ are the *sides* of the n-gon.

For small n, the following terms are used:

	n-point	n-line	n-gon
$n=3$	triangle	triangle	triangle
$n=4$	quadrangle	quadrilateral	quadrangle, quadrilateral
$n=5$	pentastigm	pentagram	pentagon
$n=6$	hexastigm	hexagram	hexagon

A triangle with vertices A, B, C and sides a, b, c is denoted ABC or abc when there is no risk of confusing it with the entire plane.

Let $\mathcal{C}_n$ be a curve of degree n defined over $\mathbf{F}_q$. Certain problems on the intersection of two curves $\mathcal{C}_n$ and $\mathcal{C}_k$ are considered, particularly when $\mathcal{C}_k$ is completely reducible and its linear components are sides of a k-line. Menelaus' Theorem gives a necessary and sufficient condition for three points, one on each side of a triangle, to be collinear. In its Euclidean form, this is expressed as a product of ratios of distances being -1. This is now generalised projectively.

Let $\mathcal{K} = \{l_1, l_2, \ldots, l_k\}$ with $l_i = \mathbf{v}(L_i)$ and L_i a linear polynomial over $\mathbf{F}_q$. Let $\mathcal{C}_k = \mathbf{v}(L_1 \cdots L_k)$. Consider the k-line $\mathcal{K}'$ that has $\mathcal{K}$ as its set of sides. Let $A_{ij} = l_i \cap l_j$, let $\mathcal{A}_i = \{A_{ij} \mid j \neq i\}$ be the set of vertices on l_i, and also let $\mathcal{A} = \{A_{ij} \mid i, j \in \mathbf{N}_k\}$ be the set of all the vertices of $\mathcal{K}'$. Consider on each l_i a multi-set G_n^i whose support is disjoint from $\mathcal{A}$, and let

$$G = \cup_{i=1}^k G_n^i.$$

For any $\mathcal{C}_k$ such that $\mathcal{C}_n$ contains no point of $\mathcal{A}$, the multi-set G consists of all common points P of $\mathcal{C}_n$ and $\mathcal{C}_k$, each counted $I(P, \mathcal{C}_n \cap \mathcal{C}_k)$ times.

Conditions are required under which a given G is the complete intersection of $\mathcal{C}_k$ with some $\mathcal{C}_n$. In fact, given G on $\mathcal{K}$, necessary and sufficient conditions are found for the set $\mathcal{M}_G = \{\mathcal{C}_n \mid \mathcal{C}_n \cap \mathcal{C}_k = G\}$ to be non-empty.

Let $\mathcal{L}$ be the linear system of all curves $\mathcal{C}_n$ through the points of G; that is, $I(P, \mathcal{C}_n, \mathcal{C}_k) \geq n_P$ for every point P with weight n_P in G. Geometrically speaking, $\mathcal{C}_n \in \mathcal{L}$ if $\mathcal{C}_n$ meets $\mathcal{K}$ at least in G. Then $\mathcal{L}$ contains, as a linear subsystem, the set $\mathcal{M}_G$ consisting of all curves $\mathcal{C}_n$ that meets $\mathcal{K}$ exactly in G.

The dimension of $\mathcal{L}$ is found, as well as the number of conditions for a plane curve $\mathcal{C}_n$ to belong to $\mathcal{L}$.

LEMMA 13.33 (i) *$\mathcal{C}_n$ meets $\mathcal{K}$ in G and some further point P if and only if all the lines l_i of $\mathcal{K}$ are components of $\mathcal{C}_n$.*

(ii) *The set $\mathcal{L}\backslash\mathcal{M}_G$ of reducible curves containing $\mathcal{C}_k$ as a component is a linear subsystem $\mathcal{L}_\mathcal{K}$ of $\mathcal{L}$ whose dimension is equal to $\mathbf{c}(n-k+2,2)-1$.*

Proof. If $P \in l_1$, then $\mathcal{C}_n$ contains $n+1$ points $\{P\} \cup G_n^1$ of l_1; so $\mathcal{C}_n$ contains l_1. However, any other line l_i meets l_1 in A_{1i}. So each l_i meets $\mathcal{C}_n$ in at least $n+1$ points. Hence $\mathcal{C}_n$ contains all l_i. Thus any curve of $\mathcal{L}$ not in $\mathcal{M}_G$ contains $\mathcal{K}$, and $\mathcal{L}_\mathcal{K}$ consists of all curves $\mathbf{v}(L_1L_2\cdots L_kF_{n-k})$, where F_{n-k} is a form of degree $n-k$; such a curve has as components the k lines l_i and $\mathcal{C}_{n-k} = \mathbf{v}(F_{n-k})$. Thus $\mathcal{L}_\mathcal{K}$ is a linear subsystem of the required dimension. □

THEOREM 13.34 *Suppose $\mathcal{L}$ is not empty, and let $\mathcal{C}'_n = \mathbf{v}(F')$ be a curve in $\mathcal{L}$.*

(i) *If $k > n$, then $\mathcal{L}_\mathcal{K} = \emptyset$ and $\mathcal{L} = \mathcal{M}_G = \{\mathcal{C}'_n\}$.*

(ii) *If $\mathcal{L}_\mathcal{K} \neq \emptyset$, then $\mathcal{M}_G$ is not a linear subsystem of $\mathcal{L}$. However, if $\mathcal{C}_n \in \mathcal{L}$ and $\mathcal{C}'_n \in \mathcal{M}_G$, there is a curve $\mathcal{C}_{n-k} = \mathbf{v}(F_{n-k})$ such that*

$$\mathcal{C}_n = \mathbf{v}(F' + L_1L_2\cdots L_kF_{n-k}).$$

(iii) *If d is the dimension of $\mathcal{L}$, then $d = \mathbf{c}(n-k+2,2)$.*

Proof. (i) If $k > n$, no $\mathcal{C}_n$ can have $\mathcal{C}_k$ as a component. So $\mathcal{L}_\mathcal{K} \neq \emptyset$. From Lemma 13.33, $\mathcal{L} = \mathcal{M}_G$ and $\mathcal{L}$ consists of the single curve $\mathcal{C}'_n$.

(ii) If $\mathcal{L}_\mathcal{K} \neq \emptyset$, then $\mathcal{L}_\mathcal{K}$ and $\mathcal{M}_G$ form a partition of $\mathcal{L}$. So $\mathcal{M}_G$ cannot be a linear subsystem. Take a point $P \in l_1$ not in G. If $\mathcal{C}_n = \mathbf{v}(F) \in \mathcal{L}$ and $\mathcal{C}'_n \in \mathcal{M}_G$, then some member of the pencil $\mathbf{v}(sF + tF')$ contains and so is in $\mathcal{L}_\mathcal{K}$. Hence $\lambda F = F' + L_1L_2\cdots L_kF_{n-k}$ for some λ and some F_{n-k}.

(iii) From (ii) and Lemma 13.33, $d = \dim \mathcal{L}_\mathcal{K} + 1 = \mathbf{c}(n-k+2,2)$. □

COROLLARY 13.35 (i) *If some curve meets $\mathcal{K}$ exactly in G, then G imposes*

$$\mathbf{c}(n+2,2) - \mathbf{c}(n-k+2,2) = \tfrac{1}{2}k(2n-k+3) - 1$$

conditions on a $\mathcal{C}_n$ to meet $\mathcal{K}$ in G.

(ii) *If $n \geq k$, then $\mathcal{K}$ imposes*

$$\mathbf{c}(n+2,2) - \mathbf{c}(n-k+2,2) = \tfrac{1}{2}k(2n-k+3)$$

conditions on a $\mathcal{C}_n$ to meet $\mathcal{K}$ in G.

THEOREM 13.36 *If $k = 1$, then $\mathcal{M}_G$ is never empty and the n conditions imposed by G are independent.*

Proof. If $\mathcal{K} = \{l_1\}$, $G = \{P_1, \ldots, P_n\}$, $P \notin l_1$, and $PP_i = \mathbf{v}(L'_i)$, all i, then $\mathbf{v}(L'_1L'_2\cdots L'_n)$ is in $\mathcal{M}_G$. □

THEOREM 13.37 *If $k = 2$, then $\mathcal{M}_G$ is never empty and the $2n$ conditions imposed by G are independent.*

Proof. If $\mathcal{K} = \{l_1, l_2\}$, $G_n^i = \{P_1^i, \ldots, P_n^i\}$, and $P_j^1 P_j^2 = \mathbf{v}(L_j')$, all j, then $\mathbf{v}(L_1' L_2' \cdots L_n')$ is in $\mathcal{M}_G$. □

Let the coordinate system in PG$(2, q)$ be fixed. If the point P lies on a side of $\mathbf{U}_0\mathbf{U}_1\mathbf{U}_2$ and is not a vertex, it is the meet of that side and one of

$$\mathbf{v}(X_1 - cX_2), \quad \mathbf{v}(X_2 - cX_0), \quad \mathbf{v}(X_0 - cX_1);$$

then c is the *coordinate* of P.

THEOREM 13.38 *Let $\mathcal{K}$ be the triangle $\mathbf{U}_0\mathbf{U}_1\mathbf{U}_2$.*

(i) *Then $\mathcal{M}_G \neq \emptyset$ if and only if*

$$\prod c = (-1)^n,$$

where the product ranges over all not-necessarily-distinct $3n$ points P of G and c is the coordinate of P.

(ii) *If $\mathcal{M}_G \neq \emptyset$, then G imposes $3n - 1$ conditions on the curves of order n to meet $\mathcal{K}$ in G; if $n = 1$ or 2, then $\mathcal{M}_G$ consists of a single $\mathcal{C}_n$.*

Proof. If $\mathcal{C}_n \in \mathcal{M}_G$ and $\mathcal{C}_n = \mathbf{v}(c_0X_0^n + c_1X_1^n + c_2X_2^n + \cdots)$, then $c_0c_1c_2 \neq 0$ since $\mathcal{C}_n$ contains no vertex of $\mathbf{U}_0\mathbf{U}_1\mathbf{U}_2$. Let

$$\begin{aligned} \mathcal{C}_n^0 &= \mathbf{v}(c_1X_1^n + \cdots + c_2X_2^n), \\ \mathcal{C}_n^1 &= \mathbf{v}(c_0X_0^n + \cdots + c_2X_2^n), \\ \mathcal{C}_n^2 &= \mathbf{v}(c_0X_1^n + \cdots + c_1X_1^n). \end{aligned}$$

Then, for $i = 0, 1, 2$,

$$G_n^i = \{n_P P \mid P \in \mathbf{u}_i;\ n_P = I(P, \mathbf{u_i} \cap \mathcal{C}_n^i)\}.$$

Hence

$$\prod c = (-1)^n c_2/c_1,\ (-1)^n c_0/c_2,\ (-1)^n c_1/c_0,$$

according as the product is taken over G_n^0, G_n^1, G_n^2. So $\prod c = (-1)^n$, where the product is taken over all points of G.

Now suppose that the condition is satisfied by the points of G; then $3n - 1$ points of G determine the remaining one, P say. Let $\mathcal{L}_P$ be the linear system of $\mathcal{C}_n$ through the points of $G \backslash \{P\}$. Then $\mathcal{L} \subset \mathcal{L}_P$ and, if d is the dimension of $\mathcal{L}_P$,

$$0 \leq \mathbf{c}(n+2, 2) - 1 - (3n - 1) = \mathbf{c}(n-1, 2) \leq d \leq \mathbf{c}(n+2, 2) - 1.$$

Since the dimension of $\mathcal{L}_\mathcal{K}$ is $\mathbf{c}(n-1, 2) - 1 < d$, so $\mathcal{L}_P \backslash \mathcal{L}_\mathcal{K} \neq \emptyset$. Now

$$\mathcal{M}_G = \mathcal{L} \backslash \mathcal{L}_\mathcal{K} \subset \mathcal{L}_P \backslash \mathcal{L}_\mathcal{K}.$$

However, if $\mathcal{C}_n' \in \mathcal{L}_P \backslash \mathcal{L}_\mathcal{K}$, it meets $\mathcal{K}$ in G', where $G' = (G \backslash \{P\}) \cup \{Q\}$ and $Q \in (\bigcup l_i) \backslash \mathcal{A}$. But G' also satisfies the given condition. So $Q = P$ and $G' = G$. Hence $\mathcal{C}_n' \in \mathcal{M}_G$ and $\mathcal{M}_G = \mathcal{L}_P \backslash \mathcal{L}_\mathcal{K} \neq \emptyset$. □

For $n = 1$, this is Menelaus' Theorem and, for $n \geq 2$, Carnot's Theorem. It is now generalised to the case that $k \geq 4$.

THEOREM 13.39 *Let $\mathcal{K} = \{l_1, l_2, \ldots, l_k\}$ be a set of k lines of* $\mathrm{PG}(2, K)$ *and let G_j be a multi-set of size n whose support is on l_j, such that no point of intersection of any two of the k lines lies in the support of any G_j. Suppose that, for each $J \subset \mathbf{N}_k$ of the following special types, there is a plane curve $\mathcal{C}_n^J$ of degree n that meets l_j in G_j exactly:*

(a) $\{l_j \mid j \in J\}$ *is a triangle;*

(b) $\{l_j \mid j \in J\}$ *is a pencil.*

Then

(i) *there is a plane curve $\mathcal{C}_n$ that meets l_j in G_j exactly, for all j;*

(ii) *if there is such a curve $\mathcal{C}_n$ and $k > n$, then $\mathcal{C}_n$ is unique.*

Proof. If all the lines are concurrent there is nothing to prove. Also the result is established for a triangle in Theorem 13.38. So assume that $k \geq 4$ and that not all the lines are concurrent; the proof is by induction.

Let $P_i = l_i \cap l_k$ for $i = 1, 2, \ldots, k-1$ and suppose that $P_1 \neq P_2$. Let $l_k = \mathbf{u}_0$ and $P_i = (0, c_i, d_i)$, $i \in \mathbf{N}_{k-1}$. By induction, there is a curve $\mathcal{C}_n^i = \mathbf{v}(F_i)$ with the correct intersection with all lines l_j, $j \neq i$.

Now $F_1(0, X_1, X_2)$ and $F_2(0, X_1, X_2)$ have degree n, and both meet l_k in G_k; hence $F_1(0, X_1, X_2)$ and $F_2(0, X_1, X_2)$ only differ by a constant factor, and may be assumed equal. Also, P_2 does not lie on the curve $\mathcal{C}_n^1$, and so $F_1(0, c_2, d_2) \neq 0$; similarly, $F_k(0, c_2, d_2) \neq 0$. Let

$$\begin{aligned}\beta' &= -F_k(0, c_2, d_2)/F_1(0, c_2, d_2),\\ F'(X_0, X_1, X_2) &= F_k(X_0, X_1, X_2) + \beta' F_1(X_0, X_1, X_2),\end{aligned}$$

and let $\mathcal{C}' = \mathbf{v}(F')$; if $F'(X_0, X_1, X_2)$ is the zero polynomial, the result is proved. Note that $\beta' \neq 0$.

The curve $\mathcal{C}'$ meets l_2 in G_2 and contains P_2; hence, by Bézout's Theorem, l_2 is a linear component of $\mathcal{C}'$. Then, for each $j = 3, 4, \ldots, k-1$, the curve $\mathcal{C}'$ meets the line l_j in G_j and also contains the point $l_j \cap l_2$; hence l_j is a linear component of $\mathcal{C}'$ for $j = 3, 4, \ldots, k-1$. Thus, if $P_i = \mathbf{v}(X_0) \cap \mathbf{v}(R_i)$ with $R_i = d_i X_1 - c_i X_2$, then

$$F'(0, X_1, X_2) = F_k(0, X_1, X_2) + \beta' F_1(0, X_1, X_2)$$

is divisible by $R_2 R_3 \cdots R_{k-1}$.

Now interchange the roles of l_1 and l_2 in the above argument. Put

$$\begin{aligned}\beta'' &= -F_k(0, c_1, d_1)/F_2(0, c_1, d_1),\\ F''(X_0, X_1, X_2) &= F_k(X_0, X_1, X_2) + \beta'' F_2(X_0, X_1, X_2),\end{aligned}$$

and let $\mathcal{C}'' = \mathbf{v}(F'')$. As before, if $F'' = 0$, the result is proved. Now

$$\begin{aligned}F'(0, X_1, X_2) - F''(0, X_1, X_2) &= \beta' F_1(0, X_1, X_2) - \beta'' F_2(0, X_1, X_2)\\ &= (\beta' - \beta'') F_1(0, X_1, X_2),\end{aligned}$$

and so $F'(0, X_1, X_2) - F''(0, X_1, X_2) = 0$ meets l_k in G_k and the point P_3, since $k \geq 4$. Hence $(\beta' - \beta'')F_1(0, X_1, X_2)$ is the zero polynomial, and so $\beta' = \beta''$. It follows that

$$\begin{aligned} F''(0, X_1, X_2) &= F_k(0, X_1, X_2) + \beta'' F_2(0, X_1, X_2) \\ &= F_k(0, X_1, X_2) + \beta' F_1(0, X_1, X_2) \end{aligned}$$

is divisible by $R_1 R_3 R_4 \cdots R_{k-1}$. Hence $F'(0, X_1, X_2) = F''(0, X_1, X_2)$ is divisible by $R_1 R_2 \cdots R_k$; here, the fact that $P_1 \neq P_2$ and so $R_1 \neq cR_2$ is used. So

$$F_k(0, X_1, X_2) + \beta' F_1(0, X_1, X_2) = R_1 R_2 \cdots R_{k-1} R.$$

Now let $l_i = \mathbf{v}(L_i)$ with $L_i = e_i X_0 + R_i$ for $i \in \mathbf{N}_{k-1}$ and define $\mathcal{C}_n = \mathbf{v}(F)$, where

$$F(X_0, X_1, X_2) = F_k(X_0, X_1, X_2) - L_1 L_2 \cdots L_{k-1} R.$$

Then $\mathcal{C}_n$ meets l_j in G_j for $j \in \mathbf{N}_{k-1}$. The curve $\mathcal{C}_n$ meets l_k where

$$F_k(0, X_1, X_2) - R_1 R_2 \cdots R_{k-1} R = 0,$$

that is, where $\beta' F_1(0, X_1, X_2) = 0$; hence $\mathcal{C}_n$ meets l_k in G_k. Hence $\mathcal{C}_n$ is the required curve.

Finally, suppose that $k > n$, and that $\mathcal{C}_n = \mathbf{v}(F)$ and $\mathcal{D}_n = \mathbf{v}(G)$ are curves satisfying the hypotheses. To obtain a contradiction, suppose that

$$H(X_0, X_1, X_2) = F(X_0, X_1, X_2) + \lambda G(X_0, X_1, X_2),$$

with $\lambda = -F(0, c_1, d_1)/G(0, c_1, d_1)$, is not the zero polynomial. Then the curve $\mathcal{E}_n = \mathbf{v}(H)$ has degree n. By Bézout's Theorem, $\mathcal{E}_n$ contains the line l_k as a linear component and so contains the points $P_1, P_2, \ldots, P_{k-1}$. Again, by Bézout's Theorem, $\mathcal{E}_n$ contains the lines l_j for $j = 1, 2, \ldots, k-1$. Hence $\mathcal{E}_n$ of degree n contains $k > n$ distinct linear components, a contradiction. Hence H is the zero polynomial, whence $\mathcal{C}_n = \mathcal{D}_n$. □

13.5 THE CONNECTION BETWEEN ARCS AND CURVES

The aim of this section is to associate a plane curve defined over $\mathbf{F}_q$ to a dual k-arc in $\mathrm{PG}(2, q)$. This shows that the dual view is useful in the study of k-arcs.

The essential result on dual k-arcs is stated and proved separately for q even and odd. As before, let $\mathcal{K}$ be a k-arc, let $\mathcal{K}'$ be a dual k-arc, and let $t = q - k + 2$. Then $\Delta(\mathcal{K}')$ stands for the set consisting of the kt points in $\mathrm{PG}(2, q)$ which lie on one but not on two of the lines in $\mathcal{K}'$.

THEOREM 13.40 *Let $\mathcal{K}'$ be a dual k-arc in $\mathrm{PG}(2, q)$ with q even. Then the tk points of $\Delta(\mathcal{K}')$ belong to a plane curve Γ_t of degree t with the following properties*:

(i) *Γ_t is unique if $k > t$, that is, $k > \frac{1}{2}q + 1$;*

(ii) *the points of $\mathcal{K}'$ distinct from those in $\Delta(\mathcal{K}')$ do not belong to Γ_t, and so no line of $\mathcal{K}'$ is a component of Γ_t;*

(iii) *if ℓ is any line of $\mathcal{K}'$ and $P \in \ell \cap \Delta(\mathcal{K}')$, then $I(P, \ell \cap \Gamma_t) = 1$; in particular, P is a non-singular, and hence an $\mathbf{F}_q$-rational point of Γ_t;*

(iv) *every linear component of Γ_t is defined over $\mathbf{F}_q$; if $k \geq 5$, this also holds for every irreducible component of Γ_t of degree two.*

Proof. By Lemma 13.29, $\prod c = 1$ for each triangle of $\mathcal{K}'$, where the product ranges over all points of $\Delta(\mathcal{K}')$ on the sides of the triangle and c is the coordinate of such a point. By Theorem 13.39, there exists a plane curve Γ_t define over $\mathbf{F}_q$, which is unique if $k > t$.

Also, by one of the hypotheses in Theorem 13.39, the points of the lines of $\mathcal{K}'$ other than those of $\Delta(\mathcal{K}')$ do not belong to Γ_t, and so no line of $\mathcal{K}'$ is component of Γ_t. Hence every line of $\mathcal{K}'$ contains at most t points of Γ_t. As each of the t points in $\Delta(\mathcal{K}')$ lying on a line ℓ of $\mathcal{K}$ is in Γ_t, none is counted twice in the intersection of ℓ and Γ_t.

If a line ℓ' is a linear component of Γ_t, then it meets every line of $\mathcal{K}'$ at a point in $\Delta(\mathcal{K}')$. Hence Γ_t contains a point in $\mathrm{PG}(2, q)$ from each line of $\mathcal{K}'$. Since $\mathcal{K}'$ consists of more than one line, Γ_t contains two distinct points of $\mathrm{PG}(2, q)$, and hence is defined over $\mathbf{F}_q$. The same argument works for a quadratic component $\mathcal{C}$ of Γ_t when $k \geq 5$, the assumption being necessary as there exist irreducible conics not defined over $\mathbf{F}_q$ which contain only four points of $\mathrm{PG}(2, q)$. □

COROLLARY 13.41 *In $\mathrm{PG}(2, q)$ with q even,*

(i) *a $(q + 1)$-arc Ω is incomplete;*

(ii) *the $q + 1$ unisecants to Ω are concurrent and Ω lies in a unique $(q + 2)$-arc.*

Proof. In the dual plane of $\mathrm{PG}(2, q)$, the set Ω is a dual $(q + 1)$-arc, and so $t = 1$. Therefore Γ_t is a line. In $\mathrm{PG}(2, q)$, the corresponding point added to Ω gives a $(q + 2)$-arc. □

COROLLARY 13.42 *If $\mathcal{K}$ is an incomplete k-arc with $3 \leq k \leq q+1$ in $\mathrm{PG}(2, q)$, q even, and if $k > \frac{1}{2}q + 1$, then*

(i) *there is a point R on no bisecant of $\mathcal{K}$;*

(ii) *if $\mathcal{K}'$ is the dual k-arc arising from $\mathcal{K}$ in the dual plane of $\mathrm{PG}(2, q)$ and Γ_t is a curve associated to $\mathcal{K}'$ via Segre's generalisation of Menelaus' Theorem, then the line ℓ corresponding to R in the dual plane is a component of Γ_t;*

(iii) *Γ_t is unique.*

Proof. (i) Since $\mathcal{K}$ is incomplete, there is such an R.

(ii) Let $\mathcal{K}'$ be the dual k-arc arising from $\mathcal{K}$ in the dual plane of $\mathrm{PG}(2, q)$. In that plane, R defines a line ℓ such that the points of ℓ cut out by the lines of $\mathcal{K}'$ are distinct. There are k such points, each lying in $\Delta(\mathcal{K}')$. By Theorem 13.40, there exists a curve Γ_t defined over $\mathbf{F}_q$ that contains these k points of ℓ. Since Γ_t has degree t and the hypothesis $k > \frac{1}{2}q + 1$ implies that $k > t$, the line ℓ must be a component of Γ_t.

(iii) Again by Theorem 13.40, since $k > t$, so Γ_t is unique. □

A result similar to Theorem 13.40 is established for q odd. Now

$$-1 = 1 \text{ for } q \text{ even}; \quad (-1)^2 = 1 \text{ for } q \text{ odd}.$$

This is the simple reason that the results for q odd are weaker than for q even; Lemma 13.29 can no longer be directly applied to Theorems 13.38 and 13.39.

Theorem 13.43 *Let $\mathcal{K}'$ be a dual k-arc in $\mathrm{PG}(2, q)$ with q odd. Then the tk points of $\Delta(\mathcal{K}')$, where $t = q + k - 2$, belong to a plane curve Γ_{2t} defined over $\mathbf{F}_q$ of degree $2t$ with the following properties:*

(i) *Γ_{2t} is unique if $k > 2t$, that is, $k > \frac{1}{3}(2q + 4)$;*

(ii) *the points of the lines of $\mathcal{K}'$ distinct from those in $\Delta(\mathcal{K}')$ do not belong to Γ_{2t} and so no line of $\mathcal{K}$ is a component of Γ_{2t};*

(iii) *if ℓ is any line of $\mathcal{K}'$ and $P \in \ell \cap \Delta(\mathcal{K}')$, then $I(P, \ell \cap \Gamma_{2t}) = 2$;*

(iv) *Γ_{2t} may contain components of multiplicity at most two but does not consist entirely of double components;*

(v) *every linear component of Γ_{2t} is defined over $\mathbf{F}_q$; if $k \geq 5$, this also holds for every irreducible component of Γ_{2t} of degree two.*

Proof. By Lemma 13.29, $\prod c = -1$ for each triangle of $\mathcal{K}'$, where the product ranges over all points of $\Delta(\mathcal{K}')$ on the sides of the triangle and c is the coordinate of such a point. So $\prod c^2 = 1$. By Theorem 13.39, there exists a plane curve Γ_{2t} define over $\mathbf{F}_q$, which is unique if $k > 2t$. Also, by one of the hypotheses in Theorem 13.39, the points of the lines of $\mathcal{K}'$ distinct those from $\Delta(\mathcal{K}')$ do not belong to Γ_{2t}, and so no line of $\mathcal{K}'$ is component of Γ_{2t}. Hence every line of $\mathcal{K}'$ contains at most $2t$ points of Γ_{2t}.

As each of the t points in $\Delta(\mathcal{K}')$ lying on a line ℓ of $\mathcal{K}'$ is in Γ_{2t} and counted twice in the intersection of ℓ and Γ_{2t}, there are no components of multiplicity three. If Γ_{2t} consisted entirely of double components, the points in $\Delta(\mathcal{K}')$ would belong to a curve of degree t and the condition $\prod d = 1$ would be satisfied for any triangle of $\mathcal{K}'$. Finally, the proof of (iv) of Theorem 13.40 also shows (v). □

Corollary 13.44 *Let $\mathcal{K}$ be an incomplete k-arc with $3 \leq k < q + 1$, q odd. If $k > \frac{2}{3}(q + 2)$, that is, $k > 2t$, then*

(i) *there is a point R on no bisecant of $\mathcal{K}$;*

(ii) *if $\mathcal{K}'$ is the dual k-arc arising from $\mathcal{K}$ in the dual plane of $\mathrm{PG}(2, q)$ and Γ_{2t} is the curve associated to $\mathcal{K}'$, then the line ℓ corresponding to R in the dual plane is a component of Γ_{2t};*

(iii) Γ_{2t} *is unique.*

Proof. The arguments used to prove Corollary 13.42 remain valid for odd q when $k > 2t$, this assumption being necessary to ensure the validity of (ii). By Theorem 13.43, there exists a plane curve Γ_{2t} defined over $\mathbf{F}_q$ such that $I(P, \Gamma_{2t} \cap \ell) \geq 2$ for every point $P \in \ell \cap \Delta(\mathcal{K}')$. Since Γ_{2t} has degree $2t$, the hypothesis $k > 2t$ implies that the line ℓ must be a component of Γ_t. □

REMARK 13.45 Segre's Theorem 13.32 can be deduced from Theorem 13.43. As $k = q + 1$, so $t = 1$. Hence, by Theorem 13.43, the points of the dual $(q+1)$-arc $\mathcal{K}'$ belong to a conic Γ_2 defined over $\mathbf{F}_q$ which does not split into two, possibly coincident, lines, as otherwise the $(q+1)$-arc $\mathcal{K}$ would be incomplete. Also, since $I(P, \Gamma_2 \cap l) = 2$ for $P \in l \cap \Delta(\mathcal{K}')$, so the lines of $\mathcal{K}'$ are tangents to Γ_2 at the points of $\Delta(\mathcal{K}')$. Since the number of $\mathbf{F}_q$-rational points of Γ_2 is exactly $q + 1$, so $\mathcal{K}'$ can be viewed as the envelope of the conic Γ_2 at its $\mathbf{F}_q$-rational points. After dualising, Theorem 13.32 follows.

EXAMPLE 13.46 As in Example 11.58, in PG(2, 13) there exists exactly one complete 12-arc, up to a projectivity, as in Table 13.2. In this case, $\Delta(\mathcal{K}')$ has size 36

Table 13.2 The 12-arc $\mathcal{K}$ and its dual $\mathcal{K}'$

	$\mathcal{K}$	$\mathcal{K}'$	
P_1	$(1,1,1)$	$X+Y+Z$	t_1
P_2	$(1,-1,1)$	$X-Y+Z$	t_2
P_3	$(-1,1,1)$	$-X+Y+Z$	t_3
P_4	$(-1,-1,1)$	$-X-Y+Z$	t_4
P_5	$(3,4,1)$	$3X+4Y+Z$	t_5
P_6	$(3,-4,1)$	$3X-4Y+Z$	t_6
P_7	$(-3,4,1)$	$-3X+4Y+Z$	t_7
P_8	$(-3,-4,1)$	$-3X-4Y+Z$	t_8
P_9	$(4,3,1)$	$4X+3Y+Z$	t_9
P_{10}	$(4,-3,1)$	$4X-3Y+Z$	t_{10}
P_{11}	$(-4,3,1)$	$-4X+3Y+Z$	t_{11}
P_{12}	$(-4,-3,1)$	$-4X-3Y+Z$	t_{12}

and its points are listed in Table 13.3.

Table 13.3 The points of $\Delta(\mathcal{K}')$

$(\pm1,\pm2,1)$	$(\pm2,\pm1,1)$	$(\pm2,\pm5,1)$
$(\pm5,\pm2,1)$	$(\pm3,\pm5,1)$	$(\pm5,\pm3,1)$
$(\pm4,\pm6,1)$	$(\pm6,\pm4,1)$	$(\pm6,\pm6,1)$

By Theorem 13.43, no plane cubic contains all points in $\Delta(\mathcal{K}')$, but there is a plane sextic that does, namely the non-singular plane curve Γ_6 defined over $\mathbf{F}_{13}$:

$$\Gamma_6 = \mathbf{v}(X_0^6 + X_1^6 + X_2^6 + 3X_0^2X_1^2X_2^2). \tag{13.4}$$

The lines of $\mathcal{K}'$ are each tangent to Γ_6 at 3 points of $\Delta(\mathcal{K}')$. In addition, Γ_6 contains 18 points on the coordinate axes; these are the points

$$B_t = (0, t, 1), \quad C_t = (t, 0, 1), \quad D_t = (t, 1, 0),$$

with $t^6 = -1$, that is, $t = \pm 2, \pm 5, \pm 6$.

These 54 points in $\mathrm{PG}(2, 13)$ are all $\mathbf{F}_{13}$-rational points of Γ_6. It should be noted that Γ_6 is optimal in that it attains the upper bound in Theorem 8.65 applied to an $\mathbf{F}_{13}$-rational linear series of dimension 2. Here, $g = 10$, $q = 13$, $n = 6$ and $S_{13} = 54$. Therefore 54 is the largest number of $\mathbf{F}_{13}$-rational points of an irreducible plane curve defined over $\mathbf{F}_{13}$ that has genus 10.

EXAMPLE 13.47 Let $f(X) = a_0X^2 + a_1X + a_2$ be a primitive polynomial of $\mathbf{F}_{q^2}[X]$; that is, f is irreducible with one root, and so the other, generating the group $\mathbf{F}_{q^4}\backslash\{0\}$. The projectivity $\mathfrak{T}$ of $\mathrm{PG}(2, q^2)$ associated to the matrix

$$\begin{bmatrix} 0 & 1 & 0 \\ 0 & 0 & 1 \\ a_2 & a_1 & a_0 \end{bmatrix}$$

is a Singer cycle; that is, $\mathfrak{T}$ acts on the set of points of $\mathrm{PG}(2, q^2)$, as well as on the set of lines of $\mathrm{PG}(2, q^2)$, as a single cycle. In particular, $\mathfrak{T}$ has order $q^4 + q^2 + 1$, and hence $\mathfrak{S} = \mathfrak{T}^{q^2+q+1}$ is a projectivity of order $q^2 - q + 1$. Also, every point orbit of $\mathfrak{S}$ is a complete $(q^2 - q + 1)$-arc $\mathcal{K}$ in $\mathrm{PG}(2, q^2)$, and every line orbit of $\mathfrak{S}$ is a complete dual $(q^2 - q + 1)$-arc $\mathcal{K}'$ in $\mathrm{PG}(2, q^2)$. For even q, the plane curve Γ_{q+1} associated to $\mathcal{K}'$ is the Hermitian curve; see Theorem 13.49. This does not hold true for odd q, the associated curve $\Gamma_{2(q+1)}$ being projectively equivalent over $\mathbf{F}_{q^6}$ to the irreducible plane curve, in affine form,

$$\mathbf{v}(Y^2 + X^2Y^{2q} + X^{2q} - 2(X^{q+1}Y^q + X^qY + XY^{q+1})).$$

However, $\Gamma_{2(q+1)}$ is birationally equivalent over $\mathbf{F}_{q^6}$ to the Hermitian curve; see Theorem 13.64.

13.6 ARCS IN OVALS IN PLANES OF EVEN ORDER

In this section, the size $m'(2, q)$ of the second largest complete arc in $\mathrm{PG}(2, q)$, q even, is obtained.

THEOREM 13.48 *In* $\mathrm{PG}(2, q)$, q *even and* $q > 4$,

(i)
$$m'(2, q) \leq q - \sqrt{q} + 1; \tag{13.5}$$

(ii) *when* q *is also square, then*

$$m'(2, q) = q - \sqrt{q} + 1. \tag{13.6}$$

Proof. Let $\mathcal{K}$ be a dual k-arc in $\mathrm{PG}(2,q)$ with $k > q - \sqrt{q} + 1$. Without loss of generality, $\mathcal{K}$ is assumed to be complete. Let Γ_t be the associated curve of degree t as in Theorem 13.40. Note that $t = q - k + 2 < \sqrt{q} + 1$.

Suppose that Γ_t has an $\mathbf{F}_q$-rational linear component; that is, there is a line ℓ of $\mathrm{PG}(2,q)$ contained in Γ_t. Then the common point of ℓ and any line in $\mathcal{K}$ is in $\Delta(\mathcal{K})$. Therefore ℓ added to $\mathcal{K}$ is a dual $(k+1)$-arc, a contradiction to $\mathcal{K}$ being complete. Thus Γ_t does not contain an $\mathbf{F}_q$-rational linear component.

Let S_q be the number of $\mathbf{F}_q$-rational points of Γ_t. From Theorem 13.40 (iii),

$$|S_q| \geq |\Delta(\mathcal{K})| = kt.$$

On the other hand, Theorem 9.67 applied to Γ_t implies that $S_q < kt$. This contradiction proves that $m'(2,q) \leq q - \sqrt{q} + 1$. If q is square, then Example 13.47 implies (13.6). □

It is still an open problem to determine all complete $(q-\sqrt{q}+1)$-arcs in $\mathrm{PG}(2,q)$ with q even and square. The following result is a step in this direction.

THEOREM 13.49 *Let $\mathcal{K}$ be a complete dual $(q^2 - q + 1)$-arc in $\mathrm{PG}(2,q^2)$ with q even. Then the associated curve Γ_t is the Hermitian curve $\mathcal{H}_q$.*

Proof. Here, $t = q + 1$. So Γ_t has genus $g \leq \frac{1}{2}(q^2 - q)$. From the proof of Theorem 13.48, the number of $\mathbf{F}_q$-rational points of Γ_t is at least $kt = q^3 + 1$. By the Hasse–Weil theorem, this is only possible when $g = \frac{1}{2}q(q-1)$ and the number of $\mathbf{F}_q$-rational points of Γ_t is exactly $q^3 + 1$. In particular, Γ_t is a non-singular model of an $\mathbf{F}_{q^2}$-maximal curve of genus $g = \frac{1}{2}q(q-1)$. Thus Γ_t is the Hermitian curve $\mathcal{H}_q$. □

13.7 ARCS IN OVALS IN PLANES OF ODD ORDER

In this section, a similar treatment to the previous section is given for planes of odd order. An upper bound is obtained for $m'(2,q)$ in the case that q is odd.

The next result is both weaker and stronger, in the sense that the arithmetic condition is stronger, as is the specification of the complete arc containing $\mathcal{K}$.

THEOREM 13.50 *Let $\mathcal{K}'$ be a dual k-arc with $3 \leq k \leq q+1$ in $\mathrm{PG}(2,q)$, q odd, and let Γ_{2t} be the associated curve. If*

(a) *Γ_{2t} has an irreducible conic $\mathcal{C}$ defined over $\mathbf{F}_q$ as a component,*

(b) $k > \frac{1}{4}(3q+5)$,

then the lines of $\mathcal{K}'$ are tangents to $\mathcal{C}$.

Proof. The lines of $\mathcal{K}'$ are of two types with respect to $\mathcal{C}$ according as the line is a tangent to $\mathcal{C}$ or not. The lines of the first type form a dual k_1-arc $\mathcal{K}'_1$, and those of the second type form a dual k_2-arc $\mathcal{K}'_2$ such that $k_1 + k_2 = k = q - t + 2$. Also, the $\mathbf{F}_q$-rational tangents to $\mathcal{C}$ distinct from the lines in $\mathcal{K}'_1$ form a dual k_3-arc $\mathcal{K}'_3$ with

$k_3 = q+1-k_1$. Since $\mathcal{C}$ meets a line at most two points, so $k_1+2k_2 \le q+1$. Hence

$$k_1 \ge q+3-2t, \quad k_2 \le t-1, \quad k_3 \le 2t-2.$$

Let ℓ be a line in $\mathrm{PG}(2,q)$ which is not a tangent to $\mathcal{C}$. At least $\frac{1}{2}(q-1)$ points of ℓ in $\mathrm{PG}(2,q)$ are the intersections of two tangents to $\mathcal{C}$ at $\mathbf{F}_q$-rational points. Also, $k > \frac{1}{4}(3q+5)$ implies that $\frac{1}{2}(q-1) > 2t-2$. As $k_3 \le 2t-2$, at least one point $P \in \ell$ must be the meet of two $\mathbf{F}_q$-rational tangents to $\mathcal{C}$ not contained in $\mathcal{K}'_3$. So both are lines in $\mathcal{K}'_1$. As ℓ is not an $\mathbf{F}_q$-rational tangent to $\mathcal{C}$, so ℓ is not in $\mathcal{K}'$. Thus $\mathcal{K}'$ consists of $\mathbf{F}_q$-rational tangents to $\mathcal{C}$. □

The next result is a key one.

THEOREM 13.51 *In* $\mathrm{PG}(2,q)$, *q odd,*

$$m'(2,q) \le q - \tfrac{1}{4}\sqrt{q} + \tfrac{7}{4}.$$

Equivalently, if $\mathcal{K}$ is a k-arc with $k > q - \frac{1}{4}\sqrt{q} + \frac{7}{4}$, that is, $\sqrt{q} > 4t-1$, then $\mathcal{K}$ is contained in the set of $\mathbf{F}_q$-rational points of a unique irreducible conic defined over $\mathbf{F}_q$.

Proof. If $k > q - \frac{1}{4}\sqrt{q} + \frac{7}{4}$, then $k \ge 5$. So, if $\mathcal{K}$ is contained in the set of $\mathbf{F}_q$-rational points of a conic, then the conic is unique.

Consider the dual plane of $\mathrm{PG}(2,q)$. Let $\mathcal{K}'$ be the dual k-arc arising from $\mathcal{K}$. Theorem 13.43 gives all the necessary information on the curve Γ_{2t} defined over $\mathbf{F}_q$ associated to $\mathcal{K}'$. For $t=1$, such a curve is an irreducible conic $\mathcal{C}$ defined over $\mathbf{F}_q$, and $\mathcal{K}'$ consists of lines tangents to $\mathcal{C}$. So it may be assumed that $t \ge 2$.

Choose an irreducible component Γ'_n of Γ_{2t}, where $n \le 2t$, and distinguish the following three cases:

(i) Γ'_n is an $\mathbf{F}_q$-rational linear component of Γ_{2t};

(ii) Γ'_n is an $\mathbf{F}_q$-rational component of degree two;

(iii) Γ'_n is an $\mathbf{F}_q$-rational component of degree at least three or Γ'_n is not defined over $\mathbf{F}_q$.

Case (i). Here, Γ'_n is an $\mathbf{F}_q$-rational line ℓ which is not in $\mathcal{K}'$. So $\mathcal{K}'_1 = \mathcal{K}' \cup \{\ell\}$ is a dual $(k+1)$-arc. If the envelope Γ'_n associated with $\mathcal{K}'_1$ is again of type (i), the process is continued until the dual arc becomes the set of all tangents to an $\mathbf{F}_q$-rational conic or the respective plane curve becomes one of the other two types.

Case (ii). Now Γ'_n is an $\mathbf{F}_q$-rational conic $\mathcal{C}$. Also,

$$k > q - \tfrac{1}{4}\sqrt{q} + \tfrac{7}{4} \Longleftrightarrow q > 4t-1 \Rightarrow q > 4t-3 \Longleftrightarrow k > \tfrac{1}{4}(3q+5).$$

So, by Theorem 13.50, $\mathcal{K}'$ consists of $\mathbf{F}_q$-rational tangents to $\mathcal{C}$.

Case (iii). Here Γ'_n is $\mathbf{F}_q$-rational with $3 \le n \le 2t$ or Γ'_n is not $\mathbf{F}_q$-rational. Suppose that Γ'_n has S non-singular and d double points in $\mathrm{PG}(2,q)$. From Theorem 9.69,

$$S + d < \tfrac{1}{2}n(q-t+2) = \tfrac{1}{2}nk. \tag{13.7}$$

By Theorem 13.43, $I(P, \Gamma_{2t} \cap \ell) = 2$ holds for each of the t points $P \in \Delta(\mathcal{K}')$ of a line ℓ in $\mathcal{K}'$. Therefore $I(P, \Gamma'_n \cap \ell) \leq 2$, and $I(Q, \Gamma'_n \cap \ell) = 0$ for $Q \notin \Delta(\mathcal{K}')$. Since

$$\textstyle\sum_{Q \in \ell} I(Q, \Gamma'_n \cap \ell) = n,$$

this implies that Γ'_n has at least $\frac{1}{2}n$, possibly double, points from ℓ in $\mathrm{PG}(2, q)$. Hence

$$S + d \geq kn/2,$$

contradicting the previous estimate (13.7).

Thus $\mathcal{K}$ is ever contained in the set of $\mathbf{F}_q$-rational points of an irreducible conic defined over $\mathbf{F}_q$. □

COROLLARY 13.52 *If $\mathcal{K}$ is a k-arc in $\mathrm{PG}(2, q)$, q odd, with $k > q - \frac{1}{4}\sqrt{q} + \frac{7}{4}$, then $\mathcal{K}$ is contained in the set $\mathcal{C}(\mathbf{F}_q)$ of all $\mathbf{F}_q$-rational points of an irreducible conic $\mathcal{C}$ defined over $\mathbf{F}_q$, and the only points for which $\mathcal{K} \cup \{Q\}$ is a $(k+1)$-arc are the $q + 1 - k$ points of $\mathcal{C}(\mathbf{F}_q) \backslash \mathcal{K}$.*

Proof. If $\mathcal{K} \cup \{Q\}$ is a $(k + 1)$-arc, then it lies in some complete arc, which is necessarily $\mathcal{C}(\mathbf{F}_q)$ by the theorem. Hence $Q \in \mathcal{C}(\mathbf{F}_q) \backslash \mathcal{K}$. □

COROLLARY 13.53 *In $\mathrm{PG}(2, q)$, q odd and $q > 49$, a q-arc is contained in the set of $\mathbf{F}_q$-rational points of a conic.*

Proof. The inequality $q > q - \frac{1}{4}\sqrt{q} + \frac{7}{4}$ holds when $q > 49$. □

REMARK 13.54 This result in fact holds true for $q \leq 49$, giving the similar result in the case that q is even.

Next, a small improvement to Theorem 13.51 is given. There the result is equivalent to $t < \frac{1}{4}(\sqrt{q}+1)$ for a k-arc that is contained in a conic, where $t = q+2-k$.

THEOREM 13.55 (Thas) *In $\mathrm{PG}(2, q)$, q odd,*

(i) *a k-arc $\mathcal{K}$ for which $t < (\sqrt{q} + 1)^2/(4\sqrt{q} + 1)$ is contained in the set of $\mathbf{F}_q$-rational points of an irreducible conic defined over $\mathbf{F}_q$;*

(ii) $m'(2, q) < q - \frac{1}{4}\sqrt{q} + \frac{25}{16}$.

Proof. (i) If $t = 1$, then $\mathcal{K}$ is a $(q + 1)$-arc, that is, an oval in $\mathrm{PG}(2, q)$, and the assertion follows from Segre's Theorem 13.32. So assume that $t > 1$. In particular, $q > 3$.

Let $\mathcal{K}'$ be the dual k-arc of $\mathcal{K}$. Assume that the plane curve Γ_{2t} defined over $\mathbf{F}_q$, associated to $\mathcal{K}'$ via Segre's generalisation of Menelaus' Theorem, is irreducible. With R the number of points of Γ_{2t} with coordinates in $\mathbf{F}_q$, (iii) of Theorem 9.57 gives that $R \leq q + 1 + (2t - 1)(2t - 2)\sqrt{q}$. So, as $\mathcal{K}$ has kt unisecants, that is, $|\Delta(\mathcal{K}')| = kt$,

$$q + 1 + (2t - 1)(2t - 2)\sqrt{q} \geq kt,$$
$$(t - 1)\{q + 1 + 2\sqrt{q} - t(1 + 4\sqrt{q})\} \leq 0.$$

Hence $t \geq (\sqrt{q}+1)^2/(4\sqrt{q}+1)$, a contradiction. Therefore Γ_{2t} is reducible.

At least one irreducible component Γ'_n has degree $n \leq t$. Consider the same three cases as in the proof of Theorem 13.51.

If Γ'_n is $\mathbf{F}_q$-rational of degree $n = 1$, then Γ'_n is a line ℓ not in $\mathcal{K}'$. So $\mathcal{K}' \cup \{\ell\}$ is a dual $(k+1)$-arc.

If Γ'_n is $\mathbf{F}_q$-rational of degree $n = 2$, then Γ'_n is an irreducible conic $\mathcal{C}$. Since $k < \frac{1}{4}(3q+5)$, that is, $t < \frac{1}{4}(q+3)$, Theorem 13.50 implies that $\mathcal{K}'$ is a set of tangents to $\mathcal{C}$ at $\mathbf{F}_q$-rational points.

Finally, suppose that Γ'_n is not $\mathbf{F}_q$-rational or is $\mathbf{F}_q$-rational with $n \geq 3$. Then

$$\{3(\sqrt{q}+1)^2/(4\sqrt{q}+1)\} - 1 \leq \sqrt{q},$$

since $q > 3$, and so $3t - 1 < \sqrt{q}$. By Theorem 9.69,

$$R < \tfrac{1}{2}n(q-t+2) = \tfrac{1}{2}nk.$$

From the proof of Theorem 13.51, Γ'_n has at least $\frac{1}{2}n$, possibly double, points from ℓ in $\mathrm{PG}(2,q)$, where ℓ is a line of $\mathcal{K}'$. Hence $R \geq \frac{1}{2}nk$, a contradiction.

Thus $\mathcal{K}'$ is contained in a dual $(k+1)$-arc. Continuing the process shows that $\mathcal{K}'$ is contained in a dual $(q+1)$-arc. Then Theorem 13.50 implies that $\mathcal{K}'$ is a set of tangents to $\mathcal{C}$ at $\mathbf{F}_q$-rational points.

(ii) The inequality $t < (\sqrt{q}+1)^2/(4\sqrt{q}+1)$ is equivalent to

$$k > q - \tfrac{1}{4}\sqrt{q} + \tfrac{25}{16} - \{\tfrac{9}{16}/(4\sqrt{q}+1)\}. \qquad \square$$

13.8 THE SECOND LARGEST COMPLETE ARC: FURTHER RESULTS

In this section similar techniques to the previous section are used, but the Stöhr–Voloch theorem is applied instead of the Hasse–Weil theorem. This makes it possible to obtain improvements on upper bounds for $m'(2,q)$.

THEOREM 13.56 *In* $\mathrm{PG}(2,p)$, *with* p *an odd prime and* $p \geq 7$,

$$m'(2,p) \leq \tfrac{44}{45}p + \tfrac{8}{9}.$$

Proof. Suppose that there exists a complete k-arc $\mathcal{K}$ with $k > \frac{44}{45}p + \frac{8}{9}$.

As in the proof of Theorem 13.51, the same three possibilities for an irreducible component Γ'_n of Γ_{2t} are considered, where Γ_{2t} is the $\mathbf{F}_q$-rational curve associated to the dual k-arc $\mathcal{K}'$.

(i) If Γ'_n is $\mathbf{F}_q$-rational and $n = 1$, then $\mathcal{K}'$ is incomplete.

(ii) If Γ'_n is $\mathbf{F}_q$-rational and $n = 2$, then Γ'_n is an irreducible conic $\mathcal{C}$ defined over $\mathbf{F}_q$. However,

$$k > \tfrac{44}{45}p + \tfrac{8}{9} > \tfrac{1}{4}(3p+5).$$

So, by Theorem 13.50, $\mathcal{K}$ is contained in the set $\mathcal{C}(\mathbf{F}_q)$ of all $\mathbf{F}_q$-rational points of $\mathcal{C}$.

(iii) Suppose that Γ'_n has S non-singular and d singular points in $\mathrm{PG}(2,q)$. Then, as in the proof of Theorem 13.51,

$$S + d \geq kn/2.$$

If Γ'_n is not $\mathbf{F}_q$-rational, then $S + d \le n^2$; so $k \le 2n \le 4t = 4(p+2-k)$. Hence $k \le \frac{4}{5}(p+2) < \frac{44}{45}p + \frac{8}{9}$ for $p \ge 5$.

Now let Γ'_n be $\mathbf{F}_q$-rational with $n \ge 3$. The linear system of all conics cuts out on Γ'_n the fixed-point-free $\mathbf{F}_q$-rational linear series L_2 that defines the Veronese map. The resulting curve is an irreducible $\mathbf{F}_q$-rational curve Γ of $\mathrm{PG}(5, K)$ of degree $2n$ where $K = \overline{\mathbf{F}}_q$. Let $\nu_0 = 0, \nu_1, \ldots, \nu_4$ be the Frobenius orders of Γ. From the fact that $\nu_i \le \deg \Gamma = 2n = 4t$ and $k > \frac{1}{4}(3p+8)$ for $p \ge 5$, it follows that $2n < p$. As $p \ge 5$, Proposition 8.50 (ii) implies that Γ is a Frobenius classical curve, that is, $\nu_i = i$ for $0 \le i \le 4$. Therefore the Stöhr–Voloch Theorem 8.65 gives that

$$\tfrac{1}{2}kn - d \le \tfrac{2}{5}n\{5(n-2)+p\} - 4d.$$

So

$$\begin{aligned} k &\le \tfrac{4}{5}\{5(n-2)+p\} \\ &\le \tfrac{4}{5}\{5(2t-2)+p\} \\ &= \tfrac{4}{5}\{10(p+1-k)+p\} \\ &= 8(p+1-k) + \tfrac{4}{5}p. \end{aligned}$$

Hence $k \le \frac{44}{45}p + \frac{8}{9}$.

If $t > \frac{1}{4}p$, then $k = p + 2 - t < \frac{3}{4}p + 2 < \frac{44}{45}p + \frac{8}{9}$, unless $p = 3$; but this is ruled out by the hypothesis. □

Now this theorem is extended to arbitrary odd powers of the characteristic.

THEOREM 13.57 *In* $\mathrm{PG}(2,q)$, $q = p^{2e+1}$ *with* p *an odd prime and* $e \ge 1$,

$$m'(2,q) \le q - \tfrac{1}{4}\sqrt{pq} + \tfrac{29}{16}p + 1.$$

Proof. Let $\mathcal{K}$ be a complete k-arc with k greater than the bound in the theorem. Let $\mathcal{K}'$ be the dual k-arc arising from $\mathcal{K}$ in the dual plane of $\mathrm{PG}(2,q)$. As in the proof of the previous theorem, it is only necessary to consider case (iii) and a component Γ'_n of Γ_{2t} of degree $n \ge 3$ that is Frobenius non-classical with respect to the linear system of conics.

If the points in $\Delta(\mathcal{K}')$ are are all double points of Γ'_n, then

$$\tfrac{1}{2}kn \le \tfrac{1}{2}(n-1)(n-2) \le \tfrac{1}{2}n^2;$$

so $k \le n \le 2t$, and hence

$$k \le \tfrac{2}{3}(q+2) < q - \tfrac{1}{4}\sqrt{pq} + \tfrac{29}{16}p + 1,$$

as required. Otherwise, some $P \in \Delta(\mathcal{K}')$ is a non-singular point of Γ'_n. If ℓ is the line in $\mathcal{K}'$ through P, then $I(P, \Gamma'_n \cap \ell) = 2$. Therefore the (P, L_1)-orders are $0, 1, 2$, where L_1 is the linear series on Γ'_n cut out by lines. By the argument in the third paragraph of Section 7.8, Γ'_n has order sequence $(0, 1, 2, 3, 4, \epsilon_5 \le 2n)$ with respect to the linear series L_2 cut out by conics. If Γ'_n is Frobenius non-classical for L_2, then the Frobenius order sequence is $(0, 1, 2, 3, \epsilon_5)$ such that $\epsilon_5 = p^m$ unless $p = 3$ and $\epsilon_5 = 6$. From the Stöhr–Voloch Theorem 8.65 applied to L_2, see also (8.65), if $d \ge 0$ points in $\Delta(\mathcal{K}')$ are double points, then

$$\tfrac{1}{2}kn - d \le \tfrac{1}{5}[(6+\epsilon_5)\{n(n-3) - 2d\} + 2n(q+5)].$$

Hence

$$k \leq \tfrac{2}{5}[(6+\epsilon_5)(n-3)+2(q+5)] \leq \tfrac{2}{5}[(6+\epsilon_5)(2t-3)+2(q+5)].$$

As $k = q+2-t$, it follows that

$$t \geq \frac{q}{4\epsilon_5+29} + \frac{6\epsilon_5+26}{4\epsilon_5+29} \geq \frac{q}{4\epsilon_5+29} + 1. \qquad (13.8)$$

If $\epsilon_5 \geq p^{e+1}$ then, as $\epsilon_5 \leq 2n \leq 4t$, it follows that

$$k \leq q+2-\tfrac{1}{4}p^{e+1} = q+2-\tfrac{1}{4}\sqrt{pq},$$

a contradiction.

Suppose, therefore, that $\epsilon_5 < p^{e+1}$. If $p \neq 3$, then ϵ_5 is a power of p; so $\epsilon_5 \leq p^e$ and, from (13.8), $t \geq q/(4p^e+29)+1$, whence

$$k > q - \tfrac{1}{4}\sqrt{pq} + \tfrac{29}{16}p + 1,$$

the required contradiction.

If $p = 3$, the same argument gives the result unless $\epsilon_5 = 6$, which also satisfies $\epsilon_5 \leq p^e$ unless $e = 1$, that is, $q = 27$. However, the result is trivially true for $q = 27$. □

THEOREM 13.58 *In* $\mathrm{PG}(2,q)$, $q = 2^{2e+1}$ *with* $e \geq 1$,

$$m'(2,q) \leq q - \sqrt{2q} + 2.$$

Proof. Assume that there exists a complete k-arc $\mathcal{K}$ with $k > q - \sqrt{2q} + 2$. As in Theorem 13.40, let Γ_t be the curve of degree $t = q+2-k$ is associated to the dual k-arc. Such a curve Γ_t is defined over $\mathbf{F}_q$ and has a non-linear irreducible component Γ'_n of degree n with $n \leq t$. Let $\nu_0 = 0, \nu_1 = \nu$ be the Frobenius orders of Γ'_n with respect to the linear series cut out by lines.

If Γ'_n is not defined over $\mathbf{F}_q$, then $kn \leq n^2$. So $k \leq n \leq t = q+2-n$; that is, $k \leq \frac{1}{2}(q+2) \leq q - \sqrt{2q} + 2$.

If Γ'_n is defined over $\mathbf{F}_q$, then, using (8.20),

$$kn \leq \tfrac{1}{2}[\nu(2g-2) + n(q+2)] \leq \tfrac{1}{2}n[\nu(n-3)+q+2].$$

Hence

$$2k \leq \nu(n-3)+q+2 \leq \nu(t-3)+q+2$$

and, since $k = q+2-n$,

$$t \geq \frac{q+2+3v_1}{\nu+2}.$$

If $\nu \leq \sqrt{q/2}$, then, for $q > 8$,

$$t \geq \frac{2(q+2)+3\sqrt{2q}}{\sqrt{2q}+4} > \left\lfloor \frac{2(q+2)+3\sqrt{2q}}{\sqrt{2q}+4} \right\rfloor = \sqrt{2q} - 1.$$

Therefore $k = q+2-t < q+3-\sqrt{2q}$. As $\sqrt{2q}$ is an integer, the theorem is proved in this case.

Since $\nu \leq n \leq t$ and ν is a power of 2, if $\nu > \sqrt{q/2}$, then $\sqrt{2q} \leq \nu_1 \leq t$; so $k \leq q+2-\sqrt{2q}$, as required.

For $q = 8$, the bound is sharp as the only complete arcs other than hyperovals are 6-arcs. □

Table 13.4 Size of the second largest complete arc in small planes

q	7	8	9	11	13	16	17	19	23	25	27	29	31	32
m'	6	6	8	10	12	13	14	14	17	21	22	24	22	24

EXAMPLE 13.59 Table 13.4 gives the known values of $m'(2,q)$ apart from those with $q \geq 64$ implicit in Theorem 13.48. Here, m' is written for $m'(2,q)$.

From Theorems 13.48 and 13.48 and Table 13.4, a q-arc in $\mathrm{PG}(2,q)$ is always incomplete.

As an application of the theorems in this section, the status of $(q-1)$-arcs can be clarified. To resolve the question of the completeness of $(q-1)$-arcs, the plane sextic Γ_6 associated to the dual $(q-1)$-arc is considered.

THEOREM 13.60 *If $q > 9$, the plane sextic Γ_6 associated to a complete $(q-1)$-arc is irreducible.*

Proof. The possibilities for Γ_6 are examined in three parts.

(i) Γ_6 *contains no $\mathbf{F}_q$-rational linear components.*

If Γ_6 did contain such a component, $\mathcal{K}$ would be incomplete.

(ii) Γ_6 *contains no multiple components.*

Suppose that $\Gamma_6 = \mathbf{v}(F)$ and $F = G_1G_2$, where $\mathcal{C}_1 = \mathbf{v}(G_1)$ has no multiple components and $\mathcal{C}_2 = \mathbf{v}(G_2)$ consists of double components. From Theorem 13.43, Γ_6 does not consist entirely of double components. By (i), it may be assumed that Γ_6 has just one double component, an $\mathbf{F}_q$-rational irreducible conic counted twice. Hence $\mathcal{C}_1$ is an irreducible conic defined over $\mathbf{F}_q$. By Theorem 13.50, the lines of $\mathcal{K}'$ are $\mathbf{F}_q$-rational tangents to $\mathcal{C}'_{2r}$, and $\mathcal{K}'$ is incomplete.

(iii) Γ_6 *is irreducible.*

If Γ_6 were reducible, then one of the following three cases would apply:

(a) Γ_6 splits into three irreducible conics defined over $\mathbf{F}_q$;

(b) Γ_6 splits into an irreducible conic and an irreducible quartic, both defined over $\mathbf{F}_q$

(c) Γ_6 splits into two irreducible, distinct cubics.

Again, by Theorem 13.50, (a) and (b) are impossible, since the fact that an irreducible conic defined over $\mathbf{F}_q$ is a component of Γ_6 implies that $\mathcal{K}$ is incomplete.

If, in case (c), Γ_6 has components Γ_3 and Γ'_3, then Γ_3 and Γ'_3 have a set $\mathcal{R}$ of at most nine common points in $\mathrm{PG}(2,q)$. Also, each line in the dual $(q-1)$-arc $\mathcal{K}'$ passes through three points of Γ_6 with intersection multiplicity two. If such a line ℓ passed through two points of Γ_3 each with multiplicity two, then Γ_3 would be reducible. Similarly, if ℓ met Γ_3 at one point not in $\mathcal{R}$ with multiplicity two and passed through two points of $\mathcal{R}$, then Γ_3 would be reducible. So ℓ must pass through three points of $\mathcal{R}$. Since no two lines in $\mathcal{K}'$ intersect on Γ_6, it follows that $q-1 \leq 3$, a contradiction. □

THEOREM 13.61 *In* $\mathrm{PG}(2,q)$, *a* $(q-1)$*-arc is incomplete for* $q > 13$ *except* *perhaps for except possibly for the fourteen values of q consisting of* $49, 81$ *and the twelve primes* $37, 41, 43, 47, 53, 59, 61, 67, 71, 73, 79, 83$.

Proof. From Table 13.4, only $q > 32$ need be considered. For q even, Theorem 13.48 shows that $m'(2,q) < q-1$ for $q \geq 8$. For q odd, Theorem 13.55 shows that $m'(2,q) < q-1$ for $q > 103$; Theorem 13.56 improves this to $q > 83$. □

REMARK 13.62 It is conjectured that the exceptions can be removed in this theorem. The corresponding irreducible sextic Γ_6 has $q-1$ triple tangents and so at least $3(q-1)$ rational points. It needs to be shown that there must exist more rational points on Γ_6 to contradict the upper bounds (8.20) or (8.25).

CONJECTURE 13.63 *In* $\mathrm{PG}(2,q)$, *with q square and* $q > 9$,

$$m'(2,q) = q - \sqrt{q} + 1.$$

Further, the corresponding arc is projectively unique.

The lowest value of q for which this conjecture is unresolved is $q = 49$; that is, it must be shown that $m'(2,49) = 43$. Related to this conjecture are Example 13.47 and the following result analogous to Theorem 13.49.

THEOREM 13.64 *Let $\mathcal{K}'$ be the dual of the cyclic (q^2-q+1)-arc in* $\mathrm{PG}(2,q^2)$ *with q odd, defined in Example 13.47. Then the associated curve $\Gamma_{2(q+1)}$ is birationally equivalent over $\mathbf{F}_{q^6}$ to the Hermitian curve $\mathcal{H}_q$.*

The proof requires some preliminary results. For a primitive (q^2+q+1)-st root of unity t in $\mathbf{F}_{q^6}$, the mapping α given by the equations,

$$\rho X_0' = tX_0, \quad \rho X_1' = t^{q^2+1}X_1, \quad \rho X_2' = X_2,$$

is a projectivity of order (q^4+q^2+1) that fixes each vertex of the fundamental triangle $\mathbf{U}_0\mathbf{U}_1\mathbf{U}_2$ of a projective coordinate system in $\mathrm{PG}(2,K)$, where $K = \overline{\mathbf{F}}_q$. Let $\mathcal{A}$ denote the collineation group generated by α.

The orbit Π of the point $\mathbf{U} = (1,1,1)$ under $\mathcal{A}$ has size q^4+q^2+1 and consists of all points $P_c = (c, c^{q^2+1}, 1)$ with $c^{q^4+q^2+1} = 1$. Note that Π lies in $\mathrm{PG}(2,q^6)$ as the (q^4+q^2+1)-st roots of unity are in $\mathbf{F}_{q^6}$.

The orbit Π may be viewed as a subgeometry of $\mathrm{PG}(2,K)$ induced by the lines meeting Π in at least two points.

PROPOSITION 13.65 *The orbit* $\Pi \cong \mathrm{PG}(2,q^2)$ *with the following properties:*

(i) *Π is a projective subplane of* $\mathrm{PG}(2,q^6)$ *lying in a non-classical position; that is,* $\Pi \neq \mathrm{PG}(2,q^2)$ *but there is a projectivity of* $\mathrm{PG}(2,q^6)$ *that maps Π to* $\mathrm{PG}(2,q)$;

(ii) *the lines of Π are* $\mathbf{v}(tX_0 + t^{q^2+1}X_1 + X_2)$, *with t running over all the (q^4+q^2+1)-st roots of unity; they form the line orbit of* $\mathbf{v}(X_0+X_1+X_2)$ *under $\mathcal{A}$.*

Proof. The line $\mathbf{v}(X_0 + X_1 + X_2)$ meets Π in exactly $q^2 + 1$ distinct points, since $u^{q^2+1} + u + 1 = 0$ implies that u is a $(q^4 + q^2 + 1)$–th root of unity. In fact, if $u^{q^4+q^2} + u^{q^2} + 1 = 0$, then $u^{q^4+q^2+1} + u^{q^2+1} + u = 0$ whence $u^{q^4+q^2+1} = 1$. Also, the polynomial $F(X) = X^{q^2+1} + X + 1$ has no multiple roots as $F(X)$ and its derivative $F'(X) = X^{q^2} + 1$ have no common root.

For any $(q^4 + q^2 + 1)$–th root $t \in \mathbf{F}_{q^6}$, the line $l_t = \mathbf{v}(tX_0 + t^{q^2+1}X_1 + X_2)$ meets Π in the points

$$P_i = (c^i, c^{(q^2+1)i}, 1) \text{ with } (c^i t)^{q^2+1} + c^i t + 1 = 0.$$

As before, there are exactly $q^2 + 1$ such points. Therefore Π equipped with all the lines l_t is a projective plane of order q^2. This is shown by a direct calculation. □

Next, intersections of Π with conics and Hermitian curves are considered.

PROPOSITION 13.66 *The conic $\mathcal{C} = \mathbf{v}(X_0^2 + X_1^2 + X_2^2)$ and the plane curve $\mathcal{D} = \mathbf{v}(X_0X_2^q - X_1^{q+1})$ have $2(q+1)$ distinct points in common, and half of these points lie in Π.*

Proof. First, $\mathbf{v}(X_2) \cap \mathcal{C} \cap \mathcal{D} = \emptyset$. Now choose any point $P = (u, v, 1)$ off the line $\mathbf{v}(X_2)$. Then P is a point of $\mathcal{D}$ if and only if $u = v^{q+1}$. In this case, P is also a point of $\mathcal{C}$ if and only if $v^{2(q+1)} + v^2 + 1 = 0$. The polynomial $f(X)$, with

$$f(X) = X^{2(q+1)} + X^2 + 1,$$

has no double roots since its derivative

$$f'(X) = 2X^{2q+1} + 2X = 2X(X^2 + 1)^q$$

has no root in common with $f(X)$.

Now $q + 1$ of the roots v of $f(X)$ give rise to points $P = (v^{q+1}, v, 1)$ lying in Π. Since

$$f(X)^q X^2 = X^{2(q^2+q+1)} + X^{2(q+1)} + X^2,$$

each root v of $f(X)$ is also a root of the polynomial $g(X) = X^{2(q^2+q+1)} - 1$. Since $f(X) = f(-X)$ and $g(X) = g(-X)$, half of the roots v of $f(X)$ are also roots of the polynomial $h(X) = X^{q^2+q+1} - 1$. This shows that the corresponding $q + 1$ points $P = (v^{q+1}, v, 1)$ lie in Π. □

PROPOSITION 13.67 *The conic $\mathcal{C} = \mathbf{v}(X_0^2 + X_1^2 + X_2^2)$ and the Hermitian curve $\mathcal{H}_q = \mathbf{v}(X_0X_1^q + X_1X_2^q + X_2X_0^q)$ have $q + 1$ distinct common points, and all lie in Π. Also, the curves $\mathcal{C}$ and $\mathcal{H}_q$ have the same tangent line at each of these points.*

Proof. From the proof of Proposition 13.66, the set Π contains $q+1$ distinct points $P = (v^{q+1}, v, 1)$ of $\mathcal{C}$, where v is a common root of $f(X)$ and $h(X)$. Each of these points also lies on $\mathcal{H}_q$, since

$$v^{2q+1} + v^{q^2+q} + v = v^{-1}(v^{q^2+q+1} + v^{2(q+1)} + v^2) = v-1(1 + v^{2(q+1)} + v^2) = 0.$$

A calculation shows that the tangent line to $\mathcal{H}_q$ at the point $P = (v^{q+1}, v, 1)$ has slope equal to $-\sqrt{v}$. The same value is obtained when $\mathcal{H}_q$ is replaced by $\mathcal{C}$. Hence

$\mathcal{C}$ and $\mathcal{H}_q$ are tangent at each of these $q+1$ points. By Bézout's Theorem 3.14, there is no other common point of $\mathcal{C}$ and $\mathcal{H}_q$, and this completes the proof. □

Now consider the family of conics $\mathcal{C}_t = \mathbf{v}(F_t)$, with

$$F_t = tX_0^2 + t^{q+1}X_1^2 + X_2^2, \tag{13.9}$$

where t is a (q^2+q+1)-st root of unity. This family consists of all conics which are images of the conic $\mathcal{C} = \mathbf{v}(X_0^2 + X_1^2 + X_2^2)$ under the action of the collineation group $\mathcal{A}$. Since $q^4+q^2+1 = (q^2+q+1)(q^2-q+1)$, the group $\mathcal{A}$ has a subgroup $\mathcal{B}$ of order q^2-q+1. The orbit of $\mathcal{C}$ under $\mathcal{B}$ consists of the conics $\mathcal{C}_t$ with t ranging over the (q^2-q+1)-st roots of unity. The next result connects the set Φ of points of $\mathcal{H}_q$ lying in Π with the sets Δ_t of points of $\mathcal{C}_t$ lying in Π.

PROPOSITION 13.68 *The sets Δ_t partition Φ into q^2-q+1 point sets, each of size $q+1$.*

Proof. It is first shown that $\mathcal{B}$ leaves $\mathcal{H}_q$ invariant. The image of $\mathcal{H}$ under α^i is the curve

$$\mathbf{v}(X_0X_1^q + u^{q(q^2-q+1)}X_1X_2^q + u^{(q+1)(q^2-q+1)}X_2X_0^q),$$

where $u = t^i$. If α^i is a generator of $\mathcal{B}$, then $i = q^2+q+1$ and $u^{q^2-q+1} = 1$, whence the result follows.

Proposition 13.65 remains valid when $\mathcal{C}$ is replaced by any conic $\mathcal{C}_t$ in the family. Therefore no point of Φ lies on more than one such conic. In fact, if both t and u are (q^2-q+1)-st roots of unity, and $P = (a,b,1)$ is a point in $\Phi\cap\Delta_t\cap\Delta_u$, Proposition 13.65 implies that the conics $\mathcal{C}_t$ and $\mathcal{C}_u$ have the same tangent at $P = (a,b,1)$. But, this can only occur when $-t^q = -u^q$, and hence $t = u$. From Proposition 13.67, each conic $\mathcal{C}_t$ in the family meets Φ in $q+1$ points. Since Φ has size q^2+q+1, the assertion follows. □

A relationship between conics $\mathcal{C}_t$ and lines l_t is given in the following result.

LEMMA 13.69 *The standard quadratic transformation τ given by the equations*

$$X_0' = X_0^2, \quad X_1' = X_1^2, \quad X_2' = X_2^2$$

leaves Π invariant. The restriction of τ to Π is a bijection and transforms Δ_t into the set Ψ_t of the common points of l_t and Π.

Proof. The invariance of Π follows from a direct calculation. The fact that the mapping $x \mapsto x^2$ is a permutation of the set of all (q^2+q+1)-st roots of unity shows that Δ_t is transformed into Ψ_t. □

The proof of Theorem 13.64 consists in showing that the transformation τ turns $\mathcal{H}_q$ into Γ_{2t}. Let Γ be the image of $\mathcal{H}_q$ under τ. From Propositions 13.67, 13.68 and 13.69, each point of the q^3+1 points of Γ lying in Π is a non-singular point of Γ, and the line l_t, where t is a (q^2-q+1)-st root of unity, is a $(q+1)$-fold tangent to Γ; that is, l_t is the tangent line to Γ at each of their common points. Such lines l_t form a cyclic dual (q^2-q+1)-arc $\mathcal{K}'$ in Π viewed as projective plane defined over $\mathbf{F}_{q^2}$. By Theorem 13.43, Γ is the curve $\Gamma_{2(q+1)}$ associated to $\mathcal{K}'$.

Proof of Theorem 13.64. It remains to show that τ maps $\mathcal{H}$ onto Γ birationally. Consider the linear system Φ of all conics,

$$\mathbf{v}(c_0X_0^2 + c_1X_1^2 + c_2X_2^2), \text{ with } \mathbf{c} = (c_0, c_1, c_2) \in \mathrm{PG}(2, K).$$

It must be shown that the linear series $g^2_{2(q+1)}$ cut out on $\mathcal{H}_q$ by Φ is simple, that is, not composed of an involution. For a point P of $\mathcal{H}_q$, the conics in Φ through P form a linear subsystem of Φ. To show that $g^2_{2(q+1)}$ is simple, it suffices to prove the existence of a point $P \in \mathcal{H}_q$ such that the conics in Φ_P have no further common point on $\mathcal{H}_q$.

It is shown that such a condition is satisfied by every point $P = (a, b, c) \in \mathcal{H}_q$ with $abc \neq 0$. The conics of Φ_P have exactly four common points in $\mathrm{PG}(2, K)$, namely,

$$P = (a, b, c), \quad P_1 = (a, -b, c), \quad P_2 = (-a, b, c), \quad P_3 = (a, b, -c).$$

None of the last three points belongs to $\mathcal{H}_q$. Thus the point $P = (a, b, c)$ is the only common point of the conics of Φ_P which lies on $\mathcal{H}_q$. This completes the proof. □

A result that improves Theorem 13.51 is the following.

THEOREM 13.70 *For $q = p^h$ with $p \geq 5$,*

$$m'(2, q) \leq q - \tfrac{1}{2}\sqrt{q} + 5.$$

Theorem 8.104 can be applied to give better bounds for $m'(2, q)$.

THEOREM 13.71 *Let $q = p^h$ with $p \geq 3$, and let $q = 3^{2e}$ when $p = 3$.*

(i) *If $q \geq 23^2$ and $q \neq 3^6$ or 5^5, then*

$$m'(2, q) \leq q - \tfrac{1}{2}\sqrt{q} + 3.$$

(ii)

$$m'(2, q) \leq \begin{cases} q - 22 & \text{when } q = 5^5, \\ q - 9 & \text{when } q = 3^6, \\ q - 9 & \text{when } q = 23^2, \\ q - 5 & \text{when } q = 19^2. \end{cases}$$

Proof. Given a k-arc $\mathcal{K}$ in $\mathrm{PG}(2, q)$ with $k > q - \frac{1}{2}\sqrt{q} + 3$, the aim is to prove that $\mathcal{K}$ is contained in an irreducible conic defined over $\mathbf{F}_q$. Let $\mathcal{K}'$ be the complete dual k-arc arising from $\mathcal{K}$ in the dual plane of $\mathrm{PG}(2, q)$.

From Theorems 13.43 and 13.50, the curve Γ_{2t} associated to $\mathcal{K}'$ does not have linear components. Further, if it has a quadratic component then $\mathcal{K}'$ is a dual $(q+1)$-arc consisting of all tangents to an $\mathbf{F}_q$-irreducible conic $\mathcal{C}$ at its points in $\mathrm{PG}(2, q)$. So assume that each component of Γ_{2t} has degree at least three. For such a component $\Gamma = \mathbf{v}(f(X, Y))$ of degree d, two cases are distinguished according as either Γ is defined over $\mathbf{F}_q$ or over a non-trivial finite extension of $\mathbf{F}_q$.

In the latter case, $f(X, Y)$ is distinct from its first conjugate $f^{(1)}(X, Y)$ and also from $cf^{(1)}f(X, Y)$ for $c \in K$; see Section 8.3. Hence Γ is distinct from its first conjugate which is the irreducible plane curve $\Gamma' = \mathbf{v}(f^{(1)}(X, Y))$ but an s-fold point P of Γ with coordinates in $\mathbf{F}_q$ is also an r-fold point of Γ'.

Let $\gamma_1, \ldots, \gamma_N$ be the branches of Γ centred at the points in $\Delta(\mathcal{K}')$, and let $\mathcal{P}_i$ be the corresponding places of $K(\mathcal{D})$. If $r_i = j_1(\mathcal{P}_i)$ denotes the order of γ_i, then the centre of γ_i is a point of Γ whose multiplicity is at least r_i. The same holds for the branches $\gamma'_1, \ldots, \gamma'_N$ of Γ' centred at the points in $\Delta(\mathcal{K}')$. By Bézout's Theorem 3.14,

$$\textstyle\sum_{i=1}^{N} r_i^2 \leq \sum_{P \in \mathcal{D}(\mathcal{K}')} I(P, \Gamma \cap \Gamma') \leq d^2 \leq d(\sqrt{q} - 2) < \frac{1}{2} d(q - \sqrt{q} + 1).$$

If M_q and M'_q, with respect to Γ, have the same meaning as in Definition 8.79, it follows that

$$2M_q + M'_q < d(q - \sqrt{q} + 1). \tag{13.10}$$

Now let Γ be an $\mathbf{F}_q$-rational irreducible component of Γ_{2t} of degree d. If (13.10) does not hold, then Theorem 8.104 applies to Γ, showing that Γ is projectively equivalent over $\mathbf{F}_q$ to the Fermat curve of degree $d = \frac{1}{2}(\sqrt{q} + 1)$; in particular, $2M_q + M'_q = d(q - \sqrt{q} + 1)$.

Thus, for every irreducible component Γ of Γ_{2t}, also

$$2M_q + M'_q \leq d(q - \sqrt{q} + 1).$$

Therefore

$$\textstyle\sum (2M_q + M'_q) \leq 2t(q - \sqrt{q} + 1), \tag{13.11}$$

where the summation is over all irreducible components of Γ_{2t}. On the other hand, Theorem 13.43 provides a lower bound for the above sum. In fact, every point $P \in \Delta(\mathcal{K}')$ is either a non-singular point or a double point of Γ. In the former case, P counts in M_q as $I(P, \Gamma_{2t} \cap \ell) = 2$ for the unique line ℓ in $\Delta(\mathcal{K}')$ which passes through P. In the latter case, P is the centre of either one branch of order 2 or of two branches of order 1. Thus P counts at least twice in $2M_q + M'_q$. Since $\Delta(\mathcal{K}')$ consists of kt points, so

$$2kt \leq \textstyle\sum (2M_q + M'_q). \tag{13.12}$$

Comparison of (13.11) and (13.12) gives $k \leq q - \sqrt{q} + 1$, contradicting the hypothesis that $k > q - \frac{1}{2}\sqrt{q} + 3$. □

Remark 13.72 Since q is odd, in the case in which q is a square, the 3 can be lowered to $\frac{5}{2}$. The best result for the case $q = 3^{2e+1}$ is that

$$m'(2, q) \leq q - \tfrac{\sqrt{3}}{4}\sqrt{q} + \tfrac{103}{16}.$$

13.9 THE THIRD LARGEST COMPLETE ARC

By Theorem 13.48, the exact value of $m'(2, q)$ is known for every even square q. The next question is to find the maximum size $m''(2, q)$ of the third largest complete arcs. The following result depending on Theorem 8.78 provides an upper bound.

THEOREM 13.73 *Let $q = 2^h$, $q \geq 16$ and q square. Then*

$$m''(2,q) \leq \begin{cases} q - 2\sqrt{q} + 6 & \textit{for } q \geq 64, \\ q - 2\sqrt{q} + 4 = 12 & \textit{for } q = 16. \end{cases} \tag{13.13}$$

Proof. Let $\mathcal{K}'$ be a complete dual k-arc in $\mathrm{PG}(2,q)$ with q as given. It must be shown that, if k is larger than the right-hand side in (13.13), then one of the following holds:

(i) $k = q - \sqrt{q} + 1$;

(ii) $k = q + 2$.

Let Γ_t be the curve defined over $\mathbf{F}_q$ associated to $\mathcal{K}'$. It must be shown that Γ_t is irreducible.

Let Γ be an irreducible component of Γ of degree n. From Theorem 13.40 (iii), Γ contains n distinct points on every line in $\mathcal{K}'$, and such points lie in $\Delta(\mathcal{K}')$.

As in the proof of Theorem 13.71, let Γ' be the first conjugate of Γ. These curves are distinct if and only if Γ is not defined over $\mathbf{F}_q$. Since every point of Γ with coordinates in $\mathbf{F}_q$ is also a point of Γ', each of the kn points in $\Delta(\mathcal{K}')$ counts in the intersection of Γ and Γ'. Since $kn = k(q - k + 2) > n^2$, Bézout's Theorem 3.14 shows that the two curves coincide, unless $n^2 \geq nk$. This inequality implies that $n \geq k$. Since $t \geq n$, also $q + 2 - k \geq k$; but this contradicts that $k > q - 2\sqrt{q} + 6$. Therefore Γ is defined over $\mathbf{F}_q$.

Next, it is shown that $\Gamma_t = \Gamma$. For, if not, then

$$2n \leq t = q + 2 - k < q + 2 - (q - 2\sqrt{q} + 6) = 2\sqrt{q} - 4$$

and $n < \sqrt{q} - 2$. Let g_n^2 be the linear series cut out on Γ by lines. It may be that Γ is Frobenius non-classical, and if this occurs then its Frobenius order $\nu_1 = q'$ is bounded above by $\frac{1}{2}\sqrt{q}$ as q' does not exceed the largest power of 2 smaller than $n < \sqrt{q} - 2$. From Theorem 8.74 applied to g_n^2,

$$nk \leq N \leq \tfrac{1}{2}\{n(n-3)q' + (q+2)n\};$$

so

$$2k \leq (n-3)q' + (q+2).$$

Now apply the conditions $k > q - 2\sqrt{q} + 6,\ n < \sqrt{q} - 2,\ q' \leq \frac{1}{2}\sqrt{q}$:

$$2(q - 2\sqrt{q} + 6) < 2k \leq (\sqrt{q} - 5)\tfrac{1}{2}\sqrt{q} + q + 2,$$

whence

$$4(q - 2\sqrt{q} + 6) < (\sqrt{q} - 5)\sqrt{q} + 2(q + 2),$$

and

$$q - 3\sqrt{q} + 20 < 0,$$

a contradiction. Hence $n = t$ and $\Gamma_t = \Gamma$.

Now Theorem 8.78 is applied. Since $k > q - 2\sqrt{q} + 6$, so

$$t = q + 2 - k < 2\sqrt{q} - 4.$$

Hence, by Theorem 8.78, the number of $\mathbf{F}_q$-rational simple points N_0 of Γ_t satisfies $N_0 \leq t(q+2-t)$. Since $\Delta(\mathcal{K}')$ has size $t(q+2-t)$, so $N_0 \geq t(q+2-t)$. Hence $N_0 = t(q+2-t)$ and again, by Theorem 8.78, Γ_t is projectively equivalent to the Hermitian curve of degree $\sqrt{q}+1$, which is case (i). □

13.10 EXERCISES

1. Show that, over $\mathbf{F}_{13}$, the sextic curve

$$\Gamma'_6 = \mathbf{v}(X_0^6 + X_1^6 + X_2^6 + X_0^3X_1^3 + X_0^3X_2^3 + X_1^3X_2^3)$$

also has $N_1 = 54$, but is not isomorphic to the curve Γ_6 in (13.4). One method is to verify that $N_2 = 162$ for Γ'_6 but $N_2 = 198$ for Γ_6.

13.11 NOTES

Several books, such as [178], [336], [350], [364], [419], [428], [460], [486], offer a detailed treatment of algebraic-geometry codes. See also the exposition by Høholdt, van Lint, and Pellikaan [227].

Algebraic curves defined over $\mathbf{F}_q$ with many $\mathbf{F}_q$-rational points have important applications to a wide variety of combinatorial problems, especially in finite geometry; see the survey paper of Szőnyi [444].

For a general account of finite projective spaces and their properties, see [216], [213], [224].

The three problems I, II, III were enunciated by Segre in his seminal paper [395].

For MDS algebraic-geometry codes, MC_k has been shown for some particular cases, namely, for codes arising from elliptic curves and curves of genus two when $q > 83$, by Munuera [340]; see also [74]. For any fixed genus g there is a q_0 such that for all $q > q_0$, the Main Conjecture holds for MDS algebraic-geometry codes arising from hyperelliptic curves defined over $\mathbf{F}_q$. This was shown by de Boer [98] using Munuera's approach; see also [48].

Theorem 13.23 is due to Glynn [175].

For surveys of further results on arcs and MDS codes, see [214], [215], [221], [222], and MacWilliams and Sloane [317, Chapter 11]. Bounds on the size of the second largest complete arcs in small planes of order $q \geq 37$ are found in [126], [97].

The formal definition of near-MDS codes is due to Dodunekov and Landjev [106]. The exact values of the maximum length of a near-MDS code were calculated by Dodunekov and Landjev [107] for $q = 2, 3, 4, 5$. Marcugini, Milani, and Pambianco considered near-MDS codes for $k = 3, 4$ and obtained a complete classification of those of maximal length for $q \leq 11$; see [323]. For more results on near-MDS codes, see Abatangelo and Larato [2], Ballico and Cossidente [36], Giulietti [169], and Landjev [290].

Sections 13.3, 13.4, 13.5 are all based on Segre's work [394], [396].

Fisher, Hirschfeld, and Thas [129], and, independently, Boros and Szőnyi [57], Cossidente [87], Kestenband [267], and Ebert [112], showed that, for $q > 2$, every point orbit of $\mathfrak{S}$ is a complete $(q^2 - q + 1)$-arc $\mathcal{K}$ in $\mathrm{PG}(2, q^2)$, and every line orbit of $\mathfrak{S}$ is a complete dual $(q^2 - q + 1)$-arc $\mathcal{K}'$ in $\mathrm{PG}(2, q^2)$. In [129] it is also shown for even q that the plane curve Γ_{q+1} associated with $\mathcal{K}'$, via Segre's generalisation of Menelaus' Theorem, is the Hermitian curve; see Theorem 13.49.

For more results on cyclic arcs, see [443], [434], [324].

Theorem 13.51 is due to Segre [398]. Theorem 13.55 is due to Thas [453]. Theorem 13.56 is due to Voloch [492]; Theorems 13.57 and 13.58 come from Voloch [495]. Theorem 13.60 is due to Bichara and Migliori [52]. For Theorem 13.70, see [217]; for Theorem 13.71, see [218].

For a proof that a q-arc is contained in the rational points of a conic, see [396].

Direct proofs that a q-arc in $\mathrm{PG}(2,q)$ is always incomplete are found in [395] for odd q and in [446] for even q.

For the classical result that $\tau_2 \leq \frac{1}{2}d(d-2)(d^2-9)$, see, for example, [396].

Complete k-arcs can be constructed from elliptic curves over $\mathbf{F}_q$. Let $\mathcal{F}$ be a non-singular plane cubic curve defined over $\mathbf{F}_q$, equipped with the group $G_{\mathcal{F}(\mathbf{F}_q)}$ isomorphic to $\mathrm{Pic}_0(\mathbf{F}_q(\mathcal{X}))$, the zero-degree $\mathbf{F}_q$-rational divisor class group of $\mathcal{F}$. Assume that $G_{\mathcal{F}(\mathbf{F}_q)}$ has even order, and let H be its subgroup of index 2. Then the the points of $\mathcal{F}$ associated to the elements of the coset $G_{\mathcal{F}(\mathbf{F}_q)}\backslash H$ form an arc in $\mathrm{PG}(2,q)$. Such an arc is either complete or may completed by adding a few points. For the proofs, see [493], [494], [167].

For Exercise 1, see [96].

Appendix A

Background on field theory and group theory

A.1 FIELD THEORY

In this section, fundamental concepts and results from field theory are listed. Apart from two generalisations of the Theorem of the Primitive Element, all the material with proofs can be found in standard textbooks.

(I) The *characteristic* of a field F is the smallest positive integer n, if it exists, such that

$$\underbrace{1+1+\cdots+1}_{n \text{ times}}=0.$$

In this case, n is prime and is denoted by p. If there is no such n, then F is of *characteristic* 0; write $p=0$.

(II) Every field has a *prime field*, that is, a subfield containing no subfield other than itself. If $p>0$, the prime field is isomorphic to $\mathbf{Z}_p=\mathbf{Z}/p\mathbf{Z}=\mathbf{F}_p$, the field of integers modulo p, while, if $p=0$, the prime field is isomorphic to $\mathbf{Q}$, the field of rational numbers.

(III) Let L be a field that contains F as a subfield. Then L is an *extension of* F, and the symbol L/F is used to express this. If L is regarded as a vector space over F, its dimension is the *degree of* L/K and denoted by $[L:K]$. If $[L:K]=n<\infty$, then L/K is a *finite extension*. If both L/F and M/L are finite extensions then M/F is also finite, and its degree is

$$[M:F]=[M:L][L:F].$$

(IV) An element $\alpha \in L$ is either *algebraic* or *transcendental* over a subfield F of L, according as there is or is not a non-constant polynomial $f(X) \in F[X]$ satisfied by α; that is, $f(\alpha)=0$.

Every algebraic element α satisfying the polynomial X^n-1 is an *n-th root of unity*. If, in addition, α does not satisfy any polynomial X^k-1 with $1 \leq k<n$, then α is a *primitive n-th root of unity*.

(V) The field L/F is an *algebraic extension*, and L is *algebraic over* F, when all elements in L are algebraic over F; otherwise, L is a *transcendental extension*. In the latter case, L has at least one transcendental element over F.

(VI) If L_1/F and L_2/F are any two extensions, then a mapping $\sigma : L_1 \to L_2$ is a *field homomorphism* if

$$\sigma(x+y) = \sigma(x) + \sigma(y), \quad \sigma(xy) = \sigma(x)\sigma(y),$$

for all $x, y \in L_1$.

(VII) A field homomorphism $\sigma : L_1 \to L_2$ is an *embedding of L_1 into L_2 over F* provided that $\sigma(x) = x$ for all $x \in F$.

Such a homomorphism must be injective, giving an isomorphism of L_1 onto the subfield $\sigma(L_1)$ of L_2. A surjective, and hence bijective, embedding of L_1 into L_2 is an *F-isomorphism.*

(VIII) A field M is *algebraically closed* if every polynomial $f(X) \in M[X]$ of positive degree has a root in M. For every field F, there exists an algebraic field extension $\bar{F}/F$ such that $\bar{F}$ is algebraically closed. Then $\bar{F}$ is uniquely determined up to an F-isomorphism, and it is *the algebraic closure of F*. If L/F is an algebraic extension, then there exists an embedding $L \to \bar{F}$ over F. There are several different embeddings but their number is at most $[L : F]$.

(IX) If $\alpha_1, \ldots, \alpha_r \in L$, then the smallest subfield M of L containing F and all elements $\alpha_1, \ldots, \alpha_r$ is denoted by $F(\alpha_1, \ldots, \alpha_r)$, and $\alpha_1, \ldots, \alpha_r$ are the *adjoined elements* of $F(\alpha_1, \ldots, \alpha_r)$. The field extension $F(\alpha_1, \ldots, \alpha_r)/F$ is algebraic if and only if every α_i is algebraic over F.

(X) If α is an algebraic element of L/F, there is a unique, monic polynomial $f(X)$ over F of minimal degree satisfied by α. This minimal degree is the *degree of α* over F. The polynomial $f(X)$ is uniquely determined, and it is the *minimal polynomial of α* over F. Other polynomials $g(X) \in F[X]$ can be satisfied by α, but this only occurs when $f(X) \mid g(X)$.

(XI) An element $\alpha \in L$ is algebraic over F if and only if $[F(\alpha) : F] = n < \infty$. Let $f(X)$ be the minimal polynomial of α over F, and put $r = \deg f(X)$. Then $[F(\alpha) : F] = r$, and a basis of $F(\alpha)/F$ is $\{1, \alpha, \ldots, \alpha^{r-1}\}$.

(XII) Conversely, let $f(X) \in F[X]$ be a non-constant polynomial; then there exists an algebraic extension field $L = K(\alpha)$ such that $f(\alpha) = 0$. If $f(X)$ is irreducible over F, this extension field is uniquely determined up to an F-isomorphism. In other words, if $L' = K(\alpha')$ is another algebraic extension field with $f(\alpha') = 0$, then there exists an F-isomorphism $\sigma : L \to L'$ with $\sigma(\alpha) = \alpha'$.

(XIII) More generally, if $f_1(X), \ldots, f_r(X) \in F[X]$ are monic polynomials of degree $d_1, \ldots, d_r$, then there is an algebraic extension M, unique up to an F-isomorphism, containing F such that all $f_i(X)$ split into linear factors over M; that is,

$$f_i(X) = \prod_{j=1}^{d_i} (X - a_{ij}), \quad a_{ij} \in M.$$

The field $F(a_{11}, \dots, a_{rd_r})$ is uniquely determined up to an F-isomorphism and is the *splitting field of* $f_1, \dots, f_r$ over F. Over the prime field of F, the decomposition

$$X^n - 1 = \prod_{d|n} F_d(X)$$

holds for every natural number n not divisible by p for $p > 0$, where

$$F_n(X) = \prod_{d|n} (X^d - 1)^{\mu(n/d)}$$

is a polynomial of degree $\varphi(n)$ whose zeros in a splitting field of $X^n - 1$ are exactly the primitive n-th roots of unity. Here $\mu(i)$ is the Möbius function:

$$\mu(i) = \begin{cases} 1 & \text{if } i = 1, \\ 0 & \text{if } i \text{ is divisible by the square of a prime }, \\ (-1)^j & \text{if } i \text{ is the product of } j \text{ distinct primes.} \end{cases} \quad \text{(A.1)}$$

and $\varphi(n)$ is the number of all primitive n-th roots of unity. The polynomial $F_n(X)$ is the *n-th cyclotomic polynomial.*

(XIV) Let $f(X) \in F[X]$ be a monic polynomial of degree $d \leq 1$. There is an algebraic extension L of K in which $f(X)$ splits into linear factors:

$$f(X) = \prod_{i=1}^{d} (X - \alpha_i).$$

The polynomial $f(X)$ is *separable* if the roots of $f(X)$ are distinct. If, however, $\alpha_i = \alpha_j$ for some $i \neq j$, then $f(X)$ is *inseparable.*

If F has zero characteristic, then all irreducible polynomials over F are separable. In the case of positive characteristic, an irreducible polynomial, $f(X) = \sum a_i X^i \in F[X]$, is separable if and only if $a_i \not\equiv 0 \pmod{p}$ for at least one index i.

(XV) The *derivative* of a polynomial $f(X) = \sum a_i X^i \in F[X]$ is the polynomial $f'(X) = \sum i a_i X^{i-1}$.

A separability criterion is that $f'(X)$ is not the zero polynomial.

(XVI) Given an algebraic extension L/K, an element $\alpha \in L$ is *separable* over F if its minimal polynomial $f(X) \in F[X]$ is a separable polynomial. The field L/K is a *separable extension* if all $\alpha \in L$ are separable over F. In zero characteristic, every algebraic extension is separable.

(XVII) If L/K is a finite extension of degree $[L : K] = n$, then L/K is separable if and only if there are n distinct embeddings $\sigma_1, \dots, \sigma_n$ of L into $\bar{F}$ over F. In this case, an element $\alpha \in L$ is in F if and only if $\sigma_i(\alpha) = \alpha$ for $i = 1, \dots, n$.

For a tower $M \supset L \supset F$ of algebraic extensions, the field M/F is separable if and only if both M/L and L/F are separable.

(XVIII) In positive characteristic, inseparable extensions exist. An element $\alpha \in L$ is *purely inseparable* over F if some power α^{p^r} of α with $r \geq 0$ lies in F. This occurs if and only if the minimal polynomial of α over F is of type $X^{p^e} - c$ with $c \in F$ and $e \leq r$. The extension L/F is *purely inseparable* if all elements $\alpha \in L$ are purely inseparable.

(XIX) Given any algebraic extension L/K, there exists a unique intermediate field S, that is, $F \subset S \subset M$, such that S/K is separable but L/S is purely inseparable. This depends on the fact that the set of all separable elements form a subfield.

(XX) If L is a field containing F as a subfield, the elements $\alpha_1, \ldots, \alpha_r \in L$ are *algebraically dependent* over F if they satisfy a non-trivial polynomial relation over F. Otherwise, $\alpha_1, \ldots, \alpha_r$ are *algebraically independent* over F.

(XXI) An infinite subset A is *algebraically independent* over F if every finite subset in A is algebraically independent; otherwise, A is *algebraically dependent.* If $\alpha_1, \alpha_2, \ldots$ are algebraically independent over F and $L/F(\alpha_1, \alpha_2, \ldots)$ is an algebraic extension, then $\{\alpha_1, \alpha_2, \ldots\}$ is a *transcendence basis of* L over F.

(XXII) If $\{\alpha_1, \ldots, \alpha_s\}$ is a finite transcendence basis of L over F, then any $s+1$ or more elements of L are algebraically dependent over F. Any two finite transcendence bases of L over F have the same number of elements, and this number is the *transcendence degree* of L over F.

Note that the function field $K(\mathcal{F})$ of an irreducible plane curve $\mathcal{F}$ is a finite extension of an algebraically closed field K of transcendence degree 1 over K. So any two elements in $K(\mathcal{F})$ are algebraically dependent over K.

(XXIII) A field F is *perfect* if all algebraic extensions L/F are separable. In zero characteristic, every field is perfect. In positive characteristic p, a field F is perfect if and only if every element of F is a p-th power of some element in F.

Important examples of perfect fields are the algebraically closed fields and finite fields.

(XXIV) An algebraic field extension L/F is *simple* if $L = F(\alpha)$ for some $\alpha \in L$. A *primitive element* of L over F is any element $\beta \in L$ such that $L = F(\beta)$.

It should be noted that the elements of a simple extension field $F(\alpha)$ may be written in the form $g(\alpha)$, where $g(X)$ is a polynomial over F of degree at most $n-1$, with $n = [F(\alpha) : F]$.

Here, $F(\alpha)$ is an algebra of rank n with basis $\{1, \alpha, \ldots, \alpha^{n-1}\}$. Using the minimal polynomial of α, say $f(X) = X^n + b_1X^{n-1} + \cdots + b_n$ with $b_i \in F$, the product of two arbitrary elements of the algebra may be calculated.

(XXV) In positive characteristic, a finite algebraic extension L/F may not be simple. A sufficient condition for a finite extension L/F to be simple is that the adjoined elements $\alpha_1, \ldots, \alpha_r$ be, with at most one exception, separable over F.

(XXVI) A complex number is *algebraic* if it is algebraic over $\mathbf{Q}$; that is, it satisfies a non-zero polynomial equation with coefficients in $\mathbf{Q}$. The set $\mathbf{A}$ of algebraic numbers is a subfield of $\mathbf{C}$. A subfield of $\mathbf{A}$ which is an algebraic extension of $\mathbf{Q}$ is an *algebraic number field.* An algebraic number is an *algebraic integer* if it satisfies a monic polynomial equation with coefficients in $\mathbf{Z}$. The set $\mathbf{B}$ of algebraic integers is a subring of $\mathbf{A}$. An algebraic integer is a rational number if and only if it is a rational integer; that is, $\mathbf{B} \cap \mathbf{Q} = \mathbf{Z}$.

The following result is an essential ingredient of other chapters.

THEOREM A.1 (Theorem of the Primitive Element) *Let F be a field containing infinitely many elements, and let $F(\alpha, \beta)$ in an algebraic extension of F. If α, β are separable over F, then there exists $\lambda \in F$ such that $\alpha + \lambda\beta$ is a primitive element of $F(\alpha, \beta)/F$.*

Proof. Let z be an indeterminate over $F(\alpha, \beta)$. Every irreducible polynomial $G(X) \in F[X]$ is also irreducible when regarded as a polynomial in $F(z)[X]$. This can be shown indirectly.

If $G(X) = U(z, X)V(z, X)/d(z)$ with $U, V \in F(z, X)$ and $d(z) \in F(z)$, then $G(X) = U(\lambda, X)(V\lambda, X)/d(\lambda)$, for any $\lambda \in F$ such that $d(\lambda) \neq 0$, is a factorisation in $F[X]$. Such a λ exists as F contains infinitely many elements. Therefore α, β are separable over $F(z)$ as well, because they satisfy the same irreducible equations over $F(z)$ as over F. Then $w = \alpha + z\beta$ is separable over $F(z)$.

Consider the simple extension $F(z)(w)/F(z)$ and denote by $g(W) \in F(z)$ the minimal polynomial of w over $F(z)$. Then $g(W)$ comes from a polynomial $G(X, W) \in F(X, W)$ in the usual way; that is, $g(W) = G(z, W)$. Since w is separable, $\partial G/\partial W \neq 0$. From the fact that

$$\deg_W \frac{\partial G}{\partial W} < \deg_W G,$$

it follows that

$$\left(\frac{\partial G}{\partial W}\right)_{W=w} \neq 0.$$

Now $G(z, \alpha + z\beta) = 0$, whence

$$\left(\frac{\partial G(z, W)}{\partial z}\right)_{W=\alpha+z\beta} + \left(\frac{\partial G(z, W)}{\partial W}\right)_{W=\alpha+z\beta} \cdot \beta = 0.$$

From this,

$$\beta = \frac{H(z, \alpha + z\beta)}{d_1(z)},$$

with $H(X, W) \in F[X, W]$ and $d_1(z) \in F[z]$. So, for $\lambda \in F\backslash\{0\}$,

$$\beta = \frac{H(\lambda, \alpha + \lambda\beta)}{d_1(\lambda)}.$$

Thus $\beta \in F(\alpha + \lambda\beta)$. Since $\alpha \in F(\alpha + \lambda\beta)$, the assertion follows. □

A major result on primitive elements is the following.

THEOREM A.2 *Every finite separable extension L/K is simple.*

This theorem holds without the hypothesis on separability provided that L has transcendence degree 1. A proof is given, for which two lemmas are required.

LEMMA A.3 *For $p > 0$, let x be a transcendental element over K, let v be the p-th root of x in the algebraic closure $\overline{K(x)}$ of $K(x)$, and put $v = x^{1/p}$. Then*

(i) $$[K(x^{1/p}) : K(x)] = p;$$

(ii) *if e is a positive integer and x^{1/p^e} is a root of x of order p^e,*
$$[K(x^{1/p^e}) : K(x)] = p^e.$$

Proof. The first assertion is a consequence of the irreducibility of the polynomial $Z^p - x \in K(x)[Z]$ over K. Applying this to $x^{1/p}$ gives the following:

$$[K(x^{1/p^2}) : K(x)] = [K(x^{1/p^2}) : K(x^{1/p}] \cdot [K(x^{1/p}) : K(x)] = p \cdot p = p^2.$$

Repeating the argument shows that $K(x^{1/p^e} : K(x)] = p^e$. □

LEMMA A.4 *Let K have the following properties:*

(a) $p > 0$;

(b) *the element u is in an extension field of $K(x)$ which is algebraic over $K(x)$, where x is transcendental;*

(c) *the element x is not separable in $K(x, u)$, where u is a root of an irreducible polynomial $F \in K(x)[U^{p^e}]$ for some $e \geq 1$;*

(d) *the polynomial F does not belong to $K(x)[U^{p^{e+1}}]$.*

Then

(i) *$K(x, u)$ contains a root x^{1/p^e} of order p^e of x;*

(ii) *u is a separable element over $K(x^{1/p^e})$.*

Proof. From the hypothesis, put $G(x, U^{p^e}) = F(x, U)$. If $\deg_V G(x, V) = d$, then $\deg_U F(x, U) = dp^e$ and $[K(x, u) : K(x)] = dp^e$. Without loss of generality, suppose that $F \in K[x, U]$ and write $G = \sum c_{ij} x^i (U^{p^e})^j$. In $\overline{K(x)}$, the element u satisfies the polynomial

$$G^{(1/p^e)} = \sum c_{ij}^{1/p^e} x^{1/p^e} U^j$$

of degree d which is the minimal polynomial of u over $K(x^{1/p^e})$. In fact, if u satisfied a polynomial $H(x^{1/p^e}, U)$ of degree $d' < d$, then u would satisfy $H^{(1/p^e)}$, which is a polynomial over $K(x)$ of degree $d'e^p < de^p$, a contradiction. Hence

$$[K(x^{1/p^e}, u) : K(x^{1/p^e})] = d$$

and, from Lemma A.3, it follows that

$$[K(x^{1/p^e}, u) : K(x)] = dp^e.$$

Thus $K(x, u) = K(x^{1/p^e}, u)$, whence $x^{1/p^e} \in K(x, u)$. Finally, (ii) is a consequence of the definition of e. □

THEOREM A.5 (Theorem of the Primitive Element for Function Fields of Transcendence Degree 1) *Let Σ be a field of transcendence degree 1 over an algebraically closed subfield K. If $x \in \Sigma$ is transcendental over K, then there is $y \in \Sigma$ such that $\Sigma = K(x, y)$.*

Proof. By Theorem A.2, take x to be not separating. Let u be a separating function of Σ. Then Theorem A.2 ensures the existence of $v \in \Sigma$ such that $\Sigma = K(u, v)$. With the notation of Lemma A.3, from (ii) in that lemma, u^{p^e} is separable over $K(x)$ for some $e \geq 0$. Similarly, there is an integer $n \geq 0$ such that v^{p^n} is separable over $K(x)$. Let $k = \min\{m, n\}$. By the sufficient condition for an extension to be simple, $K(x, u^{p^e}, v^{p^n}, x^{1/p^k}) = K(x, y)$ for some $y \in \Sigma$.

To complete the proof, it suffices to show that $u, v \in K(x, u^{p^e}, v^{p^n}, x^{1/p^e})$. In proving this for u, the hypothesis $e > 0$ may be assumed. Then x^{1/p^e} is a root of the polynomial $Z^{p^e} - x$ over $K(x)$, and hence over $K(x, u^{p^e})$. This polynomial is irreducible over $K(x, u^{p^e})$ by the following argument.

Suppose

$$Z^{p^e} - x = (Z^{m_1} + \cdots)(Z^{m_2} + \cdots), \quad 0 < m_1, m_2 < p^m.$$

In the algebraic closure $\overline{K(x, u^{p^e})}$,

$$Z^{p^e} - x = (Z - x^{1/p^e})^{p^e},$$

which shows that $Z^{m_1} + \cdots$ and $Z^{m_2} + \cdots$ must each be a power of $Z - x^{1/p^e}$. Let p^s be the largest power of p dividing m_1. Since $m_1 < p^e$, so $s < e$. Then $p^s = tm_1 + rp^s$ with t, r integers. Therefore

$$x^{1/p^{e-s}} = (x^{m_1/p^e})^s x^r \in K(x, u^{p^e}).$$

The minimal polynomial of $x^{1/p^{e-s}}$ over $K(x)$ is $Z^{p^{e-s}} - x$ contradicting the separability of $K(x, u^{p^e})$ over $K(x)$.

It follows that $[K(x, u^{p^e}, x^{1/p^e}) : K(x, u^{p^e}] = p^e$. Putting

$$d = [K(x, u^{p^e}) : K(x)],$$

this gives $[K(x, u^{p^e}, x^{1/p^e}) : K(x)] = dp^e$. Since $K(x, u^{p^e}, x^{1/p^e})$ is a subfield of $K(x^{1/p^e}, u)$ and the latter coincides with $K(x, u)$ by Lemma A.4, from $[K(x, u) : K(x)] = dp^e$ the assertion $K(x, u) = K(u^{p^e}, x^{1/p^e})$ follows. Hence $u \in K(x, u^{p^e}, x^{1/p^e})$. Similarly, $v \in K(x, v^{p^e}, x^{1/p^e})$. □

A.2 GALOIS THEORY

(I) An extension L/F is *normal* if, whenever $f(X) \in F[X]$ is an irreducible polynomial which has one root in L, then all roots of $f(X)$ lie in L. A finite extension L/F is normal if and only if L is the splitting field of a polynomial in $F[X]$. In particular, a finite normal extension is separable if and only if L is the splitting field of a separable polynomial over F.

(II) A finite, normal, separable extension L/F is a *Galois extension.* Galois theory relates field extensions to groups via the concept of an automorphism.

(III) An isomorphism of a field L onto itself is an *automorphism*. The automorphisms of L form a group, the *automorphism group of* L, written $\mathrm{Aut}(L)$. If $H \subset \mathrm{Aut}(L)$, then the set of all fixed elements of H, written L^H, is a subfield of L, the *fixed field* of H.

(IV) If L/F is finite, then $\mathrm{Aut}(L/F)$ is also finite as its order is less than or equal to $[L : F]$. If L/F is a Galois extension then equality holds, and $\mathrm{Aut}(L/F)$ is the *Galois group* of L/F. Usually, the Galois group is denoted by $\mathrm{Gal}(L/F)$.

(V) The most important result on Galois extensions is the *Galois correspondence* which is a connection between subgroups H of $\mathrm{Gal}(L/F)$ and subfields M of L intermediate between F and L. More precisely, the maps

$$M \mapsto \mathrm{Gal}(L/M),$$
$$H \mapsto H^L$$

are mutually inverse bijections between the subfields of L containing F and the subgroups of $\mathrm{Gal}(L/F)$.

(VI) (a) For every subgroup H of $\mathrm{Gal}(L/F)$,

$$[L : L^H] = |H|, \quad [L^H : F] = |\mathrm{Gal}(L/F)/|H|, \quad H = \mathrm{Gal}(L/L^H).$$

(b) For every two subgroups H, G of $\mathrm{Gal}(L/F)$,

$$H \subset G \Longleftrightarrow L^H \supset L^G.$$

(c) For every intermediate field M of L/F, there is a subgroup H of $\mathrm{Gal}(L/F)$ such that $H = \mathrm{Gal}(L/M)$.

(d) For every two intermediate fields M_1, M_2 of L/F,

$$\mathrm{Gal}(L/\langle M_1, M_2\rangle) = \mathrm{Gal}(L/M_1) \cap \mathrm{Gal}(L/M_2),$$

where $\langle M_1, M_2\rangle$ is the subfield generated by M_1 and M_2, and

$$\mathrm{Gal}(L/(M_1 \cap M_2) = \langle \mathrm{Gal}(L/M_1), \mathrm{Gal}(L/M_2)\rangle,$$

where $\langle G_1, G_2\rangle$ is the group generated by G_1 and G_2.

(e) A subgroup H of $\mathrm{Gal}(L/F)$ is a normal subgroup if and only if L^H/F is a Galois extension.

(f) If the equivalent conditions of (e) hold, then $\mathrm{Gal}(M/F)$ is isomorphic to the factor group $\mathrm{Gal}(L/F)/\mathrm{Gal}(L/M)$.

(VII) Let L/F be a finite separable extension, and let $\bar{L}$ be the algebraic closure of L. Then there exists a unique field M with following properties:

(a) $L \subset M \subset \bar{L}$;

(b) M/F is a Galois extension;

(c) if $L \subset N \subset \bar{L}$ and N/F is Galois extension, then $M \subset N$.

The field M is the *Galois closure of* L/F.

(VIII) A Galois extension L/F is *cyclic* if Gal(L/F) is cyclic. Here, two relevant special cases are considered, namely $n = [L : F]$ with n prime to p, and $[L : F] = p$.

(IX) If $\gcd(n, p) = 1$, then L/F is a *Kummer extension* if F contains all n-th roots of unity; that is, the polynomial $X^n - 1$ splits into linear factors in $F[X]$.

Every Kummer extension can be described as a simple algebraic extension in the following way. Take an element $c \in F$ which is not an m-th power in F for any divisor m of n. Then the polynomial $X^n - c$ is irreducible in $F[X]$, and L/F is the simple extension $F(v)$ with $v^n - c = 0$. The automorphisms σ in Gal(L/F) are given by $\sigma(u) = \xi u$ as ξ ranges over the n-th roots of unity in F. Conversely, If F contains all n-th roots of unity, n being prime to p, and $L = F(v)$ where $v^n - c = 0$ with c satisfying the above condition, then L/F is a cyclic extension of degree n.

(X) When $[L : F] = p$, take an element $c \in F$ that cannot be written as $a^p - a$ for any $a \in F$. Then $X^p - X - c$ is an irreducible polynomial in $F[X]$, and the simple extension $L = F(v)$ with $v^n - v - c = 0$ is an *Artin–Schreier extension* of degree p.

The automorphisms σ in Gal(L/F) are given by $\sigma(u) = u + \eta$ as η ranges over the p elements in the prime field of F. Conversely, if $L = F(v)$ is a simple extension where $v^p - v - c = 0$ and c satisfies the above condition, then L/K is cyclic of degree p.

A.3 NORMS AND TRACES

(I) Let L/F be a finite extension of degree $n = [L : F]$. Since L may be viewed as a vector space over F, every element $u \in L$ induces a linear transformation ν_u of L given by $\nu_u(x) = ux$ for $x \in L$.

(II) The *norm* of u with respect to the extension L/F is

$$\mathrm{N}_{L/F}(u) = \det(\nu_u)$$

in F.

(III) The *trace of* u with respect to L/F, written $\mathrm{T}_{L/F}(\nu_u)$, is the trace of ν_u. In other words, if $\{u_1, \ldots, u_n\}$ is a basis of L/F and

$$uu_i = \textstyle\sum_{j=1}^{n} a_{ij} u_i$$

with $a_{ij} \in F$, then

(a) $\mathrm{N}_{L/F}(u) = \det(a_{ij})$,
(b) $\mathrm{T}_{L/F}(u) = \sum_{j=1}^{n} a_{ii}$.

(IV) For $x, y \in L$ and $\alpha \in F$, with $n = [L : F]$,

(a) $\mathrm{N}_{L/F}(\alpha) = \alpha^n$;
(b) $\mathrm{T}_{L/F}(\alpha) = n \cdot \alpha$;
(c) $\mathrm{N}_{L/F}(x \cdot y) = \mathrm{N}_{L/F}(x) \cdot \mathrm{N}_{L/F}(y)$;
(d) $\mathrm{N}_{L/F}(x) = 0$ if and only if $x = 0$;
(e) $\mathrm{T}_{L/F}(x + y) = \mathrm{T}_{L/F}(x) + \mathrm{T}_{L/F}(y)$;
(f) $\mathrm{T}_{L/F}(\alpha \cdot x) = \alpha \cdot \mathrm{T}_{L/F}(x)$;
(g) for any finite extension M/L and $u \in M$,
  (i) $\mathrm{N}_{M/F}(u) = \mathrm{N}_{L/F}(\mathrm{N}_{M/L}(u))$.
  (ii) $\mathrm{T}_{M/F}(u) = \mathrm{T}_{L/F}(\mathrm{T}_{M/L}(u))$;

(V) A finite extension L/F is inseparable if and only if $\mathrm{T}_{L/F}(x) = 0$ for every $x \in L$.

(VI) L/F is separable if and only if $\mathrm{T}_{L/F} : L \to F$ is surjective.

(VII) If $f(X) = X^r + a_1X^{r-1} + \cdots + a_r \in F[X]$ is the minimal polynomial of $a \in L$ over F, and $[L : F] = n = rs$ with $s = [L : F(a)]$, then

(a) $\mathrm{N}_{L/F}(a) = (-1)^n a_r^s$;
(b) $\mathrm{T}_{L/F}(a) = -sa_1$.

(VIII) Given a separable field extension L/F of degree n, let $\sigma_1, \ldots, \sigma_n$ be the distinct embeddings of L into the algebraic closure $\bar{F}$ of F. Then, for every $a \in L$,

(a) $\mathrm{N}_{L/F}(x) = \prod_{i=1}^n \sigma_i(x)$;
(b) $\mathrm{T}_{L/F}(x) = \sum_{i=1}^n \sigma_i(x)$.

In the case of a Galois extension with Galois group $G = \mathrm{Gal}(L/F)$, these become the following:

(a) $\mathrm{N}_{L/F}(x) = \prod_{\sigma \in G} \sigma(x)$;
(b) $\mathrm{T}_{L/F}(x) = \sum_{\sigma \in G}^n \sigma(x)$.

A.4 FINITE FIELDS

Finite fields, also called Galois fields, are all of positive characteristic and have been classified. Their structure is described below.

For $q = p^n$, the splitting field $\mathbf{F}_q$ of the irreducible polynomial $X^q - X$ over the field $\mathbf{Z}/p\mathbf{Z}$ has order q; that is, it consists of q elements, and is of characteristic p.

(I) Up to isomorphism, the fields $\mathbf{F}_q$ are the only finite fields.

(II) The multiplicative group of $\mathbf{F}_q$ is a cyclic group of order $q - 1$.

(III) The additive group of $\mathbf{F}_q$ is an elementary abelian group of order q.

(IV) Every subfield of $\mathbf{F}_q$ is isomorphic to some $\mathbf{F}_{p^i}$ for a divisor i of n. Conversely, for every divisor i of n, the field $\mathbf{F}_q$ has a unique subfield of order p^i.

(V) The automorphism group of $\mathbf{F}_q$ is the cyclic group of order n, generated by $\sigma : x \mapsto x^p$.

(VI) $\mathbf{F}_q \subset \mathbf{F}_{q^m}$ for $m \geq 1$.

(VII) The extension $\mathbf{F}_{q^m}/\mathbf{F}_q$ is a Galois extension of degree m.

(VIII) The Galois group $\mathrm{Gal}(\mathbf{F}_{q^m}/\mathbf{F}_q)$ is cyclic and is generated by the Frobenius automorphism Φ of $\mathbf{F}_{q^m}$ given by $\Phi(x) = x^q$.

(IX) The norm,

$$\mathrm{N}(x) = \mathrm{N}_{\mathbf{F}_{q^m}/\mathbf{F}_q}(x) = x^{1+q+q^2+\cdots+q^{m-1}}.$$

(X) The trace,

$$\mathrm{T}(x) = \mathrm{T}_{\mathbf{F}_{q^m}/\mathbf{F}_q}(x) = x + x^q + x^{q^2} + \cdots + x^{q^{m-1}}.$$

(XI) When $x \in \mathbf{F}_{q^m}$, then $\mathrm{T}_{\mathbf{F}_{q^m}}(x) = 0$ if and only if $x = y^q - y$ for some $y \in \mathbf{F}_{q^m}$.

LEMMA A.6 (Lucas)

(i) *Let p be a prime and let the positive integers n, m have the p-adic expansions*

$$n = \sum n_i p^i, \quad m = \sum m_i p^i,$$

with $0 \leq n_i \leq p-1,\ 0 \leq m_i \leq p-1$. Then

$$\binom{n}{m} \not\equiv 0 \pmod{p}$$

if and only $n_i \geq m_i$ for all i.

(ii) *With $q = p^h$,*

$$\binom{nq}{q} \equiv n \pmod{p}.$$

Proof. (i) For an indeterminate X, consider the following expansion over $\mathbf{F}_p$:

$$\begin{aligned}(1+X)^n &= (1+X)^{n_0+n_1p+\cdots+n_rp^r}\\ &= (1+X)^{n_0}(1+X^p)^{n_1}\cdots(1+X^{p^r})^{n_r}.\end{aligned}$$

Now pick out the coefficient of X^m from both sides:

$$\binom{n}{m} = \binom{n_0}{m_0}\binom{n_1}{m_1}\cdots\binom{n_r}{m_r}.$$

This gives the result.

(ii) With $n = \sum n_i p^i$ and $q = p^h$ the p-adic expansions, compare the coefficient of X^q on both sides of

$$(1+X)^{nq} = (1+X)^{n_0p^h}\cdots(1+X)^{n_rp^{r+h}}.$$

Also, note that $n \equiv n^q \pmod{p}$ to obtain the result. □

A.5 GROUP THEORY

The purpose here is to state definitions and results from group theory which are applied in Chapter 11.

(I) The *order* of a group G, written $|G|$ or ord G, is the cardinality of the set G. If this cardinality if finite, then $|G|$ is the number of elements in G. Let p be a prime. A finite group G is a *p-group* if its order is a power of p.

(II) Two elements x, y in a group G are *conjugate* if $y = g^{-1}xg$ for some g in G. Similarly, two subgroups G, H are *conjugate* if every element in H is conjugate to some element in G.

(III) A *cyclic group* G is a group generated by a single element g; that is, G consists of all the powers of g, and is denoted $G = \langle g \rangle$. The *order of g* is the order of $\langle g \rangle$, written ord g. Hence

$$G = \{1, g, g^2, \ldots, g^{n-1} \mid g^n = 1\} \quad \text{if } G \text{ is finite,}$$
$$G = \{\ldots, g^{-2}, g^{-1}, 1, g, g^2, \ldots\} \quad \text{if } G \text{ is infinite.}$$

A *dihedral group* G is a finite group $\mathbf{D}_n$ of order $2n$ generated by two elements g and h together with the following relations:

$$\text{ord}\, g = 2, \quad \text{ord}\, h = n, \quad ghg = h^{-1}.$$

Other important groups generated by two elements g, h are the generalised quaternion groups and the semidihedral groups; both are 2-groups.

The *generalised quaternion group* of order 2^{e+1} with $e > 1$ is defined by the relations

$$\text{ord}\, g = 2^e, \quad h^2 = g^{2^{e-1}}, \quad h^{-1}gh = g^{-1}.$$

The *semi-dihedral group* of order 2^{e+1} with $e > 1$ is defined by the relations

$$\text{ord}\, g = 2^e, \quad \text{ord}\, h = 2, \quad h^{-1}gh = g^{2^{e-1}-1}.$$

(IV) An group G is *abelian* or *commutative* if $xy = yx$ for all elements x, y of G.

(V) The *centre* $Z = Z(G)$ of a group G is $Z = \{x \in G \mid xg = gx \text{ all } g \in G\}$. So a group is abelian if it coincides with its centre. Cyclic groups are abelian.

(VI) The *direct product* of the groups G and H is

$$G \times H = \{(g, h) \mid g \in G, h \in H\}$$

such that $(g_1, h_1)(g_2, h_2) = (g_1g_2, h_1h_2)$. The direct product of abelian groups is abelian.

(VII) (Lagrange's Theorem) If H is a subgroup of a finite group G, then $|H|$ divides $|G|$, and $|G|/|H|$ is the number of all cosets of H in G; in particular, if N is a normal subgroup of the finite group G, then $|G|/|N| = |G/N|$.

(VIII) The converse of Lagrange's Theorem holds for subgroups of prime power order by Sylow's Theorem. Let d be a prime dividing $|G|$. If d^n divides $|G|$ but d^{n+1} does not, then G has a subgroup of order d^n. Such a subgroup is a *d-Sylow subgroup* of G. Any two d-Sylow subgroups of G are conjugate in G and their number is congruent to 1 modulo d.

(IX) Given a subgroup of H of G, the *left coset* of an element $g \in G$ is the set $gH = \{gh \mid h \in H\}$ and the *right coset* of g is the set $Hg = \{hg \mid h \in H\}$. If G is finite, then the number of right or left cosets is the *index* of H.

A subgroup N of G is *normal*, written $N \triangleleft G$, if $gN = Ng$ for every $g \in G$. For instance, subgroups of index 2 are normal. A subgroup is normal if and only it is conjugate only to itself. The *trivial* normal subgroups of G are G and $\{1\} = I$.

If G has only trivial normal subgroups, then G is *simple*.

(X) If N is a normal subgroup of G, then the set of all left or right cosets of N in G form the *quotient* or *factor group* G/N under the operation

$$(g_1N)(g_2N) = (g_1g_2)N.$$

(XI) Given two groups A and B, an *extension* of A by B is a group G which contains a normal subgroup N such that N is isomorphic to A while G/N is isomorphic to B. If $G = NH$ and $N \cap H = \{1\}$, then the extension is *split*, or, equivalently, G is the *semidirect product* of N by H, written $G = N \rtimes H$. In this context, H is a *complement*. A sufficient condition for an extension to be split is that $|N|$ and $[G : N]$ have no common divisor. The latter condition also implies that complements are conjugate in G. If a normal subgroup S of G contains a Sylow d-subgroup S_d of G, then G=$N_G(S_d)\, S$.

(XII) A *minimal normal subgroup* of G is a normal subgroup N with $N \neq I$ which does not contain properly any other non-trivial normal subgroups of G. Every minimal normal subgroup N of a finite group G is a direct product of $N = T_1 \times \cdots \times T_k$, where the T_i are simple normal subgroups of N, pairwise conjugate under G. The *socle* of G is the subgroup generated by all minimal normal non-trivial subgroups of G, written $\mathrm{soc}(G)$.

(XIII) If $x, y \in G$, the element $x^{-1}y^{-1}xy \in G$ is the *commutator* of x and y. The subgroup G' generated by all commutators of G is the *derived* group of G. The factor group G/G' is abelian and, if N is a normal subgroup of G such that G/N is abelian, then $N \supset G'$.

(XIV) A group G is *soluble* if the sequence $G > G' > \cdots > G^{(i)} > \cdots$, in which each group $G^{(i)}$ is the derived group of the preceding, terminates with I.

(XV) A group G is *nilpotent* if it has a finite sequence $G = G_0 \triangleright G_1 \triangleright \cdots \triangleright G_i \triangleright \cdots$ such that $G_{i-1}/G_i < Z(G/G_i)$ for every $i \geq 1$. A finite group is nilpotent if and only if it is the direct product of its Sylow subgroups.

(XVI) For $g \in G$, the *centraliser* $C_G(g) = \{x \in G \mid xg = gx\}$. The intersection of all centralisers in G is the centre $Z = Z(G)$ and is a normal subgroup G. If $|G| = p^n$ with p prime and $n > 1$, then $Z(G) \neq I$, and $Z(G) \cap N$ is non-trivial for every $N \triangleleft G$. Also, a finite 2-group which contains an element of order 2 whose centraliser has order 4 is either dihedral or semidihedral or cyclic of order 4.

(XVII) For a subgroup H of G, the *normaliser* $N_G(H) = \{g \in G \mid gHg^{-1} = H\}$. Assume that H has odd order. If $N_G(H)$ has even order and $g \in N_G(H)$ has order 2, let

$$H_1 = \{h \in H \mid ghg^{-1} = h\}, \quad H_{-1} = \{h \in H \mid ghg^{-1} = h^{-1}\}.$$

Then $H = \{h_1 h_{-1} \mid h_1 \in H_1,\, h_{-1} \in H_{-1}\}$. Also, if H_1 is trivial, then H is abelian.

(XVIII) The *Frattini* subgroup of G is the intersection of all maximal subgroups of G.

(XIX) The group of all permutations of a finite set $\Omega = \{\omega_1, \ldots, \omega_n\}$ is the *symmetric group* $\mathbf{S}_n$ of Ω. Its order $|\mathbf{S}_n|$ is equal to $n!$.

The normal subgroup of $\mathbf{S}_n$ of index 2 consisting of all even permutations of Ω is the *alternating group* $\mathbf{A}_n$ of Ω.

A *permutation* group G is a group whose elements are permutations of a finite set, that is, a subgroup of a symmetric group.

(XX) If H is a finite group and G is a permutation group with a homomorphism $\rho : H \to G$, then $H/\ker\rho \cong G$ and G is the permutation group *induced by* H *on* Ω, or G is a *permutation representation of* H. If $\ker\rho$ is trivial, then this representation is *faithful*.

(XXI) Let G be a permutation group on a finite set Ω whose elements are called *points*. The *degree* of G is $|\Omega|$. The *orbit* of $\omega \in \Omega$ under G, written $G(\omega)$, consists of the images of ω under G; that is,

$$G(\omega) = \{\omega' \mid \omega' = g(\omega),\, g \in G\}.$$

The *orbit structure* of G is the list of lengths of its orbits, each counted with multiplicity.

(XXII) The *stabiliser* of a point $\omega \in \Omega$, written G_ω, is the subgroup consisting of all elements in G that fix ω; that is,

$$G_\omega = \{g \in G \mid g(\omega) = \omega\}.$$

More generally, if $\Delta \subset \Omega$, then the stabiliser of Δ under G, written G_Δ, is the subgroup consisting of all elements in G which fix each point in Δ; that is,

$$G_\Delta = \{g \in G \mid g(\omega) = \omega,\, \omega \in \Delta\}.$$

Similarly, the *global stabiliser* of Δ, written $G_{\{\Delta\}}$, consists of all elements in G which preserve Δ; that is, $G_{\{\Delta\}} = \{g \in G \mid g(\omega) \in \Delta,\, \omega \in \Delta\}$. If Δ has size n, then G_Δ is an n-point stabiliser of G.

(XXIII) The length of an orbit and the stabiliser of a point in the orbit are connected:

$$|G| = |G(\omega)|\,|G_\omega|.$$

In particular, if $G_\omega = I$, then $|G(\omega)| = |G|$ and the orbit is a *long orbit*; otherwise $G(\omega)$ is a *short orbit*.

If every orbit is long, that is, no non-trivial element in G fixes a point in Ω, then G is a *semiregular* permutation group on Ω. The *orbit structure* of G is the list of the lengths of its orbits.

(XXIV) The group G is *transitive* if it has only one orbit, and it is *regular* or *sharply transitive* if it is also semiregular. So G is transitive on Ω if and only if, for any ordered pair (ω, δ) with $\omega, \delta \in \Omega$, there exists $g \in G$ such that $\delta = g(\omega)$, and G is regular if there is exactly one such permutation $g \in G$.

(XXV) More generally, G is *n-fold transitive* on Ω if, for every n ordered pairs (ω_i, δ_i), there exists $g \in G$ such that $\delta_i = g(\omega_i)$ for $i = 1, \ldots, n$, and *sharply n-fold transitive* if there is exactly one such permutation in G.

Note that, if G is $(n-1)$-fold transitive on Ω and the n-point stabilisers are transitive on the remaining points, then G is n-fold transitive, and vice versa. Also, if G is sharply n-fold transitive, then its order is equal to

$$|\Omega|(|\Omega|-1)\cdots(|\Omega|-n+1).$$

(XXVI) Let G be a 2-transitive permutation group on a finite set Ω of size n and let N be a minimal normal subgroup of G.

(i) If N is regular, then N is elementary abelian of order $n = d^k$ for some prime number d and G is a subgroup of $\mathrm{AGL}(k, d)$.

(ii) If N is not regular, then N is a non-abelian simple group such that $N \le G \le \mathrm{Aut}(N)$.

When (i) occurs and $n = 2^k$, the group G is a subgroup of $\mathrm{A\Gamma L}(1, 2^k)$. This does not hold true for every odd n.

However, for $n = 9$, the group G is isomorphic to one of

$$\mathrm{A\Gamma L}(1,9), \quad \mathrm{AGL}(1,9), \quad \mathrm{A\gamma L}(1,9).$$

Here, $\mathrm{A\gamma L}(1,9)$ is the subgroup of $\mathrm{A\Gamma L}(1,9)$ consisting of all permutations $x \to ax^\sigma + b$ where $b \in \mathbf{F}_9$, where a is a non-zero square in $\mathbf{F}_9$ and where $\sigma \in \mathrm{Aut}(\mathbf{F}_9)$. Both $\mathrm{AGL}(1,9$ and $\mathrm{A\gamma L}(1,9)$ are sharply 2-transitive, but the 2-point stabiliser is cyclic only if $G \cong \mathrm{AGL}(1,9)$.

(XXVII) A *Frobenius* group is a transitive group which is not regular, but in which only the identity has more than one fixed point. Those elements in a Frobenius group F which have no fixed point together with the identity constitute a regular normal nilpotent subgroup N, the *kernel* of F. Also, $G = N \rtimes G_\omega$ for any $\omega \in \Omega$.

The subgroup G_ω, and every subgroup isomorphic to G_ω, is a *Frobenius complement*. If G_ω is even then it has only one element of order 2, and N is commutative. If G_ω is odd, then it is metacyclic. In any case, $Z(G_\omega)$ is non-trivial.

(XXVIII) Let G be a permutation group of Ω and d a prime number. Assume that every point of Ω is fixed by a permutation in G of order d which has no other fixed point; then G is transitive on Ω. If, in addition, $d = 2$, every point in Ω is fixed by exactly one involutory element in G, and no involutory element in G fixes more than one point of Ω then the subgroup of G consisting of all products of even number of involutory elements in G has odd order and acts transitively on Ω.

(XXIX) For $n \geq 2$, all n-transitive permutation groups are known, their classification being a corollary to that of all finite simple groups. In Chapter 11, doubly-transitive permutation groups whose 2-point stabilisers are cyclic play an important role. Below, the relevant classification theorems are stated after describing the permutation groups involved given in their usual doubly-transitive permutation representations.

(XXX) A finite group G admits a *partition* if it contains a set $\{G_1, \ldots, G_k\}$ of subgroups satisfying the following properties:

$$G = \bigcup_{i=1}^{k} G_i, \quad G_i \cap G_j = \{1\}.$$

The subgroups G_i are the *components* of the partition. A counting argument shows that $n = \sum_{i=1}^{k} |G_i| - (k-1)$.

EXAMPLE A.7 Projective linear groups

1. Let $\Omega = \mathbf{F}_q \cup \{\infty\}$. The linear fractional mappings

$$\varphi_{(a,b,c,d)}: \quad x \mapsto \frac{ax+b}{cx+d}, \quad ad - bc \neq 0,$$

with $a, b, c, d \in \mathbf{F}_q$, form a sharply triply-transitive group, the *projective general linear group* $\mathrm{PGL}(2, q)$.

2. The 2-point stabilisers are isomorphic to the multiplicative group $\mathbf{F}_q^*$ of $\mathbf{F}_q$.

3. If q is odd, then $\mathrm{PGL}(2, q)$ contains a normal 2-transitive subgroup consisting of all $\varphi_{(a,b,c,d)}$ with $ad - bc$ square in $\mathbf{F}_q$, the *projective special linear group* $\mathrm{PSL}(2, q)$; it has order $\frac{1}{2}(q^3 - q)$.

4. The 2-point stabiliser of $\mathrm{PSL}(2, q)$ is isomorphic to the subgroup of all square elements in $\mathbf{F}_q^*$.

5. Apart from the cases $q = 2, 3$, the group $\mathrm{PSL}(2, q)$ is simple.

THEOREM A.8 *For $q = p^m$, the following is a complete list of subgroups of the group* $\mathrm{PGL}(2, q)$ *up to conjugacy*:

(i) *the cyclic group of order n with $n \mid (p^m \pm 1)$;*

(ii) *the elementary abelian p-group of order p^f with $f \leq m$;*

(iii) *the dihedral group $\mathbf{D}_n$ of order $2n$ with $n \mid (q \pm 1)$;*

(iv) *the alternating group $\mathbf{A}_4$ for $p > 2$, or $p = 2$ and m even;*

(v) *the symmetric group $\mathbf{S}_4$ for $p > 2$;*

(vi) *the alternating group $\mathbf{A}_5$ for $5 \mid (q^2 - 1)$;*

(vii) *the semidirect product of an elementary abelian p-group of order p^h by a cyclic group of order n with $h \leq m$ and $n \mid (q - 1)$;*

(viii) $\mathrm{PSL}(2, p^f)$ *for $f \mid m$;*

(ix) $\mathrm{PGL}(2, p^f)$ *for $f \mid m$.*

EXAMPLE A.9 Projective unitary groups

1. Let $\mathcal{U}$ be the classical unital in $\mathrm{PG}(2, q^2)$, that is, the set of all self-conjugate points of a unitary polarity Π of $\mathrm{PG}(2, q^2)$. The *projective unitary group* $\mathrm{PGU}(3, q)$ comprises the projective transformations in $\mathrm{PG}(2, q^2)$ commuting with Π. Equivalently, $\mathrm{PGU}(3, q)$ preserves $\mathcal{U}$ and can be viewed as a permutation group on $\mathcal{U}$, since the only collineation in $\mathrm{PGU}(3, q)$ fixing every point in $\mathcal{U}$ is the identity.

2. With $\Omega = \mathcal{U}$, the group $\mathrm{PGU}(3, q)$ is a doubly-transitive permutation group on Ω and its 2-point stabiliser is isomorphic to $\mathbf{F}_q^*$.

3. $|\mathrm{PGU}(3, q)| = (q^3 + 1)q^3(q^2 - 1)$.

4. With $d = \gcd(3, q + 1)$, the group $\mathrm{PGU}(3, q)$ contains a normal subgroup $\mathrm{PSU}(3, q)$, *the special unitary group*, of index d which is still 2-transitive. If $q > 2$, the group $\mathrm{PSU}(3, q)$ is non-abelian and simple.

5. For another representation of $\mathrm{PGU}(3, q)$, let

$$M = \{m \in \mathbf{F}_{q^2} \mid m^q + m = 0\}.$$

Take an element $c \in \mathbf{F}_{q^2}$ such that $c^q + c + 1 = 0$. A homogeneous coordinate system in $\mathrm{PG}(2, q^2)$ can be chosen so that

$$\mathcal{U} = \{X_\infty\} \cup \{U = (1, u, cu^{q+1} + m) \mid u \in \mathbf{F}_{q^2},\ m \in M\}.$$

Then $\mathcal{U}$ consists of all $\mathbf{F}_{q^2}$-rational points of the Hermitian curve

$$\mathcal{H}_q = \mathbf{v}(cX_0^q X_2 + c^q X_0 X_2^q + X_1^{q+1}).$$

6. A Sylow p-subgroup S_p of $\mathrm{PGU}(3,q)$ consists of the projective transformations associated to the matrices

$$T_{a,m} = \begin{bmatrix} 1 & a^q(1+c^{q-1}) & a^{q+1}+c^{-1}m \\ 0 & 1 & a \\ 0 & 0 & 1 \end{bmatrix}$$

for $a \in \mathbf{F}_{q^2}$ and $m \in M$.

7. The set $\{T_{0,m} \mid m \in M\}$ is an elementary abelian p-group, which is $Z(S_p)$. Also, S_p is a normal subgroup of $\mathrm{PGU}(3,q)_{Y_\infty}$ and acts transitively on the other q^3 points of $\mathcal{U}$.

8. The point O is in $\mathcal{U}$, and the stabiliser T of O, Y_∞ under $\mathrm{PGU}(3,q)$ consists of the projective transformations associated to the matrices

$$\begin{bmatrix} a^{q+1} & 0 & 0 \\ 0 & a & 0 \\ 0 & 0 & 1 \end{bmatrix}$$

for $a \in \mathbf{F}_{q^2}\backslash\{0\}$.

9. The group $\mathrm{PGU}(3,q)$ is generated by S_p and T, together with the projective transformation W associated to the matrix

$$\begin{bmatrix} 0 & 0 & c \\ 0 & 1 & 0 \\ c^{-1} & 0 & 0 \end{bmatrix}$$

that interchanges the points Y_∞ and O.

10. The group $\mathrm{PGU}(3,q)$ also fixes $\mathcal{H}_q$, and hence is a K-automorphism group of $\Sigma = K(\mathcal{H}_q)$.

THEOREM A.10 *The following is the list of maximal subgroups of* $\mathrm{PSU}(3,q)$ *up to conjugacy:*

(i) *the one-point stabiliser of order* $q^3(q^2-1)/d$;

(ii) *the non-tangent line stabiliser of order* $q(q^2-1)(q+1)/d$;

(iii) *the self-conjugate triangle stabiliser of order* $6(q+1)^2/d$;

(iv) *the normaliser of a cyclic Singer group of order* $3(q^2-q+1)/d$;

further, for $q = p^k$ *with* $p > 2$,

(v) $\mathrm{PGL}(2,q)$ *preserving a conic*;

(vi) $\mathrm{PSU}(3,p^m)$, *with* $m \mid k$ *and* k/m *odd*;

(vii) *the subgroup containing* $\mathrm{PSU}(3,p^m)$ *as a normal subgroup of index* 3 *when* $m \mid k$, k/m *is odd, and* 3 *divides both* k/m *and* $q+1$;

(viii) *the Hessian groups of order* 216 *when* $9 \mid (q+1)$, *and of order* 72 *and* 36 *when* $3 \mid (q+1)$;

(ix) $\mathrm{PSL}(2,7)$ *when either* $p = 7$ *or* $\sqrt{-7} \notin \mathbf{F}_q$;

(x) *the alternating group* $\mathbf{A}_6$ *when either* $p = 3$ *and* k *is even, or* $\sqrt{5} \in \mathbf{F}_q$ *but* $\mathbf{F}_q$ *contains no cube root of unity*;

(xi) *the symmetric group* $\mathbf{S}_6$ *for* $p = 5$ *and* k *odd*;

(xii) *the alternating group* $\mathbf{A}_7$ *for* $p = 5$ *and* k *odd*;

for $q = 2^k$,

(xiii) $\mathrm{PSU}(3, 2^m)$ *with* k/m *an odd prime*;

(xiv) *the subgroups containing* $\mathrm{PSU}(3, 2^m)$ *as a normal subgroup of index* 3 *when* $k = 3m$ *with* m *odd*;

(xv) *a group of order* 36 *when* $k = 1$.

EXAMPLE A.11 Suzuki groups

1. An *ovoid* $\mathcal{O}$ in $\mathrm{PG}(3,q)$ is a point set with the same combinatorial properties as an elliptic quadric in $\mathrm{PG}(3,q)$; namely, $\mathcal{O}$ consists of q^2+1 points, no three collinear, such that the lines through any point $P \in \mathcal{O}$ meeting $\mathcal{O}$ only in P are coplanar.

2. Now assume that $q = 2q_0^2$ with $q_0 = 2^s$ and $s \geq 1$. Then $x^\varphi = x^{2q_0}$ is an automorphism of $\mathbf{F}_q$, and $x^{\varphi^2} = x^2$.

   Let $\mathcal{T}$ be the *Suzuki–Tits ovoid* in $\mathrm{PG}(3,q)$, which is the only known ovoid in $\mathrm{PG}(3,q)$ other than an elliptic quadric. The *Suzuki group* $\mathrm{Sz}(q)$, also written ${}^2B_2(q)$, is the projective group of $\mathrm{PG}(3,q)$ fixing $\mathcal{T}$. The group $\mathrm{Sz}(q)$ can be viewed as a permutation group on $\mathcal{T}$ as the identity is the only projective transformation in $\mathrm{Sz}(q)$ fixing every point in $\mathcal{T}$.

3. With $\Omega = \mathcal{T}$, the group $\mathrm{Sz}(q)$ is a 2-transitive permutation group on Ω, and its 2-point stabiliser is isomorphic to $\mathbf{F}_q^*$.

4. $|\mathrm{Sz}(q)| = (q^2+1)q^2(q-1)$.

5. The group $\mathrm{Sz}(q)$ is simple. If S_2 is a 2-Sylow subgroup of $\mathrm{Sz}(q)$, the elements of S_2 of order 2 together with the identity form an elementary abelian normal subgroup N of order q, and the quotient group S_2/N is isomorphic to N.

6. In a suitable coordinate system of $\mathrm{PG}(3,q)$ with $Z_\infty = (0,0,0,1)$,

$$\mathcal{T} = \{Z_\infty\} \cup \{(1, u, v, uv + u^{2\varphi+2} v^\varphi) \mid u, v \in \mathbf{F}_q\}.$$

A 2-Sylow subgroup S_2 consists of the projective transformations associated to the matrices

$$\begin{bmatrix} 1 & 0 & 0 & 0 \\ a & 1 & 0 & 0 \\ c & a^{\varphi} & 1 & 0 \\ ac + a^{\varphi+2} + c^{\varphi} & a^{\varphi+1} + c & a & 1 \end{bmatrix}$$

for $a, c \in \mathbf{F}_q$.

7. The subgroup S_2 is normal in $\mathrm{Sz}(q)_{Z_\infty}$ and regular on the remaining q^2 points of $\mathcal{T}$. The stabiliser $\mathrm{Sz}(q)_{Z_\infty,O}$ with $O = (1,0,0,0)$ consists of the projective transformations associated to diagonal matrices

$$\mathrm{diag}(d^{-q_0-1}, d^{-q_0}, d^{q_0}, d^{q_0+1})$$

for $d \in \mathbf{F}_{q^2}\backslash\{0\}$.

8. The group $\mathrm{Sz}(q)$ is generated by S_2 and $\mathrm{Sz}_{Z_\infty,O}$ together with the projective transformation W associated to the matrix

$$\begin{bmatrix} 0 & 0 & 0 & 1 \\ 0 & 0 & 1 & 0 \\ 0 & 1 & 0 & 0 \\ 1 & 0 & 0 & 0 \end{bmatrix}$$

that interchanges Z_∞ and O.

In the next result, $\mathrm{Sz}(2) = \mathrm{AGL}(1,5)$.

THEOREM A.12 *Up to conjugacy,* $\mathrm{Sz}(q)$ *has the following maximal subgroups:*

(i) *the one-point stabiliser of order* $q^2(q-1)$;

(ii) *the normaliser of a cyclic Singer group of order* $4(q + 2q_0 + 1)$;

(iii) *the normaliser of a cyclic Singer group of order* $4(q - 2q_0 + 1)$;

(iv) $\mathrm{Sz}(q')$ *for every* q' *such that* $q = q^m$ *with* m *prime.*

Further, the subgroups listed below form a partition of $\mathrm{Sz}(q)$:

(v) *all subgroups of order* q^2;

(vi) *all cyclic subgroups of order* $q - 1$;

(vii) *all cyclic Singer subgroups of order* $q + 2q_0 + 1$;

(viii) *all cyclic Singer subgroups of order* $q - 2q_0 + 1$.

EXAMPLE A.13 Ree groups

The Ree group can be introduced in a similar way using the combinatorial concept of an ovoid, this time in the context of polar geometries.

1. An *ovoid* in the polar space associated to the non-degenerate quadric $\mathcal{Q}$ in the space $\mathrm{PG}(6,q)$ is a point set of size q^3+1, with no two of the points conjugate with respect to the orthogonal polarity arising from $\mathcal{Q}$.

2. Now let $q = 3q_0^2$ and $q_0 = 3^s$ with $s \geq 0$. Then $x^\varphi = x^{3q_0}$ is an automorphism of $\mathbf{F}_q$, and $x^{\varphi^2} = x^3$.

   Let $\mathcal{K}$ be the *Ree–Tits ovoid* of $\mathcal{Q}$. The *Ree group* $\mathrm{Ree}(q)$, also written ${}^2G_2(q)$, is the projective group of $\mathrm{PG}(6,q)$ fixing $\mathcal{K}$.

3. The group $\mathrm{Ree}(q)$ can be viewed as a permutation group on $\mathcal{K}$ as the identity is the only projective transformation in $\mathrm{Ree}(q)$ fixing every point in $\mathcal{K}$. With $\Omega = \mathcal{K}$, the group $\mathrm{Ree}(q)$ is a 2-transitive permutation group on Ω, and its 2-point stabiliser is isomorphic to $\mathbf{F}_q^*$.

4. $|\mathrm{Ree}(q)| = (q^3+1)q^3(q-1)$.

5. $\mathrm{Ree}(3) \cong \mathrm{P\Gamma L}(2,8)$ has a normal subgroup of index 3.

6. For $q_0 > 1$, the group $\mathrm{Ree}(q)$ is simple. Let S_3 be a 3-Sylow subgroup of $\mathrm{Ree}(q)$. The elements of order 3 in $Z(S_3)$ together with the identity constitute an elementary abelian normal subgroup N of order q^2, and the factor group S_3/N is an elementary abelian group of order q.

7. In a suitable coordinate system of $\mathrm{PG}(6,q)$ with $Z_\infty = (0,0,0,0,0,0,1)$, the quadric $\mathcal{Q} = \mathbf{v}(X_3^2 + X_0X_6 + X_1X_5 + X_2X_4)$ and

   $$\mathcal{K} = \{Z_\infty\} \cup \{(1,u_1,u_2,u_3,v_1,v_2,v_3)\},$$

   with

   $$\begin{aligned} v_1(u_1,u_2,u_3) &= u_1^2u_2 - u_1u_3 + u_2^\varphi - u_1^{\varphi+3},\\ v_2(u_1,u_2,u_3) &= u_1^\varphi u_2^\varphi - u_3^\varphi + u_1u_2^2 + u_2u_3 - u_1^{2\varphi+3},\\ v_3(u_1,u_2,u_3) &= \\ &\quad u_1u_3^\varphi - u_1^{\varphi+1}u_2^\varphi + u_1^{\varphi+3}u_2 + u_1^2u_2^2 - u_2^{\varphi+1} - u_3^2 + u_1^{2\varphi+4}, \end{aligned}$$

   for $u_1, u_2, u_3 \in \mathbf{F}_q$.

8. Put

   $$\begin{aligned} w_1(u_1,u_2,u_3) &= -u_1^{\varphi+2} + u_1u_2 - u_3,\\ w_2(u_1,u_2,u_3) &= u_1^{\varphi+1}u_2 + u_1^\varphi u_3 - u_2^2,\\ w_3(u_1,u_2,u_3) &= \\ &\quad u_3^\varphi + (u_1u_2)^\varphi - u_1^{\varphi+2}u_2 - u_1u_2^2 + u_2u_3 - u_1^{\varphi+1}u_3 - u_1^{2\varphi+3},\\ w_4(u_1,u_2,u_3) &= u_1^{\varphi+3} - u_1^2u_2 - u_2^\varphi - u_1u_3. \end{aligned}$$

Then a Sylow 3-subgroup S_3 of $\mathrm{Ree}(q)$ consists of the projective transformations associated to the matrices

$$\begin{bmatrix} 1 & 0 & 0 & 0 & 0 & 0 & 0 \\ a & 1 & 0 & 0 & 0 & 0 & 0 \\ b & a^{\varphi} & 1 & 0 & 0 & 0 & 0 \\ c & b - a^{\varphi+1} & -a & 1 & 0 & 0 & 0 \\ v_1(a,b,c) & w_1(a,b,c) & -a^2 & -a & 1 & 0 & 0 \\ v_2(a,b,c) & w_2(a,b,c) & ab+c & b & -a^{\varphi} & 1 & 0 \\ v_3(a,b,c) & w_3(a,b,c) & w_4(a,b,c) & c & -b+a^{\varphi+1} & -a & 1 \end{bmatrix}$$

for $a, b, c \in \mathbf{F}_q$. Here, S_3 is a normal subgroup in $\mathrm{Ree}(q)_{Z_\infty}$ and regular on the remaining q^3 points of $\mathcal{O}$.

9. The stabiliser $\mathrm{Ree}(q)_{Z_\infty,O}$ with $O = (1,0,0,0,0,0,0)$ is the cyclic group consisting of the projective transformations associated to the diagonal matrices,

$$\mathrm{diag}(1, d, d^{\varphi+1}, d^{\varphi+2}, d^{\varphi+3}, d^{2\varphi+3}, d^{2\varphi+4})$$

for $d \in \mathbf{F}_q$.

10. The group $\mathrm{Ree}(q)$ is generated by S_3 and $\mathrm{Ree}(q)_{Z_\infty,O}$, together with the projective transformation W of order 2 associated to the matrix

$$\begin{bmatrix} 0 & 0 & 0 & 0 & 0 & 0 & 1 \\ 0 & 0 & 0 & 0 & 0 & 1 & 0 \\ 0 & 0 & 0 & 0 & 1 & 0 & 0 \\ 0 & 0 & 0 & 1 & 0 & 0 & 0 \\ 0 & 0 & 1 & 0 & 0 & 0 & 0 \\ 0 & 1 & 0 & 0 & 0 & 0 & 0 \\ 1 & 0 & 0 & 0 & 0 & 0 & 0 \end{bmatrix}$$

that interchanges Z_∞ and O.

THEOREM A.14 *Up to conjugacy,* $\mathrm{Ree}(q)$ *has the following maximal subgroups*:

(i) *the one-point stabiliser of order* $q^3(q-1)$;

(ii) *the centraliser of an involution* $z \in \mathrm{Ree}(q)$ *isomorphic to* $\langle z \rangle \times \mathrm{PSL}(2,q)$ *of order* $q(q-1)(q+1)$;

(iii) *a subgroup of order* $6(q+3q_0+1)$, *the normaliser of a cyclic Singer group of order* $q+3q_0+1$;

(iv) *a subgroup of order* $6(q-3q_0+1)$, *the normaliser of a cyclic Singer group of order* $q-3q_0+1$;

(v) *a subgroup of order* $6(q+1)$, *the normaliser of a cyclic subgroup of order* $q+1$;

(vi) $\mathrm{Ree}(q')$ *with* $q = q'^m$ *and* m *prime.*

THEOREM A.15 $\mathrm{Ree}(q)$ *has a unique conjugacy class of involutions and each involution has at least one fixed point on* $\mathcal{K}$.

THEOREM A.16 (Zassenhaus) *Let G be a 2-transitive permutation group whose 3-point stabiliser is trivial, and suppose that G has no regular normal subgroup. If G has degree $n+1$ and order $(n+1)nd$ with d even and $d \leq \frac{1}{2}(n-1)$ then G is* $\mathrm{PSL}(2,n)$ *in the usual 2-transitive permutation representation.*

THEOREM A.17 (Kantor–O'Nan–Seitz) *Let G be a permutation group whose 2-point stabiliser is cyclic. Then G has either a regular normal subgroup, or G is one of the following groups in their natural doubly-transitive permutation representations*:

$$\mathrm{PSL}(2,q),\ \mathrm{PGL}(2,q),\ \mathrm{PSU}(3,q),\ \mathrm{PGU}(3,q), \mathrm{Sz}(q),\ \mathrm{Ree}(q).$$

THEOREM A.18 (Huppert) *Assume that* (i) *occurs in* (XXVI). *If n is a power of 2, then G is a subgroup of* $\mathrm{A\Gamma L}(1,2^k)$. *This holds true for every odd n, except for* $n = 3^2, 5^2, 7^2, 11^2, 23^2, 3^4$.

REMARK A.19 In Theorem A.18, if $n = 9$ and G is sharply 2-transitive, then either $G \cong \mathrm{AGL}(1,9)$ and the 1-point stabiliser is cyclic, or $G \cong G(1,E)$ with E the unique proper nearfield of order 9 and the 1-point stabiliser the quaternion group.

A.6 NOTES

Proofs of Theorems A.1 and A.2 come from Seidenberg [400, Chapter 23].

For Lemma A.6, see Lucas [314] and Dickson [103, Vol. I, p.271].

A systematic development of the theory of finite groups is found in [239, 240, 241]; see also the survey paper [268].

Frobenius groups as in Section A.5 (XXVII) are treated in Huppert [239, Chapter I.8].

A recent textbook on permutation groups is Cameron [63], in which the classification of finite simple groups is formulated, and, as a corollary, the classification of finite 2-transitive groups is given.

For more detail on projective unitary groups, see [238, II.8], [239, II.10], [226], [268].

For more details on Suzuki groups, see [437] [439], [454], [455], [316].

For more details on Ree groups, see [454], [367], [241], [121], [304].

For more on ovoids, see [224, Appendix VI] and [362].

Theorem A.16 is found in [513]. Theorem A.17 is the main result in [257]. For a proof of Theorem A.18, see [241, Chapter XII.7].

Appendix B

Notation

$A \subset B$	A is a subset of B
${}^tM$	tranpose of the matrix M
$\mathbf{N}$	$\{1, 2, \ldots\}$, set of natural numbers
$\mathbf{N}_m$	$\{1, 2, \ldots, m\}$
$\mathbf{Z}$	ring of integers
$\mathbf{R}$	field of real numbers
$\mathbf{C}$	field of complex numbers
$\mathbf{Q}$	field of rational numbers
K	any field in Section 1.1, algebraically closed field, otherwise
p	characteristic of K, either $p = 0$ or $p \geq 2$ prime
q	a power of p (for $p \neq 0$)
$\mathbf{F}_q$	finite field with q elements
$\overline{\mathbf{F}}_q$	algebraic closure of $\mathbf{F}_q$
$\mathrm{AG}(2, K)$	affine plane over K
$\mathrm{AG}(r, K)$	r-dimensional affine space over K
$\mathrm{AG}(2, q)$	affine plane over $\mathbf{F}_q$
$\mathrm{AG}(r, q)$	affine space over $\mathbf{F}_q$
$\mathrm{PG}(2, K)$	projective plane over K
$\mathrm{PG}(r, K)$	r-dimensional projective space over K
$\mathrm{PG}(2, q)$	projective plane over $\mathbf{F}_q$
$\mathrm{PG}(r, q)$	projective space over $\mathbf{F}_q$
$\mathrm{PGL}(2, K)$	two-dimensional projective group over K
$\mathrm{PGL}(r, K)$	r-dimensional projective group over K
$\mathrm{PGL}(2, q)$	two-dimensional projective group over $\mathbf{F}_q$
$\mathrm{PGL}(r, q)$	r-dimensional projective group over $\mathbf{F}_q$
$K[X]$	polynomial ring over K in the indeterminate X
$K[X_1, X_2, \ldots, X_n]$	polynomial ring over K in $X_1, X_2, \ldots, X_n$
$K[X][Y]$	polynomial ring over $K[X]$ in the indeterminate Y
$\mathbf{F}_q[X]$	polynomial ring over $\mathbf{F}_q$ in the indeterminate X
$\mathbf{F}_q[X_1, \ldots, X_n]$	polynomial ring over $\mathbf{F}_q$ in $X_1, \ldots, X_n$
F_i	partial derivative $\partial F / \partial X_i$ for $F \in K[X_1, \ldots, X_n]$
$F^*(X_0, X_1, X_2)$	homogeneous polynomial associated to $F(X, Y) \in K[X, Y]$
$\mathcal{C}, \mathcal{F}, \mathcal{G}, \mathcal{X} \ldots \Delta, \Gamma, \ldots$	algebraic curves

$\mathbf{v}(F(X,Y))$	$\{(x_0,x_1,x_2) \in \mathrm{PG}(2,K) \mid F^*(x_0,x_1,x_2)=0\} = \mathbf{v}(F^*(X_0,X_1,X_2))$
$\mathcal{F} = \mathbf{v}(F(X_0,X_1,X_2))$	plane projective curve with homogeneous equation $F=0$
$\mathcal{F} = \mathbf{v}(F(X,Y))$	plane projective curve with affine equation $F=0$
$R(f,g)$	resultant of $f,g \in K[X]$
$R_X(f,g)$	resultant of $f,g \in K[X,Y]$ considered as polynomials in X
(F,G)	ideal generated by F and G
$I(P,\mathcal{F}\cap\mathcal{G})$	intersection number of $\mathcal{F}$ and $\mathcal{G}$ at the point P
$K[[t]]$	ring of all formal power series in t over K
$\mathrm{Aut}(K[[t]])$	K-automorphism group of $K[[t]]$
$K[[X_1,\ldots,X_n]]$	ring of all formal power series over K in $X_1,\ldots,X_n$
$K((t))$	quotient field of $K[[t]]$
$K((X_1,\ldots,X_n))$	quotient field of $K[[X_1,\ldots,X_n]]$
$\mathrm{ord}_t\, x$	order of $x \in K((t))$
X_∞, $\mathbf{U}_1$	$(0,1,0)$
Y_∞, $\mathbf{U}_2$	$(0,0,1)$
O, $\mathbf{U}_0$	$(1,0,0)$
E, $\mathbf{U}$	$(1,1,1)$
$I(P,\mathcal{F}\cap\gamma)$	intersection number of $\mathcal{F}$ and a branch γ centred at the point P
$\mathcal{O}_P$	local ring at the point P
$K(\mathcal{F})$	function field of $\mathcal{F}$
$K(\xi,\eta)$	subfield of $K(\mathcal{F})$ generated by $\xi,\eta \in K(\mathcal{F})$
Σ	field of transcendency degree 1 over K
$(\mathcal{F};(\xi,\eta))$	model of the curve $\mathcal{F}$ arising from a generic point $P=(\xi,\eta)$
Σ/Σ'	algebraic extension of Σ' to Σ
$[\Sigma:\Sigma']$	degree of Σ over Σ'
$D(\Sigma/\Sigma')$	different divisor of Σ/Σ'
$\mathcal{P},\mathcal{Q},\mathcal{R},\ldots$	places of Σ
$\mathfrak{P}(\Sigma)$	set of all places of Σ
$\mathrm{ord}_{\mathcal{P}}\, x$	order of $x \in \Sigma$
Σ^q	subfield of all q-powers in Σ
$D_t f$	derivative of $f \in K((t))$
$D_\zeta f$	derivative of f in $f \in \Sigma$ with respect to ζ
dx	differential of x
g	genus of an irreducible curve
$D_t^{(i)} f$	i-th Hasse derivative of $f \in K((t))$
$D_\zeta^{(i)} f$	i-th Hasse derivative of $f \in \Sigma$ with respect to ζ
$\mathrm{Res}_t\, y$	residue of $y \in K((t))$
$\mathrm{Res}_{\mathcal{P}}(ydx)$	residue of ydx in Σ

$D = \sum n_{\mathcal{P}}\mathcal{P}$	a divisor of Σ
Supp (D)	support of the divisor D
$A \equiv B$	equivalence of divisors A, B
$A \succ B$	divisor $A - B$ is effective
$\deg A$	degree of a divisor
div x	principal divisor of x
$\mathrm{div}(x)_0$	zero divisor of x
$\mathrm{div}(x)_\infty$	pole divisor of x
$\mathrm{Div}(\Sigma)$	divisor group of Σ
$\mathrm{Prin}(\Sigma)$	group of principal divisors of Σ
$\mathrm{Div}_0(\Sigma)$	group of zero divisors of Σ
$\mathrm{Pic}(\Sigma)$	divisor class group (Picard group) of Σ
$\mathrm{Pic}_0(\Sigma)$	zero divisor class group (Picard zero group) of Σ
$\mathcal{L}$	virtual linear series of Σ
$\dim \mathcal{L}$	dimension of $\mathcal{L}$
$\deg \mathcal{L}$	degree or order of $\mathcal{L}$
g_n^r	linear series of dimension r and order or degree n
$\|C\|$	linear series of all effective divisors equivalent to C
$\mathcal{F} \cdot \mathcal{G}$	intersection divisor cut out on $\mathcal{F}$ by $\mathcal{G}$
$\mathcal{L}(C)$	Riemann–Roch space of C
$\ell(C)$	dimension of $\mathcal{L}(C)$
i	index of speciality of a g_n^r
$\|dx\|$	canonical series
$\mathcal{K}$	canonical curve
$(j_0(P), \ldots, j_r(P))$	order sequence of Γ at P
$(\epsilon_0, \epsilon_1, \ldots, \epsilon_r)$	order sequence of Γ
$v_{\mathcal{P}}(R)$	weight of $\mathcal{P}$ in the ramification divisor R
$\mathrm{res}_{\mathcal{P}}$	residue map
$c_0(d, r)$	Castelnuovo's number
$c_1(d, r)$	Halphen's number
Φ	Frobenius map
Φ^n	n-th Frobenius map
$\mathbf{F}_q(x, y)$	$\mathbf{F}_q$-rational subfield of Σ generated by x, y
$\mathbf{F}_q(\mathcal{F})$	$\mathbf{F}_q$-rational subfield of an $\mathbf{F}_q$-rational curve $\mathcal{F}$
$\mathbf{F}_q[[t]]$	ring of formal power series over $\mathbf{F}_q$
$\mathrm{Aut}(\mathbf{F}_q[[t]])$	K-automorphism group of $\mathbf{F}_q[[t]]$
$\mathbf{F}_q((t))$	quotient field of $\mathbf{F}_q[[t]]$
$f^{(n)}$	n-th conjugate of f
$\Sigma(\mathbf{F}_q)$	set of all $\mathbf{F}_q$-rational places of Σ
$\mathcal{X}(\mathbf{F}_q)$	set of all $\mathbf{F}_q$-rational points of $\mathcal{X}$
$S_q = N_1$	number of $\mathbf{F}_q$-rational points of $\mathcal{X}$
N_i	number of $\mathbf{F}_{q^i}$-rational points of $\mathcal{X}$
R_q	number of points $P \in \mathcal{F}$ that lie in $\mathrm{PG}(2, q)$

R_q^*	number of points $P \in \mathrm{PG}(2,q)$ which are centres of $\mathbf{F}_q$-rational branches
$\widetilde{R}_q$	number of points $P \in \mathcal{F}$ in $\mathrm{PG}(2,q)$ with each r-fold point counted with multiplicity r
B_q	number of branches of $\mathcal{F}$ centred at points of $\mathrm{PG}(2,q)$
E_q	number of singular points of $\mathcal{F}$
b_P	number of $\mathbf{F}_q$-rational branches of $\mathcal{F}$ with centre at $P \in \mathrm{PG}(2,q)$
c_P	number of all branches of $\mathcal{F}$ with centre at $P \in \mathrm{PG}(2,q)$
m_P	multiplicity of a point $P \in \mathcal{F}$
$\mathcal{F}(\mathbf{F}_q)'$	set of points $P \in \mathrm{PG}(2,q)$ lying on $\mathcal{F}$
$N_g(q)$	max N_1 for given g and q
$A^+(q)$	$\limsup_{g\to\infty} N_g(q)/q$
$A^-(q)$	$\liminf_{g\to\infty} N_g(q)/q$
$\mathrm{Pic}(\mathbf{F}_q(\mathcal{F}))$	$\mathbf{F}_q$-rational divisor class group of $\mathbf{F}_q(\mathcal{F})$
$\mathrm{Pic}_0(\mathbf{F}_q(\mathcal{F}))$	$\mathbf{F}_q$-rational zero-divisor class group of $\mathbf{F}_q(\mathcal{F})$
$v_{\mathcal{P}}(S)$	weight of $\mathcal{P}$ in the Frobenius divisor S
$(\nu_0, \nu_1, \ldots, \nu_{r-1})$	Frobenius order sequence of Γ
$\zeta(\mathcal{X}, s) = Z(\mathcal{X}, t)$	zeta function of $\mathcal{X}$
$L(t) = L_q(t)$	L-polynomial of $\mathcal{X}$ over $\mathbf{F}_q$
DLS curve	Deligne–Lusztig curve of Suzuki type
DLR curve	Deligne–Lusztig curve of Ree type
$\mathcal{Z}_l$	ring of the l-adic integers
$\mathcal{J}_{\mathcal{X}}$	Jacobian variety of $\mathcal{X}$
$T_l(A)$	Tate group of A
$\mathrm{PSL}(2,q)$	two-dimensional special projective group over $\mathbf{F}_q$
$\mathrm{PGU}(3,q)$	three-dimensional projective unitary group over $\mathbf{F}_{q^2}$
$\mathrm{PSU}(3,q)$	three-dimensional projective special unitary group over $\mathbf{F}_{q^2}$
$\mathrm{Sz}(q)$	Suzuki group
$\mathrm{Ree}(q)$	Ree group
$H < G$	H is a subgroup of G
$H \lhd G$	H is a normal subgroup of G
$\mathrm{Aut}_K(\Sigma)$	K-automorphism group of Σ
G	a subgroup of $\mathrm{Aut}_K(\Sigma)$
Σ^G	fixed subfield of G
$G_{\mathcal{P}}$	stabiliser of $\mathcal{P}$ in G
$G_{\mathcal{P}}^{(i)}$	i-th ramification group of G at $\mathcal{P}$
F_i	fixed subfield of $G_{\mathcal{P}}^{(i)}$
$\mathbf{c}(n,r)$	$\binom{n}{r}$

Bibliography

[1] V. Abatangelo and G. Korchmáros, Una generalizzazione di un teorema di B. Segre sui punti regolari rispetto ad una ellisse di un piano affine di Galois, *Ann. Mat. Pura Appl.* **172** (1997), 87–102.

[2] V. Abatangelo and B. Larato, Near MDS-codes arising from algebraic curves, *Discrete Math.* **301** (2005), 5–19.

[3] M. Abdón and A. Garcia, On a characterization of certain maximal curves, *Finite Fields Appl.* **10** (2004), 133–158.

[4] M. Abdón and L. Quoos, On the genera of subfields of the Hermitian function field, *Finite Fields Appl.* **10** (2004), 271–284.

[5] M. Abdón and F. Torres, On maximal curves in characteristic two, *Manuscripta Math.* **99** (1999), 39–53.

[6] M. Abdón and F. Torres, On F_{q^2}-maximal curves of genus $\frac{1}{6}(q-3)q$, *Beiträge Algebra Geom.* **46** (2005), 241–260.

[7] S.S. Abyhankar, Coverings of algebraic curves, *Amer. J. Math.* **79** (1957), 825–856.

[8] S.S. Abhyankar, *Algebraic Geometry for Scientists and Engineers*, Mathematical Surveys and Monographs **35**, American Math. Soc., Providence, 1990, xiv+295 pp.

[9] S.S. Abyhankar, Symplectic groups and permutation polynomials. II, *Finite Fields Appl.* **8** (2002), 233–255.

[10] S.S. Abyhankar, Geometry and Galois theory, in *Advances in Algebra and Geometry*, Hindustan Book Agency, New Delhi, 2003, 109–128.

[11] R.D.M. Accola, On Castelnuovo's inequality for algebraic curves, I, *Trans. Amer. Math. Soc.* **251** (1979), 357–373.

[12] A. Aguglia and G. Korchmáros, On algebraic curves over a finite field with many rational points, *Bull. Belg. Math. Soc. Simon Stevin* **7** (2000), 333–342.

[13] A. Aguglia, G. Korchmáros and F. Torres, Plane maximal curves, *Acta Arith.* **98** (2001), 165–179.

[14] G. Ancochea, Corps hyperelliptiques abstraits de caractéristique deux, *Port. Math.* **4** (1943), 119–128.

[15] B. Angles and C. Maire, A note on tamely ramified towers of global function fields, *Finite Fields Appl.* **8** (2002), 207–215.

[16] N. Anuradha, Zeta function of the projective curve $aY^{2l} = bX^{2l} + cZ^{2l}$ over a class of finite fields, for odd primes l, *Proc. Indian Acad. Sci. Math. Sci.* **115** (2005), 1–14.

[17] N. Anuradha, Number of points on certain hyperelliptic curves defined over finite fields, *Finite Fields Appl.*, to appear.

[18] N. Anuradha and S.A. Katre, Number of points on the projective curves $aY^l = bX^l + cZ^l$ and $aY^{2l} = bX^{2l} + cZ^{2l}$ defined over finite fields, l an odd prime, *J. Number Theory* **77** (1999), 288–313.

[19] E. Arbarello, M. Cornalba, P.A. Griffiths, and J. Harris, *Geometry of Algebraic Curves*, Vol. I, Springer, New York, 1985.

[20] Y. Aubry and M. Perret, A Weil theorem for singular curves, *Arithmetic, Geometry and Coding Theory*, de Gruyter, Berlin, 1996, 1–7.

[21] Y. Aubry and M. Perret, Divisibility of zeta functions of curves in a covering, *Arch. Math. (Basel)* **82** (2004), 205–213.

[22] Y. Aubry and M. Perret, On the characteristic polynomials of the Frobenius endomorphism for projective curves over finite fields, *Finite Fields Appl.* **10** (2004), 412–431.

[23] R. Auer and J. Top, Some genus 3 curves with many points, in *Algorithmic Number Theory*, Lecture Notes in Computer Science **2369**, Springer, Berlin, 2002, 163–171.

[24] E. Bach, Weil bounds for singular curves, *Appl. Algebra Engrg. Comm. Comput.* **7** (1996), 289–298.

[25] H.F. Baker, *Abelian Functions; Abel's Theorem and the Allied Theory of Theta Functions*, reprint of the 1897 original, with a foreword by Igor Krichever, Cambridge University Press, Cambridge, 1995, xxxvi+684 pp.

[26] H.F. Baker, *Principles of Geometry, Volume V: Analytical Principles of the Theory of Curves*, Cambridge University Press, Cambridge, 1933, x+247 pp.

[27] E. Ballico, On singular curves in the case of positive characteristic, *Math. Nachr.* **141** (1989), 267–273.

[28] E. Ballico, On the general hyperplane section of a curve in char p, *Rend. Istit. Mat. Univ. Trieste* **22** (1990), 117–125 (1992).

[29] E. Ballico, Zero-dimensional projective schemes and linear general position in positive characteristic, *J. Algebra* **159** (1993), 265–274.

[30] E. Ballico, On linearly normal strange curves, *Atti Accad. Naz. Lincei Cl. Sci. Fis. Mat. Natur. Rend. Lincei Mat. Appl.* **4** (1993), 219–222.

[31] E. Ballico, Hyperplane sections of linearly normal curves, *Proc. Amer. Math. Soc.* **122** (1994), 395–398.

[32] E. Ballico, On the general hyperplane section of a projective curve, *Beiträge Algebra Geom.* **39** (1998), 85–96.

[33] E. Ballico, Space curves not contained in low degree surfaces in positive characteristic, *Note Mat.* **20** (2001), 27–33.

[34] E. Ballico, Configurations of lines and general hyperplane sections, *Beiträge Algebra Geom.* **46** (2005), 467–470.

[35] E. Ballico and A. Cossidente, On the generic hyperplane section of curves in positive characteristic, *Pure Appl. Algebra* **102** (1995), 243–250.

[36] E. Ballico and A. Cossidente, Curves in projective spaces and almost MDS codes, *Des. Codes Cryptogr.* **24** (2001), 233–237.

[37] E. Ballico, F. Giovanetti and B. Russo, On the projective geometry of strange plane curves invariant by a finite group, *Rend. Circ. Mat. Palermo Suppl.* **51** (1998), 107–114.

[38] E. Ballico and A. Hefez, Non-reflexive projective curves of low degree, *Manuscripta Math.* **70** (1991), 385–396.

[39] E. Ballico and J. Migliore, Smooth curves whose hyperplane section is a given set of points, *Comm. Algebra* **18** (1990), 3015–3040.

[40] V. Bayer and A. Hefez, Strange curves, *Comm. Algebra* **19** (1991), 3041–3059.

[41] P. Beelen, A. Garcia and H. Stichtenoth, On towers of function fields of Artin-Schreier type, *Bull. Braz. Math. Soc. (N.S.)* **35** (2004), 151–164.

[42] P. Beelen, A. Garcia and H. Stichtenoth, On ramification and genus of recursive towers, *Port. Math. (N.S.)* **62** (2005), 231–243.

[43] P. Beelen, A. Garcia and H. Stichtenoth, On towers of function fields over finite fields, *Sémin. Congr.* **11** (2005), 1–20.

[44] P. Beelen, A. Garcia and H. Stichtenoth, Towards a classification of recursive towers of function fields over finite fields, *Finite Fields Appl.* **12** (2006), 56–77.

[45] P. Beelen and R. Pellikaan, The Newton polygon of plane curves with many rational points, *Des. Codes Cryptogr.* **21** (2000), 41–67.

[46] P. Bending, A. Camina, amd R. Guralnick, Automorphisms of the modular curve, in *Progress in Galois Theory*, Developments in Mathematics **12**, Springer, New York, 2005, 25–37.

[47] H. Bennama and P. Carbonne, Courbes $X^mY^n + Y^mZ^n + Z^mX^n = 0$ et décomposition de la Jacobienne, *J. Algebra* **188** (1997), 409–417.

[48] T.G. Berry, On the nonexistence of certain geometric MDS codes, *Number-theoretic and Algebraic Methods in Computer Science*, World Science, River Edge, 1995, 19–28.

[49] E. Bertini, *Introduzione alla Geometria Proiettiva degli Iperspazi*, Principato, Messina, 1923, viii+517 pp.

[50] J. Bezerra and A. Garcia, A tower with non-Galois steps which attains the Drinfeld–Vlăduţ bound, *J. Number Theory* **106** (2004), 142–154.

[51] J. Bezerra, A. Garcia and H. Stichtenoth, An explicit tower of function fields over cubic finite fields and Zink's lower bound, *J. Reine Angew. Math.* **589** (2005), 159–199.

[52] A. Bichara and G. Migliori, Alcune osservazioni sui $(q-1)$-archi completi di $S_{2,q}$, *Boll. Un. Mat. Ital. D* **2** (1983), 21–27.

[53] R. Blache, J.P. Cherdieu and J. Estrada-Sarlabous, Some computational aspects of Jacobians of curves in the family $y^3 = \gamma x^5 + \delta$ over $\mathbf{F}_p$, *Finite Fields Appl.* **13** (2007), 348–365.

[54] E. Bombieri, Counting points on curves over over finite fields (d'après S.A. Stepanov), in *Séminaire Bourbaki* (1972/1973), Lecture Notes in Mathematics **383**, Springer, Berlin, 1974, 234–241.

[55] E. Bombieri, Hilbert's 8th problem: an analogue, *Mathematical Developments Arising from Hilbert Problems*, Proceedings of Symposia in Pure Mathematics **XXVIII**, Amer. Math. Soc., Providence, 1976, 269–274.

[56] A.I. Borevich and I.R. Shafarevich, *Number Theory*, Pure and Applied Mathematics **20**, Academic Press, London, 1966, x+435 pp.

[57] E. Boros and T. Szőnyi, On the sharpness of a theorem of B. Segre, *Combinatorica* **6** (1986), 261–268.

[58] I.I. Bouw, The p-rank of curves and covers of curves, *Courbes Semistables et Groupe Fondamental en Géométrie Algébrique*, Progress in Mathematics **187**, Birkhäuser, Basel, 2000, 267–277.

[59] R. Brandt and H. Stichtenoth, Die Automorphismengruppen hyperelliptischer Kurven, *Manuscripta Math.* **55** (1986), 83–92.

[60] E. Çakçak and F. Özbudak, Subfields of the function field of the Deligne–Lusztig curve of Ree type, *Acta Arith.* **115** (2004), 133–180.

[61] E. Çakçak and F. Özbudak, Number of rational places of subfields of the function field of the Deligne–Lusztig curve of Ree type, *Acta Arith.* **120** (2005), 79–106.

[62] E. Çakçak and F. Özbudak, Curves related to Coulter's maximal curves, *Finite Fields Appl.* **14** (2008), 209–220.

[63] P.J. Cameron, *Permutation Groups*, Cambridge Univ. Press, Cambridge, 1999, x+220 pp.

[64] A. Campillo, *Algebroid Curves in Positive Characteristic*, Lecture Notes in Mathematics **813**, Springer, Berlin, 1980, v+168 pp.

[65] P. Carbonne and T. Hénocq, Décomposition de la Jacobienne sur les corps finis, *Bull. Polish Acad. Sci. Math.* **42** (1994), 207–215.

[66] P. Carbonne, T. Hénocq, and F. Thiérry, Points de Weierstrass de deux familles de courbes, *Comm. Algebra* **27** (1999), 5235–5254.

[67] G. Cardona, J. González, J. Lario, and A. Rio, *On curves of genus* 2 *with Jacobian of* GL_2*-type*, Manuscripta Math. **98** (1999), 37–54.

[68] G. Cardona, E. Nart, and J. Pujolás, Curves of genus two over fields of even characteristic, *Math. Z.* **250** (2005), 177–201.

[69] G. Cardona and J. Quer, Curves of genus 2 with group of automorphisms isomorphic to D_8 or D_{12}, *Trans. Amer. Math. Soc.* **359** (2007), 2831–2849.

[70] M.L. Carlin and J.F. Voloch, Plane curves with many points over finite fields, *Rocky Mountain J. Math.* **34** (2004), 1255–1259.

[71] J.W.S. Cassels, Diophantine equations with special reference to elliptic curves, *J. London Math. Soc.* **41** (1966), 193–285.

[72] G. Castelnuovo, Ricerche di geometria sulla curve algebriche, *Atti R. Acad. Sci. Torino* **24** (1889), 196–223.

[73] H.C. Chang, On plane algebraic curves, *Chinese J. Math.* **6** (1978), 185–189.

[74] H. Chen and S.S.T. Yau, Contribution to Munuera's problem on the main conjecture of geometric hyperelliptic MDS codes, *IEEE Trans. Inform. Theory* **43** (1997), 1349–1354.

[75] C.C.A. Cheng, J.H. McKay, and S.S.S. Wang, Resultants of cyclotomic polynomials, *Proc. Amer. Math. Soc.* **123** (1995), 1053–1059.

[76] J.P. Cherdieu, Remarks on the zeta function of some diagonal hyperelliptic curves, *J. Pure Appl. Algebra* **190** (2004), 31–43.

[77] C. Chevalley, *Introduction to the Theory of Algebraic Functions of One Variable*, Mathematical Surveys **6**, American Math. Soc., New York, 1951, xi+188 pp.

[78] L. Chiantini and C. Ciliberto, Towards a Halphen theory of linear of series on curves, *Trans. Amer. Math. Soc.* **2** (1999), 2197–2212.

[79] L. Chiantini, C. Ciliberto and V. Di Gennaro, The genus of projective curves, *Duke Math. J.* **70** (1993), 229–245.

[80] Y. Choie and E. Jeong, Isomorphism classes of elliptic and hyperelliptic curves over finite fields $\mathbf{F}_{(2g+1)^n}$, *Finite Fields Appl.* **10** (2004), 583–614.

[81] Y. Choie and D. Yun, Isomorphism classes of hyperelliptic curves of genus 2 over $\mathbf{F}_q$, *Information Security and Privacy*, Lecture Notes in Computer Science **2384**, Springer, Berlin, 2002, 190–202.

[82] C. Ciliberto and R. Lazarsfeld, On the uniqueness of certain linear series on some classes of curves, in *Complete Intersections*, Lecture Notes in Mathematics **1092**, Springer, Berlin, 1984, 198–213.

[83] S.D. Cohen, Some function field estimates with applications, in *Number Theory and its Applications*, Lecture Notes in Pure and Applied Mathematics **204**, Dekker, New York, 1999, 23–45.

[84] P.M. Cohn, *Algebraic Numbers and Algebraic Functions*, Chapman and Hall, London, 1991, xii+192 pp.

[85] J.L. Coolidge, *A Treatise on Algebraic Plane Curves*, Dover, New York, 1959, xxiv+513 pp.; originally, Oxford University Press, 1931.

[86] J.L. Coolidge, *A History of Geometrical Methods*, Dover, New York, 1963; originally, Oxford University Press, 1940.

[87] A. Cossidente, A new proof of the existence of $(q^2 - q + 1)$-arcs in $PG(2, q^2)$, *J. Geom.* **53** (1995), 37–40; correction **59** (1997), 32–33.

[88] A. Cossidente, J.W.P. Hirschfeld, G. Korchmáros and F. Torres, On plane maximal curves, *Compositio Math.* **121** (2000), 163–181.

[89] A. Cossidente and G. Korchmáros, The Hermitian function field arising from a cyclic arc in a Galois plane, *Geometry, Combinatorial Designs and Related Structures*, London Mathematical Society Lecture Note Series **245**, Cambridge Univ. Press, Cambridge, 1997, 63–68.

[90] A. Cossidente and G. Korchmáros, The algebraic envelope associated to a complete arc, *Rend. Circ. Mat. Palermo Suppl.* **51** (1998), 9–24.

[91] A. Cossidente, G. Korchmáros and F. Torres, On curves covered by the Hermitian curve, *J. Algebra* **216** (1999), 56–76.

[92] A. Cossidente, G. Korchmáros and F. Torres, Curves of large genus covered by the Hermitian curve, *Comm. Algebra* **28** (2000), 4707–4728.

[93] A. Cossidente and A. Siciliano, Plane algebraic curves with Singer automorphisms, *J. Number Theory* **99** (2003), 373–382.

[94] A. Cossidente and A. Siciliano, The automorphism group of plane algebraic curves with Singer automorphisms, *Des. Codes Cryptogr.* **39** (2006), 33–37.

[95] R.S. Coulter, The number of rational points of a class of Artin–Schreier curves, *Finite Fields Appl.* **8** (2002), 397–413.

[96] R.N. Daskalov and M.E. Jiménez Contreras, New $(k;r)$-arcs in the plane of order thirteen, *J. Geom.* **80** (2004), 10–22.

[97] A.A. Davydov, G. Faina, S. Marcugini and F. Pambianco, Computer search in projective planes for the sizes of complete arcs, *J. Geom.* **82** (2005), 50–62.

[98] M. de Boer, MDS codes from hyperelliptic curves, in *Arithmetic, Geometry and Coding Theory*, de Gruyter, Berlin, 1996, 23–34.

[99] Y. Deng, Isomorphism classes of hyperelliptic curves of genus 3 over finite fields, *Finite Fields Appl.* **12** (2006), 248–282.

[100] Y. Deng and L. Mulan, Isomorphism classes of hyperelliptic curves of genus 2 over finite fields with characteristic 2, *Sci. China Ser. A* **49** (2006), 173–184.

[101] M. Deuring, Die Typen der Multiplikatorenringe elliptischer Funktionenkörper, *Abh. Math. Sem. Univ. Hamburg* **14** (1941), 197–272.

[102] M. Deuring, *Lectures on the Theory of Algebraic Functions of One Variable*, Lecture Notes in Mathematics **314**, Springer, Berlin, 1973, vi+151 pp.

[103] L.E. Dickson, *History of the Theory of Numbers*, Vols. I,II,III, Chelsea, New York, 1919, xii +486 pp.; 1920, v+313 pp.; 1923, xxv+803 pp.

[104] J. Dieudonné, The historical development of algebraic geometry, *Amer. Math. Monthly* **79** (1972), 827–866.

[105] J. Dieudonné, *History of Algebraic Geometry*, Chapman and Hall, New York, 1985, xii+186 pp.

[106] M.S. Dodunekov and I.N. Landjev, On near-MDS codes, *J. Geom.* **54** (1995), 30–34.

[107] M.S. Dodunekov and I.N. Landjev, Near-MDS codes over some small fields, *Discrete Math.* **213** (2000), 55–65.

[108] A. Duma, Automorfismi di corpi di funzioni algebriche, *Riv. Mat. Univ. Parma* **14** (1988), 69–89.

[109] I.M. Duursma, Monomial embeddings of the Klein curve, *Discrete Math.* **208/209** (1999), 235–246.

[110] I.M. Duursma, B. Poonen and M. Zieve, Everywhere ramified towers of global function fields, *Finite Fields and Applications*, Lecture Notes in Computer Science **2948**, Springer, Berlin, 2004, 148–153.

[111] I.M. Duursma and N. Kiyavash, The vector decomposition problem for elliptic and hyperelliptic curves, *J. Ramanujan Math. Soc.* **20** (2005), 59–76.

[112] G. Ebert, Partitioning projective geometries into caps, *Canad. J. Math.* **37** (1985), 1163–1175.

[113] A. Ebin, D. Hur, A. Katz and V. Shchogolev, A new construction of maximal curves, *Recent Progress in Arithmetic and Algebraic Geometry*, Contemporary Mathematics **386**, Amer. Math. Soc., Providence, 2005, 35–45.

[114] M. Eichler, *Introduction to the Theory of Algebraic Numbers and Functions*, translated from the German by G. Striker, Pure and Applied Mathematics **23**, Academic Press, New York, 1966, xiv+324 pp.

[115] D. Eisenbud, M. Green and J. Harris, Cayley–Bacharach theorems and conjectures, *Bull. Amer. Math. Soc.* **33** (1996), 295–324.

[116] N.D. Elkies, Explicit modular towers, in 35th Annual Allerton Conference on Common. Control and Computing, Urbana, 1997.

[117] N.D. Elkies, The Klein quartic in number theory, *The Eightfold Way*, Math. Sci. Res. Inst. Publ. **35**, Cambridge Univ. Press, Cambridge, 1999, 51–101.

[118] N.D. Elkies, Explicit towers of Drinfeld modular curves, in *European Congress of Mathematics*, Vol. II (Barcelona, 2000), Progress in Mathematics **202**, 189–198.

[119] N.D. Elkies, E.W. Howe, A. Kresch, B. Poonen, J. Wetherell and M. Zieve, Curves of every genus with many points. II. Asymptotically good families, *Duke Math. J.* **122** (2004), 399–422.

[120] A. Enge, How to distinguish hyperelliptic curves in even characteristic, *Public-key Cryptography and Computational Number Theory*, de Gruyter, Berlin, 2001, 49–58.

[121] M. Enguehard, Caractérisation des groupes de Ree, *Astérisque* **142–143** (1986), 49–139.

[122] F. Enriques, *Lezioni sulla Teoria Geometrica delle Equazioni e delle Funzioni Algebriche*, Vols. I–IV, edited by O. Chisini, Zanichelli, Bologna, 1915–1934.

[123] E. Esteves, A geometric proof of an inequality of order sequences, *Comm. Algebra* **21** (1993), 231–238.

[124] J. Estrada-Sarlabous, On the Jacobian varieties of Picard curves defined over fields of characteristic p, *Math. Nachr.* **152** (1991), 329–340.

[125] J. Estrada-Sarlabous, E. Reinaldo-Barreiro and J. Alejandro, On the Jacobian varieties of Picard curves: Explicit addition law and algebraic structure, *Math. Nachr.* **208** (1999), 149–166.

[126] G. Faina and F. Pambianco, On the spectrum of the values k for which a complete k-cap in $\mathrm{PG}(n,q)$ exists, *J. Geom.* **62** (1998), 84–98.

[127] H.M. Farkas and I. Kra, *Riemann Surfaces*, Graduate Texts in Mathematics **71**, Springer, New York, 1992, xvi+363 pp.

[128] G. Fischer, *Plane Algebraic Curves*, translated from the 1994 German original by L. Kay, Student Mathematical Library **15**, American Math. Soc., Providence, 2001, xvi+229 pp.

[129] J.C. Fisher, J.W.P. Hirschfeld and J.A. Thas, Complete arcs in planes of square order, *Combinatorics '84*, North-Holland Mathematics Studies **123**, North-Holland, Amsterdam, 1986, 243–250.

[130] G. Frey, On the structure of the class group of a function field., *Arch. Math.* **33** (1979/80), 33–40.

[131] R. Fuhrmann, Algebraische Funktionenkörper über endlichen Körpern mit maximaler Anzahl rationaler Stellen, Universität Essen, 1995.

[132] R. Fuhrmann, A. Garcia and F. Torres, On maximal curves, *J. Number Theory* **67** (1997), 29–51.

[133] R. Fuhrmann and F. Torres, The genus of curves over finite fields with many rational points, *Manuscripta Math.* **89** (1996), 103–106.

[134] R. Fuhrmann and F. Torres, On Weierstrass points and optimal curves, *Rend. Circ. Mat. Palermo Suppl.* **51** (1998), 25–46.

[135] W. Fulton, *Algebraic Curves, an Introduction to Algebraic Geometry*, Notes written with the collaboration of Richard Weiss, Advanced Book Classics, Addison-Wesley, Redwood City, 1989, xxii+226 pp.; originally, Benjamin, 1969.

[136] A. Garcia, Weights of Weierstrass points in double coverings of curves of genus one or two, *Manuscripta Math.* **55** (1986), 419–432.

[137] A. Garcia, The curves $y^n = f(x)$ over finite fields, *Arch. Math.* **54** (1990), 36–44.

[138] A. Garcia, Remarks on fixed points of automorphisms and higher-order Weierstrass points in prime characteristic, *Manuscripta Math.* **69** (1990), 301–303.

[139] A. Garcia, On higher-order Weierstrass points and the finiteness of the automorphism group, *Manuscripta Math.* **78** (1993), 413–416.

[140] A. Garcia, Some arithmetic properties of order-sequences of algebraic curves, *J. Pure Appl. Algebra* **85** (1993), 259–269.

[141] A. Garcia, Curves over finite fields attaining the Hasse–Weil upper bound, *European Congress of Mathematics*, Vol. II (Barcelona, 2000), Progress in Mathematics **202**, Birkhäuser, Basel, 2001, 199–205.

[142] A. Garcia, On curves with many rational points over finite fields, in *Finite Fields with Applications to Coding Theory, Cryptography and Related Areas*, Springer, Berlin, 2002, 152–163.

[143] A. Garcia and A. Garzon, On Kummer covers with many rational points over finite fields, *J. Pure Appl. Algebra* **185** (2003), 177–192.

[144] A. Garcia and M. Homma, Frobenius order-sequences of curves, *Algebra and Number Theory*, de Gruyter, Berlin, 1994, 27–41.

[145] A. Garcia, M.Q. Kawakita and S. Miura, On certain subcovers of the Hermitian curve, *Comm. Algebra* **34** (2006), 973–982.

[146] A. Garcia and L. Quoos, A construction of curves over finite fields, *Acta Arith.* **98** (2001), 181–195.

[147] A. Garcia and H. Stichtenoth, Elementary abelian p-extensions of algebraic function fields, *Manuscripta Math.* **72** (1991), 67–79.

[148] A. Garcia and H. Stichtenoth, A tower of Artin–Schreier extensions of function fields attaining the Drinfeld–Vlăduţ bound, *Invent. Math.* **121** (1995), 211–222.

[149] A. Garcia and H. Stichtenoth, Algebraic function fields over finite fields with many rational places, *IEEE Trans. Inform. Theory* **41** (1995), 1548–1563.

[150] A. Garcia and H. Stichtenoth, On the asymptotic behaviour of some towers of function fields over finite fields, *J. Number Theory* **61** (1996), 248–273.

[151] A. Garcia and H. Stichtenoth, On Chebyshev polynomials and maximal curves, *Acta Arith.* **90** (1999), 301–311.

[152] A. Garcia and H. Stichtenoth, A class of polynomials over finite fields, *Finite Fields Appl.* **5** (1999), 424–435.

[153] A. Garcia and H. Stichtenoth, Asymptotics for the genus and the number of rational places in towers of function fields over a finite field, *Finite Fields Appl.* **11** (2005), 434–450.

[154] A. Garcia and H. Stichtenoth, A maximal curve which is not a Galois subcover of the Hermitian curve, *Bull. Braz. Math. Soc. (N.S.)* **37** (2006), 139–152.

[155] A. Garcia and H. Stichtenoth (eds.), *Topics in Geometry, Coding Theory and Cryptography*, Algebra and Applications **6**, Springer, Dordrecht, 2007, x+200 pp.

[156] A. Garcia, H. Stichtenoth and H.G. Rück, On tame towers over finite fields, *J. Reine Angew. Math.* **557** (2003), 53–80.

[157] A Garcia, H. Stichtenoth and M. Thomas, On towers and composita of towers of function fields over finite fields, *Finite Fields Appl.* **3** (1997), 257–274.

[158] A. Garcia, H. Stichtenoth and C.P. Xing, On subfields of the Hermitian function field, *Compositio Math.* **120** (2000), 137–170.

[159] A. Garcia and F. Torres, On maximal curves having classical Weierstrass gaps, *Contemp. Math.* **245** (1999), 49–59.

[160] A. Garcia and P. Viana, Weierstrass points on certain non-classical curves, *Arch. Math.* **46** (1986), 315–322.

[161] A. Garcia and J.F. Voloch, Wronskians and linear independence in fields of prime characteristic, *Manuscripta Math.* **59** (1987), 457–469.

[162] A. Garcia and J.F. Voloch, Fermat curves over finite fields, *J. Number Theory* **30** (1988), 345–356.

[163] A. Garcia and J.F. Voloch, Duality for projective curves, *Bol. Soc. Brasil. Mat. (N.S.)* **21** (1991), 159–175.

[164] M. Gebhardt, Constructing function fields with many rational places via the Carlitz module, *Manuscripta Math.* **107** (2002), 89–99.

[165] I.M. Gel'fand, M.M. Kapranov and A.V. Zelevinsky, *Discriminants, Resultants, and Multidimensional Determinants*, Birkhäuser, Boston, 1994, x+523 pp.

[166] F. Gherardelli, L.A Rosati and G. Tomassini, *Lezioni di Geometria*, Vol. II, Milani, Padua, 1978, vii+276 pp.

[167] M. Giulietti, On plane arcs contained in cubic curves, *Finite Fields Appl.* **8** (2002), 69–90.

[168] M. Giulietti, On the number of rational points of a plane algebraic curve, *Arch. Math.* **82** (2004), 214–221.

[169] M. Giulietti, On the extendibility of near-MDS elliptic codes, *Appl. Algebra Engrg. Comm. Comput.* **15** (2004), 1–11.

[170] M. Giulietti, J.W.P. Hirschfeld and G. Korchmáros, The Desarguesian plane of order thirteen, in *Finite Geometries*, Kluwer, Dordrecht, 2001, 159–170.

[171] M. Giulietti, J.W.P. Hirschfeld, G. Korchmáros and F. Torres, Curves covered by the Hermitian curve, *Finite Fields Appl.* **12** (2006), 539–564.

[172] M. Giulietti, J.W.P. Hirschfeld, G. Korchmáros and F. Torres, Families of curves covered by the Hermitian curve, *Sémin. Cong.*, to appear.

[173] M. Giulietti and G. Korchmáros, On large automorphism groups of algebraic curves in positive characteristic, 2007, arXiv:0706.2320.

[174] M. Giulietti, G. Korchmáros and F. Torres, Quotient curves of the Deligne–Lusztig curve of Suzuki type, *Acta Arith.* **122** (2006), 245–274.

[175] D.G. Glynn, The non-classical 10-arc of $PG(4, q)$, q even, *Discrete Math.* **59** (1986), 43–51.

[176] D.M. Goldschmidt, *Algebraic Functions and Projective Curves*, Springer, New York, 2002, xvi+192 pp.

[177] J. González, Hasse–Witt matrices for the Fermat curves of prime degree, *Tohoku Math. J.* **49** (1997), 149–163.

[178] V.D. Goppa, *Geometry and Godes*, translated from the Russian by N. G. Shartse, Mathematics and its Applications (Soviet Series) **24**, Kluwer, Dordrecht, 1988, x+157 pp.

[179] R. Gow, Some TI subgroups of solvable groups, *J. London Math. Soc.* **12** (1976), 285–286.

[180] D.D. Grant, On linear forms whose coefficients are in arithmetical progession, *Israel J. Math.* **15** (1973), 204–209.

[181] S. Greco and G. Raciti, The Lüroth semigroup of plane algebraic curves, *Pacific J. Math.* **151** (1991), 43–56.

[182] L. Gruson and C. Peskine, Genre des courbes de l'espace projectif, *Algebraic Geometry*, Lecture Notes in Mathematics **687**, Springer, Berlin, 1977, 31–59.

[183] C. Guneri, A bound on the number of rational points of certain Artin–Schreier families, *Comm. Algebra* **30** (2002), 4251–4265.

[184] J. Gutierrez, D. Sevilla and T. Shaska, Hyperelliptic curves of genus 3 with prescribed automorphism group, in *Computational Aspects of Algebraic Curves*, Lecture Notes Series on Computing **13**, World Sci. Publ., Hackensack, 2005, 109–123.

[185] J. Gutierrez and T. Shaska, Hyperelliptic curves with extra involutions, *LMS J. Comput. Math.* **8** (2005), 102–115.

[186] F. Hajir and C. Maire, Asymptotically good towers of global fields, *European Congress of Mathematics*, Vol. II (Barcelona, 2000), Progress in Mathematics **202**, Birkhäuser, Basel, 2001, 207–218.

[187] J.P. Hansen, Deligne–Lusztig varieties and group codes, Lecture Notes in Mathematics **1518**, Springer, Berlin, 1992, 63–81.

[188] J.P. Hansen and J.P. Pedersen, Automorphism groups of Ree type, Deligne–Lusztig curves and function fields, *J. Reine Angew. Math.* **440** (1993), 99–109.

[189] J.P. Hansen and H. Stichtenoth, Group codes on certain algebraic curves with many rational points, *Appl. Algebra Eng. Comm. Comput.* **1** (1990), 67–77.

[190] J. Harris, The genus of space curves, *Math. Ann.* **249** (1980), 191–204.

[191] J. Harris, *Curves in Projective Space*, with the collaboration of D. Eisenbud, Séminaire de Mathématiques Supérieures **85**, Université de Montréal, Montreal, 1982, 138 pp.

[192] R. Hartshorne, *Algebraic Geometry*, Graduate Texts in Mathematics **52**, Springer, New York, 1977, xvi+496 pp.

[193] R. Hartshorne, Clifford index of ACM curves in $\mathbb{P}^3$, *Milan J. Math.* **70** (2002), 209–221.

[194] W.J. Harvey, Cyclic groups of automorphisms of a compact Riemann surface, *Quart. J. Math. Oxford Ser.* **17** (1966), 86–97.

[195] H. Hasse, Über die Kongruenzzetafunktionen, *Sitzungsber. Preuss. Akad. Wiss., Phys.-Math. Kl.* **17** (1934), 250–263.

[196] H. Hasse, Theorie der Differentiale in algebraischen Funktionenkörpern mit vollkommenem Konstantenkörper, *J. Reine Angew. Math.* **172** (1934), 55–64.

[197] H. Hasse, Theorie der höheren Differentiale in einem algebraischen Funktionenkörper mit vollkommenem Konstantenkörper bei beliebiger Charakteristik, *J. Reine Angew. Math.* **175** (1936), 50–54.

[198] H. Hasse and F.K. Schmidt, Noch eine Begründung der Theorie der höheren Differentialquotienten in einem algebraischen Funktionenkörper einer Unbestimmten, *J. Reine Angew. Math.* **177** (1937), 215–237.

[199] H. Hasse and E. Witt, Zyklische unverzweigte Erweiterungskörper vom Primzahlgrad p über einem algebraischen Funktionenkörper der Charakteristik p, *Monatshefte f. Math. und Phys.* **43** (1936), 477–492.

[200] A. Hefez, Non-reflexive curves, *Compositio Math.* **69** (1989), 3–35.

[201] A. Hefez, Irreducible plane curve singularities, *Real and Complex Singularities*, Lecture Notes in Pure and Applied Mathematics **232**, Dekker, New York, 2003, 1–120.

[202] A. Hefez and N. Kakuta, Polar curves, *J. Algebra* **181** (1996), 449–477.

[203] A. Hefez and N. Kakuta, Polars of Artin–Schreier curves, *Acta Arith.* **77** (1996), 57–70.

[204] A. Hefez and S.L. Kleiman, Notes on the duality of projective varieties, *Geometry Today*, Progress in Mathematics **60**, Birkhäuser, Boston, 1985, 143–183.

[205] A. Hefez and J.F. Voloch, Frobenius non-classical curves, *Arch. Math.* **54** (1990), 263–273.

[206] H.-W. Henn, Funktionenkörper mit grosser Automorphismengruppe, *J. Reine Angew. Math.* **302** (1978), 96–115.

[207] H.-W. Henn, Minimalanzahl von Weierstrasspunkten, *Arch. Math.* **31** (1978), 38–43.

[208] K. Hensel and G. Landsberg, *Theorie der Algebraischen Funktionen einer Variablen und ihre Anwendung auf Algebraische Kurven und Abelsche Integrale*, reprint of the 1907 original, Chelsea, New York, 1965, xvi+707 pp.

[209] C. Hering, On subgroups with trivial normalizer intersection, *J. Algebra* **20** (1972), 622–629.

[210] E.L. Hernández, A.J. Menezes and M. Muñoz, Isomorphism classes of genus-2 hyperelliptic curves over finite fields, *Appl. Algebra Engrg. Comm. Comput.* **13** (2002), 57–65.

[211] E.L. Hernández, A.J. Menezes and M. Muñoz, Isomorphism classes of genus-2 hyperelliptic curves over finite fields $\mathbf{F}_{5^m}$, *Information* **8** (2005), 837–844.

[212] H. Hilton, *Plane Algebraic Curves*, Oxford Univ. Press, Oxford, 1920, xvi+388 pp.

[213] J.W.P. Hirschfeld, *Finite Projective Spaces of Three Dimensions*, Oxford Univ. Press, Oxford, 1985, x+316 pp.

[214] J.W.P. Hirschfeld, The main conjecture for MDS codes, *Cryptography and Coding*, Lecture Notes in Computer Science **1025**, Springer, Berlin, 1995, 44–52.

[215] J.W.P. Hirschfeld, Complete arcs, *Discrete Math.* **174** (1997), 177–184.

[216] J.W.P. Hirschfeld, *Projective Geometries over Finite Fields*, second ed., Oxford Univ. Press, Oxford, 1998, xiv+555 pp.

[217] J.W.P. Hirschfeld and G. Korchmáros, On the embedding of an arc into a conic in a finite plane, *Finite Fields Appl.* **2** (1996), 274–292.

[218] J.W.P. Hirschfeld and G. Korchmáros, On the number of rational points on an algebraic curve over a finite field, *Bull. Belg. Math. Soc. Simon Stevin* **5** (1998), 313–340.

[219] J.W.P. Hirschfeld and G. Korchmáros, Arcs and curves over a finite field, *Finite Fields Appl.* **5** (1999), 393–408.

[220] J.W.P. Hirschfeld and G. Korchmáros, On the number of solutions of an equation over a finite field, *Bull. London Math. Soc.* **33** (2001), 16–24.

[221] J. W. P. Hirschfeld and L. Storme, The packing problem in statistics, coding theory and finite projective spaces, *J. Statist. Plann. Inference* **72** (1998), 355–380.

[222] J.W.P. Hirschfeld and L. Storme, The packing problem in statistics, coding theory and finite projective spaces: update 2001, *Finite Geometries*, Kluwer, Dordrecht, 2001, 201–246.

[223] J.W.P. Hirschfeld, L. Storme, J.A. Thas, and J.F. Voloch, A characterization of Hermitian curves, *J. Geom.* **41** (1991), 72–78.

[224] J.W.P. Hirschfeld and J.A. Thas, *General Galois Geometries*, Oxford Univ. Press, Oxford, 1991, xiv+407 pp.

[225] W.V.D. Hodge and D. Pedoe, *Methods of Algebraic Geometry*, Vols. I,II,III, Cambridge Univ. Press, Cambridge, 1953, viii+440 pp.; 1952, x+394 pp.; 1954, x+336 pp.

[226] A.R. Hoffer, On unitary collineation groups, *J. Algebra* **22** (1972), 211–218.

[227] T. Høholdt, J.H. van Lint and R. Pellikaan, Algebraic geometry codes, in *Handbook of Coding Theory*, Vol. I, North-Holland, Amsterdam, 1998, 871–961.

[228] M. Homma, Automorphisms of prime order of curves, *Manuscripta Math.* **33** (1980), 99–109.

[229] M. Homma, Funny plane curves in characteristic $p > 0$, *Comm. Algebra* **15** (1987), 1469–1501.

[230] M. Homma, Nonreflexive plane curves of a certain type, *Ryukyu Math. J.* **1** (1988), 22–37.

[231] M. Homma, Nonreflexive plane curves of a certain type. II, *Ryukyu Math. J.* **2** (1989), 21–28.

[232] M. Homma, A souped-up version of Pardini's theorem and its application to funny curves, *Compositio Math.* **71** (1989), 259–302.

[233] M. Homma, Linear systems on curves with no Weierstrass points, *Bol. Soc. Bras. Mat (N.S.)* **23** (1992), 91–108.

[234] M. Homma, On Esteves' inequality of order sequences of curves, *Comm. Algebra* **21** (1993), 3685–3689.

[235] M. Homma, On duals of smooth plane curves, *Proc. Amer. Math. Soc.* **118** (1993), 785–790.

[236] M. Homma, Plane curves whose tangent lines at collinear points are concurrent, *Geom. Dedicata* **53** (1994), 287–296.

[237] E.W. Howe and K.E. Lauter, Improved upper bounds for the number of points on curves over finite fields, *Ann. Inst. Fourier (Grenoble)* **53** (2003), 1677–1737; Corrigendum, to appear.

[238] D.R. Hughes and F.C. Piper, *Projective Planes*, Graduate Texts in Mathematics **6**, Springer, New York, 1973, x+291 pp.

[239] B. Huppert, *Endliche Gruppen. I*, Grundlehren der Mathematischen Wissenschaften **134**, Springer, Berlin, 1967, xii+793 pp.

[240] B. Huppert and B.N. Blackburn, *Finite Groups. II*, Grundlehren der Mathematischen Wissenschaften **242**, Springer, Berlin, 1982, xiii+531 pp.

[241] B. Huppert and B.N. Blackburn, *Finite Groups. III*, Grundlehren der Mathematischen Wissenschaften **243**, Springer, Berlin, 1982, ix+454 pp.

[242] N.E. Hurt, *Many Rational Points. Coding Theory and Algebraic Geometry*, Mathematics and its Applications **564**, Kluwer, Dordrecht, 2003, xxii+346 pp.

[243] D. Husemöller, *Elliptic Curves*, Springer, Berlin, 2004, xxi+487 pp.

[244] A. Hurwitz, Über algebraische Gebilde mit eindeutigen Transformationen in sich, *Math. Ann.* **41** (1893), 403–492.

[245] T. Ibukiyama, On rational points of curves of genus 3 over finite fields, *Tohoku Math. J.* **45** (1993), 311–329.

[246] T. Ibukiyama, T. Katsura and F. Oort, Supersingular curves of genus two and class numbers, *Compositio Math.* **57** (1986), 127–152.

[247] Y. Ihara, Some remarks on the number of rational points of algebraic curves over finite fields, *J. Fac. Sci. Tokyo* **28** (1981), 721–724.

[248] S. Irokawa and R. Sasaki, On a family of quotients of Fermat curves, *Tsukuba J. Math.* **19** (1995), 121–139.

[249] K. Iwasawa and T. Tamagawa, On the group of automorphisms of a function field, *J. Math. Soc. Japan* **3** (1951), 137–147.

[250] K. Iwasawa and T. Tamagawa, Correction: On the group of automorphisms of a function field, *J. Math. Soc. Japan* **4** (1952), 100–101.

[251] K. Iwasawa and T. Tamagawa, Correction: On the paper "On the group of automorphisms of a function field", *J. Math. Soc. Japan* **4** (1952), 203–204.

[252] J.A. Jenkins, Some remarks on Weierstrass points, *Proc. Amer. Math. Soc.* **44** (1974), 121–122.

[253] J.-R. Joly, Equations et variétés algébriques sur un corps fini, *Enseignement Math.* **19** (1973), 1–117.

[254] H. Kaji, On the tangentially degenerate curves, *J. London Math. Soc.* **33** (1986), 430–440.

[255] H. Kaji, Strangeness of higher order space curves, *Comm. Algebra* **20** (1992), 1535–1548.

[256] H. Kaneta, S. Marcugini and F. Pambianco, On arcs and curves with many automorphisms, *Mediterr. J. Math.* **2** (2005), 71–102.

[257] W.M. Kantor, M. O'Nan and G.M. Seitz, 2-transitive groups in which the stabilizer of two points is cyclic, *J. Algebra* **21** (1972), 17–50.

[258] T. Kato, Non-hyperelliptic Weierstrass points of maximal weight, *Math. Ann.* **239** (1979), 141–147.

[259] T. Kato, On criteria of $\tilde{g}$-hyperellipticity, *Kodai Math. J.* **2** (1979), 275–285.

[260] T. Kato, Conformal self-mappings of Riemann surfaces, *Sugaku* **30** (1980), 338–348.

[261] T. Kato, On Weierstrass points whose first nongaps are three, *J. Reine Angew. Math.* **316** (1980), 99–109.

[262] M.Q. Kawakita, Kummer curves and their fibre products with many rational points, *Appl. Algebra Engrg. Comm. Comput.* **14** (2003), 55–64.

[263] M.Q. Kawakita, A quotient curve of the Fermat curve of degree twelve attaining the Serre bound, *J. Algebra Appl.* **4** (2005), 173–178.

[264] M.Q. Kawakita, A quotient curve of the Fermat curve of degree twenty-three attaining the Serre bound, *J. Pure Appl. Algebra* **196** (2005), 101–110.

[265] M.Q. Kawakita and S. Miura, Quadratic forms, fibre products and some plane curves with many points, *Acta Arith.* **105** (2002), 371–386.

[266] A. Keller, Cyclotomic function fields with many rational places, in *Finite Fields and Applications*, Springer, Berlin, 2001, 293–302.

[267] B. Kestenband, Unital intersections in finite projective planes, *Geom. Dedicata* **11** (1981), 107–117.

[268] O.H. King, The subgroup structure of finite classical groups in terms of geometric configurations, in *Surveys in Combinatorics 2005*, London Mathematical Society Lecture Note Series **327**, Cambridge Univ. Press, Cambridge, 2005, 29–56.

[269] F. Kirwan, *Complex Algebraic Curves*, Cambridge Univ. Press, Cambridge, 1992, viii+264 pp.

[270] G. Kiss and G. Korchmáros, An upper bound for the number of chords of a polygon which pass through a given point in a finite plane, *Acta Sci. Math. (Szeged)* **62** (1996), 289–298.

[271] S.L. Kleiman, Tangency and duality, *Canad. Math. Soc., Conf. Proc.* **6** (1986), 163-225.

[272] S.L. Kleiman, Multiple tangents of smooth plane curves (after Kaji), *Algebraic Geometry: Sundance 1988*, Contemporary Mathematics **116**, Amer. Math. Soc., Providence, 1991, 71–84.

[273] N. Koblitz, *Introduction to Elliptic Curves and Modular Forms*, second edition, Springer, New York, 1993, x+248 pp.

[274] N. Koblitz, Hyperelliptic cryptosystems, *J. Cryptology* **1** (1989), 139–150.

[275] T. Kodama and T. Washio, Hasse–Witt matrices of Fermat curves, *Manuscripta Math.* **60** (1988), 185–195.

[276] T. Kodoma and T. Washio, A family of hyperelliptic function fields with Hasse–Witt-invariant zero, *J. Number Theory* **36** (1990), 187–200.

[277] K. Komiya, Algebraic curves with non-classical types of gap sequences for genus three and four, *Hiroshima Math. J.* **8** (1978), 371–400.

[278] A. Kontogeorgis, The group of automorphisms of the function fields of the curve $x^n + y^m + 1 = 0$, *J. Number Theory* **72** (1998), 110–136.

[279] A. Kontogeorgis, The group of automorphisms of cyclic extensions of rational function fields, *J. Algebra* **216** (1999), 665–706.

[280] G. Korchmáros and T. Szőnyi, Fermat curves over finite fields and cyclic subsets in high-dimensional projective spaces, *Finite Fields Appl.* **5** (1999), 206–217.

[281] G. Korchmáros and F. Torres, Embedding of a maximal curve in a Hermitian variety, *Compositio Math.* **128** (2001), 95–113.

[282] G. Korchmáros and F. Torres, On the genus of a maximal curve, *Math. Ann.* **323** (2002), 589–608.

[283] A. Kresch, J.L. Wetherell and M.E. Zieve, Curves of every genus with many points. I. Abelian and toric families, *J. Algebra* **250** (2002), 353–370.

[284] R.S. Kulkarni, Riemann surfaces admitting large automorphism groups, *Extremal Riemann Surfaces*, Contemporary Mathematics **201**, Amer. Math. Soc., Providence, 1997, 63–79.

[285] A.G. Kurosh, *Higher Algebra*, tenth Russian edition, translated by G. Yankovsky, Mir, Moscow, 1975, 428 pp.

[286] G. Lachaud, Sommes de Eisenstein et nombre de points de certaines courbes algébriques sur les corps finis, *C.R. Acad. Sci. Paris Sér. I Math.* **305** (1987), 729–732.

[287] G. Lachaud and M. Martin-Deschamps, Nombre de points des jacobiennes sur un corps fini, *Acta Arith.* **56** (1990), 329–340.

[288] D. Laksov, Weierstrass points on curves, *Astérisque* **87/88** (1981), 221–247.

[289] D. Laksov, Wronskians and Plücker formulas for linear systems on curves, *Ann. Sci. École Norm. Sup.* **17** (1984), 45–66.

[290] I.N. Landjev, Linear codes over finite fields and finite projective geometries, *Discrete Math.* **213** (2000), 211–244.

[291] S. Lang, *Introduction to Algebraic Geometry*, Addison-Wesley, Reading, 1972, xi+260 pp.

[292] S. Lang, *Abelian Varieties*, Springer, New York, 1983, xii+256 pp.

[293] S. Lang, *Algebra*, Graduate Texts in Mathematics **211**, Springer, New York, 2002, xvi+914 pp.

[294] K. Lauter, Improved upper bounds for the number of rational points on algebraic curves over finite fields, *C. R. Acad. Sci. Paris Sér. I Math.* **328** (1999), 1181–1185.

[295] K. Lauter, Non-existence of a curve over $\mathbf{F}_3$ of genus 5 with 14 rational points, *Proc. Amer. Math. Soc.* **128** (2000), 369–374.

[296] K. Lauter, Deligne–Lusztig curves as ray class fields, *Manuscripta Math.* **98** (1999), 87–96.

[297] K. Lauter, A formula for constructing curves over finite fields with many rational points, *J. Number Theory* **74** (1999), 56–72.

[298] K. Lauter, Geometric methods for improving the upper bounds on the number of rational points on algebraic curves over finite fields, with an appendix by J.-P. Serre, *J. Algebraic Geom.* **10** (2001), 19–36.

[299] K. Lauter, The maximum or minimum number of rational points on genus three curves over finite fields, with an appendix by J.-P. Serre, *Compositio Math.* **134** (2002) 87–111.

[300] D.B. Leep and C.C. Yeomans, The number of points on a singular curve over a finite field, *Arch. Math. (Basel)* **63** (1994), 420–426.

[301] S. Lefschetz, *Algebraic Geometry*, Princeton Univ. Press, Princeton, 1953, ix+233 pp.

[302] C. Lehr and M. Matignon, Automorphism groups for p-cyclic covers of the affine line, *Compositio Math.* **141** (2005), 1213–1237.

[303] H.W. Leopoldt, Über die Automorphismengruppe des Fermatkörpers, *J. Number Theory* **56** (1996), 256–282.

[304] V.M. Levchuk and Y.N. Nuzhin, The structure of Ree groups, *Algebra Logic* **24** (1985), 16–26.

[305] D. Levcovitz, Bounds for the number of fixed points of automorphisms of curves, *Proc. London Math. Soc.* **62** (1991), 133–150.

[306] J. Lewittes, Automorphisms of compact Riemann surfaces, *Amer. J. Math.* **85** (1963), 734–752.

[307] J. Lewittes, Places of degree one in function fields over finite fields, *J. Pure Appl. Algebra* **69** (1990), 177–183.

[308] W.C.W. Li and H. Maharaj, Coverings of curves with asymptotically many rational points, *J. Number Theory* **96** (2002), 232–256.

[309] W.C.W. Li, H. Maharaj, H. Stichtenoth, and N. Elkies, New optimal tame towers of function fields over small finite fields, in *Algorithmic Number Theory*, Lecture Notes in Computer Science **2369**, Springer, Berlin, 2002, 372–389.

[310] R. Lidl and H. Niederreiter, *Finite Fields*, Encyclopedia of Mathematics and its Applications **20**, Cambridge Univ. Press, Cambridge, 1997, xiv+755 pp.

[311] S. Ling, H. Stichtenoth and S. Yang, A class of Artin–Schreier towers with finite genus, *Bull. Braz. Math. Soc. (N.S.)* **36** (1978), 393–401.

[312] C. Lomont, Yet more projective curves over $\mathbf{F}_2$, *Experiment. Math.* **11** (2002), 547–554.

[313] D. Lorenzini, *An Invitation to Arithmetic Geometry*, Graduate Studies in Mathematics **9**, American Math. Soc., Providence, 1996, xvi+397 pp.

[314] E. Lucas, Théorie des functions numériques simplement périodiques, *Amer. J. Math.* **1**, 184–251, 289–321.

[315] I. Luengo and B. López, Algebraic curves over $\mathbf{F}_3$ with many rational points. *Algebra, Arithmetic and Geometry with Applications*, Springer, Berlin, 2004, 619–626.

[316] H. Lüneburg, *Translation Planes*, Springer, Berlin, 1980, ix+278 pp.

[317] F.J. MacWilliams and N.J.A. Sloane, *The Theory of Error-Correcting Codes*, North-Holland, Amsterdam, 1977, ix+762 pp.

[318] M.L. Madan, On a theorem of M. Deuring and I.R. Safarevic, *Manuscripta Math.* **23** (1977), 91–102.

[319] D.J. Madden and R.C. Valentini, The group of automorphisms of algebraic function fields, *J. Reine Angew. Math.* **343** (1983), 162–168.

[320] H. Maharaj and J. Wulftange, On the construction of tame towers over finite fields, *J. Pure Appl. Algebra* **199** (2005), 197–218.

[321] Y.I. Manin, The Hasse–Witt matrix of an algebraic curve, *Trans. Amer. Math. Soc.* **45** (1965), 245–246.

[322] Y.I. Manin, What is the maximum number of points on a curve over F_2? *J. Fac. Sci. Univ. Tokyo Sect. IA Math.* **28** (1981), 715–720.

[323] S. Marcugini, A. Milani, and F. Pambianco, NMDS-Codes of maximal lengths over $\mathbf{F}_q$, $8 \leq q \leq 11$, *IEEE Trans. Inform. Theory* **48** (2002), 962–966.

[324] T. Maruta, Cyclic arcs and pseudo-cyclic MDS codes, *Discrete Math.* **174** (1997), 199–205.

[325] S. Matterei, On a bound of Garcia and Voloch for the number of points of a Fermat curve over a prime field, *Finite Fields Appl.* **13** (2007), 773–777.

[326] B. Mazur, Arithmetic on curves, *Bull. Amer. Math. Soc. (N.S.)* **14** (1986), 207–259.

[327] R.J. McEliece and M.C. Rodríguez-Palánquex, Results to get maximal quasihermitian curves. New possibilities for AG codes, in *Information, Coding and Mathematics*, Kluwer Internatal Series in Engineering and Computer Science **687**, Kluwer, Boston, 2002, 55–62.

[328] A.J. Menezes (ed.), *Applications of Finite Fields*, Kluwer, Dordrecht, 1993, xi+218 pp.

[329] L. Miller, Curves with invertible Hasse–Witt-matrix, *Math. Ann.* **197** (1972), 123–127.

[330] J.S. Milne, Jacobian varieties, *Arithmetic Geometry*, Springer, New York, 1986, 167–212.

[331] T.T. Moh and W. Heinzer, On the Lüroth semigroup and Weierstrass canonical divisors, *J. Algebra* **77** (1982), 62–73.

[332] M. Moisio, A construction of a class of maximal Kummer curves, *Finite Fields Appl.* **11** (2004), 667-673.

[333] M. Moisio, On the number of rational points on some families of Fermat curves over finite fields, *Finite Fields Appl.* **13** (2007), 546–662.

[334] M. Moisio and K.O. Väänänen, A comparison of the number of rational places of certain function fields to the Hasse–Weil bounds, *Appl. Algebra Engrg. Comm. Comput.* **14** (2004), 341–359.

[335] P. Morandi, *Field and Galois Theory*, Graduate Texts in Mathematics **167**, Springer, New York, 1996. xvi+281 pp.

[336] C. Moreno, *Algebraic Curves over Finite Fields*, Cambridge Tracts in Mathematics **97**, Cambridge Univ. Press, Cambridge, 1991, x+246 pp.

[337] O. Moreno, D. Zinoviev and V. Zinoviev, On several new projective curves over $\mathbf{F}_2$ of genus 3, 4, and 5, *IEEE Trans. Inform. Theory* **41** (1995), 1643–1648.

[338] D. Mumford, *Abelian Varieties*, Tata Institute of Fundamental Research Studies in Mathematics **5**, Oxford Univ. Press, London, 1970, viii+242 pp.

[339] D. Mumford, *Curves and Their Jacobians*, Univ. of Michigan Press, Ann Arbor, 1975, vi+104 pp.

[340] C. Munuera, On the main conjecture on geometric MDS codes, *IEEE Trans. Inform. Theory* **38** (1992), 1573–1577.

[341] S. Nakajima, p-ranks and automorphism groups of algebraic curves, *Trans. Amer. Math. Soc.* **303** (1987), 595–607.

[342] S. Nakajima, On abelian automorphism groups of algebraic curves, *J. London Math. Soc.* **36** (1987), 23–32.

[343] S. Nakajima, On automorphism groups of algebraic curves, *Current Trends in Number Theory*, Hindustan Book Agency, New Delhi, 2002, 129–134.

[344] E. Nart and C. Ritzenthaler, Non-hyperelliptic curves of genus three over finite fields of characteristic two, *J. Number Theory* **116** (2006), 443–473.

[345] E. Nart and D. Sadornil, Hyperelliptic curves of genus three over finite fields of even characteristic, *Finite Fields Appl.* **10** (2004), 198–220.

[346] A. Neeman, Weierstrass points in characteristic p, *Invent. Math.* **75** (1984), 359–376.

[347] H. Niederreiter and C.P. Xing, Cyclotomic function fields, Hilbert class fields, and global function fields with many rational places, *Acta Arith.* **79** (1997), 59–76.

[348] H. Niederreiter and C.P. Xing, Towers of global function fields with asymptotically many rational places and an improvement on the Gilbert–Varshamov bound, *Math. Nachr.* **195** (1998), 171–186.

[349] H. Niederreiter and C.P. Xing, Global function fields with many rational places over the ternary field, *Acta Arith.* **83** (1998), 65–86.

[350] H. Niederreiter and C.P. Xing, *Rational Points on Curves over Finite Fields: Theory and Applications*, London Mathematical Society Lecture Note Series **285**, Cambridge Univ. Press, Cambridge, 2001, x+245 pp.

[351] Y. Noriko, On the Jacobian variety of the Fermat curve, *J. Algebra* **65** (1980), 1–35.

[352] F. Oort, Minimal p-divisible groups, *Ann. of Math.* **161** (2005), 1021–1036.

[353] G. Orzech and M. Orzech, *Plane Algebraic Curves, an Introduction via Valuations*, Monographs and Textbooks in Pure and Applied Mathematics **61**, Dekker, New York, 1981, viii+225 pp.

[354] F. Özbudak, On maximal curves and linearized permutation polynomials over finite fields, *J. Pure Appl. Algebra* **162** (2001), 87–102.

[355] F. Özbudak, A note on divisor class groups of degree zero of algebraic function fields over finite fields, *Finite Fields Appl.* **9** (2003), 129–133.

[356] F. Özbudak and H. Stichtenoth, Curves with many points and configurations of hyperplanes over finite fields, *Finite Fields Appl.* **5** (1999), 436–449.

[357] R. Pardini, Some remarks on plane curves over fields of finite characteristic, *Compositio Math.* **60** (1988), 3–17.

[358] F. Pasticci, On quotient curves of the Suzuki curve, *Ars Comb.*, to appear.

[359] J.P. Pedersen, A function field related to the Ree group, in *Coding Theory and Algebraic Geometry*, Lecture Notes in Mathematics **1518**, Springer, Berlin, 1992, 122–132.

[360] R. Pellikaan, On the gonality of curves, abundant codes and decoding, in *Coding Theory and Algebraic Geometry*, Lecture Notes in Mathematics **1518**, Springer, Berlin, 1992, 132–144.

[361] R. Pellikaan, The Klein quartic, the Fano plane and curves representing designs, in *Codes, Curves, and Signals*, Kluwer International Series in Engineering and Computer Science **485**, Kluwer, Boston, 1998, 9–20.

[362] T. Penttila and B. Williams, Ovoids of parabolic spaces, *Geom. Dedicata* **82** (2000), 1–19.

[363] M. Perret, Tours ramifiés infinies de corps de classes, *J. Number Theory* **38** (1991), 300–322.

[364] O. Pretzel, *Codes and Algebraic Curves*, Oxford Lecture Series in Mathematics and its Applications **8**, Oxford Univ. Press, Oxford, 1998, xii+192 pp.

[365] J. Rathmann, The uniform position principle for curves in characteristic *p*, *Math. Ann.* **276** (1987), 565–579.

[366] L. Rédei, *Algebra*, Pergamon, Oxford, 1967, xviii+823 pp.

[367] R. Ree, A family of simple groups associated with the simple Lie algebra of type (G_2), *Amer. J. Math.* **83** (1961), 432–462.

[368] M. Reid, *Undergraduate Algebraic Geometry*, Cambridge Univ. Press, Cambridge, 1988, viii+129 pp.

[369] B. Richter, Über einen Satz von Deuring–Safarevic–Madan, *Abh. Math. Sem. Univ. Hamburg* **52** (1982), 232–234.

[370] F. Rodriguez Villegas, J.F. Voloch, and D. Zagier, Constructions of plane curves with many points, *Acta Arith.* **99** (2001), 85–96.

[371] M.C. Rodríguez-Palánquex, L.J. García-Villalba, and I. Luengo-Velasco, Computing the genus of a class of curves, in *Applied Algebra, Algebraic Algorithms and Error-correcting Codes*, Lecture Notes in Computer Science **2227**, Springer, Berlin, 2001, 182–191.

[372] P. Roquette, Über die Automorphismengruppe eines algebraischen Funktionenkörpers, *Arch. Math.* **3** (1952), 343–350.

[373] P. Roquette, Abschätzung der Automorphismenanzahl von Funktionenkörpern bei Primzahlcharakteristik, *Math. Z.* **117** (1970), 157–163.

[374] P. Roquette, The Riemann hypothesis in characteristic p, its origin and development. I. The formation of the zeta-functions of Artin and of F.K. Schmidt, *Hamburger Beiträge zur Geschichte der Mathematik, Mitt. Math. Ges. Hamburg* **21** (2002), 79–157.

[375] M. Rosen, Some remarks on the p-rank of an algebraic curve, *Arch. Math.* **41** (1983), 143–146.

[376] M. Rosenlicht, Automorphisms of function fields, *Trans. Amer. Math. Soc.* **79** (1955), 1–11.

[377] H.-G. Rück, A note on elliptic curves over finite fields, *Math. Comp.* **49** (1987), 301–304.

[378] H.G Rück and H. Stichtenoth, A characterization of Hermitian function fields over finite fields, *J. Reine Angew. Math.* **457** (1994), 185–188.

[379] M. Rzedowski-Calderón and G.D. Villa Salvador, Automorphisms of congruence function fields, *Pacific J. Math.* **150** (1991), 167–178.

[380] G. Salmon, *A Treatise on the Higher Plane Curves*, third edition, Hodges, Foster, and Figgis, Dublin, 1879, xix+395 pp.

[381] D. Savitt, The maximum number of points on a curve of genus 4 over $\mathbf{F}_8$ is 25, with an appendix by Kristin Lauter, *Canad. J. Math.* **55** (2003), 331–352.

[382] P. Scherk, On the intersection number of two plane curves, *J. Geom.* **10** (1977), 57–68.

[383] H.L. Schmid, Über Automorphismen eines algebraische Funktionenkörpern von Primzahlcharakteristik, *J. Reine Angew. Math.* **179** (1938), 5–15.

[384] F.K. Schmidt, Analytische Zahlentheorie in Körpern der Charakteristik p, *Math. Z.* **33** (1931), 1–32.

[385] F.K. Schmidt, Zur arithmetischen Theorie der algebraischen Funktionen I, *Math. Z.* **41** (1936), 415–438.

[386] F.K. Schmidt, Die Wronskische Determinante in beliebigen differenzierbaren Funktionenkörpern, *Math. Z.* **45** (1939), 62–74.

[387] F.K. Schmidt, Zur arithmetischen Theorie der algebraischen Funktionen. II. Allgemeine Theorie der Weierstrasspunkte, *Math. Z.* **45** (1939), 75–96.

[388] W.M. Schmidt, Zur Methode von Stepanov, *Acta Arith.* **24** (1973), 347–367.

[389] W.M. Schmidt, *Equations over Finite Fields: an Elementary Approach*, second edition, Kendrick Press, Heber City, 2004, x+333 pp.

[390] R. Schoof, Nonsingular plane cubic curves over finite fields, *J. Combin. Theory Ser. A* **46** (1987), 183–211.

[391] R. Schoof, Algebraic curves and coding theory, UTM **336**, University of Trento, 1990.

[392] R. Schoof, Algebraic curves over $\mathbf{F}_2$ with many rational points, *J. Number Theory* **41** (1992), 6–14.

[393] B. Segre, Sulle curve algebriche che ammettono come trasformata razionale una curva piana dello stesso ordine, priva di punti multipli, *Math. Ann.* **109** (1933), 1–3.

[394] B. Segre, Ovals in a finite projective plane, *Canad. J. Math.* **7** (1955), 414–416.

[395] B. Segre, Curve razionali normali e k-archi negli spazi finiti, *Ann. Mat. Pura Appl.* **39** (1955), 357–379.

[396] B. Segre, Le geometrie di Galois, *Ann. Mat. Pura Appl.* **48** (1959), 1–96.

[397] B. Segre, Forme e geometrie hermitiane, con particolare riguardo al caso finito, *Ann. Mat. Pura Appl.* **70** (1965), 1–201.

[398] B. Segre, Introduction to Galois geometries, edited by J.W.P. Hirschfeld, *Atti Accad. Naz. Lincei Mem.* **8** (1967), 133–236.

[399] B. Segre, *Prodromi di Geometria Algebrica*, with an appendix by U. Bartocci and M. Lorenzani, Cremonese, Rome, 1971, vi+412 pp.

[400] A. Seidenberg, *Elements of the Theory of Algebraic Curves*, Addison–Wesley, Reading, 1968, viii+216 pp.

[401] E.S. Selmer, On the linear diophantine problem of Frobenius, *J. Reine Angew. Math.* **293/294** (1977), 1–17.

[402] J.G. Semple and G.T. Kneebone, *Algebraic Curves*, Oxford Univ. Press, Oxford, 1959, xiii+361 pp.

[403] J.G. Semple and L. Roth, *Introduction to Algebraic Geometry*, Oxford Univ. Press, Oxford, 1949, xvi+446 pp.

[404] J.-P. Serre, *Local Fields*, Graduate Texts in Mathematics **67**, Springer, New York, 1979. viii+241 pp.

[405] J.-P. Serre, Sur le nombre des points rationnels d'une courbe algébrique sur un corps fini, *C. R. Acad. Sci. Paris Sér. I Math.* **296** (1983), 397–402.

[406] J.-P. Serre, Résumé des cours de 1983–1984, Annuaire du Collège de France 1983–1984, 79–83; reprinted in Œuvres **3**, 701–705.

[407] J.-P. Serre, Rational points on curves over finite fields, unpublished notes by F.Q. Gouvea of lectures at Harvard Univ., 1985.

[408] F. Severi, *Trattato di Geometria Algebrica*, Zanichelli, Bologna, 1926, viii+358 pp.

[409] I.R. Shafarevich, On p-extensions, *Amer. Math. Soc. Transl.* **4** (1954), 59–71.

[410] I.R. Shafarevich, *Basic Algebraic Geometry*, translated from the Russian by K. A. Hirsch, Grundlehren der Mathematischen Wissenschaften **213**, Springer, Berlin, 1974, xv+439 pp.

[411] T. Shaska, Computational aspects of hyperelliptic curves, *Computer Mathematics*, Lecture Notes Series on Computing **10**, World Sci. Publishing, River Edge, 2003, 248–257.

[412] T. Shaska, Some special families of hyperelliptic curves, *J. Algebra Appl.* **3** (2004), 75–89.

[413] T. Shaska and H. Völklein, Elliptic subfields and automorphisms of genus 2 function fields, in *Algebra, Arithmetic and Geometry with Applications*, Springer, Berlin, 2004, 703–723.

[414] T. Shioda, Arithmetic and geometry of Fermat curves, in *Algebraic Geometry Seminar*, World Sci. Publishing, Singapore, 1988, 95–102.

[415] A. Siciliano, A note on rational points of curves over finite fields, *Expo. Math.* **18** (2000), 83–93.

[416] J.A. Silverman, *The Arithmetic of Elliptic Curves*, Springer, New York, 1986, xii+400 pp.

[417] B. Singh, On the group of automorphisms of a function field of genus at least two, *J. Pure Appl. Algebra* **4** (1974), 205–229.

[418] S.A. Stepanov, The number of points of a hyperelliptic curve over a finite prime field, *Izv. Akad. Nauk SSSR Ser. Mat.* **33** (1969), 1171–1181.

[419] S.A. Stepanov, *Arithmetic of Algebraic Curves*, translated from the Russian by I. Aleksanova, Consultants Bureau, New York, 1994, xiv+422 pp.

[420] S.A. Stepanov, Codes on fiber products of hyperelliptic curves, *Discrete Math. Appl.* **7** (1997), 77–88.

[421] S.A. Stepanov, *Codes on Algebraic Curves*, Kluwer/Plenum, New York, 1999, xiv+350 pp.

[422] S.A. Stepanov and M.K. Shalalfekh, Codes on fiber products of Artin–Schreier curves, *Discrete Math. Appl.* **11** (2001), 133–143.

[423] H. Stichtenoth, Über die Automorphismengruppe eines algebraischen Funktionenkörpers von Primzahlcharakteristik. I. Eine Abschätzung der Ordnung der Automorphismengruppe, *Arch. Math.* **24** (1973), 527–544.

[424] H. Stichtenoth, Über die Automorphismengruppe eines algebraischen Funktionenkörpers von Primzahlcharakteristik. II. Ein spezieller Typ von Funktionenkörpern, *Arch. Math.* **24** (1973), 615–631.

[425] H. Stichtenoth, Die Hasse–Witt–Invariante eines Kongruenzfunktionenkörpers, *Arch. Math. (Basel)* **33** (1980), 357–360.

[426] H. Stichtenoth, Zur Realisierbarkeit endlicher Gruppen als Automorphismengruppen algebraischer Funktionenkörper, *Math. Z.* **187** (1984), 221–225.

[427] H. Stichtenoth, Die Ungleichung von Castelnuovo, *J. Reine Angew. Math.* **348** (1984), 197–202.

[428] H. Stichtenoth, *Algebraic Function Fields and Codes*, Springer, Berlin, 1993, x+260 pp.

[429] H. Stichtenoth, The Fermat curve in characteristic p, in *Finite Fields: Theory, Applications, and Algorithms*, Contemporary Mathematics **225**, Amer. Math. Soc., Providence, 1999, 123–129.

[430] H. Stichtenoth and C.P. Xing, On the structure of the divisor class group of a class of curves over finite fields, *Arch. Math.* **65** (1995), 141–150.

[431] K.O. Stöhr and P. Viana, A study of Hasse–Witt matrices by local methods, *Math. Z.* **200** (1989), 397–407.

[432] K.O. Stöhr and J.F. Voloch, Weierstrass points and curves over finite fields, *Proc. London Math. Soc.* **52** (1986), 1–19.

[433] K.O. Stöhr and J.F. Voloch, A formula for the Cartier operator on plane algebraic curves, *J. Reine Angew. Math.* **377** (1987), 49–64.

[434] L. Storme and H. Van Maldeghem, Cyclic arcs in $\mathrm{PG}(2, q)$, *J. Algebraic Combin.* **3** (1994), 113–128.

[435] D. Subrao, The p-rank of Artin–Schreier curves, *Manuscripta Math.* **16** (1975), 169–193.

[436] F. Sullivan, p-torsion in the class group of curves with many automorphisms, *Arch. Math.* **26** (1975), 253–261.

[437] M. Suzuki, A new type of simple groups of finite order, *Proc. Nat. Acad. Sci. U.S.A.* **46** (1960), 868–870.

[438] M. Suzuki, On a finite group with a partition, *Arch. Math.* **12** (1961), 241–254.

[439] M. Suzuki, On a class of doubly transitive groups, *Ann. of Math.* **75** (1962), 105–145.

[440] M. Suzuki, Finite groups of even order in which Sylow 2-groups are independent, *Ann. of Math.* **80** (1964), 58–77.

[441] T. Szemberg, Automorphisms of Riemann surfaces with two fixed points, *Ann. Polon. Math.* **55** (1991), 343–347.

[442] P. Sziklai, A bound on the number of points of a plane curve over a finite field, *Finite Fields Appl.* **14** (2008), 41–43.

[443] T. Szőnyi, On cyclic caps in projective spaces, *Des. Codes Cryptogr.* **8** (1996), 327–332.

[444] T. Szőnyi, Some applications of algebraic curves in finite geometry and combinatorics, in *Surveys in Combinatorics*, London Mathematical Society Lecture Note Series **241**, Cambridge Univ. Press, Cambridge, 1997, 197–236.

[445] Y. Takizawa, Some remarks on the Picard curves over a finite field, *Tech. Report Chiba U.* **19** (2003).

[446] G. Tallini, Sui q-archi di un piano lineare finito di caratteristica $p = 2$, *Atti Accad. Naz. Lincei Rend.* **23** (1957), 242–245.

[447] G. Tallini, Le ipersuperficie irriducibili d'ordine minimo che invadono uno spazio di Galois, *Atti Accad. Naz. Lincei Rend.* **30** (1961), 706–712.

[448] G. Tallini, Sulle ipersuperficie irriducibili d'ordine minimo che contengono tutti i punti di uno spazio di Galois $S_{r,q}$, *Rend. Mat. e Appl.* **20** (1961), 431–479.

[449] J.T. Tate, Endomorphisms of abelian varieties over finite fields, *Invent. Math.* **2** (1966), 134–144.

[450] J.T. Tate, The arithmetic of elliptic curves, *Invent. Math.* **23** (1974), 179–206.

[451] O. Teichmüller, Differentialrechnung bei Charakteristik p, *J. Reine Angew. Math.* **175** (1936), 89–99.

[452] A. Temkine, Hilbert class field towers of function fields over finite fields and lower bounds for $A(q)$, *J. Number Theory* **87** (2001), 189–210.

[453] J.A. Thas, Complete arcs and algebraic curves in $PG(2,q)$, *J. Algebra* **106** (1987), 451–464.

[454] J. Tits, Les groupes simples de Suzuki et de Ree, *Séminaire Bourbaki* **6**, Soc. Math. France, Paris, 1995, 65–82.

[455] J. Tits, Ovoides et groupes de Suzuki, *Arch. Math.* **13** (1962), 187–198.

[456] J. Top, Curves of genus 3 over small finite fields, *Indag. Math. N.S.* **14** (2003), 275–283.

[457] F. Torres, Weierstrass points and double coverings of curves. With application: symmetric numerical semigroups which cannot be realized as Weierstrass semigroups, *Manuscripta Math.* **83** (1994), 39–58.

[458] F. Torres, Bounding the order of automorphisms of certain curves, *Rend. Istit. Mat. Univ. Trieste* **27** (1995), 85–116.

[459] C. Towse, Weierstrass weights of fixed points of an involution, *Math. Proc. Cambridge Philos. Soc.* **122** (1997), 385–392.

[460] M.A. Tsfasman and S.G. Vlăduţ, *Algebraic-Geometric Codes*, Mathematics and its Applications (Soviet Series) **58**, Kluwer, Dordrecht, 1991, xxiv+667 pp.

[461] M.A. Tsfasman, S.G. Vlăduţ and D. Nogin, *Algebraic Geometric Codes: Basic Notions*, Mathematical Surveys and Monographs **139**, Amer. Math. Soc., Providence, 2007, xix+338 pp.

[462] M.A. Tsfasman, S.G. Vlăduţ and T. Zink, Modular curves, Shimura curves, and Goppa codes, better than Varshamov–Gilbert bound, *Math. Nachr.* **109** (1982), 21–28.

[463] S. Tuffery, Automorphismes d'ordre 3 et 7 sur une courbe de genre 3, *Expo. Math.* **11** (1993), 159–162.

[464] S. Tuffary, Les automorphismes des courbes de genre 3 de caractéristique 2, *C. R. Acad. Sci. Paris Sér. I Math.* **321** (1995), 205–210.

[465] S. Tuffery, Déformations de courbes avec action de groupe. II, *Forum Math.* **8** (1996), 205–218.

[466] E. Ughi, On the number of points of elliptic curves over a finite field and a problem of B. Segre, *European J. Combin.* **4** (1983), 263–270.

[467] I. Vainsencher, On Stöhr–Voloch's proof of Weil's theorem, *Manuscripta Math.* **64** (1989), 121–126.

[468] R.C. Valentini, Hyperelliptic curves with zero Hasse–Witt matrix, *Manuscripta Math.* **86** (1995), 185–194.

[469] R.C. Valentini and M.L. Madan, A Hauptsatz of L.E. Dickson and Artin–Schreier extensions, *J. Reine Angew. Math.* **318** (1980), 156–177.

[470] G. van der Geer, Codes and elliptic curves, in *Effective Methods in Algebraic Geometry*, Progress in Mathematics **94**, Birkhäuser, Boston, 1991, 159–168.

[471] G. van der Geer, Curves over finite fields and codes, in *European Congress of Mathematics*, Vol. II (Barcelona, 2000), Progress in Mathematics **202**, Birkhäuser, Basel, 2001, 225–238.

[472] G. van der Geer, Coding theory and algebraic curves over finite fields: a survey and questions, in *Applications of Algebraic Geometry to Coding Theory, Physics and Computation*, NATO Science Series II Mathematics, Physics and Chemistry **36**, Kluwer, Dordrecht, 2001, 139–159.

[473] G. van der Geer and M. van der Vlugt, Kloosterman sums of the p-torsion of certain Jacobians, *Math. Ann.* **290** (1991), 549–563.

[474] G. van der Geer and M. van der Vlugt, Weight distributions for a certain class of codes and maximal curves, *Discrete Math.* **106/107** (1992), 209–218.

[475] G. van der Geer and M. van der Vlugt, Supersingular curves of genus 2 over finite fields of characteristic 2, *Math. Nachr.* **159** (1992), 73–81.

[476] G. van der Geer and M. van der Vlugt, Curves over finite fields of characteristic 2 with many rational points, *C. R. Acad. Sci. Paris Sér. I Math.* **317** (1993), 593–597.

[477] G. van der Geer and M. van der Vlugt, Fibre products of Artin–Schreier curves and generalized Hamming weights of codes, *J. Combin. Theory Ser. A* **70** (1995), 337–348.

[478] G. van der Geer and M. van der Vlugt, Quadratic forms, generalized Hamming weights of codes and curves with many points, *J. Number Theory* **59** (1996), 20–36.

[479] G. van der Geer and M. van der Vlugt, How to construct curves over finite fields with many points, in *Arithmetic Geometry*, Symposia in Mathematics **36**, Cambridge Univ. Press, Cambridge, 1997, 169–189.

[480] G. van der Geer and M. van der Vlugt, Constructing curves over finite fields with many points by solving linear equations, in *Applications of Curves over Finite Fields*, Contemporary Mathematics **245**, Amer. Math. Soc., Providence, 1999, 41–47.

[481] G. van der Geer and M. van der Vlugt, Tables of curves with many points, *Math. Comp.* **69** (2000), 797–810; update on http:// www.wins.uva.nl/ geer.

[482] G. van der Geer and M. van der Vlugt, Kummer covers with many points, *Finite Fields Appl.* **6** (2000), 327–341.

[483] G. van der Geer and M. van der Vlugt, An asymptotically good tower of curves over the field with eight elements, *Bull. London Math. Soc.* **34** (2002), 291–300.

[484] B.L. van der Waerden, *Einführung in die Algebraische Geometrie*, second edition, Grundlehren der Mathematischen Wissenschaften **51**, Springer, Berlin, 1973, xi+280 pp.

[485] B. van Geemen, K. Koike and A. Weng, Quotients of Fermat curves and a Hecke character, *Finite Fields Appl.* **11** (2005), 6–29.

[486] J.H. van Lint and G. van der Geer, *Introduction to Coding Theory and Algebraic Geometry*, DMV Seminar **12**, Birkhäuser, Basel, 1988, 83 pp.

[487] P. Viana and O.P. la Torre, Curves of genus three in characteristic two, *Comm. Algebra* **33** (2005), 4291–4302.

[488] P. Viana and J.E.A. Rodriguez, Eventually minimal curves, *Bull. Braz. Math. Soc. (N.S.)* **36** (2005), 39–58.

[489] G.D. Villa Salvador, *Topics in the Theory of Algebraic Function Fields*, Birkhäuser, Boston, 2006, xviii+652 pp.

[490] J. Voight, Curves over finite fields with many points: an introduction, in *Computational Aspects of Algebraic Curves*, Lecture Notes Series on Computing **13**, World Sci. Publ., Hackensack, 2005, 124–144.

[491] J.F. Voloch, A note on elliptic curves over finite fields, *Bull. Soc. Math. France* **116** (1988), 455–458.

[492] J.F. Voloch, Arcs in projective planes over prime fields, *J. Geom.* **38** (1990), 198–200.

[493] J.F. Voloch, On the completeness of certain plane arcs, *European J. Combin.* **8** (1987), 453–456.

[494] J.F. Voloch, On the completeness of certain plane arcs II, *European J. Combin.* **11** (1990), 491–496.

[495] J.F. Voloch, Complete arcs in Galois planes of nonsquare order, in *Advances in Finite Geometries and Designs*, Oxford University Press, Oxford, 1991, 401–406.

[496] J.F. Voloch, Jacobians of curves over finite fields, *Rocky Mountain J. Math.* **30** (2000), 755–759.

[497] R.J. Walker, *Algebraic Curves*, Princeton University Press, Princeton, 1950, x+201 pp.

[498] C.T.C. Wall, Quartic curves in characteristic 2, *Math. Proc. Cambridge Philos. Soc.* **117** (1995), 393–414.

[499] A.H. Wallace, Tangency and duality over arbitrary fields, *Proc. London Math. Soc.* **6** (1956), 321–342.

[500] W.C. Waterhouse, Abelian varieties over finite fields, *Ann. Sci. École Norm. Sup.* **2** (1969), 521–560.

[501] W.C. Waterhouse and J.S. Milne, Abelian varieties over finite fields, in *1969 Number Theory Institute*, Proceedings of Symposia in Pure Mathematics **XX**, Amer. Math. Soc., Providence, 1971, 53–64.

[502] K. Weierstrass, *Mathematische Werke, III, Abhandlungen 3*, reprint of the 1903 original, Johnson, New York, 1967, viii+362 pp.

[503] A. Weil, *Foundations of Algebraic Geometry*, American Mathematical Society Colloquium Publications **29**, American Math. Soc., New York, 1946, xix+289 pp.

[504] A. Weil, *Sur les Courbes Algébriques et les Variétés qui s'en Déduisent*, Actualités Scientifiques et Industrielle **1041**, Publications de l'Institut de Mathématique de l'Université de Strasbourg **7**, Hermann, Paris, 1948, iv+85 pp.

[505] E. Witt, Konstruktion von galoischen Körpern der Characteristik p zu vorgegebener Gruppe der Ordnung p^f, *J. Reine Angew. Math.* **174** (1936), 237–245.

[506] J. Wolfmann, The number of points on certain algebraic curves over finite fields, *Comm. Algebra* **17** (1989), 2055–2060.

[507] J. Wulftange, On the construction of some towers over finite fields. *Finite Fields and Applications*, Lecture Notes in Computer Science **2948**, Springer, Berlin, 2004, 154–165.

[508] C.P. Xing, Multiple Kummer extension and the number of prime divisors of degree one in function fields, *J. Pure Appl. Algebra* **84** (1993), 85–93.

[509] C.P. Xing and H. Stichtenoth, The genus of maximal function fields over finite fields, *Manuscripta Math.* **86** (1995), 217–224.

[510] D. Yasin, Hyperelliptic curves of genus 3 and 4 in characteristic 2, *Computational Aspects of Algebraic Curves*, Lecture Notes Series on Computing **13**, World Sci. Publ., Hackensack, 2005, 145–162.

[511] S. Yekhanin, A note on plane pointless curves, *Finite Fields Appl.* **13** (2007), 415–422.

[512] O. Zariski and P. Samuel, *Commutative Algebra*, Vols. I,II, Van Nostrand, Princeton, 1958, xi+329 pp.; 1960, vii+414 pp.

[513] H. Zassenhaus, Kennzeichnung endlicher linearer Gruppen als Permutations-gruppen, *Abh. Math. Sem. Hamburg* **11** (1936), 17–40.

[514] A. Zaytsev, The Galois closure of the Garcia–Stichtenoth tower, *Finite Fields Appl.* **13** (2007), 751–761.

[515] J.P. Zhang, On the finite group with a TI Sylow p-subgroup, *Chinese Ann. Math. Ser. B* **12** (1991), 147–151.

[516] H.J. Zhu, Hyperelliptic curves over $\mathbf{F}_2$ of every 2-rank without extra automorphisms, *Proc. Amer. Math. Soc.* **134** (2006), 323–331.

[517] H.G. Zimmer, Die Néron–Tate'schen quadratischen formen auf der rationalen Punktgruppe einer elliptischen Kurve, *J. Number Theory* **2** (1970), 459–499.

[518] H.G. Zimmer, An elementary proof of the Riemann hypothesis for an elliptic curve over a finite field, *Pacific J. Math.* **36** (1971), 267–278.

[519] T. Zink, Degeneration of Shimura surfaces and a problem in coding theory, in *Fundamentals of Computation Theory*, Lecture Notes in Computer Science **199**, Springer, Berlin, 1985, 503–511.

[520] W.A. Zúñiga-Galindo, Zeta functions of singular curves over finite fields, *Rev. Colombiana Mat.* **31** (1997), 115–124.

Index

GPSR Authorized Representative: Easy Access System Europe - Mustamäe tee 50, 10621 Tallinn, Estonia, gpsr.requests@easproject.com

www.ingramcontent.com/pod-product-compliance
Lightning Source LLC
Chambersburg PA
CBHW071954290825
31759CB00003BA/17
9780691096797